Biotechnology
Volume 2

Biotechnology

A Comprehensive Treatise in 8 Volumes

Volume 1
Microbial Fundamentals

Volume 2
Fundamentals of Biochemical Engineering

Volume 3
Biomass, Microorganisms for Special Applications,
Microbial Products I, Energy from Renewable Resources

Volume 4
Microbial Products II

Volume 5
Food and Feed Production with Microorganisms

Volume 6a
Biotransformations

Volume 6b
Special Microbial Processes

Volume 7
Enzymes in Biotechnology

Volume 8
Microbial Degradations

© VCH Verlagsgesellschaft mbH, D-6940 Weinheim (Federal Republic of Germany), 1985

Distribution:
VCH Verlagsgesellschaft, P.O. Box 1260/1280, D-6940 Weinheim (Federal Republic of Germany)
USA and Canada: VCH Publishers, 303 N.W. 12th Avenue, Deerfield Beach, FL 33442-1705 (USA)

Biotechnology

edited by H.-J. Rehm and G. Reed

Volume 2

Fundamentals of Biochemical Engineering

Volume Editor: H. Brauer

Prof. Dr. H.-J. Rehm
Institut für Mikrobiologie
der Universität
Corrensstraße 3
D-4400 Münster
Federal Republic of Germany

Dr. G. Reed
Universal Foods Corp.
Technical Center
6143 N 60th Street
Milwaukee, WI 53218
USA

Prof. Dr.-Ing. H. Brauer
Institut für Chemieingenieurtechnik
Technische Universität Berlin
Straße des 17. Juni 135
D-1000 Berlin 12

Editorial Director: Dr. Hans F. Ebel
Copy Editors: Christa Maria Schultz and Theodor C. H. Cole
Production Manager: Peter J. Biel

This book contains 551 figures and 132 tables

Deutsche Bibliothek Cataloguing-in-Publication Data

Biotechnology: a comprehensive treatise in 8 vol. / ed. by H.-J. Rehm and G. Reed. - [Ausg. in 8 Bd.]. -
Weinheim; Deerfield Beach, FL: VCH
NE: Rehm, Hans-Jürgen [Hrsg.]
Vol. 2. Fundamentals of biochemical engineering. - 1985

Fundamentals of Biochemical Engineering / vol. ed.: H. Brauer. -
Weinheim; Deerfield Beach, FL: VCH, 1985.
 (Biotechnology; Vol. 2)
 ISBN 3-527-25764-0 (Weinheim)
 ISBN 0-89573-042-1 (Deerfield Beach)
NE: Brauer, Heinz [Hrsg.]

© VCH Verlagsgesellschaft mbH, D-6940 Weinheim (Federal Republic of Germany), 1985
All rights reserved (including those of translation into other languages). No part of this book may be reproduced in any form – by photoprint, microfilm, or any other means – nor transmitted or translated into a machine language without written permission from the publishers.
Registered names, trademarks, etc. used in this book, even when not specifically marked as such, are not to be considered unprotected by law.
Compositor and Printer: Zechnersche Buchdruckerei, D-6720 Speyer
Bookbinder: Klambt-Druck GmbH, D-6720 Speyer
Printed in the Federal Republic of Germany

Preface

Biochemical Engineering is the engineers contribution to Biotechnology, an interdisciplinary field of activities, that is at present rapidly developing. Biotechnology is concerned with mass conversion by microorganisms on an industrial scale. Microbiologists, biochemists, and engineers closely cooperate to reach one of the most ambitious goals of industrial societies at the turn of the century.

Biochemical Engineering is a new engineering discipline concerned with all technical and economical aspects of microbial mass conversion. This includes the development and optimization of new technical biochemical processes and equipment as well as the planning, design, construction and operation of bioprocessing plants. The problems addressed by this new engineering discipline are not only related to microbial mass conversion in bioreactors but also to upstream and downstream processes, planning, design and operation of plants, and measurement and control.

The efficiency of bioprocesses depends heavily on the close relation between process and equipment parameters. An optimal process requires equipment fully adapted to the process. On the other hand shape and size of the equipment determines the efficiency of bioprocesses. Process and equipment are interdependent.

Microbial mass conversion in a bioreactor is a function of the laws describing the transport of momentum, heat and mass as well as bioreaction kinetics. The combination of these basic processes leads to bioreaction engineering. The establishment of this engineering field certainly benefits from the already well established field of chemical reaction engineering. The decisive difference between chemical reaction and bioreaction engineering however is the living organism. Will we be able to satisfactorily describe the properties of these organisms in engineering terms? At the present we apply the tools tested in chemical reaction engineering. We are well aware, however, that these tools may be inadequate.

In upstream and downstream processes transport phenomena play the decisive role. Especially in downstream processes, which are designed for separation of desired products of microbial mass conversion from undesired ones, transport phenomena occur in many cases under extreme conditions. This is primarily due to the small concentration of these products in the fluid and the low density difference between the microorganisms and the fluid. Conventional separation processes and equipment often cannot be applied successfully. There is a need for new and more efficient equipment and processes in downstream operations.

The scientific basis of the engineering discipline developed for the above mentioned tasks comprises transport phenomena, bioreaction kinetics, design and operation of bioreactors, upstream and downstream processes and equipment, and the fundamentals of measurement and control. It certainly is a new engineering discipline in the making. This book reflects this situation. It should be considered as a first step to the development of an engineering discipline appropriately called bioprocess engineering.

The authors of this volume are well-known experts in their field. They have made most valuable contributions to the development of bioprocess engineering and consequently to biotechnology. I want to extend my sincere thanks to them all. Thanks are also due to Dr. Hans F. Ebel and Christa Maria Schultz of VCH and the series editors Professor H.-J. Rehm and Dr. G. Reed for their most valuable support and encouragement.

Berlin, December 1984

Heinz Brauer

Editorial Advisory Board
of "Biotechnology"

Prof. Dr. S. Aiba
Osaka University
Osaka, Japan

Prof. Dr. N. Blakebrough
Department of Chemical Engineering
University of Malaya
Kuala Lumpur 22-11
Malaysia

Prof. Dr. H. Brauer
Technische Universität Berlin
Berlin (West), Germany

Prof. Dr. H. Dellweg
Institut für Gärungsgewerbe
und Biotechnologie
Berlin (West), Germany

Prof. Dr. A. L. Demain
Massachusetts Institute of Technology
Cambridge, Mass., USA

Prof. Dr. G. Durand
Institut National des Sciences Appliquées
Toulouse, France

Prof. Dr. K. Esser
Ruhr-Universität Bochum
Bochum, Federal Republic of Germany

Prof. Dr. A. Fiechter
ETH Zürich
Zurich, Switzerland

Prof. Dr. S. Fukui
Kyoto University
Kyoto, Japan

Prof. Dr. Vera Johanides
University of Zagreb
Zagreb, Yugoslavia

Prof. Dr. K. Kieslich
Gesellschaft für Biotechnologische Forschung
Braunschweig-Stöckheim
Federal Republic of Germany

Prof. Dr. H. Klaushofer
Universität für Bodenkultur
Vienna, Austria

Prof. Dr. A. Kocková-Kratochvílová
Slovak Academy of Sciences
Bratislava, Czechoslovakia

Prof. Dr. R. M. Lafferty
Technische Universität Graz
Graz, Austria

Prof. Dr. M. D. Lilly
University College London
London, United Kingdom

Prof. Dr. R. I. Mateles
Stauffer Chemical Company
Westport, Conn., USA

Prof. Dr. K. Mosbach
University of Lund
Lund, Sweden

Prof. Dr. G. Shelef
Technion – Israel Institute of Technology
Haifa, Israel

Prof. Dr. G. K. Skryabin
USSR Academy of Sciences
Moscow, USSR

Prof. Dr. H. Suomalainen †
The Finnish State Alcohol Monopoly
Helsinki, Finland

Prof. Dr. G. Terui
Osaka University
Osaka, Japan

List of Contributors

of Volume 2

Prof. Dr.-Ing. Heinz Blenke

Institut für Chemische Verfahrenstechnik
der Universität Stuttgart
D-7000 Stuttgart
Federal Republic of Germany

Dr. Alena Cejka

Schering AG
D-4619 Bergkamen
Federal Republic of Germany

Prof. Dr.-Ing. Heinz Brauer

Institut für Chemieingenieurtechnik
Technische Universität Berlin
D-1000 Berlin 12 (West)
Germany

Prof. Dr. Wolf-Dieter Deckwer

Fachbereich Chemie – Technische Chemie
Universität Oldenburg
D-2900 Oldenburg
Federal Republic of Germany

Dr. Reinhold Bronnenmeier

Linde AG
D-8021 Höllriegelskreuth
Federal Republic of Germany

Prof. dr. ir. N. W. F. Kossen

Technische Hogeschool Delft
Afdeling der Scheikundige Technologie
Bioengineering Laboratory
NL-2600 AA Delft
The Netherlands

Dr. Rainer Buchholz

Abteilung Chemietechnik
Universität Dortmund
D-4600 Dortmund
Federal Republic of Germany

Priv.-Doz. Dr. Maria-Regina Kula

Gesellschaft für Biotechnologische
Forschung mbH
D-3300 Braunschweig-Stöckheim
Federal Republic of Germany

List of Contributors

Dr.-Ing. Jürgen K. Lehmann

Gesellschaft für Biotechnologische
Forschung mbH
D-3300 Braunschweig-Stöckheim
Federal Republic of Germany

Priv.-Doz. Dr.-Ing. Herbert Märkl

Lehrstuhl B für Thermodynamik
Technische Universität München
D-8000 München 2
Federal Republic of Germany

Bernd Martin

Gesellschaft für Biotechnologische
Forschung mbH
D-3300 Braunschweig-Stöckheim
Federal Republic of Germany

Prof. Dr.-Ing. Anton Moser

Institut für Biotechnologie, Mikrobiologie
und Abfalltechnologie
Technische Universität Graz
A-8010 Graz
Austria

Prof. Dr. Ulfert Onken

Abteilung Chemietechnik
Universität Dortmund
D-4600 Dortmund
Federal Republic of Germany

Ir. N. M. G. Oosterhuis

Technische Hogeschool Delft
Afdeling der Scheikundige Technologie
Bioengineering Laboratory
NL-2600 AA Delft
The Netherlands

Dipl.-Ing. Gerd-Walter Piehl

Gesellschaft für Biotechnologische
Forschung mbH
D-3300 Braunschweig-Stöckheim
Federal Republic of Germany

Dipl.-Ing. Reimer Schultze

Gesellschaft für Biotechnologische
Forschung mbH
D-3300 Braunschweig-Stöckheim
Federal Republic of Germany

Dr. Adrian Schumpe

Fachbereich Chemie – Technische Chemie
Universität Oldenburg
D-2900 Oldenburg
Federal Republic of Germany

Dipl.-Ing. Wolfgang Sittig

Hoechst AG
Abteilung ZF 2 – Biotechnik
D-6230 Frankfurt am Main 80
Federal Republic of Germany

Werner Steven

Gesellschaft für Biotechnologische
Forschung mbH
D-3300 Braunschweig-Stöckheim
Federal Republic of Germany

Prof. Dr. Karl Heinz Wallhäusser

Hoechst AG
Pharma-Qualitätskontrolle
Gr. Biologie – Mikrobiologie
D-6230 Frankfurt am Main 80
Federal Republic of Germany

Dr.-Ing. Peter Weiland

Bundesforschungsanstalt
für Landwirtschaft
Institut für Technologie
D-3300 Braunschweig
Federal Republic of Germany

Prof. Dr. Marko Zlokarnik

Bayer AG
In AP VT
D-5090 Leverkusen
Federal Republic of Germany

Contents

1 Fundamentals of Transport Phenomena

Chapter 1

Equations for Momentum, Heat and Mass Transport, and Mass Conversion 3

by *Heinz Brauer*

Chapter 2

Differential Equations for Velocity, Temperature and Concentration Fields 23

by *Heinz Brauer*

Chapter 3

Transport Processes in Newtonian Fluids Flowing Through Tubes 33

by *Heinz Brauer*

Chapter 4

Transport Processes in Non-Newtonian Fluids Flowing Through Tubes 49

by *Heinz Brauer*

Chapter 5

Transport Processes in Fluid Flow Parallel to Plates 61

by *Heinz Brauer*

Chapter 6

Transport Processes Through the Interface of Particles 77

by *Heinz Brauer*

Chapter 7

Transport Processes in Liquid Films 113

by *Heinz Brauer*

Chapter 8

Mass Transfer Hypotheses 143

by *Heinz Brauer*

Chapter 9

Analogy of Momentum, Heat, and Mass Transfer 153

by *Heinz Brauer*

Chapter 10

Gas Solubilities in Biomedia 159

by *Adrian Schumpe*

2 Fundamentals of Microbial Reaction Engineering

Chapter 11

General Strategy in Bioprocessing 173

by *Anton Moser*

Chapter 12

Rate Equations for Enzyme Kinetics 199

by *Anton Moser*

Chapter 13

Stoichiometry of Bioprocesses 227

by *Anton Moser*

Chapter 14

Kinetics of Batch Fermentations 243

by *Anton Moser*

Chapter 15

Continuous Cultivation Techniques 285

by *Anton Moser*

Chapter 16

Special Cultivation Techniques 311

by *Anton Moser*

Chapter 17

Reaction and Mass Transfer Interactions in Microbial Systems 349

by *Anton Moser*

Chapter 18

Mechanical Stress and Microbial Production 369

by *Herbert Märkl* and *Reinhold Bronnenmeier*

3 Bioreactors

Chapter 19
Stirred Vessel Reactors 395

by *Heinz Brauer*

Chapter 20
Bubble Column Reactors 445

by *Wolf-Dieter Deckwer*

Chapter 21
Biochemical Loop Reactors 465

by *Heinz Blenke*

Chapter 22
Biological Waste Water Treatment in a Reciprocating Jet Bioreactor 519

by *Heinz Brauer*

Chapter 23
Tower-Shaped Reactors for Aerobic Biological Waste Water Treatment 537

by *Marko Zlokarnik*

Chapter 24
Modelling and Scaling-up of Bioreactors 571

by *N. W. F. Kossen* and *N. M. G. Oosterhuis*

Chapter 25
Comparative Tests for Fermentations 607

by *Jürgen K. Lehmann*

4 Selected Unit Operations

Chapter 26
Preparation of Media 629

by *Alena Cejka*

Chapter 27
Sterilization 699

by *Karl Heinz Wallhäusser*

Chapter 28
Recovery Operations 725

by *Maria-Regina Kula*

5 Measurement and Control

Chapter 29

Measurement and Instrumentation 763

by *Ulfert Onken, Rainer Buchholz,* and *Wolfgang Sittig*

Chapter 30

Control and Optimization 787

by *Ulfert Onken* and *Peter Weiland*

Index 807

1 Fundamentals of Transport Phenomena

Chapter 1

Equations for Momentum, Heat and Mass Transport, and Mass Conversion

Heinz Brauer

Institut für Chemieingenieurtechnik
Technische Universität Berlin
Berlin (West), Germany

1.1   Introduction
1.2   Momentum Transport
1.2.1   Molecular Momentum Transport in Newtonian Fluids
1.2.2   Molecular Momentum Transport in Non-Newtonian Fluids
1.2.2.1 Classification of Non-Newtonian Fluids
1.2.2.2 Momentum Transport in Viscous Non-Newtonian Fluids
1.2.3   Turbulent Momentum Transport
1.3   Heat Transport
1.3.1   Molecular Heat Transport
1.3.2   Turbulent Heat Transport
1.3.3   Ratio of Momentum and Heat Transport
1.4   Mass Transport
1.4.1   Molecular Mass Transport
1.4.1.1 Equimolar Molecular Mass Transport
1.4.1.2 Non-equimolar Molecular Mass Transport
1.4.2   Turbulent Mass Transport
1.4.3   Ratio of Momentum and Mass Transport
1.5   Convective Transport
1.5.1   Convective Transport of Heat and Mass
1.5.1.1 General Description of Convective Transport
1.5.1.2 Definition of Convective Transport Coefficients
1.5.1.3 Equations for Convective Heat and Mass Transport
1.5.2   Convective Transport of Momentum
1.6   Biochemical Mass Conversion
1.6.1   Reaction Rate Equation
1.6.2   Integration of Reaction and Transport Process
1.7   References

List of Symbols

Symbol	Units	Description
A	m²	area
a	m²/s	molecular diffusivity of heat
c_A	kmol/m³	partial mol density of component A
c_B	kmol/m³	partial mol density of component B
c_p	kJ/(kg K)	specific heat capacity of fluid
D	m²/s	diffusion coefficient
D_{AB}	m²/s	diffusion coefficient of component A in a mixture with component B
D_{BA}	m²/s	diffusion coefficient of component B in a mixture with component A
d	m	tube diameter
F	m²	cross sectional area
K	$\tau_m/(-dw/dy)^n$	Ostwald factor
M_A	kg	mass of component A
$\dot{M}_A$	kg/s	mass flux of component A
$\dot{m}_{Am}$	kg/(s m²)	molecular mass flux density of component A
$\dot{m}_{Bm}$	kg/(s m²)	molecular mass flux density of component B
$\dot{m}_{At}$	kg/(s m²)	turbulent mass flux density of component A
$\dot{m}_{Bt}$	kg/(s m²)	turbulent mass flux density of component B
$\dot{n}_{Am}$	kmol/(s m²)	molecular mol flux density of component A
$\dot{n}_{Bm}$	kmol/(s m²)	molecular mol flux density of component B
$\dot{n}_{At}$	kmol/(s m²)	turbulent mol flux density of component A
$\dot{n}_{Bt}$	kmol/(s m²)	turbulent mol flux density of component B
p	N/m²	total pressure
p_A	N/m²	partial pressure of component A
p_B	N/m²	partial pressure of component B
Q	kJ	heat
$\dot{Q}$	kJ/s	heat flux
$\dot{q}_m$	kJ/(s m²)	molecular heat flux density
$\dot{q}_t$	kJ/(s m²)	turbulent heat flux density
R	m	tube radius
R	kJ/(kmol K)	universal gas constant
$\dot{r}_A$	kmol/(s m²)	reaction flux density, reaction rate
T	K	temperature
t	s	time
w	m/s	local velocity
$\bar{w}$	m/s	mean velocity over cross sectional area
w_s	m/s	mean velocity of Stefan flow
w'_x	m/s	fluctuation velocity in x-direction
w'_y	m/s	fluctuation velocity in y-direction
x	m	length coordinate
y	m	length coordinate
z	m	length coordinate
α_z	kJ/(s m² K)	local coefficient of heat transfer
β_z	m/s	local coefficient of mass transfer
ε_m	m²/s	turbulent diffusivity of mass
ε_q	m²/s	turbulent diffusivity of heat
ε_τ	m²/s	turbulent diffusivity of momentum
η	kg/(m s)	dynamic viscosity of fluid
λ	kJ/(m s K)	heat conductivity
μ_A	kmol/kg	mol mass of component A
μ_B	kmol/kg	mol mass of component B
ν	m²/s	kinematic viscosity of fluid
ϱ	kg/m³	density of fluid
ϱ_A	kg/m³	partial density of component A
ϱ_B	kg/m³	partial density of component B
τ_m	N/m²	molecular shear stress
τ_t	N/m²	turbulent shear stress
τ_0	N/m²	yield stress of Bingham fluids
$Da \equiv \dfrac{\dot{r}_{A,max}/k}{D/R}$		Damköhler number
$Nu_z \equiv \dfrac{\alpha_z d}{\lambda}$		local Nusselt number
$Pr \equiv \dfrac{\eta c_p}{\lambda}$		"molecular" Prandtl number
$Pr_t \equiv \varepsilon_\tau/\varepsilon_q$		"turbulent" Prandtl number
$Re \equiv \dfrac{\bar{w} d}{\nu}$		Reynolds number for tube flow
$Re^+ \equiv \dfrac{\sqrt{\tau_w/\varrho}\, d}{\nu}$		shear Reynolds number
$r^* \equiv r/R$		radial coordinate

$Sc \equiv \nu/D$ "molecular" Schmidt number

$Sc_t \equiv \varepsilon_\tau/\varepsilon_m$ "turbulent" Schmidt number

$Sh_z \equiv \dfrac{\beta_z d}{D}$ local Sherwood number

$\psi_z \equiv \dfrac{dp/dz}{\varrho \bar{w}^2/2} d$ local friction coefficient

1.1 Introduction

Microbial technical processes are carried out in bioreactors. The bioreactor ist the containment for an almost unlimited number of microorganisms, each microorganism being a microreactor actually accomplishing the desired mass conversion. The contribution of an individual microorganism to the mass conversion is extremely small. Large-scale mass conversion technologies require a countless number of microorganisms, which must be contained within the smallest possible space.

Each one of the exceedingly large number of microorganisms has to be supplied with various nutrients, while at the same time conversion products have to be removed. Supply and removal are transport processes. For the prevailing conditions mass transport and mass conversion are closely linked with transport of heat and transport of momentum.

Transport of momentum, heat, and mass in bioreactors are processes with much greater consequences than in conventional chemical reactors. Insufficient transport of reactants results in substantially decreased production rates when non-microbial processes are considered. In the case of microbial processes the consequence of an inefficient mass transport process may be not only a reduced production rate but may result in an altogether different type of conversion process and conversion product.

This kind of response of an organism to malfunctions or irregularities of transport processes is one of the most fascinating aspects of the "individuality" of microorganisms. Unfortunately we do not yet know enough about the processes involved.

The less we know about the individual response of a microorganism to transport deficiencies the more we have to make sure that such deficiencies are avoided. A thorough understanding of transport processes will be helpful in the design and operation of efficient bioreactors.

Transport of momentum, heat, and mass may be achieved by conductive and convective processes. Conductive processes are due to molecular and turbulent motions, while convective transport is related to fluid motion. The basic equations for momentum, heat, and mass transport are empirically derived. There are no strictly theoretical equations available. The same is true for mass conversion processes. All equations describing chemical or biochemical conversion processes are founded upon empirical data.

The equations which will be given for transport and conversion processes, represent physical or physico-chemical laws governing the behavior of solids and fluids; they include parameters specific for the considered processes and for the particular equipment, in which the processes are carried out. This chapter is therefore devoted to a discussion of the available empirical equations describing transport of momentum, heat, and mass and the conversion of mass.

1.2 Momentum Transport

Transport of momentum is due to molecular, turbulent, and convective motions. In this order the processes of momentum transport will be discussed. Molecular and turbulent momentum transport is the subject of this section. Convective transport will be discussed in a separate section. Momentum transport occurs only in fluids. These fluids consist of two large groups: Newtonian and non-Newtonian fluids.

1.2.1 Molecular Momentum Transport in Newtonian Fluids

The majority of the conventional fluids, especially inorganic gases and liquids as well as organic gases, but also a great number of organic liquids are so-called Newtonian fluids. In these fluids molecular momentum transport is described by an empirical equation presented by Newton:

$$\tau_m = -\eta \frac{dw}{dy}. \quad (1.1)$$

This equation relates the molecular momentum flux density τ_m, which is also known as shear stress, with the velocity gradient dw/dy, which is also known as shear rate.

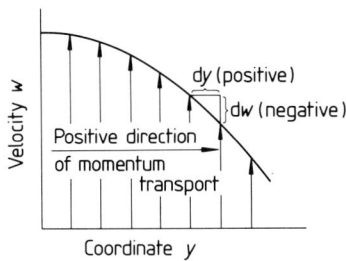

Figure 1.1. Explanation of molecular momentum transport in fluids.

The proportionality factor η is the molecular transport coefficient for momentum, better known as the dynamic viscosity of the fluid; η depends on pressure and temperature of the fluid. For all Newtonian fluids the viscosity is independent of shear stress τ_m or shear rate dw/dy. According to Eq. (1.1) even the smallest shear stress applied to Newtonian fluids will cause a fluid motion. In the absence of a gradient of the velocity $(dw/dy=0)$, there is no shear stress. Eq. (1.1) gives only one component of the stress tensor.

In Fig. 1.1 the local velocity w is plotted over the local coordinate y. According to Fig. 1.1 and Eq. (1.1) the momentum flux density τ_m is assumed to be positive in the direction of decreasing velocity.

The Newtonian equation for molecular momentum transport may also be written in the following way:

$$\tau_m = -\nu \frac{d(w\varrho)}{dy}. \quad (1.2)$$

In this equation $\nu = \eta/\varrho$ denotes the kinematic viscosity of the fluid with ϱ as density, and $w\varrho$ the momentum per unit volume of the fluid. Eq. (1.2) presents the proportionality between momentum flux density

Table 1.1. Kinematic and Dynamic Viscosity of Selected Fluids (pressure 1 bar; temperature 300 K)

	ν in m²/s	η in kg/(m s)
Mercury	$0.0112 \cdot 10^{-5}$	$151.7 \cdot 10^{-5}$
Water	$0.0681 \cdot 10^{-5}$	$85.7 \cdot 10^{-5}$
Air	$1.58 \cdot 10^{-5}$	$1.86 \cdot 10^{-5}$
Hydrogen	$11.25 \cdot 10^{-5}$	$0.89 \cdot 10^{-5}$

τ_m and gradient of momentum per unit volume. In Table 1.1 values for the dynamic and kinematic viscosity are given for a few selected fluids.

There is no absolute proof for the validity of Newton's law. Applications of this law in calculations for velocity fields are, however, in excellent agreement with extrapolated experimental results.

1.2.2 Molecular Momentum Transport in Non-Newtonian Fluids

1.2.2.1 Classification of Non-Newtonian Fluids

All fluids which do not obey Newton's law for the shear stress τ_m, given by Eq. (1.1), will be classified as non-Newtonian fluids. For this group of fluids the viscosity depends not only on temperature and pressure but also on the shear stress τ_m, time, elasticity, and other parameters.

Typical non-Newtonian fluids are melts and solutions of, e.g., polymers, thick suspensions, paints, many fermentation fluids, i.e., primarily high molecular weight fluids. Rheology, the science of properties and behavior of flowing substances, has not yet been successful in presenting a shear stress relation for all important groups of non-Newtonian fluids. Only for certain groups of non-Newtonians, such relations have been developed [1.1] to [1.5]. From a scientific point of view such shear stress relations are based on physical insight into the flow behavior of these fluids. From an engineering point of view these equations are not yet very helpful in describing fluid flow in technical equipment. For this particular problem area engineers are still forced to refer to extremely simple empirical shear stress relations with a rather narrow range of application.

There are at least two large groups of non-Newtonian fluids:

1. viscous fluids,
2. elastic fluids.

For the group of viscous non-Newtonian fluids the viscosity depends only on the shear stress, but is independent of time:

$$\eta = f_1(\tau_m) \quad \text{or} \quad \eta = f_2(dw/dy). \tag{1.3}$$

For elastic non-Newtonian fluids the viscosity is a function of time t:

$$\eta = f_3(t). \tag{1.4}$$

This implies, that elastic fluids are sensitive to distortion from an experienced or preferred shape. These fluids remember for a certain stretch of time the shape they previously possessed. When the viscosity increases with time, the behavior of the fluids is rheopectic. Thixotropic behavior implies decreasing viscosity with time.

A satisfactory mathematical description of the flow of non-Newtonian fluids in technical equipment is restricted to the groups of viscous fluids. For this group of fluids momentum transfer will be briefly discussed.

1.2.2.2 Momentum Transport in Viscous Non-Newtonian Fluids

According to the equations describing flow behavior there are three groups of viscous non-Newtonian fluids:

1. Pseudoplastic fluids,
2. dilatant fluids, and
3. Bingham fluids.

The discussion starts with the first two groups of fluids. OSTWALD and DE WAELE presented the following equation for momentum transport, known as the power law [1.6], [1.7]:

$$\tau_m = K\left(-\frac{dw}{dy}\right)^n. \tag{1.5}$$

K is the Ostwald factor, and n is the fluid index. When n=1 the Ostwald factor K is identical with the fluid viscosity η, so that Eq. (1.5) is reduced to Eq. (1.1), which has been presented for momentum transport in Newtonian fluids.

For n>1 Eq. (1.5) describes the behavior of dilatant fluids, and for n<1 of pseudoplastic fluids. In Fig. 1.2 the shear stress/shear rate relationship is given qualitatively for the three cases discussed. For Newtonian fluids with n=1 a linear relationship exists, so that the viscosity is independent of shear stress or shear rate. For dilatant and pseudoplastic fluids a non-linear relationship exists. For dilatant fluids the viscosity is reduced with increasing shear stress, while for pseudoplastic fluids the viscosity increases with shear stress.

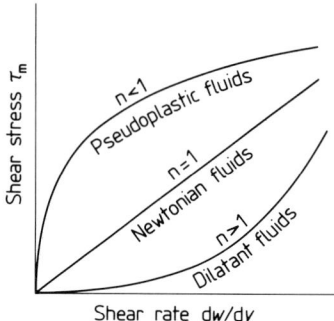

Figure 1.2. Relationship between shear stress and shear rate for Newtonian and non-Newtonian fluids.

According to Fig. 1.2 an infinitesimally small shear stress will set a dilatant fluid in motion, while for pseudoplastic fluids an infinitely large shear stress is required. This rather unrealistic behavior is expressed by the following relation:

$$\frac{d\tau_m}{d(dw/dy)} = nK\left(-\frac{dw}{dy}\right)^{n-1}. \quad (1.6)$$

$n > 1$: $-dw/dy \to 0$ $d\tau_m/d(dw/dy) \to 0$
$n < 1$: $-dw/dy \to 0$ $d\tau_m/d(dw/dy) \to \infty$

In reality the flow behavior of viscous non-Newtonian fluids asymptotically approaches that of Newtonian fluids in low and high shear stress regions. This behavior is discussed in Fig. 1.3. Curve a gives the shear stress/shear rate relationship for a pseudoplastic fluid. When this curve approaches $\tau_m = 0$, it coincides at point 1 with curve b, which represents Newtonian flow behavior in the low shear stress region. Experimental data prove that in the high shear stress region non-Newtonian flow behavior will change into Newtonian behavior. Curve a therefore ends at point 2. For higher shear stress values the flow behavior follows curve c, approaching curve d, which represents Newtonian flow behavior in the high shear stress region.

Application of the power law is limited to the range between points 1 and 2 as given in Fig. 1.3. General expressions for these limits cannot be stated. Care should be taken in the use of Eq. (1.5).

From a physical point of view application of the power law is limited to linear flows. BIRD [1.8] has proven that the power law, expressed by:

$$\tau_m = -K\left[\left(\frac{dw}{dy}\right)^2\right]^{\frac{n-1}{2}} \frac{dw}{dy}, \quad (1.7)$$

is one component of the stress tensor for Ostwald fluids. Application of this power law on other than linear fluid flows requires extreme care.

For non-Newtonian fluids a viscosity η_{n-N} can be defined as the ratio of shear stress and shear rate. From Eq. (1.7) one obtains:

$$\frac{\tau_m}{dw/dy} \equiv \eta_{n-N} = K\left[\left(\frac{dw}{dy}\right)^2\right]^{\frac{n-1}{2}}. \quad (1.8)$$

This equation shows clearly that the viscosity of Ostwald fluids is a function of the

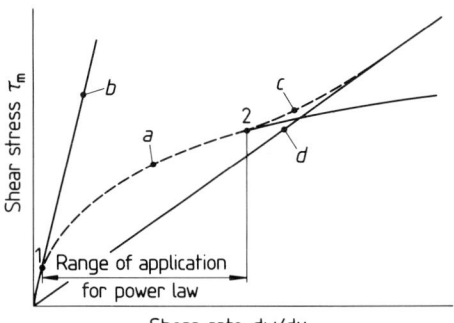

Figure 1.3. Shear stress/shear rate relationship for a pseudoplastic non-Newtonian fluid according to curve a; limiting Newtonian conditions given by curves b and d in the low and high shear stress region.

1.2 Momentum Transport

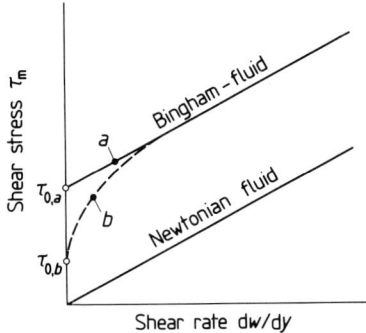

Figure 1.4. Flow behavior of a Bingham fluid.

velocity gradient. The Ostwald factor K is, however, independent of the velocity gradient.

The third group of viscous non-Newtonian fluids are called Bingham fluids [1.9]. The flow behavior of these fluids is qualitatively given by curve a in Fig. 1.4 and quantitatively by:

$$\tau_m = \tau_0 + \eta_B \left(-\frac{dw}{dy} \right). \tag{1.9}$$

The characteristic property of a Bingham fluid is the yield stress τ_0 at $dw/dy = 0$. The stress that sets a Bingham fluid into motion must exceed the yield stress. Bingham fluids do not always show a linear relationship between τ_m and dw/dy, while they do exert a non-linear behavior according to curve b in the low shear stress region.

1.2.3 Turbulent Momentum Transport

In turbulent flows molecular momentum transport is enhanced by "turbulent" momentum transport. This mode of transport is due to velocity fluctuations which are observed in the direction of the three coordinates. HINZE [1.10] therefore defines turbulence as follows: "Turbulent fluid motion is an irregular condition of flow in which the various quantities show a random variation with time and space coordinates, so that statistically distinct average values can be discerned." Turbulence can be generated by friction forces at solid walls (flow through conduits, flow past plates and bodies) or by the flow of layers of fluids with different velocities past or over one another, as stated by VON KÁRMÁN [1.11].

Turbulence will be explained here only in a very simple way. Velocity fluctuations are of statistical nature. Each component of a velocity vector has three fluctuation velocities: w'_x, w'_y, and w'_z. In Fig. 1.5 fluctuation velocity w'_y is given as a function of the time coordinate t. The time average of the fluctuation velocity is by definition zero. Fluctuation results in momentum transfer, which is explained in Fig. 1.6 as an example, between streamlines 1 and 2. The velocity of streamline 1 at a fixed point in space and time t is $w_x + w'_x$, with w_x as the time mean value of the velocity in x-direction

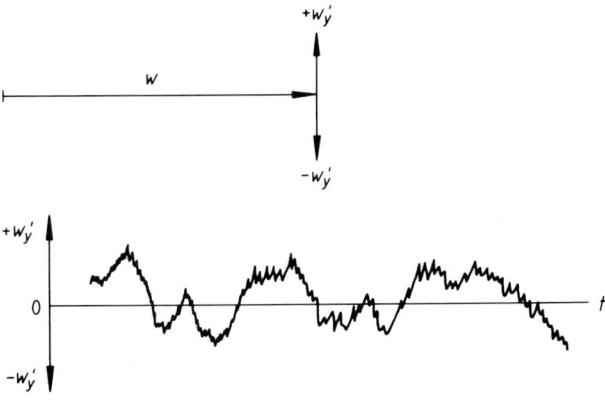

Figure 1.5. Explanation of random turbulent velocity fluctuations.

and w'_x the fluctuation velocity in the same direction. For streamline 2 the velocity is assumed to be w_x, so that $w'_x = 0$. For this condition turbulent momentum transfer is

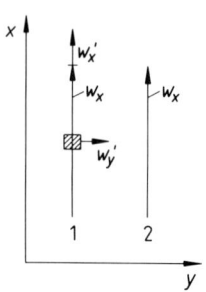

Figure 1.6. Turbulent momentum transport from streamline 1 to streamline 2.

achieved by the motion of a small fluid element from streamline 1 to streamline 2. The size of the turbulence element depends on the local conditions of turbulence. The mass transported per unit of time and area is given by $\varrho w'_y$, and the turbulent momentum per time unit transported from streamline 1 to 2 is given by:

$$\tau'_{xy} = -\varrho w'_y w'_x. \tag{1.10}$$

The average time value for the turbulent shear stress is defined as follows:

$$\tau_{xy} = -\varrho \overline{w'_y w'_x}. \tag{1.11}$$

To describe turbulent momentum transport BOUSSINESQ [1.12] introduced the simple equation:

$$\tau_t = -\varrho \varepsilon_\tau \frac{dw}{dy}, \tag{1.12}$$

which is from a mathematical point of view an analogous equation to that for molecular momentum transport as given by Eq. (1.1). The product $\varrho \varepsilon_\tau$ is the coefficient of turbulent momentum transfer, the analogous coefficient to η, which is the coefficient for molecular momentum transport. The coefficient ε_τ has the same dimension as ν, that is m^2/s.

In turbulent flow fields momentum transport is due to molecular and turbulent motion. The equation for momentum transport is therefore given by:

$$\tau = \tau_m + \tau_t = -\eta(1 + \varepsilon_\tau/\nu)\frac{dw}{dy}. \tag{1.13}$$

The ratio ε_τ/ν is a dimensionless quantity. It is a function of the properties of the turbulent flow field and local coordinates.

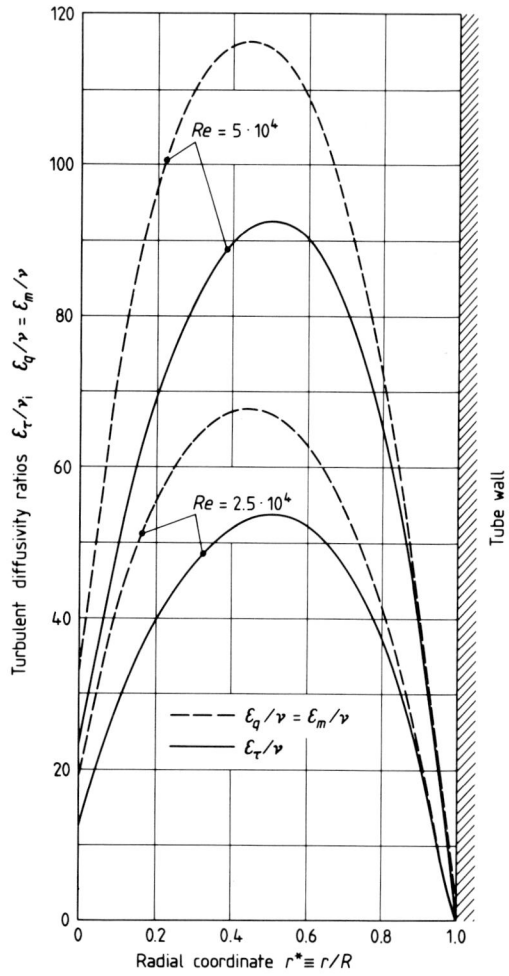

Figure 1.7. Ratio ε_τ/ν of turbulent to molecular momentum transport coefficients; ε_q/ν and τ_m/ν are presented as well as for turbulent pipe flow over the local radius r^* for two values of the Reynolds number Re.

For turbulent flow of a Newtonian fluid in circular pipes the following equation for ε_τ/v has been derived [1.1]:

$$\frac{\varepsilon_\tau}{v} = 0.5\left[\sqrt{1 + r^{*}\{2Re^{+}(0.14 - 0.08r^{*2} - 0.06r^{*4})[1 - \exp\{-Re^{+}(1-r^*)/26\}]\}^2} - 1\right] \quad (1.14)$$

with

$r^* \equiv r/R$ local radial coordinate, (1.15)

$Re^+ \equiv \dfrac{\sqrt{\tau_w/\varrho}\,R}{v}$ shear Reynolds number. (1.16)

In these equations R is tube radius, ϱ fluid density, v kinematic viscosity, and τ_w shear stress at the tube wall.

In Fig. 1.7 ε_τ/v is presented as a function of the radial coordinate r^* for two values of the Reynolds number Re for tube flow:

$$Re \equiv \frac{\bar{w}d}{v} \quad (1.17)$$

with $\bar{w}$ mean fluid velocity and d tube diameter. For tube flow turbulence starts developing at $Re = 2320$. Thus, the following relationship exists between Re and Re^+:

$$Re^+ = 0.1\,Re^{7/8}. \quad (1.18)$$

The maxima for ε_τ/v are 54 for $Re = 2.5 \cdot 10^4$ and 93 for $Re = 5 \cdot 10^4$. In these cases molecular momentum transfer is negligibly small compared with turbulent momentum transfer. The maxima occur at a value of the local radius of about $r^* = 0.5$. Towards the tube axis the ratio ε_τ/v decreases according to Eq. (1.14) and Fig. 1.7. But this part of the curve is not correct. It may be assumed that for $r^* \leq 0.5$ the maximum value of ε_τ/v applies.

Close to the tube wall ε_τ/v is proportional to the 4th power of the distance $1 - r^*$ from the wall. From Eq. (1.14) follows that:

$$\left(\frac{\varepsilon_\tau}{v}\right)_{r^*\to 1} = 2.37 \cdot 10^{-4}[Re^+(1-r^*)]^4. \quad (1.19)$$

At the tube wall ε_τ/v equals zero. This means, that very close to the wall, momentum transport in the fluid is due only to irregular molecular motion. The shear stress τ_w at the wall is given by:

$$\tau_w = \tau_m. \quad (1.20)$$

Turbulence in non-Newtonian fluid flow seems to be generally of minor importance. According to Fig. 1.3 it is to be expected, that turbulence will reduce and simplify non-Newtonian to Newtonian fluid flow.

1.3 Heat Transport

Heat transport is due to random molecular and turbulent motions. These modes of transport will be discussed in this section. Heat transport by regular or directed fluid motion, called convective heat transport, will be treated in a separate section.

1.3.1 Molecular Heat Transport

The energy is designated by the symbol Q, the Joule (J) being its unit. The energy flux $\dot{Q}$ is the energy Q transported per time unit t:

$$\dot{Q} = \frac{dQ}{dt} \quad \text{(in J/s)}. \quad (1.21)$$

By dividing $\dot{Q}$ by the area A, through which $\dot{Q}$ is transported, the energy flux density is obtained:

$$\dot{q} = \frac{\dot{Q}}{A} = \frac{d(Q/A)}{dt} \quad \text{(in J/s m}^2\text{)}. \quad (1.22)$$

Molecular heat transport occurs in solid bodies and in fluids at rest or in motion. FOURIER postulated that molecular heat transport is proportional to the temperature gradient:

$$\dot{q}_m = -\lambda \frac{dT}{dy}, \qquad (1.23)$$

T being temperature, y coordinate in the direction of considered heat transport, and λ heat conductivity or more precisely: the coefficient of molecular heat transport. The unit of λ is given by:

$$1\frac{J/s}{m\,K} = 1\frac{W}{m\,K} = 1\frac{N\,m/s}{m\,K}.$$

The heat conductivity is a measure for the quantity of heat transported in the direction of the negative temperature gradient.

In Fig. 1.8 the local temperature T is plotted over the coordinate y. The heat flux density $\dot{q}_m$ is assumed to be positive in the direction of decreasing temperature, i.e., the negative temperature gradient.

Fourier's law can also be written in the form:

$$\dot{q}_m = -a\frac{d(T\varrho c_p)}{dy}, \qquad (1.24)$$

with ϱ as density, c_p specific heat capacity of fluid, and a being the molecular heat diffusivity. This diffusivity is characteristic for the velocity with which heat is transported in the considered material. As $T\varrho c_p$ is the energy per unit volume, the gradient $d(T\varrho c_p)/dy$ gives the change of energy per unit volume in the direction of y.

For a few materials a and λ are summarized in Table 1.2.

Table 1.2. Molecular Heat Conductivity λ and Heat Diffusivity a for Various Materials at $T = 300$ K

	a in m²/s	λ in J/(m s K)
Water	$0.14 \cdot 10^{-6}$	0.61
Glass	$0.83 \cdot 10^{-6}$	1.40
Mercury	$4.33 \cdot 10^{-6}$	8.14
Steel	$14.10 \cdot 10^{-6}$	52.00
Air	$22.20 \cdot 10^{-6}$	0.026
Hydrogen	$166.73 \cdot 10^{-6}$	0.187

There is no strict evidence available for the validity of Fourier's law. Application of this law leads to results which are in excellent agreement with extrapolated experimental findings.

1.3.2 Turbulent Heat Transport

In turbulent flows heat transport due to random molecular motion is enhanced by turbulent motions. This additional mode of transport has been explained in some detail in the section on momentum transfer. The equation for turbulent heat transport is given by:

$$\dot{q}_t = -\varrho c_p \varepsilon_q \frac{dT}{dy}, \qquad (1.25)$$

with ε_q as the turbulent heat diffusivity, having the same unit as the turbulent momentum diffusivity, namely m²/s.

The sum of molecular and turbulent heat transport is described by the equation:

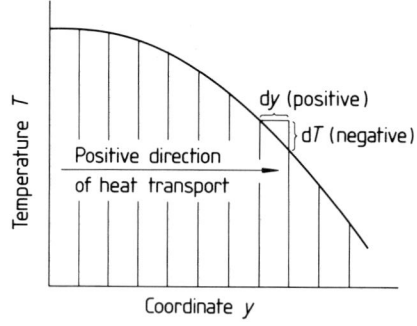

Figure 1.8. Explanation of molecular heat transport.

$$\dot{q} = \dot{q}_m + \dot{q}_t = -\lambda(1 + \varepsilon_q/a)\frac{dT}{dy}. \qquad (1.26)$$

For the ratio ε_q/a of turbulent and molecular heat diffusivity in turbulent tube flow LUDWIG [1.13] has been able to set up an equation based on results of experimental investigations:

$$\frac{\varepsilon_q}{a} = Pr(1.5 - 0.5 r^*) \frac{\varepsilon_\tau}{v}. \qquad (1.27)$$

In this equation r^* is the local radial coordinate of the tube, and Pr the Prandtl number, defined by:

$$Pr \equiv \frac{v}{a}. \qquad (1.28)$$

As ε_τ/v is well known, for instance by Eq. (1.14), ε_q/a can also be established. Eq. (1.27) may be rewritten as follows:

$$\frac{\varepsilon_q}{v} = (1.5 - 0.5 r^*) \frac{\varepsilon_\tau}{v}. \qquad (1.29)$$

In Fig. 1.7 ε_q/v is presented for two values of the Reynolds number. Close to the surface of a wall ε_q/v becomes zero. At the wall heat transport is accomplished only by molecular motion:

$$\dot{q}_w = \dot{q}_m. \qquad (1.30)$$

From the surface of a solid to a fluid the heat transport is only due to molecular conduction.

1.3.3 Ratio of Momentum and Heat Transport

In fluid flow heat transport is always combined with momentum transport. In laminar flow fields the ratio of momentum and heat flux densities is given by:

$$\frac{\tau_m}{\dot{q}_m} = Pr \frac{d(w\varrho)/dy}{d(T\varrho c_p)/dy}. \qquad (1.31)$$

This equation shows that the molecular Prandtl number Pr relates the velocity field with the temperature field in laminar flow. For turbulent transport of momentum and heat one obtains:

$$\frac{\tau_t}{\dot{q}_t} = Pr_t \frac{d(w\varrho)/dy}{d(T\varrho c_p)/dy} \qquad (1.32)$$

with

$$Pr_t \equiv \frac{\varepsilon_\tau}{\varepsilon_q} \qquad (1.33)$$

as the turbulent Prandtl number. As a first approximation it may be assumed that $Pr_t = 1$.

1.4 Mass Transport

In a solid body mass transport is only due to irregular molecular motion. In a general flow field, however, mass transport is achieved by molecular and turbulent random motions. These two cases will be discussed in this section. Convective mass transport is treated in a special section. For the sake of simplicity all discussions will be restricted to mass transfer in binary mixtures, due only to a concentration gradient. Other modes of mass transport will not be taken into consideration.

1.4.1 Molecular Mass Transport

A binary mixture consists of components A and B. It will in general suffice to consider only the transport of component A. The mass of this component is designated by M_A and the unit kg. The mass transported per time unit is given by:

$$\dot{M}_A = \frac{dM_A}{dt} \quad \text{(in kg/s)}, \qquad (1.34)$$

and the mass flux density by:

$$\dot{m}_A = \frac{d(M_A/A)}{dt} \quad \text{(in kg/m}^2\text{s)}. \tag{1.35}$$

Mass transfer occurs under conditions of equimolarity and non-equimolarity. Both cases will be discussed.

Table 1.3. Diffusion Coefficients of CO_2 in Water and in Air at a Temperature of 300 K

	D in m^2/s
Water	$1.96 \cdot 10^{-9}$
Air	$1.66 \cdot 10^{-5}$

1.4.1.1 Equimolar Molecular Mass Transport

Equimolar mass transport of component A in mixture with component B is described by Fick's law:

$$\dot{m}_{Am} = -D_{AB} \frac{d\varrho_A}{dy}, \tag{1.36}$$

with D_{AB} the coefficient of diffusion of component A in a mixture with B, and ϱ_A the partial density of component A. The unit of the diffusion coefficient is m^2/s. Fig. 1.9 indicates that the mass flux density is

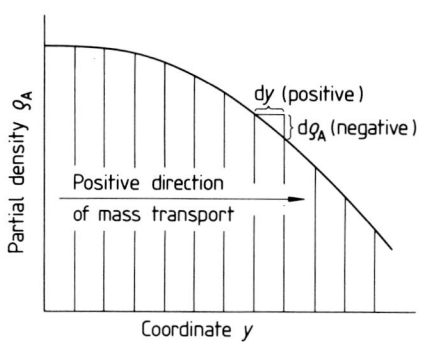

Figure 1.9. Explanation of molecular mass transport.

positive in the direction of a negative concentration gradient. In Table 1.3 two diffusion coefficients are given to indicate the value of such coefficients in liquids and in gases.
Diffusion coefficients in liquids are three to four decades smaller than in gases.

The countercurrent mass flux density for component B is given by:

$$\dot{m}_{Bm} = D_{BA} \frac{d\varrho_B}{dy}, \tag{1.37}$$

with D_{BA} as the coefficient of diffusion of component B in a mixture with A, and ϱ_B the partial density of component B. Introducing the molar densities:

$$c_A = \varrho_A/\mu_A, \quad \text{and} \tag{1.38}$$

$$c_B = \varrho_B/\mu_B \tag{1.39}$$

the mol flux densities are obtained:

$$\dot{n}_{Am} = -D_{AB} \frac{dc_A}{dy}, \quad \text{and} \tag{1.40}$$

$$\dot{n}_{Bm} = -D_{BA} \frac{dc_B}{dy}. \tag{1.41}$$

μ_A und μ_B designate the mol mass of components A and B. Replacing the partial densities for components A and B by the partial pressures p_A and p_B the diffusion equations may be written as follows:

$$\dot{m}_{Am} = -\frac{D_{AB}}{TR/\mu_A} \frac{dp_A}{dy}, \tag{1.42}$$

$$\dot{m}_{Bm} = -\frac{D_{BA}}{TR/\mu_B} \frac{dp_B}{dy}, \tag{1.43}$$

$$\dot{n}_{Am} = -\frac{D_{AB}}{TR} \frac{dp_A}{dy}, \tag{1.44}$$

$$\dot{n}_{Bm} = -\frac{D_{BA}}{TR} \frac{dp_B}{dy}. \tag{1.45}$$

An application of Dalton's law:

$$p = p_A + p_B, \qquad (1.46)$$

in which p is the total pressure of the system, is possible after differentiation of:

$$\frac{dp_A}{dy} + \frac{dp_B}{dy} = 0. \qquad (1.47)$$

With this result and from $\dot{m}_{Am} = -\dot{m}_{Bm}$ or $\dot{n}_{Am} = -\dot{n}_{Bm}$ the identity of the diffusion coefficients can be deduced as:

$$D_{AB} = D_{BA} = D. \qquad (1.48)$$

This identity is a characteristic property of equimolar diffusion.

1.4.1.2 Non-equimolar Molecular Mass Transport

The extreme case of non-equimolar molecular mass transport is uni-directional diffusion, called Stefan diffusion. For this case the transport equation will be derived.

The physical conditions of Stefan diffusion are indicated in Fig. 1.10. Due to the evaporating liquid contained in the cylindrical glass vessel a mass flux density $\dot{m}_A$ originates. For equimolar diffusion there ought to be a counterdirectional diffusion flux density $\dot{m}_B$. As the liquid surface is impenetrable for component B there must be a process that compensates the flux density $\dot{m}_B$. This compensating flow is the Stefan flow which encompasses the flow not only of component B but also of component A, that is the whole mixture of A and B. This Stefan flow density is given by:

$$\dot{m}_S = \dot{m}_{AS} + \dot{m}_{BS} = w_S \frac{p_A}{TR/\mu_A} + w_S \frac{p_B}{TR/\mu_B}, \qquad (1.49)$$

with w_S as the mean velocity of the Stefan flow, or Stefan velocity. The velocity is obtained by equating $\dot{m}_{Bm}$ and $\dot{m}_{BS}$:

$$w_S = \frac{D}{p_B} \frac{dp_B}{dy} = -\frac{D}{p_B} \frac{dp_A}{dy}. \qquad (1.50)$$

The Stefan diffusion flux density $\dot{m}_{AmS}$ is the sum of $\dot{m}_{Am}$ and $\dot{m}_{AS}$. Application of the appropriate equations leads to:

$$\dot{m}_{AmS} = -\frac{D}{TR/\mu_A} \frac{p}{p - p_A} \frac{dp_A}{dy}. \qquad (1.51)$$

Comparison with Eq. (1.42) reveals the difference between Stefan diffusion and Fick's diffusion as being the pressure factor $p/(p - p_A)$, which is always equal to or larger than one.

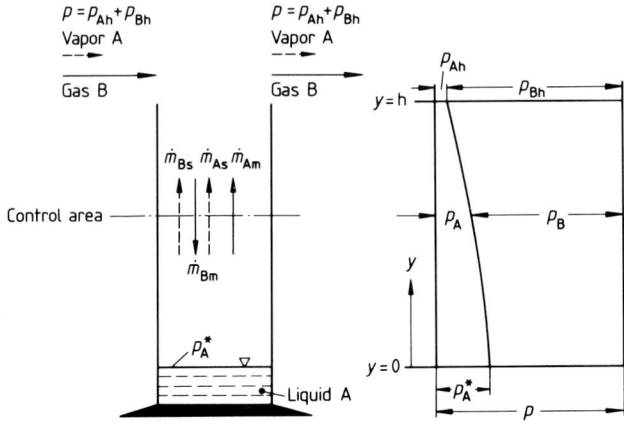

Figure 1.10. Explanation of non-equimolar mass transport; Stefan diffusion

1.4.2 Turbulent Mass Transport

In turbulent flow fields a mass transport is accomplished by random molecular and turbulent motions. Turbulent mass transport is described by the equations:

$$\dot{m}_{At} = -\varepsilon_m \frac{d\varrho_A}{dy}, \tag{1.52}$$

$$\dot{m}_{AtS} = \frac{\varepsilon_m}{TR/\mu_A} \frac{p}{p-p_A} \frac{dp_A}{dy}, \tag{1.53}$$

where ε_m is the turbulent mass diffusivity with the unit m²/s. The sum of molecular and turbulent mass transport is given by:

$$\dot{m}_A = \dot{m}_{Am} + \dot{m}_{At} = -D(1+\varepsilon_m/D)\frac{d\varrho_A}{dy}, \tag{1.54}$$

$$\dot{m}_A = \dot{m}_{AmS} + \dot{m}_{AtS} =$$
$$= -\frac{D}{TR/\mu_A}\left(1+\frac{\varepsilon_m}{D}\right)\frac{p}{p-p_A}\frac{dp_A}{dy}. \tag{1.55}$$

For ε_m the following equation applies:

$$\frac{\varepsilon_m}{\nu} = (1.5 - 0.5\,r^*)\frac{\varepsilon_\tau}{\nu}. \tag{1.56}$$

Comparison with Eq. (1.29) reveals: $\varepsilon_q/\nu = \varepsilon_m/\nu$. Fig. 1.7 gives the ratios ε_q/ν and ε_m/ν for two values of the Reynolds number.

Close to the surface of a solid body turbulent mass transport decays, so that transport occurs only by molecular motion:

$$\dot{m}_{Aw} = \dot{m}_{Am}, \tag{1.57}$$

$$\dot{m}_{AwS} = \dot{m}_{AmS}. \tag{1.58}$$

1.4.3 Ratio of Momentum and Mass Transport

In laminar flow fields the ratio of momentum and mass transfer is given by:

$$\frac{\tau_m}{\dot{m}_{Am}} = Sc\,\frac{d(w\varrho)/dy}{d\varrho_A/dy}. \tag{1.59}$$

The proportionality factor is the Schmidt number, defined by:

$$Sc \equiv \frac{\nu}{D}, \tag{1.60}$$

that relates momentum with mass transport, or the velocity field to the concentration field.

For turbulent flow fields the following equation is obtained:

$$\frac{\tau_t}{\dot{m}_{At}} = Sc_t\,\frac{d(w\varrho)/dy}{d\varrho_A/dy}. \tag{1.61}$$

The turbulent Schmidt number is defined by:

$$Sc_t \equiv \frac{\varepsilon_\tau}{\varepsilon_m}. \tag{1.62}$$

As a first approximation it may be assumed that $Sc_t = 1$.

1.5 Convective Transport

Convective transport or simply "convection" is the mode of transport of momentum, heat, and mass that occurs in a flow field. Convective transport includes molecular and turbulent transport but is decisively influenced by the participation of the flowing fluid in momentum, heat, and mass transport. Because of the similarity of heat and mass transport these two processes will be discussed together; a special section is devoted to momentum transport.

1.5.1 Convective Transport of Heat and Mass

1.5.1.1 General Description of Convective Transport

The characteristic properties of convective transport of heat and mass from the surface of a solid body to a flowing fluid will be discussed by means of Fig. 1.11. Transfer of heat or mass from the surface of the solid body to the flowing fluid is only possible by random molecular motions in the solid body and in the adjoining fluid layer. According to the distribution of the velocity w, it is assumed that the velocity is zero at the wall. This assumption has been proven to be correct for Newtonian fluids.

The conductive flux at the surface of the solid body is given by S_{m1}. The heat or mass transferred into the fluid element is split up into various streams. The molecular transport stream S_{m2} and the turbulent transport stream S_{t2} are directed through boundary 2 of the fluid element. The sum of S_{m2} and S_{t2} is of course smaller than S_{m1}, because a certain part of S_{m1} is carried away by the flowing fluid, that is by S_{c4}. The convective transport stream S_{c4} includes S_{c3} and the already mentioned fraction of S_{m1}.

The situation described remains unaltered when heat and mass are transferred from the fluid to the surface of the solid body. Momentum is always transferred from the fluid to the solid body.

1.5.1.2 Definition of Convective Transport Coefficients

The convective transport coefficient of heat is known as the heat transfer coefficient, designated by the Greek letter α. The local coefficient α_z is defined as follows:

$$\alpha_z \equiv \frac{\dot{q}_{wz}}{\Delta T}. \tag{1.63}$$

The appropriate equation for the convective local mass transfer coefficient β_z is given by:

$$\beta_z \equiv \frac{\dot{m}_{Awz}}{\Delta \varrho_A}. \tag{1.64}$$

$\dot{q}_{wz}$ and $\dot{m}_{Awz}$ are the flux densities for heat and mass at the wall and at the distance z

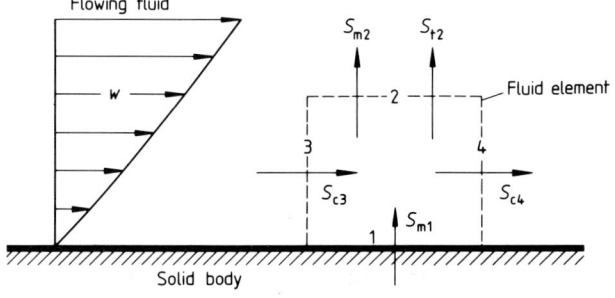

Figure 1.11. Explanation of convective transport of heat and mass.

from the entrance of the tube, given by Eqs. (1.23) and (1.36), respectively.

The convective transport coefficients given above are local quantities, that vary along the flow path of the fluid. This variation is due to the properties of the fluid and the solid surface.

For engineering purposes it is in many cases sufficient to introduce mean values of the convective transport coefficients. These coefficients will not be considered in this section.

1.5.1.3 Equations for Convective Heat and Mass Transport

The conditions for heat and mass transfer from a fluid stream to the wall of a tube are illustrated in Fig. 1.12. The fluid enters the tube at $z_1 = 0$ with the mean velocity $\bar{w}$, the mean temperature T_1, or mean partial density ϱ_{A1}. It is assumed that the mean velocity $\bar{w}$ does not vary while the fluid flows through the tube. On the way from the entrance to the exit of the tube the fluid temperature varies from T_1 to T_2 and the concentration from ϱ_{A1} to ϱ_{A2}. Assuming constant temperature at the wall, $T_w = \text{const}$, the heat flux density varies from $\dot{q}_{w1} = \infty$ at $z_1 = 0$ to $\dot{q}_{w2} = 0$ at $z_2 = \infty$. For mass transfer the concentration at the wall is constant, at the tube entrance it is $\dot{m}_{Aw1} = \infty$, and at the exit $\dot{m}_{Aw2} = 0$.

For the described conditions the local heat and mass transfer coefficients α_z and β_z, defined by Eqs. (1.63) and (1.64), will be applied. By introducing Eqs. (1.23) and (1.36) for heat and mass flux densities, $\dot{q}_{wz}$ and $\dot{m}_{Awz}$, the following equations are obtained:

$$\alpha_z = -\frac{\lambda}{\Delta T}\left(\frac{dT}{dr}\right)_{wz}, \quad (1.65)$$

$$\beta_z = -\frac{D}{\Delta\varrho_A}\left(\frac{d\varrho_A}{dr}\right)_{wz}. \quad (1.66)$$

The dimensionless quantities,

$$T^* \equiv T/\Delta T, \quad \varrho_A^* \equiv \varrho_A/\Delta\varrho_A, \\ r^* \equiv r/R = r/(d/2), \quad (1.67)$$

result in the equations for the local Nusselt number Nu_z and the local Sherwood number Sh_z:

$$Nu_z = -2\left(\frac{dT^*}{dr^*}\right)_{wz}, \quad (1.68)$$

$$Sh_z = -2\left(\frac{d\varrho_A^*}{dr^*}\right)_{wz}. \quad (1.69)$$

The Nusselt and Sherwood numbers are named after two famous German and American engineers, respectively, in honor of their outstanding contributions to the science of heat and mass transfer. The dimensionless numbers are defined by:

$$Nu_z \equiv \frac{\alpha_z d}{\lambda}, \quad (1.70)$$

$$Sh_z \equiv \frac{\beta_z d}{D}. \quad (1.71)$$

According to Eqs. (1.68) and (1.69) the local Nusselt number and Sherwood number are linear functions of the temperature or con-

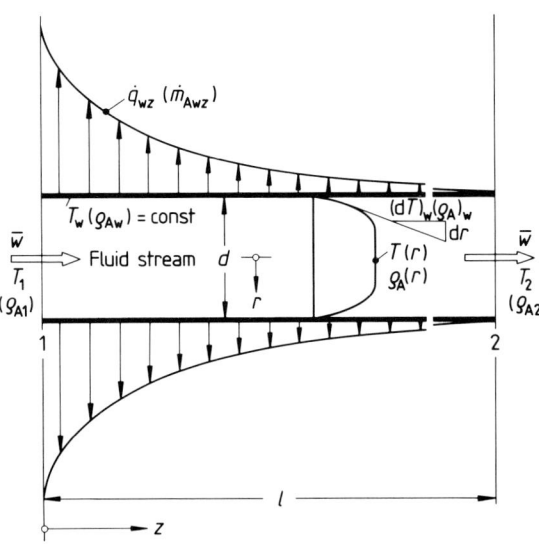

Figure 1.12. Convective heat and mass transport in tubes.

centration gradient at the wall, respectively. These gradients are obtained, as indicated in Fig. 1.12, from the temperature or concentration field. These fields are described by means of differential equations, which will be discussed in Chapter 2.

1.5.2 Convective Transport of Momentum

Momentum transport occurs in the direction of a negative velocity gradient. In a fluid flowing around a body like a plate or a spherical particle, momentum transfer to the body results in a thickening of the boundary layer in flow direction. When a

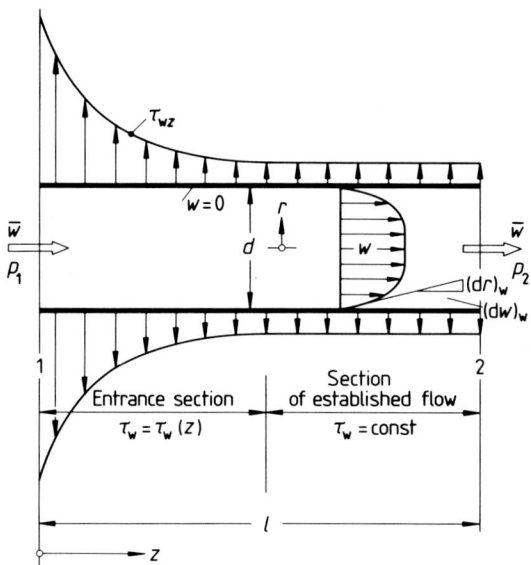

Figure 1.13. Convective momentum transport in tubes.

fluid flows through a tube, momentum transfer results in pressure drop. For this case the equation for convective momentum transfer will be considered. The physical conditions are indicated in Fig. 1.13.

It is assumed that the mean velocity $\bar{w}$ is constant. In the entrance section at $x = 0$ the velocity is independent of the radial coordinate r, so that the local shear stress is $\tau_{w0} = \infty$. Due to the action of the shear stress on the wall the velocity profile changes in such a way that the shear stress decreases and finally becomes constant. The established velocity profile is indicated in Fig. 1.13. That section of the tube, in which the velocity profile undergoes the mentioned change, is called the entrance section. This is followed by the section of established flow.

The equilibrium of shear force and pressure force is expressed by:

$$\tau_{wz} \, dA = dp \, F, \tag{1.72}$$

in which $dA = d\pi \, dz$ is the inner surface of the tube wall and $F = d^2\pi/4$ is the cross sectional area of the tube. Introducing these expressions the following relation between shear stress and pressure gradient is obtained:

$$\tau_{wz} = \frac{dp}{dz} \frac{d}{4}. \tag{1.73}$$

For Newtonian fluids the shear stress is given by Eq. (1.1), for non-Newtonian fluids Eq. (1.5) may be applied. For Newtonian fluids the following equation results:

$$\left(\frac{dw}{dr}\right)_{wz} = -\frac{dp}{dz} \frac{d}{4\eta}. \tag{1.74}$$

With the dimensionless quantities:

$$\left.\begin{array}{l} w^* \equiv w/\bar{w} \\ r^* \equiv r/R \end{array}\right\} \tag{1.75}$$

Eq. (1.74) may be rewritten into:

$$\left(\frac{dw^*}{dr^*}\right)_{wz} = -\frac{dp}{dz} \frac{d^2}{8\eta\bar{w}}. \tag{1.76}$$

The right hand side of this equation is rearranged by introducing a dimensionless momentum transfer coefficient ψ_z, which is called friction factor and defined by:

$$\psi_z \equiv \frac{dp/dz}{\varrho \bar{w}^2/2} d. \qquad (1.77)$$

Furthermore, the Reynolds number Re for tube flow, which has already been defined by Eq. (1.17) will be introduced. With ψ_z and Re Eq. (1.76) may be rewritten as follows:

$$\frac{\psi_z}{8} Re = -2 \left(\frac{dw^*}{dr^*} \right)_{wz}. \qquad (1.78)$$

The local friction factor according to this equation is proportional to the local velocity gradient at the tube wall. This gradient is obtained from the velocity field, which is described by a differential equation, that will be discussed in Chapter 2.

1.6 Biochemical Mass Conversion

Although biochemical mass conversion is treated in detail in Part II of this volume it seems to be justified to make a few remarks in this chapter of Part I. With these remarks some relations existing between microbial reactions and transport processes shall be pointed out.

There is no exact description of a biochemical reaction either occurring within or caused by an organism available. These processes are far too complicated to be fully understood today. Mathematical descriptions of microbial reactions are therefore only a very rough approximation of the real processes. All equations that have been presented for the microbial reaction rate are based on empirical results. In this respect there is a similarity with the equations presented for momentum, heat, and mass transport. Because of this similarity biochemical mass conversion will be briefly discussed in this chapter.

It is generally assumed that a biochemical mass conversion process due to microbial activities occurs within a certain fluid element. It is actually the reaction occurring within this fluid element that is considered, without referring directly to the microorganisms contained within this element. The type of empirical equation for the reaction rate may reflect to a small degree special properties of a defined species of microorganism or a population of microorganism.

All reactions, whether chemical or microbial by nature, are closely combined with transport processes. All nutrients and oxygen are transported by molecular or turbulent motions to the organisms where the conversion processes take place. It is therefore possible that the transport process is the limiting step in the overall process. On the other hand, the transport process may be so effective, that the biochemical conversion is the limiting step.

1.6.1 Reaction Rate Equation

Because of the stoichiometric relations which exist for all reactants and products it is adequate to restrict the mathematical description of a biochemical conversion to one of the components that takes part in the process. This component is designated by the capital letter A. The reaction flux is therefore designated by $\dot{R}_A$ with the unit kmol/s. The reaction flux density $\dot{r}_A$, often called reaction rate, is defined as the reaction flux per unit volume V:

$$\dot{r}_A \equiv \frac{\dot{R}_A}{V} = -\frac{dc_A}{dt} \quad \left(\text{in } \frac{\text{kmol}}{\text{m}^3 \text{ s}} \right), \qquad (1.79)$$

with c_A as the partial molar density of component A and t for time. The volumetric reaction rate $\dot{r}_A$ is identical with the variation of the concentration with time.

A well known empirical equation for the volumetric reaction rate has been presented by MONOD in [1.14]:

$$\dot r_A = \dot r_{A,\,max}\frac{c_A}{k+c_A} = \dot r_{A,\,max}\frac{c_A/k}{1+c_A/k}. \qquad (1.80)$$

$\dot r_{A,\,max}$ is the maximum of the reaction flux $\dot r_A$ specific for the considered microorganism, and k an empirical factor, which may be interpreted as the saturation constant,

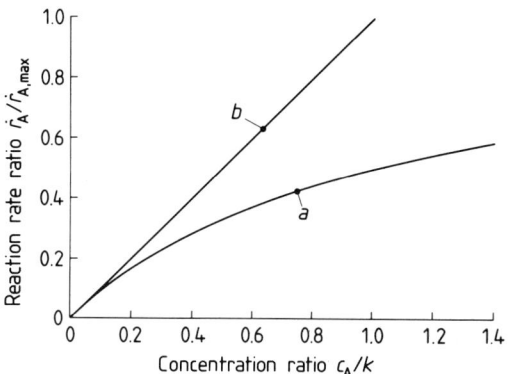

Figure 1.14. Graphical presentation of microbial conversion according to MONOD; curve b limiting case of curve a in the low concentration range.

when application of Eq. (1.80) is restricted to the low concentration range, $c_A/k \lesssim 0.1$; $\dot r_{A,\,max}$ and k must be experimentally determined.

In Fig. 1.14 the ratio $\dot r_A/\dot r_{A,\,max}$ is plotted against c_A/k; curve a represents Eq. (1.80). For $c_A/k = 0$ one obtains $\dot r_A/\dot r_{A,\,max} = 0$, and for $c_A/k = \infty$ one obtains $\dot r_A/\dot r_{A,\,max} = 1$. At low concentrations Eq. (1.80) is reduced to:

$$\frac{\dot r_A}{\dot r_{A,\,max}} = \frac{c_A}{k}. \qquad (1.81)$$

This equation is presented in Fig. 1.14 by curve b.

1.6.2 Integration of Reaction and Transport Process

In order that a biochemical reaction can take place within a microorganism the reactants must be transported to the surface of the organism. This combined process of transport and reaction is indicated in Fig. 1.15.

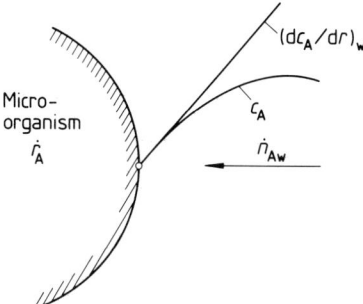

Figure 1.15. Explanation of integration of reaction and transport processes.

The physical transport of the considered reactant A is accomplished by convection and diffusion. Very close to the surface of the microorganism, transport is effected by molecular diffusion only. This step is described by the transport rate equation:

$$\dot n_{Aw} = -D\left(\frac{dc_A}{dr}\right)_w. \qquad (1.82)$$

The biochemical conversion takes place within the organism. It will be assumed that this step can be described by the equation:

$$\dot r_{Aw} = \frac{\dot r_{A,\,max}}{k} c_{Aw}, \qquad (1.83)$$

which follows from Eq. (1.81). Under steady-state conditions the reaction rate must be equal to the transport rate:

$$\dot n_{Aw} = \dot r_{Aw}. \qquad (1.84)$$

Introducing Eqs. (1.82) and (1.83) leads to:

$$\left(\frac{dc_A}{dr}\right)_w = -\frac{\dot{r}_{A,\,max}/k}{D} c_{Aw}. \qquad (1.85)$$

With the dimensionless parameters:

$$c_A^* \equiv c_A/c_{A\infty}$$
$$r^* \equiv r/R, \qquad (1.86)$$

one obtains:

$$\left(\frac{dc_A^*}{dr^*}\right)_w = -Da\, c_{Aw}^*, \qquad (1.87)$$

in which:

$$Da \equiv \frac{\dot{r}_{A,\,max}/k}{D/R} \qquad (1.88)$$

is the Damköhler number. This dimensionless number may be interpreted as the ratio of reaction velocity ($\dot{r}_{A,\,max}/k$) to transport velocity (D/R). The Damköhler number may assume the two limiting values: $Da \to 0$ and $Da \to \infty$. In the case of $Da \to 0$ the reaction velocity is the controlling step, the concentration gradient tends towards zero, and so does the biochemical conversion. In the other limiting case, $Da \to \infty$, the transport process is extremely slow and therefore is the controlling step.

When the reaction velocity controls the process, that is $Da \to 0$, the transport velocity is of minor importance. Enhancement of the biochemical process can only be achieved by improving the reaction velocity. This may necessitate the search for a more suitable microorganism. But mechanical activation of the available organism should also be taken into consideration.

On the other hand, when transport velocity controls the process, i.e., $Da \to \infty$, the reaction velocity is of minor importance. Enhancement of the biochemical process can only be achieved by improving the transport velocity. This may be done by increasing the mixing process within the reactor by increasing local convective currents and turbulence. Improvement of transport controlled processes will in many cases lead to the development of more efficient bioreactors.

1.7 References

[1.1] H. Brauer: "Stoffaustausch". Verlag Sauerländer, Aarau–Frankfurt am Main, 1971.
[1.2] H. Brauer: "Grundlagen der Einphasen- und Mehrphasenströmungen". Verlag Sauerländer, Aarau–Frankfurt am Main, 1971.
[1.3] F. Ebert: "Strömung nicht-Newtonscher Medien". Vieweg & Sohn, Braunschweig-Wiesbaden, 1980.
[1.4] H. Astarita and G. Marrucci: "Principles of Non-Newtonian Fluid Mechanics". McGraw-Hill, London, 1974.
[1.5] F. R. Eirich (ed.): "Rheology, Theory and Applications", Vol. 1–5. Academic Press, New York, 1956–1969.
[1.6] N. Ostwald: "Über die Geschwindigkeitsfunktion der Viskosität disperser Systeme". Kolloid-Z. 36 (1925), 99–117; 157–167; 248–250.
[1.7] A. de Waele: "Die Änderung der Viskosität mit der Schergeschwindigkeit disperser Systeme". Kolloid-Z. 36 (1925), 332–333.
[1.8] R. B. Bird, W. E. Stewart, and E. N. Lightfoot: "Transport Phenomena". John Wiley & Sons, New York–London, 1982.
[1.9] E. C. Bingham: "Fluidity and Plasticity". McGraw-Hill, New York–Toronto–London, 1922.
[1.10] J. O. Hinze: "Turbulence". McGraw-Hill, New York etc., 2nd Ed. 1975.
[1.11] Th. von Kármán: "Turbulence". J. R. Aeronaut. Soc. 41 (1937), 1109–1143.
[1.12] J. Boussinesq: "Theorie de l'écoulement tourbillant". Mem. Pre. par div. Sav. XXIII, Paris, 1877.
[1.13] H. Ludwig: "Bestimmung des Verhältnisses der Austauschkoeffizienten für Wärme und Impuls bei turbulenten Strömungen". Z. Flugwiss. 4 (1956), 73–81.
[1.14] A. Fiechter: "Batch and continuous culture of microbial, plant, and animal cells" in "Biotechnology", Vol. 1, "Microbial Fundamentals". Verlag Chemie, Weinheim–Deerfield Beach, Florida–Basel, 1981.

Chapter 2
Differential Equations for Velocity, Temperature and Concentration Fields

Heinz Brauer

Institut für Chemieingenieurtechnik
Technische Universität Berlin
Berlin (West), Germany

2.1 Introduction
2.2 Differential Equations
2.3 Introduction of Dimensionless Numbers
2.4 Dimensional Analysis

2.1 Introduction

In this chapter the differential equations for velocity, temperature, and concentration fields will be presented. It is not intended, however, to develop these equations on the basis of a general force balance, heat flux balance, and mass flux balance. If the reader of this book is interested in the physical and mathematical background of the presented differential equations, he should consult the relevant text books.

The differential equations will be presented for the conventional coordinate systems: rectangular, cylindrical, and spherical coordinate systems.

All equations are given in dimensional writing. In a special section of this chapter it is shown, how these equations can be transformed into dimensionless writing. This is a very safe method for determining dimensionless numbers, relevant for a defined physical situation. The last section of this chapter is devoted to dimensional analysis, another and useful method for finding the relevant dimensionless numbers. Although this section in a strict sense does not belong into a chapter on differential equations, it seems justified to include it, as both methods for determining dimensionless numbers should be explained in the same chapter.

2.2 Differential Equations

The differential equations for a velocity field result directly from a force balance on a differential volume of the flowing fluid. The forces taken into consideration are inertia force, shear force, pressure force, and gravity force. As the shear force is included, it must be decided, for which type of fluid, Newtonian or non-Newtonian, the equations will be presented. For non-Newtonian fluids there is no general equation for shear stress available. Therefore, it is only possible and rational to present the differential equations for the flow field of Newtonian fluids. The mathematical treatment of a force balance for non-Newtonian fluid flow in tubes will be given in Chapter 3 of this volume.

The differential equations for the velocity field are named after NAVIER (1827) and STOKES (1845), who made outstanding contributions to their development; they are the basis of modern fluid mechanics and will be given for laminar fluid flow. For turbulent fluid flow the equations become far more complicated and depend on the model which has been selected for mathematical description of turbulence. The differential equations for the velocity field will be supplemented by the differential equation of continuity.

The differential equations for continuity, velocity field, temperature field, and concentration field are given in Table 2.1 for a rectangular, in Table 2.2 for a cylindrical, and in Table 2.3 for a spherical coordinate system. The rectangular coordinates are x, y, and z; the cylindrical coordinates are radius r, circumferential angle θ, and length z; the spherical coordinates are radius r, circumferential angle θ, and azimutal angle φ. As indicated in Table 2.1 the differential equations consist each of several terms: The equation for the velocity field consists of momentum generation terms, the convective terms, the pressure terms, the conductive (friction) terms, and the gravity terms. The equation for the temperature field consists of the heat generation term, the convective terms, the conductive terms, and the dissipation terms. The equation for the concentration field consists of the mass generation term, the convective terms, the conductive terms, and the reaction term. The conductive terms in these equations are given by the transport equations explained in Chapter 1.

Table 2.1. Differential Equations for a Rectangular Coordinate System $(x-y-z)$

Continuity Equation	$\dfrac{\partial \varrho}{\partial t} + \dfrac{\partial}{\partial x}(\varrho w_x) + \dfrac{\partial}{\partial y}(\varrho w_y) + \dfrac{\partial}{\partial z}(\varrho w_z) = 0$	(2.1)
Equations for Velocity Field (Navier-Stokes Equations)	$\varrho \left(\dfrac{\partial w_x}{\partial t} + w_x \dfrac{\partial w_x}{\partial x} + w_y \dfrac{\partial w_x}{\partial y} + w_z \dfrac{\partial w_x}{\partial z} \right) = -\dfrac{\partial p}{\partial x} + \eta \left(\dfrac{\partial^2 w_x}{\partial x^2} + \dfrac{\partial^2 w_x}{\partial y^2} + \dfrac{\partial^2 w_x}{\partial z^2} \right) + \varrho g_x$	(2.2)
	$\varrho \left(\dfrac{\partial w_y}{\partial t} + w_x \dfrac{\partial w_y}{\partial x} + w_y \dfrac{\partial w_y}{\partial y} + w_z \dfrac{\partial w_y}{\partial z} \right) = -\dfrac{\partial p}{\partial y} + \eta \left(\dfrac{\partial^2 w_y}{\partial x^2} + \dfrac{\partial^2 w_y}{\partial y^2} + \dfrac{\partial^2 w_y}{\partial z^2} \right) + \varrho g_y$	(2.3)
	$\varrho \left(\underbrace{\dfrac{\partial w_z}{\partial t} + w_x \dfrac{\partial w_z}{\partial x} + w_y \dfrac{\partial w_z}{\partial y} + w_z \dfrac{\partial w_z}{\partial z}}_{\substack{\text{momentum generation} \\ \text{terms}}} \right) \underbrace{\vphantom{\dfrac{\partial p}{\partial z}}}_{\substack{\text{convective terms}}} = \underbrace{-\dfrac{\partial p}{\partial z}}_{\substack{\text{pressure} \\ \text{terms}}} + \eta \underbrace{\left(\dfrac{\partial^2 w_z}{\partial x^2} + \dfrac{\partial^2 w_z}{\partial y^2} + \dfrac{\partial^2 w_z}{\partial z^2} \right)}_{\substack{\text{conductive (friction)} \\ \text{terms}}} + \underbrace{\varrho g_z}_{\substack{\text{gravity} \\ \text{terms}}}$	(2.4)
Equation for Temperature Field (with Dissipation Terms)	$\varrho c_p \left(\underbrace{\dfrac{\partial T}{\partial t}}_{\substack{\text{heat generation} \\ \text{term}}} + \underbrace{w_x \dfrac{\partial T}{\partial x} + w_y \dfrac{\partial T}{\partial y} + w_z \dfrac{\partial T}{\partial z}}_{\text{convective terms}} \right) = \lambda \underbrace{\left(\dfrac{\partial^2 T}{\partial x^2} + \dfrac{\partial^2 T}{\partial y^2} + \dfrac{\partial^2 T}{\partial z^2} \right)}_{\text{conductive terms}} +$ $+ 2\eta \underbrace{\left\{ \left(\dfrac{\partial w_x}{\partial x} \right)^2 + \left(\dfrac{\partial w_y}{\partial y} \right)^2 + \left(\dfrac{\partial w_z}{\partial z} \right)^2 \right\} + \eta \left\{ \left(\dfrac{\partial w_x}{\partial z} + \dfrac{\partial w_y}{\partial x} \right)^2 + \left(\dfrac{\partial w_x}{\partial z} + \dfrac{\partial w_z}{\partial x} \right)^2 + \left(\dfrac{\partial w_y}{\partial z} + \dfrac{\partial w_z}{\partial y} \right)^2 \right\}}_{\text{dissipation terms}}$	(2.5)
Equation for Concentration Field	$\underbrace{\dfrac{\partial \varrho_A}{\partial t}}_{\substack{\text{mass generation} \\ \text{term}}} + \underbrace{w_x \dfrac{\partial \varrho_A}{\partial x} + w_y \dfrac{\partial \varrho_A}{\partial y} + w_z \dfrac{\partial \varrho_A}{\partial z}}_{\text{convective terms}} = D_{AB} \underbrace{\left(\dfrac{\partial^2 \varrho_A}{\partial x^2} + \dfrac{\partial^2 \varrho_A}{\partial y^2} + \dfrac{\partial^2 \varrho_A}{\partial z^2} \right)}_{\text{conductive terms}} + \underbrace{\dot r_A \mu_A}_{\substack{\text{reaction} \\ \text{term}}}$	(2.6)

Table 2.2. Differential Equations for a Cyclindrical Coordinate System $(r-\theta-z)$

Continuity Equation	$\dfrac{\partial \varrho}{\partial t} + \dfrac{1}{r}\dfrac{\partial}{\partial r}(\varrho r w_r) + \dfrac{1}{r}\dfrac{\partial}{\partial \theta}(\varrho w_\theta) + \dfrac{\partial}{\partial z}(\varrho w_z) = 0$	(2.7)
Equations for Velocity Field	$\varrho\left(\dfrac{\partial w_r}{\partial t} + w_r\dfrac{\partial w_r}{\partial r} + \dfrac{w_\theta}{r}\dfrac{\partial w_r}{\partial \theta} - \dfrac{w_\theta^2}{r} + w_z\dfrac{\partial w_r}{\partial z}\right) = -\dfrac{\partial p}{\partial r} +$ $+\eta\left\{\dfrac{\partial}{\partial r}\left(\dfrac{1}{r}\dfrac{\partial}{\partial r}(rw_r)\right) + \dfrac{1}{r^2}\dfrac{\partial^2 w_r}{\partial \theta^2} - \dfrac{2}{r^2}\dfrac{\partial w_\theta}{\partial \theta} + \dfrac{\partial^2 w_r}{\partial z^2}\right\} + \varrho g_r$	(2.8)
(Navier-Stokes Equations)	$\varrho\left(\dfrac{\partial w_\theta}{\partial t} + w_r\dfrac{\partial w_\theta}{\partial r} + \dfrac{w_\theta}{r}\dfrac{\partial w_\theta}{\partial \theta} + \dfrac{w_r w_\theta}{r} + w_z\dfrac{\partial w_\theta}{\partial z}\right) = -\dfrac{1}{r}\dfrac{\partial p}{\partial \theta} +$ $+\eta\left\{\dfrac{\partial}{\partial r}\left(\dfrac{1}{r}\dfrac{\partial}{\partial r}(rw_\theta)\right) + \dfrac{1}{r^2}\dfrac{\partial^2 w_\theta}{\partial \theta^2} + \dfrac{2}{r^2}\dfrac{\partial w_r}{\partial \theta} + \dfrac{\partial^2 w_\theta}{\partial z^2}\right\} + \varrho g_\theta$	(2.9)
	$\varrho\left(\dfrac{\partial w_z}{\partial t} + w_r\dfrac{\partial w_z}{\partial r} + \dfrac{w_\theta}{r}\dfrac{\partial w_z}{\partial \theta} + w_z\dfrac{\partial w_z}{\partial z}\right) = -\dfrac{\partial p}{\partial z} + \eta\left\{\dfrac{1}{r}\dfrac{\partial}{\partial r}\left(r\dfrac{\partial w_z}{\partial r}\right) + \dfrac{1}{r^2}\dfrac{\partial^2 w_z}{\partial \theta^2} + \dfrac{\partial^2 w_z}{\partial z^2}\right\} + \varrho g_z$	(2.10)
Equation for Temperature Field (with Dissipation Terms)	$\varrho c_p\left(\dfrac{\partial T}{\partial t} + w_r\dfrac{\partial T}{\partial r} + \dfrac{w_\theta}{r}\dfrac{\partial T}{\partial \theta} + w_z\dfrac{\partial T}{\partial z}\right) = \lambda\left\{\dfrac{1}{r}\dfrac{\partial}{\partial r}\left(r\dfrac{\partial T}{\partial r}\right) + \dfrac{1}{r^2}\dfrac{\partial^2 T}{\partial \theta^2} + \dfrac{\partial^2 T}{\partial z^2}\right\} + 2\eta\left\{\left(\dfrac{\partial w_r}{\partial r}\right)^2 + \left[\dfrac{1}{r}\left(\dfrac{\partial w_\theta}{\partial \theta} + w_r\right)\right]^2 + \left(\dfrac{\partial w_z}{\partial z}\right)^2\right\} +$ $+\eta\left\{\left(\dfrac{\partial w_\theta}{\partial z} + \dfrac{1}{r}\dfrac{\partial w_z}{\partial \theta}\right)^2 + \left(\dfrac{\partial w_z}{\partial r} + \dfrac{\partial w_r}{\partial z}\right)^2 + \left[\dfrac{1}{r}\dfrac{\partial w_r}{\partial \theta} + r\dfrac{\partial}{\partial r}\left(\dfrac{w_\theta}{r}\right)\right]^2\right\}$	(2.11)
Equation for Concentration Field	$\dfrac{\partial \varrho_A}{\partial t} + w_r\dfrac{\partial \varrho_A}{\partial r} + w_\theta\dfrac{1}{r}\dfrac{\partial \varrho_A}{\partial \theta} + w_z\dfrac{\partial \varrho_A}{\partial z} = D_{AB}\left\{\dfrac{1}{r}\dfrac{\partial}{\partial r}\left(r\dfrac{\partial \varrho_A}{\partial r}\right) + \dfrac{1}{r^2}\dfrac{\partial^2 \varrho_A}{\partial \theta^2} + \dfrac{\partial^2 \varrho_A}{\partial z^2}\right\} + \dot{r}_A\mu_A$	(2.12)

Table 2.3. Differential Equations for Spherical Coordinate System $(r-\theta-\varphi)$

Continuity Equation	$\dfrac{\partial \varrho}{\partial t} + \dfrac{1}{r^2}\dfrac{\partial}{\partial r}(\varrho r^2 w_r) + \dfrac{1}{r\sin\theta}\dfrac{\partial}{\partial \theta}(\varrho w_\theta \sin\theta) + \dfrac{1}{r\sin\theta}\dfrac{\partial}{\partial \varphi}(\varrho w_\varphi) = 0$	(2.13)
Equations for Velocity Field (Navier-Stokes Equations)	$\varrho\left(\dfrac{\partial w_r}{\partial t} + w_r\dfrac{\partial w_r}{\partial r} + \dfrac{w_\theta}{r}\dfrac{\partial w_r}{\partial \theta} + \dfrac{w_\varphi}{r\sin\theta}\dfrac{\partial w_r}{\partial \varphi} - \dfrac{w_\theta^2 + w_\varphi^2}{r}\right) = -\dfrac{\partial p}{\partial r} +$ $+ \eta\left(\nabla^2 w_r - \dfrac{2}{r^2}w_r - \dfrac{2}{r^2}\dfrac{\partial w_\theta}{\partial \theta} - \dfrac{2}{r^2}w_\theta \cot\theta - \dfrac{2}{r^2\sin\theta}\dfrac{\partial w_\varphi}{\partial \varphi}\right) + \varrho g_r$	(2.14)
	$\varrho\left(\dfrac{\partial w_\theta}{\partial t} + w_r\dfrac{\partial w_\theta}{\partial r} + \dfrac{w_\theta}{r}\dfrac{\partial w_\theta}{\partial \theta} + \dfrac{w_\varphi}{r\sin\theta}\dfrac{\partial w_\theta}{\partial \varphi} + \dfrac{w_r w_\theta}{r} - \dfrac{w_\varphi^2 \cot\theta}{r}\right) = -\dfrac{1}{r}\dfrac{\partial p}{\partial \theta} +$ $+ \eta\left(\nabla^2 w_\theta + \dfrac{2}{r^2}\dfrac{\partial w_r}{\partial \theta} - \dfrac{w_\theta}{r^2\sin^2\theta} - \dfrac{2\cos\theta}{r^2\sin^2\theta}\dfrac{\partial w_\varphi}{\partial \varphi}\right) + \varrho g_\theta$	(2.15)
	$\varrho\left(\dfrac{\partial w_\varphi}{\partial t} + w_r\dfrac{\partial w_\varphi}{\partial r} + \dfrac{w_\theta}{r}\dfrac{\partial w_\varphi}{\partial \theta} + \dfrac{w_\varphi}{r\sin\theta}\dfrac{\partial w_\varphi}{\partial \varphi} + \dfrac{w_\varphi w_r}{r} + \dfrac{w_\theta w_\varphi}{r}\cot\theta\right) = -\dfrac{1}{r\sin\theta}\dfrac{\partial p}{\partial \varphi} +$ $+ \eta\left(\nabla^2 w_\varphi - \dfrac{w_\varphi}{r^2\sin^2\theta} + \dfrac{2}{r^2\sin\theta}\dfrac{\partial w_r}{\partial \varphi} + \dfrac{2\cos\theta}{r^2\sin^2\theta}\dfrac{\partial w_\theta}{\partial \varphi}\right) + \varrho g_\varphi$	(2.16)
Equation for Temperature Field (with Dissipation Terms)	$\varrho c_p\left(\dfrac{\partial T}{\partial t} + w_r\dfrac{\partial T}{\partial r} + \dfrac{w_\theta}{r}\dfrac{\partial T}{\partial \theta} + \dfrac{w_\varphi}{r\sin\theta}\dfrac{\partial T}{\partial \varphi}\right) = \lambda \nabla^2 T + 2\eta\left\{\left(\dfrac{\partial w_r}{\partial r}\right)^2 + \left(\dfrac{1}{r}\dfrac{\partial w_\theta}{\partial \theta} + \dfrac{w_r}{r}\right)^2 + \left(\dfrac{1}{r\sin\theta}\dfrac{\partial w_\varphi}{\partial \varphi} + \dfrac{w_r}{r} + \dfrac{w_\theta \cot\theta}{r}\right)^2\right\} +$ $+ \eta\left\{\left[r\dfrac{\partial}{\partial r}\left(\dfrac{w_\theta}{r}\right) + \dfrac{1}{r}\dfrac{\partial w_r}{\partial \theta}\right]^2 + \left[\dfrac{1}{r\sin\theta}\dfrac{\partial w_r}{\partial \varphi} + r\dfrac{\partial}{\partial r}\left(\dfrac{w_\varphi}{r}\right)\right]^2 + \left[\dfrac{\sin\theta}{r}\dfrac{\partial}{\partial \theta}\left(\dfrac{w_\varphi}{\sin\theta}\right) + \dfrac{1}{r\sin\theta}\dfrac{\partial w_\theta}{\partial \varphi}\right]^2\right\}$	(2.17)
Equation for Concentration Field	$\dfrac{\partial \varrho_A}{\partial t} + w_r\dfrac{\partial \varrho_A}{\partial r} + w_\theta\dfrac{1}{r}\dfrac{\partial \varrho_A}{\partial \theta} + w_\varphi\dfrac{1}{r\sin\theta}\dfrac{\partial \varrho_A}{\partial \varphi} = D_{AB}\nabla^2 \varrho_A + \dot{r}_A \mu_A$	(2.18)
Nabla Operator	$\nabla^2 = \dfrac{1}{r^2}\dfrac{\partial}{\partial r}\left(r^2\dfrac{\partial}{\partial r}\right) + \dfrac{1}{r^2\sin\theta}\dfrac{\partial}{\partial \theta}\left(\sin\theta\dfrac{\partial}{\partial \theta}\right) + \dfrac{1}{r^2\sin^2\theta}\left(\dfrac{\partial^2}{\partial \varphi^2}\right)$	(2.19)

In the differential equations use is made of the following notations:

w_x, w_y, w_z (in m/s) velocity components in x, y, z-direction
w_r, w_θ, w_z (in m/s) velocity components in r, θ, z-direction
w_r, w_θ, w_φ (in m/s) velocity components in r, θ, φ-direction
g_x, g_y, g_z (in m/s^2) acceleration of gravity in x, y, z-direction
g_r, g_θ, g_z (in m/s^2) acceleration of gravity in r, θ, z-direction
g_r, g_θ, g_φ (in m/s^2) acceleration of gravity in r, θ, φ-direction
c_p (in J/kg K) specific heat capacity of system
D_{AB} (in m^2/s) diffusion coefficient of component A in a mixture with B
p (in N/m^2) total pressure in system
$\dot{r}_A$ (in kmol/s m^3) volumetric reaction flux
t (in s) time
T (in K) temperature
η (in kg/m s) dynamic viscosity of fluid mixture
μ_A (in kmol/kg) mole mass
ϱ (in kg/m^3) mean density of system
ϱ_A (in kg/m^3) partial density of component A

A solution of the above given differential equations requires the introduction of an appropriate number of boundary conditions. Analytical solutions are possible only when the differential equations can be substantially simplified. This can be done only in a very few cases of technical importance. In most cases the differential equations can only be solved by means of numerical methods and use of a fairly large computer.

2.3 Introduction of Dimensionless Numbers

It is advisable to introduce dimensionless parameters before attempting a solution of the differential equations. The applied procedure will be explained by using an example, for which Eq. (2.2) is selected:

$$\frac{\partial w_x}{\partial t} + w_x \frac{\partial w_x}{\partial x} + w_y \frac{\partial w_x}{\partial y} + w_z \frac{\partial w_x}{\partial z} =$$

$$= -\frac{1}{\varrho}\frac{\partial p}{\partial x} +$$

$$+ \frac{\eta}{\varrho}\left(\frac{\partial^2 w_x}{\partial x^2} + \frac{\partial^2 w_x}{\partial y^2} + \frac{\partial^2 w_x}{\partial z^2}\right) + g_x \quad (2.2)$$

The velocity components w_x, w_y, and w_z are reduced to dimensionless quantities by dividing through the velocity w_∞, which is, in a certain way, characteristic of the flow field. Dimensionless coordinates are obtained by dividing x, y, and z by the characteristic length l of the flow field. The dimensionless quantities are defined as follows:

$w_x^* \equiv w_x/w_\infty$; $\quad w_y^* \equiv w_y/w_\infty$; $\quad w_z^* \equiv w_z/w_\infty$

$x^* \equiv x/l$; $\qquad y^* \equiv y/l$; $\qquad z^* \equiv z/l$

Introducing these quantities into Eq. (2.2) one obtains:

$$\frac{w_\infty^2}{l}\left(\frac{l}{w_\infty}\frac{\partial w_x^*}{\partial t} + w_x^*\frac{\partial w_x^*}{\partial x^*} + w_y^*\frac{\partial w_x^*}{\partial y^*} +\right.$$

$$\left. + w_z^*\frac{\partial w_x^*}{\partial z^*}\right) = -\frac{1}{\varrho l}\frac{\partial p}{\partial x^*} +$$

$$+ \frac{\eta w_\infty}{l^2 \varrho}\left(\frac{\partial^2 w_x^*}{\partial x^{*2}} + \frac{\partial^2 w_x^*}{\partial y^{*}} + \frac{\partial^2 w_x^*}{\partial z^{*}}\right) + g_x \quad (2.20)$$

Dividing both sides by w_∞^2/l the equation assumes the following form:

$$\frac{l}{w_\infty}\frac{\partial w_x^*}{\partial t} + w_x^*\frac{\partial w_x^*}{\partial x^*} + w_y^*\frac{\partial w_x^*}{\partial y^*} + w_z^*\frac{\partial w_x^*}{\partial z^*} =$$

$$= -\frac{1}{\varrho w_\infty^2}\frac{\partial p}{\partial x^*} +$$

$$+ \frac{\eta}{w_\infty l \varrho}\left(\frac{\partial^2 w_x^*}{\partial x^{*2}} + \frac{\partial^2 w_x^*}{\partial y^{*2}} + \frac{\partial^2 w_x^*}{\partial z^{*2}}\right) + \frac{l}{w_\infty^2}g_x \quad (2.21)$$

In this equation the dimensionless parameter,

$$Re \equiv \frac{w_\infty l \varrho}{\eta}, \quad (2.22)$$

is the Reynolds number, named after the English professor for mechanical engineering, OSBORNE REYNOLDS, for his outstanding contribution to fluid mechanics and heat transfer. A second dimensionless parameter obtained is the dimensionless acceleration due to gravity:

$$g_x^* \equiv \frac{g_x}{w_\infty^2/l}. \quad (2.23)$$

The pressure p will be transformed into a dimensionless quantity by dividing through $\varrho w_\infty^2/2$, which is the dynamic pressure due to the velocity w_∞. Eq. (2.21) proposes the division of the local pressure p by ϱw_∞^2. Because of the better physical meaning the dynamic pressure $\varrho w_\infty^2/2$ is taken instead:

$$p^* \equiv \frac{p}{\varrho w_\infty^2/2}. \quad (2.24)$$

Eq. (2.21) further suggests a multiplication of the time t by the term w_∞/l. This is certainly justified. However, in many other physical situations it appeared useful to define a Fourier number:

$$Fo \equiv \frac{\eta t}{\varrho l^2} = t\frac{w_\infty}{l}\frac{\eta}{w_\infty l \varrho}. \quad (2.25)$$

The Fourier number is named after the French mathematical physicist JEAN BAPTISTE JOSEPH FOURIER (1768–1830), who made very significant contributions to the analytical treatment of heat conduction.

Making use of the Fourier number, the following expression must be introduced into Eq. (2.21):

$$\frac{tw_\infty}{l} \equiv Fo\,Re. \quad (2.26)$$

With the explained dimensionless parameters, Eq. (2.21) can be rewritten as follows:

$$\frac{1}{Re}\frac{\partial w_x^*}{\partial Fo} + w_x^*\frac{\partial w_x^*}{\partial x^*} + w_y^*\frac{\partial w_x^*}{\partial y^*} + w_z^*\frac{\partial w_x}{\partial z^*} =$$

$$= -\frac{1}{2}\frac{\partial p^*}{\partial x^*} +$$

$$+ \frac{1}{Re}\left(\frac{\partial^2 w_x^*}{\partial x^{*2}} + \frac{\partial^2 w_x^*}{\partial y^{*2}} + \frac{\partial^2 w_z^*}{\partial z^{*2}}\right) + g_x^* \quad (2.27)$$

Taking into account that there are three equations for the velocity field and the continuity equation the following general equation for the velocity component w_x^* results if there are no further dimensionless parameters due to boundary conditions:

$$w_x^* = f_1(x^*; y^*; z^*; Re; Fo; g_x^*; \partial p^*/\partial x^*) \quad (2.28)$$

The local velocity component w_x^* is a function of seven dimensionless parameters. This equation in dimensional form appears as:

$$w_x = f_2(x; y; z; l; w_\infty; \varrho; \eta; t; g_x; \partial p/\partial x) \quad (2.29)$$

The velocity component w_x is a function of ten independent parameters. Comparison with Eq. (2.28) reveals the fact that the introduction of dimensionless parameters reduces the number of parameters, in this case by three. A solution that makes use of dimensionless parameters is in general more simple or of wider applicability. This is the reason why a differential equation prior to solution, especially when numerical methods are to be applied, should be transformed into dimensionless writing. On the other hand, it is necessary to point out, that the physical interpretation of experimental or theoretical results is sometimes rendered more difficult, when use is made of dimensionless parameters.

The importance of and advantage gained by application of dimensionless numbers cannot be overestimated. Therefore, it seems to be expedient to point out still another method for the determination of dimensionless numbers. This method is based on the analysis of dimensions of the independent parameters and will be explained in the next section.

2.4 Dimensional Analysis

Each physical or physico-chemical process can be described by independent dimensional parameters. By means of dimensional analysis the dimensional parameters can be transformed into a smaller number of dimensionless parameters. The method of dimensional analysis will be explained for a simple process.

The pressure loss Δp of a fluid flowing through a horizontal tube will be investigated. A careful analysis leads to the result that Δp must be a function of diameter d and length L of the tube, the mean velocity $\bar{w}$ of the fluid, and finally of density ϱ and dynamic viscosity η of the fluid. The following relation must apply:

$$F(\Delta p; d; L; \bar{w}; \varrho; \eta) = 0. \quad (2.30)$$

Transforming the dimensional parameters into dimensionless parameters π_1 to π_n, Eq. (2.30) can be rewritten as follows:

$$F(\pi_1; \pi_2; \ldots \pi_n) = 0. \quad (2.31)$$

Number and form of the parameters π_1 to π_n will be determined.

It is always possible to arrange all independent parameters with the dependent one in such a way that only one dimensionless number results:

$$\pi = \Delta p^a d^b L^c \bar{w}^d \varrho^e \eta^f. \quad (2.32)$$

If the problem is described by more than one dimensionless number, some of the exponents a to f must become zero. In consequence of Eq. (2.32) the following relation between the dimensions of the parameters must exist:

$$\left(\frac{kg}{m\,s^2}\right)^a m^b m^c \left(\frac{m}{s}\right)^d \left(\frac{kg}{m^3}\right)^e \left(\frac{kg}{m\,s}\right)^f = 0. \quad (2.33)$$

The dimensions are those of the three basic units for mass (1 kg), length (1 m), and time (1 s). Eq. (2.33) is correct when the sum of the exponents for each basic unit becomes zero:

$$\left.\begin{array}{l} \text{mass (kg): } a \qquad\qquad\qquad + e + f = 0 \\ \text{length (m): } -a+b+c+d-3e-f=0 \\ \text{time (s): } -2a \qquad\quad -d \qquad -f=0 \end{array}\right\} \quad (2.34)$$

This system of linear equations will be solved by means of the matrix of coefficients:

$$\begin{array}{c} \\ kg \\ m \\ s \end{array} \begin{pmatrix} a & b & c & d & e & f \\ 1 & 0 & 0 & 0 & 1 & 1 \\ -1 & 1 & 1 & 1 & -3 & -1 \\ -2 & 0 & 0 & -1 & 0 & -1 \end{pmatrix}$$

As the rank r of this matrix is given by $r=3$, the number of dependent variable exponents is 3. These exponents can be freely chosen and shall be: a, b, and d. From the set of Eqs. (2.34) the following result is obtained:

$$\left.\begin{array}{l} a = -\ e - f \\ b = -\ c - f \\ d = +2e + f \end{array}\right\} \quad (2.35)$$

Introduction into Eq. (2.32) leads to the fundamental dimensionless number:

$$\pi = \Delta p^{-e-f} d^{-c-f} L^c \bar{w}^{2e+f} \varrho^e \eta^f. \quad (2.36)$$

The dimensionless number π must be split up into the relevant number i of dimensionless parameters of the problem. This number i is equal to the number of dimensional parameters n minus the rank r of the matrix. In the case under consideration this leads to:

$$\pi = F(\pi_1; \pi_2; \pi_3). \tag{2.37}$$

The definitions of these dimensionless numbers will be determined. The exponents a to f can assume any value. To arrive at the simplest expressions for π_1 to π_3 it is advisable to proceed in a first step as follows:

$$\begin{aligned} c = 1\,(e, f = 0) \quad & \pi_1 \equiv L/d \\ e = 1\,(c, f = 0) \quad & \pi_2 \equiv \frac{\varrho\,\bar{w}^2}{\Delta p} \\ f = 1\,(c, e = 0) \quad & \pi_3 \equiv \frac{\bar{w}\,\eta}{\Delta p\,d} \end{aligned} \tag{2.38}$$

In a second step the obtained dimensionless numbers will be examined with respect to their physical meaning. π_1 will be accepted as obtained. The dimensionless number π_2 will be transformed into π_5:

$$\pi_5 \equiv \frac{2}{\pi_2} = \frac{\Delta p}{\varrho\,\bar{w}^2/2}. \tag{2.39}$$

π_5 is the ratio of the pressure drop Δp and the dynamic pressure $\varrho\,\bar{w}^2/2$; this ratio is generally accepted as the best method for a dimensionless presentation of pressure drop. The dimensionless number π_3 is quite unusual. Therefore the following transformation is suggested:

$$\pi_6 \equiv \frac{\pi_2}{\pi_3} = \frac{\bar{w}\,d\,\varrho}{\eta} \equiv Re. \tag{2.40}$$

π_6 is identical with the Reynolds number Re. The result of dimensional analysis may then be written as follows:

$$\pi_5 = f(\pi_1; \pi_6). \tag{2.41}$$

According to available experimental and theoretical evidence it is justified to introduce the ratio π_5/π_1 as friction factor ψ:

$$\psi \equiv \frac{\pi_5}{\pi_1} = \frac{\Delta p}{\varrho\,\bar{w}^2/2}\,\frac{d}{L}. \tag{2.42}$$

The friction factor law for straight tubes with smooth walls may therefore be written as follows:

$$\psi = f(Re). \tag{2.43}$$

This law will be thoroughly discussed in Chapter 3.

Observation of the following rules is helpful in a rational application of dimensional analysis:

1. The physical analysis of the problem should be carried out with extreme care. Only a thorough understanding of the physical problem guarantees the correct determination of all dependent and independent dimensional parameters of the problem.
2. The independence of the dimensional parameters should be tested repeatedly. This test contributes decisively to an understanding of the problem.
3. Assuming too few independent parameters will lead to an incorrect result, too many to an unneccessary complexity of the derived equation.
4. The number of parameters consists in general of one dependent (in the example Δp) and $n-1$ independent parameters (in the example $n-1=5$). The dependent parameter should be contained in only one dimensionless parameter. This will be achieved, if the exponent of the dependent parameter is included in the group of independent exponents. In the example, the dependent exponents were assumed to be: a, b, and d; the independent exponents were c, e, and f.

Chapter 3

Transport Processes in Newtonian Fluids Flowing Through Tubes

Heinz Brauer

Institut für Chemieingenieurtechnik
Technische Universität Berlin
Berlin (West), Germany

3.1 Introduction
3.2 Momentum Transfer in Tube Flow
3.2.1 Velocity Profiles in Laminar and Turbulent Flow
3.2.2 Friction Factor Laws for Laminar and Turbulent Flow
3.3 Heat and Mass Transfer in Laminar Tube Flow
3.4 Heat and Mass Transfer in Turbulent Tube Flow
3.5 Heat and Mass Transfer in Turbulent Flow Through Rough Tubes
3.6 Mass Transfer with Biochemical Reactions
3.6.1 The Biochemical and Mathematical Problem
3.6.2 Local and Mean Concentrations
3.6.3 Mean Sherwood Number
3.7 References

List of Symbols

A	m²	area
c_A	kmol/m³	partial mol density of component A
c_{Aw}	kmol/m³	partial mol density of component A at surface of wall
c_p	kJ/(kg K)	specific heat capacity of fluid
D	m²/s	diffusion coefficient
d	m	tube diameter
F	m²	cross sectional area
k_s	m	height of roughness elements
k_w	m/s	reaction velocity factor
L	m	length of tube
$\dot{M}_A$	kg/s	mass flux of component A
$\dot{m}_A$	kg/(s m²)	mass flux density of component A
$\dot{n}_{Aw}$	kmol/(s m²)	mol flux density of component A
p	N/m²	total pressure
$\dot{Q}$	kJ/s	heat flux
$\dot{q}$	kJ/(s m²)	heat flux density
R	m	tube radius
r	m	radial coordinate
$\dot{r}_{Aw}$	kmol/(s m²)	reaction flux density of component A at surface of wall
T	K	local temperature
$\bar{T}$	K	mean temperature
T_0	K	temperature at tube entrance
T_w	K	temperature at tube wall
$\dot{V}$	m³/s	volumetric flow rate of fluid
w	m/s	local velocity
$\bar{w}$	m/s	mean velocity
w_m	m/s	maximum velocity
w_τ	m/s	shear velocity
z	m	axial coordinate
α	kJ/(s m² K)	mean coefficient of heat transfer
α_z	kJ/(s m² K)	local coefficient of heat transfer
β	m/s	mean coefficient of mass transfer
β_z	m/s	local coefficient of mass transfer
Δp	N/m²	pressure drop
η	kg/(m s)	dynamic viscosity
λ	kJ/(m s K)	heat conductivity
ν	m²/s	kinematic viscosity
ϱ	kg/m³	mean density of fluid
ϱ_A	kg/m³	local partial density of component A
$\bar{\varrho}_A$	kg/m³	mean partial density of component A
ϱ_{A0}	kg/m³	partial density of component A at tube inlet
ϱ_{Aw}	kg/m³	partial density of component A at tube wall
τ_w	N/m²	shear stress at tube wall
$Da \equiv \dfrac{k_w}{D/R}$		Damköhler number
$Nu \equiv \dfrac{\alpha d}{\lambda}$		mean Nusselt number
$Nu_z \equiv \dfrac{\alpha_z d}{\lambda}$		local Nusselt number
$Pr \equiv \dfrac{\eta c_p}{\lambda}$		Prandtl number
$r^* \equiv r/R$		radial coordinate
$Re \equiv \dfrac{\bar{w} d}{\nu}$		Reynolds number
$Sc \equiv \nu/D$		Schmidt number
$Sh \equiv \dfrac{\beta d}{D}$		mean Sherwood number
$Sh_z \equiv \dfrac{\beta_z d}{D}$		local Sherwood number
$z^* \equiv \dfrac{z/d}{Re\, Sc}$		entrance number
$\theta \equiv \dfrac{T - T_w}{T_0 - T_w}$		local temperature ratio
$\bar{\theta} \equiv \dfrac{\bar{T} - T_w}{T_0 - T_w}$		mean temperature ratio
$\xi \equiv \dfrac{\varrho_A - \varrho_{Aw}}{\varrho_{A0} - \varrho_{Aw}}$		local concentration ratio
$\bar{\xi} \equiv \dfrac{\bar{\varrho}_A - \varrho_{Aw}}{\varrho_{A0} - \varrho_{Aw}}$		mean concentration ratio
$\psi \equiv \dfrac{\Delta p}{\varrho \bar{w}^2/2}\dfrac{d}{L}$		friction coefficient

3.1 Introduction

In this chapter the transport of momentum, heat, and mass in a Newtonian fluid flowing through a straight tube will be discussed. This is one of two examples for transport processes in one-phase flow systems to be considered in detail. The second example, treated in Chapter 5, concerns the flow system surrounding a flat plate.

Momentum transport directed from the fluid to the tube wall results in pressure loss. The aim of theoretical and experimental investigations has been the determination of a general law for the pressure decrease, well known as the friction factor law. This will be presented for laminar and turbulent flow in tubes with smooth and rough walls.

Heat and mass transfer in tubes are analogous processes. A more detailed discussion, therefore, will be presented only for mass transfer. For laminar and turbulent flow in tubes with smooth and rough walls well known equations for local and mean Sherwood numbers will be given.

A special section of this chapter is devoted to mass transfer with an enzymatic biochemical reaction at the wall of the tube. This is a problem comparable to mass transfer in heterogeneous chemical reactions. Fundamentals and results of this process will be presented.

3.2 Momentum Transfer in Tube Flow

This section explains established fluid flow and velocity profiles for laminar and turbulent flow conditions and, furthermore, will describe friction factor laws.

3.2.1 Velocity Profiles in Laminar and Turbulent Flow

For laminar fluid flow in tubes with a smooth wall the velocity profile, i.e., the local velocity w as a function of the radial coordinate r, can be obtained by solving the relevant differential equation. From Eq. (2.8) the following differential equation results for established laminar and isothermal flow in a horizontal tube:

$$\eta \left\{ \frac{1}{r} \frac{\partial}{\partial r} \left(\frac{\partial w}{\partial r} \right) \right\} = \frac{\partial p}{\partial z}. \qquad (3.1)$$

The index z for the velocity w has been neglected, because for established flow there is only one velocity component. With the non-slip condition at the wall, $w=0$ at $r=R$, and the symmetry condition in the tube axis, $\partial w/\partial z = 0$ at $r=0$, the solution obtained from Eq. (3.1) is:

$$w = \frac{R^2}{4\eta} \left(-\frac{\partial p}{\partial z} \right) \left[1 - \left(\frac{r}{R} \right)^2 \right]. \qquad (3.2)$$

The negative pressure gradient for established flow equals the pressure drop Δp per unit length L of the tube:

$$-\frac{dp}{dz} = \frac{\Delta p}{L}. \qquad (3.3)$$

With this expression Eq. (3.2) can be rewritten as follows:

$$w = \frac{R^2}{4\eta} \frac{\Delta p}{L} \left[1 - \left(\frac{r}{R} \right)^2 \right]. \qquad (3.4)$$

The velocity profile is parabolic. The maximum of the velocity, w_m, occurs at the tube axis at $r=0$:

$$w_m = \frac{R^2}{4\eta} \frac{\Delta p}{L}. \qquad (3.5)$$

The mean velocity $\bar{w}$ is obtained from:

$$\bar{w} = \frac{\dot{V}}{F}, \qquad (3.6)$$

with $\dot{V}$ the volumetric flow rate and $F = d^2\pi/4$ the cross sectional area of the tube, with $d =$ tube diameter. The volumetric flow rate is given by the Hagen-Poiseuille equation:

$$\dot{V} = \int_0^R 2\pi w r \, dr = \frac{\pi R^4}{8\eta} \frac{\Delta p}{L}, \quad (3.7)$$

the mean velocity thus being:

$$\bar{w} = \frac{R^2}{8\eta} \frac{\Delta p}{L}. \quad (3.8)$$

In laminar flow the maximum velocity w_m is twice the mean velocity $\bar{w}$.

Very close to the wall, laminar flow conditions will prevail even when in the bulk of the flow turbulence has been fully established. In this "wall region" the velocity is described by an equation which may be derived from Eq. (3.4). Introducing the distance from the wall, $y = R - r$, the relation between wall shear stress τ_w and pressure drop Δp:

$$\Delta p = 4\tau_w L/d, \quad (3.9)$$

and the shear velocity w_τ, defined by:

$$w_\tau \equiv \sqrt{\tau_w/\varrho}, \quad (3.10)$$

leads to:

$$\frac{w}{w_\tau} = \frac{w_\tau y}{\nu}\left(1 - \frac{1}{2}\frac{y}{R}\right). \quad (3.11)$$

Herein, $\nu = \eta/\varrho$ is the kinematic viscosity of the fluid. Close to the wall, for $y/R \to 0$, the velocity ratio w/w_τ becomes a linear function of $w_\tau y/\nu$. The fluid layer in which this linear relationship exists is called the laminar sublayer of turbulent flow. In this sublayer most of the resistance to momentum, heat, and mass transfer is concentrated. The thickness of the sublayer is, in general, of the order of a few tenths of a millimeter. The thickness decreases, and thus the resistance to momentum, heat, and mass transfer decreases as well with increasing mean velocity of the fluid.

In turbulent flow it is assumed that the velocity ratio will remain only a function, although a non-linear function, of $w_\tau y/\nu$ over the whole cross section of the tube. This has been substantiated by experimental results. Fig. 3.1 shows the data, approximated close to the wall by curve a and at a greater distance by curve b, for which the following equation applies [3.1]:

$$\frac{w}{w_\tau} = 5.75 \log \frac{w_\tau y}{\nu} + 5.5 = 2.5 \ln \frac{w_\tau y}{\nu} + 5.5. \quad (3.12)$$

A comparison between velocity profiles in laminar and turbulent flow is shown in Fig. 3.2. In turbulent flow the ratio of maximum to mean velocity is only about 1.2, while for laminar flow it is 2.0.

3.2.2 Friction Factor Laws for Laminar and Turbulent Flow

The friction factor for established tube flow is designated by the greek letter ψ, and defined by:

$$\psi \equiv \frac{\Delta p}{\varrho \bar{w}^2/2} \frac{d}{L}. \quad (3.13)$$

For laminar flow in tubes with smooth walls one obtains from Eq. (3.8) the friction law:

$$\psi = \frac{64}{Re}, \quad (3.14)$$

range of application: $0 \leq Re \leq 2320$.

For turbulent flow in tubes with smooth walls PRANDTL [3.2] presented an equation based on Eq. (3.12), which, for the purpose of easy application, is approximated by:

$$\psi = [0.2 Re^{-0.6} + 1.3 \cdot 10^{-3} Re^{-0.2}]^{1/2} \quad (3.15)$$

range of application: $2320 \leq Re \leq \infty$.

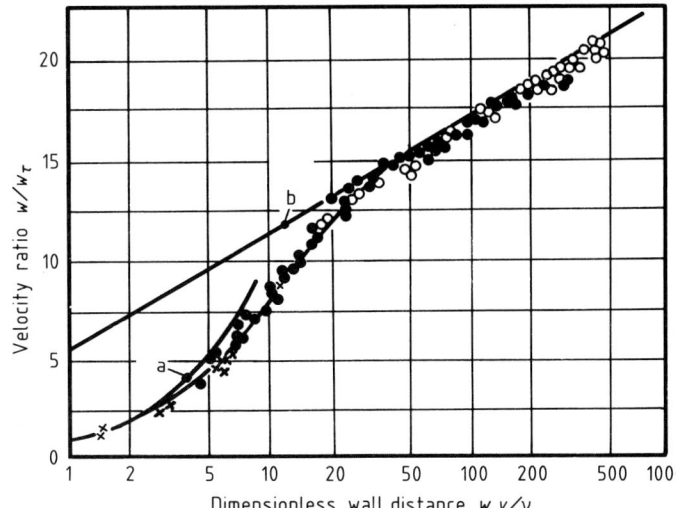

Figure 3.1. Velocity profiles for laminar and turbulent fluid flow in tubes.

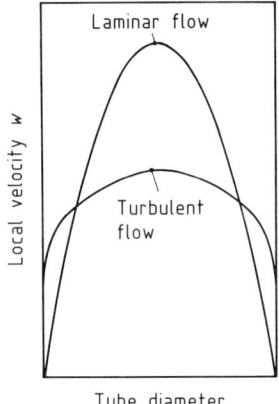

Figure 3.2. Local velocity for turbulent tube flow.

In Fig. 3.3 Eqs. (3.14) and (3.15) are presented by curves a and b, respectively. With a rather limited range of application BLASIUS [3.3] suggested the equation:

$$\psi = \frac{0.3164}{Re^{1/4}}, \qquad (3.16)$$

range of application: $2320 \leqslant Re \leqslant 10^5$.

This equation is presented in Fig. 3.3 by curve c. The transition from laminar to turbulent flow occurs at the critical Reynolds number, which was determined by SCHILLER [3.4] as $Re_{\text{crit}} = 2320$.

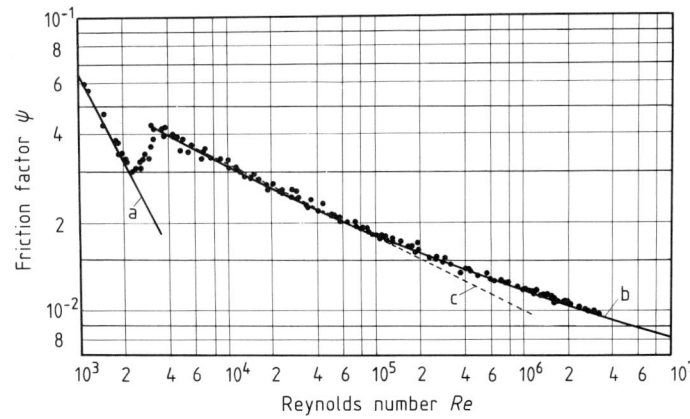

Figure 3.3. Friction factor laws for laminar and turbulent fluid flow in tubes with smooth walls; curve a after Eq. (3.14); curve b after Eq. (3.15); curve c after Eq. (3.16).

The friction factor laws mentioned so far apply to tubes with smooth walls. For rough walls the friction factor increases with the height of roughness elements. Fig. 3.4 shows the accepted plot of the friction factor for rough tubes, [3.5], [3.6]. The roughness parameter R/k_s is the ratio of tube radius R and equivalent sand roughness height k_s. When the Reynolds number Re exceeds the value given by curve d, established roughness flow prevails. Curve d represents the following equation:

$$Re_r = 396 \frac{R}{k_s} \left[2 \log \left(\frac{R}{k_s} \right) + 1.74 \right]. \quad (3.17)$$

For $Re \geqslant Re_r$ the friction factor law for established flow in rough tubes as derived by VON KÁRMÁN [3.7] is given by:

$$\psi = \frac{1}{\left[2 \log \left(\frac{R}{k_s} \right) + 1.74 \right]^2}, \quad (3.18)$$

range of application $Re_r \leqslant Re \leqslant \infty$.

In Fig. 3.4 curves a and b represent the friction factor laws for laminar and turbulent fluid flow in smooth tubes.

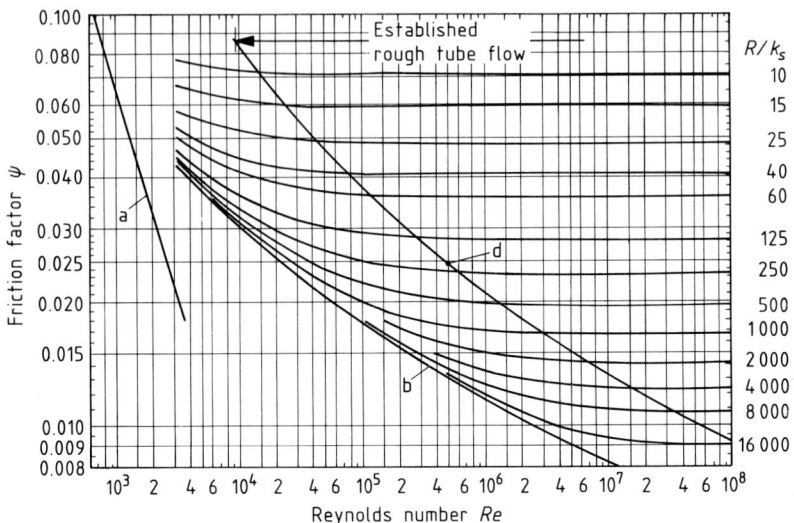

Figure 3.4. Friction factor laws for fluid flow in tubes with rough walls; curve a und b see Fig. 3.3; curve d after Eq. (3.17).

3.3 Heat and Mass Transfer in Laminar Tube Flow

Heat and mass transfer are analogous processes. The mathematical treatment, therefore, will be restricted to the mass transfer process. The final results will be presented for both processes, however.

The local concentration in a fluid is described by a differential equation obtained from Eq. (2.12). After introducing Eq. (3.4) for the local velocity w and transforming the differential equation into dimensionless writing the result is as follows [3.8]:

$$\frac{1}{2}(1 - r^{*2}) \frac{\partial \xi}{\partial z^*} = \frac{\partial^2 \xi}{\partial r^{*2}} + \frac{1}{r^*} \frac{\partial \xi}{\partial r^*}. \quad (3.19)$$

The pertaining boundary conditions are:

$z^* = 0$ and $r^* < 1$: $\xi = 1$
$z^* \geq 0$ and $r^* = 0$: $\partial \xi / \partial r^* = 1$
$z^* \geq 0$ and $r^* = 1$: $\xi = \xi_w = 0$

The dimensionless parameters introduced are defined by:

$$z^* \equiv \frac{z/d}{Re\,Sc}, \quad \text{entrance number (length coordinate)} \quad (3.20)$$

$$r^* \equiv r/R, \quad \text{radial coordinate} \quad (3.21)$$

$$\xi \equiv \frac{\varrho_A - \varrho_{Aw}}{\varrho_{A0} - \varrho_{Aw}}, \quad \text{local concentration} \quad (3.22)$$

$$Re \equiv \frac{\bar{w}\,d}{\nu}, \quad \text{Reynolds number} \quad (3.23)$$

$$Sc \equiv \frac{\nu}{D}, \quad \text{Schmidt number} \quad (3.24)$$

with z axial coordinate, r radial coordinate, R tube radius, d tube diameter, $\bar{w}$ mean fluid velocity, ν kinematic viscosity, D diffusion coefficient of component A, ϱ_A local partial density of A, ϱ_{Aw} partial density at tube wall, and ϱ_{A0} partial density at tube entrance. A comprehensive solution of Eq. (3.19) is only possible by numerical methods [3.8]. The local concentration ξ is a function of r^* and z^*:

$$\xi = f(r^*; z^*) \quad (3.25)$$

This function is presented in Fig. 3.5. At $z^* = 0$: $\xi = 1$ and for $z^* \to \infty$: $\xi \to 0$. The concentration gradient at the wall for $z^* = 0$ is $(\partial \xi / \partial r^*)_w = \infty$, and for $z^* \to \infty$: $(\partial \xi / \partial r^*)_w \to 0$.

The mean value of the concentration over the cross sectional area F of the tube is designated by $\bar{\xi}$ and determined by means of the equation:

$$\bar{\xi} = \frac{1}{\bar{w}F} \int_0^F \xi w\,dF = 4 \int_0^1 (r^* - r^{*3})\xi\,dr^*. \quad (3.26)$$

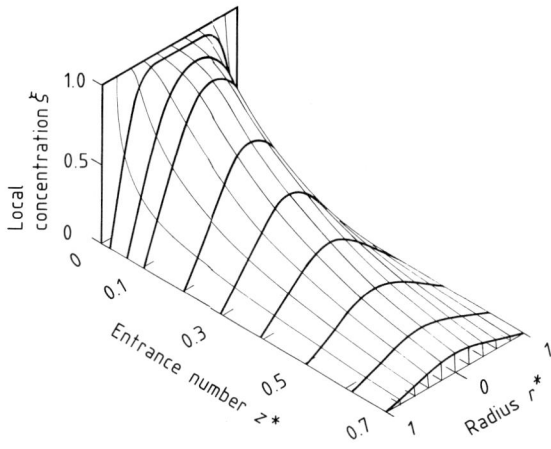

Figure 3.5. Local concentration ξ for mass transfer in laminar tube flow as a function of r^* and z^*.

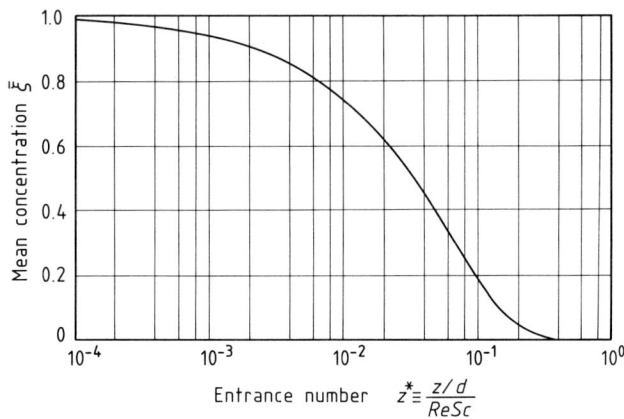

Figure 3.6. Mean concentration $\bar{\xi}$ for mass transfer in laminar tube flow as a function of z^*.

Fig. 3.6 shows the mean concentration as a function of the entrance number z^*. With the mean concentration known, the mass of component A transferred per time unit from the fluid to the wall or *vice versa* can be calculated:

$$\dot{M}_A = \dot{V}(\varrho_{A0} - \bar{\varrho}_A) = \dot{V}\varrho_{A0}\left(1 - \frac{\bar{\varrho}_A}{\varrho_{A0}}\right). \quad (3.27)$$

In dimensionless writing one obtains:

$$\frac{\dot{M}_A}{\dot{V}\varrho_{A0}} = (1-\bar{\xi})(1-\varrho_{Aw}/\varrho_{A0}). \quad (3.28)$$

In general, mass transfer calculations are not carried out by applying the mean concentration $\bar{\xi}$ but by applying the mass transfer coefficient. The local mass transfer coefficient is defined by:

$$\beta_z \equiv \frac{\dot{m}_{Aw}}{\bar{\varrho}_A - \varrho_{Aw}}. \quad (3.29)$$

Introducing Eq. (1.36) for $\dot{m}_{Aw}$ and transforming the resulting equation into dimensionless writing, leads to:

$$Sh_z = -\frac{2}{\bar{\xi}}\left(\frac{\partial \xi}{\partial r^*}\right)_w, \quad (3.30)$$

with:

$$Sh_z \equiv \frac{\beta_z d}{D} \quad (3.31)$$

as the definition of the local Sherwood number. According to Eq. (3.30) the local Sherwood number is proportional to the concentration gradient at the wall, obtained by numerical solution of Eq. (3.19), and to the reciprocal value of the mean concentration. In Fig. 3.7 curve a represents the local Sherwood number Sh_z. For $z^* \to \infty$ the Sherwood number approaches the limiting value: $Sh_\infty = 3.657$.

Besides the local Sherwood number Sh_z the mean Sherwood number Sh is important for the technical design of mass transfer equipment. This mass transfer number is defined by:

$$Sh \equiv \frac{\beta d}{D}, \quad (3.32)$$

with β as the mean mass transfer coefficient, which is derived as follows. The coefficient β is defined by:

$$\beta \equiv \frac{\dot{M}_A}{A \Delta \varrho_A}, \quad (3.33)$$

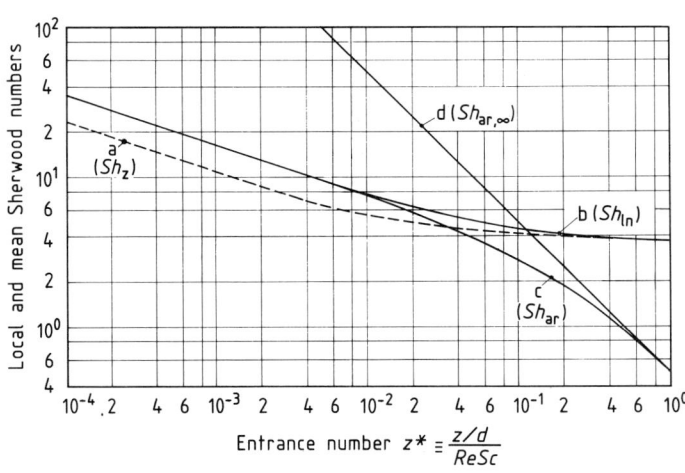

Figure 3.7. Mass transfer in laminar tube flow; curve a: local Sherwood number Sh_z; curve b: mean logarithmic Sherwood number Sh_{ln} after Eq. (3.39); curve c: mean arithmetic Sherwood number Sh_{ar} after Eq. (3.40); curve d: after Eq. (3.42).

3.3 Heat and Mass Transfer in Laminar Tube Flow

with $A = d\pi L$ being the surface area of the tube wall and $\Delta\varrho_A$ a mean value for the concentration difference. Introducing Eq. (3.27) for $\dot{M}_A$ along with dimensionless numbers results in:

$$Sh = \frac{1}{4z^*} \frac{\varrho_{A0} - \bar{\varrho}_A}{\Delta\varrho_A}. \qquad (3.34)$$

For the mean concentration difference $\Delta\varrho_A$ a mean logarithmic or a mean arithmetic value can be defined:

$$(\Delta\varrho_A)_{ln} = \frac{(\varrho_{A0} - \varrho_{Aw}) - (\bar{\varrho}_A - \varrho_{Aw})}{\ln \frac{\varrho_{A0} - \varrho_{Aw}}{\bar{\varrho}_A - \varrho_{Aw}}}, \qquad (3.35)$$

$$(\Delta\varrho_A)_{ar} = \frac{1}{2}[(\varrho_{A0} - \varrho_{Aw}) + (\bar{\varrho}_A - \varrho_{Aw})]. \qquad (3.36)$$

Here, ϱ_{A0} and $\bar{\varrho}_A$ are the mean partial densities at the entrance and the exit of the tube, respectively. With the above mean concentrations appropriate equations for the mean logarithmic and mean arithmetic Sherwood numbers are obtained:

$$Sh_{ln} = \frac{1}{4z^*} \ln(1/\bar{\xi}), \qquad (3.37)$$

$$Sh_{ar} = \frac{1}{2z^*} \frac{1 - \bar{\xi}}{1 + \bar{\xi}}. \qquad (3.38)$$

Introducing the mean concentration $\bar{\xi}$, the logarithmic and arithmetic Sherwood numbers are obtained as presented in Fig. 3.7 by curves b and c. These curves are approximated by the following equations:

$$Sh_{ln} = 3.66 + \frac{0.188 \, (Re \, Sc \, d/L)^{0.80}}{1 + 0.117 \, (Re \, Sc \, d/L)^{0.467}}, \qquad (3.39)$$

$$Sh_{ar} = \left[\frac{1}{(0.619 z^{*1/3})^5 + (2z^*)^5}\right]^{1/5}, \qquad (3.40)$$

range of application:
$0 \leqslant z^* \leqslant \infty \ (0 \leqslant Re \, Sc \, d/L \leqslant \infty)$
$0 \leqslant Re \leqslant 2320$
$0 \leqslant Sc \leqslant \infty$
$0 \leqslant L/d \leqslant \infty$.

For $z^* \to 0$ or $Re \, Sc \, d/L \to \infty$ curves b and c coincide. In this region the equation derived by LÉVÊQUE [3.9] applies:

$$Sh_{ln} = Sh_{ar} = \frac{1.615}{z^{*1/3}} = 1.615 \, (Re \, Sc \, d/L)^{1/3} \qquad (3.41)$$

range of application:
$0 \leqslant z^* \leqslant 10^{-2} \ (10^2 \leqslant Re \, Sc \, d/L \leqslant \infty)$
$0 \leqslant Re \leqslant 2320$.

For $z^* \to \infty$ or $Re \, Sc \, d/L \to 0$ the curve for Sh_{ar} approaches curve d, which represents the equation:

$$Sh_{ar, \infty} = \frac{1}{2z^*}. \qquad (3.42)$$

This is the maximum value of the arithmetic Sherwood number for all possible conditions of mass transfer in laminar and turbulent, forced and natural flow of fluids in tubes with either smooth or rough walls.

When the logarithmic or arithmetic mean Sherwood number is known the mean concentration $\bar{\xi}$ can be calculated, making use of the following equations which have been derived from Eqs. (3.37) and (3.38):

$$\bar{\xi} = e^{-(4z^* Sh_{ln})} \qquad (3.43)$$

$$\bar{\xi} = \frac{1 + 2z^* Sh_{ar}}{1 - 2z^* Sh_{ar}}. \qquad (3.44)$$

These equations can be introduced into Eq. (3.28) to calculate the transferred mass flux $\dot{M}_A$.

All equations which have been presented for mass transfer can also be applied to heat transfer when the following relations are observed:

$\varrho_A \triangleq T$ local fluid temperature
$\varrho_{A0} \triangleq T_0$ fluid temperature at tube entrance
$\varrho_{Aw} \triangleq T_w$ fluid temperature at tube wall
$\bar{\varrho}_A \triangleq \bar{T}$ mean fluid temperature

$\Delta\varrho_A \triangleq \Delta T$ mean temperature difference

$\xi \triangleq \theta \equiv \dfrac{T-T_w}{T_0-T_w}$ dimensionless fluid temperature

$\dot{M}_A \triangleq \dot{Q} = \dot{V}\varrho c_p (T_0-\bar{T})$ heat flux

$\beta_z \triangleq \alpha_z = \dot{q}/(\bar{T}-T_w)$ local heat transfer coefficient

$\beta \triangleq \alpha = \dot{Q}/(A\,\Delta T)$ mean heat transfer coefficient

$Sh_z \triangleq Nu_z = \alpha_z d/\lambda$ local Nusselt number

$Sh \triangleq Nu = \alpha d/\lambda$ mean Nusselt number

$Sc \triangleq Pr = \nu/a$ Prandtl number

Here, ϱ is the mean fluid density, c_p the specific heat capacity at constant pressure, $\dot{q}$ heat flux density, $\dot{Q}$ heat flux, λ heat conductivity, and $a = \varrho c_p/\lambda$ the heat diffusivity.

For heat as well as for mass transfer all fluid properties have to be considered at the fluid temperature:

$$T_f = \frac{1}{2}(T_0 + \bar{T}), \qquad (3.45)$$

with T_0 and $\bar{T}$ being the mean fluid temperatures at the inlet and the outlet of the tube, respectively.

All equations given in this section are in satisfactory agreement with experimental data.

Based on the results of comprehensive theoretical and experimental investigations the following mass transfer equation is suggested [3.8]:

$$Sh = \frac{Re\,Sc\,\psi/8}{1.07 + 12.7\,(Sc^{2/3}-1)\sqrt{\psi/8}}$$

$$\cdot \left(1 - \frac{180}{Re^{0.75}}\right)\left[1 + \left(\frac{d}{L}\right)^{2/3}\right], \qquad (3.46)$$

range of application: $2320 \leqslant Re \leqslant \infty$
$\qquad\qquad\qquad\qquad\;\; 0.5 \leqslant Sc \leqslant \infty$
$\qquad\qquad\qquad\qquad\;\; 2 \leqslant L/d \leqslant \infty.$

It should be noted that the value of the Schmidt number Sc ist of great importance for the influence of Schmidt and Reynolds numbers on the Sherwood number. For $Sc \to \infty$ and $Sc \to 1$ the following relations apply:

$$Sh\underset{Sc\to\infty}{} = 0.0157\,Re^{0.875}\,Sc^{1/3} \cdot$$

$$\cdot \left(1 - \frac{180}{Re^{0.75}}\right)\left[1 + \left(\frac{d}{L}\right)^{2/3}\right] \qquad (3.47)$$

$$Sh\underset{Sc\to 1}{} \sim Re^{0.75}\,Sc^{0.5}. \qquad (3.48)$$

Eq. (3.46) is in excellent agreement with experimental data. The fluid properties should be recorded at the fluid temperature T_f defined by Eq. (3.45).

3.4 Heat and Mass Transfer in Turbulent Tube Flow

The case considered here will be mass transfer in smooth tubes. The equations at the end of the previous section also apply to heat transfer when the relations between mass and heat transfer are observed.

3.5 Heat and Mass Transfer in Turbulent Flow Through Rough Tubes

Roughness of a tube wall enhances heat and mass transfer because of the significantly different flow conditions in tubes with rough walls as compared with the conditions in tubes with smooth walls. Fig. 3.8 qualitatively shows velocity and shear stress profiles in tubes with smooth and rough walls. In the case of smooth tubes the shear stress τ increases linearly from $\tau=0$ at the tube axis to $\tau=\tau_w$ at the tube wall. As with rough walls the rough elements strongly influence momentum transfer in the fluid, due to vortex formation in the wake of the roughness elements. The vortex layer extends from the distance y_0 to y_r of the wall.

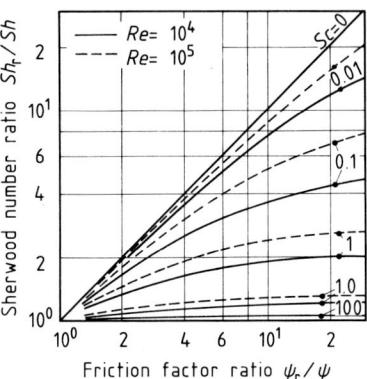

Figure 3.9. Mass transfer in turbulent fluid flow through rough tubes.

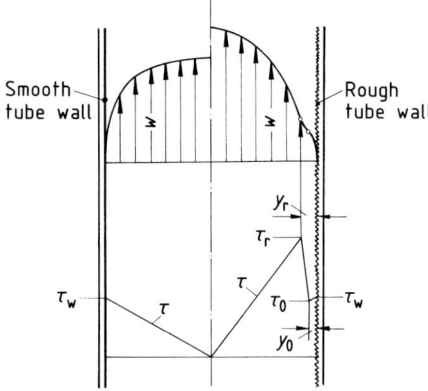

Figure 3.8. Profiles for velocity and shear stress in turbulent flow through smooth and rough tubes.

and smooth tubes over the ratio of friction factors ψ_r/ψ. This is shown in Fig. 3.9. There is a pronounced influence of roughness on mass transfer in the range of small Sc-numbers. With increasing Sc-number the roughness influence diminishes. This is due to the fact that the concentration gradient becomes steeper with increasing Schmidt number. Roughness has no chance to further increase the gradient; it has a pronounced influence on mass transfer in gases, but an almost negligible influence on mass transfer in liquids.

In this layer the shear stress increases from τ_0 to τ_r. At the maximum of the shear stress a point of inflection is observed in the velocity profile.

On the basis of the explained shear stress profile NUNNER [3.10] derived the following theoretical equation for mass transfer in turbulent fluid flow through rough tubes:

$$Sh_r = \frac{\psi}{8} \frac{Re\, Sc\, \psi_r/\psi}{1+(Sc\,\psi_r/\psi-1)1.5\,Re^{-1/8}Sc^{-1/6}}, \quad (3.49)$$

with ψ_r as the friction factor for rough and ψ that for smooth tubes. The enhancement of mass transfer by the action of roughness elements is best shown by plotting the ratio of Sherwood numbers Sh_r/Sh for rough

3.6 Mass Transfer with Biochemical Reactions

A rather simple case of a biochemical reaction will be considered to demonstrate the interactions of reaction, diffusion, and convection in a laminar flow field. The selected example involves a biochemical reaction driven by enzymes which are immobilized on the inner surface of a tube wall.

3.6.1 The Biochemical and Mathematical Problem

A reaction of immobilized enzymes at the tube wall requires transportation of the reactants to the wall and of the reaction products back into the fluid. According to Fig. 3.10 it will be assumed that the immobilized enzymes form a thin layer on the surface of the inner tube wall, and that there is no diffusional resistance within this layer. For a small element of the surface area the reaction will be considered. For small concentrations c_{Aw} of the reactant A at the wall the reaction rate $\dot{r}_{Aw}$ may be described in accordance with Eq. (1.80) by:

$$\dot{r}_{Aw} = k_w c_{Aw}. \tag{3.50}$$

Under steady state conditions the reaction rate $\dot{r}_{Aw}$ must be equal to the diffusion rate:

$$\dot{n}_{Aw} = -D \left(\frac{\partial c_A}{\partial r} \right)_w. \tag{3.51}$$

From the last two equations the concentration gradient follows as:

$$\left(\frac{\partial c_A}{\partial r} \right)_w = -\frac{k_w}{D} c_{Aw}. \tag{3.52}$$

Introducing the dimensionless radial coordinate $r^* \equiv r/R$, and dimensionless concentration leads to:

$$\bar{\xi} \equiv \frac{c_A}{c_{A0}}, \tag{3.53}$$

with c_{A0} being the concentration of component A in the fluid at the tube inlet. Eq. (3.52) may be rewritten as follows:

$$\left(\frac{\partial \xi}{\partial r^*} \right)_w = -Da\, \xi_w. \tag{3.54}$$

Here, Da is the Damköhler number, named after the German physico-chemist DAMKÖHLER, which is defined as:

$$Da \equiv \frac{k_w}{D/R}. \tag{3.55}$$

The Damköhler number may be interpreted as the ratio of reaction velocity (k_w) and diffusion velocity (D/R). Da can assume values between $Da=0$ and $Da=\infty$. With $Da \to 0$ the process is controlled by the reaction velocity, because k_w tends towards zero. The local concentration c_A becomes

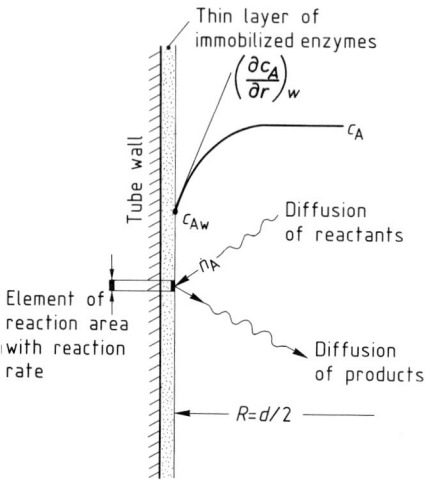

Figure 3.10. Explanation of interaction of reaction rate $\dot{r}_{Aw}$ and diffusion rate $\dot{n}_A$.

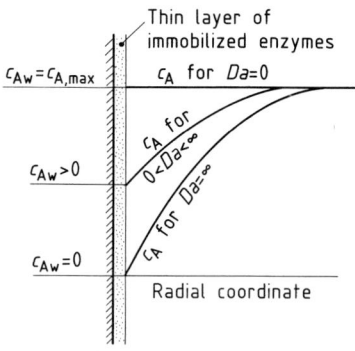

Figure 3.11. Qualitative concentration profiles for $Da=0$, $Da=\infty$, and an intermediate value of Da.

independent of the local radius as shown qualitatively in Fig. 3.11. With $Da \to \infty$ the process is controlled by the diffusion velocity D/R, which tends towards zero. In this case the concentration at the wall becomes zero, as indicated in Fig. 3.11. For any values of $0 < Da < \infty$ the concentration at the wall is given by $c_{Aw} > 0$. With increasing value of Da the concentration gradient at the wall, $(\partial c_A/\partial r)_w$, decreases steadily towards zero. For $Da = \infty$ the gradient assumes the maximum value.

The concentration field is described by the differential equation given by Eq. (3.19). While the first two boundary conditions also apply to this case, the third boundary condition has to be replaced by Eq. (3.54). The solution will be obtained by numerical methods [3.8].

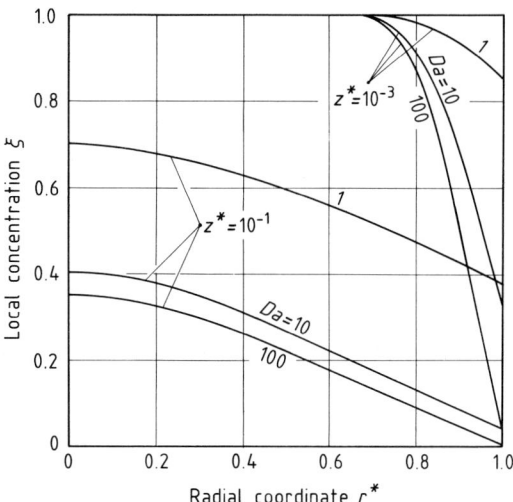

Figure 3.12. Concentration profiles for several values of z^* and Da.

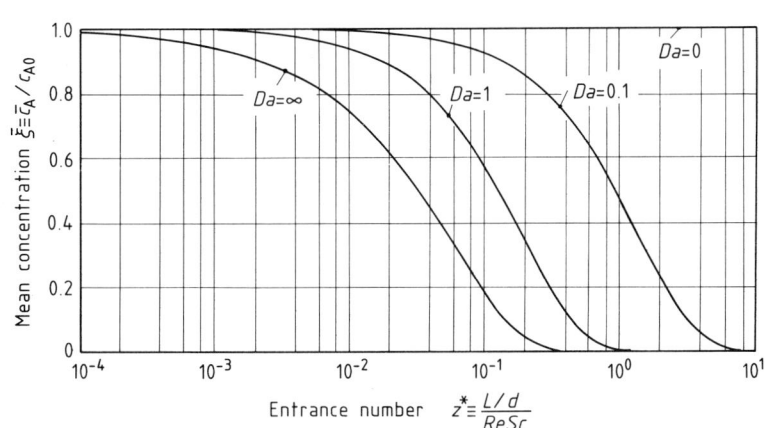

Figure 3.13. Mean concentration $\bar{\xi}$ as a function of the entrance number z^* for several values of the Damköhler number Da.

3.6.2 Local and Mean Concentrations

The local dimensionless concentration ξ is a function of three parameters:

$$\xi = f(r^*; z^*; Da). \tag{3.56}$$

In Fig. 3.12 ξ is plotted against the radial coordinate for several values of the entrance number z^* and the Damköhler number Da. For $z^* = 0$ and all values of Da the local concentration is given by $\xi = 1$. For $z^* = \infty$ and all values of Da the local concentration is given by $\xi = 0$.

The mean concentration $\bar{\xi}$ is plotted in Fig. 3.13 against the entrance number z^* for various values of the Damköhler number Da. For $Da = 0$: $\bar{\xi} = 1$, and for $Da = \infty$: $\bar{\xi} = 0$. With the mean concentration $\bar{\xi}$ the mass flux $\dot{M}_A$ can be calculated:

$$\frac{\dot{M}_A}{\dot{V} c_{A0}} = 1 - \frac{\bar{c}_A}{c_{A0}} = 1 + \bar{\xi}. \tag{3.57}$$

3.6.3 Mean Sherwood Number

Only the mean arithmetic Sherwood number Sh_{ar} will be discussed. The mean arithmetic concentration difference is in this case defined by:

$$(\Delta c_A)_{ar} = \frac{1}{2}(c_{A0} - \bar{c}_A). \qquad (3.58)$$

Introducing the mean concentration into Eq. (3.38) the result is Sh_{ar} as shown in Fig. 3.14. The theoretical results are in excellent agreement with experimental results.

The curve for $Da = \infty$ is identical with curve c presented in Fig. 3.7. For a heterogeneous biochemical reaction the Sherwood number is smaller than that for mass transfer without a chemical reaction.

With $z^* \to 0$ the Sherwood number approaches the value:

$$Sh_{ar} \atop z^* \to 0 = 2\,Da. \qquad (3.59)$$

This is the reaction controlled case. The reaction velocity due to its slowness compared with the diffusional velocity, determines the progress of the biochemical conversion. The mass transfer coefficient β_{ar} is only a function of the reaction parameter k_w, and is independent of any convection parameters:

$$\beta_{ar} \atop z^* \to 0 = k_w. \qquad (3.60)$$

For $z^* \to \infty$ the Sherwood number approaches the final value $1/2z^*$. In this convection controlled case the mass transfer coefficient β_{ar} is only a function of convection parameters:

$$\beta_{ar} \atop z^* \to \infty = \frac{1}{2}\frac{d}{L}\bar{w}. \qquad (3.61)$$

With the mean arithmetic Sherwood number Sh_{ar} known the biochemically converted mass per time unit, $\dot{M}_A$, may be calculated using Eq. (3.57). The mean concentration $\bar{\xi}$ can be obtained from Sherwood number correlations by applying Eq. (3.44), in which Sh_{ar}, as shown in Fig. 3.14, has to be introduced.

3.7 References

[3.1] H. SCHLICHTING: "Grenzschicht-Theorie". Verlag G. Braun, Karlsruhe, 1958.
[3.2] L. PRANDTL: "Führer durch die Strömungslehre". Vieweg, Braunschweig, 1949.
[3.3] H. BLASIUS: "Das Ähnlichkeitsgesetz bei Reibungsvorgängen in Flüssigkeiten". VDI-Forschungsh. 131 (1913).

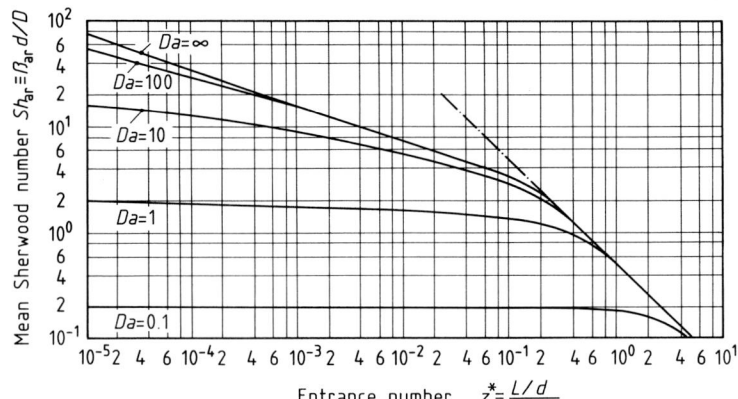

Figure 3.14. Mean arithmetic Sherwood number Sh_{ar} as a function of the entrance number z^* for several values of the Damköhler number.

[3.4] L. SCHILLER: "Experimentelle Untersuchung zum Turbulenzproblem". Z. Angew. Math. Mech. *1* (1921), 436–444.

[3.5] C. F. COLEBROOK: "Turbulent flow in pipes, with particular reference to the transition region between the smooth and rough pipe laws". J. Inst. Civ. Eng. (London) *11* (1938/39), 133–156.

[3.6] L. F. MOODY: "Friction factors for pipe flow". Trans. ASME *66* (1944) 12, 671–684.

[3.7] TH. VON KÁRMÁN: "Mechanische Ähnlichkeit und Turbulenz". Nachr. Ges. Wiss. Göttingen, Math.-Physik. Klasse 58 (1930).

[3.8] H. BRAUER: "Stoffaustausch". Verlag Sauerländer, Aarau–Frankfurt am Main, 1971.

[3.9] M. A. LÉVÊQUE: "Les lois de la transmission de chaleur par convection". Ann. Mines *12* (1928), 201–299.

[3.10] W. NUNNER: "Wärmeübergang und Druckabfall in rauhen Rohren". VDI-(Verein Deutscher Ingenieure)Forschungsh. 455, VDI-Verlag GmbH, Düsseldorf, 1956.

Chapter 4
Transport Processes in Non-Newtonian Fluids Flowing Through Tubes

Heinz Brauer

Institut für Chemieingenieurtechnik
Technische Universität Berlin
Berlin (West), Germany

4.1 Introduction
4.2 Velocity Profiles for Power Law Fluids
4.3 Friction Factors for Power Law Fluids
4.4 Mass Transfer in Power Law Fluids
4.5 Heat Transfer in Power Law Fluids
4.5.1 Heat Transfer Without Internal Frictional Heat Sources
4.5.2 Heat Transfer With Internal Frictional Heat Sources
4.6 References

List of Symbols

A	m²	area
c_p	kJ/(kg K)	specific heat capacity of fluid
D	m²/s	diffusion coefficient
d	m	tube diameter
F	m²	cross sectional area
K	$\tau/(-dw/dy)^n$	Ostwald factor
L	m	length of tube
$\dot{M}_A$	kg/s	mass flux of component A
$\dot{m}_A$	kg/(s m²)	mass flux density of component A
p	N/m²	total pressure
$\dot{Q}$	kJ/s	heat flux
$\dot{q}$	kJ/(s m²)	heat flux density
R	m	tube radius
r	m	radial coordinate
T	K	local temperature
$\bar{T}$	K	mean temperature
T_0	K	temperature at tube inlet
T_w	K	temperature at tube wall
$\dot{V}$	m³/s	volumetric flow rate
w	m/s	local velocity
$\bar{w}$	m/s	mean velocity
w_m	m/s	maximum velocity
w	m/s	shear velocity
y	m	distance from tube wall
z	m	axial coordinate
α	kJ/(s m² K)	mean coefficient of heat transfer
β	m/s	mean coefficient of mass transfer
Δp	N/m²	pressure drop
ΔT	K	temperature difference
$\Delta \varrho_A$	kg/m³	partial density difference
λ	kJ/(m s K)	heat conductivity
ϱ	kg/m³	fluid density
ϱ_A	kg/m³	local partial density of component A
$\bar{\varrho}_A$	kg/m³	mean partial density of component A
ϱ_{A0}	kg/m³	partial density of component A at tube entrance
ϱ_{Aw}	kg/m³	partial density of component A at tube wall
τ	N/m²	shear stress
τ_w	N/m²	shear stress at tube wall
n		fluid index
$Nu \equiv \dfrac{\alpha d}{\lambda}$		mean Nusselt number
$Nu_z \equiv \dfrac{\alpha_z d}{\lambda}$		local Nusselt number
$Pr \equiv \dfrac{K c_p}{\lambda} \left(\dfrac{d}{\varrho \bar{w}}\right)^{1-n} \left(\dfrac{1+3n}{4n}\right)^n$		Prandtl number
$q_F^* \equiv \dfrac{\bar{w} d^2 \Delta p/L}{\lambda (T_0 - T_w)}$		friction number
$r^* \equiv r/R$		radial coordinate
$Re_{n-N} \equiv \dfrac{\bar{w}^{2-n} d^n \varrho}{K \left(\dfrac{1+3n}{4n}\right)^n 8^{n-1}}$		Reynolds number
$Sc \equiv \dfrac{K/\varrho}{D} \left(\dfrac{d}{\varrho \bar{w}}\right)^{1-n} \left(\dfrac{1+3n}{4n}\right)^n$		Schmidt number
$Sh \equiv \dfrac{\beta d}{D}$		mean Sherwood number
$Sh_z \equiv \dfrac{\beta_z d}{D}$		local Sherwood number
$y^+ \equiv \dfrac{w^{2-n} y^n \varrho}{K}$		dimensionless distance from tube wall
$z^* \equiv \dfrac{z/D}{(Re\,Sc)_{n-N}} = \dfrac{z}{d}\dfrac{D}{\bar{w} d}$		entrance number
$\xi \equiv \dfrac{\varrho_A - \varrho_{Aw}}{\varrho_{A0} - \varrho_{Aw}}$		local concentration ratio
$\psi \equiv \dfrac{\Delta p}{\varrho \bar{w}^2/2} \dfrac{d}{L}$		friction coefficient

4.1 Introduction

Biochemical fluids may have either Newtonian or non-Newtonian properties. The most important property is the fluidity described by the relationship between shear stress and shear rate. The momentum transport in all Newtonian fluids is described by one single law, shown in Eq. (1.1), for fluids in laminar motion. For non-Newtonian fluids the situation is far more complicated. In this case there are several laws describing momentum transfer, one for each group of non-Newtonian fluids with similar properties as previously mentioned in Chapter 1. Therefore it is not possible to develop general laws for momentum, heat, and mass transfer for non-Newtonian fluids, as has been done for Newtonian fluids.

It has been pointed out that the flow properties of a large number of non-Newtonian fluids are described by the power law. For this group of fluids the velocity field in tube flow, friction factor laws, and heat and mass transfer will be discussed in this chapter. It should be borne in mind that this discussion serves as an example for non-Newtonian fluids. For other non-Newtonian fluids the laws for momentum, heat, and mass transfer may be quite different.

4.2 Velocity Profiles for Power Law Fluids

For power law fluids the shear stress/shear rate relationship is presented by the so-called power law, as given by Eq. (1.5):

$$\tau = K \left(-\frac{dw}{dr}\right)^n, \qquad (4.1)$$

with τ as shear stress, dw/dr gradient of the velocity w in the direction of the radial coordinate r, K Ostwald factor and n fluid index of this group of fluids. When $n=1$ Eq. (4.1) describes momentum transfer in Newtonian fluids; in this case the Ostwald factor is identical with the dynamic viscosity.

The velocity field for established flow of non-Newtonian fluids in tubes is described by the following equation which has been derived from Eq. (2.10):

$$\frac{dp}{dz} = \frac{\Delta p}{L} = \frac{1}{r}\frac{d}{dr}(r\tau). \qquad (4.2)$$

The boundary conditions are as follows:

$r = R: \quad w = 0$

$r = 0: \quad dw/dr = 0$

Here, $\Delta p/L$ is the pressure (p) decrease over the length L of the tube, and R the tube radius.

Integration of Eq. (4.2) gives the local velocity [4.1]:

$$w = \left(\frac{\Delta p/L}{2K}\right)^{1/n} \frac{R^{1+1/n}}{1+1/n}\left[1-\left(\frac{r}{R}\right)^{1+1/n}\right]. \qquad (4.3)$$

Introducing the relationship between wall shear stress τ_w and pressure drop Δp yields:

$$\tau_w = \frac{\Delta p}{L}\frac{R}{2}. \qquad (4.4)$$

Now, Eq. (4.3) may be rewritten as:

$$w = \frac{(\tau_w/K)^{1/n} R}{3+1/n}\left[1-\left(\frac{r}{R}\right)^{1+1/n}\right]. \qquad (4.5)$$

The mean velocity $\bar{w}$ is defined by:

$$\bar{w} = \frac{\dot{V}}{F} = \frac{8}{d^2}\int_0^R w\,r\,dr, \qquad (4.6)$$

with $\dot{V}$ the volumetric flow rate of the fluid, F cross sectional area of the tube, and d tube diameter. Introducing Eq. (4.5) for w

the integration leads to:

$$\bar{w} = \frac{(\tau_w/K)^{1/n} R}{1+1/n}.\qquad(4.7)$$

The ratio of local to mean velocity is then given by:

$$\frac{w}{\bar{w}} = \frac{1+3n}{1+n}\left[1-\left(\frac{r}{R}\right)^{1+1/n}\right].\qquad(4.8)$$

The first term on the right hand side of this equation is the ratio of maximum to mean velocity:

$$\frac{w_m}{\bar{w}} = \frac{1+3n}{1+n}.\qquad(4.9)$$

Velocity profiles for various values of n are presented in Fig. 4.1. For $n=0$ the local velocity is constant over the tube radius, so that $w/\bar{w}=1$; for $n=1$ the velocity profile is parabolic and for $n=\infty$ conical.

For turbulent flow no theoretical equations for the velocity profile have been derived. However, there are empirical equations available, which are based on a careful physical analysis of momentum transfer in non-Newtonian fluids. For power law fluids DODGE and METZNER [4.2] presented the following equations for the local velocity in turbulent tube flow:

$$\frac{w}{w_\tau} = (y^+)^{1/n} \text{ for the laminar wall layer}$$
$$\text{(laminar sublayer)}\qquad(4.10)$$

and

$$\frac{w}{w_\tau} = f_1(n)\log y^+ + f_2(n) \text{ for the tur-}\qquad(4.11)$$
$$\text{bulent core}$$

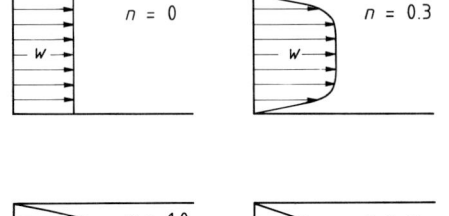

Figure 4.1. Some examples for velocity profiles of non-Newtonian power law fluids in laminar tube flow; $n=0$ and $n=0.3$ for pseudoplastic fluids; $n=1$ for Newtonian fluids; $n=\infty$ for extreme dilatant fluids.

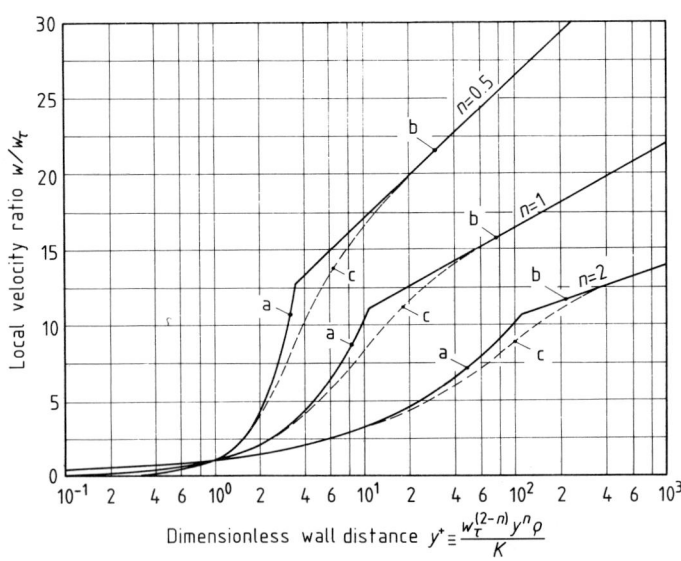

Figure 4.2. Some examples for velocity profiles of non-Newtonian power law fluids in turbulent tube flow; $n=0.5$ for pseudoplastic fluids; $n=1$ for Newtonian fluids; $n=2$ for dilatant fluids.

with:

$$f_1(n) = \frac{5.66}{n^{0.75}}, \quad (4.12)$$

$$f_2(n) = -\frac{0.40}{n^{1.2}} + \frac{2.458}{n^{0.75}}[1.960 + 1.255n - 1.628n\log(3+1/n)] \quad (4.13)$$

$$y^+ \equiv \frac{w_\tau^{(2-n)} y^n \varrho}{K} \quad (4.14)$$

$$w_\tau \equiv \sqrt{\tau_w/\varrho}. \quad (4.15)$$

For Newtonian fluids with $n=1$ Eqs. (4.10) and (4.11) are almost identical with Eqs. (3.11) and (3.12). In Fig. 4.2 Eqs. (4.10) and (4.11) are represented by curves a and b, respectively. The broken curves c give a more realistic velocity profile in the buffer layer between the linear sublayer and core. The fluid index $n=1$ represents the Newtonian case, $n=0.5$ an example for pseudoplastic fluids, and $n=2$ an example for dilatant fluids.

4.3 Friction Factors for Power Law Fluids

Friction factor laws for power law fluids have been derived on the assumption that the definition of the friction factor for Newtonian fluids may also be applied for non-Newtonians:

$$\psi \equiv \frac{4\tau_w}{\varrho \bar{w}^2/2} = \frac{\Delta p}{\varrho \bar{w}^2/2} \frac{d}{L}. \quad (4.16)$$

For laminar tube flow of power law fluids METZNER and REED [4.3] suggested the following friction factor law:

$$\psi = \frac{64}{Re_{n\text{-N}}}, \quad (4.17)$$

with $Re_{n\text{-}N}$ the Reynolds number defined by:

$$Re_{n\text{-N}} \equiv \frac{8\varrho \bar{w}^2}{\tau_w} = \frac{\bar{w}^{2-n} d^n \varrho}{K\left(\frac{1+3n}{4n}\right)^n 8^{n-1}}. \quad (4.18)$$

It has been shown that this definition for a Reynolds number coincides with that for Newtonian fluids (Re_N) multiplied with the ratio of the wall shear stress for Newtonian ($\tau_{w,N}$) and for non-Newtonian ($\tau_{w,n\text{-}N}$) fluids [4.1]:

$$Re_{n\text{-N}} \equiv Re_N \frac{\tau_{w,N}}{\tau_{w,n\text{-N}}} = \frac{\bar{w} d \varrho}{\eta} \frac{\tau_{w,N}}{\tau_{w,n\text{-N}}}. \quad (4.19)$$

Fig. 4.3 gives a comparison between Eq. (4.17), represented by the curve, and experimental data collected by METZNER and REED.

At the critical value of the Reynolds number, $Re_{n\text{-N,crit}}$, laminar fluid flow becomes unstable and turbulence starts to develop. On the basis of a stability number introduced by RYAN and JOHNSON [4.4] the following equation has been developed for

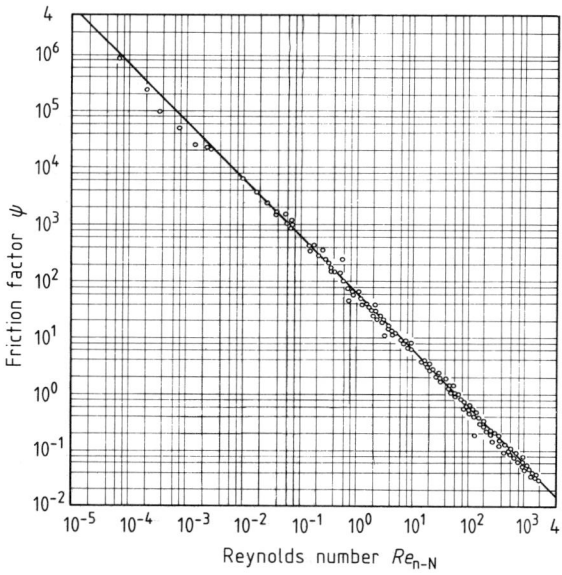

Figure 4.3. Friction factor law for laminar tube flow of pseudoplastic fluids.

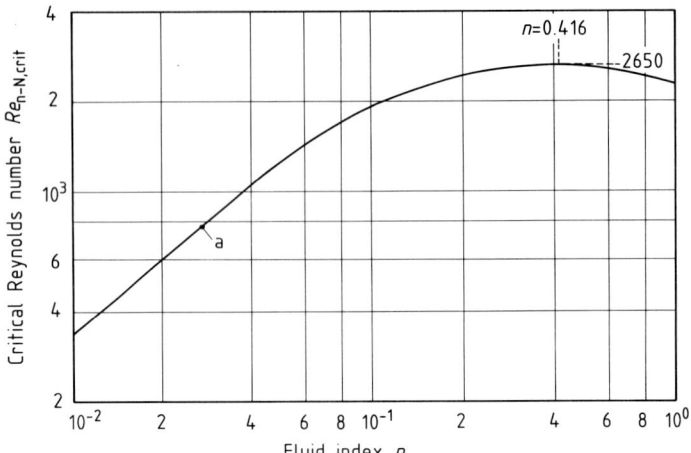

Figure 4.4. Critical Reynolds number for tube flow of pseudoplastic fluids.

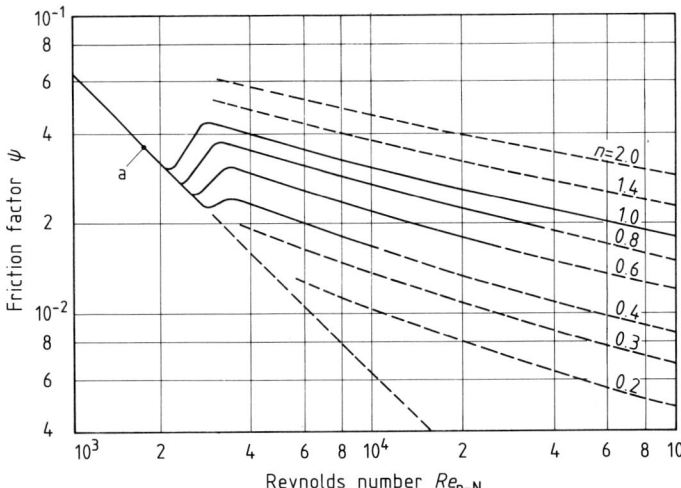

Figure 4.5. Friction factor law for turbulent tube flow of power law fluids.

the critical value of the Reynolds number [4.1]:

$$Re_{\text{n-N,crit}} = 885 \frac{8n}{(1+3n)^2} (2+n)^{\frac{2+n}{1+n}}. \quad (4.20)$$

According to this equation the critical value of Re_N for Newtonian fluids is $Re_{N,\text{crit}} = 2300$. In Fig. 4.4 $Re_{\text{n-N,crit}}$ is presented as a function of the fluid index n. For $n = 0.416$ the critical value of $Re_{\text{n-N}}$ assumes a maximum value of 2650. It therefore seems justified to assume for all values of $0.1 \leq n \leq 1$ the critical Reynolds number to be $Re_{\text{n-N}} \approx 2300$. For $n > 1$ the critical value decreases further, until it attains the value $Re_{\text{n-N,crit}} = 787$ for $n = \infty$. Experimental investigations have been carried out over the range $0.16 \leq n \leq 1$, and have been shown to be in satisfactory agreement with the assumptions made.

A friction factor law for turbulent tube flow of power law fluids has been derived by DODGE and METZNER [4.2]:

$$\sqrt{\frac{1}{\psi}} = \frac{2.0}{n^{0.75}} \log[Re_{\text{n-N}}(\psi/4)^{1-n/2}] - \frac{0.2}{n^{1.2}}. \quad (4.21)$$

The equation is presented in Fig. 4.5 for various values of the fluid index n. Those ranges of $Re_{n\text{-}N}$ and n which have been tested by experiments are given as full curves. Curve a represents the friction factor for laminar tube flow.

4.4 Mass Transfer in Power Law Fluids

A description of the basic conditions in mass transfer has been given in Sect. 1.5.1.3. The differential equation describing the concentration field for a fluid in laminar motion within a tube is obtained from Eq. (2.12). After introducing Eq. (4.3) for the local velocity w and transformation of the differential equation into dimensionless writing it follows that:

$$\frac{1+3n}{4(1+n)}(1-r^{*\,1+1/n})\frac{\partial \xi}{\partial z^*} = \frac{\partial^2 \xi}{\partial r^{*2}} + \frac{1}{r^*}\frac{\partial \xi}{\partial r^*}.$$

(4.22)

The appropriate boundary conditions are given by:

$z^* = 0$ and $r^* < 1$: $\xi = 1$

$z^* \geq 0$ and $r^* = 0$: $\partial \xi / \partial r^* = 0$

$z^* \geq 0$ and $r^* = 1$: $\xi = \xi_w = 0$

The introduced dimensionless parameters are defined by:

$$z^* \equiv \frac{z/d}{(Re\,Sc)_{n\text{-}N}} = \frac{z}{d}\frac{D}{\bar{w}d} \quad (4.23)$$

entrance number (length coordinate)

$r^* \equiv r/R$ radial coordinate   (4.24)

$$\xi \equiv \frac{\varrho_A - \varrho_{Aw}}{\varrho_{A0} - \varrho_{Aw}} \quad \text{local concentration} \quad (4.25)$$

$$Sc_{n\text{-}N} \equiv \frac{K/\varrho}{D}\left(\frac{d}{8\bar{w}}\right)^{1-n}\left(\frac{1+3n}{4n}\right)^n \quad (4.26)$$

Schmidt number

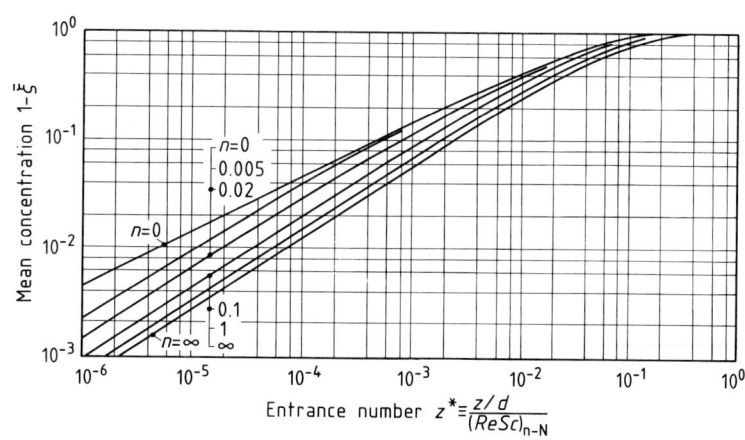

Figure 4.6. Function $1-\bar{\xi}$ of the mean concentration $\bar{\xi}$ for mass transfer in laminar tube flow of a non-Newtonian fluid as a function of z^* for various values of the fluid index n between $n=0$ and $n=\infty$.

The Reynolds number $Re_{n\text{-}N}$ has been defined by Eq. (4.19). In the latter equations z is the axial coordinate, d tube diameter, D diffusion coefficient, $\bar{w}$ mean fluid velocity, r radial coordinate, R tube radius, ϱ_A local partial density of component A, ϱ_{Aw} partial density at tube wall, ϱ_{A0} partial density at tube inlet, ϱ mean fluid density, n fluid index, and K Ostwald factor.

A comprehensive solution of Eq. (4.22) for various conditions has been presented by SCHLÜTER [4.5]. The immediate result of the integration of Eq. (4.22) is the local concentration:

$$\xi = f(r^*; z^*; n). \qquad (4.27)$$

A further integration according to Eq. (3.26) leads to the mean concentration $\bar{\xi}$. Fig. 4.6 contains the results for $\bar{\xi}$ as a function of z^* and the fluid index n. With the mean concentration known the transferred mass per time unit, M_A, can be determined according to Eq. (3.28).

The local and mean mass transfer coefficients, β_z and β, as well as the local and mean Sherwood numbers, Sh_z and Sh, are defined and calculated in the same way as in the case of Newtonian fluids. The procedure has been explained in Sect. 3.3. The local Sherwood number Sh_z is presented in

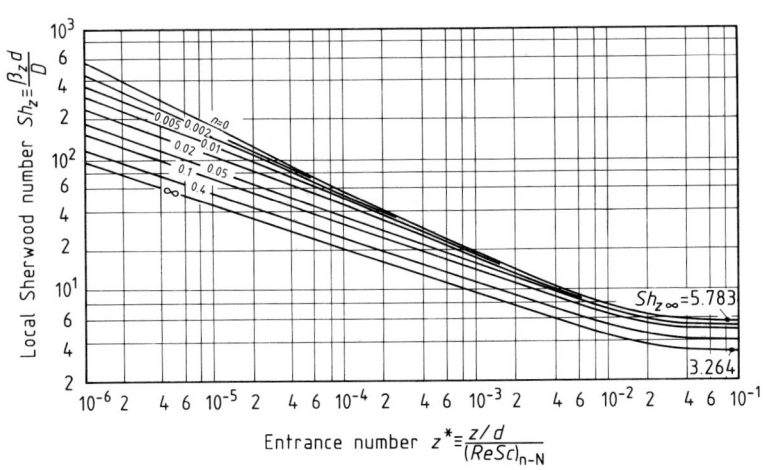

Figure 4.7. Local Sherwood number Sh_z for mass transfer in laminar tube flow of a non-Newtonian fluid as a function of z^* for various values of the fluid index from $n=0$ and $n=\infty$.

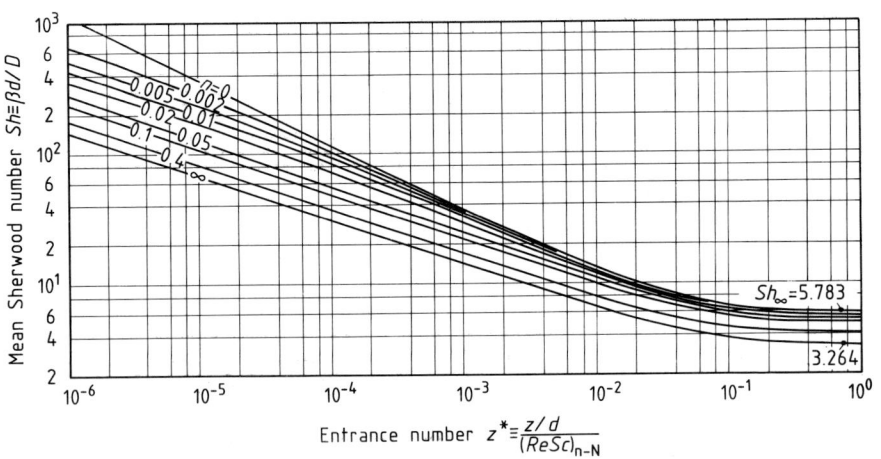

Figure 4.8. Mean Sherwood number Sh for mass transfer in laminar tube flow of a non-Newtonian fluid as a function of z^* for various values of the fluid index between $n=0$ and $n=\infty$.

Fig. 4.7 and the mean logarithmic Sherwood number Sh_{ln} in Fig. 4.8. Application of these results follows the procedure outlined in Sect. 3.3.

4.5 Heat Transfer in Power Law Fluids

There are two distinct cases of heat transfer. In the first case internal heat sources due to friction are excluded. In the second case frictional heating is included. In both cases laminar fluid flow is assumed to apply.

4.5.1 Heat Transfer Without Internal Frictional Heat Sources

With frictional heat sources excluded the considered case of heat transfer is strictly analogous to the mass transfer case considered in the preceding section. Application of the results presented for mass transfer to heat transfer requires the appropriate substitution of mass transfer quantities by the appropriate heat transfer quantities:

mass transfer quantity	$\triangleq$	heat transfer quantity	
ϱ_A	$\triangleq$	T	local temperature
ϱ_{A0}	$\triangleq$	T_0	temperature at tube inlet
ϱ_{Aw}	$\triangleq$	T_w	temperature at tube wall
$\bar{\varrho}_A$	$\triangleq$	$\bar{T}$	mean temperature
$\Delta\varrho_A$	$\triangleq$	ΔT	mean temperature difference
ξ	$\triangleq$	$T^* \equiv \dfrac{T-T_w}{T_0-T_w}$	dimensionless temperature
$\dot{M}_A$	$\triangleq$	$\dot{Q} = V\varrho c_p(T_0-\bar{T})$	heat flux

β_z	$\triangleq \alpha_z \equiv \dot{q}/(\bar{T}-T_w)$	local heat transfer coefficient
β	$\triangleq \alpha \equiv \dot{Q}/(A\Delta T)$	mean heat transfer coefficient
Sh_z	$\triangleq Nu_z \equiv \alpha_z d/\lambda$	local Nusselt number
Sh	$\triangleq Nu \equiv \alpha d/\lambda$	mean Nusselt number
Sc_{n-N}	$\triangleq Pr_{n-N} \equiv \dfrac{Kc_p}{\lambda}\left(\dfrac{d}{8\bar{w}}\right)^{1-n}\left(\dfrac{1+3n}{4n}\right)^n$	Prandtl number

Here, ϱ is the mean fluid density, c_p specific heat capacity at constant pressure, $\dot{q}$ heat flux density, $\dot{Q}$ heat flux, V volumetric flow rate of fluid, $A = d\pi L$ surface of tube, λ heat conductivity, and $a = \lambda/(\varrho c_p)$ heat diffusivity.

For heat as well as mass transfer all fluid properties have to be taken at the mean fluid temperature:

$$T_f = \frac{1}{2}(T_0 + \bar{T}) \qquad (4.28)$$

with T_0 and $\bar{T}$ as the mean fluid temperature at tube inlet and exit, respectively.

4.5.2 Heat Transfer With Internal Frictional Heat Sources

For high viscosity fluids heat production due to internal friction has to be taken into account when heat transfer is investigated. The differential equation that applies to this case is obtained from Eq. (2.11). The dissipation term has to be generalized, however, so that non-Newtonian behavior of the fluid can be properly introduced. The differential equation is as follows:

$$\varrho c_p w_z \frac{\partial T}{\partial z} = \lambda\left(\frac{\partial^2 T}{\partial r^2} + \frac{1}{r}\frac{\partial T}{\partial r}\right) + \tau\frac{\partial w_z}{\partial r}. \qquad (4.29)$$

The dissipation term $\tau \partial w_z/\partial r$ takes care of the conversion of mechanical into thermal energy.

From Eq. (4.4) the following equation for the local shear stress is obtained:

$$\tau = \frac{\Delta p}{2L} r. \qquad (4.30)$$

The velocity gradient $\partial w_z/\partial r$ is obtained from Eq. (4.3):

$$\frac{\partial w_z}{\partial r} = \bar{w} \frac{1+3n}{n} \frac{r^{1/n}}{R^{1+1/n}}. \qquad (4.31)$$

Introducing the last two equations and Eq. (4.8) for the local velocity w_z into the differential equation leads to:

$$\varrho c_p \bar{w} \frac{1+3n}{1+n} \left[1 - \left(\frac{r}{R}\right)^{1+1/n}\right] \frac{\partial T}{\partial z} =$$
$$= \lambda \left(\frac{\partial^2 T}{\partial r^2} + \frac{1}{r}\frac{\partial T}{\partial r}\right) + \frac{\Delta p \bar{w}}{L} \frac{1+3n}{2n} \left(\frac{r}{R}\right)^{1+1/n}. \qquad (4.32)$$

The following dimensionless parameters will be introduced:

$$T^* \equiv \frac{T-T_w}{|T_0-T_w|} \quad \text{dimensionless local temperature} \qquad (4.33)$$

$$r^* \equiv r/R \quad \text{radial coordinate} \qquad (4.34)$$

$$z^* \equiv \frac{z/d}{(Re\,Pr)_{n\text{-}N}} = \frac{z}{d} \frac{\lambda}{\bar{w} d \varrho c_p} \qquad (4.35)$$

entrance number (axial coordinate)

$$Re_{n\text{-}N} \equiv \frac{\bar{w}^{2-n} d^n \varrho}{K \left(\frac{1+3n}{4n}\right)^n 8^{n-1}} \qquad (4.36)$$

Reynolds number

$$Pr_{n\text{-}N} \equiv \frac{K c_p}{\lambda} \left(\frac{d}{8\bar{w}}\right)^{1-n} \left(\frac{1+3n}{4n}\right)^n \qquad (4.37)$$

Prandtl number

$$q_F^* \equiv \frac{\bar{w} d^2 \Delta p/L}{\lambda(T_0-T_w)} \quad \text{friction number} \qquad (4.38)$$

The friction number q_F^* may be interpreted as the ratio of mean frictional energy ($\bar{w} d^2 \Delta p/L$) and thermal energy on the basis of conditions at the tube inlet $\lambda(T_0-T_w)$. With the parameters given above, the following differential equation in dimensionless writing is obtained:

$$\frac{1+3n}{4(1+n)} \left[1 - \left(\frac{r}{R}\right)^{1+1/n}\right] \frac{\partial T^*}{\partial z^*} = \frac{\partial^2 T^*}{\partial r^{*2}} +$$
$$+ \frac{1}{r^*} \frac{\partial T^*}{\partial r^*} + q_F^* \frac{1+3n}{8n} r^{*(1+1/n)}. \qquad (4.39)$$

The boundary conditions are given by:

$z^* = 0$ and $r^* < 1$: $T^* = 1$

$z^* \geqslant 0$ and $r^* = 0$: $\partial T^*/\partial r^* = 0$

$z^* \geqslant 0$ and $r^* = 1$: $T^* = T_w^* = 0$

The solution of the differential equation has been presented by SCHLÜTER [4.5].

For a more profound discussion of the results it is useful to study the friction term more closely. Integration of this term over the cross section of the tube from $r^* = 0$ to r^* delivers the partial friction energy $Q_F^*(r^*)$ converted per time unit into heat:

$$Q_F^*(r^*) = \int_0^{r^*} q_F^* \frac{1+3n}{8n} r^{*1+1/n} 2\pi r^* \, dr^* =$$
$$= q_F^* \frac{\pi}{4} r^{*(3+1/n)}. \qquad (4.40)$$

The total friction energy is given by:

$$Q_F^*(r^* = 1) = q_F^* \frac{\pi}{4}. \qquad (4.41)$$

The ratio:

$$\frac{Q_F^*(r^*)}{Q_F^*(r^* = 1)} = r^{*(3+1/n)} \qquad (4.42)$$

is shown in Fig. 4.9 as a function of the radial coordinate r^* for values of n between 0 and ∞. Heat production by the work done by frictional forces is restricted to the wall region.

The immediate result of the integration of Eq. (4.39) is the local temperature:

$$T^* = f(r^*; z^*; n; q_F^*). \qquad (4.43)$$

In Fig. 4.10 the local temperature function T^*/q_F^* is plotted against the radial coordi-

4.5 Heat Transfer in Power Law Fluids

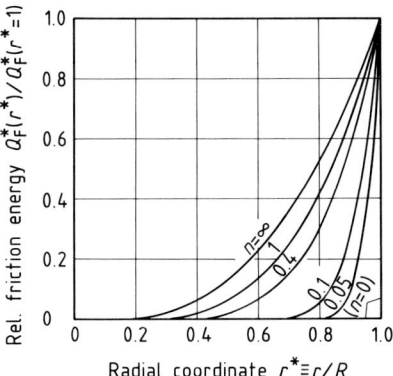

Figure 4.9. Distribution of heat production due to internal work done by frictional forces as a function of the radial coordinate r^* for various values of the fluid index n.

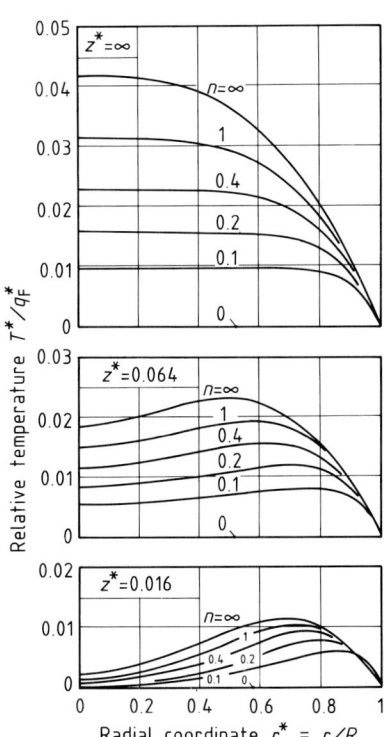

Figure 4.10. Temperature profiles due only to heat production within the fluid, as a function of the radial coordinate r^*, for three values of the axial coordinate z^* and various values of the fluid index n.

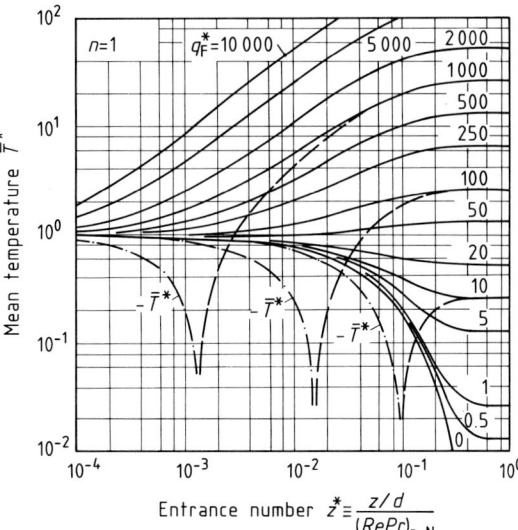

Figure 4.11. Mean temperature in a fluid with $n=1$ (Newtonian fluid) as a function of axial coordinate z^* for various values of the friction number q_F^*; full lines for cooling of the fluid; broken lines for initial heating of the fluid.

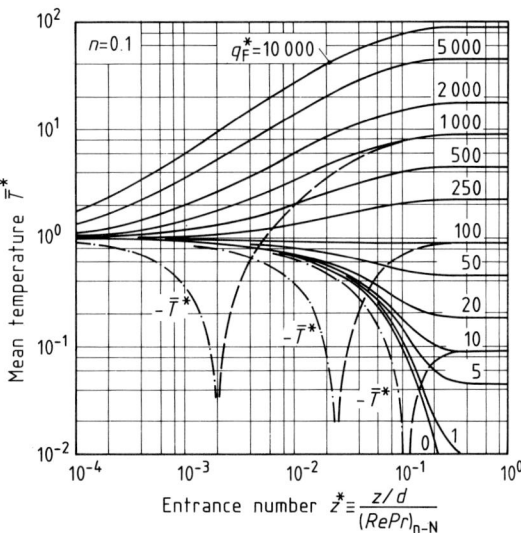

Figure 4.12. Mean temperature in a fluid with $n=0.1$ (pseudoplastic fluid) as a function of axial coordinate z^* for various values of the friction number q_F^*; full lines for cooling of the fluid; broken lines for initial heating of the fluid.

nate r^* for three values of the axial coordinate z^* and for various values of the fluid index n. These temperature profiles are only due to internal heat sources. There is no heat transfer from the wall to the fluid. Because of heat production in the wall region there is a pronounced maximum of the local temperature function close to the wall. From this maximum heat is transferred to the wall as well as towards the tube axis. For very long tubes the temperature maximum lies in the tube axis.

Further integration of the local temperature over the cross sectional area of the tube delivers the mean temperature:

$$\bar{T}^* = f(z^*; n; q_F^*). \qquad (4.44)$$

This temperature is shown in Figs. 4.11 and 4.12 for $n=1$ and $n=0.1$ as a function of the entrance number z^* for various values of the friction number q_F^*. The full lines give $\bar{T}$ for the case of cooling and the broken lines for the case of heating of the fluid in the entrance section of the tube. With very long tubes heat must always be transferred from the fluid to the tube wall, so that the case of cooling applies.

In the absence of internal heat production, $q_F^*=0$, the mean temperature approaches the limiting value $\bar{T}^*_{z^* \to \infty} = 0$ with $z^* \to \infty$. With internal heat production $\bar{T}^*_{z^* \to \infty} > 0$. In the case of heat transfer with internal heat production it is not advisable to determine the Nusselt number because of the rather complicated field of curves obtained.

4.6 References

[4.1] H. BRAUER: "Grundlagen der Einphasen- und Mehrphasenströmungen". Verlag Sauerländer, Aarau – Frankfurt am Main, 1971.
[4.2] D. W. DODGE and A. B. METZNER: "Turbulent flow of non-Newtonian systems". A.I.Ch.E. Journal 5 (1959) 2, 189–204.
[4.3] A. B. METZNER and J. C. REED: "Flow of non-Newtonian fluids – Correlation of the laminar, transition, and turbulent flow regions". A.I.Ch.E. Journal 1 (1954) 4, 434–440.
[4.4] N. W. RYAN and M. N. JOHNSON: "Transition from laminar turbulent flow in pipes". A.I.Ch.E. Journal 5 (1959) 4, 433–435.
[4.5] H. SCHLÜTER: "Berechnung des Wärmetransports in nicht-Newtonschen Flüssigkeiten bei laminarer Rohrströmung unter Berücksichtigung der Reibungsleistung". Dissertation, Technische Universität Berlin, 1969.

Chapter 5

Transport Processes in Fluid Flow Parallel to Plates

Heinz Brauer

Institut für Chemieingenieurtechnik
Technische Universität Berlin
Berlin (West), Germany

5.1   Introduction
5.2   Qualitative Description of Fluid Flow Parallel to Plates
5.3   Quantitative Description of Laminar Fluid Flow
5.3.1 Velocity Field
5.3.2 Boundary Layer Thickness
5.3.3 Local Wall Shear Stress
5.4   Friction Factor Laws for Plates
5.5   Mass Transfer at Flat Plates
5.6   Heat Transfer at Flat Plates
5.7   Mass Transfer with Biochemical Reactions
5.8   References

List of Symbols

A_p	m²	surface of plate
B	m	breadth of plate
c_p	kJ/(kg K)	specific heat capacity of fluid
D	m²/s	diffusion coefficient
k_w	m/s	reaction velocity factor
L	m	length of plate
$\dot{m}_{Awx}$	kg/(s m²)	local mass flux density at surface of plate
p	N/m²	pressure
$\dot{Q}$	kJ/s	heat flux
$\dot{q}_{wx}$	kJ/(s m²)	local heat flux density at surface of plate
T	K	local fluid temperature
T_w	K	temperature at surface of plate
T_∞	K	fluid temperature at great distance from plate
W	N	resistance force
w	m/s	local fluid velocity
w_x	m/s	local fluid velocity in x-direction
w_y	m/s	local fluid velocity in y-direction
w_∞	m/s	fluid velocity at great distance from plate
w'_∞	m/s	fluctuating velocity at great distance from plate
x	m	coordinate parallel to plate surface
y	m	coordinate perpendicular to plate surface
α	kJ/(s m² K)	mean coefficient of heat transfer
α_x	kJ/(s m² K)	local coefficient of heat transfer
β	m/s	mean coefficient of mass transfer
β_x	m/s	local coefficient of mass transfer
δ	m	thickness of velocity boundary layer
η	kg/(m s)	dynamic viscosity of fluid
λ	kJ/(m s K)	heat conductivity
ν	m²/s	kinematic viscosity
ϱ	kg/m³	mean density of fluid
ϱ_A	kg/m³	local partial density of component A
ϱ_{Aw}	kg/m³	partial density of component A at surface of plate
$\varrho_{A\infty}$	kg/m³	partial density of component A at great distance from plate
τ_w	N/m²	shear stress at surface of plate
$Da \equiv \dfrac{k_w}{D/L}$		Damköhler number
$Nu \equiv \dfrac{\alpha L}{\lambda}$		mean Nusselt number
$Nu_x \equiv \dfrac{\alpha_x L}{\lambda}$		local Nusselt number
$p^* \equiv \dfrac{p}{\varrho w_\infty^2/2}$		pressure number
$Pr \equiv \dfrac{\eta c_p}{\lambda}$		Prandtl number
$Re \equiv \dfrac{w_\infty L \varrho}{\eta}$		Reynolds number
$Sc \equiv \dfrac{\nu}{D}$		Schmidt number
$Sh \equiv \dfrac{\beta L}{D}$		mean Sherwood number
$Sh_x \equiv \dfrac{\beta_x L}{D}$		local Sherwood number
$T^* \equiv \dfrac{T - T_w}{T_\infty - T_w}$		local temperature ratio
$Tu \equiv \sqrt{w'^2_\infty/w_\infty^2}$		degree of turbulence
$w_x^* \equiv w_x/w_\infty$		local velocity ratio for x-direction
$w_y^* \equiv w_y/w_\infty$		local velocity ratio for y-direction
$x^* \equiv x/L$		coordinate ratio for x-direction
$y^* \equiv y/L$		coordinate ratio for y-direction
$\zeta \equiv \dfrac{W/A_p}{\varrho w_\infty^2/2}$		resistance factor
$\xi \equiv \dfrac{\varrho_A - \varrho_{A\infty}}{\varrho_{Aw} - \varrho_{A\infty}}$		local concentration ratio
$\tau_w^* \equiv \dfrac{\tau_w}{\varrho w_\infty^2/2}$		local wall shear stress factor

5.1 Introduction

Fluid flow parallel to plates is one of the fundamental problems of fluid mechanics. The progress made over the last decades in analyzing this type of fluid flow is largely based on the boundary layer hypothesis presented by LUDWIG PRANDTL in 1904 [5.1]. PRANDTL thus has become known as the "Father of Modern Fluid Mechanics". To understand the basic laws of fluid flow it is expedient to study the movement of a fluid parallel to a flat plate.

The boundary layer hypothesis postulates that momentum transport from a fluid to the surface of a plate is restricted to a thin layer of fluid adhering to the plate. This implies that a velocity variation occurs only in this layer, called the boundary layer. Outside the boundary layer the velocity is assumed to be constant, so that fluid flow is not influenced by the presence of the plate.

The concept of a boundary layer was developed by PRANDTL for a simplified mathematical treatment of turbulent fluid flow. For this situation the boundary layer hypothesis has maintained its fundamental importance throughout the years. The boundary layer hypothesis is one of the basic tools of fluid mechanics and will remain such for a long time to come. The lasting importance of this hypothesis is due to its inherent truth. The results of the boundary layer theory are expertly presented by SCHLICHTING [5.2].

An adequate understanding of the boundary layer concept is a necessary prerequisite to an understanding of the transport of momentum, heat, and mass. Therefore, it is important to study this concept more closely. This of course includes the study of certain insufficiencies of the boundary layer hypothesis. Soon after the hypothesis became known it was applied to laminar fluid flow. One of the important papers in this context is that by BLASIUS [5.3], published in 1908. This paper theoretically investigates laminar fluid flow parallel to a flat plate with the aid of the simplifications of the boundary layer hypothesis. Within the past years the complete differential equations for the velocity field in the vicinity of plates have been solved by numerical methods [5.4]. The results reveal the deficiencies of the hypothesis. In fact, the boundary layer hypothesis should no longer be accepted as a basis for theoretical investigations of laminar flow. For qualitative investigations, however, the boundary layer hypothesis proves to be helpful not only for turbulent but also for laminar fluid flow.

This chapter considers laminar and turbulent fluid flow parallel to flat plates as well as mass transfer. Investigations will always first be described with a mathematical treatment of transport processes in laminar fluid motion without making use of the boundary layer simplifications. The results will be compared with those obtained on the basis of the boundary layer concept. The investigations will then be extended to turbulent flow situations.

5.2 Qualitative Description of Fluid Flow Parallel to Plates

In Fig. 5.1 the flow of a fluid parallel to a flat plate is demonstrated by means of stream-lines. In any point of a stream-line the tangent gives the direction of the local velocity vector $\vec{w}$ with the components $\vec{w}_x$ and $\vec{w}_y$; $\vec{w}_x$ is the component parallel to the plate surface, $\vec{w}_y$ perpendicular to it. The velocity of the approaching fluid is given by $\vec{w}_\infty$. Long before the fluid comes in contact with the plate, the path of the stream-lines is influenced by the plate. The stream-lines are bent away from the plate, a positive velocity component, $\vec{w}_y$, develops. Towards the end of and behind the plate the stream-lines

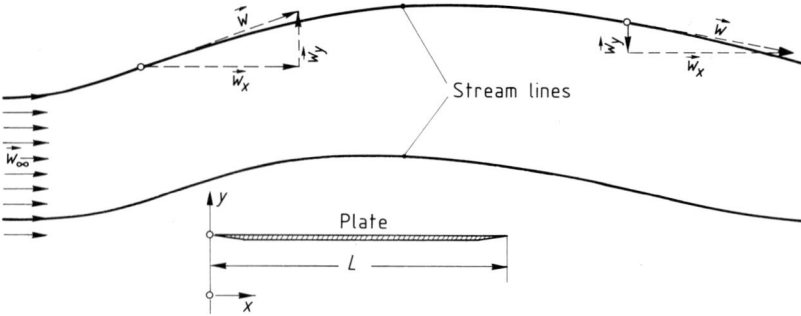

Figure 5.1. Stream-lines of real fluid flow parallel to a flat plate.

are bent toward the plate with a negative velocity component, $-\vec{w}_y$.

Detachment of the stream-lines in the vicinity of the tip of the plate at $x=0$ is caused by adhesion of the fluid at the surface of the plate. The velocity at the plate surface is zero. Reduction of fluid flow parallel to the plate must lead to a velocity normal to the surface of the plate.

A comparison between a stream-line of real fluid flow and of boundary layer flow is illustrated in Fig. 5.2. Detachment of boundary layer stream-lines from the original path occurs very suddenly when the flow has reached the tip of the plate at $x=0$. A reattachment of the stream-line does not occur. In fact, with boundary layer flow the detachment of stream-lines increases steadily even behind the plate. The boundary layer hypothesis assumes an infinitely long plate.

The result of this comparison is that the boundary layer hypothesis cannot properly describe fluid flow in the vicinity of the tip and the end of the plate. In other words, the deficiency of the boundary layer hypothesis increases with decreasing fluid velocity. This negative effect is not only observed in momentum transport but also in heat and mass transport.

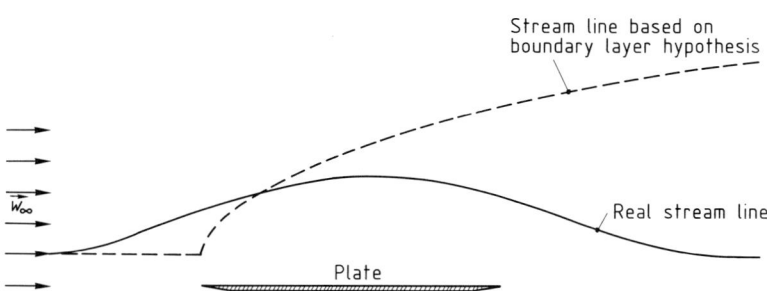

Figure 5.2. Comparison of stream-lines for real fluid flow shown by a full line and for boundary layer flow shown by a broken line.

5.3 Quantitative Description of Laminar Fluid Flow

5.3.1 Velocity Field

The velocity field is described by the equation of continuity and the appropriate Navier-Stokes equations derived from Eqs. (2.1) to (2.2). In dimensionless writing these equations with boundary conditions are as follows:

Continuity equation:

$$\frac{\partial w_x^*}{\partial x^*} + \frac{\partial w_y^*}{\partial y^*} = 0 \tag{5.1}$$

Navier-Stokes equations:

$$w_x^* \frac{\partial w_x^*}{\partial x^*} + w_y^* \frac{\partial w_x^*}{\partial y^*} =$$
$$= -\frac{\partial p^*}{\partial x^*} + \frac{1}{Re}\left[\frac{\partial^2 w_x^*}{\partial x^{*2}} + \frac{\partial^2 w_x^*}{\partial y^{*2}}\right], \tag{5.2}$$

$$w_x^* \frac{\partial w_y^*}{\partial x^*} + w_y^* \frac{\partial w_y^*}{\partial y^*} =$$
$$= -\frac{\partial p^*}{\partial y^*} + \frac{1}{Re}\left[\frac{\partial^2 w_y^*}{\partial x^{*2}} + \frac{\partial^2 w_y^*}{\partial y^{*2}}\right]. \tag{5.3}$$

Boundary conditions:

$$\left.\begin{array}{ll} x^* \to -\infty \text{ and } & 0 \leqslant y^* \leqslant +\infty \\ x^* \to +\infty \text{ and } & 0 \leqslant y^* \leqslant +\infty \\ y^* \to +\infty \text{ and } & -\infty \leqslant x^* \leqslant +\infty \end{array}\right\} \begin{array}{l} w_x^* = 1, \\ (w_y^* = 0), \end{array}$$

$$0 \leqslant x^* \leqslant 1 \text{ and } y^* = 0 \quad \left\{ \begin{array}{l} w_x^* = 0, \\ w_y^* = 0. \end{array}\right.$$

The dimensionless parameters introduced into these equations are defined as follows:

$x^* \equiv x/L$	length coordinate	(5.4)
$y^* \equiv y/L$	perpendicular coordinate	(5.5)
$w_x^* \equiv w_x/w_\infty$	velocity parallel to plate	(5.6)
$w_y^* \equiv w_y/w_\infty$	velocity normal to plate	(5.7)
$p^* \equiv \dfrac{p}{\varrho w_\infty^2/2}$	local pressure	(5.8)
$Re \equiv w_\infty L \varrho/\eta$	Reynolds number	(5.9)

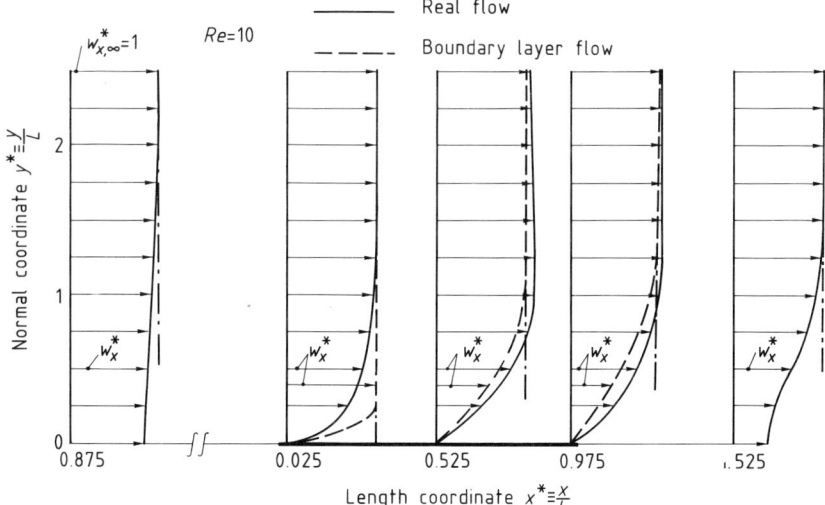

Figure 5.3. Profiles of the velocity component w_x^* parallel to the plate surface for $Re = 10$ at various positions x^*.

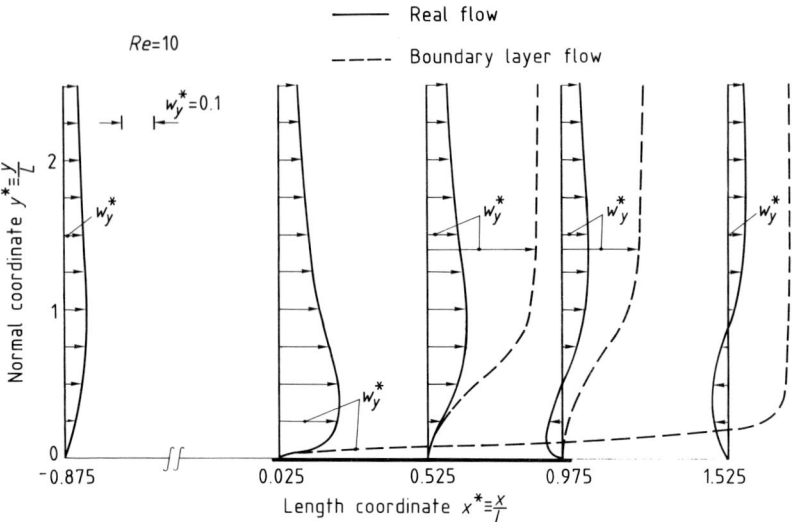

Figure 5.4. Profiles of the velocity component w_y^* normal to the plate surface for $Re = 10$ and at various positions x^*.

In these equations L is the length of the plate, ϱ fluid density, and η dynamic viscosity of the fluid.

A solution of the system of differential equations has been obtained with numerical methods by several authors, see for instance [5.4]. Some of the results are presented in Figs. 5.3 and 5.4. In Fig. 5.3 the velocity w_x^* and in Fig. 5.4 the velocity w_y^* is shown for a Reynolds number of $Re = 10$ as

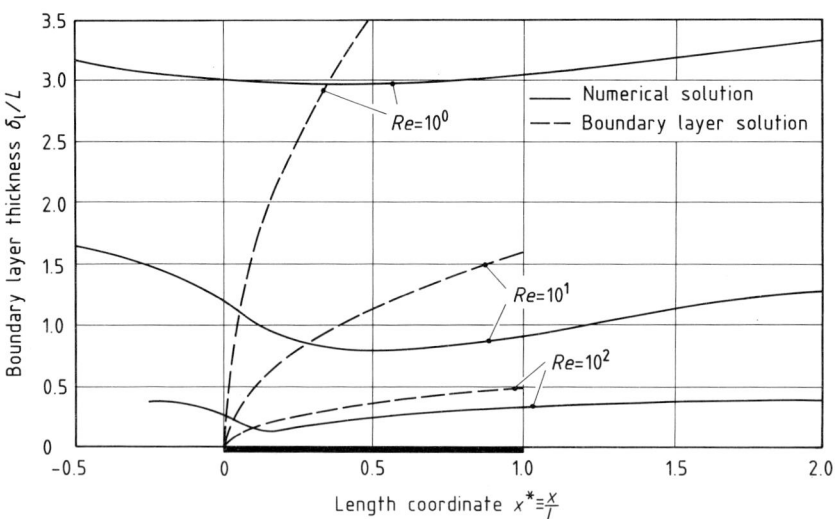

Figure 5.5. Boundary layer thickness δ/L as a function of the length coordinate x^* for real fluid flow (full lines) and for boundary layer flow (broken lines).

a function of the normal coordinate y^* for various values of the length coordinate x^*. The results obtained on the basis of the boundary layer hypothesis are included as a comparison. This proves that this hypothesis cannot adequately describe the velocity field in the low Reynolds number region.

5.3.2 Boundary Layer Thickness

The thickness of the boundary layer is designated by δ and defined by:

$$y = \delta \quad \text{when} \quad w_x = 0.99 \, w_\infty. \quad (5.10)$$

On the basis of the boundary layer hypothesis the following equation has been derived [5.2]:

$$\frac{\delta}{L} = 4.91 \frac{(x/L)^{1/2}}{Re^{1/2}}. \quad (5.11)$$

This equation is presented in Fig. 5.5 by broken lines for various values of the Reynolds number Re. For the same values of Re the full lines give the correct results obtained from the numerical solution of the complete set of differential Eqs. (5.1) to (5.3). In the low Reynolds number region the boundary layer hypothesis does not hold.

5.3.3 Local Wall Shear Stress

Momentum transfer is expressed by the local wall shear stress τ_w. For Newtonian fluids the wall shear stress is given by:

$$\tau_w = -\eta \left(\frac{\partial w_x}{\partial y} \right)_w. \quad (5.12)$$

Introducing the results obtained by the numerical solution of Eqs. (5.1) to (5.3) for the velocity gradient at the wall, $(\partial w_x/\partial y)_w$, and application of dimensionless parameters leads to the following equation:

$$\tau_w^* \, Re/2 = f(Re; x^*), \quad (5.13)$$

with:

$$\tau_w^* \equiv \frac{\tau_w}{\varrho \, w_\infty^2/2}. \quad (5.14)$$

In Fig. 5.6 the local shear stress function is presented over the length coordinate x^* for three values of the Reynolds number Re. Noteworthy are the peaks in shear stress close to the tip and the end of the plate.

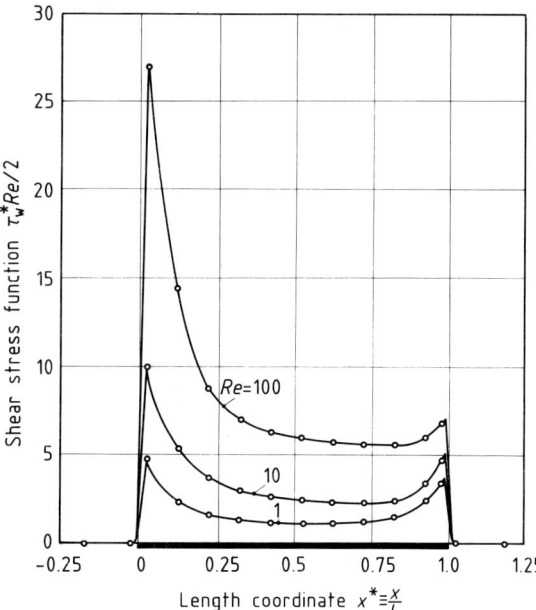

Figure 5.6. Local shear stress function for three values of the Reynolds number Re as a function of the length coordinate x^*.

5.4 Friction Factor Laws for Plates

The friction factor ζ, defined by:

$$\zeta \equiv \frac{W/A_p}{\rho w_\infty^2/2}, \qquad (5.15)$$

is a unique function of Re for laminar fluid flow, and a function of Re and the degree of turbulence for turbulent flow, defined by:

$$Tu = \sqrt{\overline{w'^2_\infty}/w_\infty^2}, \qquad (5.16)$$

with $\overline{w'^2_\infty}$ the time mean value of the square of the fluctuation velocity w'_∞ in the fluid approaching the plate. In Eq. (5.15) $A_p = L B$ is the surface area of the plate with L as the length and B the width; W is the resistance force determined by the following equation:

$$W = 2B \int_{x=0}^{x=L} \tau_w \, dx. \qquad (5.17)$$

The friction factor ζ is presented in Fig. 5.7 as a function of the Reynolds number Re. Curve a correlates the data in the laminar and curve c in the turbulent region. For laminar flow the friction factor law is given by:

$$\zeta_l = \frac{2.65}{Re^{7/8}} - \frac{1}{4Re + \dfrac{0.008}{Re}} + \frac{1.328}{Re^{1/2}}, \qquad (5.18)$$

range of application:
$10^{-2} \leqslant Re \leqslant Re_{crit} = 5 \cdot 10^5$

and for turbulent flow by:

$$\zeta_t = \frac{0.455}{(\lg Re)^{2.58}} - \frac{1}{Re} \frac{9.9 \cdot 10^3}{1 + 10^4 Tu^{1.7}}, \qquad (5.19)$$

range of application: $Re_{crit} \leqslant Re \leqslant 10^9$
$0 \leqslant Tu \leqslant 0.10$.

The transition from laminar to turbulent flow close to the surface of the plate occurs at the critical value of the Reynolds number:

$$Re_{crit} = \frac{3 \cdot 10^6}{1 + 10^4 Tu^{1.7}} \qquad (5.20)$$

range of application: $0 \leqslant Tu \leqslant 0.10$

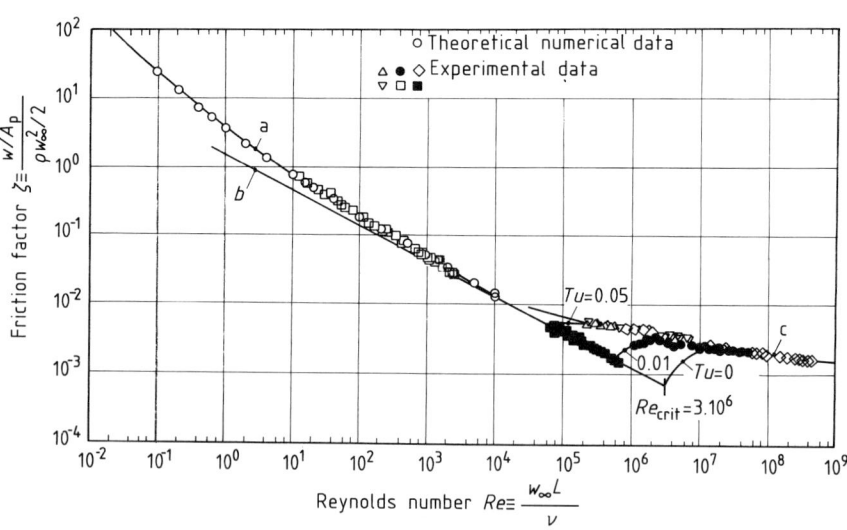

Figure 5.7. Friction factor for laminar and turbulent flow past flat plates: curve a after Eq. (5.18) for laminar real fluid flow; curve b after Eq. (5.21) for laminar boundary layer flow; curve c after Eq. (5.19) for turbulent flow.

In Fig. 5.7 Eq. (5.18) is represented by curve a, and Eq. (5.19) by curve c. For a comparison, curve b has been included representing the friction factor law, which has been derived by BLASIUS [5.3] on the basis of the boundary layer hypothesis:

$$\zeta_b = \frac{1.328}{Re^{1/2}}, \qquad (5.21)$$

range of application:
$10^3 \leqslant Re \leqslant Re_{crit} = 5 \cdot 10^5$.

This again proves the inadequacy of the boundary layer hypothesis in the low Reynolds number range.

5.5 Mass Transfer at Flat Plates

For mass transfer from the surface of the plate to the fluid it is assumed that the concentration at the plate surface is kept constant. Furthermore, it is assumed that the temperature in the considered space is constant.

The concentration field in laminar flow parallel to a flat plate is described by the differential equation derived from Eq. (2.6):

$$w_x^* \frac{\partial \xi}{\partial x^*} + w_y^* \frac{\partial \xi}{\partial y^*} = \frac{1}{Re\,Sc}\left(\frac{\partial^2 \xi}{\partial x^{*2}} + \frac{\partial^2 \xi}{\partial y^{*2}}\right). \qquad (5.22)$$

Boundary conditions:

$$\begin{array}{lll}
x^* = -\infty \text{ and } 0 \leqslant y^* \leqslant \infty: & \xi = 0 \\
-\infty \leqslant x^* \leqslant +\infty \text{ and } y^* = \infty: & \xi = 0 \\
x^* = +\infty \text{ and } 0 \leqslant y^* \leqslant \infty: & \xi = 0 \\
0 \leqslant x^* \leqslant 1 \text{ and } 0 = y^* & : \xi = \xi_w = 1 \\
-\infty \leqslant x^* \leqslant 0 \text{ and } 0 = y^* & : \partial \xi / \partial y^* = 0 \\
1 \leqslant x^* \leqslant +\infty \text{ and } 0 = y^* & : \partial \xi / \partial y^* = 0
\end{array}$$

The dimensionless parameters not yet explained, are:

$$\xi \equiv \frac{\varrho_A - \varrho_{A\infty}}{\varrho_{Aw} - \varrho_{A\infty}}, \qquad \text{local concentration} \qquad (5.23)$$

$$Sc \equiv v/D, \qquad \text{Schmidt number} \qquad (5.24)$$

with D as the diffusion coefficient, ϱ_A local partial density of component A in the fluid, $\varrho_{A\infty}$ in the approaching fluid, and ϱ_{Aw} at the surface of the plate which is assumed to be constant.

The immediate result of the numerical solution of Eq. (5.22) is the local concentration ξ and the concentration gradient $(\partial \xi / \partial y^*)_w$, which is required for the Sherwood number [5.5].

Mass transfer from the wall to the fluid or *vice versa* is described by means of the local mass transfer coefficient β_x, which is defined by:

$$\beta_x \equiv \frac{\dot{m}_{Awx}}{\varrho_{Aw} - \varrho_{A\infty}}. \qquad (5.25)$$

With Fick's law for the local mass flux density at the surface of the plate:

$$\dot{m}_{Awx} = -D\left(\frac{\partial \varrho_A}{\partial y}\right)_{wx}, \qquad (5.26)$$

and by introducing dimensionless parameters, Eq. (5.25) yields the local Sherwood number:

$$Sh_x = -\left(\frac{\partial \xi}{\partial y^*}\right)_{wx}. \qquad (5.27)$$

The local Sherwood number is defined by:

$$Sh_x \equiv \frac{\beta_x L}{D}. \qquad (5.28)$$

For one value of the Reynolds number, $Re = 40$, the local Sherwood number Sh_x is presented in Fig. 5.8 over the length coordinate x^* for various values of the Schmidt number. The highest value of Sh_x is observed at the tip of the plate, at $x^* = 0$. With increasing x^* the local Sherwood number decreases and after passing through a mini-

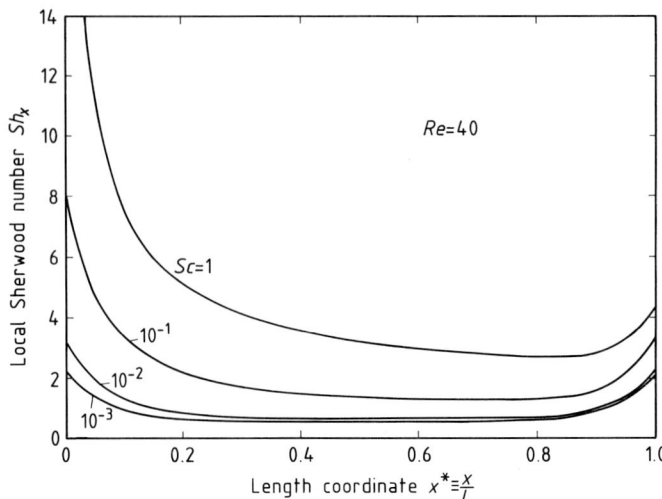

Figure 5.8. Local Sherwood number Sh_x for laminar real fluid flow past a flat plate at a constant Reynolds number but for various values of the Schmidt number Sc.

mum increases again to a second maximum at the end of the plate, at $x^* = 1$. The behavior of Sh_x qualitatively agrees with that of the local shear stress presented in Fig. 5.6.

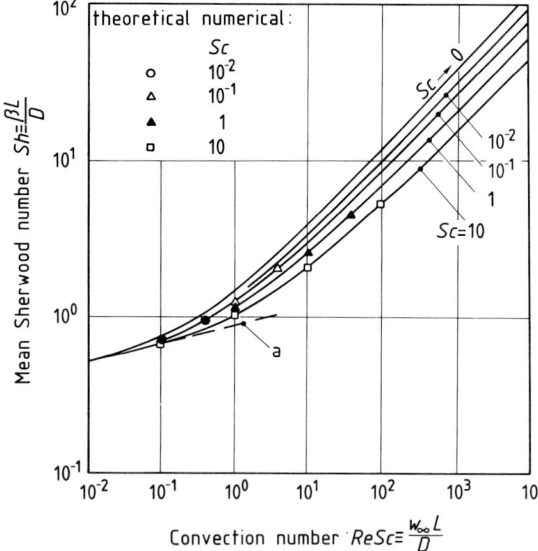

Figure 5.9. Mean Sherwood number Sh for laminar real fluid flow past a flat plate plotted against the function $Re\,Sc$ for various values of the Schmidt number Sc.

The mean Sherwood number Sh is obtained by integration of Sh_x over the length of the plate:

$$Sh = -\int_{x^*=0}^{x^*=L} (\partial \xi/\partial y^*)_{w,x}\, dx^*. \tag{5.29}$$

In Fig. 5.9 Sh is plotted against the function $Re\,Sc$ for various values of the Schmidt number. The numerical data are approximated by the following equation:

$$Sh = 0.8(Re\,Sc)^{0.1} + f_p \frac{Re\,Sc}{1+1.30(Re\,Sc)^{1/2}}, \tag{5.30}$$

with:

$$f_p = \frac{1.47}{[1+(1.67\,Sc^{1/6})^2]^{1/2}}, \tag{5.31}$$

range of application:
$$0 \leqslant Re\,Sc \leqslant \infty$$
$$0 \leqslant Re \leqslant Re_{\text{crit}} = 5 \cdot 10^5$$
$$0 \leqslant Sc \leqslant \infty.$$

From Eq. (5.30) two limiting equations may be obtained:

$$Sh = 0.80\,(Re\,Sc)^{0.1} \tag{5.32}$$

range of application:
$(Re\,Sc) \to 0$ curve a in Fig. 5.9

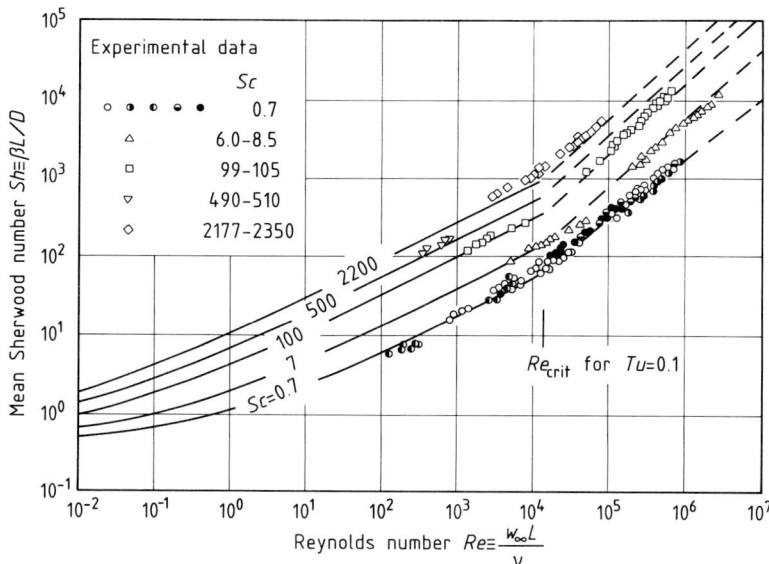

Figure 5.10. Comparison of experimental data for mass transfer in laminar and turbulent real fluid flow past a flat plate with theoretical and empirical equations (for further details see text).

$Sh = f_p \, 0.77 \, (Re \, Sc)^{1/2}$
range of application:
$Re \, Sc \to \infty$ with $Re \leqslant Re_{crit} = 5 \cdot 10^5$
$0 \leqslant Sc \leqslant \infty$.

With the additional condition, $Sc \to \infty$, $f_p = 0.88/Sc^{1/6}$ is obtained. Thus, the following equation results:

$$Sh = 0.68 \, Re^{1/2} \, Sc^{1/3} \qquad (5.33)$$

range of application:
$Re \, Sc \to \infty$ with $Re \leqslant Re_{crit}$
$Sc \to \infty$.

The last equation is based on the boundary layer hypothesis and has been presented by LÉVÊQUE [5.6]. It is a very special case of Eq. (5.30), which is of general applicability for mass transfer in laminar plate flow.

Available experimental data for mass transfer are presented in Fig. 5.10. The data are very well approximated in the low Reynolds number region by curves representing Eq. (5.30) and in the high Reynolds number region by broken curves representing the empirical equation for mass transfer in turbulent flow [5.7]:

$$Sh = \frac{0.037 \, Re^{0.8} \, Sc}{1 + 2.44 \, Re^{-0.1}(Sc^{2/3} - 1)} \qquad (5.34)$$

range of application: $Re_{crit} \leqslant Re \leqslant \infty$
$0.8 \leqslant Sc \leqslant \infty$.

To calculate the transferred mass $\dot{M}_A$ per time unit the following equation must be applied:

$$\dot{M}_A = \beta \, B L (\varrho_{Aw} - \varrho_{A0}), \qquad (5.35)$$

with BL being the surface of the plate.

All equations presented for the Sherwood number are in excellent agreement with experimental data.

5.6 Heat Transfer at Flat Plates

The heat transfer case considered in this section is analogous to the mass transfer case discussed in the previous section. The temperature T_w of the plate surface is assumed to be constant and different from the temperature T_0 of the approaching fluid. All equations which have been presented for mass transfer can also be applied to heat transfer when the following relations are observed:

$\varrho_A \triangleq T$	local temperature of fluid
$\varrho_{Aw} \triangleq T_w$	temperature of plate surface
$\varrho_{A0} \triangleq T_0$	temperature of approaching fluid
$\xi \triangleq T^* \equiv \dfrac{T-T_\infty}{T_w-T_\infty}$	dimensionless fluid temperature
$\beta_x \triangleq \alpha_x \equiv \dfrac{\dot{q}_{wx}}{T_w-T_\infty}$	local heat transfer coefficient
$\beta \triangleq \alpha \equiv \dfrac{\dot{Q}}{BL(T_w-T_0)}$	mean heat transfer coefficient
$Sh_x \triangleq Nu_x \equiv \dfrac{\alpha_x L}{\lambda}$	local Nusselt number
$Sh \triangleq Nu \equiv \dfrac{\alpha L}{\lambda}$	mean Nusselt number
$Sc \triangleq Pr \equiv \dfrac{\eta c_p}{\lambda}$	Prandtl number

In the above relations λ symbolizes the heat conductivity, c_p specific heat capacity, $\dot{Q}$ transferred heat per time unit or heat flux, and $\dot{q}_{wx} = \dot{Q}/(BL)$ heat flux density. The properties of the fluid are to be taken at the mean fluid temperature:

$$T_f = \tfrac{1}{2}(T_0 + T_w). \tag{5.36}$$

All equations for the Nusselt number obtained by the outlined procedure are in excellent agreement with experimental data.

5.7 Mass Transfer with Biochemical Reactions

The fundamentals of mass transfer with biochemical reactions at the surface of a solid have been discussed in detail in Sect. 3.6.1 and will not be repeated here.

The concentration field is described by Eq. (5.22); the boundary condition given to that equation also applies to the present case, with one exception:

$$0 \leqslant x^* \leqslant 1 \text{ and } y^* = 0: \left(\frac{\partial \xi}{\partial y^*}\right)_w = 2 Da\, \xi_w \tag{5.37}$$

in which:

$$Da \equiv \frac{k_w}{D/L}, \quad \text{Damköhler number} \tag{5.38}$$

$$\xi \equiv \frac{c_A}{c_{A\infty}}, \quad \text{local concentration} \tag{5.39}$$

$$\xi_w \equiv \frac{c_{Aw}}{c_{A\infty}}, \quad \begin{array}{l}\text{concentration at}\\\text{surface of plate}\end{array} \tag{5.40}$$

with k_w reaction velocity factor, c_A local concentration, c_{Aw} concentration at surface of the plate, and $c_{A\infty}$ concentration in the approaching fluid.

By defining the local mass transfer coefficient:

$$\beta_x \equiv \frac{\dot{n}_{Awx}}{c_{A\infty}}, \tag{5.41}$$

with:

$$\dot{n}_{Awx} = -D\left(\frac{\partial c_A}{\partial y}\right)_{wx} \tag{5.42}$$

as the local mass flux density, the following equation is obtained when dimensionless parameters are introduced:

$$Sh_x = -\left(\frac{\partial \xi}{\partial y^*}\right)_{wx}. \tag{5.43}$$

5.7 Mass Transfer with Biochemical Reactions   73

Figure 5.11

Figure 5.12

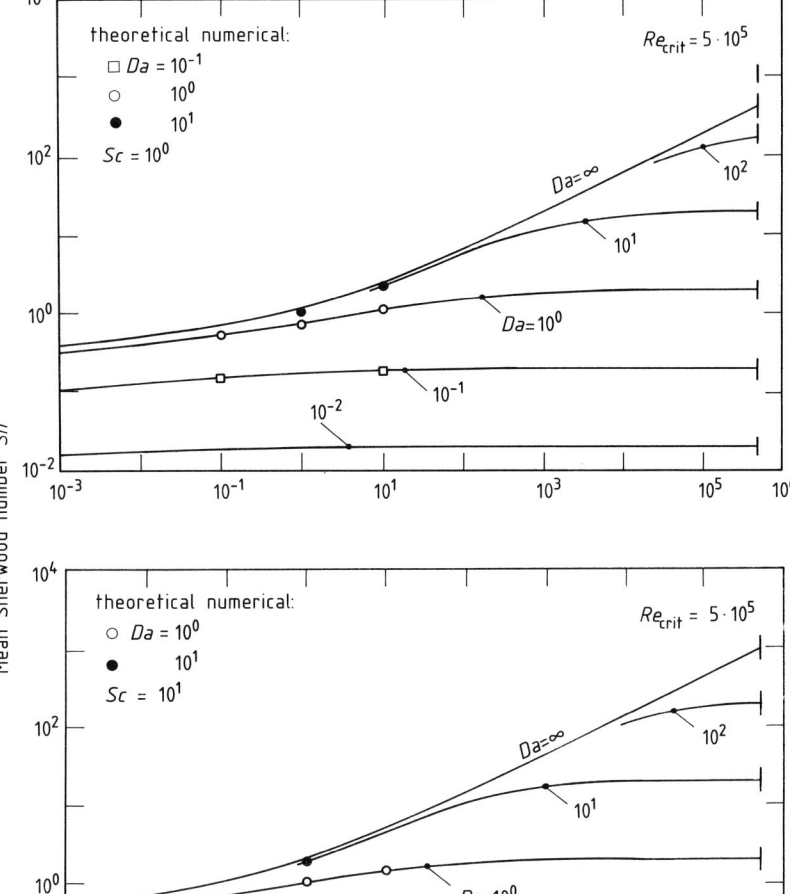

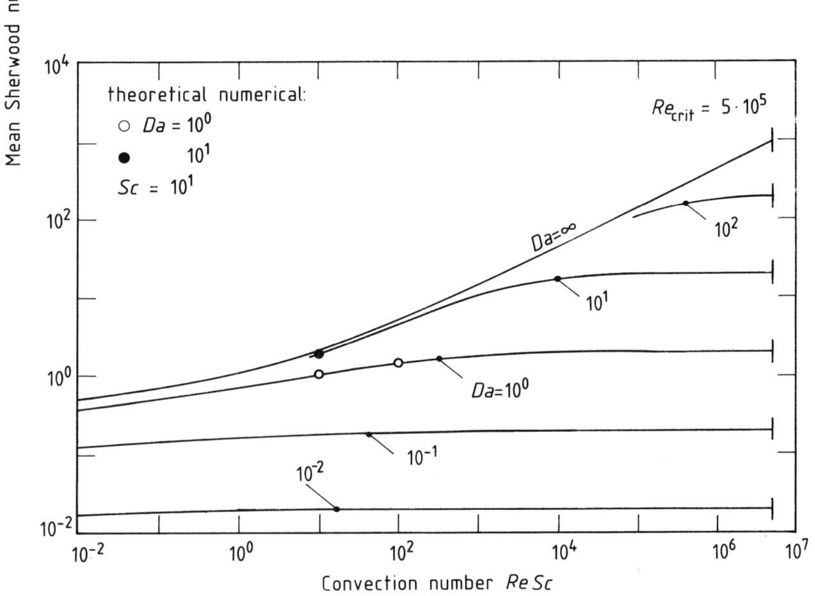

Figures 5.11 and 5.12. Mean Sherwood number for mass transfer with a biochemical reaction of first order for two values of the Schmidt number (Fig. 5.11 for $Sc = 10^0$; Fig. 5.12 for $Sc = 10^1$) plotted against the function $Re\,Sc$ for various values of the Damköhler number Da.

The local Sherwood number is defined by:

$$Sh_x \equiv \frac{\beta_x L}{D}. \quad (5.44)$$

Integration over the length of the plate gives the mean Sherwood number:

$$Sh = \int_{x^*=0}^{x^*=1} Sh_x \, dx^*. \quad (5.45)$$

The definition of Sh is given by:

$$Sh \equiv \frac{\beta L}{D}. \quad (5.46)$$

The mass of component A converted per time unit by a biochemical reaction at the surface of the plate may be calculated using the following equation:

$$M_A = \beta B L c_{A\infty}. \quad (5.47)$$

The mean transfer coefficient β is obtained from data for the mean Sherwood number [5.8]. This is shown for two values of the Schmidt number Sc in Figs. 5.11 and 5.12. Sh is plotted against the function $Re\,Sc$ for various values of the Damköhler number. The curves represent the equation:

$$Sh = \left[\cfrac{1}{\cfrac{1}{k_1(Re\,Sc)^{0.1} + f_p \cfrac{Re\,Sc}{1+k_2(Re\,Sc)^{1-k_3}}} + \cfrac{1}{(2\,Da)^2}} \right]^{1/2} \quad (5.48)$$

with:

$$k_1 = \frac{0.8}{1 + \cfrac{0.38}{Da^{0.8}}}, \quad (5.49)$$

$$k_2 = \left(1.69 + \frac{6.07}{Da}\right)^{1/2}, \quad (5.50)$$

$$k_3 = \left[\cfrac{1}{\cfrac{1}{(0.10+0.37\,Da)^2} + 4} \right]^{1/2}, \quad (5.51)$$

and f_p after Eq. (5.31). With increasing values for $Re\,Sc$ the mean Sherwood number approaches the final value $2\,Da$; this is the reaction controlled case. Diffusion controlled situations arise for $Da = \infty$. The range of application of Eq. (5.48) is given by:

$$0 \leqslant Re\,Sc \leqslant \infty$$
$$0 \leqslant Re \quad \leqslant Re_{crit} = 5 \cdot 10^5$$
$$0 \leqslant Sc \quad \leqslant \infty$$
$$0 \leqslant Da \quad \leqslant \infty$$
$$0 \leqslant n \quad \leqslant \infty.$$

5.8 References

[5.1] L. PRANDTL: "Über Flüssigkeitsbewegung bei kleiner Reibung". Verhandlungen des dritten internationalen Mathematiker-Kongresses in Heidelberg vom 8. bis 13. August 1904. Verlag B. G. Teubner, Leipzig, 1905.

[5.2] H. SCHLICHTING: "Grenzschicht-Theorie". Verlag G. Braun, Karlsruhe, 1965.

[5.3] H. BLASIUS: "Grenzschichten in Flüssigkeiten mit kleiner Reibung". Z. Math. Phys. *56* (1908), 1–37.

[5.4] H. BRAUER and D. SUCKER: "Umströmung von Platten, Zylindern und Kugeln". Chem. Ing. Tech. *48* (1976) 8, 665–671.

[5.5] H. BRAUER and D. SUCKER: "Stoff- und Wärmeübergang an umströmten Platten, Zylindern und Kugeln". Chem. Ing. Tech. *48* (1976) 9, 737–741.

[5.6] M. A. LÉVÊQUE: "Les lois de la transmission de la chaleur par convection". Ann. Mines *12* (1928), 201–299.

[5.7] V. GNIELINSKI: "Berechnung mittlerer Wärme- und Stoffübertragungskoeffizienten an laminar und turbulent überströmten Einzelkörpern mit Hilfe einer einheitlichen Gleichung". Forsch. Geb. Ingenieurwes. *41* (1975), 145–153.

[5.8] D. SUCKER and H. BRAUER: "Stoffübergang mit chemischer Oberflächenreaktion an umströmten Platten". Wärme Stoffübertrag. *12* (1979), 35–43.

Chapter 6

Transport Processes Through the Interface of Particles

Heinz Brauer

Institut für Chemieingenieurtechnik
Technische Universität Berlin
Berlin (West), Germany

6.1 Introduction
6.2 Fluid Flow around Solid Spheres
6.2.1 Velocity Field
6.2.2 Vortex Field
6.2.3 Resistance Factor
6.3 Fluid Flow around Bubbles
6.3.1 Velocity Field around Spherical Bubbles
6.3.2 Resistance Factor
6.3.3 Fluid Flow around Shapeless Bubbles
6.3.3.1 Deformation Turbulence
6.3.3.2 Velocity Field
6.3.3.3 Shear Stress
6.4 Steady-State Mass Transfer at Solid Spheres
6.4.1 Concentration Field
6.4.2 Local and Mean Sherwood Numbers
6.4.3 The Analogous Heat Transfer Case
6.5 Steady-State Mass Transfer with Biochemical Reactions at the Surface of Solid Spheres
6.5.1 The Mathematical Problem
6.5.2 Local and Mean Sherwood Numbers
6.6 Steady-State Mass Transfer at Bubbles
6.6.1 Mass Transfer through the Interface of Spherical Bubbles
6.6.2 Mass Transfer through the Interface of Non-Spherical Bubbles
6.7 Unsteady-State Mass Transfer through the Interface of Spherical Particles
6.7.1 Discussion of Mass Transfer Conditions
6.7.2 Calculation of Transferred Mass
6.7.3 Differential Equations for Concentration Fields
6.7.4 Initial, Boundary, and Interface Conditions

6.7.4.1 Initial Conditions
6.7.4.2 Boundary Conditions
6.7.4.3 Interface Conditions
6.7.5   Discussion of some Numerical Results for Mass Transfer without Fluid Motion
6.7.5.1 Mean Concentration in a Sphere
6.7.5.2 Mean Sherwood Number
6.7.6   Discussion of some Numerical Results for Mass Transfer with Fluid Motion
6.7.6.1 Mean Concentration in a Sphere
6.7.6.2 Mean Sherwood Number
6.8   References

List of Symbols

A_p	m²	surface of particle
c_A	kmol/m³	local partial mol density of component A
$c_{A\infty}$	kmol/m³	partial mol density of component A at great distance from particle
$c_{A\theta w}$	kmol/m³	local partial mol density at surface of particle
c_p	kJ/(kg K)	specific heat capacity
D	m²/s	diffusion coefficient
D_1	m²/s	diffusion coefficient in phase 1
D_2	m²/s	diffusion coefficient in phase 2
d_p	m	particle diameter
F_p	m²	cross sectional area of particle
g	m/s²	acceleration due to gravity
k_w	$\dot{r}_{A\theta w}/c_{A\theta w}^n$	reaction velocity factor
l	m	length of vortex
$\dot{M}_A$	kg/s	mass flux
$\dot{m}_{A\theta w}$	kg/(s m²)	local mass flux density at surface of particle
$\dot{n}_{A\theta w}$	kmol/(s m²)	local mol flux density at surface of particle
p	N/m²	pressure
$\dot{Q}$	kJ/s	heat flux
R	m	radius of particle
r	m	radial coordinate
T	K	local temperature
T_w	K	temperature at surface of particle
T_∞	K	temperature at great distance from particle
t	s	time
V_p	m³	volume of particle
W	N	resistance force
w_r	m/s	radial component of velocity
w_θ	m/s	tangential component of velocity
w_∞	m/s	velocity at great distance from particle
w_{r1}	m/s	radial component of velocity in phase 1
w_{r2}	m/s	radial component of velocity in phase 2
$w_{\theta 1}$	m/s	tangential component of velocity in phase 1
$w_{\theta 2}$	m/s	tangential component of velocity in phase 2
α	kJ/(s m² K)	mean coefficient of heat transfer
β	m/s	mean coefficient of mass transfer
β_θ	m/s	local coefficient of mass transfer
ε_m	m²/s	diffusivity of mass due to turbulence
η	kg/(s m)	dynamic viscosity of fluid
η_1	kg/(s m)	dynamic viscosity of phase 1
η_2	kg/(s m)	dynamic viscosity of phase 2
θ	grad	angular coordinate
θ_A	grad	angle of fluid separation
λ	kJ/(m s K)	heat conductivity
ν	m²/s	kinematic viscosity of fluid
ν_2	m²/s	kinematic viscosity of phase 2
σ	kg/s²	surface tension
ϱ	kg/m³	mean density of fluid
ϱ_A	kg/m³	local partial density of component A
ϱ_{Aw}	kg/m³	partial density of component A at surface of particle
$\varrho_{A\infty}$	kg/m³	partial density of component A at great distance from particle
ϱ_{A1}	kg/m³	partial density of component A in phase 1
ϱ_{A2}	kg/m³	partial density of component A in phase 2
ϱ_{A1p}	kg/m³	partial density of component A at interface in phase 1
ϱ_{A2p}	kg/m³	partial density of component A at interface in phase 2
ϱ_{A10}	kg/m³	partial density of component A in phase 1 at $t=0$
ϱ_{A20}	kg/m³	partial density of component A in phase 2 at $t=0$
$\varrho_{A1\infty}$	kg/m³	partial density of component A in phase 1 at $t=\infty$
$\varrho_{A2\infty}$	kg/m³	partial density of component A in phase 2 at $t=\infty$
$Da \equiv \dfrac{k_w}{D/R} c_{A\theta w}^{n-1}$		Damköhler number
$Fo_{m1} \equiv \dfrac{t}{R^2/D_1}$		Fourier number for phase 1
$Fo_{m2} \equiv \dfrac{t}{R^2/D_2}$		Fourier number for phase 2

6 Transport Processes Through the Interface of Particles

$H^* \equiv \varrho_{A1p}/\varrho_{A2p}$	Henry number
$K_F \equiv \dfrac{\varrho \sigma^3}{g \eta^4}$	fluid number
$l^* \equiv l/R$	ratio of length of vortex
$Nu \equiv \dfrac{a\, d_p}{\lambda}$	mean Nusselt number
n	order of reaction
$p^* \equiv \dfrac{p}{\varrho w_\infty^2/2}$	pressure number
$Pr \equiv \dfrac{\eta c_p}{\lambda}$	Prandtl number
$Re \equiv \dfrac{w_\infty d_p}{\nu}$	Reynolds number
$Re_2 \equiv \dfrac{w_\infty d_p}{\nu_2}$	Reynolds number for phase 2
$r^* \equiv r/R$	radial coordinate
$Sc \equiv \nu/D$	Schmidt number
$Sc_2 \equiv \nu_2/D_2$	Schmidt number for phase 2
$Sh \equiv \dfrac{\beta d_p}{D}$	mean Sherwood number
$Sh_\theta \equiv \dfrac{\beta_\theta d_p}{D}$	local Sherwood number
$Sh_1 \equiv \dfrac{\beta d_p}{D_1}$	mean Sherwood number for phase 1
$Sh_2 \equiv \dfrac{\beta d_p}{D_2}$	mean Sherwood number for phase 2
$w_r^* \equiv w_r/w_\infty$	radial velocity ratio
$w_\theta^* \equiv w_\theta/w_\infty$	tangential velocity ratio
$w_{r1}^* \equiv w_{r1}/w_\infty$	radial velocity ratio for phase 1
$w_{\theta 1}^* \equiv w_{\theta 1}/w_\infty$	tangential velocity ratio for phase 1
$w_{r2}^* \equiv w_{r2}/w_\infty$	radial velocity ratio for phase 2
$w_{\theta 2}^* \equiv w_{\theta 2}/w_\infty$	tangential velocity ratio for phase 2
$\zeta \equiv \dfrac{W/F_p}{\varrho w_\infty^2/2}$	resistance factor
$\eta^* \equiv \eta_1/\eta_2$	dynamic viscosity ratio
$\xi \equiv \dfrac{\varrho_A - \varrho_{A\infty}}{\varrho_{Aw} - \varrho_{A\infty}}$	local concentration ratio
$\xi_1 \equiv \dfrac{\varrho_{A1} - H^* \varrho_{A2\infty}}{\varrho_{A10} - H^* \varrho_{A2\infty}}$	local concentration ratio for phase 1
$\xi_2 \equiv \dfrac{\varrho_{A2} - \varrho_{A2\infty}}{\varrho_{A10}/H^* - \varrho_{A2\infty}}$	local concentration ratio for phase 2
$\bar{\varrho}_1 \equiv \dfrac{\varrho_{A1} - H^* \varrho_{A2\infty}}{\varrho_{A10} - H^* \varrho_{A2\infty}}$	mean concentration ratio for phase 1

6.1 Introduction

Transport processes through the interface of particles include momentum, heat, and mass transfer. The particles will in general have a spherical shape, and may be solids, drops, and bubbles. Momentum transfer is related to fluid flow, which will be assumed to be steady-state. Heat and mass transfer, however, may be steady-state as well as unsteady-state processes.

Almost all heat and mass transfer processes through the interface of particles are of the unsteady-state type. This is due to the fact, that the volume of the particle is in general much smaller than the volume of the environment. Because of this size relationship only a very small amount of mass can either be transferred into the particle or from the particle into the environment. The mass passing across the interface decreases with time. A further dimension is therefore introduced into the investigation of heat and mass transfer and further difficulties must be overcome in theoretical and experimental work.

The process is steady-state only in the case of mass transfer with a biochemical reaction at the surface of a particle. The majority of all publications on transport processes at spherical particles are concerned with steady-state heat and mass transfer. Although of great scientific interest, these investigations are steadily losing in importance, because they are not very useful for the solution of technical problems. The results of investigations on steady-state heat and mass transfer serve in many cases as a rough approximation for unsteady-state processes. It is, therefore, still useful to have the information on steady-state processes at hand.

This chapter is devoted to steady-state fluid flow around solid particles and drops. Steady-state mass transfer will be considered for solids and bubbles without and with chemical reactions. In the case of bubbles, deviation from the spherical shape and its influence on mass transfer will be considered. In the last section a comprehensive treatise on unsteady-state mass transfer will be presented.

6.2 Fluid Flow around Solid Spheres

It is assumed that fluid flow relative to the spherical particle is of the steady-state type. This means, for instance, that a freely falling particle has reached its final velocity. A valuable contribution to the unsteady-state flow problem has been made by HIL-PRECHT [6.1]. For steady-state conditions the velocity field, especially the vortex field in the wake of the sphere, and the friction factor will be considered.

6.2.1 Velocity Field

For steady-state conditions the velocity field in the surrounding fluid is described by differential equations derived from Eqs. (2.13) to (2.15). The notations are given in Fig. 6.1. In dimensionless writing the equations are as follows:

Continuity equation:

$$r^* \frac{\partial w_r^*}{\partial r^*} + 2 w_r^* + \frac{\partial w_\vartheta^*}{\partial \vartheta} + w_\vartheta^* \cot \vartheta = 0. \qquad (6.1)$$

Navier-Stokes equations:

$$w_r^* \frac{\partial w_r^*}{\partial r^*} + \frac{w_\vartheta^*}{r^*} \frac{\partial w_r^*}{\partial \vartheta} - \frac{w_\vartheta^{*2}}{r^*} = -\frac{1}{2} \frac{\partial p^*}{\partial r^*} +$$

$$+ \frac{2}{Re} \frac{1}{r^{*2}} \left[r^* \frac{\partial}{\partial r^*} \left(r^* \frac{\partial w_r^*}{\partial r^*} \right) + r^* \frac{\partial w_r^*}{\partial r^*} + \frac{\partial^2 w_r^*}{\partial \vartheta^2} + \right.$$

$$\left. + \cot \vartheta \frac{\partial w_r^*}{\partial \vartheta} - 2 \frac{\partial w_\vartheta^*}{\partial \vartheta} - 2 w_r^* - 2 w_\vartheta^* \cot \vartheta \right] \qquad (6.2)$$

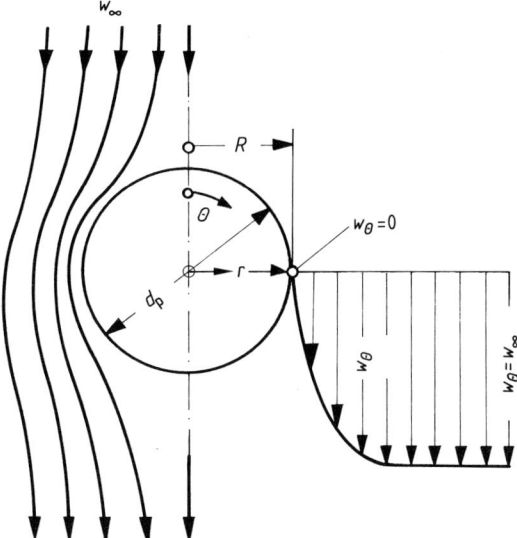

Figure 6.1. Notations for fluid flow around a solid sphere.

$$w_r^* \frac{\partial w_\vartheta^*}{\partial r^*} + \frac{w_\vartheta^*}{r^*} \frac{\partial w_\vartheta^*}{\partial \vartheta} + \frac{1}{r^*} w_r^* w_\vartheta^* = -\frac{1}{2r^*} \frac{\partial p^*}{\partial \vartheta} +$$

$$+ \frac{2}{Re} \frac{1}{r^{*2}} \left[r^* \frac{\partial}{\partial r^*} \left(r^* \frac{\partial w_\vartheta^*}{\partial r^*} \right) + r^* \frac{\partial w_\vartheta^*}{\partial r^*} + \frac{\partial^2 w_\vartheta^*}{\partial \vartheta^2} + \right.$$

$$\left. + \cot \vartheta \frac{\partial w_\vartheta^*}{\partial \vartheta} + 2 \frac{\partial w_r^*}{\partial \vartheta} - \frac{w_\vartheta^*}{\sin^2 \vartheta} \right] \qquad (6.3)$$

Boundary conditions:

$\vartheta = 0; \pi$ and $1 \leq r^* \leq \infty$: $\quad w_\vartheta^* = 0$
$\quad \partial w_r^*/\partial \vartheta = 0$

$0 \leq \vartheta \leq \pi$ and $r^* = 1$: $\quad w_\vartheta^* = 0$
$\quad w_r^* = 0$

$0 \leq \vartheta \leq \pi$ and $r^* = \infty$: $\quad w_\vartheta^* = \cos \vartheta$
$\quad w_r^* = -\sin \vartheta$

The dimensionless parameters are defined as follows:

$r^* \equiv r/R$	radial coordinate	(6.4)
$w_r^* \equiv w_r/w_\infty$	radial velocity component	(6.5)
$w_\vartheta^* \equiv w_\vartheta/w_\infty$	tangential velocity component	(6.6)

$$p^* \equiv \frac{p}{\varrho w_\infty^2/2} \quad \text{local pressure} \tag{6.7}$$

$$Re \equiv \frac{w_\infty d_\mathrm{p}}{\nu} \quad \text{Reynolds number} \tag{6.8}$$

In the above, R is the radius and d_p the diameter of the sphere, ϱ density and ν kinematic viscosity of fluid, and w_∞ velocity of approaching fluid.

Contributions to the numerical solution of the system of differential equations were especially made by JENSON [6.2], HAMIELEC and coauthors [6.3], and IHME et al. [6.4]. Some of their results will be discussed.

The flow field is presented in a very informative way by means of stream-lines, which are curves for a constant value of the dimensionless stream function ψ^*, which is related to the velocity components w_ϑ^* and w_r^* in the following way:

$$\frac{\partial \psi^*}{\partial r^*} = w_\vartheta^* r^* \sin\vartheta, \tag{6.9}$$

$$\frac{\partial \psi^*}{\partial \vartheta} = w_r^* r^{*2} \sin^2\vartheta. \tag{6.10}$$

Stream-lines around a solid spherical particle are shown in Fig. 6.2 for a Reynolds number of $Re=100$. In the wake of the sphere a ring vortex has been established, that is of fundamental importance not only for flow around the sphere, but also for mass and heat transfer.

6.2.2 Vortex Field

Due to the action of the inertia forces a vortex is established in the wake of the sphere. With increasing strength of the inertia forces, i.e., with increasing Reynolds number, the size of the vortex must increase. The size is best expressed by means of the angle of separation of the stream-lines from the sphere ϑ_A as indicated in Fig. 6.2 and the length l of the vortex. Fig. 6.3 gives the angle ϑ_A and Fig. 6.4 the relative length $l^* \equiv l/R$ as a function of the Reynolds number. Experimental and theoretical-numerical data demonstrate that vortex formation starts at a Reynolds number of $Re=20$. For the angle of separation ϑ_A and the length ratio l^* the following equations have been set up:

$$\vartheta_A = \left(\frac{\ln Re - 3}{2.3 \cdot 10^{-4}}\right)^{0.45}, \tag{6.11}$$

range of application: $20 \leqslant Re \leqslant 400$

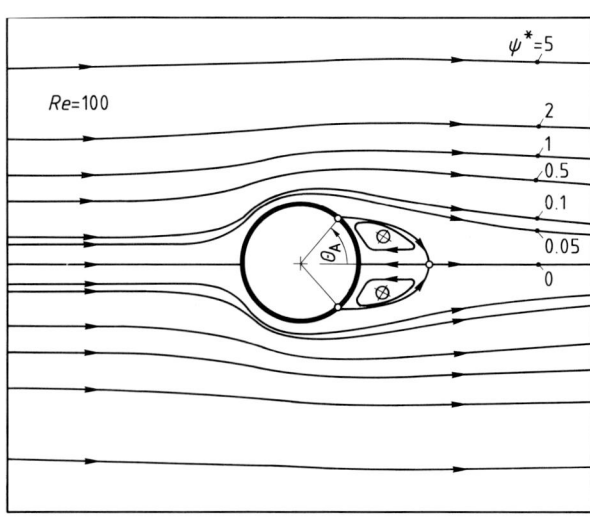

Figure 6.2. Stream-lines in the environment of a sphere at a Reynolds number of $Re=100$.

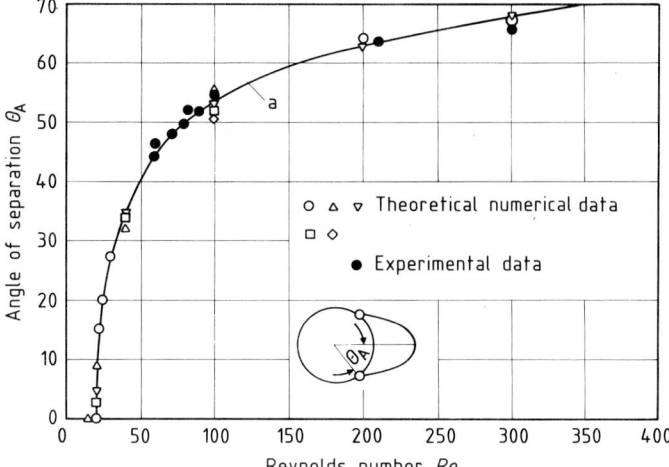

Figure 6.3. Angle of separation as a function of Re.

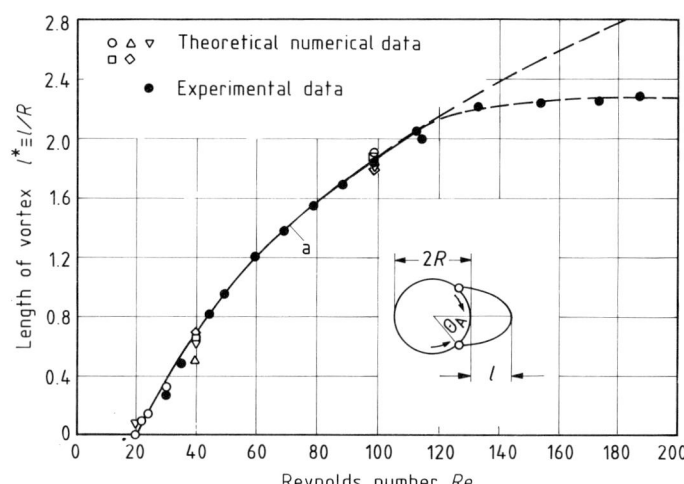

Figure 6.4. Relative length of vortex as a function of Re.

$$l^* = 1.76 \ln(Re + 22) - 6.58 \quad (6.12)$$
range of application: $20 \leqslant Re \lesssim 120$

A close examination reveals three different types of vortexes:

$20 \leqslant Re \lesssim 130$	steady-state ring vortex
$130 \lesssim Re \lesssim 450$	unsteady-state ring vortex
$450 \lesssim Re$	periodic formation and detachment of vortex elements (vortex street)

6.2.3 Resistance Factor

The resistance factor for particles is defined by:

$$\zeta \equiv \frac{W/F_\mathrm{p}}{\varrho w_\infty^2/2},$$

with W as the resistance force, $F_\mathrm{p} = d_\mathrm{p}^2 \pi/4$ the cross sectional area of the sphere, ϱ

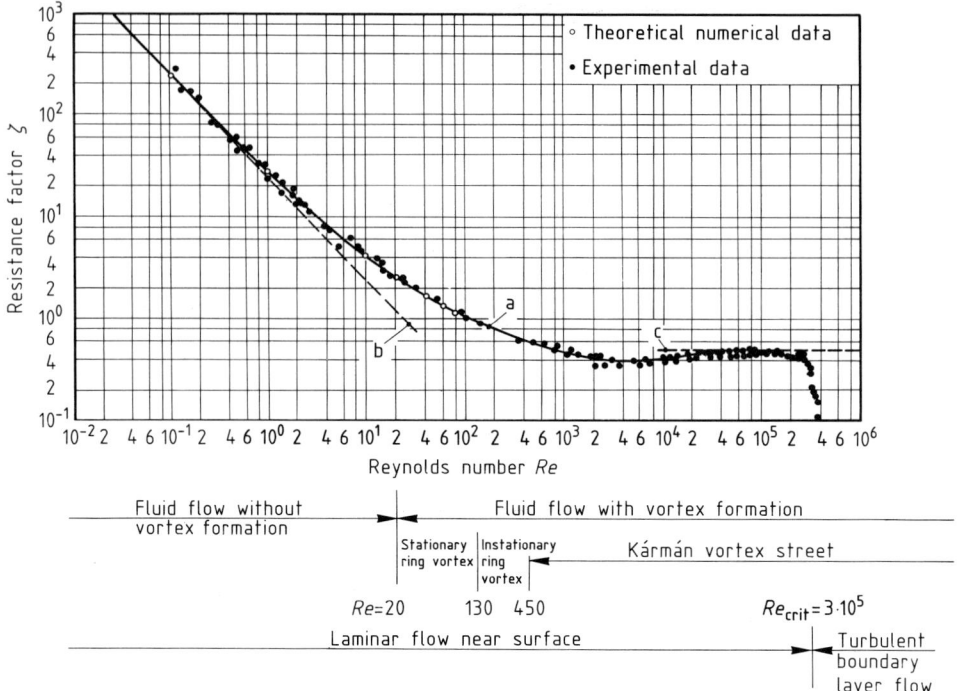

Figure 6.5. Friction factor for spherical solid particles in steady-state motion.

density, and w_∞ the velocity of the fluid approaching the particle. The resistance force is the sum of friction and pressure force, which can be obtained from the solution of the Navier-Stokes equations.

The resistance factor ζ is plotted in Fig. 6.5 against the Reynolds number Re. Experimental and numerical data are in excellent agreement. They are correlated by curve a, for which the following equation has been set up:

$$\zeta = \frac{24}{Re} + \frac{3.73}{Re^{1/2}} - \frac{4.83 \cdot 10^{-3} Re^{1/2}}{1 + 3 \cdot 10^{-6} Re^{3/2}} + 0.49 \quad (6.13)$$

range of application: $0 \leq Re \leq 3 \cdot 10^5$

In the low Reynolds number region the well known Stokes equation applies [6.5]:

$$\zeta = \frac{24}{Re} \quad (6.14)$$

range of application: $0 \leq Re \leq 0.1$

This law is represented by curve b in Fig. 6.5. In the high Reynolds number region curve c represents Newton's law:

$$\zeta = 0.49 \quad (6.15)$$

range of application: $10^4 \leq Re \leq 3 \cdot 10^5$

In Fig. 6.5 the flow regions are indicated. It should be observed that fluid flow is still laminar although it comes to vortex formation in the wake of the sphere. Whatever type of fluid flow exists far from the sphere, most probably a turbulent state, it can be stated that when Re is smaller than $3 \cdot 10^5$ fluid flow close to the surface of the sphere will be laminar. Transition to turbulence occurs at the critical value of the Reynolds number, $Re_{crit} = 3 \cdot 10^5$.

The resistance force W is the sum of friction and pressure forces. In the Stokes region, that is for $Re \leq 0.1$, the friction force is twice the pressure force. In the Newton region, that is $10^4 \leq Re \leq 3 \cdot 10^5$, the friction force is negligibly small, so that the resistance force W is only due to the pressure force.

6.3 Fluid Flow around Bubbles

Steady-state fluid flow relative to the bubble will be investigated. Small size bubbles in general have a spherical shape. Beyond a certain diameter pressure forces become much stronger than capillary forces and therefore promote aperiodic deformation of the bubbles. Spherical and non-spherical bubbles will be considered.

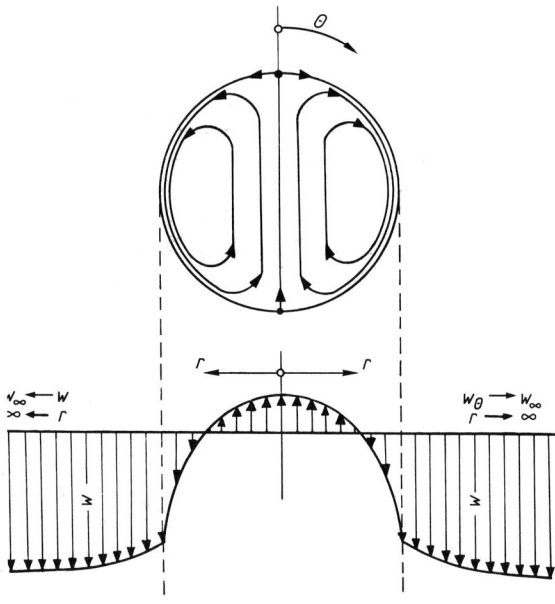

Figure 6.6. Fluid movement inside a bubble and in the surrounding liquid as observed in the equatorial plane.

6.3.1 Velocity Field around Spherical Bubbles

For steady state conditions the velocity field is described by the differential equations which have already been presented for fluid flow around solid spheres, Eqs. (6.1) to (6.3). The boundary conditions have to be reformulated. According to Fig. 6.6 fluid motion occurs not only on the outside but also on the inside of the sphere. It may be safely assumed that internal friction of the gas is negligible with regard to internal friction of the liquid. For the gas inside the sphere the assumption of potential flow conditions will therefore be justified; in the interface the shear stress must be zero [6.6].

The most important boundary conditions are those in the interface. The radial velocity in the interface is zero, because shape and size of the sphere do not change. Assuming a non-slip condition in the interface of the two fluid phases, there must be a tangential velocity w_θ. This tangential velocity

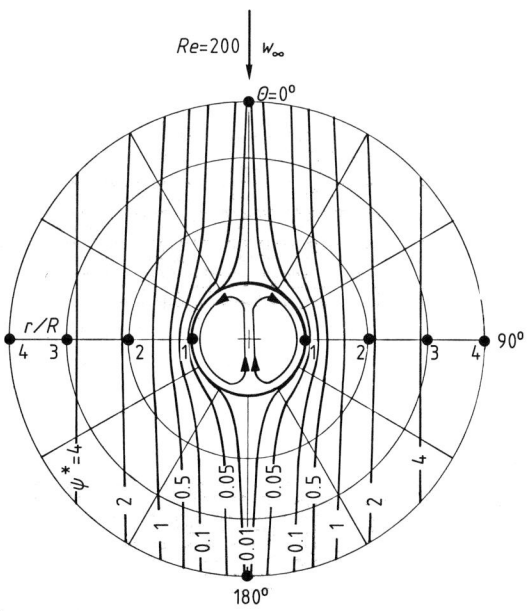

Figure 6.7. Stream-lines for fluid flow within a bubble and in the surrounding liquid.

is determined by the condition of no shear stress $\tau_{\theta r}$ in the interface:

$$\tau_{\theta r} = -\eta \left[r \frac{\partial}{\partial r} \left(\frac{w_\theta}{r} \right) + \frac{1}{r} \frac{\partial w_r}{\partial \theta} \right] = 0. \quad (6.16)$$

Introducing dimensionless quantities, one obtains:

$$\tau_{\theta r}^* = \frac{4}{Re} \left[r^* \frac{\partial}{\partial r^*} \left(\frac{w_\theta^*}{r^*} \right) + \frac{1}{r^*} \frac{\partial w_r^*}{\partial \theta} \right] = 0. \quad (6.17)$$

The dimensionless shear stress is defined as follows:

$$\tau_{\theta r}^* \equiv \frac{\tau_{\theta r}}{\varrho w_\infty^2 / 2}, \quad (6.18)$$

with $\tau_{\theta r}$ as the local shear stress. The boundary conditions for fluid flow relative to a spherical bubble can then be formulated as follows:

$\theta = 0, \pi$ and $1 \leqslant r^* \leqslant \infty$: $\quad w_\theta^* = 0$
$\qquad\qquad\qquad\qquad\qquad\quad \partial w_r^*/\partial \theta = 0$

$0 \leqslant \theta \leqslant \pi$ and $\quad r^* = 1$: $\quad w_r^* = 0$
$\qquad\qquad\qquad\qquad\qquad\quad \tau_{\theta r}^* = 0$

$0 \leqslant \theta \leqslant \pi$ and $\quad r^* = \infty$: $\quad w_\theta^* = \cos \theta$
$\qquad\qquad\qquad\qquad\qquad\quad w_r^* = -\sin \theta$

The solution of the differential equations has been obtained by applying numerical methods [6.6].

Fig. 6.7 shows stream-lines within the bubble and in the surrounding fluid. At a Reynolds number of $Re = 200$ fluid flow within the bubble and outside of the bubble comes very close to the potential flow condition. This is mainly due to the fact, that the shear stress in the interface is negligibly small.

At a first glance it may surprise that there is no fluid separation and no formation of a vortex downstream from the equator. This, too, is due to the no-shear stress condition in the interface. Separation can only take place at $\theta = 180°$.

The tangential velocity w_θ^* in the equatorial plane is shown in Fig. 6.8 versus the radial coordinate with the Reynolds number as a parameter. In the interface the velocity varies between 0.5 for creeping flow conditions and 1.5 for potential flow conditions with $Re \to \infty$. Curve a is valid for creeping flow around a solid sphere.

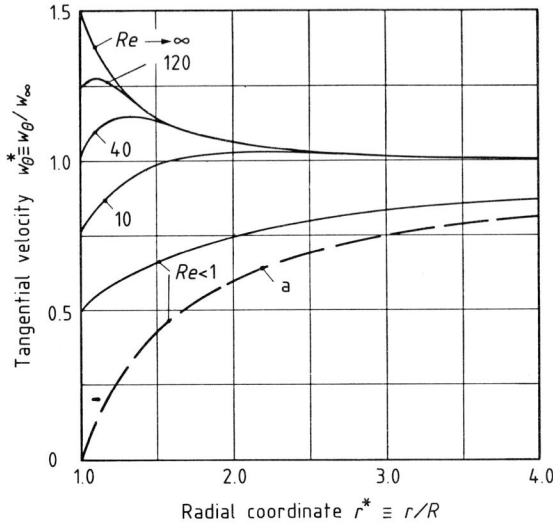

Figure 6.8. Tangential velocity w_θ^* in the equatorial plane versus the radial coordinate r^* for various values of Re; curve a for fluid flow around a solid sphere.

6.3.2 Resistance Factor

The resistance factor ζ for spherical bubbles is shown in Fig. 6.9 versus the Reynolds number. For a comparison curve e has been given, representing the resistance factor for solid spheres. The resistance factor for bubbles consists of three distinct parts. In regions a and b spherical bubbles exist. The numerical data available in these regions are approximated by the equation:

$$\zeta_{ab} = \frac{16}{Re} + \frac{14.9}{Re^{0.78}} \frac{1}{1 + 10 Re^{-0.6}}. \quad (6.19)$$

range of application: $0 \leqslant Re \leqslant Re_B$

The first term, $\zeta_a = 16/Re$, has been published by HADAMARD [6.7] and independently by RYBCZINSKI [6.8]. In the low Reynolds number region, $Re \leqslant 1$, ζ for bubbles is only 2/3 of that for solid spheres.

The upper limit for Eq. (6.19) is Re_B. At this value of Re the bubble loses its spherical shape. The resistance factor increases sharply in the transition region from Re_B to Re_C. At Re_C a new type of bubble is established that is characterized by an unstable shape. The shape of the bubble is varying in a stochastic process. In this state the resistance factor ζ_d is constant:

$$\zeta_d = 2.61 \qquad (6.20)$$
range of application: $Re \geqslant Re_C$

This value has been determined experimentally [6.9].

At the upper limiting value of Re_B local pressure forces overcome surface tension forces. The following equation for Re_B has been derived [6.10]:

$$Re_B = 3.73 \, K_F^{0.209} \qquad (6.21)$$
range of application: $Re \geqslant 10$

The dimensionless number K_F is defined by:

$$K_F \equiv \varrho \sigma^3 / (g \eta^4). \qquad (6.22)$$

In this equation ϱ is the liquid density, σ the surface tension, η the liquid viscosity, and g the gravitational acceleration. At the upper limit of the transition region the Reynolds number is given by the empirical equation:

$$Re_C = 3.1 \, K_F^{0.250}. \qquad (6.23)$$

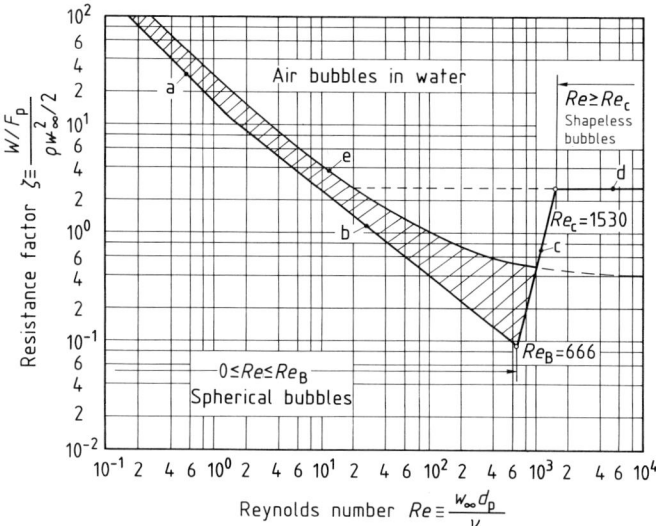

Figure 6.9. Resistance factor for spherical and non-spherical bubbles versus Reynolds number Re.

In Fig. 6.9 the transition region has been determined for air bubbles in water. In this case $Re_B = 666$ and $Re_C = 1530$.

6.3.3 Fluid Flow around Shapeless Bubbles

6.3.3.1 Deformation Turbulence

In the preceding section it has been stated, that the shape of a bubble becomes unstable when the Reynolds number exceeds the value Re_B.

A new state of bubble shape will be established. This occurs within the transition region, that extends from Re_B to Re_C. For all values of $Re \geqslant Re_C$ the shape of the bubble changes rapidly with time. The process

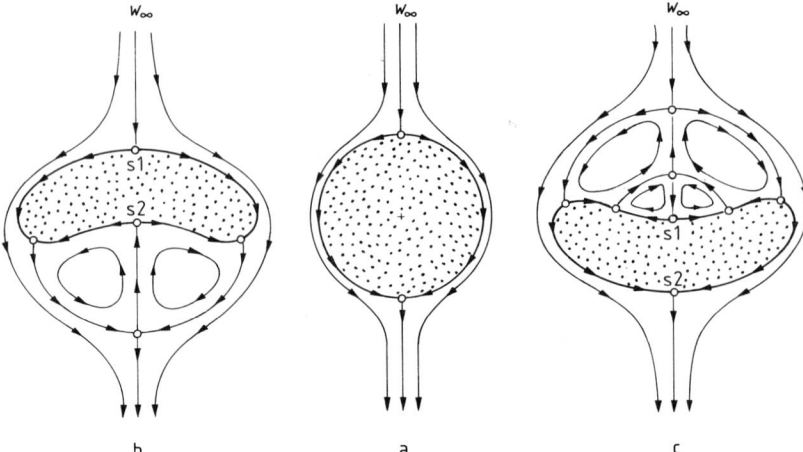

Figure 6.10. Deformation of a bubble and fluid motion.

of deformation seems to be of a stochastic nature. The consequences are stochastic motions in the surrounding fluid.

It is assumed that stochastic motions in the fluid due to deformations of the bubble have properties of conventional turbulence. This type of turbulence will be called deformation turbulence. It differs from conventional turbulence only in so far, as it has its highest value at the interface and decreases with increasing radial distance from the interface.

A very simple deformation of a bubble and the induced fluid movement in the surrounding liquid is shown in Fig. 6.10. As a basis for a mathematical description of momentum transfer under the action of deformation turbulence, it is assumed that the bubble undergoing stochastic deformation processes may be replaced by a bubble with a spherical shape, in whose vicinity a time independent turbulence function takes care of the stochastic motions within the liquid.

The turbulence function has to be determined in such a way, that calculated resistance factor and Sherwood number agree with available experimental results. Furthermore, calculated details of the velocity and concentration field must be in agreement with fundamental knowledge available for these phenomena.

6.3.3.2 Velocity Field

Figure 6.11 gives the calculated streamlines in the vicinity of a nonspherical bubble undergoing stochastic deformations. One of the striking results is the extremely large vortex behind the sphere, although ideal mobility of the interface has been assumed. The vortex is a direct consequence of the deformation process imposed on the bubble. For spherical bubbles neither vortex formation nor flow separation have been observed.

The tangential velocity $\bar{w}^*_{\partial w}$ in the interface is given in Fig. 6.12 as a function of the angle θ. The broken lines show, for purposes of comparison, the interface velocity for spherical bubbles. For non-spherical bubbles a non-symmetrical velocity profile is obtained. According to the vortex formation, the tangential velocity behind the equator is extremely small. Fig. 6.13 gives the tangential velocity in the equatorial plane at $\theta = 90°$ as a function of the radial coordinate. The broken lines are again for fluid flow around spherical bubbles and solid spheres. At first glance, it is surprising that the stochastic deformation process enforces an extremely small tangential velocity close to the interface. Of course this is due to the stochastic radial velocity, which

6.3 Fluid Flow around Bubbles

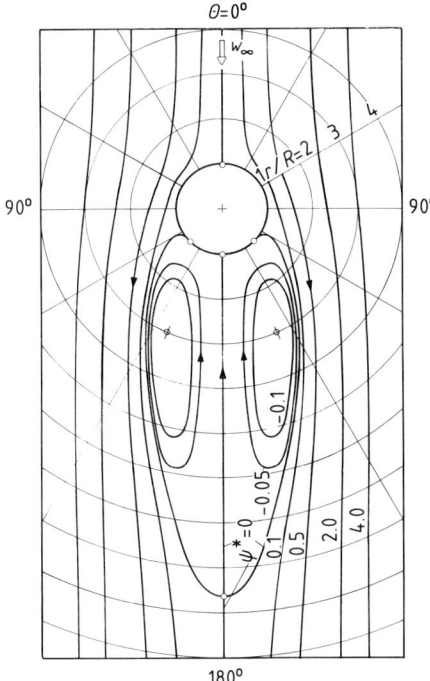

Figure 6.11. Stream-lines around a bubble undergoing stochastic deformations.

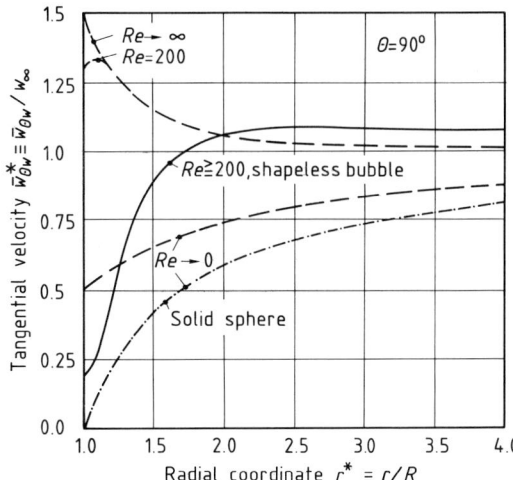

Figure 6.13. Tangential velocity in the equatorial plane versus the radial coordinate; broken curves for spherical bubbles; dotted curve for solid sphere.

has its maximum at the interface and decreases with an increasing radial coordinate. The radial motion of the interface pushes the tangential fluid motion away from the bubble interface. It is to be noted, that the tangential velocity close to the interface is even smaller than for conditions of creeping flow around spherical bubbles.

6.3.3.3 Shear Stress

Deformation turbulence causes the shear stress:

$$\tau^*_{\theta rt} \equiv \frac{\tau_{\theta rt}}{\varrho w^2_\infty/2}. \tag{6.24}$$

The turbulent shear stress varies with the radial coordinate r^* and angle θ. In Fig. 6.14 $\tau^*_{\theta rt}$ is shown as a function of r^* for two values of θ. According to the produc-

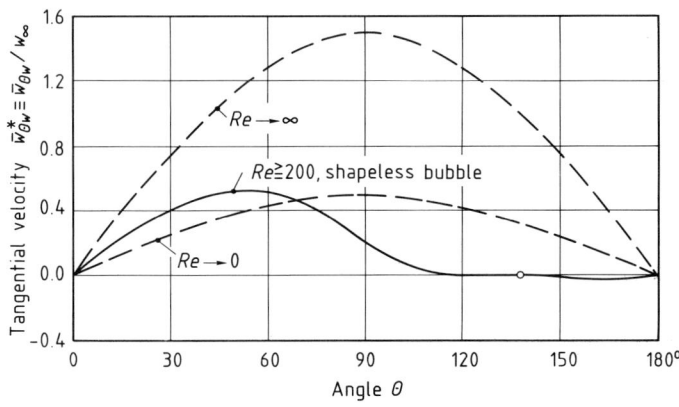

Figure 6.12. Tangential velocity $w^*_{\theta w}$ at the interface of bubbles versus the angle θ; the broken curves give the velocities for spherical bubbles for creeping flow ($Re \to 0$) and for potential flow ($Re \to \infty$).

tion of deformation turbulence the maximum of $\tau^*_{\vartheta rt}$ is at the interface of the bubble at $r^* = 1$. It may surprise, that deformation turbulence decays very rapidly. At a distance from the interface of about half a diameter, the shear stress due to deformation turbulence is zero.

Over the circumference there is a remarkable variation of the shear stress as shown in Fig. 6.15. According to vortex formation as indicated in Fig. 6.10 the minimum of the shear stress or the maximum of the tangential component of the fluctuating velocity in the interface must be close to the equatorial plane; it is exactly at $\theta = 72°$. The radial component of the fluctuating velocity is assumed to be constant over the surface of the bubble.

6.4 Steady-State Mass Transfer at Solid Spheres

Consideration will be given to the case where mass is transferred from the surface of the sphere to the fluid and *vice versa*. A chemical reaction is excluded.

6.4.1 Concentration Field

The local concentration is described by a differential equation obtained from Eq. (2.18). The coordinate system is indicated in Fig. 6.1. In dimensionless writing the differential equation is as follows:

$$w^*_r \frac{\partial \xi}{\partial r^*} + \frac{w^*_\vartheta}{r^*} \frac{\partial \xi}{\partial \theta} = \frac{2}{Re\, Sc} \frac{1}{r^{*2}} \cdot$$
$$\cdot \left[\frac{\partial}{\partial r^*} \left(r^{*2} \frac{\partial \xi}{\partial r^*} \right) + \frac{1}{\sin \theta} \frac{\partial}{\partial \theta} \left(\sin \theta \frac{\partial \xi}{\partial \theta} \right) \right].$$
(6.25)

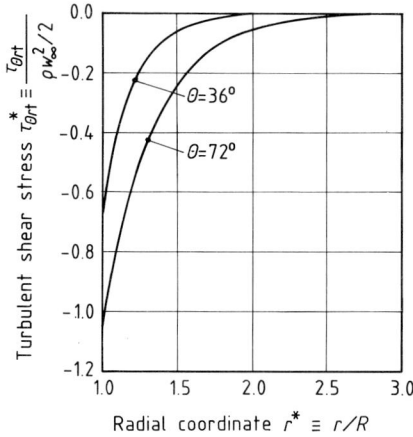

Figure 6.14. Shear stress due to deformation turbulence versus radial coordinate for two values of the angle θ.

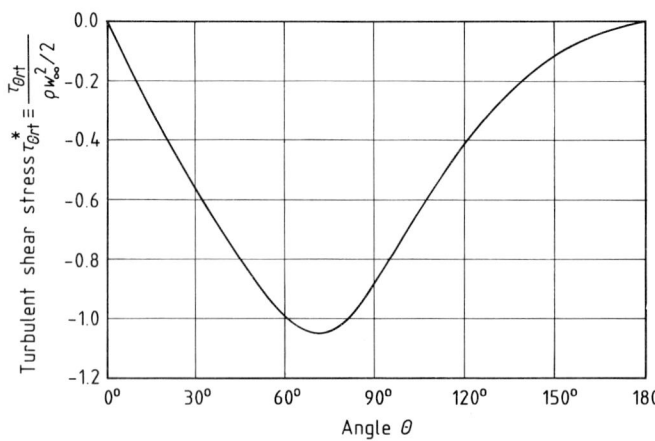

Figure 6.15. Shear stress at the interface of a bubble caused by deformation turbulence versus the angle θ.

Boundary conditions:

$\theta = 0; \pi$ and $1 \leqslant r^* \leqslant \infty$: $\partial \xi / \partial \theta = 0$
$0 \leqslant \theta \leqslant \pi$ and $r^* = 1$: $\xi = \xi_w = 1$
$0 \leqslant \theta \leqslant \pi$ and $r^* = \infty$: $\xi = \xi_\infty = 0$

The dimensionless parameters which have not yet been explained in the preceding sections are defined by:

$$\xi \equiv \frac{\varrho_A - \varrho_{A\infty}}{\varrho_{Aw} - \varrho_{A\infty}} \quad \text{local concentration} \quad (6.26)$$

$$Sc \equiv v/D \quad \text{Schmidt number} \quad (6.27)$$

Here, ϱ_A is the local partial density of component A, $\varrho_{A\infty}$ in the approaching fluid, ϱ_{Aw} at the surface of the sphere, v the kinematic viscosity of the fluid, and D the diffusion coefficient. The partial densities $\varrho_{A\infty}$ and ϱ_{Aw} are assumed to be constant.

The velocity components w_r^* and w_θ^* are obtained from the system of equations given in Sect. 6.2. A numerical solution of Eq. (6.25) has been presented by IHME et al. [6.4]. The results are the concentration fields in the surrounding fluid. Of special importance is the concentration gradient at the surface of the sphere, when the local and the mean Sherwood number are the aim of the calculations.

6.4.2 Local and Mean Sherwood Numbers

The local Sherwood number is defined by:

$$Sh_\theta \equiv \frac{\beta_\theta d_p}{D}, \quad (6.28)$$

with β_θ the local mass transfer coefficient, defined by the equation:

$$\beta_\theta \equiv \frac{\dot{m}_{A\theta w}}{\varrho_{Aw} - \varrho_{A\infty}}. \quad (6.29)$$

Introducing Fick's law for the mass flux density at the particle surface:

$$\dot{m}_{A\theta w} \equiv -D \left(\frac{\partial \varrho_A}{\partial r} \right)_{\theta w}, \quad (6.30)$$

and making use of dimensionless parameters leads to:

$$Sh_\theta \equiv -2 \left(\frac{\partial \xi}{\partial r^*} \right)_{\theta w}. \quad (6.31)$$

Integration over the surface of the sphere gives the mean Sherwood number:

$$Sh = \frac{1}{2} \int_0^\pi Sh_\theta \sin \theta \, d\theta, \quad (6.32)$$

with

$$Sh \equiv \frac{\beta d_p}{D}. \quad (6.33)$$

In Fig. 6.16 the mean Sherwood number is presented as a function of the convection number $Re\, Sc$, with the Schmidt number Sc as a further parameter [6.11]. The data given in this figure have been obtained by theoretical-numerical methods. The curves drawn through these data have been calculated by means of the following approximation equation:

$$Sh = 2 + f_k \frac{(Re\, Sc)^{1.7}}{1 + (Re\, Sc)^{1.2}}, \quad (6.34)$$

with

$$f_k = \frac{0.66}{[1 + (0.84\, Sc^{1/6})^3]^{1/3}} \quad (6.35)$$

range of application: $0 \leqslant Re \leqslant Re_{\text{crit}} = 3 \cdot 10^5$
$0 \leqslant Sc \leqslant \infty$

Eq. (6.34) is in excellent agreement with experimental data from various sources [6.12].

For potential flow conditions with $Re \to \infty$ and $Sc \to 0$ Eq. (6.34) is simplified to:

$$Sh = 2 + 0.66 \frac{(Re\, Sc)^{1.7}}{1 + (Re\, Sc)^{1.2}}. \quad (6.36)$$

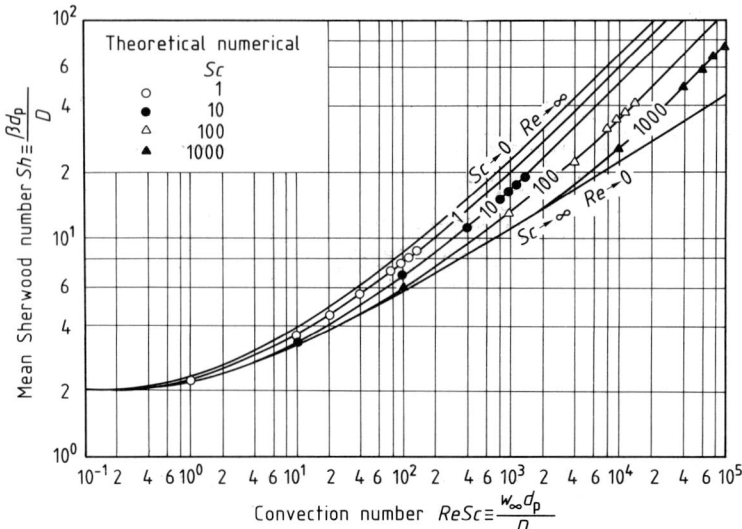

Figure 6.16. Mean Sherwood number Sh for mass transfer at solid spherical particles versus the convection number $Re\,Sc$ for various values of the Schmidt number Sc.

For boundary layer conditions with $Re \to Re_{crit} = 3 \cdot 10^5$ and $Sc \to \infty$, Eq. (6.35) gives $f_k = 0.79/Sc^{1/6}$, so that Eq. (6.34) is simplified to the well known equation:

$$Sh = 0.79\, Re^{1/2}\, Sc^{1/3}. \qquad (6.37)$$

This equation is of no practical value and proves the limitations of the boundary layer hypothesis.

Another limiting condition that has to be observed is unfortunately not included in Eq. (6.34). This case applies to creeping flow conditions: $Re \to 0$, $Sc \to \infty$. The numerical results obtained for this case have been approximated by the equation:

$$Sh = 2 + \frac{0.333\,(Re\,Sc)^{0.840}}{1 + 0.331\,(Re\,Sc)^{0.507}}. \qquad (6.38)$$

With $Re\,Sc \to \infty$ this equation is simplified to the well known Eq. (6.13):

$$Sh = 1.007\,(Re\,Sc)^{1/3}. \qquad (6.39)$$

This equation is also of no practical value.

With the mean Sherwood number according to Eq. (6.34) the mean mass transfer coefficient is known. This serves to calculate the mass $\dot M_A$ transferred per time unit from the sphere to the fluid and *vice versa*:

$$\dot M_A = \beta A_p (\varrho_{Aw} - \varrho_{A\infty}), \qquad (6.40)$$

where $A_p = d_p^2 \pi$ is the surface area of the sphere.

6.4.3 The Analogous Heat Transfer Case

Heat and mass transfer are analogous transfer processes when there is a complete formal identity of the involved dimensionless differential equations and boundary conditions. In the case of analogy, mass transfer equations may be converted into heat transfer equations by carrying out the following substitutions:

$$Sh \triangleq Nu \equiv \alpha\, d_p/\lambda \qquad \text{Nusselt number}$$
$$Sc \triangleq Pr \equiv \eta\, c_p/\lambda \qquad \text{Prandtl number}$$
$$\xi \triangleq T^* \equiv \frac{T - T_\infty}{T_w - T_\infty}$$
$$\dot M_A \triangleq \dot Q = \alpha A_p (T_w - T_\infty)$$

6.5 Steady-State Mass Transfer with Biochemical Reactions at the Surface of Solid Spheres

In these equations λ, η, and c_p are heat conductivity, dynamic viscosity, and specific heat capacity of the fluid, α the mean heat transfer coefficient, and Q heat flux.

A biochemical reaction at the surface of a solid sphere is under all circumstances a steady-state process, because the particle itself is not involved in the reaction at all. The particle offers the surface to which the reactants are transported by diffusion and convection to undergo a certain biochemical reaction. This process will be considered [6.12].

6.5.1 The Mathematical Problem

The concentration field is described by Eq. (6.25). One of the boundary conditions, however, must be reformulated, so that the biochemical reaction is taken care of. At the site of reaction the diffusion flux density:

$$\dot{n}_{A\theta w} = -D\left(\frac{\partial c_A}{\partial r}\right)_{\theta w} \qquad (6.41)$$

must be equal to the reaction flux density for which the following empirical equation will be assumed to apply:

$$\dot{r}_{A\theta w} = k_w c^n_{A\theta w}. \qquad (6.42)$$

Here, k_w is the reaction velocity constant, $c_{A\theta w}$ the molar concentration at the surface of the sphere, and n the order of the reaction. For many reactions n varies between 0 and 2. With $\dot{n}_{A\theta w} = -\dot{r}_{A\theta w}$ an equation for the concentration gradient at the surface is obtained:

$$\left(\frac{\partial c_A}{\partial r}\right)_{\theta w} = \frac{k_w}{D} c^n_{A\theta w}. \qquad (6.43)$$

With the dimensionless local concentration,

$$\xi \equiv \frac{c_A}{c_{A\infty}}, \qquad (6.44)$$

and the radial coordinate $r^* \equiv r/R$ Eq. (6.43) becomes:

$$\left(\frac{\partial \xi}{\partial r^*}\right)_{\theta w} = Da\, \xi^n_w. \qquad (6.45)$$

Da is the Damköhler number defined by:

$$Da \equiv \frac{k_w}{D/R} c^{n-1}_{A\theta w}. \qquad (6.46)$$

The Damköhler number may be interpreted as the ratio of reaction velocity and diffusion velocity; it combines the two important steps in a chemical conversion process. Da varies between two limiting values:

$Da \to 0$: reaction controlled process,
$Da \to \infty$: diffusion controlled process.

After introduction of the Damköhler number the boundary conditions can be formulated as:

$\theta = 0; \pi$ and $1 \leq r^* \leq \infty$: $\partial\xi/\partial\theta = 0$

$0 \leq \theta \leq \pi$ and $r^* = 1$: $(\partial\xi/\partial r^*)_{\theta w} = Da\, \xi^n_{\theta w}$

$0 \leq \theta \leq \pi$ and $r^* = \infty$: $\xi = \xi_\infty = 1$

The differential equation has been solved by numerical methods. The results are local concentration ξ, concentration at the surface $\xi_{\theta w}$, and the concentration gradient $(\partial\xi/\partial r^*)_{\theta w}$ at the surface of the sphere, that is required for calculation of the Sherwood number.

6.5.2 Local and Mean Sherwood Numbers

The local mass transfer coefficient for the case of an enzymatic chemical reaction at the surface of the sphere is defined by:

$$\beta_\theta \equiv \dot{n}_{A\theta w}/c_{A\infty}. \tag{6.47}$$

With $\dot{n}_{A\theta w} = -\dot{r}_{A\theta w}$ introduction of Eq. (6.42) into Eq. (6.47) after rearrangement leads to the following equation for the local Sherwood number:

$$Sh_\theta = 2 Da\, \xi_w^n. \tag{6.48}$$

The local Sherwood number is defined as:

$$Sh_\theta = \beta_\theta d_p/D. \tag{6.49}$$

Integration over the surface of the sphere leads to the mean Sherwood number:

$$Sh = \frac{1}{2} \int_0^\pi Sh_\theta \sin\theta\, d\theta. \tag{6.50}$$

The definition of Sh is as follows:

$$Sh \equiv \beta d/D. \tag{6.51}$$

Comparing Eqs. (6.47) and (6.29) it will be noticed that the definition of a local mass transfer coefficient for mass transfer with a chemical reaction does not include a concentration difference, but only the concentration $c_{A\infty}$. It does not seem advisable to define β_θ by means of the concentration difference $c_{A\infty} - c_{A\theta w}$ because $c_{A\theta w}$ varies over the surface of the sphere.

As an example of the results obtained by theoretical-numerical methods, the mean Sherwood number for a reaction order of $n=1$ is presented in Fig. 6.17 for several values of the Damköhler number. There are three special cases which will be discussed.

The first case is that of $Da \to \infty$. The process of biochemical conversion is controlled by diffusion and convective transfer only. The curve shown for $Da \to \infty$ in Fig. 6.17 is identical with the curve shown in Fig. 6.16 for $Sc \to \infty$. This curve is approximated by Eq. (6.38). The curves shown in Fig. 6.17 prove, that a catalyzed biochemical reaction reduces mass transfer. For $Da \to 0$, the Sherwood number also tends towards zero.

The second special case involves an upper limiting value for the Sherwood number, when the convection number $Re\, Sc$

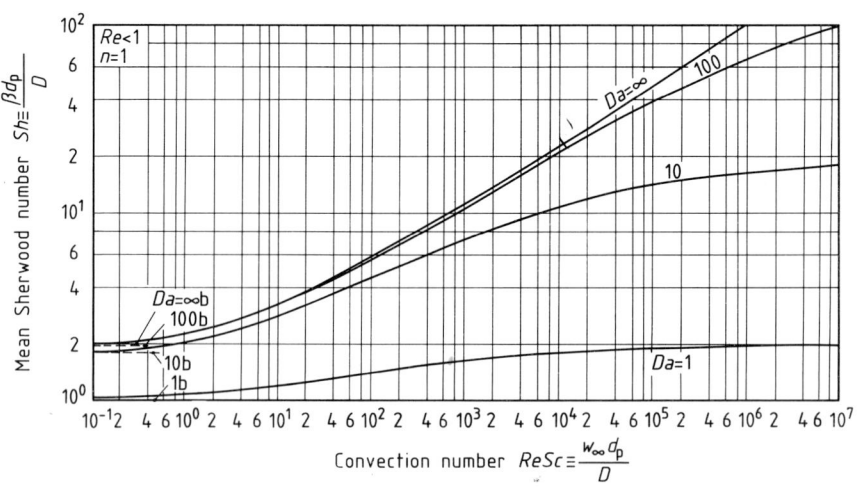

Figure 6.17. Mean Sherwood number for mass transfer with biochemical reaction at the surface of a spherical particle.

6.5 Steady-State Mass Transfer with Biochemical Reactions at the Surface of Solid Spheres

tends towards infinity. For this condition the convective transport is extremely effective, so that the concentration at the surface of the sphere is $\xi_w \to 1$. According to Eq. (6.48) the local and, consequently, the mean Sherwood number reaches the upper limiting value:

$$Sh_\infty = 2\,Da \qquad (6.52)$$

range of application: $0 \leqslant Da \leqslant \infty$
$Re\,Sc \to \infty$

For this condition, the mean mass transfer coefficient is only a function of reaction parameters:

$$\beta_\infty = k_w\, c_{A\theta w}^{n-1}. \qquad (6.53)$$

The third special case concerns the lower limiting value for the Sherwood number, when the convection number $Re\,Sc$ tends towards zero. In this case the concentration field is described by a very simple differential equation, which may be solved analytically for concrete values of the reaction order n. The obtained results are approximated by the following equation:

$$Sh_0 = \left[\left(\frac{1}{2\,Da}\right)^{1/\sqrt{n}} + \left(\frac{1}{2}\right)^{1/\sqrt{n}}\right]^{-\sqrt{n}} \qquad (6.54)$$

range of application: $0 \leqslant Da \leqslant \infty$
$0.5 \leqslant n \leqslant 2$

For $Da = \infty$, the expected value $Sh_0 = 2$ is obtained.

The available numerical data for the considered mass transfer case have been approximated by the following equation, which has been set up by ADAM [6.14]:

$$Sh = \left\{\left[\frac{1}{Sh_0 + f_k \dfrac{(Re\,Sc)^{1.7}}{1+a_1(Re\,Sc)^{1.7-a_2}}}\right]^{a_3} + \left(\frac{1}{2\,Da}\right)^{a_3}\right\}^{-1/a_3} \qquad (6.55)$$

range of application: $0 \leqslant Re\,Sc \leqslant \infty$
$0 \leqslant Re \leqslant Re_{crit} = 3\cdot 10^5$
$0 \leqslant Sc \leqslant \infty$
$0 \leqslant Da \leqslant \infty$
$0.5 \leqslant n \leqslant 2$

For Sh_0, f_k and a_1 to a_3 the following equations apply:

$Sh_0 = f_1(Da; n)$: $0 \leqslant Da \leqslant \infty$ and after Eq. (6.54)
$0.5 \leqslant n \leqslant 2$

$f_k = f_2(Sc)$: $0 \leqslant Sc \leqslant \infty$ after Eq. (6.35)

$a_1 = f_3(Da)$: $0 \leqslant Da \leqslant \infty$ and after Eq. (6.56)
$0.5 \leqslant n \leqslant 2$

$a_2 = f_4(Da)$: $0 \leqslant Da \leqslant \infty$ after Eq. (6.57)

$a_3 = f_5(Da)$: $0 \leqslant Da \leqslant \infty$ after Eq. (6.58)

The equations for a_1 to a_3 are as follows:

$$a_1 = \left[\left(\frac{2}{Da}\right)^{1/n} + 1\right]^n, \qquad (6.56)$$

$$a_2 = \left[\left(\frac{1}{8.35\cdot 10^{-4} + 1.5\cdot 10^{-2}\,Da^{1.2}}\right)^{1.75} + 128\right]^{-0.143}, \qquad (6.57)$$

$$a_3 = \left(\frac{1}{1.434\cdot 10^{-6} + 1.414\cdot 10^{-2}\,Da^{2.67}}\right)^{1/4} + 1.1. \qquad (6.58)$$

The mean deviation of data calculated by means of Eq. (6.55) and data determined by theoretical-numerical methods is $\pm 3.2\%$.

6.6 Steady-State Mass Transfer at Bubbles

Mass transfer through the interface of spherical and non-spherical bubbles will be considered. Chemical reactions will be excluded.

6.6.1 Mass Transfer through the Interface of Spherical Bubbles

Fluid motion within a bubble and in the surrounding liquid has been studied in Sect. 6.3. The concentration field is described by the differential equation presented for mass transfer at solid spheres by Eq. (6.25). The boundary conditions also apply. The local and mean Sherwood number are determined as explained in Sect. 6.4.2.

The mean Sherwood number for spherical bubbles is plotted in Fig. 6.18 against the convection number $Re\,Sc$ [6.15]. There are two limiting curves between which all others are arranged [6.12]:

$Re \to 0$ and $Sc \to \infty$; curve a in Fig. 6.18:

$$Sh_\infty = 2 + \frac{0.651\,(Re\,Sc)^{1.72}}{1+(Re\,Sc)^{1.22}} \qquad (6.59)$$

$Re \to \infty$ and $Sc \to 0$; curve b in Fig. 6.18:

$$Sh_0 = 2 + \frac{0.232\,(Re\,Sc)^{1.72}}{1+0.205\,(Re\,Sc)^{1.22}}. \qquad (6.60)$$

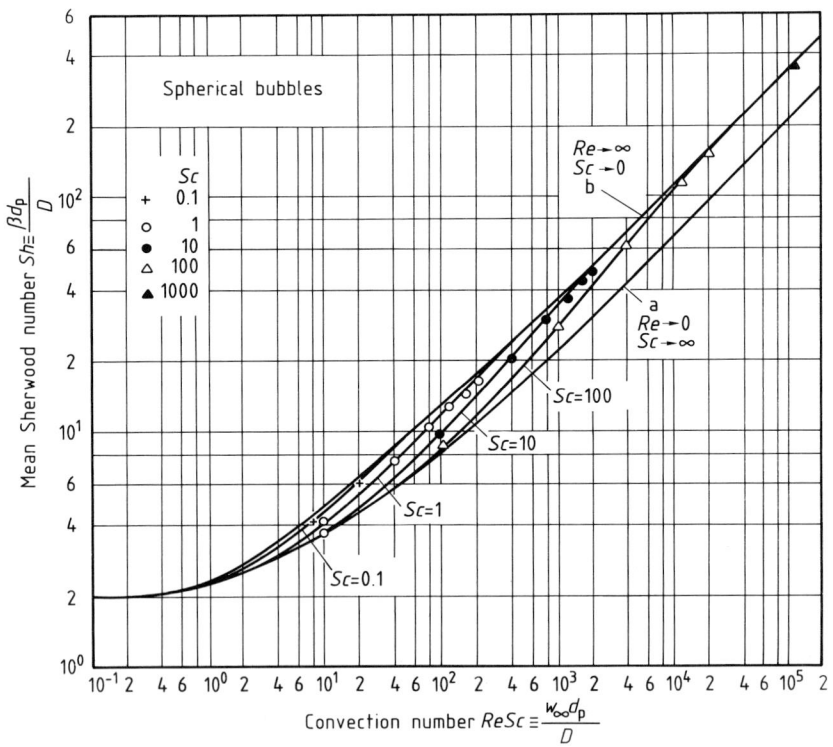

Figure 6.18. Mean Sherwood number Sh for spherical bubbles versus the convection number $Re\,Sc$ with the Schmidt number as a further parameter; curve a according to Eq. (6.59); curve b according to Eq. (6.60).

The curves for intermediary values of the Schmidt number at first follow curve a and then approach curve b. This function of the Sherwood number between the two limiting curves can be expressed by the curve presented in Fig. 6.19 and which is well approximated by the following equation:

$$\frac{Sh}{Sh_\infty} = [(1+4.33\cdot 10^{-1} Re^2)^{-1} + 4.23\cdot 10^{-5}]^{-0.055}. \qquad (6.61)$$

The Reynolds number must not exceed the value Re_B, which is given by Eq. (6.21). For $Re > Re_B$ the bubble loses its spherical shape. Introducing Eq. (6.59) into (6.61) a general equation for arbitrary values of $Re\,Sc$ and Sc is obtained:

$$Sh = \left[2 + \frac{0.651\,(Re\,Sc)^{1.72}}{1+(Re\,Sc)^{1.22}}\right] \cdot [(1+4.33\cdot 10^{-1} Re^2)^{-1} + 4.23\cdot 10^{-5}]^{-0.055} \qquad (6.62)$$

range of application:
$$100 \leq Re\,Sc \leq \infty$$
$$Re \leq Re_B \text{ [Eq. (6.21)]}$$

6.6.2 Mass Transfer through the Interface of Non-Spherical Bubbles

When the Reynolds number of bubble flow exceeds the value Re_B, which is given by Eq. (6.21), the bubble loses its spherical shape and becomes a non-spherical bubble. As explained in Sect. 6.3.3, stochastic motions of the interface of the bubble induce deformation turbulence in the surrounding liquid. In this section the influence of deformation turbulence on mass transfer will be discussed.

The local concentration is described by the differential equation:

$$w_r^* \frac{\partial \xi}{\partial r^*} + \frac{w_\theta^*}{r^*}\frac{\partial \xi}{\partial \theta} = \frac{2}{Re\,Sc}\frac{1}{r^{*2}}\left[\frac{\partial}{\partial r^*}\left\{r^{*2}\left(1+\frac{\varepsilon_m}{D}\right)\frac{\partial \xi}{\partial r^*}\right\} + \frac{1}{\sin\theta}\frac{\partial}{\partial \theta}\left(\sin\theta\,\frac{\partial \xi}{\partial \theta}\right)\right]. \qquad (6.63)$$

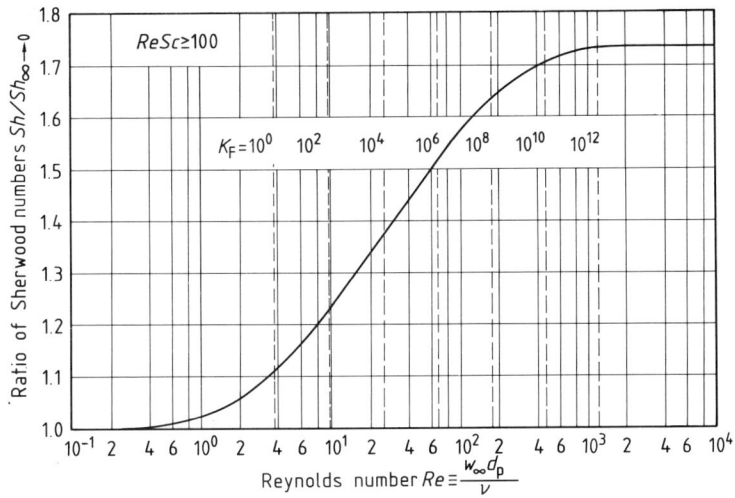

Figure 6.19. Ratio of Sherwood numbers, Sh/Sh_∞, for spherical bubbles versus the Reynolds number Re after Eq. (6.61).

Boundary conditions:

$\theta = 0; \pi$ and $1 \leqslant r^* \leqslant \infty$: $\partial \xi / \partial \theta = 0$
$0 \leqslant \theta \leqslant \pi$ and $r^* = 1$: $\xi = \xi_w = 1$
$0 \leqslant \theta \leqslant \pi$ and $r^* = \infty$: $\xi = \xi_\infty = 0$

The dimensionless quantities have been defined in Sect. 6.4.1.

The differential equation given above is an extension of Eq. (6.25) in that the molecular transport term is supplemented by a transport term that considers the action of deformation turbulence. This term is given by ε_m / D, for which the following equation has been established [6.10]:

$$\frac{\varepsilon_m}{D} = 3.34 \cdot 10^{-5} \, Re^{1.45} \, Sc \, (r^* - 1) \, 1/r^{*6}. \tag{6.64}$$

the Sherwood number is indicated in Sect. 6.4.2. The mean Sherwood number Sh is presented in Fig. 6.21 by curve d and compared with experimental data from various authors. Agreement between theory and experiment is quite satisfactory. The equation that approximates numerical and experimental data for mass transfer through the interface of non-spherical bubbles with stochastic deformations of the interface is as follows:

$$Sh = 2 + 0.015 \, Re^{0.89} \, Sc^{0.70}. \tag{6.65}$$

While the exponent of the Reynolds number is in agreement with those equations for mass transfer in turbulent fluid flow, the exponent of the Schmidt number has almost twice the accepted value.

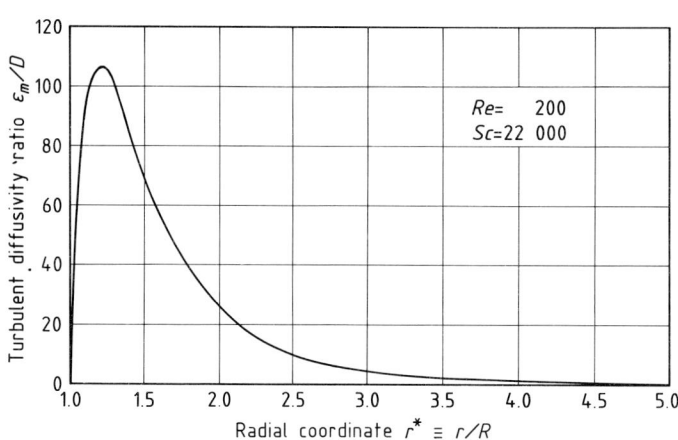

Figure 6.20. Turbulent diffusivity ratio ε_m / D around non-spherical bubbles caused by stochastic deformation of the interface.

For $Re = 200$ and $Sc = 22\,000$ this equation is presented in Fig. 6.20 versus the radial coordinate; ε_m is the turbulent diffusivity of mass and D the coefficient of molecular diffusion. At the interface ε_m / D is zero; with an increasing radial coordinate ε_m / D increases steeply, reaches a maximum with $\varepsilon_m / D = 105$ at $r^* = 1.2$, and then decreases.

The solution of the differential equation delivers the local concentration ξ and the concentration gradient $(\partial \xi / \partial r^*)_{\theta w}$ at the bubble surface, which is required for the calculation of the local and mean Sherwood number. The procedure for calculation of

Curves a and b are identical with those shown in Fig. 6.18. Curve a has been calculated for creeping flow, curve b for potential flow conditions. Symbols with two circles give numerical results. The data prove, that deformation turbulence cannot only improve mass transfer beyond curves a and b but can also lower mass transfer, so that it is poorer than for creeping flow conditions. This rather strange effect is due to the radial motion of the interface, that pushes the tangential fluid movement away from the interface. The fluid approaching this bubble does not come close enough to the bub-

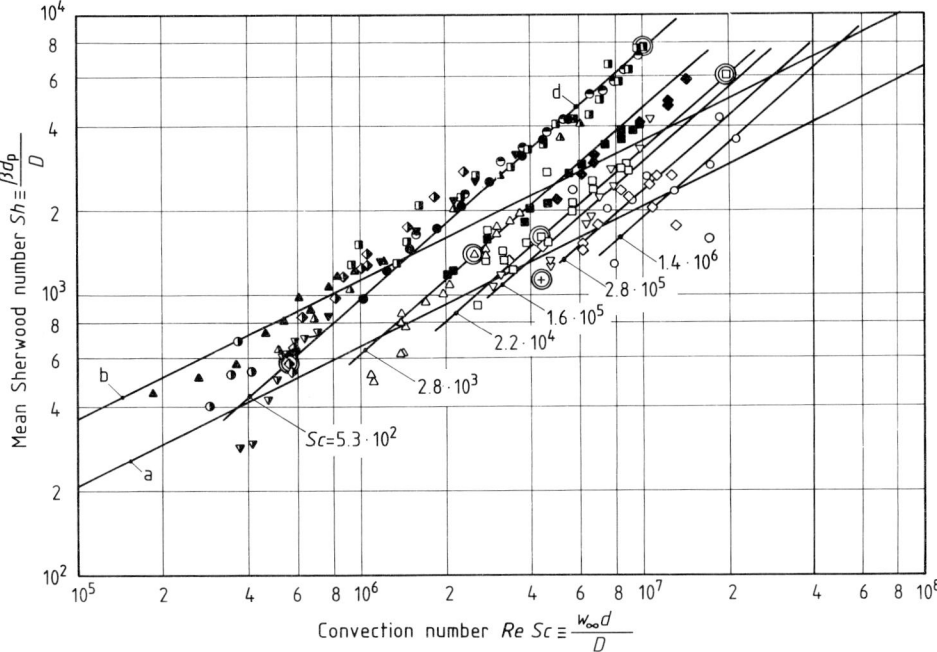

Figure 6.21. Mean Sherwood number Sh for spherical and non-spherical bubbles versus the convection number $Re\,Sc$ with the Schmidt number as further parameter.

ble surface to pick up the diffusing component when the Reynolds number is too small. In this case there is an accumulation of the diffusing component close to the surface of the non-spherical bubble. In other words, in the low Reynolds number region deformation turbulence builds up a concentration barrier around the bubble that reduces mass transfer to or from the interface.

The remarkably good agreement between theoretical-numerical and experimental data proves that the deformation turbulence hypothesis is a useful tool for a physical and mathematical description of mass transfer through the interface between two fluids subjected to stochastic deformation. According to this hypothesis, there is no analogy between momentum and mass transfer because ε_m/D is zero at the interface, while turbulent momentum transfer reaches its maximum at the interface.

6.7 Unsteady-State Mass Transfer through the Interface of Spherical Particles

Mass transfer through the interface of particles is under all conditions a time dependent process. The mathematical description of this process turned out to be extremely difficult. Therefore, investigations started with the relatively simple case of mass transfer in a quiescent surrounding fluid. In the second step, movement of the surrounding fluid was taken into account. The treatment of the unsteady-state mass transfer problem in this chapter follows this line of development.

6.7.1 Discussion of Mass Transfer Conditions

To simplify the discussion it will be assumed that the environment is at rest. For this case the concentration profiles are independent of the angular coordinate θ. For the general case of mass transfer resistance in both phases the concentration profiles are given in Fig. 6.22. Phase 1 is the sphere and phase 2 the surrounding fluid. The sphere may be a bubble, a drop, or a solid body.

Within the sphere the partial mass density ϱ_{A1} decreases from its highest value in the center of the sphere to its lowest value ϱ_{A1p} at the interface. The concentration in the environment, given by ϱ_{A2}, decreases from its highest value ϱ_{A2p} at the interface to its lowest value $\varrho_{A2\infty} = \varrho_{A2\infty}$ at a great distance from the interface. The ratio of concentrations at the interface is given by the Henry number:

$$H^* \equiv \frac{\varrho_{A1p}}{\varrho_{A2p}}. \tag{6.66}$$

For a dimensionless presentation of the concentration profiles use is made of the following definitions:

$$\xi_1 \equiv \frac{\varrho_{A1} - H^* \varrho_{A2\infty}}{\varrho_{A10} - H^* \varrho_{A2\infty}}, \tag{6.67}$$

$$\xi_2 \equiv \frac{\varrho_{A2} - \varrho_{A2\infty}}{\varrho_{A10}/H^* - \varrho_{A2\infty}}. \tag{6.68}$$

The indices 0 and ∞ designate values at time $t=0$ and $t=\infty$. The concentration difference $\varrho_{A10} - H^* \varrho_{A2\infty}$ is the maximum of

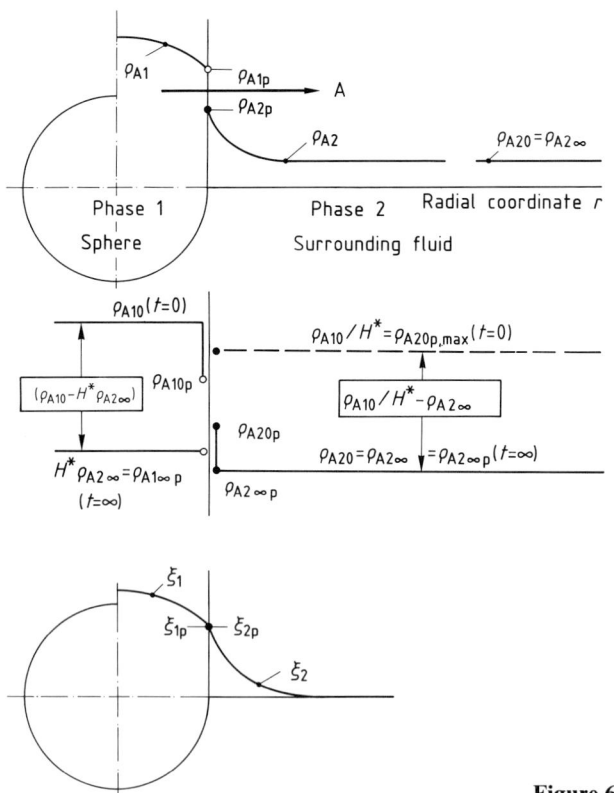

Figure 6.22. Concentration profiles and notations.

the possible change of concentration in the sphere, while $\varrho_{A10}/H^* - \varrho_{A2\infty}$ is the corresponding value in the surrounding fluid.

A dimensionless presentation of the time coordinate t is achieved by introduction of a Fourier number for each phase:

$$Fo_{m1} \equiv \frac{t}{R^2/D_1} = Fo_{m2}\frac{D_1}{D_2}, \qquad (6.69)$$

$$Fo_{m2} \equiv \frac{t}{R^2/D_2} = Fo_{m1}\frac{D_2}{D_1}. \qquad (6.70)$$

R is the radius of the sphere, D_1 the diffusion coefficient of component A in phase 1, and D_2 the diffusion coefficient in phase 2.

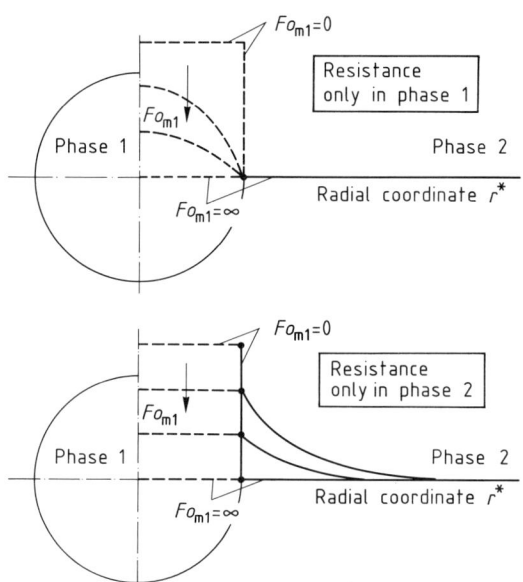

Figure 6.23. Development of concentration profiles with increasing Fourier number with resistance to mass transfer either in phase 1 or in phase 2.

There are three distinct cases of mass transfer through the interface of spherical particles:

1. General case: Resistance to mass transfer occurs in both phases. $D_1/D_2 \to 1$ and $H^* \gtrless 1$.
2. First limiting case: Resistance to mass transfer occurs in phase 1 only. $D_1/D_2 \to 0$ and $H^* < 1$; $\varrho_{A2} = \varrho_{A20} = \varrho_{A2\infty} = \varrho_{A2p} = $ const.
3. Second limiting case: Resistance to mass transfer occurs in phase 2 only. $D_1/D_2 \to \infty$ and $H^* > 1$; ϱ_{A1} is independent of radial coordinate r.

The dimensionless concentration profiles for the two limiting cases are presented in Fig. 6.23 versus the radial coordinate r^* for various values of the Fourier number.

6.7.2 Calculation of Transferred Mass

In the time interval from $t = 0$ to t the transferred mass is given by the equation:

$$M_A = \beta A_p t(\varrho_{A10} - \varrho_{A1\infty}). \qquad (6.71)$$

β is the time mean value of the mass transfer coefficient and

$$A_p = 6\frac{V_p}{d_p} \qquad (6.72)$$

the interfacial area with V_p as the volume of the particle, and d_p as the diameter.

Introducing Eqs. (6.66) and (6.72) into Eq. (6.71), the result is:

$$\frac{M_A}{V_p \varrho_{A10}} = \frac{6\beta t}{d_p}\left(1 - H^*\frac{\varrho_{A2\infty}}{\varrho_{A10}}\right). \qquad (6.73)$$

With the mean time value of the Sherwood number being:

$$Sh_1 \equiv \frac{\beta d_p}{D_1}, \qquad (6.74)$$

and the Fourier number, defined by Eq. (6.69), one obtains:

$$\frac{6\beta t}{d_p} = \frac{3}{2} Sh_1 Fo_{m1}. \qquad (6.75)$$

Eq. (6.73) may therefore be rewritten as:

$$\frac{M_A}{V_p \varrho_{A10}} = \frac{3}{2} Sh_1 \, Fo_{m1} \left(1 - H^* \frac{\varrho_{A2\infty}}{\varrho_{A10}}\right). \quad (6.76)$$

The mass M_A transferred across the interfacial area must be equal to the change of mass of component A within the sphere. The resulting equation may be written as follows:

$$\frac{M_A}{V_p \varrho_{A10}} = (1 - \bar{\xi}) \left(1 - H^* \frac{\varrho_{A2\infty}}{\varrho_{A10}}\right), \quad (6.77)$$

in which $\bar{\xi}$ is the mean concentration in the sphere. Comparison of Eq. (6.77) with (6.76) leads to the following relation:

$$1 - \bar{\xi} = \frac{3}{2} Sh_1 \, Fo_{m1} = \frac{3}{2} Sh_2 \, Fo_{m2}. \quad (6.78)$$

This shows that the mass M_A may be determined when either the mean concentration $\bar{\xi}$ or the mean Sherwood number Sh_1 are known quantities. These are the results of a theoretical-numerical treatment of the differential equations.

6.7.3 Differential Equations for Concentration Fields

The concentration field within the sphere is given by the following differential equations as obtained from Eq. (2.18):

$$\frac{\partial \xi_1}{\partial Fo_{m2}} \frac{D_1}{D_2} + \left(w_{r1}^* \frac{\partial \xi_1}{\partial r^*} + \frac{w_{\vartheta1}^*}{r^*} \frac{\partial \xi_1}{\partial \theta}\right) \frac{(Re \, Sc)_2}{2} \frac{D_2}{D_1} =$$
$$= \frac{1}{r^{*2}} \left[\frac{\partial}{\partial r^*} \left(r^{*2} \frac{\partial \xi_1}{\partial r^*}\right) + \frac{1}{\sin \theta} \frac{\partial}{\partial \theta} \left(\sin \theta \frac{\partial \xi_1}{\partial \theta}\right)\right]. \quad (6.79)$$

Similarly, the differential equation governing the concentration field in the fluid surrounding the sphere is as follows:

$$\frac{\partial \xi_2}{\partial Fo_{m2}} + \left(w_{r2}^* \frac{\partial \xi_2}{\partial r^*} + \frac{w_{\vartheta2}^*}{r^*} \frac{\partial \xi_2}{\partial \theta}\right) \frac{(Re \, Sc)_2}{2} =$$
$$= \frac{1}{r^{*2}} \left[\frac{\partial}{\partial r^*} \left(r^{*2} \frac{\partial \xi_2}{\partial r^*}\right) + \frac{1}{\sin \theta} \frac{\partial}{\partial \theta} \left(\sin \theta \frac{\partial \xi_2}{\partial \theta}\right)\right]. \quad (6.80)$$

Solution of the differential equations requires knowledge of the velocity field. For creeping flow conditions, $Re \to 0$, equations for radial and tangential velocity components in both phases have been worked out by STOKES et al. [6.16]. For the higher range of Reynolds number, $Re_2 \leqslant 150$, numerical solutions of the differential equations for the velocity field have been published [6.16].

Theoretical-numerical methods have been applied to the integration of the differential equations with due consideration of the conditions given in the next section [6.17].

6.7.4 Initial, Boundary, and Interface Conditions

The initial and boundary conditions are of a general nature, while the interface conditions depend on special mass transfer conditions.

6.7.4.1 Initial Conditions

The mass transfer process starts at time $t = 0$ with $Fo_{m1} = Fo_{m2} = 0$:

$$\left.\begin{array}{l} Fo_{m1} = Fo_{m2} = 0 \\ r^* < 1 \\ 0 \leqslant \theta \leqslant \pi \end{array}\right\} \xi_{10} = 1. \quad (6.81)$$

The initial condition is independent of mass transfer direction and distribution of mass transfer resistance in the two phases.

6.7.4.2 Boundary Conditions

Four conditions have to be observed:

1. $\left.\begin{array}{l} 0 \leqslant r^* \leqslant 1 \\ \theta = 0 \text{ and } \pi \end{array}\right\} \partial \xi_1 / \partial \theta = 0 \qquad (6.82)$

2. $\left.\begin{array}{l} r^* = 0 \\ \theta = 0 \text{ and } \pi \end{array}\right\} \partial \xi_1 / \partial r^* = 0 \qquad (6.83)$

3. $\left.\begin{array}{l} 1 \leqslant r^* \leqslant \infty \\ \theta = 0 \text{ and } \pi \end{array}\right\} \partial \xi_2 / \partial \theta = 0 \qquad (6.84)$

4. $\left.\begin{array}{l} r^* = \infty \\ \theta = 0 \text{ and } \pi \end{array}\right\} \xi_2 = 0 \qquad (6.85)$

The boundary conditions are independent of mass transfer direction and distribution of mass transfer resistance in the two phases.

6.7.4.3 Interface Conditions

A general condition is given by Henry's law, $H^* \equiv \varrho_{A1p} / \varrho_{A2p}$, that leads to:

$\left.\begin{array}{l} r^* = 1 \\ 0 \leqslant \theta \leqslant \pi \end{array}\right\} \xi_{1p} = \xi_{2p}. \qquad (6.86)$

Further interface conditions depend on the distribution of mass transfer resistance in the two phases.

First limiting case of mass transfer resistance in phase 1 only:

$\left.\begin{array}{l} r^* = 1 \\ 0 \leqslant \theta \leqslant \pi \end{array}\right\} \xi_{1p} = \xi_{2p} = 0. \qquad (6.87)$

Second limiting case of mass transfer resistance in phase 2 only:

$\left.\begin{array}{l} r^* = 1 \\ 0 \leqslant \theta \leqslant \pi \end{array}\right\} \dfrac{\partial \xi_1}{\partial Fo_{m2}} = -\dfrac{3}{2} \dfrac{1}{H^*} \int_0^\pi \left(\dfrac{\partial \xi_2}{\partial r^*}\right)_p \sin\theta \, d\theta. \qquad (6.88)$

General case of mass transfer resistance in both phases:

$\left.\begin{array}{l} r^* = 1 \\ 0 \leqslant \theta \leqslant \pi \end{array}\right\} \dfrac{D_1}{D_2} H^* = \dfrac{(\partial \xi_2 / \partial r^*)_p}{(\partial \xi_1 / \partial r^*)_p}. \qquad (6.89)$

6.7.5 Discussion of some Numerical Results for Mass Transfer without Fluid Motion

For simplification the discussion of mass transfer results will start with the simple case, in which there is no fluid motion within or outside the sphere. The discussion will be limited to the mean concentration in the sphere and the mean Sherwood number.

6.7.5.1 Mean Concentration in a Sphere

For the first limiting case of mass transfer resistance in phase 1 (sphere) only, the mean concentration $\bar{\xi}_1$ is shown in Fig. 6.24 as a function of the Fourier number Fo_{m1}. The numerical results have been approximated by the following equation [6.14]:

$$\bar{\xi}_1 = \dfrac{0.354}{1 + (2.86 \, Fo_{m1}^{0.2} + 1.82 \cdot 10^3 \, Fo_{m1}^2)^3} +$$
$$+ 0.646 \exp(-10.15 \, Fo_{m1}), \qquad (6.90)$$

range of application: $0 \leqslant Fo_{m1} \leqslant \infty$
conditions of application: $H^*(D_1/D_2)^{1/2} \to 0$
$H^* \leqslant 1$
$D_1/D_2 \leqslant 1$

For the second limiting case of mass transfer resistance in phase 2 (surrounding fluid) only, the mean concentration $\bar{\xi}_1$ is shown in Fig. 6.25 as a function of Fo_{m2}/H^*, with H^* as a parameter. The numerical results have been approximated by the following equations:

$$\bar{\xi}_1 = \exp(-b_1 Fo_{m2}^{b_2}), \qquad (6.91)$$

$$b_1 = \dfrac{1}{\left\{\left[\dfrac{H^{*0.63}}{2.977}\right]^2 + \left[\dfrac{1}{\left(\dfrac{34.27}{H^{*1.3}}\right)^3 + \left(\dfrac{3.61}{H^*}\right)^3}\right]^{2/3}\right\}^{1/2}}, \qquad (6.92)$$

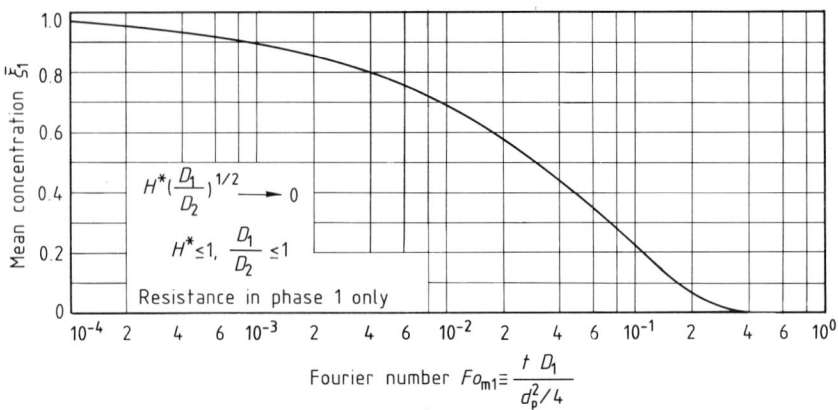

Figure 6.24. Mean concentration in the sphere for the limiting case of resistance in the sphere only.

$$b_2 = \left[\cfrac{1}{1+\left(\cfrac{1}{0.475\, H^{*\,0.12}}\right)^7}\right]^{1/7} \quad (6.93)$$

range of application: $0 \leqslant Fo \leqslant \infty$
$\qquad\qquad\qquad\qquad\quad 10^0 \leqslant H^* \leqslant 10^5$

conditions of application: $H^*(D_1/D_2)^{1/2} \to \infty$
$\qquad\qquad\qquad\qquad\qquad H^* > 1$
$\qquad\qquad\qquad\qquad\qquad D_1/D_2 > 1$

maximum relative error: 8%

For the general case of mass transfer resistance in both phases, the mean concentration $\bar{\xi}_1$ is presented in Figs. 6.26 and 6.27. The curves in Fig. 6.26 give $\bar{\xi}_1$ for $H^*(D_1/D_2)^{1/2} \leqslant 1$. In this case $\bar{\xi}_1$ is a function of Fo_{m1} and $H^*(D_1/D_2)^{1/2}$. A separate influence of H^* and D_1/D_2 has not been found. The lower limiting curve is that for $H^*(D_1/D_2)^{1/2} \to 0$; it is identical with the curve shown in Fig. 6.24. The upper limiting curve in Fig. 6.26 is that for $H^*(D_1/D_2)^{1/2} = 1$. This curve has been

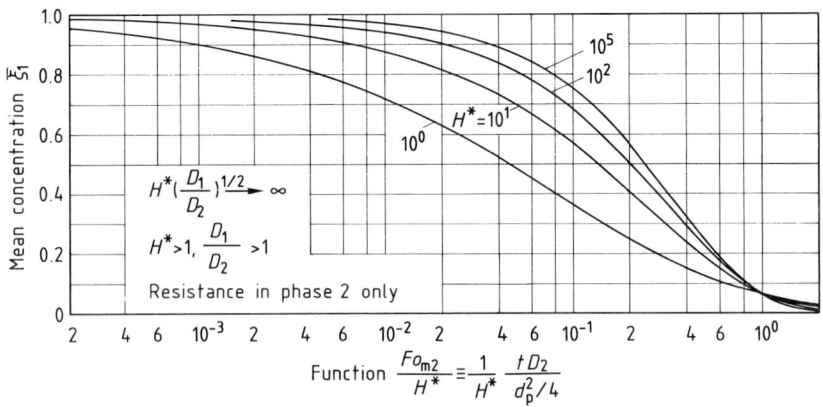

Figure 6.25. Mean concentration in the sphere for the limiting case of resistance in the surrounding fluid only.

6.7 Unsteady-State Mass Transfer through the Interface of Spherical Particles

approximated by the following equation:

$$\bar{\xi}_1 = \left[\frac{1}{(1 + 1.842\, Fo_{m1}^{0.5})^{4/3} + (14.18\, Fo_{m1}^{1.3})^{4/3}}\right]^{3/4} \quad (6.94)$$

range of application: $0 \leqslant Fo_{m1} \leqslant \infty$
conditions of application: $H^*(D_1/D_2) = 1$
$H^* \leqslant 1$
$D_1/D_2 \leqslant 1$

For all values of $H^*(D_1/D_2)^{1/2} \leqslant 1$ the mean concentration $\bar{\xi}_1$ must always be between the two limiting curves in Fig. 6.26:

$$\bar{\xi}_1 \text{ [Eq. (6.90)]} \leqslant \bar{\xi}_1 \leqslant \bar{\xi}_1 \text{ [Eq. (6.94)]}.$$

The curves in Fig. 6.27 give $\bar{\xi}_1$ for $H^*(D_1/D_2)^{1/2} \geqslant 1$. In this case, $\bar{\xi}_1$ is a function of Fo_{m2}/H^*, $H^*(D_1/D_2)^{1/2}$ and H^*. However, it has been found, that for $H^* \gg 1$, a special

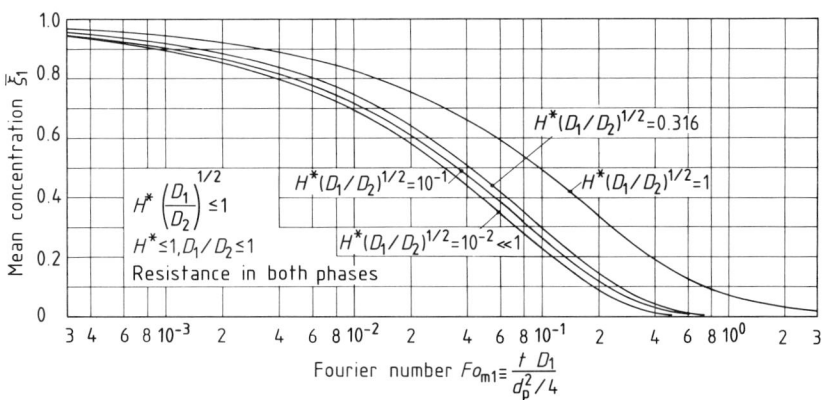

Figure 6.26. Mean concentration in the sphere for the general case of resistance in both phases with $H^*(D_1/D_2)^{1/2} \leqslant 1$.

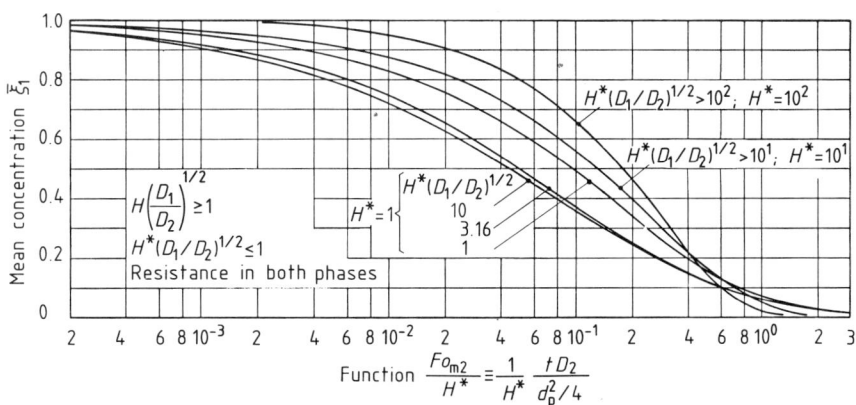

Figure 6.27. Mean concentration in the sphere for the general case of resistance in both phases with $H^*(D_1/D_2)^{1/2} \geqslant 1$.

influence of $H^*(D_1/D_2)^{1/2}$ disappears. This applies, for instance, for $H^* = 10^1$ and 10^2. These curves are identical with those shown in Fig. 6.25. The lowest possible curve is that for $H^* = 1$ and $H^*(D_1/D_2)^{1/2} \to \infty$, which is identical with the corresponding curve in Fig. 6.25.

6.7.5.2 Mean Sherwood Number

The Sherwood number Sh_1 has been defined by Eq. (6.74). It is a mean value over the interfacial area and over the time interval between $t = 0$ and t. The mathematical development of Sh_1 has been given in another publication [6.17].

For the first limiting case of mass transfer resistance in phase 1 (sphere) only, the mean Sherwood number Sh_1 is presented in Fig. 6.28. Curve a has been approximated by the equation:

$$Sh_1 = \cfrac{1}{\cfrac{Fo_{m1}^{1/2}}{4/\sqrt{\pi}} + \cfrac{\frac{3}{2} Fo_{m1}}{1 + \cfrac{0.41}{Fo_{m1}^{0.7}}}}, \tag{6.95}$$

range of application: $0 \leq Fo_{m1} \leq \infty$

conditions of application: $H^*(D_1/D_2)^{1/2} \to 0$
$$H^* \leq 1$$
$$D_1/D_2 \leq 1$$

maximum relative error: 5.4%

For the second limiting case of mass transfer resistance in phase 2 (surrounding fluid) only, the mean Sherwood number Sh_2 is shown in Fig. 6.29. Approximation equations have not been set up because of the expected complexity. If an equation for Sh_2 is of interest, it can be obtained by means of Eqs. (6.78) and (6.91).

For $Fo_{m1} \to 0$ and $\to \infty$ as well as for $Fo_{m2} \to 0$ and $\to \infty$ analytical solutions of the differential equations have been obtained:

$$Sh_1 = \frac{4/\sqrt{\pi}}{1 + H^*\left(\dfrac{D_1}{D_2}\right)^{1/2}} Fo_{m1}^{-1/2}, \tag{6.96}$$

$$Sh_2 = \frac{4/\sqrt{\pi}}{H^* + (D_2/D_1)^{1/2}} Fo_{m2}^{-1/2}, \tag{6.97}$$

range of application: $Fo_{m1}(Fo_{m2}) \to 0$
$$0 \leq H^*\left(\frac{D_1}{D_2}\right)^{1/2} \leq \infty$$

These equations apply to the general case of arbitrary distribution of mass transfer resistance in the two phases concerned.

For $Fo_{m1}(Fo_{m2}) \to \infty$ the following theoretical equations apply:

$$Sh_1 = \frac{2/3}{Fo_{m1}}, \tag{6.98}$$

$$Sh_2 = \frac{2/3}{Fo_{m2}}. \tag{6.99}$$

It is to be noted that for time $t \to 0$ and mass transfer resistance in phase 1, the mass transfer coefficient, given by:

$$\beta = \frac{2}{\sqrt{\pi}}\left(\frac{D_1}{t}\right)^{1/2}, \tag{6.100}$$

is independent of particle diameter d_p. With mass transfer resistance in phase 2 only, and $t \to 0$ one obtains from Eq. (6.97):

$$\beta = \frac{2}{\sqrt{\pi} H^*}\left(\frac{D_2}{t}\right)^{1/2}. \tag{6.101}$$

For $t \to \infty$ and arbitrary distributions of mass transfer resistance in the two phases the following equation for β is obtained:

$$\beta = \frac{1}{6}\frac{d_p}{t}. \tag{6.102}$$

The mass transfer coefficient increases linearly with particle diameter d_p.

6.7 Unsteady-State Mass Transfer through the Interface of Spherical Particles

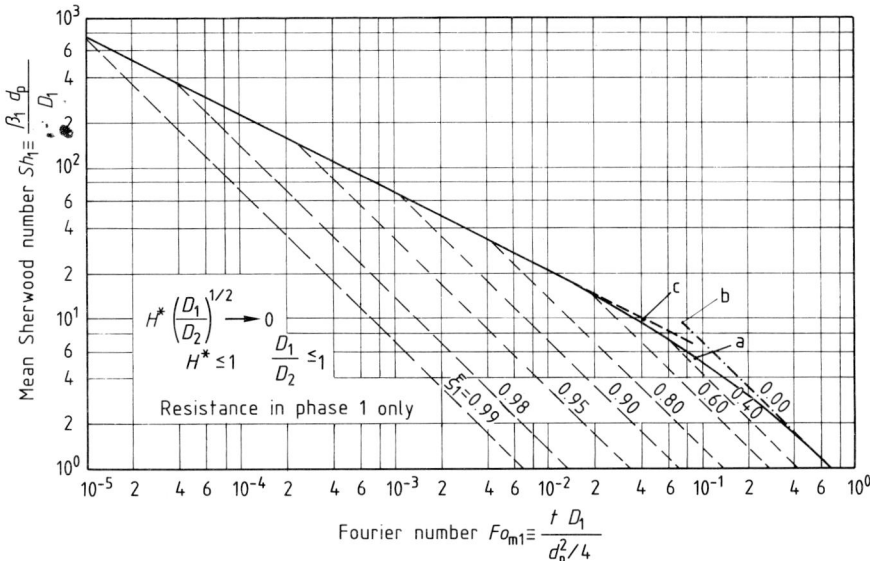

Figure 6.28. Mean Sherwood number Sh_1 for the limiting case of resistance in the sphere only.

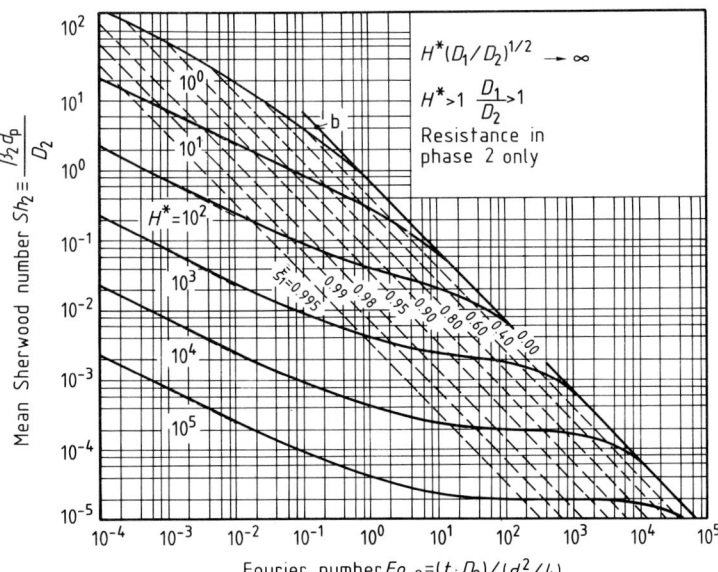

Figure 6.29. Mean Sherwood number Sh_2 for the limiting case of resistance in the surrounding fluid only.

6.7.6 Discussion of some Numerical Results for Mass Transfer with Fluid Motion

6.7.6.1 Mean Concentration in a Sphere

For the first limiting case of mass transfer resistance in phase 1 (sphere) only, the mean concentration $\bar{\xi}_1$ is shown in Fig. 6.30. There is an upper and a lower limiting curve. The upper limiting curve for $(Re\,Sc)_2/(1+\eta^*) \to 0$ is given by Eq. (6.90). The lower limiting curve for $(Re\,Sc)_2/(1+\eta^*) \to \infty$ has been approximated by the following equation:

$$\bar{\xi}_1 = \frac{0.289}{1+[(26.0\,Fo_{m1}^{0.5})^2 + (3.3\cdot 10^5\,Fo_{m1}^2)^2]^{1/2}} + 0.711\exp(-30.4\,Fo_{m1}) \qquad (6.103)$$

range of application: $0 \leq Fo_{m1} \leq \infty$
Conditions of application:
$(Re\,Sc)_2/(1+\eta^*) \to \infty$
$H^*(D_1/D_2)^{1/2} \to 0$
$H^* \leq 1$
$D_1/D_2 \leq 1$

The Eqs. (6.90) and (6.103) available for the two limiting cases $(Re\,Sc)_2/(1+\eta^*) \to 0$ and $\to \infty$, give a sound basis for mass transfer calculations when convection is either present or absent.

The symbol η^* denotes the viscosity ratio:

$$\eta^* = \eta_1/\eta_2. \qquad (6.104)$$

η_1 and η_2 are the viscosities of phase 1 fluid and phase 2 fluid.

For the second limiting case of mass transfer resistance in phase 2 (surrounding fluid) only, the mean concentration $\bar{\xi}_1$ is presented in Figs. 6.31 and 6.32 for $\eta^* = \infty$ (solid body or drop in a gaseous surrounding) and $\eta^* = 0$ (bubble).

The influence of the convection number $(Re\,Sc)_2$ increases with increasing Henry number H^* and decreasing viscosity ratio η^*. Convective mass transfer through the interface of a bubble is therefore better than through the interface of drops or solid spheres. For $(Re\,Sc)_2 \to 0$ the curves coincide with those given in Fig. 6.25, which have been approximated by Eq. (6.91).

6.7.6.2 Mean Sherwood Number

For the first limiting case of mass transfer resistance in phase 1 (sphere) only, the mean Sherwood number is shown in Fig. 6.33. The curves given for $(Re\,Sc)_2 \to 0$ and

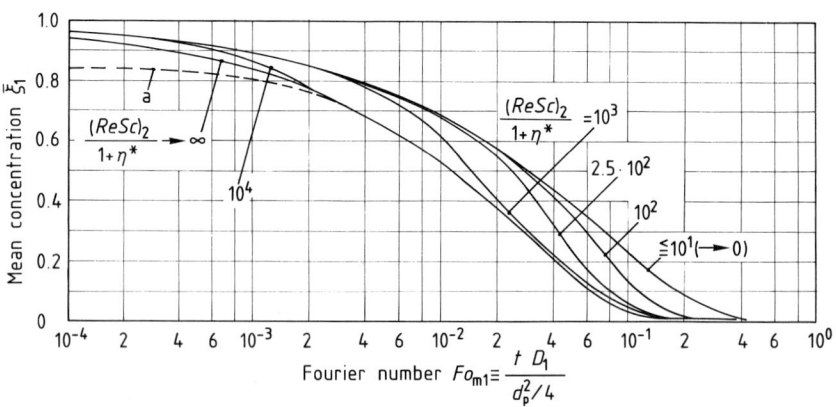

Figure 6.30. Mean concentration in the sphere for mass transfer with fluid motion in both phases for the limiting case of resistance in the sphere only.

Fig. 6.31

Fig. 6.32

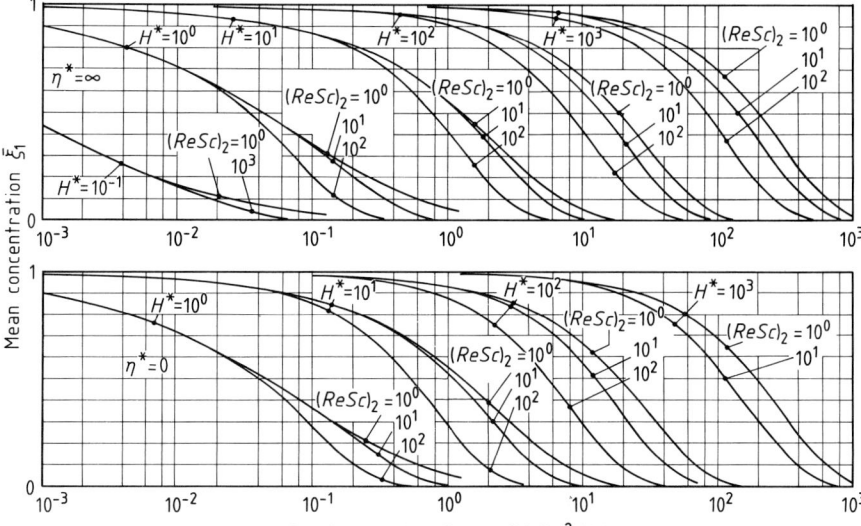

Figures 6.31 and 6.32. Mean concentration in the sphere for mass transfer with fluid motion in both phases for the limiting case of resistance in the surrounding fluid only:
Fig. 6.31: $\eta^* = \infty$, particle with rigid interface,
Fig. 6.32: $\eta^* = 0$, particle with ideal mobility of interface.

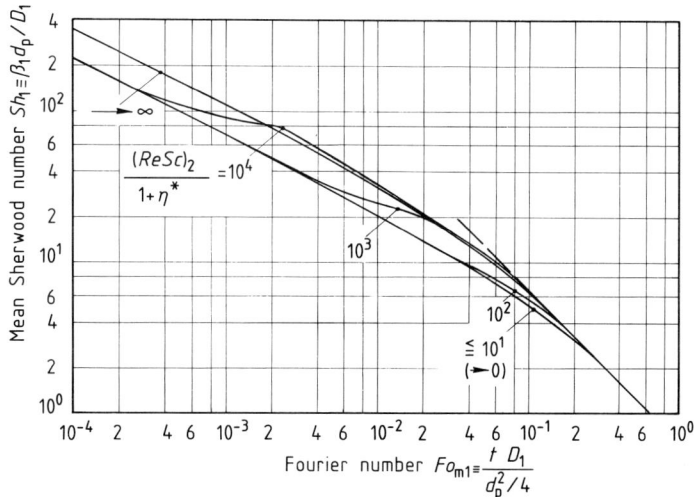

Figure 6.33. Mean Sherwood number Sh_1 for mass transfer with fluid motion in both phases for the limiting case of resistance in the sphere only.

$\to \infty$ correspond to those for the mean concentration given in Fig. 6.30. It has to be noted though, that Sh_1 may slightly exceed the Sherwood number for $(Re\,Sc)_2 \to \infty$, in a certain region. A detailed discussion of this phenomenon is to be found in [6.17].

For the second limiting case of mass transfer resistance in phase 2 (surrounding

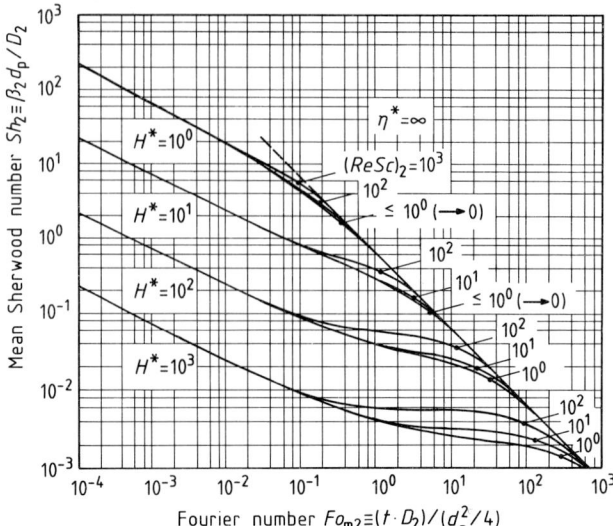

Fig. 6.34

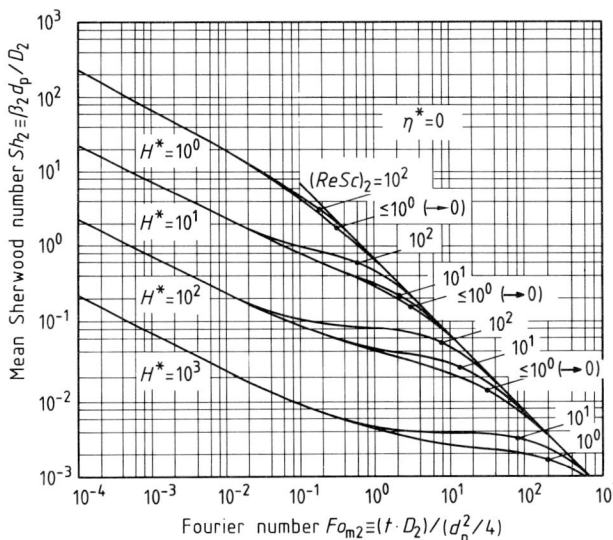

Fig. 6.35

Figures 6.34 and 6.35. Mean Sherwood number Sh_2 for mass transfer with fluid motion in both phases for the limiting case of resistance in the surrounding fluid only:
Fig. 6.34: $\eta^* = \infty$, particle with rigid interface,
Fig. 6.35: $\eta^* = 0$, particle with ideal mobility of interface.

fluid) only, the mean Sherwood number Sh_2 is presented in Figs. 6.34 and 6.35 for $\eta^* = \infty$ and $\eta^* = 0$. For $(Re\, Sc)_2 \to 0$ the curves in these two figures are identical with those shown in Fig. 6.29.

According to the numerical results presented in Figs. 6.24 to 6.35 convection will improve mass transfer when the Fourier number exceeds a certain value. There is always a starting region within which mass transfer is not influenced by convection.

6.8 References

[6.1] L. HILPRECHT: "Instationärer Impuls- und Stoffaustausch bei beschleunigter Bewegung von Einzelpartikeln". VDI-Forsch.-Heft 577, VDI-Verlag, Düsseldorf, 1976.

[6.2] V. G. JENSON: "Viscous flow round a

[6.3] A. E. Hamielec, T. W. Hoffman, and L. L. Ross (Part I); A. E. Hamielec, A. I. Johnson, and W. T. Houghton (Part II): "Numerical solution of the Navier-Stokes equation for flow past spheres"; AIChE J. *13* (1967), 212–224.

[6.4] F. Ihme, H. Schmidt-Traub, and H. Brauer: "Theoretische Untersuchung über die Umströmung und den Stoffübergang an Kugeln". Chem. Ing. Tech. *44* (1972), 306–313.

[6.5] G. G. Stokes: "On the effect of internal friction of fluids on the motion of pendulums". Trans. Cambr. Soc. Vol. 9, Part II (1851), 8–106.

[6.6] U. Haas, H. Schmidt-Traub, and H. Brauer: "Umströmung kugelförmiger Blasen mit innerer Zirkulation". Chem. Ing. Tech. *44* (1972) 18, 1060–1068.

[6.7] J. Hadamard: "Mouvement permanent lent d'une sphère liquide et visqueuse dans un liquide visqueux". C. R. Acad. Sci. Paris *152* (1911), 1735–1817.

[6.8] W. Rybszinski: Bull. Int. Acad. Sci. Cracovic A (1911).

[6.9] H. Brauer: "Grundlagen der Einphasen- und Mehrphasenströmungen". Verlag Sauerländer, Aarau – Frankfurt am Main, 1971.

[6.10] H. Glaeser and H. Brauer: "Berechnung des Impuls- und Stofftransports durch die Grenzfläche einer formveränderlichen Blase". VDI-Forsch.-Heft *581*, VDI-Verlag, Düsseldorf, 1977.

[6.11] H. Brauer and D. Sucker: "Stoff- und Wärmeübergang an umströmten Platten, Zylindern und Kugeln". Chem. Ing. Tech. *44* (1972) 5, 306–313.

[6.12] H. Brauer: "Particle/fluid transport processes". Prog. Chem. Eng. *19* (1981), 81–111.

[6.13] V. G. Levich: "Physicochemical Hydrodynamics". Prentice-Hall, Englewood Cliffs, NJ, 1962.

[6.14] U. Adam: "Approximationsgleichungen für den Stoffübergang an Kugeln", unpublished paper. Technische Universität Berlin, 1977.

[6.15] L. Oellrich, H. Schmidt-Traub, and H. Brauer: "Theoretische Berechnung des Stofftransports in der Umgebung einer Einzelblase". Chem. Eng. Sci. *28* (1973), 711–721.

[6.16] H. Brauer: "Stoffaustausch einschließlich chemischer Reaktionen". Verlag Sauerländer, Aarau – Frankfurt am Main, 1971.

[6.17] H. Brauer: "Unsteady-state mass transfer through the interface of spherical particles". Int. J. Heat Mass Transfer *21* (1978), 445–453 and 455–465.

Chapter 7
Transport Processes in Liquid Films

Heinz Brauer

Institut für Chemieingenieurtechnik
Technische Universität Berlin
Berlin (West), Germany

7.1 Introduction
7.2 Some Fluid Mechanic Properties of Liquid Films
7.3 Mass Transfer in Liquid Films with Smooth Surfaces
7.4 Mass Transfer in Liquid Films with Wavy Surfaces
7.4.1 Experimental Investigations
7.4.1.1 Tubes with a Smooth Wall
7.4.1.2 Tubes with a Wavy Wall
7.4.2 Theoretical Investigations
7.5 Liquid Film Mass Transfer with Biochemical Reactions
7.5.1 The Physico-Biochemical Problem
7.5.2 Mathematical Description of the Problem
7.5.3 Discussion of Some Results
7.5.3.1 Mass Transfer with Second Order Reactions
7.5.3.2 Mass Transfer with First Order Reactions
7.5.3.3 Comparison with Other Results
7.6 References

7 Transport Processes in Liquid Films

List of Symbols

Symbol	Units	Description
A_p	m²	interfacial area
A_w	m²	surface of tube with a wavy wall
a	m	amplitude of wave in a tube with a wavy wall
b	m	length of wave in a tube with a wavy wall
c_{Af}	kmol/m³	local partial mol density of component A
$\bar{c}_{Af}$	kmol/m³	mean partial mol density of component A
c_{Af0}	kmol/m³	partial mol density of component A at $z=0$
c_{Bf}	kmol/m³	local partial mol density of component B
c_{Bf0}	kmol/m³	partial mol density of component B at $z=0$
c_{Cf}	kmol/m³	local partial mol density of component C
D_f	m²/s	diffusion coefficient in a liquid film
D_{Af}	m²/s	diffusion coefficient of component A
D_{Bf}	m²/s	diffusion coefficient of component B
d	m	diameter of smooth tube
d_w	m	diameter of wavy tube
g	m/s²	acceleration due to gravity
k_v	(kmol s)/m³	reaction velocity factor for homogeneous reaction
k_A	(kmol s)/m³	reaction velocity factor for component A
k_B	(kmol s)/m³	reaction velocity factor for component B
L	m	length of tube
$\dot{M}_A$	kg/s	mass flux of component A
$\dot{m}_A$	kg/(s m²)	mass flux density of component A
$\dot{n}_{Az}$	kmol/(s m²)	local mol flux density of component A
R	m	radius of tube
t	s	time
U	m	circumference of tube
V_f	m³	volume of liquid
$\dot{V}_f$	m³/s	volumetric flow rate of liquid
$\dot{V}_g$	m³/s	volumetric flow rate of gas
w_f	m/s	local velocity in liquid film
$\bar{w}_f$	m/s	mean velocity of liquid film
y	m	distance from wall or free surface
β_f	m/s	coefficient of mass transfer in film for tubes with smooth walls
β_{fz}	m/s	local coefficient of mass transfer in film for tubes with smooth walls
$\beta_{f,w}$	m/s	coefficient of mass transfer in film for tubes with wavy walls
δ	m	mean film thickness
$\Delta\varrho_A$	kg/m³	concentration difference
$\varepsilon_{m,D-T}$	m²/s	diffusivity of mass due to deformation turbulence
η_f	kg/(s m)	dynamic viscosity
v_f	m²/s	kinematic viscosity
ϱ_f	kg/m³	mean density of liquid
ϱ_{Af}	kg/m³	partial density of component A in liquid film
ϱ_{Apf}	kg/m³	partial density of component A in the interface of the liquid film
$c_{Af}^* \equiv c_{Af}/c_{Af0}$		local concentration ratio for component A
$c_{Bf}^* \equiv c_{Bf}/c_{Bf0}$		local concentration ratio for component B
$D^* \equiv D_{Bf}/D_{Af}$		ratio of diffusion coefficients
$Da \equiv \dfrac{k_v c_{Bf0}\delta^2}{D_{Af}}$		Damköhler number
$H^* \equiv \dfrac{\varrho_{Apg}}{\varrho_{Apf}} = \dfrac{c_{Apg}}{c_{Apf}}$		Henry number
$K_p \equiv \dfrac{H^*}{2}\left(\dfrac{R}{\delta}-1\right)$		interface number
$Re_f \equiv \dfrac{\dot{V}_f/U}{v_f}$		Reynolds number
$Sc_f \equiv \dfrac{v_f}{D_f}$		Schmidt number
$Sh_f \equiv \dfrac{\beta_f \delta}{D_f}$		mean Sherwood number
$Sh_{fw} \equiv \dfrac{\beta_{fw}}{D_f}\left(\dfrac{3v_f^2}{g}\right)^{1/3}$		Sherwood number for mass transfer in tubes with wavy walls
$Sh_{fz} \equiv \dfrac{\beta_{fz}\delta}{D_f}$		local Sherwood number
$y^* \equiv y/\delta$		distance from wall or from free surface of film

$z^* \equiv \dfrac{z/\delta}{Re_f \, Sc_f}$ or $\dfrac{z/\delta}{Re_f \, Sc_{Af}}$ entrance number

$v^* \equiv \dfrac{v_B}{v_A} \dfrac{c_{Apf}}{c_{Bf0}}$ stoichiometric surplus factor

$\xi \equiv \dfrac{\varrho_{Afp} - \varrho_{Af}}{\varrho_{Afp} - \varrho_{Af0}}$ local concentration ratio

$\phi \equiv \beta_{fr}/\beta_f$ enhancement factor

7.1 Introduction

Mass transfer through the interface between a liquid and a gaseous phase is one of the most important operations in chemical engineering. In many cases this physical transport process is coupled with a chemical reaction, predominantly with a homogeneous reaction.

The mass flux $\dot{M}_A$ of the component A across the interfacial area A_p is described by the following equation:

$$\dot{M}_A = \beta A_p \Delta \varrho_A. \tag{7.1}$$

This equation states the fundamental problem of mass transfer in the simplest possible way. The mass flux is a linear function of the mass transfer coefficient β, the interfacial area A_p, and the difference of concentration of the transported component in the bulk of one of the fluid phases and in the interface $\Delta \varrho_A$. In order to obtain a large mass flux $\dot{M}_A$ it is necessary to produce a large interfacial area A_p in the mass transfer equipment, and to produce such fluid dynamic conditions that also a large value of the transfer coefficient β is achieved.

From a technical point of view it is advisable to consider the mass flux in relation to either the volume of the equipment V_e or the volume of one of the fluid phases V contained in the equipment. As both quantities are closely related, it does not matter which one of them is preferred. Taking the volume V_f of the liquid phase, the mass transfer equation may be written as follows:

$$\frac{\dot{M}_A}{V_f} = \beta \frac{A_p}{V_f} \Delta \varrho_A. \tag{7.2}$$

A great mass flux per unit of the liquid volume requires primarily a large interfacial area per unit of the liquid volume.

This technical aim may be achieved by one of three methods available for the production of interfacial area:

1. production of liquid films,
2. production of fluid particles,
3. production of liquid jets.

From a theoretical point of view the production of liquid films provides the greatest possible specific interfacial area A_p/V_f, while the production of liquid particles and jets results in a much smaller interfacial area. On the other hand, there are no serious technical problems involved in the production of liquid particles, while the production of jets and liquid films proves to be rather difficult. This is the reason, why in existing technical mass transfer equipment liquid particle production is by far more often used than jet and film production. With improving knowledge of the transport phenomena in all three cases the application of liquid films for mass transfer operations is steadily gaining ground.

Liquid films may be produced by the proper distribution of the liquid on the surface of vertical and horizontal tubes, of vertical and inclined flat plates, of rotating cups and disks, and on the surface of small elements like Raschig rings, for example, in packed columns. The thickness of the liquid films is in general of the order of a few tenths of a millimeter. The most important type of technical film equipment is the wetted packed column. The relatively minor application of other types of film equipment is primarily due to the fact, that it is rather difficult to generate liquid films and to ensure complete wetting of the surface. The predominant problem in film technology is therefore the development of simple and effective systems for proper distribution of the liquid on the various types of surfaces. Successful development of distribution systems will guarantee a much wider application of liquid films in mass transfer operations.

In the following sections mass transfer without and with homogeneous chemical reactions in liquid films with a smooth and wavy surface will be discussed. As a basis for the discussion, some information will be given on fluid mechanic properties of thin liquid films flowing downwards on vertical surfaces.

7.2 Some Fluid Mechanic Properties of Liquid Films

Investigations on liquid films started with a thorough theoretical analysis of liquid film flow on vertical surfaces and heat transfer by NUSSELT [7.1], [7.2]. NUSSELT assumed steady state flow conditions and a smooth free surface of the liquid films. For such conditions film flow is the result of the action of frictional and gravitational forces. The equations for the local velocity w_f, the mean velocity $\bar{w}_f$, and the film thickness δ, which NUSSELT obtained, may be written as follows:

$$w_f = \frac{g \varrho_f \delta^2}{2 \eta_f} \left[1 - \left(\frac{y}{\delta}\right)^2\right], \quad (7.3)$$

$$\bar{w}_f = \frac{1}{3} \frac{g \varrho_f \delta^2}{\eta_f} = \left(\frac{g v_f}{3}\right)^{1/3} Re_f^{2/3}, \quad (7.4)$$

$$\delta = \left(\frac{3 v_f^2}{g}\right)^{1/3} Re_f^{1/3}. \quad (7.5)$$

The symbols used in these equations have the following meaning: g acceleration due to gravitational forces, ϱ_f fluid density, η_f dynamic fluid viscosity, y distance from the film surface, $v_f = \eta_f/\varrho_f$ kinematic fluid viscosity, and

$$Re_f \equiv \frac{\bar{w}_f \delta}{v_f} = \frac{\dot{V}_f/U}{v_f} \quad (7.6)$$

the Reynolds number for film flow with $\dot{V}_f$ volumetric liquid flow rate and U circumference of a vertical tube or width of a plate.

The equations presented for film velocity and film thickness are even today assumed to be applicable, although some minor corrections are needed. The majority of theoretical investigations is still based on these equations.

In the three decades following the pioneering work of NUSSELT an increasing number of research workers was attracted to the study of liquid films. Only a few of the papers, published in these decades, and in which the entire relevant literature is listed, will be mentioned. FALLAH et al. [7.3], FRIEDMAN and MILLER [7.4], and GRIMLEY [7.5], for example, drew attention to the most important phenomenon about liquid films – the wavy structure of the film surface. Closely related to this phenomenon is the fluid mechanic instability of the liquid film and the critical Reynolds number.

In a comprehensive experimental study the author analyzed the wave structure, the area of the wavy surface, surface velocity, mean film thickness, and the value of the critical Reynolds number [7.6]. The wave

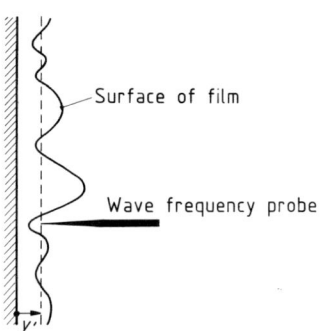

Figure 7.1. Schematic drawing of the system used for measurement of wave frequency.

structure has been described by means of the wave frequency f, which was measured with an electrical arrangement. A schematic drawing of the experimental arrangement is shown in Fig. 7.1. The frequency f measured at various values of the Reynolds number over the distance y' from the solid surface is presented in Fig. 7.2. The wave frequency increases strongly with increasing Reynolds number. The waves give evidence of a strong mechanical mixing process within the liquid film. In a paper published more than twenty years later the author interprets this mixing process as turbulence due to the deformation of the film surface [7.7]. This type of turbulence is consequently called deformation turbulence. It forms the basis for a new mass transfer theory applicable to wavy films.

One is led to assume, that the velocity profile has a non-parabolic shape on account of the wavy structure of the film surface and measured velocities [7.8]. Curve a in Fig. 7.3 shows the parabolic velocity profile of a liquid film with a smooth surface, while curve b shows the assumed profile for a film with a wavy surface. Both velocity profiles apply to the same mean velocity and mean film thickness. The result of the action of the waves on the fluid motion is an increased surface velocity and the already mentioned deformation turbulence.

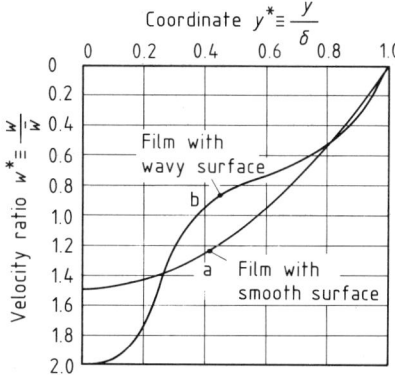

Figure 7.3. Velocity profiles in a liquid film. – Curve a for a film with smooth surface, curve b for a film with wavy surface.

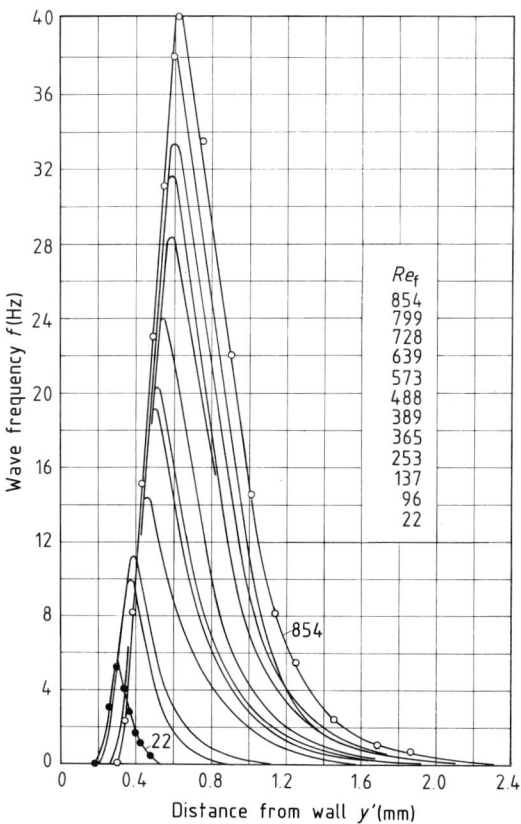

Figure 7.2. Wave frequency profiles for water films in the Reynolds number region from $Re_f = 22$ to $Re_f = 854$.

The formation of waves is the result of a fluid dynamic instability. This phenomenon has been studied theoretically by many investigators. The most comprehensive study has been carried out by GRAEF [7.9]. The theoretical results are in fair agreement with experimental data which have been correlated by the author. The resulting empirical equation is:

$$Re_{f,w} = 0.306 \left(\frac{\varrho_f \sigma^3}{g \eta_f^4} \right)^{1/10}, \qquad (7.7)$$

with σ being the surface tension. In general,

wave formation will start at a Reynolds number between 1 and 4.

Fluid dynamic instability finally leads to a turbulent state of fluid flow. This occurs at the critical value of the Reynolds number. By means of shear stress measurements the author determined the following value [7.6]:

$$Re_{crit} = 400.$$

Many other authors have also published values for Re_{crit}. In most cases Re_{crit} has been derived from experimental investigations on mean film thickness and heat transfer.

On the basis of experimental data from various sources the author derived the following equations for mean film thickness and mean velocity in the turbulent flow regime [7.6]:

$$\delta = 0.435 \left(\frac{3 v_f^2}{g}\right)^{1/3} Re_f^{8/15}, \qquad (7.8)$$

$$\bar{w}_f = 0.3315 \left(\frac{g v_f}{3}\right)^{1/3} Re_f^{7/15}. \qquad (7.9)$$

Another important result obtained by the author in the course of experimental investigations on wavy liquid films is the fact that the surface area of liquid films is increased by only a few percent due to the action of waves [7.6]. This result has been confirmed by many authors. An increase in mass transfer through the interface of wavy films can therefore not be attributed to an increased interfacial area. The observed high mass transfer rate is due to the action of deformation turbulence [7.7]. Many more authors than those already mentioned have played an important role in film research. Without going into further details, the papers published by KAPIZA [7.10], DUKLER and BERGELIN [7.11], PORTALSKI [7.12], [7.13], FEIND [7.14], COULON [7.15], HAMANN [7.16], BRAUN et al. [7.17], BAKOPOULOS [7.18], and the report by FULFORD [7.19] should at least summarily be mentioned.

7.3 Mass Transfer in Liquid Films with Smooth Surfaces

Assuming a smooth surface of the liquid film, as NUSSELT did in 1916 and 1923, a theoretical solution of the mass transfer problem poses no serious problem. The pioneer paper in this field has been published by EMMERT and PIGFORD [7.20] in 1954. The next important contribution to the problem of mass transfer in liquid films has been made by THIELE and summarized by the author [7.21].

The problem solved by THIELE will be explained by means of Fig. 7.4. It is as-

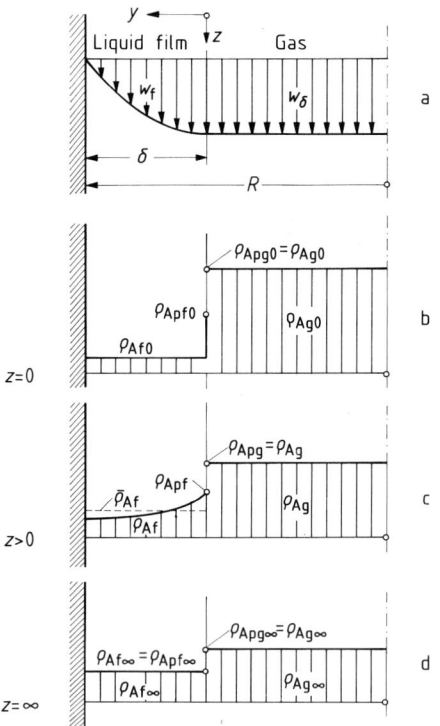

Figure 7.4. Schematic presentation of velocity and concentration profiles in a liquid film and a gas stream for cocurrent flow conditions.

7.3 Mass Transfer in Liquid Films with Smooth Surfaces

sumed that mass of component A is transferred from a gas stream into a liquid film. Both fluids move cocurrently. The velocity profile in the liquid film is parabolic, while the velocity in the gas stream is assumed to be equal to the surface velocity of the film and constant over the cross section of the flow channel.

At $z=0$ entrance conditions are given as shown in Fig. 7.4b. The concentration of component A is constant in both phases, ϱ_{Ag0} in the gas phase, and ϱ_{Af0} in the liquid phase. In the interface the concentrations are ϱ_{Apg0} and ϱ_{Apf0}. The ratio of concentration in the interface is given by the Henry number:

$$H^* \equiv \varrho_{Apg}/\varrho_{Apf}. \quad (7.10)$$

With increasing flow length z, mass is transferred from the gas stream into the liquid film. Therefore, the concentration in the gas phase decreases while it increases in the liquid phase. At $z=\infty$ equilibrium conditions are obtained.

Without going into details of the mathematical formulation of the problem and the method of numerical solution, the mean Sherwood number Sh_f will be discussed. The general equation for Sh_f is as follows:

$$Sh_f = f\left(\frac{z/\delta}{Re_f Sc_f}; K_p\right). \quad (7.11)$$

The dimensionless numbers are defined by:

$$Sh_f \equiv \frac{\beta_f \delta}{D_f} \quad \text{mean Sherwood number} \quad (7.12)$$

$$Sc_f \equiv \frac{\nu_f}{D_f} \quad \text{Schmidt number} \quad (7.13)$$

$$K_p \equiv \frac{H^*}{2}\left(\frac{R}{\delta}-1\right) = H^* \frac{2}{3} \frac{\dot{V}_g}{\dot{V}_f} \quad (7.14)$$

interface number

In these equations β_f is the mean mass transfer coefficient, D_f the coefficient of diffusion for component A, R the tube radius, and $\dot{V}_g$ the volumetric flow rate of the gas. The interface number is predominantly determined by the Henry number H^*. The ratio $\dot{V}_g/\dot{V}_f$ is related to the ratio R/δ.

In Fig. 7.5 the mean Sherwood number is shown as a function of the entrance number z^* and the interface number K_p. Sh_f increases with decreasing K_p. This is primarily due to the fact that solubility of A in the liquid film increases with a decreasing value of the Henry number H^*.

For the special case of $K_p = \infty$, that is poor solubility with $H^* \to \infty$ and for $\dot{V}_g/\dot{V}_f \to \infty$, the concentrations in the interface are independent of flow length z or en-

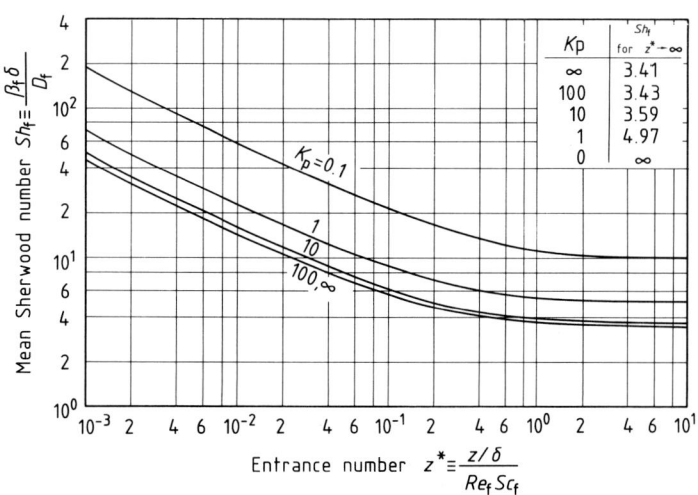

Figure 7.5. Mean Sherwood number Sh_f for mass transfer in a liquid film with smooth surface as a function of the entrance number z^* and various values of the interface number K_p.

trance number z^*. This is the case considered by EMMERT and PIGFORD [7.20]. The curve shown in Fig. 7.5 for $K_p = \infty$ may be expressed by the following correlation:

$$Sh_f = 3.4145 + \frac{0.276\, z^{*-1.2}}{1 + 0.20\, z^{*-0.7}} \qquad (7.15)$$

range of application: $0 \leqslant z^* \leqslant \infty$
$$0 \leqslant Re_f \leqslant Re_{crit} = 400$$
$$K_p \to \infty$$

This equation is in excellent agreement with experimental data.

7.4 Mass Transfer in Liquid Films with Wavy Surfaces

For the majority of industrial applications of liquid films, a wavy surface must be expected. Due to the mixing effect of the waves a substantial improvement of mass transfer has been found by many research workers. But it is not only the waves which exert a strong influence on the mass transfer process in the film. The way the film is generated and the length of the tubes or plates are important parameters for mass transfer. Many experimental and theoretical investigations have been carried out in order to obtain a better understanding of mass transfer phenomena in wavy liquid films.

7.4.1 Experimental Investigations

In order to present confirmed and for industrial purposes directly applicable results, it seems to be advisable to start with a presentation of the results of experimental investigations.

The experimental conditions consisted, in general, of a vertical tube with the liquid film at the inner surface of the tube. The gas flow was countercurrent to film flow. An effect of the gas flow on liquid flow and mass transfer in the film could not be detected.

The experiments have been carried out with two types of tubes. One type of tube had a smooth wall while the other had a wall with a wavy shape. The discussion of available results will start with the common type of tube having a smooth wall.

7.4.1.1 Tubes with a Smooth Wall

The first comprehensive and, in fact, pioneering experimental study on liquid film mass transfer has been carried out by KAMEI and OICHI, published in 1955 [7.22]. In the following decades, many investigators contributed to the wealth of knowledge available today. But not all of the obtained results have been published in such a way that they can be used for a summarizing presentation.

The mean mass transfer coefficient β_f is defined by the equation for the mass flux $\dot{M}_A$, directed from the gas to the liquid phase and crossing the interfacial area A_p:

$$\dot{M}_A = \beta_f A_p \Delta \varrho_A. \qquad (7.16)$$

The interfacial area is given by:

$$A_p = 2\pi (R - \delta) L, \qquad (7.17)$$

and the mean concentration difference by:

$$\Delta \varrho_A = \frac{(\varrho_{Af} - \varrho_{Apf})_1 - (\varrho_{Af} - \varrho_{Apf})_2}{\ln \dfrac{(\varrho_{Af} - \varrho_{Apf})_1}{(\varrho_{Af} - \varrho_{Apf})_2}}. \qquad (7.18)$$

In the last two equations R is the tube radius, δ the film thickness, L the tube length,

ϱ_{Af} the mean partial density of component A at the entrance (index 1) and exit (index 2) of the tube, and ϱ_{Apf} the partial density in the interface.

The available data [7.22] to [7.26] on mass transfer have been collected and correlated by the author [7.21] and by CARRUBA [7.23]. Fig. 7.6 shows the mean Sherwood number $Sh_{f\infty}$ divided by the square root of the Schmidt number Sc_f as a function of the Reynolds number Re_f of the liquid film. The experimental data apply to liquid films flowing down a rather long vertical tube or plate. On the basis of experimental evidence it may be safely assumed, that there will be no further decrease of the Sherwood number with increasing length of the tubes or plates. The Sherwood number shown in Fig. 7.6 is valid for the case of $L \to \infty$, L being the length of the tube.

There are three distinct regions of the Reynolds number in which different relationships between Re_f and $Sh_{f\infty}$ exist. This is the result of the variable wave effect with the Reynolds number on the mixing within the film.

The experimental data are correlated by curves a, b, and c, for which the following correlations have been determined:

$$Sh_{f\infty,a} = 2.24 \cdot 10^{-2} Re_f^{0.8} Sc_f^{0.5} \qquad (7.19)$$

range of application: $12 \leqslant Re_f \leqslant 70$

$$Sc_f \geqslant \frac{2.32 \cdot 10^4}{Re_f^{1.6}}$$

$$L/d \to \infty$$

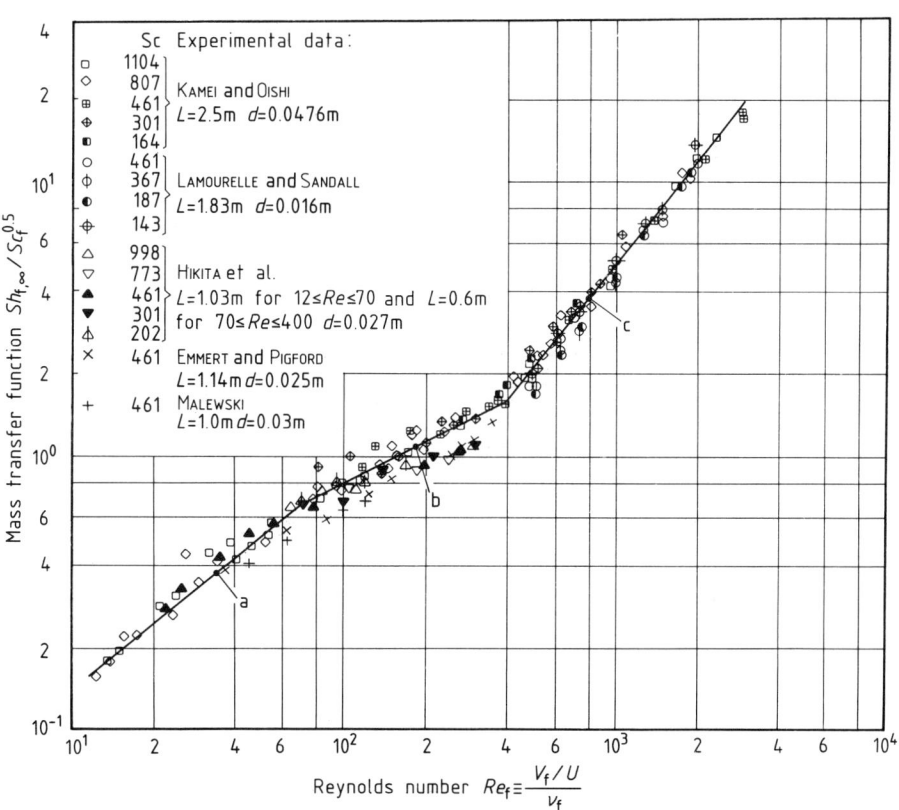

Figure 7.6. Mass transfer in wavy liquid films. Symbols represent experimental data; curve a: Eq. (7.19), curve b: Eq. (7.20), curve c: Eq. (7.21).

$$Sh_{f\infty,b} = 8.0 \cdot 10^{-2} Re_f^{0.5} Sc_f^{0.5} \quad (7.20)$$

range of application: $70 \leqslant Re_f \leqslant 400$

$$Sc_f \geqslant \frac{1.82 \cdot 10^3}{Re_f}$$

$$L/d \to \infty$$

$$Sh_{f\infty,c} = 8.9 \cdot 10^{-4} Re_f^{1.25} Sc_f^{0.5} \quad (7.21)$$

range of application: $400 \leqslant Re_f$

$$Sc_f \geqslant \frac{1.47 \cdot 10^7}{Re_f^{2.5}}$$

$$L/d \to \infty$$

The absolute minimum value of the Sherwood number is 3.4145, which applies to a film with a smooth surface and for $z^* \equiv (z/\delta)/(Re_f Sc_f) \to \infty$. The Sherwood number determined by one of the Eqs. (7.19) to (7.21) may become smaller than the limiting value 3.4145 when the Schmidt number is smaller than the limiting value given for each one of the equations. In order to prevent such a situation, the following equations should be applied:

$$Sh_{f\infty} = Sh_{f\infty,a} - \\ -(Sh_{f\infty,a}^{-10} - 3.42^{-10})^{-1/10} + 3.42 \quad (7.22)$$

range of application: $12 \leqslant Re_f \leqslant 70$
$0 \leqslant Sc_f \leqslant \infty$
$L/d \to \infty$

$$Sh_{f\infty} = Sh_{f\infty,b} - \\ -(Sh_{f\infty,b}^{-10} - 3.42^{-10})^{-1/10} + 3.42 \quad (7.23)$$

range of application: $70 \leqslant Re_f \leqslant 400$
$0 \leqslant Sc_f \leqslant \infty$
$L/d \to \infty$

$$Sh_{f\infty} = Sh_{f\infty,c} - \\ -(Sh_{f\infty,c}^{-10} - 3.42^{-10})^{-1/10} + 3.42 \quad (7.24)$$

range of application: $400 \leqslant Re_f$
$0 \leqslant Sc_f \leqslant \infty$
$L/d \to \infty$

It has been found that the tube length L has some influence on liquid film mass transfer, because shape and movement of the waves varies with tube length. Experimental investigations by BAKOPOULOS [7.18] gave results which are presented in Fig. 7.7. The influence of the tube length to diameter ra-

tio L/d is restricted to the turbulent flow region with $Re_f > 400$. Therefore, Eqs. (7.19) and (7.20) may be applied at least in the parameter range $20 \leqslant L/d \leqslant \infty$. For the turbulent region the following equation applies:

$$Sh_{f,c} = 8.9 \cdot 10^{-4} Re_f^{1.25} Sc_f^{0.5} \cdot \\ \cdot \left[1 + \left(\frac{11.3}{(L/d)^{0.5}} \right)^{20} \right]^{1/20} \quad (7.25)$$

range of application:
$400 \leqslant Re_f$
$20 \leqslant L/d \leqslant \infty$

$$Sc_f \geqslant \frac{1.47 \cdot 10^7}{Re_f^{2.5} \left[1 + \left(\frac{11.3}{(L/d)^{0.5}} \right)^{20} \right]^{1/20}}$$

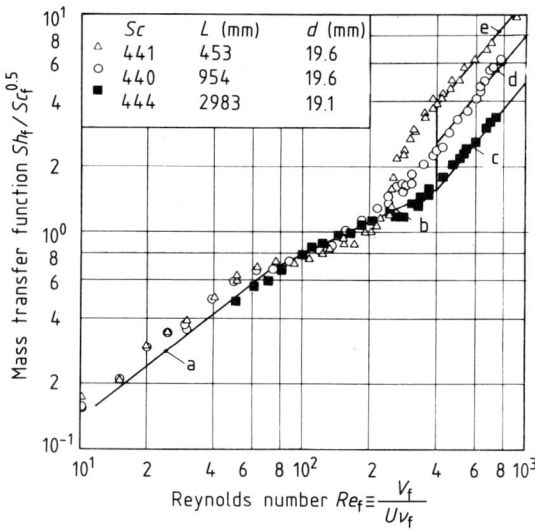

Figure 7.7. Mass transfer in wavy liquid films with variable tube length L; for curves a, b, and c see Fig. 7.6.

7.4.1.2 Tubes with a Wavy Wall

Although the application of liquid films for gas/liquid mass transfer operations will ensure high values for the mass transfer coefficient, there are many cases in which a further increase of the coefficient is of advantage. Such an increase may, for exam-

7.4 Mass Transfer in Liquid Films with Wavy Surfaces

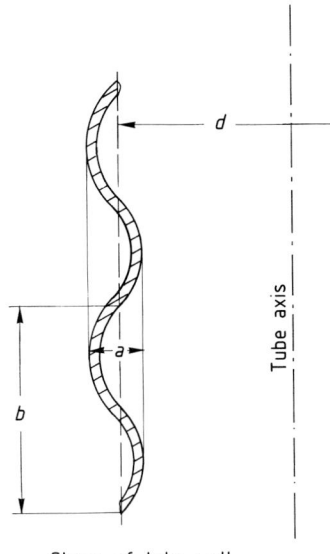

Figure 7.8. Shape of a wavy tube wall and notations.

ple, be achieved by use of tubes with a wavy wall. Fig. 7.8 gives the shape of such a tube wall. The amplitude of the wave is a, and the wave length is b. The mean diameter d_w of the wavy tube is given by:

$$d_w = \sqrt{\frac{4}{\pi} V_w / L}, \qquad (7.26)$$

with V_w as the volume of the wavy tube, and the mean circumference U_w of the tube by:

$$U_w = d_w \pi. \qquad (7.27)$$

The inner surface of the tube A_w can be determined by experiments applying the equation:

$$A_w = \frac{V}{s}, \qquad (7.28)$$

with V as the volume of liquid displaced by a given piece of tube with wall thickness s. The mass transfer experiments have been carried out using tubes with a mean diameter d_w varying from 23.7 mm to 24.9 mm. The length of the tubes varied between 210 mm and 2767 mm, the amplitude a between 0.35 mm and 2.35 mm, the amplitude ratio a/b between 0.05 and 0.53, and the tube wall surface A_w between 15 175 mm² and 213 672 mm².

The mass transfer coefficient $\beta_{f,w}$ is defined by the following equation for mass flux:

$$\dot{M}_A = \beta_{f,w} A_w \Delta \varrho_A. \qquad (7.29)$$

The mean concentration difference is defined by Eq. (7.18), and the tube surface area is given by Eq. (7.28). As there are no

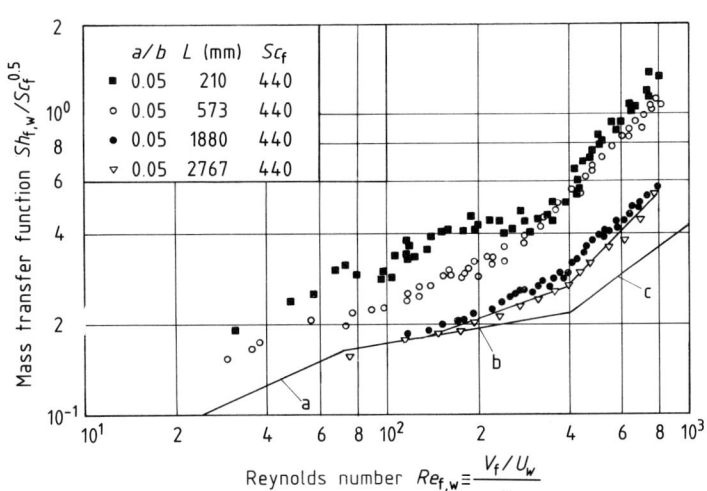

Figure 7.9. Liquid film mass transfer in vertical tubes with wavy shape of the walls as a function of $Re_{f,w}$ and tube length L.

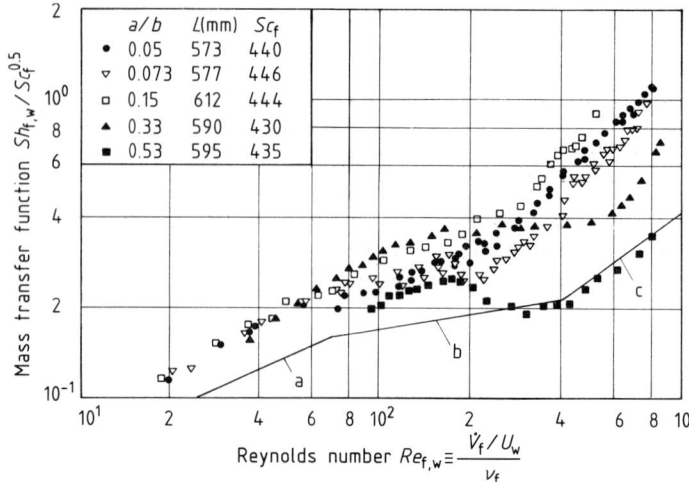

Figure 7.10. Liquid film mass transfer in vertical tubes with wavy shape of the walls as a function of $Re_{f,w}$ and amplitude ratio a/b.

equations available for the mean film thickness of wavy tubes, the dimensionless number for presentation of the mass transfer results must be defined without the use of the film thickness. The dimensionless numbers are therefore defined as follows:

$$Sh_{f,w} \equiv \frac{\beta_{f,w}}{D_f}\left(\frac{3v_f^2}{g}\right)^{1/3}, \quad (7.30)$$

$$Re_{f,w} \equiv \frac{\dot{V}_f/U_w}{v_f}. \quad (7.31)$$

Experimental results for mass transfer are presented in Figs. 7.9 and 7.10 [7.27]. In Fig. 7.9 the influence of the tube length L for a constant wave amplitude ratio a/b on mass transfer is demonstrated. Just as for smooth tubes, there is a strong influence of the tube length on mass transfer in tubes with wavy walls.

The wavy shape of the tube wall has a strong influence on liquid motion and mixing within the film. This is the reason why the relationship between $Sh_{f,w}$ and $Re_{f,w}$ is different from that for tubes with smooth

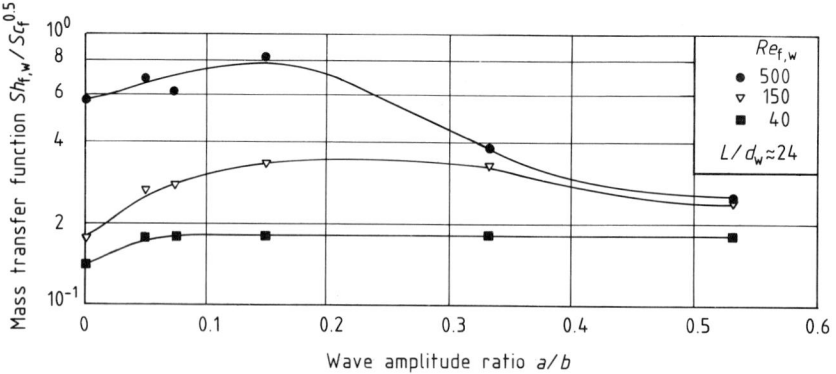

Figure 7.11. Liquid film mass transfer in vertical tubes with wavy shape of the walls as a function of the amplitude ratio a/b for three values of $Re_{f,w}$.

walls. The smooth wall case is represented by curves a, b, and c, which have been transferred from curves a, b, and c shown in Fig. 7.6. The mass transfer data presented in Fig. 7.9 may be expressed by the following correlations:

$$Sh_{f,w} = 2.90 \cdot 10^{-2} \, Re_{f,w}^{0.37} \, Sc_f^{0.5} \cdot$$
$$\cdot \left[1 + \left(\frac{3.9}{(L/d_w)^{0.30}}\right)^{20}\right]^{1/20} \quad (7.32)$$

range of application: $30 \leq Re_{f,w} \leq 400$
$8 \leq L/d_w \leq 111$
$a/b = 0.05$
$Sc_f \geq \dfrac{1.8 \cdot 10^3}{Re_{f,w}}$

$$Sh_{f,w} = 6.67 \cdot 10^{-4} \, Re_{f,w} \, Sc_f^{0.5} \cdot$$
$$\cdot \left[1 + \left(\frac{6.15}{(L/d_w)^{0.39}}\right)^{20}\right]^{1/20} \quad (7.33)$$

range of application: $400 \leq Re_{f,w}$
$8 \leq L/d_w \leq 111$
$a/b = 0.05$
$Sc_f \geq \dfrac{1.48 \cdot 10^7}{Re_{f,w}^{2.5}}$.

The influence of the wave amplitude ratio a/b on mass transfer is demonstrated by Fig. 7.10. The length ratio in this case is almost constant, $L/d_w \approx 24$. Plotting the mass transfer function against the amplitude ratio a/b for various values of the Reynolds number leads to the curves in Fig. 7.11. The optimum amplitude varies with $Re_{f,w}$ in the range from $a/b \approx 0.05$ to $a/b \approx 0.2$.

7.4.2 Theoretical Investigations

The results of numerous experimental investigations in the last decades are a sound basis for a new concept of mass transfer in wavy liquid films. The effect of the waves is a strong mixing of the liquid within the film. Any hypothesis on mass transfer in wavy films endeavors to describe the effect of the wavy film motion on mass transfer. The most advanced is the deformation turbulence hypothesis, which will be used in the following calculations.

Deformation turbulence in liquid films is the result of rapid stochastic local changes of the film thickness due to wave motions. Such changes induce stochastic variations of the local velocity with respect to its absolute value and direction. Accelerated and decelerated local fluid motions are characteristic properties of deformation turbulence.

The hypothesis of deformation turbulence in liquid films is based on two assumptions [7.7]:

1. A film with a smooth surface may be substituted for a film with a wavy surface.
2. The effect of deformation turbulence can be taken into account by an empirical mathematical equation.

Mass transfer in wavy liquid films is the result of the combined actions of molecular and turbulent motions. Turbulent motions in wavy liquid films are caused by deformation turbulence and by Reynolds turbulence. Deformation turbulence results from an instability of the interface, while Reynolds turbulence is the result of an instability of the velocity profile. The transition to deformation turbulence occurs at $Re_f = 12$ and to Reynolds turbulence at $Re_f = 400$. For $Re_f > 400$, Reynolds and deformation turbulence occur simultaneously. To simplify, the following considerations will be restricted to the Reynolds number region $Re_f < 400$, in which only deformation turbulence occurs.

The mass flux density due to deformation turbulence (D-T) is defined by the equation:

$$\dot{m}_{A,D\text{-}T} = -\varepsilon_{m,D\text{-}T} \frac{\partial \varrho_{Af}}{\partial y}, \quad (7.34)$$

with $\varepsilon_{m,D\text{-}T}$ as the appropriate diffusivity, and y the distance from the free surface of the liquid film. Experimental data for mass transfer have been used to obtain the following equations for the ratio $\varepsilon_{m,D\text{-}T}/D_f$ [7.23]:

for $12 \leq Re_f \leq 70$

$$\frac{\varepsilon_{m,D\text{-}T}}{D_f} = F_1 - (F_1^{10} + F_2^{10})^{1/10} +$$
$$+ F_2 - (F_2^{10} + F_5^{10})^{1/10} +$$
$$+ F_5(F_5^{10} - F_6^{10})^{1/10} + F_6 \qquad (7.35)$$

for $70 \leq Re_f \leq 400$

$$\frac{\varepsilon_{m,D\text{-}T}}{D_f} = F_3 - (F_3^{10} + F_4^{10})^{1/10} +$$
$$+ F_4 - (F_4^{10} + F_5^{10})^{1/10} +$$
$$+ F_5 - (F_5^{10} + F_6^{10})^{1/10} + F_6. \qquad (7.36)$$

The functions F_1 to F_6 are given by the following equations:

$$F_1 = 1.24 \cdot 10^{-3} Re_f^{1.6} Sc_f y^{*2} \qquad (7.37)$$

$$F_2 = \{[(2.13 \cdot 10^{-4} Re_f^{1.6} Sc_f)^4 + (4.94)^4]^{1/4} - 4.94\} y^{*1.5}, \qquad (7.38)$$

$$F_3 = 1.58 \cdot 10^{-2} Re_f Sc_f y^{*2},$$

$$F_4 = \{[(2.72 \cdot 10^{-3} Re_f Sc_f)^4 + (4.94)^4]^{1/4} - 4.94\} y^{*1.5}, \qquad (7.39)$$

$$F_5 = \{[(2.5 \cdot 10^{-4} Re_f^{1.5} Sc_f)^4 + (3.94)^4]^{1/4} - 3.94\} (1 - y^*)^{2.5}, \qquad (7.40)$$

$$F_6 = 7.52 \cdot 10^{-4} Re_f^{1.5} Sc_f (1 - y^*)^3. \qquad (7.41)$$

According to these equations, the diffusivity ratio $\varepsilon_{m,D\text{-}T}/D_f$ is a function of the Reynolds number Re_f, the Schmidt number Sc_f, and local coordinate $y^* \equiv y/\delta$. Fig. 7.12 gives an example for $\varepsilon_{m,D\text{-}T}/D_f$ as a function of y^* with $Re_f = 15$ and $Sc_f = 2 \cdot 10^5$. At $y^* = 0$ and $y^* = 1$ the diffusivity ratio is zero, the maximum occurs at about $y^* = 0.4$. Very close to the free surface of the liquid film the diffusivity ratio is proportional to y^{*2}:

$$\frac{\varepsilon_{m,D\text{-}T}}{D_f} \sim y^{*2} \quad \text{for} \quad y^* \to 0.$$

For mass transfer in wavy liquid films the concentration field is described by the differential equation:

$$w_f \frac{\partial \varrho_{Af}}{\partial z} = \frac{\partial}{\partial y} \left[(D_f + \varepsilon_{m,D\text{-}T}) \frac{\partial \varrho_{Af}}{\partial y} \right]. \qquad (7.42)$$

The local velocity w_f is given by Eq. (7.3), based on the assumption of a parabolic velocity distribution. This is certainly not quite correct. A better approximation of the true velocity distribution would be obtained by means of curve b presented in Fig. 7.3. Extensive investigations have shown, however, that, due to the strong influence of $\varepsilon_{m,D\text{-}T}$,

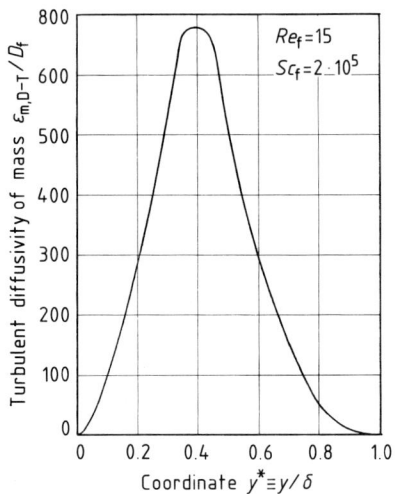

Figure 7.12. Diffusivity ratio of mass for wavy liquid films.

the shape of the velocity profile becomes less important. One is therefore justified to make use of the parabolic velocity profile in the wavy film region, $12 \leq Re_f \leq 400$. Introducing Eq. (7.3) for the local velocity w_f and dimensionless parameters leads to:

$$\frac{3}{2}(1 - y^{*2}) \frac{\partial \xi}{\partial z^*} = \frac{\partial}{\partial y^*} \left[\left(1 - \frac{\varepsilon_{m,D\text{-}T}}{D_f} \right) \frac{\partial \xi}{\partial y^*} \right]. \qquad (7.43)$$

Boundary conditions:

$z^* = 0$ and $0 \leq y^* \leq 1$: $\xi = 1$

$0 \leq z^* \leq \infty$ and $y^* = 0$: $\xi = 0$

$0 \leq z^* \leq \infty$ and $y^* = 1$: $\partial \xi / \partial y^* = 0$

The dimensionless parameters used in Eq. (7.47) are defined as follows:

$$\xi \equiv \frac{\varrho_{Afp} - \varrho_{Af}}{\varrho_{Afp} - \varrho_{Af0}} \quad \text{local concentration} \quad (7.44)$$

$$z^* \equiv \frac{z/d}{Re_f Sc_f} \quad \begin{array}{l}\text{entrance number} \\ \text{(length coordinate)}\end{array} \quad (7.45)$$

$$y^* \equiv y/\delta \quad \begin{array}{l}\text{coordinate normal} \\ \text{to fluid flow}\end{array} \quad (7.46)$$

Here, ϱ_{Af} is the local partial density of transferred component A, ϱ_{Afp} at the interface, ϱ_{Af0} at the entrance, z the length coordinate, d tube diameter, y distance from the interface, δ film thickness, Re_f and Sc_f the Reynolds and Schmidt number defined by Eqs. (7.12) and (7.13), respectively.

The solution of Eq. (7.43) has been obtained by CARRUBA [7.23].

Introducing Eq. (7.35) or (7.36) for $\varepsilon_{m,D-T}/D_f$ into Eq. (7.43) leads either to Eq. (7.19) or Eq. (7.20). This proves that the deformation turbulence hypothesis is well suited for a correct physical description of mass transfer in wavy liquid films.

7.5 Liquid Film Mass Transfer with Biochemical Reactions

Mass transfer with homogeneous biochemical reactions has attracted much attention by chemical engineers because of its technical and scientific importance. Because of the mathematical difficulties involved, however, until recently only solutions based on film- and penetration hypotheses were available. The limitations of these methods of solution are well known. In the following sections a method of solution will be presented that is free from such limitations.

7.5.1 The Physico-Biochemical Problem

Fig. 7.13 shows the concentration profiles of components A and B, reactants of a homogeneous biochemical reaction, in a liquid film. Component A is transferred from the adjacent gas stream to the liquid film. It is assumed that the concentration of component A in the gas stream is constant: $c_{Ag} = c_{Apg} = $ const. Consequently, the concentration on the liquid side of the interface is also constant: $c_{Apf} = $ const. The relation between c_{Ag} and c_{Apf} is given by Henry's law: $c_{Apf} = c_{Apg}/H^*$. This mass transfer case is identical with that discussed in Sect. 7.3 for $K_p = \infty$. The concentration c_{Af} decreases from the highest value c_{Apf} with increasing distance y from the phase boundary. Within the liquid film, component A reacts with component B, which is present in the film. As the reaction takes place close to the phase boundary, the concentration c_{Bf} decreases when the phase boundary is approached. In the phase boundary the concentration gradient $\partial c_{Bf}/\partial y$, becomes zero because of the assumption that component B does not move across the phase boundary. The result of the reaction of A and B is component C, not to be considered here.

The reaction equation is as follows:

$$\nu_A A + \nu_B B \to \nu_C C,$$

with ν_A, ν_B, and ν_C as the stoichiometric ratios. The relationship between the change of concentrations with time t is given by:

$$\frac{\partial c_{Af}}{\partial t} = \frac{\nu_B}{\nu_A} \frac{\partial c_{Bf}}{\partial t} = -\frac{\nu_C}{\nu_A} \frac{\partial c_{Cf}}{\partial t}. \quad (7.47)$$

The change of concentration with time is the volumetric reaction flux density. For components A and B the following empirical equations are assumed to apply:

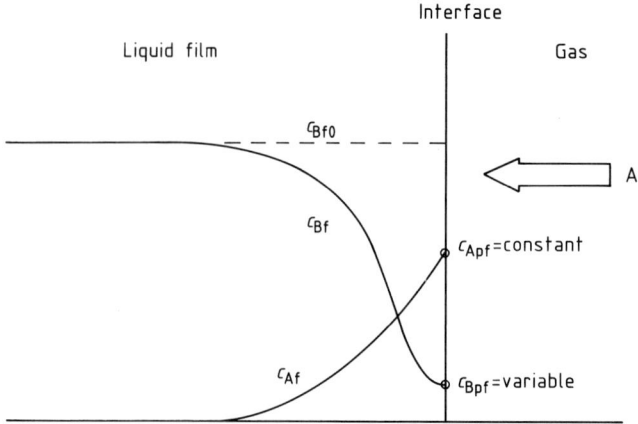

Figure 7.13. Qualitative concentration profiles for reactants A and B in a liquid film.

$$\dot{r}_A \equiv \frac{\partial c_{Af}}{\partial t} = -k_A c_{Af} c_{Bf}, \quad (7.48)$$

$$\dot{r}_B \equiv \frac{\partial c_{Bf}}{\partial t} = -\frac{\nu_B}{\nu_A} k_B c_{Af} c_{Bf}, \quad (7.49)$$

with k_A and k_B as the reaction velocity factors.

7.5.2 Mathematical Description of the Problem

The concentration field in a liquid film, in which a homogeneous biochemical reaction of second order takes place, is described by the following dimensionless differential equations and boundary conditions:

$$\frac{3}{2}(1-y^{*2})\frac{\partial c_{Af}^*}{\partial z^*} =$$

$$= \frac{\partial}{\partial y^*}\left[\left(1+\frac{\varepsilon_{m,D\text{-}T}}{D_{Af}}\right)\frac{\partial c_{Af}^*}{\partial y^*}\right] - Da\, c_{Af}^* c_{Bf}^*, \quad (7.50)$$

$$\frac{3}{2}(1-y^{*2})\frac{\partial c_{Bf}^*}{\partial z^*} =$$

$$= D^* \frac{\partial}{\partial y^*}\left[\left(1+\frac{\varepsilon_{m,D\text{-}T}}{D_{Bf}}\right)\frac{\partial c_{Bf}^*}{\partial y^*}\right] - Da\, c_{Af}^* c_{Bf}^* \nu^*. \quad (7.51)$$

Boundary conditions:

$z^* = 0$ and $y^* = 0$:
$c_{Bf}^* = 1$ and $c_{Af}^* = 1$

$z^* = 0$ and $0 < y^* \leq 1$:
$c_{Bf}^* = 1$ and $c_{Af}^* = 0$

$0 < z^* \leq \infty$ and $y^* = 0$:
$\dfrac{\partial c_{Bf}^*}{\partial y^*} = 0$ and $c_{Af}^* = c_{Apf}^* = 1$

$0 \leq z^* \leq \infty$ and $y^* = 1$:
$\dfrac{\partial c_{Bf}^*}{\partial y^*} = 0$ and $\dfrac{\partial c_{Af}^*}{\partial y^*} = 0$

The dimensionless quantities introduced are defined by:

$c_{Af}^* \equiv c_{Af}/c_{Af0}$ local concentration of A   (7.52)

$c_{Bf}^* \equiv c_{Bf}/c_{Bf0}$ local concentration of B   (7.53)

$y^* \equiv y/\delta$ distance from interface   (7.54)

$z^* \equiv \dfrac{z/\delta}{Re_f Sc_{Af}}$ entrance number   (7.55)

$Da \equiv \dfrac{k_\nu c_{Bf0}\delta^2}{D_{Af}}$ Damköhler number   (7.56)

$H^* \equiv \dfrac{c_{Apg}}{c_{Apf}}$ Henry number (7.57)

$D^* \equiv D_{Bf}/D_{Af}$ ratio of diffusion coefficients (7.58)

$v^* \equiv \dfrac{v_B}{v_A} \dfrac{c_{Apf}}{c_{Bf0}}$ stoichiometric surplus factor (7.59)

In case of a first order reaction it is assumed that component B is present in excess, so that the reaction depends only on the concentration of component A. The differential equation and boundary conditions can be simplified to:

$$\dfrac{3}{2}(1-y^{*2})\dfrac{\partial c^*_{Af}}{\partial z^*} =$$
$$= \dfrac{\partial}{\partial y^*}\left[\left(1+\dfrac{\varepsilon_{m,D\text{-}T}}{D_{Af}}\right)\dfrac{\partial c^*_{Af}}{\partial y^*}\right] - Da\, c^*_{Af} \quad (7.60)$$

$z^*=0$ and $y^*=0$: $c^*_{Af} = 1$

$z^*=0$ and $0<y^*\leqslant 1$: $c^*_{Af} = 0$

$0<z^*\leqslant\infty$ and $y^*=0$: $c^*_{Af} = c^*_{Apf}=1$

$0\leqslant z^*\leqslant\infty$ and $y^*=1$: $\dfrac{\partial c^*_{Af}}{\partial y^*}=0$

The parabolic differential equations have been solved by numerical methods. An immediate result is the local concentration ξ, from which a further integration leads to the mean concentration $\bar{c}^*_{Af}$, which is defined by:

$$\bar{c}^*_{Af} = \dfrac{1}{\bar{w}F} \int_F c^*_{Af} w_z \, dF \quad (7.61)$$

with w_z being the local and $\bar{w}$ the mean velocity of the liquid film, and F the cross sectional area of the film:

$$F = b\delta, \quad (7.62)$$

from which follows:

$$dF = b\delta\, dy^*. \quad (7.63)$$

In the last equation, b is the circumference of the tube, δ the film thickness, and $y^* \equiv y/\delta$ the local coordinate. With Eqs. (7.62) and (7.63) one obtains the following equation for the mean concentration:

$$\bar{c}^*_{Af} = \int_0^1 w^*_z c^*_{Af} \, dy^*. \quad (7.64)$$

For many applications the local mass transfer coefficient β_{fz} proves to be of use. This coefficient is defined by:

$$\beta_{fz} = \dfrac{\dot{n}_{Az}}{c_{Apf}-\bar{c}_{Af}}. \quad (7.65)$$

With the local mass flux density $\dot{n}_{Az}$, given by:

$$\dot{n}_{Az} = -D_{Af}\left(\dfrac{\partial c_{Af}}{\partial y}\right)_{y=0}, \quad (7.66)$$

the following equation for the local mass transfer coefficient results:

$$\beta_{fz} = -\dfrac{D_{Af}}{(c_{Apf}-\bar{c}_{Af})}\left(\dfrac{\partial c_{Af}}{\partial y}\right)_{y=0}. \quad (7.67)$$

Introducing the dimensionless parameters:

$\bar{c}^*_{Af} \equiv \dfrac{\bar{c}_{Af}}{c_{Apf}}$ mean concentration, (7.68)

$Sh_{fz} \equiv \dfrac{\beta_{fz}\delta}{D_{Af}}$ local Sherwood number, (7.69)

an equation for the local Sherwood number is obtained:

$$Sh_{fz} = -\dfrac{1}{(1-\bar{c}^*_{Af})}\left(\dfrac{\partial c^*_{Af}}{\partial y^*}\right)_{y^*=0}. \quad (7.70)$$

A further integration over the entrance number z^* gives the mean Sherwood number:

$$Sh_f = \dfrac{1}{z^*}\int_0^{z^*} Sh_{fz}\, dz^*, \quad (7.71)$$

with the definition of Sh_f according to Eq. (7.12).

The biochemical conversion rate increases with the Damköhler number, Da. The consequence is, that the rate of mass transfer into the liquid film must increase, too. With $Da \to \infty$ the mass transfer rate and the reaction rate approach a maximum. With $Da \to 0$, however, transfer and reaction rate approach a minimum. In this case the mass transfer process is a purely physical process; the reaction rate is zero. This case will be taken as a reference; the mass transfer coefficient is β_f. When a chemical reaction enhances mass transfer, the mass transfer coefficient is given by β_{fr}. The ratio β_{fr}/β_f is called the enhancement factor for mass transfer with chemical reactions:

$$\phi \equiv \beta_{fr}/\beta_f. \quad (7.72)$$

The enhancement factor approaches the following values:

Second order reaction

$\phi \to 1$ for $z^* \to 0$ and $z^* \to \infty$
       and $0 \leqslant Da \leqslant \infty$
               $0 \leqslant v^* \leqslant \infty$
               $0 \leqslant D^* \leqslant \infty$

$\phi \to \phi_{max}$ for $0 < z^* < \infty$
            and $0 \leqslant Da \leqslant \infty$
                   $0 \leqslant v^* \leqslant \infty$
                   $0 \leqslant D^* \leqslant \infty$

First order reaction

$\phi \to 1$ for $z^* \to 0$
       and $0 < Da \leqslant \infty$

$\phi \to \phi_{max}$ for $z^* \to \infty$
            and $0 < Da \leqslant \infty$

In the following section some of the numerical results obtained will be discussed.

7.5.3 Discussion of Some Results

The results presented in this section have been taken from investigations carried out by SPILGER [7.28] and SCHULTZ [7.29]. A few examples for mass transfer with second and first order reactions will be discussed.

7.5.3.1 Mass Transfer with Second Order Reactions

Although emphasis is placed on a discussion of the mean Sherwood number Sh and the enhancement factor ϕ, a few examples for local and mean concentration will also be described.

Discussion of Local and Mean Concentration

For the local concentrations c^*_{Af} and c^*_{Bf}, the following general functions apply:

$$\left.\begin{matrix} c^*_{Af} \\ c^*_{Bf} \end{matrix}\right\} = f_{A(B)}(y^*; z^*; Da; D^*; v^*) \quad (7.73)$$

Concentration profiles for $Da = 1000$, $D^* = 1$, and $v^* = 1$ in a smooth liquid film are presented in Fig. 7.14. The shape of the curves indicates, that most of the chemical conversion takes place close to the interface. This is even more pronounced in case of an infinitely fast reaction with $Da \to \infty$.

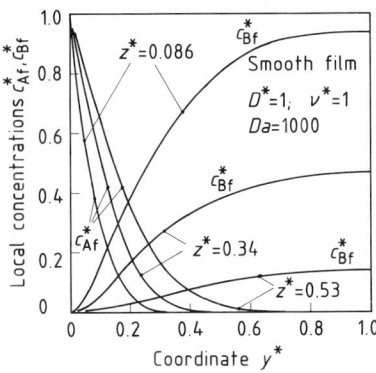

Figure 7.14. Local concentrations of components A and B for mass transfer with second order reaction in a smooth film with $Da = 1000$ for various values of z^*.

The concentration profiles are shown in Fig. 7.15 for a smooth film, and in Fig. 7.16 for a wavy film. Because of the extremely fast reaction, c^*_{Af} and c^*_{Bf} become zero at the same distance y^* from the interface. A care-

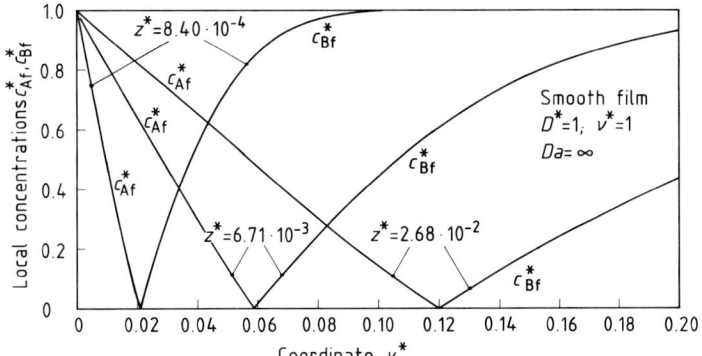

Figure 7.15. Local concentration of components A and B for mass transfer with second order reaction in a smooth film with $Da = \infty$ for various values of z^*.

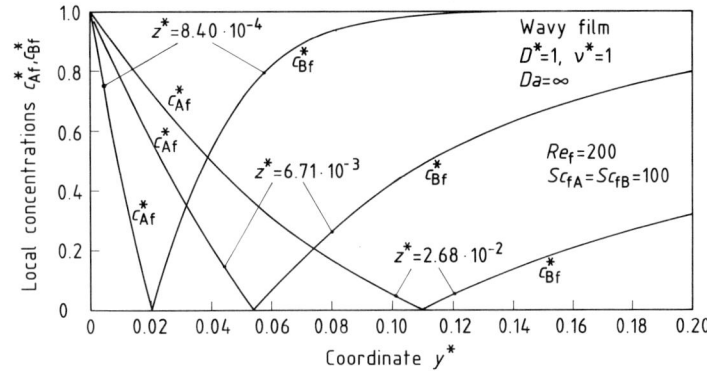

Figure 7.16. Local concentration of components A and B for mass transfer with second order reaction in a wavy film with $Da = \infty$ for various values of z^*.

ful examination of the profiles reveals the influence of deformation turbulence. In wavy films the point of $c_{Af}^* = c_{Bf}^* = 0$ is closer to the interface than in smooth films. For all values of the Damköhler number $Da < \infty$, such a singular point does not exist. The situation is indicated in Fig. 7.14.

A pronounced influence of the chemical reaction on mass transfer is only observed for relatively large values of the Damköhler

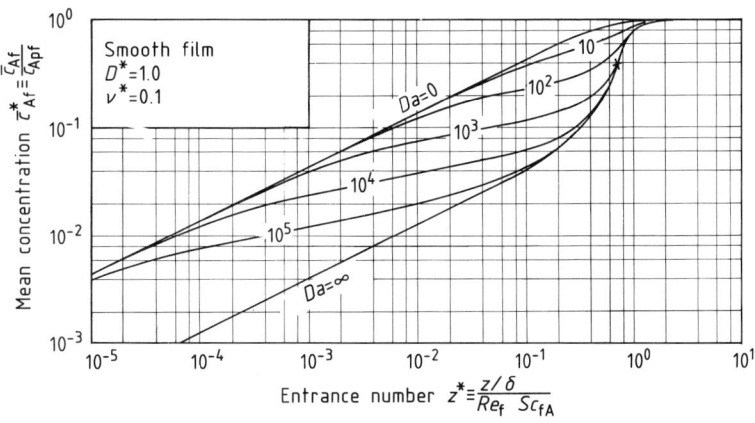

Figure 7.17. Mean concentration of component A in a smooth film for mass transfer with second order reaction as a function of entrance number z^* for various values of Damköhler number Da.

132   7 Transport Processes in Liquid Films

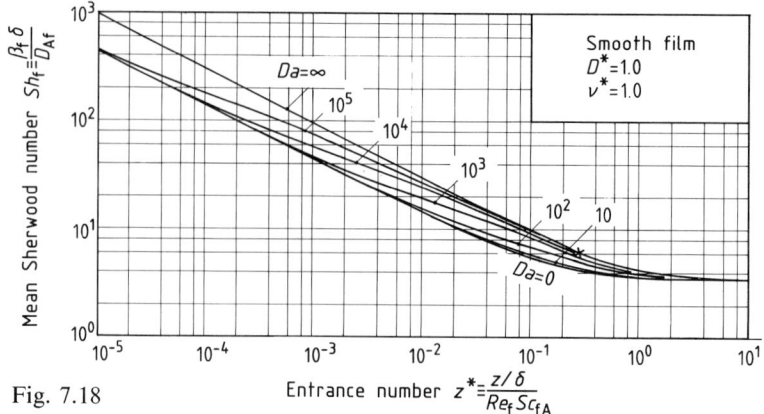

Fig. 7.18

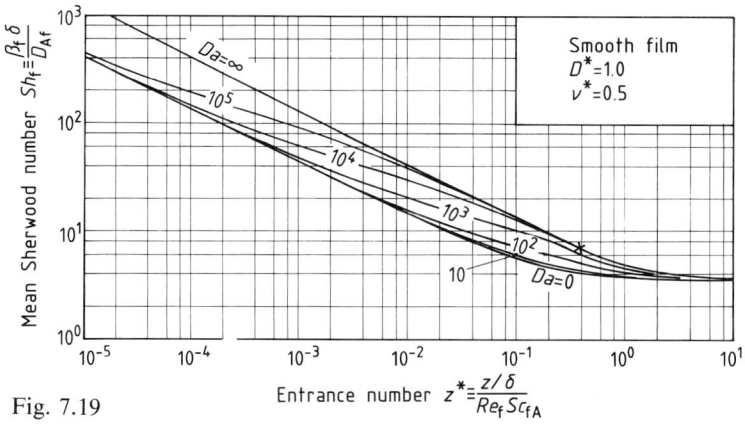

Fig. 7.19

number Da, and small values of the stoichiometric number v^*.

An example for the mean concentration $\bar{c}_A^*$ is presented in Fig. 7.17 for a smooth film. The concentration is plotted as a function of the entrance number z^* for $D^* = 1$ and $v^* = 1$, but for several values of the Damköhler number Da. With increasing z^* the mean concentration of component A approaches the limiting value of $\bar{c}_{Af}^* = 1$. In this case component B has disappeared completely. The curve for $Da \to 0$ represents mass transfer without a biochemical reaction.

Discussion of Mean Sherwood Number and Enhancement Factor

The mean Sherwood number for mass transfer with a second order reaction in a smooth film is shown in Figs. 7.18 to 7.21 as a function of the entrance number z^* and various values of the Damköhler number Da. In these presentations the ratio of diffusion coefficients $D^* \equiv D_{Bf}/D_{Af}$ has been kept constant with $D^* = 1$, while the stoichiometric number v^* has been varied in four steps from $v^* = 1$ to $v^* = 0.05$.

For all groups of curves there are two limiting curves, one for $Da = 0$ and one for

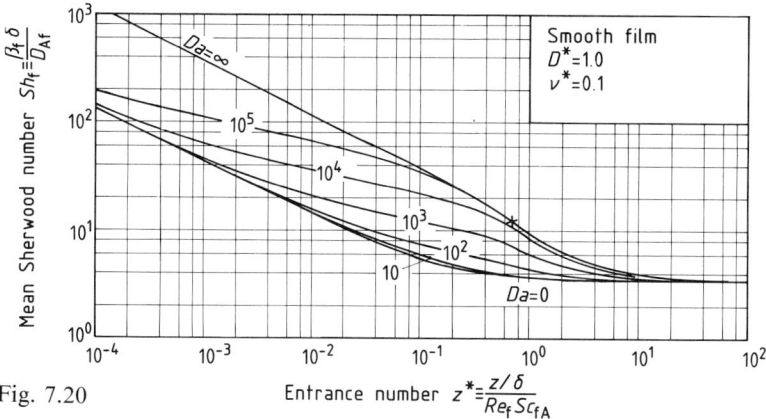

Fig. 7.20

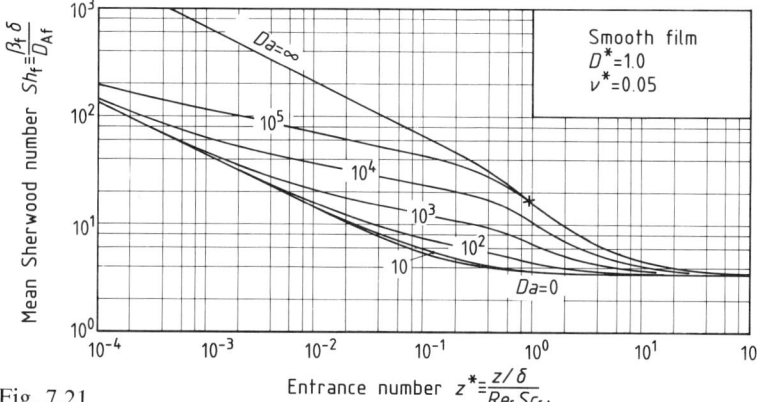

Fig. 7.21

Figures 7.18 to 7.21. Mean Sherwood number Sh_f for mass transfer with second order reaction in a smooth film as a function of the entrance number z^* for various values of Damköhler number Da, constant value of D^*, but variable stoichiometric surplus factor v^*:
Figure 7.18: $v^* = 1.0$
Figure 7.19: $v^* = 0.5$
Figure 7.20: $v^* = 0.1$
Figure 7.21: $v^* = 0.05$.

$Da = \infty$; between these all others are arranged. For all values of $Da < \infty$, the curves coincide with the lower limiting curve with $Da = 0$ and $z^* \to 0$. With increasing entrance number z^* the curves separate from that for $Da = 0$ and approach the upper limiting curve for $Da = \infty$. It should be observed, however, that all curves approach the final value $Sh_f = 3.415$ when z^* approaches infinity.

The lower limiting curve for $Da = 0$ covers the case of mass transfer in smooth liquid films without chemical reaction. This case has been discussed in Sect. 7.3. Eq. (7.15) describes the curves for $Da = 0$ in Figs. 7.18 to 7.21. The parameters D^* and v^* have no influence on the Sherwood number in this case, for which diffusion is the controlling step.

The upper limiting curve for $Da = \infty$ is a function of the stoichiometric surplus factor v^*. Decreasing values of v^* lead to an increase of the Sherwood number Sh_f, because the surplus of component A increases the reaction velocity, but of course only as long as component B is still present. With increasing z^* the mean concentration of B gradually decreases and finally becomes

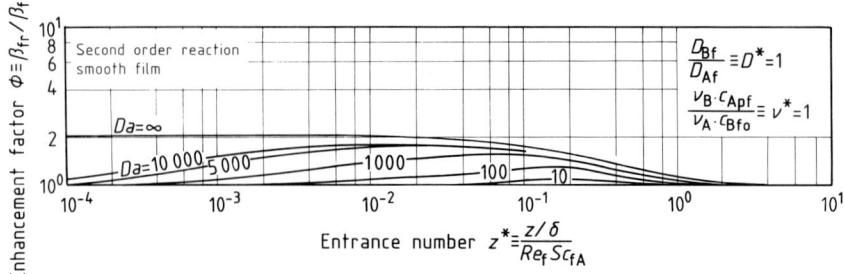

Figure 7.22. Enhancement factor ϕ for mass transfer with second order reaction in a smooth film for conditions given in Fig. 7.18.

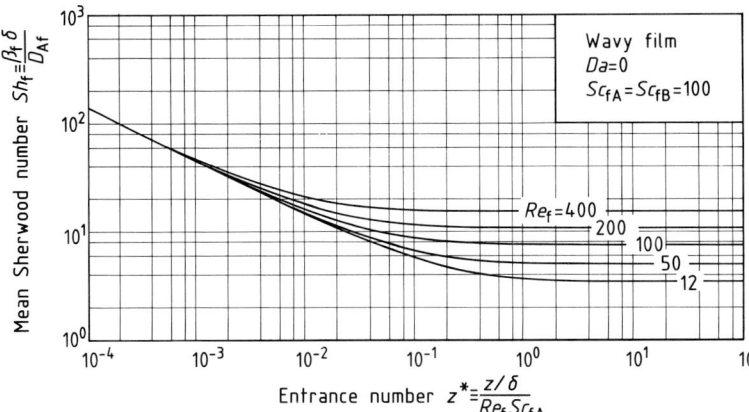

Figure 7.23. Mean Sherwood number for physical mass transfer ($Da=0$) in a wavy film as a function of the entrance number z^* and various values of Reynolds number Re_f.

zero. This occurrence is indicated by a small cross on each curve for $Da=\infty$ in Figs. 7.18 to 7.21.

For the set of parameters chosen in Fig. 7.18, the enhancement factor ϕ is presented in Fig. 7.22. This presentation makes it very clear that a substantial improvement of mass transfer by a homogeneous chemical reaction occurs only at relatively high values of the Damköhler number and in a limited range of the entrance number z^*.

For wavy films, that is, for real technical situations, mass transfer with homogeneous chemical reactions becomes a rather complex process. For a better understanding, the case of mass transfer without a reaction will be considered.

Fig. 7.23 shows the mean Sherwood number Sh_f over the entrance number z^* for $Da=0$ and $Sc_{Af}=Sc_{Bf}=100$ for several values of the Reynolds number Re_f. As deformation turbulence enhances mass transport in the liquid film, the Sherwood number increases with the Reynolds number. The influence of deformation turbulence becomes effective, however, only after the entrance number z^* has increased beyond a certain value. For $z^* \to 0$ deformation turbulence loses its effect.

Mass transfer with chemical reactions in wavy liquid films is considered by means of Figs. 7.24 to 7.26. For these figures the following parameters have been kept constant: $D^*=1.0$; $v=0.1$; $Sc_{Af}=Sc_{Bf}=100$; the variable parameter besides z^* and Da is the Reynolds number. With increasing Reynolds number, $Re_f=100$, 200, and 400, the deformation turbulence becomes more

7.5 Liquid Film Mass Transfer with Biochemical Reactions

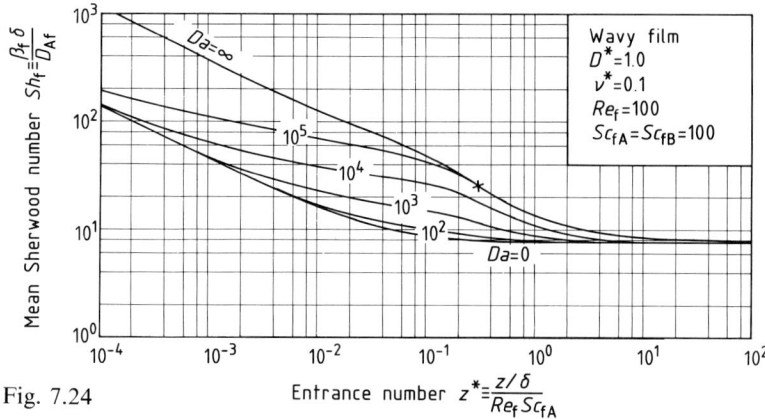

Fig. 7.24

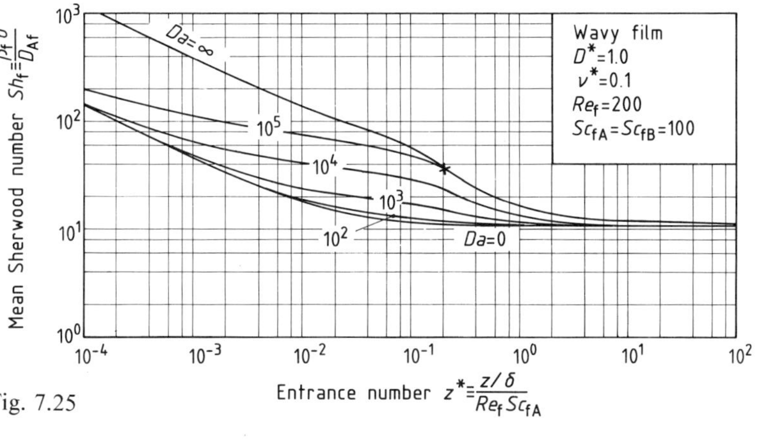

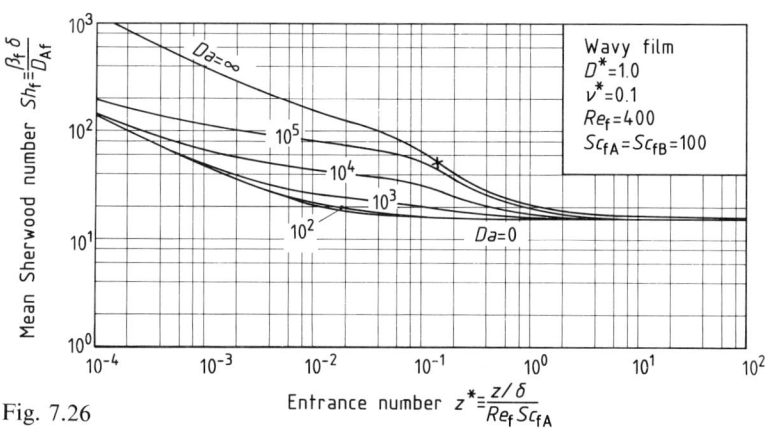

Fig. 7.25

Figures 7.24 to 7.26. Mean Sherwood number for mass transfer with second order reaction in a wavy film with $D^* = 1.0$, $v^* = 0.1$, and $Sc_{fA} = Sc_{fB} = 100$, but for variable values of the Reynolds number Re_f:
Figure 7.24: $Re_f = 100$
Figure 7.25: $Re_f = 200$
Figure 7.26: $Re_f = 400$.

Fig. 7.26

136   7 Transport Processes in Liquid Films

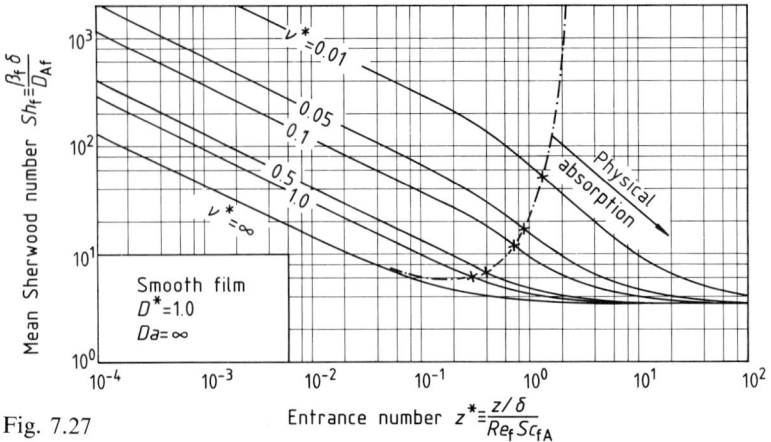

Fig. 7.27

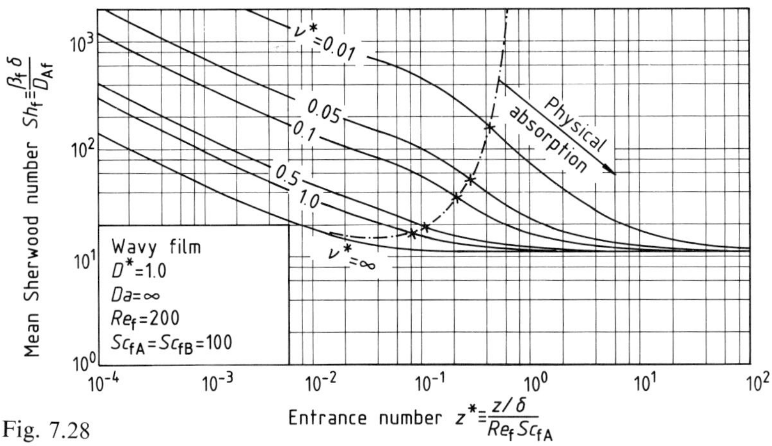

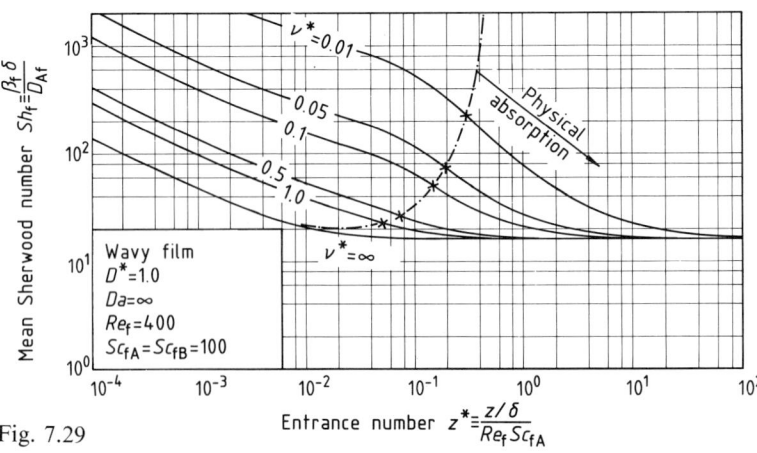

Fig. 7.28

Fig. 7.29

Figures 7.27 to 7.29:
Mean Sherwood number for mass transfer with infinitely fast second order reaction versus the entrance number z^* and for several values of the stoichiometric surplus factor v^* with $D^* = 1.0$ and $Da = \infty$.
Figure 7.27: smooth film
Figure 7.28: wavy film with $Re_f = 200$ and $Sc_{fA} = Sc_{fB} = 100$
Figure 7.29: wavy film with $Re_f = 400$ and $Sc_{fA} = Sc_{fB} = 100$.

7.5 Liquid Film Mass Transfer with Biochemical Reactions

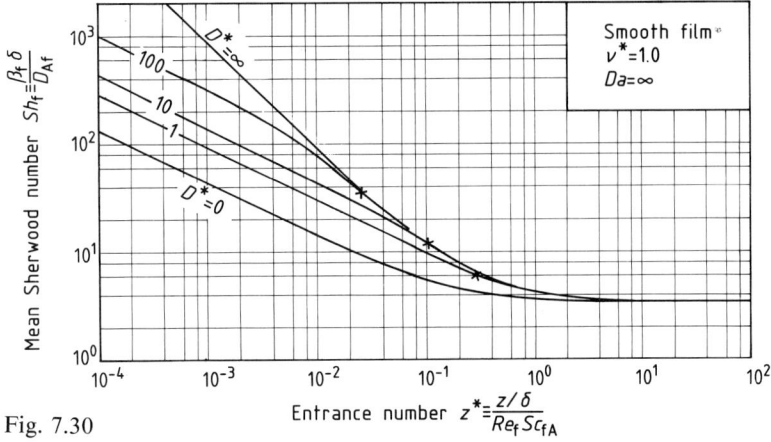

Fig. 7.30

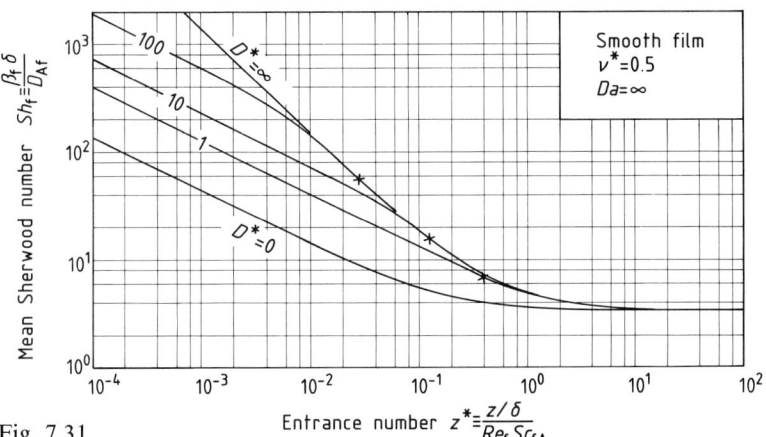

Fig. 7.31

Figures 7.30 to 7.32. Influence of the ratio of diffusion coefficients D^* on mean Sherwood number for mass transfer with an infinitely fast second order reaction in a smooth film for variable values of v^*.
Figure 7.30: $v^* = 1.0$
Figure 7.31: $v^* = 0.5$
Figure 7.32: $v^* = 0.1$.

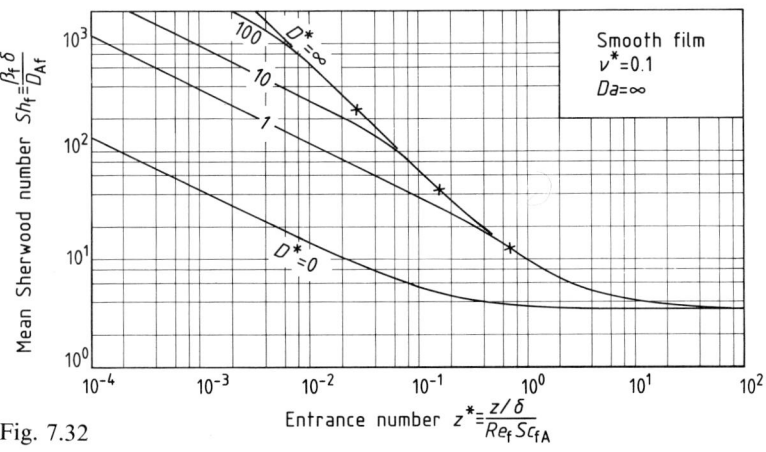

Fig. 7.32

intensive. As has been pointed out, the influence of deformation turbulence is restricted to the range of higher values of z^*. For $z^* \to 0$ the curves shown in Figs. 7.24 to 7.26 are identical with those in Fig. 7.20. For $z^* \to \infty$ the shape of the curves is primarily determined by the influence of deformation turbulence as shown in Fig. 7.23.

The influence of the stoichiometric surplus factor v^* will be discussed on the basis of Figs. 7.27 to 7.29 for an infinitely fast reaction, $Da \to \infty$. Fig. 7.27 has been prepared for mass transfer with homogeneous reactions in a smooth film, Figs. 7.28 and 7.29 for the same process in a wavy film, with the Reynolds number Re_f as a special parameter. The biochemical reaction stops when component B has been completely converted. This is the case, when the entrance number attains the value at the broken line. For higher values of z^* purely physical absorption occurs.

The lowest value of the Sherwood number is obtained for $v^* = \infty$, that is, when there is accumulation of component B. This is the case of purely physical absorption. Consequently, the mean Sherwood number increases with decreasing v^*, i.e., when the concentration of B at the entrance, c_{Bf0}, increases.

For $z^* \to 0$ waves or deformation turbulence have no influence on mass transfer, and therefore the curves in Figs. 7.27 to 7.29 are identical in this range. The influence of deformation turbulence is felt only in the region of $z^* \to \infty$.

In a last set of three figures the influence of the ratio of diffusion coefficients $D^* \equiv D_{Bf}/D_{Af}$ is exemplified. The calculations have been carried out for mass transfer with reaction in a smooth film. For $D^* = 0$ component B is not available at the location of an infinitely fast reaction. The curves for $D^* = 0$ in Figs. 7.30 to 7.32 are identical with that for $v^* = \infty$ in Fig. 7.27, which applies in case of pure physical absorption.

An upper limit for the mean Sherwood number is obtained for $D^* = \infty$. In this case the diffusion of component B occurs infinitely fast compared with that of component A. The reaction occurs very close to the interface. The Sherwood number is proportional to z^{*-1}, which is a linear function of $Re_f Sc_{fA}(z/\delta)$. The factor of proportionality increases with decreasing values of v^*, i.e., with increasing concentration of c_{Bf0} at the inlet.

7.5.3.2 Mass Transfer with First Order Reactions

From the differential equation, Eq. (7.60), it follows, that for mass transfer with a first order reaction in a smooth film the Sherwood number is a function of the entrance number z^* and Damköhler number Da. In Fig. 7.33 the enhancement factor is presented. With increasing values of z^* the Sherwood number increases steadily and approaches a constant value for $z^* \to \infty$. This value is only a function of Da. This result is obtained when convective transport becomes the controlling step. For $Da = 10^4$ the maximum value of β_{fr}/β_f is almost 28. For a wavy film the enhancement factor becomes much smaller than for a film with a smooth surface. This is evident when the results shown in Fig. 7.34 are compared with those in Fig. 7.33. The rather small enhancement is due to the fact, that the mass transfer coefficient for wavy films is already so great, that it can only be slightly increased by the homogeneous reaction. The maximum value of β_{fr}/β_f for a Damköhler number of 10^4 is 4.5. This is only 16% of the value obtained for the same process in a liquid film with a smooth surface.

7.5.3.3 Comparison with Other Results

For comparison with results from GILLILAND et al. [7.30] the curves presented in Fig. 7.20 have been recalculated and plotted in Fig. 7.35 as enhancement factor as a function of $Da\, z^*$. The enhancement factor for a second order reaction is a function of the entrance number z^*, the Damköhler number Da, the ratio of diffusivities D^*,

7.5 Liquid Film Mass Transfer with Biochemical Reactions

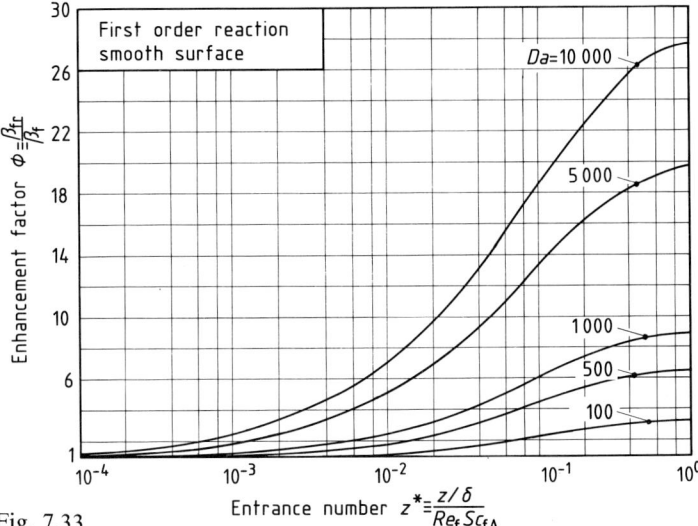

Fig. 7.33

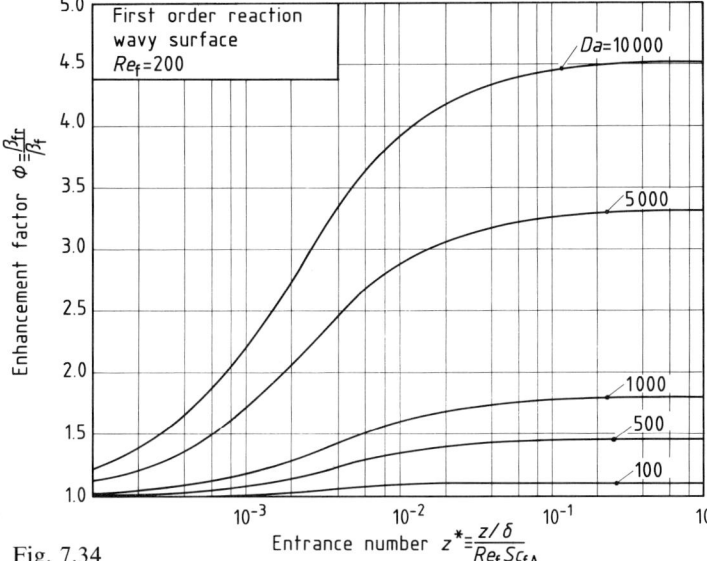

Figures 7.33 and 7.34. Enhancement factor ϕ for mass transfer with first order reaction versus entrance number z^* for various values of Damköhler number Da.
Figure 7.33: smooth film
Figure 7.34: wavy film with $Re_f = 200$.

Fig. 7.34

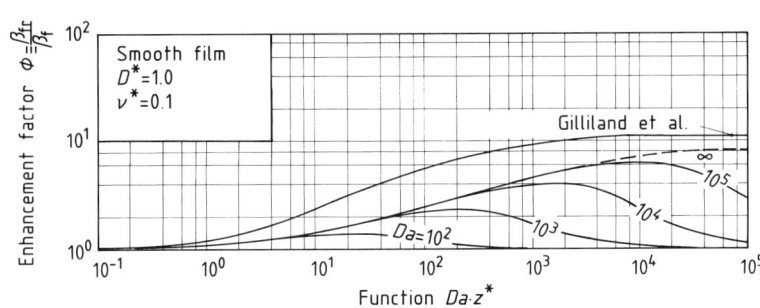

Figure 7.35. Enhancement factor ϕ for mass transfer with first order reaction in a smooth film. Curves for various values of Da have been obtained from the results presented in Fig. 7.20; the upper curve a after GILLILAND et al. [7.30].

and the stoichiometric surplus factor v^*. For the chosen values of D^* and v^* GILLILAND et al. presented a solution, that is independent of Da. These authors solved the following differential equations with boundary conditions:

$$w_\delta \frac{\partial c_{Af}}{\partial z} = D_{Af} \frac{\partial^2 c_{Af}}{\partial y^2} - k_v c_{Af} c_{Bf}, \quad (7.74)$$

$$w_\delta \frac{\partial c_{Bf}}{\partial z} = D_{Bf} \frac{\partial^2 c_{Bf}}{\partial y^2} - \frac{v_B}{v_A} k_v c_{Af} c_{Bf}. \quad (7.75)$$

$z=0$ and $y=0$: $c_{Af} = c_{Af0}$ and $c_{Bf} = c_{Bf0}$
$\phantom{z=0 \text{ and }} y>0$: $c_{Af} = 0$ and $c_{Bf} = c_{Bf0}$

$z>0$ and $y=0$: $c_{Af} = c_{Af0}$ and $\frac{\partial c_{Bf}}{\partial y} = 0$
$\phantom{z>0 \text{ and }} y=\delta$: $c_{Af} = 0$ and $c_{Bf} = c_{Bf0}$

This set of equations applies under the following conditions:

1. The velocity in the film is constant: $w_f = w_\delta$.
2. According to the boundary conditions of the solid wall, $z>0$ and $y=\delta$, component B is continuously fed into the film from the wall; this condition differs considerably from that given for Eqs. (7.50) and (7.51).

These conditions restrict the application of results to $z \to 0$. This explains the fact, that the enhancement factor ϕ is independent of the Damköhler number. From Fig. 7.20 the same result is obtained for $z^* \to 0$. Furthermore, even for $Da = \infty$ the results published by GILLILAND et al. [7.30] differ from the new results because of unrealistic boundary conditions, as pointed out above.

The results published by GILLILAND et al. are reproduced in Fig. 7.36. The enhancement factor is a function of $(Da\, z^* \pi/6)^{1/2}$ and:

$$\phi_\infty \equiv \left(\frac{\beta_{fr}}{\beta_f}\right)_\infty = 1 + \left(\frac{D_{Bf}}{D_{Af}}\right)^{1/2} \frac{v_A}{v_B} \frac{c_{Bf0}}{c_{Bpf}} =$$

$$= 1 + \frac{d^{*1/2}}{v^*}, \quad (7.76)$$

which is the enhancement factor for an infinitely fast reaction, depending on D^* and v^*. As Fig. 7.35 applies to $D^*=1$ and $v^*=0.1$, the enhancement factor for a infinitely fast reaction becomes $\phi_\infty = 11$. The curve given in Fig. 7.36 for $\phi_\infty - 1 = 10$, is the upper curve in Fig. 7.35.

The comparison of the new results with those published by GILLILAND et al. leads to the conclusion: The results of these authors are incorrect; they deliver far too high values for the enhancement factor and neglect a further influence of the Damköhler number.

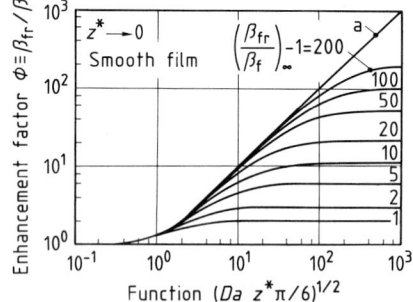

Figure 7.36. Enhancement factor ϕ after GILLILAND et al. [7.30] for mass transfer with first and second order reaction in a smooth film for $z^* \to 0$; curve a for first order reaction.

7.6 References

[7.1] W. NUSSELT: "Die Oberflächenkondensation des Wasserdampfes". Z. VDI (Verein Deutscher Ingenieure) 60 (1916), 541–546.

[7.2] W. NUSSELT: "Der Wärmeaustausch am Berieselungskühler". Z. VDI (Verein Deutscher Ingenieure) 67 (1923), 206–210.

[7.3] R. FALLAH, T. G. HUNTER, and A. W. NASH: "Isothermal flow in liquid wetted wall systems". J. Soc. Chem. Ind. (London) 53 (1934), 369T–379T.

[7.4] S. J. FRIEDMAN and C. O. MILLER: "Liquid films in the viscous flow region". Ind. Eng. Chem. *33* (1941), 885–890.

[7.5] S. S. GRIMLEY: "Liquid flow conditions in packed towers". Trans. Inst. Chem. Eng. *23* (1945), 228–235.

[7.6] H. BRAUER: "Strömung und Wärmeübergang bei Rieselfilmen". VDI-Forsch.-Heft *457* (1956) VDI-Verlag, Düsseldorf.

[7.7] H. BRAUER: "Turbulenz in mehrphasigen Strömungen". Chem. Ing. Tech. *51* (1979) 10, 934–948. See also: "Turbulence in multiphase flow". Ger. Chem. Eng. *3* (1980), 149–163.

[7.8] H. BRAUER: "Grundlagen der Ein- und Mehrphasenströmungen". Verlag Sauerländer, Aarau – Frankfurt am Main, 1971.

[7.9] M. GRAEF: "Über die Eigenschaften zwei- und dreidimensionaler Störungen in Rieselfilmen an geneigten Wänden". Mitteilungen des Max-Planck-Instituts für Strömungsforschung und der AVA, Nr. 36, Göttingen 1966.

[7.10] P. L. KAPIZA: "Wellenströmung dünner Schichten einer viskosen Flüssigkeit". J. Exp. Theor. Phys. *18* (1948), 3–18.

[7.11] A. E. DUKLER and O. P. BERGELIN: "Characteristics of flow in falling liquid films". Chem Eng. Prog. *48* (1952), 557–563.

[7.12] S. PORTALSKI: "Studies of falling liquid film flow. Film thickness on a smooth vertical plate". Chem. Eng. Sci. *18* (1963) 12, 787–804.

[7.13] S. PORTALSKI: "Eddy formation in flow down a vertical plate". Ind. Eng. Chem. Fundam. *3* (1964) 1, 49–53.

[7.14] K. FEIND: "Strömungsuntersuchungen bei Gegenstrom von Rieselfilmen und Gas in lotrechten Rohren". VDI-Forsch.-Heft *481* (1960) VDI-Verlag, Düsseldorf.

[7.15] H. COULON: "Stabilitätsverhalten bei Rieselfilmen". Chem. Ing. Tech. *45* (1973), 362–368.

[7.16] H. D. E. HAMANN: "Der wellige Rieselfilm". Dissertation, Technische Hochschule Aachen, 1972.

[7.17] D. BRAUN, H. ECKSTEIN, and J. W. HIBY: "Messung der Oberflächengeschwindigkeit von Rieselfilmen". Chem. Ing. Tech. *43* (1971) 6, 324–329.

[7.18] A. BAKOPOULOS: "Fluiddynamik und flüssigkeitsseitiger Stoffübergang in berieselten Rohrbündelkolonnen". Dissertation, Technische Universität Berlin, 1976.

[7.19] G. D. FULFORD: "The flow of liquids in thin films". In: "Advances in Chemical Engineering". Vol. 5, pp. 151–236. Academic Press, New York – London, 1964.

[7.20] R. E. EMMERT and R. L. PIGFORD: "A study of gas absorption in falling liquid films". Chem. Eng. Prog. *50* (1954) 2, 87–93.

[7.21] H. BRAUER and D. MEWES: "Stoffaustausch einschließlich chemischer Reaktionen". Verlag Sauerländer, Aarau – Frankfurt am Main, 1971.

[7.22] S. KAMEI and I. OISHI: "Mass and heat transfer in a falling film of wetted wall tower". Mem. Fac. Eng. Kyoto Univ. *17* (1955), 227–239.

[7.23] G. CARRUBA: "Stoff- und Wärmetransport im welligen Rieselfilm". Dissertation, Technische Universität Berlin, 1976.

[7.24] A. P. LAMOURELLE and D. C. SANDALL: "Gas absorption into a turbulent liquid". Chem. Eng. Sci. *27* (1972), 1934–1943.

[7.25] H. HIKITA, K. NAKANISHI, and T. NAKAOKA: "Liquid phase mass transfer in wetted wall columns". Chem. Eng. (Japan) *23* (1959), 459–466.

[7.26] W. MALEWSKI: "Zusammenhang zwischen Stoffübergang und Wellenstruktur beim welligen Rieselfilm". Chem. Ing. Tech. *37* (1965) 8, 815–825.

[7.27] B. NIELANDT: "Untersuchung des Einflusses der Rohrwelligkeit und der Rohrlänge auf den flüssigkeitsseitigen Stoffübergang in gewellten Rohren". Internal report of the Institut für Chemieingenieurtechnik, Technische Universität Berlin, 1978.

[7.28] R. SPILGER: Unpublished internal report of the Institut für Chemieingenieurtechnik, Technische Universität Berlin.

[7.29] H. SCHULTZ: "Stofftransport im Rieselfilm mit homogener chemischer Reaktion". Studienarbeit am Institut für Chemieingenieurtechnik, Technische Universität Berlin, 1982.

[7.30] E. R. GILLILAND, R. F. BADDOUR, and P. L. T. BRIAN: "Gas absorption accompanied by a liquid-phase chemical reaction". AIChE J. *4* (1958) 2, 223–230.

Chapter 8
Mass Transfer Hypotheses

Heinz Brauer

Institut für Chemieingenieurtechnik
Technische Universität Berlin
Berlin (West), Germany

8.1 Introduction
8.2 Boundary Layer Hypothesis
8.3 Film Hypothesis
8.4 Penetration Hypothesis
8.5 Deformation Turbulence Hypothesis
8.6 References

List of Symbols

A_p	m²	interfacial area
c_A	kmol/m³	local partial mol density
c_{Ap}	kmol/m³	partial mol density at the interface
c_{A0}	kmol/m³	partial mol density at $t=0$
D	m²/s	diffusion coefficient
D_1	m²/s	diffusion coefficient for phase 1
D_2	m²/s	diffusion coefficient for phase 2
L	m	length of plate
$\dot{M}_A$	kg/s	mass flux of component A
$\dot{n}_A$	kmol/(s m²)	mol flux density of component A
p	N/m²	total pressure
s_1	m	thickness of fluid layer in phase 1 close to the interface
s_2	m	thickness of fluid layer in phase 2 close to the interface
t	s	time
$\bar{w}$	m/s	mean fluid velocity
w_x	m/s	local fluid velocity in x-direction
w_y	m/s	local fluid velocity in y-direction
w_∞	m/s	fluid velocity at great distance from plate
x	m	length coordinate parallel to plate
y	m	length coordinate perpendicular to plate
β	m/s	mean coefficient of mass transfer
β_1	m/s	mean coefficient of mass transfer in phase 1
β_2	m/s	mean coefficient of mass transfer in phase 2
δ	m	thickness of velocity boundary layer and liquid film
ν	m²/s	kinematic viscosity of fluid
ϱ	kg/m³	mean fluid density
ϱ_A	kg/m³	local partial density of component A
ϱ_{Aw}	kg/m³	partial density of component A at surface of plate
$\varrho_{A\infty}$	kg/m³	partial density of component A at great distance from plate
ϱ_{A1}	kg/m³	partial density of component A in phase 1
ϱ_{A2}	kg/m³	partial density of component A in phase 2
ϱ_{A1p}	kg/m³	partial density of component A at interface of phase 1
ϱ_{A2p}	kg/m³	partial density of component A at interface of phase 2

$p \equiv \dfrac{p}{\varrho w_\infty^2/2}$	pressure number
$Re \equiv \dfrac{w_\infty L}{\nu}$	Reynolds number
$Sc \equiv \nu/D$	Schmidt number
$Sh \equiv \dfrac{\beta \delta}{D}$	Sherwood number
$w_x^* \equiv w_x/w_\infty$	fluid velocity ratio for x-direction
$w_y^* \equiv w_y/w_\infty$	fluid velocity ratio for y-direction
$x^* \equiv x/L$	length coordinate ratio for x-direction
$y^* \equiv y/L$	length coordinate ratio for y-direction
$z^* \equiv \dfrac{L/\delta}{Re\, Sc}$	entrance number
$\xi \equiv \dfrac{\varrho_A - \varrho_{A\infty}}{\varrho_{Aw} - \varrho_{A\infty}}$	local concentration ratio

8.1 Introduction

Whenever the physical and chemical details of a process are not known and understood or when a mathematical description of an understood process would be exceedingly complex and the solution too time-consuming it will be necessary to simplify matters to a point where an obtained result coincides as much as possible with the correct – but unknown – solution. The quality of such a simplified or approximated solution can only be tested by comparison with experimental results. In certain cases, however, the approximated solutions can also be compared with detailed mathematical solutions obtained by using appropriate numerical methods, by employing computers and the necessary time.

Approximate solutions of mass transfer problems are based on mass transfer hypotheses. The quality of the approximate solution will therefore depend on the quality of the hypotheses. Four mass transfer hypotheses will be discussed:

1. Boundary layer hypothesis (PRANDTL, 1904)
2. Film hypothesis (WHITMAN, 1923)
3. Penetration hypothesis (HIGBIE, 1935)
4. Deformation turbulence hypothesis (BRAUER, 1972)

The boundary layer hypothesis was suggested by PRANDTL [8.1] in 1904 for application in the high Reynolds number region. This hypothesis leads to a substantial simplification of the Navier-Stokes equations, so that solutions for the velocity field can be obtained as a precondition for the solution of the mass transfer problem.

The film hypothesis, independently suggested by JABLCZYNSKI and PRZEMYSKI [8.2] in 1912 and WHITMAN [8.3] in 1923, has become known especially through the publication by the latter author. The film hypothesis aims directly at a simplified description of mass transfer in two-phase flow systems. But soon the limitations of this hypothesis became evident. An important step in direction of a more realistic treatment of mass transfer problems seemed to be offered by the penetration hypothesis suggested by HIGBIE [8.4] in 1935. The basic assumption of both film and penetration hypothesis is the allocation of mass transfer resistance to thin layers adjacent to the interface between two fluids. However, the mathematical treatment of mass transfer across these layers follows different lines. The deficiencies of the two hypotheses are due to the fact that the mechanism of momentum, heat, and mass transport is not actually considered.

The deformation turbulence hypothesis was first suggested by BRAUER in 1972, although a detailed discussion including examples for application of this hypothesis was not published until 1979 [8.5]. This hypothesis describes the turbulent transport of momentum, heat, and mass caused by stochastic deformations of the interface between two fluids. Introduction of turbulent diffusivities into the complete differential equations for velocity, temperature, and concentration fields offers a new opportunity for a realistic description of momentum, heat, and mass transfer across the interface of a two-fluid system.

In the following sections the properties of the hypotheses and their applicability will be discussed.

In summary it may be said, that the boundary layer hypothesis is primarily applied to mass transfer problems in one-phase flow. The other hypotheses have been applied to the solution of mass transfer problems in two-phase flow, especially when wavy liquid films and nonspherical bubbles are involved. In such cases, strong mixing effects are observed, which are attributed to turbulence.

8.2 Boundary Layer Hypothesis

The most important step towards a mathematical description of momentum, heat, and mass transport in fluids by PRANDTL [8.1] in 1904 was his concept of boundary layers. This concept is based on the assumption, that resistance to momentum, heat, and mass transfer is restricted to a thin fluid layer close to the interface of a fluid/solid- or fluid/fluid-system. This thin layer, with the thickness δ, is PRANDTL's boundary layer. The boundary layer concept is based on a very careful analysis of momentum transfer in one-phase flow systems.

As PRANDTL pointed out, the applicability and usefulness of a boundary layer hypothesis increases with decreasing thickness δ of the boundary layer. The thickness δ decreases with increasing fluid velocity or Reynolds number Re. Application of the boundary layer hypothesis therefore should be restricted to the high Reynolds number region.

To describe transport processes across the boundary layer PRANDTL made use of the relevant differential equations, which he reduced to the so-called boundary layer equations. The simplifications introduced by PRANDTL will be discussed below.

The example chosen for this discussion is momentum and mass transfer in fluid flow parallel to a flat plate. A detailed discussion of this example has already been given in Sect. 5.3. Fig. 8.1 shows the flow field over a short section of the plate. The curves a, b, and c give the boundary layer thickness for three values of the approaching velocity w_∞. The boundary layer thickness δ is defined by:

$$y = \delta \quad \text{for} \quad w_x = 0.99 \, w_\infty.$$

According to Eq. (5.11) the boundary layer thickness δ is inversely proportional to the Reynolds number Re, defined by:

$$Re \equiv \frac{w_\infty L}{\nu}, \tag{8.1}$$

with L the length of the plate and ν the kinematic viscosity of the fluid. The thickness of the boundary layer decreases therefore with increasing velocity w_∞, decreasing kinematic viscosity, and decreasing length L of the plate. Within the boundary layer the velocity w_x increases from $w_x = 0$ at the wall to $w_x = 0.99 \, w_\infty$ at the outer edge of the boundary layer. For $y > \delta$ it is assumed that $w_x \approx w_\infty$. A velocity gradient, $\partial w_x / \partial y$, and therefore a shear stress τ, occurs only within the boundary layer. Outside potential flow may be assumed. Furthermore, the boundary layer concept includes the assumption that the pressure p is constant within the entire flow field within the shear as well as the potential flow region.

The flow field is correctly described by the Navier-Stokes equations, derived in this case from Eqs. (2.1) and (2.2). In the dimensionless form one obtains:

$$w_x^* \frac{\partial w_x^*}{\partial x^*} + w_y^* \frac{\partial w_x^*}{\partial y^*} = -\frac{\partial p^*}{\partial x^*} + \frac{1}{Re}\left[\frac{\partial^2 w_x^*}{\partial x^{*2}} + \frac{\partial^2 w_x^*}{\partial y^{*2}}\right] \tag{8.2}$$

$$\left\{ 1 \quad 1 \quad \delta^* \quad \frac{1}{\delta^*} \quad 0 \quad \delta^{*2} \quad 1 \quad + \quad \frac{1}{\delta^{*2}} \right\}$$

$$w_x^* \frac{\partial w_y^*}{\partial x^*} + w_y^* \frac{\partial w_y^*}{\partial y^*} = -\frac{\partial p^*}{\partial y^*} + \frac{1}{Re}\left[\frac{\partial^2 w_y^*}{\partial x^{*2}} + \frac{\partial^2 w_y^*}{\partial y^{*2}}\right] \tag{8.3}$$

$$\left\{ 1 \quad \delta^* \quad \delta^* \quad 1 \quad 0 \quad 1\delta^{*2} \quad \delta^* \quad \frac{1}{\delta^*} \right\}$$

8.2 Boundary Layer Hypothesis

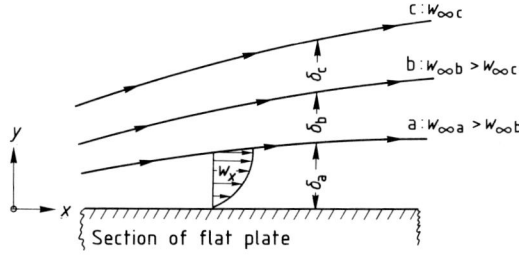

Figure 8.1. Fluid flow parallel to a section of a flat plate.

Continuity equation:

$$\frac{\partial w_x^*}{\partial x^*} + \frac{\partial w_y^*}{\partial y^*} = 0 \qquad (8.4)$$
$$\{\ 1 \qquad\quad 1\ \}$$

These equations have already been introduced by Eqs. (5.1) to (5.3). The necessary simplifications to obtain the boundary layer equations result from the following proportionalities [8.6]:

$$x^* \equiv \frac{x}{L} \sim 1 \qquad w_x^* \equiv \frac{w_x}{w_\infty} \sim 1 \qquad \delta^* \equiv \frac{\delta}{L} \ll 1$$

$$y^* \equiv \frac{y}{L} \sim \delta^* \qquad w_y^* \equiv \frac{w_y}{w_\infty} \sim \delta^* \qquad Re \sim \frac{1}{\delta^{*2}}$$

$$\frac{\partial w_x^*}{\partial x^*} \sim \frac{1}{1} = 1 \qquad \frac{\partial^2 w_x^*}{\partial x^{*2}} \sim \frac{1}{1^2} = 1 \qquad \frac{\partial^2 w_y^*}{\partial x^{*2}} \sim \frac{\delta^*}{1^2} = \delta^*$$

$$\frac{\partial w_x^*}{\partial y^*} \sim \frac{1}{\delta^*} \qquad \frac{\partial^2 w_x^*}{\partial y^{*2}} \sim \frac{1}{\delta^{*2}} \qquad \frac{\partial^2 w_y^*}{\partial y^{*2}} \sim \frac{\delta^*}{\delta^{*2}} = \frac{1}{\delta^*}$$

$$\frac{\partial w_y^*}{\partial x^*} \sim \frac{\delta^*}{1} = \delta^* \qquad \frac{\partial p^*}{\partial x^*} \approx 0$$

$$\frac{\partial w_y^*}{\partial y^*} \sim \frac{\delta^*}{\delta^*} = 1 \qquad \frac{\partial p^*}{\partial y^*} \approx 0$$

These proportionalities are indicated in Eqs. (8.2) to (8.4) below the relevant terms. Taking these proportionalities into account the continuity equation remains uneffected, while Eq. (8.3) is resolved, and Eq. (8.2) is simplified. The boundary layer equations thus appear as:

$$w_x^* \frac{\partial w_x^*}{\partial x^*} + w_y^* \frac{\partial w_x^*}{\partial y^*} = \frac{1}{Re} \frac{\partial^2 w_x^*}{\partial y^{*2}}, \qquad (8.5)$$

$$\frac{\partial w_x^*}{\partial x^*} + \frac{\partial w_y^*}{\partial y^*} = 0. \qquad (8.6)$$

From these equations, under appropriate boundary conditions, the local velocities w_x^* and w_y^* are obtained, which are required for the differential equation for the concentration field. The complete equation for the local concentration already has been given as Eq. (5.22) and will be repeated here:

$$w_x^* \frac{\partial \xi}{\partial x^*} + w_y^* \frac{\partial \xi}{\partial y^*} = \frac{1}{Re\,Sc} \left[\frac{\partial^2 \xi}{\partial x^{*2}} + \frac{\partial^2 \xi}{\partial y^{*2}} \right]$$
$$\left\{ 1 \quad 1 \ +\delta^*\frac{1}{\delta^*} \qquad \delta^{*2}\quad 1 \qquad \frac{1}{\delta^{*2}} \right\}$$
$$(8.7)$$

With the proportionalities given above and $\xi \sim 1$ the boundary layer equation for the concentration field is as follows:

$$w_x^* \frac{\partial \xi}{\partial x^*} + w_y^* \frac{\partial \xi}{\partial y^*} = \frac{1}{Re\,Sc} \frac{\partial^2 \xi}{\partial y^{*2}}. \qquad (8.8)$$

The set of boundary layer equations presented as Eqs. (8.5), (8.6), and (8.8) was first solved by POHLHAUSEN [8.7] in 1921. A slightly improved solution was obtained by LÉVÊQUE [8.8] in 1928. The result is given through Eq. (5.33). Comparison with the correct solution presented by Eq. (5.30) reveals the deficiencies of the boundary layer solution. In the low Reynolds number region the boundary layer hypothesis does not apply, as was explained in detail in Sects. 5.4 and 5.5. The low Reynolds number region includes the front end of the plate.

In the high Reynolds number region the boundary layer hypothesis does not correctly predict the influence of the Schmidt number on the Sherwood number. Furthermore, the range to which the boundary layer solution applies cannot be stated satisfactorily without a set of complete differential equations or experimental data.

The boundary layer hypothesis should thus only be used to obtain an approximate solution in the turbulent flow regime and in the laminar flow regime, when the Reynolds number approaches the critical value.

8.3 Film Hypothesis

The film hypothesis as presented by JABLCZYNSKI and PRZEMYSKI [8.2] and independently by WHITMAN [8.3] assumes that the resistance to mass transfer is concentrated in thin fluid layers on both sides of the interface between two fluids. According to this assumption concentration profiles result, which will be explained by means of Fig. 8.2.

The mass flux $\dot{M}_A$ of component A is transferred from the gas to the liquid. In the gas phase the local concentration is given by ϱ_{A2}, with ϱ_{A2p} on the gas side of the interface. The corresponding values in the liquid are ϱ_{A1} and ϱ_{A1p}. The concentrations at the interface are given by Henry's law, as explained in Sect. 6.7.1. As the diffusion coefficient for the gas phase, D_2, is much greater than that in the liquid phase, D_1, the concentration gradient in the gas phase, $(\partial \varrho_A/\partial y)_2$, must be much smaller than in the liquid phase, $(\partial \varrho_A/\partial y)_1$, according to Fick's law:

$$\dot{M}_A = -D_1 A_p \left(\frac{\partial \varrho_A}{\partial y}\right)_1 , \tag{8.9}$$

$$\dot{M}_A = -D_2 A_p \left(\frac{\partial \varrho_A}{\partial y}\right)_2 . \tag{8.10}$$

The real concentration profiles as shown in Fig. 8.2a are simplified by the film hypothesis to give profiles as shown in Fig. 8.2b. The thickness of the layer in which resistance to mass transfer occurs is s_2 for the gas phase and s_1 for the liquid phase. In these layers with mass transfer resistance the concentration varies from the value at the interface to the value in the bulk of the fluid. In

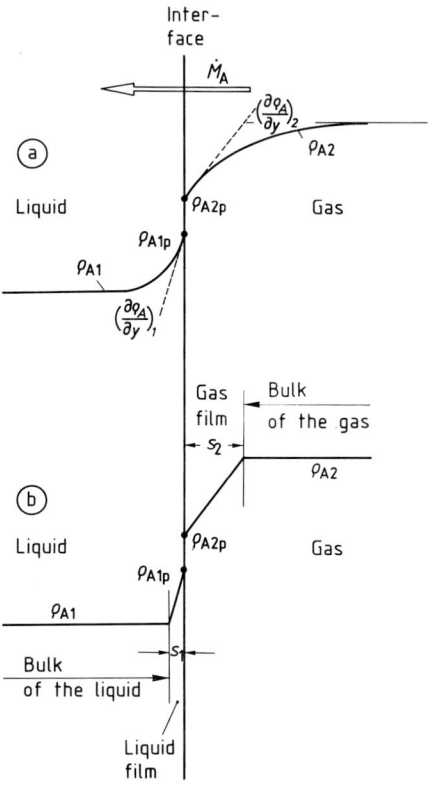

Figure 8.2. Explanation of the film hypothesis; a: concentration profiles for a realistic case; b: concentration profiles according to the assumptions of the film-hypothesis.

the bulk of the fluids the concentrations are generally assumed to be constant. This is in accordance with the assumption of complete mixing in the bulk of the fluids due to turbulence.

In the films on both sides of the interface a "stagnant fluid" is assumed, resulting in a purely molecular mass transport and a constant concentration gradient.

Mass transport through the interface can be calculated by the following two equations:

$$\dot{M}_A = \beta_2 A_p (\varrho_{A2} - \varrho_{A2p}), \quad (8.11)$$

$$\dot{M}_A = \beta_1 A_p (\varrho_{A1p} - \varrho_{A1}). \quad (8.12)$$

The mass transfer coefficients β_2 and β_1 are determined by equations expressing the fundamental statement of the film hypothesis:

$$\beta_2 \equiv \frac{D_2}{s_2}, \quad (8.13)$$

$$\beta_1 \equiv \frac{D_1}{s_1}. \quad (8.14)$$

The thickness of the films must be determined by experiment. By introducing Eqs. (8.13) and (8.14) into the relevant equations for mass flux $\dot{M}_A$ one obtains:

$$\begin{aligned}\dot{M}_A &= \frac{D_2}{s_2} A_p (\varrho_{A2} - \varrho_{A2p}), \\ \dot{M}_A &= \frac{D_1}{s_1} A_p (\varrho_{A1p} - \varrho_{A1}).\end{aligned} \quad (8.15)$$

The crucial assumption for the film hypothesis is expressed by the linear relationship between the mass transfer coefficient β or mass flux $\dot{M}_A$ and the diffusion coefficient D. This relationship has also been found experimentally. Consequently, this was accepted as proof for the validity of the film hypothesis. The thickness s of the film is a function of the properties of the fluids and the flow field, which must be determined by experiment.

A linear relationship between β and D also follows from Eq. (7.15) for $z^* \to \infty$ leading to $Sh = 3.4145$ and

$$\beta_f = 3.4145 \frac{D_f}{\delta}. \quad (8.16)$$

But this result may not be taken as proof of the film hypothesis. Eq. (8.16) applies to conditions of an extremely small concentration gradient, $\partial \varrho_A / \partial y$, thus obtaining an extremely large thickness s of the mass transfer film. This is in contradiction to the film hypothesis. Furthermore, the concept of a „stagnant fluid" within this layer is by no means justified.

In real mass transfer processes there are no situations to which the film hypothesis applies. There will always be strong arguments against the film hypothesis. In the early days of chemical engineering the film hypothesis was helpful, whereas today it must be considered as obsolete.

8.4 Penetration Hypothesis

The penetration hypothesis will be applied to the case of mass transfer in wavy liquid films. This case has been considered in Chapter 7. Fig. 8.3 shows in a simplified way the mass transfer process according to the penetration hypothesis. The behavior of a small volume of liquid will be considered.

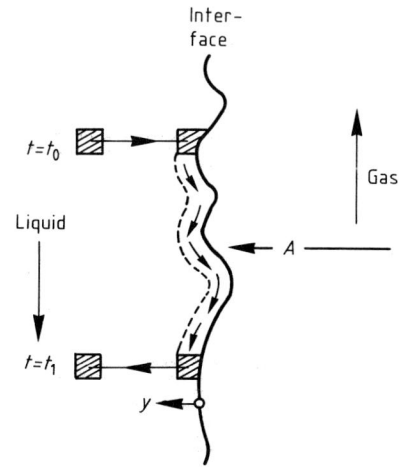

Figure 8.3. Explanation of the penetration hypothesis.

At time $t=t_0$ this fluid element leaves the bulk of the liquid and penetrates the surface of the wavy liquid film. The fluid element remains at the surface till at $t=t_1$ it returns into the bulk of the liquid.

At the surface the fluid element is exposed for some time, the residence time t_1-t_0, to the gas stream, from which component A is absorbed. This absorption is a non-steady-state process. It may be tentatively assumed that all fluid elements in the liquid film have the same velocity, implying that there is no relative velocity between fluid elements, and that the residence time is same for all elements. Based on these assumption, the following differential equation:

$$\frac{\partial c_A}{\partial t} = D \frac{\partial^2 c_A}{\partial y^2}, \qquad (8.17)$$

describes the concentration field in a fluid at rest. With the boundary conditions:

$t=t_0$ and $y>0$: $\quad c_A=c_{A0}$
$\qquad\qquad y=0$: $\quad c_A=c_{Ap}$
$\qquad\qquad y=\infty$: $\quad c_A=c_{A0}$,

the solution is:

$$\frac{c_A-c_{A0}}{c_{Ap}-c_{A0}} = \mathrm{erfc}\left(\frac{y^2}{4Dt}\right), \qquad (8.18)$$

with erfc being the complement of the error function:

$$\mathrm{erfc}(F) = 1 - \frac{2}{\sqrt{\pi}} \int_0^F e^{-F^2} dF. \qquad (8.19)$$

In these equations c_A is the local partial mol density of component A, c_{Ap} at the interface, c_{A0} at time $t=t_0$ the concentration everywhere in the liquid, y the distance from the free surface of the liquid, and D the coefficient of diffusion for component A in the liquid. Fig. 8.4 shows a qualitative graphical presentation of Eq. (8.18).

By introduction into Fick's law, which is given by:

$$\dot{n}_A = -D \left(\frac{\partial c_A}{\partial y}\right)_{y=0}, \qquad (8.20)$$

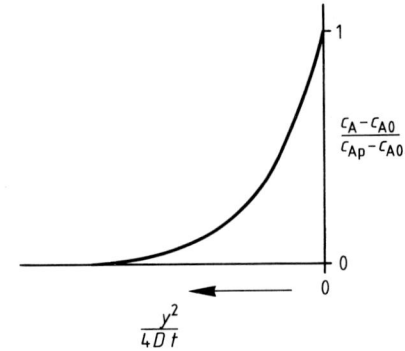

Figure 8.4. Concentration profile according to Eq. (8.18) as a result of the penetration hypothesis.

the concentration gradient obtained from Eq. (8.18), one obtains:

$$\dot{n}_A = \left(\frac{4D}{\pi \Delta t}\right)^{1/2} (c_{Ap}-c_{A0}). \qquad (8.21)$$

The first term on the right hand side of this equation may be interpreted as the mass transfer coefficient β:

$$\beta \equiv \left(\frac{4D}{\pi \Delta t}\right)^{1/2}. \qquad (8.22)$$

But this procedure is not quite correct. The case considered is mass transfer in a fluid at rest, while β is the coefficient of convective mass transfer.

According to Eqs. (8.21) and (8.22) the mass flux density $\dot{n}_A$ and the mass transfer coefficient β are functions of the square root of D. Experimental data in the high Reynolds number region confirm this relationship. It should be observed, however, that in this case a theoretical description of mass transfer in a fluid at rest is compared with experimental results for convective mass transfer. Such a comparison is not acceptable.

The attractiveness of the penetration hypothesis is due, however, to the following equation, which can be derived for the mass transfer coefficient β. Dividing the length L of the path, along which the fluid element travels at the surface of the liquid film, by the residence time Δt, one obtains a velocity. For the case of smooth liquid films the surface velocity is $1.5\bar{w}$, with $\bar{w}$ being the mean film velocity:

$$\frac{L}{\Delta t} = \frac{3}{2}\bar{w}. \tag{8.23}$$

By inserting this into Eq. (8.22) the result in dimensionless writing will be:

$$Sh = 1.382 z^{*-1/2}, \tag{8.24}$$

with

$$Sh \equiv \frac{\beta \delta}{D} \quad \text{Sherwood number} \tag{8.25}$$

$$z^* \equiv \frac{L/\delta}{Re\,Sc} = \frac{L/\delta}{\bar{w}\delta/D} \quad \text{entrance number.} \tag{8.26}$$

Eq. (8.24) can be obtained as an exact solution of the differential equation for the concentration field in a liquid film with a smooth surface [8.9]. With $z^* \to 0$ Eq. (8.24) follows from Eq. (7.15).

Essentially, it may be concluded, that the penetration hypothesis is not applicable to mass transfer in wavy liquid films, as initially assumed. It delivers correct results for mass transfer in films with a smooth surface for the special case of $z^* \to 0$. For smooth films, however, no mass transfer hypothesis is necessary, because a solution for the general case, $0 \leq z^* \leq \infty$, is available, as demonstrated in Sect. 7.3.

8.5 Deformation Turbulence Hypothesis

The deformation turbulence hypothesis fulfills some of the expectations which film and penetration hypothesis evoked. This new hypothesis is applicable to the mathematical description of momentum, heat, and mass transfer across the interface between two fluids, when this interface experiences stochastic deformations. This kind of deformation of the interface is the result of a fluid-mechanic instability of the fluid. The motions carried out by the interface are normal to the mean time position of the interface. Movement of the interface induces motions in the fluids bordering the interface.

Deformations of the interface with stochastic nature are typical for wavy liquid films, bubbles, and drops, when the diameter surpasses a certain value.

Stochastic motions induced by deformation of the interface are turbulent by nature [8.5]. This kind of turbulence has some properties in common with Reynolds turbulence and of course other unique characteristics. As the stochastic fluid motions are induced by the interface, the production of deformation turbulence has a maximum at the interface. With increasing distance from the interface deformation turbulence is damped.

Momentum, heat, and mass transfer caused by deformation turbulence can be described by means of the relevant turbulent diffusivities: ε_τ for momentum transfer, ε_q for heat transfer, and ε_m for mass transfer. Equations for these diffusivities have been presented [8.5], [8.10], [8.11]. As shown in Chapters 6 and 7, some of the equations have been applied to the solution of mass transfer problems.

The mathematical treatment of mass transfer through the interface of nonspherical bubbles may serve as an example for the application of the deformation turbulence hypothesis. As shown in Sect. 6.6.2, the turbulent mass diffusivity ε_m is introduced into the differential equation for the concentration field. The equation is solved by numerical methods. This procedure is exactly similar to that applied in the case of convective mass transfer with conventional Reynolds turbulence. The difference is only given by the equations for ε_m.

The mathematical treatment of mass transfer problems on the basis of the deformation turbulence hypothesis is the best method available for the solution of mass transfer problems when deformation of the interface causes stochastic motions in the fluids.

8.6 References

[8.1] L. PRANDTL: "Über Flüssigkeitsbewegung bei kleiner Reibung. Verhandlungen des dritten internationalen Mathematiker-Kongresses in Heidelberg vom 8. bis 13. August 1904". Verlag B.G. Teubner, Leipzig, 1905.

[8.2] K. JABLCZYNSKI and S. PRZEMYSKI: "Sur les processes des systèmes hétérogènes". J. Chim. Phys. *10* (1912), 241–288.

[8.3] W. G. WHITMAN: "The two-film theory of gas absorption". Chem. Metall. Eng. *29* (1923), 146–148.

[8.4] R. HIGBIE: "The rate of absorption of a pure gas into a still liquid during short periods of exposure". Trans. Am. Inst. Chem. Eng. *31* (1935), 365–389.

[8.5] H. BRAUER: "Turbulenz in mehrphasigen Strömungen". Chem. Ing. Tech. *51* (1979) 10, 934–948.

[8.6] H. SCHLICHTING: "Grenzschicht-Theorie". Verlag G. Braun, Karlsruhe, 1965.

[8.7] E. POHLHAUSEN: "Der Wärmeaustausch zwischen festen Körpern und Flüssigkeit mit kleiner Reibung und Wärmeleitung". Z. Angew. Math. Mech. *1* (1921), 115–121.

[8.8] M. A. LÉVÊQUE: "Les lois de la transmission de la chaleur par convection". Ann. Mines *12* (1928), 201–299.

[8.9] H. BRAUER: "Stoffaustausch einschließlich chemischer Reaktionen". Verlag Sauerländer, Aarau – Frankfurt am Main, 1971.

[8.10] G. CARRUBBA: "Stoff- und Wärmetransport im welligen Rieselfilm". Dissertation, Technische Universität Berlin, 1976.

[8.11] H. GLAESER and H. BRAUER: "Berechnung des Impuls- und Stofftransports durch die Grenzfläche einer formveränderlichen Blase". VDI (Verein Deutscher Ingenieure)-Forsch.-Heft *581* (1977) VDI-Verlag, Düsseldorf.

Chapter 9

Analogy of Momentum, Heat and Mass Transfer

Heinz Brauer

Institut für Chemieingenieurtechnik
Technische Universität Berlin
Berlin (West), Germany

9.1 Introduction
9.2 The Fundamental Equations for Analogy of Momentum, Heat, and Mass Transfer
9.3 Analogy Equation for Momentum and Heat (Mass) Transfer in Tubes
9.4 Analogy of Heat and Mass Transfer
9.5 References

9.1 Introduction

Analogy of momentum, heat, and mass transfer proved to be a useful tool for considering transport phenomena, as long as the fundamentals were not yet fully understood. With the increased understanding of transport processes, the analogy method loses its attractiveness.

The importance of the analogy method was established, when its application seemed to enable the transformation of available equations for momentum transfer into those for heat or mass transfer. But the high expectations soon led to disappointments. Application of the analogy method requires a thorough understanding of the transport processes.

The deeper insight into momentum, heat, and mass transfer processes, gained in the last decades, makes application of the analogy method, to momentum and heat as well as for momentum and mass, obsolete The correct establishment of the specific analogy equations for these cases requires the exact solution of the momentum and the heat or the mass transfer problem. With the equations for heat or mass transfer already available, there remains no necessity for applying the analogy method. Consequently, the analogy of momentum and heat and of momentum and mass has become less attractive than in former times.

The analogy of heat and mass transfer is, however, still an interesting factor in transport phenomena. But its application to the solution of technical problems is only of limited value, because mass transfer equipment is in most cases quite different from heat transfer equipment.

Although the analogy method has become of minor importance, a basic understanding seems to be desirable in order to exploit the remaining possibilities for situations in which they could be helpful in the solution of technical problems.

9.2 The Fundamental Equations for Analogy of Momentum, Heat, and Mass Transfer

Basic equations for analogy of convective transfer of momentum, heat, and mass have been developed in Sect. 1.5.1. A summary of the final equations is shown below.

Momentum transfer:

$$\frac{1}{8}\psi_z Re = -2\left(\frac{dw_z^*}{dr^*}\right)_{wz}. \qquad (9.1)$$

Heat transfer:

$$Nu_z = -2\left(\frac{dT^*}{dr^*}\right)_{wz}. \qquad (9.2)$$

Mass transfer:

$$Sh_z = -2\left(\frac{d\varrho_A^*}{dr^*}\right)_{wz}. \qquad (9.3)$$

The dimensionless parameters are defined as follows:

$$\psi_z \equiv \frac{(dp/dz)d}{\varrho\bar{w}^2/2} \quad \text{local friction factor} \qquad (9.4)$$

$$Re \equiv \frac{\bar{w}d}{\nu} \quad \text{Reynolds number} \qquad (9.5)$$

$$Nu_z \equiv \frac{\alpha_z d}{\lambda} \quad \text{local Nusselt number} \qquad (9.6)$$

$$Sh_z \equiv \frac{\beta_z d}{D} \quad \text{local Sherwood number} \qquad (9.7)$$

$$w_z^* \equiv w_z/\bar{w} \quad \text{local velocity} \qquad (9.8)$$

$$T^* \equiv \frac{T}{\Delta T} \quad \text{local temperature} \qquad (9.9)$$

$$\varrho_A^* \equiv \frac{\varrho_A}{\Delta\varrho_A} \quad \text{local concentration} \qquad (9.10)$$

In these terms dp/dz is the local gradient of the pressure p in the direction of the length coordinate z, d tube diameter, $\bar{w}$ mean fluid velocity, ϱ fluid density, ν kinematic viscosity of fluid, α_z local heat transfer coefficient, λ heat conductivity, β_z local mass transfer coefficient, D diffusion coefficient, w_z local fluid velocity, T local temperature, ΔT mean fluid temperature difference, ϱ_A local partial density of diffusing component A, and $\Delta\varrho_A$ mean concentration difference.

The above Eqs. (9.1) to (9.3) have been derived in Sect. 1.5.1 for convective transport in tubes. The form of the equations, however, does not change, when derived for other geometrical systems. The equations are generally applicable. They state that the local momentum transfer number $\psi_z Re/8$ is proportional to the dimensionless velocity gradient; the local heat transfer number Nu_z is proportional to the dimensionless temperature gradient; and the local mass transfer number Sh_z is proportional to the dimensionless concentration gradient.

When analogy between momentum and heat transfer exists the following relation must apply:

$$Nu_z = \frac{1}{8} \psi_z Re \frac{(dT^*/dr^*)_{wz}}{(dw_z^*/dr^*)_{wz}}. \quad (9.11)$$

The corresponding equation for analogy between momentum and mass transfer is:

$$Sh_z = \frac{1}{8} \psi_z Re \frac{(d\varrho_A^*/dr^*)_{wz}}{(dw_z^*/dr^*)_{wz}}. \quad (9.12)$$

The above equations may also be written in the following way:

$$\psi_z = 8 \frac{Nu_z}{Re} \frac{(dw_z^*/dr^*)_{wz}}{(dT^*/dr^*)_{wz}}, \quad (9.13)$$

$$\psi_z = 8 \frac{Sh_z}{Re} \frac{(dw_z^*/dr^*)_{wz}}{(d\varrho_A^*/dr^*)_{wz}}. \quad (9.14)$$

The analogy between heat and mass leads to the general equations:

$$Nu_z = Sh_z \frac{(dT^*/dr^*)_{wz}}{(d\varrho_A^*/dr^*)_{wz}}, \quad (9.15)$$

$$Sh_z = Nu_z \frac{(d\varrho_A^*/dr^*)_{wz}}{(dT^*/dr^*)_{wz}}. \quad (9.16)$$

These are the fundamental equations for analogy between momentum and heat, for momentum and mass, as well as for heat and mass transfer. Application requires determination of the local dimensionless gradients for velocity, temperature, and concentration for "analog conditions". As there are numerous analog conditions, there must be numerous analog equations, each one specific for the conditions of the flow, temperature, and concentration field.

Application of the fundamental equations of analogy will be discussed for a specific example in the following sections.

9.3 Analogy Equation for Momentum and Heat (Mass) Transfer in Tubes

The fundamental equations of analogy will be applied to transfer processes in tubes. It will be assumed that the velocity field is established. This means that the dimensionless velocity profile, especially the velocity gradient as indicated in Fig. 1.13, does not vary with length of the tube and with Reynolds number. With this assumption the analogy of momentum and heat (mass) requires that the temperature (concentration) field is also established. The dimensionless temperature (concentration) profile does not vary with length of tube, Reynolds and Prandtl (Schmidt) number. Analogy of momentum and mass transfer in laminar tube flow therefore exists when the friction factor law for fluid flow is given by:

$$\psi = \frac{64}{Re}, \quad (9.17)$$

and the mass transfer law by:

$$Sh = 3.66. \quad (9.18)$$

These equations have been presented in Chapter 3 as Eq. (3.14) for ψ and as Eq. (3.39) for Sh, when $z^* \to \infty$. For conditions of established profiles the gradients are obtained by numerical solution of the relevant differential equations:

$$\left(\frac{dw_z^*}{dr^*}\right)_{wz} = -4 \quad (9.19)$$

$$\left(\frac{dT^*}{dr^*}\right)_{wz} = \left(\frac{d\varrho_A^*}{dr^*}\right)_{wz} = -1.83 \quad (9.20)$$

Introducing these terms into Eqs. (9.11) and (9.12) yields:

$$\frac{Nu}{\psi Re/8} = \frac{Sh}{\psi Re/8} = \frac{-1.83}{-4.00} = 0.4575. \quad (9.21)$$

This is the specific equation for analogy between momentum and heat (mass) transfer in established laminar tube flow. Eq. (9.21) can be rewritten as:

$$Nu = Sh = \frac{0.4575}{8} \psi Re. \quad (9.22)$$

For the stated conditions the friction factor ψ is given by Eq. (3.14), $\psi = 64/Re$. From Eq. (9.22) thus follows:

$$Nu = Sh = 3.66. \quad (9.23)$$

In fact, this result has been used for the establishment of the specific analogy equations given by Eq. (9.22). The analogy equations therefore did not help to simplify the mathematical procedure for calculating the Sherwood number. The analogy of momentum and heat (mass) transfer, therefore, is of no practical value.

The analogy of momentum and heat (mass) transfer gained importance when, due to insufficient knowledge about convective transfer processes, it was assumed that there is one analogy equation applicable to various conditions. For example, Eq. (9.22) should not only be applicable to established laminar, but also to established turbulent flow conditions. For turbulent tube flow $\psi = 0.3164/Re^{1/4}$, inserted into Eq. (9.22), leads to:

$$Nu = Sh = \frac{0.4575}{8} \frac{0.3164}{Re^{1/4}} Re = 0.018 Re^{3/4}. \quad (9.24)$$

An empirical equation for the Nusselt and Sherwood number may be written as [9.1]:

$$\frac{Nu}{Pr^{0.42}} = \frac{Sh}{Sc^{0.42}} = 0.037 Re^{3/4}. \quad (9.24)$$

There is a certain similarity between the last two equations, but no agreement that could be of any practical importance. Science has devalued the concept of analogy for momentum and heat (mass) transfer.

9.4 Analogy of Heat and Mass Transfer

The fundamental equations for analogy of heat and mass transfer are Eqs. (9.15) and (9.16). The dimensionless gradients for temperature and concentration will be determined for transfer processes in laminar flow through a straight tube with a smooth wall. The differential equation and the boundary conditions are given in Sect. 3.3. For purposes of simplification it will be assumed, that the transfer processes are restricted to the entrance section of the tube, so that $z^* \to 0$. For this condition the following equations are obtained.

$$\left(\frac{d\varrho_A^*}{dr^*}\right)_{wz} = -0.5385 z_m^{*-1/3} =$$

$$= -0.5385 \left(\frac{Re\,Sc}{L/d}\right)^{1/3}, \quad (9.25)$$

$$\left(\frac{dT^*}{dr^*}\right)_{wz} = -0.5385\, z^{*-1/3} =$$
$$= -0.5385 \left(\frac{Re\, Pr}{L/d}\right)^{1/3}. \qquad (9.26)$$

Introduction into Eq. (9.15) leads to:

$$Nu_z = Sh_z \left(\frac{Pr}{Sc}\right)^{1/3}. \qquad (9.27)$$

For the considered conditions the equation for the local Sherwood number Sh_z is given by [9.1]:

$$Sh_z = 1.615 \left(\frac{Re\, Sc}{L/d}\right)^{1/3}. \qquad (9.28)$$

Introducing this into Eq. (9.27) yields:

$$Nu_z = 1.615 \left(\frac{Re\, Pr}{L/d}\right)^{1/3}. \qquad (9.29)$$

By comparing Eq. (9.28) with Eq. (9.29) the following "analogy rule" may be deduced: Transformation of a heat transfer equation into the analog mass transfer equation requires the substitution of the Nusselt number by the Sherwood number and the Prandtl number by the Schmidt number. When the equation for the Sherwood number is available, substitution of Sh_z by Nu_z and Sc by Pr delivers the heat transfer equation.

The simplicity of the "analogy rule" for analogy of heat and mass transfer guarantees its wide application, when analogy really exists. In many technical situations analogy does not exist. This is due to the fact that heat transfer equipment differs quite remarkably from mass transfer equipment, so that geometric analogy fails.

9.5 References

[9.1] H. BRAUER: "Stoffaustausch einschließlich chemischer Reaktionen". Verlag Sauerländer, Aarau–Frankfurt am Main, 1971.

Chapter 10
Gas Solubilities in Biomedia

Adrian Schumpe

Fachbereich Chemie – Technische Chemie
Universität Oldenburg
Oldenburg, Federal Republic of Germany

10.1 Introduction
10.2 Modes of Expressing Gas Solubility
10.3 Experimental Methods for Gas Solubility Determination
10.4 Estimation of Gas Solubilities
10.4.1 Pressure and Temperature Effects
10.4.2 Composition of Media
10.4.2.1 Electrolytes
10.4.2.2 Organic Solutes
10.4.2.3 Adsorption Effects
10.5 Estimated and Measured Oxygen Solubilities in Culture Media
10.6 Recommendations
10.7 References

List of Symbols

c	g/L	concentration
H_j	L/mol	ion specific constant in Eq. (10.21)
I	mol/L	ionic strength
K_s	L/mol	Sechenov constant in Eq. (10.18)
K_i	L/g	parameter defined by Eq. (10.23)
m_i	L/g	parameter defined by Eq. (10.24)
p	Pa	gas partial pressure
p_s	Pa	solvent partial pressure
R	J/(K mol)	gas constant
$T(t)$	K (°C)	temperature
V_0	cm³/mol	gas molar volume
x	–	mole fraction
X	A/(V cm)	conductivity
z	–	valency
α	–	Bunsen coefficient (defined in Sect. 10.2)
ϱ_s	g/cm³	density of solution
ϱ_{st}	g/cm³	density of solvent

Indices:
i   refers to dissolved salt or organic substance i
j   refers to ion j
0   refers to pure water

10.1 Introduction

Gas solubilities, particularly those of oxygen and carbon dioxide, are needed in biotechnology to establish mass balances, calculate yield coefficients, and evaluate volumetric mass transfer coefficients. The widely used polarographic oxygen electrodes measure only fugacity (partial pressure) of dissolved oxygen. Calculation of actual oxygen concentration requires knowledge of oxygen solubility. Gas solubilities, i.e., the concentrations of physically dissolved gas in equilibrium with a certain gas partial pressure, in aqueous media usually decrease with increasing temperature and concentration of dissolved substances (salts, sugars). Few solutes, e.g., short chain alcohols, may increase the solubility of gases. No theoretical approach applies but reliable empirical methods exist for the estimation of gas solubilities in biomedia.

10.2 Modes of Expressing Gas Solubility

The equilibrium liquid phase concentration of a gas may be expressed as:

c mass concentration (mg L^{-1}),
M molarity (mol L^{-1}),
w mass fraction (–),
x mole fraction (–),

specifying the temperature and the gas partial pressure p(Pa). As far as Henry's law holds, use of Henry's constants is convenient. They can be based on any of the various concentration measures:

$$H_M = p/M \text{ (Pa L mol}^{-1}\text{)}, \qquad (10.1)$$

$$H_x = p/x \text{ (Pa)}, \qquad (10.2)$$

$$H_L = c_g/c \text{ (–)}. \qquad (10.3)$$

H_L, the ratio of the equilibrium gas and liquid phase concentrations, equals the reciprocal of the Ostwald coefficient L at low solubilities.

$$L = \frac{1}{H_L} \qquad (10.4)$$

Ostwald coefficient (= volume of gas per unit volume of solvent).

Other coefficients are referred to a standard pressure of 101 325 Pa (1 atm), usually assuming ideal gas behavior and validity of Henry's law:

- α Bunsen coefficient $(-)$
  ($=$ volume of gas, reduced to $0\,°C$ and $101\,325$ Pa, per unit volume of solvent at a gas partial pressure of $101\,325$ Pa).
- β absorption coefficient $(-)$
  ($=$ volume of gas, reduced to $0\,°C$ and $101\,325$ Pa, per unit volume of solvent at a total pressure of $101\,325$ Pa including solvent vapor pressure).
- S Kuenen coefficient $(cm^3\,g^{-1})$
  [$=$ volume of gas (cm^3), reduced to $0\,°C$ and $101\,325$ Pa, in the quantity of solution containing 1 g of solvent. S is proportional to gas molality.]
- C_w weight solubility $(mol\,g^{-1})$
  (moles of gas per 1 g of solvent at a gas partial pressure of $101\,325$ Pa).

The conversions into the Bunsen coefficient α are:

$$\alpha = \frac{273.15}{T} L, \quad (10.5)$$

$$\alpha = \frac{101\,325}{101\,325 - p_s} \beta, \quad (10.6)$$

$$\alpha = \varrho_s (1-w) S, \quad (10.7)$$

$$\alpha = \varrho_{st} V_0 C_w, \quad (10.8)$$

where $T\,(K)$ is the temperature, p_s (Pa) the solvent vapor pressure, $V_0\,(cm^3\,mol^{-1})$ the gas molar volume at $0\,°C$ and $101\,325$ Pa, and ϱ_s (g cm^{-3}) and ϱ_{st} (g cm^{-3}) the densities of the solution and the solvent, respectively. In case of low solubility: $(1-w) \approx 1 \approx (1-x)$ and $\varrho_s \approx \varrho_{st}$. Then, the following relations are approximately valid:

$$\alpha = \frac{101.3\,V_0}{10^3 M_g p} c = \frac{101.3\,V_0}{p} M, \quad (10.9)$$

$$\alpha = \frac{\varrho_s V_0}{M_g} w = \frac{\varrho_s V_0}{M_{st}} x, \quad (10.10)$$

$$\alpha = 101.3\,V_0 H_M^{-1} = \frac{p \varrho_s V_0}{M_{st}} H_x^{-1}, \quad (10.11)$$

$$\alpha = \frac{273.15}{T} H_L^{-1}, \quad (10.12)$$

where $M_g\,(g\,mol^{-1})$ and $M_{st}\,(g\,mol^{-1})$ are the molecular weight of the gas and the solvent, respectively. For instance, the value of c for oxygen in equilibrium to moist air is related to α by:

$$c\,(mg\,O_2/L) = 2.954 \cdot 10^{-3} (P-p_s)\alpha, \quad (10.13)$$

where P (Pa) is the total pressure.

10.3 Experimental Methods for Gas Solubility Determination

Techniques of gas solubility measurement have been reviewed by MARKHAM and KOBE [10.1], BATTINO and CLEVER [10.2], and HITCHMAN [10.3]. Methods applied to biomedia have been discussed by SCHUMPE et al. [10.4].

There are chemical methods for analysis of gas saturated solutions and physical methods for determination of the amount of gas either desorbed from a saturated solution (desorption methods) or absorbed by an initially gas-free solvent (saturation methods). Chemical methods, e.g., Winkler's method for dissolved oxygen analysis, are usually not applicable to biomedia because of chemical interferences.

Desorption may be accomplished by applying vacuum, e.g., in the van Slyke apparatus, or by stripping with an inert gas. Prior desorption by evacuation is required for the saturation methods. Care has to be taken to avoid changes in composition by evaporation of volatile components. The gas-free media are equilibrated with the gas recording either the isobar volume decrease or the isochor pressure decrease. Alternatively, the increase in liquid phase oxygen fugacity may be followed with a polarographic oxygen electrode. Then, oxygen could also be introduced by decomposition of H_2O_2 as found by KÄPPELI and FIECHTER [10.5]. A technique that is based on the recording of oxygen deficiency in a continuous flow of air passed through the deaerated media, which has been applied by BABURIN et al. [10.6], is considered less accurate.

A special procedure has been applied by LIU et al. [10.7] to cultures of *Thiobacillus ferrooxidans*. The steady decrease (dp/dt) of oxygen partial pressure in the unaerated media was measured with an oxygen electrode. By relating (dp/dt) to the pseudo-steady oxygen uptake rate $[d(\text{mass } O_2)/(V_L dt)]$ of the liquid volume V_L in a Warburg-type apparatus, Henry's constant could be evaluated.

The most comprehensive and most accurate measurements of oxygen and carbon dioxide solubilities in microbial culture media have been carried out with isobar and isochor saturation methods by POPOVIC et al. [10.8], QUICKER et al. [10.9], and SCHUMPE et al. [10.4]. Since saturation could be accomplished very quickly, no or only partial inhibition of microbial respiration (e.g., by addition of formalin) was necessary. Studied media include cultures of *Candida utilis*, *Saccharomyces cerevisiae*, *Aspergillus niger*, *Penicillium chrysogenum*, *Chaetomium cellulolyticum*, *Hansenula polymorpha*, *Trichoderma reesei*, and *Escherichia coli*.

10.4 Estimation of Gas Solubilities

10.4.1 Pressure and Temperature Effects

For the pressure effect Henry's law can be assumed to be valid, i.e., the equilibrium liquid phase concentration of physically dissolved gas is proportional to the gas partial pressure.

Henry's constants depend on temperature. Usually gas solubilities in water strongly decrease with increasing temperature. For oxygen and carbon dioxide solubilities in water (α_0), HITCHMAN [10.3] and SCHUMPE et al. [10.4], respectively, corre-

Table 10.1. Coefficients of the WILHELM et al. [10.10] Correlations, Eq. (10.16), of Physical Solubilities of Gases in Water

Gas	Temperature Range (K)	A (J K^{-1} mol^{-1})	B (J mol^{-1})	C (J K^{-1} mol^{-1})	D (J K^{-2} mol^{-1})
H_2	274–339	−357.802	13897.5	52.2871	−0.0298936
N_2	273–346	−327.850	16757.6	42.8400	0.0167645
O_2	274–348	−286.942	15450.6	36.5593	0.0187662
O_3	277–293	−29.7374	3905.44		
NH_3	273–373	−162.446	2179.59	32.9085	−0.119722
H_2S	273–333	−297.158	16347.7	40.2024	0.00257153
CO	273–353	−341.325	16487.3	46.3757	
CO_2	273–353	−317.658	17371.2	43.0607	−0.00219107
CH_4	275–353	−365.183	18106.7	49.7554	−0.000285033
C_2H_6	275–353	−533.392	26565.0	74.6240	−0.00457313

lated comprehensive literature data in the temperature range of 0–50 °C:

$$\alpha_0(O_2) = 4.900 \cdot 10^{-2} - 1.335 \cdot 10^{-3} t + \\ + 2.759 \cdot 10^{-5} t^2 - 3.235 \cdot 10^{-7} t^3 + \\ + 1.614 \cdot 10^{-9} t^4, \quad (10.14)$$

$$\alpha_0(CO_2) = 1.720 - 6.689 \cdot 10^{-2} t + \\ + 1.618 \cdot 10^{-3} t^2 - 2.284 \cdot 10^{-5} t^3 + \\ + 1.394 \cdot 10^{-7} t^4, \quad (10.15)$$

where the temperature t is in °C. While Eqs. (10.14) and (10.15) are very accurate at 0–50 °C, they must not be extrapolated. For wider temperature ranges and many further gases, the equilibrium gas mole fractions x_0 in water at 101 325 Pa gas partial pressure have been correlated by WILHELM et al. [10.10] in the following form:

$$\frac{R}{4.184} \ln x_0 = A + \frac{B}{T} + C \ln T + D T, \quad (10.16)$$

where T (K) is the temperature and R (J K^{-1} mol^{-1}) is the universal gas constant. For some selected gases the coefficients are listed in Table 10.1.

10.4.2 Composition of Media

Usually electrolytes or organic solutes decrease the solubility of gases in aqueous media. Few organic solutes, e.g., short chain alcohols, may increase gas solubility. An empirical model may be used for the prediction of gas solubilities in mixed solutions of salts and alcohols (SCHUMPE and DECKWER [10.11]) and salt and sugars as well as in actual culture media (POPOVIC et al. [10.8]; QUICKER et al. [10.9]; SCHUMPE et al. [10.4]). Up to a certain concentration level the relative effects of different solutes can be treated as a "log-additive":

$$\log\left(\frac{\alpha_0}{\alpha}\right) = \sum_i \log\left(\frac{\alpha_0}{\alpha_i}\right). \quad (10.17)$$

Here (α_0/α) is the ratio of the gas solubility in water (α_0) to the one in the mixed solution (α) at the same temperature. (α_0/α_i) describes the individual effect of solute i in an equally concentrated aqueous solution of substance i alone. In Eq. (10.17) the ratios of other proportional solubility measures could be used as well (cf. Sect. 10.2). The solubilities in water (α_0) can be calculated from the correlations given in Sect. 10.4.1. The individual effects of the components may be taken from the literature, MARKHAM and KOBE [10.1], BATTINO and CLEVER [10.2], and KERTES [10.12], or estimated by the empirical methods discussed below.

10.4.2.1 Electrolytes

The salting-out of gases by electrolytes is usually well described by the empirical Sechenov equation:

$$\log\left(\frac{\alpha_0}{\alpha_i}\right) = K_{s,i} c_i, \quad (10.18)$$

where c_i is the molar concentration of the salt i and $K_{s,i}$ a slightly temperature dependent constant specific to salt i and the gas. At high values of c_i the solubilities calculated from Eq. (10.18) tend to be low. Particularly, in case of acids with concentration dependent dissociation the critical concentration can be less than 1 mol L^{-1} whereas with some salts it may be as high as 7 mol L^{-1}.

The $K_{s,i}$ salting-out constants can be estimated from empirical models which are particularly helpful if no experimental data are available. In these models the salt concentration is replaced by the ionic strength I:

$$I_i = 0.5 \sum_j c_j z_j^2, \quad (10.19)$$

where c_j and z_j are the molar concentrations and the charges of the ions j comprised in salt i, respectively. VAN KREVELEN and HOFTIJZER [10.13] suggested the following correlation:

$$\log\left(\frac{\alpha_0}{\alpha_i}\right) = (h_+ + h_- + h_g) I_i, \quad (10.20)$$

where h_+ and h_- are ion specific constants and h_g is specific to the gas and to the temperature. Comprehensive parameter sets were evaluated by DANCKWERTS [10.14] and ONDA et al. [10.15].

The van Krevelen-Hoftijzer model has been frequently applied; however, as pointed out by SCHUMPE et al. [10.16], Eq. (10.20) is physically inconsistent. Since the ion specific parameters are multiplied by the total ionic strength, the effect of a particular ion depends on the valency of the counter-ion. For mixed electrolyte solutions the solubilities predicted by Eq. (10.17) depend on the arbitrary arrangement of ions and salts required for application of Eq. (10.20). Therefore, SCHUMPE et al. [10.16] suggested the following model:

$$\log\left(\frac{\alpha_0}{\alpha_i}\right) = \sum_j H_j I_j, \quad (10.21)$$

where I_j is the ionic strength contribution of ion j:

$$I_j = 0.5 c_j z_j^2, \quad (10.22)$$

and H_j specific to ion, gas, and temperature. H_j parameter sets for oxygen (25°C, 37°C) and carbon dioxide (25°C) are listed in Tables 10.2 and 10.3. The temperature dependency is rather weak. Calculations should be based on elementary chemical principles, i.e., neutralization reactions and partial dissociation (e.g., $H_2SO_4 = H^+/HSO_4^-$; $KH_2PO_4 = K^+/H_2PO_4^-$) should be accounted for.

Table 10.2. Coefficients H_j (L mol^{-1}) of Cations for Eq. (10.21), SCHUMPE et al. [10.4]

Cation	$H_j(O_2)$, 25°C	$H_j(O_2)$, 37°C	$H_j(CO_2)$, 25°C
H$^+$	−0.776	−0.803	−0.319
Li$^+$	−0.675	−0.636	−0.178
Na$^+$	−0.568	−0.577	−0.130
K$^+$	−0.587	−0.578	−0.196
Rb$^+$	−0.618	−0.604	−0.217
Cs$^+$	−0.659	−0.612	−0.243
NH$_4^+$	−0.704	−0.681	−0.252
NEt$_4^+$		−0.709	
Mg^{2+}	−0.297	−0.321	−0.078
Ca^{2+}	−0.309	−0.316	−0.073
Ba^{2+}	−0.291	−0.299	−0.064
Mn^{2+}	−0.324	−0.325	−0.084
Fe^{2+}			−0.078
Co^{2+}		−0.317	
Ni^{2+}		−0.318	
Cu^{2+}		−0.325	−0.090
Zn^{2+}		−0.310	
Cd^{2+}		−0.320	
Al^{3+}		−0.221	−0.059
La^{3+}		−0.216	
Ce^{3+}		−0.216	
Fe^{3+}		−0.244	
Th^{4+}		−0.168	

Table 10.3. Coefficients H_j (L mol^{-1}) of Anions for Eq. (10.21), SCHUMPE et al. [10.4]

Anions	$H_j(O_2)$, 25°C	$H_j(O_2)$, 37°C	$H_j(CO_2)$, 25°C
F$^-$		0.867	
Cl$^-$	0.849	0.861	0.339
Br$^-$	0.820	0.822	0.324
J$^-$	0.784		0.309
OH$^-$	0.943	0.917	
NO$_3^-$	0.802	0.821	0.293
SCN$^-$		0.791	
BF$_4^-$		0.775	
ClO$_4^-$	0.890		
HSO$_4^-$	0.955	0.935	0.436
HSO$_3^-$			0.400
HCO$_3^-$	1.076	0.861	
H$_2$PO$_4^-$	0.997		
C$_6$H$_5$OCH$_2$COO$^-$	0.755		
SO$_4^{2-}$	0.460	0.448	0.213
S$_2$O$_3^{2-}$	0.455		0.211
CO$_3^{2-}$	0.467	0.447	
HPO$_4^{2-}$	0.477		
PO$_4^{3-}$	0.308		
Mo$_7$O$_{24}^{6-}$		0.155	

10.4.2.2 Organic Solutes

The effects of organic solutes can often be described analogously to the Sechenov equation [Eq. (10.18)]:

$$\log\left(\frac{\alpha_0}{\alpha_i}\right) = K_i c_i, \qquad (10.23)$$

where c_i (g L^{-1}) is the concentration of the organic substance i and K_i (L g^{-1}) an empirical constant specific to substance i, to the gas, and to a minor extent to the temperature. In some cases a better fit is obtained by a linear relation:

$$\left(\frac{\alpha_i}{\alpha_0}\right) = 1 - m_i c_i. \qquad (10.24)$$

For larger concentration ranges bi-parameter models may be required; or, in fact, a simple relationship may not exist at all. Unlike the salt effect, there are no empirical models for prediction of K_i or m_i values.

The coefficients for some substances often encountered in biomedia are listed in Table

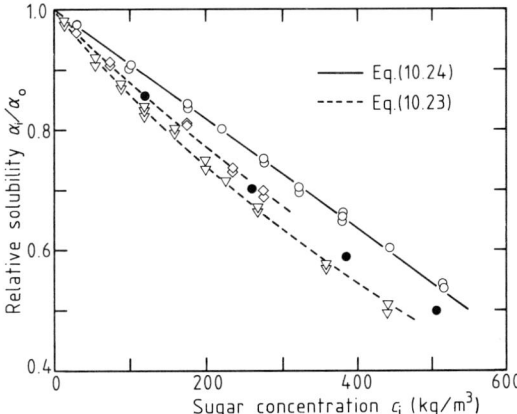

Figure 10.1. Relative oxygen solubilities at 25°C in solutions of glucose ($\triangledown$), lactose ($\diamond$) QUICKER et al. [10.9], and sucrose ($\bullet$) HIKITA et al. [10.17], ($\circ$) QUICKER et al. [10.9].

10.4. As an example, Fig. 10.1 illustrates the excellent fit of experimental relative oxygen solubilities in sugar solutions at 25°C. However, for sucrose the data of HIKITA et al. [10.17] and QUICKER et al. [10.9] do not coincide for unknown reasons.

ture. Literature data are scarce for low alcohol concentrations. Comparing oxygen solubilities reported for ethanol solutions, SCHUMPE et al. [10.4] observed large discrepancies. A recent study by KUTSCHE et al. [10.18] reported on oxygen solubilities at

Table 10.4. Coefficients K_i and m_i, Eqs. (10.23) and (10.24), for the Effects of Organic Substances on O_2 and CO_2 Solubilities, SCHUMPE et al. [10.4] and KUTSCHE et al. [10.18]

Substance	Conc. Range (g L^{-1})	Temp. (°C)	K_i (or m_i) (10^{-4} L g^{-1})	
			O_2	CO_2
Glucose	0–450	25	6.58	6.07
Lactose[a]	0–250	25	5.71	3.48
Sucrose	0–600	25	(m = 9.04)	(m = 6.87)
Molasses[b]	0–240	25	4.03	
Starch	0–200	37	6.35	
Gluconic acid	0–500	25	3.92	
Glycogen	0–250	37	6.59	
Glucosamine	0–300	37	11.23	
Glucose-(−)-phosphate	0–200	37	11.87	
ADP	0–200	37	6.35	
ATP	0–200	37	7.10	
Citric acid	0–200	25	5.09	2.68
Urea	0–300	37	3.74	
Glycerol	0–190	15	5.77	
	0–190	25	4.74	
	0–300	37	4.07	
Albumin (bovine)	0–200	37	1.81	
Albumin (chicken)	0–200	37	3.23	
Hemoglobin	0–250	37	−0.30	
α-Globulin	0–100	37	3.05	
Glycine	0–200	37	12.46	
Lysine	0–300	37	13.45	
Cysteine	0–200	37	22.82	
Caseinpeptone[b]	0– 60	30	4.3	
Meat extract[b]	0– 60	30	5.7	
Yeast extract[b]	0– 60	30	6.2	
Pharmamedia[b]	0– 80	25	1.5	

[a] technical grade, [b] measured after sterilization

Among the substances not covered in Table 10.4 are small-chain alcohols. At low temperatures, gas solubilities in aqueous alcohol solutions may exhibit maxima and minima indicating changes in water structure. Literature data are scarce for low alcohol concentrations for temperatures of 15, 25, and 37°C. The data for 37°C are shown in Fig. 10.2. The solubility rise decreases with increasing chain length of the alcohol and decreasing temperature.

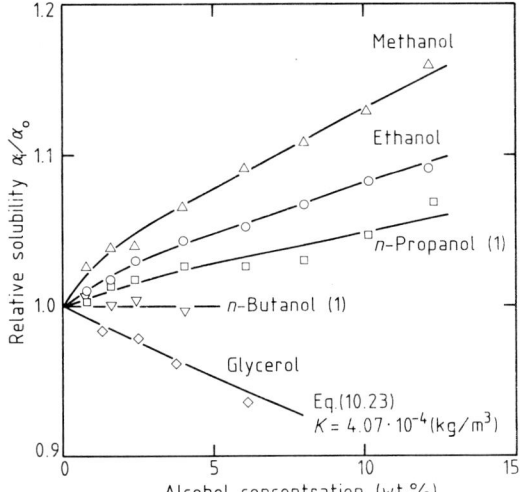

Figure 10.2. Relative oxygen solubilities at 37 °C in aqueous alcohol solutions, KUTSCHE et al. [10.18].

pylene + propylene glycol) to a culture medium and an almost threefold increase in the presence of 2 vol% sunflower oil. Adsorption of oxygen at the interface of oil drops was suggested as an explanation. However, BABURIN et al. [10.6] found no such effect in a distilled water/oil mixture. POPOVIC et al. [10.8] reported oxygen solubility to pass through a maximum at about 0.1 vol% antifoam agent but the effects were rather small.

It is even less clear how the presence of cells affects gas solubility. BABURIN et al. [10.6] suspect that cells reduce gas solubility by an effect of the membrane potential on the structure of the surrounding solution. On the other hand, QUICKER et al. [10.9] measured oxygen solubilities in culture media of *Penicillium chrysogenum* in the absence and presence of cells and found no significant differences. Cell lysis may be expected to decrease gas solubility and exhibit pH-dependent adsorption effects. These aspects deserve further investigations.

10.4.2.3 Adsorption Effects

Adsorption on solids or macromolecules may increase the overall absorption capacity of a solution. It usually does not have to be considered for an estimation of the saturation concentrations of dissolved gases, e.g., by application of Eq. (10.17). However, when measuring gas solubilities by the experimental methods discussed in Sect. 10.2 the results may be too high.

For hydrocarbon gases, the pH-dependent adsorption on aqueous albumin has been used to draw conclusions about the protein structure. For oxygen as well, an increase in absorption capacity by albumin has been reported by BABURIN et al. [10.6] whereas ZANDER [10.19], and SCHUMPE et al. [10.4] observed no such effect (cf. Table 10.4).

For hydrocarbon gases, increased absorption capacities have also been reported for micellar solutions, e.g., those of sodium dodecylsulfate. For oxygen, BABURIN et al. [10.6] reported an almost fourfold increase by addition of 4 vol% of propinole (oxypro-

10.5 Estimated and Measured Oxygen Solubilities in Culture Media

Comprehensive studies of oxygen solubilities in microbial culture media have been reported by POPOVIC et al. [10.8], QUICKER et al. [10.9], and SCHUMPE et al. [10.4]. The experimental results were compared to estimates obtained from Eq. (10.17), i.e., assuming "log-additivity" of the individual salting-out effects. Application of this concept requires that the composition is approximately known. Usually, the initial composition should be known and gross changes due to substrate con-

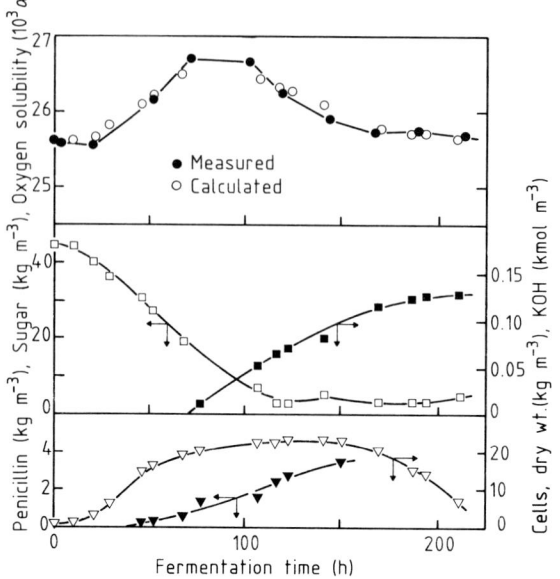

Figure 10.3. Oxygen solubilities in cultivation of *Penicillium chrysogenum*, QUICKER et al. [10.9].

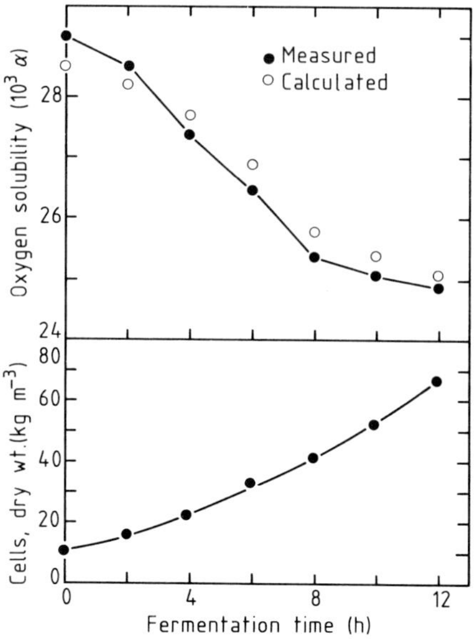

Figure 10.4. Oxygen solubilities in growing *Saccharomyces cerevisiae*, POPOVIC et al. [10.8].

Table 10.5. Culture Media Studied for Comparison of Estimated and Experimental Oxygen Solubilities

Symbol (Fig. 10.5)	Microorganism	Temp. (°C)	pH	Final Cell Dry Weight (g L^{-1})
POPOVIC et al. [10.8]:				
▲	*Candida utilis*	30	5	34
▼	*Aspergillus niger*	30	5	10
×	*Candida boidinii*[a]			
◐	*Saccharomyces cerevisiae*	33	5.5	67
▣	*Penicillium chrysogenum*	30	5	5
QUICKER et al. [10.9]:				
■ (□)[b]	*Penicillium chrysogenum*	25	6.2	10–50
SCHUMPE et al. [10.4]:				
◆ (◇)[b]	*Hansenula polymorpha*	38	5.0	5
● (○)[b]	*Saccharomyces cerevisiae*	30	5.0	9.5
▼ (▽)[b]	*Chaetomium cellulolyticum*	37	5.0	1
▲ (△)[b]	*Trichoderma reesei*[a]	30	5.0	3
(+)[b]	*Escherichia coli*	28	6.8	12–16

[a] continuous cultivation
[b] ( ) = nutrition media before inoculation

sumption and additions, e.g., of bases for pH-control, can be estimated or measured.

QUICKER et al. [10.9] tested the method successfully for batch cultivations of *Penicillium chrysogenum*. An example of the observed and predicted variation of oxygen solubility is given in Fig. 10.3. SCHUMPE et al. [10.4] presented further data on cultures listed in Table 10.5. In these studies the empirical correlations described in Sect. 10.4 were applied to estimation of the individual effects of electrolytes and organic substances.

An indirect method for assessing the effect of varying salt concentration has been suggested by POPOVIC et al. [10.8]. They correlated salting-out by electrolytes as a function of electric conductivity X by empirical equations of the form:

$$\log\left(\frac{\alpha_0}{\alpha_i}\right) = a_0 + a_1 X + a_2 X^2. \quad (10.25)$$

Then, changes in conductivity measured during bioprocesses were used as an indirect measure of changes in salting-out of oxygen by varying salt concentration. Unfortunately, the constants a_0, a_1, and a_2 are specific for the salt. Values of $\log \alpha_0/\alpha_i$ at the same conductivity may differ by more than a factor of 2. Therefore, Eq. (10.25) has to be adapted to each salt mixture encountered by independent solubility and conductivity measurements. If changes in the ionic composition are encountered the coefficients gradually lose their validity and only rough estimates can be obtained. Culture media studied by POPOVIC et al. [10.8] are listed in Table 10.5. Fig. 10.4 shows the excellent concurrence in the case of growing cell cultures of *Saccharomyces cerevisiae* where the decrease of oxygen solubility was mainly due to accumulation of ammonium sulfate.

For culture media consisting mainly of salts and carbohydrates, the direct and indirect estimation methods were applied quite successfully. Fig. 10.5 gives a parity plot for the cultures listed in Table 10.5. Estimated and measured oxygen solubilities agree within ±2%.

However, for cultures of *Escherichia coli* on nutrient media consisting mainly of proteins (yeast and meat extract, casein peptone) reproducible, drastic solubility changes were observed by SCHUMPE et al. [10.4]. These could not be understood from the available knowledge of the composition. This effect was assumed to be due to pH-dependent changes of protein structure and corresponding changes in adsorption capacity. For media of this type the methods should be applied with care.

10.6 Recommendations

Gas solubilities in mixed solutions of electrolytes and organic substances can be estimated fairly accurately by an empirical model, Eq. (10.17), that assumes "log-additivity" of the individual effects of the solutes. Relative solubilities for solutions of the individual components, which can be obtained from the literature or estimated from empirical correlations, must be introduced. For microbial culture media the composition may be only vaguely known.

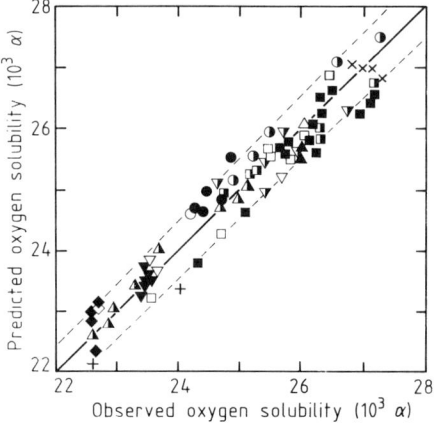

Figure 10.5. Parity plot for estimated oxygen solubilities in culture media listed in Table 10.5 (--- ±2%).

Nevertheless, the method has been applied to many cultures with a ±2% error of the solubility estimates. Care should be taken if high concentrations of dissolved protein are encountered.

10.7 References

[10.1] A. E. MARKHAM and K. A. KOBE: "Solubility of gases in liquids". Chem. Rev. 28 (1941), 519.

[10.2] R. BATTINO and H. L. CLEVER: "The solubility of gases in liquids". Chem. Rev. 60 (1966), 395.

[10.3] M. L. HITCHMAN: "Measurement of Dissolved Oxygen". John Wiley & Sons and Orbisphere Corp., Geneva–New York, 1978.

[10.4] A. SCHUMPE, G. QUICKER, and W.-D. DECKWER: "Gas solubilities in microbial culture media". Adv. Biochem. Eng. 24 (1982), 1.

[10.5] O. KÄPPELI and A. FIECHTER: "A convenient method for the determination of oxygen solubility in different solutions". Biotech. Bioeng. 23 (1981), 1897.

[10.6] L. A. BABURIN, J. E. SHVINKA, and U. E. VIESTURS: "Equilibrium oxygen concentration in fermentation fluids". Eur. J. Appl. Microbiol. Bioeng. 13 (1981), 15.

[10.7] M. S. LIU, R. M. R. BRANION, and D. W. DUNCAN: "Determination of the solubility of oxygen in fermentation media". Biotech. Bioeng. 15 (1973), 213.

[10.8] M. POPOVIC, H. NIEBELSCHÜTZ, and M. REUSS: "Oxygen solubilities in fermentation fluids". Eur. J. Appl. Microbiol. Biotechnol. 8 (1979), 1.

[10.9] G. QUICKER, A. SCHUMPE, B. KÖNIG, and W.-D. DECKWER: "Comparison of measured and calculated oxygen solubilities in fermentation media". Biotech. Bioeng. 23 (1981), 635.

[10.10] E. WILHELM, R. BATTINO, and R. J. WILCOCK: "Low-pressure solubility of gases in liquid water". Chem. Rev. 77 (1977), 219.

[10.11] A. SCHUMPE and W.-D. DECKWER: "Estimation of O_2 and CO_2 solubilities in fermentation media". Biotech. Bioeng. 21 (1979), 1075.

[10.12] A. S. KERTES: "Solubility Data Series". Pergamon Press, Oxford, 1981 (publication in progress).

[10.13] D. W. VAN KREVELEN, and P. J. HOFTIJZER: "Sur la solubilité des gaz dans les solutions aqueuses". Chimie et Industrie (1948). Numero Speciale du XXIe Congrés International de Chimie Industrielle, Bruxelles, p. 168.

[10.14] P. V. DANCKWERTS: "Gas-Liquid Reactions". McGraw-Hill, New York, 1970.

[10.15] K. ONDA, E. SADA, T. KOBAYASHI, S. KITO, and K. ITO: "Salting-out parameters of gas solubility in aqueous salt solutions". J. Chem. Eng. Jap. 3 (1970), 18.

[10.16] A. SCHUMPE, I. ADLER, and W.-D. DECKWER: "Solubility of oxygen in electrolyte solutions". Biotech. Bioeng. 20 (1978), 145.

[10.17] H. HIKITA, S. ASAI, and Y. AZUMA: "Solubility and diffusivity of oxygen in aqueous sucrose solutions". Can. J. Chem. Eng. 56 (1978), 371.

[10.18] I. KUTSCHE, G. GILDEHAUS, D. SCHULLER, and A. SCHUMPE: "Oxygen solubilities in aqueous alcohol solutions". J. Chem. Eng. Data 29 (1984), 286.

[10.19] R. ZANDER: "Der Verteilungsraum von physikalisch gelöstem Sauerstoff in wäßrigen Lösungen organischer Substanzen". Z. Naturforsch. 31c (1976), 339.

2 Fundamentals of Microbial Reaction Engineering

Chapter 11
General Strategy in Bioprocessing

Anton Moser

Institut für Biotechnologie, Mikrobiologie und Abfalltechnologie
Technische Universität Graz
Graz, Austria

11.1   Reaction Engineering and Biotechnology
11.1.1 Integrating Strategy as the Basis for a General Methodology
11.1.2 The Formal Macroapproach to Bioprocessing
11.2   The Use of Bioreactors for the Estimation of Process Kinetic Parameters
11.2.1 Bioreactors in Process Analysis and Process Development
11.2.1.1 Translation of Laboratory Culture Data to the Production Plant
11.2.1.2 Pragmatic Empiric Approach to Process Development
11.2.1.3 Systematic Empiric Approach to Process Development
11.2.2 Reactor Operation Modes and Reactor Balances
11.2.3 Pseudohomogeneity of Bioreactor Operation
11.2.4 "Perfect Bioreactors" as Model Bioreactors for Kinetic Analysis of Bioprocesses
11.3   The Use of Process Kinetics to Determine Optimum Operation of Bioreactors
11.3.1 Influence of Macromixing on Continuous Operation
11.3.1.1 Stirred Tank versus Plug Flow Reactor
11.3.1.2 Recycle Operation
11.3.1.3 Optimum Operation with Inhibition Kinetics
11.3.2 Influence of Micromixing
11.4   References

List of Symbols

Symbol	Units	Description
a	m^{-1}	specific interfacial area
C	kg	quantity of CO_2
c	$kg \cdot m^{-3}$	concentration
d	m	diameter
d_p	m	diameter of particle
D	$m \cdot s^{-1}$	diffusion coefficient
D_{eff}	$m \cdot s^{-1}$	effective dispersion coefficient
F, F_r	$m^3 \cdot s^{-1}$	liquid flow through reactor resp. by recycle
g	$m \cdot s^{-2}$	gravitational acceleration
g_c	$m^3 \cdot kg^{-1} \cdot s^{-2}$	gravitational constant ($6.671 \cdot 10^{-8}$)
H_V	J	volumetric heat of fermentation
h_v	$J \cdot m^{-3}$	heat per volume
K_S	$kg \cdot m^{-3}$	saturation constant of Monod kinetics
$k_{L1} \cdot a$	h^{-1}	specific rate coefficient of oxygen transfer
L	m	length
n	$mol \cdot m^{-3} \cdot s^{-1}$	molar flux per volume
N_j	mol	number of moles of component j
n'	$mol \cdot m^{-2} \cdot s^{-1}$	molar flux per area
O	kg	mass of oxygen
o	$kg \cdot m^{-3}$	oxygen concentration
P	kg	mass of product
P	$J \cdot s^{-1}$	power
p	$kg \cdot m^{-3}$	product concentration
p	$N \cdot m^{-2}$	pressure
q_S, q_O, q_P, q_C	$kg \cdot kg^{-1} \cdot s^{-1}$	specific rate of consumption of S and O_2 resp. formation of P and CO_2
r	$kg \cdot m^{-3} \cdot s^{-1}$	reaction rate
r_j	$kg \cdot m^{-3} \cdot s^{-1}$	rate of formation or consumption of component j
S	kg	mass of substrate
s	$kg \cdot m^{-3}$	concentration of substrate
T	K, °C	temperature
t	s	time
t_e	s	time of environmental change
t_m	s	time of mixing
t_r	s	time of reaction
t_{Tr}	s	time of transport
t_b	s	time of a batch process
$\bar{t}$	s	mean residence time
V	m^3	volume
v	$m \cdot s^{-1}$	velocity
v_{SG}	$m \cdot s^{-1}$	superficial gas velocity
X	kg	mass of biomass
x	$kg \cdot m^{-3}$	concentration of biomass

Greek letters

Symbol	Units	Description
δ	m	thickness of layers or films
ε	$m^2 \cdot s^{-3}$	energy dissipation
λ	s	life expectancy
μ	s^{-1}	specific rate of microbial growth
ϱ	$kg \cdot m^{-3}$	density

Dimensionless numbers and coefficients

Bo	Bodenstein number
Da_{II}	Damköhler number, second degree
Ha	Hatta number
N_{eq}	number of equivalent stages
Re	Reynolds number
Re_ε	Reynolds number based on energy dissipation ε
Y	yield coefficient
η_r, η_{Tr}	effectiveness factor for reaction (η_r) and transport (η_{Tr})
ν	stoichiometric coefficient
Φ	Thiele modulus
ζ	conversion

Indices

a, b, c, d; α, β, γ, δ	coefficients for stoichiometry
b	batch
crit (c)	critical
C	CO_2
eff	effective
ex	exit
G	gas
i, j	number of reaction or component
in	inlet
L	liquid
max	maximal
mm	maximum mixing
O	oxygen
0	zero

P	product
r	reaction or recycle
S	substrate or solid
ts	total segregation
Tr	transport
W	water
X	biomass
z	coordinate
*	saturation value

Abbreviations

BC	bubble column
CMMFF	completely mixed microbial film fermenter
CPFR	continuous plug flow reactor
CRR	continuous recycle reactor
CSTR	continuous stirred tank reactor
diff. R	differential reactor
DCRR	discontinuous recycle reactor
DCSTR	discontinuous stirred tank reactor
int. R	integral reactor
MPR	multi-purpose reactor
OTR	oxygen transfer rate
qss	quasi-steady-state
rds	rate determining step
RTD	residence time distribution
SCSTR	semicontinuous stirred tank reactor
STR	stirred tank reactor

11.1 Reaction Engineering and Biotechnology

11.1.1 Integrating Strategy as the Basis for a General Methodology

Microbial reaction engineering is concerned with the handling of biotechnical processes especially of microbial cells in bioreactors in order to produce microbial products on an industrial scale. These technical operations should be carried out in a reproducible manner so that the concepts of reaction engineering, well developed with chemical processes, can be successfully applied. This goal, however, can only be reached if the behavior of microorgan-

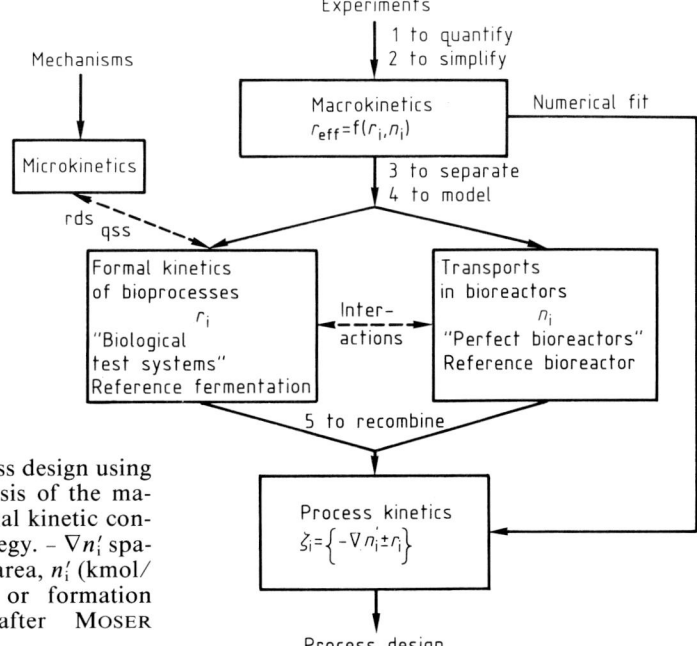

Figure 11.1. Strategies in bioprocess design using working principles 1–5 on the basis of the macroscopic principle using the formal kinetic concept as part of an integrating strategy. $-\nabla n_i'$ spatial change of mass flux through area, n_i' (kmol/m² h), r_i rate of consumption or formation (kmol/m³ h), ζ_i conversion; after MOSER [11.32].

isms is known as much as necessary for quantification of bioprocessing in reproducible operations. The nature of microbial operations, products, and techniques requires special procedures in processing, handling, and control. This, in turn, imposes special requirements on engineers and designers of the process equipment. Process and system engineering therefore play a central role in the commercialization of biotechnology. For this reason, further development of process engineering has to focus on process kinetics and bioreactor design including new biological and physical sensors and computer application. Both fields – biology and physics – have generally been treated separately, and each discipline has applied its "own methods" with a strong tendency to handle microbiological and biochemical aspects isolated from chemical engineering aspects. As a consequence, problems arise when one applies the biological data on a technical scale and the engineering data to actual bioprocessing situations. To bridge this gap in fundamental research, a clearly defined and well-developed research methodology is needed representing an integrating mode of thinking and working where biological and engineering aspects are treated. This integrating strategy, schematically shown in Fig. 11.1 after MOSER [11.1], is in agreement with the definition of biotechnology as given by the European Federation of Biotechnology:

"Biotechnology is the integrated use of biochemistry, microbiology and engineering sciences in order to achieve the technological (industrial) application of the capabilities of microorganisms, cultured tissue cells, and parts thereof".

Applying the law of conservation of mass to bioprocessing, balance equations must be written, where the overall change r_{eff}, the "macroconversion", is the result of the additive action of physical transport phenomena (n_j') and biological reactions $(r_j = v_j r)$:

$$r_{\text{eff}} \equiv \frac{\partial c_j}{\partial t} = -\nabla n_j' + v_j \cdot r. \qquad (11.1)$$

In this basic equation all activities of reaction engineering sciences are manifested: balancing, stoichiometry, kinetics of reaction rates, and physical transports. Thermodynamics are not directly included here because it emphasizes only initial and final states of the system. Due to the complexity of the additive and interactive nature of both terms in Eq. (11.1), both aspects of metabolism and bioreactor have been treated separately in the past. According to the integrating strategy, however, it is necessary to quantify bioreactors and their physical transports also with the aid of biological systems with known kinetics and *vice versa* to quantify biokinetics with the aid of bioreactors with known physical transports, using the test of pseudohomogeneity (cf. Sect. 11.2.3).

11.1.2 The Formal Macroapproach to Bioprocessing

Due to the complexity of biological and physical phenomena, simplifications must be made without losing essential information. This approach, well-known in chemical engineering, is a consequence of the macroscopic principle (GLANSDORFF and PRIGOGINE [11.2]) which was recently applied to bioprocessing by ROELS and KOSSEN [11.3] and ROELS [11.4].

Formal analogies are used in this approach to quantify bioprocesses on the basis of macroscopic process variables ("the formal macroapproach"). A metabolizing culture of microorganisms can be described on the basis of the macroscopically observed process behavior, using concentration changes of the components biomass X, substrates S_i, oxygen O, products P_j, carbon dioxide C, and heat of fermentation H_V. The overall reaction for a discontinuous process is:

$$v_S \cdot S_i + v_O \cdot O \xrightarrow{X} v_X \cdot X + v_P \cdot P + v_C \cdot C + v_{H_V} \cdot H_V. \qquad (11.2)$$

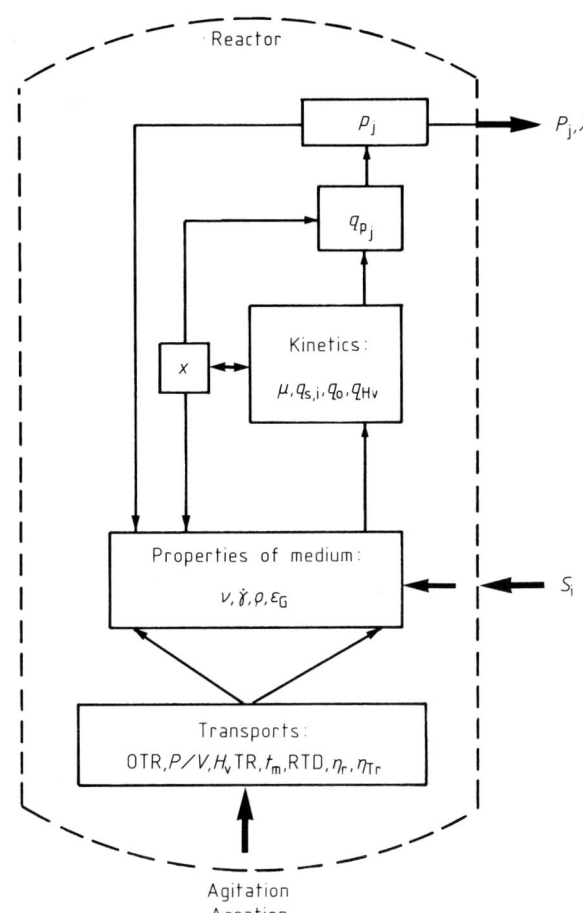

Figure 11.2. Schematic representation of the integrating strategy showing the interactions between physical transport phenomena and biokinetics in the case of a reactor of the stirred tank type; adapted from REUSS et al. [11.5]. RTD residence time distribution.

Considering the conservation of atomic species, the elementary balance equation, instead of Eq. (11.2) can be written in a general form:

$$\nu_S(C_a H_b O_c N_d) + \nu_O(O_2) + \nu_N(NH_3) \rightarrow$$
$$\rightarrow \nu_x(C_\alpha H_\beta O_\gamma N_\delta) + \nu_P(C_{\alpha'} H_{\beta'} O_{\gamma'} N_{\delta'}) +$$
$$+ \nu_C(CO_2) + \nu_W(H_2O) + \nu_{H_V} \cdot H_V \quad (11.3)$$

where $C_a H_b O_c N_d$ is a generalized carbon source, $C_\alpha H_\beta O_\gamma N_\delta$ a cell unit, and $C_{\alpha'} H_{\beta'} O_{\gamma'} N_{\delta'}$ a product. In this case, balancing is restricted to the most important elements (C, H, O, N) which is normally sufficient. In special cases the elements sul-

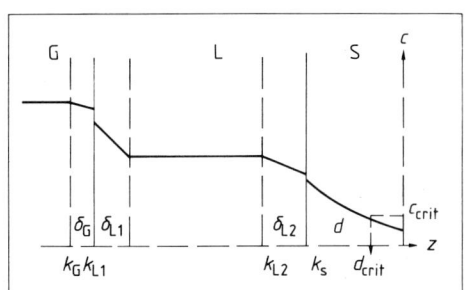

Figure 11.3. Concentration profile in heterogeneous three-phase systems with different mass transport coefficients k_G, k_{L1}, k_{L2}, and k_S in individual films with thicknesses δ_G, δ_{L1}, δ_{L2}, and d (MOSER [11.20]).

fur and phosphorus can also be significant.

The overall biosynthetic Eq. (11.3) is the net equation as a result of hundreds of metabolic reactions in the living cell. The various cycles and chains in a metabolic network, including the ATP-system and other energy handling systems, which do not result in the output of new material of cells or products, do not contribute to the net reaction. Hence, detailed knowledge of these cycles is unnecessary in a macroscopic treatment using the net stoichiometric equations. Bioprocess analysis, therefore, is considerably reduced in complexity, without loss of significant information. Microscopic processes are reflected on the macroscopic level.

Even the microbial cells suspended in the liquid phase of the bioreactor are treated as "black boxes", their macroscopic behavior, manifested in the changes of the concentrations in the liquid phase, is taken into consideration.

The result of this research strategy is a structure of the interactions according to Eq. (11.1), formulated in a sum of process kinetic equations; this is best represented by the qualitative illustration of Fig. 11.2.

Quantification for engineering purposes can adequately be achieved by calculating derived quantities like the specific rates of consumptions $q_{S,i}$, growth μ, and production $q_{P,j}$ and by formulating the interactions with the transport of mass, energy, and momentum using some other physical properties of the extracellular environment. This approach was recently used for the scale-up of penicillin production by REUSS et al. [11.5].

The generalized situation of bioprocessing including G-, L-, and S-phases can be schematically represented by the concentration profile of O_2 in Fig. 11.3 as an example of the interactions between transport and kinetics. The limiting step of physical transport from the gas into the solid phase, where microbial metabolism is occurring, is given by the hypothetical films of the gas phase δ_G and the liquid phase δ_{L1} at the G/L-interface and the liquid film at the L/S-interface δ_{L2} and the thickness of the biological catalyst δ_S. These film thicknesses represent the main resistance to mass transfer according to the two-film theory and are related to the transport coefficients of individual steps k_{L1}, k_{L2}, and k_S.

Modelling of this bioprocess therefore has to take into account the interactions be-

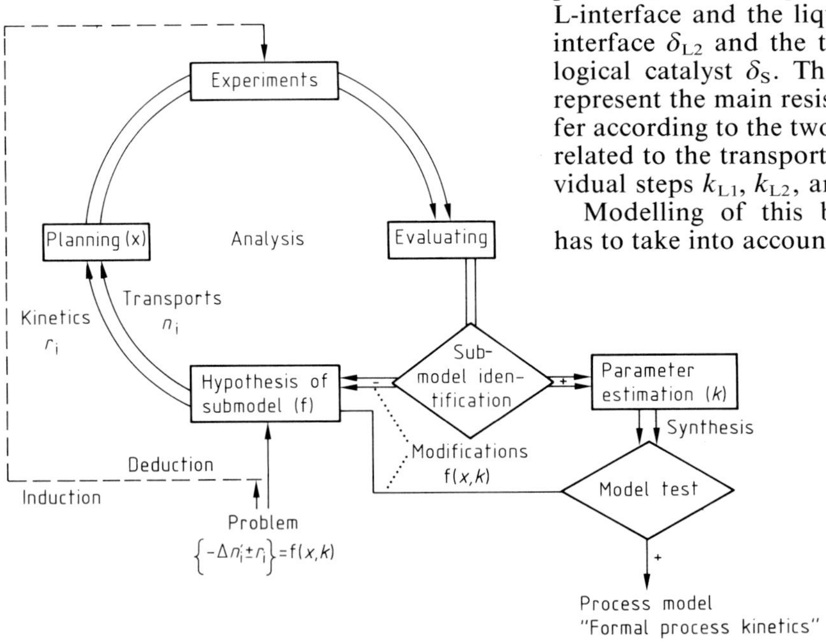

Figure 11.4. Procedure of adaptive mathematical modelling of bioprocess kinetics following a combination of deductive and inductive steps of research applying the integrating strategy according to Fig. 11.1 (MOSER [11.8]).

tween the rate constants of transports k_G, k_{L1}, k_{L2}, k_S, and the rate constant of reaction k_r. The macroscopic level operates with the yield concept where yield coefficients are given by:

$$Y_{i/j} \approx \frac{v_i}{v_j} \qquad (11.4)$$

as a substitute for true stoichiometry. The concept of yield coefficients and its relationship to stoichiometry is elaborated in more detail in Chapter 13.

As a result of this research strategy the variety of mechanisms of metabolic reactions and also the variety of mechanisms of physical transports are reduced to a simplified but adequate structure.

Process and systems engineering use mathematical models which apply a high degree of theoretical background; this is often admired but also often disapproved by biologists. However, it should be kept in mind that "without theory, practice is but routine born of habit" as LOUIS PASTEUR stated, or that "nothing has more practicability than theory, but in turn nothing is more theoretical than practice". This statement is strongly related to the deductive research methodology which is generally accepted for scientific work (POPPER [11.6]). Transferred to the situation of bioprocess analysis and design the steps of the procedure of deduction are illustrated in Fig. 11.4 (MOSER [11.7], [11.8]). While the principle of induction is primarily used only in the beginning of an investigation, the deductive method should be preferred later on, starting with a statement of a hypothesis of the investigated problem.

The effective rate, r_{eff}, can then be analyzed using mathematical model functions f with parameters k as hypotheses for biokinetics and transports based on macroscopic process variables x_i:

$$r_{eff} = f(x_i, k). \qquad (11.5)$$

As a consequence of this general strategy it can be concluded that shake flasks predominantly used in microbiological laboratories are quite useful to examine the qualitative behavior of reactants/biocatalysts (cells)/products. They can never give, however, representative data for process design or scale-up.

In this research strategy the principle of analogy is used to describe the behavior of the system under certain conditions ("if, then"). These "black-box" models, however, are not only descriptive but also predictive when applying the deductive method. Normally, only such models are thought to be of predictive nature, which are based on and derived from mechanistic assumptions. Even if some mechanistic background is advantageous for progressive interpretations of parameters with more biological significance, such an approach ("grey boxes") still remains more or less an optical illusion (ROELS and KOSSEN [11.3]) as the final subsystems are always black boxes. As a conclusion, the level to be chosen for adequate modelling is the principle of analogies, where the model parameters are to be adjusted to experiments according to HOFMANN [11.9].

However, it should be stated here, that the formal empirical approach in terms of macroscopic observables has been very fruitful by permitting chemical reactor technology to develop to a point that far surpasses the development of theoretical work in kinetics.

Sometimes kinetics are considered a subject of mystery due to the mathematical handling. But no other tool in science has such a universality and general applicability when properly applied.

Bioprocess kinetics and bioreactor design are the heart of the production of almost all industrially employed biotechnical processes. It is primarily the knowledge of kinetics and reactor design which is the basis of process engineering and distinguishes bioengineering from other branches of engineering. The selection of a reaction system which operates in the most efficient manner can be the key to the economic success or failure of the production plant.

The procedure of a systematic process kinetic analysis is schematically summarized in Fig. 11.5 (MOSER [11.8]), showing at the same time the manner in which the chapters

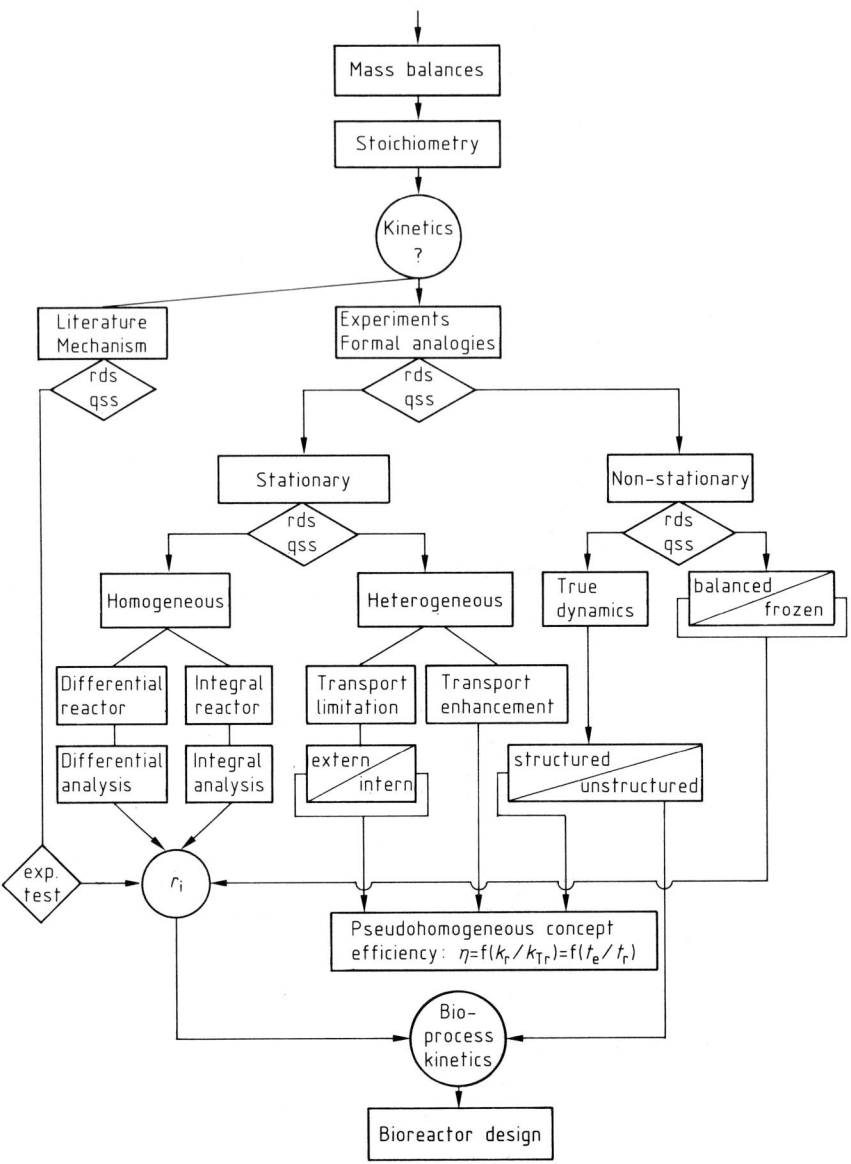

Figure 11.5. Flow sheet of the strategy of bioprocess kinetic analysis for different process situations: stationary/non-stationary, homogeneous/heterogeneous, and differential/integral reactor operations. Different steps of research are shown in order to elucidate biokinetics, which must be recombined with the mathematical model of the transports in the bioreactor (MOSER [11.8]).

in Part 2 of this volume are organized and the material presented.

A simplified treatment can be reached by using the concept of the rate-determining step ("rds") or the quasi-steady-state ("qss") and, furthermore, by introducing the concept of efficiency η, defined by:

$$\eta = r_{\text{eff}}/r_{\text{ideal}}. \tag{11.6}$$

The efficiency factor η can be interpreted as:

$$\eta \approx f(k_r/k_{\text{Tr}}), \tag{11.7}$$

resulting from the interaction between the rate constants of reaction k_r and transport k_{Tr}. Using the concept of efficiency real heterogeneous situations can be handled as pseudohomogeneous cases.

11.2 The Use of Bioreactors for the Estimation of Process Kinetic Parameters

11.2.1 Bioreactors in Process Analysis and Process Development

Bioreactors have to fulfill two requirements in bioprocessing. Beyond the use on a technical scale for industrial production, bioreactors of different size and shape serve as reaction vessels for the handling of microbial strains starting with the screening program in biological laboratories and for quantification on a bench and pilot scale.

11.2.1.1 Translation of Laboratory Culture Data to the Production Plant

The translation of laboratory culture data to production plant operation includes three steps of screening (AIBA et al., [11.10]). The objectives are: (1) using agar plating to search for better strains which grow on a given substrate and are able to metabolize the desired product, (2) using liquid substrates in shake flasks to change culture conditions and so to economize the time required for several runs in order to elucidate the qualitative or semi-quantitative behavior of the microbial reaction system, and (3) in the pilot plant, to select the best strain suitable for optimum productivity under technical conditions especially with regard to mixing and shearing on the technical scale.

11.2.1.2 Pragmatic Empiric Approach to Process Development

Process development follows predominantly a pragmatic-empirical approach, outlined by Fig. 11.6.

After an economic evaluation the microbial production pilot plant reactors are run in order to elaborate yields, some economical factors such as power consumption P/V, heat and mass transfer capabilities, and to obtain sufficient product to determine product quality. At the same time a criterion is chosen mainly pragmatically (e.g., $P/V = $ const, OTR $=$ const) for scaling-up the industrial production unit. The procedure of this pragmatic-empiric approach is simple in hierarchy and the advantage is obvious. However, it must be kept in mind, that only individual situations in process operation can be handled because the interactions between metabolisms and bioreactors are quantified simultaneously in a non-separated form.

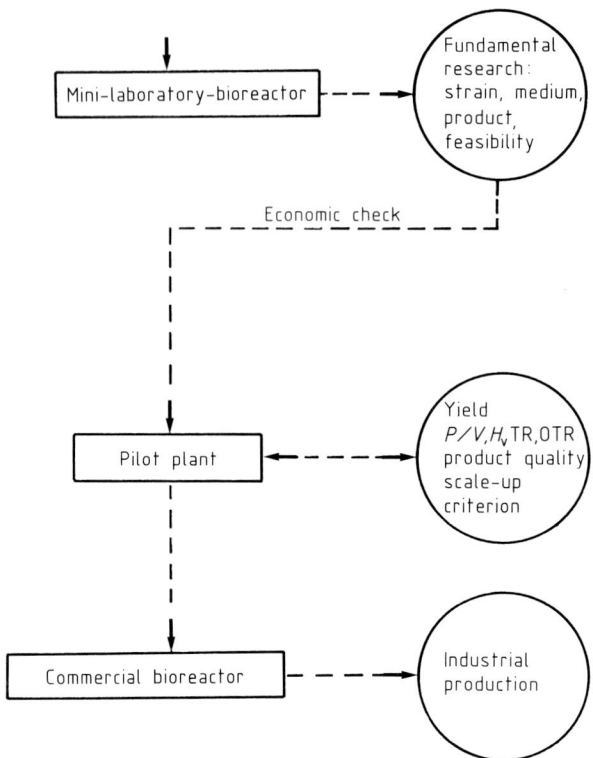

Figure 11.6. Bioreactors of different scale and their use for process design without mathematical models (pragmatic-empiric process development) (after MOSER, [11.7]). – TR Transfer rate (of H_V and oxygen, resp.).

11.2.1.3 Systematic Empiric Approach to Process Development

In agreement with the concept outlined in Sect. 11.1 a more systematic approach is to be preferred when one is interested in a generalizable scientific elucidation of bioprocessing, as indicated in more detail in Fig. 11.7 (MOSER [11.7]). The main difference to the pragmatic-empiric approach of Fig. 11.6 is that a systematic approach additionally contains a further stage of elaboration, i.e., the process kinetic analysis. The aim of this second stage in the systematic approach is to elaborate kinetics of the microbial reaction by neglecting transport phenomena. Furthermore, the bioreactor is quantified in terms of its physical transport characteristics, e.g., oxygen transfer rate OTR respecting $k_{L1}a$, mixing time, etc.

Process kinetic analysis is best carried out with the aid of mathematical models as these enable the detection of unknown factors connected with the interactions between microbial metabolism and bioreactors. Following the systematic approach, the use of pilot plants is much more essential, as the process model gained on bench scale, especially concerning the data in the "perfect bioreactor" and in the "bioreactor model", must be verified in the pilot bioreactor. As a result of the systematic-empiric approach, which is more complex than the purely empirical one, a process model on the level of a formal macroapproach is found, which is independent of scale. Process development for bioprocessing should be worked out very carefully because of the existence of many unknown biophysical phenomena. As stated in the literature (e.g., FINN and FIECHTER [11.11]; SITTIG and HEINE [11.12]; LEEGWATER et al. [11.13];

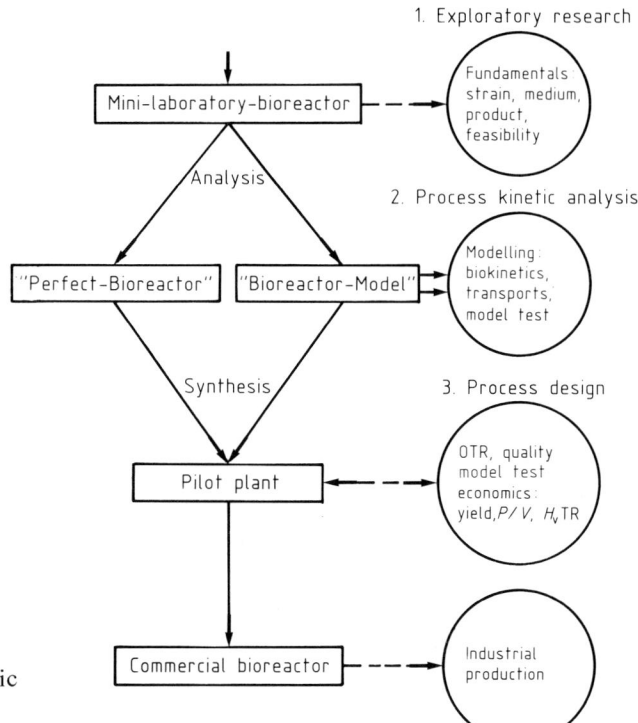

Figure 11.7. Bioreactors on different scale and their use for systematic empiric process design applying mathematical modelling (after MOSER, [11.7]).

TEMPEST and WOUTERS [11.14]; MOSER [11.15]) and recently summarized by MOSER [11.16], [11.17], the properties and performance of microorganisms are markedly influenced by the mode of operation.

11.2.2 Reactor Operation Modes and Reactor Balances

Better reliability and more secure process design is achieved by taking into account the concept of "kinetic similarity" between bioreactors (MOSER [11.7]). Here, certain rules must be followed when using bench-scale data for the design of bioreactor operation. This concept is based on the fundamental behavior of reactors shown in Fig. 11.8 for homogeneous reactors. Discontinuous (DCSTR), semi- (SCSTR), and continuous (CSTR) operational modes must be distinguished and the flow behavior of continuous reactors should be considered for limiting cases of ideally stirred tank reactors (CSTR) and plug flow reactors (CPFR).

Thus, reactor operations can be understood and characterized using the concentration profiles in time (c/t) and space (c/z). From these plots the similarity between DCSTR and CPFR as well as CPFR and NCSTR (reactor cascade with N reactors in a series) becomes evident.

As a consequence it cannot be expected that data from DCSTR fit the behavior of CSTR cultivation satisfactorily, while CPFR can be predicted on the basis of batch data.

This fact must be considered in process kinetic analysis and process design. Basically there exist two operational modes of homogeneous bioprocessing, i.e., the differ-

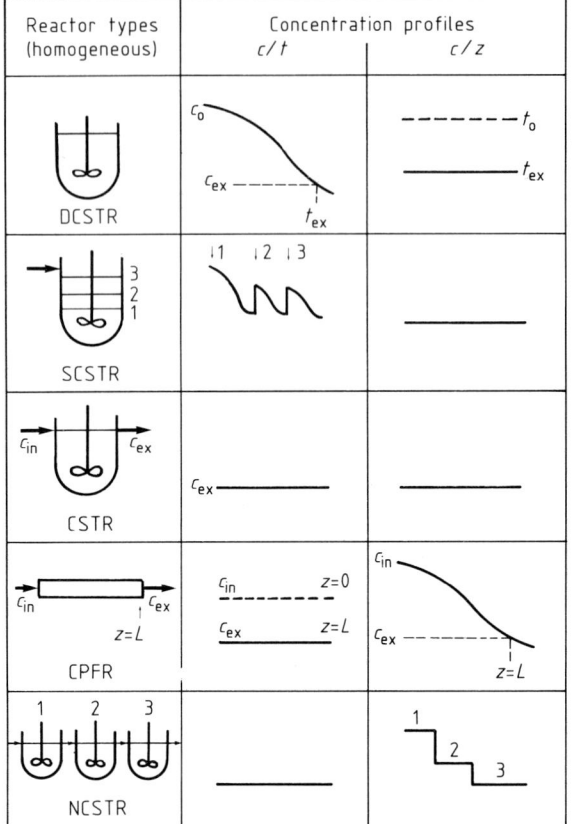

Figure 11.8. Basic reactor operational modes (homogeneous type) and their concentration profiles (MOSER [11.18]).

ential reactor and the integral reactor. A differential reactor operation is characterized by gradientless conditions, so-called "lumped parameter reactor models". Here, the reaction rate is directly related to the measured concentration:

$$r_j = \frac{c_j - c_{j,0}}{\bar{t}}. \tag{11.8}$$

This means that the reaction time t, or the mean residence time $\bar{t}$ in the case of continuous operation (in CSTR), can easily be calculated:

$$t_{\text{diff R}} \equiv \bar{t}_{\text{CSTR}} = \frac{c_j - c_{j,0}}{r_j}. \tag{11.9}$$

For integral reactor operation, Eq. (11.8) is not valid because of significant concentration gradients in the reactor configuration (so-called "distributed parameter reactor models"). Here, t can be evaluated only by integration according to:

$$t_{\text{int R}} \equiv t_b \equiv \bar{t}_{\text{CPFR}} = \int_{c_{j,0}}^{c_j} \frac{1}{r_j} dc_j, \tag{11.10}$$

where t_b is the reaction time in a batch process and $\bar{t}_{\text{CPFR}}$ the mean residence time in CPFR necessary to achieve a certain conversion.

The relationship between differential or integral reactor behavior and differential or integral kinetic analysis is schematically shown in Fig. 11.9 (MOSER [11.18]). In this scheme the drastic interactions between bioreactors and metabolism is again manifested. Integral reactor behavior, i.e., DCSTR and CPFR, require either the use

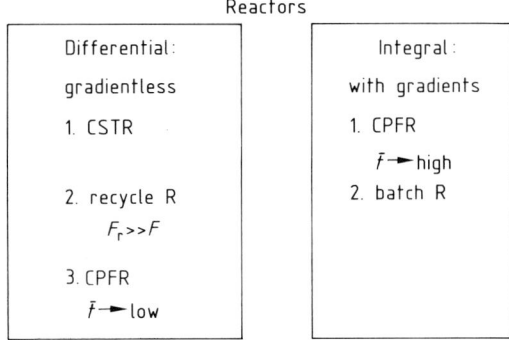

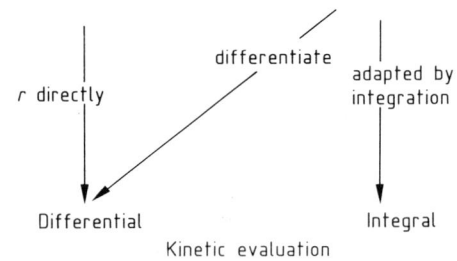

Figure 11.9. Differential and integral reactors versus differential and integral evaluation methods for kinetics. F liquid flow rate, F_r recycle liquid flow rate, R reactor.

Table 11.1. Definition of Reaction Rates or Rates of Consumption or Formation in Different Reactor Operations of the Homogeneous Type

Reactor	Reaction Rate	Eq. No.
Discontinuous stirred tank reactor (DCSTR)	$\pm r_j = \dfrac{dc_j}{dt}$	(11.13)
Semi-continuous stirred tank reactor (SCSTR) with $V(t) = V_o + F_{in} \cdot t$	$\pm r_j = \dfrac{dN_j}{V(t) \cdot dt}$	(11.14)
Continuous plug flow reactor (CPFR) (steady-state)	$\pm r_j = v_z \dfrac{dc_j}{dt}$	(11.15)
Continuous stirred tank reactor (CSTR) (steady-state) $F_{ex} = F_{in}$, V = const	$\pm r_j = \dfrac{F}{V}(c_{j,ex} - c_{j,in})$	(11.9)
Continuous recycling reactor (CRR) if $F_r \gg F$		(11.9)
Discontinuous recycling reactor (DCRR) (with V_r volume of recycle area)	$\pm r_j = \dfrac{V_r + V}{V} \cdot \dfrac{dc_j}{dt}$	(11.16)

of integral kinetic analysis in order to evaluate the biokinetics or the differentiation of reactor data by analytical, numerical, or graphical methods. Only differential reactor operations give the reaction rate directly (cf. Chapters 14 and 15).

As a first conclusion for the estimation of kinetic parameters from bioreactors it becomes evident that the correct balance equation of the mode of reactor operation must be chosen for proper kinetic analysis.

Eq. (11.1) can be rewritten for homogeneous reactors in a one-dimensional form (direction z) using the vector form:

$$\frac{\partial c_j}{\partial t} = -\text{div}(c_j \cdot v_z) + \text{div}(D \cdot \text{grad}\, c_j) + r_j \quad (11.11)$$

or:

$$\frac{\partial c_j}{\partial t} = -v_z \frac{\partial c_j}{\partial z} + D \frac{\partial^2 c_j}{\partial z^2} + r_j. \quad (11.12)$$

Here, two terms for transport phenomena appear for the two types of mechanisms of physical transport processes: convection by a flow velocity v_z and conduction by a concentration gradient, e.g., diffusion D or dispersion D_{eff}.

Eq. (11.12) can be used to derive the performance equations of basic reactor operation, summarized in Table 11.1.

In cases where significant deviations from the basic reactor models appear, more complex balance equations and models must be considered (e.g., SINCLAIR and BROWN [11.19]).

11.2.3 Pseudohomogeneity of Bioreactor Operation

In the previous section bioreactor concepts were presented using the homogeneous balances of basic reactor configurations. However, real situations of bioprocessing, where gas-, liquid-, and solid phases are present, are heterogeneous processes. This fact concerns all reaction and transfer interactions which are discussed in more detail in Chapter 17. At this point, where the use of bioreactors for the estimation of kinetic parameters is demonstrated, the problem of heterogeneity must also be considered as it heavily influences the procedure of proper kinetic analysis.

In analogy to chemical reaction engineering, heterogeneous processes can be treated and modelled as pseudohomogeneous under certain conditions. True kinetic constants k_r can be estimated, if the interaction between reaction and transport can be neglected and the transport phenomena, indicated in Fig. 11.3, do not become rate-limiting.

In general, if the transport rate constant k_{Tr} is a magnitude higher than k_r:

$$k_{Tr} \gg k_r, \tag{11.17}$$

then the efficiency factor η [cf. Eq. (11.6)] becomes unity. Sometimes in engineering science characteristic times for reaction (t_r) and transport (t_{Tr}) are used in place of the rate constants, so that Eq. (11.17) can be expressed as:

$$t_{Tr} \ll t_r, \tag{11.18}$$

where k-values are indirectly proportional to the characteristic times. An alternative to these concepts is the use of a characteristic length or diameter d instead of k_{Tr}-values. As k_{Tr} and d are related to each other according to mass transfer theories [cf. two-film theory, Eq. (8.16) in Chapter 8], these approaches are identical.

As a general conclusion for kinetic analysis of bioprocesses it becomes evident that the bioreactor used for kinetic analysis must be reexamined to determine if the above mentioned preconditions are fulfilled. This is called the test of pseudohomogeneity (MOSER [11.20], [11.18]).

Table 11.2 summarizes Eqs. (11.19)–(11.28), which are required for quantitative testing of pseudohomogeneity of three-

Table 11.2. Test of Pseudohomogeneity in Three-phase Bioprocessing with the Aid of a Set of Equations on the Basis of a Comparison of the Reaction Rate Constants k_r and that of Physical Transport k_{Tr} or of Characteristic Times t_r and t_{Tr}. – Instead of k-values characteristic thicknesses can also be applied (d_{crit}), reactor volume and temperature should be constant (MOSER [11.20])

Problem Transfer at/in Interface/Phase	in the Case of	Criteria	Eq. No.
1. G/L-Interface:	aerobic bioprocesses with flocs only	$k_{L1} \cdot a(o^* - o) \geq q_O \cdot x$ $\eta_{Tr} = 1$ if $0.3 \leq Ha$	(11.19) (11.20)
2. L-Phase: Micromixing Macromixing	STR Recycle-reactors CSTR CPFR	$t_m \leq 1/10 \cdot t_r$ $t_c \leq t_r$ $N_{eq} = 1$ $Bo_L \to \infty$	(11.21) (11.22) (11.23) (11.24)
3. L/S-Phase:	flocs and/or films	$Sh_{L2} = 2 + 0.4 \cdot Re_\varepsilon^{1/4} \cdot Sc^{1/3}$ $k_{L2} \propto v_L$	(11.25) (11.26)
4. S-Phase:	flocs and/or films 1-S-limitation	$d_p \leq d_{p,crit}$ $\eta_r = 1$ if $Da_{II} \equiv \Phi^2 \leq 0.1$ $d_{crit} = \dfrac{c^*}{K_S} \dfrac{\sqrt{1 + 2c^*/K_S}}{1 + c^*/K_S} \sqrt{\dfrac{YD_S K_S}{q_{max} \cdot \varrho}}$	(11.27) (11.28)

phase processes. The transport rates through the G/L-, L/S-interface and the L- and S-phase can be measured, calculated, and compared to the kinetic constants on the basis of these equations. External transport limitations at the L/S-interface can only be examined by means of Eq. (11.25) which represents an analogy to chemical reaction engineering. This has been verified for stirred tanks (STR) and bubble columns (BC) by relating the Reynolds number on the basis of energy dissipation ε (power input per volume) to the Sherwood number. Energy dissipation for STR can be calculated from:

$$\varepsilon_{STR} = \frac{P \cdot g_c}{V_L \cdot \varrho_L} = \frac{N_P \cdot d^5 \cdot n^3}{V_L}, \quad (11.29)$$

and for BC:

$$\varepsilon_{BC} = \frac{\Delta p \cdot F_G \cdot g_c}{V_L \cdot \varrho_L} = v_{SG} \cdot g. \quad (11.30)$$

Furthermore, internal transport limitations can be calculated by measuring and/or controlling the thickness of biofilms or diameter of bioflocs by means of Eqs. (11.27) and (11.28). Additional information can be found in Chapter 17 or in the original literature. It should be emphasized that external and internal transport limitations, especially when using STRs, cannot be excluded. As a consequence, increased values for K_S in Monod-type kinetics are found. Therefore, special bench-scale reactors are recommended for proper kinetic analysis. This will be discussed subsequently in the section on "perfect bioreactors".

In relation to the problem of pseudohomogeneity of bioprocesses it should be noted that reactions occurring inside the solid phase are almost never directly measurable. The internal fluxes of metabolic reactions can only be determined by the externally measured fluxes in the liquid medium using computer simulation. BARFORD and HALL [11.21] found that an external overall flux does not even approximately reflect the internal fluxes.

In vitro examination of an isolated part of the metabolism may be inadequate for the quantification of the coordinated and integrated biochemical control of intact living systems due to interactions between sections of metabolism *in vivo*.

11.2.4 "Perfect Bioreactors" as Model Bioreactors for Kinetic Analysis of Bioprocesses

The principal experimental difficulty associated with the determination of true biological rate coefficients in conventional bioreactors like the STR arises when measuring appropriate values for physical transports, i.e.: k_{L1}, $a_{G/L}$, k_{L2}, $a_{L/S}$, and d_p, the latter being the thickness of biofilms or the diameter of bioflocs.

As the batch process still represents the dominating operational mode in industry, process kinetic analysis must concentrate on this type. However, one of the major problems of the batch experiment lies in the difficulties involved in using it for the estimation of the K_S-value, which is, beyond μ_{max}, the second parameter in Monod-type kinetics. It is interesting to note that especially the value of K_S is strongly altered by unknown biological and physical factors (MOSER [11.18]; FIECHTER [11.22]). To separate external and internal transport limitations from biological reactions the STR is completely unsuitable. Due to a broad variation in shear rates over the reactor volume, the floc sizes exhibit a wide distribution (ATKINSON and UR-RAHMAN [11.23]) in STR, so that there is an unsatisfactorily defined situation, where internal transport phenomena falsify kinetics. Furthermore, the density of bioflocs ($\varrho_S \sim 1.05$) is about the same as that of the liquid ($\varrho_L = 1$), so that only low values of relative velocity are possible, leading to external transport limitation in STR (SHIEH [11.24]).

DECHEMA (Deutsche Gesellschaft für chemisches Apparatewesen) and the European Federation of Biotechnology (EFB) have dealt with this problem. Standardization has been proposed, e.g., for reactors the "reference bioreactor" (CRUEGER and HEINE [11.25]), for fermentations the "tests of comparison" (LEHMANN et al. [11.26]), as well as for quantification methods of bioreactors (CRUEGER [11.27]). Furthermore, some fundamental concepts are described in the literature concerning the methodology for determination of biological rate coefficients unaffected by physical transport. ATKINSON [11.28] summarizes the experimental methods, describes how to use bioreactors for this purpose, and concludes that only a biological film reactor (BFR) is suitable to estimate unfalsified biological rates as the thickness of the biofilm is uniform and can be measured easily (ATKINSON and DAOUD [11.29], KORNEGAY and ANDREWS [11.30]). However, the use of the BFR, which substantially is a sloping plane biofilm reactor, is limited to non-aseptic conditions so that ATKINSON emphasizes the use of STR together with a fluidized bed reactor (FBR), both under aseptic conditions. The microbial layer thickness in the FBR can be controlled by a combination of the shear that results from the liquid flow and an attrition caused by particle vibration. At sufficiently high flow rates, the excess of microbial growth will immediately be swept from the fermenter and contributes little to the overall performance. Under such circumstances an independent determination of the biological film thickness is not possible and the useful region of the experiment is restricted to small biological thicknesses and large thicknesses at low concentrations. The first of these conditions leads to the value of K_S as with the BFR and the latter to the ratio between μ_{max}, K_S, and the internal transport. ATKINSON and DAVIS [11.31] recommended the use of the completely mixed microbial film fermenter (CMMFF), which consists of a vertical tube containing biologically active particles. A centrifugal pump transports material from a holding vessel to the base of the tube and back to the holding vessel. The recirculation has a two-fold effect: it causes uniform concentrations throughout the reactor and also provides the necessary force to fluidize the support particles. This FBR is a method of overcoming microbial wash-out in continuous fermentation.

Recently a multi-purpose bioreactor (MPR) was shown to be a suitable concept for solving the problem of estimation of undisturbed biological rate coefficients under sterile conditions (MOSER [11.32]). This reactor configuration realized by MBR Bioreactor AG/Switzerland is characterized by use of different mixing devices in the vessel in vertical and horizontal positions. Especially cylinders are used without or with supplementary baffles and perforations at the surface to increase fluid mixing. The properties of this reactor take account of the known interfacial area for G/L- and L/S-mass transfer and theoretically calculable transfer coefficients (MOSER [11.33]; [11.15]. Furthermore, some other bench-scale reactors are described in the literature for biofilm processing. Beyond recycle reactors for carrying out heterogeneous bioreactions in analogy to chemical heterogeneous catalysis, which eliminate external transport limitations (PASPEK [11.34]; FORD et al. [11.35]) other laboratory equipment for biofilm processing is known: the totally mixed vertical rotating drum biofilm reactor (KORNEGAY and ANDREWS [11.30]) and the horizontal cylinder biofilm reactor (TOMLINSON and SNADDON [11.36]).

The proper use of bioreactors in process kinetic analysis is based on reaction engineering fundamentals. In order to estimate true biological rate coefficients it must be kept in mind, that bioreactor operation in practice includes not only homogeneous and heterogeneous processing in gradient-less reactors (e.g., CSTR) or reactors with gradients (e.g., CPFR and DCSTR) but also non-stationary modes of operation, which are very commonly used in industry. Although some efforts have been made to elucidate the dynamics of microbial populations (e.g., BAZIN [11.37]) there is still an incomplete understanding of bioreactor operation. A crude classification of the problems encountered is given in Fig. 11.10 (MOSER [11.20]). Bioprocessing as an integrating activity between biology and engineering requires criteria of classification from both fields. Beyond the reactor characteristics of heterogeneity and concentration gradients, non-stationary operation especially deals with the problem of "balanced" or "unbalanced" growth (CAMPBELL [11.38]). According to CAMPBELL's definition, growth is balanced only when all specific rates of change of process variables are constant (e.g., μ, q_O, q_C, q_S, q_P). On the other hand, steady-state growth is reached when the overall time rate of change of any metabolic variable is zero. While balanced growth is a necessary condition for steady-state growth, it is not a sufficient one. Likewise, steady-state is a sufficient condition for balanced growth but not a necessary one.

Fig. 11.10 indicates the range of bioreactor operation compared to balanced growth. Even though a mathematical solution of the variety of problems with bioreactor operations is not yet available, some concepts of general validity can be emphasized as guidelines (BARFORD et al. [11.21]; MOSER [11.16]; [11.17]; [11.39]).

Such mathematical modelling of biokinetics shows that in both cases of interactions (macroscopic transport effects and metabolic changes on the microscopic level) a drastic reduction of the complexity of models can be achieved with the aid of the principles of the rate-limiting-step concept and of the quasi-stationary concept. This requires application of the macroscopic method together with a formal kinetic analysis.

First, to allow for easy reproducibility, the reactor used for experimental investigations should be one of the basic reactor models (discontinuous and continuous stirred tank, plug flow reactor). The definition for pseudohomogeneous reaction rates in these reactor models is applicable if pseudohomogeneity is experimentally verified. Deviations due to interactions of transports can be handled using the described methodology.

Secondly, bioprocesses in non-stationary situations should be treated according to the same integrating strategy with the aid of the mentioned principles in order to yield simple but adequate kinetic models. Once again, the variety of process situations can be reduced to only a few fundamental cases. The recommendations given here can serve as guidelines for experimental work necessary to elaborate the required, relevant data and models in transient operation techniques. The dominant process in industry still is the batch process with various forms of semidis- and semicontinuous modes of operation.

For process kinetic analysis the following aspects should be observed:

1. The validity of measured kinetics on any level is restricted to a specific

Reactor operation	Biological growth	"balanced"	"unbalanced"
gradientless	steady-state	CSTR	
	quasi-steady-state	"extended culture"	
	unsteady-state	←----- CSTR, SCSTR ---------→	
	"periodic"	←---"transient operation techniques"----→	
with gradients	steady-state	←----- CPFR, NCSTR -----→	
	unsteady-state	←--- DCSTR, CPFR, NCSTR ---→	
	"periodic"	←---"transient operation techniques"----→	

Figure 11.10. Bioreactor operation techniques and their biological behavior (MOSER [11.20]).

reactor type being used for the investigations. In the case of insufficiently defined transport influences, these macrokinetics cannot be used for extrapolation to any other scale or to any other reactor.

2. Kinetic data and models should be transferred to other reactors only in the range of "kinetic similarity" (e.g., batch stirred-tank plug flow reactor), as long as there is a lack of better understanding of the interactions between environment and metabolism.

3. To some extent the kinetic models used nowadays can be used for description and simulation of bioprocessing even in the case of simple unstructured models, if the parameters of kinetic models are allowed to be adjusted to process behavior by formulating them in time-dependence [e.g., μ_{max}, K_S, and $Y = f(t)$].

4. Simple unstructured models can successfully be applied to the description of the macroscopic behavior if the models are adapted to the process (S-inhibition, 2-S-limitations, lag-, stationary-, death phase ...). Adaptive modelling according to the deductive research method is essential (MOSER [11.18]). This is also valid in balanced, frozen, and real transient situations.

5. Structured modelling is needed only in cases of real transient processing where the characteristic times of environmental change (t_e) or reaction (t_r) are comparable ($t_e \sim t_r$), if the aim of investigation is not at the level of engineering calculations of conversion but is at the microscopic level of metabolic mechanisms (TODA [11.40]). Such models should be of greater interest for engineering when rapid and secure process control is realized by developing specific sensors. Reactivity of different strains in reactors must still to be worked out.

11.3 The Use of Process Kinetics to Determine Optimum Operation of Bioreactors

Most importantly in bioprocess engineering the best suitable type of bioreactor system must be chosen and its operating conditions specified. In the past, the batch process has been predominantly used. Its design was made mainly by trial and error. Many fermentation products are still produced as "fine" chemicals, where asepsis and yield improvement by strain selection was of greater importance than any improvement in productivity due to reactor design.

In other bioprocesses such as biological waste water treatments a lack of reliable kinetic data is still a drawback in the application of process design strategies, due to the fact that significant process variables cannot be monitored. However, the process-analysis approach in analogy to chemical reaction engineering (LEVENSPIEL [11.41]; [11.42]) may contribute significantly to a systematic process development (BISCHOFF [11.43]; TOPIWALA [11.44]). The costs for separation and product purification often depend on reactor efficiency and, thus, the economics of a process may depend on the correct choice of the optimum operation of bioreactors.

In fermenter design two types of optimization problems exist, the choice of optimum productivity (i.e., maximum output) or that of optimum conversion (i.e., maximum yield). At high substrate costs, or if low concentrations in the effluent are wanted, the latter criterion becomes more important.

As will be shown later, the CSTR cannot achieve high conversions at high productivities while a CPFR is able to realize both criteria at an optimum value. However, to assess the performance of any bioreactor system, kinetics of bioprocesses must be considered in context with the bioreactor environment, i.e., rate equations have to be integrated over the reactor. In general, the performance of a reactor is the result of combining the kinetics with the balance equation of individual reactor types (cf. Sect. 11.2). Concerning the kinetics of bioprocesses a classification in analogy to chemical kinetics is very useful to show the general picture of the mentioned context between kinetics and reactor design.

11.3.1 Influence of Macromixing on Continuous Operation

Fig. 11.11 shows the dependence of reaction rates r_i (respectively, rates of formation or consumption) on concentration for vary-

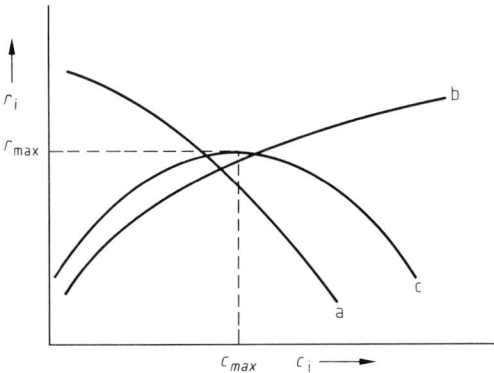

Figure 11.11. The dependence of the rate r_i of formation or consumption of a component i on the concentration for: (a) ordinary kinetics, e.g. enzyme technology; (b) autocatalytic processes, e.g. biological growth; and (c) real biotechnical processes, e.g. fermentation and waste water technology.

ing kinetics (curves a–c)). Hybrid kinetics is the dominating type in real biotechnical processes and represents a combination of an autocatalytic part with a part with ordinary kinetics at the end of the process.

Here, the reaction rate passes through a maximum at c_{max}.

Design considerations for these three cases are quite different. Also some general results appear, as will be shown.

11.3.1.1 Stirred Tank versus Plug Flow Reactor

The balance equations for reactors with and without gradients in the homogeneous case have already been used for determining reaction times needed for achieving either a certain concentration in the effluent $c_{j,ex}$ or a desirable conversion. Fractional conversion ζ of component i for constant volume systems is defined as:

$$\zeta_i = \frac{c_{i,0} - c_{i,ex}}{c_{i,0}}. \tag{11.31}$$

A graphical estimation and geometrical interpretation of the holding times t_b, $\bar{t}_{CPFR}$, and $\bar{t}_{CSTR}$ can be found according to Eqs. (11.9) and (11.10) in a plot of $1/r_i$ versus c_i. Following this concept, t_b or $\bar{t}_{CPFR}$ is the area under the curve from $c_{i,0}$ to $c_{i,ex}$, while $\bar{t}_{CSTR}$ is the area of the rectangle with sides of

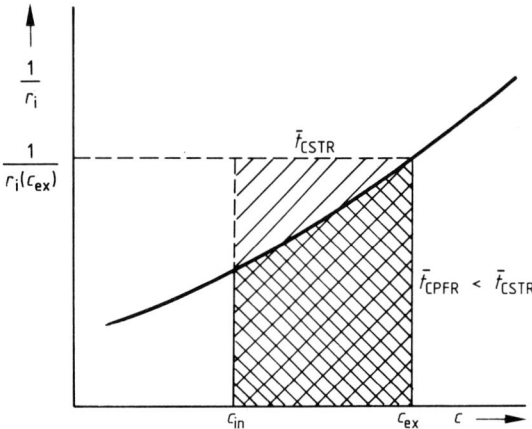

Figure 11.12. Graphical estimation of mean residence time $\bar{t}$ in a CSTR (////) or CPFR or DCSTR (\\\\) for ordinary kinetics. In this case an integral reactor is superior.

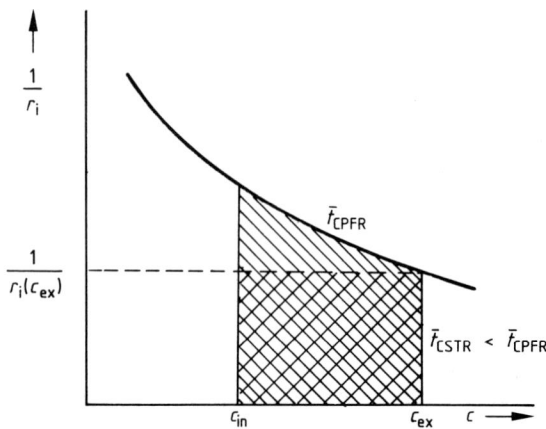

Figure 11.13. Graphical estimation of mean residence time $\bar{t}$ in a CSTR (////) or CPFR or DCSTR (\\\\) for autocatalytic processes, in which case the differential reactor is superior.

length $c_{i,ex} - c_{i,0}$ and $1/r_i$. Figs. 11.12 to 11.14 depict schematically the situations for the three basic kinetics mentioned in Fig. 11.11. Clearly the CSTR is inferior to CPFR for ordinary kinetics but requires less holding time than the DCSTR or CPFR for an autocatalytic process. In some biotechnical processes autocatalytic kinetics prevail during the exponential growth phase and its onset, while ordinary kinetics are observed as the growth rate decelerates and the stationary and death phases ensue. This suggests that optimal design of some fermentations would involve continuous operation in a CSTR combined with a CPFR (BISCHOFF [11.43]). This fact is illustrated in Fig. 11.14. Using the simple Monod-type equation for quantifying biokinetics, the holding times t_b and $\bar{t}_{CSTR}$ can be calculated by means of Eqs. (11.9) and (11.10) as follows (TOPIWALA [11.44]):

$$t_b \equiv \bar{t}_{CPFR} = \int_{s_0}^{s_{ex}} \frac{Y(K_S + s)}{\mu_{max} \cdot s \cdot x} \cdot ds \qquad (11.32)$$

or:

$$\bar{t}_{CSTR} = \frac{(s_0 - s_{ex}) \cdot Y(K_S + s)}{\mu_{max} \cdot s \cdot x}. \qquad (11.33)$$

It can be seen from Fig. 11.14 that, provided $s_0 \gg K_S$ (curve with K_{S1}), a single CSTR would be the optimum choice for all the cases except those requiring almost total conversions, where a CSTR-CPFR sequence would be optimum. The situation of curve with K_{S1} is given for biomass fermentations, where for reasons of productivity $s_0 \gg K_S$. In the case of biological waste water treatment $s_0 \approx K_S$ so that the required conversion will be reached in a CSTR-CPFR sequence (curve with K_{S2}).

By this general procedure the optimum operation of bioreactors can be estimated at varying operating conditions (LEVENSPIEL [11.42]). RICICA [11.45] applied this strategy to streptomycin production.

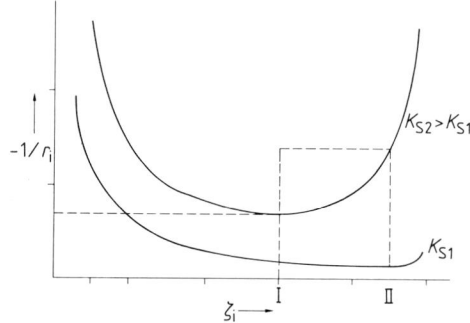

Figure 11.14. Graphical analysis of fermenter design for Monod kinetics with varying K_S-values ($K_{S1}/s_0 = 0.01$; $K_{S2}/s_0 = 1$ with $\mu_{max} = 0.8$). A certain conversion ζ_i in an optimum bioreactor system is achieved by combining a CSTR (I) with a CPFR (II) (after TOPIWALA [11.44]).

11.3.1.2 Recycle Operation

For autocatalytic reactions mixed flow (CSTR) is generally more efficient at low rates of conversion and CPFR is more efficient at high rates of conversion. However, a CPFR will not operate at all with a feed of pure reactants without cells due to the autocatalytic nature. In such a situation the feed must be continually primed with product (cell mass X), which is an ideal opportunity for using a continuous recycle reactor (CRR). When material is to be processed to some fixed conversion ζ in a CRR, reflection suggests that there exists a particular recycle ratio r which is optimal. At this point the reactor volume is minimized. The final solution is given by LEVENSPIEL [11.42] by applying $\partial x/\partial r = 0$ in the equation for recycle reactors:

$$\frac{K_S}{s_0} \cdot \ln \frac{s_0 + r \cdot s}{r \cdot s} + \ln \frac{r+1}{r} =$$
$$= \frac{r+1}{r} \cdot \frac{K_S}{s_0 + r \cdot s} + \frac{1}{r}. \qquad (11.34)$$

This equation can be solved by trial and error. The optimum r is found to be a function of K_S/s_0 and s/s_0. A graphical procedure for the determination of the optimum r is indicated in Fig. 11.15 (LEVENSPIEL [11.42]). One has to try different values of $\zeta_{i,1}$ until the two shaded areas are equal.

Comparing the CRR with single CPFR or CSTR the following conclusions can be drawn:

1. For rates of conversion smaller than the maximum rate the CSTR is superior to any CRR.
2. For rates of conversion higher than the maximum rate the CRR with the proper recycle ratio is superior to either the CPFR or CSTR.

Although Eq. (11.34) gives the best possible recycle ratio, the operational mode can still be optimized by operating at a maximum rate whenever possible. This can be achieved by an infinite recycle or by introducing proper feed at the appropriate place within the reactor.

11.3.1.3 Optimum Operation with Inhibition Kinetics

The same strategy can also be applied in the case of more complex situations in bioprocessing, where inhibition by substrates, CO_2, or products occurs. Fig. 11.16 schematically shows the best way of operating a CPFR for optimum productivity and opti-

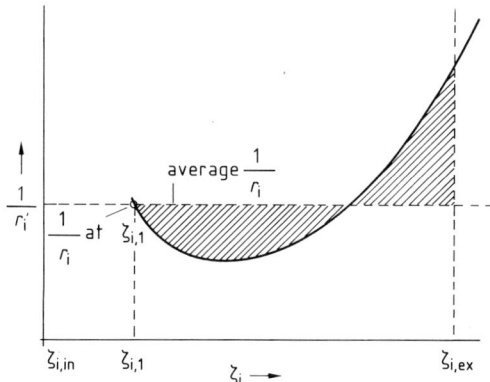

Figure 11.15. Graphical estimation of minimum residence time in a recycle reactor by adjusting the recycle ratio to an optimum value following the concept of making the shaded areas equal (after LEVENSPIEL, [11.41]).

This means that the value of rate^{-1} of the feed entering the reactor (with $c_{i,1}$) is equal to the average rate^{-1} in the reactor.

At too high recycle ratios the average rate is higher than the rate in the feed entering the reactor and *vice versa*.

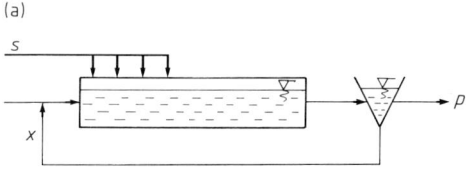

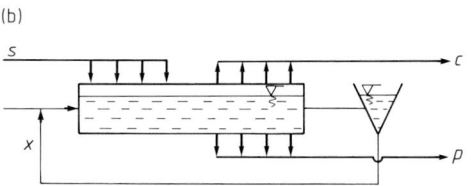

Figure 11.16. Optimum bioreactor operation with complex biokinetics in tubular reactors: (a) substrate inhibition and (b) inhibition by substrate, product, and CO_2 (after MOSER, [11.46]).

mum conversion at the same time (MOSER [11.46]). The reactor configuration (a) is the best in the case of S-inhibition and the scheme (b) is an optimal solution for inhibition by metabolites, products, and CO_2. For P-inhibition, alternatives are described in the literature: coupling fermentation with simultaneous solvent extraction (MINIER and GOMA [11.47]), or *in situ* removal under vacuum (CYSEWSKI and WILKE [11.48]).

11.3.2 Influence of Micromixing

Up to now we have considered the relation between biokinetics and design of ideal reactor types such as DCSTR, CPFR, and CSTR. It should be emphasized that the performance of reactors is also affected by other factors like interfacial transport, macro- and micromixing, properties of media, etc.

Besides transport and macromixing, which is identical with the residence time distribution (RTD), one must consider the influence of micromixing and some properties such as viscosity, shear rate, etc. Micromixing models of steady-state flow reactors distinguish totally segregated flow (ts) and maximum mixing (mm) (e.g., WEN and FAN [11.49]). While the ts-model visualizes that the feed stream enters a CPFR with side exits through which portions of the flow leave after satisfying their residence time requirement, the mm-model operates with a feed stream entering side entrances of a CPFR. Both situations are illustrated in Fig. 11.16 according to ZWIETERING [11.50]. Normally, this flow behavior is thought to be realized in ideal CSTR (mm-flow) and CPFR (ts-flow) because macro- and micromixing are connected to each other in such ideal reactor types. It should be emphasized that recycle reactors exhibit a behavior where micromixing and macromixing are functions of the recycle rate (MOSER and STEINER [11.51]) and vary between the extreme cases of ts and mm or CPFR and CSTR. For quantification of the extreme cases of micromixing a degree of segregation J was proposed by DANCKWERTS [11.52], and modified by ZWIETERING [11.50]:

$$J = \frac{\operatorname{var} \alpha_p}{\operatorname{var} \alpha} \equiv 1 - \frac{\operatorname{var} \alpha_i}{\operatorname{var} \alpha}. \qquad (11.35)$$

It is defined as the variance of the age between the "points" (var α_p) in the reactor divided by the variance of age of all molecules in the reactor or using the variance of age of "points" var α_i. A "point" was defined as a volume element, which is small compared to the reactor, but still contains a number of molecules.

The conversion of a component i in the case of the ts-flow model with $J=1$ can be calculated with the aid of the following equation presented by DANCKWERTS [11.52]:

$$c_{i,\,ts} = \int_0^\infty c_{i,\,b}(t) \cdot f(t) \cdot dt, \qquad (11.36)$$

where $c_{i,b}(t)$ is the concentration from batch kinetics and $f(t)$ a RTD-function according to the impulse method. In the case of the mm-flow model with $J=0$ a somewhat more complicated equation (ZWIETERING [11.50] results:

$$\frac{dc_i}{d\lambda} = r_i(c_i) - \frac{f(\lambda)}{1 - F(\lambda)} [c_{i,\,0} - c_i(\lambda)], \qquad (11.37)$$

where λ is the life expectation ($\lambda = 1 - \bar{t}$) and $F(\lambda)$ the RTD-step function.

Application of the concept of segregation J to bioprocessing is simplified for cases with kinetics of reaction order one. Here, Eq. (11.36) can be used, as at this reaction order Eq. (11.37) becomes identical with Eq. (11.36) (e.g., sterilization techniques). Several papers in the literature use the concept of segregation in order to computerize the effect of micromixing on microbial growth processes (TSAI et al. [11.53]), the effect of mixing on the washout and steady-state performance of continuous cultures (FAN et al. [11.54]), and the simultaneous effect of micro- and macromixing on growth (FAN et al. [11.55]).

It has been shown, that, when the number of CSTRs in a cascade is small, micromixing has a significant effect. As the number increases the micromixing effect on growth decreases. Fig. 11.17 gives a comparison of calculated exit concentrations x of cells X and s of substrate S from CSTR and CPFR (TSAI et al. [11.53]) and demonstrates that micromixing effects are very appreciable and cannot be neglected in the

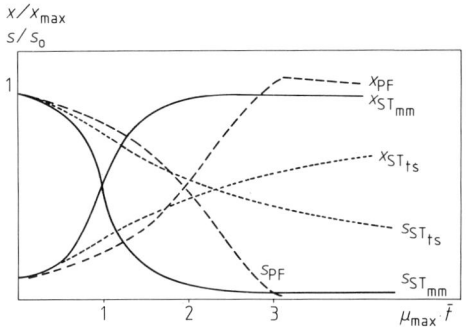

Figure 11.17. Comparison of exit concentrations for biomass x and substrate s as a function of mean residence time $\bar{t}$ for varying macromixing. - Plug flow reactor (PF) and stirred tank (ST) at different levels of micromixing; maximum mixing (mm) and total segregation (ts) (after TSAI et al. [11.53]).

design of bioreactors. Segregation may be undesirable in growth processes for any RTD. The use of a recycle type reactor as a kind of maximum mixed reactor is emphasized by SAWADA and KOJIMA [11.56]. This result is in agreement with some earlier findings of MOSER [11.57], where the behavior of "mixed plug flow" was proposed as optimum. In the search for optimum operation of bioreactors several combinations of CSTR and CPFR with and without cell recycle have been studied recently by computer simulation by TODA and DUNN [11.58], resulting in a system of two CSTR with CPFR. This system exhibits higher rates and higher productivities than a single CSTR. The importance of coupling mixing characteristics and biokinetics was also de-

monstrated by BAJPAI and REUSS [11.59]. Here circulation time distributions were measured in different reactors and used to satisfactorily predict the effect of different scales of operation upon productivities and critical dilution rates in continuous operation and fed-batch cultures of yeast. One may expect that such an approach to process analysis will lead not only to a better understanding of operation at different scales and in different reactors, but also to a more systematic strategy in bioreactor design.

11.4 References

[11.1] A. MOSER; Biotechnol. Lett. *4* (1982), 73.
[11.2] P. GLANSDORFF and I. PRIGOGINE: "Thermodynamics of Structure, Stability and Fluctuations". John Wiley, New York, 1974.
[11.3] J. A. ROELS and N. W. F. KOSSEN; Progr. Ind. Microbiol. *14* (1978), 95.
[11.4] J. A. ROELS; Biotechnol. Bioeng. *22* (1980), 2457.
[11.5] M. REUSS, R. K. BAJPAI, R. LENZ, H. NEBELSCHÜTZ, and A. PAPALEXIOU: 6th Int. Ferment. Symp., F-7.2.1. London/Ontario (Canada), 1980.
[11.6] K. POPPER: "The Logic of Scientific Discovery". Hutchinson, London, 1934, 1972.
[11.7] A. MOSER; in "Preprints 1st Eur. Congr. Biotechnol.", Interlaken/Switzerland, 1978, Part 1, p. 88.
[11.8] A. MOSER; Acta Biotechnol. *3* (1983) 3, 195.
[11.9] H. HOFMANN; Chem. Ing. Tech. *47* (1975), 191.
[11.10] S. AIBA, A. E. HUMPHREY, and N. F. MILLIS: "Biochemical Engineering". Academic Press, New York, 1976.

[11.11] R. K. FINN and A. FIECHTER; Symp. Soc. Gen. Microbiol. 20 (1979), 83.
[11.12] W. SITTIG and H. HEINE; Chem. Ing. Tech. 49 (1977), 595.
[11.13] M. P. M. LEEGWATER; J. Chem. Tech. Biotechnol. 32 (1982), 92.
[11.14] D. W. TEMPEST and J. T. M. WOUTERS; Enzyme Microbiol. Technol. 3 (1981), 283.
[11.15] A. MOSER; Biotechnol. Lett. 4 (1982), 281.
[11.16] A. MOSER; in "Proc. 3rd Austrian-Italian-Yugoslav. Conf. Chem. Eng.". 1982, Vol. 2, p. 620.
[11.17] A. MOSER: "Bioprocess Engineering". Springer Verlag, New York, 1985.
[11.18] A. MOSER: "Bioprozeßtechnik". Springer Verlag, Wien-New York, 1981.
[11.19] C. G. SINCLAIR and D. E. BROWN; Biotechnol. Bioeng. 12 (1970), 1001.
[11.20] A. MOSER; in "Proc. 2nd Intern. Conf. Waste Treatment and Utilization" (M. MOO-YOUNG et al., eds.). Vol. 2, p. 177. Pergamon Press, Oxford - New York, 1982.
[11.21] J. R. BARFORD and R. J. HALL; 7th Austrial. Conf. Chem. Eng. 21 (1977).
[11.22] A. FIECHTER; in "Biotechnology - a Comprehensive Treatise" (H. J. REHM and G. REED, eds.) Vol. 1, Chap. 7. Verlag Chemie, Weinheim-Deerfield Beach, Florida-Basel, 1982.
[11.23] B. ATKINSON and F. UR-RAHMAN; Biotechnol. Bioeng. 21 (1974), 221.
[11.24] W. K. SHIEH; Ph. D. Thesis, University of Massachusetts, Amherst, 1978.
[11.25] W. CRUEGER and H. HEINE: "Referenz Bioreaktoren" in "Arbeitsmethoden für die Biotechnologie". DECHEMA, Frankfurt am Main, 1982.
[11.26] J. LEHMANN, B. MARTIN, G. PIEHL, R. SCHULTZE, and W. STEVEN: "Vergleichstests für Fermentationen" in "Arbeitsmethoden für die Biotechnologie". DECHEMA (Deutsche Gesellschaft für chemisches Apparatewesen), Frankfurt am Main, 1982.
[11.27] W. CRUEGER: "Process Variables in Biotechnology". Working party Bioreactor Performance of EFB. DECHEMA (Deutsche Gesellschaft für chemisches Apparatewesen), Frankfurt am Main, 1984.
[11.28] B. ATKINSON: "Biochemical Reactors". Pion Ltd., London, 1974.
[11.29] B. ATKINSON and I. S. DAOUD; Trans. Inst. Chem. Eng. 48 (1970), 245.
[11.30] B. H. KORNEGAY and J. F. ANDREWS; J. Water Pollut. Control Fed. 40 (1968), 460.
[11.31] B. ATKINSON and I. J. DAVIES; Trans. Inst. Chem. Eng. 50 (1972), 208.
[11.32] A. MOSER; Adv. Ferment. 83, Suppl. to Process Biochem. (1983), 202.
[11.33] A. MOSER; Chem. Ing. Tech. 49 (1977), 612.
[11.34] St. C. PASPEK; Chem. Eng. Educ. (1980), 78.
[11.35] J. R. FORD, A. H. LAMBERT, W. COHEN, and R. P. CHAMBERS; Biotechnol. Bioeng. Symp. 3 (1972), 267.
[11.36] T. G. TOMLINSON and D. M. SNADDON; Air Water Pollut. 10 (1966), 865.
[11.37] M. J. BAZIN (ed.): "Microbial Population Dynamics", CRC Press, Boca Raton (Florida), 1982.
[11.38] A. CAMPBELL; Bacteriol. Rev. 21 (1957), 263.
[11.39] A. MOSER; Acta Biotechnol. 4 (1984), 3.
[11.40] K. TODA; J. Chem. Tech. Biotechnol. 31 (1981), 775.
[11.41] O. LEVENSPIEL: "Chemical Reaction Engineering". John Wiley, New York, 1972.
[11.42] O. LEVENSPIEL: "The Chemical Reactor Omnibook". OSU Book Stores Inc., Corvallis/OR (USA), 1979.
[11.43] K. B. BISCHOFF; Can. J. Chem. Eng. 45 (1966), 281.
[11.44] H. H. TOPIWALA; Biotechnol. Bioeng. Symp. 4 (1973), 681.
[11.45] J. RICICA; in Ferment. Adv. (D. PERLMAN, ed.) (1969), 427.
[11.46] A. MOSER; in "Proc. 33rd Canad. Chem. Eng. Conf., Toronto", October 1983, Vol. 2, p. 417.
[11.47] M. MINIER and G. GOMA; Biotechnol. Lett. 3 (1981), 405.
[11.48] G. R. CYSEWSKI and C. R. WILKE; Biotechnol. Bioeng. 19 (1977), 1125.
[11.49] C. Y. WEN and L. T. FAN: "Models for Flow Systems and Chemical Reactors". Marcel Dekker Inc., New York, 1975.
[11.50] Th. N. ZWIETERING; Chem. Eng. Sci. 11 (1959), 1.
[11.51] A. MOSER and W. STEINER; Chem. Ing. Tech. 47 (1975), 211.
[11.52] P. V. DANCKWERTS; Chem. Eng. Sci. 8 (1958), 93.
[11.53] B. I. TSAI, L. E. ERICKSON, and L. T. FAN; Biotechnol. Bioeng. 11 (1969), 181.

[11.54] L. T. FAN, L. E. ERICKSON, P. S. SHAH, and B. I. TSAI; Biotechnol. Bioeng. *12* (1970), 1019.

[11.55] L. T. FAN, B. I. TSAI, and L. E. ERICKSON; AIChE J. *17* (1971), 689.

[11.56] T. SAWADA and Y. KOJIMA; J. Ferment. Technol. *52* (1974), 848.

[11.57] A. MOSER; in "Proc. 3rd Symp. Tech. Mikrobiol.", Berlin, 1973 (H. DELLWEG, ed.), p. 61.

[11.58] K. TODA and I. J. DUNN; Biotechnol. Bioeng. *24* (1982), 651.

[11.59] R. K. BAJPAI and M. REUSS; Can. J. Chem. Eng. *60* (1982), 384.

Chapter 12

Rate Equations for Enzyme Kinetics

Anton Moser

Institut für Biotechnologie, Mikrobiologie und Abfalltechnologie
Technische Universität Graz
Graz, Austria

12.1   Introduction
12.2   Simple Enzyme Kinetics
12.2.1 Identification Plots for Simple Enzyme Kinetic Models
12.2.2 Reaction Mechanisms of Simple Enzyme Kinetics
12.3   Enzyme Kinetics with Complex Reactions
12.3.1 Reaction Mechanisms and Kinetic Equations
12.3.2 General Strategies to Derive Rate Equations for Complex Kinetics
12.4   Inhibition Kinetics of Enzyme Reactions
12.5   Effects of pH on Enzyme Kinetics
12.6   Temperature Dependence of Enzyme Kinetics
12.6.1 Arrhenius Equation
12.6.2 Enthalpy/Entropy-Compensation (Isokinetic Model)
12.7   Kinetics of Enzyme Control and Regulation
12.7.1 Physiological Regulation Phenomena
12.7.2 Kinetic Models of Enzyme Activity Control
12.7.3 Kinetic Models of the Regulation of Enzyme Synthesis
12.8   References

List of Symbols

A, B, C	kg	amounts of components
a, b, c	kg·m^{-3}	concentration of components
c	kg·m^{-3}	concentration (general term)
c_R	kg·m^{-3}	concentration of ribosomes
E	kg	amount of enzyme
e	kg·m^{-3}	concentration of enzyme
E_a	J·mol^{-1}	activation energy
$\{ES\}$	kg	amount of activated complex
$\{es\}$	kg·m^{-3}	concentration activated complex
ΔG	J·mol^{-1}	free energy of reaction
ΔH	J·mol^{-1}	reaction enthalpy
$H(H^+)$	kg	amount of hydrogen (ions)
$h(h^+)$	kg·m^{-3}	concentration of hydrogen (ions)
h	J·s	Planck constant ($6.626 \cdot 10^{-34}$ J·s)
I	kg	amount of intermediary product or inhibitor
i	kg·m^{-3}	concentration of intermediary product or inhibitor
K_I	kg·m^{-3}	inhibition constant
K_{IS}	kg·m^{-3}	constant of substrate inhibition
K_H	kg·m^{-3}	constant of Hill kinetics, Eq. (12.70)
K_m	kg·m^{-3}	constant of Michaelis-Menten kinetics, Eq. (12.7)
K_m^*	kg·m^{-3}	constant of complex enzyme kinetics, Eq. (12.32)
K_S	kg·m^{-3}	constant of Monod kinetics, Eq. (12.8)
$k_r(k_{+1}, k_{-1}, k_i)$	(m^3·mol^{-1})$^{n-1}$·s^{-1}	rate coefficient for reaction
k_B	J·K^{-1}	Boltzmann constant ($1.38 \cdot 10^{-23}$ J K^{-1})
k_{cat}	(mol P)(mol E)$^{-1}$·s^{-1}	catalytic constant, Eq. (12.28)
k_{cat}^*	(mol P)(mol E)$^{-1}$·s^{-1}	catalytic constant with complex enzyme kinetics, Eq. (12.31)
k_d	s^{-1}	rate constant of microbial death
k_E	s^{-1}	kinetic constant for enzyme synthesis, Eq. (12.72)
k_{rds}		rate constant of rate-determining step
k_∞	(m^3·mol^{-1})$^{n-1}$·s^{-1}	preexponential factor in Arrhenius equation, Eq. (12.60)
OH^-	kg	amount of OH^- ions
P	kg	amount of product
p	kg·m^{-3}	product concentration
q_S	s^{-1}	specific rate of substrate consumption
r	kg·m^{-3}·s^{-1}	rate of reaction
r_i, r_j	kg·m^{-3}·s^{-1}	rate of formation or consumption of components i, j
r_S	kg·m^{-3}·s^{-1}	rate of substrate consumption
R	J·K^{-1} mol^{-1}	universal gas constant (8.314)

List of Symbols

S	kg	amount of substrate
s	kg·m^{-3}	concentration of substrate
ΔS	J·K^{-1}·mol^{-1}	reaction entropy
T	K (°C)	temperature
T_β	K (°C)	isokinetic temperature
t	s	time
t_L	s	lag time
v	kg·m^{-3}·s^{-1}	initial rate of enzyme reactions
X	kg	amount of biomass
x	kg·m^{-3}	concentration of biomass
$Y(Y_{X/S})$	—	yield coefficient

Indices

ads	adsorption
des	desorption
denat	denaturation
cat	catalysis
E	enzyme
ES	enzyme-substrate complex
eq	equilibrium
ex	exit
i, j	components
G	gas
L	liquid
max	maximum
obs	observed
p	product
r	reaction
R	ribosome
rds	rate-determining step
S	solid or substrate
tot	total
Tr	transport
X	biomass
$\neq$	activated state-value
—	mean value
0	initial value

Greek letters

α, β	—	coefficients
μ	s^{-1}	specific rate of growth
$\zeta(\zeta_S)$	—	fractional conversion

Abbreviations

rds	rate-determining step
qss	quasi-steady-state
mRNA	messenger ribonucleic acid

12.1 Introduction

The effective rate, which is observed and measured in 3-phase experiments, is the sum of reaction and transport steps. These "macrokinetics" therefore include the external transport limitations (at G/L- and L/S-interface in the bulk of the liquid phase) and the internal transport limitations (in the S-phase). Eliminating these influences by technical manipulations, e.g., by turbulent conditions, gives the rate of change in this case of a pseudohomogeneous situation. These "microkinetics" still contain the steps of adsorption of the reactants on the active site of the catalyst, the reaction itself, and the desorption of the products. The same picture is valid in the case of heterogeneous chemical catalysis and enzymatic reactions. Microbial activities like growth and product formation can be regarded as a sequence of enzymatic reactions. On this basis PERRET [12.1] constructed a kinetic model for a growing bacterial cell population. The main pathways for major nutrients are considered together with pathways for minor nutrients and trace elements linked to each other. This metabolic network including autocatalysis, however, can be simplified with the aid of the rate-determining-step concept resulting in a "master reaction" or bottle-neck, in which the enzymatic reactions limit the total flux and the overall rate of the process (cf. Fig. 11.5).

This picture can be used to demonstrate that enzymatic reactions are the center of kinetic considerations for microbial systems, even if it is hard to directly verify such a mechanistic approach in real fermentations. According to the research strategy presented, the laws of enzyme kinetics are preferably used in analogy as first hypotheses following the deductive research methodology.

However, all comparative considerations between chemical and biological reactions are restricted by the fact that biological systems like living cells are always "open systems". They can only be compared on the formal level to open chemical systems like continuous reactors. A real "steady state" of the system is reached when at a constant reaction rate the consumed substrate and formed products are replaced by physical transports. The differences between steady state, stationary state, and transient state are illustrated in the plot of Fig. 12.1. It is interesting to note that open systems in physics and biology follow the same laws

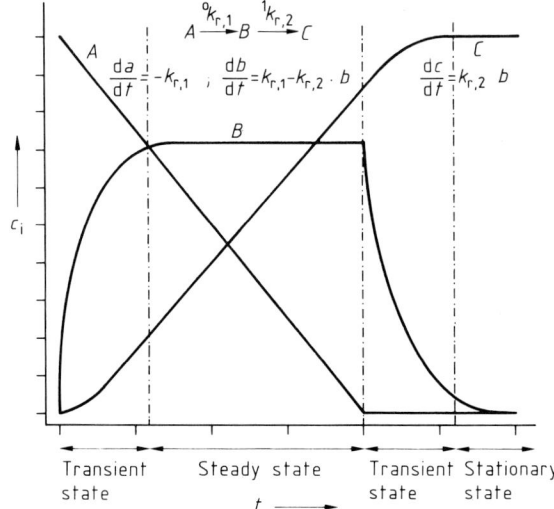

Figure 12.1. Demonstration of steady-state, stationary-state, and transient-state in reaction kinetics in the case of a consecutive chemical reaction with zero resp. first order ($^0k_{r,1}$ or $^1k_{r,2}$).

on the formal level of investigations. The biophysics of open systems were elaborated by VAN BERTALANFFY [12.2] showing some biological principles. One example chosen by VAN BERTALANFFY demonstrates the significant behavior of open systems as given in the set of reaction and transport sequences:

$$S_{ex} \xrightarrow{k_{Tr,S}} S \underset{k_{r,-1}}{\overset{k_{r,+1}}{\rightleftarrows}} I$$
$$\downarrow k_{r,+2}$$
$$P \xrightarrow{k_{Tr,P}} P_{ex} \quad . \tag{12.1}$$

In the case of steady state with $\bar{s}, \bar{i}, \bar{p}$ the following equation is valid:

$$\bar{s} : \bar{i} : \bar{p} = 1 : \frac{k_{r,+1}}{k_{r,-1}} : \frac{k_{r,+2}}{k_{Tr,P}}. \tag{12.2}$$

Some interesting conclusions can be drawn from this simple open system:

1. In steady state the ratio between components is also constant in the case of a reaction chain of arbitrary length.
2. The ratio between components depends only on system parameters like k_r and k_{Tr} and is independent of s_{ex} (principle of autoregulation).
3. Further it can be shown that:

$$\bar{s} = s_{ex} \frac{1}{1 + k_{r,+2}/k_{Tr,S}}, \tag{12.3}$$

which means that the ratio of rate constants of reaction k_r and transports k_{Tr} is governing the total flux. As a consequence a "kinetic regime" with the reaction as the rate determining step ($k_r \ll k_{Tr}$) can be distinguished from a "diffusion regime" where the transport is the rate determining step ($k_r \gg k_{Tr}$).

Some typical phenomena in biology can be explained with such simple models of open systems. An external disturbance, e.g., an increasing value of $k_{r,+2}$, results in a decrease of s, but due to an increase of transport according to $s_{ex} - s$, the open system shows a strong tendency to remain at steady state. Only in case of a constant change in k_r a new steady state is installed. For more complex open systems with feedback mechanisms this scheme can explain the appearance of oscillations.

Finally, for a proper understanding of the principles of enzyme kinetics some basic concepts from chemical kinetics should also be considered. Quantification of the rates of chemical reactions is carried out on the basis of two different approaches: 1. The reaction order or power law for homogeneous G- and L-phase reactions:

$$r = k_r \cdot c^n, \tag{12.4}$$

2. the pseudohomogeneous equation for heterogeneous chemical catalysis:

$$r = \frac{k_1 \cdot c}{1 + k_2 \cdot c}. \tag{12.5}$$

The form of Eq. (12.5) is to be preferred because it represents a generalized formulation, which serves for a variety of different situations as kinetic concept on the formal level. The symbol r is used as a general term for the reaction rate, identical with the symbol v used in the case of enzyme kinetics. Table 12.1 gives a comparison of individual concepts in different fields of science. The typical behavior of this saturation-type kinetics is shown in Fig. 12.2 with indication of enzyme kinetic parameters according to Table 12.1. Simultaneously, the plot of Fig. 12.2 shows another curve, representing Hill kinetics (see Sect. 12.7).

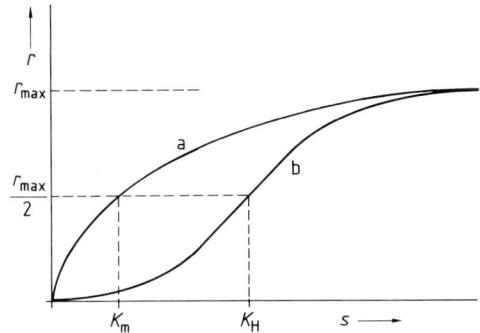

Figure 12.2. Graphical plot of basic enzyme kinetic behavior: (a) hyperbolic saturation-type kinetics (e.g., Michaelis-Menten) and (b) sigmoidal Hill kinetics.

In conclusion, Eq. (12.5) represents the generalized kinetic equation with a wide range of applicability. The appearance of this type of equation in the case of enzyme kinetics can be derived from a mechanistic background, as demonstrated in Sect. 12.2.2.

Table 12.1. Saturation Type Kinetics According to Fig. 12.2 in Different Fields of Application Compared with the General Form of Eq. (12.5) and Their Parameters

Kinetic Equations[a]	Eq. No.	Parameters
General form	(12.5)	k_1; k_2
LANGMUIR [12.3] (chemical heterogeneous catalysis) $r = k \cdot \dfrac{K \cdot p}{1 + K \cdot p}$	(12.6)	$k_1 = k \cdot K$; $k_2 = K$
HENRI [12.4], MICHAELIS-MENTEN [12.5] $r = r_{max} \dfrac{s}{K_m + s}$	(12.7)	$r_{max} = \dfrac{k_1}{k_2}$; $K_m = \dfrac{1}{k_2}$
MONOD [12.6] $q_S = q_{S,max} \dfrac{s}{K_S + s}$ $\left(\text{with } q_S = -\dfrac{1}{Y_{X/S}} \cdot \mu\right)$	(12.8)	$q_{S,max} = \dfrac{r_{max}}{x}$; $K_S = \dfrac{1}{k_2}$

[a] The symbols used in these equations are in accordance with those preferably used in the individual field of science (e.g., r, v, and q). The following relationship exists: $q_S = r_S \cdot 1/x$, with r_S = rate of substrate consumption, cf. Eq. (11.1). Often the index S is omitted for simplification as in Eq. (12.7).

12.2 Simple Enzyme Kinetics

12.2.1 Identification Plots for Simple Enzyme Kinetic Models

There are two procedures for analyzing kinetic data: The integral and the differential method.

1. In the integral method of analysis a particular form of rate equation is taken as an hypothesis, which after appropriate integration and mathematical manipulation predicts that the plot of a certain concentration function versus time should yield a straight line. Such a plot is shown in Fig. 12.3 to demonstrate the general integral method.
2. Differential methods are based on differentiation of experimental concentration versus time data in order to obtain the actual rate of reaction. Then a chosen hypothesis is tested using appropriate plots. In Fig. 12.4 this general procedure is shown for the case of differential analysis.

In both cases of analysis the model must be first identified, and only then the value of the parameter (k) can be estimated from the slope. Thus, by applying the linearization plots of Figs. 12.3 and 12.4 it is necessary to find a type of diagram where the parameter is not included in the term f(c), as realized in Figs. 12.5–12.10. If this is not possible other evaluation methods like the use of computers and non-linear regression methods must be used.

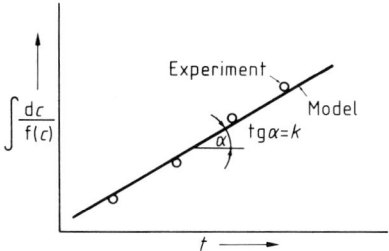

Figure 12.3. Principle of an integral method for the evaluation of the kinetic model parameter k by comparing a hypothesis (model) with experimental data in a linearization plot.

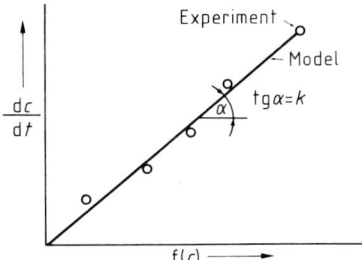

Figure 12.4. Principle of a differential method for the evaluation of the kinetic model parameter k by comparing a hypothesis (model) with experimental data in a linearization plot.

As an example these procedures are illustrated for the case of the kinetic model given in Eq. (12.5) in Figs. 12.5 to 12.10. There are advantages and disadvantages to each method. The integral method is easy to use and is recommended when testing specific mechanisms, or relatively simple rate expressions, or when the data are scattered to where one may not reliably find the derivatives needed in the differential method. The differential method is useful in more complicated cases but requires more accurate or larger amounts of data. The integral method can only test particular mechanisms or rate forms; the differential method can be used to develop or build up a rate equation to fit the data.

In general, it is suggested that integral analysis be attempted first, and, if not successful, that the differential method be tried. For complicated cases one may need to use special experimental methods which give a partial solution of the problem, or use flow reactors coupled with differential analysis.

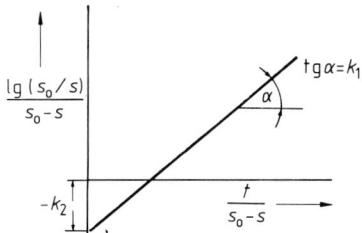

Figure 12.5. Linearization plot for the estimation of kinetic parameters k_i of saturation-type kinetics according to Eq. (12.9) following the integral method (cf. Fig. 12.3).

For simple enzyme kinetics, according to Eq. (12.5) in the case of S-utilization ($c=s$), an integral analysis is shown in Fig. 12.5. From the intercept and the slope of the linearization:

$$\frac{\lg(s_0/s)}{s_0-s} = -k_2 + \frac{k_1 t}{s_0-s}, \quad (12.9)$$

the model parameters k_1 and k_2 can be estimated. Using the Henri equation as integrated form of the Michaelis-Menten equation:

$$r_{max} \cdot t = K_m \cdot \lg \frac{s_0}{s} + (s_0-s), \quad (12.10)$$

a linearization plot of this equation has been derived as shown in Fig. 12.6, known as the Walker diagram (WALKER and SCHMIDT [12.7]), obtained by multiplying

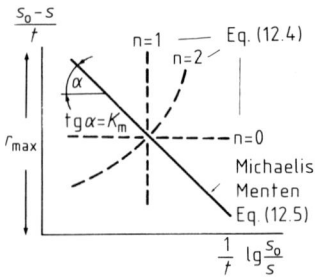

Figure 12.6. Linearization plot according to Eq. (12.11) ("Walker diagram") following the integral method for the estimation of kinetic parameters in the case of Eq. (12.4) with varying order of reaction ($n = 0, 1, 2$) and in the case of Eq. (12.5). The estimation of K_m is indicated in the plot.

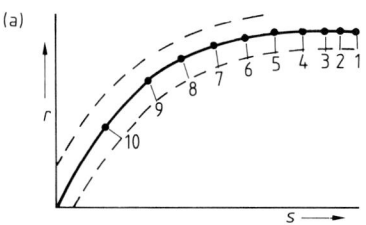

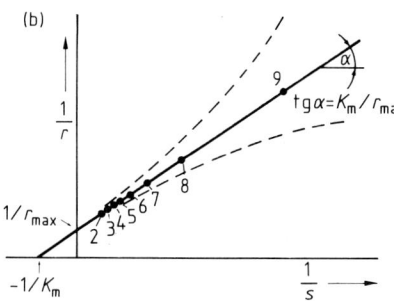

Figure 12.7. (a) Plot of biological reaction rate r vs. s in the case of saturation-type kinetics (see curve a in Fig. 12.2) with indicated points from experimental data (1–10) and (b) the consequences in parameter estimation using a double reciprocal plot according to LINEWEAVER and BURK [see Eq. (12.12)] as a case of the differential evaluation method (cf. Fig. 12.4). The dotted lines indicate the 95% confidence region. From the slope and the intercepts the parameters r_{max} and K_m can be determined.

Eq. (12.10) with $1/t$:

$$\frac{s_0 - s}{t} = -K_m \frac{1}{t} \lg \frac{s_0}{s} + r_{max}. \quad (12.11)$$

Even this plot is preferably used for kinetic analysis of biological waste water treatment, it can be generally applied advantageously because of the fact that from one single plot different kinetics can be distinguished instantaneously as shown in Fig. 12.6. The only restriction in the applicability of the Walker diagram lies in the fact that it cannot be used in the case of multiple-component kinetics like substrate utilization and simultaneous growth or product formation. This situation is explained in Chapter 14 and handled as Gates linearization.

The other evaluation method, the differential analysis, uses a number of different graphical linearizations.

The double reciprocal plot according to LINEWEAVER-BURK [12.8] is obtained after transformation of the Michaelis-Menten equation:

$$\frac{1}{r} = \frac{K_m}{r_{max}} \cdot \frac{1}{s} + \frac{1}{r_{max}}. \quad (12.12)$$

The model parameters (K_m, r_{max}) can be evaluated from the intercept and the slope as illustrated in Fig. 12.7. In this plot the region with 5% deviation is indicated by the dotted line, representing an area of 95% confidence. In conclusion it can be said that this plot is not able to give a good model identification due to the large deviation at low substrate concentrations. This is a consequence of the double reciprocal plot. Another linearization of simple enzyme kinetics is achieved with a different transformation of the Michaelis-Menten equation according to EADIE [12.9a] and HOFSTEE [12.9b]:

$$\frac{r}{s} = \frac{r_{max}}{K_m} - \frac{r}{K_m}. \quad (12.13)$$

This plot is shown in Fig. 12.8 where the estimation of the parameters is indicated together with the 95% confidence interval.

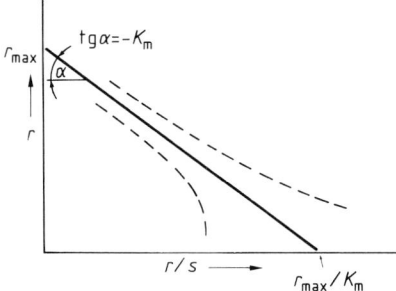

Figure 12.8. A single reciprocal plot according to EADIE and HOFSTEE [see Eq. (12.13)] as a case of the differential evaluation method (cf. Fig. 12.4) for estimation of parameters K_m and r_{max} from the slope and intercepts. Again the dotted lines indicate the 95% confidence interval.

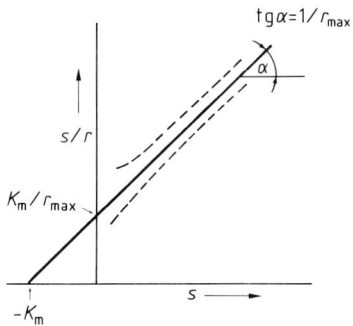

Figure 12.9. The Langmuir-plot according to Eq. (12.14) as the best diagram for a differential evaluation of saturation-type kinetics according to Eq. (12.7). The parameters r_{max} and K_m can be determined from the slope and intercepts as indicated. The dotted lines are the 95% confidence region.

Furthermore, Fig. 12.9 represents the linearization according to LANGMUIR [12.10]:

$$\frac{s}{r} = \frac{K_m}{r_{max}} + \frac{s}{r_{max}} ; \qquad (12.14)$$

this plot realizes the smallest deviation and therefore is recommended.

Last but not least, EISENTHAL and CORNISH-BOWDEN [12.11] have recently described a completely different method of plotting enzyme kinetics data. These authors showed that if the experimental s-values are plotted on a negative horizontal axis and the observed r-values are plotted on a vertical axis, then straight lines drawn through the corresponding s- and r-points intersect at $s = K_m$ and $r = r_{max}$. While Fig. 12.10b represents the conventional $r = f(s)$ plot (cf. Fig. 12.2, curve a) for better comprehensiveness, Fig. 12.10a demonstrates

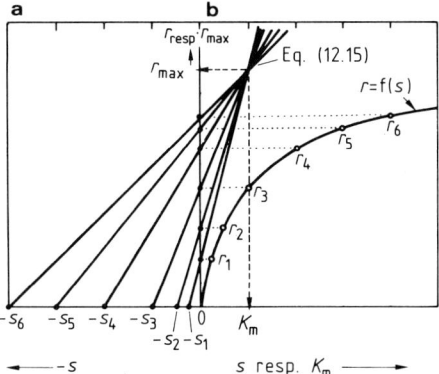

Figure 12.10. (a) The "direct linear plot" of r vs. s data and r_{max} vs. K_m according to EISENTHAL and CORNISH-BOWDEN [12.11] as a simple graphical method [see Eq. (12.15)] for the estimation of parameters r_{max} and K_m of saturation-type kinetics. Detailed explanation is given in the text. For better comprehensiveness the right-side graph (b) is added, showing how data for the conventional plot of function $r = f(s)$ according to curve a in Fig. 12.2 are used for the simple graphical procedure of the "direct linear plot" in Fig. 12.10a.

the simple graphical procedure for estimating enzyme kinetic constants, which is in agreement with the rearranged Michaelis-Menten equation, Eq. (12.11)

$$r_{max} = r + \frac{r}{s} \cdot K_m . \qquad (12.15)$$

Thus, for any value of s and r it is possible to plot r_{max} against K_m as a straight line with slope r/s, intercept $-s_i$ on the K_m-axis and intercept r_i on the r_{max}-axis. The most obvious advantage of this "direct linear plot" is that it requires no calculation.

12.2.2 Reaction Mechanisms of Simple Enzyme Kinetics

The general form of rate equations as an element of the mass balance is:

$$r_j = f(c_i, T, pH). \qquad (12.16)$$

In agreement with theory and practice this general form can be rewritten using the parameter k in the mathematical model function f as:

$$r_j = k(T) \cdot f(c_i). \qquad (12.17)$$

Kinetic analysis therefore considers both terms, temperature and concentration.

Rate equations for enzyme kinetics, "microkinetics", can be derived from postulated reaction mechanisms (cf. Fig. 11.1).

Table 12.2 summarizes four different mechanisms for simple enzyme kinetics which have been used in the past. The resulting rate equation for mechanisms 1-3 is the same and was already given in Eq. (12.7) (cf. Table 12.1). The difference between them lies in the fact that the interpretation of the model parameters r_{max} and K_m is not the same as given in Table 12.2. Mechanism No. 4 represents a more general treatment in analogy to heterogeneous chemical catalysis including reversible steps of adsorption and desorption while the reaction step is handled as irreversible.

In all cases microkinetics are derived by applying the rds-concept. The irreversible reaction is thought to be rate determining:

$$r_{tot} = r_{+2} = k_{+2} \cdot \{es\}. \qquad (12.27)$$

The unknown concentration of the active complex $\{es\}$ is then calculated by using the qss-concept. Further details can be found in text books on enzyme kinetics (e.g., SEGEL [12.16]). Both concepts, rds and qss, are widely used in the case of complex kinetic modelling.

In conclusion it can be stated that the Michaelis-Menten equation applies to many more complex mechanisms than the

Table 12.2. Reaction Mechanisms of Simple Enzyme Kinetics and Derived Kinetic Model Equations

No.	Mechanism		Interpretation of Model Parameter K_m ($r_{max} = k_{+2} \cdot e_{tot}$)
1	$E + S \underset{k_{-1}}{\overset{k_{+1}}{\rightleftharpoons}} \{ES\} \xrightarrow{k_{+2}} E + P$ (12.18) MICHAELIS-MENTEN [12.5]	$r = r_{max} \dfrac{s}{K_m + s}$ (12.7)	$K_m = \dfrac{k_{-1}}{k_{+1}}$ (12.19)
2	BRIGGS-HALDANE [12.12]		$K_m = \dfrac{k_{-1} + k_{+2}}{k_{+1}}$ (12.20)
3	$E + S \underset{k_{-1}}{\overset{k_{+1}}{\rightleftharpoons}} \{ES\} \xrightarrow{k_{+2}} E + P$ (12.21) VAN SLYKE-CULLEN [12.13]		$K_m = \dfrac{k_{+2}}{k_{+1}}$ (12.22)
4	$E + S \underset{k_{-1}}{\overset{k_{+1}}{\rightleftharpoons}} \{ES\} \xrightarrow{k_{+2}} \{EP\} \underset{k_{-3}}{\overset{k_{+3}}{\rightleftharpoons}} E + P$ (12.23) LANGMUIR-HINSHELWOOD, HINSHELWOOD [12.14] HOUGEN-WATSON [12.15]	$r = r_{max} \dfrac{K_{ads} \cdot s}{1 + K_{ads} \cdot s + K_{des} \cdot p}$ (12.24)	$K_{ads} = \dfrac{k_{+1}}{k_{-1}} \equiv \dfrac{1}{K_m}$ (12.25) $K_{des} = \dfrac{k_{-3}}{k_{+3}}$ (12.26)

Michaelis-Menten mechanism itself, but with a more complex meaning of K_m and r_{max}. In practice, therefore, it cannot be assumed that K_m (and r_{max}) can be expressed simply as in Eqs. (12.19) and (12.20). The value of r_{max} is not a fundamental property of an enzyme as it depends upon enzyme concentration. Provided that e_{tot} is known it is advantageous to define a quantity k_{cat}, the "catalytic constant" or "turnover number":

$$k_{cat} = \frac{r_{max}}{e_{tot}}. \qquad (12.28)$$

For the Michaelis-Menten mechanism, k_{cat} is identical with k_{+2} but in general the more non-committal notation k_{cat} should be preferred.

12.3 Enzyme Kinetics with Complex Reactions

12.3.1 Reaction Mechanisms and Kinetic Equations

A fundamental characteristic of enzyme catalysis is that it involves a large number of intermediate compounds formed by catalytically active groups of the enzyme with substrates and reaction products (e.g., EIGEN and HAMMES [12.17]).

Analysis of the stationary kinetics of an enzymatic process involving labile intermediates (I_i) shows that the catalytic rate constant for the process reflects the slowest rate-limiting stage of the interconversion of the intermediates (HEARON [12.18]). As indicated in Table 12.3, the general form of a rate equation as given in Eq. (12.30) remains valid also in the case of a chain of enzyme reactions but with a more complex interpretation of the model parameters [cf. Eqs. (12.31), (12.32)]. In contrast to ordinary chemical reactions possessing a virtually unlimited range of rate constants enzymes have very similar kinetic parameters ($k_{cat} \cong 10^2 \cdot s^{-1}$ and $K_m \cong 10^{-4}$ mol·L^{-1}).

Biochemical transformation of substances take place via multistage mechanisms with consecutive involvement of a large set of enzymes, presumably limited to enzymes with the k_{cat} value mentioned above.

Due to the limitations posed by the second law of thermodynamics, the extent of reaction is significantly influenced. The extent to which a chemical reaction may proceed is heavily reduced by the backward reaction. Bioprocesses with metabolizing organisms, however, very often exhibit large values of free enthalpy changes so that reverse reactions can be neglected. This holds true for aerobic as well as for anaerobic processing. The thermodynamics of equilibria, however, still remain essential in bioprocess analysis and bioreactor design in a number of enzymatic processes of carbohydrates. Significant amounts of both substrates and products are known to exist in equilibrium mixture so that in practice reversibility of reactions must be taken into account. Also, all reactions are reversible in principle. Initial rate kinetics, as treated up to now, are not representative of biotechnical processing.

It is evident, therefore, that the Michaelis-Menten mechanism is incomplete and the reversible step should be incorporated in the mechanism ($k_{-2} > 0$) as indicated in Table 12.3 [see Eq. (12.33)]. A useful form of a rate equation can be formulated applying the so-called Haldane relationship, which is an equilibrium constant [see Eq. (12.35)].

This relationship explains that in the case of reversible enzymatic reactions the kinetic parameters are dependent on each other and limited in the individual magnitude by the thermodynamic equilibrium constant of the overall reaction given in Eq. (12.35). Similar equations are obtained not only for more complicated 1-substrate or pseudo-1-substrate mechanisms but for mechanisms involving two or three reactants as well. It

Table 12.3. Reaction Mechanisms of Complex Enzyme Kinetics and Derived Kinetic Model Equations

No.	Mechanism		Model Parameters
5	$E+S \xrightleftharpoons[k_{-1}]{k_{+1}} I_1 \xrightleftharpoons[]{k_{+2}} \ldots I_i \xrightarrow{k_{i+1}} \ldots I_n \xrightarrow{k_{n+1}} E+P$ chain of irreversible reactions (HEARON [12.18]) (12.29)	$r = k_{cat}^* \cdot \dfrac{s}{K_m^* + s}$ (12.30)	$k_{cat}^* = 1 \Big/ \sum\limits_{i=2}^{n+1} 1/k_i$ (12.31) $K_m^* = \dfrac{K_m}{k_{+2}} \cdot 1 \Big/ \sum\limits_{i=2}^{n+1} 1/k_i$ (12.32)
6	$E+S \xrightleftharpoons[k_{-1}]{k_{+1}} \{ES\} \xrightleftharpoons[k_{-2}]{k_{+2}} E+P$ (12.33) reversible Michaelis-Menten-type (HALDANE [12.19]) and other types	$r = \dfrac{r_{max,S} \cdot r_{max,P} \cdot s - r_{max,S} \cdot r_{max,P} \cdot p/K_{eq}}{K_{m,S} \cdot r_{max,P} + r_{max,P} \cdot s + r_{max,S} \cdot p/K_{eq}}$ (12.34)	$K_{eq} = \dfrac{p_{eq}}{s_{eq}} = \dfrac{r_{max,S} \cdot K_{m,P}}{r_{max,P} \cdot K_{m,S}} = \dfrac{k_{+1} \cdot k_{+2}}{k_{-1} \cdot k_{-2}}$ (12.35)
7	$E+S \xrightleftharpoons[k_{-1}]{k_{+1}} \{ES\} \xrightleftharpoons[k_{-2}]{k_{+2}} \{EP\} \xrightleftharpoons[k_{-3}]{k_{+3}} E+P$ (12.36) reversible Langmuir-Hinshelwood-type		$r_{max,S} = \dfrac{k_{+2} \cdot k_{+3} \cdot e_{tot}}{k_{-2} + k_{+2} + k_{+3}}$ (12.37) $r_{max,P} = \dfrac{k_{-1} \cdot k_{-2} \cdot e_{tot}}{k_{-1} + k_{-2} + k_{+2}}$ (12.38) $K_{m,S} = \dfrac{k_{-1} \cdot k_{-2} + k_{-1} \cdot k_{+3} + k_{+2} \cdot k_{+3}}{k_{+1}(k_{-2} + k_{+2} + k_{+3})}$ (12.39) $K_{m,P} = \dfrac{k_{-1} \cdot k_{-2} + k_{-1} \cdot k_{+3} + k_{+2} \cdot k_{+3}}{(k_{-1} + k_{-2} + k_{+2}) k_{-3}}$ (12.40)
8	$\begin{array}{c} S_1 \\ E \xrightleftharpoons[k_{-1}]{k_{+1} \cdot s_1} ES_1 \\ {}_{k_{+4}} \swarrow \nwarrow {}^{k_{-4} \cdot i} \quad {}_{k_{-2}} \updownarrow {}^{k_{+2} \cdot s_2} \\ EI \xrightleftharpoons[k_{+3}]{k_{-3} \cdot P} ES_1 S_2 \\ \searrow {}^{P} \\ EIP \end{array}$ (12.41) (KING-ALTMAN [12.20])	$r = r_{max} \dfrac{s_1 \cdot s_2}{K_{S,S1} \cdot K_{m,S2} + K_{m,S2} \cdot s_1 + K_{m,S1} \cdot s_2 + s_1 \cdot s_2}$ (with $p=0$) (12.42)	r_{max}; $K_{m,S1}$; $K_{m,S2}$, and $K_{S,S1}$ = dissociation constant for S1

describes the effect of stimulation on the rate by S at the start in the complete absence of P or *vice versa* for P but also at any time during the approach to equilibrium.

There are two reasons for the decreased rate:

a) at any time, some of the P is being converted back to S and
b) at any time, some of the enzyme is combined with P so that less enzyme is available for combination with S.

Eq. (12.34) can be regarded as the general reversible form of the Michaelis-Menten equation. It has the advantage over others that it does not imply a particular mechanism and represents a purely empirical approach. There are many mechanisms more complicated than Eq. (12.33) that nonetheless generate Eq. (12.34). According to the deductive research methodology this equation can be used in model identification in practice representing a typical example of the formal kinetic approach.

The most important case of a more complex mechanism is the more realistic reversible mechanism given in Eq. (12.36) in Table 12.3.

Even if only one substrate is involved, the reaction mechanism may well contain more than one intermediate complex. HALDANE [12.19] has applied the steady-state approximation to reactions with all steps considered as reversible.

The resulting equation has the same form as Eq. (12.34) but with a different interpretation of the kinetic parameters $r_{max,S}$, $r_{max,P}$, $K_{m,S}$, and $K_{m,P}$ [cf. Eqs. (12.37) to (12.40)].

Again the Haldane relationship [K_{eq} cf. Eq. (12.35)] can be used. More complex rate equations such as those which involve several substrates, require more complex relationships but in all instances at least one relationship of this type must exist between the kinetic parameters and the equilibrium constant.

In the case of a kinetic analysis for the determination of individual values of k_i, it should be taken into consideration" that supplementary experiments are needed to estimate the equilibrium constant by special methods.

Another class of complex enzyme kinetics is the case of 2-S-mechanisms also represented in Table 12.3. The derivation of kinetic equations in this complex case can be handled easier by using the King-Altman method (see Sect. 12.3.2). As it is not normally possible to measure all of the separate rate constants, it is convenient to express the equations in coefficient form. In order to eliminate some interrelationships between the great number of coefficients, numerous methods have been used for rewriting rate equations in meaningful terms (e.g., ALBERTY [12.21]; DALZIEL [12.22]; CLELAND [12.23]). Using the concept of single enzyme kinetics with the symbols of the parameters r_{max}, K_m, and K_I for the inhibition constant of forward and backward reactions, the complicated equations can be simplified to the form of Eq. (12.42) if no product is included in the reaction mixture.

This is the equation for the initial velocity following the compulsory-order ternary-complex mechanism as a special case of bisubstrate kinetics. The meaning of all parameters is much more complex than in the case of simple enzyme kinetics and can be found in text books on enzyme kinetics (e.g. SEGEL [12.16]).

Eq. (12.42) shows that bisubstrate kinetics require a total of four model parameters: r_{max}, $K_{m,S1}$, $K_{m,S2}$, and $K_{S,S1}$, which are necessary and sufficient for a unique description of bisubstrate kinetics.

The determination of these kinetic parameters is carried out in a sequence of plots. On the basis of a certain number of experimental runs at different concentrations of s_1 and s_2 in order to measure the initial velocities, they are graphically evaluated in the manner indicated in Fig. 12.11a and b. "Primary plots" according to:

$$\frac{r_{max,1}}{r_1} = 1 + \frac{K_{m,S1}}{s_1} + \frac{K_{m,S2}}{s_2} + \frac{K_{S,S1} \cdot K_{m,S2}}{s_1 \cdot s_2} \quad (12.43)$$

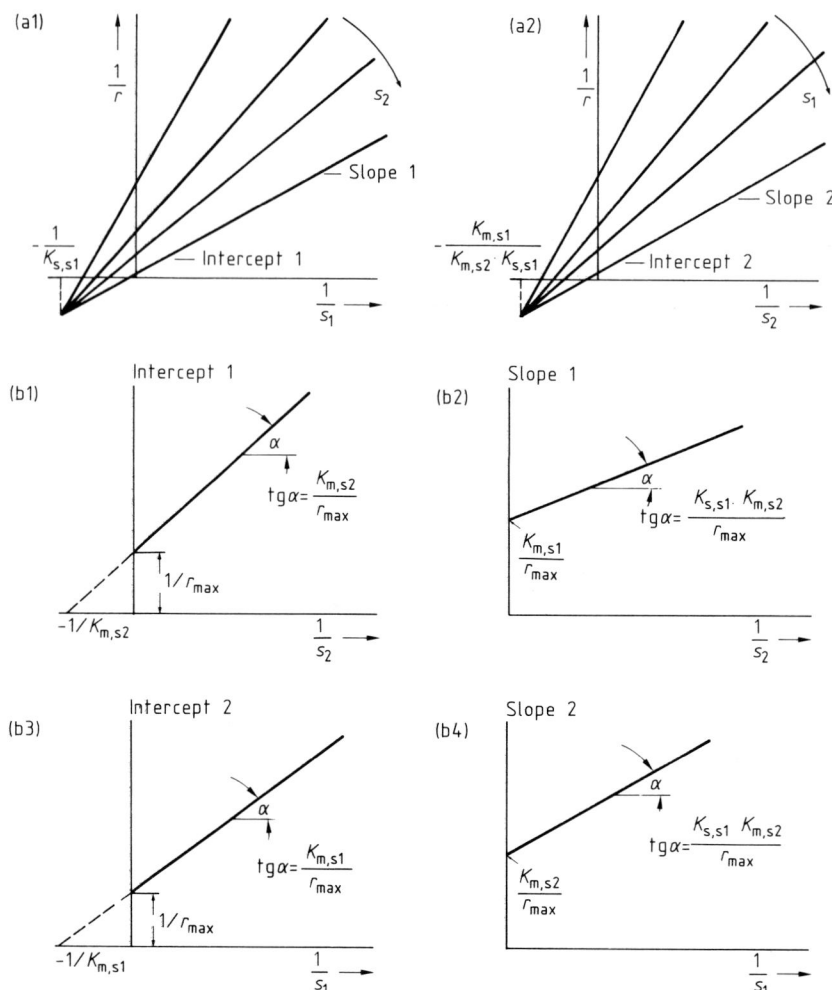

Figure 12.11. Plots for parameter estimation of 2-S-kinetics [see Eq. (12.42)]: (a) "primary plots" using experimental data: reciprocal rate r versus reciprocal variable substrate concentration s_1 (a1) and s_2 (a2) at fixed s_2 and s_1; (b) "secondary plots" using quantities from the primary plots: intercept 1 and 2 from primary plot a1 and a2, respectively, versus reciprocal fixed substrate s_1 and s_2 (graphs b1 and b3), and slope 1 and slope 2 from primary plot a1 and a2 as a function of reciprocal fixed substrate concentration (graphs b2 and b4). As outlined in the text, the parameters of Eq. (12.42), i.e., r_{max}, $K_{m,S1}$, $K_{m,S2}$, and $K_{S,S1}$ can be estimated by this sequential procedure of plots as indicated in Fig. 12.11 b1–b4.

with $1/r_i$ as a function of $1/s_1$ or $1/s_2$ give the basis for the evaluation of slopes and intercepts replotted as shown in Fig. 12.11b (1–4) to give four "secondary plots". Thus, the four replots provide the investigator with adequate equations to determine the four unknown kinetic parameters.

Concerning the variety of bisubstrate kinetics there exist only a few categories: ping-pong, ordered, random mechanism.

For more details of these mechanisms the reading of special text books on enzyme kinetics is recommended.

12.3.2 General Strategies to Derive Rate Equations for Complex Kinetics

The steady-state rate equations for any enzyme mechanism can be derived in the same way as for simple enzyme kinetics. In principle, the concepts of rds and qss must be used. In practice, however, this method is extremely laborious and liable to error for more complex mechanisms. In the case of, e.g., mechanism 7 three linear equations are obtained which must be solved for three unknowns and put into the velocity equation. This can be done by solving a third-order determinant or more simply by a direct graphic method, by KING and ALTMAN [12.20].

The method of KING and ALTMAN is applicable to any mechanism consisting of a series of reactions between different forms of one enzyme. It cannot be used for a mixture of enzymes, or to reactions containing non-enzymatic steps. Nonetheless it is applicable to most situations met in enzyme catalysis and is very useful in practice, while the principle of this method is more complex (CORNISH-BOWDEN [12.24]).

On the other hand, a general method is known from chemical heterogeneous catalysis, which can be applied to any mechanism resulting in simplified rate equations with the aid of the working principles of rds and qss (e.g., EMIG and HOFMANN [12.25]).

While simplification of rate equations according to the King-Altman method is achieved by restrictions to mechanisms with linearizable kinetics and to initial rates without the necessity to assume a rds, a general scheme can be followed in setting up rate equations in analogy to chemical heterogeneous surface reactions. Thus, simplification is obtained by assuming one reaction to be rds, while all others are in steady-state (qss-concept). A special case of application to enzyme kinetics is the ERO-mechanism (equilibrium-random) which exhibits non-linearity.

Several logical steps must be followed according to this method by EMIG and HOFMANN [12.25]:

a) Write the gross reaction.
b) Formulate an overall reaction scheme representing a simplified reaction mechanism according to the formal kinetic approach, where all steps of sorption and reaction are included.
c) Write the equations for the rates of consumptions and formations of all species.
d) Assume one reaction step of the scheme to be the rate determining step, which normally will be the reaction itself (e.g., k_2 in mechanisms 1–7).
e) Eliminate unknown magnitudes (e.g., es in mechanisms 1–5) by assuming the qss-concept for all non-rds.
f) Simplify the resulting equations by introducing equilibrium constants K.

The advantage of this general method is that real and not initial rates are the result, which is essential for biotechnical processes. The disadvantage in using the concept of rds and qss seems obvious, as it is hard to find the rds.

12.4 Inhibition Kinetics of Enzyme Reactions

A special case of inhibition has already been indirectly handled with Eq. (12.34) which can be used for the description of product inhibition under certain conditions. However, product inhibition is not always

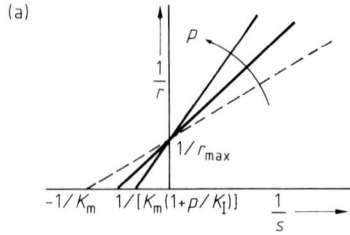
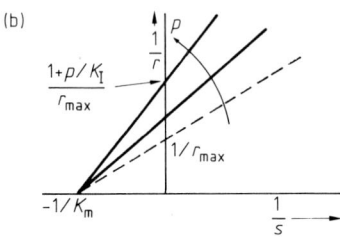
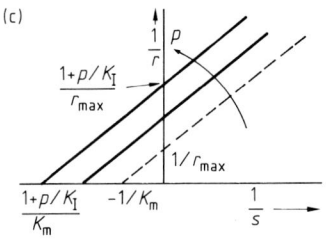
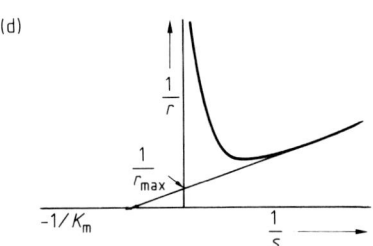

Figure 12.12. Enzyme inhibition kinetics in Lineweaver-Burk diagrams: competitive (a), non-competitive (b), uncompetitive inhibitions (c), and substrate inhibition (d). The inhibitor concentration is indicated as product concentration p (cf. Table 12.4). The parameters r_{max} and K_m in some cases can be estimated from slopes and intercepts as indicated in the graphs. For estimation of the parameter K_I see Fig. 12.13.

of this type. Moreover, there are many compounds that inhibit enzymes. Therefore, a more complete theory is required in order to account for these facts. The most important class of inhibitors reducing the rate of enzyme-catalyzed reactions is that of reversible inhibitors forming dynamic complexes with the enzyme that have different catalytic properties from those of the free enzyme.

As a basis of discussion a general scheme proposed by BOTTS and MORALES [12.26] is useful in discussing inhibitors (I) which are not reaction products:

$$E \underset{P}{\overset{S}{\rightleftharpoons}} ES$$
$$I \updownarrow \qquad \updownarrow I \qquad (12.44)$$
$$EI \underset{S}{\overset{P}{\rightleftharpoons}} EIS$$

which includes most of the simple types of inhibition:

1. competitive inhibition if EIS and the reactions involving it are missing;
2. uncompetitive inhibition if EI is missing and EIS does not break down;
3. non-competitive inhibition if the EIS complex is inactive and no effect on substrate binding occurs (E and EI have equal affinities for S).

The kinetic behavior for various inhibition types exhibits the following pattern as represented in the graphs of Fig. 12.12 a–d.

In Table 12.4 the mechanisms of these enzyme inhibitors are summarized together with the resulting kinetic model equations.

As can be seen by comparing Eqs. (12.46), (12.48), (12.50), and (12.52) the inhibited enzyme may have:

1. an increased K_m value: competitive inhibition;
2. r_{max} and K_m in a constant ratio (uncompetitive inhibition);
3. a reduced r_{max} value (pure non-com-

Table 12.4. Reaction Mechanisms of Enzyme Inhibition Kinetics and Derived Kinetic Model Equations

No.	Mechanism	Kinetic Model Equation	Parameters
9	$E+S \underset{}{\overset{K_m}{\rightleftharpoons}} \{ES\} \xrightarrow{k_{rds}} E+P$ (12.45) $+$ I $\Updownarrow K_I$ EI Competitive inhibition	$r = r_{\max} \dfrac{s}{s+K_m(1+i/K_I)}$ (12.46)	$r_{\max}, K_m, K_I$
10	$E+S \underset{}{\overset{K_m}{\rightleftharpoons}} \{ES\} \xrightarrow{k_{rds}} E+P$ (12.47) $+$ I $\Updownarrow K_I$ ESI Uncompetitive inhibition	$r = \dfrac{r_{\max}}{(1+i/K_I)} \cdot \dfrac{s}{s+K_m/(1+i/K_I)}$ (12.48)	$r_{\max}, K_m, K_I$
11	$E+S \underset{}{\overset{K_m}{\rightleftharpoons}} \{ES\} \xrightarrow{k_{rds}} E+P$ (12.49) $+$ $+$ I I $\Updownarrow K_I$ $\Updownarrow K_I$ $\{EI\} \underset{}{\overset{K_m}{\rightleftharpoons}} \{ESI\}$ Non-competitive inhibition	$r = \dfrac{r_{\max}}{(1+i/K_I)} \cdot \dfrac{s}{s+K_m}$ (12.50)	
12	$E+S \underset{}{\overset{K_m}{\rightleftharpoons}} \{ES\} \xrightarrow{k_{rds}} E+P$ (12.51) $+$ nS $\Updownarrow K_{I,S}$ ES_n Substrate inhibition	$r = r_{\max} \dfrac{s}{s+K_m+s^2/K_{I,S}}$ (12.52)	$r_{\max}, K_m, K_{I,S}$

petitive inhibition) or some combinations of these effects (mixed inhibition);
4. an unchanged rate at low s but drastically reduced velocity at high values of s (substrate inhibition).

The corresponding kinetic equation for S-inhibition is an analogous case to that for uncompetitive inhibition. The initial rate, given by Eq. (12.52), differs from the simple Michaelis-Menten type by virtue of the term s^2, significant only at high s.

Substrate inhibition plays an important role in technical bioprocesses where high yields and conversions are desired. No special linearization is known for S-inhibition. Details are discussed in Sect. 14.3.

In the preceding section several basic types of inhibition have been presented. In practice, however, mixed effects of inhibition on multisubstrate enzymes and multiple inhibitors occur so that further classification of inhibitions is needed, e.g.:

a) linear, if the replot of slope and/or intercept versus i is a straight line;
b) parabolic, if the slope and/or intercept replot is a parabola; and
c) hyperbolic, if the replot is a hyperbola (SEGEL [12.16]).

A general method of estimating the inhibition constants K_I was introduced by DIXON [12.27]. Fig. 12.13 a–c shows the Dixon plots for the main types of inhibition.

The best way for parameter estimation in all cases of inhibition are replots, where the

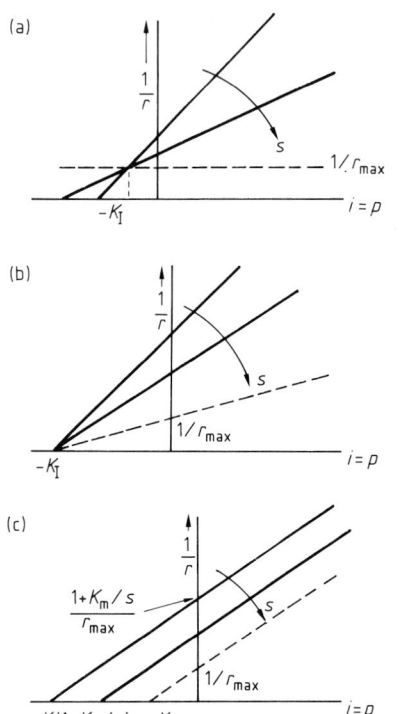

Figure 12.13. Dixon plots for enzyme inhibition kinetics for the estimation of parameter K_I together with r_{max} and K_m. The cases of competitive (a), non-competitive (b), and uncompetitive inhibition (c) are shown in diagrams of reciprocal rates r vs. inhibitor concentration i resp. p.

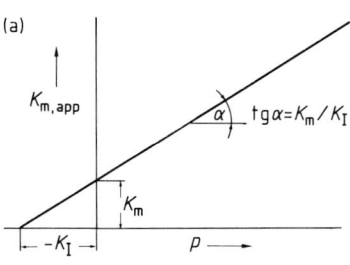

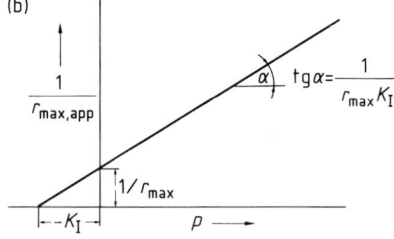

Figure 12.14. Replots of data taken from the reciprocal plot for the estimation of kinetic parameters K_m (a) and r_{max} (b) from apparent values of $K_{m,app}$ and $r_{max,app}$ in the case of competitive (a) and non-competitive inhibition (b) (compare Figs. 12.12 and 12.13).

slope and/or intercepts from Fig. 12.13 are taken. The special procedure in the case of competitive and non-competitive inhibition is represented in Fig. 12.14 a–b, indicating the values for parameters K_m and r_{max}.

Similar procedures can generally be followed using the integrated rate equation according to WALKER.

Recently a generalized enzymatic rate equation as a function of substrate conversion ζ_S [cf. Eq. (11.28)] was derived to facilitate the calculation of enzyme reactors. SIIMER [12.28] proposed a scheme, where substrate transformation is regarded formally as a two-step reaction of the type of mechanism 1 [Eq. (12.18)] but with two products P_1 and P_2, P_1 inhibiting competitively and P_2 non-competitively. The generalized form of the equation is:

$$r/r_{max} = (1-\zeta_S)/(1+\alpha\zeta_S+\beta\zeta_S^2), \quad (12.53)$$

where α and β are functions of s_0 and the dissociation constants of stimulation and inhibition according to the mechanisms proposed. Only two coefficients must be known in order to apply a calculation of enzyme reaction rates of the entire substrate concentration range. It is not necessary to know the exact reaction mechanism and the values of the kinetic (inhibition) constants, which are often not available.

structure, ionization of the substrate, and influences on binding to the enzyme and on its reactivity, only the last class is considered here in kinetic analysis of pH-effects. The simplest type of pH-effect on an enzyme, when only a single acidic or basic group is involved, is not different from the general case of hyperbolic stimulation and inhibition that was considered in Sect. 12.4.

The protonation of a basic group on an enzyme is simply a special case of the binding of a modifier at an active site. The difference, however, lies in the fact that virtually all enzymes are affected by protons, which are much smaller than any other species and have no steric effect. As a consequence, certain phenomena such as pure non-competitive inhibition are common with protons as inhibitors. Moreover, protons bind to many sites, so that it is often insufficient to consider binding at one site only.

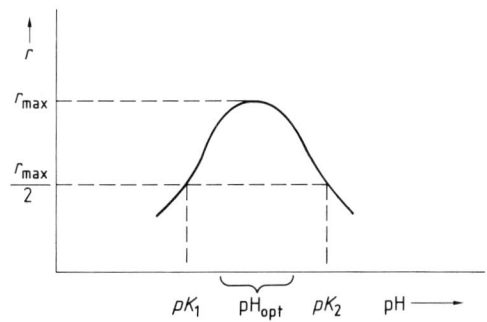

Figure 12.15. Schematic presentation of the pH-dependence of biokinetic rates in a bell-shaped curve with indication of pK_1- and pK_2-values used in formal kinetic modelling, cf. Eq. (12.57).

12.5 Effects of pH on Enzyme Kinetics

Almost all enzymes are extremely sensitive to pH. Whole cell extracts and crude enzyme preparations are well buffered but this natural buffering is lost when an enzyme is purified and must be replaced with artificial buffers. From the combined effects of irreversible damage of protein

Fortunately, however, the pH-behavior of many enzymes can be interpreted as a first approximation in terms of a simple mechanism, in which only two ionizable groups are considered. Even when every enzyme contains a large number of acidic and basic groups, the enzyme may be represented as a dibasic acid *HEH* with two non-identical acidic groups, (MICHAELIS [12.29]). For practical purposes no differentiation can be made between these groups on the basis of real experiments, because it is impossible to evaluate the corresponding "group dissociation constants". Defining "molecular dissociation constants" from combinations of them brings the practical advantage that they can be measured.

The initial velocity of enzymatic reactions frequently displays a "bell-shaped" activity profile with three distinct phases as a function of pH, shown in Fig. 12.15.

There exists the region of stimulation at low values, a region of inhibition at higher pH, and an intermediate range, the pH-optimum. Although this profile closely resembles a Gaussian curve for some values of pK_1 and pK_2 ($pK_i = -\log K_i$), it is not such a curve. Also a noticeably flat maximum can arise. In order to derive rate equations for the pH-dependence of enzyme kinetics, the bell-shaped pH activity curve can be accounted for by a simple extension of the theory for the ionization of a dibasic acid. Active sites on the free enzyme *EH* and enzyme substrate complex *SEH* are generally thought to bind and release protons and only the form *SEH* is able to react to give products. The basic mechanism is as follows:

$$S + HE^- \underset{k_{-1}}{\overset{k_{+1}}{\rightleftarrows}} HES^- \xrightarrow{k_{+2}} HE^- + P$$

with $H_2E \rightleftarrows HE^-$ ($K_{1,E}$), $HE^- \rightleftarrows E^{2-}$ ($K_{2,E}$), $H_2ES \rightleftarrows HES^-$ ($K_{1,ES}$), $HES^- \rightleftarrows ES^{2-}$ ($K_{2,ES}$). (12.54)

The scheme implies that the catalytic reaction involves only two steps as in the Michaelis-Menten mechanism. If several steps are postulated with each intermediate capable of deprotonation and protonation, the form of the final rate equation is not affected, but any chance of mechanistic interpretation of experimental results is lost.

Neglecting the fact that the scheme is an oversimplification of actual situations, this mechanism is quite useful in deriving a rate equation. The derivation is analogous to that of the Michaelis-Menten type. But instead of E and ES we have to write EH and SEH and to substitute for eh and seh. The resulting rate equation can be written in the form of an analogon to the Michaelis-Menten equation [cf. Eq. (12.7)]. The interpretation of the model parameters in this case (r^*_{max} and K^*_{max}) is much more complicated and can be expressed as:

$$r^*_{max} = r_{max}/(1 + K_{2,ES}/h^+ + h^+/K_{1,ES}) \quad (12.55)$$

$$K^*_m = K_m \frac{(1 + K_{2,E}/h^+ + h^+/K_{1,E})}{(1 + K_{2,ES}/h^+ + h^+/K_{1,ES})}. \quad (12.56)$$

Thus, the final rate equation becomes:

$$r = r_{max} \frac{1}{1 + \frac{h^+}{K_{1,ES}} + \frac{K_{2,ES}}{h^-}} \cdot \frac{s}{K_m + s}, \quad (12.57)$$

which can conveniently be applied in practice as only two parameters must be estimated for pH-dependence shown in Fig. 12.15.

Due to the high complexity it is wise to be very cautious when interpreting pH-dependent experiments. The mechanisms reflecting the influence of H^+ and also OH^- ions generally show a high diversity, where normal inhibition types like competitive, non-competitive, and mixed inhibitions are assumed or combinations with substrate ionization by H^+ and OH^- and substrate inhibition. The resulting rate equations again are taken as first hypotheses following the deductive research methodology, representing, in the case of a good fit to experimental data, a typical formal kinetic approach (ANDREYEVA and BIRYUKOV [12.30]). The pH-dependence can be modelled simply by the following polynomial fit:

$$r = r_{max}(\pm a_0 \pm a_1 \cdot pH \pm a_2 \cdot pH^2 ...), \quad (12.58)$$

where a are the coefficients to be fitted to experiments without any biological significance (cf. Chapter 14).

12.6 Temperature Dependence of Enzyme Kinetics

12.6.1 Arrhenius Equation

The effects of temperature on the rates of enzyme reactions are matters of considerable interest and importance but are fairly complicated. The experimental result normally measured, however, exhibits a simple behavior: the overall rate passes through a maximum, similar to the shape of the plot shown in Fig. 12.15 but with a very sharp decrease with increasing temperature. This figure is the result of a combined effect, where raising the temperature affects two independent processes: the enzyme-catalyzed reaction itself and the thermal inactivation of the enzyme.

It is therefore astonishing that in the majority of experimental situations both rates of enzyme catalysis and denaturation are found to obey the Arrhenius law, well known for chemical reactions. ARRHENIUS showed that over a moderate range of temperature there was a linear relation between the logarithm of the rate constant of a reaction k and $1/T$, expressed as:

$$\ln k = \ln k_\infty - \frac{E_a}{T}, \quad (12.59)$$

where T represents the absolute temperature in K and E_a the activation energy (kJ·mol^{-1}).

The estimation of the parameters E_a and k_∞ of the Arrhenius equation:

$$k = k_\infty \cdot e^{-E_a/RT}, \quad (12.60)$$

can be carried out with the aid of the Arrhenius plot shown in Fig. 12.16.

The behavior shown in this figure, where the effective rate is the additive effect of the rate of activation ($E_{a,1}$) and inactivation

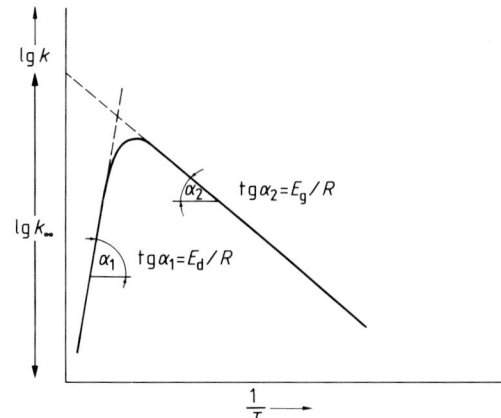

Figure 12.16. Arrhenius plot of biokinetic rate constants k: $\lg k$ vs. reciprocal absolute temperature $1/T$ resulting in two regions with two linearizations each exhibiting a formal activation energy E_a: $E_g = E_{a1}$ in case of microbial growth, $E_d = E_{a2}$ in case of microbial death [see Eq. (12.61)].

($E_{a,2}$) can be quantified approximately as follows:

$$k = k_{\infty,1} \cdot e^{-E_{a,1}/RT} - k_{\infty,2} \cdot e^{-E_{a,2}/RT}, \quad (12.61)$$

where $E_{a,2}$ will be much greater than $E_{a,1}$ (HINSHELWOOD [12.31]).

The Arrhenius equation applied to enzyme reactions is only a formal analogy, where the parameters are the result from fitting to experimental data and should be indicated as $k_{\infty,\text{obs}}$ and $E_{a,\text{obs}}$. The interpretation of these parameters, e.g., of E_a, however, under some arbitrary condition is quite useless for any deductions on reaction mechanisms. In many cases, however, the apparent values of observed parameters E_a and k_∞ give general trends (e.g., $E_{a,\text{cat}} < E_{a,\text{denat}}$). The difficulties in analyzing the function $k = k(T)$ are:

1. The effects of temperature and pH are, in general, not independent, because most ionization constants are temperature dependent.
2. Even if the overall behavior of T-dependence of rates is quantified by the Arrhenius equation in many cases, the ob-

served maximum is of no particular significance, as this value in practice varies with the experimental procedure.

Very little significance can be attached to studies on the T-dependence of v_{max} and K_m, unless the mechanistic meanings of these parameters are known.

3. Moreover, it is customary for chemical and biochemical processes to use the definition of a "water activity" a_w, rather than the water concentration itself. Water activity as a meaningful measure of water availability, determined, e.g., by the freezing point depression method, influences the rates of chemical and biochemical processes and is well known in food technology. The rates of drying, aroma loss, but also the destruction of cells and enzymes, and the yields or growth rates of microbes are affected by a_w (e.g., THIJSSEN [12.32]; LABUZA [12.33]; ESENER et al. [12.34]). Therefore, a_w should be incorporated in the Arrhenius equation in such a way, that k_∞ and E_a are treated as a_w-dependent.

4. Problems arise when applying the simple Arrhenius equation to complex reactions where the reaction mechanism contains multiple equilibria. Deviations from the Arrhenius law are often observed and appear due to enzyme denaturation and/or complex reactions.

It was emphasized that Arrhenius plots are only linear over a limited range of T and then only when a single observed reaction step is being changed with changing temperature.

One frequent result of changing conditions which change the rate of enzyme-catalyzed reactions is a change in rds.

A classic study of this type was carried out by BENDER et al. [12.35], resulting in a curved plot in an Arrhenius diagram. Considering similar cases, it becomes clear that the Arrhenius equation should certainly apply to $k_{r,\,obs}$ even if $k_{r,\,obs}$ is a composite constant. But the activation energy does not correspond to a single elementary step; it is the sum of the activation energies of the reaction steps involved.

Provided that one may compare like with like, then comparison of the activation parameters ΔH and ΔS is often much more informative than comparison of simple rate constants. Valuable information about enzyme reaction mechanisms can be obtained.

Abnormal behavior of T-dependence of an enzyme reaction is thought to include two types (TALSKY [12.36]):

a) Continuous change from $E_{a,1}$ to $E_{a,2}$ indicating a change in rds.
b) Abrupt change from $E_{a,1}$ to $E_{a,2}$ results in two separated lines in the Arrhenius plot.

Models have been proposed and discussed to account for these discontinuities in the enthalpy of activation without corresponding abrupt changes in the maximum velocity (LONDESBOROUGH [12.37]).

In the case of a changing reaction mechanism with temperature, the activation energy must vary substantially so that the Arrhenius relation has limited applicability and also the range of velocity and the influence of other factors on E_a must be considered. Functions other than the Arrhenius equation have been suggested for temperature dependence:

$$f(T) = \alpha + \beta \cdot T, \tag{12.62}$$

$$f(T) = \alpha \cdot T^\beta, \tag{12.63}$$

$$f(T) = \frac{\alpha}{\beta - T}, \tag{12.64}$$

with α and β as empirical constants (SAGUY and KAREL [12.38]). However, the range of validity is narrow and the best fit often is obtained by the hyperbolic function of Eq. (12.64), so that these functions are no real alternatives to the Arrhenius relation.

Generally the study of the temperature dependence of reaction rates gives some valuable information. Especially in the case of enzyme engineering in the food industry examples can be given on changes of enzymes during conservation, freeze-drying, and the development of stable enzymes for technical use.

12.6.2 Enthalpy/Entropy-Compensation (Isokinetic Model)

Another useful piece of information can be gained from an analysis of the temperature dependence, the enthalpy/entropy compensation (LEFFLER [12.39]) which will be discussed briefly below.

Using the activated complex (transition-state) theory according to EYRING on the basis of principles of quantum mechanics, the thermodynamical formulation of the rate constant leads to the expression:

$$k = \frac{k_B \cdot T}{h} \cdot e^{\Delta S^+/R} \cdot e^{-\Delta H^+/RT}, \quad (12.65)$$

where ΔH^+ and ΔS^+ are, respectively, the enthalpy and the entropy of formation of the activated complex (e.g., $\{ES\}$); with k_B as the Boltzmann constant and h the Planck constant.

The preexponential factor in the Arrhenius equation is thus:

$$k_\infty = \frac{k_B \cdot T}{h} \cdot e^{\Delta S^+/R}, \quad (12.66)$$

showing a deviation in the Arrhenius plot at high temperature due to f(T).

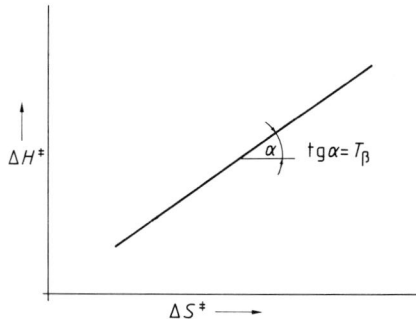

Figure 12.17. Enthalpy/entropy compensation plot according to Eq. (12.67) and the formal appearance of an "isokinetic temperature" T_β. (ΔH^+ in $kJ \cdot mol^{-1}$; ΔS^+ in $J \cdot mol^{-1} \cdot K^{-1}$).

Therefore, ΔS^+ and ΔH^+ can be calculated from the intercept and the slope in an Arrhenius plot.

The study of reaction rates in the past decades has led to the observation of a number of equilibria and rate processes in which a plotting of enthalpy changes against entropy changes (produced by a systematic variation of solvent composition, reactant molecule change, or pH) gives a straight line (LEFFLER [12.39]). The slope of such a line, as shown in Fig. 12.17 is called the "isokinetic temperature" T_β and the relationship is called enthalpy/entropy compensation.

LUMRY and EYRING [12.40] suggested that the phenomenon is a property of the proteins in the system and that such compensation is of major physiological importance for stability of proteins. They found T_β to range between 270–320 K as of this situation.

LEFFLER and GRUNWALD [12.41] show a procedure by which one can theoretically reach compensation from an analysis of thermodynamic laws for ΔH and ΔS. Thus:

$$\Delta H^+ = \Delta G^+ + T_\beta \cdot \Delta S^+, \quad (12.67)$$

meaning that a linear thermodynamic compensation of rates can occur. This would suggest an identical molecular mechanism for protein denaturation of enzymes and thermal death of microbial cells, all having a range of almost 600 kJ·mol^{-1} (LABUZA [12.33]).

If the concept of T_β were to hold true, an Arrhenius plot would intersect at a single point according to Eq. (12.65) applied to the thermal death rate of microbial cells:

$$\ln \frac{k_{d,\beta}}{T_\beta} = \left(\ln \frac{k_B}{h} + \frac{\Delta S^+}{R} \right) - \frac{\Delta H^+}{R} \cdot \frac{1}{T}, \quad (12.68)$$

where $k_{d,\beta}$ is the thermal death rate (in s^{-1}) of cells at T_β.

However, no isokinetic temperature and compensation was found by VAN UDEN [12.42]) in the case of yeast cells. Obviously the $\Delta H/\Delta S$-plot is not sufficiently sensitive to detect non-linearity over the restricted range of rates and temperatures used. The appearance of a linear compensation is not excluded when data are related to the temperature of maximum growth.

Whether true compensation occurs in bacterial death and is a property of water/protein changes still remains to be clarified. KIRK and DENNISON [12.43] found a longer negative change in ΔS with decreasing water activity a_w, so that two criteria are needed in applying $\Delta H/\Delta S$ compensation:
1. data as $f(T)$ at constant pH and a_w and,
2. data as $f(a_w)$ or $f(pH)$ at constant T.

Generally, $\Delta H/\Delta S$-analysis can lead to useful predictions of rates and activation energies from several conditions and may thus explain the change in E_a observed with changing a_w without the need to postulate a change of mechanism.

12.7 Kinetics of Enzyme Control and Regulation

The capacity of self-replication correlated with substrate consumption is not the only remarkable property of living cells. Animate systems possess an amazing degree of flexibility. When cells are exposed to new conditions they show a pronounced ability for adaption. Microorganisms adapt readily and rapidly to different kinds of environmental changes in the bioreactor and they generally do so by changing themselves structurally and functionally (see e.g., HERBERT [12.44]). The explanation of these biological phenomena offers a stimulating challenge to the principles of kinetics.

The cell depends upon a pattern of coordinated reactions. They involve in their sequences an organization in time. Moreover, the cell presents a texture of macromolecular substances like proteins, RNA, DNA, etc. in which some degree of spatial organization exists. The laws according to which the spatio-temporal organization is possible follow a pattern with a basic similarity. The network of biochemical reactions in metabolism is controlled and coordinated by an array of control and regulation pathways. As already outlined in Chapter 11, system theory is a fundamental component of biology as a science of organization. The manner in which the subsidiary processes are related and coupled in their complexity forms the content of feedback system theory.

12.7.1 Physiological Regulation Phenomena

Basically, a microorganism adapts its activities to changes in the environment by influencing the reaction pattern by the following types of mechanisms, also shown in Fig. 12.18:

1. Simple regulations by direct mass-action law. In principle these changes are generally covered by ordinary kinetics presented in Sects. 12.2 to 12.4. Rapid response is a characteristic property in this case as the time constant of these changes is small.
2. Regulation of the activity of the enzymes. The activity of enzymes depends on the structure which is influenced by small molecules. These modifiers cause changes in the conformation which can be positive or negative. These so-called allosteric controls are vital to the integration of metabolism in organisms (MONOD, CHANGEUX and JACOB [12.45]). Especially the key-enzymes in a metabolic network are affected allosterically resulting in a rapid change of the kinetic behavior (Michaelis-Menten to Hill kinetics, cf. Fig. 12.2). The time constant normally is small.
3. Regulations of the amount of enzyme, i.e., enzyme synthesis. This type of regulation is more economical as it regu-

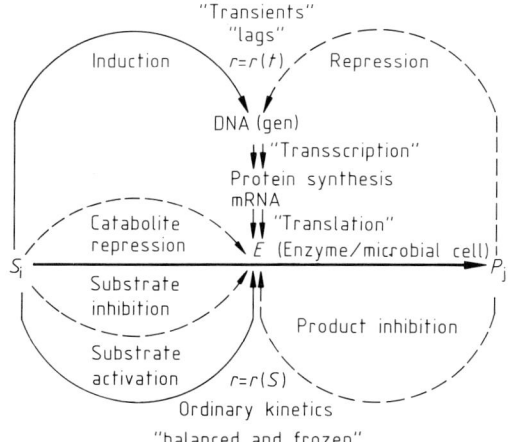

Figure 12.18. Schematic presentation of enzyme regulation principles as a basis for formal kinetic modelling (MOSER [12.70]). A detailed explanation is given in the text.

lates enzyme synthesis at the genetic level in order to control enzyme production. This is achieved by such regulatory mechanisms as induction and repression (JACOB and MONOD [12.46]), and catabolite repression (MAGASANIK [12.47]). These mechanisms, which may have drastic effects on cellular composition, have time constants in the order of minutes to hours and hence result in a much slower adaption than the mechanisms treated above. Very drastic changes are the consequence of the kinetic behavior of the cell, which may severely lag behind the changes of environmental conditions. These "lag times" are an expression of "biological inertia", and must be taken into account in mathematical modelling.

12.7.2 Kinetic Models of Enzyme Activity Control

If an enzyme molecule binds more than one molecule of S, so that the overall binding can be represented as:

$$E + nS \rightleftharpoons \{ES_n\} \rightarrow E + nP, \tag{12.69}$$

then the response of the enzyme activity to S may be sigmoid rather than hyperbolic (see Fig. 12.2). An interpretation can be found on the basis of allosteric change in conformation of the enzyme due to some inhibitors (model of MONOD, CHANGEUX and JACOB [12.45]).

As a consequence of the allosteric inhibition, a mathematic function was found by HILL [12.48], which accurately fitted all available data:

$$r = r_{max} \frac{s^{n_H}}{K_H + s^{n_H}}. \tag{12.70}$$

This equation is known as the Hill equation, with n_H being the number of substrate binding sites per molecule of enzyme, normally in the range of 1 to 3.2 (Hill coefficient) and K_H a constant comprising the intrinsic dissociation constant K_m and some interaction factors of substrate binding. The definition is analogous to K_m.

The estimation of the model parameters n_H and K_H can be carried out with the logarithmic form of the Hill equation:

$$\lg \frac{r}{r_{max} - r} = n_H \cdot \lg s - \lg K_H. \tag{12.71}$$

A Hill plot is demonstrated in Fig. 12.19.

The influence of experimental conditions on Hill parameter estimation is described by GLENDE and REICH [12.49], concluding that a non-linear regression with a minimum of 15 measuring points according to

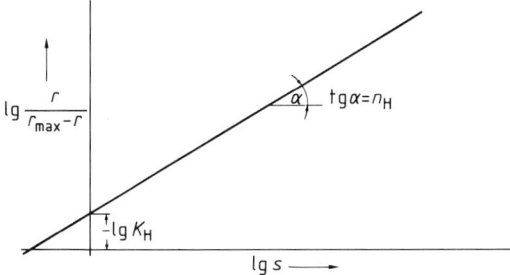

Figure 12.19. Hill plot for the estimation of kinetic parameters of Hill kinetics [see Fig. 12.2 and Eq. (12.70)], n_H and K_H according to the logarithmic form [cf. Eq. (12.71)].

REICH et al. [12.50] should be preferred as an appropriate method for an unbiased determination of parameters. The Hill model was extended by ADAIR [12.51] to allow for the existence of stable, partly liganded intermediates.

As an example of such type of kinetics the oscillations observed with glycolysis are known to be caused by feedback activation and inhibition of the regulatory enzymes phosphofructokinase and pyruvate kinase as allosteric oscillophores (HESS [12.52]).

12.7.3 Kinetic Models of the Regulations of Enzyme Synthesis

In a culture of microorganisms one is confronted with a large number of individual entities in which a large number of complexly regulated chemical reactions occurs. One of the approaches to the modelling of such a system would be to model as many of the reactions and their regulations as possible. The constructions of these so-called non-lumped structured models is most likely to be, at present, of little value.

The necessity of adequate simplifications is outlined in detail by FREDRICKSON et al. [12.53], ROELS [12.54], and ROELS and KOSSEN [12.55].

A substantial number of publications has appeared on the dynamics of enzyme synthesis in a variety of situations. Most of the models are based on more or less sophisticated versions of the operon model of JACOB and MONOD [12.46]. The role of mRNA and its stability were modelled by TERUI [12.56]. Repressor and inducer control was treated by KNORRE [12.57], IMANAKA et al. [12.58], VAN DEDEM and MOO-YOUNG [12.59], and SUGA et al. [12.60]. Allowance for dual control and catabolite repression was made by TODA [12.61]. Nearly all recent models are based on the operon model and its kinetic treatment by IMANAKA and AIBA [12.62] and BAJPAI and GHOSE [12.63].

Rather than treating these contributions in detail a simple structured model was developed showing a combination of the features of the models published (ROELS [12.54]). Recently, TODA [12.64] reviewed the effects of induction and repression of enzymes in microbial cultures as well as the modelling.

The basic concept of modelling enzyme synthesis in transient situations like change in environment, e.g., a substrate shift-up, is that the level of enzyme concentration in the cells has to increase. For this reason the structural units responsible for enzyme synthesis, i.e., mRNA in the ribosomes R must be produced initially at the level of transcription (ROELS [12.54]):

$$\frac{de}{dt} = k_E \cdot c_R \cdot x, \tag{12.72}$$

with k_E as the kinetic constant and c_R the concentration of ribosomes or mRNA; x is an indirect measure for e.

The concentration of ribosomes in a state of balanced growth is a function of the growth rate μ (TANNER [12.65]). Thus $c_R \approx \mu$.

In this sense, the changing rate of ribosomes represents a measure for "biological inertia" (lag time), which also affects the growth rate (ROMANOVSKY et al. [12.66]). This is expressed in the following differential equation:

$$\frac{dc_R}{dt} = \frac{1}{t_L}(\bar{c}_R - c_R), \tag{12.73}$$

where $\bar{c}_R$ is the stationary concentration of ribosomes, which will be reached at time t_L (lag time, $t_L \sim 1/\mu$).

In thermodynamics, the theory of incomplete systems introduces the concept of the natural times or the relaxation times of the internal processes. The system is described in terms of the externally observable variables and a number of relaxations to a change in external conditions. A small relaxation time characterizes a mechanism which adjusts quickly. The approach is more or less analogous to the transfer function approach to the dynamic behavior of systems (HIMMELBLAU and BISCHOFF [12.67]). The application of the latter approach to bioengineering systems can be found in the literature (ROELS and KOSSEN [12.55]; ROELS [12.68]).

The time constant concept provides a direct route to the choice of the degree of complexity required for the description of the behavior of a system. In principle, the behavior of a culture of organisms is described by a vast number of relaxation times resulting, among others, from the various regulatory mechanisms previously discussed. These mechanisms generally have largely different relaxation times, a highly speculative picture of which is given in Fig. 24.6. A description of the system can be simplified by basing an approach on a comparison of the relaxation times of the internal processes and those characterizing the relevant changes in external conditions (ROELS [12.69]; MOSER [12.70]).

Further applications of the kinetics of enzyme regulation are given in Chapter 16 in connection with transient modes of bioreactor operation.

12.8 References

[12.1] C. J. PERRET; J. Gen. Microbiol. 22 (1960), 589.
[12.2] L. VON BERTALANFFY: "Biophysik des Fließgleichgewichtes". Vieweg Verlag, Braunschweig, 1953.
[12.3] I. LANGMUIR; J. Am. Chem. Soc. 38 (1916), 2221.
[12.4] V. HENRI; C.R. Acad. Sci. Paris 135 (1902), 916.
[12.5] L. MICHAELIS and M. L. MENTEN; Biochem. Z. 49 (1913), 333.
[12.6] J. MONOD: "Recherches sur la Croissance des Cultures Bacteriennes". Hermann & Cie, Paris, 1942.
[12.7] A. C. WALKER and C. L. A. SCHMIDT; Arch. Biochem. 5 (1944), 445.
[12.8] H. LINEWEAVER and D. BURK; J. Am. Chem. Soc. 56 (1934), 658.
[12.9a] G. S. EADIE; J. Biol. Chem. 146 (1942), 85.
[12.9b] B. H. J. HOFSTEE; J. Biol. Chem. 199 (1952), 357.
[12.10] I. LANGMUIR; J. Am. Chem. Soc. 40 (1918), 1361.
[12.11] R. EISENTHAL and A. CORNISH-BOWDEN; Biochem. J. 139 (1974), 715.
[12.12] G. A. BRIGGS and J. B. S. HALDANE; Biochem. J. 19 (1925), 338.
[12.13] D. D. VAN SLYKE and G. E. CULLEN; J. Biol. Chem. 19 (1914), 141.
[12.14] C. N. HINSHELWOOD: "Kinetics of Chemical Change", p. 187. Clarendon Press, Oxford, 1940.
[12.15] O. A. HOUGEN and K. M. WATSON: "Chemical Process Principles", Vol. 3. J. Wiley, New York, 1947.
[12.16] I. H. SEGEL: "Enzyme Kinetics". Wiley-Interscience Publ., New York, 1975.
[12.17] M. EIGEN and G. G. HAMMES; Adv. Enzymol. 25 (1963), 1.
[12.18] J. Z. HEARON; Physiol. Rev. 32 (1952), 499.
[12.19] J. B. S. HALDANE: "Enzymes". Longmans Green, London, 1930.
[12.20] E. L. KING and C. ALTMAN; J. Phys. Chem. 60 (1956), 1375.
[12.21] R. A. ALBERTY; J. Am. Chem. Soc. 75 (1953), 1928.
[12.22] K. DALZIEL; Acta Chem. Scand. 11 (1957), 1706.
[12.23] W. W. CLELAND; Biochim. Biophys. Acta 67 (1963), 104.
[12.24] A. CORNISH-BOWDEN: "Principles of Enzyme Kinetics". Butterworths, London–Boston, 1976.
[12.25] G. EMIG and H. HOFMANN; Chem. Ing. Tech. 47 (1975), 717.
[12.26] J. BOTTS and M. MORALES; Trans. Faraday Soc. 49 (1953), 696.
[12.27] M. DIXON; Biochem. J. 55 (1953), 170.
[12.28] E. SIIMER; Biotechnol. Bioeng. 20 (1978), 1853.
[12.29] L. MICHAELIS: "Hydrogen Ion Concentration". Vol. 1. Tindall and Cox, London, 1926.
[12.30] L. N. ANDREYEVA and V. V. BIRYUKOV; Biotechnol. Bioeng. Symp. 4 (1973), 61.
[12.31] C. N. HINSHELWOOD: "The Chemical Kinetics of the Bacterial Cell". Oxford Clarendon Press, 1946.
[12.32] H. A. C. THIJSSEN; Lebensm. Wiss. Technol. 12 (1979), 308

[12.33] Th. P. Labuza; Food Technol. Febr. (1980), 67.

[12.34] A. A. Esener, G. Bol, N. W. F. Kossen, and J. A. Roels; in: Proc. 6th IFS Abstract, Adv. in Biotechnol., Vol. I, p. 339 (1980).

[12.35] M. L. Bender, F. J. Kézdy, and C. R. Gunter; J. Am. Chem. Soc. 86 (1964), 3714.

[12.36] G. Talsky; Angew. Chem. 83 (1971), 553.

[12.37] J. Londesborough; Eur. J. Biochem. 105 (1980), 211.

[12.38] I. Saguy and M. Karel; Food Technol. Febr. (1980), 78.

[12.39] J. E. Leffler; J. Org. Chem. 31 (1966), 533.

[12.40] R. Lumry and H. Eyring; J. Phys. Chem. 58 (1954), 110.

[12.41] J. E. Leffler and E. Grunwald: "Rates and Equilibria of Organic Reactions". J. Wiley, New York, 1963.

[12.42] N. van Uden, P. Abranches, and C. Cabeça-Silva; Arch. Mikrobiol. 61 (1968), 381.
N. van Uden and M. M. Vidal-Leiria; Arch. Microbiol., 108 (1976), 293.

[12.43] J. R. Kirk and D. B. Dennison; 5th Internat. Congr. Food Sci. Technol., Kyoto, Japan, 1978.

[12.44] D. Herbert; Symp. Soc. Gen. Microbiol. 11 (1961), 391.

[12.45] J. Monod, J. P. Changeux, and F. Jacob; J. Mol. Biol. 6 (1963), 306.

[12.46] F. Jacob and J. Monod; J. Mol. Biol. 3 (1961), 318.

[12.47] B. Magasanik; Symp. Quant. Biol. 26 (1961), 249.

[12.48] A. V. Hill; J. Physiol. (London) 40 (1910), 4.

[12.49] M. Glende and J. G. Reich; Acta Biol. Med. Ger. 29 (1972), 595.

[12.50] J. G. Reich, G. Wangermann, M. Falck, and K. Rohde; Eur. J. Biochem. 26 (1972), 368.

[12.51] G. S. Adair; J. Biol. Chem. 63 (1925), 529.

[12.52] B. Hess; in: Symp. Soc. Exp. Biol., p. 105. Cambridge University Press, 1973.

[12.53] A. G. Fredrickson, R. D. Megee, and H. M. Tsuchiva; Adv. Appl. Microbiol. 13 (1970), 419.

[12.54] J. A. Roels; in "Proc. 1st Eur. Congr. Biotechnol., Interlaken (Switzerland), DECHEMA 82 (1978), 221.

[12.55] J. A. Roels and N. W. F. Kossen; Prog. Ind. Microbiol. 14 (1979), 95.

[12.56] G. Terui; Pure Appl. Chem. 36 (1973), 377.

[12.57] W. A. Knorre; in "4th Symp. Cont. Cult. Microorg., Prague", p. 225. Academic Press, New York, 1968.

[12.58] T. Imanaka, T. Kaieda, and H. Taguchi; J. Ferment. Technol. 51 (1973), 423.

[12.59] G. van Dedem and M. Moo-Young; Biotechnol. Bioeng. 15 (1975), 1301.

[12.60] K. Suga, G. van Dedem, and M. Moo-Young; Biotechnol. Bioeng. 17 (1975), 185.

[12.61] K. Toda; Biotechnol. Bioeng. 18 (1976), 1117.

[12.62] T. Imanaka and S. Aiba; Biotechnol. Bioeng. 19 (1977), 757.

[12.63] R. K. Bajpai and T. K. Ghose; Biotechnol. Bioeng. 20 (1978), 927.

[12.64] K. Toda; J. Chem. Tech. Biotechnol. 31 (1981), 775.

[12.65] R. D. Tanner; Biotechnol. Bioeng. 12 (1970), 831.

[12.66] J. M. Romanovsky, N. N. Stepanova, and D. S. Chernavsky: "Kinetische Modelle in der Biophysik". VEB G. Fischer Verlag, Jena, 1974.

[12.67] D. M. Himmelblau and K. B. Bischoff: "Process Analysis and Simulation". J. Wiley, New York, 1968.

[12.68] J. A. Roels; Biotechnol. Bioeng. 22 (1980), 2457.

[12.69] J. A. Roels; J. Chem. Tech. Biotechnol. 32 (1982), 59.

[12.70] A. Moser; in "Proc. 3rd Austrian-Italian-Jugoslavian Chem. Eng. Conf.", Sept. 14–16, 1982, (F. Moser, ed.). Graz, Austria, Vol 2, p. 620.

Chapter 13
Stoichiometry of Bioprocesses

Anton Moser

Institut für Biotechnologie, Mikrobiologie und Abfalltechnologie
Technische Universität Graz
Graz, Austria

13.1 Introduction
13.2 The Concept of Yield Factors; Thermodynamic Considerations
13.2.1 The Yield Factor
13.2.2 Thermodynamic Efficiency and Degree of Reduction
13.3 Fundamentals of Stoichiometry of Complex Reaction Systems
13.3.1 Stoichiometry of Complex Reactions
13.3.2 Stoichiometry of Complex Reactor Operation
13.4 Application of Stoichiometric Concepts to Bioprocessing
13.4.1 Yield Coefficients in Fermentation Processes
13.4.2 Relationship Between Yield Factors and True Stoichiometric Coefficients
13.4.3 Integration of Elemental Balances and Kinetic Equations into Unstructured Modelling of Bioprocesses
13.4.4 Application of Balancing Methods in Bioprocessing
13.5 References

List of Symbols

C	kg	amount of CO_2 or carbon
c	$kg \cdot m^{-3}$	concentration of CO_2 (or general term)
D	s^{-1}	dilution rate
F_j	$kg \cdot s^{-1}$	mass flow of component j
H_v	$kJ \cdot m^{-3}$	volumetric heat of fermentation
ΔH_S	$kJ \cdot mol^{-1}$	heat of combustion of substrate
ΔH_X	$kJ \cdot mol^{-1}$	heat of combustion of biomass
m, m_s	s^{-1}	coefficient of maintenance
N	kg	amount of nitrogen
N_j	mol	moles of component j
$\dot{N}_j$	$mol \cdot s^{-1}$	molar flux of mass of component j
n, n_j	$mol \cdot m^{-3} \cdot s^{-1}$	mass flux per unit volume
n'	$mol \cdot m^{-2} \cdot s^{-1}$	mass flux per unit area
O	kg	amount of oxygen
P	kg	amount of product
q	$kg P \cdot kg X^{-1} \cdot s^{-1}$	specific rate of formation or consumption
r	$kg \cdot m^{-3} \cdot s^{-1}$	reaction rate (general term)
r_j	$kg \cdot m^{-3} \cdot s^{-1}$	rate of formation or consumption of component j
r_{H_v}	$kJ \cdot m^{-3} \cdot s^{-1}$	rate of formation of volumetric heat
r_C	$kg \cdot m^{-3} \cdot s^{-1}$	rate of formation of CO_2
r_O	$kg \cdot m^{-3} \cdot s^{-1}$	rate of consumption of O_2
r_P	$kg \cdot m^{-3} \cdot s^{-1}$	rate of formation of products
r_S	$kg \cdot m^{-3} \cdot s^{-1}$	rate of consumption of substrate
r_X	$kg \cdot m^{-3} \cdot s^{-1}$	rate of formation of biomass
S	kg	amount of substrate
s	$kg \cdot m^{-3}$	concentration of substrate
t	s	time (general term)
$\bar{t}$	s	mean residence time
t_m	s	mixing time
X	kg	amount of biomass
x	$kg \cdot m^{-3}$	concentration of biomass
Y	—	yield coefficient (general term)
$Y_{i/j}$	$kg_i \cdot kg_j^{-1}$	yield coefficients (cf. Table 13.2)
Y^C	—	yield based on "C-mole of biomass", Eq. (13.23)
Y_{kJ}	$kg X \cdot kJ^{-1}$	yield for heat, Eq. (13.28)

Greek letters

$\alpha, \beta, \gamma, \delta$	—	coefficients of elemental composition
γ_S	—	degree of reduction of substrate
γ_X	—	degree of reduction of biomass
η	—	energy efficiency coefficient of growth, Eq. (13.31)
η_{th}	—	thermodynamic efficiency, Eq. (13.34)
μ	s^{-1}	specific growth rate
ξ_i	mol	extent of i-th reaction
$\xi_{(t)}^*$	...	modified extent of reaction, Eq. (13.41)
ν	—	stoichiometric coefficient
$(\nu_S, \nu_X ...)$	—	stoichiometric coefficient for substrate, biomass, etc.
ν_{SX}	—	stoichiometric coefficient of reaction $S \rightarrow X$, Eq. (13.45)

Indices

ATP	adenosine triphosphate
av e$^-$	available electrons
C	CO_2 or carbon
i, j	component or reaction number
r_i	i-th reaction
max	maximum
N	nitrogen
O	oxygen
P	product
S	substrate
W	water
H_v	volumetric heat
X	biomass
t	value at time
o	spatial initial value
0	temporal initial value

Abbreviations

ATP	adenosine triphosphate
av e$^-$	available electrons
CSTR	continuous stirred tank reactor
CPFR	continuous plug flow reactor
DCSTR	discontinuous stirred tank reactor
OTR	oxygen transfer rate
Σ	sum
∇	nabla

13.1 Introduction

Stoichiometry is the science concerned with the quantitative composition of chemical compounds and quantitative conversion in chemical reactions and is one of the fundamentals of reaction engineering together with thermodynamics and kinetics. *In-vivo* processes in the past were thought to be mainly kinetically controlled without significant relevance of thermodynamic concepts.

The aspects of stoichiometry, however, have never been neglected, as yields play a central role in industry. Stoichiometric calculations, in contrast to kinetic considerations, require knowledge of the mechanism and the balance of the significant compounds ("key variables"). The principle of balancing plays a central role. As a consequence, Eq. (11.1) must be applied in all special procedures of reactor operation for macroconversions in bioprocessing.

One can distinguish between macro- and microbalances for both conservative and non-conservative properties. For conservative properties, which are not altered during the process, only macrobalances are used, for non-conservative properties microbalances also will be used (KOSSEN [13.1]; ROELS and KOSSEN [13.2]).

The general form of the principle of conservation of atomic species [cf. Eq. (11.1)] in the case of multiple reactions expressed as a sum is given by:

$$\sum_j v_j \cdot r = 0. \qquad (13.1)$$

It is difficult, however, to directly apply Eq. (13.1) to bioprocesses as several hundred components are involved in metabolism. This problem could be avoided up to now and only a fairly limited number of compounds was considered in a so-called gross stoichiometric equation of growth according to the macroscopic principle (Chapter 11). This approach was theoretically justified by ROELS [13.3].

A practical implication of Eq. (11.1), when applied to multiple reactions in steady-state, written as a sum,

$$\sum_j \nabla n'_j = -\sum_j v_j \cdot r, \qquad (13.2)$$

is that the stoichiometry of a reaction pattern can be readily studied experimentally by observations of the exchange flows ($\nabla n'_j$) with the environment in a continuous flow reactor. Most biological systems have the property that elements are conserved in all transformation processes. The balance equation for conservative properties has no production term:

$$\sum_j \nabla n'_j = 0. \qquad (13.3)$$

This equation represents an important statement of the principle of elemental balance, which is restricted to a steady state. Only the flows of matter leaving and entering the system must be considered (ROELS [13.3]). This formalism, however, results in complicated equations, which offer little advantage over the more formal equations (cf. Chapter 11). Therefore, only situations of limited complexity can be studied in more detail.

The main problem in applying stoichiometric considerations to bioprocessing, however, arises – beyond the quantification of stoichiometry in non-open reactor systems – from the complex network of metabolic reactions. Stoichiometry in simple reactions is trivial, while complex reactions can only be handled with the aid of a formal mathematical approach (SCHUBERT and HOFMANN [13.23]).

Due to the high complexity, however, it is not surprising that the historical approach in quantifying bioprocesses was much simpler, using the concept of yield factors Y. The macroscopic variable Y [cf. Eq. (11.3)], however, cannot be considered as a biological constant; this problem is discussed in Sect. 13.4. Some simple and complex cases of stoichiometry are considered in Table 13.1. This chapter concentrates mainly on stoichiometry and on some interrelations with thermodynamic and kinetic concepts.

13.2 The Concept of Yield Factors; Thermodynamic Considerations

13.2.1 The Yield Factor

The yield factor was originally defined by MONOD [13.4] in mass units by the quotient:

$$Y_{X/S} \equiv \frac{\Delta x}{\Delta s} = \frac{x_t - x_0}{s_0 - s_t} =$$

$$= \frac{r_X}{r_S} \left(\frac{\text{g cells formed}}{\text{g substrate used}} \right) \quad (13.4)$$

where Δx is the increase in biomass and Δs the amount of substrate required. This mass ratio is fully sufficient for many cases, for example, in discussing single-cell protein production from either carbohydrates or hydrocarbons (SUKATSCH and FAUST [13.5]).

In Fig. 13.1 the evaluation of the yield coefficient is shown schematically, where the value for $Y_{X/S}$ can be taken from the

Table 13.1. Problems Arising in the Stoichiometry of Technical Processes

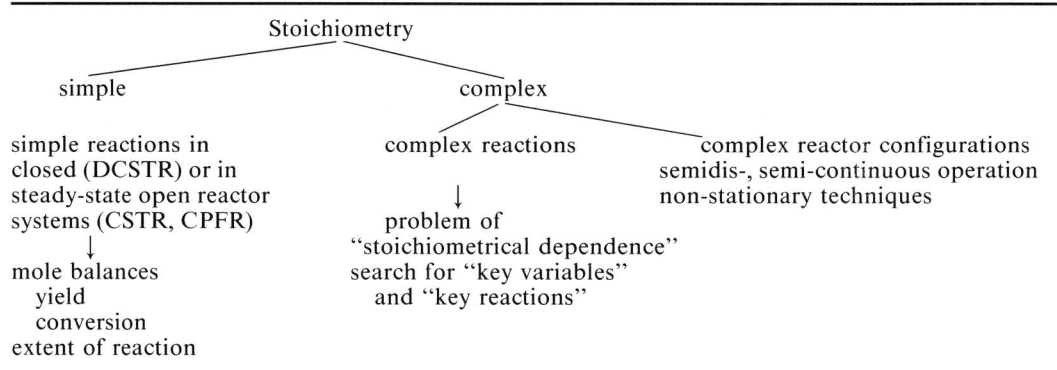

13.2 The Concept of Yield Factors; Thermodynamic Considerations

Table 13.2. Different Yield Coefficients in Bioprocessing on the Level of the Macroscopic Principle

Yield Coefficient $Y_{i/j}$	Reaction or Relation between Components	Definition	Eq. No.
$Y_{X/S}$	$S \to X$	$r_S = -1/Y_{X/S} \cdot r_X$	(13.5)
$Y_{X/O}$	$O_2 \to X$	$r_O = -1/Y_{X/O} \cdot r_X$	(13.6)
$Y_{P/S}$	$S \to P$	$r_S = -1/Y_{P/S} \cdot r_P$	(13.7)
$Y_{C/S}$	$S \to CO_2$	$r_S = -1/Y_{C/S} \cdot r_C$	(13.8)
$Y_{P/O}$	$O_2 \to P$	$r_O = -1/Y_{P/O} \cdot r_P$	(13.9)
$Y_{H_v/S}$	$S \to H_v$	$r_S = -1/Y_{H_v/S} \cdot r_{H_v}$	(13.10)
$Y_{C/O}$	$O_2 \to CO_2$	$r_O = -1/Y_{C/O} \cdot r_C$	(13.11)
$Y_{H_v/O}$	$O_2 \to H_v$	$r_O = -1/Y_{H_v/O} \cdot r_O$	(13.12)
$Y_{X/P}$	$P \sim X$	$r_P = 1/Y_{X/P} \cdot r_X = Y_{P/X} \cdot r_X$	(13.13)
$Y_{X/C}$	$CO_2 \sim X$	$r_C = 1/Y_{X/C} \cdot r_X$	(13.14)
Y_{X/H_v}	$H_v \sim X$	$r_{H_v} = 1/Y_{X/H_v} \cdot r_X$	(13.15)
$Y_{S/O}$	$O_2 \sim S$	$r_O = 1/Y_{S/O} \cdot r_S$	(13.16)
$Y_{C/P}$	$P \sim CO_2$	$r_P = 1/Y_{C/P} \cdot r_C$	(13.17)
Y_{P/H_v}	$H_v \sim P$	$r_{H_v} = 1/Y_{P/H_v} \cdot r_P$	(13.18)
Y_{C/H_v}	$H_v \sim CO_2$	$r_{H_v} = 1/Y_{C/H_v} \cdot r_C$	(13.19)

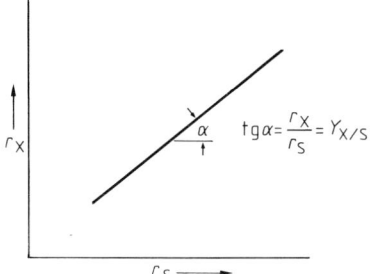

Fig. 13.1. Principle of the graphic method for the estimation of maximum yield coefficient $Y_{X/S}$.

$$\left(\frac{\text{g cells formed}}{\text{moles of substrate used}}\right) = Y_{X/S}^{\text{mol}} \quad (13.20)$$

$$\left(\frac{\text{g cells formed}}{\text{g-atoms of substrate used}}\right) = Y_{X/S}^{\text{g-at}}. \quad (13.21)$$

These units are invariable, for example, for yields of cells in g/mol of ATP formed during growth;

$$Y_{ATP} = \frac{\Delta x}{\Delta c_{ATP}}. \quad (13.22)$$

This yield factor plays an important role when considering molecular energetics of biochemical pathways. BAUCHOP and ELSDEN [13.6] found that Y_{ATP} was about 10.5 and independent of the type of organism or the environment.

There have been several suggestions of how to apply the molar concept to microorganisms. Elementary analysis performed by standard Pregl microcombustion techniques of cells revealed the following facts:

1. There is no unique empirical formula for a microorganism; the elementary composition varies with growth rate and limit-

slope of the curve, representing the data of r_{jX} vs. r_{iS}. The definition of the yield factor used by MONOD is of a purely macroscopic nature and can be applied to all compounds.

The same concept of yield can also be employed to quantify the relation between consumed or produced amounts. The definition of yield factors are summarized in Table 13.2. The estimation of yield factors can be carried out on the basis of plots similar to Fig. 13.1.

In some instances it is advantageous to express yields in mole units as follows:

ing nutrient. For yeast the following "formula" was given by HARRISON [13.7]:

$C_{3.72}H_{6.11}O_{1.95}N_{0.61}S_{0.017}P_{0.035}$

or $C_6H_{11}O_3N$.

For aerobic microorganisms the four principal elements are present in a ratio of $C_5H_7O_2N$ (HOOVER and PORGES [13.8]), which is very similar to that obtained for anaerobics (SPEECE and McCARTY [13.9]). When phosphorus is included a widely used formula is:

$C_{60}H_{87}O_{23}N_{12}P$ (McCARTY [13.10]).

2. There is no unique elementary balance equation for microbial growth but a range of such equations with different numerical coefficients.

All quantities considered, e.g., the amount of organisms containing one g-atom of total nitrogen, protein nitrogen, or DNA phosphorus, vary with growth rate and limiting substrate. It seems logical to derive the concept of a "mole" microorganism from empirical formulae which (neglecting sulfur, phosphorus, and ash) can be expressed in the general form $C_aH_bO_cN_d$. Finally, by dividing all the subscripts by a, the formula $CH_{b/a}O_{c/a}N_{d/a}$ is obtained. The definition of one "C-mole" is given by the quantity of microorganisms containing one gram-atom (12.011 g) of carbon (McLENNAN et al. [13.11]). The general use of this formula is strongly recommended for expressing quantities of microorganisms on a molar basis by HERBERT [13.12].

The concept of the "C-mole of biomass", i.e., the quantity of cells containing one gram-atom of carbon, allows yields to be expressed in molar units; this can be achieved in two ways:

$$\left(\frac{\text{g-atoms of cell-C formed}}{\text{moles of substrate used}}\right) = Y_S^C, \quad (13.23)$$

$$\left(\frac{\text{g-atoms of cell-C formed}}{\text{g-atoms of substrate-C used}}\right) = Y_C^C. \quad (13.24)$$

While both may be useful, the latter is usually to be preferred, as it is independent of the number of carbon atoms in the substrate molecule (HERBERT [13.12]).

Applying similar concepts to the important constants Y_{ATP} and Y_O, the following definitions are obtained:

$$\left(\frac{\text{g-atoms of cell-C formed}}{\text{moles of ATP formed}}\right) = Y_{ATP}^C, \quad (13.25)$$

$$\left(\frac{\text{g-atoms of cell-C formed}}{\text{g-atoms } O_2 \text{ used}}\right) = Y_O^C. \quad (13.26)$$

The quantity Y_O^C is of practical importance in the derivation of elementary balance equations and of theoretical importance in connection with oxidative phosphorylation (TEMPEST and NEIJSSEL [13.13]).

Besides the cell yields from carbon substrates and oxygen, it is suggested that cell yields from other nutrients are best expressed in the units of Eq. (13.24).

Y_N^C, Y_{Mg}^C, Y_K^C, Y_P^C, and Y_S^C

designate, respectively, the gram-atoms of cell carbon formed from one gram-atom of nitrogen, magnesium, potassium, phosphorus, and sulfur. This concept can be extended to all other essential nutrients.

Other important yield factors have been reviewed by PAYNE [13.14], BELL [13.15], and ATKINSON and MAVITUNA [13.16]. Various proposals for macroscopic efficiency measures have been put forward; they are all more or less related. The yield based on electrons available in substrates;

$$\left(\frac{\text{g of substrates}}{\text{electrons}}\right) = Y_{av\,e^-}, \quad (13.27)$$

expresses the electrons initially available in the substrate for transfer to oxygen during the combustion of this substrate. The concept of "available electrons" was first developed by MAYBERRY et al. [13.17] and later applied to biomass by MINKEWICH and EROSHIN [13.18].

A second measure proposed by PAYNE [13.14] is the yield of biomass dry weight per kJ heat of combustion of the substrate ΔH_S:

$$\left(\frac{\text{g of cell dry matter}}{\text{kJ heat of combustion of } S}\right) = Y_{kJ}. \quad (13.28)$$

This can be expressed for aerobic fermentations by:

$$Y_{kJ} = \frac{Y_{X/S}^{mol}}{\Delta H_S}\left(\frac{\text{g/mol}}{\text{kJ/mol}}\right) \quad (13.29)$$

or according to PROCHAZKA et al., [13.19] by:

$$Y_{kJ} = \frac{Y_{X/S}}{\Delta H_S - (Y_{X/S} \cdot \Delta H_X)}, \quad (13.30)$$

where ΔH_X is the heat of combustion of cell mass (in $kJ \cdot g^{-1}$).

13.2.2 Thermodynamic Efficiency and Degree of Reduction

The concept of available electrons enables us to express the energetic yield of growth as a fraction of available electrons of the substrate utilized for transfer into biomass. The energy efficiency coefficient of growth η is the part of the available electrons av e^- of the substrate transferred to biomass. $(1-\eta)$ is the energy consumed during the process of biosynthesis and released as heat.

The efficiency η is proportional to Y_{ave^-} as well as Y_{kJ} and seems to be very promising in applications, although it was suggested that the correct formulation of the second law of thermodynamics for open systems is missing (ROELS [13.3]).

The various efficiency measures have been recently reviewed by NAGAI [13.20] and are used in studying the energetics of growth and product formation (ROELS [13.21]; ERICKSON et al. [13.22]). For aerobic growth on a sole carbon and energy source without product formation the efficiency factor for oxygen of MINKEVICH and EROSHIN is shown to be related to the "degrees of reduction" of substrate γ_S and of biomass γ_X:

$$\eta \equiv Y'_{X/S} \cdot \gamma_X/\gamma_S, \quad (13.31)$$

with

$$Y'_{X/S} = \frac{Y_{X/S}}{a} = \frac{v_X}{v_S/a}, \quad (13.32)$$

where a is the number of carbon atoms in one molecule of substrate. Eq. (13.32) is a somewhat modified definition of the yield factor after Eq. (13.4) (ROELS [13.21]). $Y'_{X/S}$ expresses the moles of dry weight of biomass per carbon equivalent substrate consumed. The correlation given by Eq. (13.31) varies with the type of nitrogen source, as γ_X depends on the nitrogen source (ROELS [13.3]).

A relationship exists between $Y'_{X/S}$ and Y_{ave^-} given by the following equation (ROELS [13.3]):

$$Y_{ave^-} = \frac{Y'_{X/S}}{\gamma_S} \cdot M_X, \quad (13.33)$$

where M_X is the molecular weight of the biomass. Instead of the efficiency factor η, ROELS introduced the concept of "thermodynamic efficiency" η_{th}:

$$\eta_{th} \equiv \frac{Y'_{X/S}}{(Y'_{X/S})_{max}}, \quad (13.34)$$

where the maximum value of the yield is calculated in complete accordance with the second law. This equation has been used in the analysis of some recent yield data which have been published in the literature, represented graphically in Fig. 13.2.

Fig. 13.3 shows the same results as presented in Fig. 13.2 in a somewhat different form. The tendencies observed in Figs. 13.2 and 13.3 can be summarized as follows: For a substrate up to a degree of reduction of about 4.2, the energy content is insufficient to allow the total substrate carbon to be converted into biomass, even if the thermo-

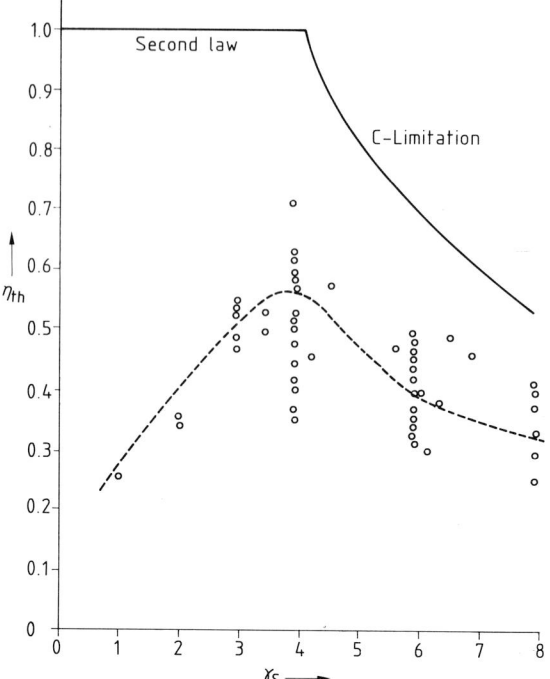

Figure 13.2. Thermodynamic efficiency η_{th} of aerobic growth with one carbon source and NH_3 as a nitrogen source as a function of the substrate's degree of reduction γ_S and theoretical limits to the efficiency (ROELS [13.3.]).

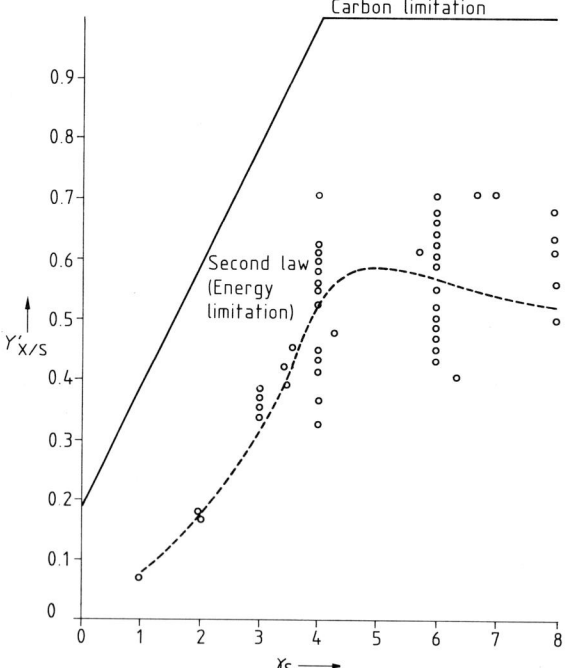

Figure 13.3. Relationship between substrate yield factor $Y'_{X/S}$ and substrate degree of reduction γ_S and the theoretical limits to $Y'_{X/S}$ (ROELS [13.3]).

dynamic efficiency is unity. For a substrate with a degree of reduction higher than 4.2, the entire carbon could be converted into biomass as far as energy requirements are concerned. The extent to which the energy can be stored in biomass becomes limiting. Hence, the thermodynamic efficiency has to decrease with growth on substrates with a high degree of reduction. For unknown reason the carbon conservation efficiency $Y'_{X/S}$ never exceeds 0.7.

A tentative model has been developed by ROELS [13.21] based on a simple microscopic assumption about the mechanism of energy transfer in a functioning organism. This model explains the reduced value of η_{th} with growth on highly reduced substrates even if it represents only a highly simplified picture of energy transformation in living organisms.

13.3 Fundamentals of Stoichiometry of Complex Reaction Systems

Bioprocessing involves a multitude of metabolic reactions even on the macroscopic level and different modes of reactor operations.

13.3.1 Stoichiometry of Complex Reactions

The stoichiometric equation in the case of complex reactions can be written in the form:

$$\sum_{i}^{M} \sum_{j}^{N} v_{ij} \cdot A_j = 0, \quad (13.35)$$

with i, the number of reactions ($1 \leq i \leq M$), j, the number of components ($1 \leq j \leq N$), A_j, the components of the reaction mixture, and v, the stoichiometric coefficients.

The differential change of the number of moles N_j of component A_j due to the reaction i is defined as:

$$(d N_j)_{r_i} = v_{ij} \cdot d \xi_i, \quad (13.36)$$

where ξ_i is the extent of reaction, defined as the change of the number of moles divided by the stoichiometric coefficient.

In complex reactions individual reaction steps are interconnected, i.e., singular components participate in different reactions causing a part of the reaction to be stoichiometrically dependent. Only the independent reactions can be determined from the change of the number of moles. A solution of the problem of stoichiometric dependency can be found with the aid of the "matrix of the stoichiometrical coefficients" with row index = number of reactions i and column index = number of components j. The number of stoichiometrically independent reactions is given by the rank of the stoichiometric coefficient matrix, R_v, which can be determined with the aid of the Gaussian method of elimination. This yields the stoichiometric coefficients of R_v and linearly independent equations, which are necessary and sufficient for calculating the conversion of the key variables and, therefore, for all other components.

Sometimes balancing is carried out without formulating the reaction equations. In this case an "element-species matrix" can be written on the basis of N species (components) with k elementary balances with row index = number of components j, and column index = number of chemical elements i, or number of elementary balances ($1 \leq i \leq k$). The number of key variables R is determined with the aid of the rank of this matrix R_β. Thus, according to SCHUBERT and HOFMANN [13.23]:

$$R = N - R_\beta \quad (13.37)$$

and

$$R_v \equiv R. \quad (13.38)$$

Normally $R_\beta = k$ in the case of N species and k elements. Hence, only $(N-k)$ net conversion rates can be chosen independently. From this reasoning it becomes clear that the number of independent kinetic equations to be postulated cannot be chosen at will; it is completely specified by the number of elementary balances k and the number of components N in the system (ROELS [13.3]). Which key variable or which kinetic equation is to be chosen strongly depends on the type of application considered. These concepts have recently been successfully applied to bioprocessing (see Sect. 13.4.4).

13.3.2 Stoichiometry of Complex Reactor Operation

The majority of stoichiometric considerations is restricted to simple reactor types such as closed reactor operations and stationary continuous stirred tanks. Balancing in the case of semi-discontinuous processing has to distinguish between components which are already present in the reactor from the beginning expressed by the number of moles $N_{j,0}$, and components which are later fed to the reactor expressed by the flux of moles $\dot{N}_{j\xi}^0$. The stoichiometric balance in the case of semi-discontinuous reactor operation is:

$$N_j = N_{j,0} + \dot{N}_j^0 \cdot t \pm \sum_i v_{ij} \cdot \xi_i, \qquad (13.39)$$

with N_j, the number of moles of component j, N_0, the temporal initial value of N, and N_j^0, the spatial initial value of N. This operation contains the time t as an independent process variable in contrast to all batch-reactor configurations.

Stoichiometry in the case of non-stationary modes of reactor operation calls for another fundamental set of equations.

The balance of component j in a stirred tank reactor is as follows:

$$\frac{dc_j}{dt} + \frac{c_j - c_j^0}{\bar{t}} = \sum_i v_{ij} \frac{d\xi_i}{dt}. \qquad (13.40)$$

Integration yields:

$$c_j = c_j^0 + (c_{j,0} - c_j^0) e^{-t/\bar{t}} + \sum v_{ij} \cdot \xi^*(t), \qquad (13.41)$$

where $\xi^*(t)$ is the modified extent of reaction.

The characteristic property of this equation is the time t as the process variable. The influence of time t can be neglected in the case of stationary conditions with $t \gg \bar{t}$ and when $c_{j,0} = c_j^0$, in a batch reactor. Therefore, the stoichiometric balance equation of a discontinuous system can be formally applied to non-stationary continuous stirred tank operation.

13.4 Application of Stoichiometric Concepts to Bioprocessing

13.4.1 Yield Coefficients in Fermentation Processes

In real fermentations yields are not constant so that the concept of the "yield-constant" must be modified. Yield factors exhibit a significant dependence on different biological and physical parameters. Although the yield factor is defined for a given strain with respect to a particular substance, it is not a function of substrate only. It also depends on the chemical nature and concentration of all other components in the medium whose actual concentrations are influenced by physical phenomena in the reactor, e.g., mixing time t_m and oxygen transfer rate OTR (MOSER [13.24], [13.25]).

Generally, the following conditions apply:

$Y = $ (strain, substrate; μ, m, S;
$\bar{t}, t_m$, OTR, C/N-, P/O-ratio, ...) (13.42)

Earlier studies have described the relationship of the yield coefficients with organism and substrate (HUMPHREY [13.26]; WANG [13.27]). These reports and others have illustrated the dependence of $Y_{X/O}$ and Y_{kJ} on $Y_{X/S}$ (MATELES [13.28]; GUENTHER [13.29]). The following function was elaborated by PIRT [13.30] to demonstrate the relationship of Y with the specific growth rate μ, realized in CSTR operating at different dilution rates D ($D = 1/\bar{t} = \mu$), and with the maintenance coefficient m:

$$\frac{1}{Y_{X/S}} = \frac{1}{Y_{X/S}^{max}} + \frac{m_s}{\mu}. \quad (13.43)$$

$Y_{X/S}^{max}$ is the value of $Y_{X/S}$ as μ approaches infinity and m_s is the specific rate of substrate consumption at a zero growth rate.

Eq. (13.43) can be used in a modified form to calculate the influence of μ and m_s on $Y_{X/S}$ (ABBOTT and CLAMEN [13.31]):

$$Y_{X/S} = \frac{\mu \cdot Y_{X/S}^{max}}{\mu + Y_{X/S}^{max} \cdot m_s}. \quad (13.44)$$

Fig. 13.4 shows the general dependence of Y on μ for $m_s = $ const.

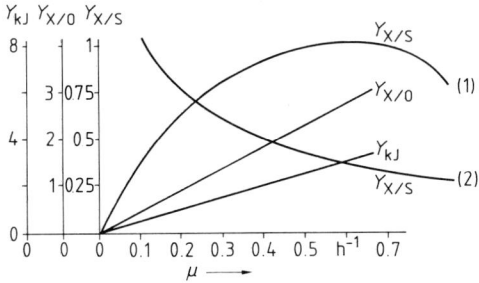

Figure 13.4. Yield coefficients as a function of specific growth rate μ, after ABBOTT and CLAMEN [13.31]. – The biomass yield $Y_{X/S}$ is shown in two curves, where $Y_{X/S}$ is the yield coefficient (1) with glucose as limiting substrate and (2) with nitrogen source as limiting substrate.

The rates at which the yield coefficients vary with μ depend on the value of the maintenance coefficient m_s. At constant μ, Fig. 13.5 shows the influence of the maintenance coefficient m_s on yields.

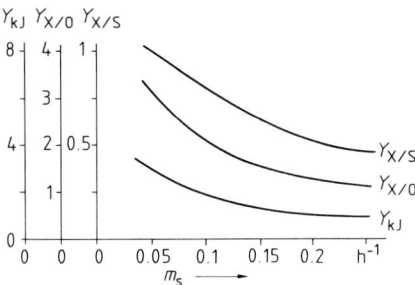

Figure 13.5. Yield coefficients as a function of specific maintenance coefficient m_S, after ABBOTT and CLAMEN [13.31].

In a number of cases this approach does not satisfy as one considers the substrate while one ought to focus on ATP. Therefore, an equation similar to Eq. (13.43) has been proposed for the specific production of ATP, where q_{ATP} is a linear function of μ.

Several attempts have been made to explain the biomass yield variation by specific growth rates based on Eq. (13.43), e.g., REUSS et al. [13.32]; ROCK et al. [13.33], but such an explanation does not appear completely justified. Eq. (13.43) cannot explain why the yield changes during the exponential phase in batch cultures where μ is constant. Furthermore, it is not true that, in the case of C_1-compounds utilizing microorganisms, $Y_{X/S}$ increases monotonically with μ or D for steady-state CSTR.

PAPOUTSAKIS and LIM [13.34] stress the carbon-flow-branching concept to illustrate the general possibility of a highly variable biomass yield only on the basis of trivial assumptions. These authors have developed an equation for the yield factor:

$$Y_{X/S} = \left(\frac{M_X}{v_{SX} \cdot M_S}\right) \frac{1}{1 + (r_1/r_2)}, \quad (13.45)$$

where r_1 and r_2 are the rates of branching metabolic pathways 1 and 2 for the carbon source, M_X and M_S are the molecular weights of biomass and substrate, respectively, and v_{SX} the stoichiometric coefficient of the reaction $S \rightarrow X$. In the case of methylotrophs there exist two distinct carbon-flow processes: assimilation (r_2) and oxidation (r_1). Cells produce only so much NADH and/or ATP through oxidation as is required for assimilation. This fine tuning of the two processes yields the maximal possible biomass. According to Eq. (13.45) the yield varies only with r_1 or r_2; this may occur by varying any concentration of the substrate or other nutrients or any of the culture conditions like T and pH.

An alternative concept has recently been proposed by AGRAWAL et al. [13.35], who achieved the desired variability of the stoichiometry by defining a yield factor, which linearly increases with substrate concentration s:

$$Y_{X/S} = a + b \cdot s, \qquad (13.46)$$

where a and b are constants. These authors incorporated the linearly increasing yield factor into Monod kinetics.

The yield factor cannot be expected to vary significantly with temperature; this has been proved by TOPIWALA and SINCLAIR [13.36]; SHILOACH and BAUER [13.37], and MOLETTA et al. [13.38]. In the case of the yield factor $Y_{X/O}$, based on BOD, a decrease with increasing temperature has been recorded by PETERS [13.39].

13.4.2 Relationship Between Yield Factors and True Stoichiometric Coefficients

On the basis of the general stoichiometric equation for microbial growth Eq. (11.2), the following balance equations for the significant elements C, H, O, and N can be developed:

starting material   products

$$\text{g-atom C:} \quad v_S \cdot a \quad = \overbrace{v_X \cdot \alpha}^{1} + v_C + v_P \cdot \alpha' \qquad (13.47)$$

$$\text{g-atom H:} \quad v_S \cdot b + 3v_N = v_X \cdot \beta + v_P \cdot \beta' + 2v_W \qquad (13.48)$$

$$\text{g-atom O:} \quad v_S \cdot c + 2v_O = v_X \cdot \gamma + v_P \cdot \gamma' + 2v_C + v_W \qquad (13.49)$$

$$\text{g-atom N:} \quad v_S \cdot d + v_N = v_X \cdot \delta + v_P \cdot \delta' \qquad (13.50)$$

In Eqs. (13.47) to (13.50) the quantities of substrate (v_S) and NH$_3$ (v_N) required to produce one C-mole of cells can be measured directly and the composition of cells and product can be determined by elementary analysis, when α, β, γ, δ and α', β', γ', δ' are given. If no nitrogen is contained in the substrate and product, then $v_N = v_X \cdot \delta$. Determination of NH$_3$ is not strictly necessary. It provides, however, a useful proof of the overall accuracy. O$_2$ and CO$_2$ can be determined with the aid of gas analyzers, giving v_O and v_C. The only quantity that cannot be conveniently measured is v_W, the moles of H$_2$O produced. This value, however, can be obtained from the difference between Eqs. (13.48) and (13.49). From Eqs. (11.2) and (13.47) to (13.50) a number of important yield constants can be derived:

$$Y_{X/S} = \frac{v_X(12\alpha + \beta + 16\gamma + 14\delta)/(1-f_a)}{v_S(12a + b + 16c + 14d)}$$

$$\left(\frac{\text{g of cells formed}}{\text{g of substrate used}}\right), \qquad (13.51)$$

where f_a is the fraction of cell mass represented by ash ($f_a \approx 0.07$–0.10) (WANG et al. [13.40]).

Other yield factors, valid for cases without product formation, can be expressed as follows (HERBERT [13.12]):

$$Y_S^C = \frac{1}{v_S} \left(\frac{\text{g-atoms of cell-C formed}}{\text{moles of substrate used}}\right),$$

$$(13.52)$$

$$Y_C^C = \frac{1}{v_S \cdot a}\left(\frac{\text{g-atoms of cell-C formed}}{\text{g-atom of substrate-C used}}\right), \quad (13.53)$$

$$Y_{CO_2}^C = \frac{v_C}{v_S \cdot a}\left(\frac{\text{moles of } CO_2 \text{ formed}}{\text{g-atoms of substrate-C used}}\right), \quad (13.54)$$

$$Y_O^C = \frac{v_X \cdot \alpha}{2 v_O} = \frac{1}{2 v_O}$$

$$\left(\frac{\text{g-atoms of all-C formed}}{\text{g-atoms } O_2 \text{ used}}\right). \quad (13.55)$$

The yield of product becomes:

$$Y_{P/S} = \frac{v_P(12\alpha' + \beta' + 16\gamma' + 14\delta')}{v_S(12a + b + 16c + 14d)}. \quad (13.56)$$

In practice it is usually convenient to calculate the values of these yield coefficients from the experimental data and to use them to obtain the values of v_S, v_O, and v_C for introduction into the cell balance equation.

MATELES [13.28] derived a useful equation for the relationship between the yields of cells on the carbon substrate and the requirement of O_2. This expression shows that O_2-demand is universally proportional to cell yield, which is a measure of the efficiency of conversion of the carbon source to cell mass.

13.4.3 Integration of Elemental Balances and Kinetic Equations into Unstructured Modelling of Bioprocesses

In order to develop a simple unstructured model for a bioprocess at least one of the reactions taking place in the culture has to be specified in kinetic terms.

Generally, a complete set of constitutive equations for each of the M chemical reactions taking place in the culture can be written in the form of a sum or of a matrix (SCHUBERT and HOFMANN [13.23]; ROELS [13.3]). The net conversion rate of each of the components follows from $r_j = v_j \cdot r$. For a system in a steady-state condition the net production rates are influenced by the flow to the system as is clear from Eq. (13.2). Furthermore, the elemental balance principle according to Eq. (13.3) specifies k relationships between the flows F_j and, hence, only $(N-k)$ net conversion rates can be

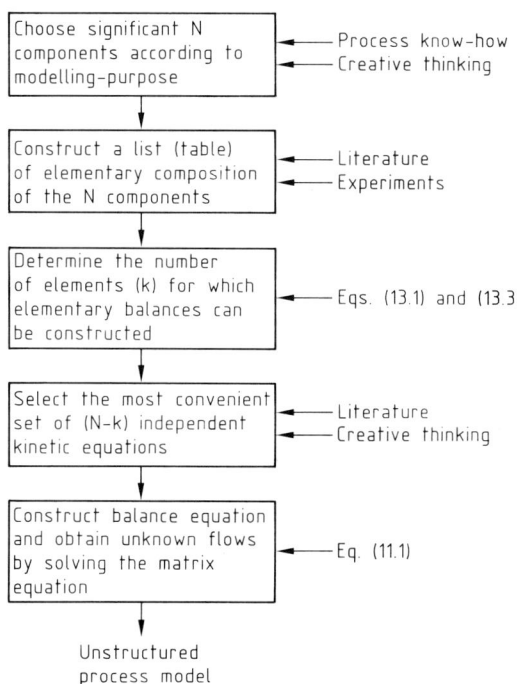

Figure 13.6. Flow sheet of the construction of unstructured models in bioprocessing by an integration of elemental balances and kinetic equations, after ROELS [13.3].

chosen independently. The number of independent kinetic equations to be postulated cannot be chosen at will, it is completely specified by the number of elemental balances (k) and the number of components in the system (N). The procedure used in forming simple unstructured process models is outlined in Fig. 13.6 (ROELS [13.3]).

13.4.4 Application of Balancing Methods in Bioprocessing

As a consequence of the increased worldwide competition in bioprocessing considerable attention has been directed towards an improvement of the efficiency of technical bioprocesses. In this situation formulation of balanced stoichiometric equations is a powerful technique.

If an elementary analysis of the cells has been carried out, so that the stoichiometric coefficients in Eq. (11.2) are known, then the mass balance equation may be set up without actually measuring the oxygen used or CO_2 formed. In this manner oxygen uptake and growth yield can be theoretically predicted as has been shown by ROELS [13.3]. A macroscopic analysis based on elemental and energy balances gives important relationships for bioprocess design. The existence of limits to the oxygen and substrate yield factor has been shown on this basis.

In the absence of information on the biosynthetic pathway an alternate approach based on reaction stoichiometry has been applied to penicillin synthesis [13.41], [13.42]. The theoretical maximum yield of penicillin from glucose was calculated and the actual conversion yield could be improved substantially by increasing and sustaining the specific rate of penicillin production and by minimizing the maintenance metabolism.

HEIJNEN et al. [13.43] developed elementary and enthalpy balances for the penicillin production process. This macroscopic analysis used carbon, hydrogen, nitrogen, oxygen, sulfur, phosphorus, and enthalpy balances for eleven relevant process variables, so that according to Eq. (13.38) at least five kinetic equations are required for a complete model. On the basis of this fermentation model the authors presented a series of simulation studies and some interesting results. They emphasize that a growth-coupled penicillin production provides an adequate description of most of the observed phenomena, which have often led to the assumption of non-growth-associated or age-dependent penicillin productivity.

Recently the impact of sugar feeding strategies has been investigated with the result that maximum productivities are more or less independent of the feeding scheme (BAJPAI and REUSS [13.44]). A strategy proposed by MOU [13.45] was employed.

Overall and instantaneous carbon balancing equations and CO_2 production data have been successfully used to calculate, on-line, the cell concentration and instantaneous growth rate by MOU and COONEY [13.46], [13.47] for penicillin production.

In an approach towards a mathematical description of activated sludge waste water treatment, biokinetic and stoichiometric relationships have led to significant insights into the process and the operation of the plant (SHERRARD [13.48]; SHERRARD and SCHROEDER [13.49]; SHERRARD [13.50]). A similar procedure of developing a reaction scheme handling kinetic expressions and reducing the system to important reactions and components has recently been presented for biological waste treatment and applied to biological nitrogen removal (IRVINE et al. [13.51]).

13.5 References

[13.1] N. W. F. KOSSEN; Proc. Soc. Gen. Microbiol. Symp. 29 (1979), p. 327, (A. T. BULL et al., eds.)

[13.2] J. A. ROELS and N. W. F. KOSSEN; Prog. Ind. Microbiol. 14 (1978), 95.

[13.3] J. A. ROELS; Biotechnol. Bioeng. 22 (1980), 2457.

[13.4] J. MONOD: "Recherches sur la Croissance des Cultures Bacteriennes". Hermann & Cie, Paris, 1942.

[13.5] D. A. SUKATSCH and U. FAUST; in: Proc. Tutzing Symp., DECHEMA Monographs No. 81 (1977), 197.

[13.6] T. BAUCHOP and S. R. ELSDEN; J. Gen. Microbiol. *23* (1960), 457.

[13.7] J. S. HARRISON; Process Biochem. *2* (1967), 41.

[13.8] S. R. HOOVER and N. PORGES; Sewage Ind. Wastes *24* (1952), 306.

[13.9] R. E. SPEECE and P. L. MCCARTY; Adv. Water Pollut. Res. *2* (1964), 305.

[13.10] P. L. MCCARTY; in: "Proc. Wastewater Reclamation Reuse Workshop", Lake Tahoe, Calif., (1970), p. 226.

[13.11] D. G. MCLENNAN, J. S. GOW, and D. A. STRINGER; Process Biochem. *8* (1973), June, 22.

[13.12] D. HERBERT; in "Continuous Culture". Vol. 6, p. 1, (A. C. R. DEAN et al., eds.), SCI, London, 1975.

[13.13] D. W. TEMPEST and O. M. NEIJSSEL; in: "Continuous Culture". Vol. 6, p. 283, (A. C. R. DEAN et al., eds.) SCI, London, 1975.

[13.14] W. J. PAYNE; Annu. Rev. Microbiol. *24* (1970), 17.

[13.15] G. H. BELL; Process Biochem. (1972) April, 21.

[13.16] B. ATKINSON and F. MAVITUNA: "Biochemical Engineering and Biotechnology, Handbook". The Nature Press and Mcmillan Publ., London, 1983.

[13.17] W. H. MAYBERRY, G. J. PROCHAZKA, and W. J. PAYNE; J. Bacteriol. *96* (1968), 1424.

[13.18] J. G. MINKEVICH and V. K. EROSHIN; Folia Microbiol. *18* (1973), 376.

[13.19] G. J. PROCHAZKA, W. J. PAYNE, and W. R. MAYBERRY; J. Bacteriol. *15* (1970), 117.

[13.20] S. NAGAI; Adv. Biochem. Eng. *11* (1979), 49.

[13.21] J. A. ROELS; Biotechnol. Bioeng. *22* (1980), 23.

[13.22] L. E. ERICKSON, I. G. MINKEVICH, and V. K. EROSHIN; Biotechnol. Bioeng. *20* (1978), 1595.

[13.23] E. SCHUBERT and H. HOFMANN; Chem. Ing. Tech. *47* (1975), 191.

[13.24] A. MOSER; Chem. Ing. Tech. *49* (1977), 612.

[13.25] A. MOSER; "Bioprocess Engineering". Springer Verlag, Heidelberg–New York, 1984.

[13.26] A. E. HUMPHREY; Process Biochem. *5* (1970), June, 19.

[13.27] D. I. C. WANG; Chem. Eng. *57* (1968), 99.

[13.28] R. I. MATELES; Biotechnol. Bioeng. *12* (1971), 581.

[13.29] K. R. GUENTHER; Biotechnol. Bioeng. *7* (1965), 445.

[13.30] S. J. PIRT; Proc. R. Soc. London Ser. B *163* (1965), 224.

[13.31] B. J. ABBOTT and A. CLAMEN; Biotechnol. Bioeng. *15* (1973), 117.

[13.32] M. REUSS, H. SAHM, and F. WAGNER; Chem. Ing. Tech. *46* (1974), 669.

[13.33] J. S. ROCK, I. GOLDBERG, and R. I. MATELES; Biotechnol. Bioeng. *20* (1978), 1557.

[13.34] E. PAPOUTSAKIS and H. C. LIM; Ind. Eng. Chem. Fundam. *20* (1981), 307.

[13.35] P. AGRAWAL, C. LEE, H. C. LIM, and D. RAMKRISHNA; Chem. Eng. Sci. *37* (1982), 453.

[13.36] H. TOPIWALA and C. G. SINCLAIR; Biotechnol. Bioeng. *13* (1971), 1975.

[13.37] J. SHILOACH and S. BAUER; Biotechnol. Bioeng. *17* (1975), 227.

[13.38] R. MOLETTA, G. GOMA, and G. DURAND; Arch. Microbiol. *118* (1978), 293.

[13.39] H. PETERS; Karlsruher Berichte zur Ingenieurbiologie *9* (1976), Ph. D. Thesis.

[13.40] D. I. C. WANG, CH. L. COONEY, A. L. DEMAIN, P. DUNWILL, A. E. HUMPHREY, and M. LILLY: "Fermentation and Enzyme Technology". J. Wiley & Sons, New York, 1979.

[13.41] CH. L. COONEY and F. ALCEVEDO; Biotechnol. Bioeng. *19* (1977), 1949.

[13.42] CH. L. COONEY; Process Biochem. *14* (1979), May, 31.

[13.43] J. J. HEIJNEN, J. A. ROELS, and A. H. STOUTHAMER; Biotechnol. Bioeng. *21* (1979), 2175.

[13.44] R. K. BAJPAI and M. REUSS; Biotechnol. Bioeng. *23* (1981), 717.

[13.45] D. G. MOU; Ph. D. Thesis, MIT Cambridge, Mass., 1979.

[13.46] D. G. MOU and CH. L. COONEY; Biotechnol. Bioeng. *25* (1983) 225.

[13.47] D. G. MOU and CH. L. COONEY; Biotechnol. Bioeng. *25* (1983), 257.

[13.48] J. H. SHERRARD; J. Water Pollut. Control Fed. *49* (1977), 1968.

[13.49] J. H. SHERRARD and E. D. SCHROEDER; J. Water Pollut. Control Fed. (1976), 742.

[13.50] J. H. SHERRARD; J. Chem. Tech. Biotechnol. *30* (1980), 447.

[13.51] R. L. IRVINE, J. E. ALLEMAN, G. MILLER, and R. W. DENNIS; J. Water Pollut. Control Fed. *52* (1980), July, 1997.

Chapter 14
Kinetics of Batch Fermentations

Anton Moser

Institut für Biotechnologie, Mikrobiologie und Abfalltechnologie
Technische Universität Graz
Graz, Austria

14.1   Introduction
14.2   Batch Fermentation Pattern and Kinetic Approaches
14.2.1 Fermentation Pattern
14.2.1.1 Morphology Pattern
14.2.1.2 Kinetic Pattern
14.2.2 Kinetic Approaches
14.2.2.1 Mechanistic Models of Microkinetics
14.2.2.2 Numerical Fitting of Bioprocess Kinetics
14.2.2.3 The Formal Kinetic Approach
14.2.3 Differential and Integral Evaluation of Batch Processes
14.3   Mathematical Models of Batch Growth Kinetics
14.3.1 Pseudohomogeneous Rate Equations for Biokinetics
14.3.1.1 Inhibition-free Substrate-limiting Model Functions
14.3.1.2 Inhibition Kinetics
14.3.1.3 Extensions of Validity of Monod-type Kinetics to All Other Growth Phases
14.3.1.4 Kinetic Modelling of Endogenous Metabolism
14.3.1.5 Kinetic Modelling of Mycelial Growth
14.3.1.6 Influence of Temperature, pH Value, Salinity, and Water Activity on Growth
14.3.2 Multi-substrate Kinetics
14.3.3 Kinetic Models for Transient Growth in Batch Processing
14.3.4 Heterogeneous Rate Equations: Macrokinetics
14.3.4.1 Deviations from Monod-type Kinetics
14.3.4.2 Macrokinetics of Linear Growth Phenomena
14.4   Mathematical Models of Product Formation Kinetics
14.4.1 Kinetic Pattern
14.4.2 Heat of Fermentation
14.4.3 Generalized and Extended Rate Equations for Microbial Product Formation
14.5   References

List of Symbols

a	m^{-1}	specific area of mass transfer
a_H	m^{-1}	specific area of heat transfer
a_i, a_P	—	stoichiometric coefficients
a, b, c, d	—	coefficients
B_x	s^{-1}	mass loading rate per unit mass
C	kg	amount of CO_2
c	$kg \cdot m^{-3}$	concentration of CO_2
c_p	$kJ \cdot mol^{-1} \cdot K^{-1}$	molar specific heat at constant pressure
c_R	$kg \cdot m^{-3}$	concentration of ribosomes
d_p	m	thickness of active zone of pellets [Eq. (14.102)]
E	kg	amount of enzyme
$ES, EP, \ldots$	kg	amount of enzyme complexes
e	$kg \cdot m^{-3}$	enzyme concentration
F	$m^3 \cdot s^{-1}$	flow rate
f_{repr}	—	factor for repression [Eqs. (14.78) to (14.82)]
H_v	$J \cdot m^{-3}$	volumetric heat of fermentation
$\Delta H_R^{(0)}$	$J \cdot mol^{-1}$	reaction enthalpy of fermentation
$\Delta H_R^{(X)}, \Delta H_R^{(S)}$	$J \cdot kg^{-1}$	heat per unit mass of biomass (X) or substrate (S)
ΔH_i	$J \cdot mol^{-1}$	heat of combustion of component i
K_d	$kg \cdot m^{-3}$	saturation-type constant in Eq. (14.67)
K_D	$kg \cdot m^{-3}$	constant for diffusional resistance in Eq. (14.17)
K_e	$kg \cdot m^{-3}$	saturation-type constant in Eq. (14.69)
$K_{I,S}$	$kg \cdot m^{-3}$	constant of substrate inhibition
$K_{I,P}$	$kg \cdot m^{-3}$	constant of product inhibition
K_S	$kg \cdot m^{-3}$	saturation constant for substrate
K_P	$kg \cdot m^{-3}$	saturation-type constant for product
K_{P1}, K_{P2}	$kg \cdot m^{-3}$	saturation constants in Eq. (14.38)
K_{repr}	$kg \cdot m^{-3}$	constant for catabolite repression [Eq. (14.82)]
K_1, K_2, K_3	$J \cdot kg^{-1} \cdot mol^{-1}$	constants in Eq. (14.128)
k_{cat}	$mol \cdot mol^{-1} \cdot s^{-1}$	catalytic enzyme constant in Eq. (12.28)
k_d	s^{-1}	rate constant of overall cell death
k_d^0	s^{-1}	rate constant of true cell death, Eq. (14.60)
k_e	s^{-1}	rate coefficient of endogenous metabolism, Eq. (14.60)
k_H	$J \cdot m^{-2} \cdot K^{-1} \cdot s^{-1}$	heat transfer coefficient, Eq. (14.123)
k_i	s^{-1}	rate coefficient of i-th reaction
k_1	$m \cdot s^{-1}$	mean relative rate of apical growth, Eq. (14.72)
k_2	$m^{-1} \cdot s^{-1}$	mean relative rate of branching, Eq. (14.73)
k_{L1}	$m \cdot s^{-1}$	mass transfer coefficient at G/L-interface
$k_{L1}a$	s^{-1}	volumetric O_2-mass transfer coefficient
k_{L2}	$m \cdot s^{-1}$	mass transfer coefficient at L/S-interface
k_P	s^{-1}	rate coefficient of production
L_H	m^{-2}	total length of hyphae per unit volume, Eq. (14.72)
M, M_i	kg	amount of mass (component i)
m, m_s, m_{H_v}	s^{-1}	maintenance coefficient (general, for substrate, volumetric heat)
N_H	m^{-3}	number of hyphal tips per unit volume, Eq. (14.72)
O	kg	amount of O_2
o	$kg \cdot m^{-3}$	O_2-concentration
P	kg	amount of product
p	$kg \cdot m^{-3}$	product concentration
p	—	power number
q, q_i	$kg_I \cdot kg_X^{-1} \cdot s^{-1}$	specific rate of bioprocess (component i)

List of Symbols

r	$kg \cdot m^{-3} \cdot s^{-1}$	rate of reaction
r_i	$kg \cdot m^{-3} \cdot s^{-1}$	rate of consumption or production of component i
S	kg	amount of substrate
S_E	kg	amount of essential substrate, Eq. (14.90)
S_I	kg	amount of growth enhancing substrate, Eq. (14.90)
s	$kg \cdot m^{-3}$	substrate concentration
s_E	$kg \cdot m^{-3}$	concentration of essential substrate, Eq. (14.90)
s_I	$kg \cdot m^{-3}$	concentration of growth enhancing substrate, Eq. (14.90)
s_c	$kg \cdot m^{-3}$	critical concentration of S
t	s	time (general term)
t_L	s	lag time
$\bar{t}$	s	mean residence time
t_{max}	s	time of maximum production
t_M	s	maturation time, Eq. (14.119)
T	K	temperature
$\overline{\Delta T}$	K	logarithmic mean temperature
V	m^3	volume
X	kg	amount of biomass
x	$kg \cdot m^{-3}$	biomass concentration
$Y, Y_{i/j}$	$kg \cdot kg^{-1}$	yield coefficient (component i / component j)
$Y^*_{S/X}$	$kg \cdot kg^{-1}$	yield coefficient relating S-consumption for biomass precursor biosynthesis, Eq. (14.120)
Y_{X/H_v}	$kg \cdot kJ^{-1}$	yield coefficient of heat related to biomass

Greek letters

Λ	s	cell age
μ	s^{-1}	specific growth rate
σ	—	ionic strength
τ	s	running time
$\varkappa$	$kg^{-1} \cdot m$	quantity defined by Eq. (14.27)

Indices

app	apparent		max	maximum
cat	catalytic		O	oxygen
c, crit	critical		0	zero, initial value
C	CO_2		obs	observed
ex	exit		P	product or particle
E	enzyme or essential for growth		R	ribosome
eff	effective		rel	relative
G	gas		repr	repression
H, H_v	(volumetric) heat		S	substrate or solid
I	growth enhancing		tot	total
in	inlet			
H_2O	water			
i, j	number of reaction or component			

Abbreviations

ATP	adenosine triphosphate
CSTR	continuous stirred tank reactor

14.1 Introduction

Batch processing or discontinuous operation of bioreactors still represents the predominant method used in industry. Even the start-up of a continuous bioprocess will be on the basis of a batch process and parameters of batch kinetic models serve as first approximating hypotheses. The most significant property of a batch process is the time-dependence of all macroscopic process variables mentioned in previous chapters (cf. Sect. 11.13). This fact leads to the classification of integral reactor behavior identical with the so-called distributed-parameter reactor, quantified with gradient-models. In consequence the definition of the reaction rate, as summarized in Table 11.1, differs from that in differential or lumped-parameter reactors quantified with non-gradient models.

The growth characteristic in a batch culture of microorganisms includes several growth phases, indicated in Fig. 14.1 (see Vol. 1, Chapter 7), each of them being of potential importance in microbial processing. Not only active cell growth but also the activities of resting and dying cells are of interest, since many bioprocesses are of commercial and environmental importance where growth has stopped. The kinetics of batch fermentations involve all these growth-phases and mathematical models of such important phenomena as microbial growth, product formation, endogenous metabolism, etc.; these will be considered for different morphological states of microbial cell populations.

14.2 Batch Fermentation Pattern and Kinetic Approaches

14.2.1 Fermentation Pattern

14.2.1.1 Morphology Pattern

There are four general microbial growth patterns, exemplified by bacteria, yeasts, molds, and viruses. Microbial viruses or phages do not follow the normal growth because they require a host organism and therefore grow exponentially but with an exponent much higher than two. Bacteria, yeasts, and molds grow exponentially in a non-limiting environment. Bacteria divide by fission, yeast by budding, and molds by chain elongation and branching. Morphology therefore deals with the appearance of the typical forms of microbial growth, including unicellular growth in the case of bacteria and some yeasts and mycelial growth in the case of molds and some yeasts. Mycelial growth can be filamentous or may be in the form of pellets, i.e., spherical agglomerates with diameters ranging from 0.1 to 10 mm of different density. This will in turn affect the local environment by changing the transport phenomena, which result in a drastic change in growth behavior. Details see Volume 1 of this Series, Chap. 7.

14.2.1.2 Kinetic Pattern

Kinetics of microbial processing has to take into account all these patterns of microbial growth. In a batch culture, the significant process variables are a variety of concentrations, each one characterized by a specific rate given in Table 14.1 by Eqs. (14.1) to (14.6).

While the specific rates are significant, the individual rates r_i are used in balance

Table 14.1. Macroscopical Concentrations in Bioprocessing and Derived Specific Rates as a Basis for Quantification and Modelling

Macroscopic Process Variable	Symbol for Mass	Concentration (in kg·m^{-3})	Absolute Rate (in kg·m^{-3}·h^{-1})	Specific Rate (in h^{-1}) Symbol Definition	Eq. No.
Biomass	X	x	r_X	$\mu = \dfrac{1}{x} \cdot \dfrac{dx}{dt}$	(14.1)
Substrate	S	s	r_S	$q_S = -\dfrac{1}{x} \cdot \dfrac{ds}{dt}$	(14.2)
Oxygen	O	o	r_O	$q_O = -\dfrac{1}{x} \cdot \dfrac{do}{dt}$	(14.3)
Product	P	p	r_P	$q_P = \dfrac{1}{x} \cdot \dfrac{dp}{dt}$	(14.4)
CO$_2$	C	c	r_C	$q_C = \dfrac{1}{x} \cdot \dfrac{dc}{dt}$	(14.5)
Heat	H_v	h_v (in kJ·m^{-3})	r_{H_v} (in kJ·m^{-3}·h^{-1})	$q_{H_v} = \dfrac{1}{x} \cdot \dfrac{dh_v}{dt}$	(14.6)

equations, expressing the absolute change in concentration.

The specific growth rate μ exhibits a typical form of dependence on time t, shown in Fig. 14.1, the behavior of which is in agreement with the shape of the x/t-curve (cf. FIECHTER [14.1], Fig. 1). The following growth phases are also indicated in Fig. 14.1: lag (1), accelerating (2), exponential (3), decelerating (4), stationary (5), and decreasing phase (6). In addition to these microbial growth phases 1–6, several regions (I–VI) are to be distinguished in Fig. 14.1 as the result of different effects on growth from a kinetic point of view (MOSER [14.2]): lag due to enzyme induction (I) or negative rate due to endogenous metabolism (m), stimulation (II), limitation by substrate shortage (III), toxicity (IV), and transport limitations (V) and inhibitions (VI).

The enormous variety of fermentation processes known can be reduced in complexity using formal kinetic concepts. With this approach, basically three types of fermentations can be distinguished (GADEN [14.3]), (Fig. 14.2); these are:

a) growth-associated product formation
b) mixed-growth-associated product formation,
c) non-growth-associated product formation.

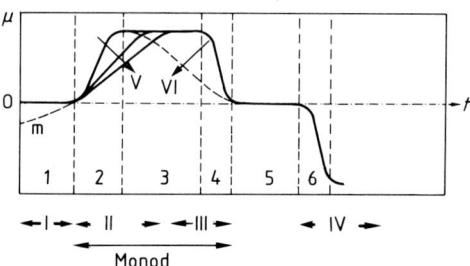

Figure 14.1. Schematic presentation of the time-dependence of the specific growth rate μ in a batch process with different ranges of growth kinetics: lag time (I) and endogenous metabolism (m), stimulation (II), limitation by substrate (III), toxicity (IV), and transport limitation in the stimulation range (V) as well as inhibitions (VI). The region of validity of the simple Monod-type kinetics is also indicated together with the microbial growth phases 1–6 according to FIECHTER [14.1], as explained in the text.

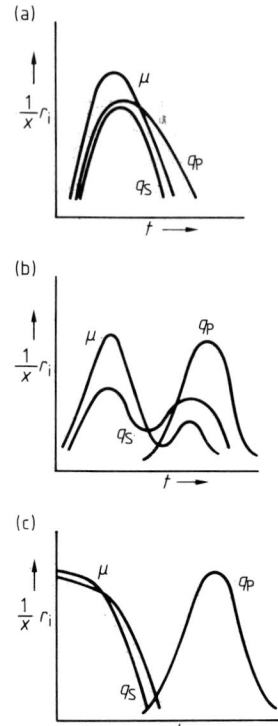

Figure 14.2. Schematic plots of the relationship of specific rates in bioprocessing (growth rate μ, substrate consumption q_S, and product formation q_P) for growth-associated (a), mixed-growth-associated (b), and non-growth-associated (c) product fermentations.

with primary metabolism. The products are historically called secondary metabolites (e.g., antibiotics and microbial toxins).

14.2.2 Kinetic Approaches

The kinetics of biological systems may be regarded at five different system levels (Fig. 14.3). Each level has a unique characteristic

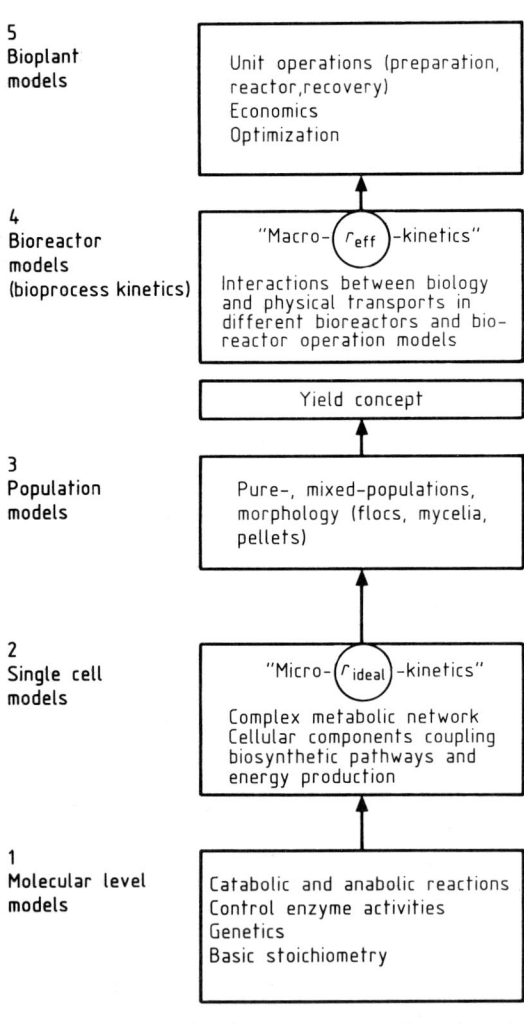

Figure 14.3. Levels of process research and modelling activities in bioprocessing.

The kinetic relationship between growth and product formation depends on the role of the product in cell metabolism, which means that microkinetics on the metabolic level determine the curves of formal kinetics. Products from type a) are usually directly engaged in a catabolic pathway, e.g., yeast fermentation with alcohol production, or they are normal intermediary metabolites, e.g., vitamins and amino acids. In the case of partly linked product formation, e.g., lactic or citric acid, the products are indirectly connected to energy production pathways and are the result of an abnormal genetically manipulated metabolism. The third group comprises fermentations in which there is obviously no association that leads to a specific kinetic treatment with some similarity to chemical reactions. In complex systems the relations among phenomena are clarified by means of system analysis (KAFAROW [14.4]). In order to control a chemical or biological conversion it is not sufficient to be informed about the underlying microkinetics but also about the physical transport phenomena (macrokinetics). The fourth level of bioreactors is the most essential one according to the macroscopic principle (see Sect. 11.1). But all other levels contribute to the development of bioreactors depending on the aim of the investigation.

The macroscopic approach using system analysis as an analogy to chemical reaction engineering is increasingly accepted and applied to bioprocess engineering problems (KAFAROW et al. [14.5]; MOSER [14.6], [14.2]; ROELS [14.7]; VOTRUBA [14.8]).

Concerning the background of biokinetic models, three different types can be distinguished: mechanistic models of microkinetics, numerical fits, and formal kinetics.

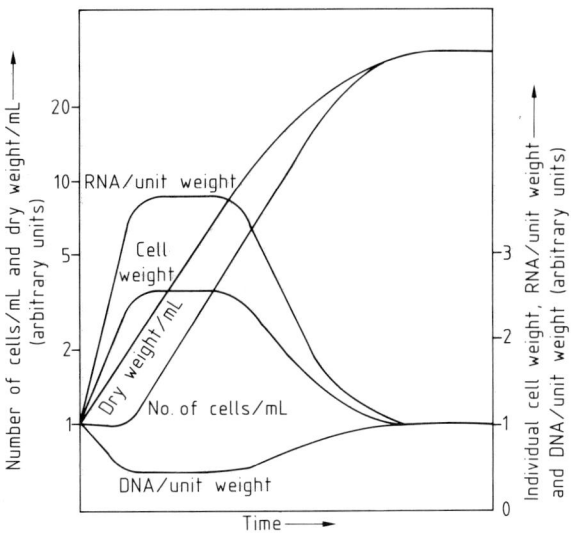

Figure 14.4. Macromolecular composition changes during a batch fermentation (HERBERT [14.12]).

14.2.2.1 Mechanistic Models of Microkinetics

Fundamental biology yields significant contributions by deriving and constructing population models from single-cell models, for which activities have been recently investigated by SHULER and DOMACH [14.11], HO and SHULER [14.10], SHULER et al. [14.11]. More complex models with a larger number of parameters are able to describe real processes better than simple models. However, due to the desired practicability of mathematical models, they should be as simple as possible. Thus, a minimum number of parameters are used for a chosen model.

Structuring a model is necessary as the composition of cell components does not remain constant under drastic environmental alterations. Fig. 14.4 shows the compositional changes of cell material during the course of a batch fermentation after HER-

BERT [14.12]. During the early and late part of the growth cycle cell composition changes significantly.

14.2.2.2 Numerical Fitting of Bioprocess Kinetics

Curve fitting can be carried out using arbitrary mathematical functions, which are flexible enough to be fitted to a set of experimental data. The advantage is, that a quantitative representation of experimental data is achieved. The validity, however, is restricted to the singular case considered and extrapolations are dangerous. The parameters of this type of model are anonymous, which means that they have nearly no biological or physical relevance. AIBA and HARA [14.13], EDWARDS and WILKE [14.14], ANDREYEVA and BIRYUKOV [14.15], and FISHMAN and BIRYUKOV [14.16] used this purely numerical approach.

In very complex fermentations where very little is known this approach is practicable (e.g., GRM et al. [14.17]). From the standpoint of chemical reaction kinetics KONO [14.18] and KONO and ASAI [14.19] to [14.21] derived equations for growth and production rate, where the inclusion of a so-called "consumption activity coefficient" ϕ renders the equation more flexible than the simple Monod relation (cf. Table 14.2).

Further information on classes of mathematical models has been presented by KOSSEN (see Chapter 24 of this volume), MOSER [14.23]; REUSS [14.25], KOSSEN [14.26], KOSSEN and ROELS [14.27], HARDER and ROELS [14.28], using some terminology from TSUCHIYA et al. [14.29].

14.2.2.3 The Formal Kinetic Approach

For complex processes the use of formal kinetics was emphasized by MOSER [14.22], [14.23], and SCHMID and SAPUNOV [14.24]. Despite the progress in the field of biology and the development of models with a mechanistic background, the mechanistic interpretation of average complex bioprocesses is a task requiring much skill. The definition of formal kinetics is (cf. Fig. 11.1) as follows:

1. They are based on real conditions which are derived experimentally and not on postulated mechanisms.
2. The mathematical function to be chosen for the model is not purely numerical but also based on analogies (see Sect. 11.1.2). Analogy means that known mathematical functions are selected, which are mainly derived from conventional mechanistic work with similar behavior, (e.g., the Monod equation as an analogy to enzyme kinetics).
3. A representative quantification of the process is obtained by adjusting the values of the model parameters to experimental data.
4. The meaning of the parameters of such formal kinetics differs from real mechanistic ones. Mechanistic interpretations can only be obtained by supplementary work. This is the main difference between formal kinetics and microkinetics; the mathematical structure used in both approaches is often the same.

14.2.3 Differential and Integral Evaluation of Batch Processes

Rate equations can be derived from experimental results applying the differential or integral evaluation method. The batch reactor exhibiting concentration-time profiles is an integral reactor. The integral evaluation method often used in microbiological laboratories with shake flask experiments, is depicted in Fig. 14.5a (MOSER and LAFFERTY [14.30]). The rate equation defining μ [see Eq. (14.1)] must be integrated. This yields an equation which can be used to determine μ in the range from t_1 to t_2:

$$\mu = \frac{\ln x_2 - \ln x_1}{t_2 - t_1}. \tag{14.7}$$

As a result, growth rates can be described as a function of initial substrate concentration $s_{0,i}$ following the integral method:

$$\mu = f(s_{0,i}). \tag{14.8}$$

On the other hand, differential evaluation of batch reactor operation requires differentiation of experimental x/t-curves with the aid of graphical, numerical, or analytical methods. The advantage of applying the differential method (see Fig. 14.5b) over the integral method is that only one experimental run is necessary at optimal conditions, but it has the disadvantage that the significant period for kinetic model identification is very short. Therefore, small time intervals must be taken as indicated in Fig. 14.5b. The problem of estimating K_s values in batch processes leads to the application of

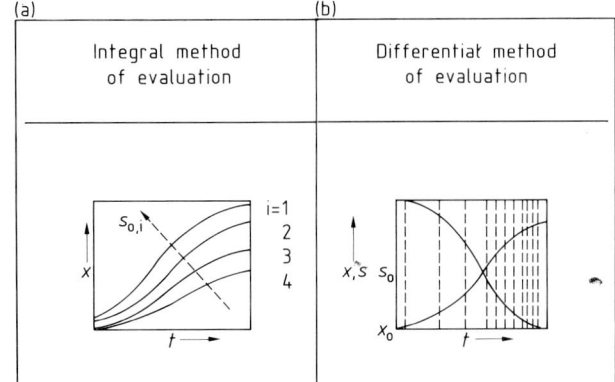

Figure 14.5. Demonstration of kinetic evaluation of the function $\mu = f(s)$ in the case of a batch process following the integral (a) and differential method (b) (MOSER and LAFFERTY [14.30]).

fed-batch cultures, where the significant period can be enlarged for safer estimation of K_s.

From the slope of the x/t-curve in batch reactors, the specific growth rate μ can be determined from measurements within a certain time interval (Δt):

$$\mu = \frac{1}{\bar{x}} \cdot \frac{\Delta x}{\Delta t}, \qquad (14.9)$$

with $\bar{x}$ as the mean value of cell concentration in the time interval Δt.

A similar procedure for integral or differential evaluation is to be followed in the case of kinetic model identification. This has already been outlined in Sect. 12.2.1 leading to identification plots by linearization of the differential or integral form of kinetic equations.

In the case of single-component systems, when catalyst concentration is constant (enzyme technology, waste water technology with constant sludge concentration), the integrated form of the Henri equation was used in the Walker diagram for parameter estimation (see Fig. 12.6).

In real multi-component systems, where both concentrations s and x vary, the problem of parameter estimation requires a simultaneous solution of both differential equations for r_x and r_s. Integration is possible by eliminating s or x with the aid of the Y-concept [see Eq. (13.5)] which yields the following equation for x:

$$\mu_{max} \cdot x_{max} \cdot t = Y \cdot K_S \cdot \ln \frac{x_{max} - x_0}{x_{max} - x}$$

$$+ (Y \cdot K_S + x_{max}) \ln \frac{x}{x_0}, \qquad (14.10)$$

and a similar equation for s:

$$q_{S, max} \cdot x_{max} \cdot t = K_S \cdot \ln \frac{s_0}{s}$$

$$+ \left(\frac{1}{Y} x_{max} + K_S\right) \ln \frac{x_{max} - Y \cdot S}{x_0}. \qquad (14.11)$$

The previously mentioned Henri equation is a special case of Eq. (14.11) with x being constant and $Y=0$. Linearization of these equations according to GATES and MARLAR [14.31] is possible by multiplying with $(1/t)$. The resulting equations are the following:

$$\frac{1}{t} \ln \frac{x}{x_0} = \frac{\mu_{max} \cdot x_{max}}{x_{max} + Y \cdot K_S}$$

$$- \frac{Y \cdot K_S}{x_{max} - Y \cdot K_S} \left(\frac{1}{t} \ln \frac{x_{max} - x_0}{x_{max} - x}\right), \qquad (14.12)$$

$$\frac{1}{t} \ln \frac{x_{max} - Y \cdot s}{x_0} = \frac{q_{S, max} \cdot Y \cdot x_{max}}{x_{max} + Y \cdot K_S}$$

$$- \frac{Y \cdot K_S}{x_{max} + Y \cdot K_S} \cdot \frac{1}{t} \ln \frac{s_0}{s}. \qquad (14.13)$$

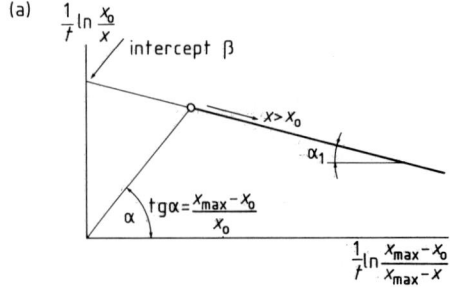

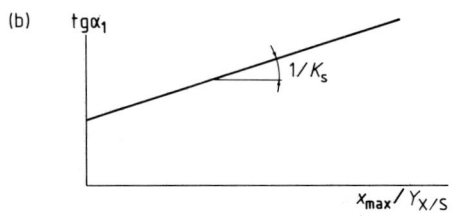

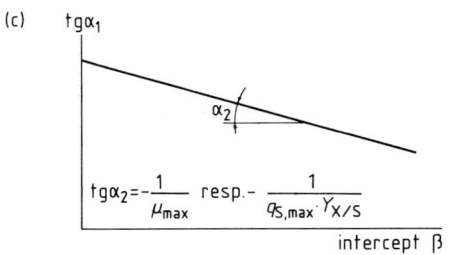

Figure 14.6. Estimation of kinetic parameters of Monod kinetics following the linearization method according to GATES and MARLAR [14.31] using a primary plot (a) and secondary plots (b and c).

An alternative to this method is a graphical trial-and-error method as shown in Fig. 14.7.

By plotting $(1/t)\ln s/s_0$ against $\ln(1+a\,d)/t$, a straight line is produced with slope c and intercept b. Since the value of a cannot be determined by direct measurement, a trial-and-error approach is used. Values of a are assumed until the best straight line is obtained. The kinetic parameters are then calculated using the values of a, b, and c from the best straight line.

A special evaluation method of batch data using the integrated form of Monod kinetics (KNOWLES et al. [14.32]) was applied by MEYRATH and BAYER [14.33]. Computer techniques facilitate the estimation of model parameters especially for batch processing (CROOKE and TANNER [14.34]).

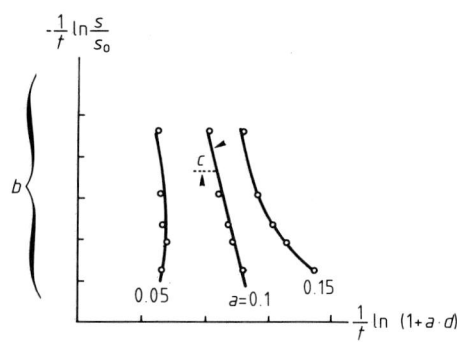

Figure 14.7. Graphical trial-and-error method from batch processes using the integrated form $\frac{1}{t}\ln\frac{s}{s_0} = c\left[\frac{1}{t}\ln(1+a\cdot d)\right] - b$, where $a = Y/x_0$, $b = \frac{\mu_{max}}{Y \cdot K_S}(x_0 + Y \cdot s_0)$, $c = 1 + \frac{x_0 + Y \cdot s_0}{Y \cdot K_S}$, and $d = s_0 - s$. For the estimation of Monod parameters:
$\mu_{max} = \frac{b}{c-1}$, $K_S = \frac{1/a + s_0}{c-1}$, and $Y = a \cdot s_0$.

The Gates linearization, however, yields different lines dependent on operational conditions, i.e., s_0, x_0, and x_{max}, so that parameter estimation (μ_{max}, Y, and K_S) requires secondary plots from the original plot according to (14.13). The slope α and intercept β of the primary plot are used in further diagrams shown in Fig. 14.6.

14.3 Mathematical Models of Batch Growth Kinetics

14.3.1 Pseudohomogeneous Rate Equations for Biokinetics

As outlined in Sect. 11.2.3 pseudohomogeneous rates of the formal kinetic type are preferably used due to their simplicity.

Fig. 14.8 gives an overview of the graphical form of kinetic rate equations encountered in this field (MOSER [14.2]).

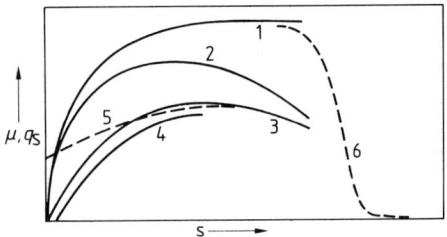

Figure 14.8. Graphical representation of kinetic model functions encountered in process kinetic analysis: μ or q_S vs. s plot in the case of simple saturation-type kinetics (1), substrate inhibition (2), product inhibition (3), endogenous metabolism (4), biosorption as a case of multi-S-kinetics (5), and transient phenomena like lag time or catabolite repression (6) (MOSER [14.2]).

14.3.1.1 Inhibition-free Substrate-limiting Model Functions

The theory of microbial growth kinetics stems from and is still dominated by MONOD's expression [14.35], [14.36] of the function $\mu = \mu(S)$, given in Eqs. (12.5) and (12.8) (Chapter 12).

This is a formal analogy to simple enzyme kinetic equations followed by a different meaning of parameters. Due to the broad application of this formulation, the behavior of Eq. (12.8) will be shown in a series of computer simulations, illustrating the influence of some variables and parameters. FIECHTER [14.1] pointed out that the shape of the growth curves depends on the initial substrate concentration. The in-

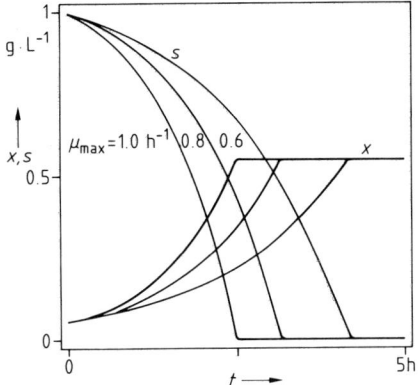

Figure 14.9. Computer simulation of a batch process with simple Monod kinetics for different values of μ_{max}: Time-dependence of biomass (x) and substrate concentration (s) in the case of $s_0 = 1$ g/L, $x_0 = 0.05$ g/L, $Y_{X/S} = 0.5$, $K_S = 0.01$ g/L (GUTKE [14.81, 14.82]).

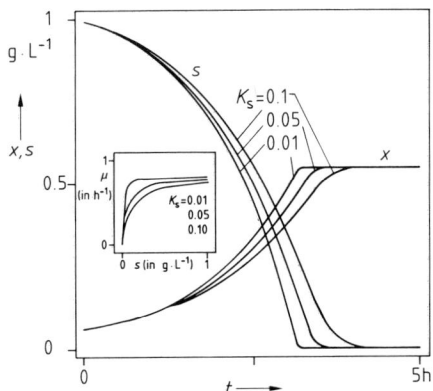

Figure 14.10. Computer simulation of a batch process with simple Monod kinetics for varying values of K_S (cf. Fig. 14.9).

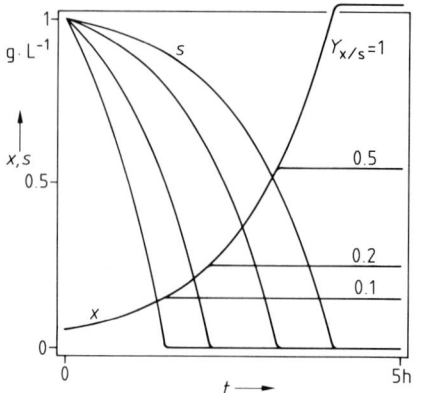

Figure 14.11. Computer simulation of a batch process with simple Monod kinetics for a varying yield coefficient $Y_{X/S}$ (cf. Fig. 14.9).

fluence of varying values of the model parameters μ_{max}, K_S, and $Y_{X/S}$ is illustrated in Figs. 14.9 to 14.11. Meanwhile, MONOD's hypothesis has been widely applied and has often proved practically satisfactory, but in a number of cases it has also failed. Several formulae as alternatives to the Monod equation are summarized in Table 14.2.

It can be shown, that the value of μ according to Monod kinetics approaches its asymptote too slowly to be a proper approximation of experimental data even in simple cases. The experimental model kinetics according to TEISSIER [14.37] and BLACKMAN [14.38] generally are thought to give a better fit. DABES et al. [14.39] made a more general approach suggesting that: (1) only the upper limit of the growth rate is fixed by a single enzymatic step; (2) at low substrate concentrations, more than one step in a series of enzyme reactions influences the growth rate.

Testing this "three-parameter equation" and BLACKMAN's kinetics against original data by MONOD for glucose limited growth

Table 14.2. Model Functions for Inhibition-free Single-substrate Limiting Kinetics

Reference	Kinetic Model	Eq. No.
MONOD [14.35]	$\mu = \mu_{max} \dfrac{s}{K_S + s}$	(12.8)
TEISSIER [14.37]	$\mu = \mu_{max} (1 - e^{-s/K_S})$	(14.14)
MOSER [14.42]	$\mu = \mu_{max} \dfrac{s^n}{K_S + s^n}$	(14.15)
CONTOIS [14.43] FUJIMOTO [14.44]	$\mu = \mu_{max} \dfrac{s}{K_S \cdot x + s}$	(14.16)
POWELL [14.40]	$\mu = \mu_{max} \dfrac{s}{(K_S + K_D) + s}$	(14.17)
BLACKMAN [14.38]	$\mu/\mu_{max} = \dfrac{1}{2} \cdot \dfrac{s}{K}$ for $s < 2K$ and $\mu/\mu_{max} = 1$ for $s > 2K$	(14.18) (14.19)
DABES et al. [14.39] "three-parameter-equation"	$s = \mu \cdot K + \dfrac{\mu - K_S}{\mu_{max} - \mu}$	(14.20)
KONO [14.18] KONO and ASAI [14.19], [14.20], [14.21]	$r_x = \mu \cdot \phi \cdot X$	(14.21)

14.3 Mathematical Models of Batch Growth Kinetics

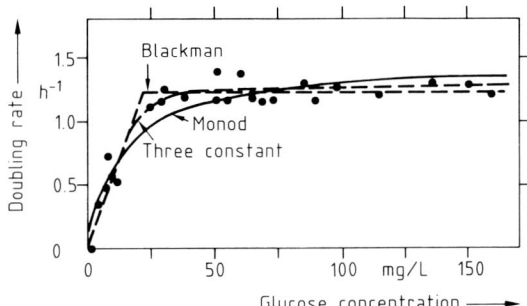

Figure 14.12. Doubling rate of *Escherichia coli* at 37°C limited by glucose concentration. Data from MONOD [14.35] as shown by DABES et al. [14.39].

of *Escherichia coli* on a synthetic medium Fig. 14.12 shows that in the concentration range between first and zero order, the Monod equation does not represent the best fit to experimental data. Although the deviations are quite significant for studies of fundamental microbial kinetics they are not as significant for bioengineering purposes in terms of practicability.

POWELL [14.40] compared different functions and elaborated another equation taking into account cell wall permeability, substrate diffusion, and cell size by a factor K_D (see Fig. 14.28). Model discriminations have shown that within a given confidence range of experimental data the competing models often cannot be distinguished (BOYLE and BERTHOUEX [14.41]). Model functions with the least number of parameters are to be preferred.

The objectives in developing a generalized model are: (1) to aid in evaluating experimental growth data and (2) to aid in determining the conceptual basis and suggest other useful forms to describe microbial growth.

According to the general behavior of growth as a function of substrate concentration, a "driving force" effecting a variation of μ with respect to s may be defined by $\mu_{max} - \mu$; the resulting rate equation is given in Table 14.3 by Eq. (14.22).

Introducing the "relative" growth rate (TEMPEST [14.45]):

$$\mu_{rel} = \mu/\mu_{max}, \qquad (14.23)$$

Eq. (14.22) can be rewritten to yield Eq. (14.24) (Table 14.3). KONAK [14.46] demonstrated that this equation reduces to the simple Monod equation for $p=2$ and to the Teissier form for $p=1$. At the same time a relationship between μ_{max} and K_S becomes apparent:

$$K_S = \frac{1}{\mu_{max} \cdot k}. \qquad (14.25)$$

Table 14.3. Generalized Specific Rate Equations for Pseudohomogeneous Single-substrate Limited Microbial Growth Kinetics in the Differential Form

KONAK [14.46]	$\dfrac{d\mu}{ds} = k(\mu_{max} - \mu)^p$	(14.22)
	$\dfrac{d(\mu_{rel})}{ds} = k \cdot \mu_{max}^{p-1} \cdot (1 - \mu_{rel})^p$	(14.24)
KARGI and SHULER [14.47]	$\dfrac{d(\mu_{rel})}{ds} = K(\mu_{rel})^m \cdot (1 - \mu_{rel})^p$	(14.26)
	with $\varkappa = d\mu_{rel}/ds$	(14.27)
	the following equation results:	
	$\ln \varkappa = \ln K + m \cdot \ln \varkappa + p \ln(1 - \varkappa)$	(14.28)
VAVILIN [14.48]	$q_S = q_{S,\,max} \dfrac{s^n}{K_S^{n-p} \cdot s_0^p + s^n}$	(14.29)

Similar generalizations of the differential form of the specific growth rate equation have been derived by KARGI and SHULER [14.47]. Eq. (14.26) is the result of this work. Values of the constants K, m, and p are given in Table 14.4.

Table 14.4. Values of the Constants of the Generalized Rate Equation (14.26) in Table 14.3

Model	K	m	p
Monod	$1/K_S$	0	2
Teissier	$1/K_S$	0	1
Moser	$n/K_S^{1/n}$	$1-1/n$	$1+1/n$
Contois	$1/K_S \cdot x$	0	2
Konak	$k \cdot \mu_{max}$	p	p

The generalized rate equation for microbial growth, therefore includes most widely used models as special cases with different constants.

It is interesting to note that VAVILIN [14.48] developed a model function independent of the reactor type used for aerobic treatment of waste waters (aeration tank, trickling filter, rotating disk) with a minimum number of coefficients. The result is Eq. (14.29) included in Table 14.3.

The analysis of a number of data has shown that $p=n-1$ is an appropriate approximation. Thus, the proposed generalized model has only three coefficients: n, K_S, and $q_{S,max}$. Larger values for n correspond to a sharper transition to components difficult to oxidize.

14.3.1.2 Inhibition Kinetics

In most cases of inhibition kinetics the equations are, like MONOD's relationship, derived from theories on the inhibition of a single enzyme.

Table 14.5. Model Equations for Substrate Inhibition Kinetics

Reference	Kinetic Model	Eq. No.
ANDREWS [14.49] to [14.51] NOACK [14.52]	$\mu = \mu_{max} \dfrac{1}{1+K_S/s+s/K_{I,S}} \approx \mu_{max} \dfrac{s}{K_S+s} \cdot \dfrac{1}{1+s/K_{I,S}}$	(14.30)
WEBB [14.58]	$\mu = \mu_{max} \dfrac{s(1+\beta \cdot s/K_S')}{s+K_S+s^2/K_S'}$	(14.31)
YANO et al. [14.59]	$\mu = \mu_{max} \dfrac{1}{1+K_S/s+\sum_j (s/K_{I,S})^j}$	(14.32)
AIBA et al. [14.60]	$\mu = \mu_{max} \dfrac{s}{K_S+s} \cdot e^{-s/K_{I,S}}$	(14.33)
Teissier-type	$\mu = \mu_{max} [\exp(-s/K_{I,S}) - \exp(-s/K_S)]$	(14.34)
WEBB [14.58]	$\mu = \mu_{max} \dfrac{s}{s+K_S(1+\sigma/K_{I,S})} \cdot e^{1.17 \cdot \sigma}$ with σ = ionic strength	(14.35)
TSENG and WAYMANN [14.55] WAYMANN and TSENG [14.56]	$\mu = \mu_{max} \dfrac{s}{K_S+s} - K_{I,S}(s-s_c)$	(14.37)
SIIMER [14.57]	$r = \dfrac{k_{cat} \cdot e_0 (1-\zeta_S)[1+\delta(1-\zeta_S)/K_S'+\varepsilon \zeta_S/K_{P2}]}{a+b\zeta_S+c\zeta_S^2}$	(14.39)

14.3 Mathematical Models of Batch Growth Kinetics

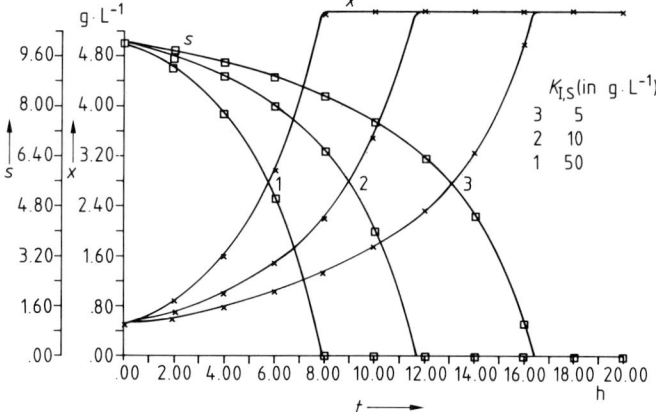

Figure 14.13. Computer simulation of a batch process with substrate inhibition kinetics at various values of the S-inhibition constant $K_{I,S}$: Time-dependence of biomass (x) and substrate concentration (s) using the following values: $Y_{X/S}=0.5$, $\mu_{max}=0.35$, $K_S=0.1$ g/L, $s_0=10$ g/L, $x_0=0.5$ g/L.

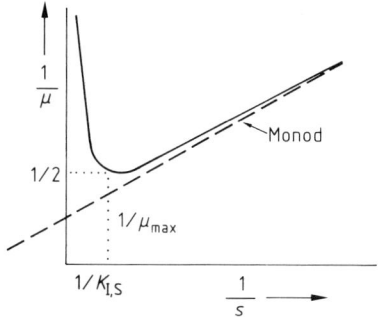

Figure 14.14. Substrate inhibition kinetics and the graphical method of parameter estimation using a double reciprocal plot according to Fig. 12.7.

constant. The meaning of the substrate inhibition constant $K_{I,S}$ is shown in Figs. 14.14 and 14.15.

Fig. 14.14 exhibits a double reciprocal plot normally used in kinetic analysis for parameter estimation. The shape of the curve is characteristic for substrate inhibition. Fig. 14.14 can be used for a rough estimation of the parameters μ_{max} and K_S by drawing the asymptote as indicated. A more accurate method of parameter estimation is shown in Fig. 14.15 (HUMPHREY [14.54]). The estimation of the three parameters can be carried out with the aid of three equations given in this figure. As a characteristic point, the maximum of the curve is taken.

Substrate Inhibition in Fermentations

ANDREWS [14.49] to [14.51] and NOACK [14.52] analyzed substrate inhibition in chemostat culture using the kinetic equation given in Table 14.5 as Eq. (14.30), in which $K_{I,S}$ is the substrate inhibition constant.

A comparison between Eq. (14.30) and Eq. (12.44) shows that the second power is absent in Eq. (14.30). EDWARDS [14.53] compared five different equations, Eqs. (14.31) to (14.35), with reference to experimental results and concluded that no model discrimination is possible. EDWARDS recommended the use of the simplest of these equivalent equations, that is Eq. (14.30). Therefore, this equation is shown in Fig. 14.13 for various values of the inhibition

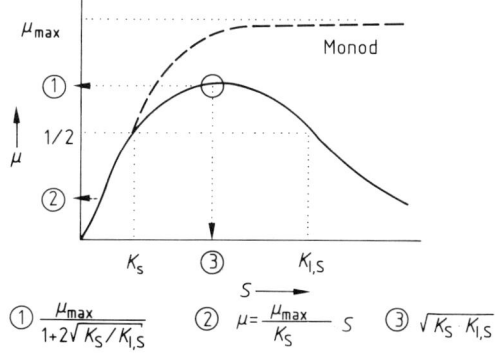

Figure 14.15. Substrate inhibition kinetics and the parameter estimation method according to HUMPHREY [14.54] using a computer simulation of Eq. (14.30).

Another method of parameter estimation of substrate inhibition kinetics is obtained from Eq. (14.30) for the case of $s \gg K_S$:

$$\frac{1}{\mu} = \frac{1}{\mu_{max}} + \frac{1}{\mu_{max} \cdot K_{I,S}} \cdot s. \qquad (14.36)$$

A quite different equation for the quantification of substrate inhibition kinetics was proposed by TSENG and WAYMANN [14.55] and WAYMANN and TSENG [14.56]. Substrate concentrations above a characteristic threshold concentration s_c linearly inhibited growth in accordance with Eq. (14.37), in which $K_{I,S}$ is the inhibition constant.

A generalized concept of a rate equation for single-substrate enzymatic reactions was proposed by SIIMER [14.57], assuming substrate S and products P_i to inhibit the reaction. The following basic reaction mechanism was suggested for the derivation of a rate equation:

[14.61] distinguishes different types of concentration-action curves, shown in Fig. 14.16, exhibiting a linear or exponential decrease or a stepwise function. Without a

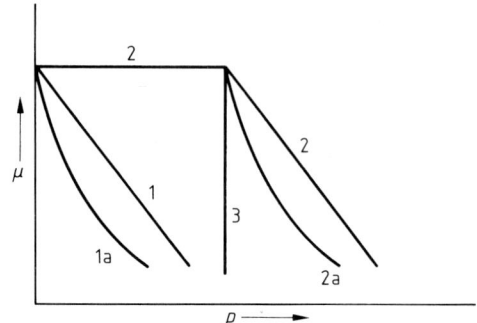

Figure 14.16. Graphical plot of the different types of product concentration (p) action on the specific growth rate μ according to HINSHELWOOD [14.61]: linear (1), exponential decrease (1 a, 2 a), or stepwise function (2, 3).

$$\begin{array}{c}
SE \underset{1/\beta\, K_s}{\rightleftharpoons} SES \xrightarrow{\delta \cdot k_{cat}} SE + P_1 + P_2 \\
\Updownarrow \beta \cdot K_s' \qquad \Updownarrow \\
EP_1 \underset{K_{P1}}{\rightleftharpoons} E \underset{K_S}{\rightleftharpoons} \{ES\} \xrightarrow{k_{cat}} E + P_1 + P_2 \qquad (14.38) \\
\Updownarrow \qquad \Updownarrow \alpha \cdot K_{P2} \qquad \Updownarrow K_{P2} \\
EP_1P_2 \underset{\gamma/\alpha\, K_{P1}}{\rightleftharpoons} EP_2 \underset{1/\alpha\, K_S}{\rightleftharpoons} EP_2S \xrightarrow{e \cdot k_{cat}} EP_2 + P_1 + P_2
\end{array}$$

According to this scheme SES and EP_2S complexes are supposed to be productive. P_1 inhibits competitively and P_2 non-competitively. Using the concept of qss with $P_1 = P_2 = P$ the rate equation as a function of substrate conversion ζ_S assumes the form of Eq. (12.45). Rewritten with parameters used in Eq. (14.38) this yields Eq. (14.39) given in Table 14.5. For a typical formal kinetic approach it is not necessary to know the exact reaction mechanism in order to use this equation in reactor calculations.

Product Inhibitions in Fermentations

If a metabolic product, formed during microbial growth, inhibits the growth rate, then the model function must take this inhibitory effect into account. HINSHELWOOD threshold concentration it is rather remarkable that the effect is often given by a simple linear relation DAGLEY and HINSHELWOOD [14.62] as shown in Eq. (14.40) in Table 14.6.

Beyond this numerical fit to product inhibition, modified by HOLZBERG et al. [14.63] with Eq. (14.41) and by GHOSE and TYAGI [14.64] in the form of Eq. (14.42) for quantifying ethanol inhibition for yeast growth, the function $\mu(s, p)$ can be modelled in analogy to enzyme kinetics. JERUSALIMSKY and NERONOVA [14.65] show the dependence of μ on P by hyperbolic or sigmoidal curves and recommend Eq. (14.44), with $K_{I,P}$ as the product inhibition constant. This model was used for a computer simulation of growth in discontinuous cul-

14.3 Mathematical Models of Batch Growth Kinetics

Table 14.6. Model Equations for Product Inhibition Kinetics

Reference	Kinetic Model	Eq. No.
DAGLEY and HINSHELWOOD [14.62]	$\mu(s,p) = \mu_{max} \dfrac{s}{K_S+s}(1-k \cdot p)$	(14.40)
HOLZBERG et al. [14.63]	$\mu = \mu_{max} - k_1(p-k_2)$	(14.41)
GHOSE and TYAGI [14.64]	$\mu = \mu_{max}(1-p/p_{max})$	(14.42)
AIBA and SHODA [14.71]	$\mu(s,p) = \mu_{max} \dfrac{s}{K_S+s} \cdot e^{-kp}$	(14.43)
JERUSALIMSKY and NERONOVA [14.65]	$\mu(s,p) = \mu_{max} \dfrac{s}{K_S+s} \cdot \dfrac{K_{I,P}}{K_{I,P}+p}$	(14.44)
BAZUA and WILKE [14.77]	$\mu_{max} = \mu_0 - k_1 \cdot \bar{p}(k_2-p)$	(14.45)
	$\mu_{max} = \mu_0(1+\bar{p}/p_{max})^{1/2}$	(14.46)
LEVENSPIEL [14.78]	$r_x = k_{obs} \cdot \dfrac{s}{K_S+s} \cdot x$	(14.47)
	with $k_{obs} = k(1-p/p_{crit})^n$	(14.48)
HOPPE and HANSFORD [14.79]	$\mu = \mu_{max} \dfrac{s}{K_S+s} \cdot \dfrac{K_{IP}}{K_{IP}+[Y_{P/S}/(s_0-s)]}$	(14.49)

ture by BERGTER [14.66] and represents the most commonly used equation (AIBA et al. [14.60], [14.67], [14.68]; PERINGER et al. [14.69]; FUKUDA et al. [14.70]). Furthermore, AIBA and SHODA [14.71] developed an empirical approach for the product inhibition pattern of yeast [cf. Eq. (14.43)].

A quite flexible class of models was described by RAMKRISHNA et al. [14.72], [14.73] using a sequence of reactions where intermediary products inactivate viable cells. With this approach oscillations of growth can also be quantified (KNORRE [14.74]).

A model of this type was used by FISHMAN and BIRYUKOW [14.16] for modelling growth in penicillin fermentation. A similar approach is used for growth analysis when there is an influence of intermediates (PETROVA et al. [14.75]; KNORRE [14.76]). These studies emphasize the importance of culture prehistory by obtaining a variety of growth curves depending on the initial conditions $P(0)$, if P is an intracellular substance.

BAZUA and WILKE [14.77] proposed a two- and a three-parameter equation, where $\bar{p}$ is the average value of P, and k_1 and k_2 are empirical constants [Eqs. (14.45) and (14.46)]. These equations show that there is

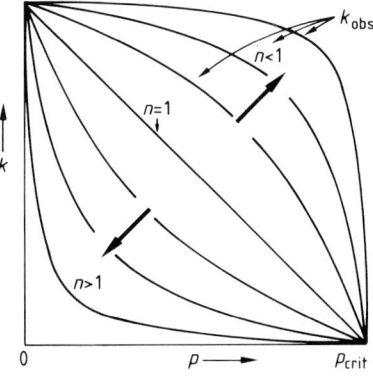

Figure 14.17. Power n shows how the observed rate constant of the Monod equation, k_{obs} [cf. Eq. (14.47)] decreases as toxic product p rises (LEVENSPIEL [14.78]).

a limiting concentration of p, beyond which the cells will not grow.

Two other equations should be mentioned: Eq. (14.47) (see also Fig. 14.17) by LEVENSPIEL [14.78] and Eq. (14.49) by HOPPE and HANSFORD [14.79]. These are included in Table 14.6.

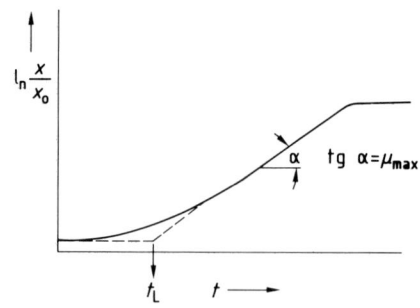

Figure 14.18b. Estimation plot for lag time t_L and maximum growth rate μ_{max}.

14.3.1.3 Extensions of Validity of Monod-type Kinetics to All Other Growth Phases

As the simple Monod-type function is only valid in the exponential and decelerating growth phase, some extensions of its validity should be introduced in order to handle the lag, stationary, and death phase as well.

Kinetic Modelling of Lag Phase

The duration t_L of a lag phase for a specific growth rate is defined by HINSHELWOOD [14.61] and LODGE and HINSHELWOOD [14.80] as being:

$$t_L = t - (2.3/\mu_{max}) \lg x/x_0. \tag{14.50}$$

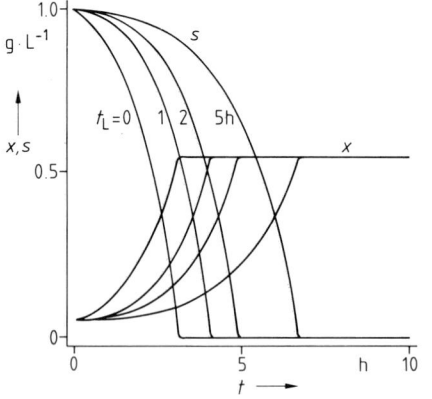

Figure 14.18a. Computer simulation of a batch process with Monod kinetics with incorporated lag time t_L for various values of t_L: Time-dependence of biomass (x) and substrate concentration (s), (cf. Fig. 14.9) (GUTKE [14.81], [14.82]).

If an inoculum is taken not from a culture in the exponential phase but in the stationary phase or from a culture with a qualitatively different medium, the synthetic apparatus, especially RNA and enzyme synthesis, must be induced. A lag phase with slowly increasing μ is the consequence. This effect is depicted in Fig. 14.18a (GUTKE [14.81], [14.82]).

It is known that the length of t_L may depend on the amount of inoculum N_0 (cf. FIECHTER [14.1]), quantified by HINSHELWOOD [14.61] by:

$$t_L = \text{const}/(N_0 - \text{const}). \tag{14.51}$$

EDWARDS [14.83] described this effect by means of the substrate concentration $\bar{s}$:

$$t_L = 1.16(\bar{s} - 0.4). \tag{14.52}$$

A simple extension of Monod-type kinetics using the lag time t_L as model parameter is given by BERGTER and KNORRE [14.84]:

$$\mu(s,t) = \mu_{max} \frac{s}{K_S + s}(1 - e^{-t/t_L}). \tag{14.53}$$

The plot in Fig. 14.18b indicates how to determine μ_{max} and t_L.

Recently a mechanistic model of lag phase was introduced by PAMMENT et al. [14.85] and discussed by BARFORD et al. [14.86]. From the observation that t_L was not dependent on the amount of inoculum, they found that t_L critically depends on enzyme availability and respiration. The lag time can be predicted with the aid of a structured model previously set up by PAMMENT et al. [14.85].

Kinetic Modelling of Stationary Phase of Growth

The logarithmic or exponential law:

$$r_x = \mu_{max} \cdot x, \qquad (14.54)$$

can be modified to incorporate the existence of a stationary growth phase. This revised form is known as the "logistic law", presented by KENDALL [14.87]:

$$r_x = \alpha \cdot x \left(1 - \frac{x}{\beta}\right), \qquad (14.55)$$

where α and β are empirical constants.

With $\alpha = \mu_{max}$ and $\beta = x_{max}$ LA MOTTA [14.88] used the logistic equation to quantify continuous growth cultures. Although this approach has been successfully applied in fitting growth curves (e.g., CONSTANTINIDES et al. [14.89] and [14.90]), the drawback is that no clear relationship between μ and s is incorporated into this type of logistic law.

Kinetic Modelling of Microbial Death Phase

Microbial death phases must be considered in bioprocesses with long residence times for example, in biological waste treatment. For this case the Monod equation must be extended by including the specific death rate:

$$k_d = -\frac{1}{x} \cdot \frac{dx}{dt}. \qquad (14.56)$$

The rate equations are:

$$r_x = (\mu - k_d) x \qquad (14.57)$$

and

$$-r_S = \frac{1}{Y_{X/S}} \cdot r_x. \qquad (14.58)$$

This model was successfully used for activated sludge processing (CHIU et al. [14.90a]).

Fig. 14.19a gives a computer simulation of these model equations for batch processes (MOSER and STEINER [14.91]) with

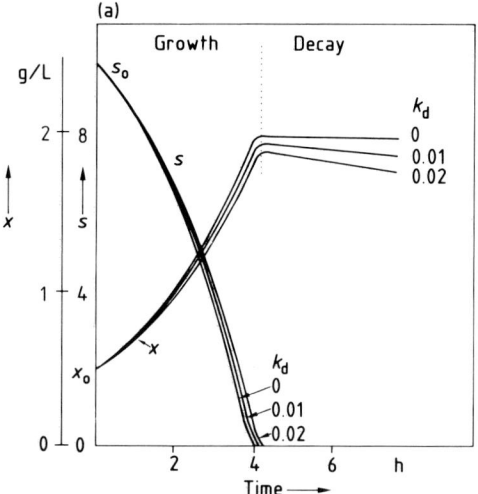

Figure 14.19a. The influence of different values of k_d from 0 to 0.02 h^{-1} on substrate concentration s and cell mass concentration x in a batch culture. The following constants were assumed: $\mu_{max} = 0.35$ h^{-1}, $K_S = 100$ mg/L, $Y = 0.15$, $s_0 = 10$ g/L, $x_0 = 0.5$ g/L (MOSER and STEINER [14.91]).

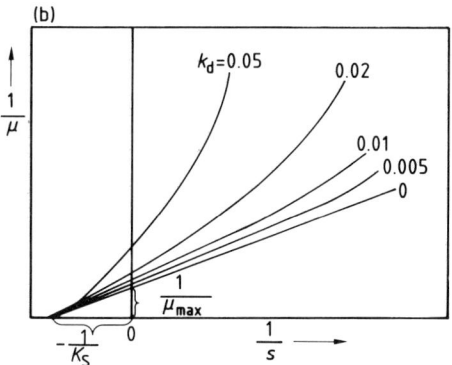

Figure 14.19b. The influence of different values of k_d from 0 to 0.05 h^{-1} on the double reciprocal plot ($1/\mu$ vs. $1/s$) of the extended Monod model for microbial growth shown in a double reciprocal plot according to Fig. 12.7 (MOSER and STEINER [14.91]).

varying values for k_d. As a true rate constant, k_d is only a function of temperature. The effect of a neglected k_d-value on the evaluation of Monod parameters is shown in Fig. 14.19b; incorrect and even negative values for K_S are the consequence (GRADY et al. [14.92]). Using the known value of k_d it is apparent that only in a plot based on the following equation the undisturbed values for μ_{max} and K_S can be estimated (MOSER and STEINER [14.91])

$$\frac{1}{\mu+k_d} = \frac{K_S}{\mu_{max}} \cdot \frac{1}{s} + \frac{1}{\mu_{max}}. \quad (14.59)$$

Physiological considerations indicate that two independent factors lead to the decline of the overall biomass. Viable cells lose mass due to endogenous metabolism and may even deteriorate. A modelling of these phenomena has been carried out by SINCLAIR and TOPIWALA [14.93], who assumed individual rates of death of viable cells, k_d^0, and endogenous metabolism, k_e. Thus:

$$k_d = k_d^0 + k_e. \quad (14.60)$$

The effect of the true rate of death k_d^0 on steady-state yields becomes significant only at very low dilution rates. Also an apparent lag can be explained in batch cultures if the inoculum has a very low viability.

14.3.1.4 Kinetic Modelling of Endogenous Metabolism

As previously outlined from a thermodynamic point of view highly structured systems such as living cells can be maintained only by dissipation of energy. The living cells which are thermodynamically characterized by a state far from equilibrium must incorporate energy-rich substances and transform the chemical energy into heat. This energy flow is necessary for maintenance of osmotic pressure and for repair processes on DNA, RNA, and other macromolecules. Thus, a source of energy must be available not only for macroscopic reactions like growth but also for maintenance of cellular structure. Therefore, the substrate balance equation must include a term m_s for maintenance (PIRT [14.94]) as shown in Eq. (14.62) in Table 14.7.

The balance for energy source utilization can be used for the estimation of the maintenance coefficient m_s, plotting r_s vs. r_x, as shown in Fig. 14.20. It is evident from this graph that:

$$m_s = k_e/Y_{X/S}, \quad (14.64)$$

indicating that the specific maintenance rate k_e may be regarded as a turnover rate of biomass which is useful for comparing the turnover rates of biomass components with maintenance energy requirements (HERBERT [14.95]; MARR et al. [14.96]). It is interesting to compare the different approaches represented in Table 14.7 in order to show the model-approaches used to quantify endogenous metabolism.

The half saturation constant is usually so small that $m_s = m_{s,max}$ during the growth process but that it is zero in the stationary phase, this effect being quite similar to the influence of Y. Therefore, both parameters are difficult to determine in batch processing. The maintenance energy after exhaustion of the energy source in the medium must be taken from the cell material itself. After utilization of an intracellular stock cells therefore start lysis at a rate constant k_e.

Logically a model is more realistic including both terms of k_d, k_e, and m_s.

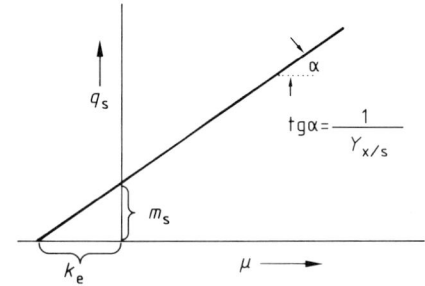

Figure 14.20. Graphical method for determination of the maintenance coefficient m_s in a plot with q_s vs. μ according to Fig. 13.1.

14.3 Mathematical Models of Batch Growth Kinetics

Table 14.7. Reaction Schemes and Kinetic Model Equations for Endogenous Metabolism

Reference	Reaction Scheme	Kinetic Equations	Parameters
PIRT [14.94]	$S \xrightarrow{\mu} X$, $S \xrightarrow{m_s}$ maintenance (14.61)	$-r_s = \dfrac{1}{Y_{X/S}} \cdot \mu \cdot x + m_s \cdot x$ (14.62) with $r_x = \mu \cdot x$ (14.63)	$m_s = -\left(\dfrac{1}{x}\dfrac{ds}{dt}\right)_m$ $m_s = \text{const}$
	$S \xrightarrow{\mu} X_v \xrightarrow{k_d} X_d$ (14.65)	$r_X = (\mu - k_d)x$ (14.57) with $-r_s = \dfrac{1}{Y_{X/S}} \cdot \mu \cdot x$ (14.58)	$k_d = \text{const}$
HUMPHREY [14.54]	$k_d \neq \text{const} = f(S)$ (14.66)	$\mu_d = \mu_{d,\max}\left(1 - \dfrac{s}{K_d + s}\right)$ (14.67)	K_d
GUTKE [14.81]	$m_s \neq \text{const} = f(S)$ (14.68)	$m_s = m_{s,\max} \dfrac{s}{K_e + s}$ (14.69)	K_e
	$S \xrightarrow{\mu} X_v \xrightarrow{k_d} X_d$, $S \xrightarrow{m_s}$ maintenance (14.70)	$r_x = (\mu - k_d)x$ (14.57) $-r_s = \dfrac{1}{Y_{X/S}} \cdot \mu \cdot x + m_s \cdot x$ (14.71)	$k_d \neq \text{const}$ $m_s \neq \text{const}$

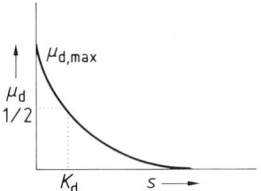

Figure 14.21. Kinetics of cell death due to substrate limitation and parameter estimation according to Eq. (14.67) (HUMPHREY [14.54]).

In summary, we have seen how a discontinuous growth rate may be quantitatively characterized by parameters like the specific growth rate, the yield coefficient, the half saturation or Michaelis-Menten-constant, the lag period, the maintenance coefficient, and the specific lysis rate.

A comparison of unstructured growth models of microorganisms is given by TAKAMATSU et al. [14.97]. These parameters are sufficient for characterization if there is no transport limitation and only one substrate is limiting which is metabolized along a single pathway. But if more substrates must be taken into account, which are sequentially or simultaneously limiting, or if a limiting substance is not metabolized but acts as a coenzyme, or if the bottleneck of metabolism is switched during the growth process the presented model has to be extended and modified.

14.3.1.5 Kinetic Modelling of Mycelial Growth

Active growth of molds is confined to the tips of the hyphae. BERGTER [14.98] presents a kinetic model which permits calculation of the relative rates of apical growth and branching of mycelial microorganisms in submerged cultures by measurement of μ

and the quotient L_H/N_H, where L_H is the total length of hyphae per unit culture volume and N_H the number of hyphal tips per unit of volume. Based on the simplest case of HINSHELWOOD's network theorem (DEAN and HINSHELWOOD [14.99]) the following equations have been suggested for describing growth of small mycelial trees of *Streptomyces hygroscopicus* on solid surfaces:

$$\frac{dL_H}{dt} = k_1 \cdot N_H, \qquad (14.72)$$

$$\frac{dN_H}{dt} = k_2 \cdot L_H, \qquad (14.73)$$

with k_1 as the relative rate of apical growth (in $\mu m \cdot h^{-1}$) and k_2 the mean relative rate of branching (in $\mu m^{-1} \cdot h^{-1}$).

It can be concluded that:

$$\mu = \sqrt{k_1 \cdot k_2}, \qquad (14.74)$$

where k_1 and k_2 exhibit a different dependence on s. With increasing s the density of the colonies, which depends on the branching rate k_2, decreases faster than the radial growth rate. This behavior can be interpreted in connection with the regulation of transport of metabolites within the hyphae. Cellular differentiation and the connection to product formation in molds was elaborated by MEGEE et al. [14.100]. Recently a review on the growth of fungi has been published by PROSSER [14.101].

14.3.1.6 Influence of Temperature, pH Value, Salinity, and Water Activity on Growth

Bioprocesses are heavily influenced by parameters like temperature, water activity, and salinity. Examples of relevant investigations have been discussed by PIRT [14.102].

The dependence of kinetic constants μ_{max}, K_S, and Y on temperature is described, e.g., by MORGAN and EDWARDS [14.103], TANNER [14.104], RYU and MATELES [14.105a], and PETERS [14.106].

The influence of a saline environment on the growth of pure and mixed cultures was recently reported and summarized by ESENER et al. [14.107]. In Fig. 14.22 typical

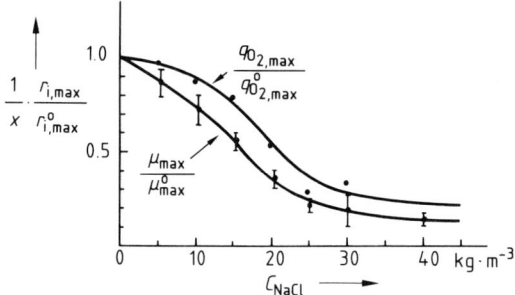

Figure 14.22. Specific rates in bioprocessing vs. NaCl concentration for microbial growth (μ_{max}) and O_2-uptake ($q_{0, max}$) in pure populations (ESENER et al. [14.107]). The rates at zero c_{NaCl} are indicated as μ^0 resp. q^0.

results are shown for the dependence of μ and q_{O_2} on NaCl concentration. Mixed culture data were taken from IMAI et al. [14.108], [14.109]. A similar decrease was found for the influence of salinity on yield factors, thermodynamic efficiency, and the observed ratio of COD to BOD in activated sludge processes (ESENER et al. [14.107]). A decrease in $Y_{0/x}$ calls for a higher aeration rate. Thus, the engineer will be faced with an optimization problem.

14.3.2 Multi-substrate Kinetics

It has become apparent over the last decade that some natural or industrial processes may be limited by more than one substrate.

For aerobic bioprocesses it is evident that, beyond any substrate, O_2 can also be rate limiting. The curve for biomass concentration increasing with time in a batch run is significantly influenced by the oxygen transfer rate quantified by the volumetric transfer coefficient $k_{L1}a$. This external transport limitation is shown in Fig. 14.31 and discussed in detail in Sect. 14.3.4.2. Kinetic modelling is successful with the aid of a double-substrate limitation function presented by REUSS and WAGNER [14.108]:

$$\mu(s, o) = \mu_{max} \frac{s}{K_S + s} \cdot \frac{o}{K_0 + o} \cdot x. \quad (14.75)$$

This equation was incorporated in the mass balance equation for O_2:

$$r_0 = k_L \cdot a(O_L^* - O_L) - \frac{1}{Y_{x/0}} \cdot \mu(s, o). \quad (14.76)$$

This equation can be used as a formal kinetic approach without assumption of a mechanism.

Situations with multi-substrate limitations are quite frequently encountered in practice. In complex industrial media in which there are not only multiple carbon sources but also a wide variety of compounds such as metabolic intermediates, amino acids, vitamins, etc. the cells pass through many transitions. A series of growth phases will result each with a successively decreasing growth rate. As growth proceeds in complex media the cells deplete the medium of useful intermediates (PARDEL [14.109]) and have to synthesise more enzymes with time. Fig. 14.23 shows one practical situation in bioprocessing (MOSER [14.2]). Classification of situations can be achieved by differentiating between sequential and simultaneous utilization with a transition case of overlapping utilization.

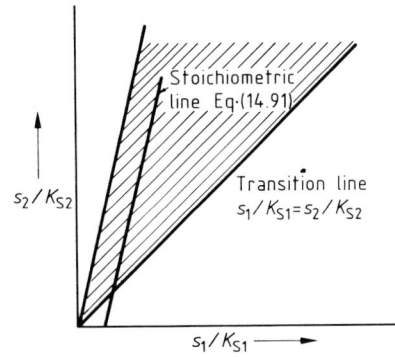

Figure 14.24. Dimensionless plot which shows the feed conditions that are required for double-substrate limitation to be possible (shaded area). It is necessary that the stoichiometric line must cross the transition line (BADER, [14.125], [14.126]).

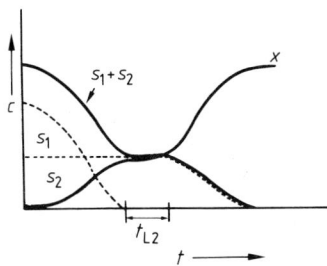

Figure 14.23. Concentration vs. time plot of substrate (s) and biomass (x) in the case of sequential substrate utilization exhibiting a diauxic lag phase t_{L2}.

While sequential or consecutive consumption of substrates are known as "diauxie" (MONOD [14.35], [14.36]) and can often be analyzed in two separated growth phases, the case of simultaneous utilization encountered in biological waste water treatment is more difficult to describe mathematically.

A general approach to consecutive substrate utilization is given by the additive function of Eq. (14.77) in Table 14.8, where $f_{repr}(s_1, t)$ is the term for the catabolic repression of S_2-utilization, which is active as long as enough s_1 is available in the medium. IMANAKA et al. [14.110] formulated Eq. (14.79) for this case which is given in Table 14.8.

BERGTER and KNORRE [14.84] used another function for this purpose, which is given by Eq. (14.80), incorporating a diauxic lag time t_{L2}. This is a formal analogy to the formulation of biological inertia, where the growth rate is related to the ri-

Table 14.8. Model Equations for Multi-substrate Kinetics

Reference	Kinetic Model	Parameters	
General form:	$\mu(s_1, s_2) = \mu(s_1) + \mu(s_2) \cdot f_{repr}$ (14.77)	$f_{repr} = f_{repr}(s_1, t)$	(14.78)
IMANAKA et al. [14.110]		$f_{repr} = \dfrac{s_2}{s_1 + s_2}$	(14.79)
BERGTER and KNORRE [14.84]		$\dot{f}_{repr}(s_1, t) = \dfrac{1}{t_{L2}}\left(\dfrac{K_S}{K_S + s} - f_{repr}\right)$	(14.80)
MOSER [14.2]; [14.22]		$f_{repr} = \dfrac{1}{1 + s_2/K_{repr}}$	(14.82)
WUHRMANN et al. [14.111]	$r_{tot} = \sum_i r_i$ (14.83)		
ATKINSON et al. [14.122] WILDERER [14.123] SHEHATA and MARR [14.112]	$q_{S, tot} = \sum_i \left(q_{S, max, i} \dfrac{s_i}{K_{S,i} + s_i}\right) x$ (14.84)	$q_{S, max, i}$ $K_{S, i}$ $q_{S, max} = \sum_i q_{S, max, i}$	(14.85)
YOON et al. [14.117] ARIS and HUMPHREY [14.118] KNORRE [14.74]	$\mu(s_1, s_2) = \mu_{max,1} \dfrac{s_1}{K_{S1} + s_1 + a_2 s_2}$ $+ \mu_{max,2} \dfrac{s_2}{K_{S2} + s_2 + a_1 s_1}$ (14.87)	$\mu_{max,1}, K_{S1}$ $\mu_{max,2}, K_{S2}$ a_1, a_2 $a_1 = 1/a_2$ $a_2 = \dfrac{k_2(k_{-1} + k_3)}{k_1(k_{-2} + k_4)}$	(14.88) (14.89)
TSAO and HANSON [14.120] TSAO and YANG [14.121]	$\mu(s_I, s_E) = \left(\mu_{max,0} + \sum_i \mu_{max,i} \dfrac{s_{I,i}}{K_{S_I,i} + s_{I,i}}\right) \prod_j \dfrac{s_{E,j}}{K_{S_E,j} + s_{E,j}}$ (14.90)		

bosomal concentration c_R, with $\bar{c}_R$ being the stationary value:

$$\frac{dc_R}{dt} = \frac{1}{t_L}(\bar{c}_R - c_R). \quad (14.81)$$

A somewhat different formal kinetic approach is reflected by Eq. (14.82), presented by MOSER [14.6], [14.22].

The case of simultaneous uptake is common in biological waste water treatment. For this case Eq. (14.83) has been suggested by WUHRMANN [14.111]. Other authors presented Eqs. (14.84) and (14.85) of Table 14.8, with $K_{S,i}$ being the constant for the i-th reaction and $q_{S, max, i}$ the contribution of the i-th reaction to the maximal rate (SHEHATA and MARR [14.112]).

A cooperative model for n interacting sites on the enzyme was derived and compared with the classical Monod model (KARGI [14.113]) resulting in a sigmoidal curve in analogy to Hill kinetics.

A strictly sequential manner of substrate utilization is to be encountered, e.g., in beer brewing, where glucose, maltose, and maltotriose are used one after the other (BUDD [14.114]). A simple model describing the kinetics of wort sugar uptake during batch fermentation of brewer's wort by yeast can be found in the literature (FIDGETT and SMITH [14.115]).

Generally, in multi-component media substrates will be utilized sequentially. Physiologically two different responses can be distinguished, known as catabolite repression and inhibition, both resulting in a diauxic growth pattern. Remarkable differences in growth behavior can be observed comparing batch and continuous operation in CSTR (HARDER and DIJKHUIZEN [14.116]), due to the existence of much lower steady-state concentrations of s_i in CSTRs so that both effects of repression and inhibition may be absent or reduced.

YOON et al. [14.117], ARIS and HUMPHREY [14.118], and also KNORRE [14.74] derived generalized Monod equations for multi-substrate systems, based on the following sequence of reactions:

$$X + a_1 S_1 \underset{k_{-1}}{\overset{k_1}{\rightleftarrows}} X'$$

$$X + a_2 S_2 \underset{k_{-2}}{\overset{k_2}{\rightleftarrows}} X''$$

$$X' \xrightarrow{k_3} 2X$$

$$X'' \xrightarrow{k_4} 2X$$

(14.86)

with X' and X'' as different intermediary states of the cells and a_1 and a_2 the stoichiometric coefficients.

Applying the steady-state approximation the result is Eq. (14.87), with a complex meaning of parameters K_{S1} and K_{S2}. The stoichiometric coefficients, given by Eqs. (14.88) and (14.89), indicate that each substrate exhibits a competitive inhibition effect. Non-competitive inhibition was considered by LEE [14.119].

Another extension of the Monod type kinetics was suggested by TSAO and HANSON [14.120] and TSAO and YANG [14.121]. They assume the existence of growth enhancing substrates S_1 and essential substrates S_E, resulting in Eq. (14.90).

The effect of growth enhancing substrates is given by a sum of Monod type expressions. The value for μ_{max} will be the sum of $\mu_{max,0}$ and all $\mu_{max,i}$. When all S_i are missing, a value of $\mu_{max,0}$ is still possible. By this approach Eqs. (14.75) and (14.76) can be explained by assuming glucose and O_2 as essential substrates. The question that generally arises is whether more than one substrate can exert control in a given system. BADER et al. [14.124], [14.125], [14.126] showed that for the growth rate to be able to switch between the substrate which is

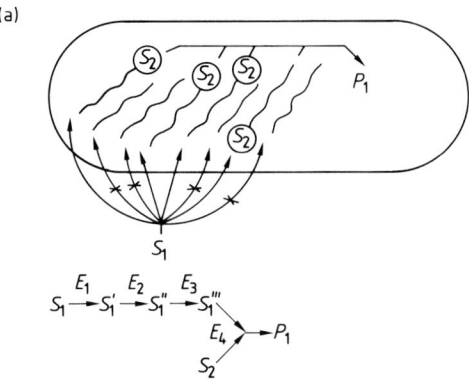

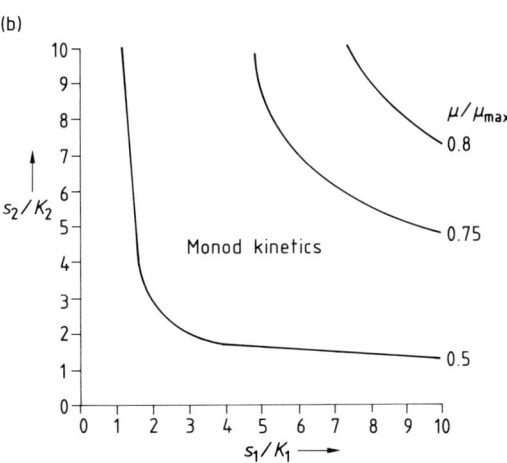

Figure 14.25. (a) Conceptual representation of the interactive model. S_1 is converted to P_1 by an enzyme which requires S_2 as a cofactor. Substrates S_1 and S_2 from two parallel pathways are combined by enzyme E to produce a product P_1 which is required for growth.
(b) Plots of lines of constant dimensionless specific growth rate μ/μ_{max} as a function of two dimensionless substrate concentrations for interactive models of the Megee-type with Monod kinetics (BADER, [14.125], [14.126]).

controlling, or to have both substrates controlling simultaneously, the stoichiometric line must intersect the transition line $\alpha = \beta$ ($s_1/K_{S1} = s_2/K_{S2}$). The stoichiometric line relates the steady-state values of the substrates which, in terms of dimensionless substrate concentrations, becomes:

$$\beta_0 - \bar{\beta} = \frac{Y_{X/S_1} \cdot K_{S1}}{Y_{X/S_2} \cdot K_{S2}} (\alpha_0 - \bar{\alpha}). \qquad (14.91)$$

This is shown in Fig. 14.24. For the intersection to occur, the following inequality must be satisfied:

$$Y_{X/S_1} \cdot K_{S1} / Y_{X/S_2} \cdot K_{S2} > \beta_0/\alpha_0 > 1. \qquad (14.92)$$

This is in agreement with the requirements proposed by SYKES [14.127] and defines the shaded region in Fig. 14.24. However, Eq. (14.92) holds only for kinetic models such as the Monod and exponential models approaching saturation asymptotically. To handle this problem it is suggested to superpose curves with constant μ as in Fig. 14.24, yielding two different models, the interactive and non-interactive model. An interactive model is based on the assumption that, if two substrates are both present in less than limiting concentrations, both substrates will affect the overall rate. The simplest model type may be constructed by simply multiplying two single-S-limited models. Fig. 14.25a is a conceptual presentation of such an approach.

A non-interactive model basically implies that μ can only be limited by one substrate at a time. Therefore, the growth rate will be equal to the lowest rate that would be predicted from the separate single-S models. For Monod-type kinetics this is written as follows:

$$\mu = \mu_{max,1} \frac{s_1}{K_{S1} + s_1} \quad \text{for} \quad \frac{s_1}{K_{S1}} < \frac{s_2}{K_{S2}} \qquad (14.93)$$

$$\mu = \mu_{max,2} \frac{s_2}{K_{S2} + s_2} \quad \text{for} \quad \frac{s_2}{K_{S2}} < \frac{s_1}{K_{S1}} \qquad (14.94)$$

Fig. 14.26a represents the concept of this approach, examples for which are to be

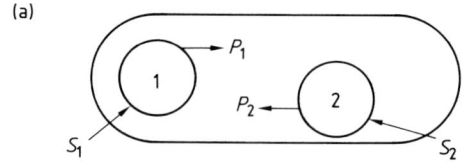

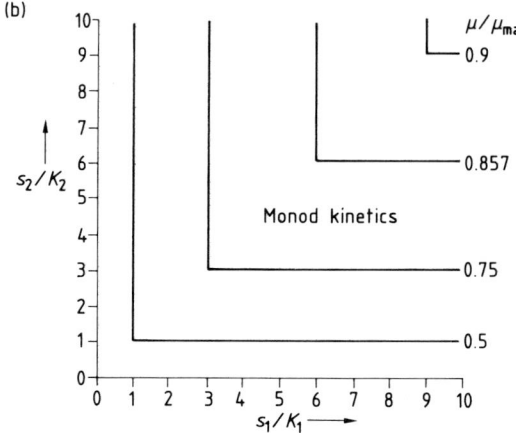

Figure 14.26. (a) Conceptual representation of the non-interactive model. Systems 1 and 2 operate independently of one another.
(b) Plots of lines of constant dimensionless specific growth rate μ/μ_{max} as a function of two dimensionless substrate concentration for non-interactive models of the Megee-type with Monod kinetics (BADER, [14.125], [14.126]).

found in the literature (SYKES [14.127]; RYDER and SINCLAIR [14.128]).

Comparisons of both approaches for Monod kinetics are shown in Figs. 14.25b and 14.26b. Non-interactive models are discontinuous functions at the transition line from one substrate limitation to another, predicting higher values of μ in the region where s_1/K_{S1} and s_2/K_{S2} are small. Interactive models are continuous functions but may yield lower values of μ when α and β both are small. It seems to be unlikely that any two cellular subsystems would be totally independent of each other even if the degree of interaction is rather small. Both types exist for certain types of substrates. Finally it was shown by BADER [14.126] that the region of simultaneous limitation by two substrates is extremely small, so that double-S-limitation can be regarded as a rare event. Nevertheless, kinetics of multi-S-systems are needed for the quantification of multilimitations occurring with a temporal shift depending on experimental conditions in one single equation.

14.3.3 Kinetic Models for Transient Growth in Batch Processing

It is convenient to group transient growth processes in two types, the relatively slow transients observed in batch cultures and the rapid changes in continuous operation. While the second type can be determined by the experimenter by changing the input, slow transients in batch processing are caused by the changing environment due to cellular activities. It was emphasized by ROELS and KOSSEN [14.129] that in batch processing non-balanced growth is dominating and therefore non-structured models fail in this practically important case. The setting up of structured models will be discussed together with the proper definition of balanced growth in Sect. 16.2. The main part of Chapter 16 will be devoted to the modelling of true transients. However, unstructured approaches to the slow transients of batch runs are also successful within a certain range of conditions. This fact can be concluded from several papers on this topic (e.g., KNORRE [14.74]; MOSER [14.6], [14.22]; BAJPAI and REUSS [14.130]; CONSTANTINIDES et al. [14.89]; KOGA et al. [14.131]; BERGTER and KNORRE [14.84]). Special formulations concerning the diauxic growth are described by ZETELAKI-HORVATH and BEKASSY-MOLNAR [14.132] using the Kono approach, BIJKERK and HALL [14.133] applying a deterministic model, and MOSER [14.134], [14.135] using the formal kinetic approach. Other approaches with a mechanistic background use the ideas of induction and repression as applied to the permease of the second substrate (VAN DEDEM and MOO-YOUNG [14.136]; MOREIRA et al. [14.137]). This model is often referred to in review articles (e.g., ROELS and KOSSEN [14.129]; PICKETT [14.138]).

The resulting equations are complex and need a large number of empirical parameters. TODA [14.139] suggested that induction and repression may be simultaneous phenomena within a "dual control" mechanism. This concept was used by IMANAKA and AIBA [14.140] in pin-pointing the difference between positive and negative controls in the dual regulatory mechanism of microorganisms. Such an approach permits a prediction of optimal control of bioprocesses in CSTR and fed-batch cultures (TODA et al. [14.141]). Recently, TODA [14.142] summarized the phenomena of induction and repression and their mathematical modelling.

The study of integrated cell control requires an understanding of a large number of interactions at different organizational levels with varying time scales and priorities. BARFORD et al. [14.86] reviewed the mechanisms of cell activities, however, most of them are still poorly understood. The concept of characteristic times or characteristic rate constants was used for the

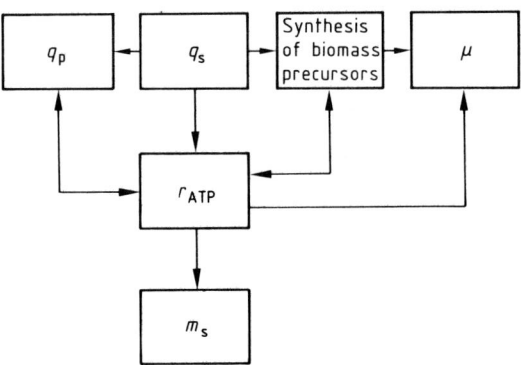

Figure 14.27. Scheme of a simple verbal model of metabolism in a block-diagram based on the rates of product synthesis ($r_p = q_p \cdot x$), substrate consumption ($r_s = q_s \cdot x$), synthesis of biomass precursors and biomass ($r_x = \mu \cdot x$), the ATP pool, and maintenance (m_s) (after ROELS and KOSSEN [14.129]).

construction of simple structured models (HARDER and ROELS [14.28]), taking into account the time regime of significant process variables according to a systematic approach. Simplicity and adequacy are needed for practiced applications. A typical block-diagram, depicted in Fig. 14.27, indicates this approach (ROELS and KOSSEN [14.129]). Substrate is used for formation of biomass, product, and ATP, which in turn is needed for growth, product synthesis, and maintenance. Similar procedures can be described, e.g., for models of diauxic growth (VAN DEDEM and MOO-YOUNG [14.136]) representing a highly structured model.

14.3.4 Heterogeneous Rate Equations: Macrokinetics

A vast number of bioprocesses can be described as homogeneous reaction systems, even when the fundamental nature is heterogeneous as is the case when several phases (G/L/S) are present. Experimental findings often can be quantified by using the pseu-

14.3.4.1 Deviations from Monod-type Kinetics

A variety of factors are known to influence Monod-type biokinetics (MOSER [14.2]; FIECHTER [14.1], i.e., the parameter K_S. Unusual or extraordinary K_S values can be the result of:

– multi-S-limitation (see Table 14.9)
– insufficient mixing in the liquid phase (see Chapter 17)
– external transport limitation (see Chapter 17)
– internal transport limitation (see Chapter 17)
– ionic strength (HORNBY et al. [14.143])
– high cell concentration (Contois kinetics, Eq. (14.16)
– endogenous metabolism (see Fig. 14.19)
– non-stationary processing (see Sect. 16.2)
– product inhibition (see Sects. 12.4 and 14.3.1.2)
– biosorption (see Sect. 14.3.4.2).

The effect of diffusion limitation on growth rates is shown in Fig. 14.28, which is

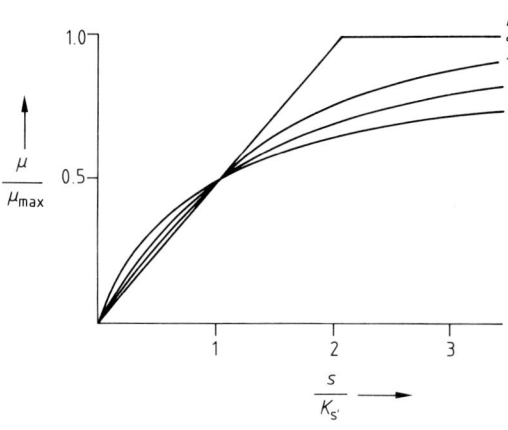

Figure 14.28. The effect of diffusion limitation (K_D) on growth rate μ in a dimensionless μ vs. s plot according to POWELL [14.40] and HUMPHREY [14.54].

dohomogeneous approach presented in Chapter 11. The empirical constants are, however, influenced by undetected transport phenomena.

based on the extended Powell equation. If K_D is only a few times the value of K_S, a Monod-type equation is obtained with an unusual K_S value according to Eq. (14.95)

14.3 Mathematical Models of Batch Growth Kinetics

Table 14.9. Macrokinetics and Explanation of Apparent Model Parameters

Reference	Macrokinetic Parameters	Eq. No.
POWELL [14.40]	$K_{S,app} = K_S + K_D$	(14.95)
HORNBY et al. [14.143]	$K_{S,app} = K_S + \dfrac{v'_{max}}{k_{L2}}$	(14.96)
SHULER and DOMACH [14.11]	$K_{S,app} = K_S + \left[\dfrac{v'_{max} \cdot K_S}{k_{L2}(s_0 + K_S)}\right]$	(14.97)
KOBAYASHI and MOO-YOUNG [14.152]	$K_{S,app} = K_S + \dfrac{3 + v'_{max}}{4 \cdot k_{L2}}$	(14.98)
MOSER [14.2]	$K_{S,1,app} = K_{S1}(1 + K_{S2}/s_2 + K_{S2} \cdot s_1/s_2 \cdot K_{S1})$	(14.99)
MONA et al. [14.148]	$q_{S,max,app} = q_{S,max} \cdot B_x$	(14.100)
	$K_{S,app} = K_S \cdot B_x$	(14.101)

in Table 14.9. HORNBY et al. [14.143] and SHULER et al. [14.9] presented Eqs. (14.96) to (14.98) for $K_{S,app}$.

Macrokinetics are described in detail by HARTMEIER [14.144], HARTMEIER et al.

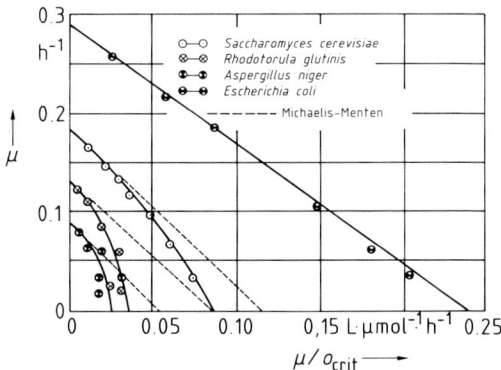

Figure 14.29. Relationship between specific growth rate μ and critical O_2-concentration o_{crit} in a semi-logarithmic plot according to Fig. 12.8 for different microbes, showing a deviation from simple Michaelis kinetics (HARTMEIER et al. [14.145]).

[14.145], and DELLWEG et al. [14.146] in studies on the respiration rate of different microorganisms. Fig. 14.29 shows a plot of respiration kinetics in an Eadie-Hofstee-diagram. Significant deviations can be observed especially in the case of mycelial growth. A theoretical interpretation will be given in Sect. 17.4. However, this kind of interaction occurring in microbial cells cannot be optimized by technical manipulation but represents a biological fact.

Multiple S-limitation can also result in higher values for K_S according to Eq. (14.99). In the case of non-stationary processing MONA [14.147] and MONA et al. [14.148] related the increased parameters $q_{S,max}$ and K_S by means of Eqs. (14.100) and (14.101) to "organic loading". These equations apply to biological waste water treatment and other non-stationary bioprocesses (e.g., AIBA et al. [14.149]).

14.3.4.2 Macrokinetics of Linear Growth Phenomena

Linear growth phases were modelled by KNORRE et al. [14.150], [14.151] assuming linearity caused by constant activity of enzymes due to a lack in concentration. Linearities indicate the presence of some limitations, although the exact nature cannot directly be deduced from this fact. In addition to a lack in nutrients, linear growth phases are the result of transport limitations.

A case of external transport limitation is demonstrated in Fig. 14.30, where linear growth is caused by an insufficient O_2-supply (REUSS and WAGNER [14.108]). Increasing the volumetric O_2-transfer coefficient $k_{L1} \cdot a$ caused the disappearance of linear growth.

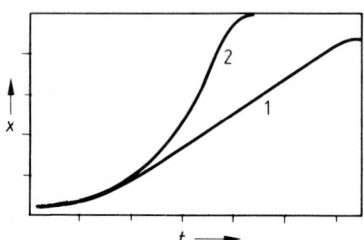

Figure 14.30. The effect of OTR (Oxygen transfer rate)-limitation on microbial growth in a plot of biomass concentration (x) vs. time during a batch process. The linear growth (curve 1) disappears by applying adequate $k_L a$-values (curve 2; exponential growth).

Another case of transport limitation concerning internal diffusion is generally encountered in the case of biofilm growth patterns. As a result of this type of internal transport limitation several equations have been developed quantifying macrokinetics of biofilm growth (cf. Sect. 16.5). Another case of a drastic change in biokinetics arises due to the "biosorption"-effect (WALTERS [14.153]; JONES [14.154]; BUSBY and ANDREWS [14.155]; MOSER [14.156]; THEOPHILOU et al. [14.157]; MOSER [14.2]).

Pellets do not always grow exponentially because of mass transfer limitations (WHITAKER and LONG [14.158]; PIRT [14.102]; METZ and KOSSEN [14.159]). Growth of fungi can occur in two morphological modes, filamentous or pulpy growth and spherical stable aggregates called pellets. In submerged cultures, growth in filamentous form follows the exponential law (METZ and KOSSEN [14.159]; VAN SUIDAM et al. [14.160]). Individual hyphae grow only at the tips at a linear extension rate. Exponential growth is maintained by continuous branching of the hyphae. This can be expressed by introducing the "hyphal growth unit", the mean length of hyphae per growing tip. The combination of a linear extension rate with a constant hyphal growth unit results in exponential growth, where the unit is independent of μ (VAN SUIDAM and METZ [14.161]).

Apart from familiar growth laws including the Monod-type and exponential growth other models used for pellet growth are summarized by VAN SUIDAM et al. [14.160], one of them being the logistic equation, given by Eq. (14.55). The most general growth model is described by the cube-root law reported by PIRT [14.94]:

$$x^{1/3} = x_0^{1/3} + k_{app} \cdot t, \tag{14.102}$$

with

$$k_{app} = k \cdot \tilde{d}_P \cdot \mu$$

and $\tilde{d}_P$ the thickness of the active zone of the pellet that contributes to growth. The rate equation thus becomes:

$$r_x = 3 \cdot k \cdot \tilde{d}_P \cdot \mu \cdot x^{2/3}. \tag{14.103}$$

The cube-root law is in fact a combination of the exponential growth law and mass transport limitation (PIRT [14.102]).

A different approach for pellet growth is given by the Gompertz's law, a formal analogy to the growth of solid tumors (CHIU and ZAJIC [14.161a]) expressed by:

$$r_x = k_1 x \cdot \exp \cdot (-k_2 t). \tag{14.104}$$

The predictions of different kinetic model equations for pellet growth are compared in their time behavior in Fig. 14.31 (VAN SUIDAM et al. [14.160]). The main objection against the models presented is that they are autonomous, i.e., the biomass rate equation is not related to the concentration of the limiting substrate and that mass transfer limitations are not considered directly in these macrokinetics. Such an integrated model for pellet growth was developed by METZ [14.162] and extended and modified by VAN SUIDAM et al. [14.160].

Another case of linear growth appears with biofilm processes, where first order kinetics in respect to substrate consumption are often found, e.g., in biological waste water treatment. ATKINSON [14.122] showed that first order kinetics can be explained by interval transport limitation oc-

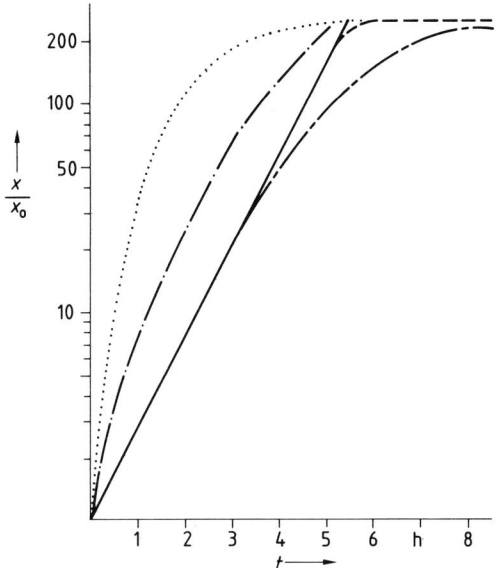

Figure 14.31. Comparison of various reported rate equations in integrated form for the biomass production in case of pellet growth. —— Exponential law; – – – Monod equation; –··– logistic law, Eq. (14.55); ····· Gompertz's law, Eq. (14.104); –·–·– cube-root law, Eq. (14.102) (VAN SUIDAM et al. [14.160]).

curring in the biofilm simultaneously with enzyme reactions. Thereby, saturation type Monod kinetics appear as a limiting case for small thicknesses of biofilms.

14.4 Mathematical Models of Product Formation Kinetics

14.4.1 Kinetic Pattern

In agreement with the fermentation pattern presented in Fig. 14.2, there are different types of kinetic modelling of product formation. On the level of formal kinetics a correlation between r_P and r_x is evident. Product formation cannot occur without the presence of cells. Thus it is expected that product formation will be coupled to growth and/or cell mass depending on the metabolic regulatory controls (GADEN [14.3]).

Furthermore, it is expected that growth and product formation are closely coupled also on a mechanistic level, whereby a causal nexus of both phenomena must exist to nutrient utilization.

The association of product formation to microbial growth, therefore, is conveniently used for mathematical modelling. In Fig. 14.32 the basic types of kinetics are demonstrated, while Table 14.10 summarizes known kinetic model equations for microbial product formation.

Growth associated production is quantified by Eqs. (14.105) to (14.107).

Substituting a Monod-type equation into Eq. (14.105) results in Eq. (14.107) which is a hyperbolic function for production in the case of growth association. Non-growth-linked product formation, described by curve II in Fig. 14.32, is more difficult to quantify, as no direct relationship to growth exists. As an alternative in this case the dependence of r_P on biomass concentration is often successfully described by Eq. (14.108) in Table 14.10. Several other equations, in Table 14.10, i.e., Eqs. (14.109) to (14.112) have been developed for various processes.

When the product accumulation rate is influenced by the product decomposition

Table 14.10. Model Equations for Pseudohomogeneous Rates of Microbial Product Formation

Reference	Kinetic Model	Eq. No.
GADEN [14.3]	$r_P = Y_{P/X} \cdot r_X$ or $q_P = Y_{P/X} \cdot \mu$ with $Y_{P/X} = \dfrac{Y_{P/S}}{Y_{X/S}}$	(14.105)
	$r_P = Y_{P/S} \cdot r_S$	(14.106)
	$r_P = q_{P,\,max} \dfrac{s}{K_S + s} \cdot x$	(14.107)
	$r_P = k_P \cdot x$	(14.108)
GIONA et al. [14.163]	$q_P = k_1 \cdot o_L + k_2$	(14.109)
GADEN [14.3]	$q_P = Y_{P/X} \cdot \mu + k_P$	(14.110)
ROWLEY and PIRT [14.164]	$q_P = q_{P,\,max} - Y_{P/X} \cdot \mu$	(14.111)
TERUI [14.165]	$q_P = q_{P,\,max} \cdot \exp\{-k_2(t - t_{max})\} + K_1 [\exp\{-k_1(t - t_{max})\} - \exp\{-k_2(t - t_{max})\}]$	(14.112)
CONSTANTINIDES et al. [14.89]	$r_P = Y_{P/X} \cdot \mu \cdot x + k_P \cdot x - k_{P,\,d} \cdot p$	(14.113)
SHU [14.166]	$r_P = \sum_i k_{1,\,i} \cdot e^{-k_{2,\,i} \cdot \Lambda}$	(14.114)
	$p = \int_0^t x(\Lambda) \int_0^\Lambda \sum_i k_{1,\,i} \cdot e^{-k_{2,\,i} \cdot \Lambda} \cdot d\Lambda \cdot d\Lambda$	(14.115)
AIBA and HARA [14.13]	$r_P = q_P(\overline{\Lambda}) \cdot x$ (14.116) $\overline{\Lambda}(t_i) = \dfrac{x_0 \cdot \Lambda_0 + \int_{t_0}^{t_i} x \cdot d\tau}{x(t_i)}$	(14.117)
BROWN and VASS [14.167]	$(r_P)_t = Y_{P/X} \cdot (r_X)_{t-t_M}$ (14.118) $P_t = Y_{P/X} \cdot x_{t-t_M}$ (14.119)	
RYU and HUMPHREY [14.105]	$q_P = q_{P,\,max} \dfrac{\varepsilon \cdot \mu_{rel}}{1 + (\varepsilon - 1) \mu_{rel}}$	(14.133)
BAJPAI and REUSS [14.130], [14.168]	$q_P = q_{P,\,max} \dfrac{s}{K_P + s(1 + s/K_{repr})}$	(14.134)
	$q_P = q_{P,\,max} \dfrac{(1 - \mu_{rel})(\mu_{rel})}{(K_P/K_S \cdot x)(1 - \mu_{rel})^2 + (1 - \mu_{rel})(\mu_{rel}) + (\mu_{rel})^2/(K_{repr}/K_S \cdot x)}$	(14.135)
KONO [14.18] KONO and ASAI [14.19] to [14.21] ASAI and KONO [14.169]	$r_P = k_{P1} \cdot \phi \cdot x + k_{P2}(1 - \phi) x$ with k_{P1} and $k_{P2} \geq 0$ and $0 < \phi < 1$	(14.136)

rate, a term must be added to account for this reaction step as in Eq. (14.113). Decomposition rates were used, e.g., for the modelling of penicillin production (CONSTANTINIDES et al. [14.89]).

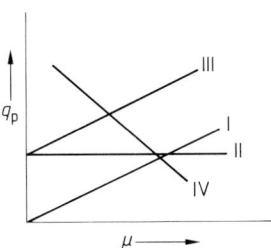

Figure 14.32. Schematic representation of the formal kinetic relationship between the specific rates of product formation (q_p) and growth (μ): growth-associated (I), non-growth-associated (II), mixed-growth-associated (III) product formation (cf. Fig. 14.2), and (IV) a case of a negative correlation.

In the group of bioprocesses where r_P is directly coupled to energy metabolism a generalized treatment seems to be possible, as ROELS and KOSSEN [14.129] pointed out.

The result is an equation for microbial production:

$$r_P = \left(\frac{1}{a_P \cdot Y_{X/S}} - \frac{Y^*_{S/X}}{a_P} \right) r_x + \frac{m_S}{a_P} x, \quad (14.120)$$

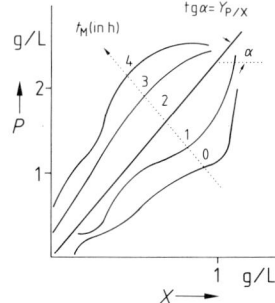

Figure 14.33. Graphic trial-and-error method of parameter estimation ($Y_{P/X}, t_M$) in the case of the maturation-time kinetic concept according to Eq. (14.119) (BROWN and VASS [14.167]).

with a_P as the stoichiometric coefficient for product formation. Eq. (14.114) states that for this case of production there exists a partly growth associated and biomass associated term.

It has been attempted to solve the difficulties in modelling especially non-growth associated product formation by introducing time-dependency even when concentration dependencies are thought to be dominant. SHU [14.166] proposed an empirical approach given by Eq. (14.114) on the basis of the assumption that r_P of individual cells is a genetically determined function of cell age Λ. Product concentration at arbitrary time, therefore, is given by Eq. (14.115). A mean cell age $\Lambda(t_i)$ was defined later on by AIBA and HARA [14.13] by Eqs. (14.116) and (14.117), with $t_0 \leq \tau \leq t_i$.

Another time concept was proposed by BROWN and VASS [14.167] introducing a maturation time t_M as indicated by Eqs. (14.118) and (14.119) and demonstrated in Fig. 14.33. Following the maturation-time concept product formation lags behind growth.

Parameter estimation can be carried out either sequentially by first determining t_M and calculating $Y_{P/X}$ (MOSER [14.2]) or simultaneously applying the graphical plot given in Fig. 14.33 as a result of a trial-and-error procedure.

14.4.2 Heat of Fermentation

Heat production is one of the macroscopic process variables considered to be significant. Experimentally, heat generation can be detected by temperature changes in the reactor, related to the volumetric heat H_v according to the Uhlich approximation:

$$r_{H_v} = \frac{dh_v}{dt} = \frac{M}{V} \cdot c_p \cdot \frac{dT}{dt}, \quad (14.121\,a)$$

with c_p as the specific heat capacity.

An alternative approach is to make an energy balance for the cooling water by measuring the flow rate F_{H_2O} and the inlet (T_{in}) and exit (T_{ex}) temperatures of the cooling water:

$$r_{H_v} = F_{H_2O} \cdot c_{p,\,H_2O}(T_{ex} - T_{in}) \tag{14.121b}$$

Eq. (14.121) implies that the whole fermenter is treated as an adiabatic calorimeter; this method is called dynamic calorimetry. Even when much of the earlier work involved the use of crude calorimeters, new techniques require relatively complicated apparatus and/or procedure for determining the heat of fermentation. Different methods are available, known as the principle of gradients, the isothermic and adiabatic principle, and the quasi-adiabatic principle: ERICKSSON [14.170]; ERICKSSON and HOLME [14.171]; JONES [14.172]; MINKEVICH and EROSHIN [14.173]; FORREST [14.174]; VAN UDEN [14.175]; IMANAKA and AIBA [14.176]; CARDOSO-DUARTE et al. [14.177]; MONK [14.178]; BAYER and FUEHRER [14.179]; KORJAGIN et al. [14.180]; MOU and COONEY [14.181]; BRONN [14.182]; WANG et al. [14.183]; LUONG and VOLESKY [14.184]; VOLESKY et al. [14.185]; SCHAUERTE [14.186].

An energy balance of a fermenter is expressed by the following equation presented by COONEY et al. [14.187]:

$$r_{H_v} = \left(\frac{dh_v}{dt}\right)_r + \left(\frac{dh_v}{dt}\right)_{ag} + \left(\frac{dh_v}{dt}\right)_{ev} + \left(\frac{dh_v}{dt}\right)_{bub} + \left(\frac{dh_v}{dt}\right)_{sur} + \left(\frac{dh_v}{dt}\right)_{Tr}, \tag{14.122}$$

which contains terms concerning reaction, agitation, evaporation, air bubbles, and heat transfer. Accurate measurements have to take into account all these terms, even though some terms can be neglected or measured prior to fermentation (agitation, surroundings, evaporation). With known heat transfer capacity of the bioreactor system, the heat production rate can be estimated by the following equation:

$$\left(\frac{dh_v}{dt}\right)_{tot} = -k_H \cdot a_H \cdot \overline{\Delta T} + r_{H_v}, \tag{14.123}$$

with $\overline{\Delta T}$ as the logarithmic mean temperature, k_H the heat transfer coefficient, and a_H the area of heat transfer. An interpretation of the kinetic term r_{H_v} can be given according to the following equation (MOU and COONEY [14.181]; COONEY et al. [14.187]) using the specific growth rate:

$$q_{H_v} = \frac{1}{x} \cdot \frac{dh_v}{dt} = \frac{1}{Y_{X/H_v}} \cdot \mu + \frac{1}{Y_{P/H_v}} \cdot q_P + m_{H_v}. \tag{14.124}$$

This type of equation was already encountered in the case of product formation in fermentations [Eq. (14.105)]. Y_{X/H_v} and Y_{P/H_v} are the thermal yields related to biomass and product formation and m_{H_v} as the maintenance coefficient. Instead of this approach an alternative formal kinetic relationship, developed by COONEY et al. [14.187], can be used:

$$q_{H_v} = (1/Y_{O/H_v}) \cdot q_O \tag{14.125}$$

with

$$1/Y_{O/H_v} = \Delta H_R^{(0)}. \tag{14.126}$$

This procedure of estimating the heat production is simpler than previous methods. The proportionally constant $1/Y_{O/H_v}$ is independent of μ, slightly dependent on the substrate, and possibly dependent on the type of organism growth. This property of $1/Y_{O/H_v}$, which is assumed to be comparable to the reaction enthalpy of fermentation, $\Delta H_R^{(0)}$, makes it extremely valuable for application to growth on complex substrates. According to MINKEVICH and EROSHIN [14.173] $\Delta H_R^{(0)}$ varies from 0.385 to 0.494, or from 0.385 for filamentous fungi to 0.565 for bacteria, after LUONG and VOLESKY [14.184], in descending order: bacteria, yeast, mold.

A quite different approach in evaluating heat generation in bioprocessing involves the combustion heat of components engaged in the conversion (IMANAKA and AIBA [14.176]). Molar enthalpies can be either taken from tables of thermodynamics or calculated on the basis of an observation by KHARASH [14.188] according to which each mole of oxygen consumed during combustion of a compound results in a release of 0.444 MJ. The following equation can be applied in the case of aerobic batch cultures of *Saccharomyces cerevisiae* for calculating the rate of heat production:

$$r_{H_v} = \left(\frac{-\Delta H_S}{M_S}\right)\frac{\mu \cdot x}{Y_{X/S}} + \left(\frac{-\Delta H_N}{M_x}\right)\mu x - \left(\frac{-\Delta H_x}{M_x}\right)\mu x - \left(\frac{-\Delta H_P}{M_P}\right)q_P \cdot x, \quad (14.127)$$

with r_{H_v} being the rate of heat of fermentation, $-\Delta H_S$, $-\Delta H_N$, $-\Delta H_x$, $-\Delta H_P$ heat of combustion of substrate, ammonia, cells, product.

Rearranging Eq. (14.127) results in an expression for the heat of fermentation per mass unit of cells produced (VOLESKY et al. [14.185]):

$$-\Delta H_R^{(X)} = K_1 + K_2/Y_{X/S} - K_3 \cdot q_P/\mu, \quad (14.128)$$

where

$$K_1 = \frac{-\Delta H_N + \Delta H_x}{M_x}, \quad (14.129)$$

$$K_2 = -\Delta H_S/M_S, \quad (14.130)$$

$$K_3 = -\Delta H_P/M_P. \quad (14.131)$$

Similarly, heat of fermentation per mass unit of substrate consumed $\Delta H_R^{(S)}$ can be expressed as follows:

$$-\Delta H_R^{(S)} = (K_1 + K_2/Y_{X/S} - K_3 \cdot q_P/\mu)Y_{X/S}. \quad (14.132)$$

BRONN [14.182] summarized data of metabolic reaction enthalpies of fermentations showing that several classes can be distinguished, such as in aerobic and anaerobic processes on carbohydrates and processes on hydrocarbons, each characterized by a more or less constant value for $\Delta H_R^{(X)}$ and $\Delta H_R^{(S)}$, independent of the strain used. Normally fermentations are exothermic.

Similar results were obtained from recent work with microcalorimetry carried out by OURA [14.189]; LOVRIEN et al. [14.190]; BRETTEL et al. [14.191; 14.192]. IMANAKA and AIBA [14.140], using the concept of heat of combustion, arrived at the same result, given by Eq. (14.126).

14.4.3 Generalized and Extended Rate Equations for Microbial Product Formation

A systematic approach of unstructured modelling of microbial production was presented by RYU and HUMPHREY [14.105]. Assuming that the conversion of an intermediate I_i to both cells and product is limited by the rate of one single enzyme in the chain, each following Monod-type kinetics, Eq. (14.133) in Table 14.10 holds. Here, $\varepsilon = K_{I,i}/K'_{I,i}$ is the ratio of the saturation constants of both enzymatic steps leading to P or X.

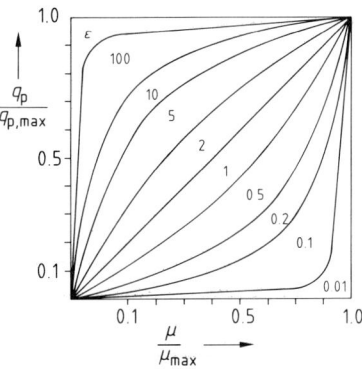

Figure 14.34. Graphical representation of Eq. (14.133) showing a generalized kinetic approach to microbial product formation due to varying values of ε (RYU and HUMPHREY [14.105], ROELS and KOSSEN [14.129]).

Fig. 14.34 represents a graphical plot of Eq. (14.133). For $\varepsilon > 1$ an equation results which is homologous to the Monod form. For $\varepsilon = 1$ the curve for Blackman kinetics results, [see Eqs. (14.18) and (14.19)] and for $\varepsilon < 1$ a parabolic curve is obtained.

Another generalization of product formation kinetics based on a mechanistic background but still using the formal kinetic approach was recently achieved by BAJPAI and REUSS [14.130], [14.193]. They proposed a substrate inhibition model as a formal analogy to catabolite repression [see Eq. (14.82)] for penicillin production as a typical case of secondary product formation. This model, along with rate equations for μ, q_S, and q_O, was successfully used to simulate experimental data. The result is Eq. (14.134) and Eq. (14.135).

14.10]. A mechanistic modelling exercise for batch penicillin fermentation was proposed by CALAM et al. [14.194] and CALAM and RUSSELL [14.195]. KREMEN [14.196] explained the differences between a high productive and a low productive situation, starting from the thermodynamics of irreversible processes.

Even when in more complex cases structured models are more promising, unstructured approaches also yield applicable results. HEGEWALD et al. [14.197] modelled antibiotic productions in batch cultures in-

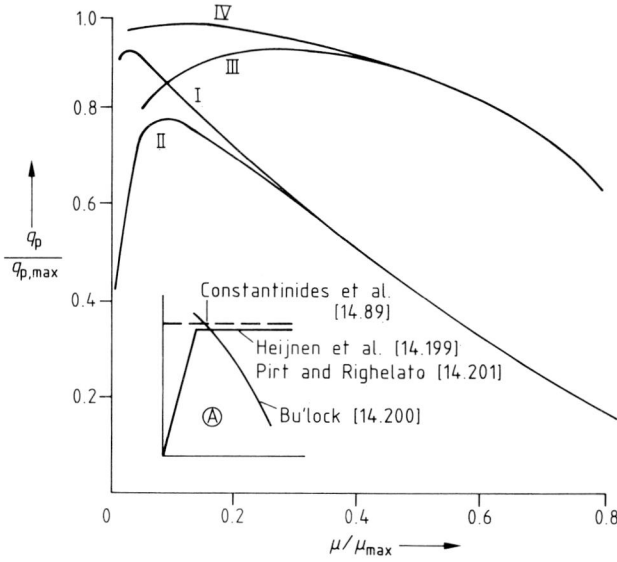

Figure 14.35. Graphical representation of Eq. (14.135) as a generalized kinetic model for microbial product formation in a plot of dimensionless specific rates q_p vs. μ. The insert A shows previously suggested patterns of penicillin production rate from the literature.

	$K_P/K_S \cdot x$	$K_{repr}/K_S \cdot x$	
I	0.00133	0.6667	
II	0.0133	0.6667	(BAJPAI and
III	0.0133	6.667	REUSS [14.193]).
IV	0.00133	6.667	

Plots of Eq. (14.135) are given in Fig. 14.35 for different sets of values of the terms ($K_P/K_S \cdot x$) and ($K_{repr}/K_S \cdot x$). The inset A in this figure compares some of the available models of secondary metabolite production showing that Eq. (14.135) does in fact represent all trends. Therefore, by enzyme kinetics a generalized form of a rate equation for microbial production was successfully developed.

The Kono approach also has been very useful in practice, as the consumption coefficient parameter ϕ can be determined easily from c/t-plots [see Eq. (14.136) in Table

corporating a term of inhibition of product formation $f(s_1)$. Instabilities of product formation of penicillin could also be explained satisfactorily (HEGEWALD et al. [14.198]).

Acknowledgement:

Most of the computer simulations were carried out by the group of Dr. W. Knorre of ZIMET at the Academy of Sciences of the German Democratic Republic in Jena based on a scientific collaboration with the author's group at the institute in Graz/Austria.

14.5 References

[14.1] A. FIECHTER: In "Biotechnology – a Comprehensive Treatise", Vol. 1, Chap. 7 (H. J. REHM and G. REED, eds.). Verlag Chemie, Weinheim – Deerfield Beach, Florida – Basel, 1981.

[14.2] A. MOSER: "Bioprozeßtechnik". Springer Verlag, New York – Wien, 1981.

[14.3] E. L. GADEN Jr.: Biotechnol. Bioeng. *1* (1959), 413.

[14.4] W. W. KAFAROW: "Kybernetische Methoden in der Chemie und Chemischen Technologie". Akademie Verlag, Berlin (GDR), 1971.

[14.5] W. W. KAFAROW, A. J. WINAROW, and L. S. GARDEJEW: "Modelling of Biochemical Reactors". Lesnaja Promyshlenost, Moskow (Russian), 1979.

[14.6] A. MOSER: In "Preprints, 1st Eur. Congress on Biotechnology", Part I, p. 88. Interlaken (Switzerland), 1978.

[14.7] J. A. ROELS: Biotechnol. Bioeng. *22* (1980), 2457.
J. A. ROELS: J. Chem. Tech. Biotechnol. *32* (1982), 59.

[14.8] J. VOTRUBA: Acta Biotechnol. *2* (1982), 2, 119.

[14.9] M. L. SHULER, S. LEUNG, and C. C. DICK: Ann. N. Y. Acad. Sci. *326* (1979), 35.

[14.10] S. V. HO and M. L SHULER: J. Theor. Biol. *68* (1977), 415.

[14.11] M. L. SHULER and M. M. DOMACH: In "ACS (Am. Chem. Soc.) – Winter Symp. on Fundamentals of Biochemical Engineering: Kinetics and Thermodynamics in Biological Systems". (H. W. BLANCH, E. T. PAPOUTSAKIS, and G. N. STEPHANOPOULUS, eds.). Boulder/Colorado (USA), 1982.

[14.12] D. HERBERT: In "Proc. 2nd Symp. Continuous Culture of Microorganisms". Vol. 12, p. 21. SCI (Society of Chemical Industry) Monogr. London, 1961.

[14.13] S. AIBA and H. HARA: J. Gen. Microbiol. *11* (1965), 41.

[14.14] H. V. EDWARDS and Ch. R. WILKE: Biotechnol. Bioeng. *10* (1968), 205.

[14.15] L. N. ANDREYEVA and V. V. BIRYUKOV: Biotechnol. Bioeng. Symp. *4* (1973), 61.

[14.16] V. M. FISHMAN and V. V. BIRYUKOV: Biotechnol. Bioeng. Symp. *4* (1974), 647.

[14.17] B. GRM, M. MELE, and M. KREMSER: Biotechnol. Bioeng. *22* (1980), 255.

[14.18] T. KONO: Biotechnol. Bioeng. *10* (1968), 105.

[14.19] T. KONO and T. ASAI: J. Ferment. Technol. *46* (1968), 391.

[14.20] T. KONO and T. ASAI: Biotechnol. Bioeng. *11* (1969), 19.

[14.21] T. KONO and T. ASAI: Biotechnol. Bioeng. *11* (1969), 293.

[14.22] A. MOSER: In "Proc. VIth International Spec. Symp. on Yeasts", (P. GALZY, A. ARNAUD, C. BIZEAU, and G. MOULIN, eds.). SI 16. Montepellier (France), 1978 b.

[14.23] A. MOSER: "Bioprocess Engineering". Springer Verlag, New York – Wien, 1985.

[14.24] R. SCHMID and V. N. SAPUNOV: "Non-Formal Kinetics", (Monographs in Modern Chemistry 14). Verlag Chemie, Weinheim – Deerfield Beach, Florida – Basel, 1982.

[14.25] M. REUSS: Fortschr. Verfahrenstech. *15 F* (1977), 549.

[14.26] N. W. F. KOSSEN: Symp. Soc. Gen. Microbiol. *20* (1979), 327.

[14.27] N. W. F. KOSSEN and J. A. ROELS: Prog. Ind. Microbiol. *14* (1978), 95.

[14.28] A. HARDER and J. A. ROELS: Adv. Biochem. Eng. *21* (1981), 56.

[14.29] H. M. TSUCHIYA, A. G. FREDRICKSON, and R. ARIS: Adv. Chem. Eng. *6* (1966), 125

[14.30] A. MOSER and R. M. LAFFERTY: In "Proc. 5th IFS (Int. Ferment. Symp.)", p. 103 (H. DELLWEG, ed.). Berlin, 1976.

[14.31] W. E. GATES and J. T. MARLAR: J. Water Pollut. Control Fed. (1968), 469.

[14.32] G. KNOWLES, A. L. DOWNING, and M. J. BARRETT: J. Gen. Microbiol. *38* (1965), 263.

[14.33] J. MEYRATH and K. BAYER: In "Proc. 3nd Symp. Techn. Mikrobiol.", p. 117 (H. DELLWEG, ed.). Berlin, 1973.

[14.34] Ph. S. CROOKE and R. D. TANNER: Appl. Math. Modelling *4* (1980), 376.

[14.35] J. MONOD: "Recherches sur la Croissance des Cultures Bacteriennes". Hermann & Cie., Paris, 1942.

[14.36] J. MONOD: Annu. Rev. Microbiol. *3* (1949), 371.

[14.37] G. TEISSIER: Ann. Physiol. Physiochim. Biol. *12* (1936), 527.

[14.38] F. F. BLACKMAN: Ann. Bot. *19* (1905), 281.
[14.39] J. N. DABES, R. K. FINN, and C. R. WILKE: Biotechnol. Bioeng. *15* (1973), 1159.
[14.40] E. O. POWELL: In "Microbial Physiology and Continuous Culture", p. 34 (E. O. POWELL, C. G. T. EVANS, R. E. STRANGE, and D. W. TEMPEST, eds.). Her Majesty's Stationary Office, Elliot Bros. & Yesman Ltd., Speke Liverpool, 1967.
[14.41] W. C. BOYLE and P. M. BERTHOUEX: Biotechnol. Bioeng. *16* (1974), 1139.
[14.42] H. MOSER: "The Dynamics of Bacterial Populations Maintained in the Chemostat". Carnegie Institution, Publ. No. 614, Washington D. C., 1958.
[14.43] D. E. CONTOIS: J. Gen. Microbiol. *21* (1959), 40.
[14.44] Y. FUJIMOTO: J. Theoret. Biol. *5* (1963), 171.
[14.45] D. W. TEMPEST: In "Continuous Culture". Chap. 6, p. 348 (A. C. R. DEAN, D. C. ELLWOOD, C. G. T. EVANS, and J. MELLING, eds.). SCI (Society of Chemical Industry), Ellis Horwood, Ltd., Chichester, 1975.
[14.46] A. R. KONAK: J. Appl. Chem. Biotechnol. *24* (1974), 453.
[14.47] F. KARGI and M. L. SHULER: Biotechnol. Bioeng. *21* (1979), 1871.
[14.48] V. A. VAVILIN: Biotechnol. Bioeng. *24* (1982), 1721.
[14.49] J. F. ANDREWS: Biotechnol. Bioeng. *10* (1968), 707.
[14.50] J. F. ANDREWS: J. Sanit. Eng. Div. (Am. Soc. Civ. Eng.) *95* (1969), 95.
[14.51] J. F. ANDREWS: Biotechnol. Bioeng. Symp. *2* (1971), 5.
[14.52] D. NOACK: "Biophysikal. Prinzipien der Populationsdynamik in der Mikrobiologie". VEB G. Thieme, Leipzig (GDR), 1968.
[14.53] V. H. EDWARDS: Biotechnol. Bioeng. *12* (1970), 679.
[14.54] A. E. HUMPHREY: ACS (Am. Chem. Soc.) Symp. Ser. (1978), 72.
[14.55] M. C. TSENG and M. WAYMAN: Can. J. Microbiol. *21* (1975), 994.
[14.56] M. WAYMANN and M. C. TSENG: Biotechnol. Bioeng. *18* (1976), 383.
[14.57] E. SIIMER: Biotechnol. Bioeng. *20* (1978), 1853.
[14.58] J. L. WEBB: "Enzyme and Metabolic Inhibitors", Vol. 1. Academic Press, New York, 1963.

[14.59] T. YANO, T. NAKAHARA, S. KAMIYAMA, and K. YAMADA: Agric. Biol. Chem. (Japanese) *30* (1966), 42.
[14.60] S. AIBA: Biotechnol. Bioeng. *10* (1968), 845.
[14.61] C. N. HINSHELWOOD: "The Chemical Kinetics of the Bacterial Cell". Clarendon Press, Oxford (UK), 1946.
[14.62] S. DAGLEY and C. N. HINSHELWOOD: J. Chem. Soc. (1938), 1942.
[14.63] I. HOLZBERG, R. K. FINN, and K. H. STEINKRAUS: Biotechnol. Bioeng. *9* (1967), 413.
[14.64] T. K. GHOSE and R. D. TYAGI: Biotechnol. Bioeng. *21* (1979), 1401.
[14.65] N. D. JERUSALIMSKY and N. M. NERONOVA: Dokl. Akad. Nauk SSSR *161* (1965), 1437 (Russian).
[14.66] F. BERGTER: "Wachstum von Mikroorganismen". VEB G. Fischer Verlag, Jena (GDR), 1972.
[14.67] S. AIBA: AIChE J. *15* (1969), 624.
[14.68] S. AIBA, M. SHODA, and M. NAGATANI: Biotechnol. Bioeng. *11* (1969), 1285.
[14.69] P. PERINGER, H. BLACHÈRE, G. CORRIEU, and A. G. LANE: Biotechnol. Bioeng. *16* (1974), 431.
[14.70] H. FUKUDA, T. SHIOTANI, W. OKADA, and H. MORIKAWA: J. Ferment. Technol. *56* (1978), 361.
[14.71] S. AIBA and M. SHODA: J. Ferment. Technol. *47* (1969), 790.
[14.72] D. RAMKRISHNA, A. G. FREDRICKSON, and H. M. TSUCHIYA: J. Ferment. Technol. *44* (1966), 210.
[14.73] D. RAMKRISHNA, A. G. FREDRICKSON, and H. M. TSUCHIYA: Biotechnol. Bioeng. *9* (1967), 129.
[14.74] W. A. KNORRE: In "Mathematische Modellbildung in Naturwissenschaft und Technik", p. 221. Akademie Verlag, Berlin (GDR), 1976.
[14.75] T. A. PETROVA, W. A. KNORRE, R. GUTKE, and F. BERGTER: Z. Allg. Mikrobiol. *17* (1977), 7, 531.
[14.76] W. A. KNORRE: In "Biophysikalische Grundlagen der Medizin", p. 132 (W. BEIER and R. ROSEN, eds.). G. Fischer Verlag, Stuttgart – New York, 1980.
[14.77] C. D. BAZUA and C. R. WILKE: Biotechnol. Bioeng. Symp. *5* (1977), 105.
[14.78] O. LEVENSPIEL: Biotechnol. Bioeng. *22* (1980), 1671.
[14.79] G. K. HOPPE and G. S. HANSFORD: Biotechn. Let *4* (1961), 39.

[14.80] R. M. LODGE and C. N. HINSHELWOOD: J. Chem. Soc. (1943), 213.
[14.81] R. GUTKE: In "Manual for UNEP/UNESCO/IORO Training Course on Theoretical Basis of Kinetics of Growth, Metabolism and Product Formation of Microorganisms", Vol. 1, p. 30 and p. 112, Academy of Sciences, Central Institute of Microbiology and Exp. Therapy (ZIMET), Jena (GDR), 1980.
[14.82] R. GUTKE: In "Manual for International Training Course on Biotechnological Fundamentals of Continuous Biomass Production", p. 39 and p. 58, Academy of Sciences, GDR, Jena (ZIMET) and Leipzig (Inst. Techn. Chemistry), 1982.
[14.83] V. H. EDWARDS: Biotechnol. Bioeng. *11* (1969), 99.
[14.84] F. BERGTER and W. KNORRE: Z. Allg. Mikrobiol. *12* (1972), 8, 613.
[14.85] N. B. PAMMENT, R. J. HALL, and J. P. BARFORD: Biotechnol. Bioeng. *20* (1978), 349.
[14.86] J. P. BARFORD: In "Microbial Population Dynamics" (M. J. BAZIN, ed.). CRC Press, Boca Raton, Florida, 1982.
[14.87] D. G. KENDALL: J. Roy. Statist. Soc. B *11* (1949), 230.
[14.88] E. J. LA MOTTA: Biotechnol. Bioeng. *18* (1976), 1359.
[14.89] A. CONSTANTINIDES, J. L. SPENCER, and E. L. GADEN Jr.: Biotechnol. Bioeng. *12* (1970), 803.
[14.90] A. CONSTANTINIDES, J. L. SPENCER, and E. L. GADEN Jr.: Biotechnol. Bioeng. *12* (1970), 1081.
[14.90a] S. Y. CHIU, L. T. FAN, I. C. KAO, and L. E. Erickson: Biotechnol. Bioeng. *14* (1972), 179.
[14.91] A. MOSER and W. STEINER: Eur. J. Appl. Microbiol. *1* (1975), 281.
[14.92] C. P. L. GRADY, L. J. HARLOW, and R. R. RIESING: Biotechnol. Bioeng. *14* (1972), 391.
[14.93] C. G. SINCLAIR and H. H. TOPIWALA: Biotechnol. Bioeng. *12* (1970), 1069.
[14.94] S. J. PIRT: Proc. R. Soc. B, *163* (1965), 224.
[14.95] D. HERBERT: In "Proc. 1st Symp. Continuous Culture of Microorganisms", p. 45. Prague (CSSR), 1958.
[14.96] A. G. MARR, E. H. NILSON, and D. J. CLARK: Ann. N. Y. Acad. Sci. *102* (1963), 536.
[14.97] T. TAKAMATSU, S. SHIOYA, and K. OKUDA: J. Ferment. Technol. *59* (1981), 131.
[14.98] F. BERGTER: Z. Allg. Mikrobiol. *18* (1978), 2, 143.
[14.99] A. C. R. DEAN and C. N. HINSHELWOOD: "Growth, Function and Regulation in Bacterial Cells." Clarendon Press, Oxford (UK), 1966.
[14.100] R. D. MEGEE III, S. KINOSHITA, A. G. FREDRICKSON, and H. M. TSUCHIYA: Biotechnol. Bioeng. *12* (1970), 771.
[14.101] J. I. PROSSER: In "Microbial Population Dynamics", Chap. 5 (M. J. BAZIN, ed.). CRC Press, Boca Raton, Florida, 1982.
[14.102] S. J. PIRT: "Principles of Microbe and Cell Cultivation". Blackwell Sci. Publ., Oxford (UK), 1975.
[14.103] M. S. MORGAN and V. H. EDWARDS: Chem. Eng. Prog. Symp. Ser. 114, *67* (1971), 51.
[14.104] R. D. TANNER: Biotechnol. Bioeng. *12* (1970), 831.
[14.105] D. Y. RYU and A. E. HUMPHREY: J. Ferment. Technol. *50* (1972), 424.
[14.105a] D. Y. RYU and R. I. MATELES: Biotechnol. Bioeng. *10* (1968), 385.
[14.106] H. PETERS: "Karlsruher Berichte zur Ingenieurbiologie", p. 9 (L. HARTMANN, ed.). Karlsruhe, 1976.
[14.107] A. A. ESENER, J. A. ROELS, and N. W. F. KOSSEN: Biotechnol. Lett. *3* (1981), 193.
[14.108] M. REUSS and F. WAGNER: In "Proc. 3rd Symp. Techn. Mikrobiol.", p. 89 (H. DELLWEG, ed.). Technische Universität Berlin, 1973.
[14.109] A. B. PARDEL: Symp. Soc. Gen. Microbiol. *11* (1961), 19.
[14.110] T. IMANAKA, K. O. HOBKA, T. YOSHIDA, and H. TAGUCHI: J. Ferment. Technol. *50* (1972), 633.
[14.111] K. WUHRMANN, F. BEUST, and T. K. GHOSE: Schweiz. Z. Hydrol. *20* (1958), 284.
[14.112] T. E. SHEHATA and A. G. MARR: J. Bacteriol. *107* (1971), 210.
[14.113] F. KARGI: J. Appl. Chem. Biotechnol. *27* (1977), 704.
[14.114] J. A. BUDD: Eur. J. Appl. Microbiol. *3* (1977), 267.
[14.115] M. FIDGETT and E. L. SMITH: J. Appl. Chem. Biotechnol. *25* (1975), 355.
[14.116] W. HARDER and L. DIJKHUIZEN: In "Continuous Culture", Vol. 6, Chap. 23 (A. C. R. DEAN, D. C. ELLWOOD, C. G. T. EVANS, and J. MELLING, eds.).

SCI (Society of Chemical Industry), London. E. Horwood Ltd., Chichester (UK), 1975.
[14.117] H. YOON, G. KLINZING, and H. W. BLANCH: Biotechnol. Bioeng. *19* (1977), 1193.
[14.118] R. ARIS and A. E. HUMPHREY: Biotechnol. Bioeng. *19* (1977), 1375.
[14.119] I. H. LEE: Ph. D. Thesis, University of Minnesota (USA), 1973.
[14.120] G. T. TSAO and Th. P. HANSON: Biotechnol. Bioeng. *17* (1975), 1591.
[14.121] G. T. TSAO and C. M. YANG: Biotechnol. Bioeng. *18* (1976), 1827.
[14.122] B. ATKINSON: "Biochemical Reactors". Pion Ltd., London, 1974.
[14.123] P. WILDERER: "Karlsruher Berichte zur Ingenieurbiologie", (L. HARTMANN, ed.), 8. Karlsruhe, 1969.
[14.124] F. G. BADER, J. S. MEYER, A. G. FREDRICKSON, and H. M. TSUCHIYA: Biotechnol. Bioeng. *17* (1975), 279.
[14.125] F. G. BADER: Biotechnol. Bioeng. *20* (1978), 183.
[14.126] F. G. BADER: In "Microbial Population Dynamics" (M. J. BAZIN, ed.). CRC Press, Boca Raton, Florida, 1982.
[14.127] R. M. SYKES: J. Water Pollut. Control Fed. *45* (1973), 888.
[14.128] D. N. RYDER and C. G. SINCLAIR: Biotechnol. Bioeng. *14* (1972), 787.
[14.129] J. A. ROELS and N. W. F. KOSSEN: In "Process in Industrial Microbiology", Vol. 14, p. 95 (M. J. BULL, ed.). Elsevier, Amsterdam, 1978.
[14.130] R. K. BAJPAI and M. REUSS: Paper at 6th IFS (Int. Ferment. Symp.) London/Ontario (Canada), 1980.
[14.131] S. KOGA, C. BURG, and A. E. HUMPHREY: Appl. Microbiol. *15* (1967), 493.
[14.132] K. ZETELAKI-HORVATH and E. BEKASSY-MOLNAR: Biotechnol. Bioeng. *15* (1973), 163.
[14.133] A. H. E. BIJKERK and R. J. HALL: Biotechnol. Bioeng. *19* (1977), 267.
[14.134] A. MOSER: Acta Biotechnol. *3* (1983), 3, 195.
[14.135] A. MOSER: In "Proc. Conference Biotech '83", p. 961. London, 1983.
[14.136] G. VAN DEDEM and M. MOO-YOUNG: Biotechnol. Bioeng. *15* (1973), 419.
[14.137] A. R. MOREIRA, G. VAN DEDEM, and M. MOO-YOUNG: Biotechnol. Bioeng. Symp. *9* (1979), 179.
[14.138] A. M. PICKETT: In "Microbial Population Dynamics", Chap. 4 (M. J. BAZIN, ed.). CRC Press, Boca Raton, Florida, 1983.
[14.139] K. TODA: Biotechnol. Bioeng. *18* (1976), 1117.
[14.140] T. IMANAKA and S. AIBA: Biotechnol. Bioeng. *19* (1977), 757.
[14.141] K. TODA: Biotechnol. Bioeng. *22* (1980), 1805.
[14.142] K. TODA: J. Chem. Tech. Biotechnol. *31* (1981), 775.
[14.143] W. E. HORNBY, M. D. LILLY, and E. M. CROOK: Biochem. J. *107* (1968), 669.
[14.144] W. HARTMEIER: Ph. D. Thesis, Technische Universität Berlin, 1972.
[14.145] W. HARTMEIER, W. K. BRONN, and H. DELLWEG: Chem. Ing. Tech. *43* (1973), 76.
[14.146] H. DELLWEG, W. HARTMEIER, and W. K. BRONN: Kemia Kemi *12* (1977), 611.
[14.147] R. MONA: Ph. D. Thesis, Eidgenössische Technische Hochschule Zürich, 1978.
[14.148] R. MONA, I. J. DUNN, and J. BOURNE: Biotechnol. Bioeng. *21* (1979), 1561.
[14.149] S. AIBA, A. E. HUMPHREY, and N. F. MILLIS: "Biochemical Engineering". Academic Press, New York, 1973.
[14.150] W. A. KNORRE, R. GUTKE, and F. BERGTER: Z. Allg. Mikrobiol. *18* (1978) 4, 255.
[14.151] W. A. KNORRE, K. MUND, and L. JAKUBIK: Z. Allg. Mikrobiol. *18* (1978) 8, 609.
[14.152] T. KOBAYASHI and M. MOO-YOUNG: J. Gen. Microbiol. *13* (1971), 893.
[14.153] C. F. WALTERS: Ph. D. Thesis, University of Illinois, Urbana (USA), 1966.
[14.154] P. H. JONES: "Adv. Water Poll. Res.", Vol. 1. Pergamon Press, New York, 1971.
[14.155] J. B. BUSBY and J. F. ANDREWS: J. Water Pollut. Control Fed. *47* (1975), 1055.
[14.156] F. MOSER: Verfahrenstechnik *11* (1977), 670.
[14.157] J. THEOPHILOU, O. WOLFBAUER, and F. MOSER: Gas Wasserfach Wasser Abwasser *120* (1979), 119.
[14.158] A. WHITAKER and P. A. LONG: Process Biochem., November (1973), 27.
[14.159] B. METZ and N. W. F. KOSSEN: Biotechnol. Bioeng. *19* (1977), 781.
[14.160] J. C. VAN SUIDAM, H. HOLS, and N. W. F. KOSSEN: Biotechnol. Bioeng. *24* (1982), 177.

[14.161] J. C. Van Suidam and B. Metz: Biotechnol. Bioeng. *23* (1981), 177.

[14.161a] Y. S. Chiu and J. E. Zajic: Biotechnol. Bioeng. *18* (1976), 1167.

[14.162] B. Metz: Ph. D. Thesis, Technische Hogeschool Delft (Netherlands), 1975.

[14.163] A. R. Giona, L. Marelli, L. Toro, and R. de Santis: Biotechnol. Bioeng. *18* (1976), 473.

[14.164] B. I. Rowley and S. J. Pirt: J. Gen. Microbiol. *72* (1972), 553.

[14.165] G. Terui: In "Microbial Engineering, 1st Intern. Symp. Adv. Microbiol. Eng.", p. 377 (Z. Sterbacek, ed.). Butterworths, London, 1972.

[14.166] P. Shu: Biotechnol. Bioeng. *3* (1961), 95.

[14.167] D. E. Brown and R. C. Vass: Biotechnol. Bioeng. *15* (1973), 321.

[14.168] A. K. Bajpai and M. Reuss: J. Chem. Tech. Biotechnol. *30* (1980), 332.

[14.169] T. Asai and T. Kono: Adv. Ferment. Suppl. Process Biochem. (1983), 212.

[14.170] R. Ericksson: In "Proc. 1st Eur. Biophys. Congress", Med. Akad. Wien (Austria), Vol. 4, p. 319 (1971).

[14.171] R. Ericksson and T. Holme: Biotechnol. Bioeng. Symp. *4* (1973), 581.

[14.172] J. M. Jones: Process Biochem., September (1973), 19.

[14.173] I. G. Minkevich and V. R. Eroshin: Folia Microbiol. *18* (1973), 376.

[14.174] W. W. Forrest: In "Methods in Microbiology", Vol. 6B, p. 285 (J. R. Norris and D. W. Ribbons, eds.). Academic Press, London – New York, 1972.

[14.175] N. Van Uden: Z. Allg. Mikrobiol. *11* (1971) 6, 541.

[14.176] T. Imanaka and S. Aiba: J. Appl. Chem. Biotechnol. *26* (1976), 559.

[14.177] J. M. Cardoso-Duarte, M. J. Marinko, and N. van Uden: In "Continuous Culture", Chap. 3 (A. C. R. Dean, D. C. Ellwood, C. G. T. Evans, and J. Melling, eds.). SCI (Society of Chemical Industry), London. E. Horwood Ltd., Chichester (UK), 1975.

[14.178] P. R. Monk: Process Biochem., December (1978), 4.

[14.179] K. Bayer and F. Fuehrer: Process Biochem. July/August (1982), 42.

[14.180] W. W. Korjagin, I. M. Cirkov, and L. I. Dowgun: In "Entwicklung von Laborfermentoren", p. 327. VI. Reinhardsbrunner Symp., Acad. Sciences of GDR. Akademie Verlag, Berlin (GDR), 1978.

[14.181] D. G. Mou and Ch. L. Cooney: Biotechnol. Bioeng. *18* (1976), 1371.

[14.182] W. K. Bronn: Chem. Ing. Tech. *43* (1971), 70.

[14.183] H. Wang, D. I. C. Wang, and C. L. Cooney: Eur. J. Appl. Microbiol. Biotechnol. *5* (1978), 207.

[14.184] J. H. T. Luong and B. Volesky: Can. J. Chem. Eng. *60* (1980), 163.

[14.185] B. Volesky, L. Yerushalmi, and J. H. T. Luong: J. Chem. Tech. Biotechnol. *32* (1982), 650.

[14.186] W. A. C. Schauerte: Ph. D. Thesis, Technische Universität München, 1981.

[14.187] Ch. L. Cooney, D. I. C. Wang, and R. I. Mateles: Biotechnol. Bioeng. *11* (1969), 269.

[14.188] M. S. Kharash: Bur. Stand. J. Res. *2* (1929), 359.

[14.189] E. Oura: In "Proc. 1st National Meet. Biophys. Biotechnol. in Finland", p. 142 (L. Patomäki and A. Kiuru, eds.). 1973.

[14.190] R. Lovrien, G. Jorgenson, M. K. Ma, and W. E. Sund: Biotechnol. Bioeng. *22* (1980), 1249.

[14.191] R. Brettel, I. Lamprecht, and B. Schaarschmidt: Eur. J. Appl. Microbiol. Biotechnol. *11* (1981), 205.

[14.192] R. Brettel, I. Lamprecht, and B. Schaarschmidt: Eur. J. Appl. Microbiol. Biotechnol. *11* (1981), 212.

[14.193] R. K. Bajpai and M. Reuss: Biotechnol. Bioeng. *23* (1981), 717.

[14.194] C. T. Calam, S. H. Ellis, and M. J. McCann: J. Appl. Chem. Biotechnol. *21* (1971), 181.

[14.195] C. T. Calam and D. W. Russell: J. Appl. Chem. Biotechnol. *23* (1973), 225.

[14.196] A. Kremen: J. Theor. Biol. *31* (1971), 363.

[14.197] E. M. Hegewald, I. Rückbeil, and W. A. Knorre: In "Proc. 7th Symp. Continuous Culture of Microorganisms", p. 717. Prague (CSSR), 1980.

[14.198] E. M. Hegewald, B. Wolleschensky, R. Gutke, M. Neubert, and W. A. Knorre: Biotechnol. Bioeng. *23* (1981), 1563.

[14.199] J. J. Heijnen, J. A. Roels, and A. H. Stouthamer: Biotechnol. Bioeng. *21* (1979), 2175.

[14.200] J. D. Bu'Lock: In "The Filamentous Fungi", Vol. 1, Chap. 3 (J. E. Smith and D. R. Berry, eds.). Edward Arnold Ltd., London, 1975.

[14.201] S. J. Pirt and R. C. Righelato: Appl. Microbiol. *15* (1967), 1284.

Chapter 15

Continuous Cultivation

Anton Moser

Institut für Biotechnologie, Mikrobiologie und Abfalltechnologie
Technische Universität Graz
Graz, Austria

15.1 Introduction
15.2 Single-stage Continuous Stirred Tank Reactor Operation
15.2.1 Estimation of Kinetic Parameters
15.2.1.1 Simple Kinetics
15.2.1.2 Complex Kinetics with Inhibitions
15.2.2 Deviations due to Maintenance, Lysis, and Incomplete Mixing
15.2.3 Product Formation Kinetics in a Single-stage Continuous Stirred Tank Reactor
15.2.3.1 Production Kinetics Without Product Inhibition
15.2.3.2 Production Kinetics With Product Inhibition
15.3 Multi-stage Continuous Culture Operation
15.3.1 Classification
15.3.2 Single-stream Multi-stage Operation
15.3.3 Multi-stream Multi-stage Operation
15.3.4 Application of Multi-stage Operation
15.4 Continuous Bioprocessing with Plug Flow Reactors
15.4.1 Performance Equations and Practical Properties
15.4.2 Real Reactor Behavior – Characterization with Residence Time Distribution
15.4.3 Application of Continuous Plug Flow Reactor Operation
15.5 Recycle Operation in Bioprocessing
15.5.1 Performance Equations and Basic Behavior
15.5.1.1 The Continuous Stirred Tank Reactor with Recycling
15.5.1.2 Cascade of Stirred Tanks with Recycling
15.5.1.3 Continuous Operation of Plug Flow Reactors with Recycling
15.5.1.4 Comparison of Performance
15.5.2 Application in Industry and Research
15.6 References

List of Symbols

A	m^2	area
Bo, Bo_L	—	Bodenstein number
c, c_i	$kg \cdot m^{-3}$	concentration (general term) of component i
D	s^{-1}	dilution rate
D_L	$m \cdot s^{-2}$	dispersion coefficient of liquid phase
Da_1	—	Damköhler number, 1st degree
F	$m^3 \cdot s^{-1}$	flow rate of liquid stream
F_r	$m^3 \cdot s^{-1}$	flow rate of recycle liquid stream
$F(t)$	—	RTD-step function
$f(t)$	—	RTD-impulse function
f_{repr}	—	factor for catabolite repression
J	—	degree of segregation
K_S	$kg \cdot m^{-3}$	saturation constant (Monod kinetics)
k	s^{-1}	rate constant (general term)
k_d	s^{-1}	rate constant of microbial death
k_P	s^{-1}	rate constant of product formation
L	m	length
m, m_S	s^{-1}	maintenance coefficient (for substrate)
N	—	number of reactor stages or number of cells
N_{equ}	—	number of equivalent stages in a reactor cascade
P	kg	amount of product
p	$kg \cdot m^{-3}$	concentration of product
q_P	$kg_P \cdot kg_X^{-1} \cdot s^{-1}$	specific rate of production
r	$kg \cdot m^{-3} \cdot s^{-1}$	reaction rate (general term)
r_P	$kg \cdot m^{-3} \cdot s^{-1}$	rate of product formation
r_S	$kg \cdot m^{-3} \cdot s^{-1}$	rate of substrate consumption
r_X	$kg \cdot m^{-3} \cdot s^{-1}$	rate of biomass growth
r	—	recycling ratio
S	kg	amount of substrate
s	$kg \cdot m^{-3}$	substrate concentration
s_i	$kg \cdot m^{-3}$	arbitrary s-value
^2s	$kg \cdot m^{-3}$	s-value for non-wash-out steady-state
s	s	spread of RTD-function, Eq. (15.44)
t	s	time
t_M	s	maturation time
$\bar{t}$	s	mean residence time
t_r	s	time of reaction
t_c	s	time of circulation
V	m^3	volume
v, v_z	$m \cdot s^{-1}$	velocity, in direction z
X	kg	amount of biomass
x	$kg \cdot m^{-3}$	concentration of biomass
^2x	$kg \cdot m^{-3}$	x-value for non-wash-out steady-state
x_i	$kg \cdot m^{-3}$	arbitrary x-value
Y	—	yield coefficient
z	—	direction, coordinate

Greek letters

β	—	cell concentration factor in sedimentation unit
μ	s^{-1}	specific growth rate
σ^2	—	variance of RTD-function (2nd moment)
ζ, ζ_s	—	fractional conversion (of substrate)

Indices

ave	average
c, crit	critical
CSTR	continuous stirred tank reactor
CPFR	continuous plug flow reactor
DCSTR	discontinuous stirred tank reactor
equ	equivalent
ex	exit
i, j	number of components or reactions or arbitrary value
int	internal
M	maturation
max	maximal
N	reactor stage
obs	observed
opt	optimal
P	product
r	recycle
repr	repression
S	substrate
tot	total
Tr	transfer
X	biomass
0	initial value

Abbreviations

CSTR	continuous stirred tank reactor
CPFR	continuous plug flow reactor
DCSTR	discontinuous (batch) stirred tank reactor
NCSTR	cascade of CSTRs with N stages
RTD	residence time distribution

15.1 Introduction

Continuous culture theory and its mathematical foundation have been presented in the basic papers of MONOD [15.1], [15.2] and NOVICK and SZILARD [15.3]. The theoretical and practical aspects are reviewed in several articles and monographs and fundamental aspects on this topic (basic behavior, stability analysis, and transients) have already been presented in Volume 1 of this series (FIECHTER [15.4]). The present chapter concentrates on the problems of reactor performances and the estimation of parameters of kinetic models not only in the case of the continuous stirred tank reactor (CSTR), but also with multi-stage systems (cascades of CSTRs), continuous plug flow reactors (CPFR), and recycle reactors.

15.2 Single-stage Continuous Stirred Tank Reactor Operation

15.2.1 Estimation of Kinetic Parameters

15.2.1.1 Simple Kinetics

The behavior of a single-stage continuous stirred tank reactor (CSTR) with simple Monod kinetics has already been discussed by FIECHTER [15.4] using a plot of steady-state concentrations, $\bar{x}$ and $\bar{s}$, versus dilution rate D. Similarly, the performance can be illustrated by a plot of $\bar{x}$ and $\bar{s}$ versus mean residence time $\bar{t}$, which is mainly used by engineers. As shown in Fig. 15.1, cell mass concentration $\bar{x}$ diminishes with reduced $\bar{t}$ until wash-out occurs at $\bar{t}_{crit}$, which value corresponds to that of the critical dilution rate D_c (FIECHTER [15.4]). The highest rates of formation of $\bar{X}$ and consumption of $\bar{S}$ are indicated in the plot at the point of maximum slope of the curves ($\bar{x}_{opt}$ and $\bar{t}_{opt}$).

To evaluate the kinetic parameters μ_{max} and K_S from a set of data from a continuous stirred tank reactor, the performance equation can be rearranged to give a linearized form with $D = 1/\bar{t}$:

$$\frac{1}{s} = \frac{\mu_{max}}{K_S} \cdot \bar{t} - \frac{1}{K_S}. \qquad (15.1)$$

A graphical representation of $1/s$ vs. $\bar{t}$ is suitable for estimating μ_{max} and K_S. This plot corresponds to the double reciprocal plot shown in Fig. 12.7b when replacing μ by the reciprocal value of $\bar{t}$ in agreement with the steady-state performance of the continuous stirred tank reactor [$D = \mu$, cf. Eq. (11.8)].

A special method for the determination of μ_{max} uses wash-out conditions ($D \gg D_c$). The balance equation of biomass for $s \gg K_S$, after transformation, becomes:

$$\ln x = \ln x_0 + (\mu_{max} - D) t. \qquad (15.2)$$

Thus, μ_{max} can easily be calculated by performing regressions on the $\ln x$ versus t data (ESENER et al. [15.5], [15.6]). The performance equation of the single-stage CSTR shows that wash-out, optimum cell concentration, and maximum production rate depend on s_0 and K_S.

In the case of CSTR with $x_0 \neq 0$, the balance equation is (LEVENSPIEL [15.7]):

$$\mu_{max} \bar{t} = \frac{(x-x_0)[Y(s_0+K_S)-(x-x_0)]}{Y \cdot s_0 \cdot x - x(x-x_0)}. \qquad (15.3)$$

Since $x_0 \neq 0$, no wash-out occurs and there is no restriction on $\mu_{max} \cdot \bar{t}$; there are always cells within the vessel.

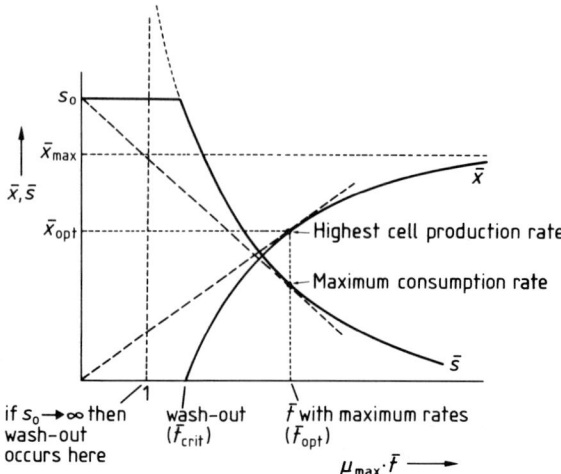

Figure 15.1. Graphical representation of concentration vs. mean residence time ($\bar{t}$) behavior in an ideally mixed, continuous stirred tank reactor for substrate-limiting, poison-free Monod kinetics with $x_0=0$. The concentration curves for biomass $\bar{x}$ and substrate $\bar{s}$ are analyzed, showing maximum and optimum values and wash-out.

15.2.1.2 Complex Kinetics with Inhibitions

The performance of CSTR-behavior is drastically changed by inhibition kinetics. The case of substrate inhibition considered by FIECHTER [15.4], exhibited more than one steady-state and more attraction domains.

The case of product inhibition and the consequences for the CSTR-performance has been investigated by LEVENSPIEL [15.8]. For $x_0=0$ the following equation is obtained:

$$\mu_{max} \cdot \bar{t} = (1-p/p_{crit})^{-n} \qquad (15.4)$$

for $\mu_{max} \cdot \bar{t} > 1$.

The properties of this performance equation can graphically be illustrated in a plot similar to that in Fig. 15.1.

To find the kinetic constants from CSTR experiments in this case one must first evaluate p_{crit} in a batch run using an excess of substrate and allow $t \to \infty$ (LEVENSPIEL [15.8]). Eq. 15.4 can be rearranged into:

$$\lg \bar{t} = -\lg \mu_{max} + n \lg[p_{crit}/(p_{crit}-p)], \quad (15.5)$$

and plotted as shown in Fig. 15.2. The slope and intercept of the best line through the data will give the kinetic parameters μ_{max} and n in the case of pure product inhibition without substrate dependence. In situations where both substrate availability and product inhibition effect the rate, the complete rate equations, Eqs. (14.47) and (14.48), must be used. The interacting effects of the four rate constants of these equations, μ_{max}, K_S, p_{crit}, and n are best found in a CSTR, running in series at different p using a cell-free feed. The performance equation [cf.

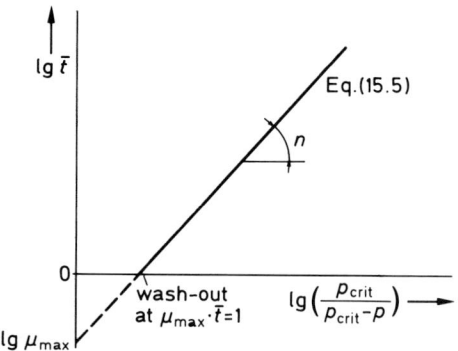

Figure 15.2. Graphical method for the evaluation of rate constants for pure product inhibition kinetics [see Eq. (14.43)] from experiments in an ideally mixed, continuous stirred tank reactor with $x_0=0$ according to Eq. (15.5) (LEVENSPIEL [15.8]).

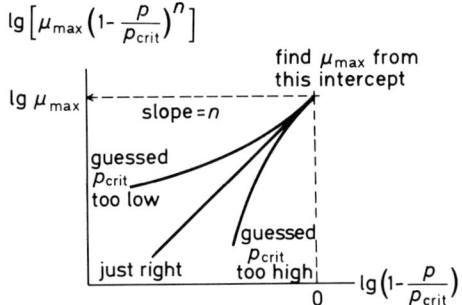

Figure 15.3. Graphical trial-and-error method for the evaluation of the toxic power number n in the generalized Monod equation [see Eq. (14.43)] according to LEVENSPIEL [15.8]. The values at the ordinate are taken from the intercept at the $\lg \bar{t}$-axis of Fig. 15.2.

Eq. (11.8)] combined with Eqs. (14.47) and (14.48), after rearrangement and with $x_0 = 0$, yields:

$$\bar{t} = 1/k_{obs} + (K_S/k_{obs})(1/s). \quad (15.6)$$

Plotting the data as in Fig. 12.12b a family of curves is obtained, each at different p. This graph gives K_S and $k(1-p/p_{crit})^n$. Then, knowing p_{crit} from previous runs, n and μ_{max} can be found using Fig. 15.3. As an alternative to the graphical evaluation, a numerical trial-and-error method can be used (LEVENSPIEL [15.8]). Due to product inhibition effects, oscillations in x and pH-value occur in CSTR also causing multiple steady-states (KNORRE [15.9]).

15.2.2 Deviations due to Maintenance, Lysis, and Incomplete Mixing

Modifications are required by the divergence between the model behavior and the experiments with CSTRs. Often one can observe that the chemostat behavior opposes a decrease in steady-state biomass with a decreasing dilution rate. This divergence can be explained by maintenance as well as by lysis effects. Including the specific lysis rate k_d and the maintenance coefficient m_S (cf. Sect. 14.3.1.4) the enlarged model is:

$$\frac{dx}{dt} = \mu(s) \cdot x - k_d x - D \cdot x, \quad (15.7)$$

$$\frac{ds}{dt} = -\frac{1}{Y_{X/S}} \cdot \mu(s) x - m_S \cdot x - D(s_0 - s), \quad (15.8)$$

which has the following non-wash-out steady-state (with $^2\bar{x}$ and $^2\bar{s}$):

$$(^2\bar{x}, {}^2\bar{s}) = \left(Y_{X/S} \frac{D}{D + k_d + m_S \cdot Y_{X/S}} \right.$$

$$\left. \cdot \left\{ s_0 - k_S \frac{D + k_d}{\mu_{max} - D - k_d} \right\}, K_S \frac{D + k_d}{\mu_{max} - D - k_d} \right). \quad (15.9)$$

The performance of a CSTR is illustrated in Figs. 15.4 and 15.5 for variations in m_S (Fig. 15.4) and k_d (Fig. 15.5) on the basis of Eqs. (15.7) and (15.8). Hence, lysis as well as maintenance causes the curve of biomass concentration against the dilution rate D to approach zero as the dilution rate tends towards zero.

Another consequence of incorporated lysis is that the wash-out point is decreased by the factor k_d:

$$D_c = \mu_{max} \frac{s_0}{s_0 + K_S} - k_d. \quad (15.10)$$

Whereas the specific lysis rate could be determined by analysis of batch experiments, this was impossible for the maintenance coefficient. This coefficient can be estimated by analysis of chemostat experiments. The theoretical basis for this is shown in Eq. (15.9). According to this relation the plot of $s_0 - \bar{s}/\bar{x}$ versus the reciprocal dilution rate $1/D$ gives a straight line:

$$\frac{1}{Y} = \frac{s_0 - {}^2\bar{s}}{{}^2\bar{x}} = \frac{1}{Y_{X/S}} + \frac{1}{D} \left(\frac{k_d}{Y_{X/S}} + m_S \right). \quad (15.11)$$

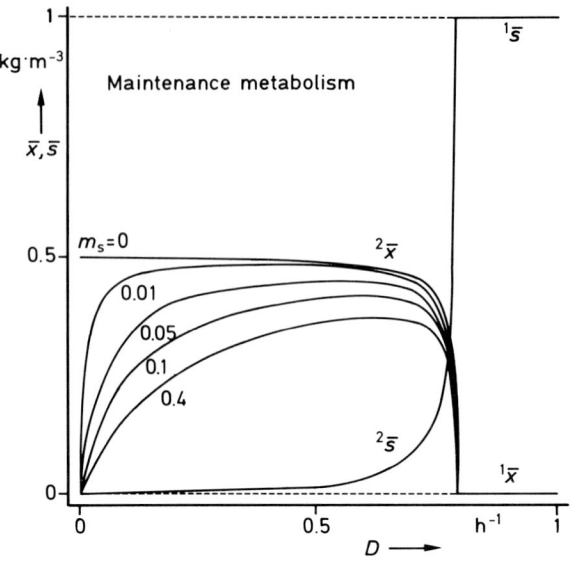

Figure 15.4. Computer simulation of the behavior of the concentrations of biomass $\bar{x}$ and substrate $\bar{s}$ under non-wash-out conditions ($^2\bar{x}$ and $^2\bar{s}$) as a function of the dilution rate D in an ideal mixed continuous stirred tank reactor with variation of the kinetic coefficient for maintenance metabolism m_S according to Eq. (15.8) (GUTKE [15.86]). Kinetic parameters: $\mu_{max} = 0.8$ h^{-1}, $K_S = 0.01$ g·L^{-1}, $Y_{X/S} = 0.5$, $s_0 = 1$ g·L^{-1}.

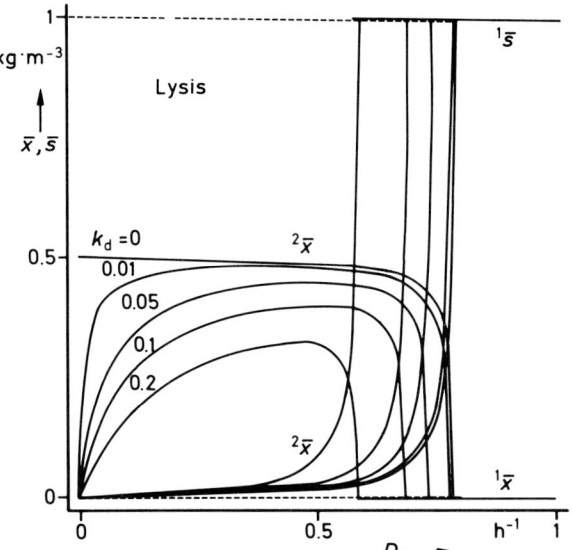

Figure 15.5. Computer simulation of the behavior of biomass $\bar{x}$ and substrate $\bar{s}$ concentration at non-wash-out conditions ($^2\bar{x}$, $^2\bar{s}$) as a function of the dilution rate D in an ideal mixed, continuous stirred tank reactor with variation of the specific lysis rate k_d according to Eqs. (15.7) and (15.9) (GUTKE [15.86]). Kinetic parameters as in Fig. 15.4 with $m_S = 0$.

We call the quotient $\bar{x}/(s_0 - \bar{s})$ the phenomenological yield coefficient Y; it equals the "true" yield coefficient, $Y_{X/S}$, if the lysis and maintenance rates fail. If for different dilution rates D the steady-state values x and s are measured with known s_0 and k_d, one can determine the "true" yield coefficient $Y_{X/S}$ as well as the maintenance coefficient m_S by linear regression on the basis of Eq. (15.11). For $k_d = 0$ the slope of the regression line is the maintenance coefficient m_S. From Eq. (15.11) we can see that the influence of k_d and m_S is quite similar. For this reason it is difficult to discriminate between both parameters. The possibility of discrimination depends on the accuracy of the estimation of the difference between wash-out point and $\mu_{max} s_0/(K_S + s_0)$, because it cannot be assumed that the specific lysis rate k_d, as determined from batch experiments, is maintained for $D \neq 0$.

SINCLAIR and TOPIWALA [15.10] modelled the behavior of CSTR considering a viability concept and also distinguished a rate of cell death k_d and endogenous metabolism of cells. On the basis of a model similar to Eqs. (15.7) and (15.8), the changes of viable, dead, and total cell mass with dilution rate approach the behavior shown in Figs. 15.4 and/or 15.5. The same authors also demonstrate the evaluation of kinetic parameters.

Recently some comments have been published stressing the complexity of maintenance by introducing the definition of "coefficient of apparently non-finalized substrate consumption" (GOMA et al., [15.11]). The difficulties in experimental verification of maintenance in the case of incorrect modelling by neglecting product formation have been stressed, e.g., during anaerobic growth (ESENER et al. [15.6]). Other interesting papers concerning maintenance have been published by KUHN et al. [15.12] and SOLOMON and ERICKSON [15.13].

Experimental data in CSTRs often deviate significantly from the behavior predicted from the normal performance due to wall growth (FIECHTER [15.4]) and imperfect mixing in the liquid phase.

Using radial flow impellers in CSTR, a model was proposed, which accounts for incomplete mixing with the aid of a two region mixing model, i.e., a maximally mixed zone near the impeller and another with total segregation (cf. Sect. 11.3.2) (SINCLAIR and BROWN [15.13a]). The authors explained the observed deviations. Similar results were obtained later by the same group using a multi-loop recirculation model previously proposed by VAN DE VUSSE [15.14]. Varying the flow rates of individual mixing modules in series, F_i, they achieved graphical plots shown in Fig. 15.6 (BROWN et al. [15.15]). On the other hand, the effect of micro-mixing on the wash-out and steady-state performance was extensively exam-

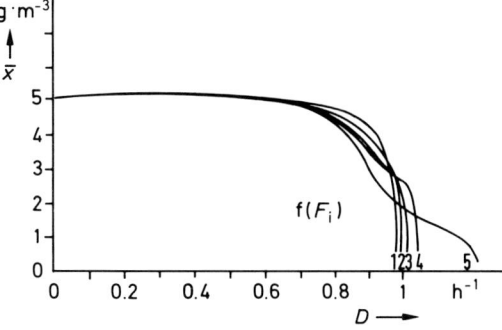

Figure 15.6. Steady-state concentration of biomass $\bar{x}$ vs. dilution rate D in a continuous stirred tank reactor with imperfect mixing. Different curves (1-5) show the effect of varying internal flow rate F_i (in $m^3 \cdot h^{-1}$) applying a multi-loop recirculation model with several mixing modules inside the reactor volume according to BROWN et al. [15.15]. $F_i = 4.05$ (1); $3.6 \cdot 10^{-3}$ (2); $2.16 \cdot 10^{-3}$ (3); $1.44 \cdot 10^{-3}$ (4); $7.2 \cdot 10^{-4}$ (5).

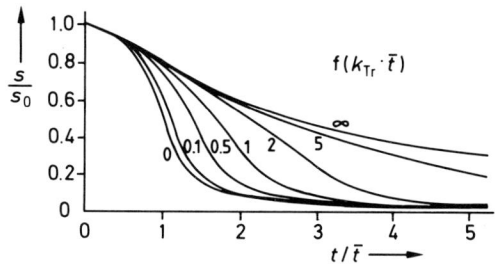

Figure 15.7. Dimensionless steady-state and exit substrate concentration $\bar{s}$ versus dimensionless mean residence time $\bar{t}$ (cf. Fig. 15.1) for a continuous stirred tank reactor with Monod kinetics in the case of $x_0 \geq 0$ using the reversed two-environmental model for quantifying incomplete mixing in the liquid phase. Different curves are drawn for varied mass transfer parameters in a dimensionless form ($k_{Tr} \cdot \bar{t}$) according to TSAI et al. [15.18].

ined by the group of FAN and coworkers (TSAI et al. [15.16], FAN et al. [15.17]; TSAI et al. [15.18]; WEN and FAN [15.19]) using a reversed two-environmental mixing model. In their studies a sequence of mixing modules (varying maximal mixing and total segregation) was used to simulate the behavior of CSTR as shown in Fig. 15.7 (TSAI et al. [15.18]).

The effect of mixing-process dynamics upon the performance of mechanically stirred bioreactors has been further studied by using a circulation-time-distribution model for fluid flow as well as a two-environmental model (BAJPAI and REUSS [15.20]). The influence of the circulation time t_c in a reactor on the bioprocess is illustrated in Fig. 15.8, showing that an increase in t_c results in a reduction of biomass yield in the aerobic cultivation of baker's yeast in a CSTR due to both a glucose- and O_2-effect.

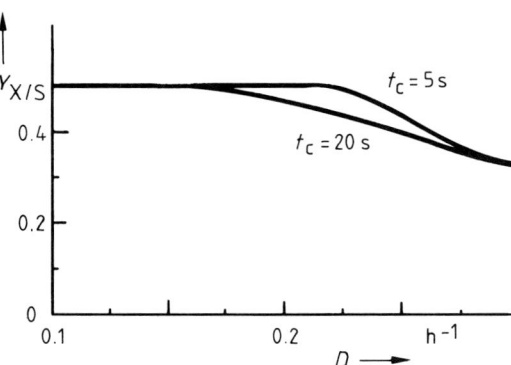

Figure 15.8. Simulation of biomass yield coefficient $Y_{X/S}$ as a function of the dilution rate D in a continuous stirred tank reactor operation of baker's yeast on glucose for different values of average circulation time t_c. In this case, the imperfect mixing of the liquid phase is quantified with the aid of the two-environmental model and using circulation time distributions of the stirred tank (BAJPAI and REUSS [15.20]).

In this sense a clear increase in D_c and s_{crit} in bioreactors with improved mixing characteristics was observed by KNOEPFEL [15.21]. Recently, performance studies of continuous culture systems have been extended by TODA and DUNN [15.22], showing that the maximum dilution rate could be elevated if a particular fermenter combination between CSTRs and CPFRs was used. Plots of x versus D were obtained, similarly as in Fig. 15.6. Imperfectly mixed bioreactor systems have recently been summarized (MOSER [15.23]).

15.2.3 Product Formation Kinetics in a Single-stage Continuous Stirred Tank Reactor

Continuous cultivation was shown to be an excellent method for research on the physiological states of growing microorganisms and especially on the regulation of metabolism. The physiological state may be defined by the quality of supplied substrates, by the dilution rate of the concentration of the limiting substrate, and – if there are multiple steady-states – by the history of the microorganisms. In discontinuous culture the different physiological states pass very quickly, whereas in a chemostat almost every possible physiological state can be maintained for any period of time. This is the advantage of continuous cultivation, especially for experimental practice in studying mathematical models of product formation kinetics.

15.2.3.1 Production Kinetics Without Product Inhibition

The product formation rate in a chemostat culture was given by PIRT [15.24] as:

$$\frac{dp}{dt} = q_P \cdot x - D \cdot p. \tag{15.12}$$

The change of product concentration in the culture over time depends on the term of synthesis ($q_p \cdot x$) and on the term of outflow of product ($-D \cdot x$).

If the product is strictly "growth-linked" [cf. Eq. (14.105)] the product concentration in the stationary state $\bar{p}$ and the output rate $D \cdot \bar{p}$ will vary with D in the same manner as the biomass concentration (cf. Fig. 15.1 according to FIECHTER [15.4]). If q_P is independent of the growth rate [cf. Eq. (14.108)], the product concentration varies inversely with dilution rate D. Then, over a wide

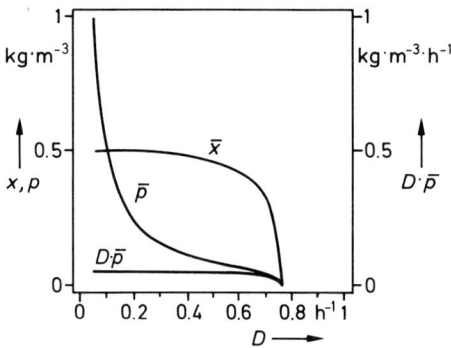

Figure 15.9. Computer simulation of the steady-state behavior of an ideally mixed, continuous stirred tank reactor with poison-free production kinetics with q_p = const. The concentrations of biomass $\bar{x}$ and product $\bar{p}$ are shown together with the productivity $D \cdot \bar{p}$ as a function of the dilution rate. Kinetic parameters are the same as given in Fig. 15.4 (HILLINGER [15.87]).

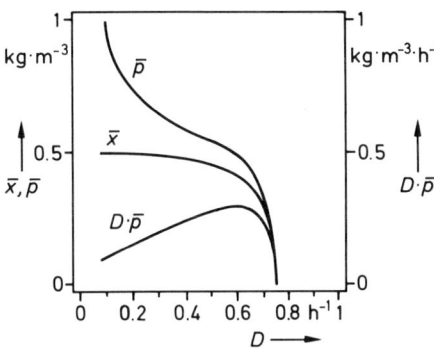

Figure 15.10. Computer simulation of the steady-state behavior of an ideally mixed, continuous stirred tank reactor with poison-free production kinetics in which q_p is partly growth-linked according to Eq. (14.95). The concentrations of biomass $\bar{x}$ and product $\bar{p}$ are shown together with productivity $D \cdot \bar{p}$ as a function of the dilution rate D with the same parameters as used in Fig. 15.9 with $k_P = 0.1\ h^{-1}$ (HILLINGER [15.87]).

range of dilution rates the output rate $D \cdot \bar{p}$ is constant, as shown in Fig. 15.9.

However, as D tends to zero eventually the assumption that k_P = const becomes invalid, either because enzyme activity decreases, because some required substrate is exhausted, because of the low flow rate, or because of regulation processes in the cell. If product formation is partly "growth-linked" and partly independent of the growth rate [cf. Eq. (14.110)], then the product concentration and the output $D \cdot \bar{p}$ in the steady-state will vary with the dilution rate D as shown in Fig. 15.10.

The product output rate, if D tends toward zero, is the "non-growth-linked" contribution k_P, multiplied by biomass concentration $\bar{x}$, expressed as:

$$D \cdot \bar{p} = (Y_{X/S} \cdot s_0) \cdot k_P. \qquad (15.13)$$

A modified kinetic approach to microbial product formation assumes that r_P is not only proportional to the actual biomass concentration but also depends on a second carbon source S_2 as HEGEWALD et al. [15.25] have pointed out:

$$r_P = q_{P,\,max} \frac{s_2}{s_2 + K_{S2}} \cdot f_{repr} - D \cdot p. \qquad (15.14)$$

This equation has been extended to supplementary O_2-limitation by BAJPAI and REUSS [15.26]. The factor f_{repr} in Eq. (15.14) represents the effect of easily metabolized sugars on the reduction of antibiotic synthesis due to repression or inhibition.

Special results on product formation in CSTRs are given by FENCL and NOVAK [15.27], RICICA [15.28], and PIRT and RIGHELATO [15.29].

15.2.3.2 Production Kinetics With Product Inhibition

Identification plots of product inhibition kinetics from CSTR-data have already been presented by FIECHTER [15.4]. The difference between batch and continuous culture with regard to product inhibition is that under conditions of continuous culturing the product is diluted, while in batch runs the product is accumulated. Therefore, in batch cultures, reaction rates generally slow down successively. In chemostat cultures, however, oscillations of x and p appear due to a periodic effect of pH-control and/or permanent inflow and outflow of fresh medium.

For a mathematical description we have to take into account the dependence of the specific growth rate μ on the substrate concentration s and on the product concentration p as shown in Fig. 14.16. Hence, in a first approximation the resulting specific growth rate $\mu = \mu(s, p)$ follows by multiplication of the Monod function with the inhibition term $f(p)$, already given in Eq. (14.44). As a further condition we take q_P as a constant assuming it to be independent of the specific growth rate. Furthermore, the influence of the pH on product formation is assumed to be negligible. On this basis the following differential equations may be derived:

$$dx/dt = \mu(s, p) \cdot x - D \cdot x, \quad (15.15)$$

$$ds/dt = D(s_0 - s) - \mu(s, p)x \cdot \frac{1}{Y_{X/S}}, \quad (15.16)$$

$$dp/dt = q_P \cdot x - D \cdot p. \quad (15.17)$$

Extensive studies on dynamic phenomena induced by product oscillations have been described by KNORRE [15.9].

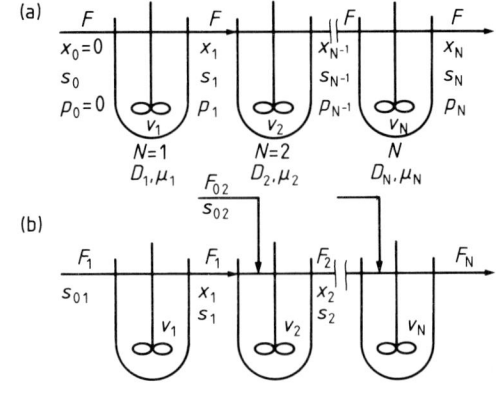

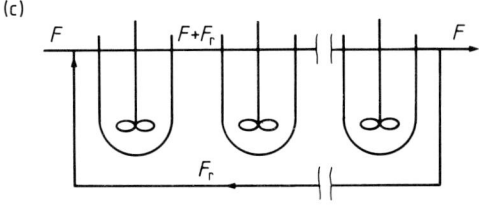

Figure 15.11. Scheme of different types of multi-stage continuous stirred tank reactor operation: single-stream multi-stage system (a), multi-stream multi-stage system (b), and multi-stage with recycling (c).

15.3 Multi-stage Continuous Culture Operation

15.3.1 Classification

Three main types of multi-stage systems may be distinguished as illustrated in Fig. 15.11 according to HERBERT [15.30]:

- single-stream multi-stage,
- multi-stream multi-stage, and
- multi-stage systems with recycling.

Single-stream systems are cascades with a single medium inflow, which is constant through all other stages. The characteristic features are:

- later stages cannot effect earlier stages,
- the first stage behaves like a CSTR,
- the dilution rate D cannot be changed in one stage without changing it in all others, as the flow is constant,
- the dilution rate in individual stages depends only on the reactor volume:

$$F = D_1 \cdot V_1 = D_2 \cdot V_2 = D_3 \cdot V_3. \quad (15.18)$$

Multi-stream systems have multiple inputs of feed and therefore are characterized as follows:

- earlier stages are independent of later stages and the first behaves like a CSTR,
- different medium feeds may be varied independently, and individual dilution rates are independent process variables.

Multi-stage systems with recycling are discussed in the next section on recycling operation.

This classification is to some extent arbitrary; it can be used to show the basic behavior. A complete mathematical analysis of every practicable type of system will assist the bioengineer in choosing the system best suitable for a particular bioprocess.

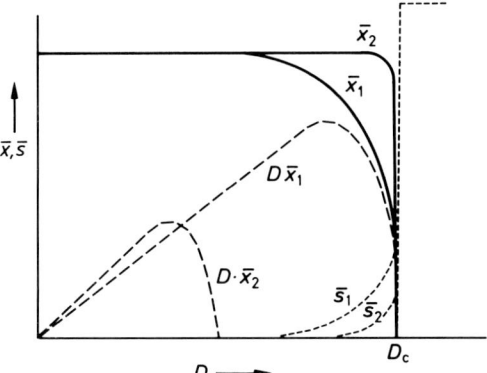

Figure 15.12. Two-stage continuous stirred tank reactor operation: Behavior of steady-state concentrations of biomass $\bar{x}$ and substrate $\bar{s}$ as a function of dilution rate D in the first stage (x_1, s_1, and Dx_1) and in the second stage (x_2, s_2, and Dx_2) according to HERBERT [15.30].

15.3.2 Single-stream Multi-stage Operation

The system to be discussed is a two-stage reactor investigated by HERBERT [15.30], MALEK and FENCL [15.31], and RICICA and NECINOVA [15.32].

The second stage is described by a more complicated balance equation. In steady-state the following condition applies:

$$x_2 = Y(s_0 - s_2), \qquad (15.19)$$

from which x_2 can be calculated when s_2 is known. This unknown concentration s_2 can be found by solving the quadratic equation:

$$(\mu_{max} - D_2)s_2^2 - \left(\mu_{max} \cdot s_0 - \frac{K_S \cdot D_1 \cdot D_2}{\mu_{max} - D_1} + K_S D_2\right)s_2 + \frac{K_S^2 \cdot D_1 \cdot D_2}{\mu_{max} - D_1} = 0, \qquad (15.20)$$

in which s_2 is expressed as a function of s_0, D_1, D_2, μ_{max}, and K_S. This quadratic equation has two positive roots and gives two values for s_2, one smaller and one greater than s_1; the first is the correct value, while the second corresponds to the biologically imaginary but mathematically real case of "reverse" growth (with conversion of cells to substrate).

Thus, from Eqs. (15.19) and (15.20) the steady-state concentrations of cells and substrate in the second stage can be obtained. The solution is graphically shown in Fig. 15.12.

Comparing the curves for $\bar{x}_1$ and $\bar{x}_2$, it will be seen that: (1) both become zero at the same critical dilution rate D_c, as stated earlier for fermenters of equal volumes, and that (2) for all dilution rates lower than D_c, x_2 is greater than x_1 (and similarly s_2 is less than s_1) because unused substrate escaping from the first fermenter is utilized for growth in the second stage.

In the second stage, utilization of substrate is almost complete even at dilution rates quite close to D_c, in other words, the used medium from the second stage is practically exhausted and little further growth would be expected. However, this can only happen under certain circumstances, as discussed by HERBERT [15.30].

This mathematical analysis can be easily extended to three or more stages; the equations for each fermenter are derived from those of the preceding one. For the N-th reactor in a cascade, the material balance equations for x and s, using symbols shown in Fig. 15.11a, are, therefore, given by:

$$V_N \frac{dx_N}{dt} = F \cdot x_{N-1} - F \cdot x_N + (\mu_N \cdot x_N) V_N, \quad (15.21)$$

$$V_N \frac{ds_N}{dt} = F \cdot s_{N-1} - F \cdot s_N - \left(\frac{\mu_N}{Y_{X/S}} x_N\right) V_N, \quad (15.22)$$

$$V_N \frac{dp_N}{dt} = F \cdot p_{N-1} - F \cdot p_N + \left(\frac{\mu_N}{Y_{P/X}} \cdot x_N\right) V_N. \quad (15.23)$$

Thus, at steady-state conditions:

$$x_N = \frac{D \cdot x_{N-1}}{D - \mu_N}, \quad (n \neq 1) \quad (15.24)$$

$$s_N = s_{N-1} - \frac{1}{D} \frac{Y_{P/X}}{Y_{X/S}} \cdot \mu_N \cdot x_N, \quad (15.25)$$

$$p_N = \frac{1}{D} (D \cdot p_{N-1} + Y_{P/X} \cdot \mu_N \cdot x_N). \quad (15.26)$$

The number of stages N and the parameters of interest for a multi-stage system can be determined by a graphical method described by, e.g., DEINDOERFER and HUMPHREY [15.33] and LUEDEKING and PIRET [15.34] in analogy to chemical process design. This procedure has been applied by FIECHTER [15.4].

In summary:

- If all fermenters contain the same volume of culture, they will all wash out at the same dilution rate D_c, which is equal to the critical dilution rate for a single stage fermenter.
- When the fermenters are of unequal volume, the whole system will wash out unless there is at least one fermenter whose dilution rate is less than D_c, though in later stages the dilution rate may be greater than D_c without these fermenters washing out.
- The critical dilution rate for the system considered as a whole is considerably smaller than that for a single fermenter.

When considering the entire system it is useful to calculate its average rate, D_{ave}, defined as the total flow F through the system divided by the total volume of culture $N \cdot V$ in all fermenters. For a single-stream chain of N fermenters this is:

$$D_{ave} = \frac{F}{N \cdot V}. \quad (15.27)$$

The wash-out rate for the whole system, i.e., the critical value of D_{ave} for any subsequent fermenter is therefore:

$$(D_{ave})_c = \frac{F_c}{N \cdot V} = \frac{D_c}{N}. \quad (15.28)$$

In fact if calculations for a third and fourth stage are made, the resulting curves are scarcely distinguishable from curve $\bar{x}_2$ in Fig. 15.12. This suggests that there will seldom be much practical advantage in using more than two stages, as far as production of cells is concerned. On the other hand, further stages might be important in obtaining cells of a desired quality. Though little growth will occur in later stages, endogenous metabolism will continue, leading to changes in the chemical composition and physiological state of the cells.

15.3.3 Multi-stream Multi-stage Operation

Compared to single-stream systems there are many more possibilities offered by multi-stream systems. These systems have multiple medium inflows which are independently variable, and which may, if desired, contain a variety of different nutrients. This can produce very complex experimental situations. We shall consider the simple case in which a single growth-limiting nutrient is

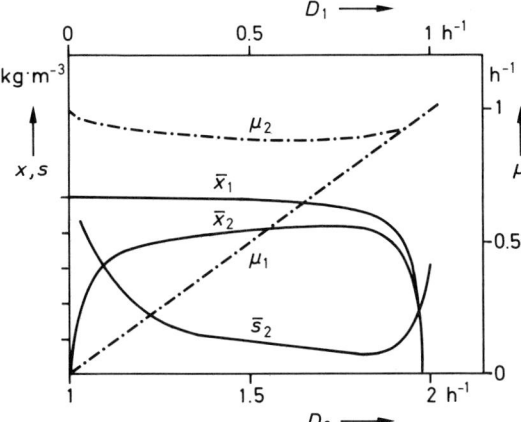

Figure 15.13. Steady-state relationship in a two-stage multi-stream continuous stirred tank reactor system where the medium inflow to the second fermenter is assumed to be held constant while the inflow to the first fermenter is varied. The curves are calculated for a constant second-stage medium inflow of $D_{02}=1.0$, and D_1 values from zero to D_c (0.98). The concentrations of biomass and substrate and the specific growth rate μ are shown for the first and second stage as a function of the dilution rates D_1 and D_2 (HERBERT [15.30]).

used in a two-stage system. Systems with different substrates fed to later systems have been described by HERBERT [15.30].

The simplest system with one substrate fed to all stages is shown in Fig. 15.11b. It is convenient to consider D_2 as the sum of two "partial dilution rates" D_{02} and D_{12}, the two inflows into fermenter 2, defined as follows:

$$D_{02}=\frac{F_{02}}{V_2}, \quad D_{12}=\frac{F_1}{V_2}, \quad D_{02}+D_{12}=D_2.$$
(15.29)

The steady-state mass balance equations for x and s result in:

$$x_2 = Y\left(\frac{D_{12}\cdot s_{01}}{D_2}+\frac{D_{02}\cdot s_{02}}{D_2}-s_2\right),$$
(15.30)

from which x_2 can be calculated when s_2 is known. To find s_2 we form Eq. (15.31) after eliminating x_2 and rearranging:

$$(\mu_{max}-D_2)s_2^2-$$
$$-\left(\frac{\mu_{max}+D_{12}\cdot s_{01}}{D_2}+\frac{(\mu_{max}-D_2)D_{02}\cdot s_{02}}{D_2}\right)-$$
$$-(D_{12}\cdot s_1+K_S\cdot D_2)s_2+K_S\cdot D_{02}\cdot s_{02}+$$
$$+K_S\cdot D_{12}\cdot s_1=0.$$
(15.31)

From this quadratic equation s_2 can be computed by inserting the value of s_1 and the steady-state behavior for any value of D_1 and D_2 can be derived; the only precondition concerns the values of kinetic parameters. The behavior of a two-stream two-stage system is illustrated in Fig. 15.13. The curves illustrate very clearly the stability of the system for operation at high second-stage growth rates, and the relative insensitivity of the second stage to changes in the first. Alternations in the first-stage dilution rate over nearly the whole of its working range ($D_1=0.15$ to 0.9 h^{-1}) produce variations of less than 10% in the second-stage cell concentration and growth rate.

The great flexibility of multi-stream systems, combined with their stability and the case of operation at high flow rates, make them superior to single-stream systems in nearly all respects. This applies also to the question of output.

15.3.4 Applications of Multi-stage Operation

Even though multi-stage systems have not been widely used in industry and research due to rather complicated handling, the potential advantages on theoretical grounds are considerable (RICICA [15.28]).

In general, multi-stage systems provide a different environment changing from stage to stage, thereby approximating the flow behavior of tubular reactors (cf. Fig. 11.8). The potentialities of multi-stage systems can be summarized as follows:

- Maximum conversion is achieved by complete utilization of the substrate

in the later stages by maximum productivity in the first stage. This is essential in the case of expensive substrates, e.g., steroid transformation or in the case of environmental problems, as for instance, waste water treatment. Predominantly, reaction orders greater than zero will give better results.
- All physiological states of microbial cultures can be maintained or run through with "matured cells" in order to obtain desirable products: e.g., secondary metabolites, intermediates, induced enzymes, spores.
- Long residence times can be maintained without reducing the dilution rates, i.e., without diminishing productivity. This fact is advantageous for bioprocesses in the stationary growth phase with complex kinetics (inhibition and/or repression) and complex media (mixture of carbon sources exhibiting polyauxia, e.g., sulfite liquors, waste water).
- High-quality products can be obtained (e.g., baker's yeast).
- Optimal operating conditions can be realized directly in gradient reactors (e.g., T-profile, changes in pH-value and p_{O_2}).

Several applications of multi-stage operation have been reported in the literature: FENCL [15.35]; GOTO et al. [15.36]; PROKOP et al. [15.37]; ERICKSON et al. [15.13]; KITAI et al. [15.38]; PACA and GREGR [15.39]; SCHÜGERL [15.40]; RICICA [15.28]; FENCL et al. [15.41]–[15.43]; TYAGI and GHOSE [15.44]; MORENO and GOMA [15.45]; IMANAKA et al. [15.46]; RYU and LEE [15.47]; FURUSAKI and MIYAUCHI [15.48]; BAHL et al. [15.49]; RYU et al. [15.50]. Generally, the advantages of multi-stage systems in terms of reaction engineering can be derived by referring to such graphical plots as those shown in Chapter 11 (see Fig. 11.14). From a practical point of view a reactor cascade facilitates automatic control of pH, temperature, p_{O_2}, supply of precursors, etc., so that their levels are optimal in the corresponding stage. The industrial application of multi-stage continuous operation, however, will always depend on a thorough economic analysis (e.g., SWARTZ [15.51]). Only the introduction of a bioprocess as a continuous process from the very beginning appears promising.

15.4 Continuous Bioprocessing with Plug Flow Reactors

The mathematical theory of continuous cultivation discussed so far concerned the well-mixed stirred tank reactor. The following section is devoted to some mathematical relations for plug flow reactors, which exhibit a flow behavior with gradients. This type of reactors is called "gradient reactor" or "reactor with distributed parameters" or "integral reactor" (MOSER [15.23], [15.53]).

15.4.1 Performance Equations and Practical Properties

An ideal case of a CPFR is represented by a narrow empty tube through which the unstirred liquid flows uniformly. The profile of concentration has been illustrated in Fig. 11.8. Since composition of the fluid varies from one position to the next along the longitudinal axis z, the material balance must be made on a differential element of fluid $dV = A \cdot dz$; where A is the cross-section area of the tube.

The steady-state mass balances for x and s in a CPFR with a recycling stream of fluid for inoculum ($r = F_r/F$) are as follows:

$$F(1+r)dx + r_X \cdot A \cdot dz = 0, \qquad (15.32)$$

$$F(1+r)ds - r_S \cdot A \cdot dz = 0. \qquad (15.33)$$

The boundary conditions for these equations are obtained by material balances around the entrance where fresh feed mixes with recycled liquids. At $z=0$, the following boundary conditions apply:

$$x_i = \frac{x_0 + r \cdot x_r}{1+r}, \quad (15.34)$$

$$s_i = \frac{s_0 + r \cdot s_r}{1+r}. \quad (15.35)$$

Eqs. (15.32) to (15.35) form a set of non-linear differential equations that are difficult to solve analytically. The equations are more readily solved by computer techniques. In certain cases, however, the equations can directly be integrated, e.g., when the amount of biomass formed by the reaction is small relative to the amount entering. Then the value of x is nearly constant over the reactor length (x_{ave}). Thus, integration of Eqs. (15.32) to (15.35) is possible, yielding the following expression with Monod kinetics:

$$(s_i - s) + K_S \frac{\ln s_i}{s} = \frac{\mu_{max} \cdot x_{ave} \cdot A \cdot z}{Y \cdot F(1+r)}. \quad (15.36)$$

The small change of x along the reactor is approximated by $(x - x_i) = Y(s_i - s)$, and substituting this expression for $(s_i - s)$ in Eq. (15.36) gives:

$$(x - x_i) = \frac{\mu_{max} \cdot x_{ave} \cdot A \cdot z}{F(1+r)} - K_S \cdot Y \cdot \ln \frac{s_i}{s}. \quad (15.37)$$

In analogy to chemical reaction engineering, POWELL and LOWE [15.54] carried out a mathematical analysis of a chain of CSTRs by computing the biomass concentration obtained in the final stage of a cascade of five chemostats with recycling in comparison with a tubular reactor. It was found that at a low rate of recycling of feed and a high recycling rate of cells more complete utilization of growth-limiting substrate can be obtained over most of the dilution rate range in the series of CSTRs than is possible with a single CSTR. This advantage increases with decreasing initial substrate concentration. GRIEVES et al. [15.55] modelled a recycling piston flow reactor and compared its performance with that of a recycling CSTR. If the objective of a process is cell mass production, the differences are slight; but if the objective is reduction in effluent substrate concentration, then a CPFR is the optimal choice. With a large recycling factor and with efficient operation of the separator, resulting in a large concentration factor, β, the differences in the critical residence time between CPFR and CSTR become much less pronounced, and the CPFR is able to provide a greater maximum production rate of microorganisms while yielding a considerably lower effluent substrate concentration than a CSTR. Fig. 15.14 illustrates this behavior by showing the performance of a CPFR in comparison with a CSTR on the basis of a concentration vs. dilution rate plot.

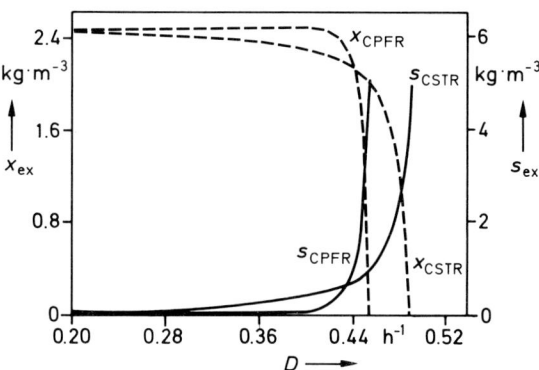

Figure 15.14. Steady-state relationship between exit biomass (x_{ex}) and substrate concentration (s_{ex}) and dilution rate D for an ideal continuous plug flow reactor (CPFR) compared to a continuous stirred tank reactor (CSTR) under the same recycling conditions ($r=0.4$ and $\beta=3$, see Fig. 15.16) according to GRIEVES et al. [15.55].

The basic difference between well-mixed flow reactors and gradient reactors such as CPFR is demonstrated in Fig. 15.15. Quantification of the flow behavior of these reactor types is shown in this graph on the basis of the residence time distribution (LEVENSPIEL [15.56]).

In continuous operation of CPFRs, the mean residence time $\bar{t}$ must be chosen according to the kinetics. Using the concept of maturation time t_M, it becomes clear that:

$$\bar{t} = t_M. \quad (15.38)$$

As all the cells leaving the system are synchronized, they will be used for dilution rates less than $1/t_M$ and the steady-state product concentration in CPFR is:

$$\bar{p} = Y_{P/X} \cdot \bar{x}. \quad (15.39)$$

In CSTRs, however, where according to the f(t) chemostat function a distribution of cell age exists in the culture, the following condition applies:

$$x(t) = \bar{x} = f(t) = D \cdot e^{-D \cdot t}. \quad (15.40)$$

Steady-state product concentration in CSTR, therefore, becomes:

$$\bar{p} = Y_{P/X} \cdot \int_{t_M}^{\infty} x(t) \cdot dt = Y_{P/X} \cdot \bar{x} \cdot e^{-D \cdot t_M}. \quad (15.41)$$

Comparing Eq. (15.41) with Eq. (15.39) shows that only at a low dilution rate the value of p in CSTRs will approximate the value obtained in CPFRs. As illustrated in Fig. 15.15, only cells with $\bar{t} > t_M$ will contribute to production.

15.4.2 Real Reactor Behavior – Characterization with Residence Time Distribution

In real reactors deviations from ideal reactor behavior occur. For CPFRs, a distribution function is normally used for quantification of the deviations from the mean value of residence time $\bar{t}$ (cf. Fig. 15.15).
Applying the law of conservation of mass without the reaction term, a dimensionless equation can be formed that includes

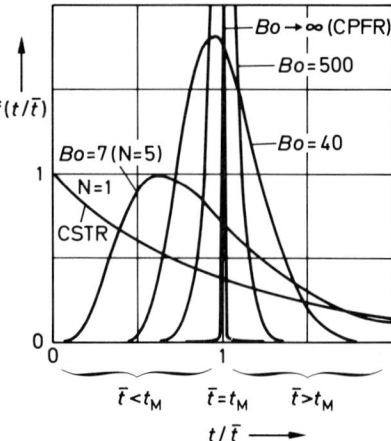

Figure 15.15. Graphical plot of the normalized residence time distribution functions following the pulse method $f(t/\bar{t})$ vs. dimensionless time $(t/\bar{t})$ for various extents of backmixing as predicted by the dispersion model using the Bodenstein number Bo as a parameter or by the tank-in-series model using the number of equivalent stages N as a parameter. Both extreme cases of continuous reactor operation, i.e., CSTR and CPFR operation are also shown.

the Bodenstein number Bo_L (LEVENSPIEL [15.56]):

$$Bo_L \equiv \frac{v_z \cdot L}{D_L}. \quad (15.42)$$

The characteristic value of the Bodenstein number (Bo) shown in Fig. 15.15 as a measure of the spread of the residence time distribution (RTD) function describes the longitudinal dispersion (D_L) in the reactor. The solution for this case is given by LEVENSPIEL [15.56]:

$$f(t/\bar{t}) = \sqrt{\frac{Bo_L}{4\pi t/\bar{t}}} \cdot \exp\left[-(1-t/\bar{t})^2 \frac{Bo_L}{4 t/\bar{t}}\right]. \quad (15.43)$$

Bo_L can be evaluated directly from the distribution function obtained by the pulse method:

$$\frac{1}{Bo_L} = \frac{1}{8}\sqrt{8 s^2/\bar{t}^2 + 1} - 1, \quad (15.44)$$

with $s^2/\bar{t}^2$ being the variance of the distribution function (σ^2 the 2nd moment, $\bar{t}$ the 1st moment), which can be estimated from experiments.

Another method of experimental verification of residence time distribution is the "step method". The resulting function $F(t)$ is related to the pulse-function $f(t)$ as follows:

$$F(t) = \int_0^t f(t) \cdot dt. \quad (15.45)$$

A second model approach to quantify RTD is the so-called tank-in-series model, which operates with the equivalent number of CSTRs, N_{equ}, in a cascade.

From the balance equation for a cascade of CSTRs [cf. Eq. (15.21) and Eq. (15.22)], the solution for RTD according to LEVENSPIEL [15.56] is:

$$f(t/\bar{t}) = \frac{N_{equ}^{N_{equ}} (t/\bar{t})^{N_{equ}-1}}{(N_{equ}-1)!} \cdot e^{-N_{equ} \cdot t/\bar{t}}. \quad (15.46)$$

With $N=1$ this equation reduces to Eq. (15.40). The model parameter N_{equ} can be estimated directly from an experimentally determined RTD-function by using the spread of distribution:

$$N_{equ} = \bar{t}^2/s^2. \quad (15.47)$$

Normally, the model with N_{equ} is used for CSTR behavior, while the dispersion model with parameter Bo is only valid for CPFR-behavior. It can be assumed that the behavior of an ideal CPFR is achieved when $Bo \to \infty$, while for real cases $N \geq 5$ and $Bo \geq 7$.

CSTR-behavior is approached only when $N \to 1$ or $Bo \to 0$. The following relationship exists between both parameters in the range of $Bo > 10$:

$$N = 1 + \tfrac{1}{2}\sqrt{Bo^2 + 1}. \quad (15.48)$$

Other models, which are valid in more complex cases of RTD containing several mixing modules, are referred to in the original literature (e.g., LEVENSPIEL [15.56]; FROMENT and BISCHOFF [15.57]).

15.4.3 Application of Continuous Plug Flow Reactor

While the CSTR-cascade is often used as a substitute for CPFR-behavior, tubular reactors are rarely found in bioprocessing. Exceptions are the waste water treatment in oxidation ponds, river analysis (METCALF & EDDY ING. [15.58], and sterilization technology, where formal kinetics follow a first-order reaction, so that conversion in CPFRs is superior to CSTRs (AIBA et al. [15.59]). In this case, the conversion can be calculated using the following equation:

$$\frac{N(L)}{N_0} = \exp\left(-Da_1 + \frac{Da_1^2}{Bo}\right), \quad (15.49)$$

with $Da_1 \equiv k_d \cdot L/v_z$ as first-degree Damköhler number and $N(L)$ the number of molecules changing with length.

Another type of application is the use of fixed bed reactors with immobilized enzymes for bioconversions, which show the potential advantages of CPFRs (LILLY and DUNNILL [15.60]).

Tubular reactors have been described by GREENSHIELDS and SMITH [15.61] and MOSER [15.23]. Their application varies from fermentation processing with microbial flocs (simple tubes or pipes, pipe-line reactors with or without degasifiers or mixing devices; see e.g., MOSER [15.62]; RUSSEL et al. [15.63]; ZIEGLER et al. [15.64]) to unconventional bioprocessing with immobilized cells, solid-substrate processing, shear sensitive tissue cell cultivation, and phototrophic organisms.

Tubular reactors offer some potential advantages over conventional CSTRs (RUSSEL et al. [15.63]; MOSER [15.23], [15.52]):

1. High productivity and optimum conversion is realized at the same time.
2. Mixing within tubular devices is more uniform by eliminating dead spaces and resulting in more secure scale-up.
3. The surface area-to-volume ratio is significantly higher resulting in facilitated transfer processes. As a consequence, mass

transfer is achieved with a comparably lower power consumption in the case of horizontal devices (MOSER [15.65]) and heat transfer is easily accomplished. This fact becomes crucial in extreme situations of bioprocessing, e.g., with solid substrates, photoreactions, and shear sensitive tissues.

4. Bioprocessing occurs with gradients of concentrations and/or temperature over the length of the tubes, so that technical manipulations for biological demands can be carried out (e.g., T-programming as a function of axial distance; substrate-dosing, CO_2-removal, etc.) in the case of inhibition and/or repression kinetics. This property is of advantage also for fundamental investigations of biological processes.

5. Horizontal configurations show that plug flow is not disturbed by CO_2 development and that hydrostatic pressure cannot become inhibiting.

6. Practical advantages exist due to the closed system of tubes (no aerosol formation in the case of waste water treatment; easy operation under pressure and/or with pure O_2). The technical arrangement is easy to handle as the basic elements are highly developed in practice (pipes, pumps, standard fittings).

15.5 Recycling Operation in Bioprocessing

As a means of resource conservation and recovery as well as for environmental problems recycling systems have become increasingly important (HAMER [15.66]). In some cases the exit stream from a fermenter can be conveniently treated (by centrifuging, filtering, or sedimentation) so that the cells are concentrated in one stream which is then returned to the fermenter. This can give improved performance, i.e., higher treatment rates, smaller vessel size, and higher conversions.

15.5.1 Performance Equations and Basic Behavior

15.5.1.1 The Continuous Stirred Tank Reactor with Recycling

In all considerations of continuous operation, recycling loops have often been included (HERBERT [15.30]). Recycling of cells from a CSTR effluent provides a means of continuous inoculation of the vessel and adds stability to the reactor.

A schematic presentation of such a recycling system is shown in Fig. 15.16a. The nomenclature is the same as previously used except for two additional parameters: the recycling ratio $r \equiv F_r/F$ and the cell concentration factor $\beta \equiv x_r/x$.

From the mass balances it follows that:

$$\mu = (1 + r - r \cdot \beta) D . \qquad (15.50)$$

The solution of the balance equations is plotted in Fig. 15.16b together with the output of cells $D \cdot x_{ex}$; for comparison, curves for cell mass concentration and output of cells in a single CSTR without recycling are indicated by dotted lines. This proves that the steady-state value of $\bar{x}$ with recycling will be greater than in a CSTR without recycling:

$$\frac{\bar{x}_{\text{recycle CSTR}}}{\bar{x}_{\text{CSTR}}} = (1 + r - r \cdot \beta) . \qquad (15.51)$$

Therefore, a dilution rate greater than the maximum growth rate may be employed, thus increasing overall productivity. The critical residence time is:

$$\bar{t}_{\text{crit}} \equiv \frac{1}{D_{\text{crit}}} = \frac{1 + r - r \cdot \beta}{\mu_{\text{max}}} . \qquad (15.52)$$

An estimate of the effect of cell mass concentration and recycling process economics may be obtained by plotting effluent substrate concentrations for several different values of x_r as shown in Fig. 15.17. Increasing x_r has the effect of minimizing s_{ex} and

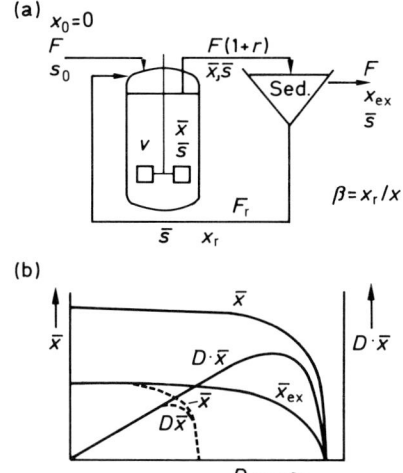

Figure 15.16. The continuous stirred tank reactor with recycling of cell mass: (a) Scheme of the reactor with the sedimentation unit (Sed.), where a recycled liquid stream F_r is applied containing concentrated cell mass x_r. (b) Steady-state concentrations of $\bar{x}$, $\bar{x}_{ex}$, and productivity $D\cdot\bar{x}$ vs. dilution rate D of recycle operation (solid lines) compared to the behavior of a single-stage CSTR (dotted lines) according to HERBERT [15.30].

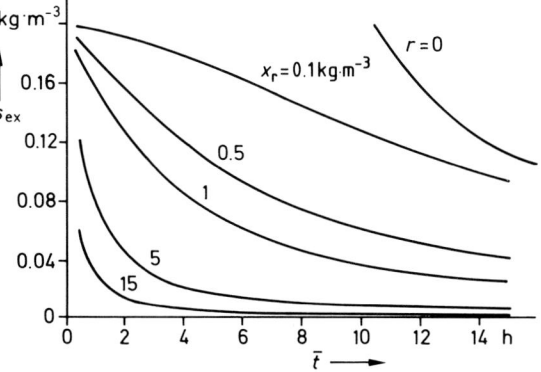

Figure 15.17. Computer simulation of steady-state effluent substrate concentration s_{ex} as a function of mean residence time $\bar{t}$ effected by recycling of cell mass concentration x_r in the case of recycling operation of a continuous stirred tank reactor according to Fig. 15.16 (ANDREWS [15.67]). Parameters: $\mu_{max}=0.2\ h^{-1}$, $K_S=0.2\ g\cdot L^{-1}$, $k_d=0.005\ h^{-1}$, $Y=0.5$, $s_0=0.2$ kg·m^{-3}, $r=0.25$.

also of making s_{ex} less sensitive or more stable with respect to variations of flow rate of constant volume reactors with residence times considerably below the wash-out value (ANDREWS [15.67]).

15.5.1.2 Cascade of Stirred Tanks with Recycling

The performance equation for this case was presented by POWELL and LOWE [15.54], as mentioned earlier; however, the equation has been shown to be analytically intractable. Nevertheless, a simple formula of the critical dilution rate D_{crit} can be given:

$$D_{crit} = \left[\frac{1-r}{(1-r\cdot\beta^{1/N})N}\right]\mu_{max}\frac{s_0}{K_S+s_0}.$$
(15.53)

This important formula shows clearly how the critical dilution rate depends on the recycling ratio r, the number of stages N, and the concentration of limiting substrate.

15.5.1.3 Continuous Operation of Plug Flow Reactors with Recycling

It has been pointed out that CPFRs containing microbial flocs can only be operated continuously on a recycling basis or with a CSTR as an inoculum reactor, otherwise wash-out will occur.

For a CPFR with recycling of non-concentrated cell mass a performance expression was given by LEVENSPIEL [15.8]. For $x_0=0$ the result is:

$$\mu_{max}\cdot\bar{t}=(r+1)\left[\frac{K_S}{s_0}\ln\frac{s_0+r\cdot s}{r\cdot s}+\ln\frac{1+r}{r}\right].$$
(15.54)

As indicated earlier (cf. Fig. 11.15), there always exists an optimal recycling ratio, at which maximum conversion can be achieved. This optimum recycling ratio with

$x_0 = 0$ is obtained by setting $dx/dr = 0$. This yields Eq. (11.31), which can be solved by trial and error. As displayed in Fig. 15.18, the optimum value, r_{opt}, is found to be a function of K_S/s_0 and s/s_0.

For a CPFR with recycling of concentrated cell mass GRIEVES et al. [15.55] derived a general equation, which can be used for computer simulation with the basic behavior:

$$-\mu_{max} \cdot \bar{t} = (1+r) \left[\frac{(1+r) K_S}{\frac{r \cdot \beta (s_0 - s)}{1 + r - r \cdot \beta} - s_0 + r \cdot s} \right.$$

$$\left. \cdot \ln \frac{s \cdot r \cdot \beta}{s_0 + s \cdot r} - \ln \frac{1+r}{r \cdot \beta} \right]. \quad (15.55)$$

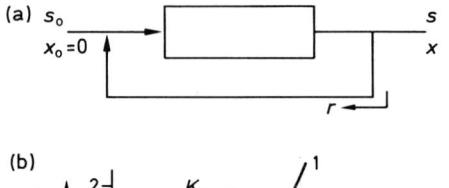

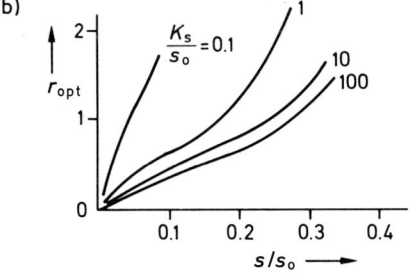

The performance calculated on the basis of Eq. (15.55) indicates the general improvement in performance with an increased recycling stream. The relationship between wash-out flow and recirculation can be deduced from Eq. (15.55) by setting s_0/s equal to unity (corresponding to zero conversion). According to ATKINSON [15.68]:

$$D_{crit} = \frac{1}{\bar{t}_{crit}} = \frac{\mu_{max}}{(1+r) \ln[(1+r)/\beta \cdot r]} \cdot \frac{s}{K_S + s}.$$
(15.56)

The improvement in performance when β is increased is shown in Fig. 15.19 where both factors, recycling ratio r and cell concentration factor β, influence the performance of a CPFR.

Figure 15.18. Recycling operation for bioprocessing: (a) scheme of recycling reactor; (b) plot of the optimum value of the recycling ratio r_{opt} as a function of the substrate concentration ratio s/s_0 with varying K_S/s_0 using a trial-and-error method for solving Eq. (15.54) (LEVENSPIEL [15.8]).

15.5.1.4 Comparison of Performance

GRIEVES et al. [15.55] compared the performance equations for CSTRs and CPFRs. The resulting curves are shown in Fig. 15.14. The degree of conversion is generally greater in the CPFR. A comparison of wash-out flow can be obtained from Eqs. (15.52) and (15.56), (ATKINSON [15.68]):

$$\frac{D_{crit, CSTR}}{D_{crit, CPFR}} = \frac{1+r}{1+r-r \cdot \beta} \cdot \ln \frac{1+r}{\beta \cdot r}. \quad (15.57)$$

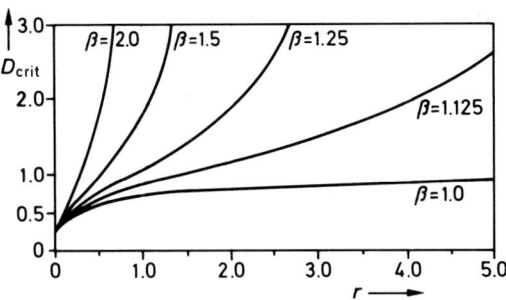

Figure 15.19. Graphical representation of the effect of cell mass concentration on the wash-out flow of continuous plug flow reactors: the critical dilution rate D_{crit} as a function of recycling ratio r with variations of cell concentration factor β (see Fig. 15.16) according to ATKINSON [15.68].

This ratio is shown in Fig. 15.20 as a function of r and β. The productivity ratio between a CSTR and a CPFR can thus be greater or less than unity depending upon the flow rate, r and β.

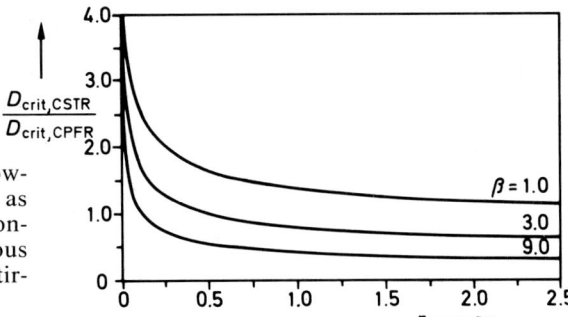

Figure 15.20. Analogous plot to Fig. 15.19 showing a comparison of the wash-out flows (D_{crit}) as a function of the recycling ratio r and cell concentration factor β in the case of a continuous plug flow reactor (CPFR) and a continuous stirred tank reactor (CSTR) (ATKINSON [15.68]).

15.5.2 Application in Industry and Research

In fermentations and other bioprocesses, process stream recycling is intimately concerned with enhanced conversion, as previously outlined. Such recycling schemes usually seek to reduce operating costs by increasing capital investment. The three major possibilities for recycling loops applied in the fermentation industry are the following: (1) processes employing gaseous substrates including pure oxygen; (2) processes employing spent medium components and/or process water reutilization; and (3) processes designed to improve fermenter productivity by the employment of elevated cell concentrations.

While the first possibility can be applied in batch and continuous processing, the second and third are only appropriate to continuous operation.

Yield, productivity, and conversion are the three most important economic factors. They are related but not necessarily compatible. Recycling operations, as indicated for the gaseous phase, the process water/medium, and cells, will not increase their compatibility, but there is no doubt that recycling will have a place in fully optimized fermentation processes designed to satisfy economic, environmental, and safety requirements (HAMER [15.66]). As advances are made in the basic understanding of microbial physiology, process systems designed to exploit these advances, possibly by exposing the cells to a sequence of appropriate conditions in a recycling stream, will be introduced (e.g., TODA and DUNN [15.22]).

In biological waste water treatment, the recycling operation of sludge has been used for a long time (e.g., METCALF & EDDY IND. [15.58]).

As a consequence, reactor stability and substrate conversion are increased. Other important applications of recycling operations are known for the optimization of process design for continuous ethanol production (e.g., CYSEWSKI and WILKE [15.69]). Generally, various systems for cell recycling are available, known as internal filtration, internal sedimentation, monostream, and external recirculation (PIRT [15.24]). Recently a computer simulation study of the ethanol fermentation with cell recycling has been reported (LEE et al. [15.70]).

Similar optimization studies have also been experimentally verified for ethanol production with *Zymomonas mobilis* (e.g., CHARLEY et al. [15.71]). However, other results show that ethanol productivity did not change markedly by an increase of cell mass recycling (MORENO and GOMA [15.45]); (DE BOKS and VAN EYBERGEN [15.72]).

Finally, several contributions in the literature suggest that the principle of recycling operations is being acknowledged in bioprocess research and development (SEIPENBUSCH and BLENKE [15.73]; BLENKE [15.74]; BULL and YOUNG [15.75]; STIEBER and GERHARDT [15.76]; CONSTANTINIDES et al. [15.77]; LIPPERT et al. [15.78]; ADLER and FIECHTER [15.79]; MOSER [15.52], [15.53]).

While the equation for CSTRs indicates that external recycling of a fluid has no effect on the performance, problems arise in the case of CPFR, where the mixing behavior is heavily influenced by recirculation. In mixing of fluids in a reactor macro- and micro-mixing have to be considered (cf. Sect. 11.3).

GILLESPIE and CARBERRY [15.80] and RIPPIN [15.81] calculated the influence of recirculation on macro-mixing of CPFRs. Evidently, plug flow behavior is lost by a successively increasing recycling ratio r. The function of the RTD of the total system $f_{tot}(t)$ is the sum of the internal RTD-functions $f_{int}(t)$ (see Eq. (15.43); MOSER and STEINER [15.82]; BLENKE [15.74]).

The basic idea of the investigations of mixing behavior with a recycling reactor is as follows. By changing r and $f_{int}(t)$ at the same time, but in such a way that $f_{tot}(t)$ remains the same (macro-mixing effect is constant or changes very little), the effects of micro- and macro-mixing can be separated. This is not possible in normal reactor configurations, where micro-mixing is coupled to macro-mixing. In this case the degree of segregation J (see Eq. [11.33]) can be expressed as a function of the recycling ratio r (DOHAN and WEINSTEIN [15.83]):

$$J = 1 - \frac{\frac{r}{r+1}\left(\sigma_{tot}^2 + 1 - \frac{r}{r+1}\right)}{\sigma_{int}^2}. \quad (15.58)$$

The relation between J and r is presented in Fig. 15.21. The curves end at a minimum value of J corresponding to the maximum value of r possible for each value of σ_{tot}^2. The curve for $\sigma_{tot}^2 = 1$, representing the single CSTR, is the envelope for all possible states of mixing.

Application of micro- and macro-mixing considerations to bioprocessing by DOHAN and WEINSTEIN [15.83] resulted in the conclusion that optimum reactor operation needs a high degree of micro-mixing (large σ_{int}^2) but at the same time a large σ_{int}^2 is not favorable to conversion. Optimum conversion clearly depends on the coupling between the two components of mixing.

As already illustrated in Fig. 11.15 an optimal recycling ratio exists for maximizing the substrate conversion. As a result of combined considerations of maximum conversion and non-wash-out conditions of cells, medium recycling is often used, e.g., for tubular and tower reactors. A loop reactor approaches the flow behavior of a CSTR at high recycling ratios ($r > 10$). This statement is valid for cells, while for substrate the behavior of the CPFR is maintained as long as $s_{ex} \approx 0$. Economic considerations suggest a decoupling of the components X, S, and O_2. Thus, there is an optimum value of r at which substrate and/or oxygen conversion is maximum. Experimental verifications are to be found in the literature (MOSER [15.62]; SCHÜGERL [15.84], [15.40]; WANDREY and FLASCHEL [15.85]). The position of the minimum of substrate concentration at the outlet (= maximum conversion ζ_S) was shown to be strongly dependent on the Damköhler number $Da_1 = \mu_{max} \cdot \bar{t}$ as well as on the K_S-value (SCHÜGERL [15.40]). Below a critical $Da_{1,crit}$ or above a critical $K_{S,crit}$ the best reactor is the CSTR.

Recycling reactors are highly suitable devices for experimental verification of mix-

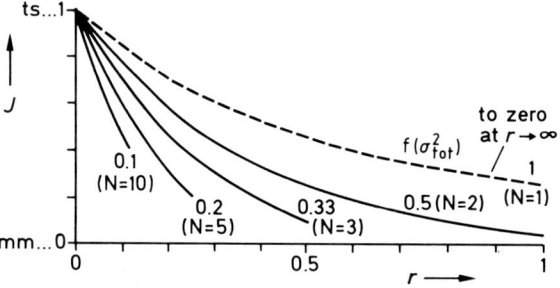

Figure 15.21. Simulation of the relation between degree of segregation J as a measure for micromixing and recycling ratio r with varied values of variance of the total residence time distribution (σ_{tot}^2). The corresponding value of the equivalent stages N are indicated in brackets (after DOHAN and WEINSTEIN [15.83]). – mm micromixing, ts total segregation.

ing models as emphasized here. Recycling reactors can be operated in both batch and continuous modes for this purpose and additionally exhibit the property of having a low cycle time distribution, which is very favorable compared to normal stirred tanks (MOSER [15.23]).

Acknowledgement

Most of the computer simulations were carried out by the group of Dr. W. Knorre of the ZIMET (Zentralinstitut für Mikrobiologie und Experimentelle Therapie) at the Academy of Sciences of the German Democratic Republic in Jena, based on a scientific collaboration with the author's institute in Graz/Austria.

15.6 References

[15.1] J. MONOD: "Recherches sur la Croissance des Cultures Bacteriennes". Hermann & Cie, Paris, 1942.
[15.2] J. MONOD: Ann. Inst. Pasteur 79 (1950), 390.
[15.3] A. NOVICK and L. SZILARD: Proc. Natl. Acad. Sci. Washington 36 (1950), 708.
[15.4] A. FIECHTER: In "Biotechnology – a Comprehensive Treatise", Vol. 1, Chap. 7 (H. J. REHM and G. REED, eds.). Verlag Chemie, Weinheim – Deerfield Beach, Florida – Basel, 1981.
[15.5] A. A. ESENER, J. A. ROELS, and N. W. F. KOSSEN: Biotechnol. Lett. 3 (1981), 15.
[15.6] A. A. ESENER, J. A. ROELS, N. W. F. KOSSEN, and J. W. H. ROOZENBURG: Eur. J. Appl. Microbiol. Biotechnol. 13 (1981), 141.
[15.7] O. LEVENSPIEL: "The Chemical Reactor Omnibook". OSU Book Stores Inc., Corvallis/Oregon (USA), 1979.
[15.8] O. LEVENSPIEL: Biotechnol. Bioeng. 22 (1980), 1671.
[15.9] W. A. KNORRE: In "Biophysikalische Grundlagen der Medizin", p. 132 (W. BEIER and R. ROSEN, eds.). G. Fischer Verlag, Stuttgart – New York, 1980.
[15.10] C. G. SINCLAIR and H. H. TOPIWALA: Biotechnol. Bioeng. 12 (1970), 1069.
[15.11] G. GOMA, R. MOLETTA, and M. NOVAK: Biotechnol. Lett. 1 (1979), 415.
[15.12] H. J. KUHN, S. COMETTA, and A. FIECHTER: Eur. J. Appl. Microbiol. Biotechnol. 10 (1980), 303.
[15.13] B. O. SOLOMON and L. E. ERICKSON: Process Biochem. Feb./March (1981), 44.
[15.13 a] C. G. SINCLAIR and D. E. BROWN: Biotechnol. Bioeng. 12 (1970), 1001.
[15.14] J. G. VAN DE VUSSE: Chem. Eng. Sci. 19 (1964), 994.
[15.15] D. E. BROWN, D. J. HALSTED, and C. G. SINCLAIR: Biotechnol. Lett. 1 (1979), 159.
[15.16] B. I. TSAI, L. E. ERICKSON, and L. T. FAN: Biotechnol. Bioeng. 11 (1969), 181.
[15.17] L. T. FAN, L. E. ERICKSON, P. S. SHAH, and B. I. TSAI: Biotechnol. Bioeng. 12 (1970), 1019.
[15.18] B. I. TSAI, L. T. FAN, L. E. ERICKSON, and M. S. K. CHEN: J. Appl. Chem. Biotechnol. 21 (1971), 307.
[15.19] C. Y. WEN and L. T. FAN: "Models for Flow Systems and Chemical Reactors". Marcel Dekker, New York, 1975.
[15.20] R. K. BAJPAI and M. REUSS: Can. J. Chem. Eng. 60 (1982), 384.
[15.21] H. P. KNOEPFEL: Ph. D. Thesis, Eidgenössische Technische Hochschule Zürich, 1972
[15.22] K. TODA and I. J. DUNN: Biotechnol. Bioeng. 24 (1982), 651.
[15.23] A. MOSER: In "Comprehensive Biotechnology", Vol. 2, Chap. 4 (M. MOO-YOUNG, ed.). Pergamon Press, London – New York, in press.
[15.24] S. J. PIRT: "Principles of Microbe and Cell Cultivation". Blackwell, Oxford (UK), 1975.
[15.25] E. M. HEGEWALD, I. RÜCKBEIL, and W. A. KNORRE: In "Proc. 7th Symp. Contin. Cult. Microorganisms", p. 717. Prague (CSSR), 1980.
[15.26] R. K. BAJPAI and M. REUSS: J. Chem. Tech. Biotechnol. 30 (1980), 332.
[15.27] Z. FENCL and M. NOVAK: Folia Microbiol. Prague 14 (1969), 314.
[15.28] J. RICICA: In "Fermentation Advances", p. 427 (D. PERLMAN, ed.). Academic Press, New York, 1969.

[15.29] S. Pirt and R. C. Righelato: Appl. Microbiol. *15* (1967), 1284.

[15.30] D. Herbert: In "Proc. Symp. Contin. Cult. Microorganisms", p. 23. Czechoslovak Academy of Sciences, Prague (CSSR), 1964.

[15.31] I. Malek and Z. Fencl: "Theoretical and Methodological Basis of Continuous Culture of Microorganisms". Academic Press, New York, 1966.

[15.32] J. Ricica and S. Necinova: Mitt. Versuchsstn. Gaerungsgewerbe Wien, *11/12* (1967), 130.

[15.33] F. H. Deindoerfer and A. E. Humphrey: Ind. Eng. Chem. *51* (1959), 809.

[15.34] R. Luedeking and E. L. Piret: Biotechnol. Bioeng. *1* (1959), 431.

[15.35] Z. Fencl: In "Proc. Symp. Contin. Cult. Microorganisms", p. 23 (I. Malek et al., eds.). Czechoslovak Academy of Sciences, Prague (CSSR), 1964.

[15.36] S. Goto, A. Kitai, and A. Ozaki: J. Ferment. Technol. *51* (1973), 582.

[15.37] A. Prokop, L. E. Erickson, J. Fernandez, and A. E. Humphrey: Biotechnol. Bioeng. *11* (1969), 945.

[15.38] A. Kitai, H. Tone, and A. Ozaki: Biotechnol. Bioeng. *11* (1969), 911.

[15.39] J. Paca and V. Gregr: Biotechnol. Bioeng. *18* (1976), 1075.

[15.40] K. Schügerl: Adv. Biochem. Eng. *22* (1982), 94.

[15.41] Z. Fencl, F. Machek, and M. Novak: In "Fermentation Advances", p. 301 (D. Perlman, ed.). Academic Press, New York, 1969.

[15.42] Z. Fencl, J. Ričica, and J. Kodešová: J. Appl. Chem. Biotechnol. *22* (1972), 405.

[15.43] Z. Fencl, E. Ujcová, F. Machek, L. Seichert, and M. Musílková: In "Proc. 7th Symp. Contin. Cult. Microorganisms", p. 49. Prague (CSSR), 1980.

[15.44] R. D. Tyagi and T. K. Ghose: Biotechnol. Bioeng. *22* (1980), 1907.

[15.45] M. Moreno and G. Goma: Biotechnol. Lett. *1* (1979), 483.

[15.46] T. Imanaka, T. Kaieda, and H. Taguchi: J. Ferment. Technol. (Japan) *51* (1973), 431.

[15.47] D. D. Y. Ryu and B. K. Lee: Process Biochem. January/February (1975), 15.

[15.48] S. Furusaki and T. Miyauchi: J. Chem. Eng. Jpn. *10* (1977) 3, 247.

[15.49] H. Bahl, W. Andersch, and G. Gottschalk: Eur. J. Appl. Microbiol. Biotechnol. *15* (1982), 201.

[15.50] Y. W. Ryu, J. M. Navarro, and G. Durand: Eur. J. Appl. Microbiol. Biotechnol. *15* (1982), 1.

[15.51] R. W. Swartz: Annu. Rep. Ferment. Proc. *3* (1979), 75.

[15.52] A. Moser: In "Proc. 33rd Canad. Engng. Conf., Toronto, October 2-5, 1983", Vol. 2, p. 417.

[15.53] A. Moser: "Bioprocess Engineering". Springer Verlag New York - Wien, 1985.

[15.54] O. Powell and J. R. Lowe: In "Proc. Symp. Contin. Cult. Microorganisms", p. 45. Czechoslovak Academy of Sciences, Prague (CSSR), 1964.

[15.55] R. B. Grieves, W. O. Pipes, W. F. Milbury, and R. K. Wood: J. Appl. Chem. *14* (1964), 478.

[15.56] O. Levenspiel: "Chemical Reaction Engineering". J. Wiley & Sons, New York, 1972.

[15.57] G. F. Froment and K. B. Bischoff: "Chemical Reactor Analysis and Design". J. Wiley & Sons, New York, 1979.

[15.58] Metcalf & Eddy Ind. (1972, 1979): "Waste Water Eng. Treatment, Disposal and Reuse". McGraw-Hill, New York, 1972, 1979.

[15.59] S. Aiba, A. E. Humphrey and N. F. Millis: "Biochemical Engineering". Academic Press, New York - London, 1973.

[15.60] M. D. Lilly and P. Dunill: Adv. Biochem. Eng. *3* (1972), 221.

[15.61] R. N. Greenshields and E. L. Smith: Process Biochem. April (1974), 11.

[15.62] A. Moser: Chem. Ing. Tech. *49* (1977), 612.

[15.63] T. W. F. Russel: Biotechnol. Bioeng. *16* (1974), 1261.

[15.64] H. Ziegler, D. Meister, I. J. Dunn, H. W. Blanch, and T. W. F. Russell: Biotechnol. Bioeng. *19* (1977), 507.

[15.65] A. Moser: Paper P-18.01301 at VIIth Int. Biotechnol. Symp., New Delhi (India), February 17-25, 1984.

[15.66] G. Hamer: Biotechnol. Bioeng. *24* (1982), 511.

[15.67] J. F. Andrews: "Kinetic and mathematical modelling", in "Ecological Aspects of Waste Water Treatment" (H. A. Hawkes and C. R. Curds,

[15.68] B. ATKINSON: "Biochemical Reactors". Pion Ltd., London, 1974.
[15.69] C. R. CYSEWSKI and CH. R. WILKE: Biotechnol. Bioeng. *20* (1978), 1421.
[15.70] J. M. LEE, J. F. POLLARD, and G. A. COULMAN: Biotechnol. Bioeng. *25* (1983), 497.
[15.71] R. C. CHARLEY, J. E. FEIN, B. H. LAVERS, H. G. LAWFORD, and G. R. LAWFORD: Biotechnol. Lett. *5* (1983), 169.
[15.72] P. A. DE BOKS and G. C. VAN EYBERGEN: Biotechnol. Lett. *3* (1981), 577.
[15.73] R. SEIPENBUSCH and H. BLENKE: Adv. Biochem. Eng. *15* (1980), 2.
[15.74] H. BLENKE: Adv. Biochem. Eng. *13* (1979), 122.
[15.75] D. N. BULL and M. D. YOUNG: Biotechnol. Bioeng. *23* (1981), 373.
[15.76] R. W. STIEBER and P. GERHARDT: Biotechnol. Bioeng. *23* (1981), 523.
[15.77] A. CONSTANTINIDES, D. BHATIA, and W. R. VIETH: Biotechnol. Bioeng. *23* (1981), 899.
[15.78] J. LIPPERT, I. ADLER, H. D. MEYER, A. LÜBBERT, and K. SCHÜGERL: Biotechnol. Bioeng. *25* (1983), 437.
[15.79] I. ADLER and A. FIECHTER: Chem. Ing. Tech. *55* (1983), 322.
[15.80] B. M. GILLESPIE and J. J. CARBERRY: Ind. Eng. Chem. Fundam. *5* (1966), 164.
[15.81] D. W. T. RIPPIN: Ind. Eng. Chem. Fundam. *6* (1967) 4, 488.
[15.82] A. MOSER and W. STEINER: Chem. Ing. Tech. *46* (1974), 695.
[15.83] L. A. DOHAN and H. WEINSTEIN: Ind. Eng. Chem. Fundam. *12* (1973) 1, 64.
[15.84] K. SCHÜGERL: Chem. Ing. Tech. *49* (1977), 605.
[15.85] C. WANDREY and E. FLASCHEL: Adv. Biochem. Eng. *12* (1979), 148.
[15.86] R. GUTKE: In "Theoretical Basis of Kinetics of Growth, Metabolism and Product Formation of Microorganisms", UNEP/UNESCO/ICRO-Training Course. ZIMET (Central Institute of Microbiology and Exp. Therapy) Academy of Sciences, Jena (GDR), 1980, 1982 and Leipzig (GDR), 1982.
[15.87] M. HILLINGER: In "Theoretical Basis of Kinetics of Growth. Metabolism and Product Formation of Microorganisms", UNEP/UNESCO/ICRO-Training Course. ZIMET (Central Institute of Microbiology and Exp. Therapy) Academy of Sciences, Jena (GDR), 1980, 1982 and Leipzig (GDR), 1982.

Chapter 16
Special Cultivation Techniques

Anton Moser

Institut für Biotechnologie, Mikrobiologie und Abfalltechnologie
Technische Universität Graz
Graz, Austria

16.1   Introduction
16.2   Transient Operation Techniques
16.2.1 Classification of Continuous Environmental Changes
16.2.2 Definitions for Balanced Growth and Steady-state Growth
16.2.3 Fed-batch Cultivation
16.2.3.1 Classification
16.2.3.2 Performance Equation for Variable-volume Continuous Stirred Tank Reactor Operation
16.2.3.3 Mathematical Modelling of Fed-batch Process Kinetics
16.2.3.4 Application of Fed-batch Cultivation
16.3   Dialysis Cultures
16.3.1 General Aspects and Application of Membrane Processing
16.3.2 Dialysis Reactor Systems
16.3.3 The Dialysis Culture Theory
16.3.3.1 Principles and Performance Equations
16.3.3.2 Implication of Theory and Comparison between Dialysis and Non-dialysis Culture Techniques
16.4   Synchronous Cultures
16.4.1 Classification
16.4.2 Scope of Synchronous Techniques
16.5   Mixed Microbial Populations
16.5.1 Classification of Types of Microbial Interactions
16.5.2 Kinetic Analysis of Microbial Interactions
16.5.2.1 Competition
16.5.2.2 Commensalism and Amensalism
16.5.2.3 Mutualism
16.5.2.4 Predation (Predator-Prey Interaction)
16.5.3 Application and Mixed Population Bioreactor Problems
16.6   Biofilm Reactor Operation
16.7   References

List of Symbols

Symbol	Units	Description
A	m^2	surface area
a	m^{-1}	specific surface area
a_{ij}, a_{ji}	—	entries of community matrix (Table 16.2)
B_x	s^{-1}	mass loading rate per unit biomass
c_μ	—	constant in Eq. (16.27)
C	—	integration constant Eq. (16.75)
D	$m \cdot s^{-1}$	diffusion coefficient
D	s^{-1}	dilution rate
d_X	m	diameter of biomass (film)
F	$m^3 \cdot s^{-1}$	flow rate
F_S, F_X	$kg \cdot m^{-3} \cdot s^{-1}$	flow rate of substrate or biomass
F'_S	$mol \cdot s^{-1}$	rate of substrate addition
G	kg	amount of G-compartment
H_v	$J \cdot m^{-3}$	volumetric heat of fermentation
K	kg	amount of K-compartment
K_{adapt}	s^{-1}	constant of adaptation in Eq. (16.33)
K_S, K_G, K_2, K_3	$kg \cdot m^{-3}$	saturation constant of Monod-type kinetics for substrate, G-compartment, or component 2 and 3
k_{L2}	$m \cdot s^{-1}$	mass transfer coefficient at L/S-interface
k_d	s^{-1}	constant of microbial death
k_r	...	reaction rate constant
k_i	...	rate constant of component i
k_P	s^{-1}	production rate constant
m_G	s^{-1}	specific turnover rate of G-compartment
m_S	s^{-1}	specific maintenance coefficient for substrate
n	—	integer, power number
n	$kg \cdot s^{-1}$	rate of permeation of mass
P_{mb}	$m \cdot s^{-1}$	permeability coefficient of the membrane
P	kg	amount of product
p	$kg \cdot m^{-3}$	concentration of product
Q	...	function of product formation activity in Eq. (16.29)
q_P, q_G	s^{-1}	specific rate of product formation or formation of G-compartment
q_S	s^{-1}	specific rate of substrate consumption
R	kg	amount of R-compartment
r_i	$kg \cdot m^{-3} \cdot s^{-1}$	rate of formation or consumption of component i
S	kg	amount of substrate
s	$kg \cdot m^{-3}$	concentration of substrate
t	s	time
t_g	s	generation time
t_p	s	period of oscillation
V	m^3	volume
w_G	—	mass fraction of component G
X	kg	amount of biomass
X_i	kg	total mass of any metabolic variable
x	$kg \cdot m^{-3}$	concentration of biomass
x_i	$kg \cdot m^{-3}$	concentration of any metabolic variable
$Y, Y_{X/S} \ldots$	—	yield coefficients

Greek letters

α, β	...	coefficients
γ_S, γ_X	—	degree of reduction
δ_i	m	thickness of i-th film
μ	s^{-1}	specific growth rate
μ_{max}	s^{-1}	maximum specific growth rate
ρ	$kg \cdot m^{-3}$	density
τ_{L1}, τ_{L2}	s	time lag in Eqs. (16.22) and (16.23)

Indices

adapt	adaptational
ave	average
charact	characteristic
crit, c	critical
ex	effluent
ferm	fermenter
G	G-compartment
i, j	components
in	inlet
K	K-compartment
L	liquid
max	maximum
mb	membrane
P	product
Q	activity function for P-formation
res	reservoir
S	solid or substrate
X	biomass
$^2x, {}^2s$	x and s-value of non-wash-out steady state
$^1x, {}^1s$	x and s-value of wash-out state
$^3x, {}^3s$	x and s-value of unstable steady state
μ	spec. growth rate
1, 2, I, II	number of species

Abbreviations

CSTR	continuous stirred tank reactor
DNA	desoxyribonucleic acid
HAc	acetic acid
RNA	ribonucleic acid

16.1 Introduction

Beyond the well-known operational methods of microbial cultivation, i. e., batch and continuous processing, other possibilities exist, which are becoming increasingly interesting for biotechnology.

Generally, there are three techniques for process improvement: strain selection and mutation, medium optimization and advanced design and operation of bioreactors. In this respect, several unconventional operation modes will be presented in this chapter: transient operation techniques, e. g., fed-batch culture, dialysis and synchronous cultivation, mixed population techniques, and biofilm operation.

16.2 Transient Operation Techniques

Transient reactor operation (PICKETT et al. [16.1]) includes all periods in which environmental conditions change other than lag-phases.

The effects of a continuously changing environment are significant in both natural and industrial microbiological systems. From an ecological point of view, studies of transients are important since naturally growing populations are subjected to a varying environment, leading to relative changes in species diversity during the course of a year.

Although continuous-culture methods are more commonly associated with physiological studies of microbial populations at the steady-state, examination of transient behavior in continuous cultures provides valuable insight into the mechanisms of regulation in biological systems (HARRISON and TOPIWALA [16.2].

Process improvement obtained with periodic operation was shown to depend on the reaction rate constants (DORAWALA and DOUGLAS [16.3]).

16.2.1 Classification of Continuous Environmental Changes

In analogy to chemical reactor operation (BAILEY [16.4]) four broad classes of periodic operation were defined as shown in Fig. 16.1. First, a process life cycle considers the long-term operation of a reactor where a cycle may be determined by poisoning of a catalyst. In all instances of bio-

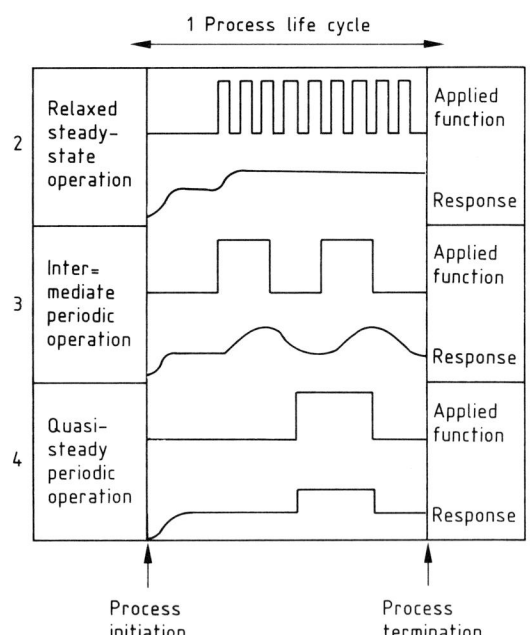

Figure 16.1. The four classes of transient (periodic) reactor operation as defined by BAILEY [16.4]. – Typical reactor output responses are shown for the four classes.

logical reactor operation this type of cycle will be short except, perhaps, where immobilized enzyme systems are involved. The other classes consider the perturbed process in more detail. Relaxed steady-state operation occurs when a system is unable to follow rapid cycling and enters what is termed a "relaxed" steady-state, a system tending towards a time invariant value (PICKETT [16.1]; [16.5]). Thus, the organism is "frozen" in its initial state and the given mechanism must not be considered as in dropwise substrate dosing (ROELS [16.6]). Relaxed steady-state, therefore, leads to a reduction in the number of differential equations in accordance with the "rate-determining-step concept" (see Sect. 11.1).

Quasi-steady periodic operation is the opposite to relaxed steady-state operation and considers a process under time-invariant conditions, i.e., when the period of the input oscillation is large compared with the system response time; the system yields a steady-state relationship with the input at any time. Here, the mechanism is fully adapted and it behaves as if the organisms were at steady-state with the particular environmental conditions. This is true for "balanced" growth in continuous culture applications; the dynamics of a given mechanism are not important. In accordance with the "quasi-steady-state concept" (see Sect. 11.1) differential equations can be reduced to algebraic equations. Intermediate periodic operation lies between both extremes and covers systems where the response time is of the same order as the imposed function cycle time so that "resonance" can happen. Here, the state of the mechanism changes dynamically and does not relate directly to the environmental conditions at the moment considered. It encompasses semi-batch periodic operation, selectivity in periodically forced CSTRs and cycled reactors with heterogeneous catalysts. Lag, overshoot phenomena, and oscillations can typically occur. Biological reactor operation will basically fall into this latter class of operation. Transient reactor operations have recently been summarized in review articles presented by PICKETT [16.5], BARFORD et al. [16.7], PICKETT et al. [16.8], DAIGGER and GRADY [16.9], SHERRARD and LAWRENCE [16.10], and CHI and HOWELL [16.11].

16.2.2 Definitions for Balanced Growth and Steady-state Growth

According to BARFORD et al. [16.7] balanced growth and steady-state growth are to be distinguished.

The concept of balanced growth was introduced by CAMPBELL [16.12] to describe the metabolic state of a culture in terms of the distributed concentration (x_i) or total mass (X_i) of any metabolic variable. According to CAMPBELL's definition, growth is balanced when the specific rate of change of all such variables is constant:

$$\frac{1}{x_i} \cdot \frac{dx_i}{dt} = \frac{1}{X_i} \cdot \frac{dX_i}{dt} = \text{const}. \quad (16.1)$$

This definition applies to a culture of microorganisms. In the case of an individual microorganism, a broader definition is required:

$$\frac{1}{x_{i,\text{ave}}} \cdot \frac{\Delta x_i}{\Delta t} = \frac{1}{X_{i,\text{ave}}} \cdot \frac{\Delta X_i}{\Delta t} = \text{const}, \quad (16.2)$$

where $t = n \cdot t_g$ with n being any integer and t_g as the generation time; $t_g = \log 2\mu$; Δx_i and ΔX_i are the changes in x_i and X_i during time Δt; $x_{i,\text{ave}}$ and $X_{i,\text{ave}}$ are the average values of x_i and X_i during the time Δt. The requirement that x_i and X_i be averaged derives from the fact that synthesis of cell components by an individual cell is not necessarily continuous and must not necessarily occur at a constant rate during the life cycle of an organism.

In addition to the growth of a single microorganism, many other microbial growth phenomena require a broader definition of balanced growth. An example is periodic behavior or sustained oscillation.

If such sustained oscillations are to be regarded as balanced growth, the definition of balanced growth may be broadened again using Eq. (16.2), but with:

$$t = n \cdot t_p, \qquad (16.3)$$

where n is any integer and t_p the period of oscillation.

These considerations are a simple example of an important concept to be developed – namely, the importance of some reference to a time scale when discussing balanced and/or steady-state growth situations (see Fig. 11.10, MOSER [16.13], and Chapter 24).

The question also has to be considered how closely balanced growth can be approached under batch growth conditions. While the ideal experimental batch curve has been referred to by HERBERT [16.14], the number of reported experimental examples of balanced growth in batch cultures is extremely small (BARFORD and HALL [16.15]).

Several important concepts are illustrated by the following observations. First, it is essential that a range of metabolic variables be considered before any well-based decision can be made as to whether balanced growth has been achieved or not. In addition, and most importantly, some metabolic processes achieve a balanced growth condition more rapidly than others. Consequently, the variable used as the criterion for attaining balanced growth must be carefully chosen and should be explicitly stated.

It is possible to consider steady-state growth as an extension of balanced growth. That is, in addition to the definitions outlined previously, a further requirement is that the overall time rate of change of the distributed concentration (x_i) or total mass (X_i) of any metabolic variable be zero. That is:

$$\frac{dx_i}{dt} = \frac{dX_i}{dt} = 0. \qquad (16.4)$$

Consequently, while balanced growth is the necessary condition for steady-state growth, it is not a sufficient condition. Likewise, steady-state is a sufficient condition for balanced growth, but not a necessary one.

Very few experimental studies precisely state what criteria were used in the determination of steady-state. It has been proposed that, given the possible experimental variations, a more precise definition of steady-state be initiated (BARFORD et al., [16.7]). While this definition is somewhat arbitrary, it is more precise than previously employed definitions.

16.2.3 Fed-batch Cultivation

16.2.3.1 Classification

Some fermentations called fed-batch cultures (YOSHIDA et al. [16.16]) utilize a continuous periodic operation technique. This transient condition assumes several forms. In "repeated" fed-batch culture (PIRT [16.17]), the complete medium is fed continuously to a batch culture; the resulting biomass is reduced by pumping out at preset intervals and the process is then repeated. In fed-batch culture, the dilution rate is therefore continually changing and represents the transient of the system. Some

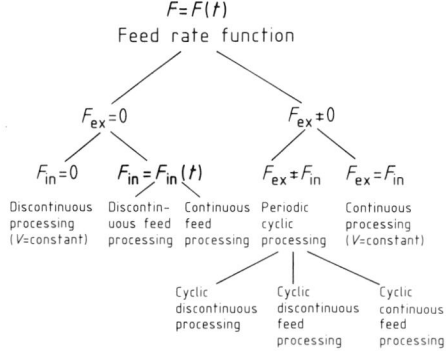

Figure 16.2. Classification scheme of bioreactor operation modes with the aid of the time dependence of flow rates in the inflow to the reactor (F_{in}) and in the outflow (F_{ex}) (MOSER [16.134]).

processes utilize a feed of the (limiting) carbon source only. This type of fed-batch culture has been termed "semi-batch" culture (YAMANE and HIRANO [16.18]). A further modification of fed-batch culture, called "extended" culture, has also been reported (EDWARDS et al. [16.19]). This technique is more controlled than other types of fed-batch culture, since environmental sensors directly measure specific parameters of culture growth and regulate the addition of the limiting carbon source to the culture. Both semi-batch and extended batch types of culture usually only cover one fed-batch "cycle" and so differ from repeated fed-batch culture in the duration of the periodicity applied to the culture.

Classification of feed culture techniques, as shown in Fig. 16.2, uses the time dependence of feed (F_{in}) and bleed (F_{ex}) streams. Discontinuous, continuous, and periodic feed techniques can be distinguished. For better comprehensiveness the graphic plot in Fig. 16.3 shows the characteristic behavior of time dependence of feed, volume, and substrate concentration during periodic operation. As can be seen, the "extended culture" is a special case where $s = $ const,

which facilitates the estimation of kinetic parameters compared to batch processes.

For quantification the following terms of $F_{in}(t)$ are commonly used [cf. Eq. (11.14)]:

linear feed: $F_{in}(t) = \alpha_1 \cdot t + \beta_1$, (16.5)

exponential feed: $F_{in}(t) = \alpha_2 \cdot e^{\beta_2 \cdot t}$, (16.6)

where α and β are constants.

16.2.3.2 Performance Equation for Variable-volume Continuous Stirred Tank Reactor Operation

The balance of biomass in the liquid phase can be formulated as:

$$\frac{d(V \cdot x_{ex})}{dt} = F_{in} \cdot x_{in} - F_{ex} \cdot x_{ex} + r_x \cdot V. \quad (16.7)$$

The balance of the limiting substrate in the liquid phase is (DUNN and MOR [16.20]):

$$\frac{d(V \cdot s_{ex})}{dt} = F_{in} \cdot s_{in} - F_{ex} \cdot s_{ex} - r_s \cdot V, \quad (16.8)$$

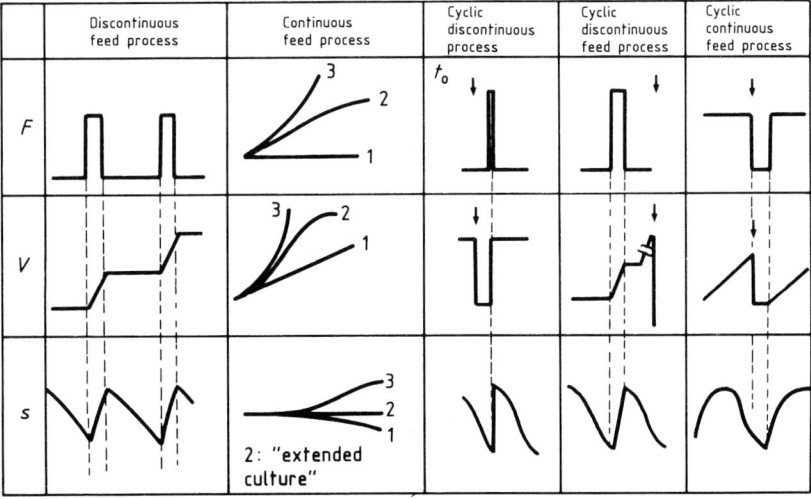

Figure 16.3. Schematic representation of the properties of discontinuous, continuous, and periodic feed bioprocessing with the aid of graphical plots of the time dependence of flow rate F, volume V, and substrate concentration s (MOSER [16.134]).

where $r_x = \mu \cdot X$ and $r_s = -1/Y \cdot r_x$ with μ according to Monod kinetics.

The variable-volume system considered here is based on several assumptions and restrictions: (1) well-mixed reactor, (2) single limiting substrate, (3) balanced growth, (4) constant yield coefficient, and (5) Monod kinetics.

The restriction that balanced cell growth must exist for the Monod relation to be valid puts a strong limitation on the possible change rates which can be considered.

A fed-batch as a special case of variable-volume CSTR-operation has been defined as a bioreactor with inflowing substrate but without outflow. For this system Eq. (16.9) applies:

$$\frac{dV}{dt} = F_{in} ; \quad (16.9)$$

here, F_{in} can be a function of time $F(t)$.
For $x_{in} = 0$ Eq. (16.7) becomes:

$$\frac{d(V \cdot x_{ex})}{dt} \equiv \left[\frac{dV}{dt} \cdot x_{ex} + \frac{dx_{ex}}{dt} V\right] = \mu \cdot x_{ex} \cdot V.$$
$$(16.10)$$

Combining Eqs. (16.9) and (16.10) yields:

$$\frac{dx_{ex}}{dt} = -\frac{F_{in}}{V} x_{ex} + \frac{\mu_{max} \cdot s_{ex}}{K_S + s_{ex}} x_{ex} . \quad (16.11)$$

The substrate balance, Eq. (16.8), reduces for the fed-batch reactor in a similar way to:

$$\frac{ds_{ex}}{dt} = (s_{in} - s_{ex}) \frac{F_{in}}{V} - \frac{\mu_{max} \cdot s_{ex}}{K_S + s_{ex}} \cdot \frac{x_{ex}}{Y} . \quad (16.12)$$

Equations (16.9), (16.11), and (16.12) provide the mathematical model for the fed-batch reactor.

It is instructive to compare the fed-batch model with the model for a constant-volume chemostat [see Eq. (11.8) with $D = \mu$]. These equations are formally identical with Eqs. (16.11) and (16.12), but it is important to note that the physical meaning of the terms is not identical. Comparing the term $-F_{in} \cdot x_{ex}/V$ in Eq. (16.11) with $D \cdot \bar{x}$ shows that the origin of this term for the fed-batch is the expression $x_{ex} \cdot dV/dt$; it thus represents a decrease in cell concentration due to the volume change, which arises from the inlet flow rate F_{in}. On the other hand, $-F_{in} \cdot x_{ex}$ in the case of the chemostat is a wash-out term, expressing the mass flow rate of cells leaving the reactor with the outgoing stream.

Mathematically the two models are not identical. The dilution rate F/V in the chemostat is constant for constant flow, whereas in a fed-batch fermentation continuous feeding causes a volume increase and a corresponding decrease in dilution rate, F_{in}/V. A fed-batch can be compared to a constant-volume chemostat whose feed rate is decreasing slowly. Because the mathematical form of the fed-batch model in Eqs. (16.11) and (16.12) is identical to that of the chemostat, it can be concluded that the fed-batch will behave analogously. A dynamic steady-state will be achieved for sufficiently low flow rates, such that the specific growth rate is exactly maintained equal to the dilution rate F_{in}/V. This phenomenon has been previously identified and the dynamic steady-state has been termed a "quasi-steady-

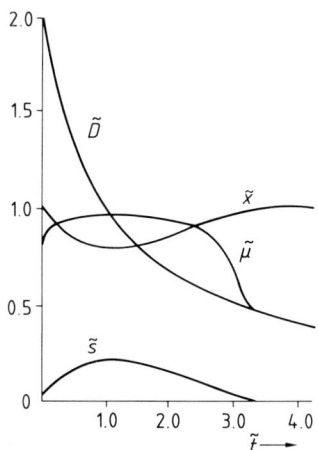

Figure 16.4. Computer simulation of the fed-batch process based on Eqs. (16.9) to (16.12) in dimensionless form ($\tilde{x} = x/Y \cdot s_{in}$; $\tilde{s} = s/s_{in}$; $\tilde{D} = D/\mu_{max}$; $\tilde{F} = F_{in}/V \cdot \mu_{max}$; $\tilde{\mu} = \mu/\mu_{max}$; and $\tilde{t} = t \cdot \mu_{max}$). As can be seen, a quasi-steady-state with $\tilde{D} = \tilde{\mu}$ is attained during a certain period of time (DUNN and MOR [16.20]).

state". This is characterized by a constant value of x_{ex}, which must exist because $\mu = F_{in}/V$. Since the volume is increasing steadily, μ must be maintained by a decrease in s_{ex}; therefore, ds_{ex}/dt is not zero. Computer simulations obtained by solving Eqs. (16.9), (16.11), and (16.12) numerically show that the phenomena do indeed occur as described. A computer simulation of fed-batch cultivation is shown in Fig. 16.4, calculated on the basis of dimensionless variables (DUNN and MOR [16.20]).

As can be seen, S steadily decreases during the quasi-steady-state, $\mu = D$, in order to maintain the quasi-steady-state as the volume increases.

When the quasi-steady-state is achieved, the equality of μ and F_{in}/V leads to the following relationship for s_{ex}:

$$s_{ex} = \frac{F_{in}}{V} \cdot \frac{K_S}{\mu_{max} - \frac{F_{in}}{V}}. \qquad (16.13)$$

For the special case of constant feed rate, F_{in}, the fed-batch has some further properties that are worth noting.

a) From Eq. (16.10) under the quasi-steady-state conditions with $\mu = (F_{in}/V)$, it follows that:

$$\frac{d(V \cdot x_{ex})}{dt} = F_{in} \cdot x_{ex} = \text{const}. \qquad (16.14)$$

The rate of change of total biomass is constant.

b) From Eq. (16.9) the volume must increase linearly with time:

$$V = V_0 + F_{in} \cdot t. \qquad (16.15)$$

c) The rate of change of μ during a quasi-steady-state period is:

$$d\left(\frac{F_{in}}{V}\right) = -\frac{F_{in}^2}{V^2} = -\frac{F_{in}^2}{(V_0 + F_{in} \cdot t)^2}; \qquad (16.16)$$

at low t when $V = V_0$ this becomes:

$$\frac{d\mu}{dt} = -\frac{F_{in}^2}{V_0^2}, \qquad (16.17)$$

at higher values of t when $V \gg V_0$:

$$\frac{d\mu}{dt} = -\frac{1}{t^2}. \qquad (16.18)$$

Thus, it is seen that μ decreases most rapidly during the first period of the fed-batch, the rate of decrease becoming slower with time. For the purpose of obtaining a particular desired rate of change of μ, the flow rate F_{in} could be changed progressively during the course of fermentation.

6.2.3.3 Mathematical Modelling of Fed-batch Process Kinetics

Unstructured Models

In general, unstructured models can be considered as a suitable approximation for two distinct cases. These cases arise when the composition of the organisms is not relevant to the aspects of the system described by the model, or when it is independent of time, i.e., in balanced growth. Thus, unstructured models are adequate for relaxed steady-state and quasi-steady periodic operation as shown in Fig. 16.1, while structured approaches are to be preferred in the case of intermediate periodic operation. However, even in this case the use of unstructured but modified models for the kinetics is possible and advantageous in some cases, as unstructured models contain fewer process variables to be measured but more kinetic parameters to be evaluated (MOSER [16.13]).

Unstructured but modified modelling of process kinetics has been successfully applied to biological waste-water treatment, where non-stationary behavior occurs as a consequence of fluctuations in feed-stream or shock-loading. MONA et al. [16.21] modified the Monod-type kinetics and adapted them to real situations by introducing load-dependent kinetic parameters:

$$q_{S,\,max} = k_1 \cdot B_x, \qquad (16.19)$$

$$K_S = k_2 \cdot B_x, \qquad (16.20)$$

with k_1 and k_2 as empirical constants and B_x as the mass loading rate per unit of biomass (sludge):

$$B_x = \frac{F \cdot s_0}{V \cdot x} \,. \qquad (16.21)$$

Further, these authors introduced time lags τ_{L1} and τ_{L2} responsible for time dependence:

$$\hat{q}_{S,\,max} + \tau_{L1} \frac{d\hat{q}_{S,\,max}}{dt} = k_1 \cdot B_x \qquad (16.22)$$

and

$$\hat{K}_S + \tau_{L2} \frac{d\hat{K}_S}{dt} = k_2 \cdot B_x \,, \qquad (16.23)$$

with $\hat{q}_{S,\,max}$ and $\hat{K}_S$ as the deviation variables defined as:

$$\hat{q}_{S,\,max} = q_{S,\,max}(t) - \bar{q}_{S,\,max} \qquad (16.24)$$

and

$$\hat{K}_S = K_S(t) - \bar{K}_S \,. \qquad (16.25)$$

The basis of this formal kinetic modelling is shown in Fig. 16.5 together with the definitions of τ_L and $\bar{q}_{S,\,max}$.

A formal kinetic approach to product formation under transient conditions has been presented by GUTKE et al. [16.22]:

$$k_P = k_P(\mu; \dot{\mu}) \,. \qquad (16.26)$$

As the physiological state can be characterized by the specific growth rate μ, the non-stationary state is quantified by the time derivation $\dot{\mu}$, which is assumed to be constant (c_μ):

$$\dot{\mu} = -c_\mu \,. \qquad (16.27)$$

The specific growth rate $\mu(t)$ is then given by:

$$\mu(t) = \mu_0 - c_\mu \cdot t \,. \qquad (16.28)$$

Another attempt to develop a generalized rate equation was recently made by HARDER and ROELS [16.23] by application of the relaxation time concept. A model de-

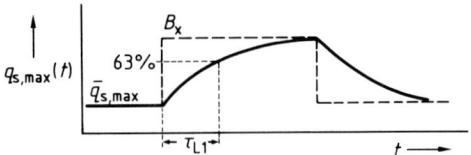

Figure 16.5. Illustration of the non-stationary behavior of a bioprocess as a consequence of a stepwise change in feed stream (dotted line represents the organic loading B_x) and an unstructured model approach to non-stationary kinetics using load depending kinetics [cf. Eqs. (16.19) and (16.20)] together with time lags (τ_L) for q_S as indicated (MONA et al. [16.21]).

scribing the dynamics of product formation was derived using Eq. (14.10b) together with Monod-type kinetics for r_s incorporating endogenous metabolism with the aid of Eq. (14.62) which results in an expression for steady-state:

$$q_P = \frac{Q}{Y_{X/S}} \mu + Q \cdot m_S \,, \qquad (16.29)$$

with Q as the product formation activity function, which is related in steady-state $(\bar{Q})$ to μ:

$$\bar{Q} = f(\mu) \,. \qquad (16.30)$$

Extension to dynamic situations is achieved by the formulation of a rate of adaptation of Q to environmental conditions with the aid of the intrinsic balance equation:

$$\dot{Q} = \frac{1}{x} (r_Q - r_x \cdot Q) \,, \qquad (16.31)$$

in which r_Q is the rate of synthesis of Q.

Combining Eqs. (16.30) and (16.31) with an equation for the quantification of the difference between actual rate of synthesis of Q and the steady-state rate by using a constant of adaptation (K_{adapt}), which can be estimated from shift-down or shift-up experiments, finally results in the following expression for the rate of change of Q:

$$\dot{Q} = -(K_{adapt} + \mu)(Q - \bar{Q}) \,. \qquad (16.32)$$

Fig. 16.6 shows the apparent relationship between Q and μ according to Eq. (16.32) in a dynamic situation. When K is large, the steady-state relationship is obtained.

The relaxation time t_{charact} for adaptation to a new steady state during a shift in continuous culture is given by:

$$t_{\text{charact}} = \frac{1}{(\mu + K_{\text{adapt}})} = \frac{1}{\mu(1 + K_{\text{adapt}}/\mu)} \cdot \quad (16.33)$$

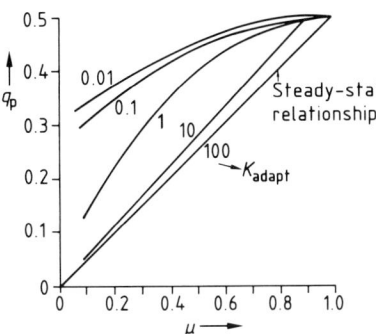

Figure 16.6. Steady-state and dynamic relationship between specific rate of product formation q_P and specific growth rate μ for various values of the constant of adaptation K_{adapt} according to Eq. (16.33). The steady-state relationship $Q = 0.5 \cdot \mu$, determined from continuous culture, is shown as limiting case (HARDER and ROELS [16.23]).

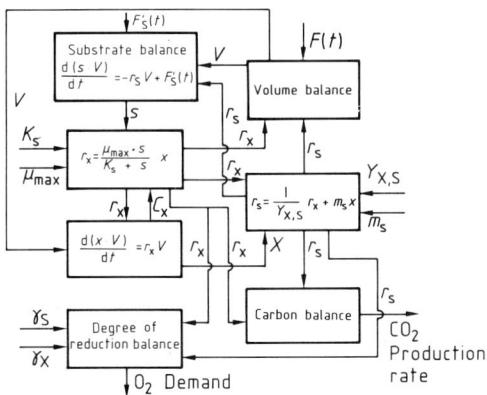

Figure 16.7. A block diagram of unstructured modelling for fed-batch growth kinetics according to ROELS [16.6].

If K is large, a new steady state will be obtained almost instantaneously without lag of the organism. If K_{adapt} is small, the time constant for adaptation will be equal to $1/\mu$, i.e., dilution through growth controls the adaptational process.

ESENER et al. [16.24] applied unstructured models to fed-batch processing on the basis of Eqs. (16.11) and (16.12) for x and s. If such an unstructured model is combined with the elemental and energy-balance principles, the time dependence of O_2-consumption and CO_2- or H_v-production can be calculated for batch, continuous, and fed-batch cultivation. A block diagram of such an approach is shown in Fig. 16.7, taken from ROELS [16.6].

The same investigation of fed-batch growth (ESENER et al., [16.24]) has shown that in those cases where detailed kinetics play a role, i.e., in the shift from a process controlled by the maximum specific growth rate of the organism to a situation where growth is limited by the addition rate of the substrate, the model only provides a poor fit to the phenomena actually observed. In fact, such deviations between the predictions of a simple unstructured model and the much more complex reality are easily understood if the internal functioning of an organism is studied in somewhat more detail. Basically a microbial system adapts its activities to changes in the environment by the types of mechanisms illustrated in Fig. 24.6. Structured models, therefore, should be more adequate for a better description of the system during highly transient periods.

Structured models

In developing structured models one has to select those parameters, which are most relevant for the description of the physiological state of the organism. Information from molecular chemistry, microbiology, and biochemistry is required for this purpose. A logical first choice would be to select DNA, RNA, carbohydrate, and protein contents of the cells to describe their physiological state.

All of these variables can be experimentally determined and their dependence on the steady-state growth rate is well established. It has to be noted, however, that even the consideration of these four relevant components is not sufficient to fully describe the activities and qualities of the organisms; e.g., no information can be obtained about the geometrical structure of the cells nor about the dependence of diffusional processes on such structures (KOSSEN [16.25]). Thus, a structured model should be formed in such a way that it only provides information about the most relevant processes and variables.

Rather simple models can be obtained by considering a few variables as an extension of the unstructured models, which consider one biotic variable (ESENER et al. [16.26]). These models, in which the activity of biomass is specified by more than one and up to three or four variables, are called compartmental models. They have moderate mathematical complexity and are easier to verify experimentally.

aspects of structured modelling. Recent reviews on structured models and critical literature surveys can be consulted for further information (FREDRICKSON [16.32]; ROELS [16.29]; HARDER and ROELS [16.23]).

Diagrams of a two-compartment model are shown in Fig. 16.8 and of a three-compartment model in Fig. 16.9. Here, the external limiting substrate is assumed to be converted to K, from which G forms with constant yield. A third reaction involves the turnover of macromolecules, i.e., G material, back to K material (small molecules). In this manner the maintenance requirements for turnover should be incorporated into the model. The maintenance process is accounted for only by the depolymerization reaction of the macromolecules without

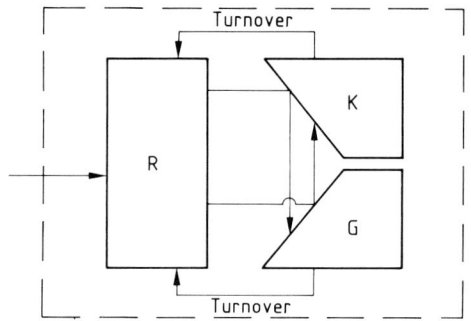

Figure 16.9. Schematic representation of a three-compartment model according to WILLIAMS [16.27], including a G-, K-, and R-compartment.

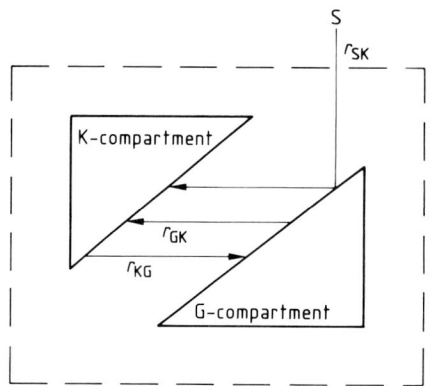

Figure 16.8. Schematic representation of a two-compartment model according to WILLIAMS [16.27], containing the G- and the K-compartment (x_G and x_K). The following rates are included: r_{SK}, r_{KG}, and r_{GK} (see text).

The first formal compartmental model was developed by WILLIAMS [16.27], [16.28]. Some inconsistencies of this model were later corrected by ROELS and KOSSEN [16.33] and ROELS [16.29]. RAMKRISHNA et al. [16.30] and FREDRICKSON et al. [16.31] have contributed greatly to the theoretical

mass loss in the biotic phase. In the presented model, RNA, carbohydrate, and other small cellular molecules are lumped into one compartment, which will be referred to as the K-compartment. Since RNA is the main constituent, the amount of K may be expected to be proportional to the RNA content. The other compartment is called the G-compartment and contains the genetic material and all the rest of the cell constituents, i.e., proteins, DNA, structural material, etc. This approach more or less assumes the RNA concentration (and thus synthesis) to be the bottleneck, and since RNA plays a central role in the synthesis of proteins, this may be assumed to be a reasonable approach to structured modelling.

Thus, the following reactions, essentially based on the work of WILLIAMS [16.27] as well as on modifications and extensions by ROELS and KOSSEN [16.33], and ROELS [16.34], can be formulated:

$$S \xrightarrow{G} Y_{SK} K \quad \text{rate } r_{SK}, \quad (16.34)$$

$$K \xrightarrow{K} Y_{KG} G \quad \text{rate } r_{KG}, \quad (16.35)$$

$$G \longrightarrow K \quad \text{rate } r_{GK}. \quad (16.36)$$

A computer simulation of the Williams model is shown in Fig. 16.10, illustrating the basic behavior of the two-compartment model.

The next step is the postulation of rate expressions for r_{SK}, r_{KG}, and r_{GK}.

For a description of the state of the culture as a function of time, the approach advocated by HARDER and ROELS [16.23] can be used without complications when the kinetics and the stoichiometry of the process involved are defined, such as:

a) the conversion of the substrate to the K-compartment, (r_{SK}),

b) the transformation of the K- into the G-compartment, (r_{KG}), and
c) the turnover of the compartments of the biomass, (r_{GK}).

The balance equations for the rate of change of the substrate concentration, the biomass concentration, and the fraction of the G-compartment are now obtained by application of the formalism treated by HARDER and ROELS [16.23]. The resulting equations are:

$$\frac{ds}{dt} = -q_{S,\max} \frac{s}{K_S+s} \cdot \frac{w_G}{K_G+w_G} x + F_S, \quad (16.37)$$

with w_G being the mass fraction of component G and F_S the flow of substrate to the system (in $kg \cdot m^{-3} \cdot h^{-1}$);

$$\frac{dx}{dt} = Y_{SK} \cdot q_{S,\max} \frac{s}{K_S+s} \cdot \frac{w_G}{K_G+w_G} x + (Y_{KG}-1) f(w_G) x + F_X, \quad (16.38)$$

with F_X as the net flow of unstructured biomass dry weight to the system (in $kg \cdot m^{-3} \cdot h^{-1}$); and

$$\frac{dw_G}{dt} = -Y_{SK} \cdot q_{S,\max} \frac{s}{K_S+s} \cdot \frac{w_G}{K_G+w_G} \cdot w_G + f(w_G) \cdot \{w_G + Y_{KG}(1-w_G)\} - m_G w_G, \quad (16.39)$$

with m_G as the specific turnover rate of compartment G (maintenance rate). These equations contain the transport contributions F_S and F_X, which depend on the mode of operation and are given in the following table:

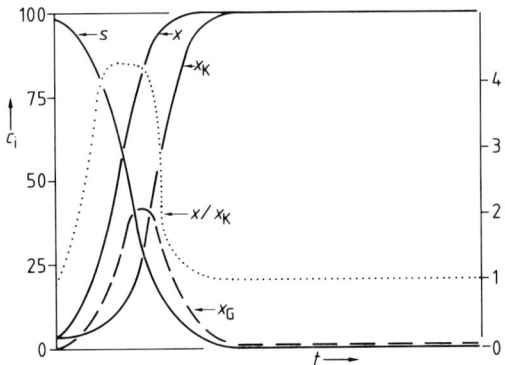

Figure 16.10. Representation of a batch-culture growth based on the two-compartment model by a graphical plot of concentrations versus time for substrate s, total biomass x, G-compartment biomass x_G, and K-compartment biomass x_K as well as the ratio between x and x_K (WILLIAMS [16.27], [16.28]).

	F_S	F_X
for batch operation	0	0
for CSTR with feedback of fraction $(1-w_D)$ of the biomass leaving the system	$D(s_{in}-s_{ex})$	$-w_D \cdot D \cdot x$
for fed-batch culture	$F_S(t)$	0

A most important feature of Eqs. (16.37) to (16.39) is the fact that the intrinsic balance equation is independent of the mode of operation.

The necessity for a model of such simplicity may not be obvious although sufficient knowledge is available for the construction of a model more closely describing reality. Moreover, two factors should be considered:

- A number of regulatory mechanisms at the level of energy generation and consumption operate with such small relaxation times that a pseudo-steady-state hypothesis with respect to these mechanisms is justified. Hence, the introduction of these details seems unnecessary.
- A minimum of complexity is desirable because the complex model often proves very difficult to verify and may fit experimental results without having any relationship to the behavior of the organism (ROELS and KOSSEN [16.33]). Only after obtaining experimental evidence that the simple model can be rejected because of an insufficient fit of the data or unrealistic parameter values, should additional data be introduced. An additional "hidden" variable must be specified.

If the obtained relations for r_S are compared, the structured model can be seen to have a maintenance term dependent on the growth rate:

$$m_S = \frac{k_{KG}}{Y_{SK}} \cdot \frac{D + m_G}{(k_{KG} \cdot Y_{KG})^2} \cdot$$
$$\cdot [k_{KG} Y_{KG} - (D + m_G)](1 - Y_{KG}). \tag{16.40}$$

The two-compartment model exhibits many features observed in batch and continuous culture experiments. The model must be considered as a preliminary proposal because many of the kinetic assumptions do not rest on solid biochemical facts about the internal regulation of the cell. A more thorough study of known regulatory phenomena and an empirical study of transient situations, for example, in continuous culture, is needed.

16.2.3.4 Application of Fed-batch Cultivation

For fundamental research of process kinetics a fed-batch process is of advantage, as it permits maintenance of low S-concentrations over a long period, which is favorable for the estimation of kinetic parameters: K_S in Monod kinetics, maintenance coefficients, and bioenergetics (ESENER et al. [16.24]), and also ordinary kinetics (see e.g., LEE and YAN [16.35]). Several computer simulations have been carried out in order to demonstrate productivities compared to the CSTR at constant volume (KELLER and DUNN [16.36]) and also to show the influence of different errors, e.g., fluctuations in volume and S-feed (KELLER and DUNN [16.37]). Computed results by sensitivity analysis also provide very useful information on the degree of precision of parameter values in process equations (KISHIMOTO et al. [16.38]). Furthermore, the fed-batch process provides the same information as the CSTR without requirement for an outflow, without volume control, and without the necessity of shifting steady-states.

Fed-batch cultures have the unique feature of realizing transient conditions with the growth rate under control between fixed values. There is evidence that the maximum rates of some processes can be achieved only transiently. Examples are the production of secondary metabolites (PIRT [16.17]; BAJPAI and REUSS [16.39], [16.40]; GUTKE and KNORRE [16.41] and [16.42]), and processes with complex kinetics like catabolite repression and/or inhibitions as in yeast technology with the glucose effect (AIBA et al. [16.43], DAIRAKU et al. [16.44]), and waste water treatment.

A review of some fermentation processes, in which batch-feed or continuous-feed has been used or tested including a description of methods of nutrient addition was published by WHITAKER [16.45].

16.3 Dialysis Cultures

16.3.1 General Aspects and Application of Membrane Processing

In several microbial processes economic efficiency can be drastically increased when a technical solution for the discharge of accumulated inhibitory or toxic products or metabolites is available. Examples are the production of ethanol from carbohydrates (KOSARIC et al. [16.46]; CHARLEY et al. [16.47]; ROGERS et al. [16.48]), the utilization of whole or deproteinized whey for the production of food yeasts, nitrogenous feed supplement for ruminants (GERHARDT [16.49]; GERHARDT and GALLUP [16.50]; LANE [16.51]), the production of salicylic acid from naphthalene, the threonine biosynthesis (ABBOTT and GERHARDT [16.52]), and the continuous aseptic production of phytoplankton (MARSOT et al. [16.53]).

A number of different approaches have been described as technical solutions for the mentioned problems by MAIORELLA et al. [16.54], HAMER [16.55], and LILLY [16.56]: vacuum, membrane, retractive and extractive fermentation, biphasic and ion-exchange resin culture techniques.

Although membrane processes have been studied for more than a century, they have only recently become of interest for industrial separations. This is due to the fact that these processes allow separation of dissolved materials from one another or from a solvent without phase change. In contrast to evaporation or crystallization processes for product recovery, membrane processes do not require the energy represented by the latent heat of vaporization or crystallization. Since energy costs represent a sizable portion of the total operating costs for most separations, the possibility of using membrane processes is attractive (e.g., RAUTENBACH and ALBRECHT [16.57]; PYE and HUMPHREY [16.58]).

Among the various membrane techniques, dialysis appears desirable for the initial study, because of its simplicity and the existence of a theoretical basis for its application to bacterial cultures (SCHULTZ and GERHARDT [16.59]).

In dialysis, i.e., the separation of solute molecules by their unequal diffusion through a semi-permeable membrane based on a concentration gradient, the small molecular products are removed from the immediate environment of the bacterial cell (and ultimately from the intracellular enzyme site), thus relieving the feedback inhibition by a product that normally regulates its production. As more product is withdrawn by dialysis, more substrate is consumed and more product is obtained; i.e., the fermentation becomes more efficient.

16.3.2 Dialysis Reactor Systems

Dialysis processes were first developed as a batch technique using a simple dialysis flask (GERHARDT and GALLUP [16.50]). A simple internal filtration unit is schematically shown in Fig. 16.11a. The process is readily adapted to continuous operation as illustrated in Fig. 16.11b. A culture of active cells is maintained in a confined zone of the vessel. Substrate enters the reaction zone by diffusing through a dialysis membrane from the medium zone. Reaction then takes place and product diffuses back through the membrane into the medium zone where it is recovered in an overflow. Cells cannot escape the reaction zone and extremely high cell densities can be achieved.

The configurations of dialyzer-dialysis fermenter systems are graphically shown in Fig. 16.11c.

Last not least, Fig. 16.11d represents a scheme of a laboratory-scale filter fermenter with internal separation of cells by means of filtration or dialysis recently developed (DOSTALEK and HÄGGSTROM [16.60]). A filtration-fermenter with a rotat-

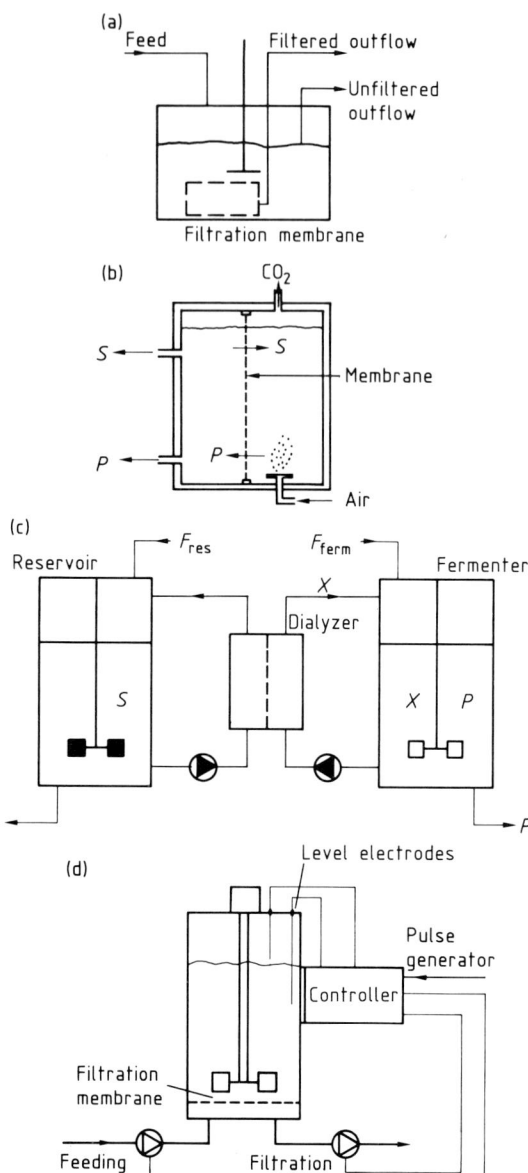

Figure 16.11. Schemes of different types of dialysis bioreactors: (a) simple internal filtration unit (GERHARDT and GALLUP [16.50]); (b) simple continuous dialysis fermenter (MAIORELLA et al. [16.54]); (c) a reactor system consisting of a fermenter, a reservoir, and a dialyzer (SCHULTZ and GERHARDT [16.59]); (d) continuous culture with feedback using a filtration membrane on the bottom of the vessel (DOSTALEK and HÄGGSTRÖM [16.60]).

ing microfiltration unit has been described by SORTLAND and WILKE [16.61].

The problem of membrane fouling has been overcome in the rotor fermenter (Fig. 16.12) (MARGARITIS and WILKE [16.62]). This is a continuous pressure dialysis reactor first suggested by BHAGAT and WILKE [16.63]. The fixed membrane is replaced by a rapidly rotating membrane cylinder. Feed and air are pumped into the annular reaction zone. This zone is maintained under pressure (115–170 kPa), so that filtrate is continuously forced through the membrane. Because the membrane is rotating, a strong centrifugal force is developed at the membrane surface and large molecules impinging on the surface are thrown back into the annular zone. Only a very thin steady-state cake is developed. Using a laboratory-scale rotor fermenter, cell densities of 50 g·L^{-1} and ethanol productivities of 36 g·L^{-1}·h^{-1} have been achieved.

The rotor fermenter appears attractive in terms of its high productivity. It fails, however, when rated on mechanical and operating simplicity.

A high cell density plug fermenter has been described, which is a particularly interesting version of the dialysis fermenter (BAKER and KIRSOP [16.64]). A dense plug of yeast is maintained between two support plates. The medium is pumped through the plug under high pressure (120–1100 kPa) and rapid reaction takes place. Another approach to achieving a high rate dialysis process involves the use of hollow fiber reactors (KAN and SHULER [16.65]). Hollow fiber reactors provide extremely large membrane surface areas, so that rapid substrate diffusion and high reaction rates are possible. The hollow fiber reactor is arranged like a shell and tube heat exchanger. The tubes are fine hollow (0.5 mm) fibers of membrane material. A bundle of 1000 of these fiber tubes can be packed into a single shell of 3 cm diameter.

Hollow fiber reactors have been used on a laboratory scale to produce several high-value biological products (KAN and SHULER [16.66]; WEBSTER [16.67]), for ethanol production (see MAIORELLA et al. [16.54]), and generally for bioprocessing

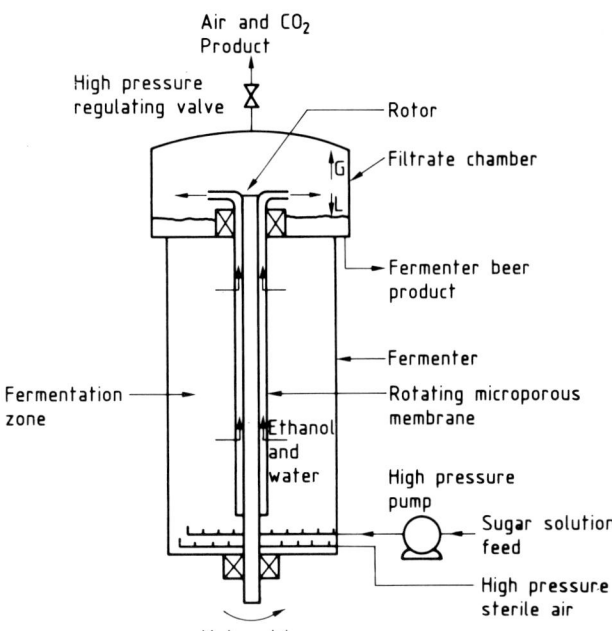

Figure 16.12. Schematic diagram of the rotor-fermenter assembly as a continuous pressure dialysis reactor using a rotating stainless steel membrane cylinder (MARGARITIS and WILKE [16.62]). G Gas, L Liquid.

with enzymes and/or immobilized cells (e.g., MESSING [16.68]; VENKATASUBRAMANIAN [16.69]). The hollow fiber reactor has potential to achieve yeast cell densities approaching the maximum cell packing density (200 g·L^{-1}) and correspondingly high ethanol productivities should be possible. Venting of carbon dioxide gas may pose problems, however (SHIOTANI and YAMANE [16.70]; CHO et al. [16.71]), and membrane plugging may also be a problem, so that the hollow fiber reactor will not be important as an industrial ethanol reactor.

Recently, an optimization study of process design for continuous alcohol production by means of *Zymomonas* sp. was carried out with a two-stage CSTR system; a hollow fiber ultrafiltration unit was used for cell recycling after the second stage, assuring continuous operation for extended periods. The effluent ethanol concentration was increased up to > 100 g/L and a 60% increase of the volumetric productivity was observed compared to the normal single-stage CSTR without cell recycling (CHARLEY et al. [16.47]). Similar reactors with ultrafiltration membranes have increasingly been reported in the literature (e.g., HONG et al. [16.72]). In order to reduce the severe flux losses during processing, self-cleaning membranes with membrane-immobilized enzymes have been developed (VELICANGIL and HOWELL [16.73]).

16.3.3 The Dialysis Culture Theory

16.3.3.1 Principles and Performance Equations

Four basic modes are generally possible in dialysis culture operation shown in Fig. 16.11c: (1) continuous reservoir and continuous fermenter with F_{res} and $F_{ferm} > 0$; (2) batch reservoir and batch fermenter with F_{res} and $F_{ferm} = 0$; (3) batch reservoir and continuous fermenter with $F_{res} = 0$ and

$F_{\text{ferm}} > 0$; and (4) continuous reservoir and batch fermenter with $F_{\text{res}} > 0$ and $F_{\text{ferm}} = 0$ (SCHULTZ and GERHARDT [16.59]).

Mass transport through the membrane depends strongly on the membrane permeability coefficient P_{mb}, which includes several unknown factors and which must be measured experimentally:

$$P_{\text{mb}} = \frac{n}{A_{\text{mb}} \cdot \Delta S}, \quad (16.41)$$

with n as the rate of permeation, A_{mb} the membrane area, and ΔS the substrate concentration difference across the membrane. The permeability coefficient reflects the overall mass transport resistance, which can be quantified using the 2-film theory:

$$\frac{1}{P_{\text{mb}}} = \frac{1}{P_{\text{mb}}^0} + \sum_i \frac{\delta_i}{D}, \quad (16.42)$$

where P_{mb}^0 is the true permeability, D the diffusion coefficient, and δ_i the thickness of transport limiting films in the liquid.

Using Eq. (16.41) for membrane permeation, the basic-balance equations for substrate and product are as follows:

$$V_{\text{res}} \frac{ds_{\text{res}}}{dt} = F_{\text{res}} \cdot s_{\text{res}}^0 - F_{\text{res}} \cdot s_{\text{res}} +$$

$$+ P_{\text{mb,S}} \cdot A_{\text{mb}} (s_{\text{ferm}} - s_{\text{res}}), \quad (16.43)$$

$$V_{\text{res}} \frac{dp_{\text{res}}}{dt} = F_{\text{res}} \cdot p_{\text{res}}^0 - F_{\text{res}} \cdot p_{\text{res}} +$$

$$+ P_{\text{mb,P}} \cdot A_{\text{mb}} (p_{\text{ferm}} - p_{\text{res}}). \quad (16.44)$$

A similar balance can be established in the fermenter for substrate, biomass, and product in the general form:

$$V_{\text{ferm}} \frac{ds_{\text{ferm}}}{dt} = F_{\text{ferm}} (s_{\text{ferm}}^0 - s_{\text{ferm}}) -$$

$$- P_{\text{mb,S}} \cdot A_{\text{mb}} (s_{\text{ferm}} - s_{\text{res}}) - r_S \cdot V_{\text{ferm}}, \quad (16.45)$$

$$V_{\text{ferm}} \frac{dx_{\text{ferm}}}{dt} = F_{\text{ferm}} \cdot x_{\text{ferm}}^0 -$$

$$- F_{\text{ferm}} \cdot x_{\text{ferm}} + V_{\text{ferm}} \cdot r_x, \quad (16.46)$$

$$V_{\text{ferm}} \frac{dp_{\text{ferm}}}{dt} = F_{\text{ferm}} (p_{\text{ferm}}^0 - p_{\text{ferm}}) -$$

$$- P_{\text{mb,P}} \cdot A_{\text{mb}} (p_{\text{ferm}} - p_{\text{res}}) \cdot r_P \cdot V_{\text{ferm}}. \quad (16.47)$$

These five mass balance equations plus model equations for kinetics (r_x, r_S, and r_P, see Chapter 14) are mathematically sufficient to determine dialysis culture operations (SCHULTZ and GERHARDT [16.59]; COULMAN et al. [16.74]).

16.3.3.2 Implication of Theory and Comparison between Dialysis and Non-dialysis Culture Techniques

Case 1 of completely continuous dialysis operation is the only one which is essentially steady-state in nature, so that Eqs. (16.43) to (16.47) can be simplified, as time derivatives are all equal to zero. Solutions are fully described in original papers together with the behavior of x, s, and the critical dilution rate D_c as a function of dialysis. Using this solution a comparison between dialysis and non-dialysis continuous cultures can be realized as shown in Fig. 16.13 a–c. The comparison is carried out with respect to cell concentration (a), productivity (b), and efficiency of cell production (c), which is defined as the actual production rate divided by the production rate equivalent to complete utilization of the substrate supplied.

The most striking and important difference between non-dialysis and dialysis continuous cultures is the much higher cell concentration attainable in dialysis culture, especially at low dilution rates. Further, it can be seen that a maximum production rate in continuous dialysis cultures is achieved at a lower dilution rate than in non-dialysis continuous cultures, which is reached, however, at the expense of lower efficiencies in converting substrate to cells.

Case 2 (batch operation) is still predominantly used in process analysis. Mathemati-

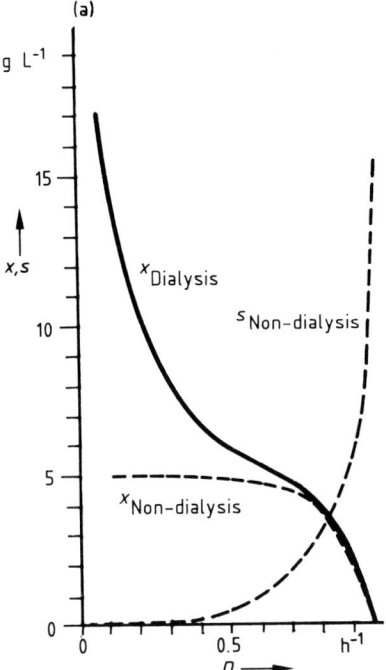
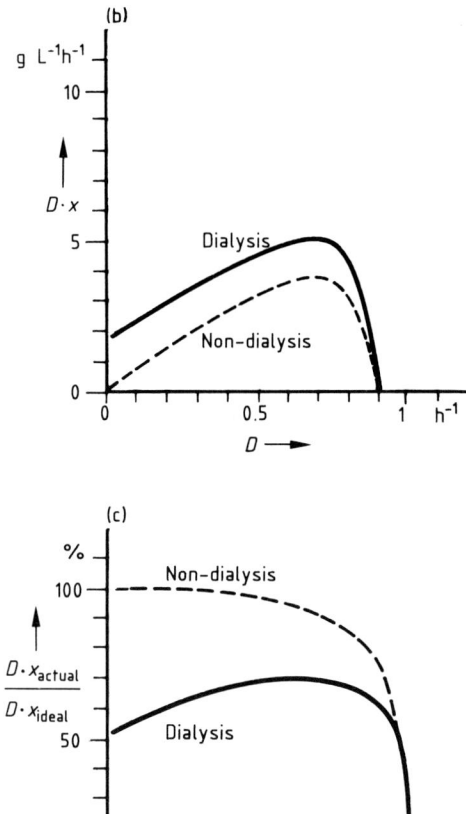

Figure 16.13. Calculated comparison of complete continuous dialysis and non-dialysis continuous culture as a function of the dilution rate D with respect to (a) cell concentration x, (b) cell mass productivity, and (c) cell production efficiency. The following conditions were assumed: $V_{ferm} = 1$ L, $\mu_{max} = 1$ h^{-1}, $K_S = 2 \cdot 10^{-4}$ g·mL^{-1}, $Y_{X/S} = 0.5$; $s^0_{res} = s^0_{ferm} = 10^{-2}$ g·mL^{-1}; $F_{res} = 500$ mL·h^{-1}; $P_{mb} \cdot A_{mb} = 420$ cm^3·h^{-1} (after SCHULTZ and GERHARDT [16.59]).

cally, the values of F_{res} and F_{ferm} in Eqs. (16.43) to (16.47) become zero. The expected growth pattern for the full batch dialysis culture is best evaluated by computer and shown in Fig. 16.14.

Case 3, the batch fermenter/continuous reservoir dialysis culture, can be evaluated similarly. The behavior is also shown in Fig. 16.14. Linear growth can be maintained as S_{res} is constant due to membrane diffusion. Thereby, the limitation of dialysis fermentation becomes evident. The inherently slow process of diffusion of both nutrients and metabolic products through the membrane is the rate determining step. In this way, inhibitory and/or toxic substances can be accumulated, which limit maximum cell concentration.

In the case of product formation in dialysis cultures, the only additional information needed is the permeability characteristic of the product for the membrane material and also the kinetic model for product formation. The intricacies of fitting a mathematical model to the kinetics of product formation was illustrated for steroid conversions by CHEN et al. [16.75], where phenomena such as substrate solubility, coprecipitation, feedback and substrate inhibition had to be taken into account.

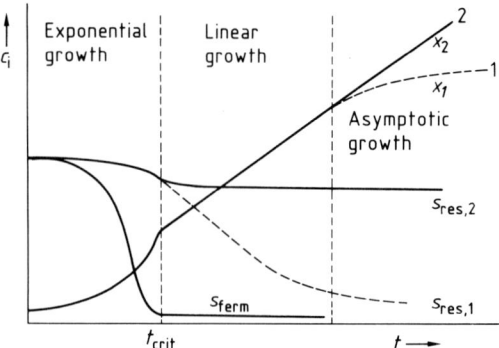

Figure 16.14. Calculated changes in time of cell concentration x and limiting substrates s_{ferm} and s_{res} for dialysis culture systems operated (1) with batch fermenter and batch reservoir, and (2) with batch fermenter and continuous reservoir. The critical time t_{crit}, when substrate diffusion through membrane becomes limiting, is indicated in the plot (after SCHULTZ and GERHARDT [16.59]).

Some analytical solutions are possible for fully continuous operation. The final concentration of the product, if it cannot diffuse across the membrane, will be $p = P/V_{ferm}$. However, if the product is diffusible ($P_{mb,P} > 0$), then it will be distributed between the fermenter and the reservoir, so that $p = P/(V_{ferm} + V_{res})$; thus:

$$p_{res} = \frac{p_{ferm}}{\dfrac{F_{res}}{P_{mb,P} \cdot A_{mb}} + 1}. \qquad (16.48)$$

In order to obtain a large fraction of product in the reservoir effluent free from cells, both the permeability $P_{mb,P} \cdot A_{mb}$ and F_{res} must be large in comparison to F_{ferm}. This type of operation results in a low concentration of product and creates difficulties in product recovery.

Experimental tests for dialysis culture processes are described in the literature for the ammonium-lactate fermentation of whey (STIEBER et al. [16.76]), and an improved mathematical model was developed thereafter by STIEBER and GERHARDT [16.77], incorporating P-inhibition. Recently, the mathematical model was again modified to incorporate a second feed stream of cells, substrate, and product into the fermenter. The behavior of this dialysis culture system with a cell-feed flow from a prefermenter or from cell-recycling was examined and the process has been improved (STIEBER and GERHARDT [16.78]).

16.4 Synchronous Cultures

For many fundamental investigations it is of great importance to study a synchronous culture of microorganisms. Such situations arise in mutation experiments but are of general interest with respect to some inadequacy of the continuous culture theory. In recent years the advent of cell synchrony as a refinement in continuous techniques has added new dimensions to the study of microbial growth (ZEUTHEN [16.79]; CAMERON and PADILLA [16.80]). The potential advantage of synchronous cultures is related directly to the fact that in this technique the cells of the population are all in the same stage of cell development (physiological state). The population serves as an amplification of the cells and, thus, it becomes possible to study the behavior of the cells by observing the behavior of the population. Different methods are known and have been described as synchronous/synchronized pulsed and phased culture techniques (DAWSON [16.81]).

16.4.1 Classification

Growth synchrony can be established by different methods, e.g., shocks or other sudden changes in growth conditions (LARK and MAALOE [16.82]; CUTLER and

EVANS [16.83]), especially with inhibition effects using X-rays (SPOERL and LOONEY [16.84]; HILZ and ECKSTEIN [16.85]), with periodic feeding and substrate exhaustion (WILLIAMSON and SCOPES [16.86]), and by separating cells with different diameters (MITCHISON and VINCENT [16.87]) or with density-gradient-centrifuges (WIEMKEN et al. [16.88]). According to DAWSON [16.89], cell synchrony can be also achieved by periodic substrate feeding. This has later been used in continuous synchronous cultures of microorganisms (KJAERGAARD and JOERGENSEN [16.90]). Two general classes of methods exist for attaining synchrony in batch cultures (MAALOE [16.91]): methods whereby cells at the same stage of division are selected or separated from a randomly dividing population to serve as an inoculum for synchronous cultures, and secondly, methods in which the entire population is manipulated by applied constraints of the environment (which may be nutritional, physical, physiological, or growth inhibitory) to align the cells and produce the inoculum for a synchronized culture. The two methods are sometimes alternatively referred to as selection or induced synchrony, respectively (JAMES [16.92]).

Recently, continuous methods of growth synchronization have been developed. These can be considered as developments or variations of the chemostat method and retain the same advantages, namely those of constantly reproducible growth at any chosen rate and for any desired period of operation.

1. The temporary oscillations observed in the steady-state of the chemostat, when the flow rate has been altered either by "step-up" or "step-down" changes, have been used by VON MEYENBURG to effect a synchronization of yeast cells. This synchrony generally lasts for two or three generations, but on occasion has been stabilized for much longer periods (VON MEYENBURG [16.93]). For t-steps see ZEUTHEN [16.79].

2. An alternative method for synchronizing the cell population in a continuous culture uses pulsed additions of limiting

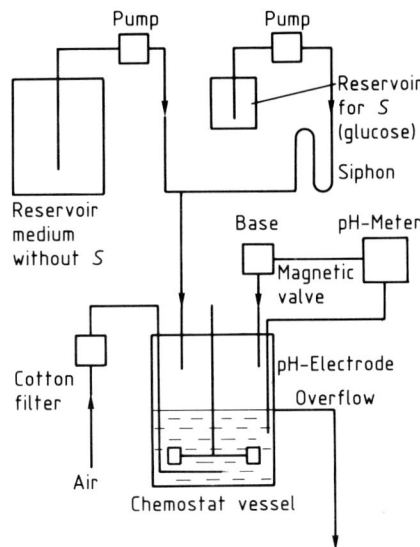

Figure 16.15. Scheme of the apparatus for synchronous culture based on a chemostat applying periodic substrate feeding with time intervals as indicated in Eq. (16.49) (KJAERGAARD and JOERGENSEN [16.90]).

nutrient, at doubling time intervals, to a chemostat culture growing on a medium deficient in that nutrient. HANSCHE [16.94] and GOODWIN [16.95] have both used this method. A modified method proposed by KJAERGAARD and JOERGENSEN is probably better than the one presented by GOODWIN, who used phosphate-limited cultures, because cells may contain different amounts of phosphorus depending on growth rate or growth conditions (TEMPEST et al. [16.96]). The principle of the apparatus shown in Fig. 16.15 is that of a normal chemostat with the exception that the substrate fed with the dilution rate is without a carbon source, and that this carbon source is fed automatically to the vessel at certain intervals in volumes that are negligible compared to the reactor volume.

The time between two additions of carbon source is calculated as follows:

$$\frac{\ln 2}{\mu} + \left(\frac{1}{D} - \frac{1}{\mu}\right)\ln 2 = \frac{\ln 2}{D}. \quad (16.49)$$

This system will give a continuous synchronous culture, in which the cell growth rate is independent of the dilution rate (ln 2/D) as long as $D<\mu$. A new generation of cells is produced every ln 2/D.

3. Another method of synchronization is that of continuous phased culturing, which enables the perspectives of cell behavior to be delineated still further (DAWSON and KURZ [16.97]).

In this technique a single total addition of nutrients is made, at intervals of a chosen doubling time, to a cell population growing "in synchrony". This procedure may be repeated continually and maintains the synchrony.

In Fig. 16.16 the arrangement used for producing a synchronous (phased) culture is diagrammatically compared with that for obtaining continuous asynchronous cultures (chemostat) or batch cultures.

In continuous phased culture, fresh medium is added repeatedly at doubling time intervals to an equal volume of culture; after each addition, one half of the diluted culture is removed to avoid an otherwise exponential increase required in the size of the culture system. The cells adjust their growth rate to the dosing interval and become synchronized to it: by changing the time interval, the doubling time can be changed (DAWSON [16.81], [16.98], [16.99]).

In Table 16.1 the main characteristics of the various techniques now available for cultivating microbes are summarized. It may be noted that the relative merits of constancy and changeability, which HERBERT showed to characterize "open" and "closed" systems in asynchronous cultures apply equally well to cultures with cells growing "in synchrony".

From this classification it becomes clear that GOODWIN [16.95] and VON MEYENBURG [16.93] established real continuous, synchronous cultures in chemostats, whereas the phased culture described by DAWSON [16.81] can hardly be classified as a chemostat culture since half of the volume is exchanged every doubling time. However, this technique is an extension of the continuous culture theory, enabling continually synchronized growth to proceed in a continuous culture as a research tool for conducting studies of the cell cycle. Thus, continuous

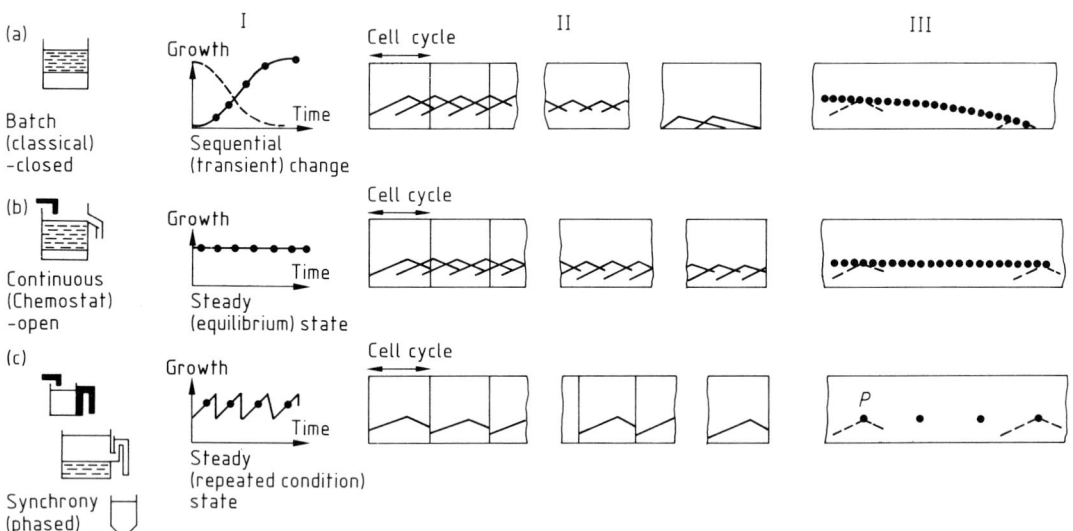

Figure 16.16. Arrangements for obtaining asynchronous cultures in batch (a) or continuous chemostats (b), and synchronous (phased) culture (c); corresponding cell mass concentration versus time plots (I); illustration of the changes with time in terms of the cell cycle (II); a particular event (P) in the cycle (III) (DAWSON [16.81] and [16.99]).

Table 16.1. Main Characteristics of Techniques Available for Cultivating Microbes (DAWSON [16.81])

Type	Method	System	Growth Rate	Technique
Asynchronous	Batch	Closed	Changing	Traditional Bactogen
	Continuous	Open	Constant	Chemostat Turbidostat
Synchrony	Batch	Closed	Changing	Synchronous Synchronous
	Continuous	Open	Constant	Pulsed Pulsed

phased cultures permit a systematic study of the cell cycle under constant conditions: some examples of the method are described in the literature (DAWSON [16.81], [16.100]).

16.4.2 Scope of Synchronous Techniques

Contemporary knowledge of cell metabolism has been mainly acquired using material from asynchronous populations of poorly defined origin and from *in vitro* and not *in vivo* studies.

In batch culture, randomized cell cycle activities are manifest throughout the entire growth sequence, but post-cycle activity only during the final or terminal generation.

In continuous cultures in the chemostat the cell cycle represents, when randomized, the possibilities present in the first propagator stage; and the post-cycle, the range of activity possible in subsequent stages. However, it should be remembered that the multi-stage chemostat will have randomly distributed populations of different age structure in each stage, and these will undergo further randomization between stages.

In continuous phased cultures, however, cell cycle growth takes place and is followed in the same (second stage) vessel by post-cycle development. The cells may be harvested at any point on a basis of elapsed time, much as a tube fermenter might be harvested by length of transit, but with the practical advantage in phased cultures of having a fully mixed and, if required, aerated condition present in the culture vessel throughout the cultivation period (DAWSON [16.101]).

By considering growth and metabolism in the culture in terms of cell cycle and post-cycle activities of the cells, it becomes possible to relate presently incompatible areas in batch and continuous cultures. For instance, if we consider the culture in terms of the cell, the growth curve can be rearranged to yield a possible connection between the areas of growth and secondary metabolism, which BU'LOCK et al. [16.102], BU'LOCK [16.103] describes as the "tropho- and idiophase" (see DAWSON [16.81]).

In phased cultures the cell population is synchronized, and doubling time, nutrient supply per cell, population numbers, and temporal activities become empirical parameters involved in the operation of the technique.

DAWSON [16.81], [16.99] also reviews some of the earlier findings with phased cultures on the basis of cell cycle and post-cycle activities, which the experimental dimensions of the chemostat cannot approach. Because of the need to evaluate parameters both qualitatively and quantitatively in depth in such studies, electronic (Coulter) counting and sizing of cell populations have been used for examining the cultures instead of the compound microscope employed in earlier studies.

16.5 Mixed Microbial Populations

The terms "mixed microbial populations", "heterogeneous populations", and "natural populations" all have been used to denote systems in which natural selection occurs and determines which organisms can survive and which will predominate in the ecosystem. Shifts in species may occur due to natural interactions between species or in response to an external stimulus such as a change in environment. Multiple microbial species are interacting not only in natural ecosystems such as the biological treatment of water (GAUDY and GAUDY [16.104]) and wastes (rumen, soil) but also in technical manufacturing of, e.g., cheese, wine, sauerkraut. Some species found in nature are very difficult to grow as pure laboratory cultures, stressing the importance of interactions between singular species of microbial communities. This section concentrates on the analysis of microbial interactions, starting with the characterization of two-species interactions.

16.5.1 Classification of Types of Microbial Interactions

Various terms are used to denote various types of interactions; however, there is much overlap in the meaning of these terms so that they do not fit all categories of interaction patterns (BUNGAY and BUNGAY [16.105]; NOACK [16.106]; MEERS [16.107]). Table 16.2 shows a classification scheme of all possible combinations of interactions known, i.e., competition, commensalism, amensalism, mutualism, and predation. The influence of species i on species j is determined by defining an element a_{ij} in the so-called community matrix (BAILEY and OLLIS [16.108]). If a_{ij} is positive, species j has a positive effect on growth of species i, while an inhibitory effect is described by a negative value of a_{ij} or a neutral effect by a zero a_{ij} and *vice versa* for a_{ji}.

Competition occurs when a community of two or more species are mutually limiting because of their joint dependence on a common factor external to them.

Commensalism is the case, where the growth of one species is promoted by the presence of a second species in a population, the growth of this second species being unaffected by the presence of the first.

Mutualism is similar to the situation given with commensalism, but both organisms grow faster in the presence of the other than they do separately. It could be caused by the production of growth factors or products which serve as nutrients.

Synergism is a third type of mutualism, in which the formation of specific products is greater in mixed than in pure cultures.

Predation occurs when an organism totally engulfs and digests another organism which thereby loses the ability to reproduce.

Amensalism is the situation, where the growth of one species is repressed because of the presence of a toxic substance produced by another and represents the opposite to commensalism.

Neutralism, which is relatively rare, means that both species have no observable effect on one another.

Symbiosis is again similar and occurs if the mutualistic partnership is necessary for survival of one species.

Parasitism, which is hard to be distinguished as a microbial interrelationship from predation, occurs when one organism feeds or reproduces at the expense of tissues or body fluids of another.

Supplementary to this classification scheme a distinction is often made between open and closed environments, i.e., continuous and discontinuous reactors. An open environment leads to population stability (homeostasis), although oscillations may be observed due either to periodic fluctuations in the environment or to interactions between microbial species. The factors influencing the survival of given species are

Table 16.2. Classification of Pairwise Interactions Between Microbial Populations Based on the Signs of the Entries a_{ji} and a_{ij} from the Community Matrix A (i=j) – adopted from MAY [16.107a].

		Effect of Species j on Species i (sign of a_{ij})		
		−	0	+
Effect of Species i on Species j (sign of a_{ji})	−	− − Competition	− 0 Amensalism	− + Predation
	0	0 − Amensalism	0 0 Neutralism	0 + Commensalism
	+	+ − Predation	+ 0 Commensalism	+ + Mutualism

different in closed and open systems (MEERS [16.107]).

16.5.2 Kinetic Analysis of Microbial Interactions

16.5.2.1 Competition

With the simplest model of two species (x_1, x_2) competing for the same limiting substrate (s) the principle of selection can be demonstrated ("survival of the fittest" organism with the greater specific growth rate). The corresponding reaction scheme and kinetics are listed in Table 16.3.

Fig. 16.17 shows the steady-state values of this model as a function of D. Evidently, no coexistence is possible. However, for lower dilution rates $(D<D_S)$ the first species survives, whereas for a middle range of D $(D_S<D<D_c)$ the second species survives. Above the critical dilution rate D_c both organisms wash out. The plot of the kinetics, $D=\mu_1(^2s)$ and $D=\mu_2(^3s)$, indicates that the switch from the first surviving species to the second is due to the intersection of both μ-characteristics (GUTKE [16.109]).

The principle factors governing the kinetic behavior are the differences in μ and K_S of both species (SIKYTA et al. [16.110]). Thus, with known values of μ_{max} and K_S,

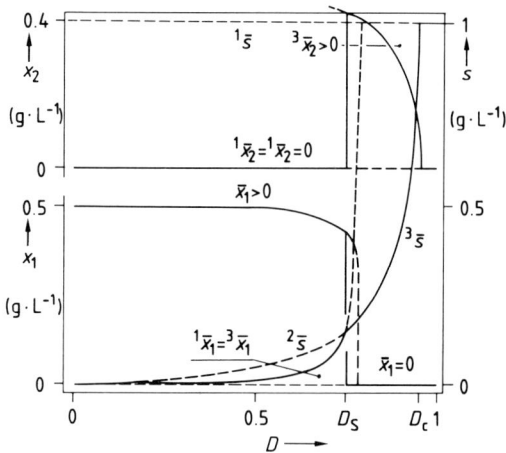

Figure 16.17. Computer simulation of the steady-state concentrations of two microbial species x_1 and x_2 and of the limiting commonly used substrate s as a function of dilution rate D according to Eqs. (16.51) to (16.53). The kinetic constants are as follows: $\mu_{max,1}=0.8$ h^{-1}; $K_{1S}=0.01$ g·L^{-1}; $\mu_{max,2}=1$ h^{-1}; $K_{2S}=0.05$ g·L^{-1}; $Y_{x2/s}=0.5$. No coexistence of both species is possible in the steady-state. For lower dilution rates $(D<D_S)$ x_1 survives, whereas for $D_S<D<D_c$ x_2 dominates. Above D_c, both x_1 and x_2 will wash out (GUTKE [16.109]).

Fig. 16.18 can serve for the identification of a given trend and for the prediction of population changes. The dilution rate can thus be used as a controlling factor. Growth of two competing species in a closed environ-

Table 16.3. Types of Microbial Interactions in Mixed Populations and the Corresponding Reaction Schemes with Kinetic Equation

Type	Reaction Scheme	Process Kinetic Equations for CSTR	
Competition	$S \begin{smallmatrix}\nearrow X_1 \\ \searrow X_2\end{smallmatrix}$ (16.50)	$r_{x_1} = \mu_1(s)x_1 - Dx_1$	(16.51)
		$r_{x_2} = \mu_2(s)x_2 - Dx_2$	(16.52)
		$r_s = -\dfrac{1}{Y_{X_1/S}}\mu_1(s)x_1 - \dfrac{1}{Y_{X_2/S}}\mu_2(s)x_2 + D(s_0 - s)$	(16.53)
Commensalism and Amensalism REILLY [16.119]	$S_I \to X_1 \to P$ (16.54) $S_{II} \to X_2$	$r_{x_1} = (\mu_1 - D)x_1$	(16.55)
		$r_{x_2} = (\mu_2 - D)x_2$	(16.56)
		$r_{s_1} = D(s_{10} - s_1) - \dfrac{\mu_1 \cdot x_1}{Y_{X1}}$	(16.57)
		$r_{s_2} = -Ds_2 + \dfrac{\mu_1 \cdot x_1}{Y_{X1}} - \dfrac{\mu_2 \cdot x_2}{Y_{X2}}$	(16.58)
Mutualism	$S_I \to X_1 \begin{smallmatrix}P_2 \\ \searrow \\ P_1\end{smallmatrix} X_2 \leftarrow S_{II}$ (16.59)	$r_{xi} = \mu_i x_i - Dx_i$	(16.60)
		$r_{Si} = -\dfrac{1}{Y_{X_i}}\mu_i x_i + D(s_0 - s)$	(16.61)
		$r_{P1} = q_{pi} \cdot x_1 - \dfrac{1}{Y_{2/P1}}\mu_2 x_2 - D \cdot p_1$	(16.62)
		$r_{P2} = q_{P2} \cdot x_2 - \dfrac{1}{Y_{1/P2}}\mu_1 x_1 - D \cdot p_2$	(16.63)
		resp.	
		$r_s = D(s_0 - s) - \mu_1 x_1/Y_{1S} - \mu_2 x_2/Y_{2S}$	(16.64)
		$r_P = -D \cdot p + q_P x_2 - \mu_1 x_1/Y_{X/P}$	(16.65)
Predation LOTKA [16.126] VOLTERRA [16.111] BUNGAY and BUNGAY [16.105] CURDS [16.127]	$S \to X_1 \to X_2 \dashrightarrow X_i$ (16.66)	$r_{x_1} = \mu_1 x_1 - k_1 x_1 \cdot x_2$	(16.67)
		$r_{x_2} = k_2 x_1 \cdot x_2 - k_3 x_2$	(16.68)
		$r_{x_1} = \mu_1 x_1 - Dx_1 - \dfrac{\mu_2}{Y_{X_1/x_2}}x_2$	(16.69)
		$r_{x_2} = \mu_2 x_2 - Dx_2$	(16.70)
		$r_s = -\dfrac{\mu_1}{Y_{X/S1}}x_1 + D(s_0 - s)$	(16.71)
		with $\mu_1 = f(s)$ according to Monod kinetics and	
		$\mu_2 = f(x_1) = \mu_{max,2}\dfrac{x_1}{K_2 + x_1}$	(16.72)

ment was analyzed by VOLTERRA [16.111], who established many of the fundamentals of mathematical ecology. POWELL [16.112] and MOSER [16.113] have provided mathematical analyses of the fate of contaminants or of mutants in CSTRs.

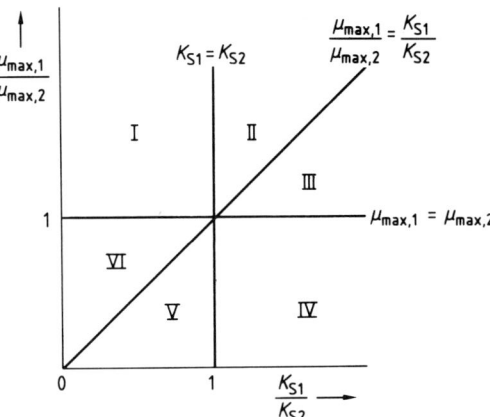

Figure 16.18. Generalized plot of theoretical dependences of specific growth rates on limiting substrate concentrations for a population including two strains (mutants 1 and 2) according to SIKYTA et al. [16.110]. Six different cases are illustrated in this plot, where correspondingly denoted areas reflect the magnitude of the ratios $\mu_{max,1}/\mu_{max,2}$ and K_{S1}/K_{S2}.

Type of mixed cultures:	Predominant organism:
I	1
II	1
III	
$D < \mu_c$	2
$D > \mu_c$	1
IV	2
V	2
VI	
$D < \mu_c$	1
$D > \mu_c$	2

Other reported trends are the computer aided analysis of mixed cultures (OHTAGUCHI et al. [16.114]), the mathematical description of the competition between two and three species under dual S-limitation in a CSTR (GOTTSCHAL and THINGSTAD [16.115]), and the analysis of a two-stage CSTR-system as an attractive alternative (STEPHANOPOULOS and FREDRICKSON [16.116]). The advantageous use of multi-stage systems had been recommended earlier by VELDKAMP and JANNASCH [16.117]. The often observed complexity of microbial interactions, however, needs a much more elaborated kinetic modelling including other categories than just competition. By varying the concentration of the medium components, various interacting systems can be created (e.g., TSENG and PHILLIPPS [16.118]).

16.5.2.2 Commensalism and Amensalism

Many microbes produce substances that promote the growth of other species, which is probably the most widespread form of commensalism [cf. Eq. (16.54) in Table 16.3]. The combinations of possible commensal relationships, however, are legion and the interactions may well have been competitive as well as commensalistic, but studies have rarely been very detailed.

Kinetic model equations for pure commensalism without inhibition employ Monod-type relations for $\mu_i = f(s_i)$ as well as the assumption that the nutrient for the dependent species is produced without loss of yield from the nutrient for the independent species. The problem of a commensalistic model with feedback inhibitions and activation by excreted metabolites has been treated by REILLY [16.119]. Inhibition was assumed to be competitive and activation to be additive: This complicated commensalistic system is less stable than the non-inhibited system, with a limited cycle response occurring after dilution rate changes.

Similar results have been obtained by SHEINTUCH [16.120] who examined the dynamics of commensalistic systems with self- and cross-inhibition. Multiplicity of steady-states was observed as well as oscillatory states. Population dynamics and stability analyses for commensalism and other interactions are widespread in the literature (e.g., LEE et al. [16.121]; MIURA et al.

[16.122]; TSENG and PHILLIPPS [16.118]). It has often been concluded that mixed cultures are very complex and that the interacting mechanisms may not be as simple as commensalism and/or competition.

16.5.2.3 Mutualism

Mutualism, sometimes called proto-cooperation, is not obligatory for survival, and could be caused by the production of growth factors or products P, which serve as nutrients, as indicated by Eq. (16.59) in Table 16.3.

Following this scheme, the specific growth rates (μ_1 and μ_2) can be expressed as follows:

$$\mu_1 = \mu_{max,1} \frac{s_1}{K_{S1}+s_1} \cdot \frac{p_2}{K_{P2}+p_2}, \quad (16.73)$$

and

$$\mu_2 = \mu_{max,2} \frac{s_2}{K_{S2}+s_2} \cdot \frac{p_1}{K_{P1}+p_1}, \quad (16.74)$$

and substituted into the balance equations for CSTR operation as given by Eqs. (16.60) to (16.63).

A somewhat more complex case, where x_1 grows on S, which is toxic at high concentrations to x_2 and where x_2 produces a nutrient for x_1, may be modelled by using the same biomass balance as before but different S-balances with kinetic equations for S-inhibition and co-utilization of P [cf. Eq. (16.74)] (MEYER et al. [16.123]) as shown in Eqs. (16.64) and (16.65) in Table 16.3.

The model behavior, determined with parameters taken from the pure cultures, demonstrates the coexistence of both species at lower dilution rates above which the singular point of coexistence could not be obtained. It seems generally valid that a mutualistic interaction in a CSTR is established only in a limited range of initial conditions of X_i and S (MEYER et al. [16.123]; MIURA et al. [16.122]; TAGUCHI et al. [16.124]).

16.5.2.4 Predation (Predator-Prey Interaction)

A food chain of mixed populations indicated by the scheme of Eq. (16.66) contains bacteria in aquatic and terrestrial environments while protozoa and ciliates represent other members of a biocoenosis. A prey (x_1) is consumed by the predator (x_2).

GAUSE [16.125] systematically studied the interactions between ciliates and their prey in a closed laboratory environment; such studies were also carried out extensively by VOLTERRA [16.111]. Together with LOTKA [16.126] this author developed early mathematical models. The original Lotka-Volterra analysis considered μ_1 and μ_2 to be constants, but normally they would depend on their respective substrates.

The model equations for simple prey-predator interactions initially developed by LOTKA and VOLTERRA and later refined by BUNGAY and BUNGAY [16.105] are given by Eqs. (16.67) and (16.68) in Table 16.3, in which k_1 is the killing efficiency constant based upon encounters, k_2 the predator growth constant, proportional to the yield coefficient $Y_{x1/x2}$, and k_3 the specific death rate.

The process kinetic model equations for CSTR operation have been developed by CURDS [16.127], as shown by Eqs. (16.69)–(16.72). A computer simulation of this set of predator model equations is illustrated in Fig. 16.19, which can also be used for stability analysis. Three possible states may exist in this system, depending upon growth parameters: stable oscillations (limit cycles) or damped oscillations, or an asymptotic approach to the steady-state value.

If Eq. (16.67) is divided by Eq. (16.68) and integrated, Eq. (16.75) is obtained:

$$-k_3 \lg x_1 + k_2 x_1 - \mu_1 \lg x_2 + k_1 x_2 = C. \quad (16.75)$$

Each value of the integration constant C gives a closed cycle for the relationship between x_1 and x_2 on a x_1/x_2 plane. This phase plane analysis, shown in Fig. 16.20, demonstrates that the populations of x_1 and x_2 oscillate around a vortex point, with the

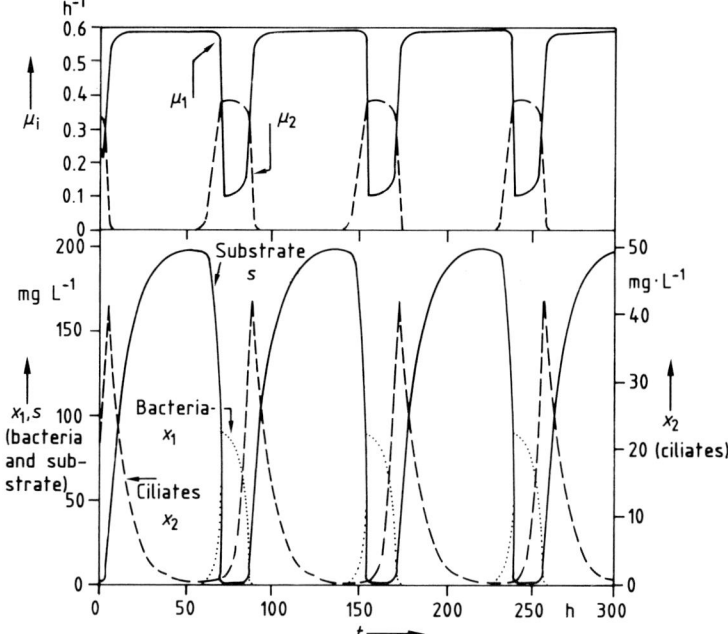

Figure 16.19. Computer analysis of predation interactions for mixed populations of bacteria (x_1) and ciliates (x_2) exhibiting stable oscillations in x_1, x_2, limiting substrate s, and specific growth rates μ_1 and μ_2 (CURDS [16.127]). The kinetic constants are as follows: $s_{in} = 200$ mg·L^{-1}; $D = 0.1$ h^{-1}; $\mu_{max,1} = 0.6$ h^{-1}; $\mu_{max,2} = 0.43$ h^{-1}; $K_{S1} = 4$ mg·L^{-1}; $K_{S2} = 12$ mg·L^{-1}; $Y_{X/S1} = 0.45$; $Y_{X/S2} = 0.54$.

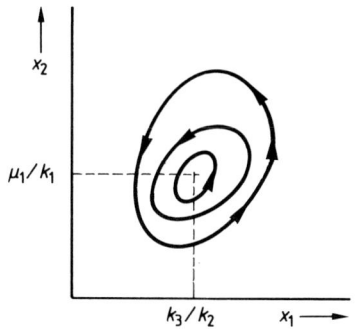

Figure 16.20. Closed-cycle relationship between predator (x_2) and prey (x_1) in predator-prey phase space from the classical theory as a solution of Eq. (16.75) (LOTKA [16.126]; VOLTERRA [16.111]; CANALE [16.128]). Different closed cycles ("trajectories") appear as a consequence of varied constants C in Eq. (16.75).

equilibrium point at the x_1- or x_2-axis given by the ratio of μ_1/k_1 or k_3/k_2 (CANALE [16.128]). Biologists and mathematicians have shown considerable interest in the Lotka-Volterra model. Cyclic population data have then been compared to models and showed some discrepancies, e.g., two singular points.

In this situation SUDO et al. [16.129] proposed a more sophisticated model by distinguishing between two kinds of bacterial concentration: bacterial food unavailable to the protozoon due to flocs and available because of dispersion, respectively. With this concept of microbial floc formation and disintegration, a better agreement was reported. On the other hand, TSUCHIYA et al. [16.130] and JOST et al. [16.131] demonstrated that under certain conditions (combinations of dilution rate and S-concentration data) it is also possible that predator-prey systems in a CSTR do not exhibit oscillatory behavior.

Stability analyses for such cases result in an operating diagram for the general dynamic features according to JOST et al. [16.131]. This shows regions of different characteristics for given values of kinetic constants when plotting $1/D$ vs. s_0. However, deviations from this model behavior have been observed in experiments leading to an important model, that incorporates the multiple saturation predation rate:

$$\mu_2 = \mu_{max,2} \frac{x_1^2}{(K_2+x_1)(K_3+x_1)}. \quad (16.76)$$

This equation reduces to the Monod-type equation, Eq. (16.72); with x_1 much larger than K_2 and K_3, and μ_2 varies as x_1^2. As a result, it can be shown that sustained oscillations disappear when t increases. It is commonly accepted by ecologists that environmental heterogeneity exerts a stabilizing influence on predator-prey interactions. This fact led to investigations of combined wall growth in a CSTR with predation (RATNAM et al. [16.132]). Data obtained showed that bacteria but not protozoa were attached strongly to the walls and that wall growth had a significant effect on the dynamics.

For further details on mathematical modelling of mixed culture kinetics the reader is referred to recent reviews (BAZIN [16.133], MOSER [16.134b]).

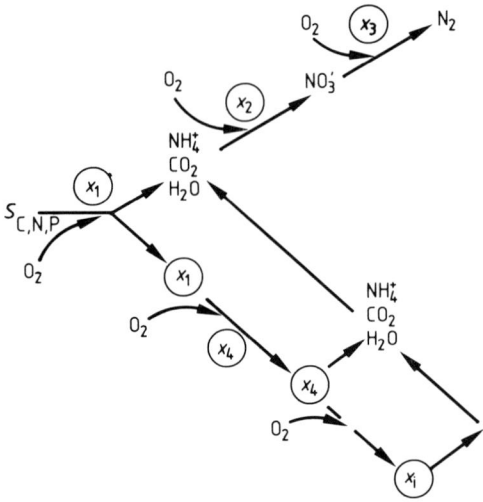

Figure 16.21. Reaction sequences in typical aerobic waste water treatment systems using activated sludge as a mixed population ("biocoenosis"). Mixed substrate containing carbon, nitrogen, and phosphorus is supplied ($S_{C,N,P}$), which then is consumed by saprophytic bacteria (X_1), followed by a "fret chain" ($X-X_i$). Under certain conditions, ammonia is consumed by nitrifiers (x_2) followed by denitrification through x_3 (WILDERER and HARTMANN [16.162]).

16.5.3 Application and Mixed Population Bioreactor Problems

Typical examples of mixed culture application are biological waste water treatment, aerobic and anaerobic sludge digestion (biogas production) and also corn, cheese, and yoghurt production, the human skin, the alimentary canal, and generally aquatic environments such as rivers, and soils. In natural self-purifying systems, e.g., rivers, a succession of populations occur: chemo-organotrophic bacteria, ciliates, nitrifiers, algae, and fishes. In biological waste water treatment reactor systems, a multitude of consecutive reactions are known as a result of activities of the biocoenosis, shown in Fig. 16.21.

Another case of mixed population interactions is the rumen fermentation (e.g., CZERKAWSKI [16.135]), where mathematical models have also been developed for a better understanding and for technical manipulations.

In a situation where the optimal strategy for the production of ethanol from sugars involves batch fermentation of high concentrated sugar solutions, the use of dual organisms, possessing different substrate and product inhibition characteristics, can bring about improved ethanol productivity. The extent of any improvement depends crucially on the functional form of the inhibition relationships and on the values of the kinetic parameters (e.g., JONES and GREENFIELD [16.136]).

Mathematical modelling in this and other cases of mixed populations is a completely unstructured approach. Handling the bioreactor as a black box, is often and even predominantly used: e.g., biological waste water treatment (GAUDY and GAUDY [16.104]), and biogas fermentation (ANDREWS [16.137]; CHEN and HASHIMOTO

[16.138]), resulting in arbitrary values of kinetic parameters of simple kinetic models. However, in accordance with the formal macro-approach some improvements can be achieved.

Nevertheless, in cases where evidently different populations are significant, as in biogas production, where the three steps of fermentative, acetogenic, and methanogenic reactions with strains x_i take place following the scheme:

$$S_i \xrightarrow{X_1} \text{fatty acids} + H_2 + CO_2 \longrightarrow$$

$$\xrightarrow{X_2} HAc + H_2 + CO_2 \longrightarrow$$

$$\xrightarrow{X_3} CH_4 + CO_2, \qquad (16.77)$$

adequate kinetic modelling must follow a more structured approach, incorporating different strains and substrates with stimulating and inhibiting properties.

A similar but simpler situation occurs with nitrification or denitrification of biological waste water. Both have been described as a two-step microbial process (TANAKA et al. [16.139]; EGGERS and TERLOUW [16.140]). Beyond this field of fascinating possibilities of application and research activities some bioreactor operation problems should be referred to as typical for mixed populations. An infinite number of stable steady-states can occur in the case of phenol degradation (HUMPHREY and YANG [16.141]), which is thought to be of general importance to biology (HUMPHREY [16.142]).

This infinitely variable problem in bioreactor operation is still waiting to be solved. There have been relatively few systematic tests of the principles of interactions of mixed populations. Experimental work on the basis of kinetic model theories should be stimulated by means of a CSTR, a CSTR-cascade, or a continuous plug flow fermenter.

16.6 Biofilm Reactor Operation

Biofilm operation is the oldest form of bioprocessing; biofilms generally occur in nature. Due to the adhesion of microbes, biofilm formation has to be taken into ac-

Table 16.4. Biofilm Operation Compared to Floc Bioprocessing (MOSER [16.134a])

Criteria	Flocs	Biofilms
Mode of operation	Discontinuous Continuous (Wash-out!)	Continuous (No wash-out!)
Process control	Multiple, difficult	Simple
Product recovery	Expensive	Easy and cheaper
Kinetics	Homogeneous Heterogeneous	Heterogeneous and pseudohomogeneous
Transport problems		
G/L	yes	yes
L	yes	yes
L/S	yes, difficult to handle	yes, easier to handle
S	yes, difficult to handle	yes, easier to handle
Particle size	Size distribution uncontrolled in STRs	Film thickness uncontrolled but also possibilities for controlled and uniform film growth

count in river as well as in chemostat analysis. The significance of microbial films in fermenters was extensively studied and reviewed by ATKINSON and FOWLER [16.143]; ATKINSON [16.144]; ATKINSON and KNIGHTS [16.145]; HARREMOËS [16.146] and CHARAKLIS [16.147].

Properties of biofilms are summarized in Table 16.4. Accordingly, process intensification using cell support systems are promising (ATKINSON et al. [16.148]). The potential advantages are as follows:

- As wash-out is not possible, highest throughputs can be achieved in reactors. Biomass hold-up is, therefore, independent of throughput.
- Due to the adhesion of cells on support surfaces, highest biomass concentrations can be realized.
- Due to the adhesion to solid particles, the principles of fluidization can be applied to bioreactor operation with the advantage of having no restrictions in transport processes especially at the L/S-interface (turbulence) and in the S-phase (abrasion or scouring).
- The appearance of transport limitations and concentration gradients may be advantageous in some cases:
  a) With mixed populations different species may occur at various depths within the particle, e.g., nitrifiers near the surface and denitrifiers towards the center (EGGERS and TERLOUW [16.140]).
  b) When using cells, which must be "mature" before producing the product (cf. Sect. 14.4) their retention within a particle may be beneficial, e.g., citric acid and secondary metabolite production.
  c) When using S-inhibiting media, the most favorable S-concentration may occur towards the center of the film.

Performance is independent of film thickness when operating with thick films. However, in thick biofilms S-limitations occur generally and result in a loss of cell viability. Thereby, the adhesive bond to the support surface is weakened so that sloughing occurs, which disturbs reactor performance. Therefore, control of biofilm thickness is of central importance. The thickness can be controlled by means of mechanical scraping, abrasion, or friction and self-regulation at higher hydrodynamic stress.

Higher relative velocities can be realized. As a result, external transport limitation can be excluded and even easily studied simultaneously with internal transport limitations in the case of uniform and controlled biofilm thickness. Therefore, biofilm reactors are superior to the conventional stirred tank reactors.

Last but not least, pellet processing must be mentioned here (METZ and KOSSEN [16.149]). Usually the investigators consider pellet formation an undesirable feature, leading to inhomogeneous mycelia. Nevertheless, controlled pellet formation offers the advantage of a much lower viscosity of pellet suspensions in comparison to a filamentous broth (pulp) with the result, that pellet processes are energetically more efficient than a traditional pulp-like process. The activity loss in the center of a pellet is very well compensated for on a macroscopic level by the reduced power requirement for such a process (ROELS and VAN SUIDAM [16.150]).

Industrial-scale fermentations using biofilm operation, schematically shown in Fig. 16.22 a–c, are represented by:

- trickling filters (fixed bed reactors) and rotating disk fermenters in biological waste water treatment. An old type of percolating filter is the "quick-vinegar process", which uses packings of beech-wood chips as support material;
- fluidized bed bioreactors, recently used in biological waste water treatment and as an enzyme reactor;
- trays used for animal tissue culturing, where cells are growing and adhering on a surface in the presence of unstirred layers of medium;
- bacterial leaching of ores for the recovery of metals by natural adhesion on solid materials;

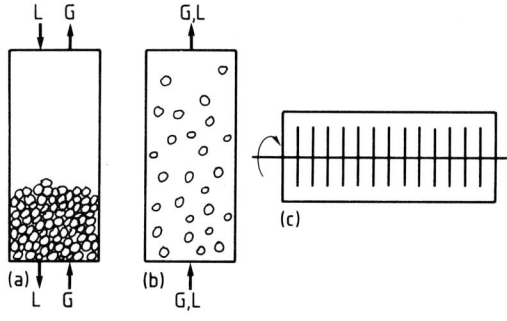

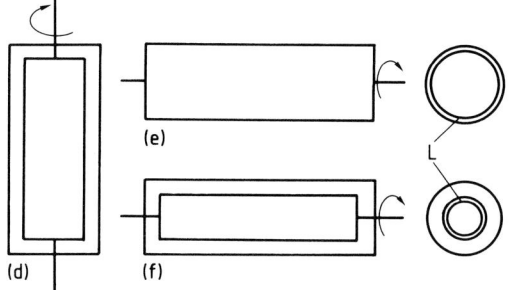

Figure 16.22. Schematic representation of biofilm reactors in industry (a)–(c) and bioprocess research (d)–(f). – The fixed bed reactor or percolating filter (a); the fluidized bed reactor or completely mixed microbial film fermenter (b); the rotating biodisk reactor (c); the annular biofilm reactor in vertical position (d) (KORNEGAY and ANDREWS [16.153]); the horizontal rotating tube with the biofilm on the inner wall (e) (TOMLINSON and SNADDON [16.152]); the thin-layer-fermenter (f) (GORBACH [16.154]; MOSER [16.155]–[16.158]).

- conventional stirred tanks also show biofilm formation at the wall, which is, however, often ignored.

Furthermore, biofilms are emerging as a most critical factor affecting natural aquatic systems, water distribution systems, waste water processing, heat exchanger operation, velocity of ships, and even human disease. The most common method of controlling biofilm accumulation in practice is chlorination of water; the toxicity, however, limits its use. Thus, research in biofilm processing has been stimulated (CHARAKLIS [16.147]). Fig. 16.22 d–f represents several biofilm reactor configurations mentioned in the literature, which are suitable in bench-scale process research.

A sponge fermenter, not shown in Fig. 16.22, was used by FREEMAN [16.151] for microbial growth in a cellulose sponge. More adequate configurations for process research are the horizontal rotating tube reactor (TOMLINSON and SNADDON [16.152]), the annular biofilm reactor in vertical position (KORNEGAY and ANDREWS [16.153]), and the thin-layer film fermenter (GORBACH [16.154]) used as a horizontal device with special potential advantages for a systematic bioprocess analysis (MOSER [16.155]–[16.158]). Another model biofilm reactor is the sloping plane reactor (MAIER et al. [16.159]), which was extensively examined as a biological film reactor by ATKINSON and his group (ATKINSON and DAOUD [16.160]) together with the completely mixed microbial film fermenter as a type of fluidized bed. All these reactors with a simple geometry exhibit ideal properties of hydrodynamics, which facilitates modelling of transfer processes.

16.7 References

[16.1] A. M. PICKETT: Process Biochem. *14* (1979), Nov. 10.
[16.2] D. E. F. HARRISON and H. H. TOPIWALA: Adv. Biochem. Eng. *3* (1974), 167.
[16.3] T. G. DORAWALA and J. M. DOUGLAS: AIChE J. *17* (1971), 974.
[16.4] J. E. BAILEY: Chem. Eng. Commun. *1* (1973), 111.
[16.5] A. M. PICKETT: In "Microbial Population Dynamics", Chap. 4 (M. BAZIN, ed.). CRC Press, Boca Raton, Florida, 1982.
[16.6] J. A. ROELS: J. Chem. Tech. Biotechnol. *32* (1982), 59.

[16.7] J. P. BARFORD, N. B. PAMMENT, and R. J. HALL: In "Microbial Population Dynamics", Chap. 3 (M. J. BAZIN, ed.). CRC Press, Boca Raton, Florida, 1982.

[16.8] A. M. PICKETT, M. J. BAZIN, and H. H. TOPIWALA: Biotechnol. Bioeng. 21 (1979), 1043.

[16.9] G. T. DAIGGER and C. P. L. GRADY: Biotechnol. Bioeng. 24 (1982), 1427.

[16.10] J. H. SHERRARD and A. W. LAWRENCE: J. Water Pollut. Control Fed. 47 (1975), 1848.

[16.11] C. T. CHI and J. A. HOWELL: Biotechnol. Bioeng. 18 (1976), 63.

[16.12] A. CAMPBELL: Bacteriol. Rev. 21 (1957), 263.

[16.13] A. MOSER: Acta Biotechnologica 4 (1984), 1.

[16.14] D. HERBERT: Symp. Soc. Gen. Microbiol. 11 (1961), 391.

[16.15] J. P. BARFORD and R. J. HALL: J. Gen. Microbiol. 114 (1979), 267.

[16.16] F. YOSHIDA, T. YAMANE, and K.-I. NAKAMOTO: Biotechnol. Bioeng. 15 (1973), 257.

[16.17] S. J. PIRT: J. Appl. Chem. Biotechnol. 24 (1974), 415.

[16.18] T. YAMANE and S. HIRANO: J. Ferment. Technol. 55 (1977), 156.

[16.19] V. H. EDWARDS, M. J. GOTTSCHALK, A. Y. NOOJIN III, L. B. TUTHILL, and A. L. TANNAHILL: Biotechnol. Bioeng. 17 (1970), 975.

[16.20] I. J. DUNN and J. R. MOR: Biotechnol. Bioeng. 17 (1975), 1805.

[16.21] R. MONA, I. J. DUNN, and J. R. BOURNE: Biotechnol. Bioeng. 21 (1979), 1561.

[16.22] R. GUTKE, G. GIRA, K. MUND, and W. A. KNORRE: Biotechnol. Lett. 2 (1980), 315.

[16.23] A. HARDER and J. A. ROELS: Adv. Biochem. Eng. 22 (1982), 56.

[16.24] A. A. ESENER, J. A. ROELS, and N. W. F. KOSSEN: Biotechnol. Bioeng. 23 (1981), 1851.

[16.25] N. W. F. KOSSEN: Symp. Soc. Gen. Microbiol. 29 (1979), 327.

[16.26] A. A. ESENER, T. VEERMAN, J. A. ROELS, and N. W. F. KOSSEN: Biotechnol. Bioeng. 24 (1982), 1749.

[16.27] F. M. WILLIAMS: J. Theor. Biol. 15 (1967), 190.

[16.28] F. M. WILLIAMS: In "Systems Analysis and Simulation Theory", Vol. 1, Chap. 3, p. 197 (B. C. PATTEN, ed.). Academic Press, New York, 1971.

[16.29] J. A. ROELS: "Energetics and Kinetics in Biotechnology". Elsevier Biomedical, Amsterdam, 1983.

[16.30] D. RAMKRISHNA, A. G. FREDRICKSON, and H. M. TSUCHIYA: Biotechnol. Bioeng. 9 (1967), 129.

[16.31] A. G. FREDRICKSON, R. D. MEGEE III, and H. M. TSUCHIYA: Adv. Appl. Microbiol. 13 (1970), 419.

[16.32] A. G. FREDRICKSON: Biotechnol. Bioeng. 18 (1976), 1481.

[16.33] J. A. ROELS and N. W. F. KOSSEN: Prog. Ind. Microbiol. 14 (1978), 95.

[16.34] J. A. ROELS: In "Proc. 1st Eur. Congr. Biotechnol.", Interlaken (Switzerland). DECHEMA Monographien, Vol. 82, p. 221. Verlag Chemie, Weinheim – Deerfield Beach, Florida – Basel, 1978.

[16.35] H. H. LEE and B. D. YAN: Chem. Eng. Sci. 36 (1981), 483.

[16.36] R. KELLER and I. J. DUNN: J. Appl. Chem. Biotechnol. 28 (1978), 508.

[16.37] R. KELLER and I. J. DUNN: J. Appl. Chem. Biotechnol. 28 (1978), 784.

[16.38] M. KISHIMOTO, T. YAMANE, and F. YOSHIDA: J. Ferment. Technol. 54 (1976), 891.

[16.39] R. K. BAJPAI and M. REUSS: J. Chem. Tech. Biotechnol. 30 (1980), 332.

[16.40] R. K. BAJPAI and M. REUSS: Biotechnol. Bioeng. 23 (1982), 717.

[16.41] R. GUTKE and W. A. KNORRE: Biotechnol. Bioeng. 23 (1981), 2771.

[16.42] R. GUTKE and W. A. KNORRE: Biotechnol. Bioeng. 24 1983), 2129.

[16.43] S. AIBA, S. NAGAI, and Y. NISHIZAWA: Biotechnol. Bioeng. 18 (1976), 1001.

[16.44] K. DAIRAKU, Y. YAMASAKI, K. KUKI, S. SHIOYA, and T. TAKAMATSU: Biotechnol. Bioeng. 23 (1981), 2069.

[16.45] A. WHITAKER: Process Biochem. 15 (1980), April, May, 10.

[16.46] N. KOSARIC, I. RUSSELL, D. C. M. NG, and G. S. STEWART: Adv. Appl. Microbiol. 26 (1980), 147.

[16.47] R. C. CHARLEY, J. E. FEIN, B. H. LAVERS, H. G. LAWFORD, and G. R. LAWFORD: Biotechnol. Lett. 5 (1983), 169.

[16.48] P. L. ROGERS, K. J. LEE, M. L. SKOTNICKI, and D. E. TRIBE: In "Proc. 2nd Symp. Bioconversion and Biochemical Eng.", Vol. 2, p. 359 (T. K. GHOSE, ed.). Pramodh Kapur, Raj Bandhu Ind. Co, New Delhi, 1980.

[16.49] P. GERHARDT: US Pat. 3186917 (1965).

[16.50] P. GERHARDT and D. M. GALLUP: J. Bacteriol. 86 (1963), 919.

[16.51] A. G. LANE: J. Appl. Chem. Biotechnol. *27* (1977), 165.

[16.52] B. J. ABBOTT and P. GERHARDT: Biotechnol. Bioeng. *12* (1970), 577 and 603.

[16.53] P. MARSOT, R. FOURNIER, and C. BLAIS: Biotechnol. Lett. *3* (1981), 689.

[16.54] B. MAIORELLA, Ch. R. WILKE, and H. W. BLANCH: Adv. Biochem. Eng. *20* (1981), 43.

[16.55] G. HAMER: Biotechnol. Bioeng. *24* (1982), 511.

[16.56] M. D. LILLY: J. Chem. Tech. Biotechnol. *32* (1982), 162.

[16.57] R. RAUTENBACH and R. ALBRECHT: Chem. Ing. Tech. *54* (1982), 229.

[16.58] E. K. PYE and A. E. HUMPHREY: Interim Report to US Dept. of Energy (1979), p. 79.

[16.59] J. S. SCHULTZ and P. GERHARDT: Bacteriol. Rev. *33* (1969), 1.

[16.60] M. DOSTALEK and M. HÄGGSTROM: Biotechnol. Bioeng. *24* (1982), 2077.

[16.61] L. SORTLAND and Ch. R. WILKE: Biotechnol. Bioeng. *11* (1969), 805.

[16.62] A. MARGARITIS and Ch. R. WILKE: Biotechnol. Bioeng. *20* (1978), 709, 727.

[16.63] A. K. BHAGAT and C. R. WILKE (1966) – Lawrence Radiation Lab. Report UCRL-16574.

[16.64] D. BAKER and B. H. KIRSOP: J. Inst. Brew. *79* (1973), 487.

[16.65] J. K. KAN and M. L. SHULER: AIChE Symp. Ser. *172* (1976), 31.

[16.66] J. K. KAN and M. L. SHULER: Biotechnol. Bioeng. *20* (1978), 217.

[16.67] I. A. WEBSTER: J. Chem. Tech. Biotechnol. *31* (1981), 178.

[16.68] P. A. MESSING: "Immobilized Enzymes for Industrial Reactors". Academic Press, New York, 1975.

[16.69] K. VENKATASUBRAMANIAN (ed.): "Immobilized Microbial Cells". ACS (Am. Chem. Soc.) Symp. Ser. *106*, Washington D.C., 1979.

[16.70] T. SHIOTANI and T. YAMANE: Eur. J. Appl. Microbiol. Biotechnol. *13* (1981), 96.

[16.71] G. H. CHO, Ch. Y. CHOI, Y. D. CHOI, and M. H. HAN: J. Chem. Tech. Biotechnol. *32* (1982), 959.

[16.72] J. HONG, G. T. TSAO, and P. C. WANKAT: Biotechnol. Bioeng. *23* (1981), 1501.

[16.73] O. VELICANGIL and J. A. HOWELL: Biotechnol. Bioeng. *23* (1981), 843.

[16.74] G. A. COULMAN, R. W. STIEBER, and P. GERHARDT: Appl. Environ. Microbiol. *34* (1977), 725.

[16.75] J. W. CHEN, F. J. HILLS, H. J. KOEPSELL, and W. O. MAXON: Ind. Eng. Chem. Process Des. Dev. *4* (1965), 421.

[16.76] R. W. STIEBER, G. A. COULMAN, and P. GERHARDT: Appl. Environ. Microbiol. *34* (1977), 733.

[16.77] R. W. STIEBER and P. GERHARDT: Appl. Environ. Microbiol. *37* (1979), 487.

[16.78] R. W. STIEBER and P. GERHARDT: Biotechnol. Bioeng. *23* (1981), 523 and 535.

[16.79] E. ZEUTHEN (ed.): "Synchrony in Cell Division and Growth". Interscience, New York, 1964.

[16.80] I. L. CAMERON and C. M. PADILLA (eds.): "Cell Synchrony". Academic Press, New York, 1966.

[16.81] P. DAWSON: J. Appl. Chem. Biotechnol. *22* (1972), 79.

[16.82] K. G. LARK and O. MAALOE: Biochim. Biophys. Acta *21* (1956), 448.

[16.83] R. G. CUTLER and J. E. EVANS: J. Bacteriol. *91* (1966), 469.

[16.84] E. SPOERL and D. LOONEY: Cell Tissue Res. *17* (1959), 320.

[16.85] H. HILZ and H. ECKSTEIN: Biochem. Z. *340* (1964), 351.

[16.86] D. H. WILLIAMSON and A. W. SCOPES: Cell Tissue Res. *20* (1962), 338.

[16.87] J. M. MITCHISON and W. S. VINCENT: Nature *205* (1965), 987.

[16.88] A. WIEMKEN, K. VON MEYENBURG, and P. MATILE: In "Proc. 2nd Symp. Protoplasts". Brno (Czechoslovakia) 1968.

[16.89] P. DAWSON: Can. J. Microbiol. *11* (1965), 893.

[16.90] L. KJAERGAARD and B. B. JOERGENSEN: Biotechnol. Bioeng. *21* (1979), 147.

[16.91] O. MAALOE: In "The Bacteria", Vol. 4, Chap. 1 (I. GUNSALUS and R. Y. STANIER, eds.). Academic Press, New York, 1963.

[16.92] T. W. JAMES: In "Cell Synchrony", p. 1 (J. L. CAMERON and G. M. PADILLA, eds.). Academic Press, New York, 1966.

[16.93] K. VON MEYENBURG: Vierteljahresschr. Naturforsch. Ges. Zurich *114* (1969), 113.

[16.94] P. E. HANSCHE: J. Theor. Biol. *24* (1969), 335.

[16.95] B. C. GOODWIN: Eur. J. Biochem. *10* (1969), 511.

[16.96] D. W. TEMPEST, J. W. DICKS, and J. R. HUNTER: J. Gen. Microbiol. *45* (1966), 135.

[16.97] P. DAWSON and W. KURZ: Biotechnol. Bioeng. *11* (1969), 843.

[16.98] P. DAWSON: In "Proc. 4th IFS (Int. Ferment. Symp.)", p. 121, Kyoto, Jap., 1973.

[16.99] P. DAWSON: In "Proc. 2nd Symp. Bioconversion and Biochem. Eng.", Vol. 2, p. 275 (T. K. GHOSE, ed.). Pramodh Kapur, Raj Bandhu Ind. Co, New Delhi, 1980.

[16.100] P. DAWSON: In "Proc. 6th IFS (Int. Ferment. Symp.)", Vol. 31. London/Ontario (Canada), 1980.

[16.101] P. DAWSON: In "Proc. 6th IFS (Int. Ferment. Symp.)", Vol. 32, p. 191. London/Ontario (Canada), 1980.

[16.102] J. D. BU'LOCK, D. HAMILTON, M. A. HULME, A. J. POWELL, D. SHEPHERD, H. M. SMALLEY, and G. N. SMITH: Can. J. Microbiol. *11* (1965), 765.

[16.103] J. D. BU'LOCK: "The Biosynthesis of Natural Products", Chap. 1. McGraw-Hill, New York, 1965.

[16.104] A. F. GAUDY Jr. and E. T. GAUDY: Adv. Biochem. Eng. *2* (1972), 97.

[16.105] H. R. BUNGAY III and M. L. BUNGAY: Adv. Appl. Microbiol. *10* (1968), 269.

[16.106] D. NOACK: "Biophysikalische Prinzipien der Populationsdynamik in der Mikrobiologie". VEB G. Thieme Verlag, Leipzig (GDR), 1968.

[16.107] J. L. MEERS: Crit. Rev. Microbiol. *2* (1973), 139.

[16.107a] R. M. MAY: "Stability and Complexity in Model Ecosystems", p. 25. Princeton University Press, Princeton N.J., 1973.

[16.108] J. E. BAILEY and D. F. OLLIS: "Biochemical Engineering Fundamentals", p. 642. McGraw-Hill, New York, 1977.

[16.109] R. GUTKE: In "Manual for UNEP/UNESCO/ICRO Training Course on Theoretical Basis of Kinetics of Growth, Metabolism and Product Formation of Microorganisms", Vol. 1, p. 112, Academy of Science, Central Institute of Microbiology and Exp. Therapy (ZIMET), Jena (GDR), 1980 (or pp. 39, 58 in Ed. 1982).

[16.110] B. SIKYTA, M. NOVAK, and P. DOBERSKY: In "FEMS (Federation of European Microbiological Societies) Symp.", Vol. 4, p. 119 (J. MEYRATH and J. D. BU'LOCK, eds.), Academic Press, New York, 1977.

[16.111] V. VOLTERRA: "Leçons sur la Theorie Mathematique de la Lutte pour la Vie" Gauthier-Villars, Paris, 1931.

[16.112] E. O. POWELL: J. Gen. Microbiol. *18* (1958), 259.

[16.113] H. MOSER: Quant. Biol. *22* (1957), 121.

[16.114] K. OHTAGUCHI, I. ENDO, and I. INOUE: Int. Chem. Eng. *19* (1979), 313, 591.

[16.115] J. C. GOTTSCHAL and T. F. THINGSTAD: Biotechnol. Bioeng. *24* (1982), 1403.

[16.116] G. STEPHANOPOULOS and A. G. FREDRICKSON: Biotechnol. Bioeng. *21* (1979), 1491.

[16.117] H. VELDKAMP and H. W. JANNASCH: J. Appl. Chem. Biotechnol. *22* (1972), 105.

[16.118] M. TSENG and C. R. PHILLIPPS: Biotechnol. Bioeng. *23* (1981), 1639.

[16.119] P. J. REILLY: Biotechnol. Bioeng. *16* (1974), 1373.

[16.120] M. SHEINTUCH: Biotechnol. Bioeng. *22* (1980), 2557.

[16.121] I. H. LEE, A. G. FREDRICKSON, and H. M. TSUCHIYA: Biotechnol. Bioeng. *18* (1976), 513.

[16.122] Y. MIURA, H. TANAKA, and M. OKAZAKI: Biotechnol. Bioeng. *22* (1980), 929.

[16.123] J. S. MEYER, H. M. TSUCHIYA, and A. G. FREDRICKSON: Biotechnol. Bioeng. *17* (1975), 1065.

[16.124] H. TAGUCHI, T. YOSHIDA, and K. NAKATANI: J. Ferment. Technol. *56* (1978), 158.

[16.125] G. F. GAUSE: "The Struggle for Existence". Williams & Wilkins, Baltimore/Maryland (USA), 1934.

[16.126] A. J. LOTKA: "Elements of Physical Biology". Williams & Wilkins Co., Baltimore/Md. (USA), 1925.

[16.127] C. R. CURDS: Water Res. *5* (1971), 793.

[16.128] R. P. CANALE: Biotechnol. Bioeng. *11* (1969), 887.

[16.129] R. SUDO, K. KOBAYASHI, and S. AIBA: Biotechnol. Bioeng. *17* (1975), 167.

[16.130] H. M. TSUCHIYA, J. F. DRAKE, J. L. JOST, and A. G. FREDRICKSON: J. Bacteriol. *110* (1972), 1147.

[16.131] J. L. JOST, J. F. DRAKE, A. G. FREDRICKSON, and H. M. TSUCHIYA: J. Theor. Biol. *41* (1973), 461.

[16.132] D. A. RATNAM, St. PAVLOU, and A. G. FREDRICKSON: Biotechnol. Bioeng. *24* (1982), 2675.

[16.133] M. J. BAZIN: In "Mixed Culture Fermentation", Chap. 2 (M. E. BUSHELL and J. H. SLATER, eds.). Academic Press, London – New York, 1981.

[16.134a] A. MOSER: "Bioprozeßtechnik", p. 56, Springer Verlag, Wien – New York, 1981.

[16.134b] A. MOSER: "Bioprocess Engineering", p. 8, Springer Verlag, Vienna – New York, 1985.

[16.135] J. W. CZERKAWSKI: Process Biochem., October (1973), 25.

[16.136] R. P. JONES and P. F. GREENFIELD: Biotechnol. Lett. *3* (1981), 225.

[16.137] J. F. ANDREWS: Biotechnol. Bioeng. Symp. *2* (1971), 5.

[16.138] Y. R. CHEN and A. G. HASHIMOTO: Biotechnol. Bioeng. Symp. *8* (1978), 269.

[16.139] H. TANAKA, S. UZMAN, and I. J. DUNN: Biotechnol. Bioeng. *23* (1981), 1683.

[16.140] E. EGGERS and T. TERLOUW: Water Res. *13* (1979), 1077.

[16.141] A. E. HUMPHREY and R. D. YANG: Biotechnol. Bioeng. *17* (1975), 1211.

[16.142] A. E. HUMPHREY: Chem. Eng. Prog., May (1977), 85.

[16.143] B. ATKINSON and H. W. FOWLER: Adv. Biochem. Eng. *3* (1974), 221.

[16.144] B. ATKINSON: "Biochemical Reactors". Pion Ltd., London, 1974.

[16.145] B. ATKINSON and A. J. KNIGHTS: Biotechnol. Bioeng. *17* (1975), 1245.

[16.146] P. HARREMOES: In "Water Pollution Microbiology", Vol. 2, Chap. 4, John Wiley, New York, 1978.

[16.147] W. G. CHARAKLIS: Biotechnol. Bioeng. *23* (1981), 1923.

[16.148] B. ATKINSON, G. M. BLACK, and A. PINCHES: Process Biochem. *15* (1980), APRIL/MAY, 24.

[16.149] B. METZ and N. W. F. KOSSEN: Biotechnol. Bioeng. *19* (1977), 781.

[16.150] J. A. ROELS and J. C. VAN SUIDAM: Biotechnol. Bioeng. *22* (1980), 463.

[16.151] R. R. FREEMAN: Biotechnol. Bioeng. *3* (1961), 339.

[16.152] T. G. TOMLINSON and D. H. SNADDON: Int. J. Air Water Pollut. *10* (1968), 865.

[16.153] B. H. KORNEGAY and J. F. ANDREWS: J. Water Pollut. Control Fed. *40* (1968) R, 460.

[16.154] G. GORBACH: Monatsschr. Brau. *22* (1968), 49.

[16.155] A. MOSER: Chem. Ing. Tech. *45* (1973), 1313.

[16.156] A. MOSER: Chem. Ing. Tech. *49* (1977), 612.

[16.157] A. MOSER: Biotechnol. Lett. *4* (1982), 281.

[16.158] A. MOSER: "Advances in Fermentation '83", Symp. (Proc. Biochem.) London, September 1983, 202.

[16.159] W. J. MAIER, V. C. BEHN, and C. D. GATES: J. ASCE, SED *93* (1967), SA 4, 91.

[16.160] B. ATKINSON and J. S. DAOUD: Trans. Inst. Chem. Eng. *48* (1970), 245.

[16.161] N. P. HARRIS and G. S. HANSFORD: Water Res. *10* (1976), 935.

[16.162] P. WILDERER and L. HARTMANN: München, Beitr. Abwasser Fisch. Flußbiol. *29* (1978), 9.

Chapter 17

Reaction and Mass Transfer Interactions in Microbial Systems

Anton Moser

Institut für Biotechnologie, Mikrobiologie und Abfalltechnologie
Technische Universität Graz
Graz, Austria

17.1   Introduction
17.2   Problems and Concept of Solution
17.2.1   Experimental Phenomena with Microbial Processes
17.2.1.1 External Transport Limitation
17.2.1.2 Internal Transport Limitation
17.2.1.3 Transport Enhancement
17.2.1.4 Dynamic Response Lag Time of Measuring Electrodes
17.2.2   Concept for the Solution of the Problems of Interactions
17.3   External Transport Limitation
17.3.1   Theory
17.3.2   Practical Application of Theory
17.3.2.1 Transport Limitation at the Liquid/Solid Interface
17.3.2.2 Liquid-phase Micromixing
17.3.2.3 Transport Limitation at the Gas/Liquid Interface in Case of Oxygen Transfer
17.4   Internal Transport Limitation
17.5   Combined External and Internal Mass Transfer Effects
17.6   Transport Enhancement in Bioprocessing
17.7   References

List of Symbols

a	m^{-1}	specific area
c	$kg \cdot m^{-3}$	concentration (general term)
D	$m^2 \cdot s^{-1}$	diffusion coefficient
Da_{II}	—	Damköhler number, 2nd degree
d	m	diameter (general term)
d_P	m	diameter of particle
Ha	—	Hatta number
K_S, K_O	$kg \cdot m^{-3}$	Monod saturation constant for substrate and oxygen
k	—	rate coefficient (general term)
k_r	—	reaction rate coefficient
k_{Tr}	—	transport rate coefficient
k_E	s^{-1}	electrode response coefficient
k_1, k_2, k_3	—	biological rate equation coefficients
k_r^L, k_r^S	—	reaction rate coefficients in liquid or solid phase
k_G	$m \cdot s^{-1}$	transport rate coefficient in G-phase
k_{L1}, k_{L2}	$m \cdot s^{-1}$	transport rate coefficient in L-phase
k_S	$m \cdot s^{-1}$	transport rate coefficient in S-phase
$k_L a$	s^{-1}	volumetric rate coefficient of oxygen transfer
k_{mix}	s^{-1}	coefficient of mixing time
m_S, m_O	s^{-1}	specific rate coefficient of maintenance of substrate and O_2
n, n_i	$kg \cdot m^{-3} \cdot s^{-1}$	mass flux per volume
n_i'	$kg \cdot m^{-2} \cdot s^{-1}$	mass flux through area
O	kg	amount of oxygen
o	$kg \cdot m^{-3}$	concentration of oxygen
P	kg	amount of product
P	$J \cdot s^{-1}$	power
P_G	$J \cdot s^{-1}$	power in case of gassed liquid
p	$kg \cdot m^{-3}$	product concentration
q_S, q_O	s^{-1}	specific rate of consumption of substrate or oxygen
q_P	s^{-1}	specific rate of production
r	m	radius
r_i	$kg \cdot m^{-3} \cdot s^{-1}$	rate of formation or production of component i (S, X, O, P)
r_i'	$kg \cdot m^{-2} \cdot s^{-1}$	surface-based reaction rate
r_i^L	$kg \cdot m^{-3} \cdot s^{-1}$	reaction rate in liquid phase
Sh	—	Sherwood number
St	—	Stanton number
S	kg	amount of substrate
s	$kg \cdot m^{-3}$	substrate concentration
t	s	time (general term)
$\bar{t}$	s	mean residence time
t_e	s	characteristic time of environmental change
t_m	s	mixing time
t_r	s	characteristic time of reaction
t_{Tr}	s	characteristic time of transport
V	m^3	volume
v	$m \cdot s^{-1}$	velocity
x	$kg \cdot m^{-3}$	biomass concentration
z	—	coordinate
α	—	shape parameter in Eq. (17.23)
γ	—	diffusivity parameter in Eq. (17.23)
$\dot{\gamma}$	s^{-1}	shear rate

δ	m	thickness, depth
$\varepsilon(\bar{\varepsilon})$	m$^2 \cdot$s^{-3}	(mean) energy dissipation per unit mass
Φ	—	Thiele modulus
ϕ^2	—	modulus based on observed reaction rate (Fig. 17.10)
η	—	effectiveness factor (general term)
$\eta_r, \eta_{r,V}$	—	(volume-based) effectiveness factor for reaction rate
$\eta_{r,A}$	—	surface-based effectiveness factor
$\hat{\eta}_r$	—	overall effectiveness factor
η_{Tr}	—	effectiveness factor for transport (enhancement factor)
τ	s	characteristic time
ρ	kg$\cdot$m^{-3}	density

Indices

app	apparent
A	area
B	bulk
crit	critical
E	electrode
e	environment
eff	effective
G	gas
i, j	component
ideal	ideal condition
L, L1, L2	liquid (1: film at G/L-, 2: film at L/S-interface)
m	mixing
max	maximum
O	oxygen
P	particle
r	reaction
rel	relative
S	substrate
S	solid
T	tank
Tr	transport
V	volume
$^-$	mean value
$^\wedge$	dimensionless value
*	value under saturation conditions at the interfacial film

Abbreviations

BRE	biological rate equation
CSTR	continuous stirred tank reactor
OTR	oxygen transfer rate
STR	stirred tank reactor

17.1 Introduction

Microorganisms are able to adapt to various types of environmental conditions. Research concerning these phenomena of adaptation has been very scarce. The cells generally appear to adapt themselves by changing structurally and functionally (LEEGWATER et al. [17.1]; TEMPEST and WOUTERS [17.2]). As a consequence, organisms in a CSTR (Continuous Stirred Tank Reactor) are expected to behave significantly different, structurally and functionally, from those growing in batch culture. This fact was the basis of the integrating strategy outlined in Chapter 11. In the present chapter, bioprocess operation is considered for the case, that one phase in the three-phase reaction system (G/L/S) significantly determines the overall rate of the process. In general this will be the solid phase. Processes based on biofilms, pellets, and bioflocs beyond a critical minimum diameter require a strategy that takes into account the interactions between biological reactions and physical phenomena of mass transfer.

Similar concepts of interactions are of practical significance in all experimental situations where rate data from bioreactor operations are analyzed. Generally, all experiments must be understood as complex interrelations not only between bioreactions and physical transports but also as interrelations with the measuring system. Each type of these interrelations will be elaborated in more detail concerning different kinds of intra- and extracellular phenomena and different transport mechanisms. For all situations of reaction and transport occurring simultaneously or in series and a concept for solution of general validity will be presented in Sect. 17.2.2 and outlined for each situation in Sects. 17.3–17.6. Further phenomena even complicate the simple pseudo-homogeneous approach. As indicated in Fig. 11.2, significant environmental changes can be produced by the cells themselves by causing an increased viscosity in the fermentation medium. Many interrelations exist between the fields of knowledge presented in Chapters 1 through 16. It must be realized, however, that the fundamentals of physical mass transfer and chemical reactions, as presented in Chapters 1 to 9, have not yet been fully integrated into microbial processes. The mathematical description of these processes, especially their physical part, is still based on very rough empirical assumptions and approximations as in the case of biological reactions. This chapter deserves special attention and engagement of all biotechnologists in order to finally succeed in an exact description of microbial processes. With this in mind, the following sections should be studied and accepted as a challenge to the reader for further improvement. Even though the basic principles of transport processes were extensively outlined in Part I of this volume (BRAUER [17.25]) and biokinetics in previous chapters, there is still a necessity to demonstrate some essential and accepted concepts to the problem of interactions between biology and physics in order to facilitate the reader's approach. However, this chapter concentrates on the influence of transport phenomena on kinetics and the evaluation of model parameters.

17.2 Problems and Concept of Solution

17.2.1 Experimental Phenomena with Microbial Processes

In bioprocessing experiments several problems arise as a consequence of different types of interactions between reaction and mass transfer. Fig. 17.1 summarizes all these situations in a general schematic diagram.

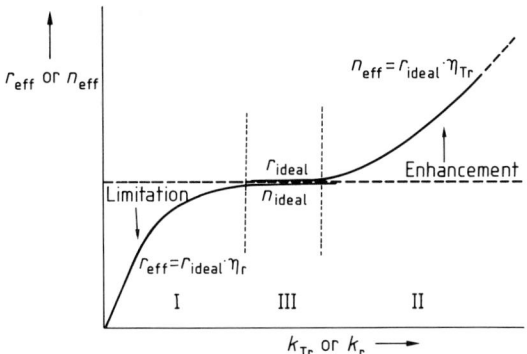

Figure 17.1. Schematic representation of the effect of interactions between reaction and mass transfer showing the region of transport limitations (I) at low values of transport rate coefficients k_{Tr}, and the region of transport enhancement (II) at high values of homogeneous reaction rate coefficients k_r. Only in the medium region (III), the macro-conversion r_{eff} becomes identical with the ideal reaction rate r_{ideal}.

where reaction and transport occur in series (external limitation) or parallel (internal limitation).

17.2.1.1 External Transport Limitation

External transport limitation prevails when the liquid-film resistance at the surface of the solid phase (cells) becomes rate limiting, i.e., k_{L2} in Fig. 11.3. In this case the oxygen uptake rate, representative for the macroscopic formal approach, is significantly dependent on the fluid velocity in a stirred tank bioreactor (NGIAN and LIN [17.3] and CASTALDI and MALINA [17.4]). Increasing the rotational speed in stirred tank reactors (LA MOTTA [17.5]) or recycle flow in tubular reactors may finally lead to

Table 17.1. Classification of the Different Types of Interactions Between Microbial Metabolism, Kinetics, and the Physical Transport Phenomena in Bioreactors

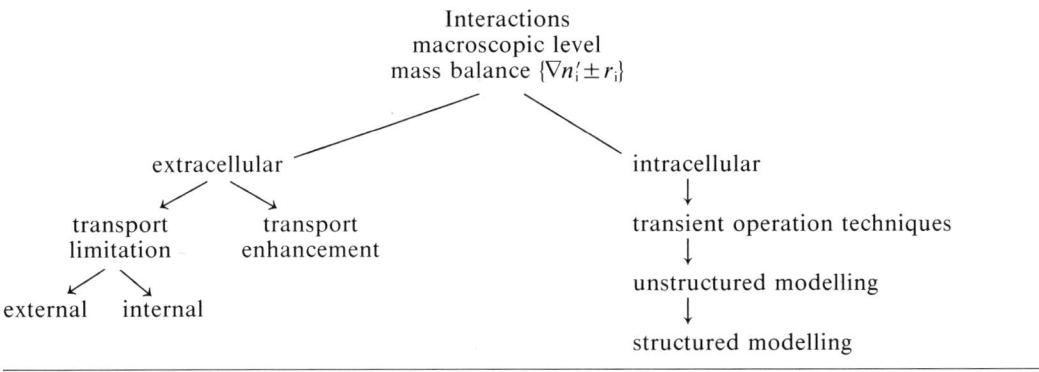

The different experimental situations can be classified according to the distinction between limitation of reaction rate by transport and enhancement of transport rate by reaction both occurring on the macroscopic, extracellular level, as indicated in Table 17.1.

Systematization in more detail can be achieved by subdividing into so-called external and internal transport limitations, the result that the rate becomes insensitive to the fluid velocity as indicated in Fig. 17.1. Similarly, the importance of sufficient micromixing in a fermentation with two immiscible fluids (such as hydrocarbon processing) was drastically illustrated by EINSELE [17.6]. By increasing the rotational speed of an impeller-type CSTR, productivity of cell mass can be increased by a factor of about 3.

Another case of external transport limitation is observed when the limitation of the growth rate is due to insufficient oxygen supply. As demonstrated in Fig. 14.30, linear growth occurs as a result of $k_{L1}a$-limitation (REUSS and WAGNER [17.7]).

17.2.1.2 Internal Transport Limitation

Strong deviations from a simple pseudo-homogeneous treatment of bioprocesses are observed for internal limitations of reaction rates due to diffusional resistance inside microbial agglomerations (k_S in Fig. 11.3). Such cases have been presented in Figs. 14.29 and 14.32.

Typical examples have been described by ATKINSON and UR-RAHMAN [17.8], LA MOTTA [17.9], NGIAN et al. [17.10], HARREMOËS [17.11], and KORNEGAY and ANDREWS [17.12]. For internal limitation, the macrokinetics still exhibit saturation-type kinetics but with a drastically increased value of K_S due to increased and over-critical diameters of bioflocs, biofilms, or pellets (BHAVARAJU and BLANCH [17.13]; MIURA [17.14]).

17.2.1.3 Transport Enhancement

An experimentally quite different situation arises when the physical transport, e.g. OTR, is enhanced by the presence of a biological or biochemical process. This phenomenon was observed, for example, in the case of chemical reactions by DANCKWERTS [17.15], where it plays a central role for the design of absorbers for fast reactions. The problem of OTR-enhancement in bioprocessing was first encountered in the application of sulfite oxidation and the glucose-oxidase method for the quantification of the volumetric transfer coefficient $k_{L1} \cdot a$ (REITH [17.16] and HSIEH et al. [17.17]). Higher values of $k_{L1} \cdot a$ are obtained as would be expected by pure physical absorption, which is questionable with microbial processes (e.g., MOSER [17.18]).

17.2.1.4 Dynamic Response Lag Time of Measuring Electrodes

The measurement of bioprocessing data, such as the oxygen transfer coefficient $k_{L1} \cdot a$ in fermentation media with the dynamic oxygen-electrode method, requires modelling of the complete dynamics of the system. The following steps must be included in handling the dynamics of the bioreactor: liquid-phase dynamics ($k_{L1} \cdot a$), gas-phase dynamics (k_G and $\bar{t}_G$), electrode dynamics (k_E and τ_E) including the resistances of internal electrolyte, membrane, and external liquid film on the outer surface of the membrane electrode. These problems have been elaborated in detail by LEE and TSAO [17.19] and RUCHTI et al. [17.20]. Incorrect values of $k_{L1} \cdot a$ are obtained when these influences are neglected.

17.2.2 Concept for the Solution of the Problems of Interactions

Much theoretical work exists in the literature of chemical engineering and heterogeneous catalysis concerning the coupling of reaction and transport processes within porous solids along with other internal and external interactions. Such work is directly applicable to the analysis of the interactions within microbial aggregates such as flocs, films, or pellets. The main distinction between the laws of chemical kinetics and biokinetics have already been illustrated in Fig. 12.18 and indicated in Table 17.1.

Table 17.2 summarizes the concept for solving the problem of interactions. The introduction of an effectiveness factor η, which is defined as the ratio of the observed rate (macro-conversion r_{eff}) to the rate which would occur in the absence of any interaction, is very useful. Two different definitions for the effectiveness factor are given by Eqs. (11.5) and (17.4); η_r relates to the ideal reaction rate and η_{Tr} to the ideal

Table 17.2. Modelling of Reaction and Mass Transfer Interactions in Microbial Systems: General Concept for the Solution

General Problem:	$r_{eff} = f(k_r, k_{Tr})$ or $f'(t_r, t_{Tr}$ or $t_e)$		(17.1)

Special Situation:
- Extracellular level: $f(k_r, k_{Tr})$
- Intracellular level: $f'(t_r, t_e)$ — see Sect. 12.7 and Sect. 16.2

Transport limitation / Transport enhancement

	Transport limitation	Transport enhancement	
General Concept:	$r_{eff} = r_{ideal} \cdot \eta_r$ (17.2)	$n_{eff} = n_{ideal} \cdot \eta_{Tr}$ (17.3)	
Definition of η:	$\eta_r = \dfrac{r_{eff}}{r_{ideal}}$ (11.5)	$\eta_{Tr} = \dfrac{n_{eff}}{n_{ideal}}$ (17.4)	
Interpretation of η:	$\eta_r = f(k_r/k_{Tr})$ (17.5)	$\eta_{Tr} = f(k_{Tr}/k_r)$ (17.6)	
Special Case:	external: $\eta_r = f(k_r/k_{Tr})$ see Sect. 17.3 internal: $\eta_r = f(\sqrt{Da_{II}}$ or $\Phi)$, $0 \le \eta_r \le 1$ (17.7) (17.9) see Sect. 17.4	$\eta_{Tr} = f(Ha)$, $\eta_{Tr} \ge 1$ (17.8) (17.10) see Sect. 17.6	

combined effects: see Sect. 17.5

transport rate. In all cases, the effectiveness factor can be interpreted as the functional relationship between k_r and k_{Tr} [cf. Eqs. (17.5) and (17.6)].

In this concept, the rate constants k_r and k_{Tr} are compared. Similarly, values for characteristic time (t_r, t_{Tr} and t_e), for changes in reaction, transport, and environment can also be used ([17.21], [17.22], [17.23]).

Theoretical solutions for some cases will be given in the following sections, handling separately the case of reaction and transport in series or next to each other.

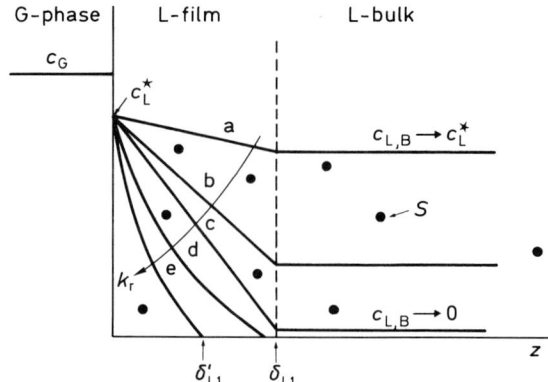

Figure 17.2. Pseudo-homogeneous model approach to heterogeneous G/L/S-processes, where the solid particles are suspended in the liquid phase: concentration/space profiles for different ratios between reaction rate and transport rate coefficient (MOSER [17.22]).
a: "kinetic regime" $k_r < k_{Tr}$
b: interaction regime $k_r \sim k_{Tr}$
c: "diffusion regime" $k_r > k_{Tr}$
d, e: transport enhancement interaction regime.

17.3 External Transport Limitations

17.3.1 Theory

If the chemical or biological reaction is slow, we may assume that the reaction occurs in the liquid bulk and that no reaction takes place in thin liquid layers at the interface, where, according to the simple two-film theory, the resistance to mass transfer can be assumed to be concentrated. This is illustrated in Fig. 17.2. Comparing Fig. 17.2 with Fig. 11.3 it becomes evident that external transport limitation in 3-phase bioprocessing can take place at the G/L- and L/S-interface. In the case of small solid particles the pseudo-homogeneous treatment of the solid phase can be applied ($d_S < \delta_L$).

With the assumption that the rate constants k_r and k_{Tr} are known, which, however, is never the case, the following empirical balance equation may be applied (qss):

$$r_{eff} = k_r \cdot c^n = k_{Tr}(c^* - c) . \quad (17.11)$$

Eliminating the unknown bulk concentration c the following result for $n=1$ is obtained:

$$r_{eff} = k_r c^* \frac{1}{1 + k_r/k_{Tr}} . \quad (17.12)$$

Two other identical forms may be given as:

$$r_{eff} = k_{Tr} c^* \frac{1}{1 + k_{Tr}/k_r} , \quad (17.13)$$

$$r_{eff} = c^* \frac{1}{\dfrac{1}{k_{Tr}} + \dfrac{1}{k_r}} . \quad (17.14)$$

Keeping in mind that for a G/L-process $k_{Tr} = k_{L1} \cdot a_L$ and for a L/S-process $k_{Tr} = k_{L2} \cdot a_S$, this equation can be used to illustrate the fluctuations in the interrelating steps of reaction and transport.

In a slow reaction, i.e., in the case of the "kinetic regime" with $k_r/k_{Tr} \ll 1$, Eq. (17.14) reduces to:

$$r_{eff} \equiv r_{ideal} = k_r \cdot c^* , \quad (17.15)$$

while in the case of the "transport or diffusion regime" $k_r/k_{Tr} \gg 1$, Eq. (17.14) becomes:

$$r_{eff} \equiv n_{ideal} = k_{Tr} \cdot c^*, \qquad (17.16)$$

which is the equation for pure physical absorption.

Thus, comparing Eq. (17.13) with the concept of an effectiveness factor, $\eta_r = 1$ for conditions of kinetic control [Eq. (17.15)]. For external transport limitation the following equation for η_r results:

$$\eta_r = \frac{1}{1 + k_r/k_{Tr}}. \qquad (17.17)$$

This result is in formal agreement with Ohm's law in electrical engineering for resistances in series. All phenomena in series behave in this way.

Therefore, for a 3-phase heterogeneous bioprocess, the equation for the effective rate in the S-phase will represent a steady-state condition and is given by:

$$r_{eff,S} = c^* \frac{1}{\dfrac{1}{k_{L1} \cdot a_L} + \dfrac{1}{k_{mix}} + \dfrac{1}{k_{L2} \cdot a_S} + \dfrac{1}{k_{r,i}^S}}, \qquad (17.18)$$

with k_{mix} as the formal rate constant for fluid mixing, which is indirectly proportional to the mixing time t_m and can be neglected for turbulent conditions. An overall effective rate constant k_{eff}, can be formulated as follows:

$$1/k_{eff} = 1/k_{L1} \cdot a_L + 1/k_{L2} \cdot a_S + 1/k_r. \qquad (17.19)$$

17.3.2 Practical Application of Theory

Three different situations in external transport situations can be distinguished in 3-phase bioprocessing, i.e., limitations in the k_{L2}-film, in the k_{L1}-film, or in the liquid (t_m). When considering pseudo-homogeneous bioprocessing, however, only k_{L1}-limitation can be handled. The influence of liquid mixing is then formally a parallel phenomenon while the k_{L2}-limitation cannot be elucidated at all in stirred-tank-type reactors.

17.3.2.1 Transport Limitation at the Liquid/Solid Interface

This classical situation in bioprocessing is complicated by the complex fluid dynamics for multi-phase systems. A useful approach to mass transfer problems is therefore based on the application of dimensionless groups which has been thoroughly explained in Chapters 1 to 9. The numbers of Sherwood Sh, Schmidt Sc, Reynolds Re, and Grashof (or Archimedes) Gr (or Ar) are relevant for external mass transfer (MOO-YOUNG and BLANCH [17.24]). The most advanced treatment of mass transfer without and with chemical reactions has been presented by BRAUER [17.25] in Chapters 1 to 9.

A typical example of a solution for the case of biokinetics [see Eq. (12.8)] has been given by ATKINSON and UR-RAHMAN [17.8] using Eq. [17.14] and replacing the rate constant k_r by an equation which takes also into account internal transport limitation (η_r):

$$k_{r,app} = q_{S,max} \frac{s}{K_S + s} \cdot \eta_r(s/K_S, k_3 \cdot d). \qquad (17.20)$$

They conclude that with

$$k'_{L2} = k_{L2}/(\rho_S \cdot a_S) \qquad (17.21)$$

the limitation of the liquid-phase diffusion will only have a small effect on the overall rate of substrate uptake:

$$k'_{L2}/k_{r,app} \geq 10. \qquad (17.22)$$

Finally these authors derive an equation by substituting Eqs. (17.20) and (17.21) into Eq. (17.22) leading to

$$Sh_{L2} \frac{\alpha^2}{\gamma} = \frac{60 \cdot \eta_r \left(k_3 \cdot \alpha \cdot \dfrac{1}{a_S}\right)^2}{1+s/K_S}, \quad (17.23)$$

which is plotted in Fig. 17.3 in terms of $Sh_{L2} \cdot (\alpha^2/\gamma)$ vs. s/K_S for various values of $k_3 \cdot \alpha \cdot a_S$, k_3 being a coefficient of the "biological rate equation" defined in Eq. (17.26).

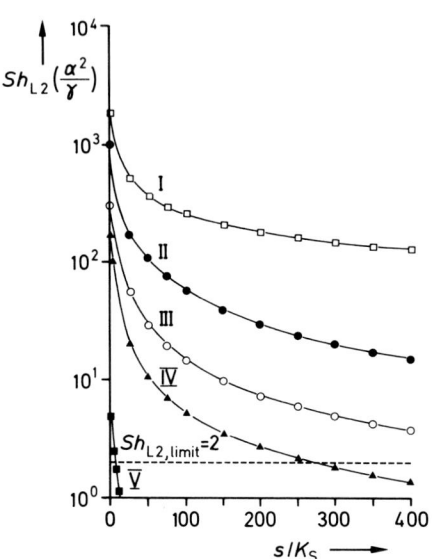

Figure 17.3. Sherwood number Sh_{L2}, required for a minimum limitation of liquid-phase diffusion according to Eq. (17.23) with varied values of $k_3 \cdot \alpha \cdot d_S$: (I) 30; (II) 10; (III) 5; (IV) 3; (V) 0.5. The limiting Sherwood number $Sh_{L2}(\alpha^2/\gamma) = 2$ is indicated by a dotted line (ATKINSON and UR-RAHMAN [17.8]).

The curves represent the value of Sh_{L2} required to achieve a minimum L-phase diffusional limitation for bioflocs of given size $(k_3 \cdot a_S)$ exposed to a given concentration. Clearly the larger the particle and the smaller the concentration, the larger the k_{L2}-value required. Since γ, relating diffusivity within biomass D_{eff} to that of the free solution D, is probably only slightly less than unity and α, which is a shape parameter, only slightly greater than unity (1.06 for cylindrical and 1.16 for spherical flocs), it follows that the smallest value of Sh_{L2} (α^2/γ) is approximately two, sufficient only for the smallest flocs as shown in Fig. 17.3. Substituting this value into Eq. (17.26) for the conditions most sensitive to a k_{L2}-limitation, i.e., $c \to 0$, assuming that (with $d_S = 1/a_S$) $k_3 \cdot \alpha \cdot d_S < 0.5 \,(\eta_r \sim 1)$, leads to the criterion, where there will be no k_{L2}-limitation even under quiescent conditions (ATKINSON and UR-RAHMAN [17.8]):

$$k_3 \cdot \alpha \cdot d_S \leq 0.182 \quad (17.24)$$

17.3.2.2 Liquid-phase Micromixing

Concerning the influence of micromixing in the L-phase there is a lack of adequate data due to the absence of a methodology. Quantification of micromixing in the L-phase is achieved with the aid of the terminal mixing time t_m, which however, can only be applied to stirred tank reactors and reactors with circulation (BRYANT [17.26]). Better elucidation of the physics of micromixing was achieved by direct measurement of the glucose uptake in cells with a fluorometer, carried out by EINSELE et al. [17.27]. The comparison of these measurements with conventional t_m values suggested that bulk liquid mixing is only important when it is related to the response time of cells (ca. 4 s). At present, the stirred tank is the dominant type of bioreactors, where nearly complete mixing is assumed to be attained, which is questionable at large scale.

17.3.2.3 Transport Limitation at the Gas/Liquid Interface in Case of Oxygen Transfer

This field of bioengineering is widely elaborated in textbooks and articles, as the problem of OTR plays a central role in bioprocessing where aerobic fermentations dominate. A sufficient oxygen supply is often recognized as the factor determining bioreactor productivity especially for growing cells with high oxygen demand, high cell concentrations, or high viscosities. Major efforts, as in L-micromixing, have been made to eliminate O_2 limitations in industrial fermentation, and have stimulated the design of new bioreactors with unconventional aerators. Nevertheless, the problem of O_2 transport-rate (OTR) limitation still remains and requires a continuous control of the dissolved O_2-concentration in the medium.

Clearly, Eq. (17.12) must also apply to external transport limitation in pseudo-homogeneous bioprocesses, when steady-state conditions are considered ([17.28], [17.29]).

17.4 Internal Transport Limitation

In bioprocessing the limiting mass transfer step possibly shifts from the external case (G/L- or L/S-interfaces) to the interior of the cells, when compact cell aggregates such as microbial flocs, biofilms, cellular tissues, immobilized cells and enzymes, mycelia and mold pellets are considered. Here, simultaneous diffusion and reaction takes place in the solid phase. In analogy to chemical heterogeneous catalysis the theory of "mass transport with simultaneous reaction" can be applied as outlined by BRAUER [17.25].

The situation of mass transfer through a layer of biofilm, where simultaneously reaction occurs, is shown in Fig. 17.4. Several cases of concentration profiles represent the combined action of external and internal transport limitation and reaction in the biofilm.

A steady-state solution with a Monod-type for the reaction term, with $r_{max} = q_{S, max} \cdot x$, is based on

$$D_{eff} \frac{d^2 s}{dz^2} = a_f \cdot r_{max} \frac{s}{K_S + s}. \tag{17.25}$$

Integration of this equation with appropriate boundary conditions gives the concentration gradient ds/dz in the biofilm. Numerous workers have presented analytical solutions taking various types of kinetics into account (HAUG and MCCARTY [17.30]; HARREMOËS [17.11]; ATKINSON and DAOUD [17.31]; ATKINSON and DAVIES [17.32]; RITTMANN and MCCARTY [17.33], and WILLIAMSON and CHUNG [17.34]).

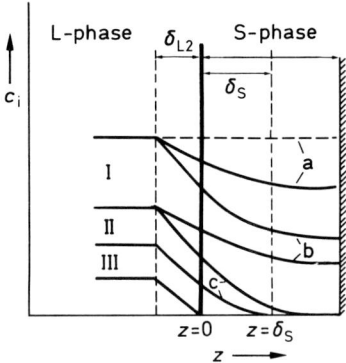

Figure 17.4. Concentration/space profile in case of a heterogeneous liquid-solid bioprocess (partially aerobic biofilm model) with (a) fully penetrated biofilm, (b) shallow biofilm, and (c) deep biofilm (incomplete penetration). Profiles I, II, and III represent the situations of combined effects of external and internal limitations.

The solution of Eq. (17.25) has been given in numerical form as the "biological rate equation" (BRE according to ATKINSON [17.35]), which is in general form

$$r'_i = f(k_2 \cdot s_L, k_3 \cdot d), \tag{17.26}$$

with k_2 and k_3 together with k'_1, called the BRE-coefficients, and s_L the concentration of the liquid medium bathing the particle. While k_1 and k_2 have the same meaning as in the case of pseudohomogeneous rates [cf. Eq. (12.7)] of biokinetics, generally written as rate r_i (in $kg \cdot m^{-3} \cdot h^{-1}$):

$$r_i = f(k_1, k_2, s_L) \tag{17.27}$$

the third coefficient contains internal mass transport limitation:

$$k_3 = \sqrt{k_1'/D},\qquad (17.28)$$

with

$$k_1' = a_S \cdot k_1.\qquad (17.29)$$

The BRE can also be written as

$$r_1' \text{ resp. } \eta_r = f(s_L/K_S, \Phi)\qquad (17.30)$$

using the Thiele modulus Φ.

Fig. 17.5 shows the graphical representation of the BRE in a conventional Monod-kinetic plot for better comparability as dependence of specific rate q_S versus substrate concentration s_L with variation of the thickness d of biological catalyst such as microbial flocs. As can be seen, the Monod-equation is a limiting case of the BRE at low values of d.

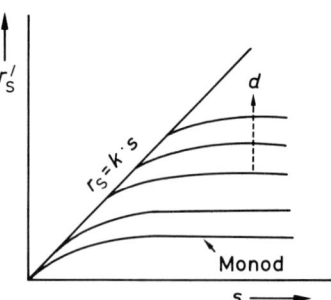

Figure 17.6. Graphical representation of the biological rate equation (BRE) for biofilms according to ATKINSON [17.35] in a diagram of surface-based consumption rate r_S' versus substrate concentration s at varied biofilm thickness d. A pseudo-first-order reaction apparently appears as an asymptotic case as indicated.

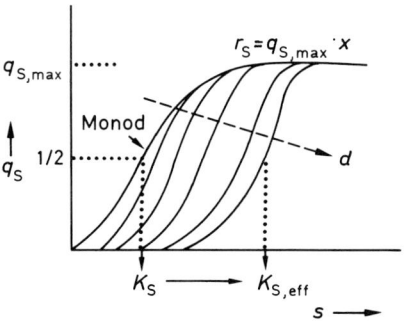

Figure 17.5. Graphical presentation of the biological rate equation (BRE) for microbial flocs according to ATKINSON [17.35] in a diagram of specific consumption rate of substrate q_S versus substrate concentration s at varied diameters of bioflocs d. The Monod-type kinetics are the limiting case for small d, while $K_{S,\text{eff}}$ increases with increasing thickness.

Fig. 17.6 shows the graphical solution of the BRE for biofilm processing in a plot of surface-related rate r_S' versus s_L with variation of film thickness d. The appearance of a pseudo-first-order kinetic approach, often encountered in this field (cf. Sects. 14.3.4, 16.6.2), can easily be seen on this plot.

For better practicability several simplifications of the biological rate equation according to ATKINSON can be derived, which are summarized in Table 17.3 for microbial flocs [Eqs. (17.31)–(17.33)] and in Table 17.4 for biofilms [Eqs. (17.34) to (17.36)].

The corresponding effectiveness factors in the pseudo-homogeneous rate equation,

$$r_{S,\text{eff}} = q_{S,\text{max}} \frac{s}{K_S + s} x \cdot \eta_r,\qquad (17.37)$$

are defined in detail in the literature (ATKINSON [17.35]).

Using the concept of η_r, a graphical plot representing a solution of the BRE is illustrated in Fig. 17.7 as dependence of η_r on s/K_S at varying values of $k_3 \cdot \delta$.

The significance of internal mass transfer resistance on the interpretation of the kinetic data of suspended growth systems (cf. Fig. 14.29) will be explained quantitatively. The biological rate equation usually is transformed for parameter estimation into the following equations: the Langmuir plot (cf. Fig. 12.9)

$$\frac{r_{\max}}{r_{\text{eff}}} = \frac{K_S}{s} \cdot \frac{1}{\eta_{r,V}} + \frac{1}{\eta_{r,V}}\qquad (17.38)$$

17.4 Internal Transport Limitation

Table 17.3. Equations Derived from the Biological Rate Equation (BRE) (ATKINSON [17.35]) which can be used for the quantification of different cases of "biofloc processing"

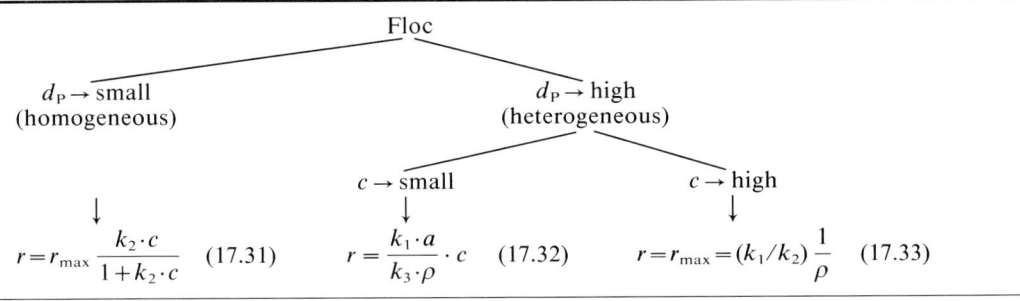

Table 17.4. Equations Derived from the Biological Rate Equation (BRE) ATKINSON [17.35]) which can be used for the quantification of different cases of "biofilm processing"

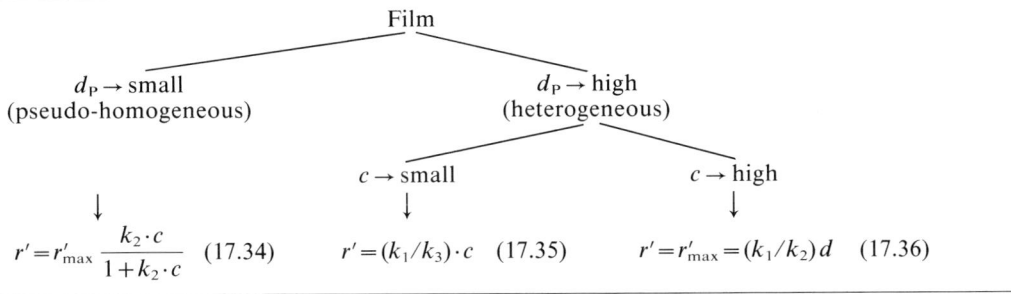

and the Eadie-Hofstee plot (cf. Fig. 12.8)

$$\frac{r_{eff}}{r_{max}} = \eta_{r,V} - \frac{r_{eff} \cdot K_S}{s \cdot r_{max}} \quad (17.39)$$

resp. the Lineweaver-Burk plot (cf. Fig. 12.7)

$$\frac{r_{max}}{r_{eff}} = \frac{1}{\eta_{r,V}} + \frac{K_S}{\eta_{r,V}} \cdot \frac{1}{s}. \quad (17.40)$$

The effect of internal mass transfer resistances can be analyzed on these plots as illustrated in Figs. 17.8 to 17.10 (e.g., SHIEH [17.36]). The linearity of all three linearized plots is significantly distored under the influence of internal mass transfer quantified by ϕ^2 especially with high particle diameters. Larger K_S-values are obtained, while r_{max} ($= q_{S,max} \cdot x$) is not affected provided the S-concentration is reasonably high. If the S-concentration is not sufficiently high, then both K_S and r_{max} would be affected.

The same principles of the theory of mass transfer with simultaneous reaction were used to characterize the diffusional limitation in suspensions of filamentous microorganisms (REUSS et al. [17.37], [17.38]). If mycelial broth can be considered to consist of small spheres of filaments with the coupled phenomena the net specific uptake rate of oxygen in the broth can be calculated by Eq. (17.37) with appropriate values for η_r according to ATKINSON [17.35] using the Thiele modulus Φ. If the assumptions concerning the hypothetical geometry of mycelium structure are reasonable, then Φ should be constant at given fluid dynamics,

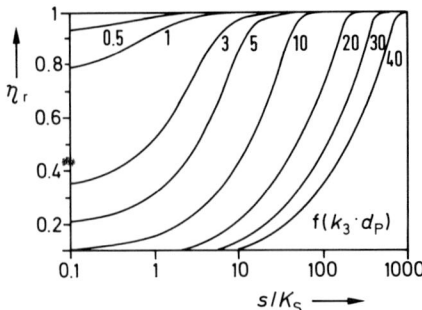

Figure 17.7. Numerical solution corresponding to the biological rate equation [Eq. (17.26)] in a plot of volume-based effectiveness factor η_r versus dimensionless substrate concentration s/K_S at varied values of dimensionless biofilm thickness $k_3 \cdot d$ (ATKINSON [17.35]).

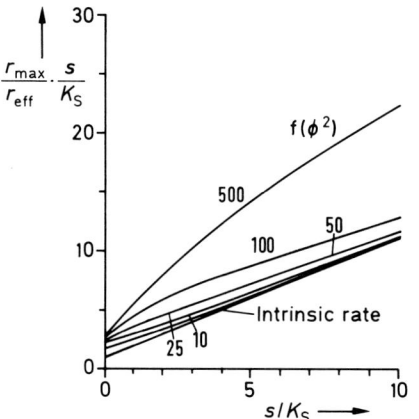

Figure 17.8. Graphical illustration of the effect of internal transport limitation on the Langmuir plot (cf. Fig. 12.9) at varied values of the modulus ϕ^2 (SHIEH [17.36]).

where

$$\varepsilon_{max} = 0.5 \cdot \bar{\varepsilon} \left(\frac{d_T}{d_i}\right)^3 \tag{17.42}$$

with d_T tank diameter, d_i impeller diameter and $\bar{\varepsilon}$ average energy dissipation rate per unit mass. Eq. (17.41) means that for elements large enough to correspond to eddies of inertial subrange, the maximum stable diameter is only a function of ε and is independent of broth viscosity.

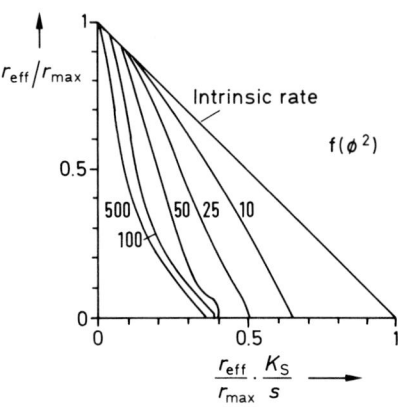

Figure 17.9. Graphical representation of the effect of internal mass transfer resistance on the Eadie-Hofstee plot (cf. Fig. 12.8) with varied values of the modulus ϕ^2 (SHIEH [17.36]).

which fact was experimentally verified. The effect of agitator speed can be satisfactorily accounted for by Φ. Applying the statistical theory of turbulence (KOLMOGOROFF, [17.39] it could be concluded that the mean size of particles is controlled by turbulent shearing action equivalent to the energy dissipation ε:

$$\bar{d}_P = c \cdot \varepsilon_{max}^{-0.25}, \tag{17.41}$$

Thus, it is evident from Eq. (17.41) resp. (17.42) substituted into the equation for Φ that Φ should correlate with power input with an exponent of (-0.25), which fact has also been verified in experiments (REUSS et al., [17.38]). Eq. (17.41) also provides an opportunity to verify the theory for the type of experiments in which macrokinetics have been found, where O_2 uptake rates of the mycelium varied with variation of impeller/tank diameter ratio (e.g., STEEL and MAXON [17.40]).

Finally the practicability of the η_r-concept will be illustrated for stirred tank reactors. Here, it is known that due to the gradients in shear rates, a given floc size distribution results. Therefore, the estimation of the BRE-coefficient k_3 is not possible in stirred tanks but only in biofilm reactors (see Fig. 16.22). Nevertheless, the η_r-concept remains a useful approach, as it was shown that a mean floc size $\bar{d}_P$, is sufficient to characterize a given distribution function (ATKINSON and UR-RAHMAN [17.8]). The mean floc size closely corresponds to the "surface" mean floc size ("Sauter diameter")

$$\bar{d}_P = \frac{\sum_i n_i \cdot d_i^3}{\sum_i n_i \cdot d_i^2} \quad (17.43)$$

with n_i as the number of flocs with diameter d_i. This mean diameter can then be used in the graphical plot of the BRE to determine the effective rate to be expected in the experimental situation.

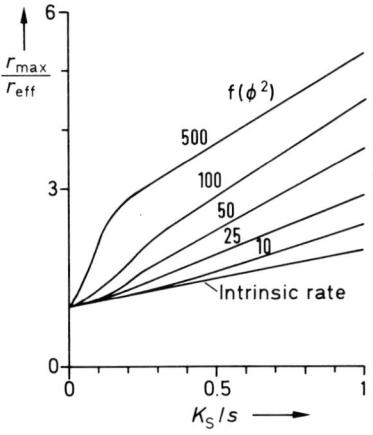

Figure 17.10. Graphical demonstration of internal mass transfer resistance on the Lineweaver-Burk plot (cf. Fig. 17.7) with variation of the modulus ϕ^2 (SHIEH [17.36]).

The use of a surface-based effectiveness factor $\eta_{r,A}$ instead of volume-based factors previously used ($\eta_r \equiv \eta_{r,V}$) has been suggested by FUJIE et al. [17.41].

17.5 Combined External and Internal Mass Transfer Effects

In practical investigations the phenomena of external and internal mass transport cannot be completely separated. Experimentally, it is possible to attain a condition at which the effect of external transport is negligible by increasing the relative fluid velocity and also the effect of internal limitation by reducing particle size. Generally, however, both phenomena must be taken into consideration by checking the experimental conditions.

The typical situation of experiments where both phenomena are significant is shown in Fig. 17.4. In most aerobic biofilm systems O_2 does not fully penetrate the whole biofilm (FUJIE et al. [17.41]). Therefore it is convenient to divide the overall biofilm system into four regions: bulk liquid, diffusion layer δ_{L2} according to two-film mass transfer theory, active or aerobic biofilm δ_S, and anaerobic or inactive biofilm. Increasing the relative fluid velocity v_{rel} at the L/S-interface results in a deeper active biofilm. Different curves of the concentration profile are obtained for different ratios of k_{L2} and k_S.

An analytical solution of this problem of combined external and internal transports is possible only for simple reaction rate equations. A numerical solution can be achieved by assuming first-order approximations (RONY [17.42]). It is also possible to solve the problem without a numerical solution of the differential equation by using the result of the computations that have been performed for the case in which only internal transport has been considered. HORVATH and ENGASSER [17.43] show, by using Eadie-Hofstee plots, that external and internal transports can be best identified.

In Eq. (17.44) the combined effects of external and internal transport limitations are quantified, showing the type of interaction.

$$-r_{\text{eff}} = \left(\frac{1}{\frac{1}{k_r \cdot \eta_r} + \frac{1}{k_{L2}}}\right) \cdot s \cdot x = k_{\text{app}} \cdot s \cdot x \quad (17.44)$$

where $s \equiv c_{L,B}$.

The behavior of this type of equation can graphically be shown by plotting k_{app} against $s_{L,B}$ in a log/log-diagram, Fig. 17.11 (WATANABE et al., [17.44]).

thickness and case II is characterized by a half-order reaction rate. A series of nitrification experiments has been carried out to verify the model.

Coupled external and internal transports were also considered by REUSS [17.45] showing a graphical solution of the problem, assuming saturation type kinetics for respiration of pellets and defining an overall transfer coefficient [cf. Eq. (17.19)].

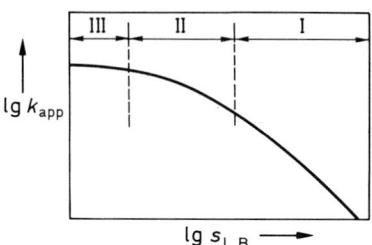

Figure 17.11. Relationship between overall rate coefficient k_{app} and bulk substrate concentration $s_{L,B}$ in a double logarithmic plot according to Eq. (17.44), representing a formal kinetic approach to the problem of combined external and internal mass transfer limitation. The three regions (I, II, III) belong to the concentration profiles I, II, and III in Fig. 17.4 as referred to in the text (WATANABE et al. [17.44]).

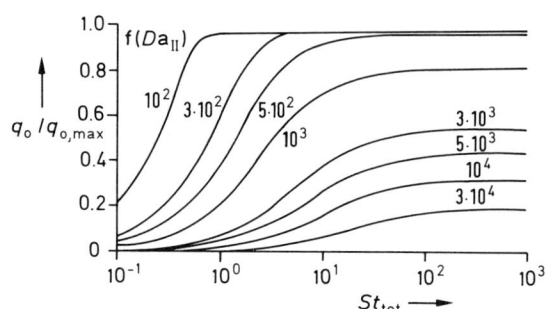

Figure 17.12. Graphical plot of the normalized effective oxygen consumption rate $q_O/q_{O,\text{max}}$ versus a modified Stanton number ($St_{\text{tot}} = k_{\text{Tr,tot}} \cdot r_P^2/D_{\text{eff}}$) representing the ratio between external and internal transport rate as a function of the Damköhler number ($Da_{II} = q_{O,\text{max}} \cdot r_P^2/K_S \cdot D_{\text{eff}}$) as a measure for the ratio between maximum reaction rate and internal transport rate, showing the influence of external O_2 transport rate on effective reaction rate with pellets (REUSS [17.45]).

For external limitation (see profile III in Fig. 17.4), c_S is equal to zero and Eq. (17.11) reduces to

$$-r_S = k_{L2} \cdot c_{L,B} \, . \quad (17.45)$$

This means that a bulk first-order reaction occurs at extremely low values of $c_{L,B}$ which is in accordance with enzyme kinetics. If L/S-mass transfer is negligible, i.e., I and II in Fig. 17.11,

$$c_{L,B} \cong c_S \quad \text{and} \quad (17.46)$$

$$k_{\text{app}} \cong k_r , \quad (17.47)$$

then Eq. (17.44) is simplified to obtain the concentration profiles I and II in Fig. 17.4. Case I is characterized by a zero-order reaction rate where δ_O is the oxygen penetration

With this boundary condition Eq. (17.25) was numerically solved and plotted as shown in Fig. 17.12. Calculations demonstrate to which extent r_{eff} of O_2-utilization of pellets increases by increasing external transports. O_2-limitation still remains a function of pellet size, maximum respiration rate, and diffusion coefficient (Da_{II}) and can only be avoided by an increase of external O_2-concentration (cf. profile a in Fig. 17.4). On the basis of these calculations the potential influence of process variables in experiments, e.g., agitation, aeration, can be predicted, resp. efficiency can be increased.

Finally, an overall effectiveness factor $\hat{\eta}_r$ was proposed as more valuable when L-film resistance of mass transfer cannot be neglected. Since $\hat{\eta}_r$ approaches $\eta_{r,V}$ when L/S-mass transport becomes infinite, the latter can be regarded as a limiting case of the former (YAMANE, [17.46]).

17.6 Transport Enhancement in Bioprocessing

When a gas is absorbed into a liquid and reacts there, the concentration profiles are disturbed by the chemical reaction in the bulk (cf. Fig. 17.4, curves d and e). Further increases of the reaction rate at constant mass transfer lead to a further increase in the gradient in the L-film at the interface, with the result that reaction and transport occur simultaneously at this place. This results in a decrease of the hypothetical film thickness from δ_{L1} to δ'_{L1}. The enhancement of mass transfer is due to the chemical or biological reaction.

The possible rate enhancement of G/L-oxygen transfer by viable respiring microbial cells was attained to some extent in various bioreactors (e.g., TSAO et al. [17.47]) but was later on challenged by experiments (YAGI and YOSHIDA [17.48]; LINEK et al. [17.49]).

Furthermore, TSAO et al. [17.47] even found a more pronounced effect of OTR-enhancement due to the respiration of viable cells or in a glucose oxidase system and explained the difference between theory and experiment by the accumulation of microorganisms in the surface region (so-called "two-zone model").

The full pattern of observed or challenged η_{Tr} can only be understood by taking into account all interactions between physical and biological parameters. Thus, enhancement depends on a variety of factors (MOSER [17.18]):

$$\eta_{Tr} = f(k_{L1}, n_O, n_S, q_{O,max}, x, K_O, m_S, K_S, a) \quad (17.48)$$

assuming double-substrate limitation kinetics (glucose and O_2) and including a term for maintenance, m_S.

This systematic study shows that bio-enhancement would significantly contribute to OTR, if it exists at all, only in aeration systems with low k_{L1}-values at high x and high $q_{O,max}$, S-limitation and at low m_S in the region of $n_O = 1$. Recent model verification experiments led to the conclusion that the effect of OTR-enhancement is difficult to observe experimentally because microbial cells rapidly adapt to the environment. As a consequence the needed preconditions, e.g., highest respiration rate at O_2-limiting conditions, cannot be realized with known biological test fermentation systems (KÜNG and MOSER [17.50]).

MERCHUK [17.51] explained the discrepancies of the "extra-enhancement" with the aid of a simple model considering linear, constant, and exponential cell distribution in the L-film at the G/L-interface.

This result is supported by recent investigations using the absorption system carbon dioxide phosphate buffer-carbonic anhydrase for the measurement of k_{L1} and a indicating that absorption enhancement is completely in line with ordinary theory, and bio-enhancement does not require any new special theory such as the 2-zone model, (ALPER et al., [17.52]).

17.7 References

[17.1] M. P. M. LEEGWATER, O. M. NEIJSSEL, and D. W. TEMPEST: J. Chem. Tech. Biotechnol. *32* (1982), 4, 92.
[17.2] D. W. TEMPEST and J. T. M. WOUTERS: Enzyme Microbiol. Technol. *3* (1981), 283.
[17.3] K. F. NGIAN and S. H. LIN: Biotechnol. Bioeng. *18* (1976), 1623.
[17.4] F. J. CASTALDI and J. F. MALINA: J. Water Pollut. Control Fed. *54* (1982), 261.
[17.5] E. J. LA MOTTA: Biotechnol. Bioeng. *18* (1976), 1359.
[17.6] A. EINSELE: Ph. D. Thesis 4900, Eidgenössische Technische Hochschule Zürich, 1972.
[17.7] M. REUSS and F. WAGNER: In "Proc. 3rd Symp. Tech. Mikrobiol.", p. 89 (H. DELLWEG, ed.). Institut für Gärungsgewerbe und Biotechnologie, Berlin, 1973.
[17.8] B. ATKINSON and F. UR-RAHMAN: Biotechnol. Bioeng. *21* (1979), 221.
[17.9] E. J. LA MOTTA: Environ. Sci. Technol. *10* (1976), 765.
[17.10] K. F. NGIAN, W. R. B. LIN, and W. R. B. MARTIN: Biotechnol. Bioeng. *19* (1977), 1773.
[17.11] P. HARREMOËS: Vatten *2* (1977), 122.
[17.12] B. H. KORNEGAY and J. F. ANDREWS: In "Proc. 24th Ind. Waste Conf.", p. 1398. Purdue (USA), 1969.
[17.13] S. M. BHAVARAIU and H. W. BLANCH: J. Ferment. Technol. *6* (1975), 413.
[17.14] Y. MIURA: Adv. Biochem. Eng. *4* (1976), 3.
[17.15] P. V. DANCKWERTS: "Gas-Liquid Reactions". McGraw-Hill, New York, 1970.
[17.16] T. REITH: Ph. D. Thesis, Technische Hogeschool Delft (The Netherlands), 1968.
[17.17] D. P. H. HSIEH, R. S. SILVER, and R. I. MATELES: Biotechnol. Bioeng. *11* (1969), 1.
[17.18] A. MOSER: In "Proc. 2nd Symp. Bioconversion and Biochemical Eng.", Vol. 2, p. 253 (T. K. GHOSE, ed.). Pramodh Kapur, Raj Bandhu Ind. Co., New Delhi, 1980.
[17.19] Y. H. LEE and G. T. TSAO: Adv. Biochem. Eng. *13* (1979), 35.
[17.20] G. RUCHTI, I. J. DUNN, and J. R. BOURNE: Biotechnol. Bioeng. *23* (1981), 277.
[17.21] I. A. ROELS: J. Chem. Tech. Biotechnol. *35* (1982), 59.
[17.22] A. MOSER: "Bioprozeßtechnik". Springer Verlag, Wien – New York, 1981.
[17.23] A. MOSER: Acta Biotechnol. *4* (1984), 3.
[17.24] M. MOO-YOUNG and H. W. BLANCH: Adv. Biochem. Eng. *19* (1981), 1.
[17.25] H. BRAUER: In "Biotechnology", Vol. 2, Part I (H. J. REHM and G. REED, eds.). Verlag Chemie, Weinheim – Deerfield Beach, Florida – Basel, 1984.
[17.26] J. BRYANT: Adv. Biochem. Eng. *5* (1977), 101.
[17.27] A. EINSELE, D. L. RISTROPH, and A. E. HUMPHREY: Biotechnol. Bioeng. *20* (1978), 1487.
[17.28] H. TAGUCHI and A. E. HUMPHREY: J. Ferment. Technol. *44* (1966), 881.
[17.29] N. D. DANG, D. A. KARRER, and I. J. DUNN: Biotechnol. Bioeng. *19* (1977), 853.
[17.30] R. T. HAUG and P. L. MCCARTY: Tech. Rep. Civil Eng. Dept. Stanford Univ. No. 149, February 1971.
[17.31] B. ATKINSON and I. S. DAOUD: Trans. Inst. Chem. Eng. *46* (1968), T 19.
[17.32] B. ATKINSON and I. J. DAVIES: Trans. Inst. Chem. Eng. *52* (1974), 248.
[17.33] B. E. RITTMANN and P. L. MCCARTY: Biotechnol. Bioeng. *22* (1980), 2343, 2359.
[17.34] K. J. WILLIAMSON and T. H. CHUNG: "Proc. 49th Nat. Meet. Am. Inst. Chem. Eng.", Houston/Texas (USA), 1975.
[17.35] B. ATKINSON: "Biochemical Reactors". Pion Ltd., London, 1974.
[17.36] W. K. SHIEH: Water Res. *14* (1980), 695.
[17.37] M. REUSS, R. K. BAJPAI, R. LENZ, H. NIEBELSCHÜTZ, and A. PAPALEXIOU: "Proc. 6th IFS (Int. Ferment. Symp.)". London/Ontario (Canada), 1980, F-7.2.1.
[17.38] M. REUSS: J. Chem. Tech. Biotechnol. *32* (1982), 81.
[17.39] A. N. KOLMOGOROFF: C. R. Acad. Sci. USSR *30* (1941), 301; ibid. *32* (1941), 16.
[17.40] R. STEEL and W. D. MAXON: Biotechnol. Bioeng. *8* (1966), 97.
[17.41] K. FUJIE, T. FURUYA, and H. KUBOTA: J. Ferment. Technol. *57* (1979), 99.
[17.42] P. R. RONY: J. Chem. Tech. Biotechnol. *13* (1971), 431.
[17.43] C. HORVATH and J.-C. ENGASSER: Biotechnol. Bioeng. *16* (1974), 909.
[17.44] Y. WATANABE, M. ISHIGURO, and K.

NISHIDOME: Prog. Water Technol. *12* (1980), 233.
[17.45] M. REUSS: Fortschr. Verfahrenstech. *14F* (1976), 551.
[17.46] T. YAMANE: J. Ferment. Technol. *59* (1981), 375.
[17.47] G. T. TSAO, A. MUKERJEE, and Y. Y. LEE: In "Proc. 4th IFS (Int. Ferment. Symp)", p. 65. Kyoto (Japan), 1972.
[17.48] H. YAGI and F. YOSHIDA: Biotechnol. Bioeng. *17* (1975), 1083.
[17.49] V. LINEK, M. SOBOTKA, and A. PROKOP: Chem. Eng. Sci. *29* (1974), 637.
[17.50] W. KÜNG and A. MOSER: Biotechnol. Lett., in press.
[17.51] J. C. MERCHUK: Biotechnol. Bioeng. *19* (1977), 1885.
[17.52] E. ALPER, B. WICHTENDAHL, and W.-D. DECKWER: Chem. Eng. Sci. *35* (1980), 217, 1264.

Chapter 18
Mechanical Stress and Microbial Production

Herbert Märkl
Lehrstuhl B für Thermodynamik
Technische Universität München
Garching, Federal Republic of Germany

Reinhold Bronnenmeier
Linde AG
Höllriegelskreuth, Federal Republic of Germany

18.1 Mechanical Forces Affecting Technical Cultures of Microorganisms
18.1.1 Morphological Variations
18.1.2 Release of Intracellular Material
18.1.3 Changes of Metabolism, Growth Rate, and Production Rate
18.2 Experimental Methods and Definitions
18.2.1 Definition of Damage
18.2.2 Damage Mechanisms
18.2.3 Experimental Procedure
18.2.4 Standardization of Experiments
18.3 Experimental Data
18.3.1 Microorganisms Growing as Single Cells
18.3.1.1 *Tetrahymena pyriformis* (Protozoa)
18.3.1.2 Green Algae and Cyanobacteria
18.3.2 Filamentous Microorganisms
18.3.2.1 *Spirulina platensis*
18.3.2.2 Molds
18.4 Impact on Reactor Scale-up
18.5 References

List of Symbols

D	m	diameter of agitator
E_{260}	—	extinction at 260 μm
$\dot{E}_{260}$	—	release rate of intracellular substances
$k_L a$	s^{-1}	oxygen transfer capacity
n	s^{-1}	number of revolutions
N	W	dissipated energy
v_{tip}	m s^{-1}	tip velocity of agitator
V	L	fermenter volume
V_{prod}	L	volume produced
VCC	—	viable cell counts
μ	h^{-1}	growth rate

18.1 Mechanical Forces Affecting Technical Cultures of Microorganisms

Due to the need of heat and mass transfer, technical cultures of microorganisms have to be mixed by mechanical, hydraulic, or pneumatic means. In each case mechanical forces are generated which affect cultures of microorganisms in multiple ways.

18.1.1 Morphological Variations

Hydromechanical forces influence the morphology of microorganisms. Much experimental work concerning the influence of mechanical forces on the morphology is done with molds because of their use in a number of important industrial fermentation processes.

The form of mold mycelia can be classified into two categories: the filamentous and the pellet form. The most important process in which pellet growth is being widely used is the production of citric acid. METZ and KOSSEN [18.1] published a literature review on "the growth of molds in the form of pellets". According to these authors the citric acid production process is started with an inoculum consisting of pellets of 0.2–0.5 mm diameter with a loose character. At the production level an optimum pellet concentration of $2.8 \cdot 10^8$ pellets/m^3 is achieved. Toward the end of the fermentation the pellets reach a diameter of 1.0–2.0 mm and have a much smoother surface.

The pellet size is influenced to a large extent by the level of agitation applied to the suspension. Two mechanisms of changing the size of a pellet were found by TAGUCHI et al. [18.2]. One of these is the decrease in diameter of the pellet by chipping off pellicles from the surface of pellets, and the other is the direct breakup of the pellet structure at higher hydrodynamic loads. Strong agitation results in smaller and more compact pellets.

A literature review and experimental results of the morphology of filamentous molds have been reported by METZ [18.3], VAN SUIJDAM and METZ [18.4], and METZ et al. [18.5]. Longest hyphae are observed when mycelium is grown under shake flask conditions or on solid media. In stirred reactors the hyphae become shorter when stirring speed is increased. The length of main hyphae ranges from a minimum of 80 μm at high agitation intensity up to more than 300 μm at moderate agitation.

At high agitation intensity hyphae are highly branched and thick, forming small compact units according to the results presented in Fig. 18.1.

In addition to this effect of stirrer speed, a change in length of the main hyphae is observed with varying growth rates (Fig. 18.2). At higher growth rates longer hyphae are observed. VAN SUIJDAM and METZ [18.4] developed a model describing the influence of growth rate and shear stress in the fermenter on the morphology of filamentous molds. The main concept of this

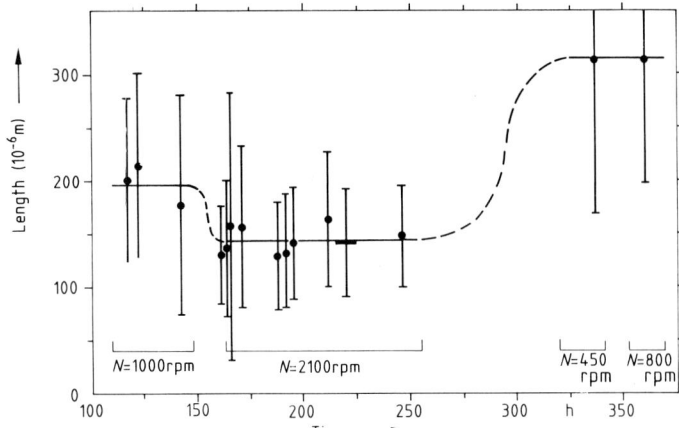

Figure 18.1. Variation in the length of the main hyphae during a continuous experiment at varying stirrer speeds (N) with *Penicillium chrysogenum*, dilution rate 0.055 h^{-1}, 5 L fermenter, impeller diameter 50 mm; METZ [18.3].

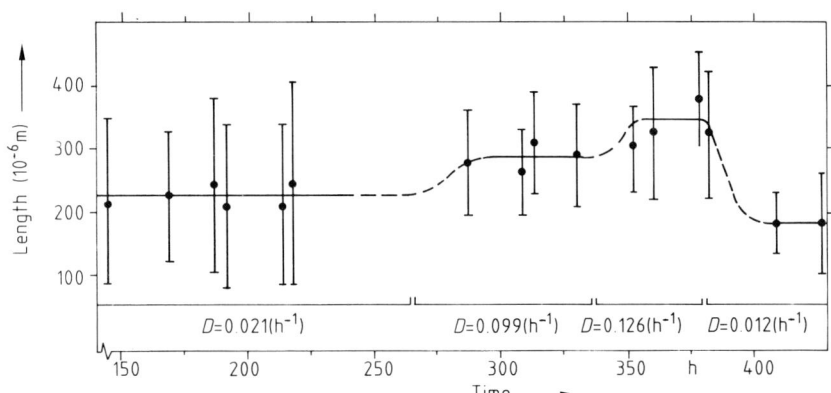

Figure 18.2. Variation in the length of the main hyphae during a continuous experiment with varying growth rates with *Penicillium chrysogenum*. Stirrer speed $N = 1000$ rpm, 5 L fermenter, impeller diameter 50 mm; METZ [18.3].

model is the dynamic equilibrium between growth and breakup of the hyphae. Although there was a qualitative agreement between predictions of the model and experimental results, it was concluded that quantitatively the model was not valid. The authors have made it clear that neglect of the variation of the tensile strength of the hyphae with age and culturing conditions could have been one of the causes of disagreement.

Changes of the morphology of the blue-green alga *Spirulina platensis* were observed by BRONNENMEIER and MÄRKL [18.6]. Details will be given in Sect. 18.3.

18.1.2 Release of Intracellular Material

Agitated cultures of molds are well known to release intracellular material to the fermentation broth due to mechanical forces. TANAKA et al. [18.7] report the result of extensive experiments with mycelial suspensions of *Mucor javanicus* and *Rhizopus javanicus*. Agitation of defined mold suspensions resulted in leakage of intracellular substances consisting of RNA-related nucleotides, mostly mononucleotides with a

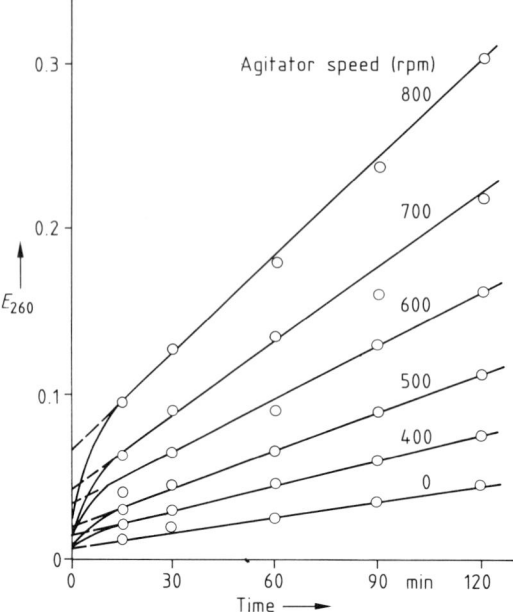

Figure 18.3. Effect of agitator speed on amount of nucleotides (extinction of cell-free liquid at 260 nm) leaked from a model suspension of *Rhizopus* mycelia. – 5 L jar fermenter, $D_{Impeller}/D_{Fermenter} = 0.5$. Preparation of mycelial model suspension: inoculum 3×10^5 spores/mL; cultivation in reciprocal shaker (115 oscillations/min), 30 °C, 24 h; pulp-form mycelia were filtered and washed 3 times with distilled water; the washed mycelia were suspended in distilled water to a concentration of 1 g dry weight per L; TANAKA et al. [18.7].

maximum absorption at 260 nm. This phenomenon was not accompanied by fracture of mycelia. The composition of the nucleotides was the same regardless of culture apparatus and culture age. The rate of observed leakage was dependent on the agitator speed. At any constant agitator speed the nucleotides leaked from mycelia to water in direct proportion to agitation time (Fig. 18.3).

Eighteen strains of filamentous microorganisms, which included species of *Mucor*, *Rhizopus*, *Aspergillus*, *Penicillium*, *Neurospora*, *Piricularia*, *Cephalosporium*, and *Streptomyces*, were examined by TANAKA et al. [18.8]. It was found that the leakage of low molecular nucleotides in stirred mycelial suspensions is a common phenomenon in filamentous microorganisms. The rate of leakage of intracellular substances from mycelia by mechanical forces depends on the culture conditions and the age of the mycelia used.

Adaption phenomena are observed in growing cultures. During cultivation of *Aspergillus niger* S 59, UJCOVÁ et al. [18.9] found a lower leakage of intracellular material at higher stirring speeds. The possible reason for this observation is suggested by MUSÍLKOVÁ et al. [18.10]. High agitation leads to a morphologically compact and strong mycelium, and, in addition, the cell wall provides a greater resistance to the action of hydrolases. This effect was shown in experiments with two *Aspergillus niger* mutants by measuring the quantity of protoplasts released after a 2-hour action of snail gastric juices. The quantity of protoplasts released from hyphae of *A. niger* is smaller after cultivation at higher impeller speed. Hyphae cultivated at the lowest impeller speed were the most sensitive to lytic enzymes.

18.1.3 Changes of Metabolism, Growth Rate, and Production Rate

SITTIG and HEINE [18.11] reported a drastic change of metabolism during fermentation of *Methylomonas clara* in a loop reactor after changing the mixing system in the reactor. During continuous fermentation, mixing was provided alternately by an air-lift pump (low shear forces in the fermentation broth) and by free jet propulsion (high shear forces). Although the production rate of the single cell protein with free jet was the same, the consumption of the substrate was higher by 50% (Fig. 18.4).

At the same time the production of CO_2 almost doubled after changing from the air-lift pump to free jet propulsion. By using free jet propulsion the transfer of oxygen increased, as was expected, but the better

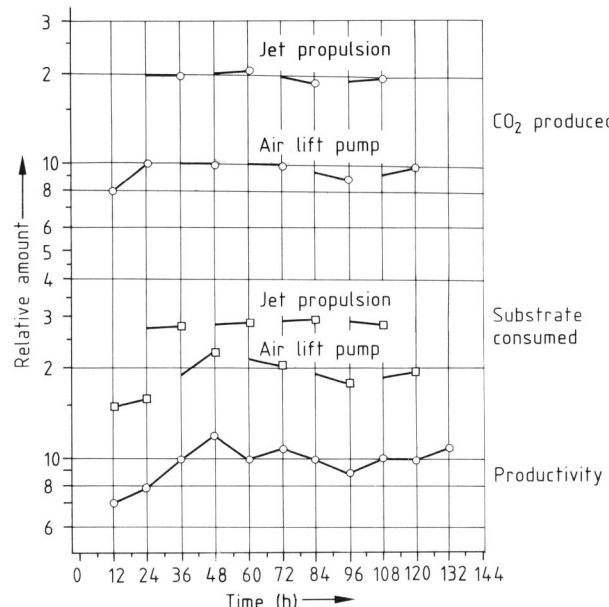

Figure 18.4. Change in metabolism due to mechanical forces; SITTIG and HEINE [18.11].

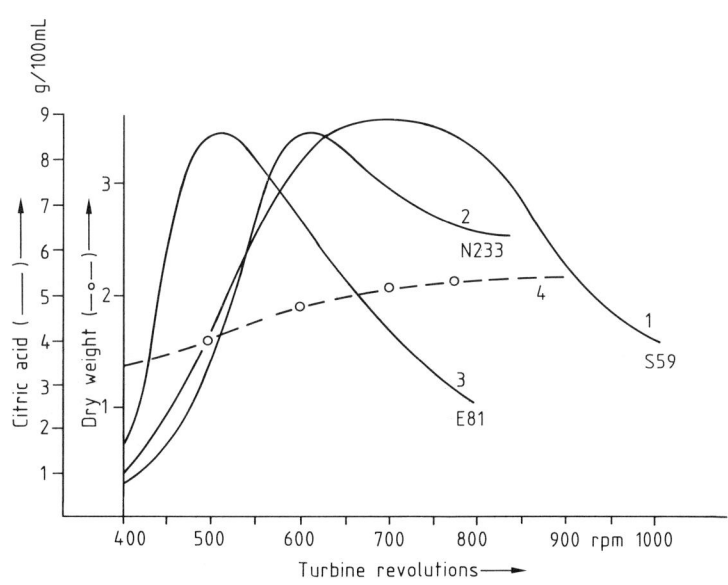

Figure 18.5. Production of citric acid after seven days of batch fermentation in various high-production mutants of *Aspergillus niger* as a function of turbine revolutions. – Curves 1–3: production of citric acid in mutants S 59, N 233, and E 81. Curve 4: dependence of the biomass dry weight of strain S 59 on mixing. Fermenter contained 1 liter of fermentation medium. Turbine stirrer. Aeration 1 L/min; UJCOVÁ et al. [18.9].

availability of oxygen only resulted in higher production of CO_2 but not of microbial protein.

Similarly the production rate of a desired product or the growth rate of microbial biomass can be affected by hydrodynamic forces as demonstrated by experiments of UJCOVÁ et al. [18.9].

The production of citric acid and the growth of mycelia was observed during batch fermentations of different strains of *Aspergillus niger*. A maximum citric acid production level was reached in a diffusely growing mycelium, the production level being highly dependent on the speed of the agitator.

As shown in Fig. 18.5, optimum production was observed at different stirring speeds for each single strain. The E 81 and N 233 mutants have a relatively narrow interval of mixer speeds optimal for citric acid production; this interval is broader for the S 59 mutant. The low productivity in the range of low mixer speeds most probably is attributed to the shortage of dissolved oxygen. Curve 1 represents the production of citric acid and curve 4 the biomass growth. Maximum growth is reached at a higher stirrer speed than the maximum of citric acid production.

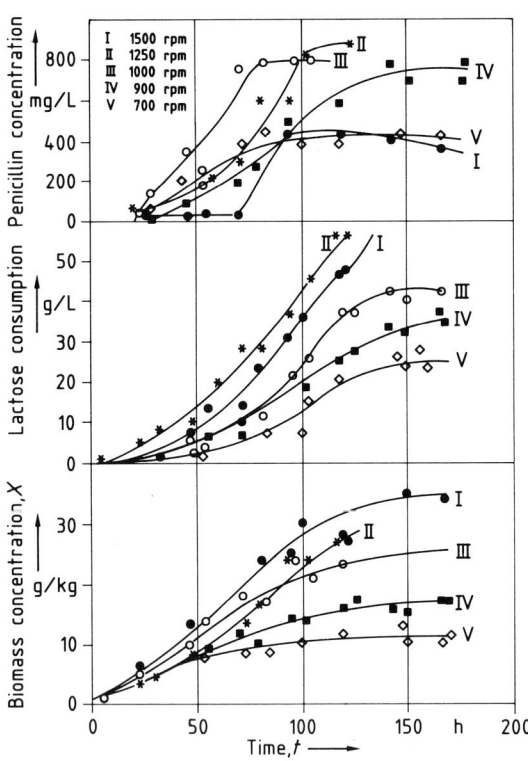

Figure 18.6. Penicillin concentration, lactose consumption, and biomass concentration as a function of the cultivation time for strain H 613 grown in a draught tube with a propeller in a tank of 24 L volume. Parameter: impeller speed (after KÖNIG et al. [18.11a]).

Another example is given by KÖNIG et al. [18.11a]. They studied the effect of stirring speed on the growth of *Penicillium chrysogenum* and the production of penicillin. The result is shown in Fig. 18.6. It is clearly demonstrated that at high stirring speed (1500 rpm, curve I) the penicillin concentration remains low although at this impeller speed the highest biomass concentration is reached. From these findings the following conclusions are drawn:

a) Mixing of technical cultures of microorganisms favors the transport of nutrients and oxygen to the single organism and the transport of metabolic products away from it. On the other hand, mixing may result in damage of organisms at higher mixing rates. Both effects overlap in technical systems.

b) The mechanisms which influence product formation by stirring may be different from those affecting the growth of microbial biomass.

c) Optimal conditions either for the output of a desired product or for the growth of microbes vary for different organisms and even for different strains of the same species to a large extent.

18.2 Experimental Methods and Definitions

For the scale-up of microbial production systems and for the rational design of new bioreactor configurations the capacity of microorganisms for hydrodynamic stress should be known quantitatively. As shown by the above mentioned experiments (Fig. 18.5) of UJCOVÁ et al. [18.9] such information is necessary for each single mutant of one species because the characteristic data may vary considerably from one strain to the other.

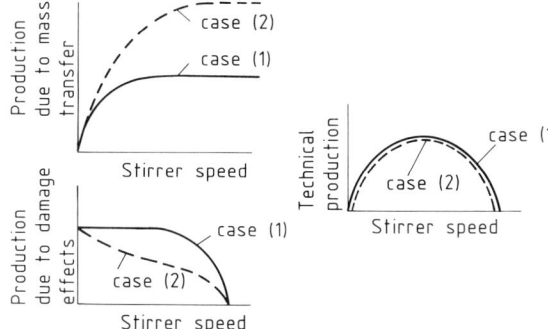

Figure 18.7. Overlapping of mass transfer and damage effects.

There is a lack of information of this kind in the literature. Even for industrial processes, working for years, explicit data about damage are often not available. This is linked to the above discussed overlapping of transport and damage effects. Let us take Fig. 18.5 as an example: The production of citric acid runs through an optimum with increasing stirrer speed. From the data given one cannot distinguish between the following possible causes for this phenomenon as demonstrated in Fig. 18.7: (1) The resulting "Technical Production" simply is the addition of two phases. At low stirrer speed the production is dominated by mass transfer effects, at high stirrer speed by damage effects. (2) Overlapping takes place over a wide range of stirrer speed. Already at low stirrer speed, when O_2-transport is still highly limiting, mechanical effects do not allow a higher production rate. In the first case the theoretical maximum production rate is almost reached. In the second case, due to damage of organisms, the culture cannot be supplied sufficiently with oxygen in this type of reactor.

18.2.1 Definition of Damage

The terms "transport" and "damage" are used rather global in this argumentation. It is necessary to take a closer look at these terms. Microbial production is associated with the supply of different substances and the withdrawal of metabolic products. One of the most important examples is the supply of oxygen to aerobic organisms. It is well known that with an increase in mixing the transport of oxygen from a gas bubble to a single microorganism is accelerated: gas-liquid interfacial area is increased, boundary layers both at the gas bubble and the organism are smaller, convective transport mechanisms within the liquid are influenced positively. In the case of pellet or cell-aggregate forming organisms oxygen also has to be transported from outside of the structure to a single cell located inside the aggregate. Mixing affects this transport indirectly because at different levels of mixing different sizes of pellets or cell-aggregates are growing. Big pellet structures are disrupted and in that way both the interfacial area between liquid and pellet is enlarged and transport paths inside the pellet become smaller.

REUSS [18.12] developed a mathematical model for the supply of cell-aggregates and filamentous organisms with oxygen, coupling molecular diffusion inside the structure with the reaction. It was shown that the model may also be useful for the analysis of pulp-like filamentous microbial systems although it presumes the presence of homogeneous spherical elements such as homogeneous pellet structures in which diffusion and reaction takes place.

The word "damage" is defined in this chapter in the sense that all mechanically generated effects shall be summarized which affect the aim of a technical process in a negative sense.

Let us explain this definition with two examples: Although filamentous, aggregate or pellet forming organisms may be ruptured at a special level of mixing intensity, we may not call this "damage" as long as no negative effects on the process are observed (the consequences could even be positive as was shown above). On the other hand, it is clearly a case of damage when the formation of a desired product at high stirring speeds is suppressed although the organism itself does grow well, as is the case for the production of citric acid with *Aspergillus niger* S 59 (Fig. 18.5).

18.2.2 Damage Mechanisms

How is hydrodynamic stress generated? Regarding the flow around a turbine blade (Fig. 18.8) as an example, a stagnant area is found at the front and a strongly accelerating stream is produced at the tips. At the back of the blade there is a low pressure zone with eddies moving away from it.

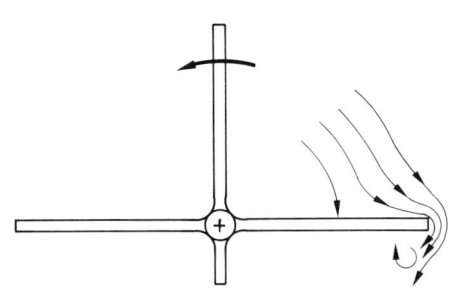

Figure 18.8. Flow around a turbine blade.

In this stream two destructive mechanisms are possible:

1. Rapid pressure change over the blade accompanied by acceleration of the microorganisms.
2. High shear stresses at the turbine tips.

In all practical flow systems both mechanisms are present. Filamentous, floc, or pellet forming organisms are possibly damaged more easily by the second mechanism when, due to shear flow, a critical tensile stress of one single filament is exceeded. Single cell organisms with a compact shape may be more sensitive to rapid pressure change or high accelerations which are of oscillating nature in turbulent flow. Which part of an organism is hurt, whether internal membranes, the cytoplasmatic membrane, or other structures, is not clear and may differ from one case to another.

Macroscopically observed damage most probably represents a dynamic statistical equilibrium of a number of single effects: 1. A particular organism may be hurt passing a zone of high energy dissipation and later the same organism will recover again to a certain extent. Thus, a balance is established between damage and recovering events, the equilibrium of which will be determined by the probability that one organism will be exposed to a damage situation, to the degree of damage, and to the recovery rate. 2. Fatigue phenomena may play a role in some cases. Possibly an organism will not be hurt having a contact to stress conditions for the first time. Only steadily recurring hydrodynamic stress may result in a damage effect in this case (TAGUCHI et al. [18.2]). 3. It is well known that some organisms are able to adapt to a special level of mechanical stress if sufficient time is provided for this process. 4. Different cell age results in different sensitivity to hydromechanical forces.

18.2.3 Experimental Procedure

Systematic experiments are the only way to provide quantitative information on the influence of mechanical forces on microbial production. Although these experiments may be rather difficult in one or the other case, the general strategy should be clear:

a) The key parameter measured for quantification of mechanical effects should be closely related to the aim of the technical process under investigation: production rate of a product, quality of product, yield coefficients, and so on.

Secondary parameters like release of intracellular material or microscopically obtained morphological data may be very helpful but they can be misleading when their relation to the direct process aim is not investigated.

b) Separation of the influence of damage effects and transport effects on microbial production is desirable. One method applicable in some cases, is the variation of inlet concentration of the consumed gas (e.g., oxygen or CO_2 for aerobic or photosynthetic cultures, respectively).

In other cases mathematical modelling may be helpful for analyzing experimental data.

c) In stirred photosynthetic cultures of microorganisms, the reaction rate may depend on the stirring speed due to the so-called flash light effect as was demonstrated by MÄRKL [18.13].

In a cylindrical glass fermenter, illuminated from the outside, the light intensity in the center will be smaller than in the outer regions of the fermenter due to absorption of light. Therefore, mixing results in a dynamic light-dark pattern "seen" by a single alga when moving through different light regions.

The results of five experiments, performed with increasing density of algae, are shown in Fig. 18.9. Experiments No. 1-3 are carried out with the illumination of fluorescent tube lamps (55 W/m^2), in experiment 4 and 5 a Xenon lighting system (530 W/m^2) is used. The CO_2 concentration in the gas bubble, serving as the CO_2 supply for the culture, is established at a value of as high as 1 percent. It can be proved that in this case the rate of the reaction is not limited by the supply of CO_2.

At an algal density of 0.17 g/L there are almost no light gradients in the culture. As a consequence mixing has no influence on the photosynthetic reaction rate.

Higher densities of algae result in an enhancement of photosynthetic activity with increasing stirring speed. The influence of the "flash light effect" increases with higher light gradients within the culture and with increasing light intensity.

When testing the hydrodynamic stress capacity of microalgae by variation of stirring speed in a fermenter, one has to use thin cultures of algae (compare Exp. 1 in Fig. 18.9). This avoids overlapping with the flash light effect.

18.2.4 Standardization of Experiments

In the literature different types of experimental procedures are described. In most cases the conventional stirred tank reactor is used. Different values of stress are generated by different stirring speeds.

These experiments are rather limited for two reasons. First, the range of mechanical stress that can be generated is limited by the available range of stirring speed and, second, standardization of experimental procedure is almost impossible due to the high variety of stirrer and reactor vessel geometry. In addition to this, data are required not only for stirred tank reactors but also for the development of production in new unconventional reactor and aeration systems.

For these reasons a new test apparatus was developed by MÄRKL [18.14], in which a free jet (Fig. 18.10) generates a defined hydrodynamic stress in a fermenter.

A free jet is a relatively simple and clearly defined flow system in which both pressure change and shear stress are pres-

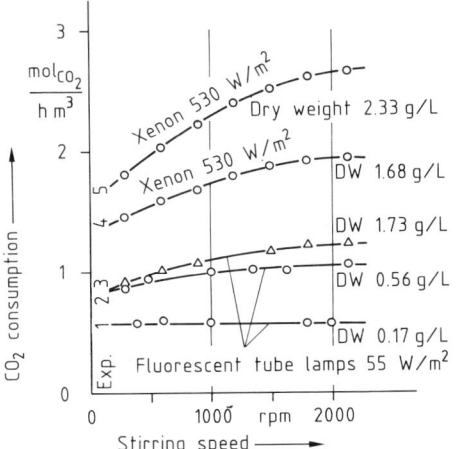

Figure 18.9. Net photosynthetic reaction rate as influenced by mixing. 2 L fermenter illuminated from outside, impeller-diameter 50 mm, experiments with *Chlorella vulgaris*, $T = 27\,°C$, pH = 6.5; MÄRKL [18.13].

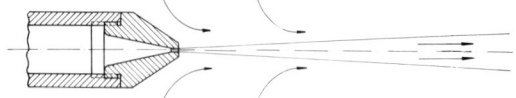

Figure 18.10. Hydrodynamic stress in a free jet.

ent. The latter can be estimated from basic research work, especially that of PRANDTL [18.15] and REICHARDT [18.16].

The working principle of the test apparatus, shown in Fig. 18.11, is as follows: A fraction of the turbidostatically controlled culture (57%) is forced out of the fermenter vessel by a small increase of pressure inside the fermenter into a second vessel, where a defined pressure is adjusted by sterile nitrogen. The constant pressure level can be varied from very low values up to 100 bar.

This pressure forces the culture back into the fermenter by flow through a nozzle of 0.35 mm diameter. At this nozzle a free jet is formed. The system operates under sterile conditions and the nozzle is always submerged. As there are no pumps and only ball valves are used in the system, no uncontrolled damage of the organisms can occur. The cultivation of the test organisms in the fermenter is performed at such stirring conditions that no damage of the organisms by hydrodynamic stress is to be expected.

After application of stress the response of the culture, that is, the metabolic activity, the growth rate, or production rate, can be measured. Because the stress is applied to a continuous culture representing a well balanced system, even very small damage effects can be determined. The stress in the free-jet method is generated during a limited period of time. This procedure enables one to measure not only damage effects but also the recovery behavior of the culture.

MIDLER and FINN [18.17] proposed a constant-shear device resembling a Couette viscometer to produce a well defined hydrodynamic stress in the laminar region (Fig. 18.12).

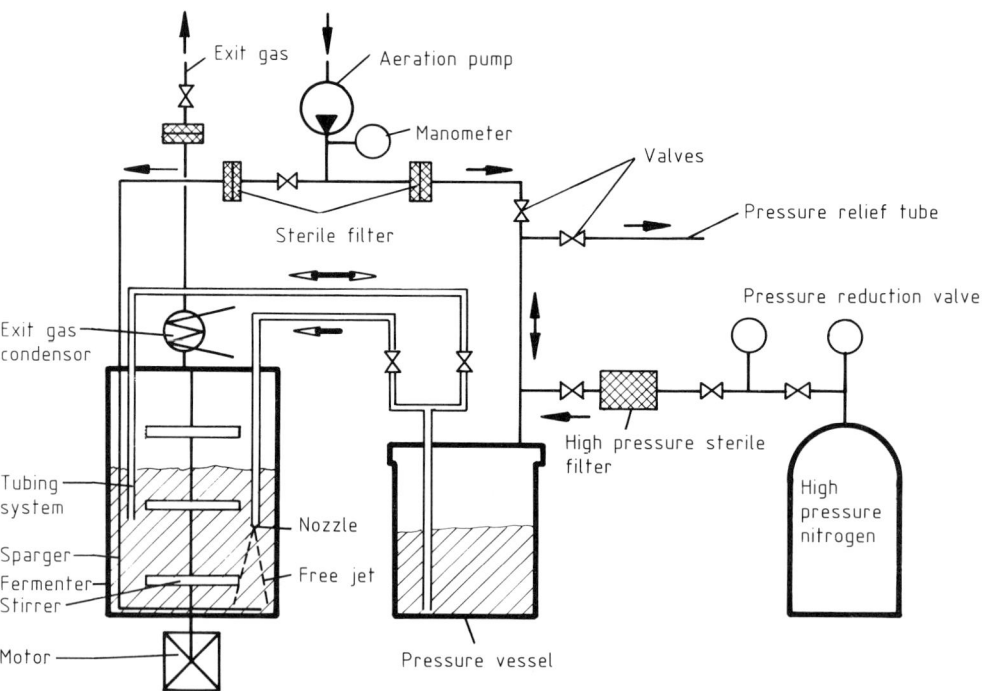

Figure 18.11. Test apparatus for free-jet experiments. ▷ Culture flow, ▶ gas flow.

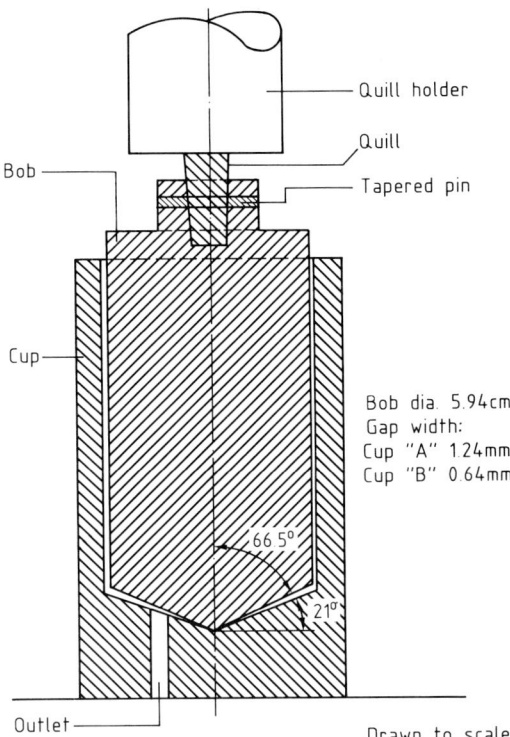

Figure 18.12. Laminar shear device (after MIDLER and FINN [18.17]).

Two different gap widths of 0.64 and 1.24 mm can be used. Experiments up to a shear rate of 1970 s^{-1} have been reported.

18.3 Experimental Data

The experimental data reported here cover the following subjects:

- A model system for evaluating hydromechanical stress, where the bursting of protozoa is taken as a measure of stress intensity.
- Investigations showing the growth of different algae and cyanobacteria as a function of hydromechanical stress.
- Influence of mechanical forces on growth and productivity of mold cultures.

18.3.1 Microorganisms Growing as Single Cells

18.3.1.1 *Tetrahymena pyriformis* (Protozoa)

MIDLER and FINN [18.17] investigated the hydromechanical stress capacity of this protozoon (Fig. 18.13) which has a size of about 80 μm. They exposed suspensions of cells of *Tetrahymena pyriformis* to different hydromechanical stress conditions in turbulent agitated vessels with changing vessel and impeller geometry and in a laminar shear device shown in Fig. 18.12.

The number of bursted cells during or after the stress experiments served as a measure for the degree of damage. The number of undamaged cells was determined microscopically.

The cultures showed a greater sensitivity with increasing age and the salt concentration of the suspending media. In experiments with the laminar shear device (Fig.

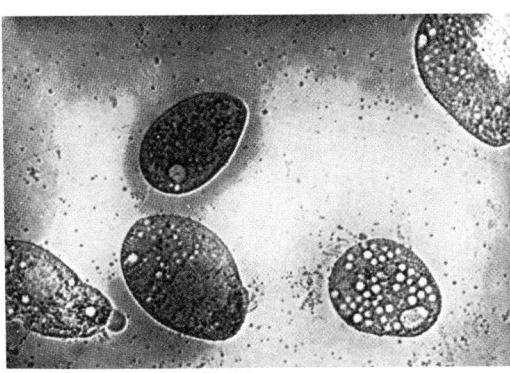

Figure 18.13. *Tetrahymena pyriformis* (after MIDLER and FINN [18.17]).

18.14a) at constant shear rates damage was stronger at higher viscosity. When mechanical stress was generated in a stirred tank reactor by agitators with varying diameters, the damage rate could be well correlated with the impeller tip velocity (Fig. 18.14b).

In Fig. 18.14b it can be seen that no damage of cells occurred below a defined value of impeller tip speed (0.8 m/s; critical hydrodynamic stress). During the course of each stress experiment in agitated vessels

18.3.1.2 Green Algae and Cyanobacteria

Several single-cell-type green algae and one kind of cyanobacterium (Fig. 18.15) were investigated for their hydromechanical stress capacity by BRONNENMEIER and MÄRKL [18.6].

These test organisms differ considerably in shape, size, cellular organization, and

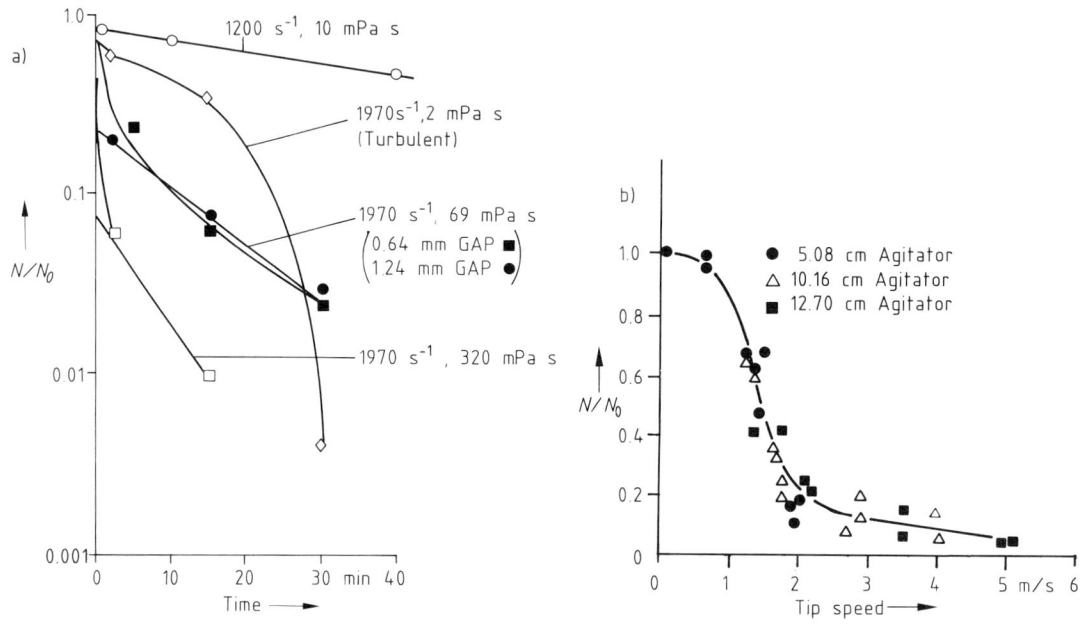

Figure 18.14. a) Fractional survival of *Tetrahymena* in the laminar shear device as a function of time (parameter: shear rate, viscosity). b) Fractional survival of *Tetrahymena* in agitated vessels as a function of agitator tip speed. – N/N_0 = number of not bursted cells to total cell number (after MIDLER and FINN [18.17]).

two different phases of cell destruction were observed. Immediately after the beginning of stress a rapid "primary" disruption of cells occurred. It was followed by a phase of slow "secondary" disruption of the remaining cells at a constant bursting rate.

The authors show that at turbulent flow conditions damaging mechanisms are of higher efficiency than at laminar flow conditions.

cell wall structure from one another. While the wild-type strain of *Chlamydomonas reinhardii* possesses a three-layered cell wall of glycoproteins, the middle layer being submicroscopically structured (SCHLÖSSER et al. [18.18]), the cell-wall-defect mutant strain CW 15 nearly has no cell wall. The cell wall of *Chlorella vulgaris* is formed by two layers of pectin and cellulose, respectively, into which reticular fibers of cellulose are embedded (DAWES [18.19]). The

18.3 Experimental Data

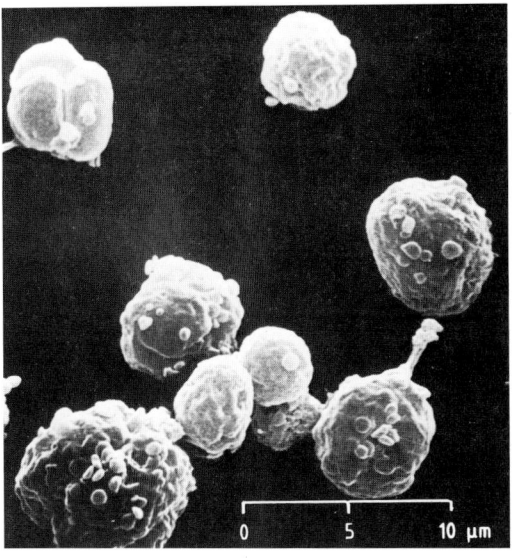

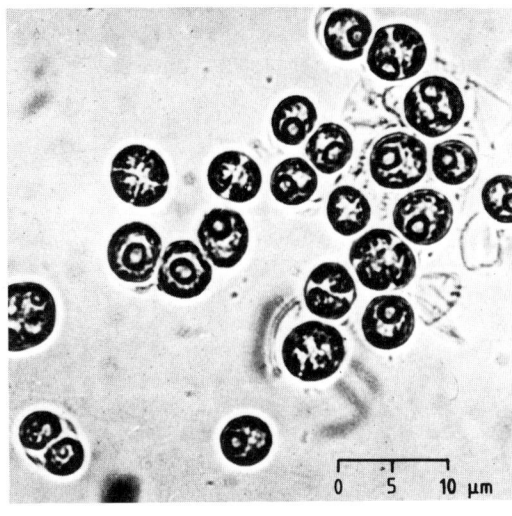

Figure 18.15. Testorganisms
a) *Chlamydomonas reinhardii*
   wild-type strain 11/32-89 of the Sammlung von Algenkulturen, Pflanzenphysiol. Institut der Universität Göttingen (green alga) and
   Chlamydomonas reinhardii
   cell-wall-defect mutant strain CW 15, isolated by DAVIES and PLASKITT [18.21] (green alga).
b) *Chlorella vulgaris* (green alga).
c) *Anacystis nidulans* (cyanobacterium).

rod-shaped cyanobacterium *Anacystis nidulans* has a cell wall composition resembling to that of Gram negative bacteria (peptidoglycan, outer membrane).

The morphology of cultures of *Chlamydomonas reinhardii* is rather heterogeneous, as flagellated and not flagellated cells occur at different growth stages in the generation cycle as described by ESSER [18.20].

Free Jet Experiment

These microorganisms were exposed to hydrodynamic stress by means of the free jet test apparatus (Figs. 18.10 and 18.11). The standard test for the described procedure consists in the circulation of 1 L of the fermenter liquor through the pressure nozzle. This cycle is performed 5 times. As an exception *Chlorella vulgaris* and *Anacystis*

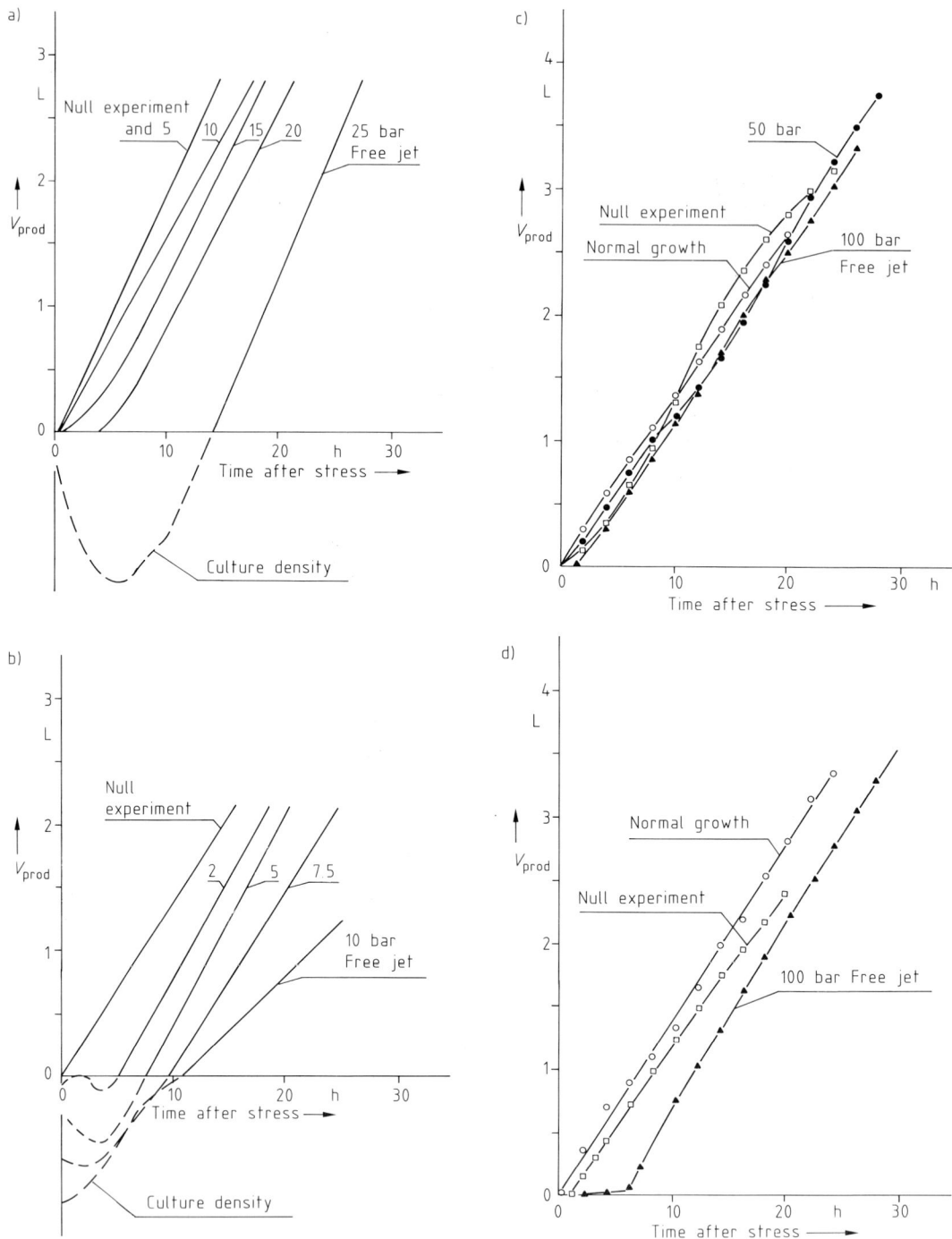

Figure 18.16. Results of free jet experiments with different single-cell-type microorganisms (detailed explanation see text). – a) *Chlamydomonas reinhardii* wild-type strain 11/32–89, b) *Chlamydomonas reinhardii* cell-wall-defect mutant, strain CW 15, c) *Chlorella vulgaris*, d) *Anacystis nidulans*.

nidulans were exposed to 20 stress cycles because these organisms are well known for their high resistance against mechanical stress. Performance of null experiments was as follows: one liter of the culture was pumped to the pressure vessel and then back to the fermenter without passing through the nozzle. Thus, the effect caused by changing culture conditions was determined.

The results are shown in Fig. 18.16. The time after the free jet exposure of the turbidostatically controlled test cultures is plotted at the abscissa of all figures. The ordinates show the volume of the culture, produced by the turbidostat after mechanical stress application. The slope of the curves is a direct measure for the growth rate of the culture, because the culture-density in the fermenter is kept constant by turbidostatic control. When the production reaches the mark at 1.74 L, the culture has doubled once. The parameter labelling the curves denotes the pressure which was applied to press the culture through a nozzle, thus producing a free jet.

In experiments a) and b) the strongest mechanical stress values have been generated by a pressure of 25 and 10 bar, respectively, while in experiments c) and d) even 100 bar were applied to generate the free jet. The experiments with both *Chlamydomonas reinhardii* strains (Fig. 18.16 a, b) indicate that an increase of mechanical stress results in increasing damage of the culture. In the case of the wild-type strain (Fig. 18.16 a) a nozzle pressure of more than 10 bar causes growth of the turbidostatically controlled culture to stop for some time. After the application of 25 bar for generation of the free jet, the time needed for the recovery of the culture has lengthened to 15 h. The reason for this delay is the decrease of the culture density, marked by a dotted line. This line represents the measurement of the turbidostatic probe (photosensor) inside the fermenter in arbitrary units. It can be seen that the density has still decreased 6 h after the 25 bar experiment. Various reasons may be suggested, for example, a progressive reduction of pigments or destruction of organisms. Contrary to the disintegration of heavily damaged cells, a recovery process is observed, balancing the destructive effect after 6 h. The recovery can be attributed to the growth of the lesser or not damaged part of the culture population. As soon as the culture density has reached the initial value, the operation of the turbidostat is restarted.

The critical value of hydrodynamic load below which no essential damage occurs, can be detected directly. In this case it was concluded that a hydrodynamic load produced by a nozzle pressure up to nearly 15 bar had no effect. If these data are compared with Fig. 18.16 b, one can see at once, that the cell-wall-defect mutant strain CW 15 (of the same alga) is much more sensitive to hydromechanical stress: the critical stress limit is evidently below 2 bar free jet stress. Already at this value a measurable decrease of the culture density (dotted line) and a recovery time of the turbidostat of about 5 h was observed. This strain was completely killed in a free jet, generated by 15 bar (experiment not shown in the figure), a stress level where the wild-type strain was only slightly damaged.

Despite the much higher free jet stress exposure, caused by 100 bar pressure in the case of the experiments with *Chlorella vulgaris* (Fig. 18.16 c) and *Anacystis nidulans* (Fig. 18.16 d), no strong damage effects could be observed. The recovery time of the culture of *C. vulgaris* was only about one hour, that of *A. nidulans* about 6 hours. In both cases, neither a release of intracellular substances nor a decrease of viable cell counts was observed. The critical stress values in the free-jet experiments are scarcely below 100 bar for *C. vulgaris* and clearly below 100 bar, but not known exactly, for *A. nidulans*.

In free-jet experiments with *Chlamydomonas reinhardii*, a release of intracellular substances, measured as extinction at 260 nm of the centrifuged, cell-free culture broth, could be observed. In the case of the wild-type strain (Fig. 18.17 a), up to the 15 bar free jet, only a slight increase occurred. Beyond 15 bar the release of UV-absorbing material increased considerably and was continued for 6 hours after stress

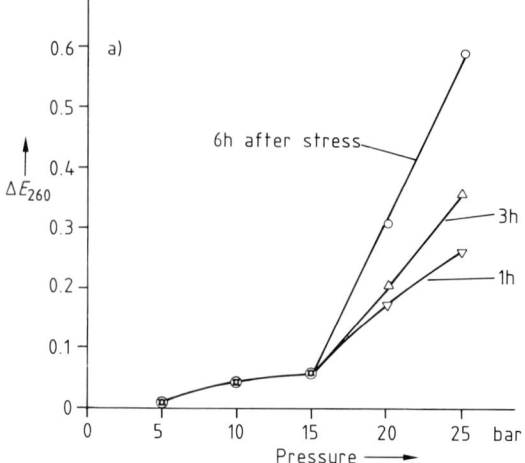

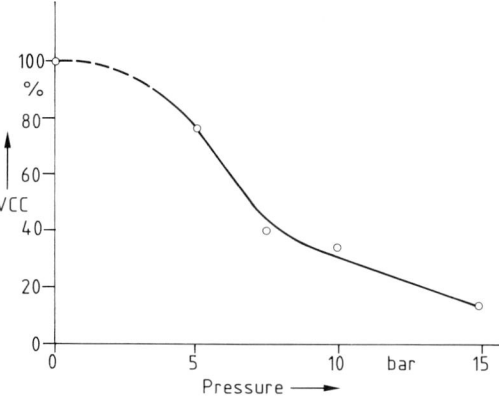

Figure 18.18. Decrease of viable cell counts (VCC) caused by free-jet stress. *Chlamydomonas reinhardii*, cell-wall-defect mutant CW 15.

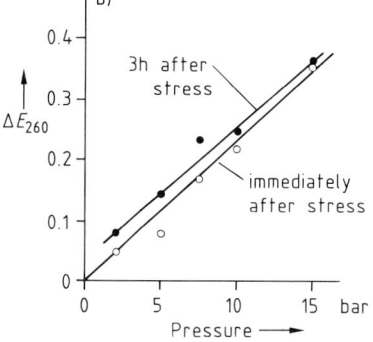

Figure 18.17. Release of intracellular substances caused by free-jet stress. a) *Chlamydomonas reinhardii* wild-type strain. b) *Chlamydomonas reinhardii* cell-wall-defect mutant CW 15.

application. This result correlates with the gradual decrease of the turbidity of the culture, shown in Fig. 18.16a. With the cell-wall-defect mutant strain a linear increase of extinction was found with an increase of applied pressure. Three hours after the end of stress exposure no further release of intracellular material could be observed. In addition to these effects, a decrease of the viable cell counts during the stress time was found with the cell-wall-defect mutant strain as shown in Fig. 18.18. It can be stated that, for example, after the 7.5 bar experiment, 40% of the cells remained alive. If the time required for 40% of cells to grow again to 100% is calculated, taking into account the normal growth rate of the culture, a time of 10 h is obtained. This time is identical with the recovery time of the turbidostat shown in Fig. 18.16b.

Stirring Experiments

In these experiments hydrodynamic stress was generated by a fast rotating, three stage turbine stirrer (diameter 5 cm, 2 L fermenter). Stirring experiments were performed with both strains of *Chlamydomonas reinhardii* but not with *Chlorella* and *Anacystis* because the latter were well known not to be affected by stirring (up to the available stirrer speed of 3000 rpm). Aeration was performed with a mixture of air and 1% CO_2. This CO_2 concentration is not limiting even at a low stirrer speed (300 rpm) as was proved earlier by MÄRKL [18.22]. Culture density was kept low in order to avoid overlapping with flash light effects (see Fig. 18.9).

Stirring experiments were performed with turbidostatically controlled continuous cultures. At each stirring speed the experiment was continued until either a stationary state was reached or growth stopped. The observed steady state represents a dynamic equilibrium including both, damage and recovery effects. The results of these experiments are plotted in Fig. 18.19.

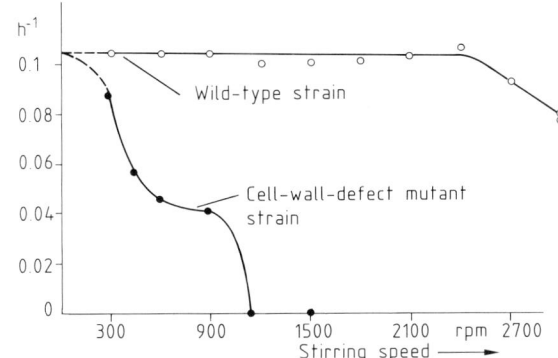

Figure 18.19. Growth of *Chlamydomonas reinhardii*, wild-type strain, and the cell-wall-defect mutant strain CW 15 at different stirring speeds in the turbidostat.

It can be seen clearly that the wild-type strain is not damaged at a stirring speed up to 2400 rpm.

The growth rate of the cell-wall-defect mutant strain, however, is already strongly restricted at small stirring speeds. At a stirring speed of 1200 rpm no growth is observed at all. The critical range in stirring is beyond 2400 rpm for the wild-type strain and probably already at 300 rpm for the cell-wall-defect mutant strain.

With both strains a linear dependence of the release-rate of intracellular substances on the stirring speed could be observed as shown in Fig. 18.20. The actual release rate of intracellular substances during continuous cultivation of *Chlamydomonas reinhardii* was calculated by taking the dilution effect of substrate flow into the fermenter into account.

18.3.2 Filamentous Microorganisms

18.3.2.1 *Spirulina platensis*

Spirulina platensis (Fig. 18.21), which forms long helical cell aggregations, was investigated by BRONNENMEIER and MÄRKL [18.6] in the same manner as described for *Chlamydomonas reinhardii* in Sect. 18.3.1.2. *Spirulina platensis* is a relatively large prokaryotic microorganism. One cell has an average diameter of ca. 15 µm. The helical structure is formed by numerous cells, connected by plasmatic filaments. This type of cellular organization is called a trichome. In a trichome each cell can double during the growth phase, whereas only apical growth occurs in the mycelium of molds. A trichome is able to grow as large as several millimeters in length. It is covered by a sheath.

Free-jet Experiments
Fig. 18.22 shows the results of free-jet experiments. The graph is typical for a microorganism highly sensitive to hydrodynamic stress. The critical load is in the range between 2 and 4 bar. The extreme pressure of 15 bar causes a recovery time of about 14 hours. Measurements of the culture density during the recovery time indicate a different type of damage and recovery from that

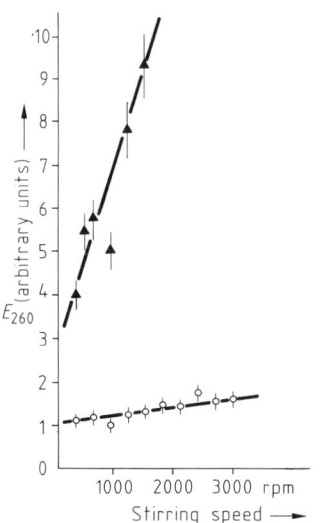

Figure 18.20. Release rate of intracellular substances over stirring speed (stationary values), during continuous cultivation of *Chlamydomonas reinhardii* wild-type strain (○) and cell-wall-defect mutant strain CW 15 (▲).

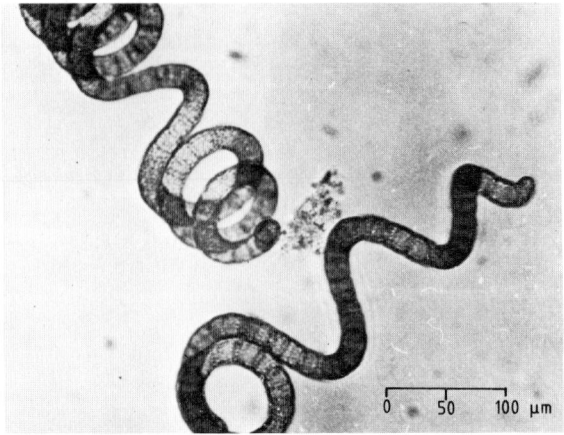
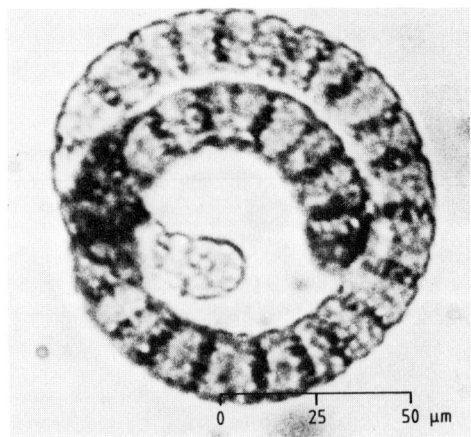

Figure 18.21. Different enlargements of *Spirulina platensis*.

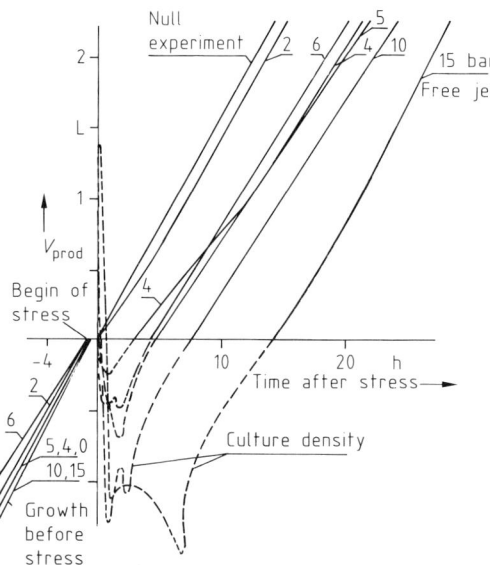

Figure 18.22. Free-jet experiments with *Spirulina platensis*. – (—) production of the fermenter, (---) density of the culture (turbidity probe).

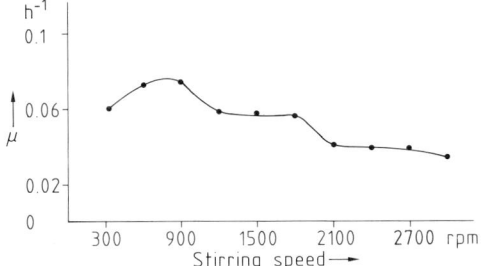

Figure 18.23. Specific growth rate of *Spirulina platensis* at different stirring speeds in the turbidostat.

observed in experiments with the unicellular *Chlamydomonas reinhardii* strains.

Stirring Experiments

Results of stirring experiments with *Spirulina platensis* are shown in Fig. 18.23. In the range from 300–3000 rpm stationary states of the continuous culture could be established. The growth rate strongly depends on the stirring speed. An optimum stirring speed of 900 rpm was observed, and the growth rate at higher speeds decreased in small increments. At 3000 rpm only half of the maximum growth rate was observed. During these experiments no significant increase of UV-absorbing substances in the culture liquid could be measured.

Microscopic examination showed that, in free-jet experiments and in stirring experiments, hydrodynamic load has a strong influence on the morphology of this organism. Some photographs, taken during stirring experiments, are shown in Fig. 18.24.

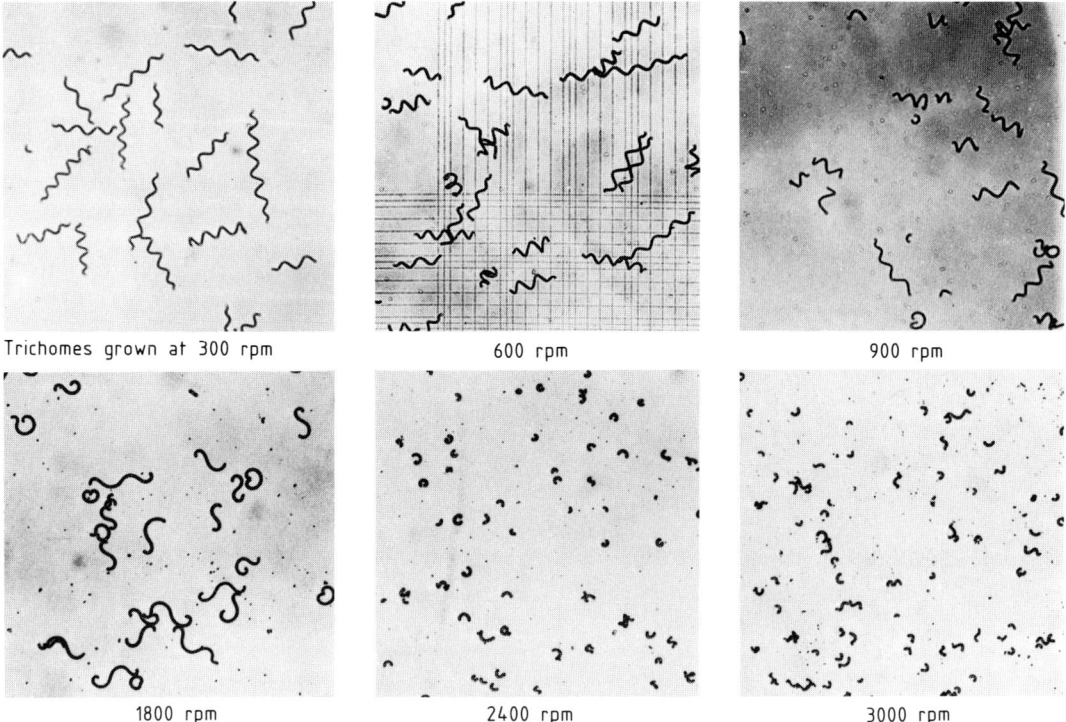

Figure 18.24. Influence of stirring speed on trichome length of *Spirulina platensis*.

From the stirring experiments with *Spirulina platensis* the question arose whether its growth rate is determined by the length of the trichomes or directly by hydrodynamic stress.

To find an answer to this question the following experiments were performed: A steady state with a high stress level (e.g., 3000 rpm) was established. At this stirring speed, only very short trichomes were present in the culture. Then, the stirring speed was suddenly reduced to 300 rpm with the result that only short segments of trichomes were in the culture. At this condition the typical growth rate at 300 rpm was measured after a short relaxation time although only short trichomes were present. As a result it can be stated that the growth rate of *Spirulina platensis* does not depend on the trichomal length but depends on the actual level of applied hydrodynamic stress.

18.3.2.2 Molds

Many publications are available, which hint that mechanical stress can lead to lower product yields. But only a few publications deal with experimental work showing the direct influence of mechanical stress on growth or productivity of mold cultures. An example is the investigation of UJCOVÁ et al. [18.9], already discussed in detail in Sect. 18.1. One of the often cited publications on this topic are the experimental investigations of TANAKA et al. [18.7], [18.8], [18.23], [18.24]. They show that mycelial growth in submerged fermenter cultures is inhibited with increasing mechanical agitation as shown in Fig. 18.25.

TANAKA and his coauthors [18.7], [18.8], [18.23] additionally found that many molds release intracellular low molecular RNA-related nucleotides from their mycelia at cer-

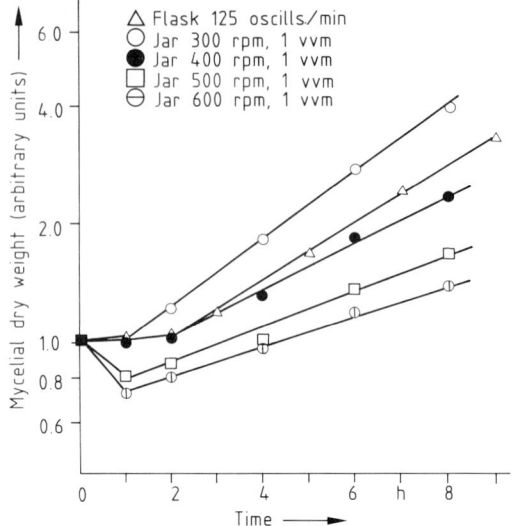

Figure 18.25. Influence of agitation conditions on mycelial growth (dry weight) of *Mucor javanicus* NI 1051 (after TANAKA and UEDA [18.23]).

was varied, the release rate was lower at higher viscosities, see Fig. 18.26. ΔE_{260} indicates the amount of intracellular nucleotides leaked for one hour at a standard concentration of mycelial suspension as compared to unstirred suspensions. The preparation of the investigated mycelial model suspensions is described in the legend of Fig. 18.3.

If model suspensions of *Mucor javanicus* NI 1051 and *Rhizopus javanicus* Takeda were agitated in fermenters with different

tain culture conditions although no visible damage of the mycelium occurs.

Eighteen different mold species and strains differing widely in their mycelial diameters (0.8–16 μm) have been investigated by TANAKA et al. [18.8]. The authors found no correlation between the diameter of the mycelial threads and the rate of release of low molecular weight nucleotides.

In experiments with *Mucor javanicus* where the viscosity of the suspending liquid

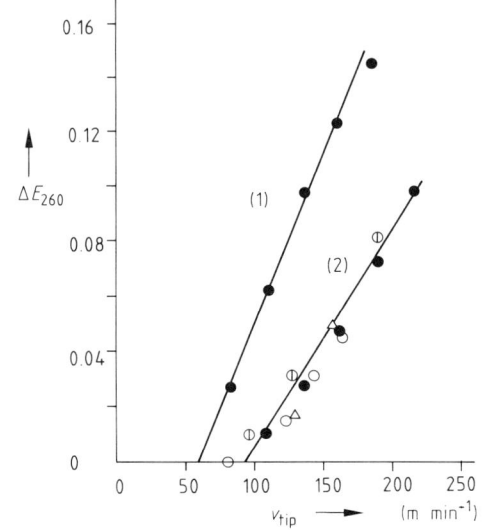

Figure 18.27. Correlation between ΔE_{260} and impeller tip velocity.
(1) *M. javanicus* NI 1051,
(2) *R. javanicus* Takeda
5 L jar D_i/D_T 10 L jar D_i/D_T
   ○ 0.38   △ 0.50
   ● 0.50
   ⊙ 0.60
D_i impeller diameter, D_T tank diameter (after TANAKA [18.25]).

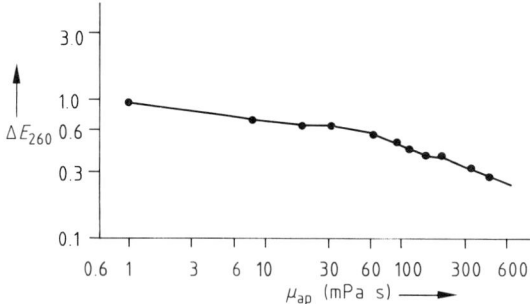

Figure 18.26. Effect of viscosity of liquid used for suspension on ΔE_{260} with *Mucor javanicus* (after TANAKA et al. [18.8]).

stirrers and geometrical dimensions, these authors found nearly linear correlations between low molecular nucleotide release rates and impeller tip velocity (Fig. 18.27) or $k_L a$ (oxygen transfer rate) values (Fig. 18.28).

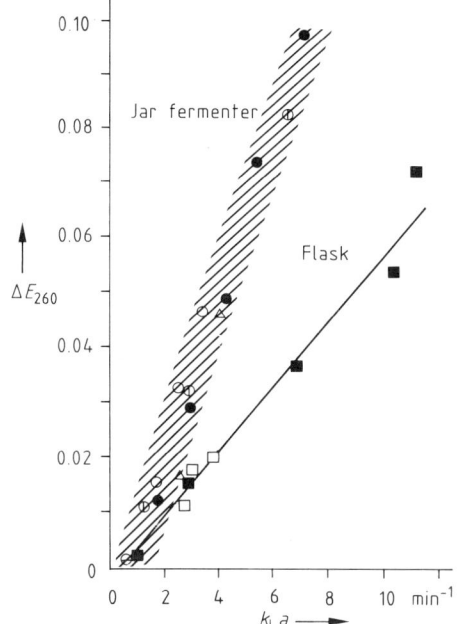

Figure 18.28. Correlation between ΔE_{260} and $k_L a$.

5 L jar D_i/D_T Air flow rate (vvm)
○ 0.38   1
● 0.50   1
◐ 0.60   1
— 0.50   0~2
10 L jar D_i/D_T
△ 0.50   1
□ 500 mL shaking flask (reciprocal shaker)
■ 500 mL Erlenmeyer flask (rotary shaker)
(after TANAKA [18.25]).

The rheological characteristics of fermenter broths change greatly during batch cultivation of molds (MORRIS et al. [18.26]), due to the increasing biomass concentration, morphological changes of the mold, consumption of substrates and the release of products. Additionally the rheology of mold cultures is strongly influenced by the mechanical agitation as was demonstrated for a penicillin fermentation by KÖNIG et al. [18.11a]: If the culture was grown at low impeller speeds, filamentous mold prevailed, leading to high apparent viscosities even at low biomass concentrations. At faster impeller speeds, small pellet-like agglomerates were formed. Their size diminished and their density increased with further ascending impeller speed, resulting in a less viscous broth.

When the *Penicillium chrysogenum* culture was grown at a medium impeller speed ranging from 900 to 1000 rpm (draught tube reactor with a propeller in a tank of 24 L volume) filamentous interconnections between small mycelia pellets were formed at higher biomass concentrations and caused a sharp increase of the apparent viscosity. At the high stirrer speed of 1850 rpm small fragments of hyphae indicated significant cell damage.

A change in rheological properties can influence the productivity of a process as was shown in several cases, when the viscosity of culture media was artificially increased through the addition of pseudoplastic polymer additives.

For example, MOO-YOUNG et al. [18.27] found a four-fold increase in enzyme titer and biomass in cultures of *Aspergillus oryzae* and *Aspergillus niger* when they added Polyol[1] or Carbopol[2] to the cultures; BOTRE et al. [18.28] reported a 38% increase in penicillin production with the addition of PVP[3] (a thickening agent) to the fermentation media, and many more similar effects, summed up in the review by METZ and KOSSEN [18.1], are known. The reason for the higher productivity of mold cultures at higher viscosities is not yet quite clear. This may be due to better dispersion of spores before the sporulation, which leads to a lower pellet diameter and better mass transfer as well as a breaking of big pellets into smaller units, followed by a change to more filamentous and dispersed growth of pellets with better mass transfer, according to MOO-YOUNG et al. [18.27], ELMAYERGI and MOO-YOUNG [18.29], and ELMAYERGI et al. [18.30]. But other mechanisms cannot be excluded, for example, a change in surface tension, better supplementation with essential metal ions in the case of addition of anionic polymers through a polymeric film around the pellets (ELMAYERGI et al.

---
[1] Polyalcohol, [2] acrylic acid polymers, [3] polyvinylpyrrolidone.

[18.29]), or, last but not least, a diminishing effect on high local turbulence intensity and dispersion forces (Kobayashi and Suzuki [18.31]). Even higher mass transfer rates at the gas bubble surface in pseudoplastic fluids is proposed as a possible positive influence on metabolic activity by Moo-Young et al. [18.27].

18.4 Impact on Reactor Scale-up

It is commonly assumed that reduction in metabolic activity of microorganisms is caused by local shear forces resulting in destruction of organisms. But the disruption of an organism does not always lead to a lower growth or production rate of the culture as was shown in stirring experiments with *Spirulina platensis* (Sect. 18.3.2.1). From this it can be stated that not the disruption of the trichomes, which may be related to the applied shear forces, but other mechanisms do affect the growth of this cyanobacterium.

There is no general analytic theory for scale-up procedures of technical flow systems with respect to hydrodynamic stress. There is some lack of knowledge of the basic damage mechanism. But there are quite a few different types of flow patterns which may cause disintegration (Hinze [18.32]), and there is a large variety of rheological properties of fermentation fluids (Taguchi [18.33]).

Until now, empirically developed strategies for scale-up are based on only few experimental data. Midler and Finn showed (Fig. 18.14b) that the fractional survival of *Tetrahymena* cells measured in agitated vessels with different impeller diameters (5.08 cm; 10.16 cm; 12.7 cm) and varying stirrer speed can be well correlated using the agitator tip velocity as a parameter.

Scale-up at a constant agitator tip speed is also suggested by the data shown in Fig. 18.27 (Tanaka [18.25]). The leakage of nucleic-acid-related substances from the mold *Rhizopus javanicus* is correlated linearly with impeller tip velocity. These measurements were performed in a 5 L and a 10 L fermenter, with three different impeller diameters in the 5 L fermenter, and variation of the agitator speed.

Leakage data of *Rhizopus javanicus* also yield a good linear correlation with the measured oxygen transfer capacity ($k_L a$) as shown in Fig. 18.28. This means that in this system the oxygen transfer characteristic can only be improved when the mechanical stress to the organism is raised in the same proportion (taking the leakage of intracellular material as a measure).

The use of the agitator tip velocity as a scale parameter is not possible when the stress situation in a reactor should be estimated in air-lift, jet aeration, or other novel aeration and propulsion systems. Bronnenmeier and Märkl [18.6] published stress experiments, performed in a conventionally stirred fermenter and in a free-jet system as shown in Sect. 18.3, Figs. 18.16 to 18.24. The critical stress values for jet and stirring experiments for three organisms are shown in Table 18.1. The organisms are not damaged at stress values lower than those shown in the table.

The energy data, given in Table 18.1, for the free jet experiments are calculated from the applied pressure by which the algal suspension is pressed through the nozzle. In the stirring experiments at all levels of agitation the dissipated energy is measured. The volumetric dissipated energy is given in parentheses, taking the active fermenter volume as a reference in both cases. It is very obvious that the dissipated energy at the critical level of hydrodynamic load, both in the free jet experiment and in the stirring experiment, is more or less of the same order of magnitude. In addition it is found that for every test organism the critical dissipated energy in the free jet is about nine times lower than in the stirring experiment, which may be seen in Fig. 18.29. This correlation can possibly be explained by the

Table 18.1. Critical Stress Values in Free Jet and Stirring Experiments. – The values in parentheses represent the volumetric dissipated energy. The reference volume is the active fermenter volume, $V = 1.74$ L.

Testorganism	in Free-jet Experiment	in Stirring Experiments
1. *Chlamydomonas reinhardii* CW 15 mutant strain	0–2 bar (0–0.22 W/L)	300 rpm (0.28 W/L)
2. *Chlamydomonas reinhardii* wild-type strain	15 bar (4.54 W/L)	2400 rpm (41 W/L)
3. *Spirulina platensis*	2–4 bar (0.22–0.63 W/L)	900 rpm (3.89 W/L)

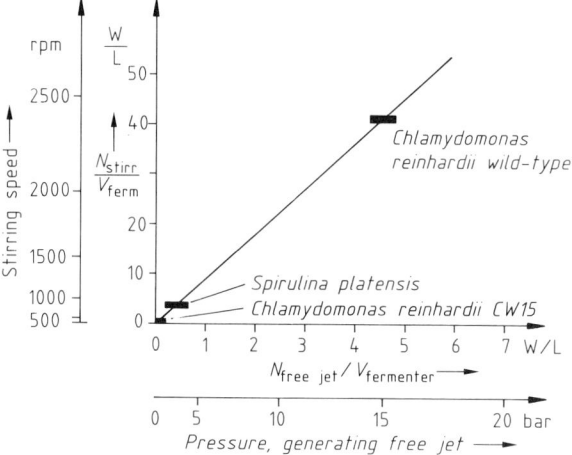

Figure 18.29. Critical dissipated energy in free jet and stirring experiments. N/V = dissipated energy per active fermenter volume.

fact that flow energy in the free jet is dissipated in a considerably smaller volume.

It is not easy to decide whether dissipated energy or linear velocity should be preferred as scale parameter for the estimation of the mechanical stress situation in fermentation systems on the basis of the experimental data available. This may be demonstrated by taking a turbulent stirred vessel as an example. In this case the dissipated volumetric energy (N/V) as a function of agitator diameter (D) and the number of revolutions (n) is well known to be:

$$\frac{N}{V} \sim n^3 D^2 \qquad (18.1)$$

or

$$\sqrt[3]{\frac{N}{V}} \sim n D^{2/3}. \qquad (18.2)$$

The tip speed velocity (v_{tip}):

$$v_{\text{tip}} \sim n D. \qquad (18.3)$$

Comparison of Eqs. (18.2) and (18.3) reveals that there is no significant difference between the two concepts for moderate values of D.

Acknowledgement

We are grateful to Professor D. VORTMEYER for many helpful discussions. We thank Professor H. TANAKA and Professor R. K. FINN for going through the original version of our paper and for many valuable comments.

18.5 References

[18.1] B. Metz and N. W. F. Kossen; Biotechnol. Bioeng. *19* (1977), 781.
[18.2] H. Taguchi, T. Yoshida, Y. Tomita, and S. Teramoto; J. Ferment. Technol. *46* (1968), 814.
[18.3] B. Metz: "From Pulp to Pellet". PhD Thesis, Delft University of Technology, The Netherlands, 1976.
[18.4] J. C. van Suijdam and B. Metz; Biotechnol. Bioeng. *23* (1981), 111.
[18.5] B. Metz, E. W. de Bruijn, and J. C. van Suijdam; Biotechnol. Bioeng. *23* (1981), 149.
[18.6] R. Bronnenmeier and H. Märkl; Biotechnol. Bioeng. *24* (1982), 553.
[18.7] H. Tanaka, J. Takahashi, and K. Ueda; J. Ferment. Technol. *53* (1975), 18.
[18.8] H. Tanaka, T. Mizuguchi, and K. Ueda; J. Ferment. Technol. *53* (1975), 35.
[18.9] E. Ujcová, Z. Fencl, M. Musílková, and L. Seichert; Biotechnol. Bioeng. *22* (1980), 237.
[18.10] M. Musílková, E. Ujcová, J. Placek, Z. Fencl, and L. Seichert; Biotechnol. Bioeng. *24* (1981), 441.
[18.11] W. Sittig and H. Heine; Chem. Ing. Tech. *49* (1977), 595.
[18.11a] B. König, Ch. Seewald, and K. Schügerl; Eur. J. Appl. Microbiol. Biotechnol. *12* (1981), 205.
[18.12] M. Reuss, R. K. Bajpai, and W. Berke; J. Chem. Tech. Biotechnol. *32* (1982), 81.
[18.13] H. Märkl, in "Algae Biomass" (G. Shelef and C. J. Soeder, eds.), pp. 361-383. Elsevier/North-Holland Biomedical Press, Amsterdam-New York-Oxford 1980.
[18.14] H. Märkl; Fifth Int. Ferment. Symp. Berlin *4* (1976), 79.
[18.15] L. Prandtl: "Strömungslehre". Verlag Friedrich Vieweg & Sohn, Braunschweig 1942.
[18.16] H. Reichardt; VDI-Forschungsheft 414 (1942), 13.
[18.17] M. Midler and R. K. Finn; Biotechnol. Bioeng. *8* (1966), 71.
[18.18] U. G. Schlösser, H. Sachs, and D. G. Robinson; Protoplasma *88* (1976), 51.
[18.19] C. J. Dawes; J. Sci. *66* (1966), 317.
[18.20] K. Esser: "Kryptogamen", pp. 117-128. Springer Verlag, Berlin-Heidelberg-New York 1976.
[18.21] D. R. Davies and A. Plaskitt; Genet. Res. Camb. *17* (1971), 33.
[18.22] H. Märkl; Biotechnol. Bioeng. *19* (1977), 1851.
[18.23] H. Tanaka and K. Ueda; J. Ferment. Technol. *53* (1975), 27.
[18.24] H. Tanaka and K. Ueda; J. Ferment. Technol. *53* (1975), 143.
[18.25] H. Tanaka; J. Ferment. Technol. *54* (1976), 818.
[18.26] G. G. Morris, R. N. Greenshields, and E. L. Smith; Biotechnol. Bioeng. Symp. *4* (1973), 535.
[18.27] M. Moo-Young, T. Hirose, and K. H. Geiger; Biotechnol. Bioeng. *11* (1969), 725.
[18.28] C. Botre, L. Cieri, and F. Giordeni; Farmaco Ed. Prat. *19* (1964), 507.
[18.29] H. Elmayergi and M. Moo-Young; Biotechnol. Bioeng. Symp. Ser. *4* (1973), 507.
[18.30] H. Elmayergi, J. M. Scharer, and M. Moo-Young; Biotechnol. Bioeng. *15* (1973), 845.
[18.31] H. Kobayashi and H. Suzuki; J. Ferment. Technol. *50* (1972), 835.
[18.32] J. O. Hinze; AIChE J. *1* (1955), 289.
[18.33] H. Taguchi; in "Adv. Biochem. Eng. 1" (T. K. Ghose and A. Fiechter, eds.), pp. 1-30. Springer Verlag, Berlin-Heidelberg-New York 1971.

3 Bioreactors

Chapter 19
Stirred Vessel Reactors

Heinz Brauer

Institut für Chemieingenieurtechnik
Technische Universität Berlin
Berlin (West), Germany

19.1   Introduction
19.2   General Description of Stirred Vessel Reactor
19.3   Fluid-Mechanic Properties of Fermentation Fluids
19.4   Description of Stirrers
19.4.1   Functions of Stirrers
19.4.2   Shape of Stirrers and Arrangement in the Vessel
19.5   Mechanics of Fluid Flow in Stirred Vessels
19.5.1   Stirrer Model
19.5.2   Primary and Secondary Fluid Flow
19.5.3   Volumetric Flow Rates and Energy Contents of Primary and Secondary Flows
19.5.4   Energy Transfer from the Stirrer to the Fluid
19.5.4.1   Energy Transfer Curve
19.5.4.2   Theoretical Analysis of Energy Transfer
19.5.4.3   Semi-empirical Equations for Energy Transfer to Newtonian Fluids
19.5.4.4   Calculation of Energy Transfer to Non-Newtonian Fluids
19.6   Heat Transfer in Stirred Vessels
19.6.1   Introduction
19.6.2   Physical Phenomena of Heat Transfer in Stirred Vessels
19.6.2.1   Mathematical Problem of Heat Transfer
19.6.2.2   Temperature Field in a Quiescent Fluid
19.6.2.3   Temperature Field in a Fluid in Motion but Without Dissipation of Mechanical Energy
19.6.2.4   Temperature Field in a Fluid in Motion With Dissipation of Mechanical Energy for $Br = +10$
19.6.2.5   Temperature Field in a Fluid in Motion With Dissipation of Mechanical Energy for $Br = -10$
19.6.3   Empirical Equations for Heat Transfer
19.7   Gas Dispersion in Stirred Vessels
19.7.1   Physical Analysis of the Dispersion Process

19.7.2 Bubble Generation by Means of Various Dispersion Elements
19.7.3 Bubble Movement in a Stirred Vessel
19.7.4 Gas Content and Flooding
19.7.5 Energy Transfer in the Dispersion Process
19.8   Mass Transfer in Stirred Vessels
19.9   References

List of Symbols

A	m²	overall heat transfer area
A_i	m²	inner surface of vessel
A_o	m²	outer surface of vessel
A_m	m²	mean area of vessel wall
A_p	m²	interfacial area
a	m²/s	molecular heat diffusivity
a_p	m²/m³	specific interfacial area
b	m	thickness of blade and paddle
c_p	kJ/(kg K)	specific heat capacity
D	m	vessel diameter
d_{gl}	m	equivalent stirrer diameter
d_k	m	center line diameter of disk stirrer
d_L	m	hole diameter of disk stirrer
d_r	m	stirrer diameter
d_s	m	disk diameter of turbine stirrer
d_w	m	diameter of stirrer shaft
E_p	N m	energy of primary liquid flow
E_s	N m	energy of secondary liquid flow
g	m/s²	acceleration due to gravity
H	m	height of liquid in vessel
h_b	m	distance from bottom of vessel
h_p	m	pitch of propeller
h_r	m	height of stirrer, paddle, blade
h_s	m	disk thickness of turbine stirrer
h_w	m	height of wetted stirrer shaft
k	$\tau/(-dw/dy)^m$	consistency (Ostwald) factor
k	kJ/(s m² K)	overall heat transfer coefficient
k	m	baffle thickness
l	m	length of blade
M	N m	torque of stirrer
$\dot{M}_A$	kg/s	mass flux
N	N m/s	power
n	1/s	number of revolutions of stirrer
$\dot{Q}$	kJ/s	heat flux
$q_{b,R}$	kJ/(s m³)	dissipation energy
R_1	m	radius of stirrer model
R_2	m	radius of vessel
r	m	radial coordinate
s	m	radial length of baffle
T	K	local fluid temperature in vessel
T_0	K	fluid temperature at $t=0$
T_w	K	temperature of vessel wall
$\bar{T}_i$	K	mean fluid temperature within the vessel
$\bar{T}_o$	K	mean fluid temperature in vessel jacket
t	s	time
u	m/s	radial velocity
v	m/s	tangential velocity
V	m³	volume of vessel
V_f	m³	volume of liquid in vessel
V_g	m³	volume of gas in vessel
$\dot{V}_g$	m³/s	volumetric flow rate of gas
$\dot{V}_p$	m³/s	volumetric flow rate of primary flow
$\dot{V}_s$	m³/s	volumetric flow rate of secondary flow
w	m/s	axial velocity
y	m	length coordinate
α_i	kJ/(s m² K)	coefficient of heat transfer at inner surface of vessel
α_o	kJ/(s m² K)	coefficient of heat transfer at outer surface of vessel
β	m/s	coefficient of mass transfer
$\dot{\gamma}$	1/s	shear rate, velocity gradient
$\Delta\varrho_A$	kg/m³	mean concentration difference
η	kg/(s m)	dynamic viscosity of fluid
λ	kJ/(s m K)	heat conductivity of fluid
λ_w	kJ/(s m K)	heat conductivity of vessel wall
ν	m²/s	kinematic viscosity of fluid
ϱ	kg/m³	density of fluid
σ	kg/s²	surface tension
τ	N/m²	shear stress
ω	1/s	angular velocity of stirrer
$b^* \equiv b/d$		blade thickness ratio
$Br \equiv \dfrac{\omega^2 R_1^2 \eta}{\lambda(T_w - T_o)}$		Brinkman number
c		number of blades and paddles
$D^* \equiv D/d_r$		diameter ratio
$d_{gl}^* \equiv d_{gl}/D$		ratio of equivalent diameter of stirrer
$d_r^* \equiv d_r/D$		stirrer diameter ratio

$d_s^* \equiv d_s/D$	plate diameter ratio	
$d_w^* \equiv d_w/D$	shaft diameter ratio	
$E^* \equiv \dfrac{E_p + E_s}{\varrho \omega^2 R_1^2 R_2^3}$	total energy of flow	
$E_p^* \equiv \dfrac{E_p}{\varrho \omega^2 R_1^2 R_2^3}$	energy of primary flow	
$E_s^* \equiv \dfrac{E_s}{\varrho \omega^2 R_1^2 R_2^3}$	energy of secondary flow	
$Fo \equiv \dfrac{ta}{R_2^2}$	Fourier number	
$Fr \equiv \dfrac{n d_r^2}{g}$	Froude number	
$Ga \equiv \dfrac{d_r^3 g \varrho^2}{\eta^2}$	Galilei number	
$H^* \equiv H/D$	liquid height ratio	
$h_b^* \equiv h_b/D$	ratio of distance from bottom of vessel	
$h_r^* \equiv h_r/D$	ratio of height of stirrer	
$h_s^* \equiv h_s/D$	plate thickness ratio	
$h_w^* \equiv h_w/D$	wetted height of stirrer shaft ratio	
i	number of baffles	
$k^* \equiv k/D$	baffle thickness ratio	
$l^* \equiv l/D$	blade length ratio	
m	fluid index	
$Nu \equiv \dfrac{\alpha_i D}{\lambda}$	Nusselt number	
$Ne \equiv \dfrac{N}{\varrho n^3 d_r^5}$	Newton number	
$Pr \equiv \dfrac{\eta c_p}{\lambda}$	Prandtl number	
$q_{t,R}^* \equiv \dfrac{q_{t,R}}{\lambda(T_w - T_o)/R_2}$	dissipation number	
$r^* \equiv r/R_2 = r/(D/2)$	radial coordinate	
$Re \equiv \dfrac{n d_r^2}{v}$	Reynolds number	
$Re_{12} \equiv \dfrac{R_1 R_2 \omega}{v}$	Reynolds number	
$Re_m \equiv \dfrac{n^{2-m} d_r^2 \varrho}{k}$	Reynolds number for power law fluids	
$T^* \equiv \dfrac{T - T_o}{T_w - T_o}$	local temperature ratio	
$T^+ \equiv \dfrac{T^* - T_{min}^*}{T_{max}^* - T_{min}^*}$	dimensionless local temperature	
$u^* \equiv \dfrac{u}{\omega R_1}$	radial velocity ratio	
$u_+ \equiv u^*/Re$	generalized radial velocity	
$v^* \equiv \dfrac{v}{\omega R_1}$	tangential velocity ratio	
$\dot{V}_g^* \equiv \dfrac{\dot{V}_g}{n d_r^3}$	gas flow rate number	
$\dot{V}_p^* \equiv \dfrac{\dot{V}_p}{\omega R_1 R_2^2}$	dimensionless primary flow rate	
$\dot{V}_s^* \equiv \dfrac{\dot{V}_s}{\omega R_1 R_2^2}$	dimensionless secondary flow rate	
$w^* \equiv \dfrac{w}{\omega R_1}$	axial velocity ratio	
$w_+ \equiv w^*/Re$	generalized axial velocity	
$We \equiv \dfrac{n^2 d_r^3 \varrho}{\sigma}$	Weber number	
z	number of holes in disk stirrer	
$z^* \equiv z/R_2$	axial coordinate	

19.1 Introduction

The stirred vessel is still the most widely applied type of bioreactor in fermentation technology. This bioreactor consists of a cylindrical vessel with height H and diameter D, their ratio being of the order of 1 to 3. In the vertical axis of the vessel a rotating stirrer is arranged. The stirrer forms gas bubbles within the fluid and keeps microorganisms in motion.

The stirred vessel of the conventional design is best suited for batch operation, and fermentation of small charges. But the charges may have quite different fluid-mechanical and biological properties. In this respect the conventional stirred vessel reactor is of almost universal applicability. It is the optimum design for a great number of small charge batch fermentation processes.

Due to the complexity of the nature of the fermentation fluids, the velocity-, temperature-, and concentration fields, local and overall microbial reaction rate, gas content, energy transfer from the stirrer to the fluid, and heat transfer for heating and cooling purposes cannot yet be described by general physical and physico-chemical laws. Therefore, recourse must be had to empirical equations. It will, however, be endeavored to elucidate the scientific background as much as possible.

One of the important properties of fermentation fluids is their viscosity, ranging over several decimals. Low viscosity fluids, for example, may be aqueous solutions; high viscosity fermentation fluids are thick suspensions. The situation becomes even more difficult when this property is described as a function of fermentation time. Viscosity and other fluid properties may be strongly time dependent. The consequence is, that energy transfer, oxygen transfer, and all other steps of the fermentation process must be described as a function of time. This is, at least in a strict sense, not yet possible. Modelling of fermentation processes demands important simplifications. Fermentation fluids are generally of non-Newtonian nature.

The viscosity of these fluids does not only depend on temperature and concentration of the various components but also on the shear stress that is related to the movement of the fluid. The stress dependence on viscosity is not a universal function valid for all non-Newtonian fluids. There are many groups of fermentation fluids, each one with a specific stress dependence on viscosity. In general, the viscosity and the stress relationship has to be determined by experiment for each fluid.

Bubble formation, especially bubble diameter, depends on the local viscosity. This local fluid property is unknown and will probably remain unknown because of the complexity of the three-dimensional fluid flow in a bioreactor of conventional design. Therefore, no sound physical basis for a rational description of bubble generation and movement in a fermentation fluid is available. The consequence is, that gas hold-up, gas flooding of the stirrer, and mass transfer from the bubble to the fluid can only be described by empirical equations.

As far as energy transfer from the stirrer to Newtonian as well as non-Newtonian fluids is concerned, the situation is much better. Attempts to model this transfer process have been quite successful. The energy transfer process is well understood and physically sound equations are available for calculation of energy transfer for various stirrers.

In the following sections of this chapter a short description of the conventional design of stirred vessel fermenters and fermentation fluid properties will be given along with a detailed discussion of commonly applied stirrers, of fluid flow within vessels, of energy transfer from stirrer to fluid, and of heat transfer, gas dispersion, and mass transfer.

19.2 General Description of Stirred Vessel Reactors

A simple type of stirred fermenter is shown in Fig. 19.1. The fermenter consists of a cylindrical vessel with a diameter D

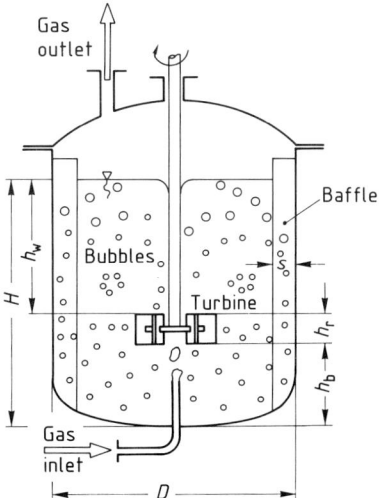

Figure 19.1. General layout of a conventional stirred vessel reactor.

equal to the height H of the fermentation fluid. The vessel is normally equipped with four baffles. The rotating stirrer, in this case it is a turbine, is mounted on a shaft that is arranged in the vertical center-line of the fermenter. The gas is introduced through the bottom of the vessel. The gas rises freely until it contacts the bottom side of the circular disk of the turbine. Here, the gas is picked up by the liquid, that, due to centrifugal forces, moves in a radial direction from the center to the wall of the vessel. The gas dispersion takes place within the active region, i.e., the rotational volume of the stirrer. The details of the dispersion process will be explained in a later section.

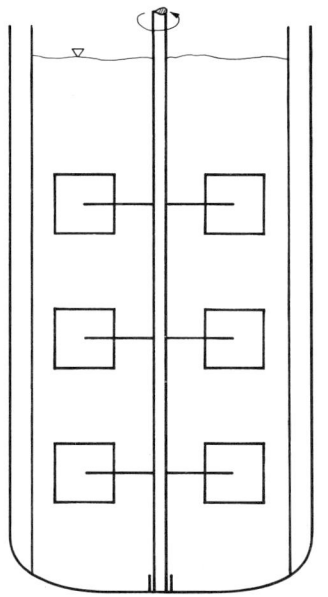

Figure 19.2. General layout of a slender fermenter with a three-stage stirrer arrangement.

Fig. 19.2 shows a fermenter that consists of a rather slender vessel, for which height/diameter ratios of 2:1 and 3:1 are quite common [19.1], [19.2]. Several stirrers are mounted onto the rotating shaft in order to achieve an energy distribution in the vessel favorable to the fermentation process.

19.3 Fluid-Mechanic Properties of Fermentation Fluids

The fermentation fluid commonly consists of a liquid and small gas bubbles dispersed therein. The liquid contains the microorganisms and necessary nutrients for their biological activity as well as metabolic products. The most important property of fermentation fluids is the viscosity. This viscosity is always that of the liquid. Extreme care has to be taken in order to avoid the presence of gas bubbles, when the viscosity has to be determined experimentally. However, it will be rather difficult to remove fine bubbles, especially from highly viscous liquids. The presence of fine gas bubbles in the liquid will lead to a reduction of its viscosity. Furthermore, care has to be taken during measurement so that no sedimentation of the microorganisms takes place.

Fluids exhibit either Newtonian or non-Newtonian flow behavior determined by their viscosity. Some remarks on fluid viscosity are also included in Chapters 1 and 4 of this volume.

The viscosity η is defined for Newtonian and non-Newtonian fluids as follows:

$$\eta \equiv -\tau/\dot{\gamma}, \qquad (19.1)$$

where τ is the local shear stress and $\dot{\gamma}$ the local shear rate. In a linear flow field the shear rate $\dot{\gamma}$ is the local gradient of the velocity w:

$$\dot{\gamma} = dw/dy, \qquad (19.2)$$

where y is the coordinate normal to the direction of the fluid velocity. Eq. (19.1) shows that the viscosity depends on the local properties of the velocity field and on the shear stress.

For the simple case of Newtonian fluids the shear stress is described by the empirical equation:

$$\tau = -\eta \frac{dw}{dy}. \qquad (19.3)$$

Introducing τ into Eq. (19.1) leads to $\eta = \eta$, that is, η is a constant. The viscosity of a Newtonian fluid is independent of the properties of the velocity field; η is independent of shear stress and shear rate.

For a certain class of non-Newtonian fluids the shear stress may be described by the following empirical equation as has been suggested by OSTWALD and DE WAELE [19.3], [19.4]:

$$\tau = -k[\dot{\gamma}]^m, \qquad (19.4)$$

where k is the consistency factor and m the fluid index. Introducing Eq. (19.4) into Eq. (19.1) one obtains:

$$\eta = k[\dot{\gamma}]^{m-1}. \qquad (19.5)$$

In this case the viscosity depends on the fluid properties k and m and on the local properties of the velocity field. In the case of non-Newtonian fluids the viscosity is a function of the shear rate. This is true for all groups and types of non-Newtonian fluids, whether their flow behavior may be described by Eq. (19.4) or not.

All non-Newtonian fluids with a flow behavior described by the so-called power law, given by Eq. (19.4), are called power law fluids. The fluid index m can vary from zero to infinity. In the special case of $m=1$, the power law turns into Newton's law. Fig. 19.3a shows the influence of the fluid index on the relationship between shear stress and shear rate, while in Fig. 19.3b the viscosity η is shown as a function of shear stress. Fluids with an index of $m<1$ are called pseudoplastic fluids, those with $m>1$ dilatant fluids. Fermentation fluids with non-Newtonian behavior are primarily of the pseudoplastic type.

There are other groups of non-Newtonian fluids for which flow behavior is described by other relationships between shear stress and shear rate [19.5] to [19.11]. But these relationships are generally far more complicated than the one given by Eq. (19.4), especially when elastic properties of the fluid have to be taken into account.

In flow fields of non-Newtonian fluids the viscosity is a local property, generally

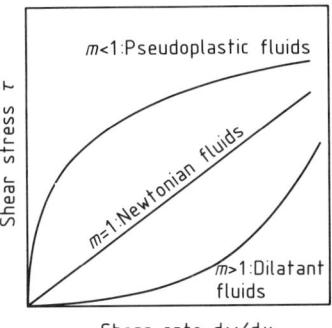

Figure 19.3a. Shear stress/shear rate relationship for Newtonian and non-Newtonian fluids.

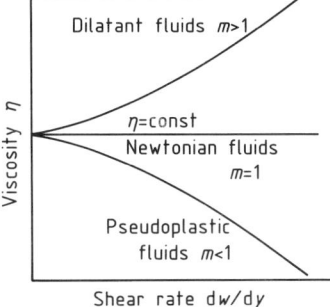

Figure 19.3b. Viscosity of Newtonian and non-Newtonian fluids as a function of shear stress.

varying with the local coordinates. Furthermore, with progressing microbial reaction the viscosity varies [19.12]. It is this kind of variation of the property of fermentation fluids that makes it extremely difficult to predict fluid flow in a fermenter.

19.4 Description of Stirrers

Next to the shape of the vessel it is the shape of the stirrer, its rotational speed, and its arrangement in the vessel that is of utmost importance to fluid flow and therefore to the fermentation process. Consequently, the stirrer deserves special attention to its design and process engineering.

19.4.1 Functions of Stirrers

The shape of a stirrer and the arrangement of stirrers within a vessel should reflect the functions which the stirrer has to meet. These functions are prescribed by the process carried out in the vessel. The four fundamental functions of stirrers are the following [19.13]:

1. Energy transfer to the fluid,
2. dispersion of gas in the liquid,
3. separation of gas and liquid,
4. mixing of all components of the fermentation fluid.

The stirrer is at the same time an energy transfer element, a gas dispersion element, a separation element, and a mixing element. Due to the combination of the four functions shape, speed, and arrangement of the stirrer will be the result of a compromising process found mostly by practical experience.

The energy transferred by the stirrer to the fluid is required for a suitable fluid movement in the vessel. Fluid movement is always connected with energy dissipation. Mechanical energy is thereby converted into thermal energy. The so-called loss of mechanical energy must be balanced by mechanical energy constantly transferred to the fluid by the stirrer. The energy is necessary for maintaining a fluid movement that is best suited for the process. This fluid movement should be achieved with a minimum of energy. The function of a stirrer as an energy transfer element may be described as the property *to achieve and maintain a prescribed fluid movement within a vessel with a minimum of energy.*

Energy is not only required for fluid movement but for gas bubble generation as well. However, this energy is a very small fraction of the overall energy transferred and may be neglected. The gas dispersion process consists of two parts:

1. Gas bubble generation,
2. gas bubble distribution in the fluid.

The second part of the process depends first of all on the fluid movement achieved by the stirrer and will not be discussed at this point. The all important function of a stirrer as an element for gas dispersion is the generation of gas bubbles. The mass flow rate across the interface between a gas and a liquid is proportional to the interfacial area. Increased interfacial area leads to an increased mass transfer rate. A large interfacial area requires the generation of small bubbles from a given gas flow rate. Optimum mass transfer conditions may require prescription of bubble diameter. The function of a stirrer as a gas dispersion element may be described as the property *to achieve a prescribed interfacial area between gas and liquid and its distribution in the vessel.*

In many cases surface-active agents may concentrate in the interface thereby hindering the diffusional process across the interface. Generation of interfacial area must therefore include the demand for its periodic renewal. Although bubble generation requires energy, this energy is generally more than compensated by energy reduction due to the partial contact of the stirrer with the low viscosity gas instead of the high viscosity liquid.

The third function of a stirrer is concerned with the separation of gas and liquid. The separation process is often more complicated than the dispersion process.

Big bubbles separate easily from the liquid, while the separation of very small bubbles may be extremely difficult. The separation process, on the other hand, depends heavily on the fluid flow in the vessel and the shape of the vessel. The function of a stirrer as a separation element may be described as the property *to achieve conditions for bubble diameter and bubble movement in the vessel which enable an easy separation of gas and liquid.*

The third function of the stirrer is closely related to the first two functions.

The fourth function of a stirrer is generally considered as the most important one. All components of the fermentation fluid or the biological suspension should be completely mixed if possible. This includes the liquid, the living microorganisms, and the gas bubbles. When all components are mixed in an ideal way the consequence is a constant biochemical reaction in all volumetric elements of the biological suspension. Concentrations and temperature will be the same for all volumetric elements. But this is never possible, because the quality of the mixture depends on the fluid flow in the vessel, which does not permit an ideal mixture. Research has furthermore established that the quality of mixture in vessels with conventional rotating mixing elements is restricted by the imposed rotational movement of the fluid. After these critical remarks the function of a stirrer as a mixing element may be described as the property *to achieve a state of mixture that is favorable to the biochemical reactions taking place in the biological suspension with a minimum of energy and without inflicting any harm on the microorganisms.*

Most of the stirrers used in technical equipment have not been designed on the basis of the described functions. This is first of all due to the fact that research has only lately brought some light into the extremely complicated phenomena of stirred gas/liquid systems. The stirrers will be described in the following section.

19.4.2 Shape of Stirrers and Arrangement in the Vessel

Shape and arrangement of the stirrers in the bioreactor exert a strong influence on fluid flow within the reactor and, consequently, on the microbial processes. Although a great number of different types of stirrers is available, only a few have gained importance and are widely used. These stirrers are the following:

1. Turbine stirrer with disk (Fig. 19.4),
2. turbine stirrer without disk (Fig. 19.5),
3. paddle stirrer (Fig. 19.6),
4. propeller stirrer (Fig. 19.7),
5. disk stirrer (Fig. 19.8).

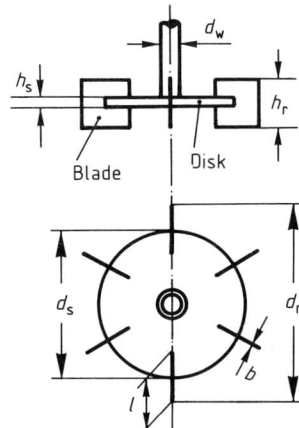

Figure 19.4. Shape and dimensions of turbine stirrer with disk (Rushton turbine).

The disk stirrer is not yet widely used. But it has some exceptional properties that may bring the disk stirrer the importance it deserves.

The dimensions of the stirrers are the following:

d_r stirrer diameter
h_r stirrer height, blade height, paddle height

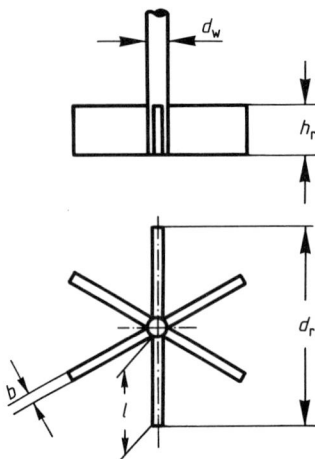

Figure 19.5. Shape and dimensions of turbine stirrer without disk.

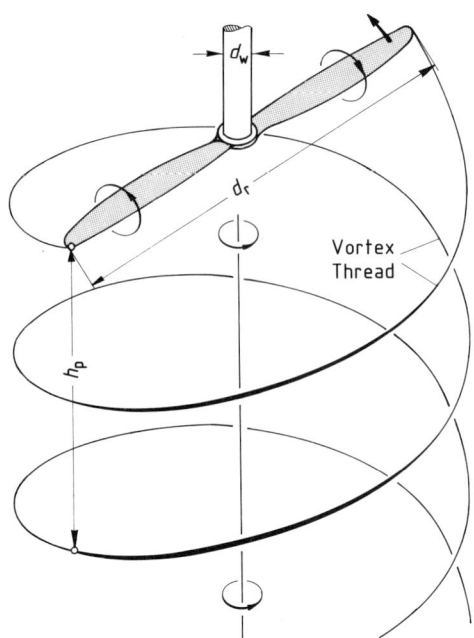

Figure 19.7. Shape and dimensions of two-blade propeller stirrer.

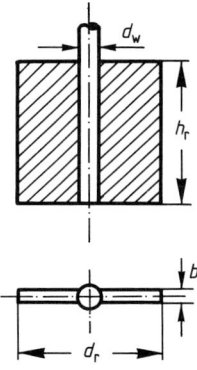

Figure 19.6. Shape and dimensions of paddle stirrer.

- d_s disk diameter of turbine stirrer
- h_s disk thickness of turbine stirrer
- h_p pitch of propeller
- l length of blade
- b thickness of blade and paddle
- d_L hole diameter of disk stirrer
- d_k center line diameter of disk stirrer
- c number of blades of turbine, paddle and propeller stirrer
- z number of holes in disk stirrer

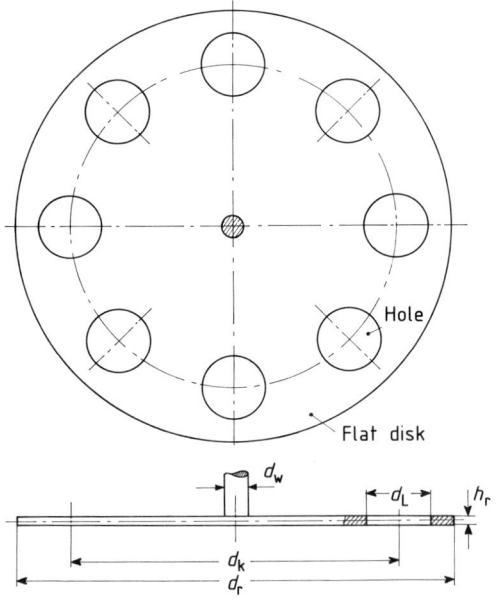

Figure 19.8. Shape and dimensions of holed disk stirrer.

The arrangement of a stirrer in a vessel is shown in Fig. 19.1. The vessels may not be equipped with baffles. The important dimensions are the following:

D vessel diameter
H liquid height in vessel
h_b distance from bottom of vessel
h_w height of rotating axis in liquid
s radial length of baffle

By arranging the stirrers and baffles in conventional vessels as shown in Fig. 19.1 the following ratios of geometrical dimensions are normally observed:

$H/D \approx 1$ for all types of stirrers
$s/D \approx 1/10$ for turbine and paddle stirrers
$h_b/D \approx 1/3$ for all types of stirrers
$d_r/D \approx 1/3$ for turbine and paddle stirrers
$\approx 1/2$ to $2/3$ for propeller and disk stirrers

Application of stirrers depends strongly on the viscosity of the fluid as indicated in Fig. 19.9.

19.5 Mechanics of Fluid Flow in Stirred Vessels

The stirrer produces a very complicated three-dimensional flow field. Until a few years ago, neither experimental nor theoretical methods were available for the investigation of fluid flow and related transport processes in stirred vessels. In the last few years, however, appropriate experimental and theoretical methods have been developed and successfully tested [19.5], [19.14] to [19.20]. Theoretical numerical methods proved to be more helpful than any other method for gaining deeper insight and understanding into flow phenomena in stirred reactors. These investigations are based on a relatively simple model of conventional stirrers. The results obtained help to explain the characteristics of primary and secondary fluid flow and energy transfer in stirred vessels.

19.5.1 Stirrer Model

Observations by various workers on the fluid flow in mixing vessels have led to the conclusion that, in the laminar flow regime, the fluid contained in the cylindrical volume generated by a rotating stirrer, behaves almost like a solid body [19.21] to [19.25]. Therefore, the stirrer may be replaced by a cylindrical element of the same rotational volume as the stirrer. Fig. 19.10 shows a model equivalent to a paddle stirrer.

Both stirrer and model are of the same height h_r. The diameter of the stirrer is d_r and that of the model d_{gl}. The diameter d_{gl} has been selected in such a way that the model and the stirrer transfer the same energy to the liquid. According to comprehensive experimental investigations on laminar flow fields in stirred vessels, d_r and d_{gl} differ only by a few percent, with d_{gl} being slightly larger. This proves that the model

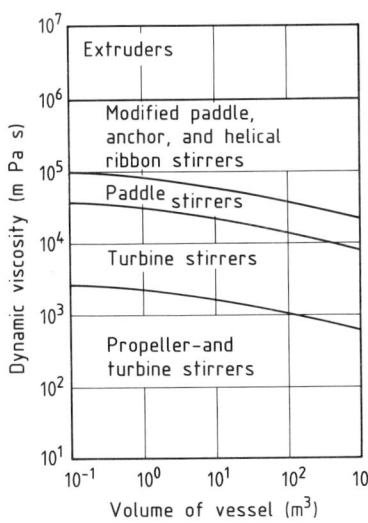

Figure 19.9. Application of type of stirrer depending on viscosity of fluid.

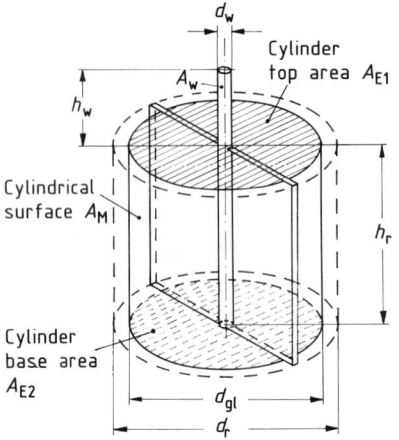

Figure 19.10. Stirrer model presented for a paddle stirrer.

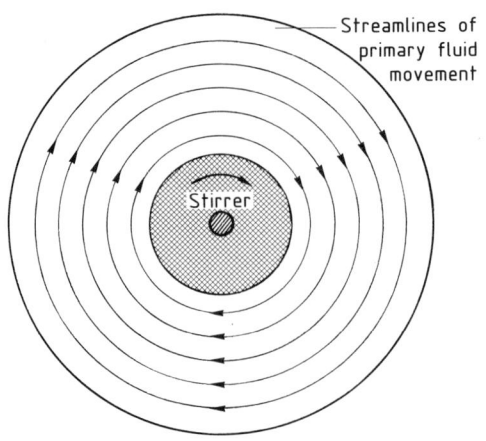

Figure 19.11. Streamlines of primary fluid movement in a stirred vessel fermenter.

represents an acceptable substitute for the stirrer, and that it may be used as a basis for the theoretical investigation of three-dimensional liquid flow in the vessel [19.21], [19.26]. Some results of the theoretical numerical investigations, pertaining to primary and secondary liquid flows and their energy content, will be discussed in detail.

19.5.2 Primary and Secondary Fluid Flow

By transferring energy to the liquid, the revolving stirrer produces a rotational motion, which will be referred to as primary liquid flow and, owing to centrifugal forces, also a secondary liquid flow. The primary flow is illustrated in Fig. 19.11 and the secondary in Fig. 19.12. In the horizontal plane of the stirrer, the centrifugal forces drive the liquid outwards in a radial direction. At the vessel wall the liquid flows either upwards or downwards thus forming two vortex rings.

For two values of the Reynolds number, i.e., $Re_{12}=1$ and 1000, the tangential velocity representing the primary flow and the secondary liquid flow is shown in Figs.

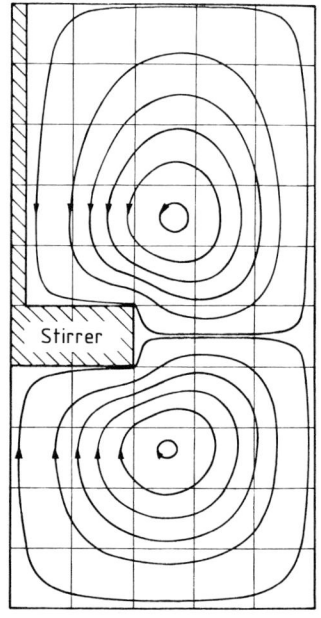

Figure 19.12. Streamlines of secondary fluid movement in a stirred vessel fermenter.

19.13 to 19.16. The Reynolds number Re_{12} is defined as follows:

$$Re_{12} \equiv \frac{R_1 R_2 \omega}{\nu}, \qquad (19.6)$$

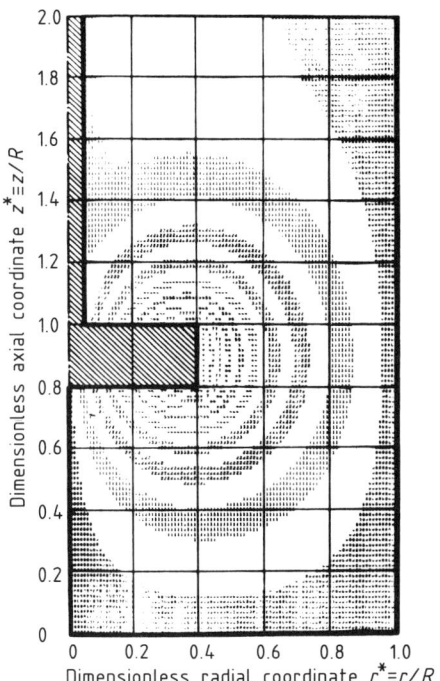

Figure 19.13. Field of tangential fluid velocity v^+ in a stirred vessel at $Re_{12}=1$.

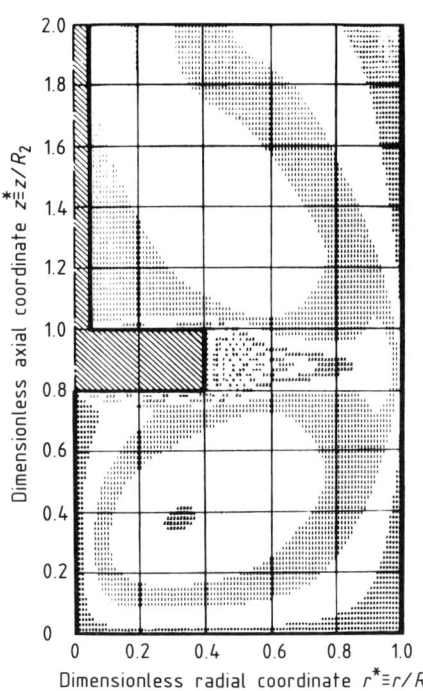

Figure 19.15. Field of tangential fluid velocity v^+ in a stirred vessel at $Re_{12}=10^3$.

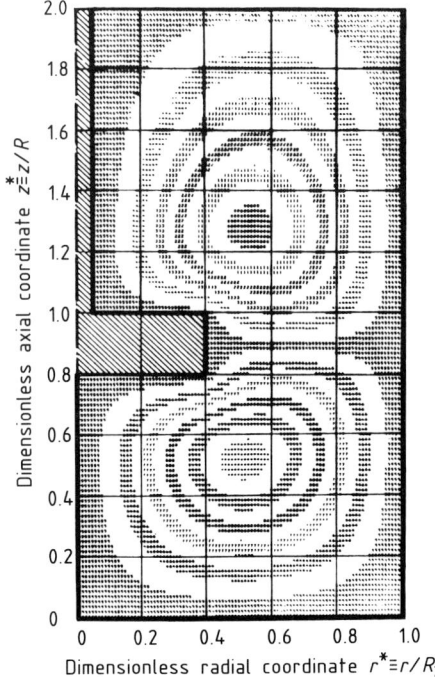

Figure 19.14. Field of streamlines ψ^+ in a stirred vessel at $Re_{12}=1$.

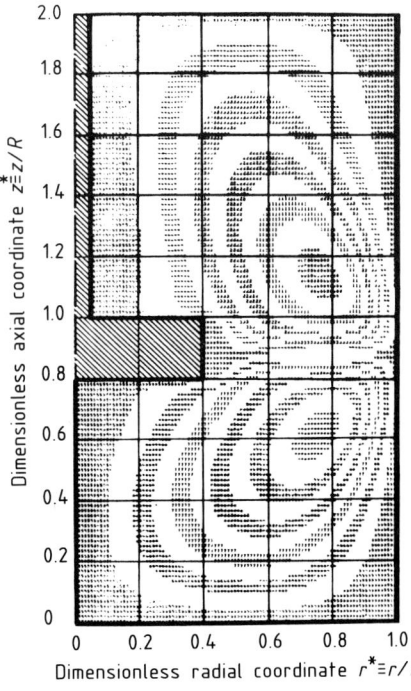

Figure 19.16. Field of streamlines ψ^+ in a stirred vessel at $Re_{12}=10^3$.

where $\omega = 2\pi n$ denotes the angular velocity, n the stirrer speed, $R_1 = d_r/2$ the stirrer radius, $R_2 = D/2$ the vessel radius, and ν the kinematic viscosity of the liquid. The Reynolds number Re_{12} is related to the conventional Reynolds number Re by the following equation:

$$Re = Re_{12} \frac{2}{\pi} \frac{d_r}{D}. \tag{19.7}$$

Hence, Re is defined by:

$$Re \equiv \frac{n d_r^2}{\nu}. \tag{19.8}$$

In a suitable dimensionless form [19.27], which will not be explained in this chapter for $Re_{12} = 1$ or $Re = 0.255$, Fig. 19.13 shows the tangential flow field, represented by lines of constant tangential velocity v^+, whereas Fig. 19.14 shows the secondary flow field, represented by lines of constant stream function ψ^+. In the alternating blank and dotted areas, v^+ and ψ^+ vary only within narrow limits. In the range of low Reynolds numbers the curves of constant v^+ are elliptical. The streamlines shown in Fig. 19.14 are almost circular, at least those near the center of the vortices.

For $Re_{12} = 1000$ or $Re = 255$ the corresponding flow fields are shown in Figs. 19.15 and 19.16. With increasing Reynolds number the centrifugal force and hence its effect on liquid flow in the vessel increase substantially. This is shown by the deformation of the flow field, i.e. by the dislocation of the center of the vortices of secondary flow with increasing Reynolds number. Furthermore, for $Re_{12} = 1$, the maximum radial velocity is about 0.27% of the maximum tangential velocity, but for $Re_{12} = 1000$ the corresponding figure is 26%.

Heat and mass transfer in mixing vessels are strongly dependent on the secondary flow, while the primary flow has no influence at all. However, there is no possibility to suppress the primary flow, since this constitutes the energy source for the secondary flow.

A better understanding of heat and mass transfer processes in mixing vessels will be provided by the knowledge of volumetric flow rates of primary and secondary liquid flows as well as their energy contents.

19.5.3 Volumetric Flow Rates and Energy Contents of Primary and Secondary Flows

The volumetric flow rates of primary and secondary flows are designated by $\dot{V}_p$ and $\dot{V}_s$, respectively. The dimensionless flow rates $\dot{V}_p^*$ and $\dot{V}_s^*$ are defined as follows:

$$\dot{V}_p^* \equiv \frac{\dot{V}_p}{\omega R_1 R_2^2}, \tag{19.9}$$

$$\dot{V}_s^* \equiv \frac{\dot{V}_s}{\omega R_1 R_2^2}. \tag{19.10}$$

The primary liquid flow is defined as that flow which passes a vertical cross-section of the vessel in a tangential direction. The secondary flow passes across the cylindrical surface linking the centers of the two vortex rings.

Fig. 19.17 shows the flow rates $\dot{V}_p^*$ and $\dot{V}_s^*$ for the primary and secondary flows as well as their ratio $\dot{V}_s^*/\dot{V}_p^*$ plotted against the Reynolds number Re_{12}. Dimensions of the vessel, for which the theoretical investigations have been carried out, are as follows:

$d_r/D = 0.40 \qquad d_w/D = 0.05$
$h_r/D = 0.10 \qquad H/D = 1.00$
$h_b/D = 0.40$

According to Fig. 19.17 the dimensionless flow rate of the primary flow $\dot{V}_p^*$ is practically independent of the Reynolds number Re_{12}. Hence, $\dot{V}_p$ is roughly a linear function of angular velocity $\omega = 2\pi n$ of the stirrer. In the case of the secondary flow rate, however, the relation is quite different. The secondary liquid flow rate $\dot{V}_s^*$ increases with Re_{12} and approaches an almost constant value at $Re_{12} \geq 100$. In the range of low Reynolds numbers the rate $\dot{V}_s$ is proportional to ω^2.

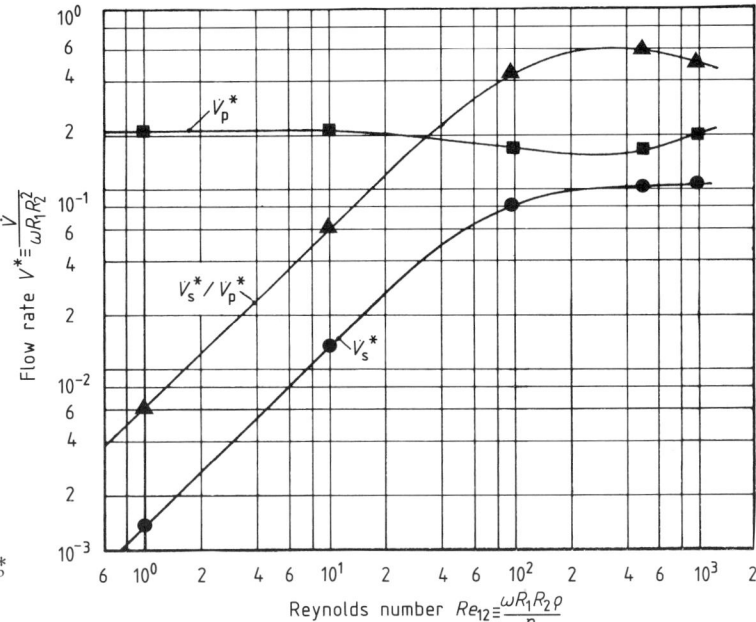

Figure 19.17. Volumetric flow rates of primary and secondary fluid flows; $\dot{V}_p^*$ and $\dot{V}_s^*$, and the ratio $\dot{V}_s^*/\dot{V}_p^*$ for a stirrer with $d_r/D = 0.4$ and $h_r/D = 0.1$.

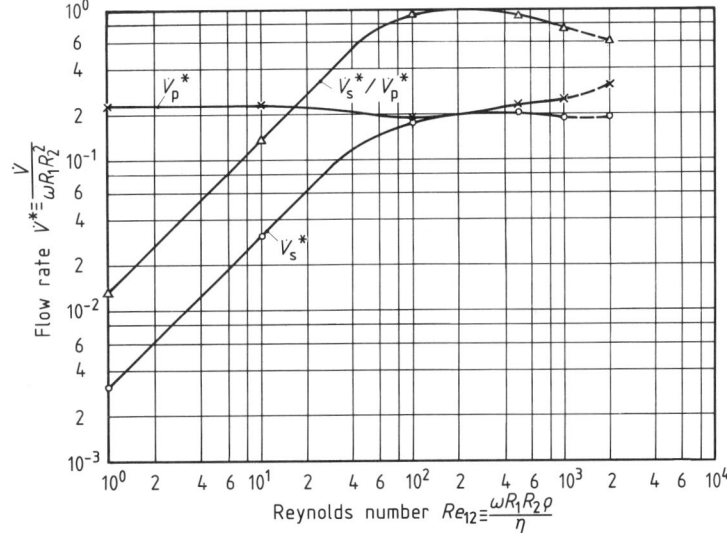

Figure 19.18. Volumetric flow rates of primary and secondary fluid flows; $\dot{V}_p^*$ and $\dot{V}_s^*$, and the ratio $\dot{V}_s^*/\dot{V}_p^*$ for a stirrer with $d_r/D = 0.633$ and $h_r/D = 0.031$.

The curve for $\dot{V}_s^*/\dot{V}_p^*$ is nearly parallel to that for $\dot{V}_s^*$.

At $Re_{12} = 1$ the secondary flow rate $\dot{V}_s^*$ is only 0.6% of the primary flow rate $\dot{V}_p^*$. Since it is the secondary flow rate which influences the mixing process and not the primary flow rate, the stirrer exerts only a very weak influence on the mixing process at this low Reynolds number. This is the reason why in this Reynolds number range the helical stirrer is preferred, since it produces stronger axial liquid motion.

At $Re_{12} = 1000$ the secondary flow has already increased to a remarkable intensity, so that $\dot{V}_s^*$ is about 50% of $\dot{V}_p^*$.

Fig. 19.18 presents the flow rates for another set of vessel dimensions:

$d_r/D = 0.633$ $d_w/D = 0.067$
$h_r/D = 0.031$ $H/D = 1.000$
$h_b/D = 0.469$

The curves are very similar to those shown in Fig. 19.17. The primary flow rate $\dot{V}_p^*$ has almost the same value. The secondary flow rate $\dot{V}_s^*$ is, however, higher than that shown in Fig. 19.17. At $Re_{12} \geqslant 100$ the secondary flow is as large as the primary flow.

For the second set of geometrical data, the kinetic energy contained in the two liquid streams has been determined. The dimensionless energies are defined by:

$$E_p^* \equiv \frac{E_p}{\varrho \omega^2 R_1^2 R_2^3}, \quad \text{energy of primary liquid flow} \quad (19.11)$$

$$E_s^* \equiv \frac{E_s}{\varrho \omega^2 R_1^2 R_2^3}, \quad \text{energy of secondary liquid flow} \quad (19.12)$$

$$E^* \equiv \frac{E_p + E_s}{\varrho \omega^2 R_1^2 R_2^3} \quad \text{total energy of flow.} \quad (19.13)$$

These energy ratios are plotted in Fig. 19.19 as functions of the Reynolds number Re_{12}. At $Re_{12} \geqslant 200$ the ratio E_s^*/E_p^* has reached its maximum value, i.e., 0.2. The maximum energy content of the secondary flow amounts to only 20% of the energy content of the primary fluid flow.

With decreasing Reynolds number the energy content of the secondary flow decreases very rapidly. At $Re_{12} = 1$, for example, the ratio E_s^*/E_p^* is only $3 \cdot 10^{-5}$ or 0.003%. In this range of Re_{12} the energy E_s is proportional to ω^4.

19.5.4 Energy Transfer from the Stirrer to the Fluid

The energy required for the fermentation process is of great interest to the bioengineer. The physical fundamentals of energy transfer and equations for the energy transfer factor, the so-called Newton number, will be presented for Newtonian and non-Newtonian fluids. The discussion on energy transfer will be introduced by an explanation of the energy transfer factor curve.

19.5.4.1 Energy Transfer Curve

The energy transfer curve, or power curve, gives the relation between the energy transfer factor, the so-called Newton num-

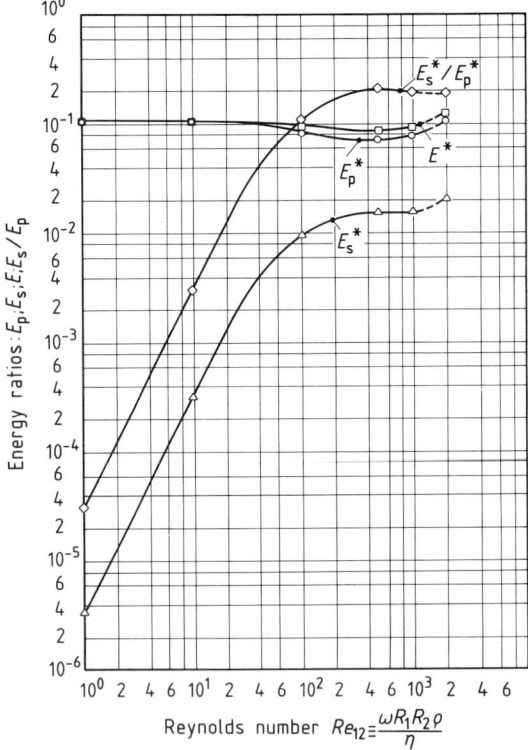

Figure 19.19. Energy of primary and secondary fluid flows; E_p^* and E_s^*, as well as the ratio E_s^*/E_p^* and the total energy $E^* = E_p^* + E_s^*$ for a stirrer with $d_r/D = 0.633$ and $h_r/D = 0.031$.

ber Ne and the relevant dimensionless numbers:

$$Ne \equiv \frac{N}{n^3 d_r^5 \varrho}. \qquad (19.14)$$

N is the energy transferred by the stirrer to the fluid, n number of revolutions of the stirrer per time unit, d_r diameter of stirrer, and ϱ density of liquid. In most cases the Newton number is presented as a function of the Reynolds number, which has already been defined by Eq. (19.8).

There is no method available that permits a straightforward theoretical calculation of the Newton number, and it is not to be anticipated that such a method will be available in the foreseeable future. The functional relationship between the Newton number and all dimensionless parameters can only be determined by experiment. But the intuition of an engineer such as BÜCHE [19.28] has been necessary to present the experimental results in the best possible and most rational way. Results obtained for the relationship between Newton number Ne and Reynolds number Re are shown in Fig. 19.20. These results will be discussed in some detail.

The data have been obtained for a six-bladed turbine stirrer arranged in a vessel without baffles. Curve a fits the data closely. At values of the Reynolds number Re below 10 the Newton number Ne is proportional to Re^{-1}. With increasing values of Re the slope of the curve changes. Beyond $Re \approx 10^2$ the Newton number is, in this special case, proportional to $Re^{-0.28}$. According to experimental evidence available, the energy transfer curve consists of three distinct parts that reflect well defined flow regions:

1. Region of laminar fluid flow in the vessel.
   Reynolds number range: $0 \leqslant Re \leqslant 10$
   Ne-Re relationship: $Ne \sim Re^{-1}$
2. Region of transition from laminar to turbulent fluid flow.
   Reynolds number range:
   $10^1 \leqslant Re \leqslant 10^2$
   Ne-Re relationship: changing from $Ne \sim Re^{-1}$ to $Ne \sim Re^{-0.28}$
3. Region of turbulent fluid flow in the vessel.
   Reynolds number range:
   $10^2 \leqslant Re \leqslant \infty$
   Ne-Re relationship: $Ne \sim Re^{-0.28}$

For different types of stirrers the transition region may shift slightly to either lower or higher values of the Reynolds number and

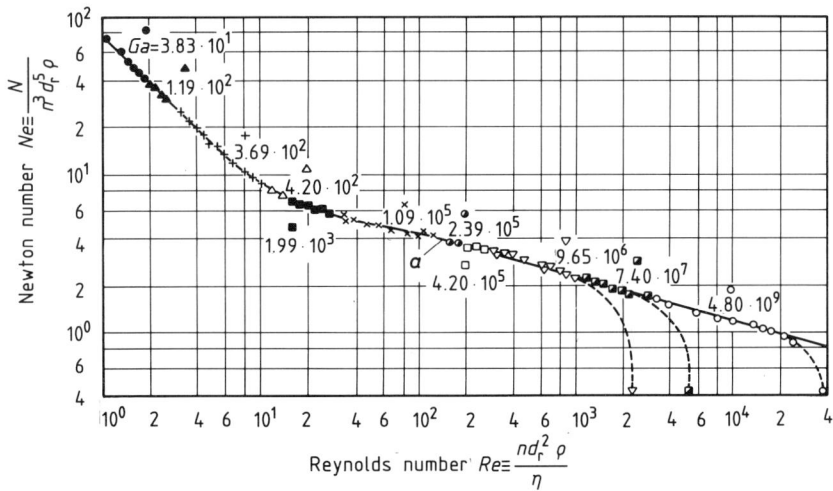

Figure 19.20. Energy transfer curve for a turbine stirrer in a vessel without baffles.

the exponent of the Reynolds number in the Ne-Re relationship may change slightly as well.

A close inspection of the experimental data, shown in Fig. 19.20, reveals that there are eleven groups of data. Each group of data covers a relatively small range of the Reynolds number. This range has a well defined upper limit of the Reynolds number, that is given by Re_1 for vessels not equipped with baffles. When the Reynolds number exceeds this limiting value, operation of the stirrer becomes unstable. The transition from stable to unstable operation starts with a strong curl-like deformation of the fluid surface around the rotating stirrer shaft. When the curl reaches the stirrer the energy is transferred intermittently. At the same time the energy transfer to the fluid is, in the mean, reduced and gas is dispersed in the liquid. Some of the curves, given by broken lines in Fig. 19.20, show this situation. It is advisable to stay well below the upper limiting Reynolds number Re_1 in order to have stable operating conditions.

The lower limiting value of the Reynolds number is given from a theoretical point of view for every group of data by $Re = 0$. For usual experimental conditions the value of the Reynolds number Re_0 at the lower limit is, according to the data shown in Fig. 19.20, roughly equal to $Re_1/3$. This value depends on the sensitivity of the instrument that measures the torque in the rotating shaft of the stirrer.

The groups of data are obtained in the same experimental unit with different viscosity fluids. Theoretical investigations [19.23], [19.29] have disclosed the fact, that the transition to unstable operation conditions is due to the situation already described and is a function of another dimensionless number, the Galilei number:

$$Ga \equiv \frac{d_r^3 g \varrho^2}{\eta^2}. \qquad (19.15)$$

In this equation g is the gravitational acceleration. The upper limiting value of the Reynolds number must be a function of the Galilei number besides other parameters. The energy transfer factor Ne must be a function of Re and Ga and further geometrical parameters:

$$Re_1 = f(Ga; \text{ geometrical parameters}) \qquad (19.16)$$

$$Ne = f(Re; Ga; \text{ geometrical parameters}). \qquad (19.17)$$

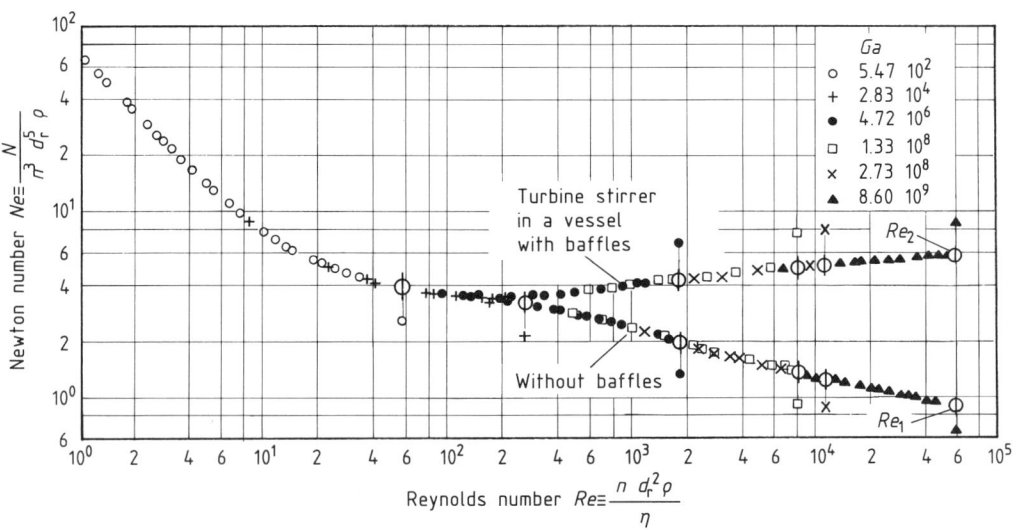

Figure 19.21. Energy transfer curve for a turbine stirrer in a vessel with and without baffles.

These equations hold only for stirrer operation in pure fluids. In the case of liquid/gas systems some further parameters must be introduced, that take into account the gas flow rate and the gas properties.

The data for the energy transfer factor shown in Fig. 19.20 have been obtained in vessels without baffles. In Fig. 19.21 data are shown for energy transfer in vessels with and without baffles. An effect of the baffles on the Newton number is observed in the turbulent flow region only for the chosen experimental conditions. At a value of the Reynolds number of 10^4 the Newton number for vessels with baffles is roughly three times that for vessels without baffles. It is therefore advisable to use vessels without baffles, if possible.

Baffles are still quite often assumed to prevent the formation of the already described curl-like deformation of the fluid surface. But experiments have proven this assumption to be wrong. In vessels with baffles the curl-like deformation of the fluid surface is also observed although in quite a different way. Duration of the curl-like deformation is limited in the case of vessels with baffles and is unlimited in the case of vessels without baffles. Furthermore, the described deformations wander around in vessels with baffles and stay in a fixed position around the rotating stirrer shaft in vessels without baffles.

19.5.4.2 Theoretical Analysis of Energy Transfer

In order to maintain the desired fluid movement in a vessel mechanical energy must be continuously transferred to the fluid. For steady-state conditions this energy is continuously converted into heat. Energy is transferred from the entire wetted surface A of the stirrer to the fluid. This surface consists of four parts, as indicated in Fig. 19.10:

$$A = A_M + A_{E1} + A_{E2} + A_W, \quad (19.18)$$

with A_M as the cylindrical surface, A_{E1} and A_{E2} the cylinder top and cylinder base area, and A_W the wetted surface of the rotating shaft. According to this energy transfer surface area the transferred energy is given by:

$$N = N_M + N_{E1} + N_{E2} + N_W. \quad (19.19)$$

Each fraction of the overall energy has to be determined. From a differential element of the surface area dA of the stirrer the differential energy:

$$dN = 2\pi n\, dM \quad (19.20)$$

is transferred to the liquid, with n as the rotational speed of the stirrer and dM the differential torque:

$$dM = r\tau\, dA. \quad (19.21)$$

In this equation τ is the local shear stress at the radius r. Introduction into Eq. (19.20) leads to:

$$dN = 2\pi n r \tau\, dA. \quad (19.22)$$

This equation has to be applied for each fraction of the local energy given in Eq. (19.19). For each fraction the appropriate expressions for the local values for shear stress τ, radius r, and surface area dA must be determined and introduced. For conditions of laminar flow of Newtonian fluids this has been done by THIELE [19.21]. The resulting equations are as follows:

$$Ne_M = 2\pi^3 \frac{d_{gl}^*}{Re} \int_{h_b^*}^{h_b^*+h_r^*} \left[-\frac{\partial}{\partial r^*}\left(\frac{v^*}{r^*}\right)\right]_{(r^*=d_{gl}^*/2)} dz^*, \quad (19.23)$$

$$Ne_{E1} = \frac{\pi^3}{2}\frac{d_{gl}^*}{Re}\frac{1}{d_{gl}^{*3}} \int_{d_w^*}^{d_{gl}^*/2} r^{*2}\left[-\frac{\partial v^*}{\partial z^*}\right]_{(z^*=h_b^*+h_r^*)} dr^*, \quad (19.24)$$

$$Ne_{E2} = \frac{\pi^3}{2}\frac{d_{gl}^*}{Re}\frac{1}{d_{gl}^{*3}} \int_0^{d_{gl}^*/2} r^{*2}\left[+\frac{\partial v^*}{\partial z^*}\right]_{(z^*=h_b^*)} dr^*, \quad (19.25)$$

$$Ne_W = 2\pi^3 \frac{d_{gl}^*}{Re}\frac{d_w^{*3}}{d_{gl}^{*3}} \int_{h_b^*+h_r^*}^{h^*} \left[-\frac{\partial}{\partial r^*}\left(\frac{v^*}{r^*}\right)\right]_{(r^*=d_w^*/2)} dz^*. \quad (19.26)$$

The four Newton numbers are defined by Eq. (19.14) with N the appropriate fraction according to Eq. (19.19); Re is the Reynolds number defined by Eq. (19.8). The other dimensionless parameters are defined as follows:

$d_{gl}^* \equiv d_{gl}/D$	equivalent diameter	(19.27)
$d_w^* \equiv d_w/D$	diameter of rotating shaft	(19.28)
$h_b^* \equiv h_b/D$	distance from bottom of vessel	(19.29)
$h_r^* \equiv h_r/D$	stirrer height	(19.30)
$H^* \equiv H/D$	liquid height	(19.31)
$r^* \equiv r/(D/2)$	radial coordinate	(19.32)
$v^* \equiv v/(\omega d_{gl}/2)$	tangential velocity	(19.33)
$z^* \equiv z/(D/2)$	length coordinate	(19.34)

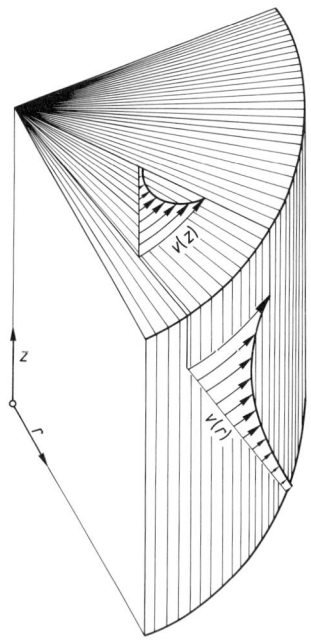

Figure 19.22. Schematic presentation of the tangential velocity as a function of the radial and axial coordinates r and z close to the cylindrical and the circular surface of the stirrer model.

The geometric dimensions of vessel and stirrer assembly are given in Figs. 19.1, 19.4 to 19.8, and 19.10. The coordinates z and r are indicated in Fig. 19.22. In the same figure, the tangential velocities $v(r)$ and $v(z)$ are shown on the cylindrical and the circular surface of the stirrer model.

It follows from Eqs. (19.23) to (19.26) that the energy transfer is only a function of the gradients of the tangential velocities in radial and axial direction, $\partial v/\partial r$ and $\partial v/\partial z$. These velocity gradients are obtained as a result of a theoretical numerical solution of the appropriate differential equations for the velocity field in stirred vessels. The solution has been worked out by THIELE [19.21]. By comparing theoretical and experimental results for energy transfer, it has been possible to determine the equivalent diameter d_{gl}. It has been found, that in the laminar flow regime d_{gl} is only a few percent smaller than the real stirrer diameter d_r. This proves the quality of the stirrer model in the laminar flow region. Due to this success the model concept was also applied in the turbulent flow regime. The equations for energy transfer presented in the following section are based on this concept.

19.5.4.3 Semi-empirical Equations for Energy Transfer to Newtonian Fluids

The equations presented in the following subsections for energy transfer to Newtonian fluids are based on the model concept explained in the preceding section. Empirical expansions have been necessary for application of the equations in the entire relevant region of the Reynolds number.

The General Equations

The general form of the energy transfer equation is as follows:

$$Ne = f(Re, d_r^*; d_s^*; d_w^*; h_r^*; h_s^*; h_w^*; l^*; b^*; H^*; s^*; k^*; c; i) \quad (19.35)$$

with $Ga \to \infty$.

The energy transfer equations have been set up by IHME [19.22], who made use of prior findings by BRAUER [19.23], BRAUER and THIELE [19.30], and THIELE [19.21]. The general form of the equation for the upper limiting value of the Reynolds number Re_1 in vessels without baffles and Re_2 in vessels with baffles is as follows:

$$\left.\begin{array}{l} Re_1 \\ Re_2 \end{array}\right\} = \left.\begin{array}{l} f_1 \\ f_2 \end{array}\right\} \begin{array}{l} (Ga; Re; d_r^*; d_s^*; d_w^*; \\ h_r^*; h_s^*; h_w^*; l^*; b^*; H^*; \\ s^*; k^*; c; i) \end{array} \quad (19.36)$$

The equations for Re_1 and Re_2 have been set up by IHME [19.22]. The definitions of the dimensionless numbers and quantities will be repeated here:

$Ne \equiv \dfrac{N}{\varrho n^3 d_r^5}$	Newton number	(19.14)
$Re \equiv \dfrac{n d_r^2 \varrho}{\eta}$	Reynolds number	(19.8)
$Ga \equiv \dfrac{d_r^3 g \varrho^2}{\eta^2}$	Galilei number	(19.15)
$d_r^* \equiv d_r/D$	stirrer diameter ratio	(19.37)
$d_s^* \equiv d_s/D$	plate diameter ratio	(19.38)
$d_w^* \equiv d_w/D$	rotating shaft diameter ratio	(19.28)
$h_r^* \equiv h_r/D$	stirrer height ratio	(19.30)
$h_s^* \equiv h_s/D$	plate thickness ratio	(19.39)
$h_w^* \equiv h_w/D$	rotating shaft height ratio	(19.40)
$H^* \equiv H/D$	liquid height ratio	(19.31)
$l^* \equiv l/D$	blade length ratio	(19.41)
$b^* \equiv b/D$	blade thickness ratio	(19.42)
$s^* \equiv s/D$	baffle length ratio	(19.43)
$k^* \equiv k/D$	baffle thickness ratio	(19.44)
c	number of blades and paddles	
i	number of baffles	

For dimensions of turbine and paddle stirrers see Figs. 19.1 and 19.4 to 19.6

Energy Transfer Equations for Turbine and Paddle Stirrers in Vessels Without Baffles

The energy transfer equation is given by:

$$Ne = \frac{C_1}{Re} + \left(\frac{1}{C_2 + C_3 Re} - \frac{1}{C_2 + C_4 Re^2}\right) + \frac{Re}{C_5 + C_6 Re^{1.113}}. \quad (19.45)$$

The first term on the right hand side applies to the laminar region; it is based on theoretical analysis. The second term applies to the transition region and the third term to the fully turbulent region. The last two terms are empirical terms based on all relevant experimental data available in the literature until 1974 [19.22].

C_1 to C_6 are functions of geometric ratios. The most important one is C_1:

$$C_1 = 4\pi^3 \left(\frac{d_{gl}^*}{d_r^*}\right)^3 \left[\frac{\{(0.6 h_r^{*0.6})^5 + (1.3 h_r^*)^5\}^{1/5}}{d_{gl}^*(1 - d_{gl}^{*2})} + \frac{0.95 h_w^{*1.4}}{d_w^*(1 - d_w^{*2})}\left(\frac{d_w^*}{d_{gl}^*}\right) + \frac{6}{\pi^3} d_{gl}^{*1/3}\left\{2 - \left(\frac{d_w^*}{d_{gl}^*}\right)^4\right\}\right]. \quad (19.46)$$

d_{gl} is the equivalent diameter of the stirrer model. The equation for the two types of turbine stirrers is as follows:

$$d_{gl}^* = 1.09 \, d_r^* \, b^{*0.015} \cdot$$

$$\cdot \left[\left\{1 + \frac{1}{(167 + 0.00167 \, c^4)^{0.5} \, d_r^{*1.375}}\right\}^{-4} + \frac{1.417}{d_r^{*0.36} \, c^{0.44}}\right]^{-1/2} \cdot$$

$$\cdot \left[1 + \frac{1}{d_r^{*0.75} \left\{3.4324 \left(\dfrac{l^*}{d_r^*}\right)^{0.24} + 1.4 \cdot 10^6 \left(\dfrac{l^*}{d_r^*}\right)^6\right\}^{5/3}}\right]^{-1/5} \quad (19.47)$$

For paddle stirrers the following equation applies:

$$d_{gl}^* = 1.09 b^{*0.015} \cdot$$

$$\cdot \left[17.5 h_r^{0.004} \left(\frac{0.0708}{d_r^{*1.15}} - 1\right) + 17.5\right]^{-1} \cdot$$

$$\cdot \left[1 + \frac{1}{d_r^{*0.75}\left\{3.4\left(\frac{l^*}{d_r^*}\right)^{0.24} + 5\cdot 10^5\left(\frac{l^*}{d_r^*}\right)^6\right\}^{5/3}}\right]^{-1/5} \quad (19.48)$$

The functions C_2 to C_6 apply to both types of turbine stirrers and to the paddle stirrer:

$$C_2 = \left[\left\{\left(0.024\left[\frac{d_r^*}{h_r^*}\right]^{2.13}\right)^4 + (0.55\, d_r^{*2.35})^4\right\} \cdot \right.$$
$$\cdot \left\{\left(\frac{7.18}{c^{1.1}}\right)^4 + (0.001\, c^2)^4\right\}\right]^{1/4} \cdot$$
$$\cdot \left[\left(0.138\, \frac{d_r^*}{l^*}\right)^{6.4} + 1\right]^{1/8} \quad (19.49)$$

$$C_3 = \frac{0.0046\, d_r^{*0.735}}{c^{0.4}} \left[\frac{0.0009}{h^{*2}} + 0.36\right] \cdot$$
$$\cdot \left[\left(0.069\, \frac{d_r^*}{l^*}\right)^{4.8} + 1\right]^{1/3} \quad (19.50)$$

$$C_4 = 0.056\, \frac{h_r^{*0.1}}{c^{2.42}} \quad (19.51)$$

$$C_5 = \left[\frac{1}{c^3}\left(\frac{2.6}{h_r^{*3}} + 1.5 \cdot 10^4\right)\right]^{1/3} \quad (19.52)$$

$$C_6 = 3.35\, d_r^{*1.1} \left[\left\{\frac{2.77 \cdot 10^{-6}}{c^{2.56} h_r^{*4}} + \frac{8.64 \cdot 10^{-3}}{c^{0.96} h_r^{*0.52}} \cdot \right.\right.$$
$$\left.\cdot \left(1 + \frac{0.45\,[b^{*0.005} - 0.97]^4}{c^{0.6}[0.004 + h_r^{*2.1}]}\right)\right\}^{1.25} +$$
$$+ \frac{0.0014}{c^{0.5}}\right]^{1/5} \left[\left(0.042\, \frac{d_r^{*0.5}}{l^*}\right)^4 + 1\right]^{1/4} \cdot$$
$$\cdot [1 + 6 \cdot 10^{-6} c^4]^{0.1}. \quad (19.53)$$

The range of application for Eqs. (19.45) to (19.53) as given below has been determined by a large number of experiments carried out by various authors:

$$0 \leqslant Re \leqslant 5 \cdot 10^5$$
$$Re \leqslant Re_1$$
$$0.25 \leqslant d_r^* \leqslant 0.9$$
$$0.025 \leqslant h_r^* \leqslant 0.9$$
$$0.0175 \leqslant l^* \leqslant 0.45$$
$$0.002 \leqslant b^* \leqslant 0.04$$
$$2 \leqslant c \leqslant 12$$
$$0.05 \leqslant h_w^* \leqslant 0.7$$
$$0.2 \leqslant h_b^* \leqslant H^* - (h_r^* + h_w^*)$$
$$H^* = 1$$

It is advisable to make use of a computer when the given equations are applied. For the most important parameters – h_r^*, d_r^*, and c – Fig. 19.23 shows the space of application. Two examples have been taken for a comparison between calculated and measured values for Ne from the paper published by IHME [19.22] (Figs. 19.24 and 19.25). These contain data for turbine and paddle stirrers. The maximum deviation between calculated and measured values of Ne occurs in the transition region and may reach up to ±10%.

In Sect. 19.5.4.1 it has been discussed, that the Reynolds number Re must not ex-

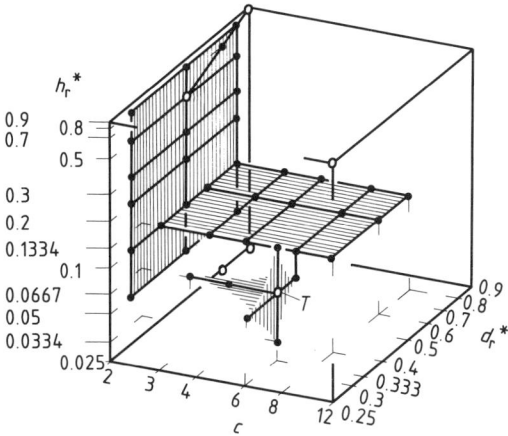

Figure 19.23. Space of application for Eq. (19.45) with respect to the three most important parameters for turbine and paddle stirrers in vessels without baffles.

19.5 Mechanics of Fluid Flow in Stirred Vessels 417

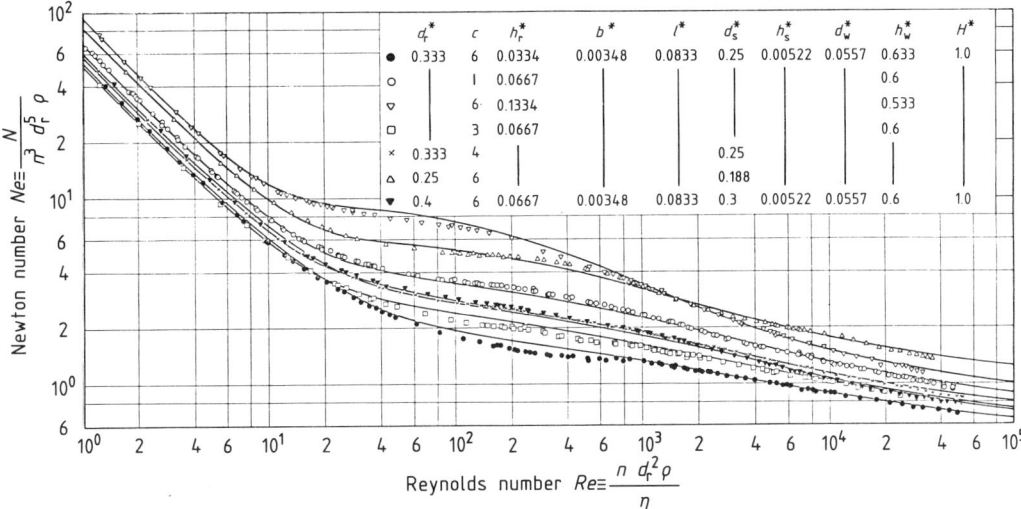

Figure 19.24. Comparison between measured and calculated *Ne*-data for turbine stirrers in vessels without baffles. The curves have been calculated by means of Eq. (19.45).

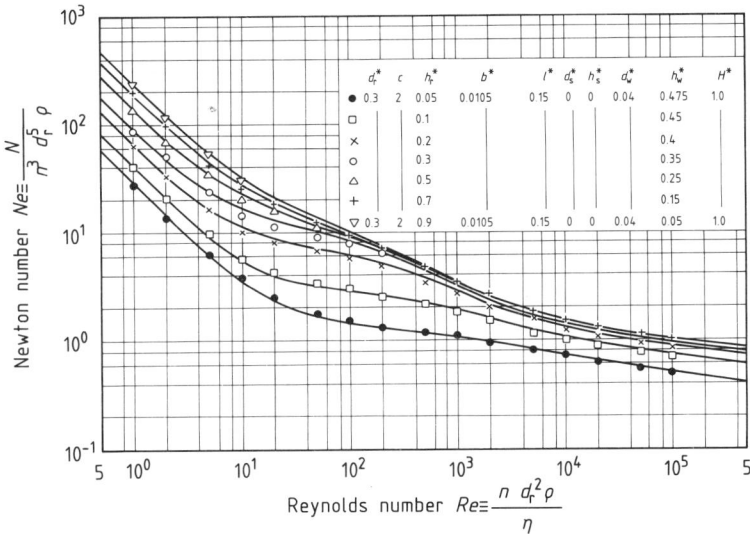

Figure 19.25. Comparison between measured and calculated *Ne*-data for paddle stirrers in vessels without baffles. The curves have been calculated by means of Eq. (19.45).

ceed an upper limiting value given by Re_1. On the basis of experimental data, an equation has been developed:

$$\frac{Re_1^2}{Ga} = 0.24 h_w^* \left[\frac{1}{d_r^{*6}} + 57\right]^{1/6}$$

$$\cdot \left[\left(\frac{1}{h_r^{*6}} + \frac{6.87 \cdot 10^5}{h_r^{*1.5}}\right)^{1/3} \frac{h_w^{*2}}{c \, d_r^* \, Ga^{0.3}} +\right.$$

$$\left. + \frac{0.8}{c^{0.18}} \left(\frac{5.8 \cdot 10^{-4}}{h_r^{*5.6}} + 1\right)^{0.05}\right]. \quad (19.54)$$

Range of tested application:

$10^4 \leq Ga \leq \infty$
$0.25 \leq d_r^* \leq 0.9$
$0.025 \leq h_r^* \leq 0.9$
$0 \leq h_w^* \leq 0.7$
$2 \leq c \leq 12$

$\left.\begin{array}{l} b^* = 0.00348 \\ l^*/d_r^* = 0.25 \\ H^* = 1 \end{array}\right\}$ parameters of less influence

When the Reynolds number Re proves to be greater than Re_1 either n or d_r must be reduced.

Energy Transfer Equations for Turbine and Paddle Stirrers in Vessels With Baffles

The vessels are assumed to be equipped with four baffles, $s^* = 0.1$. The energy transfer equation is given as follows:

$$Ne = \frac{C_1}{Re} + \left[\left\{\left(\frac{C_7}{Re}\right)^2 + \left(\frac{C_8}{Re^{0.1}}\right)^2\right\}^5 + C_9^{10}\right]^{-1/10} \quad (19.55)$$

The term C_1/Re applies to the laminar flow region with C_1 given by Eq. (19.46). The other two terms in Eq. (19.55) apply to the transition and fully turbulent region. The functions C_7 to C_9 are given below:

$$C_7 = \frac{0.195}{[(0.21 h_r^{*0.35})^2 + (20.8 h_r^{*2})^2]^{1/2}}$$

$$\cdot \left[\left(\frac{4.7}{c}\right)^2 + 1\right]^{1/2} \frac{d_r^{*1.1}}{b^{*0.2}(0.19 + 7.6 h_s^*)}, \quad (19.56)$$

$$C_8 = \frac{C_9 \left[0.27 + 0.45 \left(\frac{h_s^*}{h_r^*}\right)^{0.25}\right]}{\dfrac{0.13}{\left[1 + \left(4 \dfrac{d_r^*}{c}\right)^5\right]^{1/5}} + 0.089 \left(\frac{h_s^*}{h_r^*}\right)^{0.5} + 60 \left(\frac{h_s^*}{h_r^*}\right)^5} \quad (19.57)$$

$$C_9 = b^{*0.11} d_r^{*1.25} \left[\frac{7.1 \cdot 10^{-6}}{h_r^{*4.6}} + \left(\frac{0.22}{h_r^{*7}} + 80\right)^{0.2}\right]^{0.625}$$

$$\cdot \left(\frac{342}{c^4} + 0.098\right)^{0.25} \cdot$$

$$\cdot \left[\frac{0.13}{\left\{1 + \left(5.5 \dfrac{d_r^*}{c}\right)^5\right\}^{1/5}} + 0.089 \left(\frac{h_s^*}{h_r^*}\right)^{0.5} + 60 \left(\frac{h_s^*}{h_r^*}\right)^5\right] \cdot$$

$$\cdot \left[\left(\frac{0.22 \, d_r^{*0.75}}{l^{*0.75}}\right)^{10} + 1\right]^{1/10}. \quad (19.58)$$

The range of application has been determined through experiments and is given as follows:

$0 \leq Re \leq \infty$
$Re \leq Re_2$
$0.25 \leq d_r^* \leq 0.6$
$0.03 \leq h_r^* \leq 0.6$
$0.05 \leq l^* \leq 0.3$
$0.002 \leq b^* \leq 0.03$
$2 \leq c \leq 12$
$0.025 \leq h_s^*/d_r^* \leq 0.25$ and $h_s^* = 0$
$0.2 \leq h_w^* \leq 0.7$
$0.2 \leq h_b^* = H^* - (h_r^* + h_w^*)$

$\left.\begin{array}{l} H^* = 1 \\ s^* = 0.1 \\ k^* = 0.0174 \\ i = 4 \end{array}\right\}$ baffle properties

The baffle thickness ratio $k^* \equiv k/D$ is a parameter of lesser importance. The equations may be applied safely to other values of k^*.

For application of Eq. (19.55) the use of a computer is advised. A field of application with respect to the three most important parameters is given in Fig. 19.26. Two examples have been taken for a comparison between calculated and measured values

19.5 Mechanics of Fluid Flow in Stirred Vessels   419

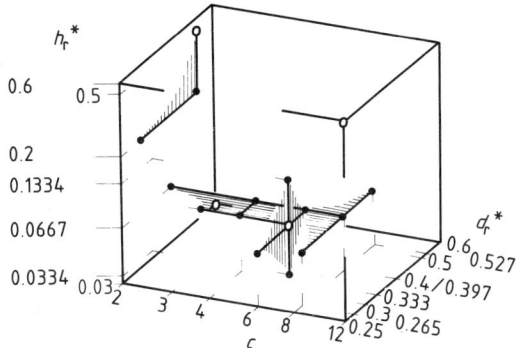

Figure 19.26. Space of application for Eq. (19.55) with respect of the three most important parameters for turbine and paddle stirrers in vessels with baffles.

for Ne from the paper published by IHME [19.22] (Figs. 19.27 and 19.28). They contain data for turbine and paddle stirrers from various authors.

In Sect. 19.5.4.1 it has been discussed, that the Reynolds number must not exceed the upper limiting value Re_2 when the vessels are equipped with baffles. On the basis of experimental data an equation has been developed:

$$Re_2 = Re_1 \left[\frac{0.376}{(h_r^{*0.9} + 7.5 h_r^{*4})^{0.5} c^{0.22}} \cdot \left(\frac{4.8 \cdot 10^{-3}}{d_r^{*3}} + 1 \right) \right]^{\frac{1}{2m}} \quad (19.59)$$

with

$$m = 1 + \frac{2.8 \cdot 10^6}{Ga} . \quad (19.60)$$

Range of application tested:

$10^4 \leq Ga \leq \infty$
$0.03 \leq h_r^* \leq 0.6$
$0.25 \leq d_r^* \leq 0.6$
$2 \leq c \leq 12$
$b^* = 0.00348$
$h_s^* = 0.00522$
$l^*/d_r^* = 0.25$
$H^* = 1$
$s^* = 0.1$
$k^* = 0.0174$
$i = 4$

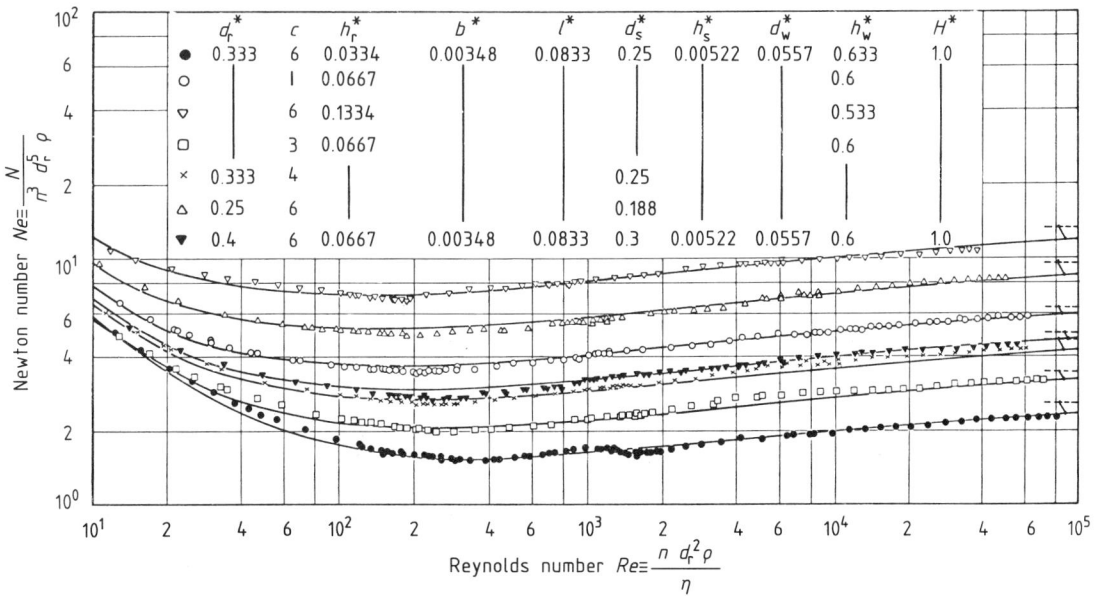

Figure 19.27. Comparison between measured and calculated Ne-data for turbine stirrers in vessels with baffles. The curves have been calculated by means of Eq. (19.55).

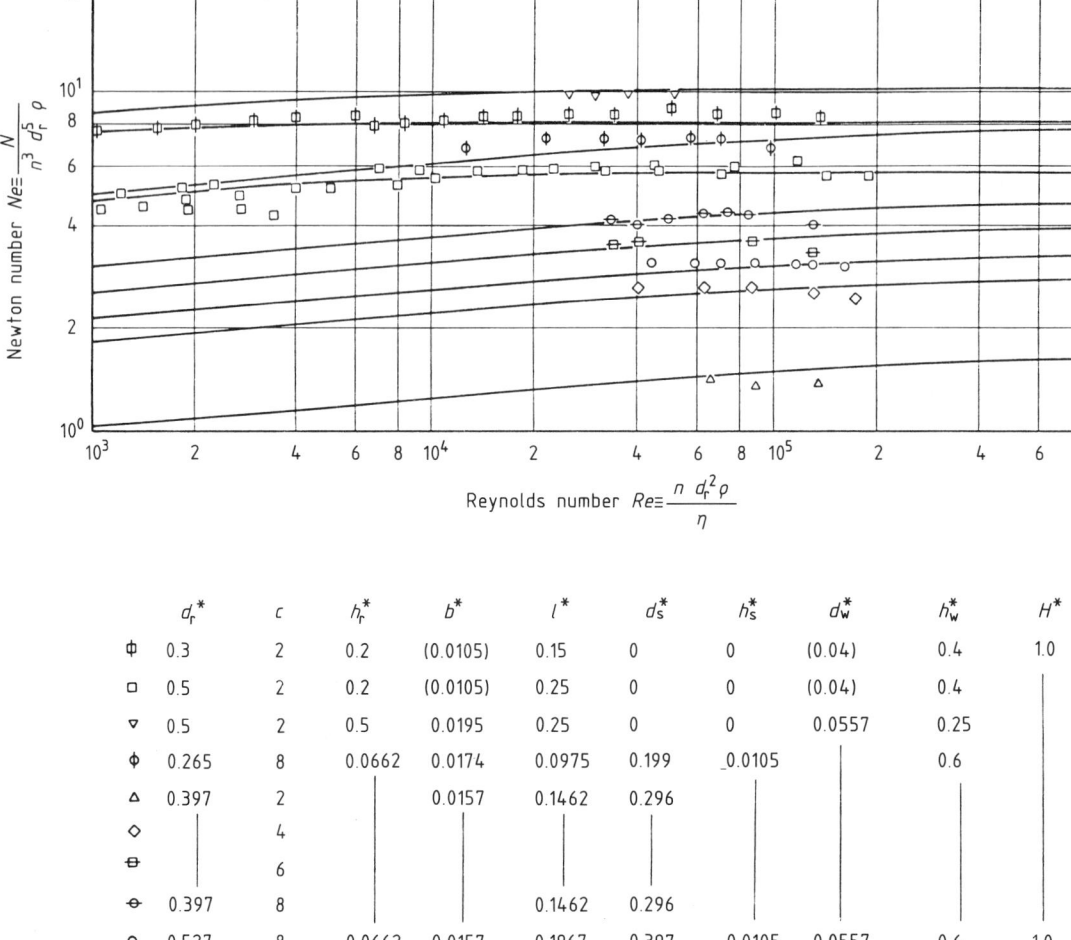

Figure 19.28. Comparison between measured and calculated Ne-data for turbine and paddle stirrers in vessels with baffles. The curves have been calculated by means of Eq. (19.55).

	d_r^*	c	h_r^*	b^*	l^*	d_s^*	h_s^*	d_w^*	h_w^*	H^*
ϕ	0.3	2	0.2	(0.0105)	0.15	0	0	(0.04)	0.4	1.0
□	0.5	2	0.2	(0.0105)	0.25	0	0	(0.04)	0.4	
▽	0.5	2	0.5	0.0195	0.25	0	0	0.0557	0.25	
ϕ	0.265	8	0.0662	0.0174	0.0975	0.199	0.0105		0.6	
△	0.397	2		0.0157	0.1462	0.296				
◇		4								
⊖		6								
⊕	0.397	8			0.1462	0.296				
○	0.527	8	0.0662	0.0157	0.1967	0.397	0.0105	0.0557	0.6	1.0

For Re_1 Eq. (19.54) has been given. The difference between Re_1 and Re_2 is relatively small. When Re_1 turns out to be greater than Re_2 either n or d_r must be reduced.

Discussion of the Influence of the Three Most Important Parameters for Energy Transfer

The most important parameters for energy transfer by turbine and paddle stirrers are the stirrer diameter ratio d_r^*, the stirrer height ratio h_r^*, and the number of blades or paddles c. In Fig. 19.29 Ne is presented as a function of these three parameters for vessels without (o) and with (m) baffles. When the Newton number is given as a function of h_r^*, then d_r^* and c are kept constant. The values for the constant parameters are given in the graph. Fig. 19.29 shows a remarkable influence of the stirrer height h_r^* on the Newton number when the vessel is equipped with baffles. As the desired fluid flow in the vessel is to be achieved with the lowest possible energy transfer, common practice in the use of baffles deserves reexamination.

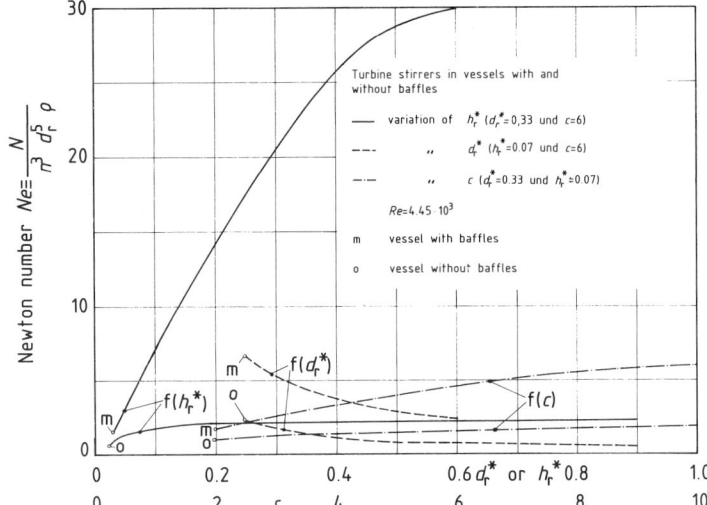

Figure 19.29. Influence of the most important parameters on the Newton number Ne for turbine stirrers in vessels without and with baffles.

Energy Transfer Equations for Propeller and Disk Stirrers

The information available on energy transfer by propeller and holed disk stirrers is still rather scarce. The results of experimental investigations carried out by GLAESER et al. [19.31] are shown in Figs. 19.30 and 19.31. Shape and dimensions of these stirrers are given in Figs. 19.7 and 19.8. The vessel had been equipped with four baffles.

Fig. 19.30 shows the results for 1-, 2-, 3-, and 4-bladed propellers in the high turbulent flow region. The arrow indicates the upper limiting Reynolds number Re_2. In the high turbulent flow region the Newton number Ne is independent of Re. The results are correlated by the following empirical equation:

$$Ne = 0.026\, c, \tag{19.61}$$

with c being the number of propeller blades. For a 4-bladed propeller $Ne = 0.104$, that is about 2% of the Newton number for turbine stirrers.

In Fig. 19.31 the experimental data for disk stirrers are presented. The number of holes near the outer edge of the disk has been varied between $z=0$ and $z=12$. In one case there were four more holes arranged on an inner circle. The diameter of the

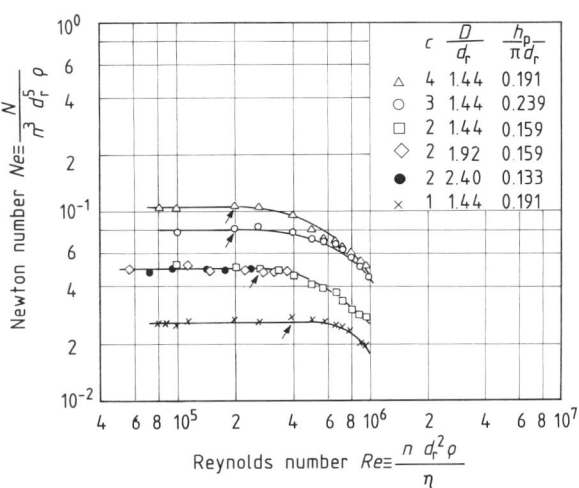

Figure 19.30. Newton number Ne for 6-propeller stirrers in baffled vessels as a function of the Reynolds number Re in the high turbulent region.

holes was 21 mm, the diameter ratio d_r/D was 0.5. Within the tested range the data are well correlated by the equation:

$$Ne = 0{,}04 + 0.0075\, z. \tag{19.62}$$

With $z = 12$ the Newton number becomes $Ne = 0.13$. This value is less than 3% of that for turbine stirrers.

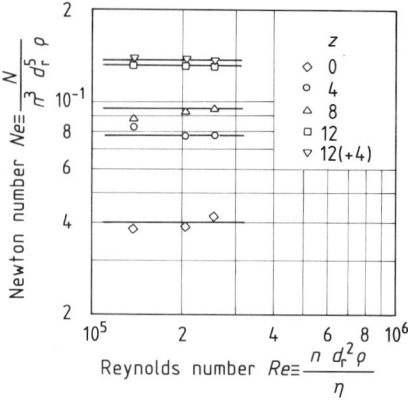

Figure 19.31. Newton number Ne for 5-holed disk stirrers in baffled vessels as a function of Reynolds number Re in the high turbulent region.

19.5.4.4 Calculation of Energy Transfer to Non-Newtonian Fluids

The viscosity of the majority of non-Newtonian liquids is relatively high. As a consequence, the stirrer operations take place predominantly in the laminar flow region, so that the Reynolds number is in most cases below a value of roughly 10.

For immediate application, information on energy transfer is available for power law fluids. For other fluids, especially when viscoelastic effects are present, it is suggested to rely on specific laboratory tests.

Energy Transfer Calculation Method according to Metzner-Otto-Schilo

METZNER and OTTO [19.32] developed and tested a rough but simple method with which the energy transfer in non-Newtonian liquids can be calculated when information is available on energy transfer in Newtonian liquids for the same type of stirrer and vessel. The method consists of introducing a characteristic viscosity η_c of the non-Newtonian liquid in the vessel under stirring conditions. Using this viscosity in the Reynolds number of equations available for Newtonian fluids, these equations may be used to calculate Ne for non-Newtonians.

For purely viscous non-Newtonian fluids, the so-called power law fluids, the characteristic viscosity η_c is given by Eq. (19.5). For the mean velocity gradient or shear rate $\dot{\gamma}$ divided by the number of revolutions per time unit n METZNER and OTTO found a value between $\dot{\gamma}/n = 11.5$ to 13 for turbine stirrers and $\dot{\gamma}/n = 10$ for a propeller stirrer. In the experiments carried out by METZNER and OTTO the fluid index varied from $m = 0.2$ to 1.5 and $m = 0.16$ to 1.0, respectively. The simple method suggested by METZNER and OTTO has been found satisfactory for engineering purposes [19.2], [19.33], [19.34]. CALDERBANK and MOO-YOUNG [19.35] found that $\dot{\gamma}/n$ is a function of the stirrer diameter ratio d_r^*. In a still later paper SCHILO [19.24] found that for paddle stirrers $\dot{\gamma}/n$ becomes independent of the diameter ratio d_r^*, when d_r^* is smaller than 0.7 and is a function of the fluid index m. The results obtained by SCHILO may be expressed by the empirical equation:

$$\frac{\dot{\gamma}}{n} = 4.4 \, m^{-1.625}. \tag{19.63}$$

tested range: $0.488 \leq m \leq 0.81$
$\qquad\qquad\qquad d_r^* \leq 0.7$

It is suggested to make use of Eq. (19.63) for turbine and paddle stirrers. In the case of propeller and disk stirrers Eq. (19.63) should be used with caution. However, the mistake that may be involved by applying Eq. (19.63) is not very serious, as the energy transfer by propeller and disk stirrers is much smaller than that by turbine and paddle stirrers.

The procedure for calculation of the transferred energy per time unit is as follows:

1. When the fluid index m and the number of revolutions n of the stirrer are known, $\dot{\gamma}$ follows from Eq. (19.63).
2. With the known shear rate $\dot{\gamma} = dw/dy$, the characteristic viscosity η_c is obtained from a $\eta_c/(dw/dy)$-curve as shown in Fig. 19.4. This curve must be determined

by means of a viscosimeter for the non-Newtonian fluid in question.

3. The characteristic viscosity η_c is applied to determination of the Reynolds number:

$$Re \equiv \frac{n d_r^2 \varrho}{\eta_c}.$$

4. Calculation of the upper limiting Reynolds number Re_1 or Re_2. The Reynolds number Re must be smaller than Re_1 or Re_2. If this condition is not fulfilled, the number of revolutions n of the stirrer or the stirrer diameter d_r must be reduced.

5. With the Reynolds number Re known the Newton number can be determined by means of one of the appropriate equations presented in the preceding sections: Eqs. (19.45), (19.55), (19.61), or (19.62).

6. Determination of the transferred energy per time unit N from:

$$N = Ne \varrho n^3 d_r^5.$$

Energy Transfer Calculation Method according to Schilo

Based on the concept of the cylinder model discussed in Sect. 19.5.1, SCHILO [19.24] derived a theoretical equation for the Newton number. Comparison with experimental results made it necessary to introduce an empirical correction. The equation presented by SCHILO holds for the following conditions:

1. power law fluids with $0 \leq m \leq 1$,
2. paddle stirrers, and
3. laminar flow region.

The equation is given as follows:

$$Ne = 0.7 \pi^2 \left(\frac{4\pi}{m}\right)^m \frac{(h_r/d_r) D^{*2}}{(D^{*2/m} - 0.75)^m} \frac{1}{Re_m^{0.9}}. \quad (19.64)$$

The Reynolds number Re_m and D^* are defined as follows:

$$Re_m \equiv \frac{n^{2-m} d_r^2 \varrho}{k}, \quad (19.65)$$

$$D^* \equiv 1/d_r^* = D/d_r. \quad (19.66)$$

In Eq. (19.65) k is the consistency factor of the shear stress/shear rate relationship defined by Eq. (19.4). Experimental data obtained by SCHILO for paddle stirrers in power law fluids are shown in Fig. 19.32. The curve drawn through the data has been calculated according to Eq. (19.64).

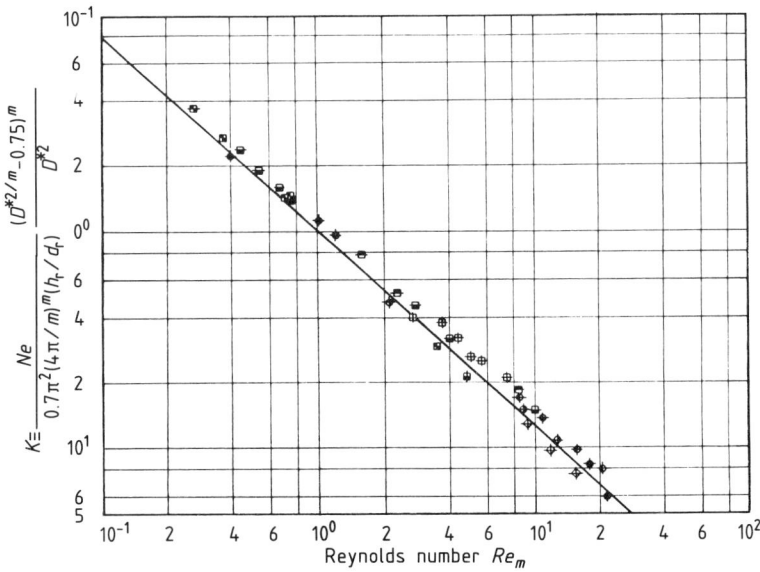

Figure 19.32. Newton number Ne for paddle stirrers in power law fluids; curve after Eq. (19.64).

19.6 Heat Transfer in Stirred Vessels

19.6.1 Introduction

The aim of the stirring process in a batch fermenter must be to obtain constant conditions for microbial reactions in the entire volume of the fermenter. This requires constant temperature of the fluid and constant concentrations for all components of the fermentation fluid within the reactor. Due to the fluid flow conditions in a stirred vessel the temperature cannot be constant under any conditions. Instead, the temperature will in general exhibit a strong variation not only with the space coordinates but also with time.

Heat transfer in stirred vessels is an unsteady-state process. Because of the complexity of this process experimental investigations have been carried out for relatively simple steady-state conditions. Therefore, the results obtained can be considered only as a very rough approximation to the real process.

All mechanical energy transferred to the fluid in order to maintain fluid movement is dissipated into heat. The dissipation process is proportional to the gradients of the velocity components, primarily of the tangential velocity. The consequence is, that heat production takes place in a space close to the surface of the stirrer. Whatever the initial conditions of heat transfer are, after a certain time, the dissipated mechanical energy must be transferred from the fluid through the walls of the vessel to an outside medium.

In the following subsections it is intended to transmit to the reader a fairly accurate picture of the real heat transfer process by means of graphical presentations of the temperature field. After the physical phenomena have been understood, empirical equations describing steady-state heat transfer will be discussed.

19.6.2 Physical Phenomena of Heat Transfer in Stirred Vessels

The physical phenomena of heat transfer will be discussed on the basis of results obtained by a comprehensive theoretical-numerical study of the problem [19.27], [19.36].

19.6.2.1 Mathematical Problem of Heat Transfer

The temperature field in a quiescent fluid and in a fluid in laminar motion, contained in a stirred vessel of the type shown in Fig. 19.1 with a stirrer modelled according to Fig. 19.10, is described by the following differential equation:

$$\frac{\partial T^*}{\partial Fo} + Re_{12}^2 Pr \left[u_+ \frac{\partial T^*}{\partial r^*} + w_+ \frac{\partial T^*}{\partial z^*} \right] =$$

$$= \left[\frac{1}{r^*} \frac{\partial}{\partial r^*} \left(r^* \frac{\partial T^*}{\partial r^*} \right) + \frac{\partial^2 T^*}{\partial z^{*2}} \right] + q_{i,R}^*. \quad (19.67)$$

On the left hand side the first term predicts the variation of temperature with time, while the second term predicts temperature variation due to fluid motion. On the right hand side the first term predicts temperature variation due to molecular heat conduction, while the second term is the so-called dissipation term, which describes the dissipation of mechanical energy into heat. This term is given by:

$$q_{i,R}^* = Br \left\{ 2 Re_{12}^2 \left[\left(\frac{\partial u_+}{\partial r^*} \right)^2 + \left(\frac{u_+}{r^*} \right)^2 + \right. \right.$$

$$+ \left(\frac{\partial w_+}{\partial z^*} \right)^2 + \frac{1}{2} \left(\frac{\partial w_+}{\partial r^*} + \frac{\partial u_+}{\partial z^*} \right)^2 \right] +$$

$$+ \left[\left(\frac{\partial v^*}{\partial z^*} \right)^2 + \left(r^* \frac{\partial (v^*/r^*)}{\partial r^*} \right)^2 \right] \right\}. \quad (19.68)$$

The first term represents the contribution of radial and axial velocities, u_+ and w_+, and the second term of tangential velocity v^* to the dissipation heat. The dimensionless quantities of the last two equations are defined as follows:

$$Br \equiv \frac{\omega^2 R_1^2 \eta}{\lambda(T_w - T_o)} =$$
$$= \pi^2 \frac{n^2 d_{gl}^2 \eta}{\lambda(T_w - T_o)} \quad \text{Brinkman number} \quad (19.69)$$

$$Fo \equiv \frac{ta}{R_2^2} = \frac{4t\lambda}{\varrho c_p D^2} \quad \text{Fourier number} \quad (19.70)$$

$$Pr \equiv \frac{v}{a} = \frac{\eta c_p}{\lambda} \quad \text{Prandtl number} \quad (19.71)$$

$$q_{t,R}^* \equiv \frac{q_{t,R}}{\lambda(T_w - T_o) R_2^2} \quad \text{local dissipation energy} \quad (19.72)$$

$$Re_{12} \equiv \frac{\omega R_1 R_2 \varrho}{\eta} =$$
$$= \frac{\pi}{2} \frac{1}{d_{gl}^*} Re \quad \text{Reynolds number} \quad (19.6)$$

$$Re \equiv \frac{n d_{gl}^2 \varrho}{\eta} \quad \text{Reynolds number} \quad (19.8)$$

$$T^* \equiv \frac{T - T_o}{T_w - T_o} \quad \text{local temperature} \quad (19.73)$$

$$u^* \equiv \frac{u}{\omega R_1} = \frac{u}{\pi n d_{gl}} \quad \text{local radial velocity} \quad (19.74)$$

$$v^* \equiv \frac{v}{\omega R_1} = \frac{v}{\pi n d_{gl}} \quad \text{local tangential velocity} \quad (19.33)$$

$$w^* \equiv \frac{w}{\omega R_1} = \frac{w}{\pi n d_{gl}} \quad \text{local axial velocity} \quad (19.75)$$

$$u_+ \equiv \frac{u^*}{Re} \quad \text{relative radial velocity} \quad (19.76)$$

$$w_+ \equiv \frac{w^*}{Re} \quad \text{relative axial velocity} \quad (19.77)$$

$$r^* \equiv r/R_2 = 2r/D \quad \text{radial coordinate} \quad (19.78)$$

$$z^* \equiv z/R_2 = 2z/D \quad \text{axial coordinate} \quad (19.79)$$

For purposes of simplification the geometrical ratios of the stirrer system have been kept constant:

$$d_{gl}^* \equiv d_{gl}/D = 0.40$$
$$d_w^* \equiv d_w/D = 0.05$$
$$h_b^* \equiv h_b/D = 0.40$$
$$h_r^* \equiv h_r/D = 0.10$$
$$h_w^* \equiv h_w/D = 0.50$$
$$H^* \equiv H/D = h_b^* + h_r^* + h_w^* = 1.0$$

The symbols used in the last equations which have not yet been explained are: R_1 the radius of the stirrer model and R_2 of the vessel, λ heat conductivity, c_p specific heat capacity, $a = \lambda/(\varrho c_p)$ heat diffusivity, t time, T local temperature at arbitrary time t, T_o temperature at $t = 0$, and T_w wall temperature of the vessel.

The Brinkman number Br is one of the most important dimensionless parameters for this heat transfer problem; Br may be interpreted as the ratio of the time independent frictional heat $\omega^2 R_1^2 \eta$ and the heat $\lambda(T_w - T_o)$ either transferred from the vessel wall to the fluid or from the fluid to the wall in the initial condition $t = 0$.

It should be noted that, according to Eq. (19.67), the convective heat transport is only due to radial and axial fluid motion. The tangential fluid velocity v does not support heat transport within the fluid. This means that the rotational or primary fluid motion is of no importance to heat transport, although this part of the fluid motion contains the major portion of mechanical energy. The transportation of heat in the fluid is only supported by secondary fluid motion. In heat generation by internal friction, however, it is the tangential velocity that delivers the greatest portion.

The energy equation has been solved for the following conditions:

Initial condition:

$$Fo = 0; \quad T^* = 0. \quad (19.80)$$

Boundary conditions:

$$\left.\begin{array}{lll}
1.\ z^*=0 & \text{and}\ 0\leqslant r^*\leqslant 1 & \text{bottom of vessel} \\
2.\ z^*=2H^* & \text{and}\ d_w^*\leqslant r^*\leqslant 1 & \text{free surface of liquid} \\
3.\ r^*=1 & \text{and}\ 0\leqslant z^*\leqslant 2H^* & \text{cylindrical surface of vessel}
\end{array}\right\} T^*=1$$

$$\left.\begin{array}{lll}
4.\ r^*=0 & \text{and}\ 0\leqslant z^*\leqslant 2h_b^* & \text{axis of vessel} \\
5.\ r^*=d_{gl}^* & \text{and}\ 2h_b^*\leqslant z^*\leqslant 2(h_r^*+h_b^*) & \text{cylindrical surface of stirrer model} \\
6.\ r^*=d_w^* & \text{and}\ 2(h_r^*+h_b^*)\leqslant z^*\leqslant 2H^* & \text{cylindrical surface of stirrer shaft}
\end{array}\right\} \dfrac{\partial T^*}{\partial r^*}=0 \quad (19.81)$$

$$\left.\begin{array}{lll}
7.\ z^*=2h_b^* & \text{and}\ 0\leqslant r^*\leqslant d_r^* & \text{bottom area of stirrer model} \\
8.\ z^*=2(h_r^*+h_b^*) & \text{and}\ d_w^*\leqslant r^*\leqslant d_r^* & \text{top area of stirrer model}
\end{array}\right\} \dfrac{\partial T^*}{\partial z^*}=0$$

A numerical method for the solution of the differential equation for the temperature field in the stirred vessel has been developed by THIELE [19.21]. The result obtained by applying this method is the local temperature:

$$T^*=f(Fo;\ Re_{12}^2 Pr;\ Re_{12};\ Br;\ r^*;\ z^*).$$

For some selected conditions the temperature field will be discussed.

19.6.2.2 Temperature Field in a Quiescent Fluid

When the fluid is at rest ($Re_{12}^2 Pr=0$), there is no dissipation of mechanical energy into heat ($Br=0$). For this condition a very simple structure of the temperature field is obtained. The temperature is only a function of the Fourier number Fo and the coordinates. For $Fo=1.043\cdot 10^{-2}$ the temperature field is given in Fig. 19.33.

For $Fo=0$ the dimensionless temperature $T^*=0$ exists in the entire volume of the liquid. With progressing time heat starts to penetrate into the liquid, so that the temperature rises steadily. The isotherms pre-

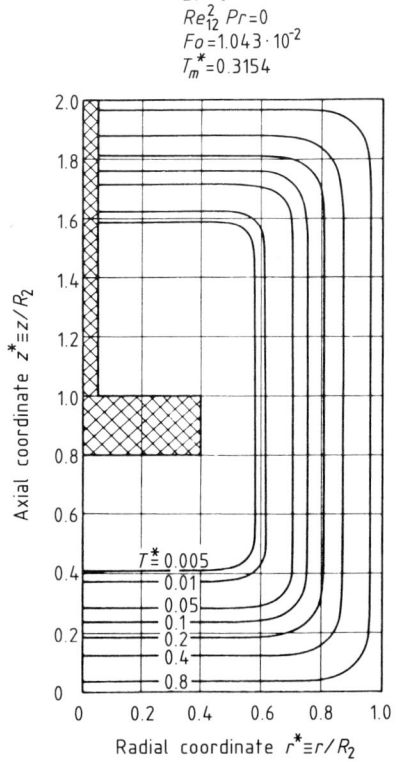

Figure 19.33. Temperature field in a quiescent fluid for $Fo=1.043\cdot 10^{-2}$.

sented in Fig. 19.33 are parallel to the outer surface of the liquid over wide areas. For $Fo \to \infty$ the temperature asymptotically approaches the final value $T^* = 1$, in the entire volume of liquid. At the selected Fourier number of $Fo = 1.043 \cdot 10^{-2}$ the mean temperature is only $T_m^* = 0.3154$, and therefore still rather low compared to the final value $T_m^* = 1$.

19.6.2.3 Temperature Field in a Fluid in Motion but Without Dissipation of Mechanical Energy

When the fluid is set into motion it is the secondary flow that exerts a strong influence on the structure of the temperature field. This is exemplified by Figs. 19.34 to 19.36. With increasing time, i.e., Fourier number Fo, the temperature field reflects the characteristic properties of the secondary flow field, that is the vortex ring above and below the mean stirrer plane.

With the assumption that there is no dissipation of mechanical energy into heat ($Br = 0$), the temperature field is due only to the heat intruding into the liquid from the outside. At the outer boundary the temperature is $T^* = 1$. With progressing time the temperature of the fluid is elevated above the initial temperature $T^* = 0$. For $Fo \to \infty$ the temperature of the entire fluid approaches the final value $T^* = 1$.

It should be noted, that the structure of the temperature field becomes more pronounced with increasing Fourier number, but at the same time the local temperature

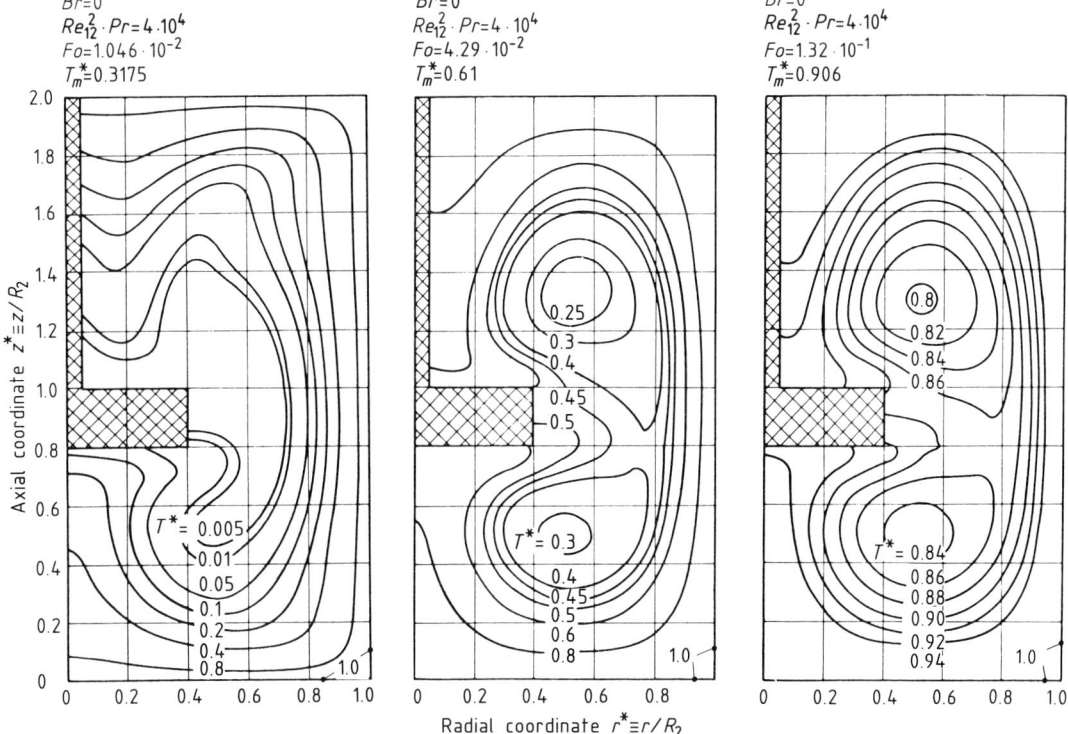

Figures 19.34 to 19.36. Temperature fields in a fluid in motion but without dissipation of mechanical energy.
Figure 19.34: $Fo = 1.046 \cdot 10^{-2}$, Figure 19.35: $Fo = 4.29 \cdot 10^{-2}$, Figure 19.36: $Fo = 1.32 \cdot 10^{-1}$.

19.6.2.4 Temperature Field in a Fluid in Motion With Dissipation of Mechanical Energy for $Br = +10$

When dissipation of mechanical energy is taken into account, heat is transferred into the fluid from the outer boundary and at the same time from the surface of the stirrer. In this case the temperature field is built up from the outer and the inner boundary of the fluid. This is the characteristic starting condition when the temperature field is built up for the case of heat transfer with dissipation of mechanical energy. This situation is clearly shown in Fig. 19.37.

Figures 19.37 to 19.39 demonstrate the development of the temperature field with changing Fourier number Fo. For the heat transfer case under consideration it proved to be advisable not to present the dimensionless temperature T^* but the reduced dimensionless temperature T^+ which is defined as follows:

$$T^+ \equiv \frac{T^* - T^*_{min}}{T^*_{max} - T^*_{min}}. \tag{19.82}$$

T^*_{min} and T^*_{max} are minimum and maximum values of T^* in the temperature field for the given value of Fo. The advantage is the variation of T^+ between $T^+_{min} = 0$ for $T^* = T_{min}$ and $T^+_{max} = 1$ for $T^* = T^*_{max}$. T^*_{min} varies according to the prescribed boundary conditions from $T^*_{min} = 0$ and $T^*_{min} = 1$. This variation is independent of the Brinkman number. The range of T^*_{max}, however, has the lower limit at $T^*_{max} = 1$ and the upper limit at $T^*_{max} > 1$, depending on the value of the Brinkman number.

The temperature fields shown in Figs. 19.37 to 19.39 have been calculated for a value of the Brinkman number of $Br = +10$. In this case the heat of dissipation is about ten times the heat transferred at the beginning of the process from the outside to the fluid. The structure of the temperature field is therefore determined by dissipation of mechanical energy in the range of low values of Fo and, because of the relatively high value of $Re_{12}^2 Pr = 4 \cdot 10^4$ with $Re_{12} = 20$, by the secondary fluid motion in the range of high values of Fo.

The heat currents moving into the fluid from the outer boundary and from the inner boundary, i.e., the surface of the stirrer, do not yet influence each other at $Fo = 5.40 \cdot 10^{-3}$ in Fig. 19.37. At $Fo = 1.06 \cdot 10^{-2}$ in Fig. 19.38, however, the two temperature fields grow together. With further increasing Fourier number the heat current from the outer boundary loses its influence on the temperature field. In Fig. 19.39 the temperature field is already dominated by the heat of dissipation and at the outer boundary a heat current is directed from the fluid to the wall of the vessel, that is, the direction of the heat current has been inverted. This phenomenon is indicated by the minimum value of T^*, at $T^*_{min} = 1$, occurring at the outer boundary. The temperature of the fluid is now higher than $T^* = 1$ throughout.

The dissipation of mechanical energy into heat occurs primarily, as all figures from the temperature field indicate, at the cylindrical surface of the stirrer model. In the center line of this surface the highest value of the temperature with $T^+ = 1$ is observed. The contribution of the circular top and bottom areas of the stirrer model is naturally very small.

The final structure of the temperature field has almost been attained at $Fo = 4.33 \cdot 10^{-1}$, as shown in Fig. 19.39. With further increasing Fourier number steady-state conditions for heat transfer have been reached. All the heat originating from the dissipation process, which is the total mechanical energy transferred from the stirrer to the fluid, must be transferred through the vessel walls to the surroundings.

The distance between two isotherms is proportional to the local temperature gradient. At the vessel wall this gradient is proportional to the local heat transfer coeffi-

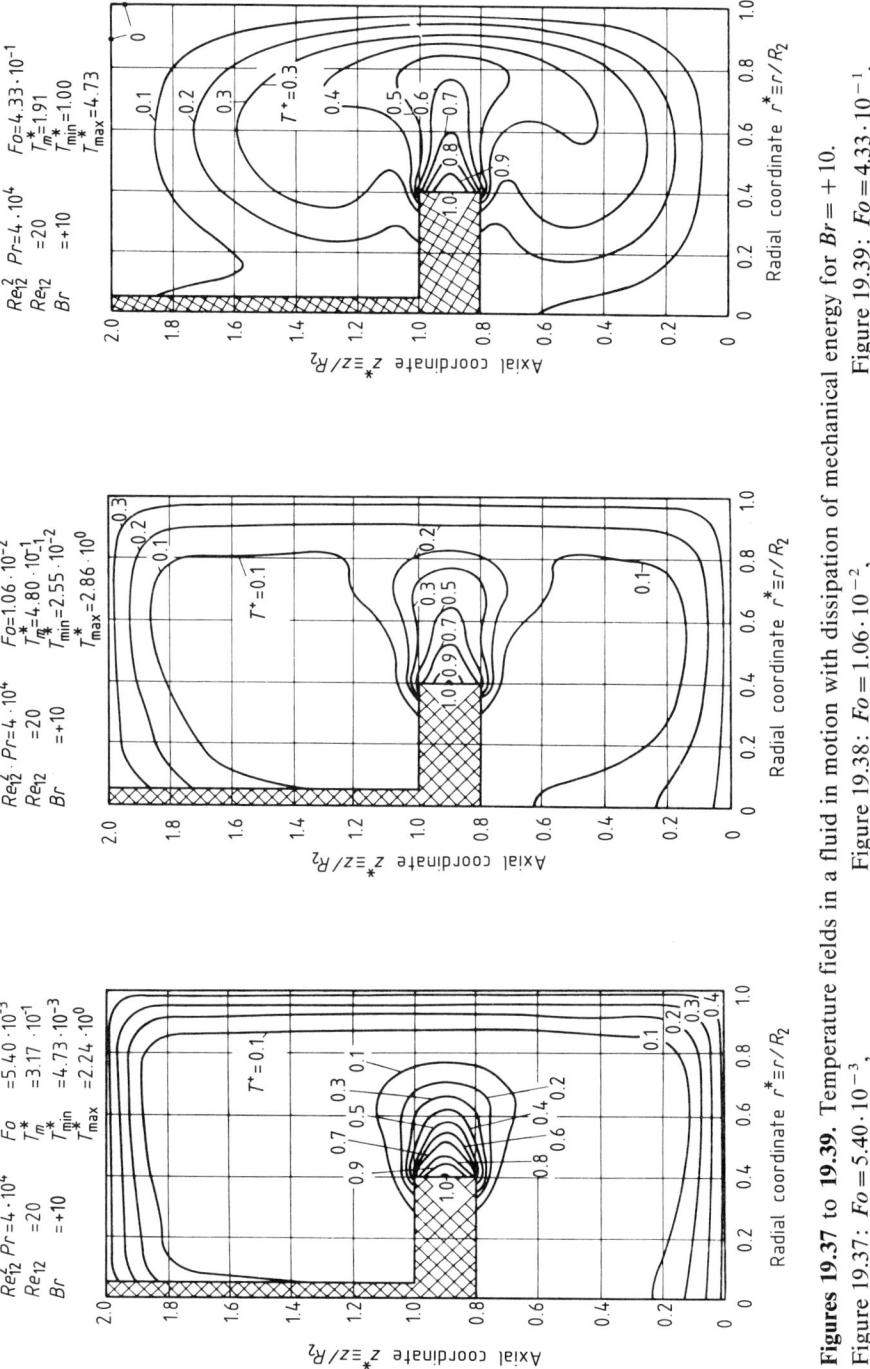

Figures 19.37 to 19.39. Temperature fields in a fluid in motion with dissipation of mechanical energy for $Br = +10$.
Figure 19.37: $Fo = 5.40 \cdot 10^{-3}$; Figure 19.38: $Fo = 1.06 \cdot 10^{-2}$; Figure 19.39: $Fo = 4.33 \cdot 10^{-1}$.

cient. Fig. 19.39 shows that the heat transfer coefficient varies quite remarkably at different areas of the wall of the vessel. The highest value of the heat transfer coefficient is observed in the mean stirrer plane, the lowest values in the corners and at the bottom of the vessel.

19.6.2.5 Temperature Field in a Fluid in Motion With Dissipation of Mechanical Energy for $Br = -10$

With a negative Brinkman number heat is withdrawn from the fluid right from the start of the process. In this case the temperature of the fluid T_o at $t = 0$ is higher than the temperature of the vessel wall T_w.

Approaching steady-state conditions for heat transfer at higher values of the Fourier number, it is irrelevant whether the Brinkman number has a negative or a positive value. In both cases the mechanical energy transferred to the fluid must be withdrawn as heat. The temperature fields are therefore the same for $Br = +10$ and $Br = -10$ with $Fo \to \infty$. This is quite obvious when one compares the temperature fields for $Br = -10$ as depicted in Fig. 19.40, and for $Br = +10$ as in Fig. 19.39.

It has been mentioned that for $Fo \to \infty$ the structure of the temperature field is strongly influenced by the secondary fluid flow. The strength of the secondary flow, however, decreases rapidly with decreasing

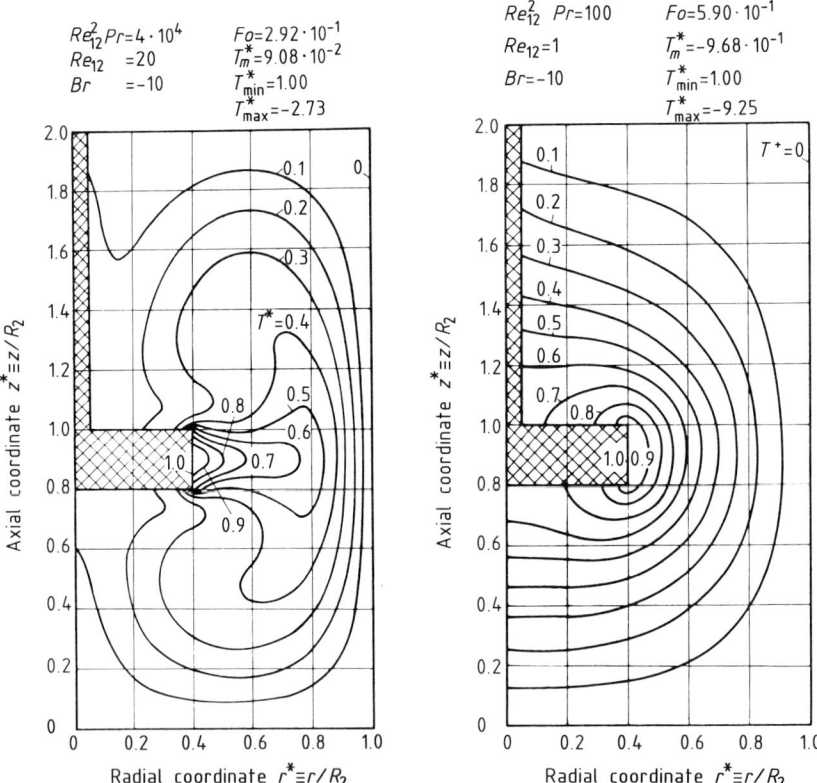

Figures 19.40 and 19.41. Temperature fields in a fluid in motion with dissipation of mechanical energy for $Br = -10$.
Figure 19.40: $Re_{12}^2 Pr = 4 \cdot 10^4$,   Figure 19.41: $Re_{12}^2 Pr = 10^2$.

Reynolds number. This can be deduced from Figs. 19.17 and 19.18, which present the volumetric flow rates. It is therefore only natural that, for $Re_{12}^2 Pr = 100$ with $Re_{12} = 1$, the temperature field presented in Fig. 19.41 does not show an effect of the secondary flow. This temperature field is typical for heat transfer with dissipation of mechanical energy with negligible secondary flow and approaching steady-state conditions.

19.6.3 Empirical Equations for Heat Transfer

The basis of heat transfer calculations for technical purposes is the assumption of steady-state conditions. In this case the following equation for the heat flux $\dot{Q}$ applies:

$$\dot{Q} = kA \Delta T_m, \quad (19.83)$$

with k as the overall heat transfer coefficient defined by:

$$\frac{1}{kA} = \frac{1}{A_i \alpha_i} + \frac{s_w}{A_m \lambda_w} + \frac{1}{A_o \alpha_o}. \quad (19.84)$$

In this equation α_i and α_o are the heat transfer coefficients on the inner and the outer surface of the vessel wall, A_i and A_o the relevant surface areas, s_w the thickness of the wall, λ_w the heat conductivity coefficient of the wall material, and

$$A_m = \frac{A_o - A_i}{\ln(A_o/A_i)} \quad (19.85)$$

the mean wall area. The temperature difference ΔT_m is the logarithmic mean value:

$$\Delta T_m = \frac{T_i - T_o}{\ln(T_i/T_o)}, \quad (19.86)$$

with T_i being a mean value of the liquid in the vessel and T_o a mean temperature of the fluid outside the vessel, for instance, in a jacket of the vessel.

The above equations may also be applied when the heat exchanger is a tube coil arranged inside the vessel. In this case A_o and A_i are the inner and the outer surface of the tube, and T_o and T_i the temperatures of the fluids within the tube and within the vessel.

According to Eqs. (19.83) and (19.84) heat transfer calculations require equations for the two heat transfer coefficients α_i and α_o. For heat transfer on the outer surface the conventional equations for heat transfer in tubes may be applied. For heat transfer on the inner surface of the vessel special equations are required, for which the limitations and shortcomings have already been mentioned.

In a comprehensive study POGGEMANN et al. [19.37] collected all available equations for heat transfer on the inner surface of the vessel from the international literature. Comparison shows that the equations differ by a factor of about 2. Within this uncertainty, an influence of the type of stirrer and heat exchanger can hardly be eliminated. It therefore seems to be justified to present only one equation that should be applicable to heat transfer calculations, of course within the range of the given uncertainty. This equation is formulated as:

$$Nu = 0.464 \, Re^{2/3} Pr^{1/3} (\eta/\eta_w)^{0.14}. \quad (19.87)$$

The dimensionless numbers are defined by:

$Nu \equiv \alpha_i D/\lambda$ Nusselt number (19.88)

$Re \equiv n d_r^2 \varrho/\eta$ Reynolds number (19.8)

$Pr \equiv \eta c_p/\lambda$ Prandtl number (19.11)

In Eq. (19.87) η and η_w are the dynamic viscosity of the fluid in the vessel at a mean fluid temperature and wall temperature, respectively. The other fluid properties have to be determined for the mean fluid temperature.

Eq. (19.87) is within the mentioned range of uncertainty applicable to heat transfer in vessels with and without baffles, to turbine, paddle, propeller, and anchor stirrers, and to pure Newtonian and non-Newtonian fluids. For non-Newtonian fluids the characteristic viscosity η_c has to be introduced and determined according to the procedure outlined in Sect. 19.5.4.4. In case of aerated vessels the results obtained by Eq. (19.87) should be considered as maximal, so that a reduction by about 20% seems to be advisable.

19.7 Gas Dispersion in Stirred Vessels

Many microbial processes require oxygen, which must be transferred into the fermentation fluid from dispersed gas bubbles. Gas dispersion is therefore an important part of microbial processes. Generation and dispersion of bubbles in a fermentation fluid is accomplished by the rotating stirrer. It has already been pointed out, that the stirrer is a multi-purpose element, which has to transfer energy to the fluid in order to maintain fluid motion, and which has to disperse gas at the same time. In the following subsections a physical analysis of the dispersion process will be presented; the properties of various dispersion elements, bubble movement within the fermentation fluid, gas holdup and flooding conditions, and finally the influence of the dispersion process on energy transfer will be discussed.

19.7.1 Physical Analysis of the Dispersion Process

The dispersion process consists of two distinct parts: Generation of bubbles and distribution of the bubbles within the fermentation fluid. Bubble generation requires size reduction of a certain volume of gas. This process will be analyzed. Distribution of the bubbles within the fluid volume requires an appropriate fluid motion, which also has to be produced by the stirrer.

Bubble generation is accomplished by rotating stirrers in several steps:

1. Formation of a gas reservoir in the vessel from which bubble generation can be initiated. This reservoir is of decisive importance for the entire process. Its volume should exceed a minimum, especially when very small bubbles shall be generated. In some cases the reservoir is established in a region of low local pressure from which the gas cannot easily escape. In other cases the gas is entrapped by an appropriate design of the stirrer.

2. Transport of the gas to the site of bubble generation. In most cases this is achieved by fluid currents due to the action of centrifugal forces in a rotating body of fluid. Shear stress is acting in the interface of the gas/liquid system, but is not yet strong enough to separate small elements from the bulk of the gas.

3. Generation of low pressure regions in the bulk of the liquid by the rotating stirrer, where high shear stresses become effective. Such conditions are observed in the wake of blunt bodies, from which whirling filaments extend into the bulk of the liquid. Depending on the pressure conditions many whirling filaments may be combined to an areal site. The high shear stress is due to the rotational speed of the filaments. Bubble generation, due to the action of the shear stress, is enhanced with increasing length and decreasing diameter. Dispersion elements therefore should be designed in such a way, that filamental or areal sites of bubble generation are achieved.

Bubble movement in the liquid is of great importance for residence time of the bubbles and consequently mass transfer across the gas/liquid interface. Thus, the dispersion element must generate an appropriate three-dimensional liquid flow in the vessel. The primary, that is, the tangential or rotational fluid motion, will pick up the bubbles and help to increase the residence time of the bubbles in the bulk of the liquid. Primary fluid motion, however, will not redirect the bubbles into the bubble generation process for bubble or interface renewal. The renewal process can only be accomplished by help of the secondary flow. Of course secondary flow will also help to extend the residence time. This again proves the importance of the secondary fluid flow. Dispersion elements should be designed in such a way, that strong secondary currents are produced.

19.7.2 Bubble Generation by Means of Various Dispersion Elements

The process of bubble generation by means of dispersion elements, which are of technical importance, will be considered on the basis of the physical phenomena explained in the preceding subsection. Fundamental research on the dispersion of gas in stirred vessels has been carried out by BIESECKER [19.38], VAN'T RIET [19.39], GLAESER et al. [19.31], and many others [19.40] to [19.45].

The first dispersion element described will be the Rushton turbine shown in Fig. 19.42. For purposes of simplification the turbine stirrer is shown with only one blade. The gas is supplied from below and collects in the reservoir situated around the center of the bottom side of the disk on which the blade is mounted. The disk prevents the gas from easily escaping the reservoir in a vertical direction. The gas can only leave the reservoir in a radial direction. In this direction shear stresses, exerted by the liquid, transport the gas to the sites of bubble generation. These sites are whirling filaments in the wake of blades as indicated in Fig. 19.42. With increasing distance from the blade the diameter of the filaments decreases. Because of the reduced pressure in these filaments, there is a constant strong flow of the gas from the blade to the tip of the filament. The rotational motion of the

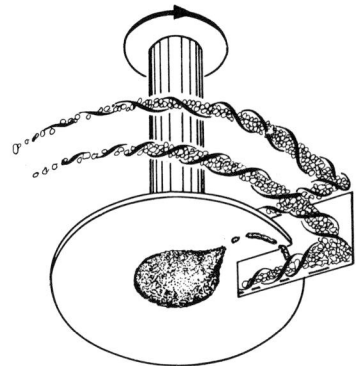

Figure 19.42. Bubble generation in vortex threads in the wake of a turbine stirrer.

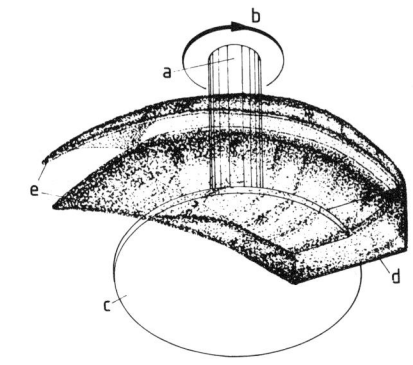

Figure 19.43. Bubble generation in vortex sheets in the wake of a turbine stirrer.

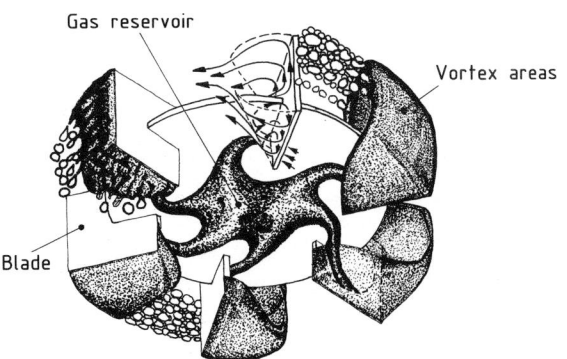

Figure 19.44. Gas supply to the vortex sheets in the wake of the blades of a turbine stirrer.

filaments supports the action of the shear stress by which the formation of small bubbles is achieved.

With increasing flow rate the gas intrudes one whirling filament after another. It is therefore possible, that only a few of the whirling filaments behind the blades are filled with gas. When the gas flow rate is still further increased, vortex areas are generated as indicated in Fig. 19.43 for a one-blade and in Fig. 19.44 for a six-blade turbine. In general, the bubble diameter increases and the diameter distribution becomes broader when the gas flow rate leads to the formation of area sites.

When the gas flow rate is further increased a large gas-filled space builds up behind the blades. For high viscosity fluids this already happens at relatively low gas flow rates. In this situation rather large bubbles are generated.

The final stage of gas dispersion is reached when flooding occurs. In this case the gas reservoir extends to the edge of the disk of the turbine stirrer. The reservoir overflows by discharging very big bubbles,

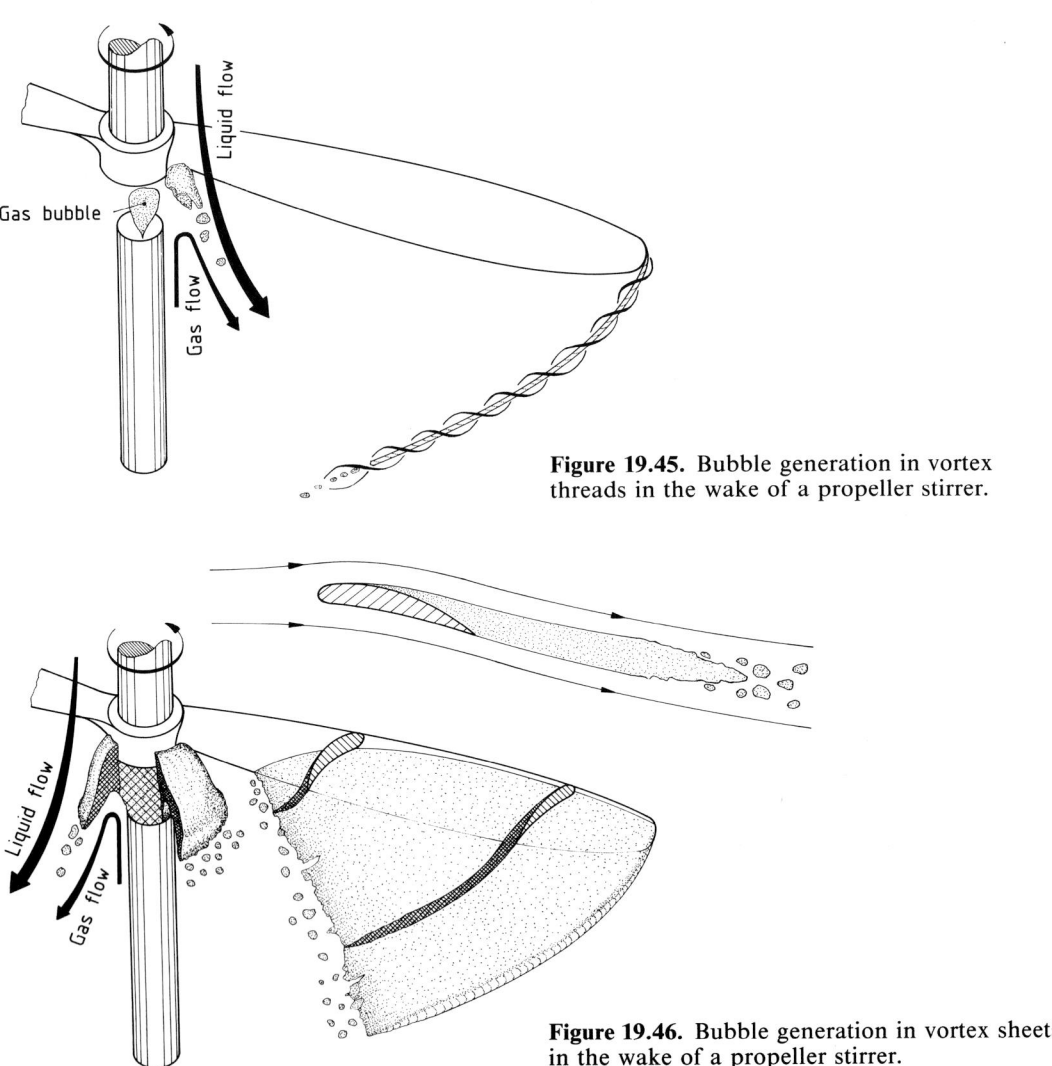

Figure 19.45. Bubble generation in vortex threads in the wake of a propeller stirrer.

Figure 19.46. Bubble generation in vortex sheets in the wake of a propeller stirrer.

which, due to the buoyancy forces, rise directly to the surface of the liquid without intruding the whirling filaments. Flooding should be avoided by all means, because the interfacial area is quite remarkably reduced.

Turbine stirrers without a central disk are not well suited for gas dispersion, as a gas reservoir cannot be built up properly.

Gas dispersion by means of propeller stirrers is shown in Figs. 19.45 and 19.46. The propeller must produce a liquid flow that is directed downwards to the bottom of the vessel. This fluid flow prevents the big bubbles emerging from the gas tube to rise in the center of the vessel. The liquid flow breaks up the big gas bubbles. The smaller bubbles are generated in this manner carried away by the liquid flow and finally enter the whirling filaments or vortex areas.

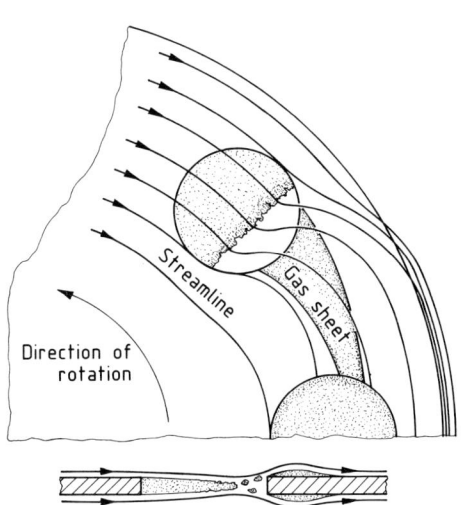

Figure 19.47. Bubble generation in gas sheets of a disk stirrer.

Flooding occurs when the gas flow rate and the initial bubbles are so great that the liquid flow can not carry the bubbles away.

The holed disk stirrer, shown in Fig. 19.47, is an excellent dispersion element. The energy transferred to the liquid is extremely small. Excellent conditions are offered for the formation of a gas reservoir. Reduced pressure regions are built up within the holes and in the wake of the holes. In both cases very thin gas sheets come into existence from which shear forces generate very small bubbles. Flooding occurs when the gas reservoir extends over the outer edge of the disk.

With other rotating gas dispersion elements the process of bubble generation is more or less the same.

19.7.3 Bubble Movement in a Stirred Vessel

Very small bubbles follow the primary and secondary movement of the liquid. Big bubbles, however, will determine liquid movement in the vessel. This is the case when flooding occurs [19.38]. Movement of small bubbles is indicated in Fig. 19.48, while Fig. 19.49 shows bubble movement for flooding conditions.

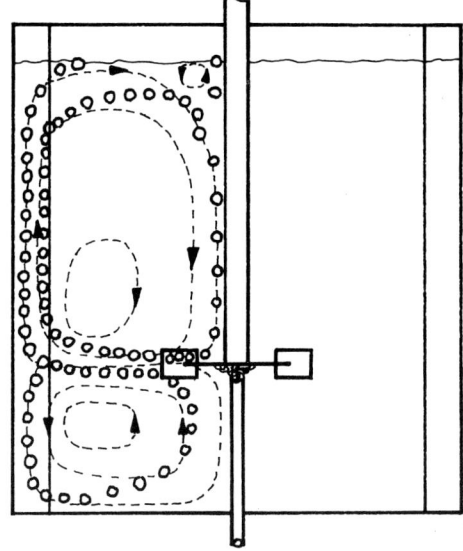

Figure 19.48. Movement of gas bubbles in a vessel at low gas flow rates.

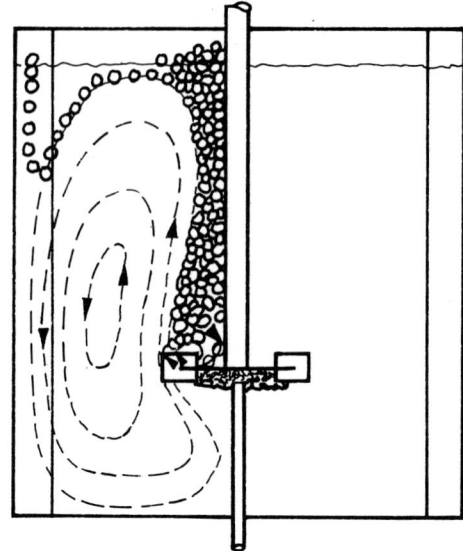

Figure 19.49. Movement of gas bubbles in a vessel at high gas flow rates.

The general relation for the gas content V_g/V_f is given by:

$$V_g/V_f = f(\dot{V}_g^*; Re; Fr; We; \text{geometrical parameters}) \quad (19.89)$$

with

$$\dot{V}_g^* \equiv \dot{V}_g/(n d_r^3), \quad \text{gas flow rate number} \quad (19.90)$$

$$Fr \equiv \frac{n^2 d_r}{g}, \quad \text{Froude number} \quad (19.91)$$

$$We \equiv \frac{n^2 d_r^3 \varrho}{\sigma}. \quad \text{Weber number} \quad (19.92)$$

$\dot{V}_g$ is the flow rate, g the gravitational acceleration, and σ the surface tension. The Reynolds number Re has been defined by Eq. (19.8). Data will be presented for vessels with turbine, propellers, and holed disk stirrers, so that the geometrical parameters

With normal operational conditions prevailing in a fermenter the bubble movement is still determined by the liquid movement. Primary and secondary movement are of equal importance. The rotational primary movement will carry the bubbles along and increase the residence time of the bubbles in the vessel. This increases the gas hold-up which again increases the turbulence in the liquid. A high degree of turbulence will increase the mass transfer in the liquid phase.

19.7.4 Gas Content and Flooding

For presentation of gas content data in a stirred vessel, use is made of the dimensionless ratio V_g/V_f, with V_f as the volume of the liquid and V_g the volume of the gas in the gas/liquid volume. The gas is in general introduced into the vessel at the bottom side of the stirrer.

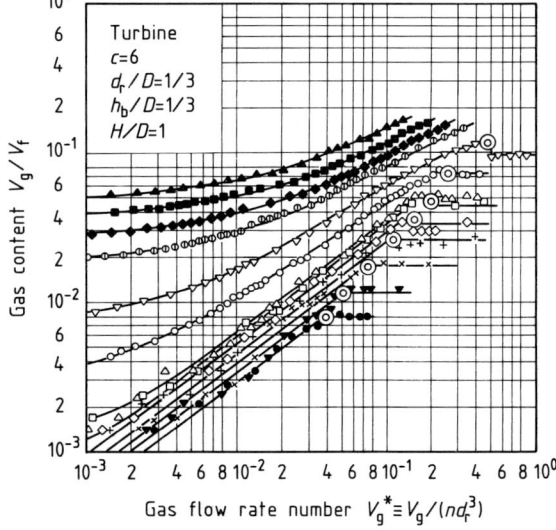

Figure 19.50. Gas content V_g/V_f over the gas flow rate number $\dot{V}_g^*$ and various values of the Froude or Reynolds number for a baffled vessel with a turbine stirrer.

have been held constant for each set of data presented.

In Fig. 19.50 some of the available data on the gas content in vessels with turbine stirrers are presented as an example. The majority of data indicates that the gas content in a stirred vessel is of the order of only a few percent or even less. This is quite typical and does not vary appreciably with type of stirrer and liquid.

The gas content increases with gas flow rate number $\dot{V}_g^*$ and Reynolds or Froude number. For Reynolds numbers below $Re = 5.94 \cdot 10^4$ the gas content varies according to straight parallel lines drawn through the data. For values of Re beyond $5.94 \cdot 10^4$ the gas content approaches a constant value when the gas flow rate number approaches $\dot{V}_g^* = 0$. This behavior discloses the two mechanisms by which gas is introduced into the liquid. The gas flow rate $\dot{V}_g$, that is, the gas that is forced into the liquid, is responsible for the so-called volumetric aeration. Another part of the gas content is due to surface aeration, that is, a gas stream that enters the liquid through the free surface. In general, this is achieved by curl-like deformations of the free surface which, when extending to the stirrer plane, supply this region of low pressure with the gas to be dispersed. The effectiveness of such a whirlpool, with respect to gas supply to the liquid, increases with rotational motion of the liquid mass. Surface aeration is therefore proportional to the Reynolds number Re, as clearly shown in Fig. 19.50.

For each value of Re the gas content reaches an upper limiting value at a certain gas flow rate, $\dot{V}_{max}^*$. Flooding occurs when $\dot{V}^* > \dot{V}_{max}^*$. For gas dispersion in vessels by means of Rushton turbines BIESECKER [19.38] derived the following equation:

$$\dot{V}_{max}^* = Fr \frac{d_r/D}{H/D - h_b/D}. \qquad (19.93)$$

The geometric ratios given in Fig. 19.50 lead to $\dot{V}_{max}^* = Fr/2$. Similar results have been obtained by HENZLER [19.42] and ZLOKARNIK and JUDAT [19.43].

Further results for gas content in vessels with propeller and holed disk stirrers are shown in Figs. 19.51 and 19.52. The curves for gas dispersion with propellers show two distinct breaks. At the first break, indicated by arrows, it occurs for the first time, that a small amount of gas is not properly dispersed. A small fraction of the introduced gas, contained in large bubbles, rises close to the rotating axis directly to the free surface of the liquid. This amount of non-dispersed gas increases until the parameter $(\dot{V}_g^* Re)$ reaches the value $(\dot{V}_{g,max}^* Re)$, which is indicated by a double circle. Fig. 19.51 shows the wide range in which flooding gradually develops. The following equa-

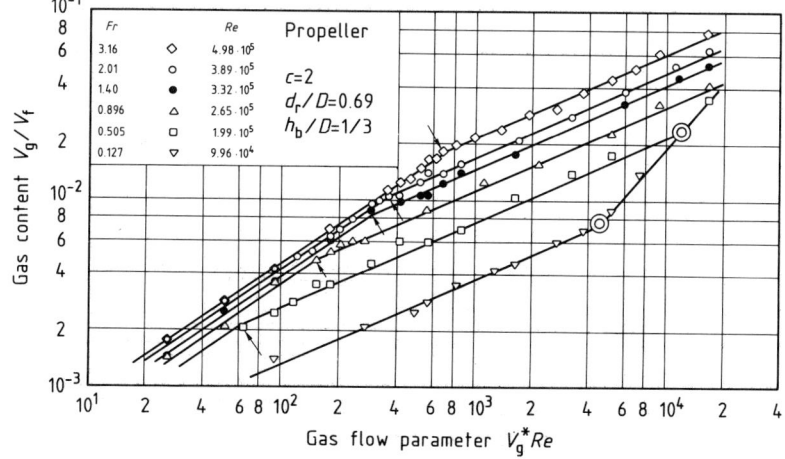

Figure 19.51. Gas content V_g/V_f over the gas flow parameter $\dot{V}_g^* Re$ and various values of the Froude or Reynolds number for a baffled vessel equipped with a propeller stirrer.

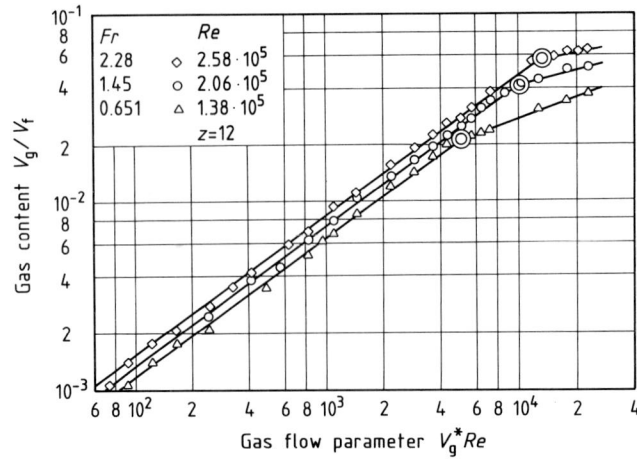

Figure 19.52. Gas content V_g/V_f over the gas flow parameter $\dot{V}_g^* Re$ and various values of the Froude or Reynolds number for a baffled vessel equipped with a holed disk stirrer.

tions have been set up [19.31]:

$$\dot{V}_{g,\max}^* = 4.45 \cdot 10^{-2} c^{0.23} (d_r/D)^{0.28} \quad (19.94)$$
$$\text{for } d_r/D \leq 0.5$$

$$\dot{V}_{g,\max}^* = 4.63 \cdot 10^{-3} c^{0.23} (d_r/D)^{-3} \quad (19.95)$$
$$\text{for } d_r/D \geq 0.5$$

The maximum for the gas flow parameter for holed disks is indicated in Fig. 19.52 by double circles. From all the available data the following correlation has been obtained [19.31]:

$$\dot{V}_{g,\max}^* = 47.2 \frac{Re^{(0.0246 z^{1.66} - 1)}}{e^{(0.0695 z^2 + 0.813 z - 6.405)}} \quad (19.96)$$

It should be observed that, although propeller and holed disk stirrers transfer only a fraction of the energy transferred by the Rushton turbine, the gas content is in all of these cases of the same order of magnitude.

19.7.5 Energy Transfer in the Dispersion Process

Energy transfer in aerated systems is an extremely complicated physical process and has not yet been sufficiently analyzed. There are no equations of general impor-

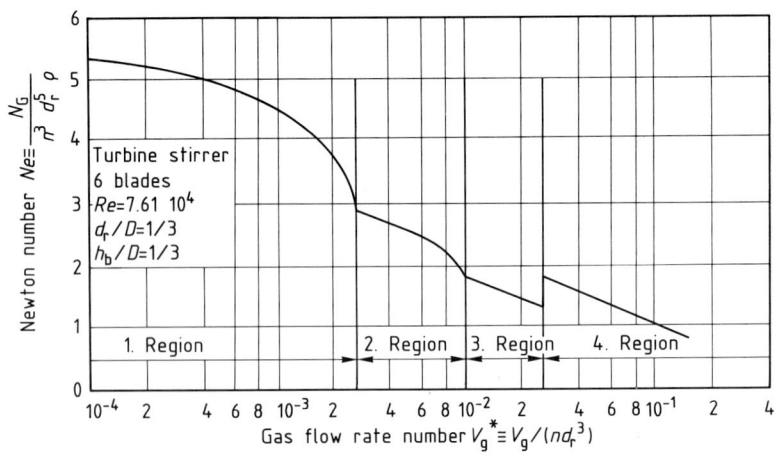

Figure 19.53. Newton number Ne for a six-bladed turbine stirrer in a baffled vessel as a function of the gas flow rate number $\dot{V}_g^*$.

tance available for practical application. However, the information which has been gathered by excellent experimental investigation by BIESECKER [19.38], VAN'T RIET [19.39] and GLAESER et al. [19.31] may serve to understand some of the phenomena involved.

Fig. 19.53 gives the Newton number of a 6-bladed turbine stirrer in a baffled vessel under conditions of aeration as a function of the gas flow rate number $\dot{V}_g^*$. With increasing gas flow rate, while all other parameters are kept constant, the energy transfer from the stirrer to the two-phase fluid decreases, because an increasing portion of the stirrer surface comes in contact with the less viscous gas. There are four distinct regions to be observed.

In the 1. region at very low gas flow rates the gas collects beneath the stirrer. A gas reservoir is built up from which single gas bubbles are torn away, due to liquid movement, in direction of the periphery of the stirrer by the action of centrifugal forces. Observations revealed the fact, that gas dispersion starts in the wake of one blade. An increased gas flow rate leads to gas dispersion in the wake of two blades. At the end of the 1. region all blades take part in the gas dispersion process. In Sect. 19.7.2 it already has been explained that gas dispersion starts in the vortex threads.

In the 2. region further increase of the gas flow rate leads to the formation of vortex sheets. The procedure of sheet formation is the same as that for vortex thread

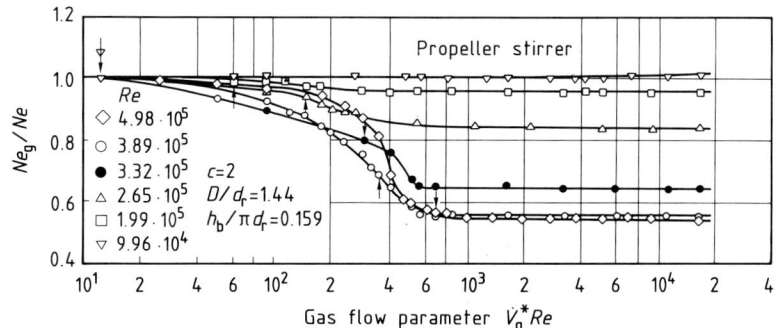

Figure 19.54. Ratio of Newton number Ne_g/Ne for a propeller stirrer in a baffled vessel over the gas flow parameter $\dot{V}_g^* Re$ for various values of Re.

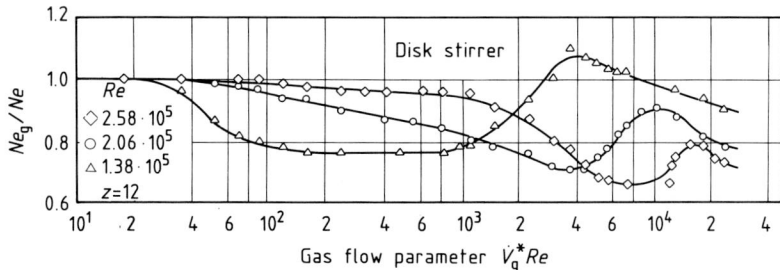

Figure 19.55. Ratio of Newton number Ne_g/Ne for a holed disk stirrer in a baffled vessel over the gas flow parameter $\dot{V}_g^* Re$ for various values of Re.

formation in the 1. region. At the end of the 2. region vortex sheets have been built up behind all blades.

In the 3. region there are stable vortex sheets behind all blades of the stirrer region. This is the region with highest gas dispersion efficiency. The energy transfer to the two-phase fluid has been reduced to about 1/3.

In the 4. region the increased gas flow rate leads to flooding conditions for the stirrer. This is characterized by the formation of very big bubbles. The gas dispersion process becomes ineffective.

Similar observations have been made when other types of stirrers are used as dispersion elements. Some further examples are given in Figs. 19.54 and 19.55 for propeller and holed disk stirrers. In these figures the ratio of Newton numbers Ne_g/Ne is given as a function of the gas flow parameter ($\dot{V}_g^* Re$). Ne_g is the Newton number for the aerated and Ne for the non-aerated system.

On the basis of available evidence it seems safe to assume that energy transfer is reduced to:

60% for turbine stirrers,
60% for paddle stirrers,
70% for propeller stirrers,
90% for disk stirrers.

It appears justified to accept these figures not only for dispersion processes in Newtonian fluids but also in non-Newtonian fluids. Further investigations are necessary, especially with non-Newtonian fluids [19.34]. But, expectations should not be too high. Energy transfer in aerated non-Newtonian systems is one of the most difficult problems in biochemical engineering.

19.8 Mass Transfer in Stirred Vessels

Mass transfer in microbial systems may in many cases be considered as a stationary process, mostly by absorption but in certain cases also by desorption. In aerobic processes the oxygen must be transferred from the gas bubbles to the surrounding liquid. But the stationary process for the liquid is a non-stationary process for the dispersed gas phase. It is the stationary process of mass transfer in the liquid which will be considered.

Reactions of microorganisms may be interpreted as homogeneous reactions. In this case physical mass transfer from a bubble to the microorganisms is enhanced, depending on the reaction velocity. The enhancement increases with the reaction velocity, which in general, however, is unknown. This is the reason, why mass transfer in fermentation fluids is not yet well understood. Equations, presented so far, do not take reactions of microorganisms into account, and consequently deliver the minimum values for the mass transfer coefficients.

If A is the transferred component, the mass transfer rate $\dot{M}_A$ is given by the general equation:

$$\dot{M}_A = \beta A_p \Delta \varrho_A, \qquad (19.97)$$

with β as a mean mass transfer coefficient, A_p the interfacial area in the volume V of the vessel, and $\Delta \varrho_A$ the logarithmic mean value of the concentration difference:

$$\Delta \varrho_A = \frac{(\varrho_{Ap1} - \varrho_{A1}) - (\varrho_{Ap2} - \varrho_{A2})}{\ln \dfrac{\varrho_{Ap1} - \varrho_{A1}}{\varrho_{Ap2} - \varrho_{A2}}}, \qquad (19.98)$$

with ϱ_A being the partial density of component A at some distance from the interface and ϱ_{Ap} the partial density of the liquid side of the interface, which may be assumed to be the saturation value. Index 1 denotes the partial densities in the stirrer plane and index 2 in the free surface of the fermentation fluid. Because of the mixing effect of the stirrer, the partial densities ϱ_{A1} and ϱ_{A2} generally will have the same value.

It has become a widely accepted practice to devide the mass transfer rate $\dot{M}_A$ by the volume V of the reactor, so that Eq. (19.97) may be rewritten as:

$$\frac{\dot{M}_A}{V} = \beta a_p \Delta \varrho_A , \quad (19.99)$$

with

$$a_p = A_p/V \quad (19.100)$$

as the specific interfacial area. This procedure tacitly implies the assumption, that the transfer rate $\dot{M}_A$ is a linear function of the reaction volume V. This, however, is not quite correct.

Application of Eq. (19.99) for design work or process evaluation requires detailed information on the mass transfer coefficient β and the specific area a_p. Due to experimental difficulties it is not yet possible to determine β and a_p separately with desired accuracy. Therefore, the product βa_p is determined, for which dimensionless correlations have been presented by various authors [19.34], [19.41], [19.46].

Based on a large amount of data for mass transfer across the interface of a gas and a Newtonian liquid published by various authors [19.47] to [19.55], JUDAT [19.46] presented the following correlation:

$$(\beta a_p)^* = 9.8 \cdot 10^{-5} (N/V)^{*0.40} / \\ /(B^{-0.6} + 0.81 \cdot 10^{-0.65/B}) . \quad (19.101)$$

The dimensionless parameters used in this equation are defined as follows:

$$(\beta a_p)^* \equiv \beta a_p (v/g^2)^{1/3} \quad (19.102)$$

mass transfer parameter,

$$(N/V)^* \equiv (N/V)/[\varrho(vg^4)^{1/3}] \quad (19.103)$$

power parameter,

$$B = (\dot{V}_g/D^2)/(vg)^{1/3} \quad (19.104)$$

gasification parameter.

In these parameters $v = \eta/\varrho$ is the kinematic viscosity with η as the dynamic viscosity and ϱ the density of the liquid, g acceleration due to gravity, N energy transferred per time unit from the stirrer to the fluid, V_g gas flow rate, and D vessel diameter. The maximum deviation of the experimental data

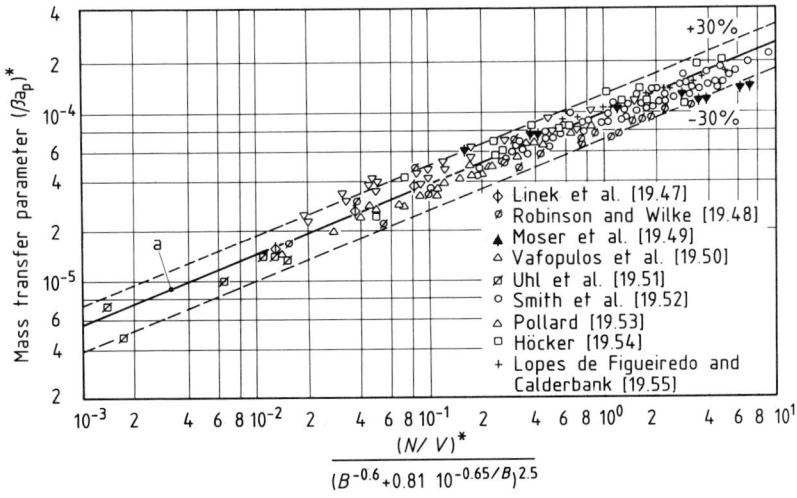

Figure 19.56. Mass transfer in stirred vessels; comparison of available data with Eq. (19.101) represented by curve a.

from Eq. (19.101) is ±30%. A comparison between Eq. (19.101) and the available data is shown in Fig. 19.56.

According to investigations by HENZLER [19.34] it should be justifiable to apply Eq. (19.101) also to mass transfer in non-Newtonian liquids, when the characteristic viscosity is determined according to the procedure outlined in Sect. 19.5.4.4. But caution is advisable in any case.

19.9 References

[19.1] K. KIPKE: "Offene Probleme in der Rührtechnik". Chem. Ing. Tech. 54 (1982), 5, 416–425.

[19.2] H.-J. HENZLER: "Verfahrenstechnische Auslegungsunterlagen für Rührbehälter als Fermenter". Chem. Ing. Tech. 54 (1982) 5, 461–467.

[19.3] N. OSTWALD: "Über die Geschwindigkeitsfunktion der Viskosität disperser Systeme". Kolloid-Z. 36 (1925) 2, 99–117; 157–167; 248–250.

[19.4] A. DE WAELE: "Die Änderung der Viskosität mit der Schergeschwindigkeit disperser Systeme". Kolloid-Z. 36 (1925), 332–333.

[19.5] H. BRAUER: "Grundlagen der Einphasen- und Mehrphasenströmungen". Verlag Sauerländer, Aarau-Frankfurt am Main, 1971.

[19.6] F. R. EIRICH: "Rheology, Theory and Applications", Vol. 1 (1956), Vol. 2 (1958), Vol. 3 (1960). Academic Press, New York.

[19.7] A. G. FREDRICKSON: "Principles and Applications of Rheology". Prentice Hall, Englewood Cliffs (N.Y.), 1964.

[19.8] A. B. METZNER: "Non-Newtonian technology". In "Advances in Chemical Engineering", Vol. 1, pp. 77–153. Academic Press, New York, 1956.

[19.9] A. ASTARITA and G. MARUCCI: "Principles of Non-Newtonian Fluid Mechanics". McGraw-Hill (UK), London, 1974.

[19.10] D. EBERT: "Strömung nicht-Newtonscher Medien". Vieweg u. Sohn, Braunschweig-Wiesbaden, 1980.

[19.11] G. BÖHME: "Strömungsmechanik nicht-Newtonscher Fluide"; Teubner Studienbücher Mechanik. Teubner, Stuttgart, 1981.

[19.12] H. TAGUCHI: "The nature of fermentation fluids". In "Advances in Biochemical Engineering", Vol. 1, pp. 1–30. Springer Verlag, New York – Heidelberg – Berlin, 1971.

[19.13] H. BRAUER: "Power consumption in aerated stirred tank reactor systems". In "Advances in Biochemical Engineering", Vol. 13, pp. 87–119. Springer Verlag, New York-Heidelberg-Berlin.

[19.14] H. THOMAE: "Das turbulente Strömungsfeld in einem Rührkessel". Chem. Ing. Tech. 42 (1970) 5, 317–322.

[19.15] A. MERSMANN, W.-D. EINENKEL, and M. KÄPPEL: "Auslegung und Maßstabsvergrößerung von Rührapparaten". Chem. Ing. Tech. 47 (1975) 23, 953–964.

[19.16] V. NOVÁK and F. RIEGER: "Homogenization efficiency of helical ribbon and anchor agitators". Chem. Eng. J. 9 (1975) 1, 63–70.

[19.17] V. NOVÁK and F. RIEGER: "Power consumption scale-up in agitating non-Newtonian fluids". Chem. Eng. Sci. 29 (1974) 11, 2229–2234.

[19.18] J. THYN, V. NOVÁK, and P. POCK: "Effect of the measured volume size on the homogenization time". Chem. Eng. J. 12 (1976), 211–217.

[19.19] V. V. CHAVAN and J. ULBRECHT: "Efficiency of a screw mixer". Chem. Ing. Tech. 46 (1974) 1, 35 (Synopsis).

[19.20] D. D. KALE, R. A. MASHELKAR, and J. ULBRECHT: "High speed agitation of non-Newtonian fluids: influence of elasticity and fluid inertia". Chem. Ing. Tech. 46 (1974) 2, 69 (Synopsis).

[19.21] H. THIELE: "Strömung und Leistungsbedarf beim Rühren Newtonscher Flüssigkeiten mit Anker-, Blatt- und Turbinenrührern im laminaren Bereich". Dissertation, Technische Universität Berlin, 1972.

[19.22] F. IHME: "Leistungsbedarf und Verformung der Flüssigkeitsoberfläche beim Rühren Newtonscher Flüssigkeit mit Turbinen-, Blatt- und Ankerrührern". Fortschr.-Ber. VDI-Z. Reihe 7, Nr. 44.

[19.23] H. BRAUER: "Ansatz zur theoretischen Berechnung des Leistungsbedarfs und des Wärmeübergangs beim Rühren". Chem. Ing. Tech. *39* (1967) 5/6, 209–217.

[19.24] D. SCHILO: "Leistungsbedarf beim Rühren nicht-Newtonscher Flüssigkeiten". Dissertation, Technische Universität Berlin, 1968.

[19.25] S. NAGATA, K. YAMAMOTO, K. HASCHIMOTO, and Y. NARUSE: "Studies on the flow patterns of liquids in a cylindrical mixing vessel, over a wide range of Reynolds number". Mem. Fac. Eng. Kyoto Univ. *22* (1960), 68–85.

[19.26] H. BRAUER: "Development and performance of a stagnation jet mixer". Ger. Chem. Eng. *4* (1981), 144–154.

[19.27] H. BRAUER: "Wärmetransport in Rührgefäßen". Waerme Stoffuebertrag. *13* (1980), 105–113.

[19.28] W. BÜCHE: "Leistungsbedarf von Rührwerken". VDI-Z. *81* (1937) 37, 1065–1069.

[19.29] H. BRAUER and D. MEWES: "Einfluß von Strombrechern auf die Rührerleistung". Chem. Ing. Tech. *45* (1973) 7, 461–467.

[19.30] H. BRAUER and H. THIELE: "Grundlagen für die Dimensionierung von Rührern und Rührgefäßen". DECHEMA Monographien Bd. 66, pp. 69–98, Verlag Chemie, Weinheim, 1971.

[19.31] H. GLAESER, B. BIESECKER, and H. Brauer: "Begasung von Flüssigkeiten mit Propeller- und Lochscheibenrührern". Verfahrenstechnik *7* (1973) 2, 31–40.

[19.32] A. B. METZNER and R. E. OTTO: "Agitation of non-Newtonian fluids". A.I.Ch.E. Journal *3* (1957) 1, 3–10.

[19.33] K. KIPKE and E. TODTENHAUPT: "Rühren von nicht-Newtonschen Flüssigkeiten". Verfahrenstechnik *16* (1982) 6, 497–503.

[19.34] H.-J. HENZLER: "Begasen höherviskoser Flüssigkeiten". Chem. Ing. Tech. *52* (1980) 8, 643–652.

[19.35] P. H. CALDERBANK and M. B. MOO-YOUNG: "The power characteristics of agitation for the mixing of Newtonian and non-Newtonian fluids". Trans. Inst. Chem. Eng. *39* (1961), 337–347.

[19.36] H. BRAUER and H. THIELE: "Leistungsbedarf und Wärmeübertragung beim Rühren im laminaren Strömungsbereich". Verfahrenstechnik *5* (1971), 420–428, 448–452.

[19.37] R. POGGEMANN, A. STRIFF, and P.-M. WEINSPACH: "Flüssigkeiten". Chem. Ing. Tech. *51* (1979) 10, 948–959.

[19.38] B. O. BIESECKER: "Begasen von Flüssigkeiten mit Rührern". VDI-Forsch.-Hefte *554*. VDI-Verlag, Düsseldorf, 1972.

[19.39] K. VAN'T RIET: "Turbine Agitator Hydrodynamics and Dispersion Performance". Dissertation, Technische Hogeschool Delft, 1975.

[19.40] H. JUDAT: "Zum Dispergieren von Gasen". Dissertation, Universität Dortmund, 1976.

[19.41] H. JUDAT: "Begasen niedrigviskoser Flüssigkeiten". Chem. Ing. Tech. *51* (1979) 7, 710–716.

[19.42] H.-J. HENZLER: "Begasen höherviskoser Flüssigkeiten". Chem. Ing. Tech. *52* (1980) 8, 643–652.

[19.43] M. ZLOKARNIK and H. JUDAT: "Rohr- und Scheibenrührer – zwei leistungsfähige Rührer zur Flüssigkeitsbegasung". Chem. Ing. Tech. *39* (1967) 20, 1163–1168.

[19.44] M. ZLOKARNIK: "Scale-up of surface aerators for waste water treatment". In "Advances in Biochemical Engineering", Vol 11, pp. 157–180. Springer Verlag, New York–Heidelberg–Berlin, 1979.

[19.45] H. HÖCKER and G. LANGER: "Zum Leistungsverhalten begaster Rührer in Newtonschen und nicht-Newtonschen Flüssigkeiten". Rheol. Acta *16* (1977), 400–412.

[19.46] H. JUDAT: "Stoffaustausch gas/flüssig im Rührkessel – eine kritische Bestandsaufnahme". Chem. Ing. Tech. *54* (1982) 5, 520–521 (Synopsis).

[19.47] V. LINEK, J. MAYRHOFEROVÁ, and J. MOSNEROVÁ: "The influence of diffusivity on liquid phase mass transfer in solutions of electrolytes". Chem. Eng. Sci. *25* (1970), 1033–1045.

[19.48] C. W. ROBINSON and C. R. WILKE: "Oxygen absorption in stirred tanks: A correlation for ionic strength effects". Biotech. Bioeng. *15* (1973), 755–782.

[19.49] A. MOSER, V. EDLINGER, and F. MOSER: "Sauerstofftransport und Austauschflächen in Öl-Wasser-Dispersionen". Verfahrenstechnik *9* (1975) 11, 553–565.

[19.50] J. VAFOPULOS, K. SZTATESCNY, and F. MOSER: "Der Einfluß des Partial- und Gesamtdruckes auf den Stoffaustausch". Chem. Ing. Tech. *47* (1975) 16, 681 (Synopsis).

[19.51] V. W. UHL, R. L. WINTER, and E. L. HEIMARK: "Mass transfer in large secondary treatment aerators". AIChE-Symposium Ser. *73* (1977) 167, 33–41.

[19.52] J. M. SMITH, K. VAN'T RIET, and J. C. MIDDLETOOS: Scale-up of agitated gas-liquid reactors for mass transfer". Proc. Second Europ. Conf. Mixing, Cambridge 1977, F4-51/66.

[19.53] G. J. POLLARD: "Flooding and aeration efficiency in standard stirred vessels". Proc. Intern. Symp. Mixing, Mons 1978, C4-1/C4-16.

[19.54] H. HÖCKER: "Untersuchung zum Leistungsbedarf und Stoffübergang in Rührreaktoren". Dissertation, Universität Dortmund, 1979.

[19.55] M. H. LOPES DE FIGUEIREDO and P. H. CALDERBANK: "The scale-up of aerated mixing vessels for specified oxygen dissolution rates". Chem. Eng. Sci. *34* (1979), 1333–1338.

Chapter 20
Bubble Column Reactors

Wolf-Dieter Deckwer

Fachbereich Chemie, Fachgebiet Technische Chemie
Universität Oldenburg
Oldenburg, Federal Republic of Germany

20.1 Introduction
20.2 Hydrodynamics
20.2.1 General Remarks
20.2.2 Flow Regimes
20.2.3 Pressure Drop
20.2.4 Energy Input
20.2.5 Gas Holdup
20.2.6 Bubble Properties and Bubble Swarm Dynamics
20.2.7 Mixing
20.2.7.1 Liquid Phase Dispersion
20.2.7.2 Gas Phase Dispersion
20.3 Mass Transfer
20.3.1 General Considerations
20.3.2 Liquid-Solid Mass Transfer
20.3.3 Gas-Liquid Mass Transfer
20.3.3.1 Liquid Side Mass Transfer Coefficient
20.3.3.2 Volumetric Mass Transfer Coefficient
20.4 Heat Transfer
20.5 Models of Bubble Columns
20.6 References

List of Symbols

Symbol	Units	Description
a	m^{-1}	specific interfacial area, defined by Eq. (20.12)
c	$kg\ m^{-3}$	concentration
c_o	$kg\ m^{-3}$	concentration of tracer when distributed over entire reactor volume
c_p	$m^2\ K^{-1}\ s^{-2}$	specific heat capacity
Δc	$kg\ m^{-3}$	driving concentration difference
d_B	m	mean bubble diameter
d_p	m	particle diameter
d_S	m	volume-to-surface (Sauter) mean diameter
D	$m^2\ s^{-1}$	diffusivity
D_{eff}	$m^2\ s^{-1}$	effective diffusivity
D_G	$m^2\ s^{-1}$	gas phase dispersion coefficient
D_L	$m^2\ s^{-1}$	liquid phase dispersion coefficient
D_N	m	nozzle or hole diameter
D_R	m	reactor diameter
D_S	$m^2\ s^{-1}$	solid (biomass) dispersion coefficient
g	$m\ s^{-2}$	acceleration due to gravity
G	$kg\ s^{-1}$	oxygen mass flow rate, Eq. (20.28)
h	$kg\ K^{-1}\ s^{-3}$	heat transfer coefficient
k	$kg\ s^{n-2}\ m^{-1}$	fluid consistency index, Eq. (20.39)
k	$kg\ m\ K^{-1}\ s^{-3}$	thermal conductivity
k_L	$m\ s^{-1}$	liquid side mass transfer coefficient at gas/liquid interface
k_s	$m\ s^{-1}$	liquid/solid mass transfer coefficient
L	m	reactor length
M_R	kg	mass present in reactor
Δp	$kg\ m^{-1}\ s^{-2}$	total pressure drop
Δp_L	$kg\ m^{-1}\ s^{-2}$	pressure drop due to hydrostatic head
Δp_S	$kg\ m^{-1}\ s^{-2}$	pressure drop caused by gas sparger
P	$kg\ m^{-1}\ s^{-2}$	pressure
P_T	$kg\ m^{-1}\ s^{-2}$	pressure at reactor top
P_G	$kg\ m^2\ s^{-3}$	power input by gas phase, Eq. (20.4)
P_L	$kg\ m^2\ s^{-3}$	power input by liquid phase, Eq. (20.5)
q	$m^3\ s^{-1}$	gas throughput
R	$m^2\ s^{-2}\ K^{-1}$	specific gas constant
R	m	reactor radius
r	m	radial coordinate
T	K	temperature
$u_{B\infty}$	$m\ s^{-1}$	rise velocity of single bubbles
U_{BS}	$m\ s^{-1}$	bubble slip velocity, Eq. (20.14)
u_C	$m\ s^{-1}$	liquid circulation velocity, Eq. (20.17)
u_G	$m\ s^{-1}$	linear gas velocity
u_{Go}	$m\ s^{-1}$	gas velocity in sparger holes
$\bar{u}_G$	$m\ s^{-1}$	mean linear gas velocity
u_G^*	$m\ s^{-1}$	bubble rise velocity in swarm
u_L	$m\ s^{-1}$	linear liquid velocity
u_L'	$m\ s^{-1}$	interstitial liquid velocity
u_S	$m\ s^{-1}$	settling velocity of particles
$\dot{V}_G$	$m^3\ s^{-1}$	gas throughput
$\dot{V}_{LN}$	$m^3\ s^{-1}$	liquid flow rate through nozzle
x	m	axial coordinate
$\dot{\gamma}$	s^{-1}	shear rate
δ	m	distance between tracer injection and measuring point
ε	$m^2\ s^{-3}$	specific energy dissipation rate
ε_G		relative gas holdup
$\bar{\varepsilon}_G$		mean gas holdup
μ	$kg\ m^{-1}\ s^{-1}$	dynamic viscosity
μ_{eff}	$kg\ m^{-1}\ s^{-1}$	kinematic viscosity
ν_{eff}	$m^2\ s^{-1}$	effective kinematic viscosity
ν_t	$m^2\ s^{-1}$	turbulent kinematic viscosity
ϱ	$kg\ m^{-3}$	density
σ	$kg\ s^{-2}$	surface tension
τ	s	mixing time
Bo		Bond number, Eq. (20.34)
De		Deborah number, Eq. (20.37)
Fr		Froude number, Eq. (20.36)
Ga		Galilei number, Eq. (20.35)
Pe		Peclet number, Fig. 20.11
Sc		Schmidt number, Eq. (20.33)
Sh		Sherwood number, Fig. 20.11
X		Power number, Eq. (20.29)
Y		Sorption number, Eq. (20.28)
α		ratio of maximum hydrostatic head to pressure at top of reactor, Eq. (20.3)

Subscripts

B	biomass or bubble
G	gas
L	liquid
S	solid

20.1 Introduction

Bubble column reactors are vessels for contacting gas and liquid media. The liquid often contains suspended solids like solid reactants, catalysts, or biomass. Because of their simple construction, ease of maintenance, excellent mixing, and excellent heat and mass transfer properties bubble columns are widespread in the chemical industry. Of course in biotechnology, the stirred vessel is the most frequently employed type of reactor. However, other and novel reactors like bubble columns become increasingly important. Such reactors may provide more advantageous technical and economical conditions, and hence improve overall reactor performance. The main field of potential industrial applications of bubble columns and their various modifications are certainly large-scale bioprocesses carried out continuously. Examples include SCP production at ICI in Billingham (50000 t/a), the Hoechst-Uhde SCP demonstration plant (1000 t/a) and various large-scale units for waste water treatment. However, bubble columns are also under investigation for penicillin production and cultivation of animal cells.

In biotechnological applications, i. e., primarily aerobic fermentations, bubble columns are usually denoted as tower fermenters, tower bioreactors, tower loop reactors, etc. The basic construction of a bubble column in its simplest form is schematically shown in Fig. 20.1. It is characteristic that there are no moving parts, therefore, no sealing problems occur, which facilitates sterile operation of the unit. The gas (air) is distributed at the reactor bottom by means of various kinds of spargers, and rises upwards in form of bubbles concurrently or countercurrently to the liquid. As the bubbles, rising faster than the surrounding liquid, carry liquid upwards in their wake, a liquid circulation pattern is established which causes high mixing intensity in the liquid phase. The diameter of the cylindrical reactor is usually enlarged at the top to fa-

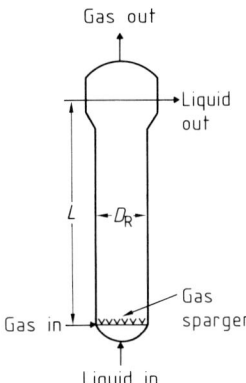

Figure 20.1. Conventional bubble column.

cilitate bubble disengagement and foam destruction.

With regard to their performance, bubble columns have a high degree of flexibility which can additionally be altered and matched to special tasks by installing various devices. Some of the more frequent modifications are shown in Fig. 20.2. For instance, the simple (single staged) bubble columns as shown in Fig. 20.1 can be modified by introducing trays and channels. Also packings can be inserted as, for example, motionless mixers which provide for high mass transfer rates. The liquid circulation flow pattern with stochastic and random-like character can be ordered and structured by insertion of a central tube of appropriate dimensions. Bubble columns with internal or external loops are used to recycle and uniformly distribute biomass over the reactor volume. Liquid flow recirculating by loops gives a more uniform liquid velocity profile (WEILAND and ONKEN [20.1]). An external loop can also be used for heat removal. Other devices for efficient heat transfer in bubble columns are summarized in Fig. 20.3.

The choice of the gas distribution system can have a great effect on the performance of bubble column type bioreactors. Even in tall reactors, where the bubbles have a long rise time, the mode of the gas distribution and the construction of the sparger influences the interfacial area and hence the oxygen mass transfer rate. Of course, this is

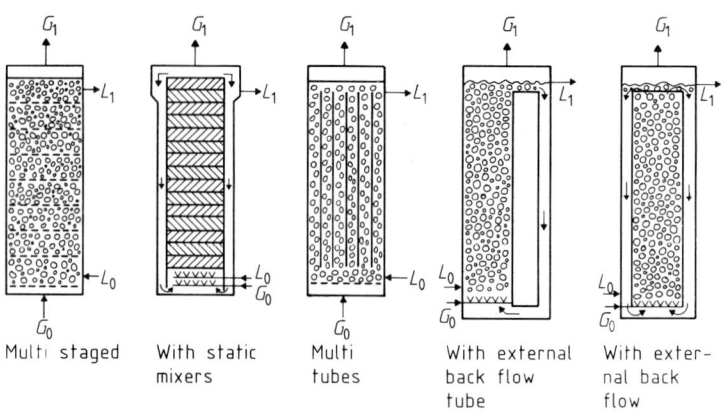

Figure 20.2. Modified bubble columns.

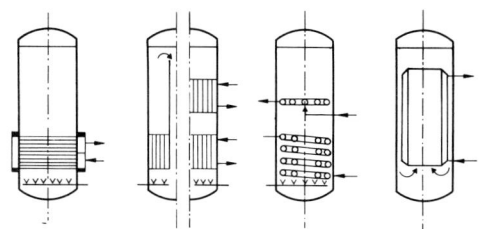

Figure 20.3. Arrangements for heat exchange.

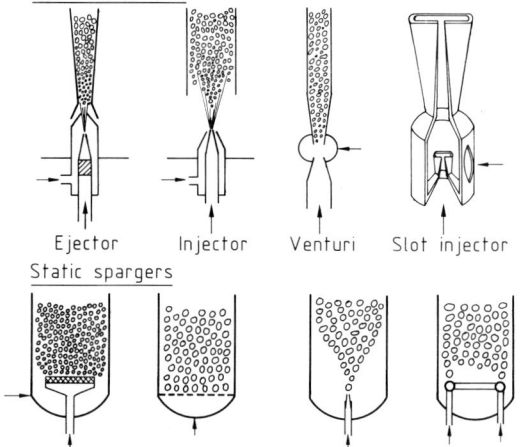

Figure 20.4. Static and dynamic gas spargers.

particularly important for fermentation media which hinder coalescence of bubbles. Typical spargers used in bubble columns and bioengineering applications are shown in Fig. 20.4. Spargers which disperse the gas flow by means of pores and holes such as sintered and perforated plates, one-hole nozzles and ring spargers are called static distributors while two-phase (gas-liquid) nozzles are dynamic spargers. Here, the gas is distributed by the kinetic energy of the liquid jet giving a fine and rather uniform bubble size distribution. Because of the danger of clogging sintered plates are not recommended for industrial use. However, promising results have been obtained with sintered plate spargers in several yeast and bacterial fermentations in lab-scale bubble columns. Like sintered plates, perforated rubber plates also lead to high interfacial areas, and it is claimed that due to their periodic deformations clogging does not occur.

20.2 Hydrodynamics

20.2.1 General Remarks

To characterize the multi-phase flow in bubble columns a number of macroscopic quantities can be used as, for instance, pressure drop, gas holdup, liquid velocity distribution, dispersion and mixing properties. The pressure drop corresponds to the hydrostatic head of the fluid medium in the reactor and is needed to calculate the energy input by the gas flow. Together with the bubble diameter the gas holdup decisively determines the gas-liquid interfacial area which is available for oxygen mass transfer. A further important point of information is the radial gas holdup distribution which can be used for calculating the liquid circulation velocity. Particularly in bioprocesses, knowledge of dispersion and mixing properties is important. Often, aeration can not be completely uniform over the entire reactor volume. Then mixing supplies the biomass with oxygen and other substrates.

For an understanding and description of the behavior of bubble columns it would be desirable to start from more fundamental phenomena, for instance, bubble formation, fluid flow around single rising bubbles, interaction of bubbles in swarms, etc. Such approaches have been successful only for few special situations. SCHÜGERL [20.2] has proposed various turbulence properties to characterize two-phase flow in biological systems. Knowledge of quantities such as intensity and scale of turbulence, autocorrelation functions, turbulence power, and energy spectra may improve understanding and provide for a better insight in such systems. But it is presently not well understood how the above quantities do influence reactor performance and how they can successfully be employed for estimations of macroscopic properties such as mixing times or mass transfer coefficients. Further fundamental work is needed to elucidate the connection between microscopic phenomena such as structure of turbulence and macroscopic design parameters.

In spite of the simple construction of bubble columns and their modifications, their design and scale-up do not present an easy task as SHAH and DECKWER [20.3] have pointed out. This fact has to be attributed to the complex fluid mechanic phenomena and the great sensitivity of the gas-in-liquid dispersion to liquid properties. Therefore, most correlations for bubble column design parameters are purely empirical. Hence, extrapolation to other operating conditions and application to other systems may be dangerous. This is also a particular difficulty when dealing with biological culture media. Here, trace impurities and small amounts of metabolites may exert a strong effect on gas holdup, foam formation, and mass transfer properties. Care and discretion is therefore always recommended when applying empirical correlations to such culture media.

20.2.2 Flow Regimes

At low gas velocities, i.e., $u_G \leq 5$ cm/s, the bubble sizes are rather uniform and their rise velocities are approximately equal. In this case, there is hardly any interaction between the bubbles in the swarm which rise upward in a uniform manner without significant coalescence. This flow regime is called bubbly or "pseudohomogeneous" flow. In this flow regime, which is most often desirable, only a certain maximum amount of gas can be transported through the column. The appropriate gas throughput can be estimated from flooding point calculations, the slip velocity concept, and drift flux theory. The amount of gas transportable in the bubbly flow regime can be considerably increased by increasing the concurrent liquid flow rate. Thus the gas holdup is decreased, so that bubbly flow is stabilized. In loop bubble columns with high liquid circulation bubbly flow can be maintained at gas velocities higher than 5 cm/s.

If the gas flow increases above the flooding point, larger bubbles of high buoyancy are formed and transition to heterogeneous or churn-turbulent flow takes place. As shown in Fig. 20.5, large bubbles move predominantly close to the center line in a churn-like rotating motion coexisting with small bubbles typically found in bubbly flow. The large rising bubbles cause circulation of liquid within the column. Therefore, the heterogeneous flow is also called

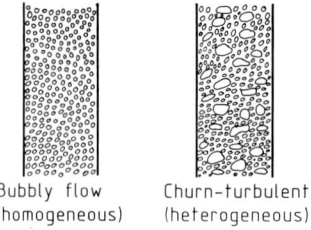

Figure 20.5. Major flow regimes in bubble columns.

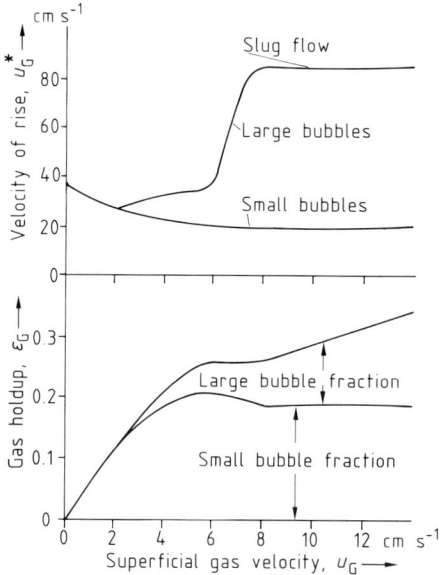

Figure 20.6. Fractional holdup and rise velocities of small and large bubbles.

recirculating flow regime. Fig. 20.6 shows holdup data of small and large bubbles and their rise velocities in a 10 cm diameter bubble column for the air-water system (BEINHAUER [20.4]). Transition from bubbly to heterogeneous flow can be recognized easily by the sharp increase of the bubble rise velocity (u_G^*), cf. Fig. 20.6 or a plot of $u_G^* = u_G/\varepsilon_G$ vs. u_G. Though the fraction of big bubbles is relatively small their rise velocities are high. Therefore, at high gas velocities, most of the gas is transported through the column by large bubbles and bubble clusters which are inefficient for mass transfer. Thus, the gas phase conversion and the oxygen transfer efficiency in reference to the power input decreases significantly.

A peculiar situation occurs in small diameter reactors in which large bubbles are stabilized to form slugs. In Fig. 20.6 a constant rise velocity u_G^* of about 80 cm/s at gas velocities above 8 cm/s is the result of slug formation. Slug flow only occurs in small diameter tower reactors and is not typical for bubble columns of industrial size. However, for highly viscous Newtonian and non-Newtonian culture media bubble slugs can be found even if the column diameter is as large as 30 cm.

The type of flow regime depends not only on the gas flow rate but also on such parameters as sparger design, physicochemical properties, and liquid velocity. Also the presence of suspended solids can influence the transition from one flow regime to the other. In highly viscous media large bubbles can be formed at gas velocities of considerably less than 5 cm/s. Porous spargers with mean pore sizes of less than about 200 μm commonly lead to bubbly flow-up to gas velocities of about 5 to 8 cm/s. On the other hand, if other static spargers like perforated plates or single and multi-hole distributors with openings larger than 1 mm in diameter are used, homogeneous flow can only be realized at very low gas velocities.

20.2.3 Pressure Drop

In vertically sparged reactors the pressure drop due to liquid inertia and wall friction can be neglected. The pressure drop is therefore composed of the drop exerted by the gas sparger (Δp_S) and the hydrostatic head of the liquid (Δp_L):

$$\Delta p = \Delta p_S + \Delta p_L. \tag{20.1}$$

Usually, $\Delta p_L \gg \Delta p_S$. The pressure profile along the column is advantageously used for gas holdup determinations and is given by:

$$P(z) = P_T[1 + \alpha(1-z)], \qquad (20.2)$$

where P_T is the pressure at the reactor top, z the dimensionless axial coordinate ($z \equiv x/L$), and α the ratio of maximum hydrostatic head to P_T:

$$\alpha = \frac{\varrho_L(1-\varepsilon_G)gL}{P_T}. \qquad (20.3)$$

Here, g is the gravitational constant, ε_G the gas holdup, ϱ_L the liquid density, and L the dispersion height.

20.2.4 Energy Input

The power input in a tower reactor is an important quantitiy for varying the operating conditions. In addition, the power input is frequently used to correlate hydrodynamic and mass transfer parameters as, for instance, oxygen transfer efficiencies. In the case of static spargers, the energy transferred into the reactor is the kinetic energy of the gas flow and the compression energy to overcome the pressure drop. Hence, the power input by the gas flow is given by:

$$P_G = \dot{V}_G \varrho_G [RT \ln(1+\alpha) + u_{Go}^2/2]. \qquad (20.4)$$

Here, $\dot{V}_G$ is the volumetric gas flow rate, ϱ_G gas density, R specific gas constant, T absolute temperature, and u_{Go} gas velocity in the sparger holes. α is given by Eq. (20.3). In the case of dynamic spargers, one also has to consider the power input by the liquid jet:

$$P_L = \frac{8}{\pi^2} \frac{\dot{V}_{LN}^3}{D_N^4} \varrho_L, \qquad (20.5)$$

where D_N is the nozzle diameter and $\dot{V}_{LN}$ the liquid flow rate. In recent developments it has been found that KOLMOGOROFF's theory of isotropic turbulence is a useful tool to correlate various parameters of tower fermenters. In this theory the specific energy dissipation rate (referred to mass) plays an important role. The specific energy dissipation rate ε is approximately given by:

$$\varepsilon = \dot{V}_G \Delta p / M_R, \qquad (20.6)$$

which, for bubble columns, reduces to:

$$\varepsilon = u_G g. \qquad (20.7)$$

20.2.5 Gas Holdup

The fractional gas holdup ε_G is an important factor to characterize gas-in-liquid dispersions. Because of the relation $\varepsilon_G = 1 - \varepsilon_L$, the gas holdup determines the liquid volume available for a starting biological reaction. In addition, ε_G is of utmost importance for the gas-liquid interfacial area. ε_G depends mainly on gas throughput, sparger design, and physico-chemical properties. If the column diameter is large compared to the bubble diameter, say larger by a factor of about 40, the column diameter has no significant effect. This is commonly valid for $D_R \geq 10$ cm. The influence of the gas velocity u_G on ε_G can be simply expressed by:

$$\varepsilon_G \propto u_G^n. \qquad (20.8)$$

At low gas velocities, and if porous spargers or other efficient gas distributors are used bubbly flow prevails. Then the exponent n may vary from 0.7 to 1.2. In churn-turbulent flow, which occurs at higher gas velocities, and if single and multi-nozzle spargers ($D_N \geq 1$ mm) are used, n is in the range of 0.5 to 0.7. Fig. 20.7 shows data for water and various aqueous systems including results measured in large-scale equipment with a diameter of up to 5.5 m. The flow is churn-turbulent throughout, and the exponent in Eq. (20.8) is 0.6.

The gas holdup is a quantity which can be measured easily. Therefore, numerous data are available from the literature and

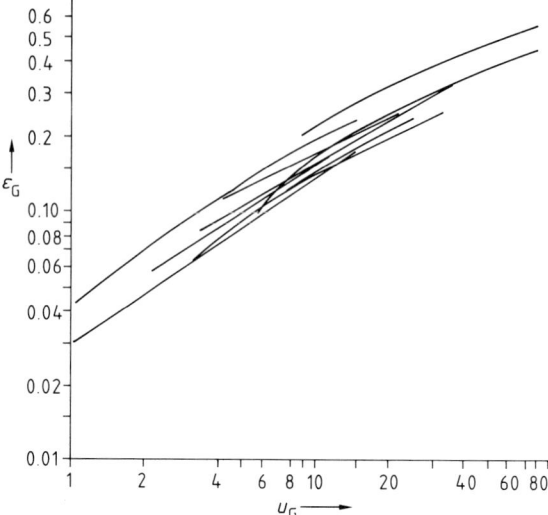

Figure 20.7. Gas holdup vs. gas velocity for churn-turbulent flow ($0.2\,\text{m} \leq D_R \leq 5.5\,\text{m}$).

many correlations for ε_G have been proposed. A summary can be found in a report by SHAH et al. [20.5]. A correlation which can be recommended as a conservative estimate is that of AKITA and YOSHIDA [20.6]:

$$\frac{\varepsilon_G}{(1-\varepsilon_G)^4} = 0.2 \left(\frac{g D_R^2 \varrho_L}{\sigma}\right)^{1/8} \left(\frac{g D_R^3}{v_L^2}\right)^{1/12} \frac{u_G}{\sqrt{g D_R}}.$$

(20.9)

In this correlation the column diameter D_R is only included for presenting the measured dependencies in dimensionless form. It has to be pointed out that considerably higher holdup values than those predicted by Eq. (20.9) have been reported for nutrition and fermentation media (SCHÜGERL et al. [20.7]; SCHÜGERL [20.2]), particularly if more effective spargers such as porous plates and two-phase nozzles were used. However, it has not yet been possible to summarize the data measured for such systems into applicable correlations. Gas holdup correlations for aqueous carboxymethyl cellulose (CMC) solutions, which are often useful for simulating high viscous non-Newtonian fermentation media, have been reported by SCHUMPE et al. [20.8].

20.2.6 Bubble Properties and Bubble Swarm Dynamics

Bubble diameter, bubble rise velocity, bubble size distribution, and bubble velocity profiles have a direct influence on bubble column performance. Bubble collectives are characterized either by the mean bubble diameter:

$$d_B = \Sigma n_i d_{Bi} / \Sigma n_i, \qquad (20.10)$$

or better by volume-to-surface (Sauter) mean diameter:

$$d_S = \Sigma n_i d_{Bi}^3 / \Sigma n_i d_{Bi}^2. \qquad (20.11)$$

The latter quantitiy is directly connected to the specific interfacial area a by:

$$a = 6 \varepsilon_G / d_S. \qquad (20.12)$$

Again, a number of correlations for d_S can be found in the literature (SHAH et al. [20.5]). However, it is hardly possible to recommend any of these correlations for biological media. In general, one has to assume that the bubble size distributions, which by the way are often bimodal, depend on liquid phase properties like σ_L, μ_L, and ϱ_L, gas velocity, and sparger design. The complexity of this dependence can be discerned from the comprehensive data on d_S and mass transfer reported by SCHÜGERL et al. [20.7]. These authors used liquid model media for yeast fermentation, i.e., solutions of alcohols and glucose in the presence of salt mixtures typical in yeast fermentations. Due to the large variability of their findings SCHÜGERL et al. could not propose correlations but discussed their results with respect to coalescence promoting and hindering properties of the liquid media and on the basis of a bubble size stability diagram derived theoretically.

Bubble coalescence and breakup are important phenomena in understanding bubble swarm dynamics. In particular, bubble coalescence has a detrimental effect on the scale-up of tall and non-staged tower bioreactors in which all the gas is distributed from the bottom. Coalescence predominates at high gas velocities, i.e., for high bubble densities. It is understood that the transition from bubbly flow to churn-turbulent flow is the result of a coalescence process. CALDERBANK et al. [20.9] suggest that coalescence sets in if the ratio of the distance between two bubble centers and the Sauter mean diameter is less than 2. Though the mechanism of coalescence in bubble swarms is not yet well understood, it is assumed that coalescence rates are mainly affected by the composition of the liquid phase and its physical properties.

In highly viscous media coalescence rates are high, indeed spherical cap bubbles and bubble slugs can be observed even at very low gas flow rates. Rapid coalescence does occur by an "overtaking" mechanism in highly viscous media. In liquids of low viscosity, as, for instance, solutions of alcohols and electrolytes (nutrition media), surface tension forces and the size of the primary bubbles generated at the sparger play an important role. It is believed that coalescence takes place more readily in pure liquids. In this case, the interface is mobile and the liquid film between bubbles can be narrowed quickly to the point of rupture by a process of stretching. In solutions of alcohols, electrolytes, etc., and liquids with impurities, film stretching is opposed by interfacial tension gradients which lead to a reduced film thinning rate. The effect of various substances including antifoaming agents is discussed in more detail by SCHÜGERL et al. [20.7] and SCHÜGERL [20.2]. A brief summary can also be found in SHAH et al. [20.5].

Tall bubble columns without additional stages within the column certainly should not be used in fermentations during which bubbles tend to coalesce rapidly. Coalescence can be suppressed to some extent by increasing the liquid flow rate by means of an external or internal loop (WEILAND and ONKEN [20.1]). However, the use of a staged column or the introduction of mechanical energy by moving parts like stirrer will be more effective.

The rise velocity u_G^* of the bubbles in the swarm is given by:

$$u_G^* = u_G/\varepsilon_G, \quad (20.13)$$

while the relative (slip) bubble velocity is found from:

$$u_{BS} = \frac{u_G}{\varepsilon_G} - \frac{u_L}{1-\varepsilon_G}, \quad (20.14)$$

for concurrent flow of gas and liquid. The terminal rise velocity of single bubbles is also often used as a correlating parameter. This velocity can be estimated by the method of CLIFT et al. [20.10].

20.2.7 Mixing

The global mixing effects in tower bioreactors can conveniently be described by dispersion coefficients which for each phase are defined in analogy to Fick's law for diffusive transport.

20.2.7.1 Liquid Phase Dispersion

Dispersion in a liquid phase has been investigated by many authors. All relevant contributions have been reviewed by SHAH et al. [20.5], [20.11]. The empirical correlation of DECKWER et al. [20.12] is recommended to calculate the liquid phase dispersion coefficient D_L:

$$D_L = 0.678\, u_G^{0.3} D_R^{1.4}. \quad (20.15)$$

Here, as well as in the following correlations, the unit of the dispersion coefficient is m²/s, the unit of velocities is m/s, and that of D_R is m. On the basis of the circulation cell model (see Sect. 20.5) JOSHI [20.13] proposed the relation:

$$D_L = 0.33\,(u_C + u_L)\,D_R, \quad (20.16)$$

where the circulation velocity u_C is calculated by means of the equation:

$$u_C = 1.31 \left[g D_R \left(u_G - \frac{\varepsilon_G}{1-\varepsilon_G} u_L - \varepsilon_G u_{B\infty} \right) \right]^{1/3} \quad (20.17)$$

with $u_{B\infty}$ as the terminal rise velocity of single bubbles, which is usually in the range of 0.2 to 0.25 m/s. Both Eqs. (20.15) and (20.16) describe experimental data equally well and do also apply to large-scale industrial bubble columns with a diameter of 3.2 m and a height of 19 m. Liquid phase dispersion coefficients in staged and packed bubble columns are reviewed by SHAH et al. [20.5].

through a maximum value. KÖNIG et al. interpret their results with the Sauter diameters of the bubbles which were measured simultaneously. For very small diameters ($d_S \leq 1$ mm) most of the liquid is attached to the gas-liquid interphase and carried upward with the bubbles which causes an underpressure and leads to violent eddies. At medium values of d_S (1 to 3 mm) bubbly flow prevails which yields extremely low values of D_L. For larger bubble diameters the flow is churn-turbulent giving again high values of D_L.

In fermentation technology one is often more interested in the mixing time than in the dispersion coefficient. Both mixing time and dispersion coefficient are directly correlated by the transient solution of the dispersion model. However, the mixing time must be referred to a certain degree of homogeneity. For the case of a pulse tracer in-

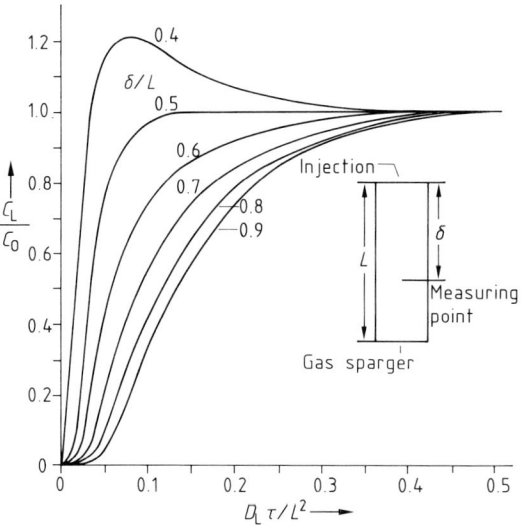

Figure 20.8. Homogeneity vs. dimensionless dispersion group $A = D_L \tau / L^2$.

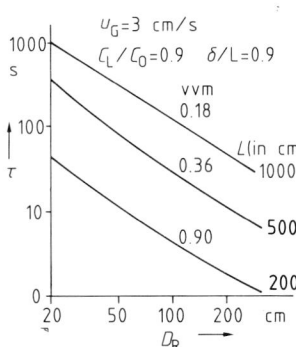

Figure 20.9. Mixing time vs. column diameter.

jection the solution curves of the dispersion model are given in Fig. 20.8 for various values of δ/L, i.e., the dimensionless distance between tracer injection and measuring point. For a given value of δ/L and a desired degree of homogeneity C_L/C_o the value of:

$$A(C_L/C_o, \delta/L) = \frac{D_L \tau}{L^2} \quad (20.18)$$

can be read from the abscissa in Fig. 20.8. Hence, by introducing this value in an ap-

An unusual behavior of liquid phase dispersion was observed by KÖNIG et al. [20.14] who reported on measurements in bubble columns equipped with porous spargers and employing diluted solutions of alcohols as liquid phase. The findings of these authors may be relevant to biological systems. The dependency of D_L on u_G reveals steep, unusual changes and D_L passes

propriate correlation for D_L as, for example, Eq. (20.16), the mixing time τ (in s) is given by:

$$\tau = \frac{A}{0.678} \frac{L^2}{u_G^{0.3} D_R^{1.4}}. \qquad (20.19)$$

In Fig. 20.9 τ is plotted vs. tower diameter D_R for three different values of height L. It can be seen that it is particularly the ratio of L to D_R which influences the mixing time.

20.2.7.2 Gas Phase Dispersion

Data on gas phase dispersion are rare and subject to considerable scatter. Careful measurements were carried out by MANGARTZ and PILHOFER [20.15]. The findings of these authors indicate that besides column diameter it is mainly the rise velocity of the bubbles in the swarm which influences gas phase dispersion. MANGARTZ and PILHOFER recommend the following empirical correlation for the gas phase dispersion coefficient:

$$D_G = 50 \, D_R^{1.5} \, (u_G/\varepsilon_G)^3, \qquad (20.20)$$

which was confirmed by other investigations for larger bubble columns.

Gas phase dispersion coefficients are high, particularly for larger diameter columns and often considerably larger than those of the liquid phase. However, the impact of gas phase dispersion on bubble column performance has been scarcely taken into account. Because of their large scale, and, if gas phase conversion (oxygen uptake rates) is high, dispersion in the gas phase may have a significant influence on tower bioreactor performance.

20.3 Mass Transfer

20.3.1 General Considerations

An important problem in a variety of aerobic fermentations is the transport of oxygen from the air bubbles to the site of reaction, i.e., the biomass phase suspended in the culture media. One of the major reasons that oxygen transfer can play an important role in many biological processes is certainly the limited oxygen capacity of the fermentation broth due to the low solubility of O_2. The oxygen transfer path in a typical three-phase biological system is schematically shown in Fig. 20.10. Of course, the situation is completely equivalent to that encountered in catalytic slurry reactors. Before the oxygen can be consumed by biomass particles several physical resistances have to be overcome. In terms of the simple film theory possible resistances can be:

- diffusion through the liquid films around the bubble and the biomass particles (pellets, flocculant, fibrous, or filamentous material)
- effective internal diffusion simultaneously with consumption by the reaction in the interior of the biomass particles.

The latter phenomenon can practically not be influenced by any external or engi-

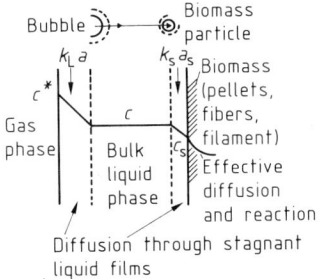

Figure 20.10. Mass transfer resistances in biological reactor.

neering measures. The interaction of internal diffusion and reaction in biological systems can be treated in analogy to heterogeneous catalysis by introducing the concept of effectiveness factors (KOBAYASHI et al. [20.16]; ATKINSON [20.17]; BAILY and OLLIS [20.18]). On the other hand, operational conditions and hydrodynamic flow behavior has a marked influence on transfer of oxygen through the liquid films around both the bubble and the biomass particle.

20.3.2 Liquid-Solid Mass Transfer

Mass transfer coefficients, k_s, for transport from the bulk of the liquid phase to the external surface of solid particles suspended in bubble columns have been measured for various conditions. It is assumed that these data also apply to mass transfer through the liquid film around biomass particles. An appropriate correlation for the liquid-solid mass transfer coefficient k_s which extends over a wide range of Schmidt numbers ($137 \leq Sc = v/D \leq 50000$) and includes data for higher viscous media has been given by SÄNGER and DECKWER [20.19]:

$$\frac{k_s d_p}{D} = 2 + 0.545 \left(\frac{v}{D}\right)^{1/3} \left(\frac{\varepsilon d_p^4}{v^3}\right)^{0.264}. \quad (20.21)$$

Here, d_p is the diameter of the solid particles, D the diffusivity, v the kinematic viscosity, and ε the specific energy dissipation rate which is calculated from Eq. (20.7).

20.3.3 Gas-Liquid Mass Transfer

For calculating the gas-liquid mass transfer rate in tower bioreactors it is sufficient to know the liquid side volumetric mass transfer coefficient ($k_L a$) which is the product of the liquid side mass transfer coefficient k_L and the specific interfacial area a. For gases relevant to biotechnology gas side resistances for mass transfer are negligibly small. As absorption enhancement due to fast reaction of the dissolved gaseous reactant can not be expected for biological reactions (ALPER et al. [20.20]) there is no need to separate the product $k_L a$ into its individual constituents. Nevertheless, it is often useful to know the value of the liquid side mass transfer coefficient k_L for the investigated system.

20.3.3.1 Liquid Side Mass Transfer Coefficient

The mass transfer coefficient k_L strongly depends on the kind of bubble surface; small bubbles behave like rigid spheres which result in low values of k_L. The k_L values for such small bubbles ($d_B < 2.5$ mm) can be estimated from the empirical correlation of CALDERBANK and MOO-YOUNG [20.21]:

$$k_L = 0.31 \left(\frac{(\varrho_L - \varrho_G)\mu_L g}{\varrho_L^2}\right)^{1/3} \left(\frac{\mu_L}{D\varrho_L}\right)^{-2/3}. \quad (20.22)$$

For larger bubbles the surface is not rigid but deformable and partly oscillating. Therefore, higher k_L values are observed. For $d_B > 2.5$ mm they can be calculated by means of the following equation [20.21]:

$$k_L = 0.42 \left(\frac{(\varrho_L - \varrho_G)\mu_L g}{\varrho_L^2}\right)^{1/3} \left(\frac{\mu_L}{D\varrho_L}\right)^{-1/2}. \quad (20.23)$$

In Eqs. (20.22) and (20.23) k_L is in cm/s. From Eq. (20.23) it can be seen that the k_L values for large bubbles follow the penetration theory. This has also been confirmed by HALLENSLEBEN [20.22] who studied mass transfer from single bubbles. In particular it was shown that k_L values determined from single bubble measurements can be applied to bubble swarms in bubbly flow.

AKITA and YOSHIDA [20.23] also proposed an empirical, dimensionless relation for estimating k_L. This correlation is given by:

$$\frac{k_L d_S}{D} = 0.5 \left(\frac{\nu_L}{D}\right)^{1/2} \left(\frac{g d_S^3}{\nu_L^2}\right)^{1/4} \left(\frac{g d_S^2 \varrho_L}{\sigma}\right)^{3/8}. \quad (20.24)$$

The correlations presented by CALDERBANK and MOO-YOUNG [20.21] as well as by AKITA and YOSHIDA [20.23] are recommended for k_L estimation in media of relatively low viscosity. In bubble columns, the bubble size varies usually between 1 and 5 mm and therefore, the k_L values show considerable variation, i.e., k_L is frequently in the range from 0.01 to 0.03 cm/s. This is also demonstrated by Fig. 20.11 where re-

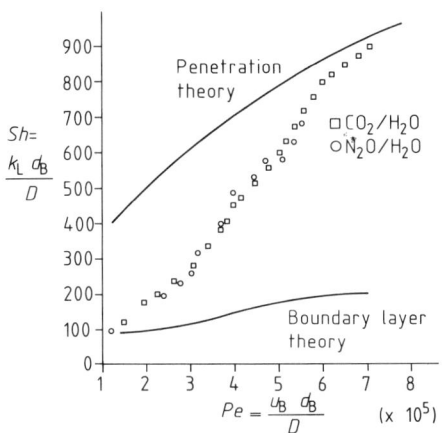

Figure 20.11. Sherwood number, $Sh(k_L d_B/D)$, as a function of the Peclet number, $Pe(u_B d_B/D)$, for single bubbles.

sults from single bubble measurements are plotted in a dimensionless form (HALLENSLEBEN [20.22]). For the range of Peclet numbers encountered in bubble columns the Sherwood number varies over a wide range. For highly viscous liquids with non-Newtonian behavior (CMC solutions) SCHUMPE and DECKWER [20.24] proposed to estimate k_L data from the empirical equation:

$$k_L = 4.5 \cdot 10^{-3} \bar{u}_G^{0.08} \mu_{\text{eff}}^{-0.32}, \quad (20.25)$$

where the mean gas velocity $\bar{u}_G$ is in cm/s and the effective dynamic viscosity μ_{eff} in Pa s.

20.3.3.2 Volumetric Mass Transfer Coefficient

Many investigations and a huge amount of data on $k_L a$ can be found. However, often the data reported for the same or for similar conditions differ widely. The reason for these discrepancies can be manifold as, for instance, presence of trace impurities, use of insufficient experimental techniques and inappropriate methods for data evaluation. The matter has been briefly discussed by SHAH et al. [20.5] and in more detail by DECKWER [20.25]. Volumetric mass transfer coefficients depend on the gas velocity and the gas sparger used. In addition, they are very sensitive to physico-chemical properties, particularly to those which promote or prevent coalescence. The column diameter may also exert some influence as long as it is small, say $D_R \leq 15$ cm. Some authors also found an influence of the dispersion height and the liquid velocity. It has, however, been shown that these dependencies are only the result of applying an insufficient model for data evaluation (DECKWER et al. [20.26]). No dependency of $k_L a$ on liquid flow rate ($u_L \leq 10$ cm/s) can be observed when applying the dispersion model which correctly interprets the concentration jump at column inlet. The simplest form to empirically correlate $k_L a$ data is by:

$$k_L a = b u_G^n, \quad (20.26)$$

where the exponent n depends on the flow regime. For bubbly flow and developing churn-turbulent flow, i.e., up to gas velocities of about 8 cm/s, n is approximately 0.8. The regression coefficient b depends mainly on sparger design and liquid properties. For water and aqueous electrolyte solutions aerated by 1 mm holes b is 0.467 (u_G in m/s, $k_L a$ in s^{-1}). For aeration with a sintered plate b increases to 1.2 to 1.4.

AKITA and YOSHIDA [20.6] developed a dimensionless empirical correlation for $k_L a$ which is given by:

$$\frac{k_L a D_R^2}{D} = 0.6 \left(\frac{v_L}{D}\right)^{0.5} \left(\frac{g D_R^2 \varrho_L}{\sigma}\right)^{0.62} \left(\frac{g D_R^3}{v_L^2}\right)^{0.31} \varepsilon_G^{1.1}. \quad (20.27)$$

Fig. 20.12 shows that this correlation applies well to large-scale equipment (reactors up to 5.5 m in diameter). Of course, for large size reactors the influence of the diameter disappears. For $D_R > 60$ cm the calculations are carried out with $D_R = 60$ cm. Use of Eq. (20.27) can be recommended for liquid media of lower viscosity (≤ 21 mPa s) and static spargers. However, it should be pointed out that the application of the correlation of AKITA and YOSHIDA [20.6] usually gives a conservative estimate.

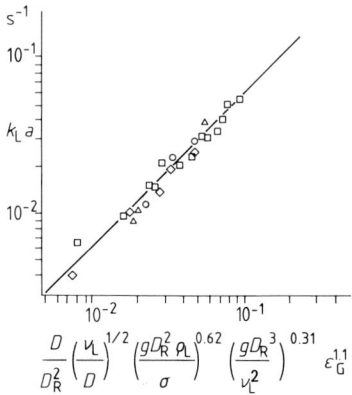

Figure 20.12. Volumetric mass transfer coefficient $k_L a$ from the Akita-Yoshida correlation. – Comparison with data from other authors in large-scale equipment (0.2 m $\leq D_R \leq$ 5.5 m).

SCHÜGERL et al. [20.7] and SCHÜGERL [20.2] have reported on $k_L a$ values in simulated culture media for SCP fermentations. Considerably larger values than those predicted by Eq. (20.27) have been found. It was not possible to present the various findings in single- and multistaged columns in form of mathematical correlations. The results are qualitatively discussed on the basis of the coalescence properties, see Sect. 20.2.6. ZLOKARNIK [20.27], [20.28] reported on comprehensive mass transfer measurements in large-scale bubble columns for waste water treatment. The results obtained with two-phase nozzles as spargers are represented by introducing a dimensionless sorption number Y which is defined by:

$$Y = \frac{G}{Lq\Delta c}\left(\frac{v^2}{g}\right)^{1/3} = \frac{k_L a}{u_G}\left(\frac{v^2}{g}\right)^{1/3}, \quad (20.28)$$

where G is the oxygen mass flow rate through the interface (in kg/s), L the contactor height (in m), q the gas throughput (in m^3/s), and Δc the mean driving concentration difference. If one assumes as a rough approximation $k_L a \propto u_G$ then Y should be constant for a given system if porous and perforated plates are used to disperse the gas. If the gas is sparged by mechanical agitation or by the energy of the liquid jet in two-component nozzles then Y is a function of the power number X which is given by:

$$X \equiv \frac{P_L/q}{\varrho(vg)^{2/3}}. \quad (20.29)$$

Here, P_L is the power of the liquid jet. The sorption number Y presents a reasonable measure to characterize mass transfer properties of tower bioreactors. Table 20.1 summarizes experimental studies in bubble columns of various geometry and gives the sorption number calculated for $u_G = 5$ cm/s. Besides physico-chemical properties it is particularly the gas distributor which affects the sorption number. This can clearly be discerned from Table 20.1. For single and multiorifice (perforated plates, spider type) spargers, with hole diameters of 1 mm or larger Y varies from 2.4 to 5.4, the mean value being about 4. In the case of porous plate distributors Y is about 10 for water, but for non-coalescing media considerably larger values can be obtained, especially if the pore diameter is small. The high $k_L a$

Table 20.1. Typical Experimental Values of Sorption Number of Various Columns (evaluated for $u_G = 5$ cm/s)

Gas Sparger (mean orifice or pore diameter in mm)	Column Sizes D_R (in cm)	L (in cm)	Liquid System	Sorption Number $10^5 Y$	Reference
Multi-orifice, 30	550	900	Tap water	2.4	KATAOKA et al. [20.29]
Multi-orifice, 1.6	100	120	Sulfite soln.	3.7	KASTANEK et al. [20.30]
Single- and multi-orifice, 3.8–8.6	8.6, 183 762	600–1600	Tap water	5.4	JACKSON and SHEN [20.31]
Multi-orifice, 1	20	720	Tap water Salt solns.	3.6 4.3	DECKWER et al. [20.12]
Multi-orifice, 0.5 (perforated plate, 180 bores)	14	400	Demineralized water solns. of alcohols (0.5–2%) and salts	8.7 10.5	SCHÜGERL et al. [20.7]
Porous plate, 0.150	15	440	Tap water Salt solns.	9.9 15.4	DECKWER et al. [20.12]
Porous plate, 0.0175	14	400	Demineralized water solns. of alcohols (0.5–2%) and salts	11.2 >33.0	SCHÜGERL et al. [20.7]
Two-phase nozzle	30	230	Sulfite soln.	9–18[a]	NAGEL et al. [20.32]
Slot injector (two-phase)	280	700	Tap water	8–16[a]	ZLOKARNIK [20.27], [20.28]

[a] Dependent on energy provided by liquid jet

values and, hence, sorption numbers attainable in such liquids have to be entirely attributed to small bubble sizes. High sorption numbers can be obtained by application of two-phase nozzles, of course, where additional energy is introduced by the liquid jet.

Oxygen mass transfer into highly viscous solutions of glycerol, carboxymethyl cellulose (CMC), and polyacrylamide (PAA) has been measured by SCHÜGERL and co-workers in single- and multistage tower reactors.

The principal results have been summarized and reviewed by SCHÜGERL [20.33]; there are, however, some doubts about the reliability of the data. For instance, $k_L a$ values measured by other authors (NAKANOH and YOSHIDA [20.34]; DECKWER et al. [20.35]) in CMC solutions and in single-stage bubble columns differ by a factor of 5 to 10. Also, the observed dependencies on u_G and effective viscosity v_{eff} are different. The correlation of HENZLER [20.36] which is based on the data of BUCHHOLZ et al. [20.37] for CMC solutions is as follows:

$$k_L a \sim u_G^{0.1} v_{\text{eff}}^{-0.87}. \quad (20.30)$$

The correlation of NAKANOH and YOSHIDA [20.34] is given by:

$$k_L a \propto u_G v_{\text{eff}}^{-0.28}. \quad (20.31)$$

The complete dimensionless correlation of NAKANOH and YOSHIDA is given by:

$$k_L a' \frac{D_R^2}{D} =$$
$$= 0.09 \, Sc^{0.5} \, Bo^{0.75} \, Ga^{0.39} \, Fr (1 + c \, De^m)^{-1}. \quad (20.32)$$

The dimensionless numbers are defined as follows:

$Sc = v_{\text{eff}}/D \qquad$ Schmidt number $\quad (20.33)$

$Bo = g D_R^2 \varrho_L / \sigma$ Bond number   (20.34)

$Ga = g D_R^3 / v_{eff}^2$ Galilei number   (20.35)

$Fr = u_G / \sqrt{g D_R}$ Froude number   (20.36)

$De = \dfrac{u_G (1 + \varepsilon_G) \lambda}{\varepsilon_G d_S}$ Deborah number (20.37)

The Deborah number De accounts for elastic properties of the solutions. For polyacrylate solutions (up to 0.1%) NAKANOH and YOSHIDA suggest for the constants c and m the values: $c = 0.13$ and $m = 0.55$. For inelastic liquids c is zero. Eq. (20.32) describes experimental data for Newtonian (sucrose solution $\leq 50\%$) and non-Newtonian (CMC $\leq 1\%$, polyacrylate $\leq 0.1\%$) liquids up to effective viscosities of about 0.05 Pa s. The flow regime is assumed to be churn-turbulent at low gas velocities and exhibits slug flow at higher flow rates.

$k_L a$ values were also taken in a 14 cm diameter column with CMC concentrations up to 1.6% corresponding to viscosities of about 0.1 Pa s (DECKWER et al. [20.35]). In this investigation a remarkable effect of the gas sparger on $k_L a$ was observed at low gas velocities. A typical result is shown in Fig. 20.13. It can be seen that by using a rubber plate and a sintered plate high $k_L a$ values can be obtained at gas velocities of about 1 cm/s. When a perforated plate is used such high $k_L a$ values can only be obtained at a threefold gas flow rate in case of the sintered plate or even higher flow rates in case of the rubber plate. The reason for the high $k_L a$ values at low gas velocities with rubber and sintered plates is that bubbly flow can be realized.

For aeration with perforated and sintered plates slug flow prevails at gas velocities above 2 cm/s. The $k_L a$ data measured in the 14 cm diameter reactor under slug flow conditions have been correlated by DECKWER et al. [20.35]:

$$k_L a = 2.08 \cdot 10^{-4} \bar{u}_G^{-0.59} \mu_{eff}^{-0.84}. \qquad (20.38)$$

Here, $k_L a$ is in s^{-1}, $\bar{u}_G$ is the mean gas velocity in cm/s, and the effective viscosity is in Pa s. The effective viscosity in Eq. (20.38) as well as in the correlation of NAKANOH and YOSHIDA [20.32] has been calculated from the Ostwald-de Waele power law:

$$\mu_{eff} = k \dot{\gamma}^{n-1}. \qquad (20.39)$$

The shear rate $\dot{\gamma}$ prevailing in bubble columns depends on the aeration rate and was calculated from the relation:

$$\dot{\gamma} = 50 u_G, \qquad (20.40)$$

where $\dot{\gamma}$ is in s^{-1} and u_G in cm/s. Eq. (20.40) was proposed by NISHIKAWA et al. [20.38].

20.4 Heat Transfer

Wall-to-dispersion heat transfer coefficients h in two-phase and three-phase (slurry) column reactors can be obtained from an equation presented by DECKWER [20.39] and DECKWER et al. [20.40]:

$$\frac{h}{\varrho c_p u_G} = 0.1 \left[\frac{u_G^3}{v \dot{g}} \left(\frac{v \varrho c_p}{k} \right)^2 \right]^{-1/4}. \qquad (20.41)$$

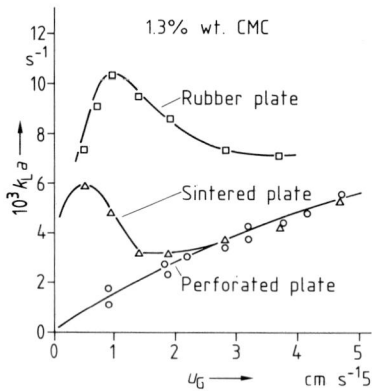

Figure 20.13. Effect of a gas sparger on the volumetric mass transfer coefficient in solutions of CMC.

The heat transfer coefficients obtained from this equation are in good agreement with the findings reported by NISHIKAWA et al. [20.38] for media of lower and higher viscosity including CMC solutions. NISHIKAWA et al. [20.38] have proposed valuable correlations for the average shear rate in bubble dispersions as a function of the gas velocity, for instance, Eq. (20.40) by assuming that heat transfer data in aerated Newtonian and non-Newtonian fluids follow the same dependencies. With the shear rate known, the effective viscosity of non-Newtonian media in bubble columns can be obtained from the shear stress vs. shear rate curve. Use of Eq. (20.41) is also recommended for bubble column bioreactors. However, one should consider the possibility of slime formation (adherence of biomass particles) on the heat transfer surfaces which might reduce the heat transfer rate considerably.

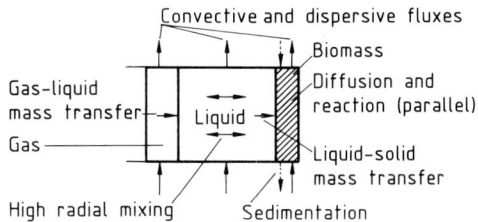

Figure 20.14. Schematic volume element of a bioreactor.

20.5 Models of Bubble Columns

For design and scale-up of reactors mathematical models and computational optimization procedures are employed. The use of mathematical models requires introduction of the microscopic picture of mass transfer and biological reaction into the governing macroscopic balance equations which describe fluid mechanics appropriately. In tower reactors the length to diameter ratio is usually large. Therefore, the assumption of complete mixing is not justified. It is now generally accepted that the axial dispersed plug flow model is a pertinent approach to mathematically model bubble column bioreactors. The differential equations of this model are obtained in the usual way by balancing over a volume element under consideration of those phenomena which are assumed to be of influence. In view of Fig. 20.14 the steady-state balance equations are as follows:

Gas phase:

$$\frac{d}{dx}\left(\varepsilon_G D_G \frac{dc_G}{dx}\right) - \frac{d}{dx}(u_G c_G) - k_L a(c_L^* - c_L) = 0$$
(20.42)

Liquid phase:

$$\frac{d}{dx}\left(\varepsilon_L D_L \frac{dc_L}{dx}\right) - u_L \frac{dc_L}{dx} + k_L a(c_L^* - c_L) - k_S a_S(c_L - c_S) = 0$$
(20.43)

Biomass phase (external surface):

$$k_S a_S(c_L - c_S) - R(c_S, c_C, c_B, D_{eff} \ldots) = 0.$$
(20.44)

Of course, the rate term R in the oxygen balance depends additionally on the local concentrations of the carbon source and the biomass (c_C and c_B), and equivalent balances have to be formulated for both of them. In the case of biomass it may be necessary to take into account sedimentation which gives the following balance equation:

$$\frac{d}{dx}\left(\varepsilon_L D_S \frac{dc_B}{dx}\right) + (u_S - u_L)\frac{dc_B}{dx} + R' = 0$$
(20.45)

where R' is the generation term for biomass. u_S is the settling velocity of the biomass particles in the swarm. The biomass dispersion coefficient D_S can be assumed to be equal to the liquid phase dispersion coeffi-

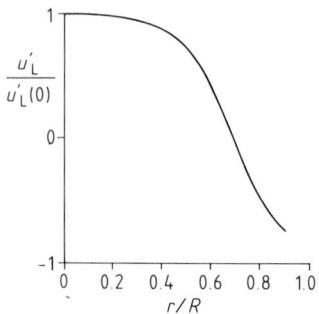

Figure 20.15. Dimensionless radial liquid velocity profile in bubble columns.

cient D_L. A detailed dynamic model for biomass production which is based on the dispersed plug flow model has been given by LUTTMANN et al. [20.41]. The assumption of a plug flow profile modified by dispersive flow can only be justified for multistage reactors and bubble columns with a high recirculation rate of liquid flow (WEILAND and ONKEN [20.1].) In single-stage bubble columns pronounced radial liquid velocity profiles exist. Fig. 20.15 shows such a dimensionless velocity profile which has been calculated from an empirical relation given by RIQUARTS [20.42].

On the basis of a force balance and with knowledge of the radial gas holdup profile UEYAMA and MIYAUCHI [20.43] developed a recirculation flow model which presents a fundamental approach to the description of the hydrodynamic behavior of bubble columns in churn-turbulent flow. By means of this model the liquid velocity profile and some other quantities like bubble slip velocity and turbulent viscosity can be calculated. However, the mathematical equations for calculating the profile are complex which makes the computation cumbersome. By neglecting the effect of wall friction on the pressure gradient RIQUARTS [20.42] arrived at a much simpler equation. According to RIQUARTS, the profile of the interstitial liquid velocity is given by:

$$u'_L(r/R) = u'_L(0) - \frac{\bar{\varepsilon}_G g D_R^2}{32 v_t}\{1 - [1-(r/R)^2]^2\}, \quad (20.46)$$

where the center line velocity follows from:

$$u'_L(0) = \frac{1 - 0.75 \bar{\varepsilon}_G}{1 - \bar{\varepsilon}_G} \cdot \frac{\bar{\varepsilon}_G g D_R^2}{48 v_t}. \quad (20.47)$$

Here, $\bar{\varepsilon}_G$ is the average gas holdup, and the turbulent viscosity v_t can be calculated either from the relation given by UEYAMA and MIYAUCHI [20.43] or from RIQUARTS [20.44]:

$$\frac{u'_L(0) D_R}{v_t} = 19. \quad (20.48)$$

JOSHI and SHARMA [20.45] extended the energy balance method introduced by WHALLEY and DAVIDSON [20.46] and proposed a multiple cell circulation model. As indicated in Fig. 20.16 many circulation cells persist in axial direction. The height of each circulation cell approximately equals the column diameter, irrespective of gas flow rate, dispersion height, and column diameter. The main result of JOSHI's and SHARMA's treatment is an explicit expression for the liquid circulation velocity which is useful for correlating liquid dispersion coefficients, see Sect. 20.2.7.1 and Eqs. (20.16) and (20.17), and heat transfer coefficients.

The correlation of liquid phase dispersion coefficients with the average liquid circulation velocity, as proposed by JOSHI and SHARMA [20.45], apparently presents a contradiction. The liquid dispersion coefficient is a measure of the irregular stochastic mix-

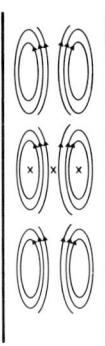

Figure 20.16. Liquid circulation cells in axial direction (JOSHI and SHARMA [20.45]).

ing processes. In contrast, the circulation velocity derived from the energy balance is deterministic in its nature. This problem is discussed in detail by RIQUARTS [20.44]. The author has analyzed the mass balance equations by taking into account a simplified liquid velocity distribution, i. e., by dividing the entire column into a central upflow and an annular downflow region. RIQUARTS [20.44] finally obtained a balance equation which is equivalent to the one-dimensional dispersed plug flow model, and concludes that the liquid phase dispersion coefficient is, in fact, a measure of stochastic mixing processes even when the circulation flow is chiefly responsible for the axial mass transfer. One can therefore conclude that the apparent discrepancy between the deterministic and stochastic approach is immaterial. Both physical concepts are equally capable of representing transfer phenomena in the liquid phase of bubble columns and are probably limiting cases of a more general hydrodynamic model.

20.6 References

[20.1] P. WEILAND and U. ONKEN; Ger. Chem. Eng. 4 (1981), 174.
[20.2] SCHÜGERL, K.; Adv. Biochem. Eng. 22 (1982), 94.
[20.3] Y. T. SHAH and W.-D. DECKWER: "Fluid-fluid reactors" in "Scale-up in the Chemical Process Industries" (A. BISIO and R. KABEL, eds.). Wiley & Sons, New York 1984.
[20.4] R. BEINHAUER: "Dynamische Messung des relativen Gasgehalts in Blasensäulen mittels Absorption von Röntgenstrahlen". Thesis, Technische Universität Berlin, Berlin 1971.
[20.5] Y. T. SHAH, B. G. KELKAR, S. P. GODBOLE, and W.-D. DECKWER; AIChE J. 28 (1982), 353.
[20.6] K. AKITA and F. YOSHIDA; Ind. Eng. Chem. Process Des. Dev. 12 (1973), 76.
[20.7] K. SCHÜGERL, J. LÜCKE, and U. OELS; Adv. Biochem. Eng. 7 (1977), 1.
[20.8] A. SCHUMPE, K. NGUYEN-TIEN, and W.-D. DECKWER; Chem. Ing. Tech. 53 (1981), 886.
[20.9] P. H. CALDERBANK, M. B. MOO-YOUNG, and R. BIBBY; Proc. 3rd Eur. Symp. React. Eng., Amsterdam 1964.
[20.10] R. CLIFT, J. R. GRACE, and M. E. WEBER: "Bubbles, Drops and Particles". Academic Press, New York 1978.
[20.11] Y. T. SHAH, G. J. STIEGEL, and M. M. SHARMA; AIChE J. 24 (1978), 369.
[20.12] W.-D. DECKWER, R. BURCKHART, and G. ZOLL; Chem. Eng. Sci. 29 (1974), 2177.
[20.13] J. B. JOSHI; Trans. Inst. Chem. Eng. 58 (1980), 155.
[20.14] B. KÖNIG, R. BUCHHOLZ, J. LÜCKE, and K. SCHÜGERL; Ger. Chem. Eng. 1 (1978), 199.
[20.15] K.-H. MANGARTZ and TH. PILHOFER; Verfahrenstechnik 14 (1980), 40.
[20.16] T. KOBAYASHI, G. VAN DEDEM, and M. MOO-YOUNG; Biotech. Bioeng. 15 (1973), 27.
[20.17] B. ATKINSON: "Biochemical Reactors". Pion Ltd., London 1974.
[20.18] J. E. BAILEY and D. F. OLLIS: "Biochemical Engineering Fundamentals". McGraw-Hill, New York 1977.
[20.19] P. SÄNGER and W.-D. DECKWER; Chem. Eng. J. 22 (1981), 179.
[20.20] E. ALPER, Y. SERPEMEN, and W.-D. DECKWER; in "Advances in Biotechnology" (M. MOO-YOUNG, C. W. ROBINSON, and C. VEZINA, eds.), Vol. 1, p. 511. Pergamon Press, Toronto 1980.
[20.21] P. H. CALDERBANK and M. MOO-YOUNG; Chem. Eng. Sci. 16 (1961), 39.
[20.22] J. HALLENSLEBEN: "Simultaner Stoffaustausch von CO_2 und Sauerstoff an Einzelblasen und in Blasenschwärmen". Dissertation, Universität Hannover, Hannover 1980.
[20.23] K. AKITA and F. YOSHIDA; Ind. Eng. Chem. Process Des. Dev. 13 (1974), 84.
[20.24] A. SCHUMPE and W.-D. DECKWER; Ind. Eng. Chem. Process Des. Dev. 21 (1982), 706.
[20.25] W.-D. DECKWER: "Reaktionstechnik in Blasensäulen-Reaktoren". Verlag Sauerländer, Aarau–Frankfurt am Main 1985.
[20.26] W.-D. DECKWER, K. NGUYEN-TIEN, B. G. KELKAR, and Y. T. SHAH; AIChE J. 29 (1983), 915.
[20.27] M. ZLOKARNIK; Chem. Eng. Sci. 34 (1979), 1265.

[20.28] M. ZLOKARNIK; Korrespondenz Abwasser *27* (1980), 194.

[20.29] H. KATAOKA, H. TAKEUCHI, K. NAKAO, H. YAGI, T. TADAKI, T. OTAKE, T. MIYAUCHI, K. WASHIME, K. WATANABE, and F. YOSHIDA; J. Chem. Eng. Jpn. *12* (1979), 105.

[20.30] F. KASTANEK, J. KRATOCHVIL, and M. RYLEK; Collect. Czech. Chem. Commun. *42* (1977), 3549.

[20.31] M. L. JACKSON and C.-C. SHEN; AIChE J. *24* (1978), 63.

[20.32] O. NAGEL, B. HEGNER, and H. KÜRTEN; Chem. Ing. Tech. *50* (1978), 934.

[20.33] K. SCHÜGERL; Adv. Biochem. Eng. *19* (1981), 72.

[20.34] M. NAKANOH and F. YOSHIDA; Ind. Eng. Chem. Process Des. Dev. *19* (1980), 190.

[20.35] W.-D. DECKWER, K. NGUYEN-TIEN, A. SCHUMPE, and Y. SERPEMEN; Biotech. Bioeng. *24* (1982), 461.

[20.36] H. J. HENZLER; Chem. Ing. Tech. *52* (1980), 643.

[20.37] H. BUCHHOLZ, R. BUCHHOLZ, J. LÜCKE, and K. SCHÜGERL; Chem. Eng. Sci. *33* (1978), 1061.

[20.38] M. NISHIKAWA, H. KATO, and K. HISHIMOTO; Ind. Eng. Chem. Process Des. Dev. *16* (1977), 133.

[20.39] W.-D. DECKWER; Chem. Eng. Sci. *35* (1980), 1341.

[20.40] W.-D. DECKWER, Y. LOUISI, A. ZAIDI, and M. RALEK; Ind. Eng. Chem. Process Des. Dev. *19* (1980), 699.

[20.41] R. LUTTMANN, M. THOMA, R. BUCHHOLZ, and K. SCHÜGERL; Comp. Chem. Eng. *7* (1983), 43, 51.

[20.42] H.-P. RIQUARTS; Chem. Ing. Tech. *53* (1981), 60.

[20.43] K. UEYAMA and T. MIYAUCHI; AIChE J. *25* (1979), 258.

[20.44] H.-P. RIQUARTS; Ger. Chem. Eng. *4* (1981), 18.

[20.45] J. B. JOSHI and M. M. SHARMA; Trans. Inst. Chem. Eng. *57* (1979), 244.

[20.46] P. B. WHALLEY and J. F. DAVIDSON; Proc. Symp. Multiphase Flow Systems. Inst. Chem. Eng. Symp. Series No. 38, London 1974.

Chapter 21
Biochemical Loop Reactors

Heinz Blenke

Institut für Chemische Verfahrenstechnik
Universität Stuttgart
Stuttgart, Federal Republic of Germany

21.1 Principles and Types of Loop Reactors (LR)
21.1.1 Flow Paths and Arrangements
21.1.2 Types of Circulation Drives
21.1.3 Differences from Other Jet Reactors
21.1.4 Some Construction Examples
21.1.4.1 Loop Reactors
21.1.4.2 Loop Reactor Parts
21.2 Demands on Bioreactors
21.2.1 Defined Distribution
21.2.2 Fine Dispersion
21.2.3 No Cell Damage
21.2.4 Large Specific Heat Transfer
21.2.5 Simple Construction
21.2.6 Simple Mode of Operation
21.2.7 Easy Adjustment
21.2.8 Good Computability
21.3 Fluid Dynamics of Loop Reactors (LR)
21.3.1 Fluid Dynamic Design of LR for Aqueous (Quasi-)Homogeneous L-Systems
21.3.1.1 Fundamental Fluid Dynamic Parameters
21.3.1.2 Circulation Intensity in JLR with (Quasi-)Homogeneous Aqueous L-Systems
21.3.1.3 Experimental Determination of ζ_U, ζ_U^* and Re_m/Re_1
21.3.1.4 Power and Efficiency of the Circulation Drive (JLR)
21.3.1.5 Application of Previous Results for JLR to PLR
21.3.2 (Quasi-)Homogeneous Highly Viscous L-Systems
21.3.2.1 Experimental Data
21.3.2.2 Theoretical Relations for Newtonian Fluids
21.3.2.3 Some Results for Newtonian Fluids
21.3.2.4 Comparison between PLR with Newtonian and Non-Newtonian (Pseudoplastic) Fluids

21.3.3 Flow Behavior of Heterogeneous Solid-Liquid-Systems (S-L-Systems or Suspensions) in JLR
21.4 Fluid Dynamics of Loop Reactors with Aerated Gas-Liquid-Systems (G-L-Systems)
21.4.1 Coalescence Behavior of G-L-Systems
21.4.2 Volumetric Gas Hold-up of G-L-Systems
21.4.2.1 Definitions
21.4.2.2 Air-Water-System at $w_G \lesssim 10$ cm s^{-1}
21.4.2.3 Sulfite-System at $w_G \lesssim 60$ cm s^{-1}
21.4.2.4 Summary of Gas Hold-up
21.4.3 Circulation Flow of G-L-Systems in JLR and ALR
21.4.3.1 Methods for the Determination of Liquid Circulation Velocity in G-L-Systems
21.4.3.2 JLR and ALR with Air-Water-System at $w_G \lesssim 10$ cm s^{-1}
21.4.3.3 ALR with Sulfite-System at $w_G \lesssim 60$ cm s^{-1}
21.5 Mixing Behavior of LR
21.5.1 Basic Principles
21.5.2 Mixing Behavior
21.6 Residence Time Behavior of LR
21.7 Mass Transfer with (Bio-)Chemical Reaction in G-L-Systems
21.7.1 Mass Transfer
21.7.2 Combination of Diffusion (Film Theory) and Chemical Reaction
21.7.3 Determination of $k_L \equiv \beta_{ph}$ and a_R
21.7.4 Determination of $\bar{c}_o$ in LR
21.8 Possibilities for Improving Productivity of LR
21.9 Some Examples of Biotechnical Applications of LR
21.9.1 Raw Materials and Energy
21.9.2 Food
21.9.3 Biological Waste Water Purification
21.10 Outlook
21.11 References

List of Symbols

a	m	distance of draft tube from liquid jet nozzle
a	m^{-1}	specific interfacial area
a_a	m^{-1}	specific G-L (gas-liquid) interfacial area in annulus
a_i	m^{-1}	specific G-L interfacial area in inner space
a_L	m^{-1}	specific G-L interfacial area ref. to liquid volume
a_R	m^{-1}	specific G-L interfacial area ref. to reactor volume
A	m^2	G-L interfacial area
A_a	m^2	cross-sectional area of annulus
A_a	m^2	G-L interfacial area in annulus
A_i	m^2	G-L interfacial area in inner space
A_o	m	distance of draft tube from surface
A_u	m	distance of draft tube from bottom
A_D	m	distance of jet nozzle from upper edge of draft tube
c	kmol m^{-3}; kg m^{-3}	concentration
c_a		volumetric solid particle concentration in annulus
c_o	kmol m^{-3}; kg m^{-3}	liquid side oxygen concentration at G-L interfacial area
$\bar{c}_o$	kmol m^{-3}	mean oxygen concentration over G-L interfacial area
$\bar{c}_o^M$	kmol m^{-3}	mean concentration over G-L interfacial area according to complete mixing model
$\bar{c}_o^S$	kmol m^{-3}	mean concentration over G-L interfacial area according to segregation model
c_r		relative concentration
c_G	kmol m^{-3}	molar oxygen concentration in G-phase
$\bar{c}_{Ga}^S$	kmol m^{-3}	mean O$_2$ concentration of G-phase in annulus according to segregation model
$\bar{c}_{Gi}^S$	kmol m^{-3}	mean O$_2$ concentration of G-phase in inner space according to segregation model
c_L	kmol m^{-3}	molar oxygen concentration in L-phase
c_{Lr}		relative oxygen concentration in L-phase ref. to $\bar{c}_o$
c_{LS}	kmol m^{-3}	molar oxygen concentration at L-S interfacial area
c_{O_2}	kmol m^{-3}	oxygen concentration
c_{So}		volumetric solid concentration at uniform distribution throughout reactor volume
c_{Sub}	kmol m^{-3}	substrate concentration
c_τ	kg m^{-3}	total tracer concentration in circulation system at time τ
c_τ	kg m^{-3}	tracer concentration in the outlet flow at time τ
Δc	kmol m^{-3}	driving concentration difference
C		constant
C_n		integration constant
D	m	reactor diameter
D_E	m	mean diameter of draft tube
D_I	m	diameter of dispersing tube
D_L	m^2 s^{-1}	diffusivity of gas (O$_2$) in liquid
D_M	m	diameter of the measuring ball
D_P	m	diameter of propeller
D_S	m	diameter of a solid particle
D_{eff}	m^2 s^{-1}	effective longitudinal diffusivity
D_1	m	diameter of liquid jet nozzle
F_{ALR}	N	air-lift driving force
g	m s^{-2}	gravity constant
$\dot{G}$	m^3 s^{-1}	gas volume flow
h		inhomogeneity
$\hat{h}$		desired maximal inhomogeneity
H	m	reactor height
H_G	m	settling height
H_L	m	height of liquid
H_P	m	delivery height of propeller
He		Henry coefficient

Symbol	Units	Description
$H(\tau)$		distribution function
i_U		number of circulations within time t
i_{UM}		number of circulations to obtain required degree of mixing
k		constant
k_s		correction factor for effect of slenderness ratio
k_n	$kmol^{1-n} \cdot m^{3(n-1)} s^{-1}$	nth order reaction rate constant
k_L	$m\, s^{-1}; m\, h^{-1}$	mass transfer coefficient
k_1		correction factor for liquid jet effect
k_2	$m^3\, kmol^{-1}\, s^{-1}$	2nd order reaction rate constant
K	$N\, m^{-2}\, s^n$	consistency factor
K_M	$kg\, m^{-3}; g\, L^{-1}$	Michaelis constant
L	m	length of measuring section
L	m	flow path
L_E	m	length of draft tube
L_I	m	length of dispersing tube
L_U	m	length of one circulation
$\dot{L}$	$m^3\, s^{-1}$	liquid volume flow
$\dot{m}_{O_2}$	$kg\, m^{-3}\, h^{-1}$	specific oxygen mass flow
M	kg	mass
M_R	kg	reaction mass
$\dot{M}$	$kg\, s^{-1}$	mass flow
$\dot{M}_G$	$kg\, s^{-1}$	gas mass flow
$\dot{M}_L$	$kg\, s^{-1}$	liquid mass flow
$\dot{M}_{O_2}$	$kg\, s^{-1}$	oxygen mass flow
$\dot{M}_{L1}$	$kg\, s^{-1}$	liquid mass flow through liquid nozzle (in heterogeneous systems)
$\dot{M}_1$	$kg\, s^{-1}$	mass through flow
$\dot{M}_2$	$kg\, s^{-1}$	circulation mass flow
$\dot{M}_3$	$kg\, s^{-1}$	total mass flow
n		flow index
n		order of reaction
n_{eq}		equivalent number of consecutive equal volume ideal STR
n_P	$s^{-1}; rpm$	propeller rotational speed
n_U		circulation number
$\dot{n}$	$kmol\, m^{-2}\, h^{-1}$	molar flow density
$\dot{n}_{O_2}$	$kmol\, m^{-2}\, h^{-1}$	oxygen molar flow density
$\dot{n}_{ph}$	$kmol\, m^{-2}\, h^{-1}$	molar flow density (pure physical diffusion)
N	kmol	molar amount of oxygen
$\dot{N}$	$kmol\, h^{-1}$	molar flow
$\dot{N}_a$	$kmol\, h^{-1}$	transferred molar flow of oxygen in annulus
$\dot{N}_i$	$kmol\, h^{-1}$	transferred molar flow of oxygen in inner space
$\dot{N}_{O_2}$	$kmol\, h^{-1}$	molar flow of oxygen
p_P	$N\, m^{-2}$	pressure head produced by propeller
Δp_U	$N\, m^{-2}$	pressure drop of one circulation
P	kW	total power input
P_L	kW	liquid jet power input
P_P	kW	pumping power of propeller
P_U	kW	circulation power
$\dot{q}$	$kW\, m^{-3}$	specific total caloric power
r_U	$s^{-1}; h^{-1}$	circulation rate
s		slenderness ratio
s_E	m; mm	thickness of draft tube wall
s_I	m; mm	thickness of dispersing tube
t	s	time
$\bar{t}$	s	mean residence time
t_a	s	time of measuring ball to pass measuring section
$\bar{t}_G$	s	mean residence time of gas
t_M	s	mixing time
$\bar{t}_U$	s	mean circulation time
T	K; °C	temperature
ΔT_m	K; °C	mean temperature difference
T_R	K; °C	reaction temperature
u_{O_2}		oxygen conversion rate
V	m^3	volume
V_F	m^3	liquid film volume
V_G	m^3	gas volume
V_L	m^3	liquid volume
V_R	m^3	reactor volume; reaction volume
$\dot{V}$	$m^3\, s^{-1}$	volume flow
$\dot{V}_G$	$m^3\, s^{-1}$	gas volume flow in G-L-system
$\dot{V}_{G\alpha}$	$m^3\, s^{-1}$	gas volume through flow
$\hat{\dot{V}}_G$	$m^3\, s^{-1}$	maximal receptivity of gas flow (air sparging at top of reactor)
$\dot{V}_L$	$m^3\, s^{-1}$	liquid volume flow in G-L-system
$\dot{V}_P$	$m^3\, s^{-1}$	volume flow from propeller
$\dot{V}_1$	$m^3\, s^{-1}; L\, s^{-1}$	liquid volume through flow
$\dot{V}_2$	$m^3\, s^{-1}$	liquid volume circulation flow
$\dot{V}_3$	$m^3\, s^{-1}$	total liquid volume flow
w	$m\, s^{-1}$	velocity
w_a	$m\, s^{-1}$	liquid velocity in annulus

Symbol	Units	Description
w_i	m s^{-1}	mean liquid velocity after momentum transfer in outlet of draft tube
w_m	m s^{-1}	mean liquid circulation velocity
w_G	m s^{-1}; cm s^{-1}	superficial gas velocity
w_M	m s^{-1}	velocity of measuring ball
w_S	m s^{-1}	free-settling velocity of single particle resp. measuring ball
w_{La}	m s^{-1}	liquid velocity in annulus
w_{Lm}	m s^{-1}	mean liquid velocity in heterogeneous systems
w_{SS}	m s^{-1}	free-settling velocity of particle clouds
w_{L1}	m s^{-1}	liquid velocity in liquid nozzle in heterogeneous systems
w_1	m s^{-1}	liquid velocity in liquid nozzle
w_2	m s^{-1}	velocity of circulation liquid entering draft tube
x		coordinate
x_U		numbers of completed circulations for any concentration element
X_o		dimensionless upper distance of draft tube
X_u		dimensionless lower distance of draft tube
y		coordinate
z		coordinate

Greek letters

Symbol	Units	Description
β_{ch}	m s^{-1}	chemical mass transfer coefficient
β_{ph}	m s^{-1}	physical mass transfer coefficient
$\dot{\gamma}$	s^{-1}	shear rate
δ	m; µm	film thickness
ε		(mean) volumetric gas hold-up
ε_a		volumetric gas hold-up in annulus
ε_i		volumetric gas hold-up in inner space
ε_l		local volumetric gas hold-up
ζ_a		resistance number for external tubular flow
ζ_i		resistance number for internal tubular flow
ζ_o		resistance number for flow around upper edges of draft tube
ζ_u		resistance number for flow around lower edges of draft tube
ζ_S		resistance coefficient of solid particle
ζ_U		circulation resistance number for JLR
ζ_W		resistance coefficient of measuring ball
ζ_U^*		circulation resistance number for LR without liquid jet
η	Pa s	dynamic viscosity
η_s	Pa s	apparent dynamic viscosity
η_L	Pa s	dynamic viscosity of liquid
η_U		efficiency of circulation
ν	m^2 s^{-1}	kinematic viscosity
ν_m	m^2 s^{-1}	mean kinematic liquid viscosity of total flow
ν_1	m^2 s^{-1}	kinematic viscosity of liquid entering through jet nozzle
ϱ	kg m^{-3}	density
ϱ_m	kg m^{-3}	mean liquid density of total flow
$\bar{\varrho}_a$	kg m^{-3}	mean density of G-L-system in annulus
ϱ_M	kg m^{-3}	density of measuring ball
ϱ_S	kg m^{-3}	density of solid particles
ϱ_1	kg m^{-3}	density of liquid entering through jet nozzle
τ	N m^{-2}	shear stress
τ		relative residence time
τ_M		relative mixing time related to $\bar{t}_U$
τ_U		relative time related to $\bar{t}_U$
Φ		chemical acceleration

Dimensionless numbers

B	volumetric number
Bo	Bodenstein number
Eu	Euler number
Fr_G	Froude number ref. to gas
Fr_S	Froude number ref. to solid particles
Ha	Hatta number
Ne	Newton number
N_V	volume flow number
Re_m	Reynolds number ref. to circulation liquid
Re_G	Reynolds number ref. to gas
Re_P	Reynolds number ref. to propeller
Re_S	Reynolds number ref. to solid particles
Re_P^+	modified Reynolds number ref. to propeller
Re_1	Reynolds number for liquid nozzle

Symbols

∞	infinite
$\wedge$	maximum
$\vee$	minimum
—	mean
Δ	difference
Σ	summation

Abbreviations

ALR	air-lift loop reactor
BCR	bubble column reactor
BOD	biological oxygen demand
CMC	carboxymethylcellulose
FJR	free jet reactor
G	gas
JLR	jet loop reactor
JNR	jet nozzle reactor
L	liquid
LR	loop reactor
PLR	propeller loop reactor
S	solid
SCP	single cell protein
STR	stirred tank reactor

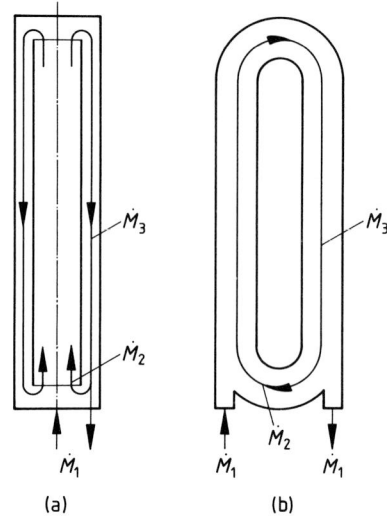

Figure 21.1. Principles of Loop Reactors (LR). (a) with internal circulation, (b) with external circulation. $\dot{M}_1$ through flow, $\dot{M}_2$ circulation flow, $\dot{M}_3$ total flow.

21.1 Principles and Types of Loop Reactors (LR)

21.1.1 Flow Paths and Arrangements

LR are chemical and bioreactors for fluids and fluidized systems, in which, as shown in Fig. 21.1 at least one definitely directed circulation flow occurs [21.1]. The circulation flow $\dot{M}_2$ can be superimposed by a through flow $\dot{M}_1$ resulting in the total flow $\dot{M}_3 = \dot{M}_1 + \dot{M}_2$. In this way the flow pattern of a loop is formed; hence the name.

Typical arrangements are slim tower-like apparatuses with internal guiding parts, which direct the flow of the "internal circulation" (in this case a concentric "draft

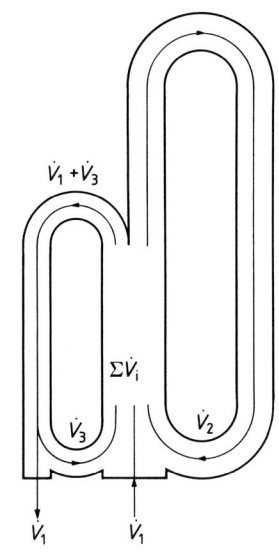

Figure 21.2. Twin-LR with two external circulations. $\dot{V}$ volume flow.

tube"), as illustrated in Fig. 21.1a, or as shown in Fig. 21.1b LR-types having an "external circulation". LR with internal circulation are also operated with inversed

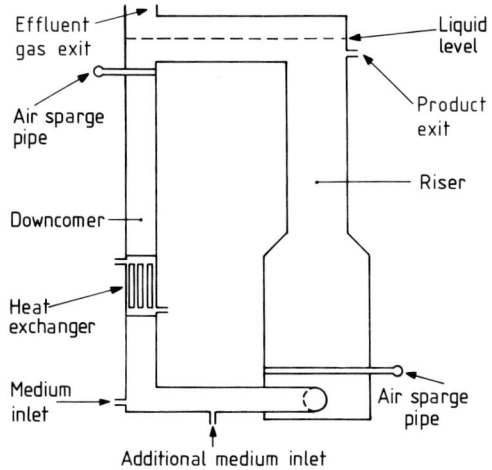

Figure 21.3. ICI Pressure Cycle Fermenter with external circulation.

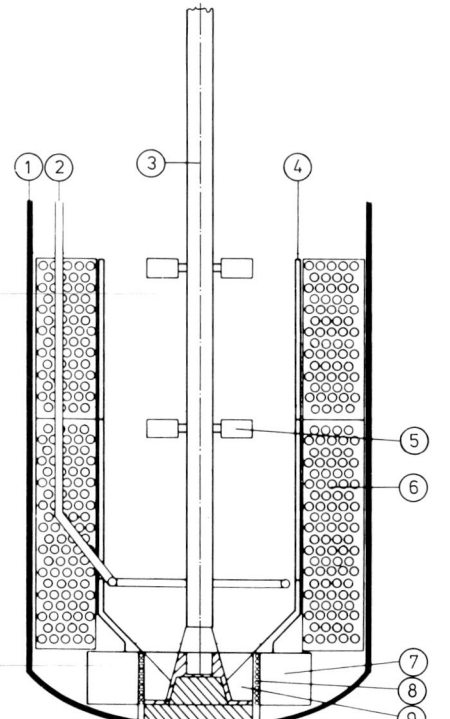

Figure 21.4. Immobilization of microorganisms on glass in a combination of LR and STR.
1 reactor wall, 2 air sparger, 3 stirrer shaft, 4 draft tube, 5 blade stirrer, 6 carrier material (e.g., glass balls), 7 guide device, 8 emulsifier, 9 turbine stirrer.

flow direction (see e.g. Figs. 21.10 and 21.16). Fig. 21.2 shows the principle of a twin-LR with two external circulations, in which the flow velocities and thus the mean residence times can be selected independently.

Fixed parts can be arranged in the flow-through zones of all types of LR, e.g., according to Fig. 21.3 heat exchangers and air spargers or as in Fig. 21.59 baffle plates to redisperse coalesced gas bubbles or packing as carrier material for immobilized cells or enzymes and blade stirrers for redispersion in the inner space as illustrated in Fig. 21.4 [21.2] or "static mixers" in the upflow which effect redispersion and uniform distribution of the gas bubbles [21.3].

21.1.2 Types of Circulation Drives

One classifies loop reactors not only according to flow path and type of design, but also by the mode of circulation drive:

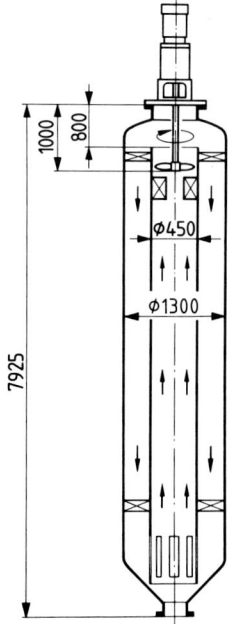

Figure 21.5. Propeller Loop Reactor (PLR).

- *Propeller Loop Reactors* (PLR) with hydromechanic propeller flow drive, as shown in Fig. 21.5 [21.4].
- *Air-lift Loop Reactors* (ALR) with hydrostatic circulation drive caused by different densities (especially on account of different gas hold-up) in communicating spaces, for example as shown in Fig. 21.3 with external and in Fig. 21.6a with internal circulation.

sary, liquid nozzles allow very large reactor volumes with smaller construction heights as shown in Fig. 21.7 [21.7], [21.8].

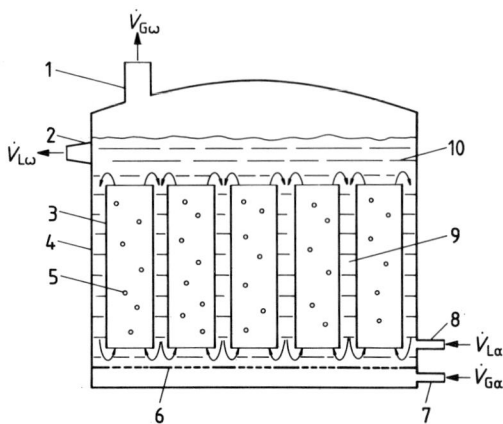

Figure 21.7. Bio-ALR with several parallel arranged draft tubes, each of which has an individual gas sparger at the perforated bottom plate.
1 gas exit, 2 liquid outlet, 3 draft tubes, 4 reactor wall, 5 rising flow with gas bubbles within the draft tubes, 6 perforated plate for gas sparging, 7 gas feed, 8 liquid inlet, 9 outer space between the draft tubes, 10 G-L-separation zone.

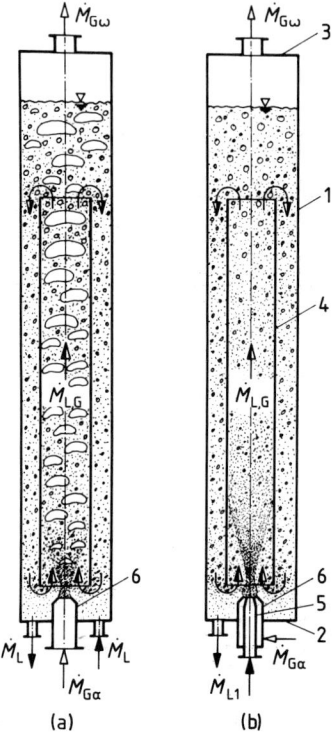

Figure 21.6. Schemes of Air-lift Loop Reactor (ALR), left (a), and Jet Loop Reactor (JLR), right (b), with G-L-system.
1 jacket, 2 bottom, 3 lid, 4 draft tube, 5 liquid nozzle, 6 gas sparger.

- *Jet Loop Reactors* (JLR) with hydrodynamic jet flow drive, as shown in principle in Fig. 21.6b [21.6]. The liquid jet $\dot{M}_{L1}$ is injected through at least one liquid nozzle with high velocity $w_{L1} \gtrsim 20$ m s^{-1}.

Several draft tubes *parallel* to one another with individual gas feed and, if neces-

21.1.3 Differences from Other Jet Reactors

As shown in Fig. 21.8, one can operate basically different G-L-reactors with the same mode of gas and liquid feeds – in this case a central liquid jet $\dot{V}_L$ with gas feed $\dot{V}_{G\alpha}$ through a concentric ring nozzle [21.9].

The design in Fig. 21.8a is the abovementioned *"jet loop reactor"* JLR, characterized by the draft tube of length L_E, which effects the definitely directed flow and consequent low-loss inter-mixing in the *whole* reaction space. The draft tube with its rela-

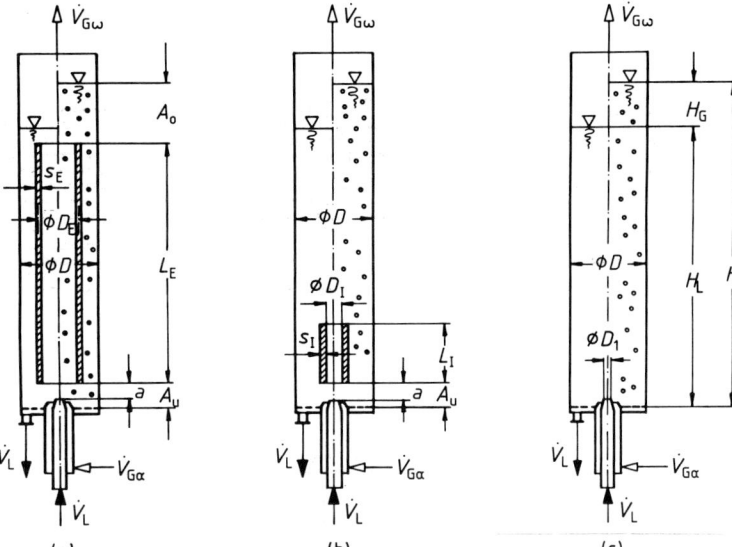

Figure 21.8. Different types of jet reactors. (a) Jet Loop Reactor (JLR), (b) Jet Nozzle Reactor (JNR), (c) Free Jet Reactor (FJR).

tively large diameter D_E compared with that of the liquid jet nozzle D_1 has hardly any influence on the dispersion caused by the liquid jet. This occurs practically just as with the free jet reactor (FJR) shown in Fig. 21.8c. The function of the draft tube in the JLR is thus not the intensification of dispersion, but a definite distribution of all components throughout the whole reactor content.

On the other hand the *"jet nozzle reactor"* (JNR) shown in Fig. 21.8b forms a space, with its comparatively small dispersing tube, in which the gas is very finely dispersed by intensive impulse transfer and high energy-dissipation density [21.10], [21.11].

The *"free jet reactor"* (FJR) as shown in Fig. 21.8c has no internal parts at all to influence distribution or dispersion. It represents a bubble column reactor (BCR) with jet gas sparging at the bottom.

The author and his co-workers investigated almost all reactor types cited above – i.e., ALR, JLR, JNR, FJR and combinations – in order to compare them with the same G-L-system under the same operational and geometrical conditions.

21.1.4 Some Construction Examples

21.1.4.1 Loop Reactors

Fig. 21.9 shows some constructions of PLR and gives an impression of the reactor size; for example in the type shown in Fig. 21.9b the active reaction volume is approximately $V_R \approx 200$ m^3. That is a remarkable volume for a PLR. Some examples of ALR with internal and external circulation are shown schematically in Fig. 21.10 [21.12].

Fig. 21.11 represents a JLR series made of glass. This model series includes reactor diameters $D = 100$ to 400 mm; heights $H = 1100$ to 2700 mm; draft tube diameters $D_E = 51$ to 211 mm; reactor volumes $V_R = 4$ to 210 L. The liquid jet is injected at the bottom in the center. A side-tube inlet through the reactor wall and the lower end of the draft tube allows introduction of liquid or gas to the liquid jet directly above the liquid nozzle. The liquid leaves the reactor through the side-tube at the bottom around the liquid inlet tube.

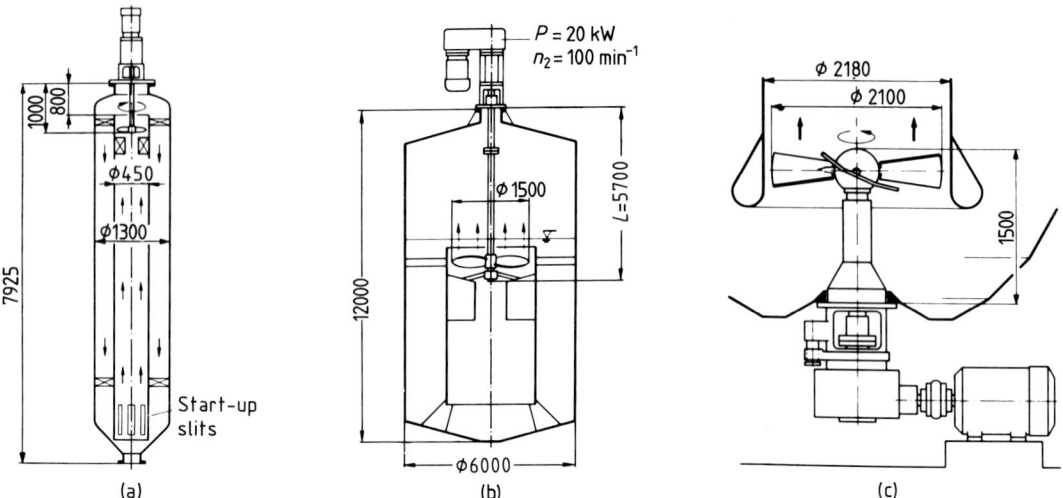

Figure 21.9. Types of construction of loop reactors mainly used with chemical L-systems but in principle also appropriate for G-L-biosystems.
(a) slim tower type PLR, (b) compact PLR, (c) propeller with bottom drive in PLR.

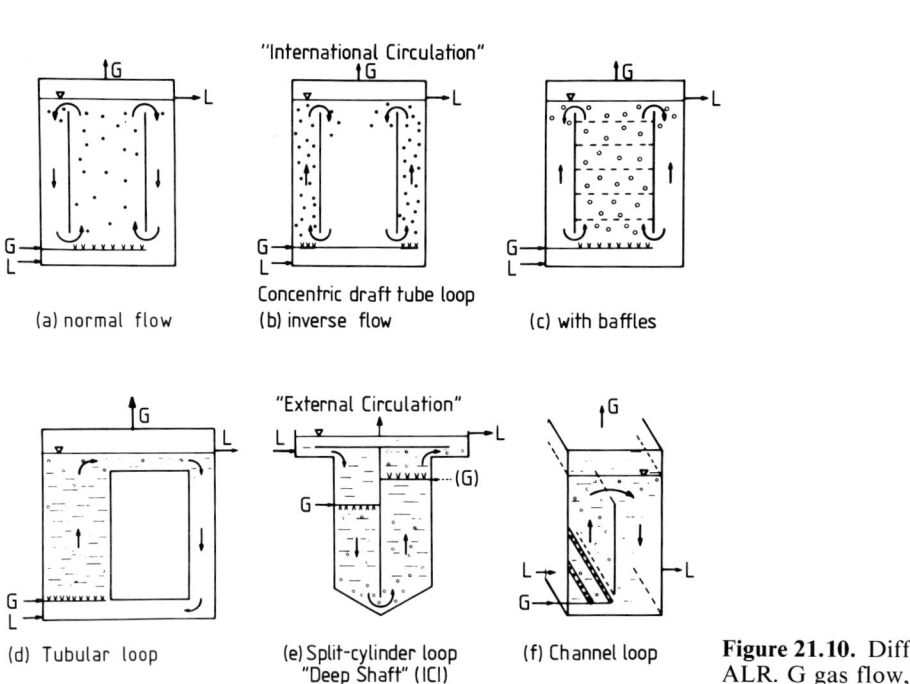

Figure 21.10. Different types of ALR. G gas flow, L liquid flow.

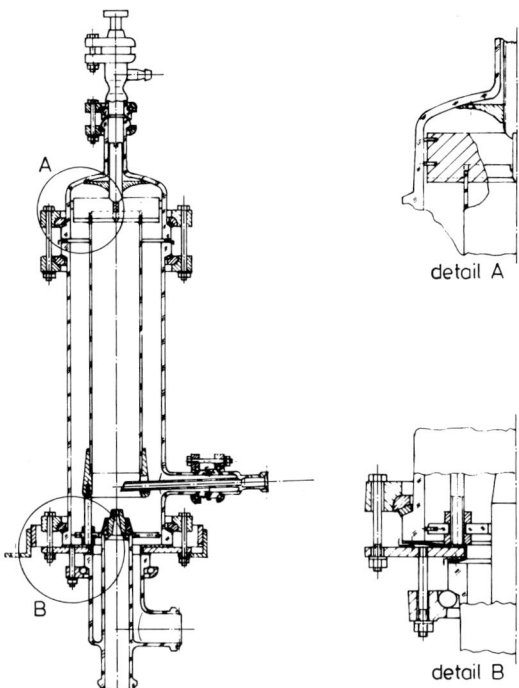

Figure 21.11. Series of laboratory JLR made of glass.

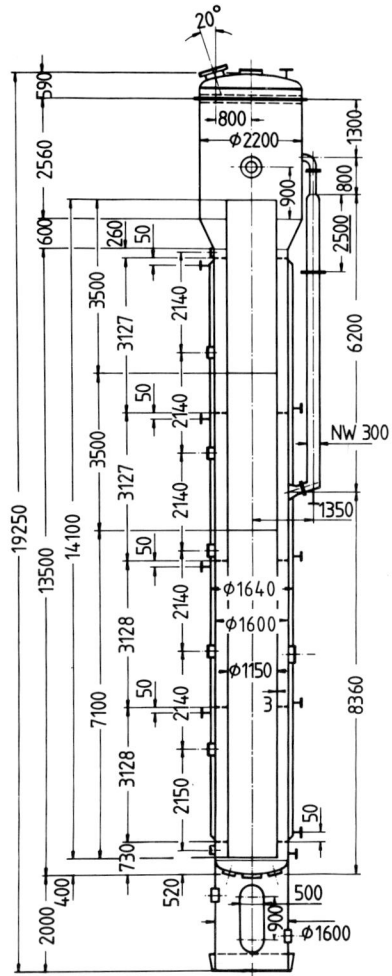

Figure 21.12. Design of a pilot plant LR, which can be operated as ALR or JLR.

A pilot plant fermenter for the SCP process of Hoechst-Uhde, which can be operated as ALR or as JLR is shown in Fig. 21.12 with all dimensions [21.13]. With a reactor volume $V_R \approx 40$ m^3 it produces about 1000 t/a SCP.

In the JLR examples given up to now the submerged gas sparging always took place – either in the inner space or in the annulus – at the bottom of the reactor in the rising liquid flow zone. Experiments were naturally also performed, and commercial LR developed in which submerged air-sparging takes place into the downward-directed liquid flow in the upper part of the reactor.

The principle of such an experimental apparatus is indicated in Fig. 21.13a with a G-L-ring nozzle. Gas deflectors according to Fig. 21.13b considerably increase $\hat{V}_G$ and therewith the maximal superficial gas velocity $\hat{w}_G = 4 \cdot \hat{V}_G / \pi D^2$, because they reduce the recirculation of gas at the top into the downflow and thus increase the air-lift effect, as well as the jet drive of the circulation flow. Fig. 21.14 shows that with increasing distance A_D from the upper edge of the draft tube at the same specific power input P_L/V_R of the liquid jet the maximal values of $\hat{V}_G$ and $\hat{w}_G$ are considerably increased, because in this way the circulation drive is increased. Examples of gas through flows which can be so achieved with the air-water system are presented in Fig. 21.15. They show that apparently with

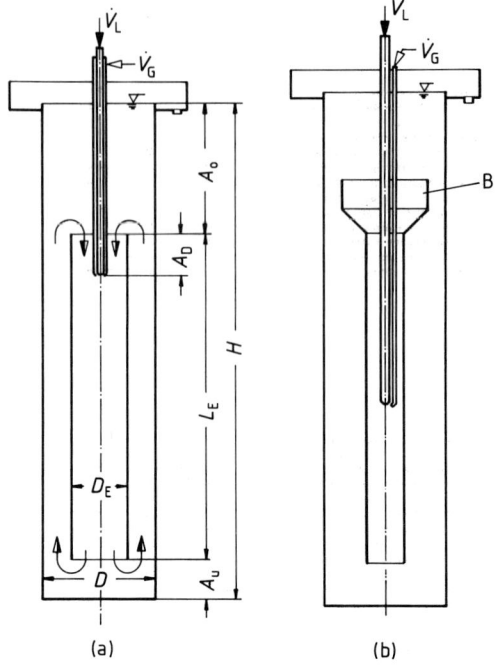

Figure 21.13. Sketch of the model reactor JLR 630, (a) without, (b) with gas deflector B.

gas sparging from top to bottom, as shown in Fig. 21.16. The flow behavior corresponds in principle to that already described.

A fundamentally different type of gas feed, from above the liquid surface, is used in the "plunging liquid jet reactor" [21.16], whose mode of function is shown schematically in Fig. 21.17 [21.17]. The plunging jet hits the liquid surface as a G-L-jet and transports its own gas content and additional gas sucked out of the reactor head into the liquid phase; at the same time this effects a mechanical defoaming.

The "emulsion biological reactor" is also charged from above with a G-L-phase as shown in Fig. 21.18, which represents a JLR with external circulation [21.18]. It contains a fixed bed of packings onto which a G-L-mixture ("emulsion") is sprayed as uniformly as possible by the recirculating pump IV, via the heat exchanger V, the venturi nozzle III, and a distributor VI. The oxygen transfer rate per unit power input is given by $\dot{M}_{O_2}/P \approx 1.6$ to 2.5 kg/kWh (see Table 21.1).

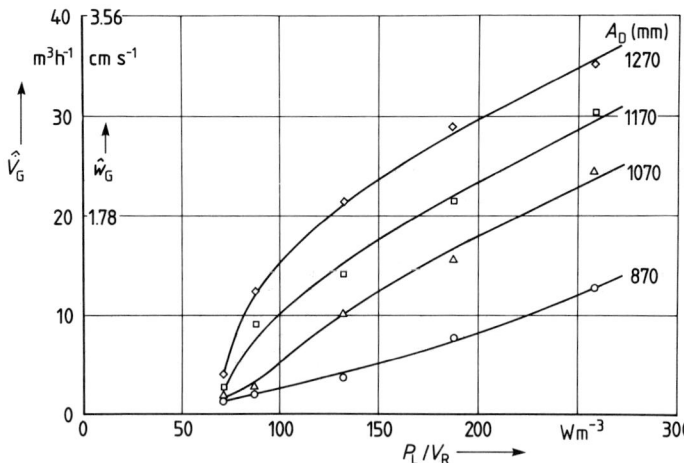

Figure 21.14. Maximal gas through flow $\hat{\dot{V}}_G$ with gas deflector B and liquid nozzle diameter $D_1 = 10$ mm. $\dot{V}_G = 10$ m^3 h^{-1} corresponds to $w_G = 0.89$ cm s^{-1}.

this mode of operation only relatively low superficial gas velocities (in this case $w_G \lesssim 2.5$ cm s^{-1}) can be achieved [21.14].

The "compact reactor" [21.15] is mentioned as a further example of submerged

As a conclusion to the selection of typical construction types, the most important problems to be solved in designing reactors for heterogeneous G-L-biosystems are given in Fig. 21.19 [21.19]. The left side rep-

21.1 Principles and Types of Loop Reactors (LR)   477

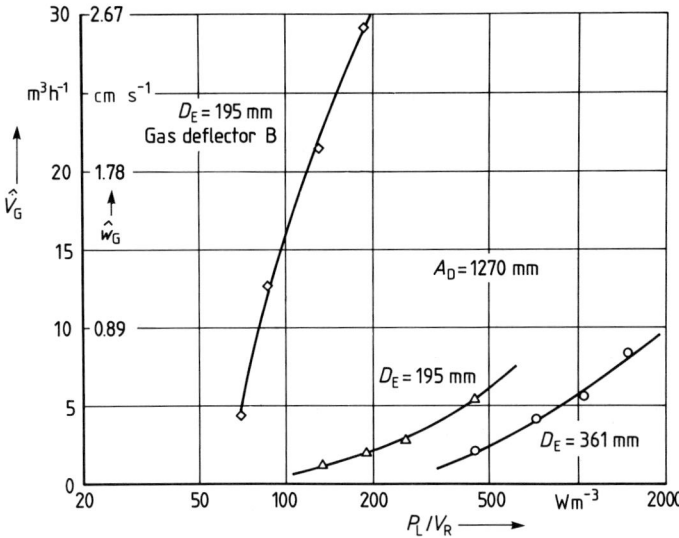

Figure 21.15. Maximal gas through flow $\dot{V}_G$ with inclined aeration tube according to Fig. 21.21b (right). Lower curves without gas deflector; upper curve with gas deflector B. D_E and A_D according to Fig. 21.13. $\dot{V}_G = 10$ m^3 h^{-1} corresponds to $w_G = 0.89$ cm s^{-1}.

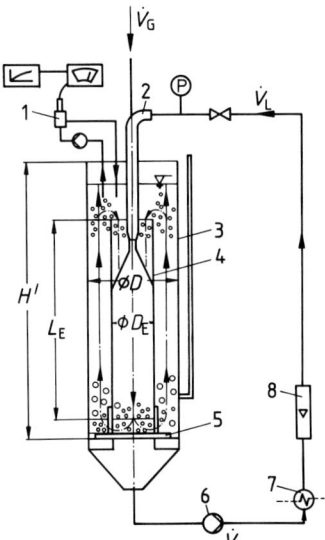

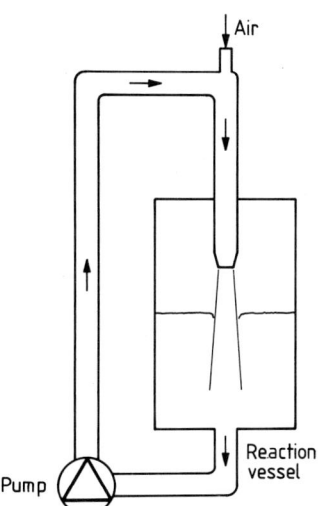

Figure 21.16. "Compact reactor" with circulation flow as shown in Fig. 21.13.
1 O$_2$ probe, 2 suction of the circulating G-L-system and fresh air, 3 reactor wall, 4 draft tube, 5 reflecting plate, 6 liquid pump, 7 heat exchanger, 8 flow meter.

Figure 21.17. Sketch of the plunging liquid jet reactor.

Table 21.1. Comparison of the Interfacial Area and the Oxygen Transfer Rate of Various Reactor Types

Gas Liquid Reactor Type	Interfacial Area per unit power input A/P (in m^2/kW)	Oxygen Transfer Rate per unit power input $\dot{M}_{O_2}/P$ (in kg/kWh)
Air-lift loop reactor ALR	300–500	1.2–2.0
Bubble column reactor BCR	200–600	0.8–2.4
Stirred tank reactor STR	300–600	1.2–2.4
Jet loop reactor JLR	1000–1200	4.0–6.0

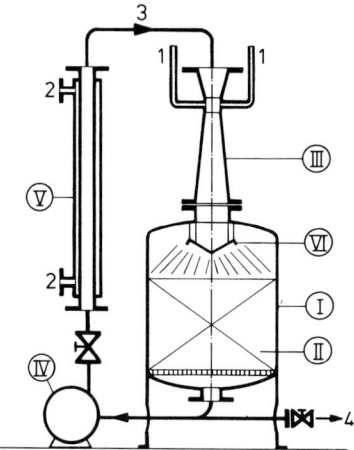

Figure 21.18. Example of an "emulsion biological reactor".
I reactor, II packing of fixed bed, III venturi, IV recirculating pump, V heat exchanger, VI internal distributor, 1 gas in, 2 cooling water in and out, 3 liquid in, 4 liquid out.

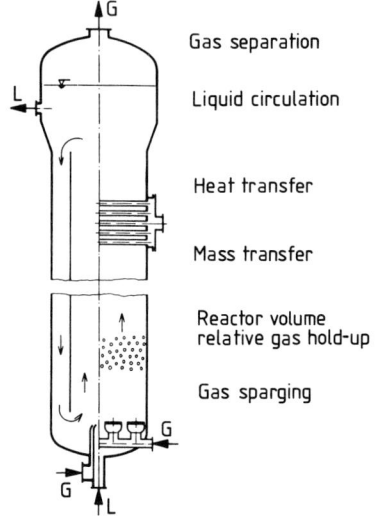

Figure 21.19. General problems in reactor design for G-L-systems. The left side represents a JLR, the right side a bubble column.

resents a JLR with internal circulation and gas feed through a ring nozzle at the bottom; the right side shows a bubble column with various internal parts and functional details, which are valid in principle for all bioreactors with G-L-systems (see further details in Sect. 21.2).

21.1.4.2 Loop Reactor Parts

Examples of gas sparging for ALR and for bubble columns are presented schematically in Fig. 21.20 [21.19]. The "single hole plate", in which gas and liquid phases flow through a large central hole, represents a transition stage between a sieve plate and a nozzle. Various gas sparging nozzles in combination with a centrally injected liquid jet for JLR are shown in Fig. 21.21 [21.1], [21.20], [21.21]. Fig. 21.22 shows the mode of operation of the "radial flow nozzle", which has proved itself in the "BIOHOCH Reactor" of Hoechst for biological waste water purification [21.22].

Since most G-L-biosystems have a tendency to foam, mechanical, hydraulic, or chemical defoamers are often necessary.

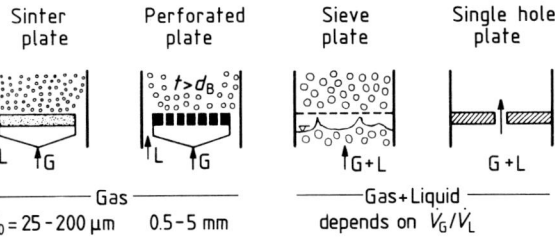

Figure 21.20. Examples of bubble formation by pores and holes.

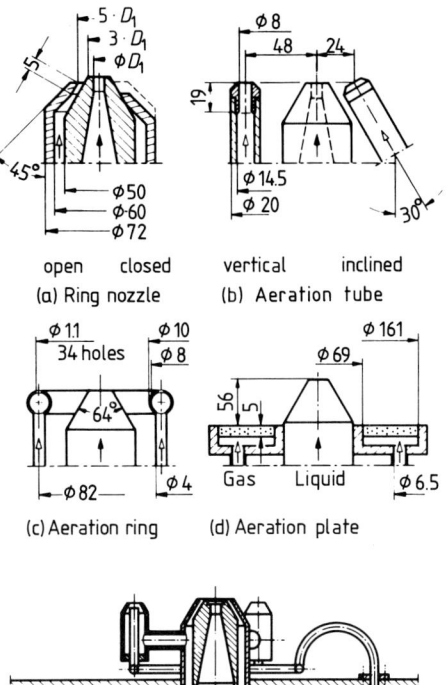

(a) Ring nozzle (b) Aeration tube

(c) Aeration ring (d) Aeration plate

(e) Central G-L-nozzle, surrounded by another three G-L-nozzles mounted on a central ring

Figure 21.21. Examples of G-L-nozzle arrangements for JLR.

But the easiest way to solve a problem is to avoid it, as shown in the example of Fig. 21.23.

The only free surface of the system occurs here under the influence of strong centrifugal and shear forces, which avoid bubble formation at gas exit *in statu nascendi*. This hydrodynamic foam suppression requires no shaft arrangements like mechanical defoamers, which facilitates the maintenance of sterility in the reaction system. Also it has no influence on the reaction system, unlike chemical foam inhibitors or other defoamers. Furthermore it allows the complete utilization of the whole reactor space [21.23].

Since it is most important from a biological and economic point of view to maintain the sterility of all bioreactors, tight seals are essential for all shafts. In the case of small fermenters, propeller drive from below is often used so that the lid is reserved for other purposes (nutrient input, inoculation, inspection, defoaming ...). Such a drive arrangement is illustrated in Fig. 21.9c. For larger PLR one usually prefers to use an arrangement with the shaft passing overhead through the lid, e.g., as shown in Fig. 21.9a and b.

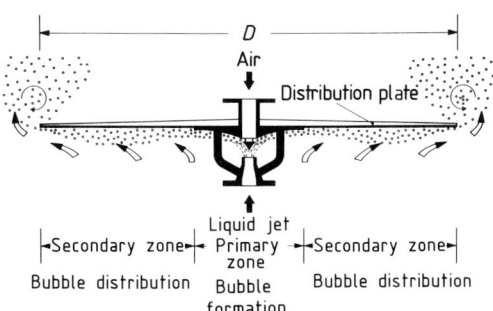

Figure 21.22. "Radial flow nozzle" in "BIO-HOCH Reactor" of Hoechst AG.

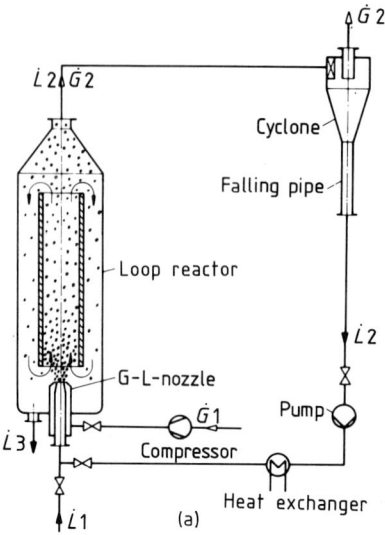

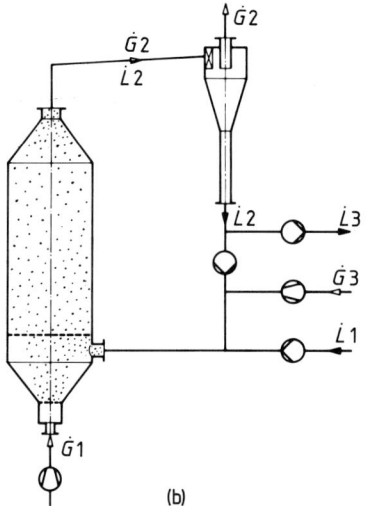

Figure 21.23. Loop reactors as examples for equipment with G-L-systems and foam preventing operation.
(a) Loop reactor with internal circulation.
(b) Loop reactor with external circulation and additional gas input $\dot{G}3$ into the downflow.

21.2 Demands on Bioreactors

With respect to this topic reference shall be made first of all again to Fig. 21.19 and the problems indicated there for the design of G-L-bioreactors. Fundamentally the following demands are essentially valid for all types of bioreactors with heterogeneous G-L-systems [21.1], [21.24], [21.25].

21.2.1 Defined Distribution

A clearly defined distribution of all components throughout the reaction space V_R is required. Thereby the flow, mixing, and residence time behavior are influenced. Thus for example in the SCP process of Hoechst-Uhde for growth of *Methylomonas clara* on methanol, the microorganisms, substrate, and oxygen must be distributed as uniformly as possible throughout the liquid phase so that the limits $1‰ \lesssim c_{Sub} \lesssim 1\%$ and $c_{O_2} \lesssim 1$ ppm are not exceeded anywhere for long periods of time. In addition, no stable foam layer should be formed in which microorganisms might be flotationally enriched, damaged, or might even die. Also sedimentation of reaction components, e.g., cell-agglomerates should be avoided.

All LR-types fulfill these requirements very well due to their specifically directed adjustable flows and concentrations throughout the whole reaction space, defoaming, and swirling of sediments, especially with the corresponding design of the reactor bottom.

21.2.2 Fine Dispersion

A fine dispersion of all disperse phases (G, L, S) is an important requirement for successful operation of any bioreactor. This especially influences the local and overall gas hold-up, specific interfacial area, and mass transfer. This demand is best fulfilled by PLR and JLR due to their intensive primary dispersion in strong impact and shearing fields and additional effective redispersion in the forced repeated passage through the propeller or jet zone. With respect to primary dispersion, this is also valid for the combination of the LR with a jet nozzle according to Fig. 21.8b, at low gas through flows and with weakly coalescing systems.

21.2.3 No Cell Damage

Cell damage should be avoided at high power inputs with sensitive organisms, especially with plant cell fermenters. All types of ALR are especially suited for this, but also JLR which do not have very high injected liquid jet velocities ($w_{L1} \lesssim 30$ m s^{-1}). The necessary circulation can be achieved under these conditions according to Eq. (21.22) by using correspondingly large nozzle diameters D_1.

21.2.4 Large Specific Heat Transfer

Required is a large heat transfer area arranged *in* the reactor, which is not overgrown by microorganisms. In the case of highly exothermic bioreactions, e.g., in the SCP process, a specific caloric power of $\dot{q} \approx 30$ to 40 kW m^{-3} has to be removed at a very small mean temperature difference $\Delta T_m \lesssim 10$ K between the biosystem and cooling medium, since, as microorganisms are used, the reaction temperature as a rule lies close to 37 °C. With all types of LR with internal circulation, double-jacketed draft tubes can be used to advantage as additional internal cooling equipment without flow impairment and with minimal volume requirements.

21.2.5 Simple Construction

Simplicity of design and the realization of large bioreactors of several 1000 m^3 with low construction costs is desirable. ALR and JLR fulfill this demand very well since they only consist of a tower or tank with one, or several, draft devices and several nozzles, as shown in Fig. 21.7.

ICI bioreactors have volumes $V_R \approx 2000$ m^3, the "BIOHOCH Reactor" of Hoechst at Kelheim has an activated volume of $V_R \approx 3000$ m^3. In this respect bubble columns are also very favorable as illustrated by the Bayer AG "tower reactors" with $V_R \approx 13\,000$ m^3.

21.2.6 Simple Mode of Operation

Good sterilizability and maintenance of sterility, low mechanical maintenance, low power requirements for the necessary mixing and dispersion effects, as well as the feasibility of large sizes should be provided. ALR and JLR fulfill these demands very well, as later discussions – especially of the specific power requirement – will show.

21.2.7 Easy Adjustment

Operating conditions should be easily adjustable, e.g., in batch operation over wide ranges of temperature, concentration, viscosity, etc., or to production changes in continuous processes. All types of LR fulfill these demands very well.

21.2.8 Good Computability

Design and operation behavior should be well computable and suitable for scale-up with the help of model systems and model equipment.

In this respect the development and scientific investigation of LR lie behind that of the STR and BCR which have been investigated and operated much longer. Therefore corresponding R&D work is of particular importance for potential applications and biotechnical, as well as economic improvements.

21.3 Fluid Dynamics of Loop Reactors (LR)

21.3.1 Fluid Dynamic Design of LR for Aqueous (Quasi-)Homogeneous L-Systems

Although in biotechnology LR are especially suitable and used in heterogeneous G-L-systems, the basic principles and the most important fluid dynamic parameters and relations will first be demonstrated briefly using the fundamental homogeneous liquid system of drinking water. In principle these considerations are approximately also valid for multi-phase systems with such fine dispersion und uniform distribution of all components that the systems can be treated as quasi-homogeneous.

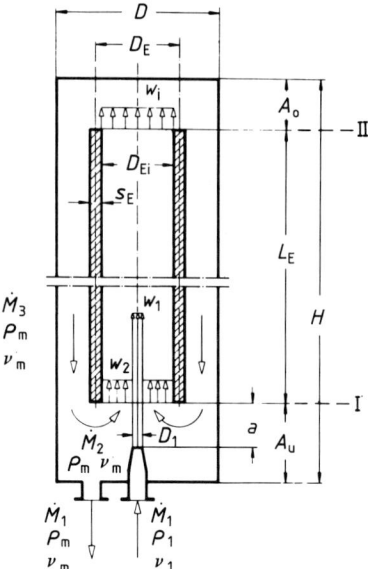

Figure 21.24. Schematic drawing of a JLR and its function showing the most important fluid dynamic parameters.
D internal diameter of loop reactor, H filling height of reaction system, D_E mean diameter of draft tube, D_{Ei} internal diameter of draft tube, L_E length of draft tube, A_u distance of draft tube from bottom, A_o distance of draft tube from surface, a distance of draft tube from liquid nozzle, D_1 diameter of liquid nozzle and liquid jet, $\dot{M}_1$ liquid mass through flow, $\dot{M}_2$ liquid circulation mass flow, $\dot{M}_3$ total mass flow of liquid.

As fluid dynamics are most instructive in JLR and the results can easily be modified for ALR and PLR, we shall consider as a basic example first of all the JLR shown in Fig. 21.24 [21.1], [21.26], [21.27].

21.3.1.1 Fundamental Fluid Dynamic Parameters

Important parameters for fluid dynamic design are the following:

- height-to-diameter ratio of the reactor ("slenderness ratio"):

$$s \equiv \frac{H}{D} ; \qquad (21.1)$$

- effective reactor volume:

$$V_R = \frac{\pi}{4} D^2 H = \frac{\pi}{4} s D^3 ; \qquad (21.2)$$

- reaction mass:

$$M_R = \varrho_m V_R = \varrho_m \frac{\pi}{4} s D^3 ; \qquad (21.3)$$

- circulation number:

$$n_U \equiv \frac{\dot{M}_3}{\dot{M}_1} = \frac{\dot{M}_1 + \dot{M}_2}{\dot{M}_1} = 1 + \frac{\dot{M}_2}{\dot{M}_1} ; \qquad (21.4)$$

- mean circulation velocity:

$$w_m \equiv \frac{8 \dot{V}_3}{\pi D^2} = \frac{8 \dot{M}_3}{\varrho_m \pi D^2} = \frac{8 \dot{M}_1}{\varrho_m \pi D^2} n_U \sim n_U ; \qquad (21.5)$$

- circulation rate from Eqs. (21.3) and (21.5) with Eq. (21.1):

$$r_U \equiv \frac{\dot{M}_3}{M_R} = \frac{\dot{V}_3}{V_R} = \frac{w_m}{2H} \equiv \bar{t}_U^{-1} ; \qquad (21.6)$$

- mean circulation time:

$$\bar{t}_U \equiv \frac{M_R}{\dot{M}_3} = \frac{2H}{w_m} \equiv r_U^{-1} ; \qquad (21.7)$$

- mean residence time of the continuous through flow $\dot{M}_1$ in V_R for liquid systems:

$$\bar{t} \equiv \frac{M_R}{\dot{M}_1} = \frac{\dot{M}_3 M_R}{\dot{M}_1 \dot{M}_3} = n_U \bar{t}_U = \frac{n_U}{r_U} ; \qquad (21.8)$$

- liquid velocity (for JLR) at nozzle outlet:

$$w_1 = \frac{4 \dot{V}_1}{\pi D_1^2} = \frac{4 \dot{M}_1}{\varrho_1 \pi D_1^2} ; \qquad (21.9)$$

- nozzle Reynolds number with Eq. (21.9):

$$Re_1 \equiv \frac{w_1 D_1}{\nu_1} = \frac{4 \dot{M}_1}{\nu_1 \varrho_1 \pi D_1} \sim D_1^{-1} ; \qquad (21.10)$$

- mean circulation Reynolds number with Eqs. (21.4) and (21.5):

$$Re_m \equiv \frac{w_m D}{\nu_m} = \frac{8 \dot{V}_3}{\nu_m \pi D} = \frac{8 \dot{M}_1 n_U}{\nu_m \varrho_m \pi D} \sim \frac{n_U}{D} ; \qquad (21.11)$$

- circulation Reynolds ratio with Eqs. (21.10) and (21.11):

$$\frac{Re_m}{Re_1} \equiv 2 \frac{\nu_1 \varrho_1 D_1}{\nu_m \varrho_m D} n_U \sim \frac{D_1}{D} n_U ; \qquad (21.12)$$

or for constant geometrical and material parameters:

$$\left(\frac{Re_m}{Re_1}\right)_{\frac{D_1}{D}} \sim n_U . \qquad (21.13)$$

21.3.1.2 Circulation Intensity in JLR with (Quasi-)Homogeneous Aqueous L-Systems

It is obvious that the contents of LR are the more intensely mixed, the larger the mean circulation velocity w_m. The aim of this fluid dynamic consideration, which is demonstrated here for a JLR as shown in Fig. 21.24, is to achieve the fastest possible circulation flow with a given liquid jet power input P_L, Eq. (21.24).

The intensity of the circulation flow is a function of the resistance number ζ_U (resp. the Euler number Eu_U [21.28]):

$$\zeta_U \equiv \frac{2 \Delta p_U}{\varrho_m w_m^2} \equiv Eu_U . \qquad (21.14)$$

Δp_U is the pressure drop due to one circulation of the fluid. For the resistance number ζ_U the following equation has been derived which may be safely applied to technical scale reactors:

$$\zeta_U \approx \frac{\varrho_m}{2 \varrho_1} \left(\frac{D_1}{D} n_U\right)^{-2} \left(\frac{D_E}{D}\right)^{-2} . \qquad (21.15)$$

Substituting $(D_1/D) n_U$ by an expression obtained from Eq. (21.12) it follows:

$$\frac{Re_m}{Re_1} \approx \frac{\nu_1}{\nu_m} \left(2 \frac{\varrho_1}{\varrho_m}\right)^{\frac{1}{2}} \zeta_U^{-\frac{1}{2}} \left(\frac{D_E}{D}\right)^{-1} . \qquad (21.16)$$

According to the laws of fluid dynamics, ζ_U can be expressed as the sum of the consecutive partial resistances, ζ_i and ζ_a, for the internal and external tubular flow and ζ_o and ζ_u for the flow around the upper and lower edges of the draft tube, in this way defined as:

$$\zeta_U^* \equiv \zeta_i + \zeta_a + \zeta_o + \zeta_u. \qquad (21.17)$$

ζ_i and ζ_a can be calculated from well-known fluid dynamic laws. However, unfortunately we could not find similar calculation methods for ζ_o and ζ_u as $f(D_E/D; s_E; A_o; A_u$ and draft tube shape). Thus we were forced to perform experiments in order to determine ζ_U and ζ_U^*.

21.3.1.3 Experimental Determination of ζ_U, ζ_U^* and Re_m/Re_1

There are two different possibilities of determining ζ_U resp. ζ_U^*. Both have been realized in two parallel experimental ways [21.26], [21.29] also for comparison and control.

- In a first group of experiments the circulation numbers n_U were determined by measuring the flow velocity w_a in the annulus of a model JLR made of acrylic glass

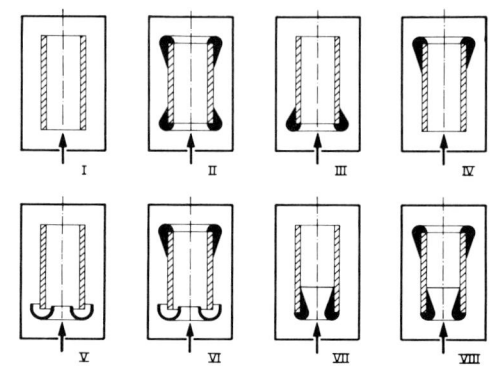

Figure 21.25. Types of draft tubes used in the fluid dynamic experiments with model JLR, which led to the results shown in Figs. 21.26 and 21.27.

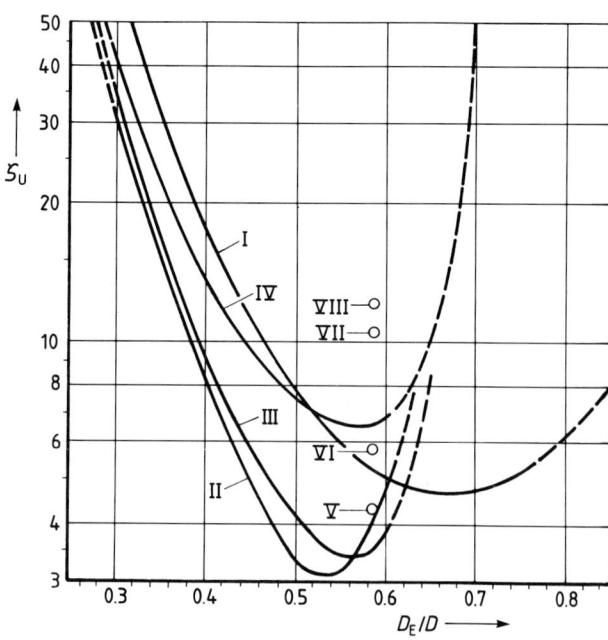

Figure 21.26. Circulation resistance number ζ_U, obtained by Eq. (21.15) and experiments in model JLR as a function of D_E/D and draft tube shape according to Fig. 21.25.

with $D = 290$ mm and $s = 5$ using drinking water as L-system [21.26].

In these experiments the following parameters have been kept constant, because in preliminary test runs they had been found to be optimal:

$$L_E \approx 7.5 D_E; \quad a \approx 0.1 D; \quad \frac{A_o}{D} \approx 0.2 \frac{D}{D_E};$$

$$\frac{A_u}{D} \approx 0.15 \frac{D}{D_E}. \qquad (21.18)$$

Over the whole range investigated, [21.26], [21.27], [21.29], from

$$2 \cdot 10^4 \leq Re_1 \leq 6 \cdot 10^5 \qquad 0 < \dot{M}_1 \leq 1.4 \text{ kg} \cdot \text{s}^{-1}$$

$$29 \leq \frac{D}{D_1} \leq 159 \qquad 0.3 \leq \frac{D_E}{D} \leq 0.82$$

$$27 \leq \frac{D}{s_E} \leq 100 \qquad X_o \equiv 4 \frac{D_E}{D} \frac{A_o}{D} \approx 0.82$$

$$X_u \equiv 4 \frac{D_E}{D} \frac{A_u}{D} \approx 0.58$$

$$(21.19)$$

it was found for each given geometry for highly turbulent flow:

$$\zeta_U = \text{const}. \qquad (21.20)$$

That means according to Eq. (21.15):

$$\frac{D_1}{D} n_U = \text{const}, \qquad (21.21)$$

and according to Eq. (21.16) or (21.12) for each geometry and fluid system:

$$\frac{Re_m}{Re_1} = \text{const}. \qquad (21.22)$$

The experimental results from the model JLR ($D = 290$ mm, $s = 5$) with the various draft tubes shown in Fig. 21.25 are shown

Table 21.2. Correction Factor k_1 for Varying Values of D_E/D

D_E/D	0.3	0.4	0.5	0.6	0.7	0.8
k_1	1.40	1.32	1.24	1.16	1.08	1.00

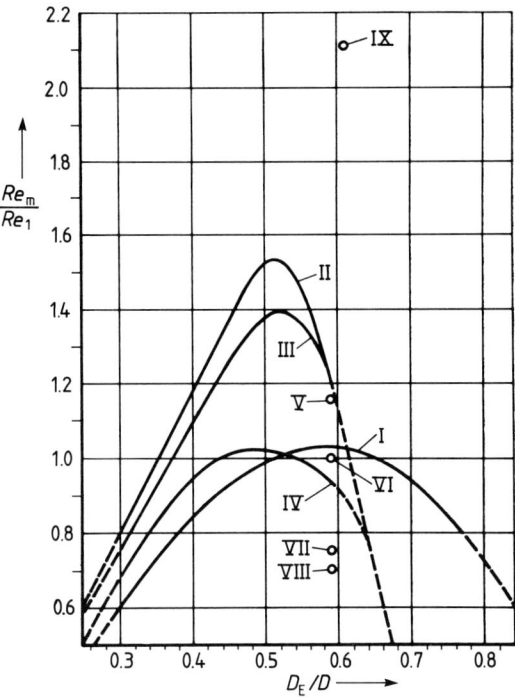

Figure 21.27. The circulation Reynolds ratio Re_m/Re_1 obtained by Eq. (21.16) and experiments in model JLR as a function of D_E/D and draft tube shape according to Fig. 21.25. Draft tube IX is shown in Fig. 21.28.

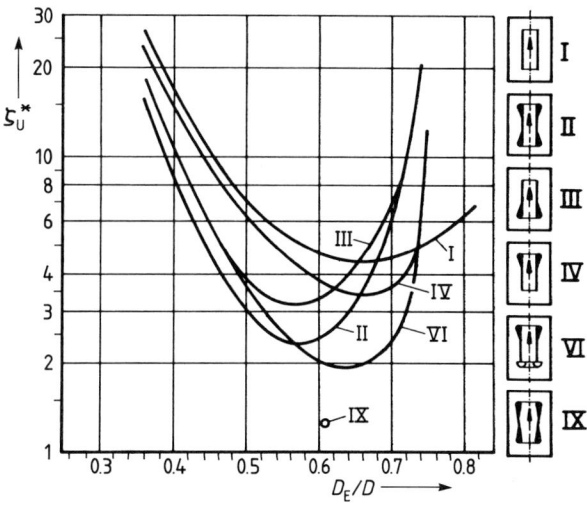

Figure 21.28. Circulation resistance number ζ_U^* obtained by Eq. (21.17) and experiments.

in Fig. 21.26 for ζ_U and in Fig. 21.27 for Re_m/Re_1 including draft tube type IX from Fig. 21.28 considering k_1, Eq. (21.23), from Table 21.2.

- In the second series of experiments we measured ζ_o and ζ_u in order to determine ζ_U^* of Eq. (21.17). The results are shown in Fig. 21.28.

The ζ_U^*-curves are in very good agreement with the ζ_U-curves in Fig. 21.26 as found with the model JLR. The results of both investigations lead to exact mutual agreement when a correction factor k_1 is applied to ζ_u in Eq. (21.17), which takes into account the effect of the liquid jet (which only occurs in the model JLR), giving:

$$\zeta_U = \zeta_i + \zeta_a + \zeta_o + k_1 \zeta_u . \quad (21.23)$$

The k_1-values resulting from the matching of both sets of experiments are given in Table 21.2 [21.27].

- The ζ_U- and ζ_U^*-values given in Figs. 21.26 and 21.28 are valid for a slenderness ratio $s = H/D = 5$. Then the momentum of the liquid jet in the JLR is completely transferred inside the draft tube [21.30]. Should

21.3.1.4 Power and Efficiency of the Circulation Drive (JLR)

In JLR the liquid jet power input P_L is given by:

$$P_L = \dot{V}_1 p_1 = \frac{\pi}{8} \varrho_1 D_1^2 w_1^3 = \frac{\pi}{8} \varrho_1 v_1^3 Re_1^3 D_1^{-1}$$

$[\text{N m s}^{-1} \equiv \text{J s}^{-1} \equiv \text{W}].\quad (21.24)$

Under steady-state conditions the liquid jet power input P_L must cover the circulation power P_U:

$$P_U = \dot{V}_3 \Delta p_U \,[\text{N m s}^{-1} \equiv \text{J s}^{-1} \equiv \text{W}]. \quad (21.25)$$

With Eqs. (21.14) and (21.4) one obtains:

$$P_U = \dot{V}_3 \frac{\varrho_m}{2} w_m^2 \zeta_U = 0.5 \dot{M}_3 w_m^2 \zeta_U =$$
$$= 0.5 \dot{M}_1 w_m^2 \zeta_U n_U , \quad (21.26)$$

or with Eqs. (21.5) and (21.11):

$$P_U = 0.5 \varrho_m \dot{V}_3 w_m^2 \zeta_U = \frac{\pi}{16} \varrho_m D^2 w_m^3 \zeta_U =$$
$$= \frac{\pi}{16} \varrho_m v_m^3 Re_m^3 D^{-1} \zeta_U . \quad (21.27)$$

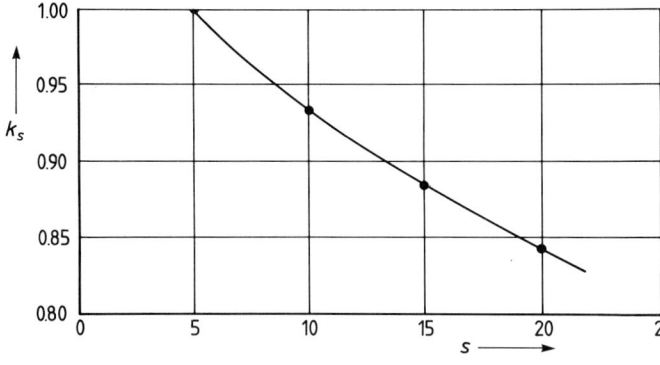

Figure 21.29. Correction factor k_s, with which Re_m/Re_1 (Fig. 21.27) has to be multiplied, if $s \geq 5$.

the loop reactor be designed with higher s, then at constant ζ_o and ζ_u only ζ_i and ζ_a increase. This can be allowed for by multiplying the Re_m/Re_1-values calculated with Eq. (21.12) or (21.16) from n_U, ζ_U or ζ_U^* at $s=5$ by a correction factor k_s, given in Fig. 21.29 for other values of s [21.26], [21.27], [21.31].

With Eqs. (21.24) and (21.27) the efficiency of the circulation generation in the JLR is given by:

$$\eta_U \equiv \frac{P_U}{P_L} = 0.5 \frac{\varrho_m}{\varrho_1} \left(\frac{v_m}{v_1}\right)^3 \left(\frac{Re_m}{Re_1}\right)^3 \frac{D_1}{D} \zeta_U . \quad (21.28)$$

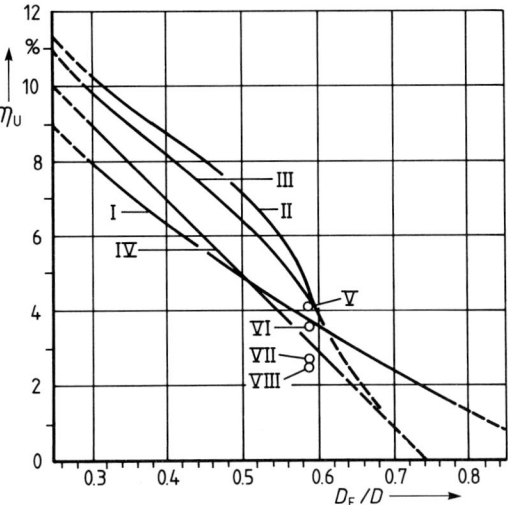

Figure 21.30. Efficiency η_U, Eqs. (21.28) and (21.29), of circulation drive in JLR.

With Eq. (21.16) for ζ_U Eq. (21.28) may be rewritten as

$$\eta_U \approx \frac{v_m}{v_1} \frac{Re_m}{Re_1} \left(\frac{D_E}{D}\right)^{-2} \frac{D_1}{D}. \qquad (21.29)$$

An analysis of the equations for η_U reveals that:

η_U increases with decreasing D_E/D,

η_U increases with increasing D_1/D, and

η_U increases with decreasing ζ_U.

In Fig. 21.30 η_U is plotted versus D_E/D.

21.3.1.5 Application of Previous Results for JLR to PLR

For propeller loop reactors (PLR) $\zeta_U = \zeta_U^*$ is valid. The pressure drop Δp_U, due to one total circulation, equals the pressure head produced by the propeller:

$$\Delta p_U = \zeta_U^* \frac{\varrho_m}{2} w_m^2 = \varrho_m g H_P = p_P. \qquad (21.30)$$

Thus for PLR one can determine w_m with Eq. (21.30); Re_m with Eq. (21.11); r_U with Eq. (21.6), and $\bar{t}_U$ with Eq. (21.7), directly for known values of ϱ_m, H_P, and ζ_U^*.

The propeller power is:

$$P_P = \dot{V}_P \cdot p_P \quad [W], \qquad (21.31)$$

with the volume flow from the propeller $\dot{V}_P$ [m^3 s^{-1}] and the pressure head p_P [N m^{-2} ≡ Pa].

21.3.2 (Quasi-)Homogeneous Highly Viscous L-Systems

21.3.2.1 Experimental Data

Some results of experiments with (quasi-)homogeneous highly viscous systems in PLR and JLR will be indicated here briefly [21.32], [21.33].

As Newtonian fluids glucose/water mixtures were used with dynamic viscosities in the range:

$\eta = 10^{-3}$ to 5 Pa s   for the PLR,

$\eta = 10^{-3}$ to 0.4 Pa s   for the JLR.

For Newtonian fluids the well-known relation between the shear stress τ [N m^{-2} ≡ Pa] and the shear rate $\dot{\gamma} = -dw_x/dy$ [s^{-1}] applies:

$$\tau = \eta \cdot \dot{\gamma}, \qquad (21.32)$$

with the dynamic viscosity η.

Carboxymethylcellulose (CMC) in water was used as non-Newtonian fluid. For such non-Newtonian viscous fluids the law of OSTWALD and DE WAELE may be applied:

$$\tau = K \dot{\gamma}^n, \qquad (21.33)$$

with consistency factor K and flow index n.

With Eq. (21.33) an apparent viscosity can be defined:

$$\eta_s \equiv \frac{\tau}{\dot{\gamma}} = K \dot{\gamma}^{n-1}. \qquad (21.34)$$

$n < 1$ holds for pseudoplastic (structure viscous) fluids, $n > 1$ for dilatant media and $n = 1$ for Newtonian fluids, for which K is identical to the dynamic viscosity η. From Eq. (21.33) it follows that with increasing $\dot{\gamma}$ the apparent viscosity η_s decreases for pseudoplastic fluids, while it increases for dilatant media, and remains constant for Newtonian fluids with $\eta_s \equiv \eta \equiv K$.

All LR investigated were made of acrylic glass and had the following dimensions:

Internal diameter $D = 190$ mm,

slenderness ratio $s = 5$ or 10,

diameter ratio $D_E/D = 0.67$ for PLR,
$\qquad\qquad\qquad\quad = 0.6$ for JLR.

The speed of the propeller in PLR was varied from $n_P = 100–1700$ rpm and the jet volume flow in JLR from $\dot{V}_1 = 0.1–20$ L s^{-1} with nozzle diameters $D_1 = 1.8–10$ mm.

The following parameters were measured: The circulation volume flow $\dot{V}_P$ (PLR) and $\dot{V}_3$ (JLR) using the maximal velocity $\hat{w}_a$ in the annulus; furthermore the power input P_P (PLR) and P_L (JLR).

21.3.2.2 Theoretical Relations for Newtonian Fluids

For geometrically and hydrodynamically similar designs, characterized by the diameter D_P and the speed n_P of the propeller, the following similarity numbers can be used for scale-up from models to large-scale units.

The "volume flow number"

$$N_V \equiv \frac{\dot{V}_P}{n_P D_P^3} \qquad (21.35)$$

is a constant characteristic number for a certain propeller, pump or agitator.

The propeller Reynolds number is another important parameter:

$$Re_P \equiv \frac{n_P D_P^2}{\nu} = \frac{\varrho n_P D_P^2}{\eta}. \qquad (21.36)$$

For PLR the total circulation flow $\dot{V}_P$ can be characterized by the mean circulation Reynolds number:

$$Re_m \equiv \frac{8 \dot{V}_P}{\nu \pi D} = \frac{8 n_P D_P^3}{\nu \pi D} N_V. \qquad (21.37)$$

Similarly a circulation Reynolds ratio can be defined for the PLR as:

$$\frac{Re_m}{Re_P} = \frac{8 D_P}{\pi D} N_V \sim \frac{D_P}{D} N_V, \qquad (21.38)$$

or for constant geometrical parameters:

$$\left(\frac{Re_m}{Re_P}\right)_{\frac{D_P}{D}} \sim N_V. \qquad (21.39)$$

For propellers (as for agitators) the driving power P_P can be characterized by the "Newton number":

$$Ne \equiv \frac{P_P}{\varrho n_P^3 D_P^5}. \qquad (21.40)$$

21.3.2.3 Some Results for Newtonian Fluids

a) It is seen from Fig. 21.31 that for JLR in the turbulent range ($Re_1 \gtrsim 5000$) $(Re_m/Re_1)_t = 1.03 = $const; this confirms Eq. (21.22). In the range of higher viscosity Re_m/Re_1 decreases as can be expected because in Eq. (21.16) the resistance number ζ_U, Eq. (21.14), increases. In the laminar range the proportionality $(Re_m/Re_1)_l \sim Re_1$ approximately holds. At higher values of s, ζ_U increases due to greater partial resistances ζ_i and ζ_a, and at equal Re_1 consequently Re_m/Re_1 decreases. The measured data in Fig. 21.31 can be correlated by the equation:

$$\frac{Re_m}{Re_1} = \left(\frac{C_1}{Re_1^{C_2}} + C_3\right)^{-1}, \qquad (21.41)$$

with the empirical constants C_1, C_2, and C_3.

b) A comparison of Figs. 21.31 and 21.32 confirms that, according to Eq.

21.3 Fluid Dynamics of Loop Reactors (LR)   489

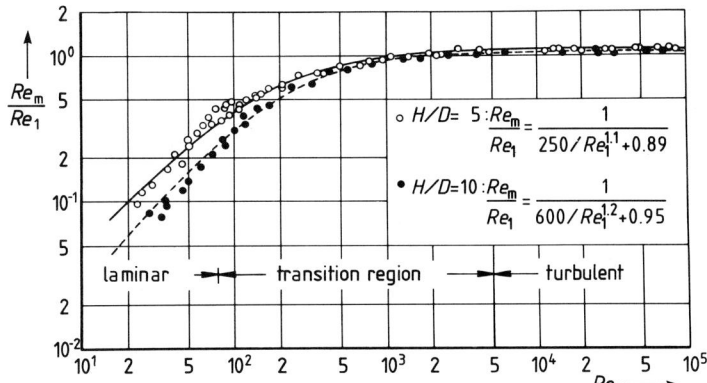

Figure 21.31. Circulation Reynolds ratio Re_m/Re_1, Eq. (21.41), of JLR as a function of the liquid jet Reynolds number Re_1 for Newtonian fluids.

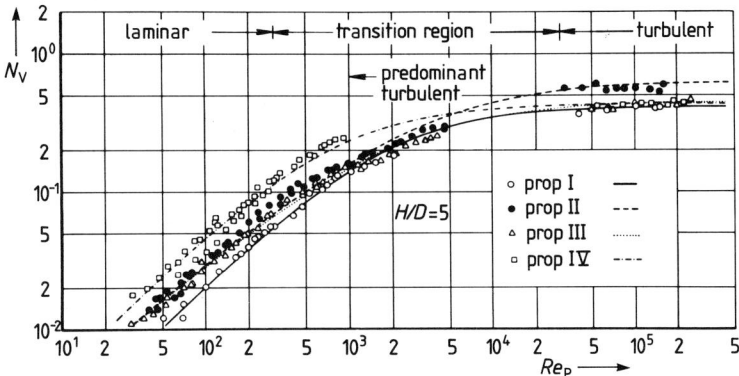

Figure 21.32. Volume flow number N_V of PLR, Eq. (21.42), as a function of the propeller Reynolds number Re_P, Eq. (21.36); compare with Fig. 21.31.

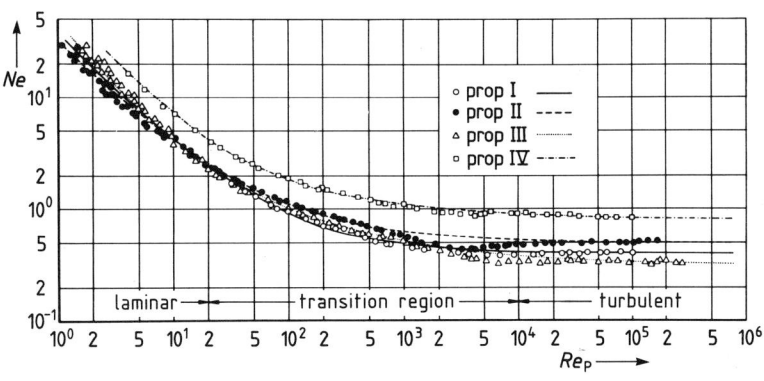

Figure 21.33. Newton number Ne as a function of the propeller Reynolds number Re_P for the propeller types of Fig. 21.34.

(21.39), $N_V \sim Re_m/Re_P$ of a given PLR corresponds to Re_m/Re_1 of a similar JLR.

Thus the measured points in Fig. 21.32 can be correlated with the same type of empirical equation:

$$N_V = \left(\frac{C_1}{Re_P^{C_2}} + C_3\right)^{-1}. \tag{21.42}$$

The limiting values are for $Re_P \to 0$:

$$N_V \approx Re_P^{C_2} C_1^{-1}, \tag{21.43}$$

and for $Re_P \to \infty$:

$$N_V \approx C_3^{-1} \approx \text{const}. \tag{21.44}$$

Figure 21.34. Types of propellers used in PLR.

c) According to Fig. 21.33 the curves for the Newton number Ne over Re_P for the propellers of Fig. 21.34 correspond to that of all investigated agitator types [21.34], [21.35].

21.3.2.4 Comparison between PLR with Newtonian and Non-Newtonian (Pseudoplastic) Fluids

For non-Newtonian fluids the apparent viscosity η_s according to Eq. (21.34) depends on the shear rate $\dot{\gamma}$. The discussion will be restricted to the PLR using the empirical relation found for agitators [21.36]:

$$\bar{\dot{\gamma}} = k n_P, \tag{21.45}$$

where k is a constant for a certain agitator type. This leads, with the apparent viscosity, Eq. (21.34), to the following definition of a Reynolds number of the propeller:

$$Re_P^+ \equiv \frac{\varrho n_P D_P^2}{K(k n_P)^{n-1}} = \frac{\varrho n_P^{2-n} D_P^2}{K k^{n-1}}. \tag{21.46}$$

For Newtonian fluids with $n=1$ and $K \equiv \eta$ Re_P^+ is identical with Re_P, Eq. (21.36).

With the substitution of Re_P by Re_P^+ the Newton number Ne for agitation of non-

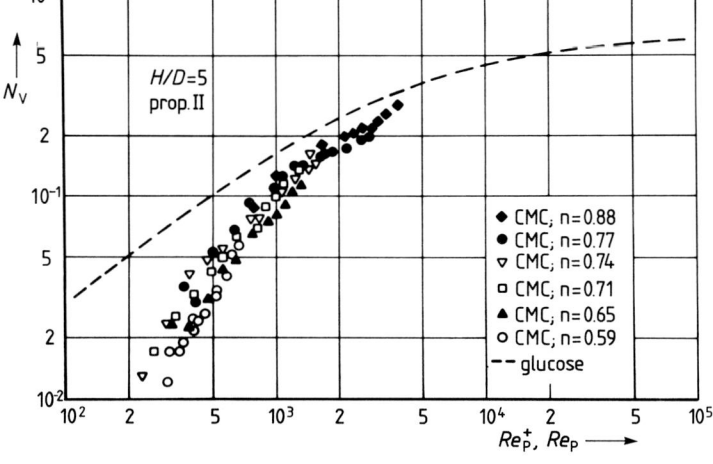

Figure 21.35. Volume flow number N_V as a function of the modified propeller Reynolds number Re_P^+, Eq. (21.46), for pseudoplastic fluids, compared with Re_P, Eq. (21.36), for Newtonian fluids.

Newtonian fluids can be taken from Fig. 21.35.

The dependence of the volume flow number N_V on Re_P^+ measured with CMC deviates, however, significantly from that of Newtonian fluids, as shown in Fig. 21.35.

Corresponding relationships for JLR are discussed in [21.32].

21.3.3 Flow Behavior of Heterogeneous Solid-Liquid-Systems (S-L-Systems or Suspensions) in JLR

In biosystems it is often essential to avoid settling (sedimentation) of particles, such as cell agglomerates, on the bottom of the reactor. We therefore carried out theoretical and experimental studies on the fluidization and distribution of solid particles as well as the fluid dynamics of suspensions in JLR. The material and geometric parameters of the solid particles were varied according to Table 21.3. Their concentration c_a in the annulus was measured using a newly developed photometric method incorporating a He-Ne gas laser [21.37]. This method, however, restricts the volumetric solid concentration to $c_{So} \lesssim 1\%$ at uniform distribution throughout the reactor volume V_R.

First it is of interest to know from which values of $Re_1 = \check{R}e_1$ onwards particles are carried from the inside of the JLR into the annulus, thus the start of the transition from a fluidized bed in the inner space to a circulation of the particles. At $Re_1 < \check{R}e_1$ a fluidized bed inside the draft tube is maintained. This mode of operation represents a fluidized bed LR with internal circulation. Correlation of the experimental data resulted in the relationship for the minimum Reynolds number [21.37]:

$$\check{R}e_1 = Fr_S^{0.5} \cdot 10^{(4.08 + 10.38 D_S/D)} + \\ + c_{So} \cdot 10^{(6.0 + 0.15 Fr_S - 15.5 D_S/D)}. \quad (21.47)$$

Here the Froude number Fr_S is defined with the settling velocity w_S [see Eq. (21.50)] of a single particle, its diameter D_S and the gravity constant g:

$$Fr_S \equiv \frac{w_S^2}{g D_S}. \quad (21.48)$$

From $\check{R}e_1$ onwards the suspended particles are transported first of all in irregular clouds from the inner space to the annulus. Uniform particle discharge and particle circulation starts at $\hat{R}e_1$:

$$\hat{R}e_1 \equiv \frac{w_{L1} \hat{D}_1}{v_1} = \\ = 23 \cdot 10^4 \, Fr_S^{0.38} \left(\frac{\varrho_S}{\varrho_L}\right)^{0.62} \left(\frac{D_S}{D} c_{So}\right)^{0.26}. \quad (21.49)$$

For conventional operation $\check{R}e_1 \lesssim Re_1 \lesssim \hat{R}e_1$ holds.

A further investigation [21.38] extended the solid particle concentration up to about 10%. As solids only small glass balls were used, which are described in Table 21.4. In the range of higher particle concentrations the free-settling velocity w_S of single par-

Table 21.3. Material and Geometric Parameters of Solid Particles

Serial Number	Material	Density (in g/cm³)	Particle Form	Measurements (in mm)	Tolerance (in μm)
1	PA[a]	1.13	Pellet	Diameter 1.50/2.50/ 3.175/4.00	20
2	POM[b]	1.45	Pellet		
3	Aluminum	2.70	Pellet		
4	PVC	1.41	Cube	Edge length $4 \times 2.5 \times 1.5$	

[a] PA Polyamide   [b] POM Polyoxymethane (Hostaform)

Table 21.4. Particle Diameter and Free-settling Velocity of the Glass Balls

Smallest particle diameter D_S (in mm)	0.15	0.42	0.8	2.0
Largest	0.25	0.59	1.2	
Mean free-settling velocity w_S (in m s^{-1})	0.014	0.085	0.17	0.28
Solid density ϱ_S (in kg m^{-3})		2500		

ticles must be replaced by that of particle clouds w_{SS}. In [21.37] we referred to [21.39] for single particles:

$$w_S = \left[\frac{4}{3}\frac{D_S g}{\zeta_S}\left(\frac{\varrho_S}{\varrho_L}-1\right)\right]^{\frac{1}{2}}, \quad (21.50)$$

with

$$\zeta_S = \frac{24}{Re_S} + \frac{4}{Re_S^{0.5}} + 0.4, \quad (21.51)$$

and

$$Re_S \equiv \frac{w_S D_S}{v_L}. \quad (21.52)$$

For the JLR the free-settling velocity w_{SS} of the particle cloud can be determined by the approximation:

$$\frac{w_{SS}}{w_S} \approx e^{-4.5 c_{So}}. \quad (21.53)$$

With some simplifications the following approximation results:

$$w_1 D_1 \equiv Re_1 v_1 =$$
$$= \frac{D}{2}\left(4\frac{\Delta\varrho}{\varrho_m}g H c_{So}\frac{w_{SS}}{w_m} + \zeta_U w_m^2\right)^{\frac{1}{2}}, \quad (21.54)$$

with $\Delta\varrho = \varrho_S - \varrho_L$ (21.55)

and ζ_U according to Fig. 21.26.

Eq. (21.54) yields the required values of w_1 and Re_1 in order to obtain a mean velocity of circulation w_m at given parameters.

The minimum values $\check{R}e_1(\check{w}_1)$, at which circulation of solids starts were determined in this investigation to be:

$$(w_1\check{}D_1) \equiv \check{R}e_1 v_1 =$$
$$= 1.5^{\frac{1}{2}}\left(2\zeta_U^{\frac{1}{2}}\frac{\Delta\varrho}{\varrho_m}g H c_{So} w_{SS}\right)^{\frac{1}{3}} D_E. \quad (21.56)$$

The investigations presented in [21.37] and [21.38] can be applied to or specifically supplemented for given dimensions and characteristic data of biosystems, e.g., movable carriers for fixed enzymes or for agglomerates of microorganisms, especially pellet forming kinds. Eq. (21.49) should certainly also be valid over the range of smaller particles and free-settling velocities – e.g. for cell agglomerates.

21.4 Fluid Dynamics of Loop Reactors with Aerated Gas-Liquid-Systems (G-L-Systems)

21.4.1 Coalescence Behavior of G-L-Systems

An important parameter for fluid dynamics, mixing and residence time behavior, as well as for mass transfer and utilization of the reaction space in G-L-systems is the volumetric gas hold-up ε, which essentially depends on the bubble size distribution.

In pure liquids, gas bubbles tend to unite. This coalescence occurs because of the thinning of the liquid lamella between neighboring gas bubbles. This takes place very rapidly in pure liquids so that even with finest primary dispersion quite uniform stable bubble sizes between about 2 and 5 mm depending on the degree of turbulence in the space in question are formed after leaving the dispersing shearing fields.

In solutions (e.g. of salts) or in mixtures (e.g. with alcohol) which tend to foam, coalescence can be strongly impeded. For example, practically no coalescence occurs in an aqueous 1.0 N Na_2SO_4 solution (71 g salt per L), and stable bubble sizes of between 0.3 and 0.5 mm are formed. This results in a much larger G-L interfacial area and consequently in higher oxygen transfer. Also much more gas is sucked into the downflow. Thus coalescence behavior must be considered when comparing experimental results obtained with different G-L-systems and transferring results from a model to large-scale designs [21.40].

21.4.2 Volumetric Gas Hold-up of G-L-Systems

21.4.2.1 Definitions

Gas hold-up in loop reactors is described by the following dimensionless parameters:

- mean volumetric gas hold-up (see Fig. 21.36):

$$\varepsilon \equiv \frac{V_G}{V_R} = \frac{V_G}{V_G + V_L} = \frac{V_R - V_L}{V_R} = 1 - \frac{V_L}{V_R} = \frac{H_G}{H};$$
(21.57)

- volumetric gas hold-up in the inner space:

$$\varepsilon_i \equiv \frac{V_{Gi}}{V_{Ri}} = 1 - \frac{V_{Li}}{V_{Ri}};$$
(21.58)

- volumetric gas hold-up in the annulus:

$$\varepsilon_a \equiv \frac{V_{Ga}}{V_{Ra}} = 1 - \frac{V_{La}}{V_{Ra}};$$
(21.59)

- local gas hold-up:

$$\varepsilon_l \equiv \frac{V_{Gl}}{V_{Rl}}.$$
(21.60)

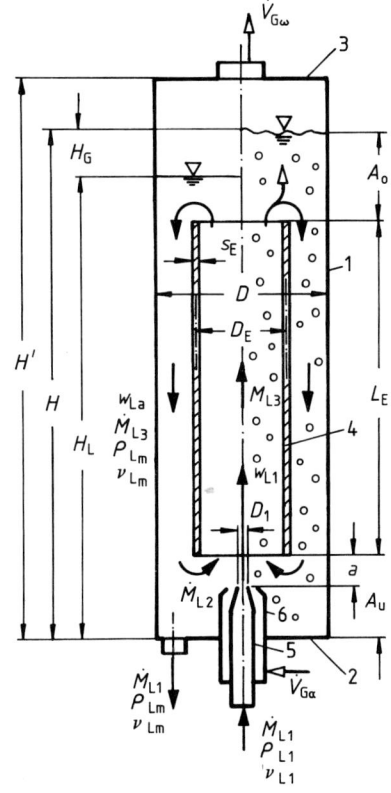

Figure 21.36. JLR with a gas-liquid-system (right), and without aeration (left) showing the main parameters of optimization of fluid dynamics and mass transfer.
1 Reactor wall, 2 bottom, 3 top plate, 4 draft tube, 5 liquid-jet nozzle, 6 air sparger.

21.4.2.2 Air-Water-System at $w_G \lesssim 10$ cm s^{-1}

The gas hold-up ε_a was measured by means of a radioactive density method [21.44], [21.20].

The liquid velocity w_{La} in the annulus is measured using temperature impulses from a heating element arranged in the annular space. Their transition time across the measuring section beginning 75 mm below the heating element, is recorded electronically by sensors 100 mm apart. The mean gas hold-up ε is determined by the height difference $H - H_L = h - h_L = H_G$ in a standpipe. The dimensions of the loop reactors used in the experiments are $D = 290$ mm; $D_E/D = 0.59$; $s = H/D = 6$ to 22. Although all gas spargers shown in Fig. 21.21 were investigated, this discussion is limited to the ring nozzle.

Fig. 21.37 shows experimental data of ε_a for various ε-values at "superficial gas velocities" $w_G \approx 2$ to 8 cm s^{-1}. w_G is based on the *total* reactor cross-section:

$$w_G \equiv \frac{4 \dot{V}_G}{\pi D^2}. \qquad (21.61)$$

It is surprising that for $\varepsilon \lesssim 20\%$ in JLR ($Re_1 > 0$) $\varepsilon_a > \varepsilon$, which is due to local gas bubble motion in opposite direction to the liquid jet drive. In general, however, $\varepsilon_a < \varepsilon$, which is the main basis of the air-lift drive.

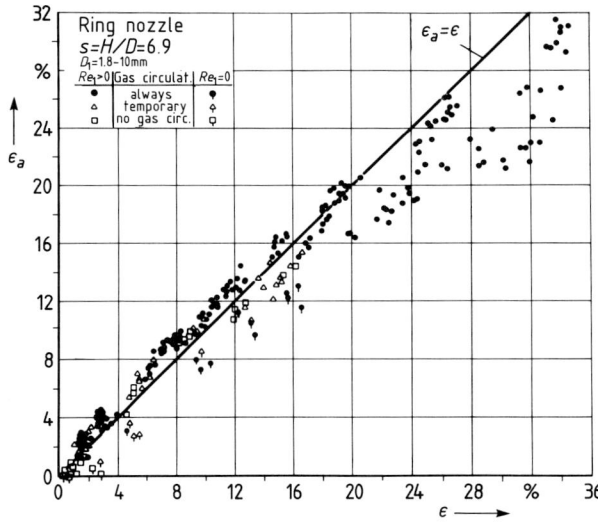

Figure 21.37. Gas hold-up ε_a in the annulus and mean gas hold-up ε in the total reaction system for ALR ($Re_1 = 0$) and JLR ($Re_1 > 0$) with air-water-system.

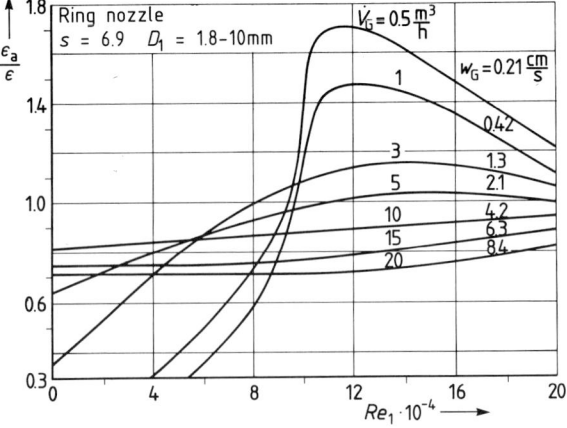

Figure 21.38. Gas hold-up ratio $\varepsilon_a/\varepsilon$ as a function of Re_1 and w_G.

Fig. 21.38 shows the correlation $\varepsilon_a/\varepsilon = f(Re_1, w_G)$. Thus desired $\varepsilon_a/\varepsilon$-values can be systematically selected or undesired values avoided. To ensure continual gas circulation at $0.7 \lesssim \varepsilon_a/\varepsilon \lesssim 1.3$ the following conditions must be fulfilled:

- ALR ($Re_1 = 0$): $w_G \gtrsim 2$ cm s^{-1}
- JLR ($Re_1 > 0$): $Re_1 \gtrsim 10^5$ at $w_G \gtrsim 2$ cm s^{-1}.

21.4.2.3 Sulfite-System at $w_G \lesssim 60$ cm s^{-1}

These investigations aiming mainly at oxygen transfer, were carried out using the oxidation of sodium sulfite to sodium sulfate in aerated aqueous solutions with cobalt as catalyst.

At high sulfite concentrations this system is practically non-coalescent. For this reason the small gas bubbles last a long time with intensive primary dispersion and accordingly many are also sucked into the annulus. For this system the gas hold-up ε has been determined. In Fig. 21.39, it is shown that the higher Re_1 is in JLR, the more rapidly ε increases; initially linearly with increasing gas flow rate $\dot{V}_G$ (w_G).

For the ALR a further investigation [21.41] resulted in the gas hold-up correlation as shown in Fig. 21.40:

$$\varepsilon = f(Re_G \cdot Fr_G \cdot s), \qquad (21.62)$$

which is well confirmed by the experimental results plotted for ALR having $D = 290$, resp. 630 mm and $H = 1.74$ to 6.96 m. The dimensionless parameters are defined as follows:

$$Re_G \equiv \frac{w_G \cdot D}{\nu_G}, \quad \text{gas Reynolds number,} \qquad (21.63)$$

$$Fr_G \equiv \frac{w_G^2}{gH}, \quad \text{gas Froude number.} \qquad (21.64)$$

Furthermore the investigation [21.41] revealed an approximately linear relationship between ε_a in the annulus and ε_i in the inner space with ε according to Fig. 21.41.

The equation for ε_a is:

$$\varepsilon_a \approx 0.916\,\varepsilon, \qquad (21.65)$$

with a mean deviation of $\pm 10\%$.

For the ratio $D_E/D = 0.64$ used in the experiments it follows:

$$\varepsilon_i \approx 1.134\,\varepsilon, \qquad (21.66)$$

with a mean deviation of $\pm 20\%$.

For the ratio $\varepsilon_i/\varepsilon_a$ the following equation has been obtained:

$$\frac{\varepsilon_i}{\varepsilon_a} \approx 1.238. \qquad (21.67)$$

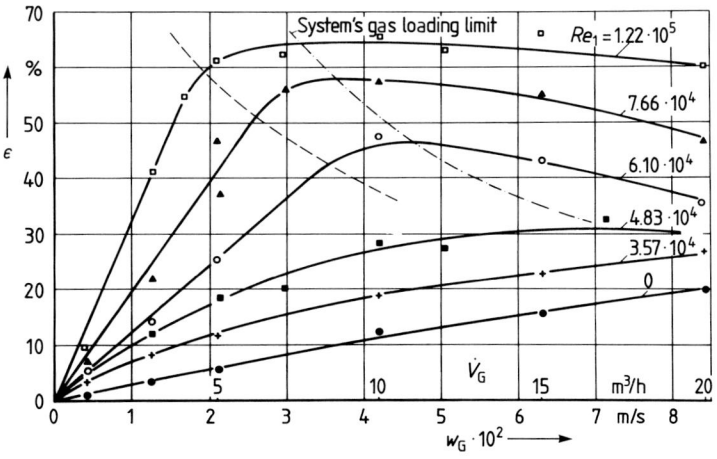

Figure 21.39. Mean gas hold-up ε in ALR ($Re_1 = 0$) and JLR ($Re_1 > 0$; $D_1 = 3.5$ mm) with sulfite-system depending on $\dot{V}_G$ (w_G).

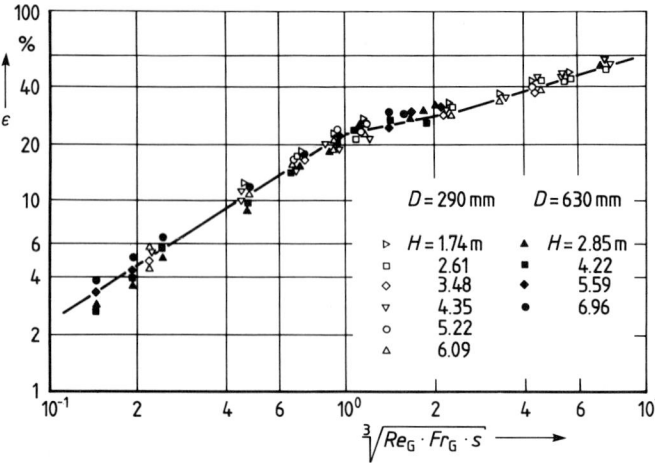

Figure 21.40. Correlation of the mean gas hold-up ε with $(Re_G \, Fr_G \, s)^{1/3}$ for ALR at various reactor diameters D and mixture heights H (sulfite-system).

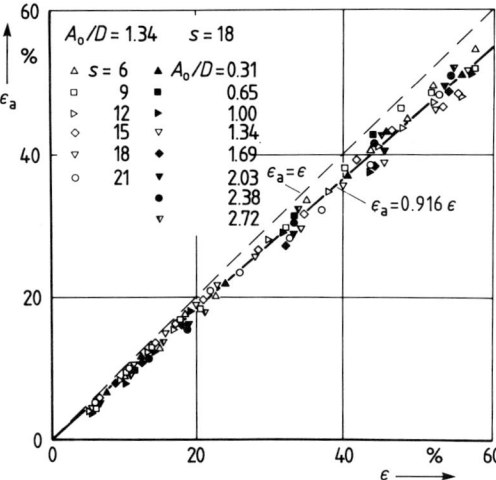

Figure 21.41. Correlation of the gas hold-up ε_a in the annulus with the mean gas hold-up ε in ALR with sulfite-system.

$$\bar{t}_G \equiv \frac{V_G}{\dot{V}_G} = \varepsilon \frac{V_R}{\dot{V}_G}. \qquad (21.68)$$

- Small values of ε result in large values of liquid volume, $V_L = (1-\varepsilon) V_R$, in which the reaction occurs.

21.4.3 Circulation Flow of G-L-Systems in JLR and ALR

21.4.3.1 Methods for the Determination of Liquid Circulation Velocity in G-L-Systems

In principle all methods for the determination of the circulation velocity in LR are based on length and time measurements, for instance the *"temperature-, salt-, and acid-base-impulse method"*.

Another method is the *"measuring ball method"* [21.42], which can be considered as a development of the "sponge method" [21.26]. In the meantime it has often been successfully used in various G-L-systems. By means of an "electronic stop watch", the time t_a is measured, which a circulating

21.4.2.4 Summary of Gas Hold-up

The mean and local volumetric gas hold-up ε, ε_a, and ε_i, is important for the following reasons:

- With increasing gas hold-up the mean residence time increases according to the equation:

measuring ball of about 10–30 mm diameter requires to pass through the measuring section L – in this case between the two ring-antennae arranged around the reactor wall. The velocity of the measuring ball is:

$$w_M = \frac{L}{t_a}. \qquad (21.69)$$

The measuring ball transmits to, or receives from, the antennae high-frequency signals in the MHz range. The moment when the ball passes the antenna plane is marked by a change of field strength.

At all operating conditions with G-L-systems in LR the inaccuracy of w_{La} lay in the range of ±10%; the inaccuracy of the acid-base-impulse method is in the range of ±5%.

The accuracy of w_{La} seems to be poor but it must be considered that this method records all fluctuations of the fluid dynamic system during the long-time measurement.

into account. It is given accurately enough in the range of $10^3 \lesssim Re < 10^5$ with its resistance coefficient ζ_w by the equation [21.39]:

$$w_S = \left[\frac{4}{3}\frac{D_M g}{\zeta_w}\left(\frac{\varrho_M}{\bar{\varrho}_a} - 1\right)\right]^{\frac{1}{2}} =$$

$$= 54.5 \left(D_M \frac{\varepsilon_a}{1-\varepsilon_a}\right)^{\frac{1}{2}} \text{ (in cm s}^{-1}\text{)}, (21.70)$$

with the volumetric gas hold-up ε_a and the mean density $\bar{\varrho}_a$ of the G-L-system in the annulus; the density $\varrho_M = \varrho_L$, and the diameter of the measuring object D_M (in cm). The downwards flow velocity in the annulus is given by:

$$w_{La} = w_M - w_S. \qquad (21.71)$$

Fig. 21.42 shows that even at high gas hold-up, $\varepsilon_a \approx 40\%$, w_{La}, Eq. (21.71), is in very good agreement with the directly measured w_{La} values using the acid-base-impulse

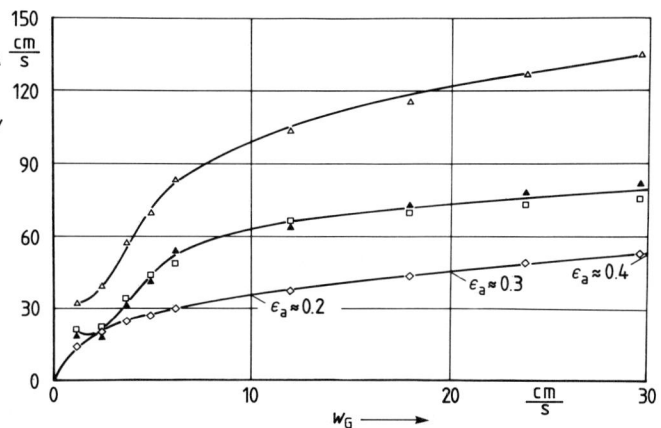

Figure 21.42. Values of the flow velocity w_{La} in the annulus of an ALR with sulfite-system as a function of the superficial gas velocity w_G.
△ measured values of w_M with "measuring ball method", ◊ calculated values of w_S, Eq. (21.70), ▲ resulting values w_{La}, Eq. (21.71), with "measuring ball method", □ measured values of w_{La}^* with "acid-base-impulse method", $D_M = 15$ mm.

It must be remembered that in homogeneous systems $w_{La} = w_M$ is valid if the density of the measuring ball is $\varrho_M = \varrho_L$. In determining w_{La} from the experimentally determined w_M for G-L-systems, the settling velocity of the measuring ball must be taken

method. This method is very suitable for fully automatic plant operation, also for non-transparent reactors and systems. The "measuring ball method" is however not suitable for highly viscous fluids with a distinct laminar flow profile [21.43].

21.4.3.2 JLR and ALR with Air-Water-System at $w_G \lesssim 10$ cm s^{-1}

These investigations [21.44], [21.20] were carried out in an experimental pilot plant with a LR comprising a ring nozzle sparger at the bottom, as Fig. 21.21a shows in detail. For all acrylic glass reactors the dimensions, indicated in Fig. 21.36, were as follows:

- internal diameter of the reactor $D = 290$ mm,
- diameter ratio $D_E/D = 0.59$,
- slenderness ratio of the reaction space $s = H/D = 6$ to 22.

Re_m is defined by:

$$Re_m \equiv \frac{w_{Lm} D}{v_L} = \frac{8 \dot{V}_L}{\pi v_L (1-\varepsilon) D}, \quad (21.72)$$

with the total volumetric gas hold-up ε and the mean liquid circulation velocity w_{Lm}:

$$w_{Lm} = \frac{8 \dot{V}_L}{\pi D^2 (1-\varepsilon)}. \quad (21.73)$$

From Fig. 21.43 it follows that:

- Operation with a liquid jet, but without gas sparging ($\dot{V}_G = 0$), i.e., JLR with homogeneous L-system, is characterized by the linear correlation of Eq. (21.22).
- Operation with gas sparging through the ring nozzle, but without a liquid jet ($Re_1 = 0$), i.e., ALR with G-L-system, is characterized by the operating points on the ordinate.

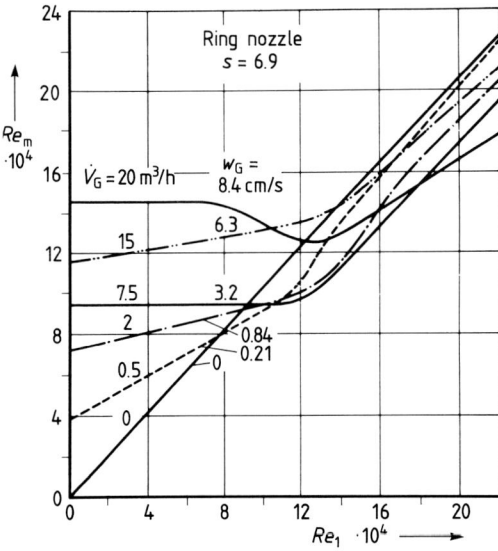

Figure 21.43. Liquid circulation Reynolds number Re_m in ALR and JLR depending on gas through flow $\dot{V}_G$ and liquid nozzle Reynolds number Re_1 in air-water-system.

In Fig. 21.43 the characteristic "flow chart" is shown for heterogeneous G-L-systems in JLR. It gives the Reynolds number Re_m for liquid circulation as a function of the nozzle Reynolds number Re_1 (liquid jet), with gas flow rate $\dot{V}_G$ and superficial gas velocity w_G as parameters.

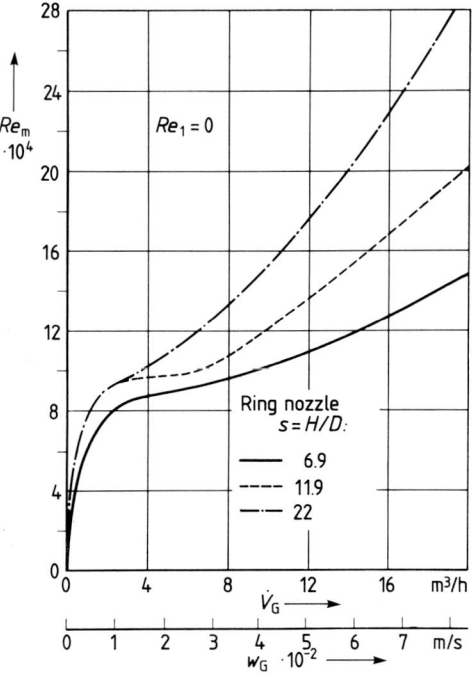

Figure 21.44. Liquid circulation Reynolds number Re_m in ALR depending on $s = H/D$ and $\dot{V}_G (w_G)$ in air-water-system.

21.4 Fluid Dynamics of Loop Reactors (G-L-Systems)

- In operation with gas sparging and liquid jet, i.e., JLR with G-L-system, 2 ranges can be clearly distinguished:
  - $Re_1 \lesssim 1.2 \cdot 10^5$ in which the circulation is mainly due to air-lift drive.
  - $Re_1 \gtrsim 1.2 \cdot 10^5$ in which the circulation is mainly due to liquid jet drive.

From this one can conclude that in G-L-systems the liquid jet induces an effective circulation, characterized by Re_m, only at $Re_1 \gtrsim 1.2 \cdot 10^5$; but even then Re_m is always smaller (up to about 25%) than in homogeneous L-systems. This is mainly due to the fact that a considerable part of the L-jet power input is now used for gas dispersion, which lowers the driving effect.

Fig. 21.44 shows clearly that the circulation flow of the coalescent air-water-system in ALR ($Re_1 = 0$) rapidly increases with increasing slenderness ratio s, that means at constant diameter D with increasing height H. This is plausible since the air-lift driving force F_{ALR} increases directly with s, because it is proportional to the height H of the communicating spaces and to the density difference $\Delta \varrho$, thus $F_{ALR} \sim H \Delta \varrho \sim s \Delta \varrho \sim s \Delta \varepsilon$.

21.4.3.3 ALR with Sulfite-System at $w_G \lesssim 60$ cm s^{-1}

Fig. 21.45 shows that $(D_E/D)_{opt} = 0.64$ is obtained for G-L-systems with maximum circulation velocity w_{La} at $w_G \lesssim 12$ cm s^{-1}. The circulation rate r_U, given for this heterogeneous system by:

$$r_U \equiv \frac{\dot{V}_L}{V_L} = \frac{w_{Lm}}{2H} = \bar{t}_U^{-1} \approx$$

$$\approx \frac{w_{La}}{H} \frac{1-\varepsilon_a}{1-\varepsilon}\left[1-\left(\frac{D_E}{D}\right)^2\right], \quad (21.74)$$

has a maximum at $(D_E/D)_{opt} = 0.59$.

The measured velocity values $w_{La} = f(w_G, s)$ [21.41] are shown in Fig. 21.46. The circulation velocity w_{La} increases rapidly with s, reaches a maximum at $w_G \approx 20$ cm s^{-1}, then it decreases again slightly.

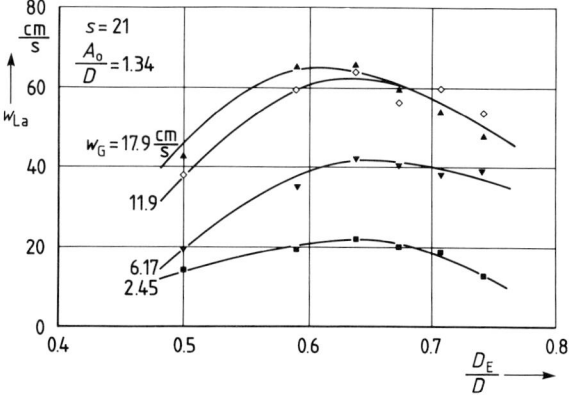

Figure 21.45. Influence of the draft tube diameter ratio D_E/D on the liquid velocity w_{La} in the annulus of ALR with sulfite-system.

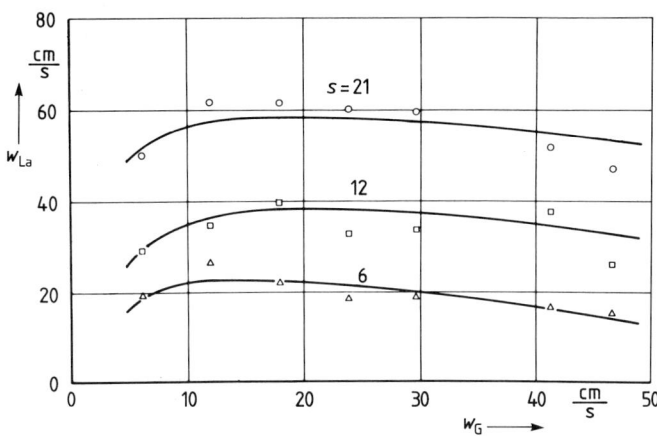

Figure 21.46. Influence of the superficial gas velocity w_G and the slenderness ratio s on the liquid velocity w_{La} in the annulus of ALR with sulfite-system.

21.5 Mixing Behavior of LR

21.5.1 Basic Principles

The importance of the mixing behavior of bioreactors has already been discussed in Sect. 21.2.1. In all types of LR two fundamentally different mixing effects are superimposed [21.1].

1. *Longitudinal mixing* occurs in each circulation due to the velocity profile, turbulence, dead spaces, and molecular diffusion. It can be mathematically formulated according to two models:

- In the *diffusion model* the mentioned effects are considered according to Fick's laws of diffusion by summing them all up in the effective longitudinal diffusion coefficient D_{eff}. The diffusion coefficient is characterized by the Bodenstein number Bo:

$$Bo \equiv \frac{w_m L_U}{D_{eff}} = \frac{2 w_m H}{D_{eff}}, \qquad (21.75)$$

with $L_U = 2H$, the length of the circulation path for the fluid. As explained in [21.45] and [21.46], for LR with homogeneous aqueous systems the following values for Bo are obtained:

$$Bo \approx 50 - 200. \qquad (21.76)$$

- In the *tanks in series model* the real tube reactor is replaced by a series of stirred tank reactors (STR) with the equivalent number n_{eq} of consecutive equal volume ideal STR, which results in the same longitudinal mixing effect.

When $Bo > 8$, both models can be combined by the relation [21.47]:

$$n_{eq} = 1 + \frac{1}{2}(Bo^2 + 1)^{\frac{1}{2}} \approx 1 + \frac{Bo}{2}. \qquad (21.77)$$

Thus with Eq. (21.76) it follows that:

$$n_{eq} \approx 25 - 100. \qquad (21.78)$$

From this it is clear that, under turbulent flow conditions, plug flow practically occurs in loop reactors, which hardly contributes to the mixing effect.

2. *Backmixing* due to recycling of the directed circulation flow (Fig. 21.1), which characterizes the LR, is decisive for its mixing effect in turbulent flow.

Imagining oneself in the situation of a stationary observer at a point in the LR, at which a tracer is injected into the flow as an impulse at time $t=0$ (impulse tracing), then after each circulation the observer will detect a widening concentration distribution of the tracer (impulse response) as shown in Fig. 21.47 for $Bo = 200$. The relative time τ_U plotted on the abscissa refers the actual time t to the mean circulation time $\bar{t}_U$:

$$\tau_U \equiv \frac{t}{\bar{t}_U} = t r_U = \frac{t}{\bar{t}} n_U = \tau n_U \equiv i_U. \qquad (21.79)$$

τ_U is identical to the number of circulations i_U at the mean velocity w_m, Eq. (21.5), in time t.

The measured concentration of the tracer is referred to its mean value c_∞ which

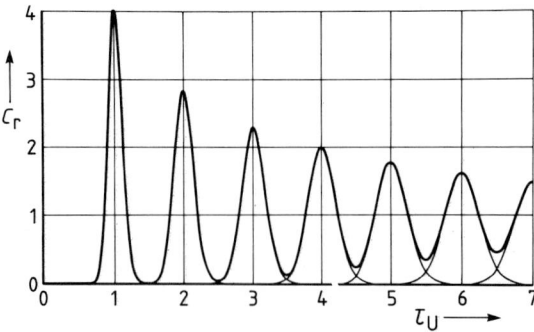

Figure 21.47. Impulse response c_r, Eq. (21.80), in the real circulation flow of JLR ($Bo = 200$) spreads with relative time τ_U, Eq. (21.79), after the impulse injection.

would result if it were uniformly distributed throughout the whole reactor volume V_R. Thus the relative concentration is:

$$c_r \equiv \frac{c}{c_\infty}. \tag{21.80}$$

The total concentration c_τ in the circulation system at time τ is given by the sum of all the concentration elements which have circulated a different number of times ($x_U = 1, 2, 3 \ldots$):

$$x_U = \frac{L}{L_U}, \tag{21.81}$$

with L the flow path in t and $L_U = 2H$ the length of one circulation.

For the calculation only noteworthy concentrations in the range $x_U = \tau_U \pm 2$ need be considered. Starting with [21.48] the derivations in [21.49] resulted in the following correlation for the impulse distribution:

$$c_r = \sum (c_r)_{x_U} = \sum_{x_U=\tau_U-2}^{x_U=\tau_U+2} \left(\frac{Bo}{4\pi\tau_U}\right)^{\frac{1}{2}} \cdot \exp\left[-\frac{(x_U-\tau_U)^2}{4\tau_U} Bo\right]. \tag{21.82}$$

21.5.2 Mixing Behavior

Usually mixers are judged by the mixing time required to achieve a certain degree of mixing throughout the mixer space. Comparisons are only of value when based on the same degree of mixing, which is defined in [21.49] as follows:

$$h \equiv \frac{c - c_\infty}{c_\infty} = c_r - 1. \tag{21.83}$$

In Fig. 21.48 the maxima and minima of the impulse distribution, calculated with Eq. (21.82) are connected by envelope curves, which are approximately exponential functions. The desired maximal inhomogeneity $\hat{h}$ requires the mixing time t_M. Using the relative mixing time,

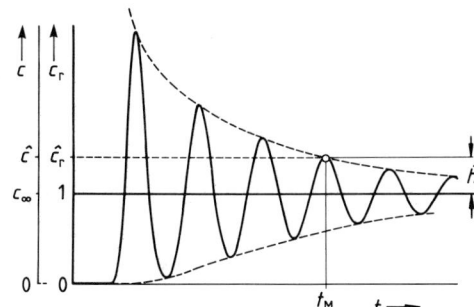

Figure 21.48. Determination of the inhomogeneity $\hat{h}$ characterizing the degree of mixing by the upper envelope curve connecting the maxima of the impulse distribution function.

$$\tau_M \equiv \frac{t_M}{\bar{t}_U} \equiv i_{UM}, \tag{21.84}$$

which is identical with the number of circulations in LR necessary to achieve $\hat{h}$, the following empirical equation gives a good approximation to the experimental results [21.49]:

$$\tau_M = Bo \frac{0.692 - \ln \hat{h}}{39.48}. \tag{21.85}$$

On the basis of Eqs. (21.84) and (21.85) and an expression for the circulation time $\bar{t}_U$ the following equation for the mixing time t_M has been derived:

$$t_M = \frac{(0.692 - \ln \hat{h})(s - 0.24) D}{19.74} \frac{Bo}{w_m}; \tag{21.86}$$

after introduction of the Reynolds number, defined by Eq. (21.11), one obtains:

$$t_M = \frac{(0.692 - \ln \hat{h})(s - 0.24) D^2}{19.74 \, v_m} \frac{Bo}{Re_m}. \tag{21.87}$$

From Eq. (21.86) resp. (21.87) the following conclusions can be drawn:

- the larger the maximal allowable inhomogeneity $\hat{h}$, the shorter the required mixing time t_M;
- the larger the reactor height H resp. the slenderness ratio s, the longer t_M, because at the same w_m fewer circu-

lations per unit time reduce the most effective backmixing;
- the larger w_m resp. Re_m, the shorter t_M, because more frequent circulations improve the most effective backmixing;
- Bo should be small, but it is in fact rather high in turbulent flow – especially in G-L-systems.

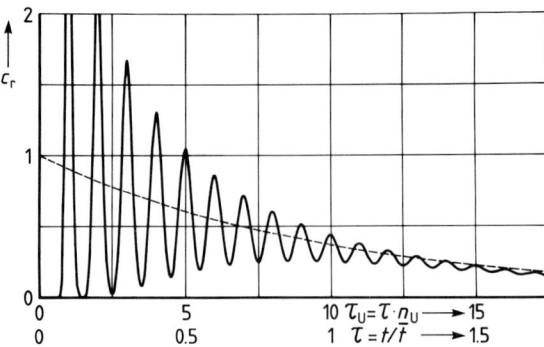

Figure 21.49. Impulse response or distribution function of a LR with continuous operation ($Bo = 200$; $n_U = 10$).

21.6 Residence Time Behavior of LR

When in continuous operation a through flow is superimposed on the circulation flow of the closed system (batch operation), then impulse responses as shown in Fig. 21.49 result. But in this case also the same relative mixing times τ_M are needed for a certain specified inhomogeneity $\hat{h}$.

After each circulation, tracer mass within the reactor decreases in the ratio:

$$\frac{\dot{M}_2}{\dot{M}_3} = \frac{n_U - 1}{n_U}, \quad (21.88)$$

and thus after x_U completed circulations by:

$$\left(\frac{n_U - 1}{n_U}\right)^{x_U}. \quad (21.89)$$

This is the term, by which for continuous operation Eq. (21.82) has to be multiplied to give the impulse distribution [21.45]:

$$c_r = \sum (c_r)_{x_U} = \sum_{x_U = \tau_U - 2}^{x_U = \tau_U + 2} \left(\frac{n_U - 1}{n_U}\right)^{x_U} \left(\frac{Bo}{4\pi \tau_U}\right)^{\frac{1}{2}}$$
$$\cdot \exp\left[-\frac{(x_U - \tau_U)^2}{4\tau_U} Bo\right]. \quad (21.90)$$

Replacing τ_U according to Eq. (21.79) by the relative residence time τ of the reactor, i.e. $\tau_U = n_U \cdot \tau$.

Eq. (21.90) is identical with the "distribution function":

$$H(\tau) \equiv \frac{c_r}{c_\infty},$$

which, as an impulse response, characterizes the residence time behavior of LR as demonstrated in Fig. 21.49.

21.7 Mass Transfer with (Bio-)Chemical Reaction in G-L-Systems

21.7.1 Mass Transfer

In addition to *distribution* of all reaction components in heterogeneous G-L reaction systems *dispersion* is very important. It determines the specific G-L interfacial area and consequently the mass transfer between the phases. As a rule, (bio-)chemical reactions take place in the liquid phase. Gaseous reaction partners (e.g. O_2) must be transported from the gas phase into the liq-

21.7 Mass Transfer with (Bio-)Chemical Reaction in G-L-Systems

uid phase. Conversely the gaseous reaction products formed (e.g. CO_2) must also be transferred out of the liquid phase into the gas phase. For the sake of simplicity, this discussion is limited to the sulfite-system, which is often used to simulate aerobic biosystems [21.50].

The sulfite oxidation simulates the O_2 consumption of the microorganisms. The O_2 transfer from the gas bubble to the surface of the microorganisms is decisively important for the determination of the gas dispersion. However, some biological questions are still unanswered: permeation inhibition at individual microorganisms due to possible surrounding envelopes of excretions, as shown in Fig. 21.50 [21.51]; uncontrolled cell agglomerations; the mechanism of the O_2 introduction into the cell; and the metabolism kinetics within the cell. COHEN and MONOD showed in 1957 that these complex effects can be generally described by the equation:

$$r_1 = -\frac{dc_1}{dt} = \frac{k \cdot c_1'}{K_M + c_1'}. \quad (21.91)$$

This relationship is also known as the Michaelis-Menten equation, with k reaction rate constant, c_1' molar concentration of the observed reactant (e.g. O_2) at the entrance to the permease zone, and K_M Michaelis constant. In the diffusion zone the first diffusion law of FICK is valid.

Physical and operating data are taken from the SCP process with *Methylomonas clara* on methanol.

The main process-engineering problem is to bring about the required specific O_2 intake,

$$\dot{m}_{O_2} \equiv \dot{m}_{V_L} \equiv \frac{\dot{M}_{O_2}}{V_L} = 8 \text{ kg m}^{-3} \text{ h}^{-1} =$$
$$= 2.2 \text{ g m}^{-3} \text{ s}^{-1}, \quad (21.92)$$

as economically as possible; that means with as low a specific power input P/V_L as possible and high O_2 conversion:

$$u_{O_2} = \frac{\dot{N}_{O_2}}{\dot{V}_{G\alpha} c_{G\alpha}}. \quad (21.93)$$

From the molar flow balance of the reactor the O_2 molar flow is experimentally determined as:

$$\dot{N}_{O_2} = \dot{N}_{O_2\alpha} - \dot{N}_{O_2\omega} \text{ [kmol h}^{-1}\text{]}, \quad (21.94)$$

which is related to the mass flow as follows:

$$\dot{N}_{O_2} = \frac{\dot{M}_{O_2} \text{ [kg h}^{-1}\text{]}}{32 \text{ [kg kmol}^{-1}\text{]}}. \quad (21.95)$$

Combining mass transfer (O_2) and chemical reaction the *"molar flow density"* referred to the unknown G-L inerfacial area A is *theoretically determined* as:

$$\dot{n}_{O_2} \equiv \frac{\dot{N}_{O_2}}{A} \equiv \dot{n} \text{ [kmol m}^{-2} \text{ h}^{-1}\text{]}. \quad (21.96)$$

From the measured values of $\dot{N}_{O_2}$ and the calculated values of $\dot{n}_{O_2}$ the G-L interfacial area results:

$$A = \left(\frac{\dot{N}}{\dot{n}}\right)_{O_2}, \quad (21.97)$$

and therewith the specific interfacial area:

$$a_R = \frac{A}{V_R}, \quad (21.98)$$

resp. with Eq. (21.57):

$$a_L = \frac{A}{V_L} = \frac{a_R}{1-\varepsilon}. \quad (21.99)$$

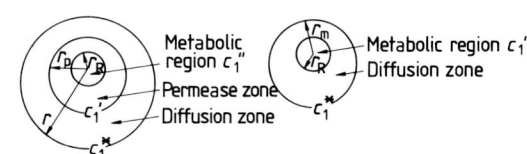

Figure 21.50. Models for a single microorganism.

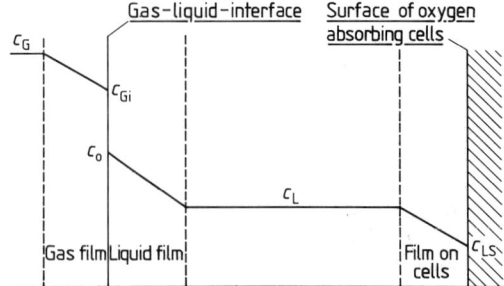

Figure 21.51. Oxygen transfer from an air bubble to the surface of a cell within the liquid phase according to film theory.

Considering the oxygen path from the bulk of a gas bubble to the cell surface, as is shown in Fig. 21.51 based on the *film theory*, the following assumptions can be made:

- *Diffusion* in the gas film can be neglected, because of the high turbulence in the gas bubbles; i.e., $c_{Gi}=c_G$.
- Henry's law is valid for the *absorption* of O_2 from the gas phase into the liquid phase according to which:

$$c_o = \frac{c_G}{He}. \tag{21.100}$$

- The pure *physical diffusion* in the liquid film follows Fick's first law of diffusion (Fig. 21.52):

$$\dot{n}_{ph} = -D_L \frac{dc}{dx} = \frac{D_L}{\delta}(c_o - c_L) =$$
$$= \beta_{ph}(c_o - c_L) = \beta_{ph}\Delta c. \tag{21.101}$$

The physical mass transfer coefficient is, in biotechnology, usually designated by:

$$k_L \equiv \beta_{ph} \equiv \frac{D_L}{\delta}. \tag{21.102}$$

- The *convection* in the bulk liquid occurs so intensively in turbulent systems that $c_L=$ const.

- The diffusion resistance in the film at the surface of the cells can be neglected because the L-S interfacial area of all individual cells is 10^5 to 10^6 times larger than the G-L interfacial area A.

In consequence the rate determining step is concentrated on the diffusion in the liquid film as shown in Fig. 21.52.

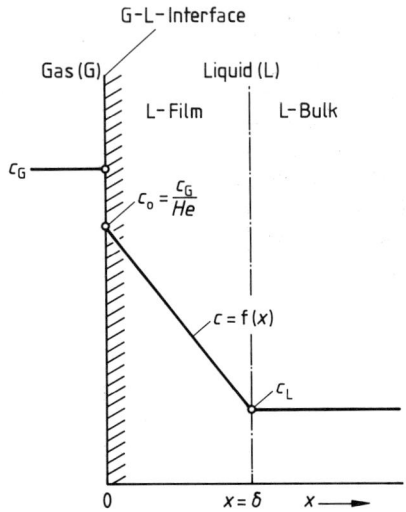

Figure 21.52. Oxygen concentration profile in the liquid film for diffusion with slow chemical reaction according to the film model.

21.7.2 Combination of Diffusion (Film Theory) and Chemical Reaction

The change of O_2 concentration c in a small volume element of the L-film at the G-L-interface is given by Fick's second law of diffusion (D):

$$\left(\frac{dc}{dt}\right)_D = D_L \frac{d^2c}{dx^2}. \tag{21.103}$$

In the sulfite-system considered the O_2 reaction is of the order $n=2$ in the range investigated; thus the concentration change by reaction (R) is also given by:

$$\left(\frac{dc}{dt}\right)_R = -k_n c^n = -k_2 c^2 . \quad (21.104)$$

At steady-state conditions,

$$\frac{dc}{dt} = \left(\frac{dc}{dt}\right)_D + \left(\frac{dc}{dt}\right)_R = 0 , \quad (21.105)$$

the following equation for the local concentration c results:

$$D_L \frac{d^2 c}{dx^2} = k_n c^n . \quad (21.106)$$

By means of this equation the following solution has been derived for the oxygen rate of consumption:

$$\dot{n} = \dot{n}_{O_2} = \frac{\dot{N}}{A} = \beta_{ch} \bar{c}_o (1 + C_n)^{\frac{1}{2}} =$$
$$= \beta_{ph} \bar{c}_o Ha (1 + C_n)^{\frac{1}{2}} , \quad (21.107)$$

with the "chemical mass transfer coefficient":

$$\beta_{ch} \equiv \left(\frac{2}{n+1} D_L k_n \bar{c}_o^{n-1}\right)^{\frac{1}{2}} , \quad (21.108)$$

the Hatta number [21.51]:

$$Ha \equiv \frac{\beta_{ch}}{\beta_{ph}} = \frac{\left(\frac{2}{n+1} D_L k_n \bar{c}_o^{n-1}\right)^{\frac{1}{2}}}{D_L/\delta} , \quad (21.109)$$

the integration constant C_n and $\bar{c}_o$ the mean value of the concentration across the whole G-L interfacial area A (see Fig. 21.52).

In the highly turbulent sulfite-system at $T_R = 25\,°C$ with approximate values of $D_L \approx 2 \cdot 10^{-9}$ m^2 s^{-1} and $\beta_{ph} \equiv k_L \approx 4.6 \cdot 10^{-4}$ m s^{-1} the imaginary thickness δ of the liquid film in Fig. 21.52 is according to Eq. (21.102) $\delta \approx 4$ μm.

In nearly all technical cases for slow to rapid reactions the approximation of REITH [21.52] is of sufficient accuracy:

$$\dot{n} \approx (\beta_{ph}^2 + \beta_{ch}^2)^{\frac{1}{2}} (\bar{c}_o - c_L) \approx$$
$$\approx \beta_{ph} \bar{c}_0 (1 - c_{Lr})(1 + Ha^2)^{\frac{1}{2}} \approx$$
$$\approx \beta_{ch} \bar{c}_o (1 - c_{Lr})(1 + Ha^{-2})^{\frac{1}{2}} , \quad (21.110)$$

with the relative bulk concentration:

$$c_{Lr} \equiv \frac{c_L}{\bar{c}_o} . \quad (21.111)$$

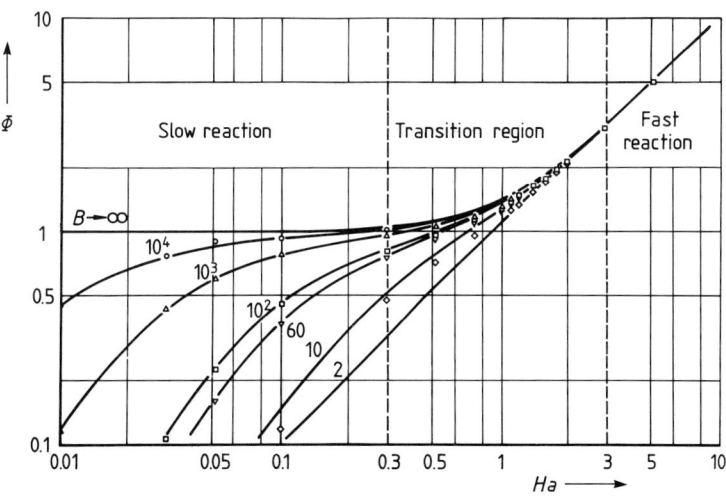

Figure 21.53. Chemical acceleration Φ, Eq. (21.112), with $k_L \equiv \beta_{ph}$, Eq. (21.102), as a function of the Hatta number Ha, Eq. (21.109), and the volumetric number B, Eq. (21.113), for a second order reaction.

It is in full agreement with the theoretical derivation for $n=1$ according to the penetration theory [21.53].

The chemical acceleration Φ is defined by:

$$\Phi \equiv \frac{\dot{n}}{\dot{n}_{ph}} = \frac{Ha(1+C_n)^{\frac{1}{2}}}{1-c_{Lr}} \approx (1+Ha^2)^{\frac{1}{2}}. \tag{21.112}$$

If Φ with the bulk concentration $c_L=0$ is plotted against Ha, using the "volumetric number" [21.54]

$$B \equiv \frac{V_L}{V_F} = \frac{V_L}{A\delta} = \frac{V_L k_L}{AD_L} = \frac{k_L V_R(1-\varepsilon)}{AD_L} =$$

$$= \frac{k_L(1-\varepsilon)}{a_R D_L}, \tag{21.113}$$

as a parameter, Fig. 21.53 [21.72] results. It shows especially clearly for $B \to \infty$, i.e., very small liquid film volume V_F compared with liquid volume V_L:

- $Ha \lesssim 0.3$: range of slow reaction, because $\dot{n} = \dot{n}_{ph}$; with $\beta_{ch} \ll \beta_{ph}$,
- $Ha \gtrsim 3$: range of rapid reaction with

$$\dot{n} \approx Ha \, \dot{n}_{ph} \approx \beta_{ch} \bar{c}_o. \tag{21.114}$$

This result is obtained by Eq. (21.110) with $\beta_{ph} \ll \beta_{ch}$ and $c_L \approx 0$.

- $0.3 \lesssim Ha \lesssim 3$: transition range for which $\dot{n}$ according to Eq. (21.110) applies with maximal deviation of $+5\%$ from the exact solution given by Eq. (21.107).

In the exact solution given by Eq. (21.107) the integration constant:

$$C_n = c_{Lr}^{n+1} \left\{ \left[\frac{n+1}{2} Ha(B-1) \right]^2 c_{Lr}^{n-1} - 1 \right\}, \tag{21.115}$$

includes the relative bulk concentration c_{Lr} defined by Eq. (21.111).

21.7.3 Determination of $k_L \equiv \beta_{ph}$ and a_R

Starting with Eq. (21.110), squaring, expanding with $a_R^2 = (A/V_R)^2$ and using Eq. (21.97) results in the relationship:

$$\left[\frac{\dot{N}}{V_R(\bar{c}_o - c_L)} \right]^2 = a_R^2 \beta_{ch}^2 + a_R^2 \beta_{ph}^2 \tag{21.116}$$

leading to the "Danckwerts plot" in Fig. 21.54. It was determined here in a JLR [21.54] using the sulfite-system, at various catalyst concentrations and thus reaction

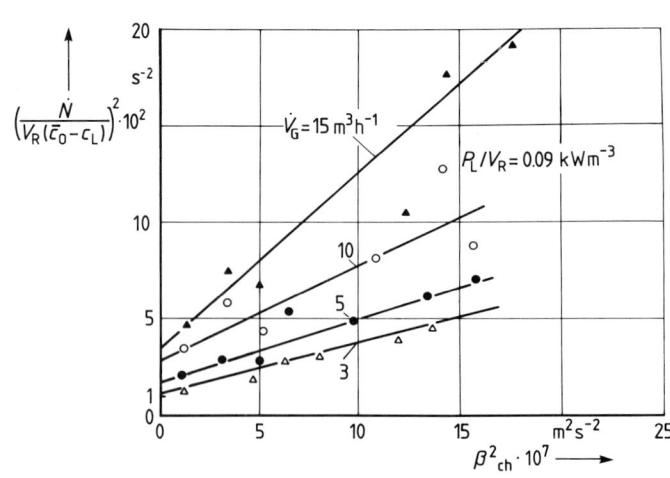

Figure 21.54. "Danckwerts plot" according to Eq. (21.116) from which one can graphically determine k_L and a_R.

rate constants, and various gas through flows $\dot{V}_G$ (w_G).

Now all the related coordinate values are known and from the diagram one can graphically derive:

- a_R^2 as the slope of the straight lines in Fig. 21.54
- $(a_R \beta_{ph})^2$ as the point of intersection at the ordinate
- from these two values $\beta_{ph} \equiv k_L$.

21.7.4 Determination of $\bar{c}_o$ in LR

In all discussions in Sect. 21.7 the mean concentration $\bar{c}_o$ across the whole interfacial area plays a decisive role. Unfortunately it is especially difficult to determine it in LR because completely different conditions exist in the inner space than in the annulus, and the local determination of c_o and its assignment to the related partial volume causes great difficulties. A fundamental step forward in the solution of this problem for LR is the new method of measurement of local O_2 concentration in the gas phase [21.41]. The results clearly show a very unfavorable separation effect of the gas bubbles by the flow reverse around the upper edge of the draft tube in LR: Large O_2-rich bubbles leave the G-L-system, small O_2-poor bubbles are sucked into the annulus and recycled, thus lowering decisively the mean O_2 concentration $\bar{c}_o$ in LR.

As long as local O_2 concentrations in the LR could not be measured, one could only use the inlet and outlet concentrations $c_{o\alpha}$ and $c_{o\omega}$ to determine the mean concentration $\bar{c}_o$. For this the following two theoretical models were used:

(1) The *segregation model* considers the LR to be an ideal tube reactor without any mixing of gas bubbles. Accordingly they pass through the reactor completely segregated. The mean concentration is then given by [21.54]:

$$\bar{c}_o^S = \frac{c_{o\alpha} - c_{o\omega}}{\ln \frac{c_{o\alpha}}{c_{o\omega}}}. \tag{21.117}$$

It is the highest mean concentration which can occur with given values of $c_{o\alpha}$ and $c_{o\omega}$.

(2) The *complete mixing model* considers the LR to be an ideal STR, for which complete backmixing is valid, thus leading to

$$\bar{c}_o^M = c_{o\omega}. \tag{21.118}$$

It results in the lowest mean concentration, if no other negative separation effects occur, such as discussed above for the LR.

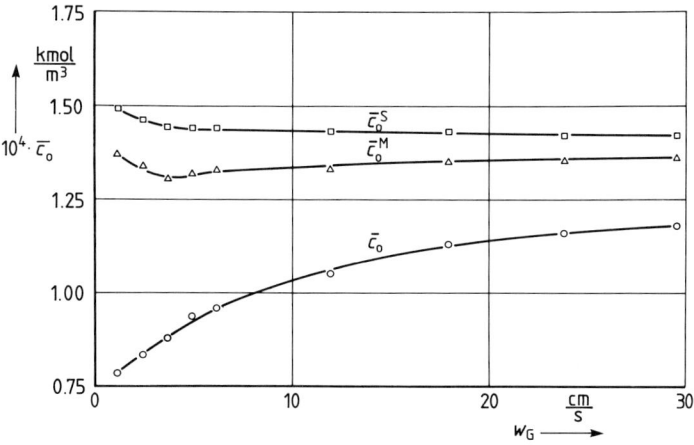

Figure 21.55. Comparison of $\bar{c}_o$-values for the discussed models plotted against the superficial gas velocity w_G.

To consider this unique characteristic of LR, a special *"recirculation model"* was developed for this reactor type [21.41]. It considers the inner and annular spaces separately with respect to the relevant local parameters. The investigations showed that for each of the two partial spaces the segregation model holds approximately. Then the total molar amount of O_2 in the reactor is given by:

$$N = N_i + N_a = V_{Gi}\bar{c}_{Gi}^S + V_{Ga}\bar{c}_{Ga}^S = V_G \bar{c}_G. \quad (21.119)$$

With

$$V_G = \varepsilon V_R; \quad V_{Gi} = \varepsilon_i V_{Ri}; \quad V_{Ga} = \varepsilon_a V_{Ra} \quad (21.120)$$

and $\bar{c}_o = \bar{c}_G/He$, Eq. (21.100), from Eq. (21.119) the mean O_2 concentration in the whole reaction space according to the recirculation model is derived as

$$\bar{c}_o = \frac{\varepsilon_i V_{Ri} \bar{c}_{Gi}^S + \varepsilon_a V_{Ra} \bar{c}_{Ga}^S}{He \varepsilon V_R}. \quad (21.121)$$

Fig. 21.55 shows a comparison of the $\bar{c}_o$ values for the three models considered here plotted against the superficial gas velocity w_G for the ALR [21.41] with $s = 21$ and sulfite-system. As to be expected $\bar{c}_o^M$ lies below $\bar{c}_o^S$ over the whole range. But, because of the negative separation effect, $\bar{c}_o$, which is most accurately found using the recirculation model, lies again substantially below $\bar{c}_o^M$.

In spite of this inherent disadvantage of all conventional LR-types they have already been proven to be very efficient. But having recognized this disadvantage it is a self-evident task for engineers to overcome this and to further improve the productivity of LR. Only two examples may indicate directions for such development.

21.8 Possibilities for Improving Productivity of LR

A quite simple arrangement is a "degassing head" [21.1], [21.41] as shown in Fig. 21.56 and applied as Fig. 21.12 shows. It reduces the suction of smaller O_2-poor bubbles into the downwards flow, and thus the recirculation of these "useless" bubbles. Evidently the same effect can be achieved

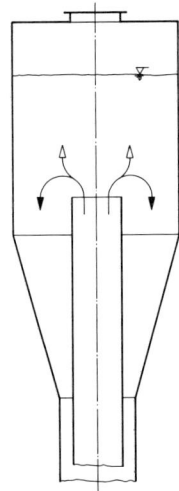

Figure 21.56. Scheme of a degassing head.

with a gas deflector as shown in Figs. 21.13b and 21.62 in the case of reverse flow direction.

Another possibility is shown in Fig. 21.57 [21.73]. From the feed gas $\dot{M}_{Gin}$ (open arrows) introduced at the bottom of the inner space, $\dot{M}_{G2}$ is sucked into the surrounding annulus. Before it can recirculate around the lower edge of the draft tube 2 into the inner space and lower the O_2 concentration there, it enters a second external annulus and rises there due to its buoyancy.

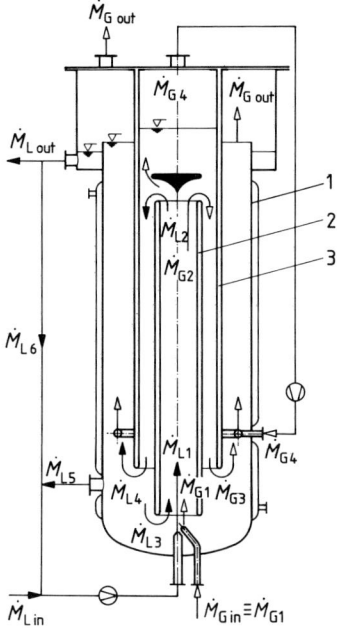

Figure 21.57. Design of a modified JLR with recirculation of L-phase (full arrows) and direct through flow of the G-phase (open arrows) forming a combination of a modified LR with a surrounding bubble column in one apparatus.
1 reactor wall, 2 inner draft tube, 3 outer draft tube.

21.9 Some Examples of Biotechnical Applications of LR

Without doubt the cooperation of natural science and engineering in biotechnology is of fundamental importance for the future of mankind [21.55]. The UNO estimates optimistically that in 100 years from now the world population could stabilize itself at about 11 billion people, i.e. 2½ times as many as today. Of these, probably 9 billion (today 3.2) would live in currently developing countries. On account of the growing numbers and the increasing demands of the world population, it is unavoidable that nature will have to provide more and more food, raw materials, and energy, and more and more material and energy waste has to be disposed of. One of the main future responsibilities of technology, and especially of biotechnology, is to help solve these global tasks economically, ecologically, and humanely. This may be briefly indicated by three examples in which LR probably can contribute decisively to progress.

21.9.1 Raw Materials and Energy

Nature annually produces nearly 200 billion t of renewable biomass with an energy value of about 100 billion t "coal units". Currently only about 2–3% of this is used for food, energy, and fiber production. This inexhaustible, because continually renewed, biomass must be utilized by biotechnology for economic industrial use, e.g., by hydrolysis of cellulose, semi-cellulose and starch to fermentable sugar [21.56], [21.57].

An example is the Hoechst-Uhde process, shown in Fig. 21.58, for the continuous production of ethanol starting from raw materials containing sugar, starch, and cellulose [21.58], [21.59]. The injected air not only effects the air-lift circulation drive in the ALR, but also the oxygen supply to the yeasts. With a mean residence time of 1–5 hours, 95% alcohol yields can be achieved [21.57]. The defined flow and the backmixing of the ALR are especially advantageous here.

Bioalcohol (ethanol) obtained from regenerative, systematically grown and cultivated biomass (agricultural and forest products) can substitute for fossil raw materials as high-octane fuel for use in vehicles, and the valuable by-products can, among others, be used as chemical raw materials. Good prospects exist for this in our more temperate climate [21.56].

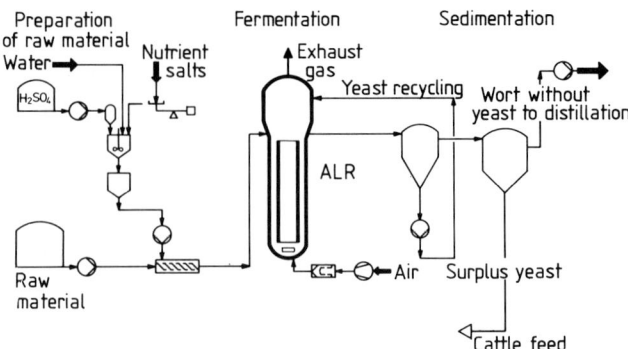

Figure 21.58. Flow sheet of the continuous Hoechst-Uhde ethanol production process on sugar, starch, and cellulose media with ALR as bioreactor.

21.9.2 Food

Today about 500 million people are starving; the FAO estimates that by 1985 it could be 750 million. Apart from this about 40% of the world population suffers from protein deficiency – especially in Africa, the Near- and Far-East.

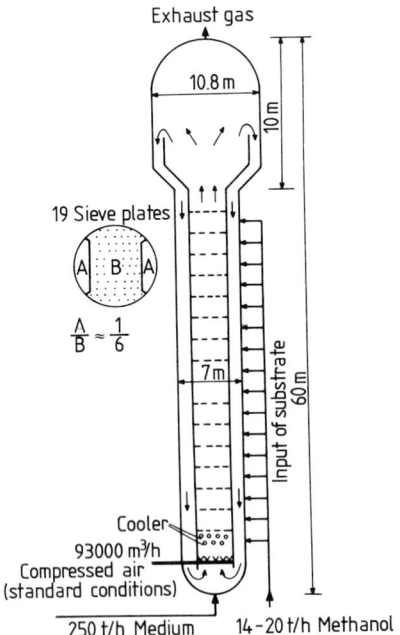

Figure 21.59. A large scale ICI Pressure Cycle Fermenter with internal circulation for SCP production; see also Fig. 21.3.

Industrial production of microbial bioprotein, so-called single cell protein SCP [21.6], [21.60], [21.61] can help to solve some of the acute problems. The large-scale development of the SCP process began with *Pseudomonas methylotrophus* on methanol early in the seventies in the ICI Pressure Cycle Fermenter. Fig. 21.3 shows a pilot plant LR with external circulation for 1000 t SCP per year. ICI have also developed an ALR arrangement with internal circulation, according to Fig. 21.59. In the meantime both LR-types are in operation with $V_R \approx 2000$–3000 m^3, especially in SCP plants of about 70 000 t per year (in operation since 1980) [21.62]. Fig. 21.59 shows a new large-scale arrangement of an ICI Pressure Cycle Fermenter [21.63]. In order to reduce the high circulation velocity of the ALR due to its large filling height of about 55 m and to improve the conversion of atmospheric oxygen, 19 sieve plates are arranged in the draft tube, which simultaneously effect a redispersion of coalesced gas bubbles. Since it is very difficult to rapidly and uniformly distribute the substrate (methanol) throughout the 2000 m^3 large reactor, it is added at around 1000 sites along the vertical axis of the reactor.

The Mitsubishi Gas Chemical Company, Inc. (MGC), one of the largest Japanese methanol producers, also uses an ALR as SCP bioreactor in a 500 t per year pilot plant with $V_R \approx 20$ m^3 [21.64]. The cell productivity is claimed to be 5 kg m^{-3} h^{-1}. After comparison tests with a STR fermenter, the ALR was preferred, because for sim-

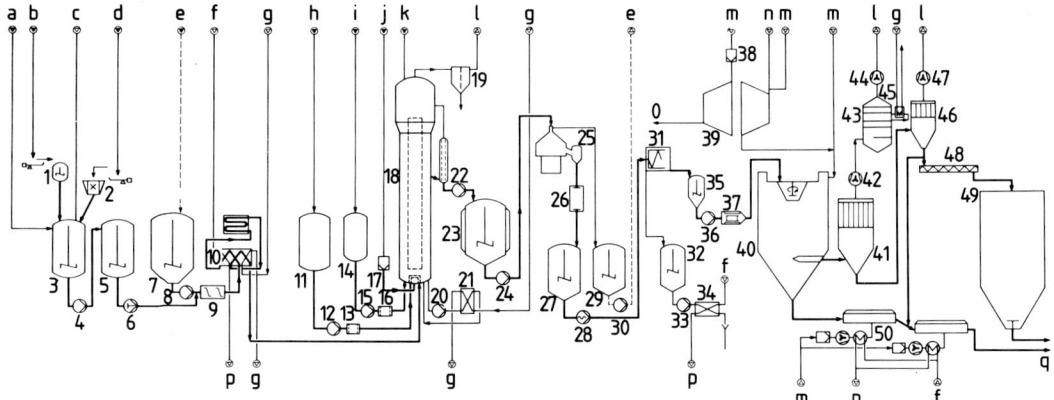

Figure 21.60. A flow sheet of the Hoechst-Uhde process for SCP production; pilot plant with LR (JLR or ALR) as bioreactor designed for 1000 t per year.
1 mixing vessel for trace elements, 2 crushers, 3 mixing vessel for nutrient solution, 4 pump, 5 nutrient solution reservoir, 6 pump, 7 recycle water reservoir, 8 pump, 9 mixing device, 10 thermal sterilization section, 11 methanol reservoir, 12 pump, 13 sterilizing filter, 14 NH_3 reservoir, 15 pump, 16 sterilizing filter, 17 sterilizing filter, 18 fermenter, 19 separator, 20 pump, 21 cooler, 22 pump, 23 harvest tank, 24 pump, 25 separator, 26 conditioning section, 27 concentrate vessel, 28 pump, 29 clear-liquid vessel, 30 pump, 31 decanter, 32 waste water tank, 33 pump, 34 waste water conditioning section, 35 concentrate vessel, 36 pump, 37 thermolysis facilities, 38 filter, 39 air compressor, 40 drier, 41 filter, 42 ventilator, 43 scrubber/cooler, 44 ventilator, 45 cooler, 46 solids separator, 47 ventilator, 48 screw conveyer, 49 silo, 50 granulating facilities.
a H_3PO_4, b trace elements, c H_2O, d nutrient salts, e recycle water, f steam, g cooling water, h methanol, i NH_3, j process air, k inoculant, l waste air, m air, n gas, o process air to 18, p condensate, q bagging, storage, dispatch.

ilar specific oxygen inputs it allows simpler large-scale design and lower power input.

Fig. 21.60 [21.65] shows a flow sheet of the SCP process being developed by Hoechst-Uhde using *Methylomonas clara* on methanol with a 40 m³ LR, as shown in Fig. 21.12. This bioreactor can be operated as ALR or as JLR, and is designed for 1000 t SCP per year. Pilot plants for SCP production with ALR and *inverse* flow direction also proved to be successful.

Investigations using a JLR showed that with *Candida lipolytica* on *n*-paraffin a definite increase in productivity with increasing mean circulation velocity w_m could be achieved. Considerations in [21.66] show that this significant productivity increase is clearly due to the more frequent forced passages through, and the consequent redispersion of, cell agglomerates in the shearing fields of the liquid jets.

21.9.3 Biological Waste Water Purification

The best way to reduce material and energy wastes is without doubt recycling and processing to the greatest degree possible. This "process-integrated" environmental protection simultaneously spares the limited resources and the environment. Unfortunately it is subject to scientific-technical and economic limits, which however, using biotechnology – e.g., the enzyme technology – can be considerably influenced. Nevertheless, residual material (and energies) remain, which can be returned to the cycle of nature as harmlessly as possible [21.67].

A pre-eminent example of this is biological waste water purification. Until about 1970 biological purification was performed in flat "activated basins" with surface aeration, as shown in Fig. 21.61. The air (and thus oxygen) required by the aerobic microorganisms which consume the remaining impurities, was "beaten" into the waste water across its surface. In fact this equipment lends itself to efficient energy dissipation,

currently being constructed or started-up at several sites – in several cases in connection with public waste water purification. There is a parallel arrangement of several draft tubes, each with its own circulation, with radial flow nozzles (Fig. 21.22) and gas deflectors (Fig. 21.13b). This arrangement

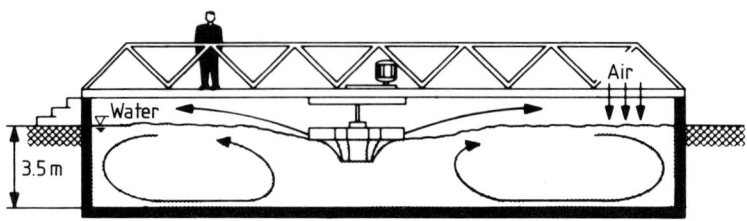

Figure 21.61. Flat concrete basin with surface aeration in biological sewage purification; technological standard about 1970.

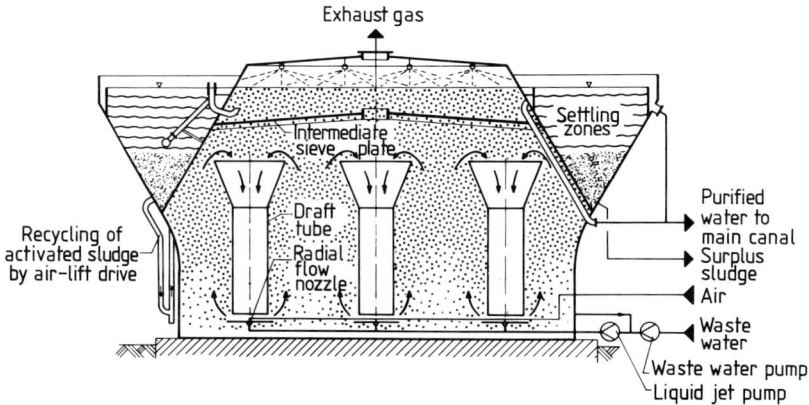

Figure 21.62. Scheme of a large scale "BIOHOCH Reactor" with several parallel draft tubes, corresponding to the arrangement of Fig. 21.7, each draft tube with a gas deflector and a radial flow nozzle as shown in Fig. 21.22.

as well as smell and noise production; the whole plant requires a great deal of space.

The next step resulted in covered basins with medium-height and submerged aeration. The operation was less noisy and less smelly; the energy and space requirements were lower [21.68], [21.69].

But in fact it was the transition from basin biology to "tower-" or "high-biology" which brought about decisive progress.

Fig. 21.62 [21.70] shows the design of such a "BIOHOCH Reactor", which are

corresponds in principle to that of Fig. 21.7. This type of "BIOHOCH Reactor" is currently being built or operated with volumes of up to 8000 m³. The main advantages are: 50 to 80% less energy; stable purification process; low outlet values; no problems with smell and noise; rational above-ground construction with prefabricated elements; location without disturbance and with minimum space requirements in factories or natural landscape and residential areas (Fig. 21.63).

An extreme case of height resp. of depth of such LR-construction is represented undoubtedly by the ICI Deep Shaft Pressure Cycle Fermenter according to Fig. 21.64. Here the activated tank is similar to a well construction and is sunk up to 200 m deep

Figure 21.63. View of a large scale "BIOHOCH Reactor" as shown in Fig. 21.62. It demonstrates that it can be located without disturbance in natural landscape as well as in residential areas. - Data: diameter: 23 to 45 m, filling height: 20 m, activated volume: 8000 m³ per reactor, capacity: 15 000 kg/d BOD_5, air feed: up to about 11 000 m³/h.

into the earth [21.71]. In principle the operation is the same as that for the SCP Pressure Cycle Fermenter except that here most of the air is introduced into the downflow tube. The circulation is started up and stabilized in operation by the further injection of compressed air at relatively shallow depth in the upflow side, as Fig. 21.64 indicates.

Bubble contact time, high pressure and high turbulence combine to yield good oxygen transfer characteristics of about 3 kg m^{-3} h^{-1}; but normally less than this is required, even for high strength industrial waste water. Oxygen efficiencies of the order of 90% are achievable for depths greater than 100 m with energy requirements around 3 kg O_2 per kWh. For instance, an 8 m diameter by 100 m deep shaft suffices to treat the sewage from a population of 500 000 or for more concentrated industrial waste waters up to 2 000 000 population equivalent. Conventional shallow basins for these duties would occupy several hectares [21.71]. The adaption of such deep shaft ALR-types with huge capacities to all environments is optimal, because they are completely hidden in the earth.

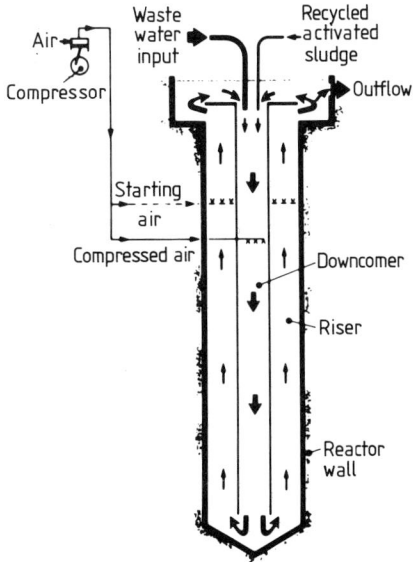

Figure 21.64. ICI Deep Shaft Pressure Cycle Fermenter (PCF) for biological waste water purification. The adaptation to any kind of surroundings is optimal, because this large bioreactor is constructed below ground level.

21.10 Outlook

As this treatise shows, LR of various types are suitable for many applications in biotechnology. Compared with gas-sparged stirred tanks and bubble column fermenters, they have only been under scientific research and development for the relatively short time of about 20 years and consequently there are still many biotechnological problems, technical improvements and possibilities of application to be investigated and solved. But if LR were not expected to prove useful as bioreactors the editors would scarcely have selected them as emblem of this series of books. Nevertheless, they can – like all other reactor types – never, even after further developments, become *the* optimal universal bioreactor. The extraordinary variety of biotechnological systems and processes will always require various types of bioreactors.

Acknowledgements

I express cordial thanks to my cooperators Dipl.-Ing. Ursula Gross, Dipl.-Ing. Andreas Hermann, and Dipl.-Ing. Werner Sternad for thoroughly reading and correcting the whole manuscript and arranging the nomenclature, Ing. Gerhard Friedrich for accurately producing or completing the numerous figures, and my secretary Luise Preiss for accurate typing of the difficult text and patiently supporting its repeated modifications.

Furthermore I thank the publishers and editors for indulgently enduring my time lag, and especially the volume editor for his liberal generosity.

21.11 References

[21.1] H. BLENKE: "Loop Reactors." Adv. Biochem. Eng. *13* (1979), 121–214.

[21.2] H.-J. REHM: "Probleme bei der Entwicklung von Apparaten in der Biotechnologie." Verfahrenstechnik *17* (1983) 4, 238–243.

[21.3] R. EHRAT: "Ein neuer Oxidationsreaktor." Chemie-Technik *5* (1976), 467–469.

[21.4] K. KIPKE: "Strömungstechnische Auslegung von Leitrohrpropellern für Schlaufenreaktoren." Maschinenmarkt *88* (1982) 43, 881–884.

[21.5] J. D. LITTLEHAILES: Lecture at 1. Symp. Mikrobielle Proteingewinnung, Braunschweig-Stöckheim 1975, Proceedings pp. 43–48.

[21.6] H. BLENKE: "Industrielle Eiweißproduktion – eine Nahrungsquelle der Zukunft. Brennpunkte der Forschung." Deutsche Verlagsanstalt, Stuttgart, 1981.

[21.7] P. G. COOPER, and R. S. SILVER: "Basin fermenter for single cell protein." Chem. Eng. Prog. *71* (1975) 9, 85–88.

[21.8] J. C. JOHNSON: "Yeasts for Food and other Purposes." Noyes Data Corporation, New Jersey, USA, 1977.

[21.9] H. BLENKE and W. HIRNER: "Stoffübergang in Gas/Flüssigkeits-Strahlreaktoren und Blasenkolonnen." VDI (Verein Deutscher Ingenieure)-Berichte *218* (1974), 549–578. Chem. Ing. Tech. *46* (1974) 8, 353.

[21.10] O. NAGEL, H. KÜRTEN, and R. SINN: "Strahldüsenreaktoren." Chem. Ing. Tech. *42* (1970) 7, 474–479 and *42* (1970) 14, 921–926.

[21.11] O. NAGEL, H. KÜRTEN, and R. SINN: "Stoffaustauschfläche und Energiedissipationsdichte als Auswahlkriterium für Gas/Flüssigkeits-Reaktoren." Chem. Ing. Tech. *44* (1972) 6, 367–373 and *44* (1972) 14, 899–903.

[21.12] U. ONKEN and P. WEILAND: "Airlift fermenters: Construction, behavior and uses." Adv. Biotech. Processes *1* (1983), 67–95.

[21.13] UHDE GmbH, Dortmund: Fermenter, Konstruktionszeichnung, 1976.

[21.14] H. VOGEL: "Untersuchung neuartiger Varianten des Schlaufenreaktors." Diplomarbeit, Institut für Chemische Ver-

fahrenstechnik, Universität Stuttgart, 1978.
[21.15] N. RÄBIGER and A. VOGELPOHL: "Der Kompaktreaktor, ein neuentwickelter Schlaufenreaktor mit hoher Stoffaustauschleistung." Chem. Ing. Tech. 55 (1983) 6, 486–487.
[21.16] K. SCHREIER: "Neuer Hochleistungsfermenter nach dem Tauchstrahlverfahren." Chem. Ztg. 99 (1975), 328–331.
[21.17] R. LAFFERTY, A. MOSER, W. STEINER, A. SARIA, and J. WEBER: "Gas/Flüssigkeitsstrahl-Schlaufenreaktor." VDI (Verein Deutscher Ingenieure)-Berichte 315 (1978). Chem. Ing. Tech. 50 (1978) 5, 401.
[21.18] C. LABBI, J. M. LEBAULT, A. ZOULALIAN, and G. BESSON: "Réacteur air/liquide à mélange externe." Extrait d'Informations Chimie 212 (1981) Avril.
[21.19] H. GERSTENBERG: "Blasensäulen-Reaktoren." Chem. Ing. Tech. 51 (1979) 3, 208–216.
[21.20] K. BOHNER and H. BLENKE: "Gasgehalt und Flüssigkeitsumwälzung im Schlaufenreaktor." Verfahrenstechnik 6 (1972) 2, 50–57.
[21.21] R. SEIPENBUSCH: "Wachstum der Candida lipolytica auf n-Paraffin im Strahl-Schlaufenreaktor." Dissertation, Universität Stuttgart, 1981.
[21.22] H. G. MÜLLER and G. SELL: "Der Radialstrombegaser – Ein leistungsfähiger Belüfter für biologische Abwasserreinigungsanlagen." Hoechst AG, Frankfurt, 1983.
[21.23] H. BLENKE: "Verfahren und Vorrichtung zur Entgasung von Gas-Liquid-Systemen ohne Schaumbildung." Ger. Pat. DOS P 33 31 993.6, 5 September, 1983.
[21.24] H. BLENKE: "Schlaufenreaktoren – Bioreaktoren." BMFT-Statusseminar "Bioverfahrenstechnik" Braunschweig-Stöckheim, 1977. Proceedings, Ed. DFVLR, pp. 5–44.
[21.25] K. SCHÜGERL: "Neue Bioreaktoren für aerobe Prozesse." Chem. Ing. Tech. 52 (1980) 12, 951–965.
[21.26] H. BLENKE, K. BOHNER, and S. SCHUSTER: "Beitrag zur optimalen Gestaltung chemischer Reaktoren." Chem. Ing. Tech. 37 (1965) 3, 289–294.
[21.27] H. BLENKE, K. BOHNER, and W. PFEIFFER: "Hydrodynamische Berechnung von Schlaufenreaktoren für Einphasensysteme." Chem. Ing. Tech. 43 (1971) 1/2, 10–17.
[21.28] P. GRASSMANN: "Physikalische Grundlagen der Chemie-Ingenieur-Technik", 2nd Ed. Verlag Sauerländer, Aarau-Frankfurt, 1970.
[21.29] H. BLENKE, K. BOHNER, and W. HIRNER: "Druckverlust bei der 180°-Strömungsumlenkung im Schlaufenreaktor." Verfahrenstechnik 3 (1969) 10, 444–452.
[21.30] H. BLENKE, K. BOHNER, and E. VOLLMERHAUS: "Untersuchungen zur Berechnung des Betriebsverhaltens von Treibstrahlförderern." Chem. Ing. Tech. 35 (1963) 3, 201–208.
[21.31] W. STEIN: "Beitrag zur Berechnung von Schlaufenreaktoren." Dissertation, Universität Stuttgart, 1968.
[21.32] R. MARQUART: "Umwälzung mittel- bis hochviskoser, Newtonscher und nicht-Newtonscher Flüssigkeiten in Propeller- und Strahl-Schlaufenreaktoren." Dissertation, Universität Stuttgart, 1977.
[21.33] R. MARQUART and H. BLENKE: "Umlauf mittel- bis hochviskoser, Newtonscher und nicht-Newtonscher Flüssigkeiten in Propeller-Schlaufenreaktoren." Verfahrenstechnik 12 (1978) 11, 721–726.
[21.34] M. ZLOKARNIK: "Rührtechnik." In "Ullmanns Encyklopädie der technischen Chemie," 4th Ed., Vol. 2, pp. 259–281. Verlag Chemie, Weinheim, 1972.
[21.35] M. ZLOKARNIK: "Modellübertragung in der Verfahrenstechnik." Chem. Ing. Tech. 55 (1983) 5, 363–372.
[21.36] A. B. METZNER and R. E. OTTO: "Agitation of non-Newtonian fluids." AIChE J. 3 (1957) 1, 3–10.
[21.37] W. PFEIFFER, H. BLENKE, and E. MUSCHELKNAUTZ: "Fluiddynamik von Suspensionen in Strahl-Schlaufenreaktoren." Verfahrenstechnik 11 (1977) 2, 95–101.
[21.38] P. ZEHNER: "Suspendieren von Feststoffen im Strahl-Schlaufenreaktor." Chem. Ing. Tech. 52 (1980) 11, 910–911.
[21.39] H. BRAUER: "Grundlagen der Einphasen- und Mehrphasenströmungen." Verlag Sauerländer, Aarau-Frankfurt, 1971.
[21.40] M. ZLOKARNIK: "Koaleszenzphänomene im System gasförmig/flüssig und deren Einfluß auf den O_2-Eintrag bei der biologischen Abwasserreinigung." Korrespondenz Abwasser 27 (1980) 11, 728–734.
[21.41] W. SCHUMM: "Örtliche Sauerstoffkonzentrationen und Phasengrenzflächen im

Schlaufenreaktor." Dissertation, Universität Stuttgart, 1982. – W. SCHUMM and H. BLENKE: "Örtliche Sauerstoffkonzentrationen im Schlaufenreaktor und ihre Bedeutung für die Berechnung der Phasengrenzfläche." Verfahrenstechnik 14 (1980) 12, 797–803. Chem. Ing. Tech. 53 (1981) 6, 444–446.

[21.42] K. G. SCHMIDT and H. BLENKE: "Meßmethode zur Bestimmung der Flüssigkeitsgeschwindigkeit von heterogenen Gas-Liquid-Systemen in Schlaufenapparaten." Verfahrenstechnik 17 (1983) 10/11, 593–597.

[21.43] B. VON DER LINDEN: "Untersuchung der Begasung höherviskoser Newtonscher Flüssigkeiten in Schlaufenreaktoren." Dissertation, Universität Stuttgart, 1983.

[21.44] K. BOHNER: "Gasgehalt und Flüssigkeitsumwälzung im Schlaufenreaktor." Dissertation, Universität Stuttgart, 1971.

[21.45] J. LEHNERT: "Berechnung von Mischvorgängen in schlanken Schlaufenreaktoren." Verfahrenstechnik 6 (1972) 2, 58–64.

[21.46] W. STEIN: "Zur Berechnung von Schlaufenreaktoren." Chem. Ing. Tech. 40 (1968) 17, 829–837.

[21.47] J. PAWLOWSKI: "Reaktionsrohr mit Cellarströmung und seine Verweilzeit-Eigenschaften." Chem. Ing. Tech. 34 (1962) 9, 628–631.

[21.48] O. LEVENSPIEL and W. K. SMITH: "Notes on the diffusion-type model for the longitudinal mixing of fluids in flow." Chem. Eng. Sci. 6 (1957) 227–233.

[21.49] J. LEHNERT: "Berechnung von Mischvorgängen in schlanken Schlaufenapparaten." Dissertation, Universität Stuttgart, 1972.

[21.50] H. BLENKE, W. REULE, and W. SCHUMM: "Beitrag zur Entwicklung wirtschaftlich optimaler Schlaufenreaktoren für begaste wäßrige Biosysteme." In "Bioreaktoren," 2. BMFT-Statusseminar Bioverfahrenstechnik, Jülich, 1979, Proceedings (J. GARTZEN, ed.), pp. 273–298.

[21.51] B. ATKINSON: "Biochemical Reactors." Pion Ltd., London, 1974.

[21.52] T. REITH: "Physical aspects of bubble dispersions in liquids." Ph.D. Thesis, Technische Hogeschool Delft, 1968.

[21.53] P. V. DANCKWERTS: "Significance of liquid film coefficients in gas absorption." Ind. Eng. Chem. 43 (1951) 6, 1460–1467.

[21.54] W. HIRNER: "Stoffübergang und Stoffübergangsfläche in Gas/Flüssigkeits-Strahlreaktoren." Dissertation, Universität Stuttgart, 1973. – W. HIRNER and H. BLENKE: "Gasgehalt und Phasengrenzfläche in Schlaufen- und Strahlreaktoren." Verfahrenstechnik 11 (1977) 5, 297–303.

[21.55] H. BLENKE: "Der Mensch und sein Lebensraum – Eingriff und Wandel." In "Verhandlungen der Gesellschaft Deutscher Naturforscher und Ärzte, 109. Versammlung," pp. 5–13. Springer Verlag, Berlin, 1978.

[21.56] M. DAMBROTH: "Industrierohstoffe vom Acker." Umschau 82 (1982) 11, 359–362.

[21.57] H. MENRAD and A. KÖNIG: "Alkoholkraftstoffe." Springer Verlag, Wien–New York, 1982.

[21.58] U. FAUST and R. KNECHT: "Novel continuous fermentation process for ethyl alcohol." Third International Symposium of Alcohol Fuels Technology, Asilomar, California, May 28–31, 1979.

[21.59] UHDE GmbH, Dortmund: "Der neue kontinuierliche Ethanol-Prozeß," 1980.

[21.60] H. BLENKE, M. SCHLINGMANN, and W. SITTIG: "The Industrial Production of Single Cell Protein (SCP)." Int. Coll. C.P.C.I.A., Paris, 1977.

[21.61] H. BLENKE: "Single Cell Protein (SCP) – Ein industriell produzierter Rohstoff für Nahrungsmittel." VDI (Verein Deutscher Ingenieure)-Berichte 277 (1977), 127–136.

[21.62] A. CRUEGER and W. CRUEGER: "Lehrbuch der angewandten Mikrobiologie." Akademische Verlagsgesellschaft, Wiesbaden, 1982.

[21.63] K. SCHÜGERL: "Apparatetechnische Aspekte der Kultivierung von Einzellern in Turmreaktoren." Chem. Ing. Tech. 55 (1983) 2, 123–134.

[21.64] M. KURAISHI: "Single Cell Protein Process Development with Methanol as Substrate." DECHEMA Monogr. 83 (1979), 111–124.

[21.65] UHDE GmbH, Dortmund: "Anlagen zur Erzeugung von Bioprotein," 1982.

[21.66] R. SEIPENBUSCH and H. BLENKE: "The loop reactor for cultivating yeast on n-paraffin substrate." Adv. Biochem. Eng. 15 (1979), 1–40.

[21.67] H. BLENKE: "Saubere Umwelt – Wünsche, Möglichkeiten, Grenzen." Chem. Ing. Tech. *54* (1982) 4, 277–278.

[21.68] G. LIPPHARDT: "Sauberes Wasser – technische Möglichkeiten, wirtschaftliche Konsequenzen." Chem. Ing. Tech. *54* (1982) 4, 279–286.

[21.69] M. ZLOKARNIK: "Eignung und Leistungsfähigkeit von Belüftungsvorrichtungen für die biologische Abwasserreinigung." Chem. Ing. Tech. *52* (1980) 4, 330–331.

[21.70] VERBAND DER CHEMISCHEN INDUSTRIE e. V. (VCI), Frankfurt: "Chemie und Umwelt – Wasser," 1982.

[21.71] D. A. HINES: "The Large Scale Pressure Cycle Fermenter Configuration." DECHEMA Monogr. *82* (1978), 55–64.

[21.72] W. REULE: "Zur Stoffübertragung im Strahl-Schlaufenreaktor mit dem Natriumsulfit-System." Dissertation, Universität Stuttgart, 1983.

[21.73] H. BLENKE: "Verfahren und Vorrichtung zur Durchführung von (bio-)chemischen Reaktionen in einem Gas/Flüssigkeitssystem." EP (European Pat.) 0 007 489, 10 November, 1982.

Chapter 22

Biological Waste Water Treatment in a Reciprocating Jet Bioreactor

Heinz Brauer

Institut für Chemieingenieurtechnik
Technische Universität Berlin
Berlin (West), Germany

22.1 Introduction
22.2 Development of Bioreactors for Waste Water Treatment
22.3 Properties of the Reciprocating Jet Bioreactor
22.4 Reaction of Bacteria to Mechanical Stress
22.5 Analysis of Oxygen Transport in the Biosuspension
22.6 Description of a Pilot Plant for Biological Waste Water Treatment
22.7 Properties of the Waste Water
22.8 Parameters of the Purification Process
22.9 Experimental Conditions
22.10 Microbial Purification of Waste Water in the Reciprocating Jet Bioreactor
22.11 Application of the Reciprocating Jet Bioreactor
22.12 References

List of Symbols

a	m	amplitude of reciprocating movement
D	m	diameter of bioreactor
D_s	m²/s	diffusion coefficient of oxygen in waste water
d_b	m	diameter of holes in sieve plates
E	K	reaction specific temperature
f	s^{-1}	frequency of reciprocating movement
H	m	height of bioreactor
h_p	m	distance between sieve plates
M_B	kg/m³	specific biomass in bioreactor
R_s	$\frac{kg/m^3 \, h}{kg/m^3}$	specific conversion rate, Fig. 22.13
T	K	temperature of biosuspension
t_b	m	spacing of holes in sieve plates
$\dot{V}_g$	m³/s	volumetric flow rate of air
$\dot{V}_r$	m³/s	volumetric recycle flow rate
$\dot{V}_z$	m³/s	volumetric flow rate of waste water
η	kg/m s	dynamic viscosity of waste water
η_g	kg/m s	dynamic viscosity of air
ϱ	kg/m³	density of water
ϱ_g	kg/m³	density of air
ϱ_{gs}	kg/m³	partial density of oxygen in air
ϱ_a	kg/m³	concentration of pollutant (TOC) in waste water at exit of bioreactor
ϱ_z	kg/m³	concentration of pollutant (TOC) in waste water at entrance of bioreactor
ϱ_n	kg/m³	concentration of pollutant (TOC) in waste water that cannot be further reduced by action of microorganisms
σ	kg/s²	surface tension in the interface between air bubble and waste water
M_B^*	—	biomass/pollutant ratio, Eq. (22.7)
T^*	—	reduced reaction temperature, Eq. (22.8)
t_v^*	—	dimensionless residence time of waste water in bioreactor, Eq. (22.9)
$\dot{V}_r^*$	—	recycle ratio, Eq. (22.5)
ϱ_z^*	—	pollutant concentration ratio, Eq. (22.6)
φ	—	efficiency of microbial conversion, Eq. (22.3)

22.1 Introduction

The only high-efficiency bioreactor available for aerobic biological waste water treatment is the reciprocating jet bioreactor developed by the author. This new bioreactor consists essentially of a relatively slender cylindrical vessel, in which a package of sieve plates carries out a reciprocating motion. By this motion the biosuspension, consisting of the polluted water, air bubbles, and bacteria, is forced jet-like through the holes in the sieve plates. The reciprocating motion of the package of sieve plates reduces the size of the agglomerates of bacteria, ultimately to individual bacteria, substantially enhancing oxygen transfer from the air bubbles into the water, biochemical activity of the bacteria, and flocculation velocity of the bacteria as soon as the biosuspension leaves the bioreactor and flows into the sedimentation tank.

The aerobic biological waste water treatment bacteria are applied as microreactors for the conversion of harmful organic and inorganic substances. For organic compounds the conversion is as follows (HABECK-TROPFKE [22.1]):

$$\text{organic compounds in aqueous solution} + O_2 \xrightarrow{\text{bacteria}} CO_2 + H_2O + \text{energy} + \text{cellular matter}$$

According to this simple equation oxygen is required for the microbial conversion of harmful compounds. In most cases the composition of the mixture and the concentration of the harmful organic and inorganic compounds contained in the water is unknown. The conversion of such a mixture of compounds requires a very large population of bacteria. The individual species of microorganisms are generally unknown. The most suitable population of bacteria will develop in the bioreactor during its full operation. No seeding is necessary.

A survey of the development of biological waste water treatment over the last hundred years will be helpful in a correct estimation of the properties of the reciprocating jet bioreactor.

22.2 Development of Bioreactors for Waste Water Treatment

Conventional biological waste water treatment is carried out in large, centralized purification plants, installed far away from the townships. The first generation of such plants are the so-called sewage farms. The water was spread in a relatively thin layer over a large area. The primary aim of such a system was mechanical purification. Biological treatment was not yet seriously considered. The second generation of biological waste water treatment plants consists of large tanks made of concrete. Their depth is such as to satisfy the process capacity requirement. The volume may be of the order of several thousand or even ten thousand cubic meters. In most cases these tanks are open so that the typical unpleasant smell is easily emitted. In order to reduce the emittance of odorous substances the tanks are lately supplied with a cover. The main properties of the biological tanks of the second generation are as follows:

> The motion of the biosuspension in large volume tanks can neither be controlled nor predicted. Flow conditions in such tanks are characterized by large dead volumes. Furthermore, oxygen transfer into the biosuspension is very poor. The consequence is a rather low volume-based biological purification efficiency, which steadily decreases with increasing volume of the tank.

Large volume tanks in general have a substantial site requirement. In case of open tanks, the existence of a large interface between the sewage and ambient air supports the emission of the typically bad smell, normally associated with waste water treatment plants.

Conventional waste water treatment plants are not suitable for factories located in populated areas. In such cases the waste water has to be transported over relatively long distances to treatment plants located in unpopulated areas so that emissions are not objectionable. The alternative is a new type of purifying plant which may safely be operated even in densely populated areas and which is not too expensive. A new technical system consisting of a high-efficiency bioreactor and sedimentation tank, and which will satisfy these requirements has been under development in the last few years. This system is ready to leave the laboratory stage. It may be called "3rd generation of waste water treatment plants".

The general requirements for high-efficiency biological waste water treatment plants of the third generation may be formulated as follows (BRAUER and SUCKER [22.2]):

1. Small volume of bioreactor and sedimentation tank
2. Small site requirement
3. Closed, emission proof, bioreactor and sedimentation tank
4. No scaling-up problems
5. High operational flexibility
6. High volume-based purification efficiency of the bioreactor, at least 10 times that of conventional bioreactors
7. High volume-based sedimentation efficiency of the sedimentation tank
8. Oxygen supply for the biosuspension by means of natural air
9. High utilization of the oxygen contained in the air
10. Possibility of incorporation into production processes

These requirements are fully met by the reciprocating jet bioreactor and the high-efficiency sedimentation tank.

22.3 Properties of the Reciprocating Jet Bioreactor

The reciprocating jet bioreactor consists of a slender cylindrical vessel with a height/diameter ratio of about 4:1. The height of the pilot reactor was 120 cm and the diameter 30 cm. The active volume of the bioreactor was about 70 L. The most important element of the bioreactor is the reciprocating package of sieve plates. Fig. 22.1 shows a photograph of such a package. The sieve plates are arranged on a central shaft. The diameter of the holes in the plates is 12 mm, the spacing of the holes 27 mm. The distance of the plates is 50 mm, the amplitude of the reciprocating movement 100 mm, and the frequency $f = 1$ Hz. Diameter and spacing of the holes, distance between plates, as well as amplitude and frequency of the reciprocating motion are kept constant, independent of the size of the bioreactor.

The reciprocating element enforces, as indicated in Fig. 22.2, a jet-like motion of the biosuspension through the holes in the plates followed by an intensive vortex flow. The space available to one jet with the surrounding vortex is one elementary unit of the bioreactor. The volume of the reciprocating jet bioreactor consists of a large number of such elements, arranged in parallel and in sequence. As the volume of one elementary unit is $V_e = 31.6$ cm^3, the number of such units per m^3 of the bioreactor is 31 676.

When the reciprocating motion of the package of sieve plates changes direction,

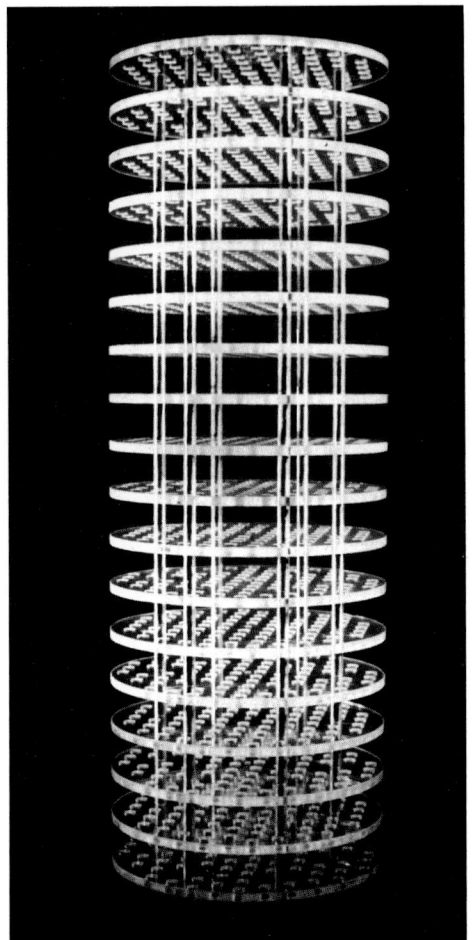

Figure 22.1. Photograph of a package of sieve plates.

the fluid motion in the jets and vortices also changes its direction. The alternating upward and downward motion of the biosuspension in the jets is part of the secondary motion. The primary motion is the time-independent axial flow of the biosuspension through the bioreactor. It is the secondary motion that is responsible for a thorough mixture within the biosuspension. Part of the biosuspension contained in one elementary unit is withdrawn, while the same volume of biosuspension is added to this unit. By this process of periodical exchange of fluid between neighboring elementary units, the intensive mixing is achieved. Consequently, the conditions for microbial reactions are identical for all elementary units.

The reciprocating motion of the package of sieve plates induces very strong shear stresses within the biosuspension, which are proportional to the power input. For the reciprocating jet bioreactor the power input N transfer from the fluid into the bacteria.
2. Enhancement of the biochemical activity of the individual bacterium.
3. Enhancement of the flocculation velocity after the bacteria leave the space of high shear stress in the bioreactor and get into the sedimentation tank.

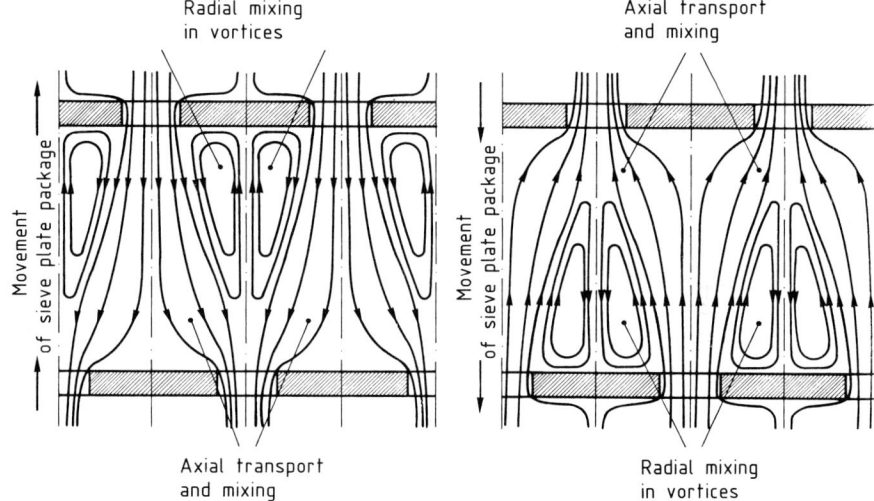

Figure 22.2. Fluid motion in the space between two sieve plates.

per unit volume V is of the order of $N/V = 1000$ to 2000 W/m^3. In comparison, conventional bioreactors require a power input of about $N/V = 50$ W/m^3. The energy dissipation in the reciprocating jet bioreactor and, consequently, the strength of the resulting shear stress is 20 to 40 times that of conventional bioreactors.

The decisive and unique properties of the reciprocating jet bioreactor are the result of the extremely strong shear stresses present in the biosuspension:

1. Generation of very small agglomerates of bacteria, ultimately to individual bacteria. For such conditions the largest possible surface area of the mass of bacteria is available for mass

4. Generation of very small air bubbles and periodic renewal of the bubble surface twice in one reciprocating cycle.
5. Very intensive mixing of the biosuspension in the radial and axial direction.

The generation of individual bacteria or very small agglomerates as well as that of small air bubbles is achieved while the biosuspension is forced to flow through the holes in the sieve plates and at the edge of the free jet that penetrates the space between two sieve plates. The consequences of the action of the mechanical stress on the bacteria will be discussed in the following section.

22.4 Reaction of Bacteria to Mechanical Stress

The enhancement of the chemical activity of the surface of a solid body due to mechanical stress is a well known effect, which is called the "mechano-chemical effect". A theoretical description of this effect has been offered by SCHÖNE [22.3].

An appreciable mechano-biochemical effect has been observed during investigations of biological waste water treatment in the reciprocating jet reactor [22.2]. The mechanical stress must be applied so that none of the bacteria are destroyed. Shear stresses within the biosuspension seem to be especially well suited for action on the bacteria.

In 1977 SITTIG and HEINE [22.4] reported already on the variation of the metabolism of *Methylomonas clara* under strong local shear stresses. Due to the design of the loop reactor in which the fermentation process was carried out, only a fraction of the microorganisms could be exposed to the mechanical stress. Therefore, the increase of metabolism was not very strong. Similar results have been observed by UJCOVÁ et al. [22.5]. From these two investigations it follows, that mechanical stress enhances metabolism and that the mechanical stress should be available in the whole volume of the bioreactor.

The strong mechano-biochemical effect observed in the reciprocating jet bioreactor is due to the fact, that in every elementary unit the mechanical stress is of the same strength and that there are no dead spaces. All microorganisms contained in the bioreactor are exposed to the same mechanical stress.

The action of the mechanical stress on the bacteria yields the following results: Small shear stress leads to a breakup of agglomerates, so that finally individual bacteria will result. By this process more bacteria are made available for the biochemical conversion process. In other words, the biochemical conversion capacity of the bacteria is fully exploited by making the surface of the bacteria available to mass transfer across this surface. With increasing shear stress the biochemical activity of the individual bacterium is increased. This activation process is closely related to phenomena occurring outside the bacterial cell wall. HUBERT and WERNER [22.6] as well as WEHMEYER et al. [22.7] reported on an increase of the electrical surface charge of the bacteria with increasing mechanical stress. Furthermore, these authors observed an increased biochemical conversion and flocculation of the bacteria as a function of the surface charge. The result of mechanical stress exerted on microorganisms is therefore threefold:

1. Generation of individual bacteria or small agglomerates of bacteria enhancing the biochemical conversion capacity.
2. Increasing the biochemical activity of the individual bacterium.
3. Increasing the flocculation velocity of the bacteria when these leave the region of high mechanical stress.

The consequences of the immediate results of mechano-biochemical activation are:

a) Reduction of the volume of the bioreactor because of a substantial increase of the biochemical conversion rate of the bacteria.
b) Reduction of the volume of the sedimentation vessel because of a substantial increase of the flocculation velocity of the bacteria.

The sediment of the bacteria obtained in the reciprocating jet bioreactor can be seen in Fig. 22.3, which has been obtained by means of an electron microscope. The bacteria have the shape of spheres, the diameter is about 0.5 μm. With this information the number of bacteria contained in an elementary unit of the reciprocating jet bioreactor may be determined. The volume of such a unit is 31.6 cm^3. Assuming the very

Figure 22.3. Electromicrograph of the sediment of mechanically activated bacteria.

22.5 Analysis of Oxygen Transport in the Biosuspension

low volumetric concentration of 0.5%, the number of bacteria in this elementary unit is about 10^{12}, or one million times one million bacteria. Each one of the bacteria must be supplied not only with organic or inorganic material but also with oxygen. The process of oxygen supply is rather difficult, not only because of the low solubility of oxygen in water, but also because of limitations occurring in the physical transfer process in the biosuspension. This transfer process will be considered in the following section.

In aerobic biological waste water treatment organic compounds contained in the water are microbially converted by means of oxygen, which must be transferred into the water. The most economic transfer is achieved when the oxygen contained in natural air can be used. In this case air must be introduced and dispersed in the waste water. The transport of oxygen from an air bubble to the bacteria will be analyzed.

The transport of oxygen in the biosuspension includes six steps according to Fig. 22.4 [22.2]:

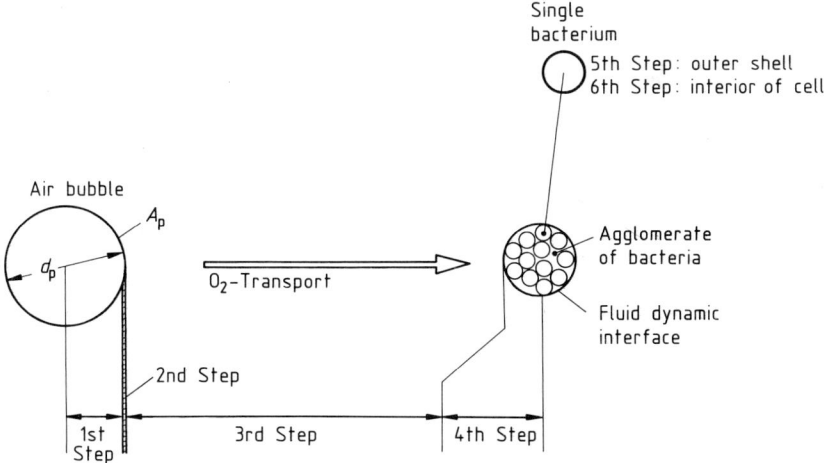

Figure 22.4. Analysis of oxygen transport in waste water from the air bubble to the interior of a bacterium.

1. Transport from the interior to the bubble surface
2. Transport across the interface between bubble and water
3. Transport in the waste water from the interface of the bubble to the fluid dynamic interface of the agglomerate of bacteria
4. Transport within the agglomerate to the individual bacterium
5. Transport across the outer 'shell' of the bacterium
6. Transport within the bacterium

The analysis of the oxygen transport will be restricted to these aspects, which are important for the transport rate. Within the bubble any resistance to oxygen transport may be neglected. Because of effective molecular and convective transport the oxygen concentration will be nearly constant within the volume of the bubble, however, decreasing with time. An important resistance to oxygen transport may exist in the interface between bubble and water due to the assemblage of surface active molecules. This diffusion barrier will of course be disrupted by the strong shear stress acting on the bubbles, which will periodically enforce disintegration and renewal of bubbles.

A rather strong resistance to oxygen transport may occur in step 3, primarily because of the distance between the interfaces of air bubble and agglomerates of bacteria. The intensive convective motion within the liquid enforced by the reciprocating motion of the package of sieve plates will effectively reduce this resistance. This is due to two effects: The high degree of mixing supports molecular diffusion, while the disruption of large bubbles reduces the distance between the bubbles and the agglomerates of bacteria.

A further important resistance occurs within the agglomerates. But this may also be effectively reduced by the disruption of agglomerates and generation of single bacteria by the action of mechanical stress.

In step 5 the oxygen has to move through the cell wall. This process is by no means well understood because of incomplete knowledge of the structure and functions of the elements of the cell wall and membrane. Experimental investigations of electrokinetic properties of bacteria indicate that the action of mechanical forces influences these properties, including the permeability of the wall for oxygen and other materials (HUBERT and WERNER [22.6]; WEHMEYER et al. [22.7]). Nothing is known about the resistance to oxygen transport within the bacterium.

The analysis of oxygen transport within the biosuspension revealed the important fact that the same mechanical stress, which enhances biochemical conversion of the bacteria, also enhances the necessary oxygen transport.

22.6 Description of a Pilot Plant for Biological Waste Water Treatment

Since 1978 comprehensive experimental investigations on aerobic biological waste water treatment have been carried out. The pilot plants available for these investigations have been operated in biological one-stage and two-stage processes. Fig. 22.5 gives a schematic drawing of the two-stage plant, which will be described in detail.

The two-stage pilot plant consists of two independent biological cycles. Such systems enable the successful purification of even heavily polluted waters, for which conventional systems cannot be used unless the load of the waste water is substantially reduced by adding clean water.

From the waste water storage tanks **1** and **2** the raw water flows through pipe **3** to the mixing unit **4**. For start-up operations fresh water is added at this point to the waste water. The fresh water is supplied by the public system, flows through pipe **5** and metering devices **6** to the mixing unit **4**. From here, pipe **7** leads the water to heat

22.6 Description of a Pilot Plant for Biological Waste Water Treatment

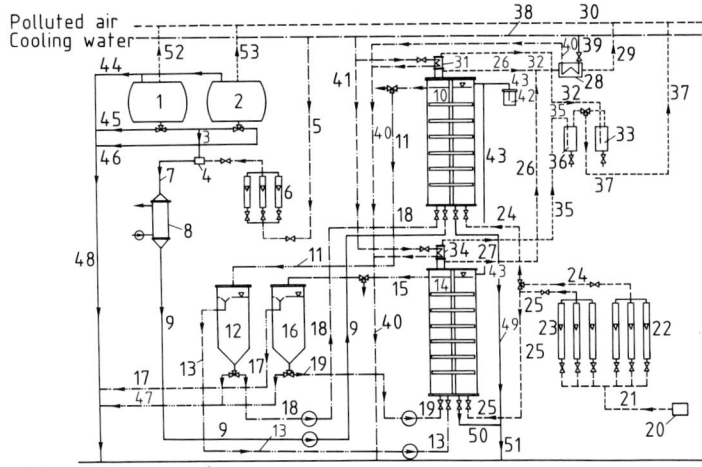

Figure 22.5. Schematic drawing of a pilot plant for aerobic biological waste water treatment (for details see text).

exchanger **8**, in which the water is brought to an optimal operation temperature, which is about 35 to 37 °C. Pipe **9** connects the heat exchanger with the bottom plate of bioreactor **10** of the first biological stage. After successful biological treatment of the waste water in the reciprocating jet bioreactor the water flows through pipe **11** to the sedimentation tank **12**, where the biomass is separated from the pretreated waste water. This water is directed through pipe **13** to bioreactor **14** of the second biological stage for final treatment. The resulting biosuspension flows through pipe **15** to the second sedimentation vessel **16**, where a complete separation of the biomass from the purified water is carried out. The purified water is discharged through pipe **17** into the public system. Recycling of biomass is achieved by means of pipes **18** and **19**.

The microbial conversion of the organic compounds contained in the waste water requires the addition of oxygen. For this reason both bioreactors are supplied with natural air. By means of compressor **20** air is directed via pipe **21**, metering and regulation systems **22** and **23**, and via pipe **24** to bioreactor **10**, and via pipe **25** to bioreactor **14** of the second stage. The air supplies the water with oxygen and picks up carbon dioxide, one of the products of microbial reactions. The air leaves the bioreactors **10** and **14** through pipes **26** and **27** and is directed to condenser **28**, where some of the water vapor is condensed. From there the air flows through pipe **29** to the air discharge system for purification if necessary.

The oxygen content of the air leaving the two bioreactors is determined. For this purpose air is withdrawn from bioreactor **10** and taken through condenser **31** and pipe **32** to the measuring device **33**. From bioreactor **14** air is taken via condenser **34** and pipe **35** to the measuring device **36**. The air leaving both measuring devices flows through pipe **37** into the discharge pipe **30**.

The condensers **28**, **31**, and **34** are supplied with cooling water from the central supply line **38**. Condenser **28** is connected with this supply line by pipe **39**. After leaving condenser **28** the cooling water is discharged through pipe **40**. The condensers **31** and **34** are connected with supply line **38** by pipe **41**. The water leaves the condensers by pipe **40**.

If foaming occurs in the bioreactors **10** and **14** antifoam agent is supplied from system **42** via pipe **43**. When the pilot plant has to be emptied, the water is withdrawn through pipes **44** to **51**. The supply tanks **1** and **2** are connected with the air discharge system by pipes **52** and **53**.

22.7 Properties of the Waste Water

In earlier experiments various industrial and synthetic waste waters have been tested. The results of some of these experiments have been published by BRAUER and SUCKER [22.2]. The reputation of the reciprocating jet bioreactor has been founded on these results. Furthermore, these results gave encouragement to a more fundamental investigation. The properties of the waste water used in these investigations will be discussed.

The waste water had been obtained from the public waste water treatment plant in Marienfelde in Berlin. This waste water results from a thermal conditioning process carried out at a temperature of about 300°C. It has a dark brown color. The pollutant concentration is of the order of 3500 to 4000 g TOC per m³ waste water, or 12075 to 14800 g COD per m³, or 5541 to 6333 g BOD$_5$ per m³. TOC is the Total Organic Carbon of the organic compounds contained in the waste water; COD is the Chemical Oxygen Demand required for chemical conversion of the organic compounds contained in the water; BOD$_5$ is the Biochemical Oxygen Demand in 5 days required for microbial conversion of organic compounds contained in the waste water. The relation between these units differs from one waste water to another. For the waste water used in the mentioned experiments the relation between COD and TOC is given in Fig. 22.6. From these results the following relation has been derived:

$$\frac{COD}{TOC} \approx 3.45.$$

The relation between BOD$_5$ and TOC is given by:

$$\frac{BOD_5}{TOC} \approx 1.58,$$

and that for COD and BOD$_5$ by:

$$\frac{COD}{BOD_5} \approx 2.18.$$

The pollutant contained in the waste water is in general a mixture of a great number of unknown compounds. The concentration of individual compounds is known only in exceptional cases. This is the reason why the pollutant concentration is given in the units which have been explained above.

Microbial conversion of a large fraction of the organic compounds contained in the waste water used for the experiments is possible, as SARFERT [22.8] as well as FRENZEL and SARFERT [22.9] have established. For a smaller fraction of the organic compounds microbial conversion seems to be rather difficult or even impossible. These specific

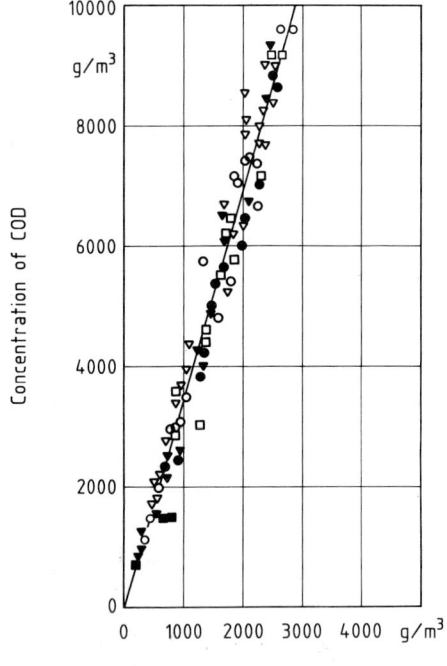

Figure 22.6. Relation between pollutant concentration for the waste water used in the experiments, expressed as g COD per m³ and as g TOC per m³.

compounds have been analyzed by FRENZEL and SARFERT [22.10]. According to tests carried out in the pilot plant, the ratio of unconvertible (TOC_R) and convertible (TOC_0) pollutants has been found to be:

$$\frac{TOC_R}{TOC_0} = \frac{1232}{3850} = 0.32.$$

This is a figure found in many other cases.

22.8 Parameters of the Purification Process

The result of a microbial purification process is described by the difference between the pollutant concentration in the waste water at the inlet (ϱ_z) and at the outlet (ϱ_a) of the bioreactor. For this concentration difference the following equation has been suggested (GRÜGER [22.11]):

$$\varrho_z - \varrho_a = f(\dot{V}_z; \dot{V}_r; \varrho; \eta; \varrho_z; \varrho_n; D_s;$$
parameters of waste water
$D; H; d_b; t_b; h_p; a; f;$
parameters of bioreactor
$M_B; T; E;$
parameters of bacteria
$\dot{V}_g; \varrho_g; \eta_g; \varrho_{gs}; \sigma)$ (22.1)
parameters of air

The parameters are: $\dot{V}_z$ volumetric flow rate of waste water, $\dot{V}_r$ volumetric recycle flow rate, ϱ density and η dynamic viscosity of waste water, ϱ_z pollutant concentration of waste water entering the bioreactor, ϱ_n pollutant concentration of waste water that cannot be reduced by microorganisms, D_s diffusion coefficient of oxygen in waste water, D diameter and H height of bioreactor, d_b diameter and t_b spacing of holes in the sieve plates, h_p distance between plates, a amplitude and f frequency of reciprocating movement of the package of sieve plates, M_B specific biomass in bioreactor, T temperature of biosuspension in bioreactor, E reaction specific temperature of bacteria, $\dot{V}_g$ volumetric flow rate of air, ϱ_g density and η_g dynamic viscosity of air, ϱ_{gs} partial density of oxygen in air, and σ surface tension between air bubble and waste water.

Because of the great number of parameters – 23 have been listed – one may not expect to find a solution of Eq. (22.1) on the basis of experiments. Far reaching simplifications must be introduced. From the parameters of the waste water the following may be safely neglected:

$\varrho; \eta; D_s.$

In most technical applications these parameters will vary only slightly. From the parameters of the bioreactor the following may be neglected:

$d_b; t_b; h_p; a.$

These parameters are kept absolutely constant. Of the parameters of the bacteria none can be neglected. The reaction specific temperature E is not yet known, however, and thus cannot be introduced into an empirical equation. The parameters of the air will all be neglected:

$\dot{V}_g; \varrho_g; \eta_g; \varrho_{gs}; \sigma.$

With the exception of $\dot{V}_g$ the parameters will be constant, because natural air will, in general, be introduced into the bioreactor. The volumetric flow rate $\dot{V}_g$ is kept constant at a level, where the oxygen content of the waste water in the bioreactor will not become the limiting factor in the microbial reaction.

For further simplification of Eq. (22.1) the volume of the bioreactor V is introduced, which is a function of diameter D and height H.

With these simplifications the following equation is obtained:

$$\varrho_z - \varrho_a = $$
$$= f(\dot{V}_z; \dot{V}_r; \varrho_z; \varrho_n; V; f; M_B; T; E) \quad (22.2)$$

In this equation dimensionless parameters will be introduced:

$$\varphi \equiv \frac{\varrho_z - \varrho_a}{\varrho_z - \varrho_n} \quad \text{efficiency of microbial conversion} \quad (22.3)$$

$$t_v^* \equiv t_v f = \frac{V}{\dot{V}_z + \dot{V}_r} f \quad \text{dimensionless residence time} \quad (22.4)$$

$$\dot{V}_r^* \equiv \frac{\dot{V}_r}{\dot{V}_z + \dot{V}_r} \quad \text{recycle ratio} \quad (22.5)$$

$$\varrho_z^* \equiv \frac{\varrho_z}{\varrho_n} \quad \text{pollutant concentration ratio} \quad (22.6)$$

$$M_B^* \equiv \frac{\dot{M}_B}{\varrho_z - \varrho_n} \quad \text{biomass/pollutant ratio} \quad (22.7)$$

$$T^* \equiv T/E \quad \text{reduced reaction temperature} \quad (22.8)$$

The final equation, describing the efficiency of microbial conversion of the pollutants contained in the waste water, is as follows:

$$\varphi = f(t_v^*; \dot{V}_r^*; \varrho_z^*; M_B^*; T^*). \quad (22.9)$$

This is the basis for evaluation of experimental results.

22.9 Experimental Conditions

The experiments have been carried out in the pilot plant described in Sect. 22.6. The pilot plant had been arranged for one-stage operation. The amplitude of the reciprocating package of sieve plates was 100 mm, which is twice the distance between two sieve plates.

The volumetric flow rate $\dot{V}_g$ of the air varied between $\dot{V}_g = 0.8$ and $2.5 \, \text{m}^3/\text{h}$ in order to insure a minimum oxygen content of $\varrho_s = 0.2 \, \text{mg/L}$ in the biosuspension. The specific biomass in the bioreactor M_B depends on the specific biomass M_{Br} in the recycle flow rate $\dot{V}_r$, and on the flow rate of the waste water $\dot{V}_z$. The relation is as follows:

$$M_B = \dot{V}_r^* M_{Br}. \quad (22.10)$$

The specific biomass in the recycle flow depends strongly on the effectiveness of the sedimentation system. With the high efficiency sedimentation system, the specific biomass in the recycle flow assumed values between $M_{Br} = 20$ to $30 \, \text{kg/m}^3$. With a conventional value for the recycle ratio of $\dot{V}_r^* = 0.15$, the specific biomass in the bioreactor varied between 10 and $15 \, \text{kg/m}^3$.

22.10 Microbial Purification of Waste Water in the Reciprocating Jet Bioreactor

For easy application of the results it is advisable to present the efficiency φ of microbial conversion as a function of conventionally applied parameters:

$$\varphi = f(T; t_v; \varrho_z; f; \dot{V}_r^*). \quad (22.11)$$

The results discussed in this section have been obtained by GRÜGER [22.11].

In Fig. 22.7 the conversion efficiency φ is shown over the reaction temperature T for three values of the residence time t_v. The concentration of pollutant in the waste water at the entrance of the bioreactor has been either $\varrho_z = 600 \, \text{g}$ TOC per m^3 waste water or 2000 (2200) g TOC per m^3. The values for the residence time have been 15,

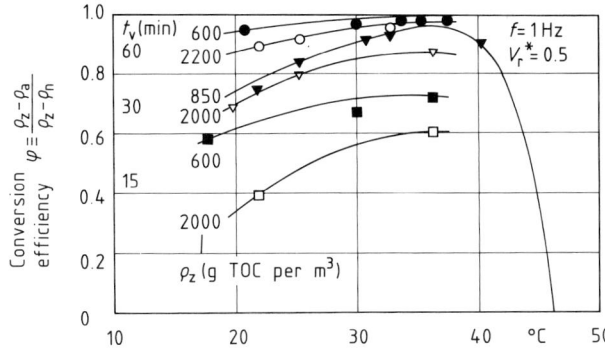

Figure 22.7. Conversion efficiency φ as a function of reaction temperature T for various values of the residence time t_v and pollutant concentration ϱ_z at inlet of bioreactor.

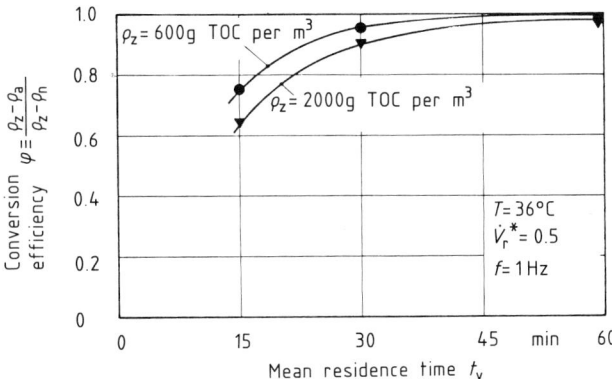

Figure 22.8. Conversion efficiency φ as a function of residence time t_v for two values of the pollutant concentration ϱ_z at inlet of bioreactor for optimum temperature $T=36\,°C$.

30, and 60 min. The frequency of the reciprocating motion with $f = 1$ Hz and the recycle ratio with $\dot{V}_r^* = 0.5$ have been constant parameters.

According to Fig. 22.7 the conversion efficiency φ increases steadily with increasing temperature T, assuming a maximum value at $T \approx 36\,°C$, sharply decreasing thereafter. The optimum temperature for aerobic biological waste water treatment is $T_{\text{opt}} = 36\,°C$. At this optimum temperature the conversion efficiency reaches values between 90 and 98%, when the residence time of the waste water in the bioreactor is either $t_v = 30$ or 60 min. The influence of temperature decreases with increasing residence time t_v. These results prove the extreme efficiency of the reciprocating jet bioreactor. The efficiency is at least 30 times better than conventional purification systems. Operation of the bioreactor at the elevated temperature of $T = 36\,°C$ poses no problems in industry, as most waste waters are available at even higher temperatures. On the other hand, heat losses can be easily avoided by insulation of the reactor. Because of the extremely small volume and closed design of the reciprocating jet bioreactor, insulation can be achieved at minimum costs.

The conversion efficiency φ at optimum temperature $T = 36\,°C$ is shown in Fig. 22.8 as a function of the residence time t_v for two values of the pollutant concentration ϱ_z. This figure again demonstrates a decreasing influence of increasing residence time t_v on the pollutant concentration ϱ_z. For the two waste waters tested the conversion efficiency is quite satisfactory for a residence time of $t_v \geq 30$ min.

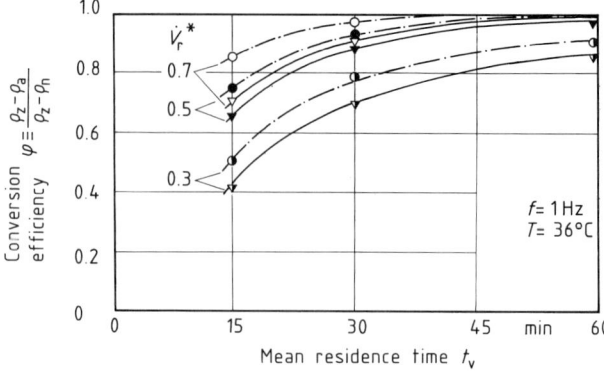

Figure 22.9. Conversion efficiency φ as a function of residence time t_v for three values of the recycle ratio $\dot{V}_r^*$ and two values of the pollutant concentration ϱ_z at inlet of bioreactor.
— · — $\varrho_z = 600$ g TOC per m^3,
——— $\varrho_z = 2000$ g TOC per m^3.

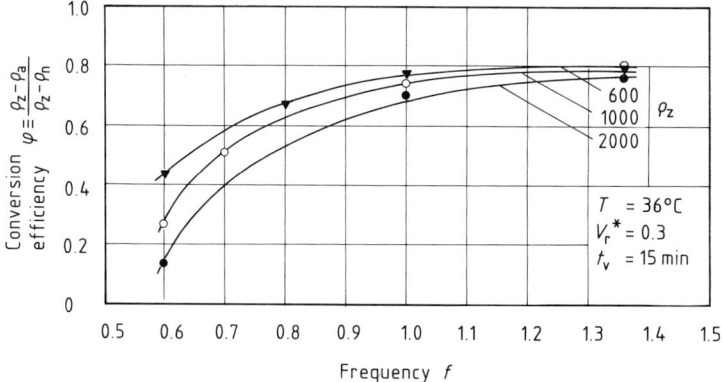

Figure 22.10. Conversion efficiency φ as a function of frequency f of the reciprocating package of sieve plates for three values of the pollutant concentration ϱ_z at inlet of bioreactor (g TOC per m^3) and for a residence time $t_v = 16$ min of the waste water in the bioreactor.

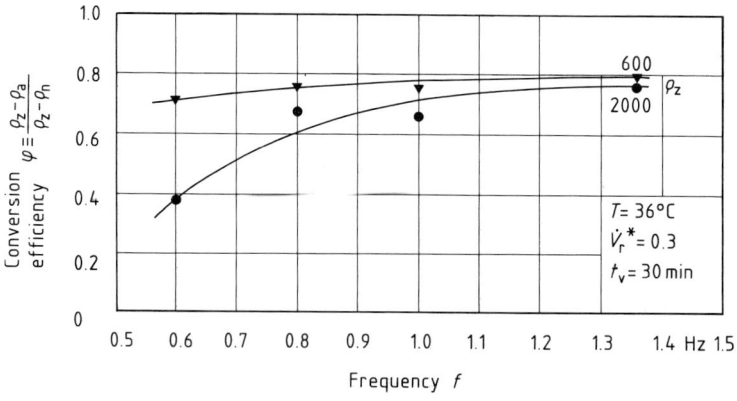

Figure 22.11. Same as Figure 22.10 but for a residence time $t_v = 30$ min, ϱ_z given in g TOC per m^3.

Fig. 22.9 shows the conversion efficiency φ as a function of the mean residence time t_v for three values of the recycle ratio $\dot{V}_r^*$. For the tested waste water the recycle ratio $\dot{V}_r^*$ should not be smaller than $\dot{V}_r^* = 0.5$, because the conversion efficiency will decrease rather strongly. Increasing the recycle ratio beyond $\dot{V}_r^* = 0.5$ does not substantially increase the conversion efficiency φ. It is therefore suggested that the reciprocating jet bioreactor be operated with a recycle ratio of $\dot{V}_r^* = 0.5$. This is also the accepted value for conventional purification systems.

In Figs. 22.10 and 22.11 the conversion efficiency φ is plotted against the frequency f of the reciprocating package of sieve plates for three values of the pollutant concentration ϱ_z at the entrance of the bioreactor. It should be noted that these results have been obtained for the unfavorable value of $\dot{V}_r^* = 0.3$. The results in Fig. 22.10 have been obtained for a mean residence time of $t_v = 15$ min and those in Fig. 22.11 for $t_v = 30$ min. The results presented in these figures indicate a steady increase of the conversion efficiency with increasing frequency f up to a maximum at $f \approx 1$ Hz. Therefore, the reciprocating jet bioreactor should not be operated at a frequency beyond $f = 1$ Hz. Depending on mean residence time and pollutant concentration ϱ_z the frequency f may be reduced. A reduction of the frequency f has considerable practical consequences, as the power input increases with the cube of the frequency.

The volumetric power input for the reciprocating motion of the sieve plate package, which is the power N divided by the volume V of the bioreactor, is shown in Fig. 22.12 as a function of frequency f. For example, decreasing the frequency from $f = 1$ to 0.8 Hz results in a decrease of the volumetric power input from 2800 W/m³ to 1400 W/m³, that is by 50%.

In Fig. 22.13 the specific conversion rate R_s is presented as a function of the pollutant concentration ϱ_a in the water at the exit of the bioreactor. The specific conversion rate is the converted pollutant mass (kg TOC) per time unit and per unit of dry biomass (kg DB). This quantity is a measure of the conversion rate of a single bacterium. It is therefore suited for comparison of effec-

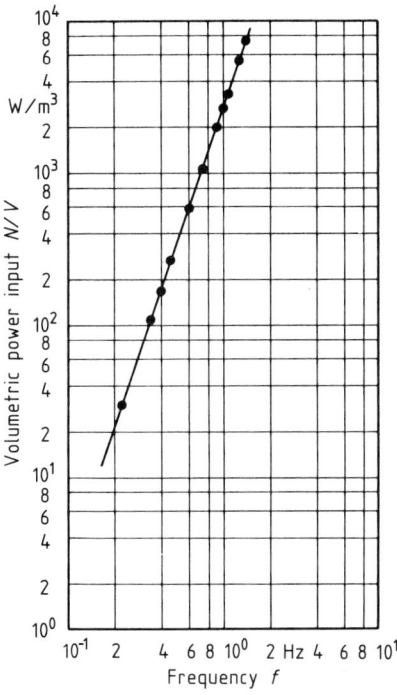

Figure 22.12. Volumetric power input N/V as a function of frequency f of reciprocating package of sieve plates.

Figure 22.13. Specific conversion rate R_s (given in kg of converted pollutant mass TOC per hour and m³ water, divided by kg dry biomass DB per m³ water) as a function of the pollutant concentration ϱ_a (g TOC per m³ water) at exit of bioreactor. – Abbreviations see text.

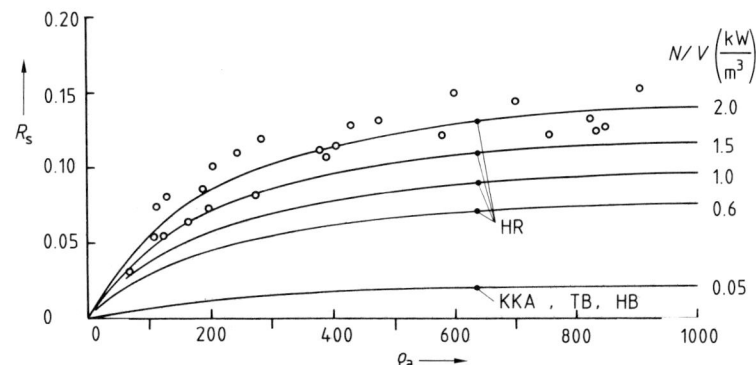

tiveness of different types of bioreactors. In Fig. 22.13 HR indicates the specific conversion rate of the reciprocating jet bioreactor for various values of the volumetric power input N/V, while KKA, TB, and HB indicate R_s for public waste water plants and industrial tower reactors. Circles indicate experimental data obtained for $N/V = 2$ kW/m³. For this curve the specific conversion rate is 7 times that of conventional plants.

Taking into account that the specific biomass in the reciprocating jet bioreactor is up to ten times that of conventional plants, the reciprocating jet bioreactor is superior to conventional plants by a factor of up to 70. The volume of this new bioreactor is therefore only a fraction of that of conventional bioreactors. It should be mentioned at this point, that the same progress is achieved with the new sedimentation system.

22.11 Application of the Reciprocating Jet Bioreactor

Application of the reciprocating jet bioreactor depends on the unique properties of this novel bioreactor, which may be summarized as follows:

1. Extremely high efficiency of the microbial conversion of organic and inorganic pollutants per unit volume of the bioreactor, due to mechanical activation of the bacteria. The consequences are:
   1.1 Very short residence time of the waste water in the bioreactor, which results in a very small volume of the bioreactor.
   1.2 Very small site requirement which results in great freedom of plant installation.
   1.3 High operational flexibility which results in very fast adaptation to varying waste water properties.
   1.4 Very short start-up time.
   1.5 Possibility of incorporation of waste water treatment into production plants in which the waste water is being produced.
2. Favorable design of the bioreactor, based on the concept of elementary units as microbial reaction space. The consequences are:
   2.1 Simplicity of design work with complete freedom of arranging the elementary units in parallel or in sequence, so that diameter or height may be freely chosen.
   2.2 No scaling-up problem, which results in greater reliability of the design work.
   2.3 Closed, emission-proof system which results in great freedom of site selection for installation.
3. High efficiency of the oxygen supply procedure based on the concept of periodic renewal of air/water interface and reversal of fluid flow in the vortices of the elementary units. The consequences are:
   3.1 Very high mass transfer rates across the air/water interface.
   3.2 Oxygen supply from natural air that is dispersed in the liquid. Application of oxygen enriched air or pure oxygen is not required, even in cases of highly polluted water.
   3.3 Low air flow rates.
4. Very high sedimentation velocity for the microorganisms. The consequences are:
   4.1 Very short residence of the waste water in the sedimentation vessel, which results in a very small volume of the vessel.
   4.2 Very dense sediment of microorganisms in the sedimentation vessel, which results in a very high biomass content in the recycle flow and in the bioreactor.

The unique properties of the high-efficiency reciprocating jet bioreactor in combination with a high-efficiency sedimentation process offer new possibilities for microbial purification of industrial and public waste waters. Waste water treatment can be carried out directly at or close to the place of waste water production. The waste water treatment plant can be incorporated into industrial production plants. Production and purification of waste water are under the same responsibility. This will certainly result in a substantial reduction of waste water production and further improvement of the purification process.

The new waste water purification system can also be incorporated in densely populated areas.

Furthermore, the reciprocating jet bioreactor may not only be used for the waste water purification process but also for various other biotechnological processes.

22.12 References

[22.1] H. H. HABECK-TROPFKE: "Abwasserbiologie". Werner-Verlag, Düsseldorf 1980.
[22.2] H. BRAUER and D. SUCKER: "Biological waste water treatment in a high-efficiency reactor"; Ger. Chem. Eng. 2 (1979) 2, 77–86.
[22.3] R. SCHÖNE: "Eine Modellvorstellung über die Steigerung der chemischen Aktivität von Kristalloberflächen durch Prallbeschuß mit Feststoffkörnchen". Dissertation, Technische Universität Berlin, 1968.
[22.4] W. SITTIG and H. HEINE: "Erfahrungen mit großtechnisch eingesetzten Bioreaktoren"; Chem. Ing. Tech. 49 (1977) 8, 595–605.
[22.5] E. UJCOVÁ, Z. FEND, M. MUSILKOVÁ, and L. SEICHERT: "Dependence of release of nucleotides from fungi on fermentor turbine speed"; Biotechnol. Bioeng. 22 (1980), 237–241.
[22.6] M. HUBERT and U. WERNER: "Die Oberflächenladung von Mikroorganismen – eine neue Meßgröße zur Führung biotechnologischer Prozesse"; Verfahrenstechnik 17 (1983) 1, 19–22.
[22.7] F. WEHMEYER, M. HUBERT, and U. WERNER: "Electrokinetic measurements during continuous cultivation of Saccharomyces cerevisiae: pH-dependency of surface charge"; not yet published.
[22.8] F. SARFERT: "Beschaffenheit von "Filtrat"-Wässern in Abhängigkeit von der jeweils ausgeschalteten thermischen Konditionierung und deren Beurteilung hinsichtlich ihrer Einflüsse auf Kläranlagen". Lecture in Haus der Technik, Essen, April 1971.
[22.9] H.-J. FRENZEL and F. SARFERT: "Untersuchungen zur Abbaubarkeit von "Filtrat"-Wässern aus der thermischen Schlammkonditionierung gemeinsam mit häuslichem Abwasser in Schlammbelebungsanlagen"; Kommunalwirtschaft 9 (1971), 329–332.
[22.10] H.-J. FRENZEL and F. SARFERT: "Untersuchungen über die Natur einiger in "Filtraten" thermisch konditionierter Klärschlämme enthaltenen Stoffe"; Gas Wasserfach Wasser/Abwasser 114 (1973) 7, 330–333.
[22.11] W. GRÜGER: "Untersuchung der biologischen Abwasserreinigung im Hubstrahlreaktor". Dissertation, Technische Universität Berlin, 1983.

Chapter 23

Tower-Shaped Reactors for Aerobic Biological Waste Water Treatment

Marko Zlokarnik

Bayer AG
Leverkusen, Federal Republic of Germany

23.1 Introduction
23.1.1 Abstract
23.1.2 Background
23.2 Choice of Aeration System
23.3 Design Data for Injectors
23.3.1 Sorption Characteristic of an Injector
23.3.1.1 Definition of $k_L a$
23.3.1.2 Formulation According to the Theory of Similarity
23.3.1.3 Allowance for Variation of $Y(X)$ with Temperature
23.3.2 Pressure Drop Characteristic of an Injector
23.3.2.1 Definition of Individual Δp Values
23.3.2.2 Formulation According to the Theory of Similarity
23.4 Description of Bayer Injectors
23.4.1 Bayer 8/14 Injector
23.4.2 Bayer Slot Injector
23.4.3 Sorption Characteristics of Both Injector Types in Relation to the Degree of Coalescence of the System
23.4.4 Efficiency
23.5 Coalescence Phenomena
23.5.1 Effect of Material Parameters on Bubble Coalescence
23.5.2 Effect of Process Parameters on Bubble Coalescence
23.5.3 Effect of Geometric Parameters on Bubble Coalescence
23.5.4 Implications for Practical Work
23.6 Bayer Tower Biology®
23.6.1 Technical Principle
23.6.2 Plants so far Constructed
23.7 BIOHOCH Reactor® of Hoechst AG
23.7.1 Process Engineering Concept
23.7.2 Radial Flow Nozzle
23.7.3 Survey of Plants Already Completed or Under Construction
23.8 References

List of Symbols

a	m^2/m^3	volume-related interfacial area
c'	kg/m^3	O_2 saturation concentration at gas inlet
c''	kg/m^3	O_2 saturation concentration at gas outlet
$\mathbb{D}$	m^2/s	diffusivity of O_2 in water
d_1	m	diameter of gas flow channel in nozzle
d	m	diameter of liquid flow channel in nozzle
E	kg/kWh	oxygen transfer efficiency
g	m/s^2	acceleration due to gravity
G	kg/h	oxygen transfer rate
h	m	injector clearance
H	m	height of liquid
k_L	m/s	mass transfer coefficient
p_1	N/m^2	pressure of ambient
p_2	N/m^2	pressure of gas at inlet of reactor
P	W	compressor power
P_L	W	pump power
q	m^3/s	gas flow rate
q_L	m^3/s	liquid flow rate
V	m^3	volume of reactor
Δc_m	kg/m^3	mean logarithmic concentration difference
Δp	N/m^2	pressure drop of gas in nozzle
Δp_L	N/m^2	pressure drop of liquid in nozzle
ν	m^2/s	kinematic viscosity of gas
ν_L	m^2/s	kinematic viscosity of liquid
ϱ	kg/m^3	density of gas
ϱ_L	kg/m^3	density of liquid
σ	N/m	surface tension
$Eu \equiv \dfrac{\Delta p\, d_1^4}{\varrho q^2}$		Euler number for gas flow
$Eu_L \equiv \dfrac{\Delta p_L d^4}{\varrho_L q_L^2}$		Euler number for liquid flow
$Re_L \equiv \dfrac{q_L}{\nu_L d}$		Reynolds number for liquid flow in nozzle
$X \equiv \dfrac{P_L/q_2}{\varrho(\nu g)^{2/3}}$		gas dispersion number
$Y \equiv \dfrac{G}{H q_1 \Delta c_m}\left(\dfrac{\nu^2}{g}\right)^{1/3}$		gas sorption number

23.1 Introduction

23.1.1 Abstract

The aeration of bacterial cultures in activated sludge tanks is the main operation in aerobic biological waste water purification with which process engineers have concerned themselves over the past ten years. The study of this process has shown that efficient utilization of atmospheric oxygen can only be achieved by means of relatively tall liquid columns and a correspondingly long residence time of the gas phase. This consideration, along with the need for space-saving plant designs and odorless, noiseless operation, led in Britain and Canada to shaft-like designs for the activated sludge tank (ICI – "deep shaft process"). Parallel developments in the Federal Republic of Germany resulted – simply because of the soil conditions – in tower-shaped reactors. Bayer AG, Leverkusen, for example, developed the concept of "Tower Biology", while Hoechst AG, Frankfurt, came up with the "Biohigh Reactor".

This chapter is devoted mainly to the *process engineering* research carried out in order to develop tower-shaped reactors. Details are given of the criteria by which the aeration devices (injectors) were developed and the factors that led to their optimization. It was essential here that special attention should be paid to the physical, process-related and geometric factors affecting the process of coalescence, which runs counter to dispersion of the gas. The high efficiency of the technical-scale injectors developed is indicated by the fact that, under the coalescence-promoting conditions encountered with the usual type of effluent, they achieve an efficiency $E \approx 3.8$ kg O_2 per kWh and 80% O_2 utilization (4% by volume O_2 in the off-gas) with a liquid height of 17 meters, calculated on standard conditions.

Finally, a description is given of biological waste water treatment plants, either completed or still under construction,

23.2 Choice of Aeration System

which have been designed according to the principle of the tower-shaped activated sludge tank.

23.1.2 Background

The aeration of a bacterial culture ("activated sludge") in a sewage treatment unit (activated sludge basin or tank) so far has been regarded as the most important engineering operation process in aerobic biological waste water treatment, especially as it may account for up to 30% of the running costs. The function of the aeration devices is to ensure that the bacterial culture is supplied with sufficient oxygen while at the same time maintaining an adequate flow velocity in the treatment unit so that no biomass is deposited on the bottom, where it might putrefy. For decades, activated sludge units were normally constructed in the form of flat, open basins; these take up a large space and – particularly when surface aeration is employed – are a constant source of troublesome noise and unpleasant odors (aerosol emission).

It was obvious that, if the activated sludge units were to be built in the form of a tower or shaft, not only would the aeration efficiency improve but the plant would also take up much less space and the odor emission could be considerably reduced. Such a unit could easily be covered over and, in view of the height of the liquid column above the aeration devices, oxygen from the gaseous phase (air) would be absorbed by the liquid under a higher system pressure. These ideas were put into practice in the seventies in Britain and Canada in the form of the ICI "Deep Shaft Process", while in the Federal Republic of Germany tower-shaped activated sludge units (Bayer AG's Tower Biology®, Hoechst AG's BIO-HOCH Reactor®) were developed.

This chapter describes the problems associated with the concept of the tower-shaped reactor and the solutions which finally made their industrial-scale realization possible.

The central feature of a tower-shaped activated sludge tank is its aeration system, which has to ensure that the biomass is supplied efficiently with oxygen. This involves two requirements: (1) A high efficiency E (in kg/kWh) of the O_2 uptake must be guaranteed, combined with a minimum gas throughput in order to prevent foaming problems and to reduce the cost of any off-gas treatment (thermal or biological). This means (2) that a high degree of utilization of the atmospheric oxygen must be achieved, which is then reflected in the small O_2 molar fraction x'' in the off-gas.

These aims can be achieved through careful selection of process engineering parameters (shape of the aerator and its operating parameters) and of the height H of the liquid column, a geometric parameter. There can be no doubt that only what are known as volume aerators are suitable in tower-shaped reactors; stirrers cannot be used for obvious reasons. The only possibilities here are gas dispersers (porous aerators, static mixers) or two-phase nozzles (injectors). It should be remembered that the job of these aeration devices is not only to produce fine gas bubbles but, even more importantly, to distribute them rapidly and evenly over the entire cross-section of the activated sludge tank so that they cannot collide with each other as they rise to the surface and "coalesce" to form larger bubbles. It must also be noted that the swarms of gas bubbles in the tower have the function of maintaining an intensive liquid circulation that prevents precipitation of the flocs while at the same time ensuring sufficient back-mixing of the liquid over the height of the tower. This is the only way to obtain an O_2 concentration gradient over the entire height that ensures intensive O_2 absorption near the bottom and prevents the escape of dissolved O_2 from the upper layers of the liquid column.

The aeration device chosen by Bayer AG was the injector. There were two reasons for this:

a) Good results had already been obtained in the 60s with a small type of injector (8/14 injector), which was used in basins 4 meters deep at the Dormagen factory.
b) In view of the *two* freely selectable process parameters (gas throughput *and* liquid throughput), it was reasonable to expect that injectors would be more effective than gas dispersers in mixing the gas bubbles quickly into the liquid volume and distributing them over the cross-section of the tower, thus reducing bubble coalescence.

Hoechst AG, too, opted for the use of two-phase nozzles right from the start, cf. Sect. 23.7.1.

23.3 Design Data for Injectors

Injectors are two-phase nozzles, in which the kinetic energy of the liquid propulsion jet is used to break up the gas into very fine bubbles. They are positioned just above the floor of the treatment tank, the height of the liquid column in the tank thus becoming a freely selectable parameter which can therefore be optimized. This arrangement means that the rising gas bubbles have to traverse the full height of the liquid column. If an injector is to be designed ideally for the required O_2 uptake, data must be obtained which permit the necessary process conditions – as far as mass transfer is concerned – to be calculated in advance and also indicate the power consumption. Sorption and pressure drop characteristics must therefore be compiled for the injector. These are representations of relevant test results based on the theory of similarity, the "sorption characteristics" serving to describe both the relationship between the absorption rate and the two process parameters (gas and liquid throughputs) and to allow for the effect of the coalescence behavior of the system on mass transfer. The term "system" here includes both the physical properties of the liquid and the position of the injector. Fine primary bubbles in a coalescence-promoting system do, of course, merge rapidly to form larger bubbles. This process also depends greatly on geometrical parameters.

23.3.1 Sorption Characteristic of an Injector [23.1]

23.3.1.1 Definition of $k_L a$

Mass transfer in a gas/liquid system is generally described in terms of the gross absorption rate equation:

$$G/V = k_L a \Delta c_m, \qquad (23.1)$$

which serves as a definition equation for the volume-related sorption coefficient $k_L a$:

$$k_L a \equiv \frac{G}{V \Delta c_m}, \qquad (23.2)$$

where

G/V is volume-related mass transfer through the interface (in kg/m³ s),
k_L liquid-side mass transfer coefficient (in m/s),
a volume-related interfacial area (in m²/m³), and
Δc_m mean logarithmic concentration difference (in kg/m³).

Δc_m is defined as:

$$\Delta c_m \equiv \frac{c' - c''}{\ln \frac{c' - c}{c'' - c}}. \qquad (23.3)$$

This assumes that the actual conditions are sufficiently far away from equilibrium, that the back-mixing of the gas phase is negligible and that back-mixing of the liquid is complete.

c' and c'' are the O_2 saturation concentrations corresponding to the O_2 content of the gas at the gas inlet and outlet, respectively; c is the O_2 concentration in the liquid.

The saturation concentration at the gas inlet,

$$c' = c_s x'(1 + 0.1 H), \qquad (23.4)$$

with $x' = 0.21$ for air, depends only on the temperature and the height H (in m) of the liquid above the injector. But the saturation concentration at the gas outlet:

$$c'' = c_s x'', \qquad (23.5)$$

is also dependent on the O_2 molar fraction x'' in the off-gas, which is calculated from the absorption rate G and the air throughput q using the formula:

$$x'' = (qx' - G/\varrho_{O_2})/(q - G/\varrho_{O_2}). \qquad (23.6)$$

The gas throughput q (in m³/s) and O_2 gas density ϱ_{O_2} (in kg/m³) relate here to identical conditions (e.g., standard conditions).

The relationship between the O_2 saturation concentration c_s and the temperature can be obtained from the relevant tables.

23.3.1.2 Formulation According to the Theory of Similarity

When formulating the absorption characteristics according to the theory of similarity, it must be remembered that the definition of $k_L a$ as a volume-related intensive quantity implies three consequences:

1. Independence of geometric parameters (in view of the assumption that the gas/liquid system is quasi-uniform).
2. Independence of the material parameters of the gas (in view of the assumption $k_G \gg k_L$).
3. Formulation of the process parameters as intensive quantities.

The following material parameters of the liquid phase have to be taken into account:

ϱ density of liquid,
ν kinematic viscosity of liquid,
$\mathbb{D}$ diffusivity of O_2 in the liquid,
σ surface tension of the liquid, and
S_i material parameters that describe the coalescence behavior of the system and whose number and nature are as yet unknown.

With injectors, the gas throughput q and the liquid throughput q_L occur as (extensive) process parameters. Since the hydrodynamic behavior of bubble columns is given by the superficial velocity of the gas, $v = q/A$ with A as the cross-sectional area, this quantity can be used to advantage as a reasonably intensively, formulated process parameter. The liquid throughput q_L tells us nothing about its dispersing effect on the gas continuum. Instead of q_L we therefore select the power P_L of the propulsion jet throughput:

$$P_L = \Delta p_n q_L, \qquad (23.7)$$

where Δp_n is the pressure drop in the propulsion jet nozzle. Selection of P_L does, however, require a knowledge of the pressure drop characteristic, and thus we arrive at the intensive quantity P_L/q. The third process parameter is acceleration due to gravity g which, because of the extreme density differences within the gas/liquid system, must greatly affect the hydrodynamics of the process.

The quantity in question, $k_L a$, is therefore given by the following relationship:

$$k_L a = f(\underbrace{\varrho, \nu, \mathbb{D}, \sigma, S_i}_{\text{Material}}; \underbrace{v, P_L/q, g}_{\text{Process}}) \qquad (23.8)$$

Target parameter   Material parameters   Process parameters

This relationship between nine dimensional quantities can be reduced by dimensional analysis to a relationship between $9-3=6$ numbers, because 3 basic units (mass, length, time) occur in their dimensions. With regard to the parameters designated S_i, which describe the coalescence behavior of gas bubbles in solutions, it should be noted that they are always converted by means of the parameters ϱ, ν, and g – regardless of their dimensions – to numbers based solely on physical properties. It follows that:

$$(k_L a)^* = \Psi[v^*, (P_L/q)^*, Sc, \sigma^*, S_i^*]. \quad (23.9)$$

The dimensionless numbers are defined as follows:

$$(k_L a)^* \equiv k_L a (\nu/g^2)^{1/3} \quad (23.10)$$

$$v^* \equiv v(g\nu)^{-1/3} \quad (23.11)$$

$$(P_L/q)^* \equiv (P_L/q)/[\varrho(\nu g)^{2/3}] \quad (23.12)$$

$$Sc \equiv \nu/\mathbb{D} \text{ (Schmidt number)} \quad (23.13)$$

$$\sigma^* \equiv \sigma/[\varrho(\nu^4 g)^{1/3}] \quad (23.14)$$

$S_i^* \equiv$ dimensionless numbers allowing for the coalescence parameters S_i

When absorption takes place in the *same* material system and at constant temperature, the numbers Sc, σ^*, S_i^* remain constant and the relationship is reduced to:

$$(k_L a)^* = \Psi_1[v^*, (P_L/q)^*], \quad (23.15)$$

with Sc, σ^*, $S_i^* = $ const.

In many research projects on bubble columns with gas dispersers (porous or perforated plates), it has been found that $k_L a \propto v$ or $k_L a/v = $ const (gas disperser). This is an obvious finding from the physical point of view, if the definition equations for $k_L a$ and v are called to mind:

$$\frac{k_L a}{v} = \frac{G}{V \Delta c_m} \cdot \frac{A}{q} = \frac{G}{H q \Delta c_m}, \quad (23.16)$$

with the column volume $V = HA$. The mass transfer rate G is proportional to the liquid height H, the gas throughput q, and the mean logarithmic concentration difference Δc_m.

We can assume that this also applies to injectors, when $P_L/q = $ const (identical gas bubble size distribution). Thus, the three-parameter function Eq. (23.15) is reduced by one parameter to give:

$$\frac{(k_L a)^*}{v^*} = \Psi_1\{(P_L/q)^*\} \text{ or} \quad (23.17)$$

$$\frac{k_L a}{v}\left(\frac{\nu^2}{g}\right)^{1/3} = \Psi_1\left\{\frac{P_L/q}{\varrho(\nu g)^{2/3}}\right\}. \quad (23.18)$$

With high bubble columns, more accurate allowance can probably be made for the influence of the gas throughput q (in $v = q/A$) if this is related to the mean system pressure:

$$q_1 = q/(1 + 0.05\, H), \quad (23.19)$$

with H in m. The power P_L of the propulsion jet, however, affects a gas throughput that is subject to the system pressure at the level of the injector:

$$q_2 = q/[1 + 0.1(H-h)], \quad (23.20)$$

with h in m being the injector clearance.

Taking this into account, the two numbers must be defined as follows:

$$Y \equiv \frac{G}{H q_1 \Delta c_m}\left(\frac{\nu^2}{g}\right)^{1/3}, \quad \text{sorption number} \quad (23.21)$$

$$X \equiv \frac{P_L/q_2}{\varrho(\nu g)^{2/3}}. \quad \text{dispersion number} \quad (23.22)$$

The object of the tests will be to determine the functional relationship:

$$Y = \Psi_1(X). \quad (23.23)$$

This is referred to in the following as the *sorption characteristic* of the injector.

23.3.1.3 Allowance for Variation of $Y(X)$ with Temperature

A separate investigation of the influence of temperature on mass transfer in a water/air system within a temperature range of $\vartheta = 17\text{--}45\,°\text{C}$ has shown that the relationship $Y = f(X)$ makes full allowance for this effect; the Schmidt number $Sc \equiv \nu/\mathbb{D}$ need not, therefore, be used [23.2].

23.3.2 Pressure Drop Characteristic of an Injector

23.3.2.1 Definition of Individual Δp Values

If an injector is positioned just above the floor of a tall treatment tank ("tower") to serve as a gas disperser, it cannot normally function as an ejector. The gas is fed to it via a separate compressor, which has to work against a pressure composed of the hydrostatic pressure of the liquid column, $H\varrho g$, and the gas-side pressure drop Δp of the injector. The adiabatic power P of the compressor is determined from the following relationship:

$$P = \frac{\kappa}{\kappa - 1} q p_1 \left[\left(\frac{p_2}{p_1}\right)^{\frac{\kappa - 1}{\kappa}} - 1\right] / \eta_c, \qquad (23.24)$$

with

$\kappa = 1.4$,

and

$$p_2 = \Delta p(q) + H\varrho g + p_1. \qquad (23.25)$$

The efficiency of a compressor is normally taken as $\eta_c = 0.60$.

The injector does, however, have to be supplied – via a liquid pump – with liquid from the treatment tank which, in the form of a propulsion jet, ensures the dispersion of the gas continuum. Therefore, the liquid throughput suffers a pressure loss Δp_L. The pump power P_L is determined as follows:

$$P_L = \Delta p_L q_L / \eta_L, \qquad (23.26)$$

where q_L (in m^3/s) is the liquid throughput and η_L the efficiency of the pump ($\eta_L \approx 0.75$).

If both powers, P and P_L, needed to achieve a given O$_2$ uptake, G (in kg/h), are known, the efficiency of the O$_2$ uptake is given by:

$$E \equiv \frac{G}{P + P_L} \quad (\text{in kg/kWh}). \qquad (23.27)$$

The efficiency E of the O$_2$ uptake is the parameter by which the operating conditions of the injectors are optimized.

23.3.2.2 Formulation According to the Theory of Similarity

Fig. 23.1 shows a schematic cross-section through an injector: this comprises the propulsion jet nozzle and the mixing chamber. d and d_1 are the characteristic diameters of the respective parts of the injector at their narrowest points; Δp_L is the pressure drop of the liquid throughput within the propulsion jet nozzle.

If we consider an injector of given geometry, Δp and Δp_L will depend on the characteristic linear dimensions mentioned, the throughputs q and q_L, as well as on the material parameters of the two fluids, with the densities ϱ and ϱ_L and kinematic viscosities ν and ν_L:

$$\Delta p, \Delta p_L = f(d, d_1; \varrho, \varrho_L, \nu, \nu_L; q, q_L). \qquad (23.28)$$

By dimensional analysis these two relationships are reduced to:

$$Eu \equiv \frac{\Delta p\, d_1^4}{\varrho q^2} = f_1\left(\frac{d_1}{d}, \frac{\varrho}{\varrho_L}, \frac{\nu}{\nu_L}, \frac{q_L}{\nu_L d}, \frac{q}{q_L}\right) \qquad (23.29)$$

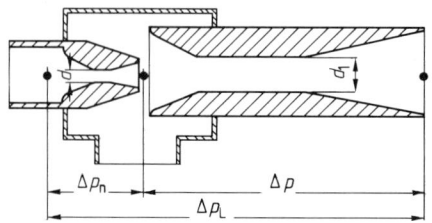

Figure 23.1. Schematic cross-section of an injector showing the pressure drop ranges.

and

$$Eu_L \equiv \frac{\Delta p_L d^4}{\varrho_L q_L^2} = f_2\left(\frac{d_1}{d}, \frac{\varrho}{\varrho_L}, \frac{v}{v_L}, \frac{q_L}{v_L d}, \frac{q}{q_L}\right) \quad (23.30)$$

The Euler numbers Eu and Eu_L depend on only two process numbers: $Re_L \equiv q_L/(v_L d)$ and q/q_L, if the material system and geometry are given:

$$Eu = f_1(Re_L, q/q_L), \quad (23.31)$$

$$Eu_L = f_2(Re_L, q/q_L). \quad (23.32)$$

The Euler number Eu_n, which is used in determining Δp_n, is now given by the following relationship:

$$Eu_n = \frac{\Delta p_n d^4}{\varrho_L q_L^2} = Eu_L - \left(\frac{d}{d_1}\right)^4 \frac{\varrho}{\varrho_L}\left(\frac{q}{q_L}\right)^2 Eu \quad (23.33)$$

The determination of the pressure drop characteristics will not be described in more detail since measurement of Δp and Δp_L presents no difficulty. It should be pointed out, however, that for this measurement it is best to position the injector in a bubble column and cover it with a liquid layer of known height, which is then taken into account when the results are analyzed.

23.4 Description of Bayer Injectors [23.1], [23.2]

23.4.1 Bayer 8/14 Injector

The Bayer 8/14 injector [23.3] has been used successfully for years in many waste water treatment plants; its design is similar to that of the ejectors, cf. Fig. 23.2. It is injection-molded in polypropylene, cf. Fig. 23.3, and is therefore corrosion-resistant. Its smooth, water-repellent surface remains largely free of incrustation. These injectors are usually arranged in clusters of four about 3/4–1 meters above the bottom, and

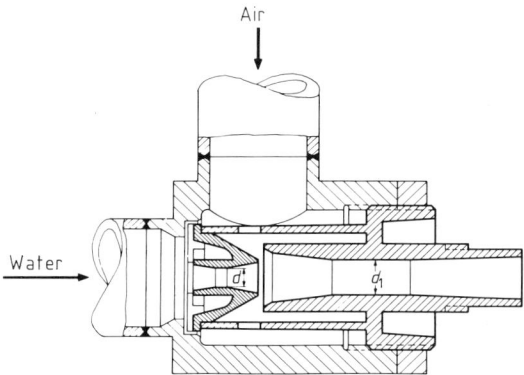

Figure 23.2. Cross-section through the Bayer 8/14 injector.

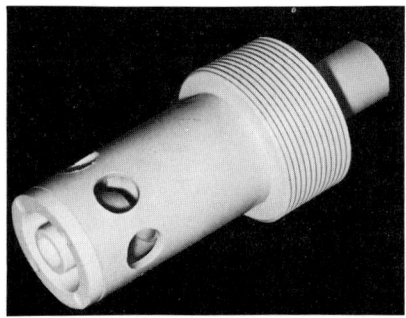

Figure 23.3. Photograph of the Bayer 8/14 injector. – Its total length is 13 cm.

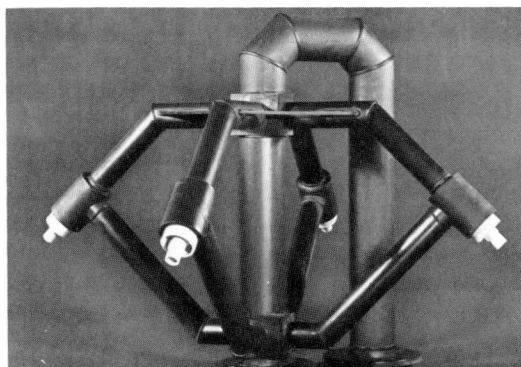

Figure 23.4. Injector cluster.

the free jet is directed towards the floor at an angle of 45°, cf. Fig. 23.4. This ensures liquid circulation at the bottom, with the result that precipitation of solids is prevented. This arrangement also means, however, that the free jet of the gas/liquid dispersion fans out immediately above the floor of the activated sludge tank to produce a swarm of gas bubbles, which can then rise through the entire height of the liquid column. This gives the longest possible gas residence time in the activated sludge unit. The characteristic dimensions of this type of injector are given by the smallest diameter of the propulsion jet nozzle ($d = 8$ mm) and the smallest diameter of the mixing chamber ($d_1 = 14$ mm).

A geometrically similar scale-up of this injector with the aim of removing any risk of clogging and reducing the number of injectors in the tank automatically involves a drastic decline in its efficiency. The liquid propulsion jet does most of its gas dispersion work at its circumference u, which in the case of a geometrically similar enlargement increases only linearly ($u = \pi d$), whereas its cross-sectional area A increases with the square of the diameter ($A = \pi d^2/4$). As the injector is enlarged, a growing proportion of the propulsion jet therefore leaves the mixing chamber without doing work. This problem, which occurs with all axially operating injectors, has been successfully solved with the development of the slot injector.

23.4.2 Bayer Slot Injector

The special design feature of the slot injector [23.4] is the transition of the mixing chamber cross-section (viewed in the direction of flow) from circular to slot-shaped without any change in the cross-sectional area of the mixing chamber, Fig. 23.5. This has a two-fold effect. First, the mixing chamber has predominantly converging faces along which the shear rate increases without the pressure drop increasing to the same degree as in accelerated flow. This means that a fine gas dispersion is produced in the mixing chamber but the free jet still retains a high kinetic energy. Secondly, the free jet leaves the injector as a relatively flat ribbon which mixes comparatively quickly with the surrounding liquid, thus helping to suppress bubble coalescence. The shape of the propulsion jet nozzle is adapted to that of the mixing chamber: it, too, tapers down to a slot. This ensures that even the corners of the slot-shaped outlet are completely filled with the gas/liquid dispersion.

This injector, too, is made from polypropylene by the injection molding process, Fig. 23.6. Since its outlet is slot-shaped, un-

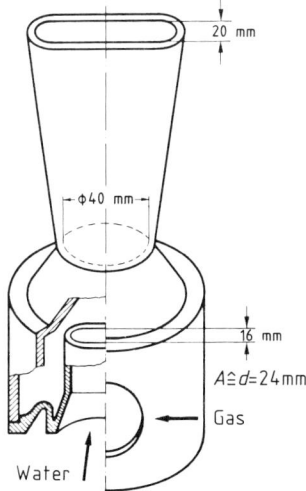

Figure 23.5. Partly sectioned side elevation of the Bayer slot injector.

like the circular outlet of the 8/14 injector, its characteristic dimensions are as follows:

Propulsion jet nozzle:
Slot width 16 mm; cross-sectional area of slot equals that of a circle of diameter $d = 24$ mm.

Mixing chamber:
Slot width 20 mm; cross-sectional area of slot equals that of a circle of diameter $d_1 = 40$ mm.

Slot injectors are normally arranged not in clusters but singly. They are positioned about 1 meter above the floor so that they point towards the bottom of the tank at an angle of 30° to the horizontal, cf. Fig. 23.7. They are installed at about the same intervals as the 8/14 injector clusters (one slot injector per $8-10$ m^2).

23.4.3 Sorption Characteristics of Both Injector Types in Relation to the Degree of Coalescence of the System

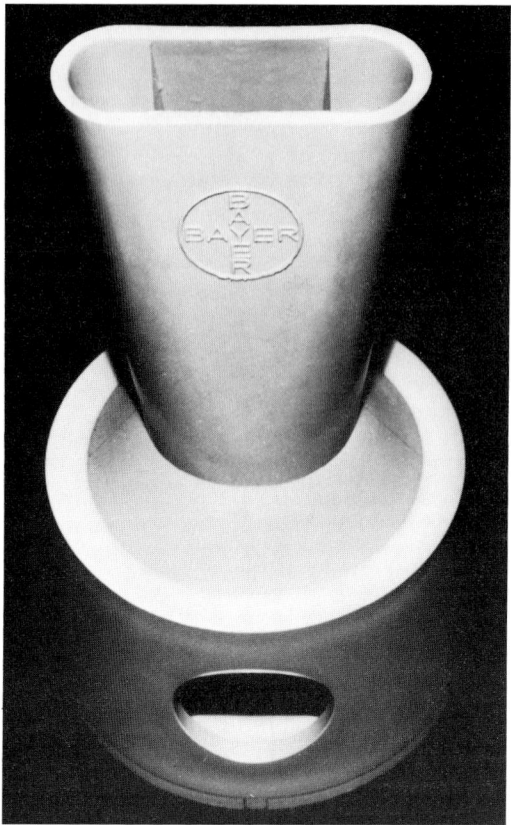

Figure 23.6. Photograph of the Bayer slot injector. – Its total length is 19 cm.

Figure 23.7. Photograph of the piping with slot injectors inside the tower. – The bottom pipe is for water supply, the middle one for air and the top one for sludge recycle.

Both injectors were fixed to the wall of the respective tanks 1 meter above the floor so that the free jet was at an angle of 45° (8/14 injector) or 30° (slot injector) to the horizontal. The tank diameter was 1.6 meters for the 8/14 injector but 2.8 meters for the slot injector; in both cases the height H of the liquid above the injector was 7 meters (height of liquid above the floor = 8 m).

The mass transfer behavior of both injectors was measured under steady state test conditions, hydrazine being used to remove (chemically bind) the absorbed oxygen (hydrazine method) [23.5]. The advantage of this method over non-steady state test conditions is that the measurements are more accurate and easier to evaluate; it also ensures that the chemical and physical properties of the system do not change in the course of the measurements. When pure

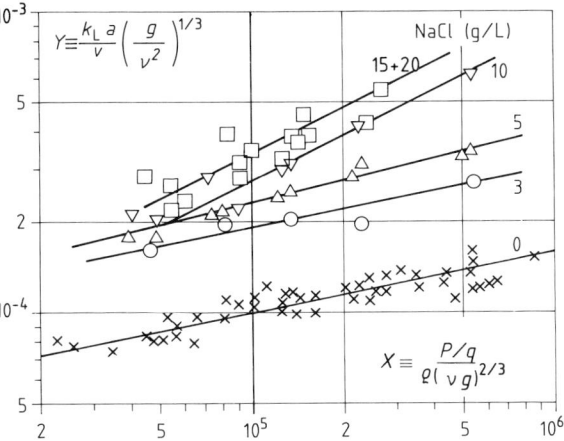

Figure 23.8. Sorption characteristics for the 8/14 injector as a function of the degree of coalescence of the system, expressed in terms of the NaCl concentration.

water is used, the salt concentration is always $\ll 1$ g/L, resulting in a material system which promotes coalescence. The characteristic concentration gradient under steady-state test conditions is given by the mean logarithmic concentration difference Δc_m, which takes into account the change in the O_2 saturation concentration along the height of the liquid (variation in hydrostatic pressure and in the O_2 molar fraction as a result of absorption). For details of the test evaluation see Sect. 23.3.1.2.

It is well-known that the coalescence of primary fine gas bubbles in aqueous solutions of certain substances is greatly suppressed, resulting in a much faster absorption rate than with the coalescence-promoting pure water/air system, cf. Sect. 23.5. In the sorption measurements the degree of coalescence of the system was varied by adding common salt and is expressed in terms of the NaCl concentration. Later (in Sect. 23.5) it is shown that the coalescence behavior of any liquid mixture can be described in terms of the *equivalent* NaCl *concentration;* this can be done by fairly simple means using the "manometric method" [23.5]. In this way it is possible to determine the sorption characteristic of the injector

that applies to mass transfer within the material system concerned.

Fig. 23.8 shows the sorption characteristic of the 8/14 injector as a function of the degree of coalescence of the system. It can be seen from the increase in Y with X that 3 g NaCl per L is sufficient to suppress the coalescence of the gas bubbles in the bubble swarm to the point where the absorption rate is almost doubled. If the salt concentration is increased further, the coalescence of the gas bubbles in the free jet is impeded and the power of the propulsion jet utilized even more: the exponent with which the gas dispersion number X influences the sorption characteristic increases. The following process relationships apply:

water: $Y = 1.0 \cdot 10^{-5} X^{0.2}$
water + NaCl(3 g/L): $Y = 1.9 \cdot 10^{-5} X^{0.2}$
water + NaCl(5 g/L): $Y = 1.3 \cdot 10^{-5} X^{0.25}$
water + NaCl(10 g/L): $Y = 1.2 \cdot 10^{-6} X^{0.5}$
water + NaCl(20 g/L): $Y = 1.5 \cdot 10^{-6} X^{0.5}$

Fig. 23.9 shows the corresponding sorption characteristics for the slot injector. If the $Y(X)$ relationship for pure water is compared for the two types of injector, the two sorption characteristics are seen to be almost identical. This proves how successfully the aeration density – and therefore the bubble coalescence in the swarm – has been

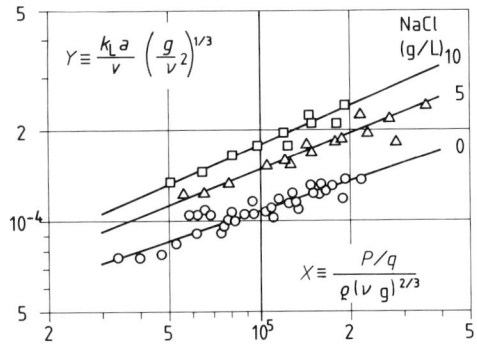

Figure 23.9. Sorption characteristics of the slot injector as a function of the degree of coalescence of the system, expressed in terms of the NaCl concentration.

reduced by the fanning-out of the gas/liquid free jet from the slot injector compared to the 8/14 injector, despite an eight- to nine-fold increase in the characteristic cross-sections. If the $Y(X)$ relationship in salt solutions is compared, however, it is apparent that in material systems in which coalescence is inhibited, mass transfer is not accelerated as much as was the case with the 8/14 injector. The only possible explanation for this is that the slot injector gives rise to larger primary gas bubbles than the 8/14 injector.

For the slot injector the following process relationships apply:

water: $Y = 2.4 \cdot 10^{-6} X^{0.33}$
water + NaCl (5 g/L): $Y = 1.65 \cdot 10^{-6} X^{0.39}$
water + NaCl (10 g/L): $Y = 1.25 \cdot 10^{-6} X^{0.43}$

The difference in the extent to which the degree of coalescence affects mass transfer for the two injectors can best be illustrated by plotting the multiplying factor m for physical absorption:

$$m = (k_L a)_{\text{solution}} / (k_L a)_{\text{solvent}} \triangleq Y_{\text{solution}} / Y_{\text{water}} \tag{23.34}$$

as a function of the salt concentration (cf. Fig. 23.10).

This graph shows the relationship $m = f(c_{\text{NaCl}})$ for a mixing vessel with (self-aspirating) pipe stirrer [23.5] as well as for both types of injectors. It is immediately apparent here that with the stirrer, the relationship $m = f(c_{\text{NaCl}})$ does not vary at all with the gas dispersion number X; with the slot injector the degree of dependence is only slight, whereas with the 8/14 injector the above relationship varies greatly according to the gas dispersion number X. The reason for this is as follows: In the mixing vessel, equilibrium exists at all times between the dispersing effect of the stirrer and the rate of coalescence in the gas/liquid dispersion; more intimate mixing increases both the dispersing effect and the coalescence rate. The injectors, on the other hand, are capable of exerting only a relatively slight influence on the hydrodynamics in the bubble swarm. Here, the gas bubbles are generated only once; after that, they can only become larger. Since they rise up in parallel in the bubble swarm, and are not continuously back-mixed as in the mixing vessel, coalescence in the bubble column can be successfully suppressed using much smaller salt concentrations than in the mixing vessel. If, therefore, the difference between the two injectors as regards the relationship $m = f(c_{\text{NaCl}})$ is as great as indicated in Fig. 23.10, this can only be explained by the fact that the slot injector produces much larger primary gas bubbles than the 8/14 injector.

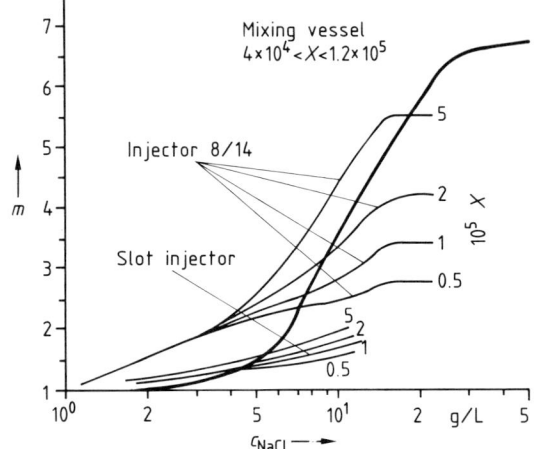

Figure 23.10. Relationship between the multiplying factor m and the NaCl concentration c for the vessel with pipe stirrer (heavy line) [23.5] and for both types of injector.

On the basis of what has been said so far, it may be concluded that, with the development of the slot injector, the task of designing a large, clog-free yet efficient injector for coalescing material systems has been successfully accomplished (both injectors having virtually the same sorption characteristics in pure water). For the aeration of material systems in which coalescence is to some degree inhibited, however, preference should be given, where possible, to the 8/14 injector.

23.4.4 Efficiency [23.2]

If the design data for both types of injector are available in the form of sorption characteristics (Figs. 23.8 and 23.9) and pressure drop characteristics (the latter not being explicitly described here), conclusions can be drawn as to the efficiency E (in kg/kWh) of the O_2 uptake that can be achieved under given conditions. The object of these calculations is therefore the quantity:

$$E = G/\Sigma P. \qquad (23.35)$$

It is important to explain first of all which powers P are taken into account and how this is done.
- a) To calculate the adiabatic compressor power, both the hydrostatic pressure of the liquid column and the gas-side pressure drop in the injector are used. The pressure drop of the gas throughput in the pipe system cannot be included in the calculation because it is different for each plant. The efficiency of the compressor is taken to be $\eta = 0.6$.
- b) Calculation of the pump power is based only on the liquid-side pressure drop in the injector and a pump efficiency of $\eta = 0.75$. This means that we proceed as if the propulsion jet throughput were removed from the treatment tank at the level of the injector and recirculated via pump and injector. Thus, the pump power for raising the effluent stream to be treated to the surface of the liquid in the treatment tank is not taken into account, for the following three reasons:
  1. We do not wish to specify here whether the treatment tank is to take the form of a deep shaft or a high tower.
  2. The effluent stream flowing through the treatment plant is not directly proportional to the propulsion jet throughput, because the O_2 uptake is a function of the organic load and not of the quantity of effluent.
  3. It is possible for the kinetic energy of the purified waste water stream leaving a tall treatment plant to be at least partially utilized.

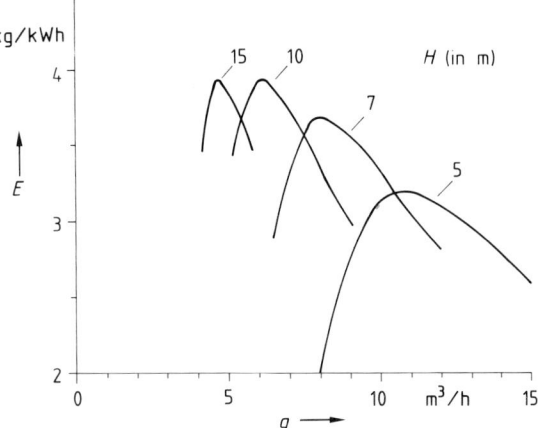

Figure 23.11. Relationship between oxygen transfer efficiency E (in kg/kWh) and gas flow rate q (in m³/h) at different liquid heights H (in m) under standard conditions for the 8/14 injector. $G/V = 0.1$ kg/m³ h of oxygen, one injector per 2 m², temperature 10°C, $c = 0$ mg/L O_2.

The relationship $E(q, H)$ determined in this way for the 8/14 injector under given conditions is shown in Fig. 23.11. Since injectors have two freely selectable process parameters (gas and liquid throughputs), a given O_2 uptake can be achieved with any number of pairs of gas throughput and liquid throughput values, but there is *only one* pair of values at which this O_2 uptake is achieved with a minimum of power consumption, i.e., where $E = $ max. (For each gas throughput value q there exists therefore a certain liquid throughput necessary to achieve the desired oxygen uptake G, though this is not shown in Fig. 23.11 for the sake of simplicity.)

It can be seen from Fig. 23.11 that the efficiency of mass transfer increases with the liquid height H, the gas throughput at which $E = $ max declines simultaneously.

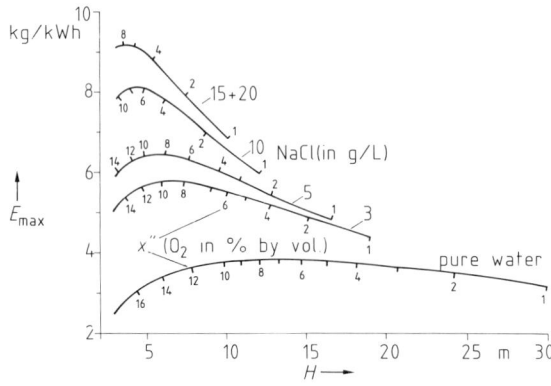

Figure 23.12. Relationship $E_{max}(H, x'')$ for the 8/14 injector under the same conditions as in Fig. 23.11.

Both these relationships have to do with the fact that a higher liquid column prolongs the residence time of the gas bubbles, thus increasing the utilization of the atmospheric oxygen; less air is therefore required to achieve the same O_2 uptake.

If the E_{max} values are now plotted against the height of the liquid, we obtain the graph in Fig. 23.12; it can be seen, first of all, that in the case of pure water (coalescing material system) the highest possible E_{max} value, $E_{max} = 3.8$ kg O_2 per kWh, is reached with a liquid height of $H = 12-14$ m, the O_2 concentration in the off-gas being $x'' = 6-8\%$ by vol. The curve for $E_{max}(H)$ is so flat at this point that at $H = 20$ m the amount of off-gas can be further reduced ($x'' \approx 3$ vol-%) without a significant drop in E ($E_{max} = 3.7$ kg/kWh at $H = 20$ m).

Fig. 23.12 also shows the relationship $E_{max}(H, x'')$ for various degrees of coalescence (= sodium chloride concentrations). The tremendous influence of the coalescence behavior of the liquid on the obtainable E_{max} value and on the optimum liquid height is apparent. For a liquid whose coalescence behavior corresponds to that of a salt solution of 15 g NaCl per L, for example, a maximum possible E_{max} value of 9.2 kg O_2 per kWh is already reached at $H = 4$ m, whereas when $H = 10$ m the off-gas contains only 1 vol-% O_2 and in this case $E_{max} = 6.8$ kg/kWh.

In view of the outstanding importance of the coalescence behavior for the O_2 uptake in the case of volume aeration with injectors, this quantity must be determined exactly before the height of the liquid in the treatment tank can be decided or the injectors can be designed. For this purpose the coalescence behavior of the effluent as purified in a pilot plant (at the outlet!) should be monitored over a long period by the "pressure gauge method" [23.5]. For further details see Sect. 23.5.4.

Fig. 23.13 corresponds to Fig. 23.12 and shows the relationship $E_{max}(H, x'')$ for the slot injector. A comparison of these relationships for pure water confirms the equivalence of the two injectors from an energetic point of view. Even with a degree of coalescence corresponding to that of a solution of 5 g NaCl per L, however, the maximum oxygen transfer efficiency of the slot injector is only $E_{max} = 4.6$ kg/kWh, that is 70% of the value obtainable with the 8/14 injector.

The disadvantage of the so-called "standard conditions" ($\vartheta = 10$°C, $c = 0$ mg O_2 per L) is that the bacterial culture is not viable under these conditions. It is now known, however, that with volume aeration a dissolved oxygen concentration of 1 mg O_2 per L is sufficient. The relationship $E_{max}(H, x)$ for the slot injector, where $c = 1$ mg O_2 per L and $\vartheta = 20$ and 40°C, is therefore illustrated in Fig. 23.14. It is apparent from this

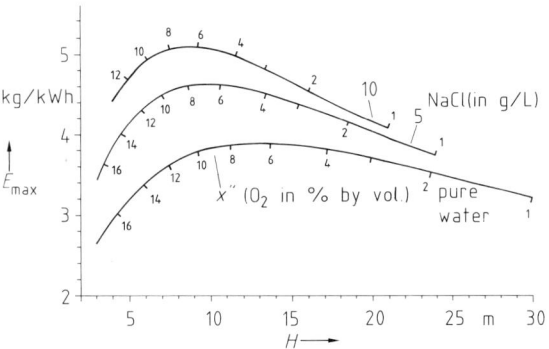

Figure 23.13. Relationship $E_{max}(H, x'')$ for the slot injector. – $G/V = 0.1$ kg/m³ h O_2, one injector per 10 m², standard conditions.

that the optimum liquid height is now between 15 and 20 m, where $x'' = 5$–8 vol-% and $E_{max} = 3.2$ kg O_2 per kWh.

It may seem surprising that the 40°C curve is (slightly) above the 20°C curve, since O_2 solubility and, therefore, the concentration difference Δc_m decrease rapidly with increasing temperature. The reason is that with increasing temperature the kinematic viscosity ν of the liquid also declines,

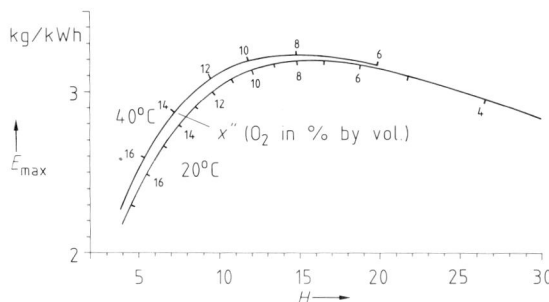

Figure 23.14. Relationship $E_{max}(H, x'')$ for the slot injector where $c = 1$ mg/L O_2 and $\vartheta = 20$ and 40°C. Other conditions as in Fig. 23.13.

whereas the diffusivity $\mathbb{D}$ of O_2 in water is greatly enhanced. Since this greatly increases k_L, and therefore $k_L a$ (and thus Y), it more than compensates, within this temperature range, for the disadvantage of decreasing O_2 solubility.

23.5 Coalescence Phenomena

Fine gas bubbles dispersed in a pure liquid (e.g., pure water) tend to join to form larger ones. This process, known as bubble coalescence, is caused by the fact that the liquid film (liquid lamella) between two adjacent gas bubbles becomes thinner and thinner until it eventually ruptures ("drainage"). The process is extremely rapid in pure liquids but extremely slow in solutions with a tendency to foam: the capacity for foaming directly reflects the coalescence behavior of the system.

The reason for the thinning of the liquid film is the pressure difference between the film and the bulk of the liquid. These are caused by the different capillary pressures which are, in turn, due to the different radii of curvature and to van der Waals' forces. These pressure differences Δp form, in each case, an equilibrium of forces with the difference in the surface tension $\Delta \sigma$:

$$\delta \Delta p = 2 \Delta \sigma; \qquad (23.36)$$

with δ film thickness.

23.5.1 Effect of Material Parameters on Bubble Coalescence

We should remember at this point that for solutions, an increase in surface area leads to a change in the surface concentration, which, in turn, alters the surface tension. We know from Gibbs' rule on the dependence of the surface tension of a solution on its concentration that with an increase in surface area the surface concentration Γ_i of the solute i is altered in a quite specific manner:

$$\Gamma_i = - \frac{c_i}{RT} \frac{d\sigma}{dc_i}. \qquad (23.37)$$

When the term $d\sigma/dc$ is positive, there is a deficiency of the dissolved substance at the surface; when it is negative, there is an excess. Despite this, however – according to MARUCCI's theory of coalescence [23.6] – the considerable increase in surface area during the individual process of coalescence always involves an appreciable increase in surface tension.*⁾ Thus, the quantity $d\sigma/dc$ greatly affects the dynamics of the film and should therefore form a suitable basis for the prediction of the coalescence behavior of solutions. (For pure liquids, on the other hand, $d\sigma/dc = 0$; there is therefore no increase in the surface tension of the thinning lamella which could slow down or suppress the process of coalescence.)

Fig. 23.15 shows two bubble columns ($D = 600$ mm, $H = 1800$ mm) of acrylic glass,

Figure 23.15. Optical demonstration of coalescence phenomena in two bubble columns of 0.6 m diameter and 1.8 m liquid height operated under identical conditions (see text). Left column with 20 g NaCl per L water, right column with tap water.

both equipped with the 8/14 injector and operated under identical conditions ($q = 15$ m³/h, standard conditions, $q_f = 3$ m³/h). While the right-hand bubble column is filled with drinking water, the one on the left contains 20 g NaCl per L. It can be seen from the picture that the free jet of the gas/liquid dispersion, after it disintegrates into a bubble swarm, forms much larger gas bubbles in the drinking water than in the saline solution. Stable gas bubbles in pure water in fact have a diameter of 3–5 mm, whereas a 20 g/L NaCl solution suppresses bubble coalescence almost entirely. Here, the gas bubbles retain virtually the same size as the primary bubbles (approx. 0.3–0.5 mm in the case of the 8/14 injector), resulting in a considerable increase in the interfacial area, especially as fine gas bubbles also rise to the surface more slowly than larger ones and thus increase the gas content in the liquid (gas hold-up).

In order to obtain quantitative data regarding the effect of inorganic salts (electrolytes) and of normal aliphatic alcohols on the suppression of bubble coalescence in aqueous solutions, absorption measure-

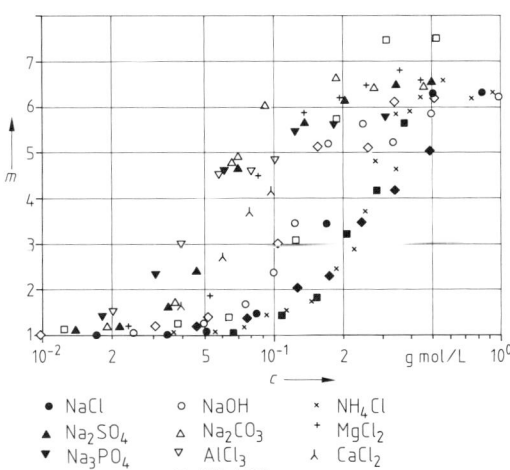

- NaCl   ○ NaOH   × NH₄Cl
- Na₂SO₄ △ Na₂CO₃ + MgCl₂
- Na₃PO₄ ▽ AlCl₃   λ CaCl₂
- KBr   □ NH₄HCO₃
- KNO₃   ◇ (NH₄)₂CO₃

Figure 23.16. Multiplying factor m as a function of the type of salt and salt concentration. Measurements by the pressure gauge method. (Pipe stirrer of 60 mm ∅ at 1800 rpm; liquid volume 60 L). From [23.28].

---

* According to MARUCCI [23.6], the lamella thins out to a few hundred Å before it ruptures; values for the surface area per unit volume may be as high as 10^6 cm⁻¹.

ments were carried out in the pure N_2/solution system by what is known as the "pressure gauge method"; this technique, by which measurements are possible in any liquid, will be discussed at greater length later on. In Figs. 23.16 and 23.17 the results of these measurements are represented in the form $m = f(c)$, where m is the enhancement or multiplying factor by which the sorption coefficient $k_L a$ for pure water must be multiplied to obtain the sorption coefficient in the solution in question. This

the water structure in the aqueous solution. It was possible to conclude from this that the term $d\sigma/dc$ reflects the structure of water in aqueous solutions.

Unfortunately, the findings obtained with inorganic salts and normal aliphatic alcohols are not universally applicable; later measurements with aqueous solutions of benzene and its derivatives show complex $m(c)$ dependences which cannot be correlated solely with the corresponding values of $d\sigma/dc$.

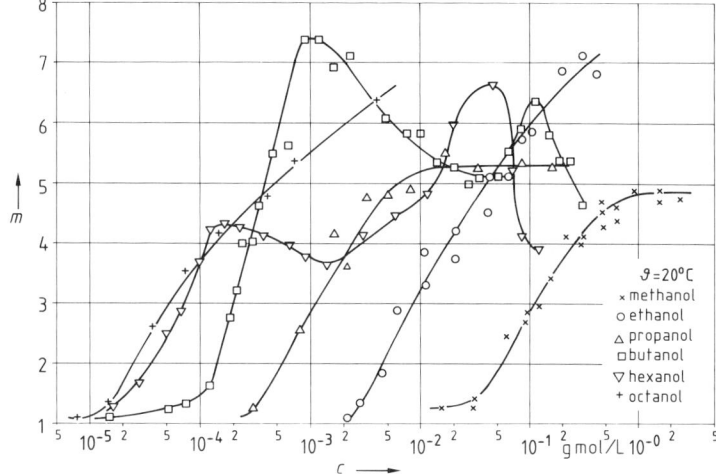

Figure 23.17. Multiplying factor m as a function of the concentration c of various normal aliphatic alcohols. Test conditions as in Fig. 23.16. From [23.28].

enhancement factor m in chemical process engineering corresponds therefore to the α factor in effluent treatment technology:

$$\alpha \equiv m \equiv (k_L a)_{\text{solution}} / (k_L a)_{\text{pure water}}. \quad (23.34)$$

It is apparent from Figs. 23.16 and 23.17 that with intensive aeration (very fine primary gas bubbles) m may have a value of 6–7, different concentrations being necessary to achieve this depending on the type of salt or alcohol. It has been demonstrated [23.7] that the results obtained with the two classes of material can be clearly correlated with the corresponding $d\sigma/dc$ values; the most significant aspect of the above publication, however, was the proof that this correlation can also be established using material parameters which describe the state of

However, there are also some compounds which, even in very low concentrations in water, greatly accelerate coalescence. Fig. 23.18 shows data obtained with so-called non-ionic surfactants [23.1] normally used as antifoams.

It can be seen from this graph that as little as 3 ppm of certain antifoams is sufficient to reduce the O_2 uptake by 50% compared to pure water. It is therefore clear that non-ionic surfactants should not be used as foam suppressants in activated sludge tanks.

Little attention has so far been devoted to the coalescence-promoting effect of the activated sludge; the flocs of the activated sludge, as "crystallization nuclei", may in fact be expected to promote bubble coalescence, i.e., to adversely affect the O_2 up-

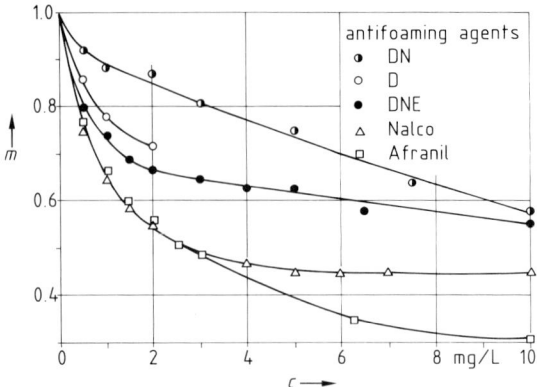

Figure 23.18. Multiplying factor m as a function of the concentration of various types of antifoam. Test conditions: bubble column ∅ 600 × 1800 mm; Bayer 8/14 injector, $q = 18$ m³/h, $q_L = 3$ m³/h. From [23.1].

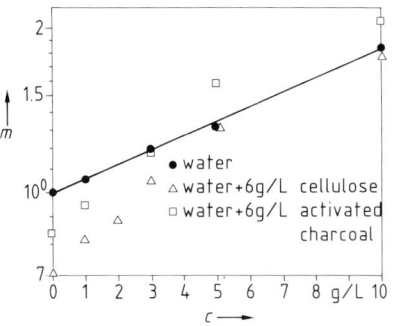

Figure 23.19. Multiplying factor m as a function of NaCl concentration and of solids. Test conditions as in Fig. 23.18. From [23.1].

23.5.2 Effect of Process Parameters on Bubble Coalescence

It is obvious that bubble coalescence will be especially pronounced when extremely fine primary gas bubbles are produced and that it cannot possibly occur with large, stable gas bubbles; significant changes in the m value will therefore take place only in the case of "fine-bubble aeration".

The injectors produce very fine primary gas bubbles and, with the high-energy free jet of the gas/liquid dispersion, also cause the gas bubbles from the free jet to intermix rapidly with the surrounding liquid, thus delaying bubble coalescence to a certain extent even in coalescence-promoting material systems. This means that the primary gas bubble is particularly well "preserved" in material systems in which coalescence is inhibited. This is apparent, for example, from the sorption characteristics $Y(X)$ of the two Bayer injectors (see Figs. 23.8 and 23.9). It can be seen that, with both injectors, a relatively low "salt load" of 3–5 g/L NaCl is sufficient to bring about an m value of 1.5–2. The 8/14 injector generates particularly fine primary gas bubbles and is therefore especially suitable for use in material systems where coalescence is inhibited.

take. In tests on activated sludge [23.1] from a treatment tank for the purification of chemical effluent it was shown that, even after repeated washing with pure water, the activated sludge was still so heavily loaded with organic solvents that the result was $m \approx 0.5$. On the other hand, absorption tests [23.1] in pure water with 6 g/L cellulose or activated carbon showed that the finely dispersed solid does not, on its own, promote coalescence to a great extent and that 3 g/L NaCl is sufficient to eliminate its coalescence-promoting effect, cf. Fig. 23.19.

23.5.3 Effect of Geometric Parameters on Bubble Coalescence

The greater the tendency of a material system towards bubble coalescence, the more it is affected by various geometric parameters, because these parameters also determine the aeration density and the hydrodynamics of the system. Two of the most important geometric parameters are discussed briefly below:

a) Liquid height above the injector.

If the aeration device produces relatively large gas bubbles, or if the fine primary gas bubbles coalesce immediately above the gas disperser to form larger ones, then the height of the liquid above the gas disperser has no significant effect on bubble coalescence. In the case of the injectors discussed above, however, that generate fine or extremely fine gas bubbles which, through the free jet, are mixed into the surrounding liquid, the situation is quite different. Tests [23.1] with slot injectors have shown that in this case coalescence in pure water is only complete after a rise of 3 meters. When tests are carried out to obtain design data for injectors to be used as aeration devices, the height of the liquid must therefore be $\geqslant 3$ m; the sorption characteristics for Bayer injectors described in Sect. 23.4.3 were determined at $H = 7$ m.

b) Angle of inclination of the free jet of injectors.

Since injectors emit a free jet of the gas/liquid dispersion several meters in length, which sucks in the surrounding liquid and sets it in motion, there are certain angles of inclination at which the surrounding liquid induces a constriction of the free jet, thus accelerating coalescence.

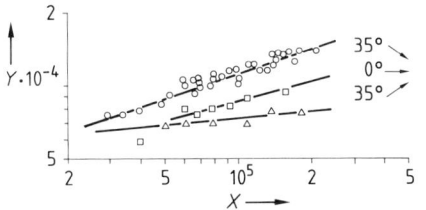

Figure 23.20. Sorption characteristic $Y(X)$ for the Bayer slot injector (technical scale), as a function of angle of inclination of the jet. From [23.1].

It can be seen from Fig. 23.20 that, with the technical-scale slot injector – all other operating conditions being identical – the O_2 uptake is increased by 50% (!) if the injector is directed towards the bottom at an angle of 35° to the horizontal rather than upwards at the same angle to the horizontal.

To sum up, it may be said that all three classes of parameters (physical, process-related, and geometric) govern the coalescence of gas bubbles in liquids and greatly affect the efficiency of injectors. The coalescence conditions of the system as a whole result from the interaction between the coalescence behavior of the liquid and the hydrodynamic behavior of the free jet and the bubble swarm. The two last mentioned depend in turn on the type of injector and the flow pattern of the surrounding liquid (H/D). *Due to the antagonism between the dispersing action of the injector and the bubble coalescence in the gas/liquid dispersion, laboratory measurements cannot provide reliable design data for injectors for use in material systems that promote coalescence.* Measurements on a technical scale are therefore essential.

23.5.4 Implications for Practical Work

a) Design data for aeration devices have to be determined in pure water. The tests must be carried out with technical-scale equipment; suitably large tanks or basins are therefore required. The results of laboratory tests are completely meaningless in this regard, since both the gas dispersion process and the counteractive process of bubble coalescence are heavily dependent on scale.

b) For aeration devices producing fine primary gas bubbles, design data (sorption characteristics) must also be drawn up for degrees of coalescence different from that of water. These measurements should also be carried out on a technical scale as under a); the best test substance to use here is common salt (NaCl) in concentrations of 1, 2, 3, 5, 10, and 15 g/L.

c) The only other parameter of which an exact knowledge is required for practical purposes is the coalescence behavior of the effluent concerned. This does not, however, require any further tests on a technical or semi-technical scale. The necessary data can be obtained in laboratory tests using any aeration device that produces fine primary gas bubbles and for which the relationship $m = f(c_{NaCl})$ has been determined in advance. If measurements on biologically treated waste water are carried out with this laboratory equipment, the m value determined can be used to calculate the "corresponding NaCl concentration" having the same coalescence behavior as the biologically treated waste water. On this basis the optimum operating conditions for the aeration device can be calculated using that sorption characteristic of the desired gas disperser which is applicable to the degree of coalescence of the effluent.

The laboratory tests described under c) must be carried out over a long period with effluent that has already been purified biologically: because biological sewage treatment takes place in vessels with almost ideal back-mixing, the resulting BOD_5 concentration corresponds to that at the outlet and is well below the concentration at the inlet.

Fig. 23.10 gives a practical guide to the tests mentioned under c) above. The heavy line represents the $m(c)$ relationship found for sodium chloride solutions of various concentrations in a 70 L test vessel with a self-aspirating pipe stirrer. Let us now suppose that tests carried out in this vessel with the biologically treated effluent has yielded an m value of $m = 2.0$. According to Fig. 23.10 this would mean that the coalescence behavior of this effluent corresponds to that of a NaCl solution of 6 g/L. The design data used to size and optimize the desired aeration device should therefore have been attained in a sodium chloride solution of ≈ 5 g/L.

The "corresponding salt concentration" giving the same coalescence behavior as the effluent in question can be determined using any aeration device provided that this produces very fine primary gas bubbles. In this test, measurements are made with an O_2 measuring electrode and evaluated in the form $k_L a = \Delta \ln \Delta c / \Delta t$.

In view of the attempts being made to recycle process waste water, the possibility cannot be excluded that industrial effluents will, in future, often have a coalescence behavior differing greatly from that of pure water. They will sometimes also contain substances that will prevent determination of the degree of coalescence by means of Δc measurement with O_2 electrodes. In this case the so-called "pressure gauge method" can be employed; this permits the accurate determination of $k_L a$ in any liquid [23.5].

The basic procedure here is as follows: in a closed vessel, under isobaric test conditions, a pure gas (e. g., high-purity nitrogen) is introduced into the liquid from the gas space above the liquid, the shrinkage of this space due to absorption being compensated by feeding in a corresponding quantity of gas from a pressurized storage container. The pressure drop in the storage container with time is registered by means of a pressure transmitter and a recorder, which permits accurate determination of $k_L a$.

23.6 Bayer Tower Biology®

23.6.1 Technical Principle

The following process engineering requirements led to the development of the Bayer Tower Biology® principle: The demand for a high efficiency E of the O_2 uptake and for a minimum production of off-gas so that thermal treatment of the latter could, where necessary, be carried out economically.

We have already seen from the relationships E_{max}, $x'' = f(H)$ for the two injectors described (Figs. 23.12 and 23.13) that, with the usual effluent, a high degree of utilization of atmospheric oxygen under coalescence-promoting conditions can only be achieved with relatively high liquid columns and therefore correspondingly long residence times of the gas bubbles in the liquid. Fig. 23.14 shows clearly that the optimum liquid height, as far as the oxygen uptake efficiency alone is concerned, is about 15 m, the O_2 concentration in the off-gas still amounting to roughly 8% by volume (degree of utilization 62%). Higher levels of oxygen utilization are only possible with even higher water columns; these are then achieved at the price of a slight drop in efficiency. It is also apparent from Fig. 23.14 that a reduction to about 4% by volume of oxygen in the off-gas (O_2 utilization 81%) is not possible with a liquid height of less than 26 m; the efficiency of the O_2 uptake here, however, is still about 3.0 kg/kWh.

The idea for the Bayer Tower Biology® concept [23.8] is to construct the treatment unit (activated sludge tank) in the form of a tall vessel, the oxygen being supplied by means of injectors. There were three important requirements here: the injectors had to be only a relatively short distance (½–1 m) above the bottom, positioned so that they point towards the bottom, and they also had to be equidistant from each other. The aim of the first requirement is to give the gas bubbles the greatest possible vertical distance to travel before they reach the surface; the second is to ensure that there is a vigorous flow of liquid at the bottom of the tank which prevents deposits.

The purpose of the third requirement is to promote the formation of so-called "aeration chimneys" above the injectors to ensure uniform back-mixing over the height of the liquid; this is essential for good O_2 absorption (Δc should be as high as possible and positive throughout).

In the original design, secondary clarifiers (sedimentation funnels or "Dortmund wells") are designed to separate the biomass from the purified waste water. As most of the sedimented biomass is returned to the activated sludge tank ("recycled sludge"), one way of avoiding an increase in the power requirement of the pump was to arrange the sedimentation funnels so that the liquid level in them largely corresponds with that in the main tank. In the case of the Leverkusen Tower Biology, this resulted in a "collar" around the towers, whereas the other Tower Biology units so far constructed have separate secondary clarifiers (sedimentation funnels or „Dortmund wells") at the top of each tower. This concept is illustrated in the sketch in Fig. 23.21.

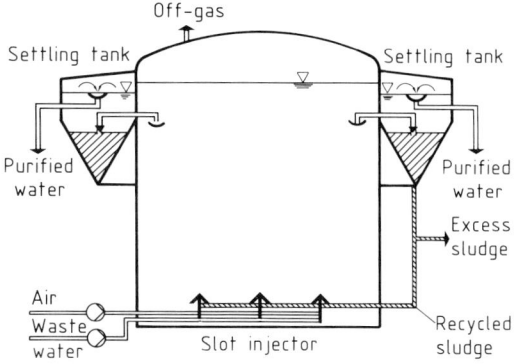

Figure 23.21. Schematic cross section of a Bayer Tower Biology® reactor.

23.6.2 Plants so far Constructed

a) Bayer AG, Leverkusen Works (Germany) [23.9]–[23.11]

The industrial effluent from more than 100 chemical production facilities at Bayer AG's Leverkusen works is conveyed through a double sewer more than 2.5 km long to the "joint sewage treatment plant" in the Leverkusen suburb of Bürrig. A "basin biology" plant with surface aeration has been in operation there since 1971. This first treatment facility built at Bürrig was able to handle part of the industrial effluent from the Bayer factory as well as the municipal sewage from the Wupper water authority area. The Bayer Tower Biology® fa-

Figure 23.22. Photograph of the Tower Biology® plant at Bayer's main factory in Leverkusen, on stream since December 1980.

cilities were erected during the second phase of construction which began in 1976, involving a capital expenditure of DM 135 million. This facility went on stream in December 1980 (cf. Fig. 23.22). Since then it has served as the first stage in the biological treatment of the entire effluent from Bayer's Leverkusen works; the outlet leads to the basin biology, the second stage in biological treatment, which deals not only with the pretreated Bayer effluent but also with the municipal sewage from the Wupper water authority area (dry weather flow).

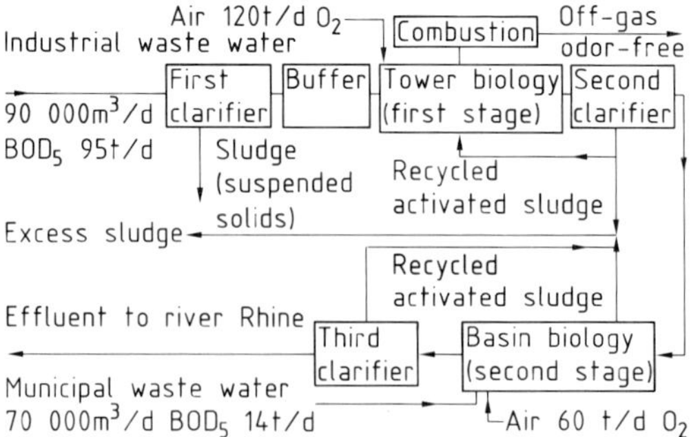

Figure 23.23. Flow diagram of the Leverkusen plant.

The design data for the whole plant (cf. sketch in Fig. 23.23) are:

Intake:
Industrial effluent: 90000 m³/d with 95 t/d BOD_5; plus municipal sewage: 70000 m³/d with 14 t/d BOD_5.

Primary clarification:
Residence time 2.5 h.

Tower Biology:
4 rubber-lined steel tanks (Ø 26 × 30 m) in parallel; liquid height 26.5 m, tank volume 13600 m³, total volume 54400 m³. Aeration via 4 × 72 slot injectors. O_2 uptake 120 t/d, 22000 m³/h air (standard conditions) being required for an O_2 utilization of up to $x''=4\%$ O_2 by volume. Volume load 1.8 kg/m³ d BOD_5, sludge load 0.32 kg BOD_5 per kg dry mass per day, residence time 14.5 h.

Off-gas purification:
The off-gas is pre-heated in a countercurrent to about 580 °C and mixed with approximately 160 m³/h of natural gas. "Non-flaming combustion" is induced by infrared radiation. This oxidation takes place in a reaction chamber at 750 °C. The

flue gas is subsequently cooled in a counter-current heat exchanger to 200°C, and the HCl resulting from the oxidation of chlorinated hydrocarbons is removed in a dry absorption column (burnt lime, CaO). The spent lime is separated by a filter bag and used afterwards for waste water neutralization. The off-gas purified in this way is discharged into the atmosphere through a short chimney at a temperature of about 180°C.

Intermediate clarification:

Each tower has a "collar" of 16 sedimentation funnels of 9 m Ø with a total volume of 16×420 m^3 = 6720 m^3 per tower. Residence time 7 h, superficial velocity 0.95 m/h.

Basin biology
(second stage in the biological treatment of Bayer effluent):

2 flat basins in series ($H = 3.5$ m) with respective volumes of 11 500 and 25 000 m^3. Residence time 3.4 + 7.5 h, O$_2$ uptake 60 t/d.

Secondary clarification
(sedimentation of biomass):

Basin with scraper ($\tau = 3$ h) and sedimentation funnels ($\tau = 6.3$ h).

b) Bayer AG, Brunsbüttel Works (West Germany) [23.12], [23.13]

Following several years' experience with the 8 m-high biotanks (Fig. 23.24, foreground), the second phase of construction was completed in 1979. Six tanks then went into operation (Ø 10×15 m, volume 1200 m^3) having a total volume of 7200 m^3; these carry out the biological purification of 5600 m^3/d of waste water with a BOD$_5$ load of 4.2 t/d. This requires an O$_2$ uptake of 7.2 t/d. The secondary clarifiers (sedimentation funnels) are situated between the towers and are also covered.

c) Bayer Factory at Thane (India)

Tower Biology in concrete towers (2 tanks of Ø 15×17 m, water level 15 m) with adjacent sedimentation funnels. Waste wa-

Figure 23.24. Aerial view of the Bayer Tower Biology® reactors (far right) at Bayer's Brunsbüttel works.

ter throughput 3600 m³/d with a BOD₅ load of 2 t/d; O₂ uptake 4 t/d. This treatment plant commenced operation in 1981.

d) Königsbacher Brewery, Koblenz (West Germany) [23.14]

One of the main reasons for the choice of Tower Biology in this case was the lack of space available. The treatment plant consists of a tower ($\varnothing$ 20 × 20 m) with an adjacent funnel-shaped secondary clarifier. 2000 m³/d of effluent are treated; the BOD₅ load is 3 t/d. O₂ uptake is 4.2 t/d. The organisms' one-sided diet (carbohydrates) encourages the formation of filamentous activated sludge, which tends to rise to the surface during deaeration of the Tower Biology® outlet. To deal with this, a new type of flotation unit was installed, consisting of a single flotation cell ($D = 2.5$ m; $V = 30$ m³) with a funnel-shaped nozzle (induced air flotation) [23.15]. This gives a recycled sludge with about 10 g/L dry matter and a water-clear outflow of the purified waste water. The Tower Biology went on stream in 1981.

e) Petrochemical Plant in Wesseling, North Rhine-Westphalia (West Germany) [23.16]

Here, there is 8000 m³ of waste water to be treated daily, containing an ammonium (NH_4^+) load of 8 t/d as well as a BOD₅ load of 6 t/d. Nitrification and denitrification processes therefore have to be integrated into the biological treatment for purposes of nitrogen elimination. Following extensive trials on a semi-technical scale, the industrial-scale plant shown in the sketch in Fig. 23.25 was designed. The plant comprises a two-step nitrification, the first step also incorporating the aerobic oxidation of carbon as well as 70% nitrification (from 800–1000 ppm to 100–150 ppm). In the second step, the remaining NH_4^+ is degraded to 5–30 ppm. The nitrites and nitrates thus formed then have to be reduced to nitrogen with the aid of hydrogen donors (e.g., methanol). To save on methanol, the outflow from the second nitrification step is divided into two streams; 40% of the total is fed to a denitrification unit upstream from the first nitrification unit, and 60% passes to the downstream denitrification unit to which is added an effluent free of N but rich in hydrocarbons (methanol). An activated sludge tank is then required downstream from the denitrification unit to degrade the residual BOD₅ resulting from these additional streams.

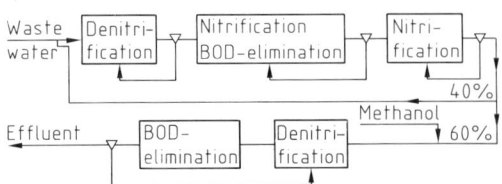

Figure 23.25. Flow diagram of combined BOD elimination and nitrification/denitrification.

Since the nitrification unit has a high O₂ demand (4.57 kg oxygen per kg of ammonia nitrogen), the entire plant was designed in the form of a multi-step Tower Biology; the O₂ uptake is 50 t/d. It comprises seven towers of between 24 and 17 m in height plus six sedimentation funnels, see Fig. 23.26. These commenced operation in July 1981 and have proved entirely satisfactory.

f) Dynamit Nobel, Lülsdorf Works (West Germany) [23.17]

In this case the effluent (15 000 m³/d) carries not only the BOD₅ load (14 t/d) but also the NO_3^- load (3 t/d) from the nitration process. The plant consists of two parallel lines, each containing a denitrification unit ($\varnothing$ 12 × 22 m; $V = 2500$ m³) and an activated sludge unit ($\varnothing$ 17 × 20 m, $V = 4500$ m³). This treatment plant went on stream in July 1983.

g) Lehrter Zucker AG, Lehrte, near Hannover (West Germany) [23.18]

The special feature of sugar factories as far as their effluents are concerned is that effluent only occurs during a particular season (after the sugar beet harvest) and also that it contains a high carbohydrate concentration and is therefore best pretreated in an anaerobic unit. This is also true for Lehrter

Figure 23.26. Bayer Tower Biology® facility at a refinery operating according to the principle illustrated in Fig. 23.25.

Zucker AG. Here, only the downstream aerobic treatment unit takes the form of a Tower Biology plant ($\varnothing$ 15 × 18 m; $V = 3200$ m^3); the O$_2$ uptake is 5–7 t/d (max.). This plant handles 3000 m^3 of waste water per day, which contains 7–10 g/L of COD and 4.5–7.2 g/L of BOD$_5$. The plant went into operation in the autumn of 1982.

h) Bitburger Brewery, Bitburg (West Germany) [23.19]

The excellent results obtained with the combination of Tower Biology and induced air flotation at the Königsbacher brewery in Koblenz prompted the Bitburger brewery to choose the same system. The plant at Bitburg comprises a tower ($\varnothing$ 20 × 24 m, $V = 7300$ m^3) and a downstream flotation unit consisting of two parallel flotation cells, each with a volume of 50 m^3. The plant commenced operation in January 1984 and treats 7200 m^3 of effluent daily.

23.7 BIOHOCH Reactor® of Hoechst AG*)

23.7.1 Process Engineering Concept

The incentive to develop a tower-shaped reactor at Hoechst AG came chiefly from the need to save space and reduce the emission of off-gas. The design of this facility can be seen as a further stage in the devel-

---

* This section was kindly contributed by H. G. MÜLLER, Hoechst AG, Frankfurt am Main, Federal Republic of Germany.

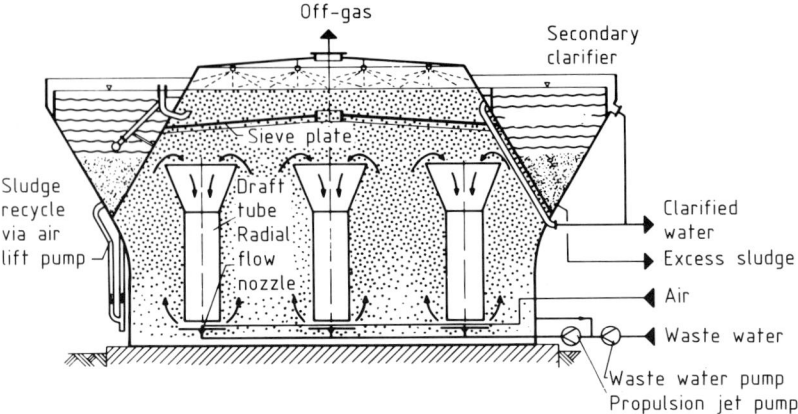

Figure 23.27. BIOHOCH Reactor® of Hoechst AG (multi-loop reactor).

opment that began with the low-level design (1967) with basin depths of 3.5 m, continued with the medium-level unit with water depths of 10 m and finally led to the BIOHOCH reactor [23.20].

The BIOHOCH reactor consists basically of an almost cylindrical activated sludge tank and a conical secondary clarification unit forming a ring around the activated sludge tank, cf. Fig. 23.27. The activated sludge tank is divided by a sieve plate into two chambers. This results in a reduction in the gases (N_2, CO_2) dissolved in the sludge-water mixture in the upper part of the reactor. It improves the settling characteristics of the activated sludge in the secondary clarifier.

The biomass precipitated from the treated waste water flows back from the secondary clarifier into the aeration tank via air lift pumps. The quantity of sludge recycled is controlled by the air fed into the sludge recycle pipe.

In order to improve sludge sedimentation, rotating rakers are installed in the secondary clarifier. The function of the draft tubes installed in parallel in the aeration tank is to improve vertical back-mixing of the activated sludge. The difference in density between the aerated liquid outside the draft tubes and the non-aerated liquid inside gives rise to a strong circulatory current. The water emerges horizontally from the lower end of the draft tubes, thus preventing the deposition of solids on the bottom; the raw sewage added is mixed quickly and evenly with the contents of the reactor.

The activated sludge is aerated by radial flow nozzles on the bottom, through which charging of the reactor with pretreated, neutralized effluent also takes place.

In order to improve the bubble distribution, the radial flow nozzles are arranged beneath the draft tubes.

The operation of a bubble column or loop reactor is chiefly influenced by the arrangement and mode of operation of the aerators. This applies in particular to the large-surface aeration of the activated sludge stage in a waste water treatment plant.

In the technology of waste water treatment, aerators have to meet a number of requirements:

- high oxygen input
- optimum utilization of the oxygen content of the aeration air and thus low off-gas quantities
- thorough back-mixing of the waste-sludge mixture
- noiseless operation
- reliability of operation, in particular freedom from clogging.

On the basis of these criteria, a wide variety of aeration systems were examined, the most favorable results being obtained with the two-phase nozzles [23.21], [23.22]. With these nozzles, the air is dispersed by the kinetic energy of a water jet (propulsion jet).

23.7.2 Radial Flow Nozzle

Extensive tests resulted in the development of the radial flow nozzle, Fig. 23.28. With this nozzle, the propulsion jet is converted with the aid of a deflecting element into a radially spreading fan jet. The air is introduced into the liquid through an annular slot behind the deflecting element [23.23]. The gas distributor plate prevents rapid rising of the bubbles.

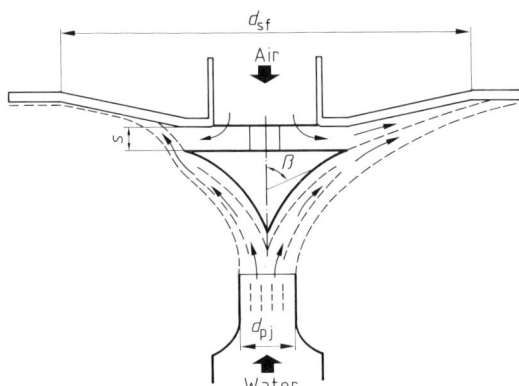

Figure 23.28. Radial flow nozzle. d_{sf} shear field diameter, d_{pj} propulsion diameter, β angle of deflecting cone, s slot width.

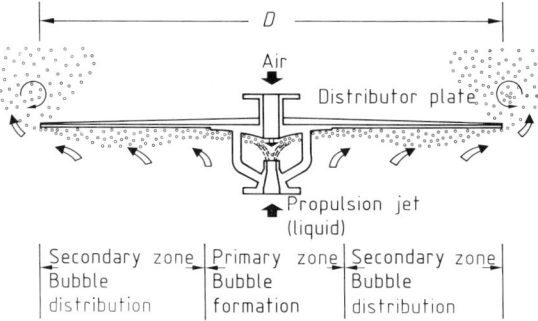

Figure 23.29. Bubble formation and bubble distribution using a radial flow nozzle.

The turbulent flow of the bubble/water mixture is compelled to flow radially outwards and to mix with the secondary water drawn in. The gas content declines with increasing distance from the bubble formation zone, thus reducing bubble coalescence. A vortex of gas bubbles forms above the distributor plate. Fig. 23.29 shows this nozzle in cross-section.

The re-shaping of the propulsion jet to form a fan means that when the jet diameter is increased, the shearing area and the cross-sectional area of the jet both increase with the square of the jet diameter. The geometric parameters such as the deflection angle, the diameter or the gap width s have in this respect a considerable influence on bubble formation.

The radial flow nozzles are made from ordinary steel or high-grade steel, the plug, which is subject to high stress, is made from a special alloy (see Fig. 23.30). Fig. 23.31 shows the bubble pattern of a radial flow nozzle.

Tests performed with model nozzles in smaller units by means of fixed measuring devices employing the hydrazine method or sulfide oxidation method yielded no significant, utilizable results. No more meaningful information was obtained when using the mobile input tests commonly employed in waste water treatment technology. They were not suitable as a basis for designing the aeration system of the BIOHOCH reactor®.

With all these methods it is assumed that in the volume examined, the oxygen transfer rate coefficient $k_L a$ is constant.

However, this does not apply to aeration tanks with high water levels. In this case, both the $k_L a$ value and the concentration difference are locally dependent values.

An additional factor is that the coalescence conditions present in the waste water are completely different from those in a test solution, for example. Studies have shown that the oxygen transfer factor α, which represents the ratio of oxygen uptake in the waste water to oxygen uptake in the treated effluent, does not only depend on the quality of the waste water, but also on geometric and hydraulic parameters, cf. Sect. 23.5.3.

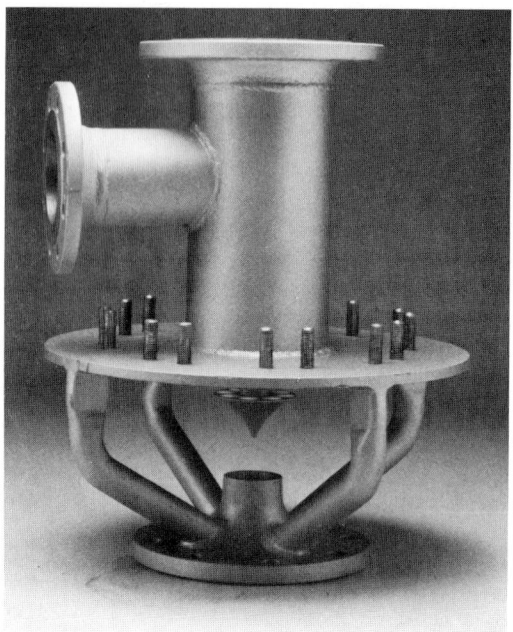

Figure 23.30. Radial flow nozzle with 50 mm propulsion jet diameter.

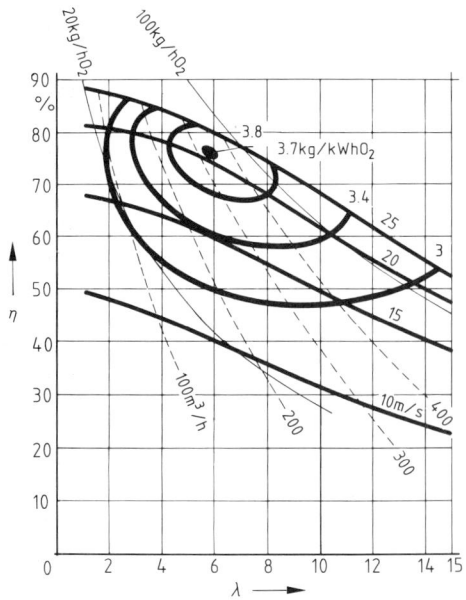

Figure 23.32. Working diagram for a radial flow nozzle (type 30/1500) for coalescence conditions $m = 0.9-1.6$ and standard conditions ($c = 0$ mg/L O_2). λ air/water ratio (m³ air per m³ water), η oxygen utilization (%), $H = 17.5$ m.

Figure 23.31. Bubble pattern of a radial flow nozzle.

In order to obtain the most realistic results possible, the nozzles were tested solely in waste water/sludge suspensions. The radial flow nozzles used were of the same design as those later employed in industrial scale plants. The test column was 18 m high and 5 m in diameter.

For the purpose of determining the degree of oxygen utilization, the amount of air and off-gas was established, as well as the oxygen concentration in the water and off-gas. Oxygen uptake was obtained from the mass balance, as were also the values for degree of oxygen utilization and efficiency.

The various operating states (water and air throughput etc.) and the corresponding efficiency values can be jointly represented by means of characteristic curves, an example of which is shown in Fig. 23.32 for a nozzle with 32 mm diameter at a water level of 17.5 m [23.24].

The optimum condition under which this nozzle achieves its maximum efficiency is a propulsion jet velocity of 21 m/s and an air/water ratio of $\lambda = 6$ (m³ air per m³ water). At this point the data are as follows:

Degree of oxygen
utilization $\qquad \eta = 76\%$

Efficiency $E = 3.8$ kg/kWh
Oxygen uptake $G = 72$ kg/h
Air throughput
(standard conditions) $q = 320$ m^3/h
O_2 concentration in
the off-gas $x'' = 5\%$ by volume

In order to be able to apply the results to industrial scale plants – e.g., for the purpose of arranging the radial flow nozzles in the BIOHOCH reactor® – it was necessary to obtain an insight into the local mass transfer. The oxygen concentration profiles were therefore determined as a function of height and diameter. During the test, the oxygen concentrations were measured simultaneously in the liquid and gas phase by means of adjustable probes.

In Fig. 23.33 the oxygen concentration is plotted against the liquid height. The left-hand curve shows the oxygen concentration in the ambient liquid, the right-hand curve the oxygen concentration in the liquid at the phase boundary of the air bubbles. The velocity of the propulsion jet in these experiments was 19 m/s.

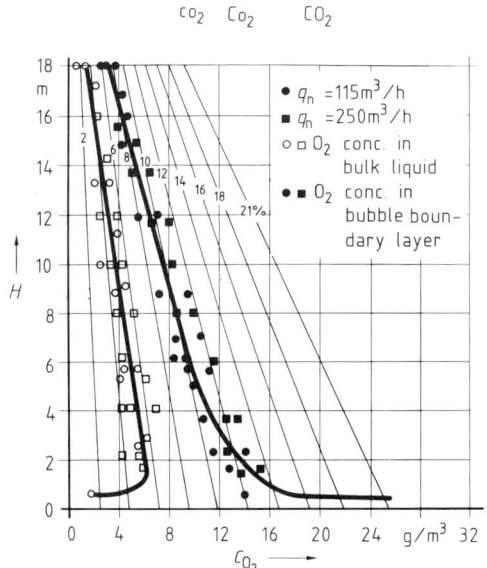

Figure 23.33. Oxygen concentration profile in waste water (type 30/1500). $m = 0.9-1.4$; propulsion velocity $v = 19$ m/s.

From the oxygen profiles the local mass transfer rate coefficients were determined. These values are plotted in Fig. 23.34 against liquid height H. In the lower section, in the area of activity of the nozzles, $k_L a$ values were obtained, which after a few meters dropped to the levels associated with bubble columns [23.25], [23.26], [23.27].

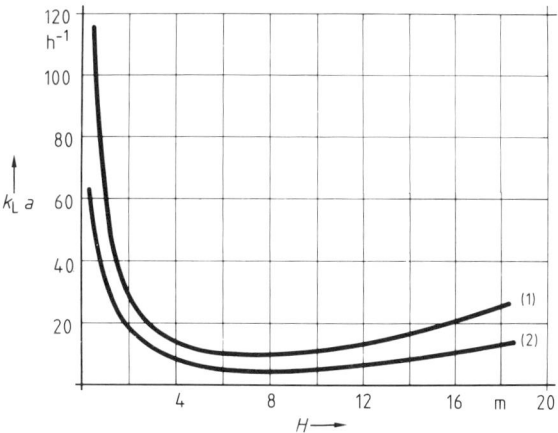

Figure 23.34. Mass transfer rate coefficient $k_L a$ as a function of height (type 30/1500). – Waste water $\alpha = 0.6-1.2$, propulsion velocity $v = 19$ m/s, (1) $q_n = 250$ m^3/h, (2) $q_n = 115$ m^3/h.

Radial flow nozzles with propulsion jet diameters of 30–70 mm are in operation in a number of waste water treatment plants within the Hoechst group and elsewhere. Studies carried out at plants in operation have fully confirmed the test results.

23.7.3 Survey of Plants Already Completed or Under Construction

a) Hoechst AG, Kelheim (Donau) Works (West Germany)

Kelheim is the site of the first of a number of BIOHOCH reactors that have been

Figure 23.35. Photograph of the BIOHOCH Reactor® at the Kelheim works of Hoechst AG.

built. It has been in operation since October 1981 (Fig. 23.35).

The reactor has a liquid height of 15 m and an activated sludge volume of 3000 m^2; its diameter is 16–24 m.

In the activated sludge tank, seven radial flow nozzles are installed beneath the draft tubes. The propulsion jet is 30 mm in diameter. At a degradation capacity of 8 t/d COD, the BOD$_5$ concentration averages 16.5 mg/L. 65–85% oxygen utilization is achieved (O$_2$ concentration in the off-gas 3–7%).

b) Hoechst AG, Griesheim Works (West Germany)

The special feature of the Griesheim effluent is its high content of substances formed during the manufacture of basic products and intermediates for dyes, pharmaceuticals etc., some of which have low degradability or inhibit degradation. In order to improve the degradation efficiency, a modified aeration process is used, the special feature of which is the addition of activated carbon in powder form to the activated sludge. The plant was started up in 1983 (Fig. 23.36).

2 BIOHOCH reactors (steel tanks)
diameter 23–45 m
liquid height 20 m
aerated liquid volume 8000 m^3
 per reactor
oxygen utilization approx. 60–70%
waste water throughput 60 000 m^3/d
treatment capacity 15 000 kg/d BOD$_5$
air throughput 3000–10 800 m^3/h

c) Cassella AG/Hoechst AG, Offenbach Works (West Germany)

Space saving was the decisive reason for building this joint plant. The mechanically and chemically pretreated effluent from the Offenbach works passes through a pipeline some 3 km in length to the treatment plant, where it is mixed with the pretreated effluent from Cassella AG immediately prior to the activated sludge stage.

Design data for the activated sludge unit:

2 BIOHOCH reactors (with integrated secondary clarification)
steel tanks (epoxy resin coating)
diameter 20–39 m
liquid height 20 m

Figure 23.36. Photograph of the BIOHOCH Reactor® to be used at the Griesheim works of Hoechst AG.

aerated liquid volume	6000 m³ per reactor
treatment capacity	24 000 kg/d BOD₅
air input (standard conditions)	3000–10 800 m³/h

d) Hoechst AG, Hoechst Works (West Germany) (treatment plant under construction)

The treatment capacity of the central waste water treatment plant had to be increased due to rising organic load in the effluent and the desire for more efficient degradation. The capacity of the three parallel waste water treatment units will then be 160 000 kg/d BOD₅.

The design data for the 3rd unit are as follows:

2 BIOHOCH reactors
steel tanks/epoxy resin coating
diameter	32.1–45.5 m
liquid height	20 m
aerated liquid volume	15 000 m³ per reactor
treatment capacity	60 t/d BOD₅
air input (standard conditions)	6000–21 600 m³/h

e) E. Merck, Darmstadt (West Germany)

The waste water treatment plant at the Darmstadt works of E. Merck is being enlarged in order to handle the biological treatment of industrial effluent.

The design data for two BIOHOCH reactors were determined by Merck in the course of long-term trials under operating conditions in a pilot plant.

The experimental data obtained led to the construction of *one* large-scale unit, which came on stream in 1983.

Design data:
1 BIOHOCH reactor (steel tanks/rubber lined)
diameter	18.8–33.4 m
liquid height	18 m
aerated liquid volume	5000 m³
air input (standard conditions)	3500–8000 m³/h
oxygen utilization	60–70%
maximum COD load	30 000 kg/d

f) American Hoechst Corp., Baton Rouge Works (USA)

The waste water treatment plant at the Baton Rouge works went on stream in mid-

Figure 23.37. Photograph of the BIOHOCH Reactor® at the Baton Rouge works of the American Hoechst Corporation.

April 1983. Here, the BIOHOCH reactor principle is employed in a modified form. The activated sludge tank – a cylindrical vessel – is installed separately from the secondary clarifier – a Dortmund tank. A draft tube is installed inside the tank. The aeration system comprises a radial flow nozzle RS 50/3500*), a propulsion jet pump and a screw compressor unit. The air volume is max. 560 m³/h. The surface of the water in the activated sludge tank is sprayed with an antifoam.

The design data are as follows:
diameter/height   8 m/14 m
water volume   240–320 m³/d
treatment capacity   1500 kg/d BOD

Fig. 23.37 shows an over-all view of the plant.

g) Riedel de Haen, Seelze (West Germany)
At Riedel de Haen, the biological waste water treatment plant has been enlarged by a second unit, which went into operation in the spring of 1983. The plant was built for the biological treatment of industrial effluent from the batchwise production of laboratory chemicals. The activated sludge tank, made of concrete, has a diameter of 20 m and a liquid height of 8.75 m. The activated sludge volume is approx. 2700 m³.

Radial flow nozzles are used for aeration of the tank.

Design data:
waste water throughput   200 m³/h
oxygen uptake   approx. 250 kg/h
air input
  (standard conditions)   3300 m³/h
oxygen utilization   40%–50%

23.8 References

[23.1] M. ZLOKARNIK; Chem. Ing. Tech. *50* (1978) 9, 715; Chem. Eng. Sci. *34* (1979) 10, 1265–1271.

[23.2] M. ZLOKARNIK; Verfahrenstechnik *13* (1979) 7/8, 601–604.

---

* 50 mm diameter of the propulsion jet nozzle, 3500 mm diameter of the distribution plate

[23.3] DOS (Ger. Pat. Appl.) 24 08 064 of September 4, 1975; DE (Ger. Pat.) 24 08 064 of December 29, 1983.

[23.4] DOS (Ger. Pat. Appl.) 26 34 494 of February 2, 1978; DE (Ger. Pat.) 26 34 494 of April 14, 1983; US Pat. 4 162 970 of July 31, 1979.

[23.5] M. ZLOKARNIK; Adv. Biochem. Eng. *8* (1978), 133–151.

[23.6] G. MARUCCI; Chem. Eng. Sci. *24* (1969), 975–985.

[23.7] M. ZLOKARNIK; Preprints Discussion Paper 1/3; First European Congress on Biotechnology, Interlaken, Switzerland, September 1978.

[23.8] DOS (Ger. Pat. Appl.) 25 12 815 of October 7, 1976; 25 16 914 of October 21, 1976; 27 05 243 of August 17, 1978.

[23.9] G. DIESTERWEG, H. FUHR, and P. REHER; Industrieabwasser (1978) 6, 5–13.

[23.10] E. SCHELLMANN; Prog. Water Technol. *12* (1980) 215–232.

[23.11] P. LINGEN; Journées Information Eaux, October 2/3, 1980, Poitiers, France.

[23.12] G. DIESTERWEG and I. PASCIK; Hydrocarbon Process. (Nov. 1980), 191–195.

[23.13] I. PASCIK and G. DIESTERWEG; Umwelt *4* (1982), 231–233.

[23.14] E. SCHMIDT; Wasser Luft Betr. (1983) *10,* 15–22.

[23.15] M. ZLOKARNIK; Chem. Ing. Tech. *53* (1981) 8, 600–606; Ger. Chem. Eng. *5* (1982) 2, 109–115.

[23.16] I. PASCIK; Hydrocarbon Process. (Oct. 1982) 80–84; Chem. Ing. Tech. *54* (1982) 12, 1190–1192.

[23.17] ANONYMOUS; Wasser Luft Betr. (1983) 10, 9.

[23.18] H. J. DANKERT and I. PASCIK; Zuckerindustrie *108* (1983), 847–852.

[23.19] ANONYMOUS; Wasser Luft Betr. (1984) 1/2, 24.

[23.20] A. BAUER and G. LEISTNER; Chem. Tech. *7* (1978) 31–65.

[23.21] H. G. MÜLLER and G. SELL; DECHEMA Monographien Vol. 86/II, No. A75-1801, pp. 589–596. Verlag Chemie, Weinheim–New York 1980.

[23.22] M. ZLOKARNIK; Korrespondenz Abwasser *27* (1980), 194–209.

[23.23] H. G. MÜLLER and G. SELL; Eur. Pat. Appl. No. 0044498 of July 13, 1981.

[23.24] H. G. MÜLLER and G. SELL; Chem. Ing. Tech. *56* (1984), 5.

[23.25] W. D. DECKWER, R. BURCKHARD, and G. ZOLL; Chem. Eng. Sci. *29* (1974) 11, 2177–2188.

[23.26] K. B. WANG and L. T. FAN; Chem. Eng. Sci. *33* (1978) 7, 945–952.

[23.27] R. BURCKHARD and W. D. DECKWER; Verfahrenstechnik *10* (1976) 6, 429–433.

[23.28] M. ZLOKARNIK; Korrespondenz Abwasser *27* (1980) 11, 728–734.

Chapter 24

Modelling and Scaling-up of Bioreactors

N. W. F. Kossen
N. M. G. Oosterhuis

Technische Hogeschool Delft
Afdeling der Scheikundige Technologie
Bioengineering Laboratory
Delft, The Netherlands

24.1 Introduction
24.2 Modelling of Bioreactors
24.2.1 The Model Cycle
24.2.2 Kinds of Models
24.2.3 Complexity of the Model
24.2.4 Solving the Equations
24.2.5 Parameter Sensitivity
24.2.6 Experimental Design/Parameter Optimization/Testing of the Model
24.2.7 Conclusions
24.3 Scaling-up of Bioreactors
24.3.1 Introduction
24.3.2 Scale-up Methods in Use
24.3.2.1 Fundamental Method
24.3.2.2 Semi-fundamental Method
24.3.2.3 Dimensional Analysis
24.3.2.4 Rules of Thumb
24.3.2.5 Trial and Error
24.3.3 Dimensional Analysis
24.3.3.1 Mechanistic Background of DA
24.3.3.2 The Use of Dimensionless Groups for Scaling-up
24.3.3.3 Heterogeneous Systems
24.3.3.4 The Generation of the Dimensionless Groups
24.3.3.5 Some Examples
24.3.4 Regime Analysis
24.3.4.1 Introduction
24.3.4.2 Methods for Regime Analysis
24.3.5 Scaling-down of Parts of the Process
24.4 References

List of Symbols

Symbol	Units	Description
a	m^{-1}	specific surface concentration
C	$kg\ m^{-3}$	concentration
C_{in}	$kg\ m^{-3}$	concentration in inlet
C_m	$kg\ m^{-3}$	concentration of microorganisms
C_{out}	$kg\ m^{-3}$	concentration in outlet
C_s	$kg\ m^{-3}$	concentration of substrate
D	m	stirrer diameter
$\mathbb{D}$	$m^2\ s^{-1}$	diffusion coefficient
$\mathbb{D}_e$	$m^2\ s^{-1}$	dispersion coefficient
d_i	m	internal diameter
d_p	m	particle diameter
Fr	—	Froude number (see Table 24.5)
g	$m\ s^{-2}$	acceleration due to gravity
H	m	height of column
H_s	m	height of sludge bed
k	s^{-1}	1st order reaction rate constant
k_l	$m\ s^{-1}$	mass transfer coefficient
$k_l a$	s^{-1}	mass transfer parameter
K_s	$kg\ m^{-3}$	Monod constant
L	m	length
N	s^{-1}	stirrer speed
N_m	m^{-3}	number of microorganisms per volume
P_1	$N\ m^{-2}$	pressure at the bottom
P_2	$N\ m^{-2}$	pressure at the top
P	W	power
P_g	W	gassed power
P_i	—	parameter sensitivity
pO_2	$N\ m^{-2}$	oxygen pressure
Q, Q_{gas}	$m^3\ s^{-1}$	volumetric gas flow
$\tilde{Q}$	$kg\ s^{-1}$	accumulation term
r, r_c	$kg\ m^{-3}\ s^{-1}$	reaction rate
R	$N\ m\ kmol^{-1}\ K^{-1}$	gas constant
Re	—	Reynolds number (see Table 24.5)
t	s	time
t_c	s	time constant for conversion
t_{circ}	s	circulation time
$t_\mathbb{D}$	s	time constant for diffusion
t_f	s	time constant for flow
t_m	s	mixing time
t_{mt}	s	time constant for mass transfer
T	°C, K	temperature
v	$m\ s^{-1}$	velocity
v_p	$m\ s^{-1}$	velocity of particle
v_s	$m\ s^{-1}$	superficial velocity
v_{tip}	$m\ s^{-1}$	tip speed of stirrer
V	m^3	volume
x	m	distance
X	—	parameter value
Y	—	value of output variable
α	—	growth parameter
η	$N\ s\ m^{-2}$	dynamic viscosity
κ	—	dimensionless distance
μ	s^{-1}	specific growth rate
μ_{max}	s^{-1}	maximum specific growth rate
ξ	—	dimensionless concentration
Π	$kg\ s^{-1}$	conversion term
Π_n	—	dimensionless group
ϱ	$kg\ m^{-3}$	density
$\Delta\varrho$	$kg\ m^{-3}$	density difference
ϱ_l	$kg\ m^{-3}$	liquid density
σ	$N\ m^{-1}$	surface tension
τ	s	residence time
τ_0	$N\ m^{-2}$	yield stress
Φ	$kg\ s^{-1}$	transport term
Φ_{circ}	$m^3\ s^{-1}$	circulating flow
Φ_m	$kg\ s^{-1}$	mass flow
Φ_p	$m^3\ s^{-1}$	pumping capacity of stirrer
Φ_v	$m^3\ s^{-1}$	volumetric flow

Subscripts

l	liquid
M	model scale
p	particle
P	prototype (production) scale

For the definition of the dimensionless numbers see Table 24.5.

24.1 Introduction

Bioreactors for the production of fermented foods and beverages have been in use for at least 4000 years. Their size has increased slowly but continuously. By using a trial and error procedure these bioreactors have attained a remarkable level of sophistication. Obviously mathematical modelling and science-based scale-up procedures are not a necessity to develop bioreactors. A few disadvantages of the trial and error procedure should be kept in mind, however:

1. Generally, trials have to be performed at production scale. This is an expensive procedure, especially if the size of production increases;
2. the errors can have (and often have had) disastrous effects;
3. the procedure is very time consuming. Innovations had a time constant of ages in the past.

In the 19th and 20th century the rise of the natural and technical sciences had two important effects:

1. Insight in the mechanisms governing the fermentation process (metabolism, kinetics, transport phenomena, thermodynamics), resulting in better understanding, and
2. the incorporation of these mechanisms into mathematical models resulting in better control of the process.

This enabled the designer to use the mathematical model of the process instead of the process itself to examine his ideas (design and scale-up, modification, optimization, etc.). This method is cheap, fast, and relatively harmless; however, when used to extrapolate results obtained with the model (to other scales or process conditions), the method can only be applied under two conditions:

1. The model should be of a mechanistic nature (i.e., it should contain all the relevant mechanisms, contrary to so-called black-box models containing only formal mathematical relations between process variables without a physical background).
2. Rigorous experimental testing of the model and its parts on a laboratory scale.

If well tested mechanistic models are available, extrapolation is possible, at least in principle.

One of the definitions of a model is:
"A model is a representation of a part of reality. Manipulation of the model gives the information we need about that part of reality to serve a specified goal".
This goal can be:

- design/scale-up/modification
- start-up/operation/shut-down
- process control
- optimization.

The types of models we need to serve these goals can be quite different. For the design of a system we need a rather detailed mechanistic model. Usually these models are non-linear and are based on the assumption of steady-state conditions. For the control of a system we can use rather simple models. Usually they are linearized over the area under consideration and describe the transient (non-steady-state) behavior of the system for small perturbations of the process variables (DORF [24.1]). For the start-up or shut-down of process equipment mechanistic models describing the behavior of the system under larger perturbations should be available. This chapter is divided into two parts: modelling and scaling-up. In the part about modelling aspects not directly relevant for scaling-up will also be treated. The emphasis will be on mathematical modelling.

24.2 Modelling of Bioreactors

24.2.1 The Model Cycle

The development of a (mathematical) model is a cyclic event. This is represented schematically in Fig. 24.1 (KOSSEN [24.2]).

A mathematical model is always set up on the basis of a well formulated verbal model. The goals of the model should be defined explicitly. Many kinds of mathematical models can be set up for a particular process. This is the subject of Sect. 24.2.2.

To avoid unnecessary computational efforts the mathematical model has to be reduced to a state where it describes the relevant phenomena only. Methods for this procedure will be mentioned in Sect. 24.2.3.

After solving the equations (Sect. 24.2.4) one should raise the question how sensitive the model is for the values of the different parameters involved (Sect. 24.2.5).

Because experiments are usually very expensive, efficient experimentation is a necessity. The purpose of the experiments is to validate the model. Usually this asks for adjustment of the model by optimization of the parameters and separate experiments to test the model. This is the subject of Sect. 24.2.6.

Quite often a realistic parameter optimization does not produce a proper fit between model and experiment. In that case the structure of the model has to be changed. This is done in the verbal stage, whereafter the whole modelling procedure starts again.

Properly speaking, a model can never be validated but only refuted (POPPER [24.3]). Here, validation means to determine if the model describes reality accurately enough to serve its purpose.

24.2.2 Kinds of Models

Several kinds or classes of models exist. A survey is given in Fig. 24.2. For a detailed description the reader is referred to the literature (FREDERICKSON et al. [24.4], ROELS and KOSSEN [24.5]).

The distinction between black-box models and gray-box models is essential. Black-box models only give a formal description of a phenomenon and are of a non-mechanistic (phenomenological) nature. They do not allow for extrapolation (e.g., scale-up). When studying a system it initially always turns out to be a black box. Gradually its structure is unravelled, whereby this black-box nature changes into a gray-box one. However, this is an optical illusion. The gray box is a collection of smaller black boxes, the "elements" of the system, arranged in what is called the 'structure' of the system (the whole of black boxes with their interconnections) (HOFSTADTER [24.6]; see also Fig. 24.3).

An important question has to be raised now: if a gray-box (or mechanistic) model is nothing more but a collection of black-box models, how can a gray-box model then be used for extrapolation? The answer to this question is: a particular system is a

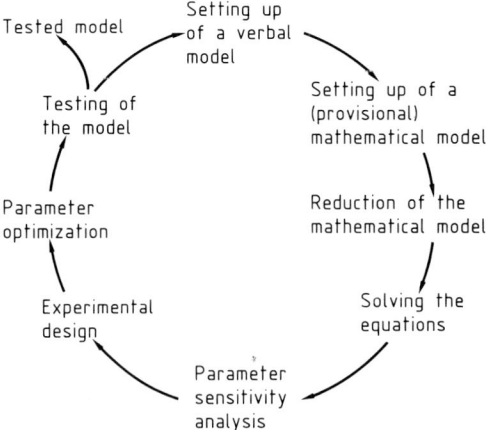

Figure 24.1. The "modelling cycle".

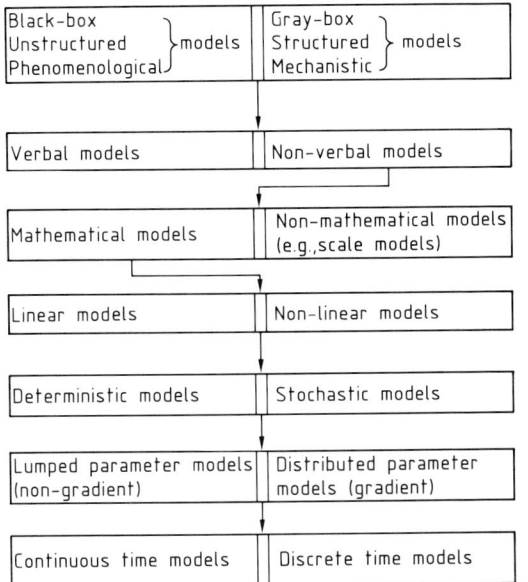

Figure 24.2. Classification of models (I).

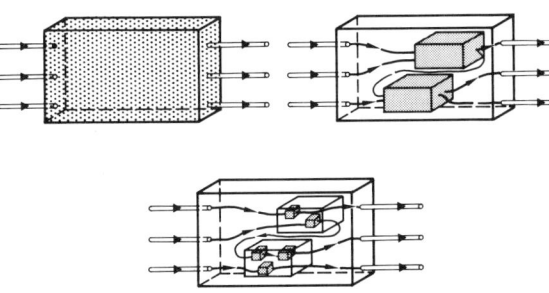

Figure 24.3. Black-box vs. gray-box models.

unique combination of non-unique elements (the constituent black boxes). In bioreactor design these 'elements' can be kinetic equations, transport equations, etc.

Generally, each of these elements has been already validated, but only for a limited range of conditions. *If all the relevant elements are present in the model, and if, during extrapolation, all these elements stay within their range of validity, extrapolation is allowed, otherwise not.*

This problem is by no means philosophical. It is the essence of engineering activi-ties, like design and scaling-up. If one of the relevant elements is not present, or fails during extrapolation, the model as a whole can give erroneous results. If one of the elements has not yet been validated, this should be done on the laboratory scale, taking into account the range of conditions this element will meet after extrapolation (after scaling-up, for example). This procedure is called 'scaling-down' and will be dealt with more extensively in Sect. 24.3.

A few remarks about the different kinds of mathematical models follows based on Fig. 24.2. Most biological phenomena result in *non-linear* mathematical models. The Monod equation is one of the simplest examples:

$$\mu = \mu_{max} \frac{C_s}{K_s + C_s}. \quad (24.1)$$

However, various mathematical techniques have been developed for *linear* equations (OGATA [24.7]). For non-linear systems different kinds of non-linearities require different kinds of techniques and these are limited in number. Fortunately, the range of process conditions is often sufficiently limited to allow linearization. Sometimes this enables analytical solutions of the equations. Otherwise, the calculations must be performed with a computer.

Deterministic models give a precise description of the behavior of the system. An example is the equation:

$$\frac{dC_m}{dt} = \mu C_m, \quad (24.2)$$

which after integration results $(C_m|_{t=0} = C_{m_0})$ in:

$$C_m = C_{m_0} \exp(\mu t). \quad (24.3)$$

With this equation one value for C_m is found for every value of t.

Our knowledge of birth and death processes of individual cells is very fragmentary. As a consequence, these processes appear to be of a stochastic nature. If small amounts of microorganisms are present in a system (during sterilization, for example)

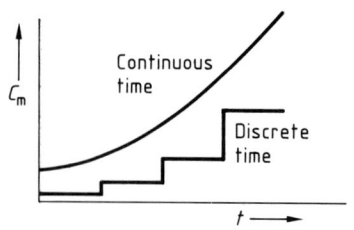

Figure 24.4. Continuous time vs. discrete time models.

we therefore have to use *stochastic models*. These models, as distinct from deterministic models, only predict a probability for the behavior of a system. Monte Carlo techniques are used frequently for the modelling of these systems (VERKOOYEN [24.8]; see also FREDERICKSON [24.9] and GOEL and RICHTER-DYN [24.10]).

When concentration- or temperature gradients are absent in a system *lumped parameter* (or non-gradient) *models* can be used. These models are simple when compared with *distributed parameter* (or gradient-) *models* that have to be used when gradients cannot be neglected. This subject will be discussed in more detail when dealing with balance equations at the end of this section.

Continuous time models describe phenomena of a non-abrupt nature like the growth of biomass when large amounts of microorganisms are present:

$$C_m = C_{m_0} \exp(\mu t). \tag{24.4}$$

C_m as a function of t is a smooth curve. For a synchronized culture the situation is different. Between short periods of division of cells long periods exist where the number of cells is constant. These phenomena are described with *discrete time models* like Eq. (24.5):

$$N_{t+\Delta t} = N_t + \alpha N_t ; \tag{24.5}$$

see also Fig. 24.4.

When small amounts of microorganisms are present discrete time models have to be used (as well as stochastic models as mentioned earlier). The classification of mathematical models as given above is not the only one; other classifications are possible. A very useful one is given in Fig. 24.5.

Constitutive Equations

Kinetic, transport, and thermodynamic equations are dealt with in other chapters of this volume. Only a few remarks will be made here.

A general problem with the existing kinetic equations is their black-box character. They are seldomly sufficiently structured to describe transient phenomena, although such phenomena occur frequently in bioreactors (microorganisms circulating through areas with different concentrations). However, structuring the kinetic model soon results in models with too many parameters. Testing these models is very laborious. Also the measurements are generally not sufficiently accurate to discriminate between models. In the authors' opinion it is more useful to simulate the production-scale process conditions on a laboratory scale and to subject the microorganisms to these conditions. This results in a black-box description that is, however, as a consequence of the set-up of the model experiment, useful for scaling-up.

Biokinetics are generally of little importance for continuous and fed-batch processes (ROELS [24.11]) because the bioreactor is then usually controlled by transport processes instead of kinetic processes.

Balance Equations

Balance equations will also be dealt with in a brief manner because they are described elsewhere in this book.

For the design of bioreactors the models should contain mechanistic descriptions of

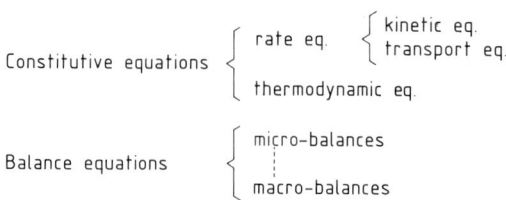

Figure 24.5. Classification of models (II).

conversion and transport phenomena, collected in so-called balance equations (mass-, heat-, and momentum balances). The general form of such an equation is:

$$\dot{Q} = \pi + \Phi \qquad (24.6)$$

$\dot{Q}$ ist the accumulation term, π the conversion term, and Φ the transport term.

Balance equations can only be set up for extensive properties (biomass, amount of substrate, enthalpy, etc.). Extensive properties are proportional to the total mass of the system. No balance equation can be made for intensive properties (for example, temperature and pressure).

Several kinds of balances can be used, depending on the properties of the bioreactors. If the bioreactor is perfectly mixed, we can use a macro-balance of the bioreactor (i.e., the balance is made over the reactor as a whole) [Eq. (24.7)]. This is a non-gradient model. If the reactor is not perfectly mixed we have to use, as a start for our calculations, a balance over a part of the reactor taken so small that gradients in that element can be neglected. This is a micro-balance [Eq. (24.8)]. Integration of this balance over the size of the reactor results in an equation describing the behavior of the reactor as a whole, including the influence of gradients. We then have a gradient model:

$$\Phi_v (C_{in} - C_{out}) - Vr_c = V \frac{dC_{out}}{dt}, \qquad (24.7)$$

$$-v \frac{\partial C}{\partial x} - r_c = \frac{\partial C}{\partial t}. \qquad (24.8)$$

Intermediate kinds of balances between macro and micro are possible (HIMMELBLAU and BISCHOFF [24.12]). Going from macro- to micro-balances the details provided by the model increase, but the effort in handling the equations sometimes becomes insurmountable (KOSSEN [24.13]).

For conservative properties like total mass, total energy, and amount of elements, the balance equations contain no conversion term:

$$\dot{Q} = \Phi. \qquad (24.9)$$

Any of the members of the second group of models can belong to one or more classes of the group of mathematical models. A constitutive equation like the kinetic equation, for example, can be a non-linear, deterministic, non-gradient continuous time model. In fact, the majority of kinetic equations belong to these classes.

24.2.3 Complexity of the Model

Mathematical models can range from very simple to extremely complicated. *The art of model building is to find the right optimum between simplicity and reliability in order to serve the goal of the model.*

Simple models are easy to handle but may not provide enough information to describe the process. Complicated models sometimes give a better description of the process, but may give rise to insurmountable problems in obtaining the right values for the many parameters present and in the handling of the model.

In the authors' experience the most efficient way of modelling is to start as simple as possible, taking into account only the most important mechanisms. The complexity of models can also quite often be reduced considerably by comparing the time constants of the different mechanisms (ROELS [24.11]). Mechanisms with time constants much larger than the process itself can be omitted, mechanisms with time constants much smaller than the process itself are at equilibrium or at a steady state (see Fig. 24.6).

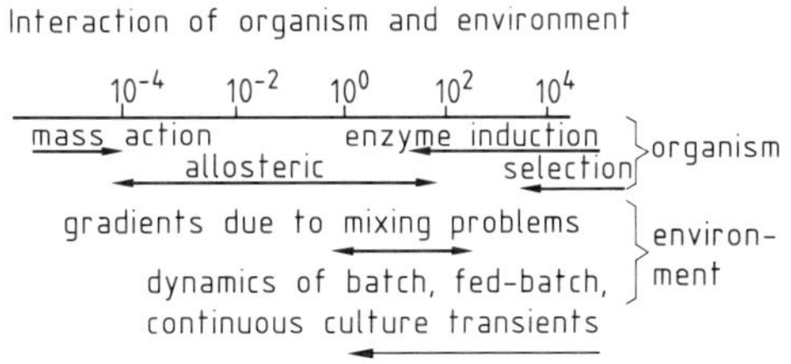

Interaction of organism and environment

| 10^{-4} | 10^{-2} | 10^{0} | 10^{2} | 10^{4} |

mass action — enzyme induction } organism
allosteric — selection

gradients due to mixing problems } environment
dynamics of batch, fed-batch, continuous culture transients

Figure 24.6. Time constants.

If, for example, the mixing time of a process is much smaller than the conversion- and mass transfer times, a non-gradient model (macro-balance) can be used. In this way considerable reductions of the complexity of the models are possible.

FREDERICKSON et al. [24.4] and ROELS [24.11] describe the need for structured kinetic models. The incorporation of more than two or three compartments (elements) in the structured model is generally useless.

One of the main problems is the modelling of flow in bioreactors, particularly in stirred vessels. Very little is known about flow patterns and circulation times on the production scale (JANSEN [24.14], EINSELE [24.15], [24.16]). Semi-empirical models (mixed vessels in series, plug flow with dispersion, etc.) often have been proposed but evidence for the validity of these models on a production scale is scarce. These models are often based on residence time distribution (RTD) measurements (LEVENSPIEL [24.17]). A problem with RTD measurements is that they indicate that different fluid elements spend different times in the reactor but not where (for example, in regions of high or low concentration of oxygen). With modified RTD techniques this can be elucidated (tracer injection in different areas of the system) but this is very time-consuming (DANCKWERTS [24.18]).

24.2.4 Solving the Equations

"Solving" equations generally means that one or more of the variables are obtained in an explicit form – as for example, $C = f(v, x, t,$ kinetic constants) instead of the implicit formulation of Eq. (24.8). There is a strong tendency to "run to the computer" once a mathematical model has been formulated and must be solved.

Analytical solutions, however, give much more insight into the behavior of a system than numerical solutions obtained with digital computers. Unfortunately, analytical solutions are often difficult to obtain.

FRANKS [24.19] gives a survey of the different kinds of mathematical models and the possibilities to solve them analytically.

Analytical solutions exist mainly for linear equations. There are, however, very interesting approximate solutions possible, even for non-linear partial differential equations (HIMMELBLAU and BISCHOFF [24.12]; FINLAYSON and SCRIVEN [24.20]; MERCER and O'DRISCOLL [24.21]; RAMACHANDRAN [24.22]).

Quite often the model has to be solved by numerical methods with a computer. Nowadays, user-orientated computer languages exist like CSMP (Continuous System Modelling Program) that can be mastered to a large extent within a few days once some basic knowledge about programming is available. Other languages (Fortran, Pascal) are used very often. There is ample literature on programming: MCCRACKEN [24.23], GOTTFRIED [24.24], ZAKS [24.25], JENSEN and WIRTH [24.26].

There are some inherent dangers in the use of computers, however. As long as no violation of the syntax occurs (the syntax is the set of rules and statements of a particular computer language), the computer will accept any model, right or wrong. Because the computer is used for the more complicated models a check on the correctness of the computer program is not easy. Nevertheless, this check is an absolute necessity. To some extent one can rely on one's intuition: "Does the model show what I expected intuitively?". However, this can be dangerous because many systems show a "counterintuitive behavior". Therefore, a theoretically better founded method is necessary. One such method is shown in Fig. 24.7.

Most models in use in biotechnology can be modified to such an extent that an analytical solution for this modified version is possible. This can be done by looking at the steady state ($t \to \infty$) or by lumping (perfect mixing assumed). The results of this procedure can be compared with the results of the computer program obtained under similar conditions. Other possibilities are: linearization of the model or the use of one of the other approximate analytical tech-

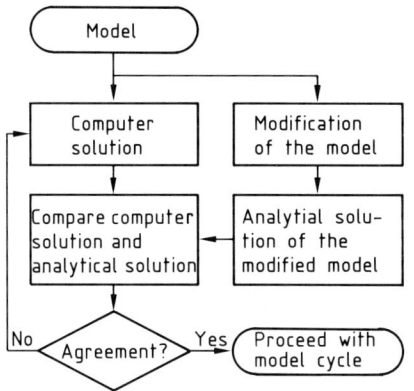

Figure 24.7. Check of the computer model.

niques mentioned earlier. These approximations can be compared again with the results of the computer program. These procedures give a reasonable idea about the correctness of the computer program.

24.2.5 Parameter Sensitivity

Before a comparison can be made between the mathematical model and the experimental results, the right values of the different parameters in the model have to be known. Obtaining these parameters can be very time-consuming, particularly in biotechnology where the systematic compilation of parameter values is still in its infancy. This refers to kinetic parameters (K_s, μ_{max}, Y, maintenance, etc.), transport parameters ($\mathbb{D}$, $k_1 a$, etc.) and thermodynamic parameters (e.g., solubilities of different components).

The influence of the various parameters on model behavior can be very different. Parameters having a large influence should be known accurately, for others a span of an order of magnitude is often sufficient. Activities to obtain an insight into the influence of parameters are called parameter sensitivity analyses. The purpose of these analyses is to avoid unnecessary efforts in obtaining accurate values of less relevant parameters.

One of the simplest ways for this analysis is presented below:

1. Obtain a first estimate of the values of the parameters involved;
2. introduce these parameter values into the (computer) model;
3. vary the values of these parameters one by one, e.g., 10% and register their influence on the most important output variable of the model (e.g., a

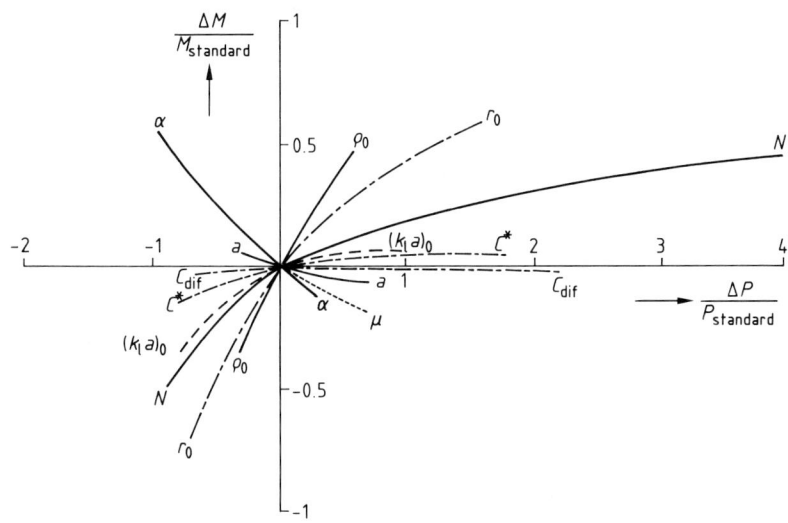

Figure 24.8. Parameter sensitivity of maximum biomass concentration.

concentration, a temperature, or a reaction rate);
4. the parameter sensitivity P_i can now be defined as:

$$P_i = \frac{\Delta Y / Y}{\Delta X_i / X_i}, \qquad (24.10)$$

- Y being the value of the important output variable at the original value of the parameter,
- ΔY the variation of this variable due to the variation of the parameter,
- X_i the original value of the parameter,
- ΔX_i the variation of this parameter.

Presentation in a graphical form rapidly shows the relative importance of the different parameters involved. In Fig. 24.8 an example is given for the influence of different parameters on the maximum biomass (M) due to growth of mycelial pellets (METZ [24.27]). The exact meaning of the different parameters is not relevant here, but the picture shows that ϱ_0 has a very pronounced influence on M, while C^* has virtually no influence at all. Therefore, an approximate estimate of C^* is sufficient, but ϱ_0 should be measured very accurately.

24.2.6 Experimental Design/ Parameter Optimization/ Testing of the Model

Experiments are necessary to optimize the parameters in the model and to test the model. Therefore, these three subjects are dealt with in one section.

It is possible to obtain a (black-box) model of a system directly by comparison of input/output data obtained by experiments (stimulus/response techniques like frequency response analysis). These methods can be of great value for systems where no *a priori* knowledge of the system structure and the involved mechanisms are available. They are often used in process control. Surveys are presented among others by SEINFELD and LAPIDUS [24.28], OGATA [24.7], and BECKEY and YAMASHIRO [24.29]. The application of these methods to biotechnological problems is described by DANG [24.30]. When combined with the scaling-down approach this certainly is a valuable technique that deserves further attention in biotechnology.

The discussion for the rest of this section will be restricted to mechanistic (gray-box) models.

It is essential to make a clear distinction between parameter optimization and model testing. During parameter optimization the values of the parameters in the model are altered in such a way that an optimal fit is obtained between model and experiments. The procedure is presented schematically in Fig. 24.9.

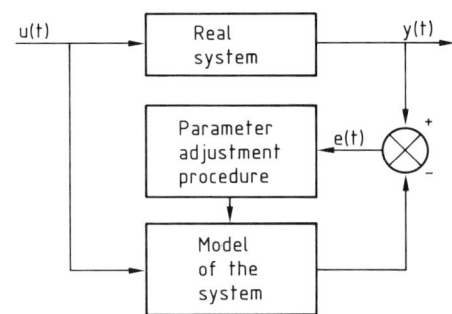

Figure 24.9. Parameter optimization.

After this optimization of parameters the critical evaluation of the value of the model (the testing of the model) still must be accomplished. *This has to be done by new experiments, independent of the ones used for the optimization of the parameters, and preferably under different conditions of temperature, concentration, etc.* Only then can the usefulness of the model be estimated, as well as the presence of all relevant mechanisms in the model, and its capability to predict the behavior of the system under different conditions (extrapolation). The misuse of models usually is a consequence of the unawareness of the clear difference between parameter optimization and model testing (see the publication of WEI [24.31] entitled: "The Least Square Fitting of an Elephant").

When this distinction is made, parameter optimization is a useful and valuable technique, provided that:

(1) The amount of adjustable parameters is limited to two or three. In biological systems the accuracy of the measurements is usually too poor to allow for a meaningful optimization of more than three parameters. REN DER YANG and HUMPHREY [24.32] and KROUWEL [24.33] have shown, for example, that a distinction between different models for microbial growth is already virtually impossible with 3-parameter models.
(2) The values of the optimized parameters stay within physically realistic ranges.

The reasoning given under (1) applies to *adjustable* parameters only, and not to many parameter models as such. A model of a fermentation plant, for example, can easily contain 30 or 40 parameters.

The values of these parameters have all been obtained, however, by literature study or by detailed separate experiments and are not used to fit the model. Therefore, such a model can be very valuable, despite its many parameters.

Techniques to obtain and optimize parameter values are described among others by SEINFELD and LAPIDUS [24.28] and, for biokinetic parameters, by JOHNSON and BERTHOUEX [24.34], [24.35]. Also, experimental design as such is described in detail in these publications. On-line optimization of parameters with a computer is possible.

See also BOX et al. [24.36], DRAPER and SMITH [24.37], and HIMMELBLAU [24.38], for proper experimental design and the use of statistical techniques necessary to test the model as a whole.

24.2.7 Conclusions

1. A model is set up for some purpose or goal. Different goals ask for different models.
2. Many kinds or classes of models exist. A model usually belongs to several classes. The distinction between black-box models and gray-box models is essential. The kind of model is a function of the system under investigation.
3. Extrapolation with the help of a model is only possible if all the relevant elements are present and if they stay within their range of validity during extrapolation.
4. Mathematical modelling is a cyclic event. The best procedure is to start as simple as possible and to gradually increase the complexity, if necessary. Comparison of time constants is a valuable tool for the set-up and reduction of a model.
5. Computer calculations always have to be checked by using modified versions of the model or by using approximate solutions that allow for an analytical solution.
6. Proper experimental design is extremely important. Parameter optimization and testing of the model have to be clearly distinguished.

24.3 Scaling-up of Bioreactors

24.3.1 Introduction

Commonly, the term scaling-up is used in two different, albeit related, senses. First of all scaling-up is used to indicate the general tendency in society to develop new systems that are larger than the existing ones. These systems can be ships, airplanes, administrations, factories, etc. The 20th century is to some extent characterized by this kind of scale-up.

The term scale-up is also used for a procedure whereby a large-scale system (the prototype) is designed and built on the basis of the results of experiments with small-scale models. The difference with the first definition lies in the basis of comparison (already existing system vs. small-scale experimental systems).

In this section scale-up is used in the second sense only. M will be used for small-scale (model) systems and P for the production-scale (prototype) system.

The first question regarding scaling-up problems is why they do exist at all. Why can't we simply build an industrial-scale fermenter, geometrically similar to the laboratory fermenter, and with equal aeration rate (vvm), stirrer speed, or power per unit volume (P/V)? To answer this question one has to realize that, for the design of bioreactors, three different kinds of phenomena are important:

– thermodynamic phenomena ⎫ (scale-
– micro-kinetic phenomena ⎬ independent)
– transport phenomena (scale-dependent)

Of these phenomena the first and second are independent of scale (if measured properly). A typical thermodynamic property like the solubility of oxygen in a broth, for example, is not dependent on the scale of the bioreactor (fermenter) nor is the micro- (or intrinsic) kinetic behavior of single microorganisms (growth and product formation as a function of the *local* environmental conditions like nutrient concentration, pH, and temperature).

The actual oxygen concentration and kinetic behavior of microorganisms in a bioreactor are, however, generally speaking rather dependent on scale. The reason is that oxygen and other nutrients involved in the conversion processes are consumed constantly and have to be supplied by transport processes.

Furthermore, the microorganisms are always subjected to (turbulent) shear phenomena, which is also a kind of transport process (momentum transport). Shear phenomena can either influence the microorganism itself (damage) (VAN SUIJDAM [24.39]) or influence the formation of agglomerates (flocs, pellets) of microorganisms (MIDLER and FINN [24.40]; TAGUCHI et al. [24.41]).

Transport processes are very dependent on scale (generally speaking, the time constants[*] for transport processes increase with scale). This is virtually the only reason that scale-up problems do exist at all. Because transport to microorganisms is essential in (industrial) fermentations, scale-up problems can always be expected (although more for aerobic processes than for anaerobic ones and more for continuous- and fed-batch processes than for batch processes).

The fact that the magnitude of transport processes differs at different scales has important consequences for the environment of the microorganisms. At small scales gradients of concentration and pressure are very small (well mixed system). The turbulent shear is relatively low. At production scale large gradients can exist. When the microorganism travels through the bioreactor it meets constantly changing nutrient and oxygen concentrations and pressures. Furthermore, the turbulent shear is higher. Often these phenomena have a profound influence on the behavior of the microorganisms, see TAGUCHI et al. [24.42]; WANG and FEWKES [24.43]; HATTORI et al. [24.44]; KASZAB et al. [24.45]. Therefore, the transport phenomena are important enough to deserve more attention.

Transport processes in bioreactors are governed by two transport mechanisms:

• flow (convection)
• diffusion (conduction)

---

[*] For the use of time constants see Sect. 24.2.

The time constant for flow-transport is:

$$t_f = L/v. \quad (24.11)$$

This means that liquid velocities should increase proportionally with scale in order to keep t_f constant [basically the behavior of a bioreactor does not change during scale-up if the ratios of the time constants for transport (t_f and $t_{\mathbb{D}}$) and for conversion (t_c = concentration/reaction rate) are kept constant]. For stirred vessels, where $v \sim N \cdot L$, this means that N should be kept constant during scaling-up. This results in an increase in power per unit volume (P/V) proportional to the square of the size of the reactor ($P/V \sim L^2$), which is generally not acceptable. Usually P/V is kept more or less constant during scale-up, resulting in $t_f \sim L^{2/3}$. The rationale behind this will be given later (Sect. 24.3.2).

The time constant for diffusion transport is:

$$t_{\mathbb{D}} = L^2/\mathbb{D}. \quad (24.12)$$

It is quite obvious after inspection of this simple equation that, in case of diffusion limited processes, an increase of scale under conditions of geometrical similarity has dramatic effects on the transport time.

It will be clear that the transport times (t_f and $t_{\mathbb{D}}$) can increase considerably during scaling-up while the conversion time t_c remains more or less constant. This results in an increasing influence of transport phenomena on reactor behavior during scaling-up. It is often observed that at a small scale a process is determined by kinetic phenomena (kinetic regime) and at a large scale by transport phenomena (transport regime); see TAGUCHI et al. [24.42]; WANG and FEWKES [24.43]; BROWN [24.46]. Such a change in regime whereby $t_c > t_f$ or $t_{\mathbb{D}}$ for M and $t_c < t_f$ or $t_{\mathbb{D}}$ for P has severe consequences for scaling-up.

Phenomena directly related to flow and diffusion are:

- shear
- mixing
- mass transfer ($k_l a$)
- heat transfer
- macro-kinetics (a form of apparent kinetics as a result of a combination of micro-kinetics and diffusion: occurring in immobilized systems).

Therefore, we can expect these phenomena to be liable to change during scaling-up, depending on the scale-up criteria used. Because the kinetic behavior of microorganisms is dependent upon their local environmental conditions and their variations, and because these are determined to a great extent by mixing phenomena, one can also expect that the apparent kinetic behavior can change with scaling-up (BLAKEBROUGH [24.47]).

The arguments given above are not restricted to bioreactors. For the same reasons the scale-up of chemical reactors is often a problem as well.

In bioreactors, however, an additional complication, compared with chemical reactors, is due to properties of microorganisms such as growth, adaptation, decay, and shear-sensitivity. Therefore, the scale-up of bioreactors can be very difficult indeed.

Theoretically the design of a bioreactor could be set up as given below:

- The behavior (the kinetics of growth and product formation) of a selected strain is determined under a wide variety of environmental conditions like nutrient and dissolved oxygen concentrations and shear.
- From these results the optimal conditions for growth and product formation are selected.
- The kinetics are inserted in the mass balance. The mass-, heat-, and momentum micro-balances are solved, resulting in a detailed model that gives the relation between environmental conditions in the bioreactor and primary operating variables (stirrer speed, rate of aeration and of substrate addition, etc.). With this model the values of the primary operating variables necessary to meet the optimal conditions in the bioreactor are calculated. This bioreactor

can be new (the exception) or of an existing type (the rule).

For several reasons this procedure does not work. The three most important reasons are:

- Extensive experiments are needed to determine the kinetics in sufficient detail. This is rather time-consuming.
- Solving the micro-balance equations for anything but the simplest geometry and flow conditions is practically impossible.
- Optimal environmental conditions often ask for conflicting values of the primary operating variables (for example, a high power per unit volume for mass transfer and a low one for cells, due to shear).

In practice, productive strains are selected in simple laboratory-scale bioreactors (usually petri dishes and shake flasks) under conditions that bear absolutely no resemblance to conditions that can be realized on a production scale (OLDSHUE [24.48]). Then, gradually the strain is tested in a number of existing bioreactors of increasing scale. By trial and error the process conditions (stirrer speed, medium composition, etc.) are found under which the strain gives an acceptable production. Rules of thumb are used as a rough guideline (AIBA et al. [24.49]; BARTHOLOMEW [24.50]; EINSELE [24.16]).

The disadvantage of this procedure is obvious. The behavior of microorganisms (growth and product formation) depends on their phenotype, which is a result of their genotype and the environmental conditions. So, for a given genotype, the environmental conditions completely determine their behavior. In the above mentioned procedure these conditions at laboratory scale differ widely from the conditions at production scale (mixing, aeration, shear, kind of substrate). Therefore, the results of small-scale experiments are often of little use for scale-up. The present screening practice has been described by OLDSHUE [24.48] as: "Looking for potential football players at a baseball game". The great advantage, however, is that the use of petri dishes and shake flasks allows for massive screening tests.

There is a third way to design a bioreactor. It is called *scaling-down* and is essentially an "environmental approach" (YOUNG [24.51]). The experiments at the small scale are performed under environmental conditions comparable to the conditions that can be realized (physically and economically) at the production scale (OOSTERHUIS and KOSSEN [24.52]; OLDSHUE [24.53]). These conditions are among others: OTR (oxygen transfer rate), mixing times, shear, and substrate composition. In Fig. 24.10 this procedure is presented schematically.

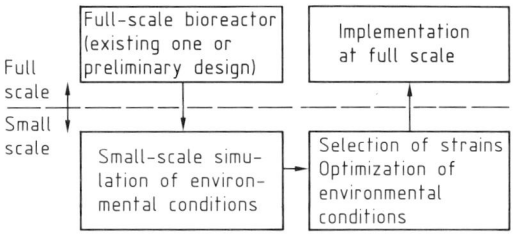

Figure 24.10. Scaling-down.

In fact, this method is a kind of kinetic research under realistic circumstances. It is particularly useful for the general situation when new strains have to be used in old fermenters or if the operational conditions in existing fermenters have to be optimized. It can also be used, however, when a new fermenter has to be designed. First, a preliminary design is made, based on general and incomplete information about the strain to be used. This preliminary design allows for an educated guess of the range of environmental conditions to be expected and is simulated on a small scale (scale-down of conception).

The advantage of this method is obvious: screening and other small-scale experiments under conditions representative for the production scale. The main disadvantage is that until now no cheap screening tests are available in which large-scale environmental conditions can be met.

24.3.2 Scale-up Methods in Use

A number of scale-up methods exist:

1. Fundamental method (solving the micro-balances for momentum-, mass-, and heat transfer)
2. Semi-fundamental method (solving simplified balances)
3. Dimensional analysis (including regime analysis)
4. Rules of thumb
5. Trial and error

These are archetypes, often used in combination with each other. The scale-up methods mentioned will be dealt with in more detail in this section. Dimensional analysis will be treated more extensively in a separate section (24.3.3).

First, it should be mentioned that in the fermentation industry quite often an increase of production is not realized by building larger fermenters, but by building duplicates of existing ones. This is the safest way to increase production and results in a very flexible set-up of fermenters that can be used for the production of different products at the same time, if necessary. Furthermore, the use of pilot plants is widespread in biotechnology (BARTHOLOMEW [24.50]; KAROW et al. [24.54]). This clearly shows the existence of uncertainties in scaling-up. A good pilot plant is a scaled-down plant, not a scaled-up lab experiment (CONN [24.55]). It is meant for the gathering of information for the design of the full-scale plant but also for other aspects of production:

- long term effects (corrosion, accumulation during recycling, etc.),
- process control,
- production of test batches for marketing,
- process demonstration.

24.3.2.1 Fundamental Method

When using the fundamental method micro-balances are set up for momentum-, mass-, and heat transfer. Examples of these balances have been shown in Chapter 1 of this volume (Fundamentals of transport phenomena). An extensive treatment is given by BIRD et al. [24.56]. A number of complications arise when these balances are used for bioreactor design.

In the first place, the balances, if applied to a stirred vessel, have to contain terms for transport in three directions. Furthermore, the boundary conditions are very complicated.

In the second place, the balances are coupled: the solution of the momentum balance gives the flow components that have to be used in the mass- and heat balances.

In the third place, the momentum balances are usually set up for a homogeneous fluid which is not very realistic for an aerated fermentation broth. WALLIS [24.57] gives suggestions how this problem can be solved but it remains an extra complicating factor.

The main problem is the impossibility to solve the micro-momentum-balance. As a result the fundamental method is used only for very simple systems:

- well defined flow conditions (e.g., laminar flow of a liquid film),
- absence of flow (e.g., diffusion of nutrients in a stagnant layer of microorganisms).

The last mentioned application of micro-balances is important for immobilized microbial systems. If the behavior of a thin layer is known, the influence of an increase of layer thickness can be predicted.

It has to be mentioned that generally temperature gradients in bioreactors are very small, which allows for a simplified set-up of the heat balance. This, however, does not change much in the overall picture for this scale-up method. The main problem is and remains the impossibility to find solutions for the micro-momentum-balances for anything but the simplest flow conditions.

24.3.2.2 Semi-fundamental Method

Here, simplified equations are used, particularly for the flow, so that problems related to the solution of the micro-momentum-balance are avoided. For stirred vessels as well as for bubble columns a wide variety of flow models is available in the literature. Surveys are given among others by LEVENSPIEL [24.17]; SHAH [24.58]; SHAFTLEIN and RUSSEL [24.59]; NORWOOD and METZNER [24.60]; KHANG and LEVENSPIEL [24.61]; HEYNEN and VAN'T RIET [24.62]; TOWELL and ACKERMAN [24.63]. Most models refer to bulk flow and do not give information about flow near important regions such as stirrer blades, cooling coils, and vessel walls. The flow models used are generally one of the following three types:

- plug flow,
- plug flow with dispersion (dispersion is a diffusion process of fluid elements),
- well mixed: one tank,
    series of tanks.

Combinations of these models are used as well.

The well mixed one-tank model is by far the most popular one, mainly because of its simplicity. The flow models result in transport terms that are used in (micro-)mass balances. For a one-directional flow with dispersion and for a situation of tanks in series the mass balances for steady-state conditions are given below:

$$-v \frac{dC}{dx} + \mathbb{D}_e \frac{d^2 C}{dx^2} - r = 0 \qquad (24.13)$$

(plug flow with dispersion)

$$\Phi_v (C_n - C_{n+1}) = V_{n+1} r_{n+1} \qquad (24.14)$$

(tank No. $n+1$)

These balances can be solved quite easily numerically and sometimes also analytically, depending on the form of the conversion term r.

Residence time distribution (RTD) measurements are also important to learn about the bulk flow behavior in bioreactors (OVASKAINEN et al. [24.64]; POPOVIĆ et al. [24.65]; LAINE and KUOPPAMÄKI [24.66]). In essence, this is a method whereby the response of an output signal to an input signal is observed, resulting in a black-box model of the flow in the system. The signals are tracer concentrations (dyes, acids, radio-active tracers, etc.).

The semi-fundamental method is, together with the use of rules of thumb, probably the most widespread method for the design of bioreactors. However, the bulk flow models mentioned above usually result from observations in small-scale apparatus (10 to 100 liter). With a few exceptions like JANSEN et al. [24.14]; MIDDLETON [24.67]; BRYANT and SADEGHZADEH [24.68]; and MANN and MAVROS [24.69] very little has been published about bulk flow in industrial-scale reactors. This makes scale-up based on these flow models rather risky. One of the reasons why large-scale data are lacking, is the problem of measurement of flow at a large scale. With the use of the flow-follower technique (MANN and MAVROS [24.69]; BRYANT and SADEGHZADEH [24.68]) this problem can be overcome. The circumstances allowing for extrapolation of the flow models mentioned here to a larger scale will be dealt with in Sect. 24.3.3.

24.3.2.3 Dimensional Analysis

This method is very powerful although it has severe limitations. As an introduction it is mentioned here that dimensional analysis (DA) is a technique whereby dimensionless groups of parameters are kept constant during scaling-up. The physical meaning of these groups is a ratio of time constants for the different mechanisms involved. Therefore, keeping the dimensionless groups constant means that the relative importance of the mechanisms involved does not change during scaling-up.

The limitations are fourfold:

- Often it is not possible to keep all the dimensionless groups constant during scaling-up. One then has to determine the most important groups, neglecting the rest (regime analysis). If a change in regime takes place during scaling-up, however, the formal DA method breaks down completely.
- The formal application of DA sometimes leads to technically unrealistic situations (power consumption, stirrer speed, etc.).
- Some systems are rather autonomous: the bubble size in coalescing systems, for example, is almost independent of scale and process conditions. This violates the basic concept of formal DA: geometric similarity.
- The choice of parameters involved in DA is not always obvious and sometimes rather arbitrary.

Nevertheless, the method offers a considerable potential and will be treated in a separate section.

24.3.2.4 Rules of Thumb

EINSELE [24.16] mentions scale-up rules used in a number of European fermentation industries. The result is mainly the application of a number of rules of thumb (see Table 24.1).

These criteria are all closely related and refer mainly to oxygen transfer (pO_2 is a function of $k_l a$ which in turn is a function of P/V, provided $v_{tip} > 3$ m s^{-1}). Correlations are available to calculate the values of these criteria at different scales (MOO-YOUNG and BLANCH [24.70]). They are presented below for stirred vessels as well as for bubble columns, in both cases for low viscous aqueous systems only.

Table 24.1. Scale-up Criteria in Fermentation Industries

% of Industries	Scale-up Criterion Used
30	constant P/V
30	constant $k_l a$
20	constant v_{tip}
20	constant pO_2

Stirred vessel (with one standard Rushton-turbine impeller; see among others VAN'T RIET [24.71]).

See the list of symbols at the beginning of this chapter for the meaning and the dimensions of the different symbols used.

The difference between a coalescing and a non-coalescing system is not absolute; it depends on the fluid properties, but also on the size of the reactor. In large reactors bubbles have more time to coalesce before they leave the system than in small reactors. Therefore, a system that is non-coalescing at a small scale tends to become coalescing at a large scale. If the equations for $k_l a$ in stirred vessels are written in a more general form, then:

$$k_l a = c \left(\frac{P_g}{V}\right)^a (v_s)^b. \qquad (24.15)$$

Table 24.2. Influence of Scale on the Exponents of Eq. (24.15)

Vessel Size in m³	a	b	
0.005	0.95	0.67	
0.5	0.6–0.7	0.67	(BARTHOLOMEW [24.50])
50	0.4–0.5	0.50	
0.002–2.6	0.4	0.5	(VAN'T RIET [24.71] coalescing systems)

Power input	$P = 6\varrho N^3 D^5$	(ungassed)
	$P_g = 0.4 P$	(gassed)
	$P_g = 0.0312\, P Fr^{-0.16} Re^{0.064} \left(\dfrac{Q}{ND^3}\right)^{-0.38} (T/D)^{0.8}$	
		(gassed, REUSS [24.72])
Mass transfer	$k_l a = 2.6 \cdot 10^{-2} (P_g/V)^{0.4} (v_s)^{0.5}$	(coalescing systems)
	$k_l a = 2.0 \cdot 10^{-3} (P_g/V)^{0.7} (v_s)^{0.2}$	(non-coalescing systems)
	(see comment at the end of this "table")	
Mixing/ circulation	$\left.\begin{array}{l} t_{circ} = V/\Phi_{circ} \\ \Phi_{circ} = 2\Phi_p \\ \Phi_p = 1.3\, ND^3 \end{array}\right\}\ Nt_{circ} = \dfrac{V}{2.6\, D^3}$ $(Re > 5 \cdot 10^3)$	
	$t_m = 4\, t_{circ}$	
Shear	$\sim \eta N$	(laminar flow)
	$\sim \varrho (ND)^2$	(turbulent flow)
	$\sim N/D$	(WANG [24.43])
	$\sim \varrho (ND)^2 N \sim P/V$	(turb. flow and fatigue)
	(for more details on shear see comments below)	

Some common values of process variables at production scale are:

- P/V 2–4 kW/m^3
- v_{tip} >3 m/s
- Q/ND^3 0.1–0.15

Bubble column (see HEYNEN [24.62])

Power input	$P = v_s \varrho_1 g V$	($H < 2$ m)
	$P = \dfrac{\Phi_{m,gas} R T}{M_{gas}} \ln\left(\dfrac{P_1}{P_2}\right)$	($H > 2$ m)
Mass transfer	$k_l a = 0.32\, v_s^{0.7}$	
	$\left.\begin{array}{l} 0.55\% \text{ decrease of oxygen} \\ \text{concentration in air} \\ \text{per meter column} \end{array}\right\}$	(for coalescing systems)
	For non-coalescing systems $k_l a$ depends entirely on the initial bubble size.	
Mixing	$\mathbb{D}_e^l = 0.35\, (v_s g T^4)^{1/3}$	
	$t_m \sim H^2/\mathbb{D}_e^l$	
Shear	In bubble columns shear is usually not important.	

Results of BARTHOLOMEW [24.50] show that for increasing vessel size a and b approach the values for coalescing systems (see Table 24.2).

For the influence of scaling-up on shear the situation is not very clear. An increased shear can have negative effects (damage to microorganisms) as well as positive effects (decrease of floc size). Under conditions of laminar flow the shear stress is proportional to ηN. At production scale laminar flow is not very likely to occur, but at laboratory scale, when working with viscous broth, laminar flow conditions are more common than most people realize. Scaling-up then results in a change of flow regime which is rather unpleasant if one is not prepared for it. Under conditions of turbulent flow the shear stress is proportional to $\varrho (ND)^2$. This results in constant values of ND as a scale-up criterion. If not only the turbulent shear stress is important but also the frequency of exposure to this shear stress (due to fatigue phenomena or time for flocs to grow) one should expect $\varrho (ND)^2 N$ to be important. This has the same effect as keeping P/V

constant. WANG and FEWKES [24.43] suggest stirrer shear divided by stirrer flow ($N^2D^2/ND^3 = N/D$) as a criterion in situations where mycelia tend to agglomerate. The situation appears rather confusing. Fortunately many microorganisms are not very sensitive for the condition of shear in fermenters. An exception, however, are those microorganisms that tend to form agglomerates (REUSS et al. [24.73]).

The relations for stirred vessels and bubble columns as given above are very useful for the calculation of the order of magnitude of the different phenomena involved in the performance of the bioreactor. They can be used for the design of scale-down experiments to find an approximation of the conditions the microorganisms will meet on a production scale.

The scale-up criterion generally used is a similar P/V in order to obtain similar $k_l a$ values for model and prototype (VAN'T RIET [24.71]). As one can see, however, this scale-up criterion results in a considerable increase of t_m (factor 4.5) and turbulent shear (factor 4.6). If mixing can be expected to cause complications on the production scale [compare t_m with the time constant for substrate consumption t_c ($=C/r$)] then extra measures have to be taken to solve this problem (BRYANT [24.74]) (for example, by introducing the substrate at a number of different points in the bioreactor; see SENIOR [24.75]). If shear sensitive microorganisms are used, an increase of the turbulent shear of the order of magnitude mentioned could have a rather negative effect on the productivity at production scale. If the ex-

Table 24.3. Different Scale-up Criteria and Their Consequences

Scale-up Criterion	Value at 10 m³ Scale (V_M = 10 liter)					
	P	P/V	N (or t_m^{-1})	ND	Re	N/D
Equal P/V	10^3	1	0.22	2.15	21.5	0.022
Equal N (or t_m^{-1})	10^5	10^2	1	10	10^2	0.1
Equal tip speed	10^2	0.1	0.1	1	10	10^{-2}
Equal Re number	0.1	10^{-4}	10^{-2}	0.1	1	10^{-3}
Equal shear to flow ratio	10^8	10^5	10	10^2	10^3	1

They also allow for the calculation of process variables (P/V or t_m for example) once a particular scale-up criterion has been chosen (for example, equal tip-speed of the impeller). One can of course also choose a constant t_m as a scale-up criterion and observe the consequences for tip-speed. The results of such an exercise are presented in Table 24.3. In this table the values in every row give, for a particular scale-up criterion, the ratio of the values for prototype and model of the variables mentioned at the top of the table. The calculations have been set up for geometrically similar systems with V_M = 10 liter and V_P = 10 m³, which results in a linear scale-up factor of 10. As one can see, different scale-up criteria result in entirely different process conditions on a production scale.

periments at M-scale had been performed according to the principle of scaling-down, a much more realistic picture of the behavior of the microorganism would have been obtained from the experiments.

A few other remarks should be added. First of all, something should be said about the change of the aeration rate during scaling-up. One of the rules is to keep the vol-

Table 24.4. Consequences of Scale-up Rules for Aeration

Aeration Rule	$(v_s)_P/(v_s)_M$	Q_P/Q_M
Q/ND^3 constant	$(D_P/D_M)^{1/3}$	$(D_P/D_M)^{7/3}$
v_s constant	1	$(D_P/D_M)^2$
vvm constant	D_P/D_M	$(D_P/D_M)^3$

ume of air per volume of fermenter per minute (vvm, a quantity close to one) constant. Other rules are: a constant value for the ratio of the air flow rate (Q) and the pumping capacity (ND^3) or a constant value for the superficial gas velocity v_s. These rules result in values for the aeration rate that differ widely as is shown in Table 24.4. For the calculation of v_s from the rule $Q/ND^3 =$ const, it has been assumed that P/V is constant.

As can be seen from Table 24.4, Q_P/Q_M always increases during scaling-up. The same holds for $(v_s)_P/(v_s)_M$ except when v_s is constant, of course. The use of $v_s =$ const has the advantage that the holdup does not change during scaling-up. The use of a constant vvm during scale-up is usually rather unsatisfactory.

Some additional remarks should also be made about the influence of viscous systems on performance during scaling-up. For viscosities of 1 Pa·s (1000 cP), not an unusual figure for a mycelium or a xanthan broth, the Reynolds number at the small scale (1 liter, $D = 3.5 \cdot 10^{-2}$ m, $N = 10$ s^{-1}) is $Re = 12.25$. This means that the flow in the reactor is laminar. Therefore, the reactor is not very well mixed. For the large scale (100 m^3, $D = 2$ m, $N = 2$ s^{-1}) $Re = 8 \cdot 10^3$, which results in turbulent conditions and, therefore, much better mixing. For extreme, but not unusual, situations like this the prediction of large-scale behavior from small-scale experiments is impossible. Therefore, model experiments for viscous systems should always be set up with great care.

24.3.2.5 Trial and Error

In Sect. 24.3.1 it has already been mentioned that for the longest time of the existence of the human race, trial and error was the only method available for process improvement. This is no longer true. The 'method' is still in use for optimizing processes gradually, but for scaling-up it is of no use.

From the aforementioned a few conclusions can be drawn:

1. Design of bioreactors 'from first principles' (the fundamental method) is not possible.
2. The semi-fundamental method and the rules of thumb method give right orders of magnitude only.
3. From conclusions 1 and 2 it is clear that there is a need for other methods in order to describe more accurately the influence of a change of scale on the performance of microorganisms.

It will be shown that dimensional analysis opens a number of possibilities. Finally, scaling-down as a way of thinking about problems concerned with changes of scale, combined with elements of this chapter and the ones about dimensional analysis, will be shown to be the best possibility for the time being.

24.3.3 Dimensional Analysis

Dimensional analysis (DA) has long been applied for the design of scale-up experiments. Originally it was mainly used for flow problems, later it was extended to other problems, also in process engineering. JOHNSTONE and THRING [24.76] presented a survey of the application of the technique in process technology. Many books on the subject have been published, among others by PAWLOWSKI [24.77], ZLOKARNIK [24.78], and BECKER [24.79]. The method has been used successfully in many situations, yet it has its limitations as mentioned earlier. In this section the pros and cons of the method will be given.

24.3.3.1 Mechanistic Background of DA

The background of the DA can be shown to be rather mechanistic. When the momentum, mass, and heat balances, as well as their boundary- and initial conditions, are written in a dimensionless form, the dimensionless numbers, which are the backbone of the method, appear automatically.

As an example, the mass balance for a plug flow with dispersion system is given by:

$$-v\frac{dC}{dx} + \mathbb{D}_e\frac{d^2C}{dx^2} - r = 0, \qquad (24.16)$$

$$\left.\begin{array}{l} C|_{x=0} = C_0 \\ \left.\dfrac{dC}{dx}\right|_{x=L} = 0 \end{array}\right\} \text{boundary conditions.}$$

Assuming a first order reaction ($r = kC$), the differential equation is brought into a dimensionless form by dividing C and x by proper constants of equal dimensions (C_0 and L, for example) followed by dividing all coefficients in the equation by one of them:

$$-\frac{vC_0 d(C/C_0)}{L d(x/L)} + \frac{\mathbb{D}_e C_0 d^2(C/C_0)}{L^2 d(x/L)^2} - kC_0(C/C_0) \qquad (24.17)$$

or:

$$-\frac{vL}{\mathbb{D}_e}\frac{d(C/C_0)}{d(x/L)} + \frac{d^2(C/C_0)}{d(x/L)^2} - \frac{kL^2}{\mathbb{D}_e}(C/C_0) \qquad (24.18)$$

$$C|_{x=0}/C_0 = 1 \quad \frac{d(C_{x=L}/C_0)}{d(x/L)} = 0 \qquad (24.19)$$

(boundary conditions)

Generally, the new dimensionless variables (C/C_0 and x/L) are represented by Greek letters ($\xi = C/C_0$ and $\kappa = x/L$), resulting in:

$$-\frac{vL}{\mathbb{D}_e}\frac{d\xi}{d\kappa} + \frac{d^2\xi}{d\kappa^2} - \frac{kL^2}{\mathbb{D}_e}\xi = 0 \qquad (24.20)$$

$$\xi|_{\kappa=0} = 1 \quad \left.\frac{d\xi}{d\kappa}\right|_{\kappa=1} = 0. \qquad (24.21)$$

From observation of the last equations it will be clear that the solution of these equations will be of the general form:

$$\xi = f\left(\frac{vL}{\mathbb{D}_e}, \frac{kL^2}{\mathbb{D}_e}, \kappa\right). \qquad (24.22)$$

If, therefore, M and P are geometrically similar and if the dimensionless groups ($vL/\mathbb{D}_e$ and $kL^2/\mathbb{D}_e$) and the dimensionless boundary conditions are the same for M and P, then the solution of the balance equation in its dimensionless form is also the same for M and P. This means that $\xi(\kappa)$ is equal for M and P. In other words:

Equal dimensionless groups of parameters for M and P result in geometrically similar dimensionless profiles (of velocity, concentration, and temperature) for M and P, provided M and P have geometrically similar shapes.

The non-dimensionless profiles can be quite different. This is shown in Fig. 24.11 where profiles are given for $\xi(\kappa)$ and for $C(x)$ both for M and P.

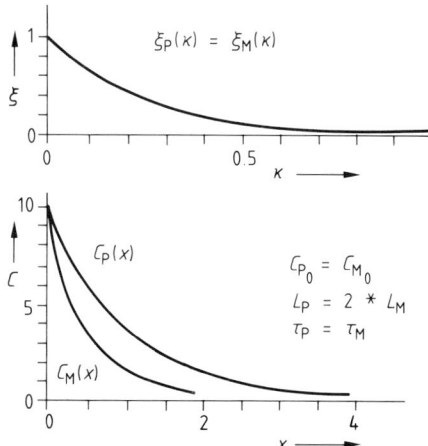

Figure 24.11. Concentration profiles.

It thus can be concluded that:

1. Geometrically similar profiles in M and P are only possible if the geometrical shapes of M and P are similar as well. Furthermore, it is clear from observation of the mass balance that geometrically similar concentration profiles are only possible if the flow profiles are similar as well.

2. The dimensionless groups of parameters can be regarded as ratios of time constants:

$$\frac{vL}{\mathbb{D}_e} = \frac{v}{L} * \frac{L^2}{\mathbb{D}_e} = \frac{1}{t_f} * t_{\mathbb{D}_e} \qquad (24.23)$$

and

$$\frac{kL^2}{\mathbb{D}_e} = \frac{1}{t_c} * t_{\mathbb{D}_e}. \qquad (24.24)$$

Therefore, the constancy of these groups for M and P results in equal ratios of the characteristic times (and, therefore, equal relative importance of the different mechanisms involved like flow, dispersion, and conversion).

Table 24.5. Dimensionless Numbers

Momentum	Reynolds	inertia forces / viscous forces	$Re \equiv \frac{\varrho v D}{\eta}$	$\left(Re^{b)} \equiv \frac{\varrho N D^2}{\eta}\right)$
	Froude	inertia forces / gravitational forces	$Fr \equiv \frac{v^2}{gL}$	$\left(Fr^{b)} \equiv \frac{N^2 D}{g}\right)$
	Weber	inertia forces / surface forces	$We \equiv \frac{\varrho v^2 d_p}{\sigma}$	$\left(We^{b)} \equiv \frac{\varrho N^2 D^2 d}{\sigma}\right)$
	Power number	total dissip. power / power due to inertia	$P_0 \equiv \frac{P}{\varrho N^3 D^5}$	
Mass	Sherwood	total mass transfer / mass transfer by diffusion	$Sh \equiv \frac{kD}{\mathbb{D}}$	
	Schmidt	(hydrodynamic boundary layer / mass transfer boundary layer)³	$Sc \equiv \frac{v}{\mathbb{D}}$	
	Péclet	(mass transfer by convection / mass transfer by diffusion)	$Pe \equiv \frac{vL}{\mathbb{D}}$	
	Fourier	(process time / diffusion time)	$Fo \equiv \frac{\mathbb{D} t}{D^2}$	
	Biot	(external mass transfer / internal mass transfer)	$Bi \equiv \frac{k d_p}{\mathbb{D}}$ [a]	
Heat	Nusselt	(total heat transfer / heat transfer by conduction)	$Nu \equiv \frac{\alpha D}{\lambda}$	
	Prandtl	(hydrodynamic boundary layer / thermal boundary layer)³	$Pr \equiv \frac{v}{a}$	
	Péclet$_h$	(heat transfer by convection / heat transfer by conduction)	$Pe_h \equiv \frac{vL}{a}$ $[a = \lambda/(\varrho c_p)]$	
	Fourier$_h$	(process time / heat conduction time)	$Fo_h \equiv \frac{at}{D^2}$	
	Biot$_h$	(external heat transfer / internal heat transfer)	$Bi_h \equiv \frac{\alpha d_p}{\lambda}$ [a]	
Chem. React.	Damköhler I	(chemical reaction rate / mass transport by convection)	$Da_I \equiv \frac{rL}{vC}$	
	Damköhler II	(chemical reaction rate / mass transport by diffusion)	$Da_{II} \equiv \frac{rL^2}{\mathbb{D} C}$	
	Thiele Modulus	(chemical reaction rate in particle / diffusion in particle)^½	$\Phi = R\sqrt{\frac{r}{\mathbb{D} C}}$	

[a] indicating that D or λ are related to the dispersed phase; [b] for stirred vessels

3. The solution of the balance equations for heat, mass, or momentum transfer can always be presented in the form:

$$\Pi_1 = f(\Pi_2, \Pi_3, \ldots) \qquad (24.25)$$

where Π_1, Π_2, etc. are the dimensionless groups. Representations of the form:

$$\Pi_1 = C \Pi_2^\alpha \Pi_3^\beta \ldots \qquad (24.26)$$

as often found in literature, are not a logical consequence of DA, although in a number of situations they work quite well.

tum, mass, and heat balances are presented in Table 24.5.

Some dimensionless groups (*We* and *Bi*) typically refer to two-phase systems. Some groups appear for mass transfer as well as for heat transfer. All of them can be interpreted as a ratio of time constants.

The groups used in the mass balance [Eq. (24.20)] were $\dfrac{vL}{\mathbb{D}_e} \equiv Pe$ and $\dfrac{rL^2}{\mathbb{D}_e} \equiv Da_{II}$. In practice it is often quite difficult to obtain the dimensionless groups. There are many situations where a micro-balance equation

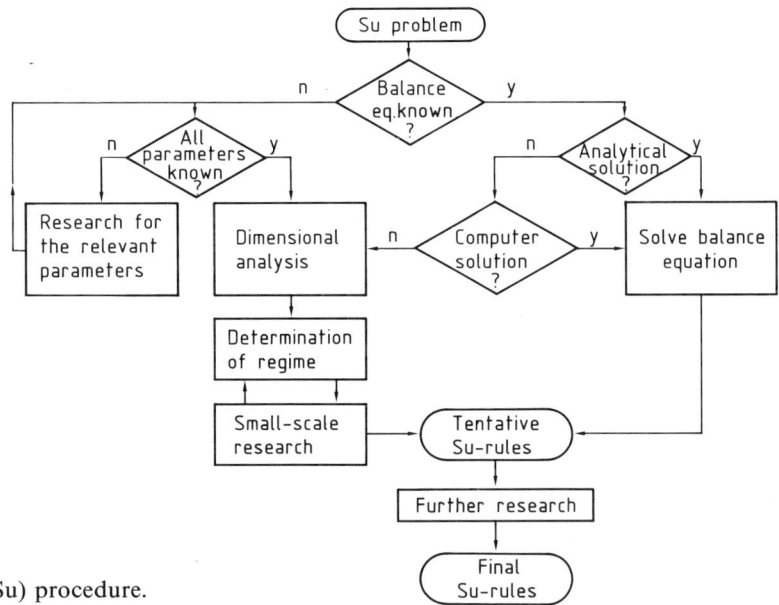

Figure 24.12. Scaling-up (Su) procedure.

4. The fact that ξ is only a function of three dimensionless quantities:

$$\xi = f\left(\frac{vL}{\mathbb{D}_e}, \frac{kL^2}{\mathbb{D}_e}, \kappa\right) \qquad (24.27)$$

means that, to check the influence of v, L, $\mathbb{D}_e$, and k on $\xi(\kappa)$ experimentally, only the two dimensionless numbers containing these parameters have to be varied, and not the four parameters separately.

The most important dimensionless groups, resulting from DA of the momen-

cannot be set up. Fig. 24.12 gives an idea of how the DA procedure is connected with the procedure of setting up balances.

24.3.3.2 The Use of Dimensionless Groups for Scaling-up

Once the dimensionless groups have been obtained their use for the proper set-up of scaling-up or -down experiments is, at least in principle, rather simple: equal values of these groups for M and P. In practice, the situation is more complicated. For flow problems in homogeneous systems the method works well. For situations where mass and/or heat transfer problems play a role as well, and for flow in heterogeneous systems the method becomes more complicated. As an example, the flow of a homogeneous system in a stirred vessel will be presented.

In a stirred vessel with a homogeneous fluid, geometrically similar flow patterns ask for constant values of Re and Fr for M and P:

$$Re_M = Re_P : \left(\frac{\varrho N D^2}{\eta}\right)_M = \left(\frac{\varrho N D^2}{\eta}\right)_P \quad (24.28)$$

$$Fr_M = Fr_P : \left(\frac{N^2 D}{g}\right)_M = \left(\frac{N^2 D}{g}\right)_P . \quad (24.29)$$

If the same liquid is used for M and P, the conditions $ND^2 = \text{const}$ and $N^2 D = \text{const}$ cannot be matched at different scales. One can, however, choose different liquid systems for M and P in order to match the equality of Re and Fr. This is the case if:

$$\frac{\varrho_M}{\eta_M} = \frac{\varrho_P}{\eta_P} * \left(\frac{D_P}{D_M}\right)^{3/2}, \quad (24.30)$$

which results in a much lower value of η_M compared with η_P if $\varrho_M \sim \varrho_P$. When water is the fluid at P-scale this is not easy to realize. Fortunately for baffled systems the flow is almost independent of Fr so that the only requirement for geometrically similar flow patterns at M and P is: $Re_M = Re_P$. For flow patterns the situation is clear, however, for bioreactors one is also interested in mixing times (t_m) and power per unit volume (P/V), because P/V largely determines the value of $k_l a$ as previously shown. It has been explained in Sect. 24.3.2.4 that equal Re numbers for M and P result in extremely low values for P/V and very high values of t_m at P-scale. This has disastrous effects on the behavior of the bioreactor on a production scale. Fortunately, the bulk-flow pattern in a stirred vessel is virtually independent of Re as long as $Re > 10^4$. This fact usually supplies sufficient possibilities for proper experimental set-up. For example, equal P/V as a scale-up criterion results for $D_P/D_M = 10$ in $Re_P = 21.5 \cdot Re_M$ (see Table 24.3). However, t_m is still too large. If $Re_M > 10^4$, then the flow pattern in the prototype is similar to that of the model. If, however, $Re_M < 10^3$ and $Re_P > 10^4$ complications arise (viscous systems).

As we have seen for the simple situation mentioned here, the use of the DA technique is well possible although some pitfalls exist. For systems with (bio)chemical reactions and/or heat transfer the situation is more complicated, as already mentioned. The number of dimensionless groups to be kept constant is then about 4 or 5. This is generally not possible. One then has to find out which mechanisms control the system at M and P (regime analysis).

If, for example, at M- and P-scale the mixing time is much shorter than the conversion time the flow profile is obviously not very critical. The system can be considered to be under a chemical regime. One then only has to keep those dimensionless numbers constant that contain the time constants for the relevant conversion processes. Regime analysis will be dealt with in Sect. 24.3.4.

24.3.3.3 Heterogeneous Systems

For heterogeneous systems another problem arises. During scaling-up one is usually not interested in geometric similarity for the dispersed particles present in the heterogeneous system. Generally, the size of flocs, pellets, air bubbles, droplets, and crystals should not grow proportionally with the scale of the apparatus but should remain constant during changes of scale. This is possible, but it is a violation of the geometric similarity. The size of particles in dispersions is determined by two opposing processes: dispersion and coalescence. Both these processes are a function of flow properties and surface properties. If the surface properties do not change during scaling-up, the flow properties determine the particle size. Particularly the scale of the energy dissipating eddies is important for the dispersion phenomena (SHINNAR and CHURCH [24.80]; KOLMOGOROV [24.81]). This size is determined by the energy input per unit volume (P/V). Thus, if P/V is kept constant during scaling-up, the size of the eddies does not change. If, however, Re is kept constant the size of the eddies changes proportionally with scale. Therefore, in the literature $P/V=$ const is the general scale-up rule for dispersion processes (VAN'T RIET [24.71]). It should be mentioned here, however, that P/V is not a dimensionless group. Some authors, among others ZLOKARNIK [24.82], use P/V as a parameter in DA analysis.

The theory behind the principle of constant P/V values during scaling-up is not used correctly when applied to a stirred vessel. The theory (KOLMOGOROV [24.81]) assumes isotropic turbulence which is more or less the case for bubble columns but definitely not for stirred vessel. For experiments at one scale, particle sizes correlate well with P/V, see among many others YOSHIDA [24.83]; CALDERBANK [24.84]; SKELLAND and SEKSARIA [24.85]. For changing scales the time available for particles to coalesce increases with increasing scale. We have already seen that this causes the power of P/V in the correlation for $k_l a$ to change during scaling-up. More mechanistic models are needed here (see OOSTERHUIS and KOSSEN [24.52]; KOSSEN [24.13]).

To keep the particles in a heterogeneous system (flocs, pellets) or to eliminate them finally (air bubbles), the linear velocities in M and P must be the same if the particle size in M and P is the same. The advantage of $P/V=$ const as a scale-up criterion is that the linear velocities change only moderately with scale:

$$v_P/v_M = (D_P/D_M)^{1/3}. \quad (24.31)$$

As mentioned earlier, however, t_m increases considerably. It is important to realize that for non-coalescing systems (or better formulated systems with limited coalescing properties) the initial particle size is very important. This initial particle size is determined largely by the flow conditions near the stirrer or the sparger. For strongly coalescing systems the initial particle size has a limited influence on the average particle size which is to a large extent determined by the equilibrium size of the particle and the flow conditions that enhance coalescence. These flow conditions are mainly determined by the flow at some distance from the stirrer. The equilibrium size of the particles (bubbles or droplets) can be estimated with a simple balance of forces:

$$\tfrac{1}{4}\pi d_p^2 \varrho v_p^2 = \pi d_p \sigma, \quad (24.32)$$

which results in:

$$d_p = \frac{4\sigma}{\varrho v_p^2}. \quad (24.33)$$

For air bubbles in water $v_p = 0.25$ m s^{-1} over a large range of diameters, $\varrho = 10^3$ kg m^{-3}, and $\sigma = 72 \cdot 10^{-3}$ N m^{-1}. This results in $d_p = 4.6 \cdot 10^{-3}$ m, a value close to the one observed (5–6 mm) (see HEYNEN and VAN'T RIET [24.62]).

24.3.3.4 The Generation of the Dimensionless Groups

There are many ways to generate the dimensionless groups of parameters. One method has already been presented, starting with the micro-balances. Often it is not possible to set up a micro-balance. The first and main problem that has to be solved then is to determine the relevant parameters. This inventory is facilitated if different groups of parameters are distinguished:

- geometrical parameters (D, H, d_p)
- fluid/solid/gas properties (ϱ, η, σ)
- process variables (N, P, v)
- dimensional constants (g, R)

The inventory is a matter of experience and physical intuition. If too many parameters are collected, including non-relevant ones, too many groups are generated that have to be kept at a constant value for M and P. If a relevant parameter is overlooked, an important group will be missing, resulting in an incomplete description of the system.

In practice, non-relevant groups can be eliminated with regime analysis. Missing important groups usually are detected during experimentation when it appears not to be possible to describe the behavior of the system with the existing groups.

Once the relevant parameters have been obtained, their collection in different dimensionless groups is a well established standard procedure. PAWLOWSKI [24.77] gives an extensive treatment of the theories behind this procedure. The number of groups can be calculated from the number of parameters (n) and the number of primary quantities (m). The primary quantities used often are mass, length, time, and temperature although other choices are possible. The Buckingham theory states that the complete (smallest) set of dimensionless groups that describes the system is $n-m$.

For the formation of the groups several methods exist:

- Gauss-Jordan reduction method,
- Rayleigh's method,
- Gukhman's method.

The Gauss-Jordan reduction method and Rayleigh's method will be demonstrated next, and some examples will be given. A variation of the method of GUKHMAN has been published recently by QURAISHI and FAHIDY [24.86].

A point to mention here is the way the groups are presented. As an example (taken from the following section) the flow of a fluid jet through a bed of microbial aggregates is used. One of the dimensionless groups appears to be a kind of Froude number: $Fr \equiv v_1^2/(gH)$. This group does not reveal its physical meaning until written in a different way, which is also correct from a point of view of DA because multiplication of a dimensionless group with another dimensionless group is always allowed:

$$Fr \equiv \frac{\varrho_1 v_1^2}{\Delta \varrho g H}. \qquad (24.34)$$

Now the group is equal to the ratio of the pressure exerted by the jet $(\varrho_1 v_1^2)$ and the pressure needed to lift the bed of aggregates $(\Delta \varrho g H)$. It follows that, if $Fr \gg 1$, this means that $\varrho_1 v_1^2 \gg \Delta \varrho g H$ and therefore the jet can break through the bed.

24.3.3.5 Some Examples

Two examples will be presented here. The first one stems from a publication of IRVING et al. [24.87]:

In a large sludge digester (Fig. 24.13), microbial flocs have to be kept in suspension due to a circulating flow induced by gas lift pumps. A gas lift pump is a kind of a draught tube placed in the digester with a gas sparger at the bottom.

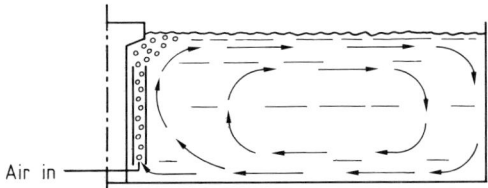

Figure 24.13. Sludge digester.

It became clear from some preliminary runs with the prototype that the pumps were not capable of maintaining the required concentration of solids. Because experiments on the P-scale are very expensive it was decided to scale the system down and to look for a solution at the M-scale.

The relevant parameters are:

- characteristic dimension H (m)
- liquid velocity in the tank v_1 (m s^{-1})
- density of the liquid ϱ_1 (kg m^{-3})
- density difference between the liquid in the tank and the gas/liquid mixture in the gas lift pump $\Delta \varrho$ (kg m^{-3})
- acceleration due to gravity g (m s^{-2})

According to the Buckingham theory five parameters and three dimensional quantities should result in two dimensionless groups. The method of RAYLEIGH will be used to obtain the groups. It is assumed that the solution of the problem can be written as:

$$v_1 = H^\alpha \varrho_1^\beta \Delta\varrho^\gamma g^\delta. \tag{24.35}$$

By the use of dimensions this relation can be written as:

$$\frac{L}{t} = L^\alpha \left(\frac{M}{L^3}\right)^\beta \left(\frac{M}{L^3}\right)^\gamma \left(\frac{L}{t^2}\right)^\delta. \tag{24.36}$$

Because every physical meaningful relation ought to be homogeneous in its dimensions we come to the following conclusions for the dimensions M, L, and t:

M: $0 = \beta + \gamma$
L: $1 = \alpha - 3\beta - 3\gamma + \delta$
t: $-1 = -2\delta$.

We now have four unknown powers ($\alpha, \beta, \gamma, \delta$). However, we have three equations, therefore three unknown powers can be eliminated:

The first equation thus leads to: $\beta = -\gamma$
The third equation yields: $\delta = \frac{1}{2}$
The second equation leads to the conclusion that: $\alpha = \frac{1}{2}$.

The powers in the original equation can now be rewritten as:

$$v_1 = H^{\frac{1}{2}} \varrho_1^{-\gamma} \Delta\varrho^\gamma g^{\frac{1}{2}}. \tag{24.37}$$

Comprising equal powers gives:

$$\frac{v_1}{(gH)^{\frac{1}{2}}} = \left(\frac{\Delta\varrho}{\varrho_1}\right)^\gamma, \tag{24.38}$$

or more generally:

$$\frac{v_1}{(gH)^{\frac{1}{2}}} = f\left(\frac{\Delta\varrho}{\varrho_1}\right). \tag{24.39}$$

Groups of dimensionless variables remain equally useful when multiplied with each other. To obtain groups with a clear physical meaning we therefore take the square of the first group and multiply this with the reverse of the second group. The result is:

$$\frac{\varrho_1 v_1^2}{\Delta\varrho g H} = f\left(\frac{\Delta\varrho}{\varrho_1}\right). \tag{24.40}$$

The physical interpretation of the first group is: resistance due to the flow of the circulating liquid relative to the buoyancy of the gas/liquid mixture in the gas lift pump.

If both dimensionless groups are kept at a constant value the result is:

$$v_{1M} = v_{1P}\left(\frac{H_M}{H_P}\right)^{\frac{1}{2}}, \tag{24.41}$$

$$\left(\frac{\Delta\varrho}{\varrho_1}\right)_M = \left(\frac{\Delta\varrho}{\varrho_1}\right)_P. \tag{24.42}$$

This $\Delta\varrho$ in the gas lift pump is determined by the gas holdup in the pump. This gas holdup is proportional to:

$$\frac{v_s}{v_b} = \frac{Q_{gas}}{H^2 v_b}. \tag{24.43}$$

Because the rising velocity of air bubbles (v_b) is roughly constant for bubble diameters larger than 2 mm, this means that, in order to keep $\Delta\varrho$ constant, Q_{gas} must be proportional to H^2 or:

$$Q_{gas\,M} = Q_{gas\,P}\left(\frac{H_M}{H_P}\right)^2. \tag{24.44}$$

The model was built according to these scaling-down criteria. Improvements were found by experimentation with the model. These improvements appeared to be successful after application in the prototype.

The second example describes a similar approach as the first one. Now the Gauss-Jordan reduction method will be used to obtain the dimensionless groups. The problem was worked out by the authors (BOLLE et al. [24.88]). The performance of a large waste water installation appeared to be less than expected. Flow problems (short circuiting) were assumed to be the cause. RTD measurements on the P-scale confirmed this assumption. To solve the problem a model was built with a linear scale ratio of 1/20. Fig. 24.14 schematically shows a relevant part of the system.

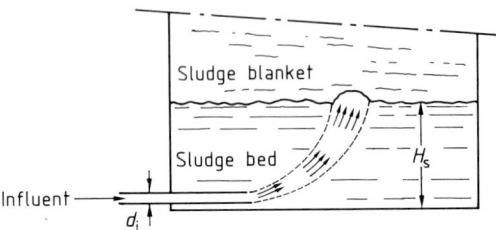

Figure 24.14. Schematic picture of the flow system.

The problem could be localized as short circuiting of the sludge bed. The relevant parameters selected were:

- liquid: v_1, ϱ_1, η_1
- bed: $d_p, \Delta\varrho_p, \tau_0$
- gas: v_s
- reactor: H_s, d_i
- general: g, t

The dimensionless groups will be obtained by writing the variables and the dimensions in one matrix. The left part of this matrix has to be transformed into a diagonal matrix. The groups are then easily obtained, as will be shown. The method is described in detail by PAWLOWSKI [24.77].

	ϱ_1	H_s	t	g	d_i	v_s	d_p	$\Delta\varrho_p$	τ_0	v_1	η_1
M	1	0	0	0	0	0	0	1	1	0	1
L	−3	1	0	1	1	1	1	−3	−1	1	−1
t	0	0	1	−2	0	−1	0	0	−2	−1	−1

Addition of three times the first row to the second row results in:

	ϱ_1	H_s	t	g	d_i	v_s	d_p	$\Delta\varrho_p$	τ_0	v_1	η_1
M	1	0	0	0	0	0	0	1	1	0	1
L	0	1	0	1	1	1	1	0	2	1	2
t	0	0	1	−2	0	−1	0	0	−2	−1	−1

The dimensionless groups are now obtained as follows: g is divided by H_s and multiplied by t^2. ϱ_1 does not appear in this group because at the place of the column of g that corresponds with ϱ_1 in the left matrix, a zero is found. The first dimensionless group now is $\Pi_1 \equiv \dfrac{g t^2}{H_s}$.

In the same way the other groups are found:

$$\Pi_1 \equiv \frac{gt^2}{H_s} \quad \Pi_2 \equiv \frac{d_i}{H_s} \quad \Pi_3 \equiv \frac{v_s t}{H_s} \quad \Pi_4 \equiv \frac{d_p}{H_s}$$

$$\Pi_5 \equiv \frac{\Delta\varrho_p}{\varrho_1} \quad \Pi_6 \equiv \frac{\tau_0 t^2}{\varrho_1 H_s^2} \quad \Pi_7 \equiv \frac{v_1 t}{H_s} \quad \Pi_8 \equiv \frac{\eta_1 t}{\varrho_1 H_s^2}$$

For the sake of better physical interpretation the following rearrangements are made:

Π_7, Π_8, and Π_9 are recombined into

$$\Pi_8' \equiv \frac{\varrho v_1 d_i}{\eta_1}$$

Π_1, Π_5, and Π_7 are recombined into

$$\Pi_1' \equiv \frac{\varrho_1 v_1^2}{\Delta\varrho_s g H_s}.$$

Table 24.6 shows the final set of dimensionless numbers, together with their values at P- and M-scale.

As presently one was not especially interested in the effect of time, the groups Π_3 and Π_7 were neglected. The group Π_6 is so small that it is very unlikely that this group is of any importance (the turbulent shear stress is much larger than the yield stress of the bed). Group Π_6 was therefore also neglected. Π_2 and Π_4 were kept constant (geometric similarity). The most important group was Π_1. The physical interpretation of this group has been mentioned in a previous section. As a consequence of keeping Π_1 constant, the value of the Re number decreased considerably (Π_8). Because it was clearly observed that despite this low Re_M number the jet at M-scale was still turbulent, this low Re number was no limitation for the procedure.

Table 24.6. Dimensionless Numbers

	P	M
$\Pi_1 \equiv \dfrac{\varrho_1 v_1^2}{\Delta \varrho_p g H_s}$	2.88	2.88
$\Pi_2 \equiv \dfrac{d_i}{H_s}$	$1.9 \cdot 10^{-2}$	$1.9 \cdot 10^{-2}$
$\Pi_3 \equiv \dfrac{v_s t}{H_s}$	$9.1 \cdot 10^{-2}$	–
$\Pi_4 \equiv \dfrac{d_p}{H_s}$	10^{-3}	10^{-3}
$\Pi_5 \equiv \dfrac{\Delta \varrho_p}{\varrho_1}$	10^{-1}	10^{-1}
$\Pi_6 \equiv \dfrac{\tau_0}{\varrho_1 v_1^2}$	$2.9 \cdot 10^{-3}$	–
$\Pi_7 \equiv \dfrac{v_1 t}{H_s}$	$1.1 \cdot 10^{-3}$	–
$\Pi_8 \equiv \dfrac{\varrho_1 v_1 d_i}{\eta_1}$	$9.1 \cdot 10^4$	1029

Improvements of the flow system were performed in the model for $\Pi_1 = $ const. The solutions proved to be successful when translated backwards to the prototype.

24.3.4 Regime Analysis

24.3.4.1 Introduction

The word 'regime' is used for the dominance of a particular mechanism in the performance of a system. A kinetic (or chemical) regime for a bioreactor, for example, means that the performance of the bioreactor is dominated by the kinetic processes (growth and product formation), and not by transport processes.

A regime can be pure (only one mechanism dominates) or mixed (two or more mechanisms have a comparable influence on the system).

Regime analysis is important when DA is used for the design of scale-up experiments because, as we have seen, not all dimensionless groups can be kept at a constant value during scaling-up. The importance of regime analysis is, however, by no means restricted to DA. It is an important activity (or at least it should be one) in the set-up of every experiment as it gives an answer to the question: "Do I measure the mechanism I want to, or do other mechanisms influence my results as well?". Regime analysis is closely related to the parameter sensitivity analysis mentioned in Sect. 24.2.5.

Three important questions have to be answered by regime analysis:

- Is one regime rate determining?
- Which regime is this?
- Does this regime change during a change of scale?

24.3.4.2 Methods for Regime Analysis

Regime analysis is possible in a number of ways. They can be subdivided into experimental and theoretical methods. Table 24.7 gives a survey.

Table 24.7. Methods for Regime Analysis

1. Experimental methods	change of velocity change of concentration change of particle size change of temperature
2. Theoretical methods	analytical methods: – time constants – dimensionless numbers numerical methods: parameter sensitivity analysis

Experimental Methods

The basis of the experimental method of regime analysis is the variation of a quantity that has a pronounced influence on one of the mechanisms involved in the behavior of the system.

Velocity is a quantity that can generally be varied easily, at least on the laboratory scale (increase of stirrer speed, for example). If this results in an increased conversion rate, then transport processes are important, although not necessarily the sole rate determining process. One can think of transport processes like bulk mixing, transfer of oxygen from air to liquid, transport of nutrients to particles containing immobilized microorganisms, decrease of floc size, etc.

Concentration changes can also be applied easily. If an increase of the glucose concentration has no influence on the conversion rate, for example, then a different factor will be rate determining, unless the concentration is already at its saturation value ($C \gg K_s$). This last possibility also indicates a different factor being rate determining, but this could be an enzyme concentration in the microorganism and not a transport process. A well known example of the influence of the concentration on the conversion is the influence of a change in oxygen concentration on the OUR (Oxygen Uptake Rate) of a mycelial broth as mentioned among others by REUSS [24.73].

At low oxygen concentrations transport resistance is important, at high oxygen concentrations the uptake rate is determined by kinetics only (pure kinetic regime). The influence of impeller speed is obvious (influence on floc size). This example also shows that the regime is not only dependent on scale but also on the operating conditions. This is a general rule. Comparison of the operating conditions for M and P therefore is always necessary.

Particle size can be varied by using sieve fractions, for example. A change in the conversion rate due to a change in particle size is an indication of the influence of internal diffusion in the particles on the conversion process.

Temperature changes generally have a much larger influence on kinetics than on transport processes. The amount of increase of the conversion rate with temperature therefore can be used as an indication of the rate determining process.

The experimental method gives a clear answer if the variation of the quantity has no influence on the conversion process as a whole. If there is an influence the regime can still be either mixed or pure. Combinations of experimental techniques (change in velocity followed by a change in concentration) can solve this problem. It is also possible to do some simple calculations to find out whether or not the response of the system to the change of the particular quantity is as expected for a pure regime. LEVENSPIEL [24.17] gives good examples of the use of the experimental techniques.

The experimental technique can give an answer to the first two questions mentioned (i.e., is one regime rate determining and which regime is this?). It does not tell us whether the regime will change during scaling-up or not. Therefore, we need theoretical methods.

Theoretical Methods

The theoretical methods can be subdivided into methods based on an analytical solution (usually of an approximate nature) and methods based on the numerical solution of the balance equations, followed by parameter sensitivity analysis. The analytical method can be based on a number of different principles (comparison of time constants, of transfer- and conversion steps, of forces or pressures, or the use of the dimensionless groups). Because the dimensionless groups can be seen as ratios of time constants, forces, etc., their use is comprised in the use of time constants and the like.

The use of time constants is particularly illustrative for mass transfer- and conversion processes. Comparison of the time constants for conversion, mixing, flow, diffusion, mass transfer, and residence gives a good idea which regime is present. If, for example, t_m is large compared to t_c and the other time constants, mixing (flow) is the dominating regime. If at the same time $1/k_l a > t_c$, then the supply of oxygen is a problem as well.

The different time constants can be calculated from the correlation given in Sect. 24.3.2.4, the kinetic, geometrical, and physical parameters involved, and the relations for the time constants given below:

mixing time $\quad t_m \equiv \dfrac{V}{2.6 N D^3}$

($Re > 5 \cdot 10^3$, Rushton turbine)

conversion time $\quad t_c \equiv \dfrac{C}{r}$

diffusion time $\quad t_D \equiv \dfrac{L^2}{\mathbb{D}}$

residence time $\quad \tau \equiv \dfrac{V}{\Phi_v} \text{ or } = \dfrac{L}{v}$

mass transfer time $\quad t_{mt} \equiv \dfrac{1}{k_l a}$.

To compare the different flow mechanisms, time constants can be used as well but, in the authors' experience, comparison of the different pressures, tensions and stresses involved give a better insight. The most important ones are given below:

shear stress (laminar)	$\dfrac{v}{\eta_L}$	N m^{-2}
shear stress (turbulent)	ϱv^2	N m^{-2}
buoyancy	$\Delta \varrho g L$	N m^{-2}
yield stress	τ_0	N m^{-2}
surface tension	σ	N m^{-1}

Note the different dimension of the surface tension.

As we have seen, the different dimensionless groups can be composed from the time constants and the pressures given above: Re = (turbulent shear stress/laminar shear stress), Pe = (residence time/diffusion time), etc. The use of the pressures or stresses is similar to the use of time constants. If, for example, the laminar shear stress is very small compared with the turbulent shear stress, the turbulent shear stress dominates. If τ_0 is much smaller than ϱv^2 the influence of τ_0 on the process can be neglected (this was the case in the second example of the last section). A very high or very low dimensionless group is also an indication that one particular mechanism predominates. An exception is the Re number for pipe flow. Here, laminar shear stresses predominate if $Re < 2000$ and turbulent shear if $Re > 10^4$. This is a consequence of an unfortunate definition of Re for pipe flow.

24.3.5 Scaling-down of Parts of the Process

In the preceding sections a number of points related to changes of scale have been presented:

1. The main scale-up problems are transport problems.
2. The environmental approach (scaling-down) is essential to obtain reliable results at M-scale on the behavior of microorganisms at P-scale.
3. Geometrical similarity between M and P is useful if one particular part of the process (bulk flow, shear, mass transfer, or mixing) is studied. It is generally not possible to find one particular set of operating variables that results in similar behavior at M and P for more than one part.
4. Two-phase systems generally show a roughly equal size of the dispersed phase at M and P instead of a size proportional to the size of the reactor.

The points given above (particularly 2 and 3) reveal that *the effects of changes of scale can only be investigated if the different parts (mechanisms) are studied separately*. The consequences of this conclusion will be discussed now.

1. Growth and production

Growth and production have to be studied at M-scale under conditions similar to the full scale (scaling-down). Once a general idea exists of the size, shape, and auxiliary equipment (stirrers, cooling coils, etc.) of the bioreactor on the production scale, the likely environmental conditions at this scale can be estimated, among others, by means of the equations presented in Sect. 24.3.2.4 (mixing times, variations in dissolved oxygen concentration, shear, etc.). Regime analysis and the time constant concept are useful tools here. In a small-scale stirred vessel these conditions can be simulated by varying the air flow rate or oxygen concentration by adding the substrate pulsewise and by choosing $(ND)^2 = $ const as a scale-down criterion if an influence of shear is expected. There is no need for geometric similarity except for the stirrer if shear is important.

The procedure given above cannot be used for the general screening of microorganisms when large numbers of strains have to be inspected. However, as soon as a potential candidate for production appears the procedure mentioned above should be applied. The substrates and additives (antifoam) used should be the same as at P-scale. This is another important point that is often overlooked.

2. Flow and mixing

Flow and mixing problems have to be studied under conditions of geometrical similarity, although sometimes a geometrically similar element of an apparatus can be used as well (for example, a liquid entrance system plus its immediate surroundings). The Re number is not very critical, provided it is in the same region for M and P (either the laminar or the turbulent region). For transport through boundary layers (for example, heat transport from the bulk of the liquid to cooling coils) Re numbers play an important role, however; see, among others, BEEK and MUTZALL [24.89]. DA is a very powerful tool here.

3. Aeration and other dispersion processes

For aeration and other dispersion processes geometrical similarity is desirable, although not always for the system as a whole. Stirrers should always be geometrically similar, however. $P/V = $ const is, as far as the literature is concerned, a reasonable scale-up rule, although many other relations exist. However, very little is known about the validity of this rule for large-scale ratios. In the authors' opinion there is a great need for structured models. DA is a useful tool for this kind of research.

The procedure presented here can be summarized as scaling-down of parts of the process. It asks for a good communication between microbial physiologists and process engineers and can therefore be considered to be biotechnology in the true sense of the word.

24.4 References

[24.1] R. C. DORF: "Modern Control Systems". Addison-Wesley, Reading, Mass., 1980.
[24.2] N. W. F. KOSSEN: "Mathematical modelling of fermentation processes: scope and limitations"; Proc. Microbial Technology (BULL, ELLWOOD, and RATLEDGE, eds.), Symp. 29. Society for General Microbiology, Cambridge University Press, Cambridge 1979.
[24.3] K. R. POPPER: "Conjectures and Refutations". Routledge & Kegan Paul, London 1972.
[24.4] A. G. FREDERICKSON, R. D. MEGEE, and M. M. TSUCHIYA: "Mathematical models for fermentation processes"; Adv. Appl. Microbiol. 13 (1970), 419–465.
[24.5] J. A. ROELS and N. W. F. KOSSEN: "On the modelling of microbial metabolism"; Progr. Ind. Microbiol. 14 (1978), 95–203.
[24.6] D. R. HOFSTADTER: "Gödel, Escher, Bach". The Harvester Press, Stanford Terrace 1979.
[24.7] K. OGATA: "Modern Control Engineering". Prentice Hall, Englewood Cliffs, N. J., 1970.
[24.8] A. H. M. VERKOOYEN: "A Stochastic Model for the Growth of Yeast on Li-

[24.9] A. G. FREDERICKSON: "Stochastic models for sterilization"; Biotech. Bioeng. *8* (1966), 167–182.

[24.10] N. S. GOEL and N. RICHTER-DYN: "Stochastic Models in Biology". Academic Press, New York 1974.

[24.11] J. A. ROELS: "Mathematical models and the design of biochemical reactors"; J. Chem. Tech. Biotechnol. *32* (1982), 59–72.

[24.12] D. M. HIMMELBLAU and K. B. BISCHOFF: "Process Analysis and Simulation: Deterministic Systems". Wiley Interscience, New York 1968.

[24.13] N. W. F. KOSSEN: "Models in bioreactor design"; Proc. 3rd Int. Conf. on Computer Appl. in Ferment. Technology, Soc. Chem. Ind. London, 1981.

[24.14] H. JANSEN, S. SLOT, and H. GÜRTLER: "Determination of mixing times in large-scale fermenters using radioactive isotopes"; Prepr. First Eur. Congr. on Biotechnology Part II: Poster Papers, pp. 80–83, Dechema, Frankfurt 1978.

[24.15] A. EINSELE: "Charakterisierung von Bioreaktoren durch Mischzeiten"; Chem. Rundsch. *29* (1976), 53–55.

[24.16] A. EINSELE: "Scaling-up bioreactors"; Process Biochem. *13* (1978), 13–14.

[24.17] O. LEVENSPIEL: "Chemical Reaction Engineering". Wiley & Sons, New York 1972.

[24.18] P. V. DANCKWERTS: "L'utilisation des traceurs dans les recherches industrielles"; Génie Chim. *83* (1960), 75–80.

[24.19] R. G. E. FRANKS: "Modelling and Simulation in Chemical Engineering". Wiley Interscience, New York 1972.

[24.20] B. A. FINLAYSON and L. E. SCRIVEN: "The method of weighted residuals and its relation to certain variational principles for the analysis of transport processes"; Chem. Eng. Sci. *20* (1965), 395–404.

[24.21] D. G. MERCER and K. F. O'DRISCOLL: "Kinetic modelling of a multiple immobilized enzyme system I: Development and testing of the model"; Biotech. Bioeng. *23* (1981), 2447–2464.

[24.22] P. A. RAMACHANDRAN: "Solution of immobilized enzyme problems by collocation methods"; Biotech. Bioeng. *17* (1975), 211–226.

[24.23] D. D. MCCRACKEN: "A Guide to FORTRAN IV Programming". Wiley & Sons, New York 1972.

[24.24] B. S. GOTTFRIED: "Programming with FORTRAN IV". Quantum Publishers, New York 1972.

[24.25] R. ZAKS: "An Introduction to Pascal"; Sybex, Berkeley, Calif. 1980.

[24.26] K. JENSEN and N. WIRTH: "Pascal User Manual and Report". Springer Verlag, New York 1974.

[24.27] B. METZ: "From Pulp to Pellet". Dissertation, Delft 1976.

[24.28] J. H. SEINFELD and L. LAPIDUS: "Mathematical Models in Chemical Engineering", Vol. 3. Prentice Hall, New York 1974.

[24.29] G. A. BECKEY and S. M. YAMASHIRO: "Parameter estimation in mathematical models of biological systems"; Adv. Biomed. Eng. *6* (1976), 1.

[24.30] N. D. P. DANG: "Modelling of dynamic biological processes with empirical transfer functions"; J. Ferment. Technol. *53* (1975), 885–894.

[24.31] J. WEI: "The least square fitting of an elephant; Chemtech (1975), 128–129.

[24.32] REN DER YANG and A. E. HUMPHREY: "Dynamic and steady-state studies of phenol biodegradation in pure and mixed cultures"; Biotech. Bioeng. *17* (1975), 1211–1235.

[24.33] P. G. KROUWEL: "Immobilized Cells for Continuous Solvent Production". Dissertation, Delft 1982.

[24.34] D. B. JOHNSON and P. M. BERTHOUEX: "Using multiresponse data to estimate biokinetic parameters"; Biotech. Bioeng. *17* (1975), 571–583.

[24.35] D. B. JOHNSON and P. M. BERTHOUEX: "Efficient biokinetic experimental design"; Biotech. Bioeng. *17* (1975), 557–570.

[24.36] G. E. P. BOX, W. G. HUNTER, and J. S. HUNTER: "Statistics for Experimenters". Wiley & Sons, New York 1978.

[24.37] N. DRAPER and H. SMITH: "Applied Regression Analysis". Wiley & Sons, New York 1966.

[24.38] D. M. HIMMELBLAU: "Process Analysis by Statistical Methods". Wiley & Sons, New York 1968.

[24.39] J. C. VAN SUIJDAM: "Mycelial Pellet Suspensions". Dissertation, Delft 1980.

[24.40] M. MIDLER and R. K. FINN: "A model system for evaluating shear in the design of stirred fermenters"; Biotech. Bioeng. *8* (1966), 71–84.

[24.41] H. TAGUCHI, T. YOSHIDA, Y. TOMITA, and S. TERAMOTO: "The effects of agitation on the disruption of the mycelial pellets in stirred fermenters" (in Japanese); J. Ferment. Technol. *46* (1968), 814–822.

[24.42] H. TAGUCHI, T. IMANAKA, S. TERAMOTO, M. TAKATSU, and M. SATO: "Scale-up of glucoamylase fermentation by *Endomyces* sp." (in Japanese); J. Ferment. Technol. *46* (1968), 823–828.

[24.43] D. I. C. WANG and R. C. J. FEWKES: "Effect of operating variables and geometric parameters on the behavior of non-Newtonian, mycelial, antibiotic fermentations"; Dev. Ind. Microbiol. *18* (1977), 39–56.

[24.44] K. HATTORI, S. YOKOO, and O. IMADA: "Scale-up of L-glutamic acid fermentation from hydrocarbons"; J. Ferment. Technol. *52* (1974), 132–139.

[24.45] I. KASZAB, I. HÓGYE, S. KOMÓCSI, and J. SZILÁGYI: "Possible side effect of scale-up: extra-cellular accumulation of asparagine in Tobramycin fermentation"; Process Biochem. (1981), Febr./March, 38–39/49.

[24.46] D. E. BROWN: "Industrial-scale operation of microbial processes"; J. Chem. Tech. Biotechnol. *32* (1982), 34–46.

[24.47] N. BLAKEBROUGH: "Mixing effects in biological systems"; Chem. Eng. (1972) Febr., 58–63.

[24.48] J. Y. OLDSHUE: "Let's understand mixing"; Chemtech (1981) Sept., 554–561.

[24.49] S. AIBA, A. E. HUMPHREY, and N. F. MILLIS: "Biochemical Engineering" (Chap. 7). University of Tokyo Press, Tokyo, 1973.

[24.50] W. H. BARTHOLOMEW: "Scale-up of submerged fermentations"; Adv. Appl. Microbiol. *2* (1960), 289.

[24.51] T. B. YOUNG: "Fermentation scale-up: industrial experience with a total environmental approach"; Ann. N. Y. Acad. Sci. *326* (1979), 165–180.

[24.52] N. M. G. OOSTERHUIS and N. W. F. KOSSEN: "Scale-up of bioreactors"; Proc. 2nd Eur. Conf. on Ind. Microbiol., Milan, Sept. 15–17 1982.

[24.53] J. Y. OLDSHUE: "Fermentation mixing scale-up techniques"; Biotech. Bioeng. *8* (1966), 3–24.

[24.54] E. O. KAROW, W. H. BARTHOLOMEW, and M. R. SFAT: "Oxygen transfer and agitation in submerged fermentation"; J. Agric. Food Chem. *1* (1953), 302–306.

[24.55] A. L. CONN: "Experiences in transition from lab to plant"; Chemtech (1975), 154–159.

[24.56] R. B. BIRD, W. E. STEWART, and E. N. LIGHTFOOT: "Transport Phenomena". Wiley & Sons, New York 1960.

[24.57] G. B. WALLIS: "One-dimensional Two-phase flow". McGraw-Hill, New York 1969.

[24.58] Y. T. SHAH: "Gas-liquid-solid Reactor Design". McGraw-Hill, New York 1979.

[24.59] R. W. SHAFTLEIN and T. W. F. RUSSEL: "Two-phase reactor design, tank-type reactors"; Ind. Eng. Chem. *60* (1968), 12–27.

[24.60] K. W. NORWOOD and A. B. METZNER: "Flow patterns and mixing rates in agitated vessels"; A. I. Ch. E. J. *6* (1960), 432–437.

[24.61] S. J. KHANG and O. LEVENSPIEL: "New scale-up and design method for stirrer agitated batch mixing vessels"; Chem. Eng. Sci. *31* (1976), 569–577.

[24.62] J. J. HEYNEN and K. VAN'T RIET: "Mass transfer, mixing, and heat transfer phenomena in low-viscous bubble column reactors"; IVth Eur. Conf. on Mixing, April 27–29 1982, Noordwijkerhout (The Netherlands).

[24.63] G. D. TOWELL and G. H. ACKERMAN: "Axial mixing of liquid and gas in large bubble reactors"; Chem. React. Eng. *B 3* (1972), 1–12.

[24.64] P. OVASKAINEN, R. LUNDELL, and P. LAIHO: "Engineering of fermentation plants, Part 2: Fermenter design and scale-up". Process Biochem. *11* (1976), 37–39/55.

[24.65] M. POPOVIĆ, A. PAPALEXIOU, and M. REUSS: "Gas residence time distribution in stirred tank reactors"; Proc. VIth Int. Ferment. Symp. London, Ontario, Canada, 1980, paper F-7.1.10.

[24.66] J. LAINE and R. KUOPPAMÄKI: "Development of the design of large-scale fermenters"; Ind. Eng. Chem. Process Des. Dev. *18* (1979), 501–506.

[24.67] J. C. MIDDLETON: "Measurement of circulation within large mixing vessels"; IIIrd Eur. Conf. on Mixing, April 4–6 1979, York, U. K., paper A 2.

[24.68] J. BRYANT and S. SADEGHZADEH: "Terminal mixing times in stirred vessels"; IVth Eur. Conf. on Mixing, April 27–29 1982, Noordwijkerhout (The Netherlands), paper B 4.

[24.69] R. MANN and D. MAVROS: "Analysis of unsteady tracer dispersion and mixing in a stirred vessel using interconnected networks of ideal flow zones"; IVth

Eur. Conf. on Mixing, April 27–29 1982, Noordwijkerhout (The Netherlands), paper B3.

[24.70] M. MOO-YOUNG and H. W. BLANCH: "Design of biochemical reactors; mass transfer criteria for simple and complex systems"; Adv. Biochem. Eng. *19* (1981), 1–71.

[24.71] K. VAN'T RIET: "Review of measuring methods and results in non-viscous gas-liquid mass transfer in stirred vessels"; Ind. Eng. Chem. Process Des. Dev. *18* (1979), 367–375.

[24.72] M. REUSS, R. K. BAJPAI, R. LENZ, H. NIEBELSCHÜTZ, and A. PAPALEXIOU: "Scale-up strategies based on the interaction of transport and reaction"; VIth Int. Ferment. Symp., July 20–25 1980, London, Ontario, Canada, paper F-7.2.1.

[24.73] M. REUSS, R. K. BAJPAI, and W. BERKE: "Effective oxygen-consumption rates in fermentation broth with filamentous organisms"; J. Chem. Tech. Biotechnol. *32* (1982), 81–91.

[24.74] J. BRYANT: "The characterization of mixing in fermenters"; Adv. Biochem. Eng. *5* (1977), 101–123.

[24.75] P. J. SENIOR and J. WINDASS: "The ICI single cell protein process"; 13th Int. TNO Conference 1980 (TNO, P.O. Box 297, The Hague, The Netherlands), pp. 97–101.

[24.76] R. E. JOHNSTONE and M. W. THRING: "Pilot Plants, Models, and Scale-up Methods in Chemical Engineering". McGraw-Hill, New York 1957.

[24.77] J. PAWLOWSKI: "Die Ähnlichkeitstheorie in der physikalisch-technischen Forschung". Springer Verlag, Berlin 1971.

[24.78] M. ZLOKARNIK: "Ähnlichkeitstheorie in der Verfahrenstechnik". Bayer, Leverkusen 1974.

[24.79] H. A. BECKER: "Dimensionless Parameters; Theory and Methodology". Applied Science Publishers, London 1976.

[24.80] R. SHINNAR and J. M. CHURCH: "Predicting particle size in agitated dispersions"; Ind. Eng. Chem. *52* (1960), 253–256.

[24.81] A. N. KOLMOGOROV: C. R. Acad. Sci. USSR *30* (1941), 301; *32*, 16.

[24.82] M. ZLOKARNIK: "Sorption characteristics for gas-liquid contacting in mixing vessels"; Adv. Biochem. Eng. *8* (1978), 134–151.

[24.83] F. YOSHIDA and T. YAMADA: "Dispersion of oil in water and gas-bubble columns and agitated vessels"; Chemeca 1970 (Butterworth, Australia).

[24.84] P. H. CALDERBANK: "Physical rate processes in industrial fermentation", part I: "The interfacial area in gas-liquid contacting with mechanical agitation"; Trans. Inst. Chem. Eng. *36* (1958), 443–463.

[24.85] A. H. P. SKELLAND and R. SEKSARIA: "Minimum impeller speeds for liquid-liquid dispersion in baffled vessels"; Ind. Eng. Chem. Process Des. Dev. *17* (1978), 56–61.

[24.86] M. A. QURAISHI and T. Z. FAHIDY: "A simplified procedure for dimensional analysis employing SI units"; Can. J. Chem. Eng. *59* (1981), 563–566.

[24.87] S. IRVING, R. KING, and D. BULL: "Mixing studies in a large-scale sludge digester"; Chem. Eng. (1978) Nov., 831–837.

[24.88] W. BOLLE, J. VAN BREUGEL, G. C. VAN EYBERGEN, and N. W. F. KOSSEN: "Fluid flow in up-flow reactors used for anaerobic treatment of waste water"; 2nd Int. Congr. on Anaerobic Digestion, Travemünde (Germany) 1981.

[24.89] W. J. BEEK and K. M. K. MUTZALL: "Transport Phenomena". John Wiley, London 1975.

Chapter 25

Comparative Tests for Fermentations

Jürgen K. Lehmann

with the assistance of *Bernd Martin, Gerd-Walter Piehl, Reimer Schultze,* and *Werner Steven*

Gesellschaft für Biotechnologische Forschung mbH,
Braunschweig-Stöckheim, Federal Republic of Germany

25.1 Introduction
25.2 Reference Reactor and Test Conditions
25.3 Reference Conditions
25.3.1 Process Conditions
25.3.1.1 Batch Process
25.3.1.2 Fed-Batch Process
25.3.2 Documentation
25.3.3 Control Parameters
25.4 Process Monitoring
25.4.1 Volume-related Oxygen Uptake Rate
25.4.2 Volume-related CO_2 Evolution Rate
25.4.3 Respiratory Quotient
25.5 Comparative Tests
25.5.1 Baker's Yeast
25.5.1.1 Cultivation
25.5.1.2 Analysis
25.5.1.3 Results
25.5.2 Biosynthesis of Antibiotics
25.5.2.1 Pentalenolactone Cultivation
25.5.2.2 Analysis
25.5.2.3 Results
25.5.3 Production of Xanthan Polysaccharide
25.5.3.1 Cultivation
25.5.3.2 Analysis
25.5.3.3 Results

25.5.4 Biotransformation of Gluconic Acid
25.5.4.1 Cultivation
25.5.4.2 Analysis
25.5.4.3 Results
25.6 Addresses of Strain Collections
25.7 Equipment List
25.8 Reagents Required
25.9 References

List of Symbols

BDM	—	biological dry mass
D	s^{-1}	shear velocity
p_{O_2}	% air saturation	oxygen tension
p	bar	pressure
Q_{O_2}	mg/L·h	volume-related oxygen uptake rate
Q_{CO_2}	mg/L·h	volume-related CO_2 evolution rate
q_{O_2}	mg/g BDM·h	specific organism-related oxygen uptake rate
q	mg/g BDM·h	metabolic quotient
total O_2	g/L	integral of the volume-related oxygen uptake rate
t	°C	temperature
RQ	mol CO_2/mol O_2	respiratory quotient
rpm	min^{-1}	rotary speed
vvm	L air/L liquid·min	aeration rate
$\dot{V}_G$	N m³/h	gas flow rate
W_S	cm/s	tip velocity of stirrer
X	g/L	cell density, concentration
x	mol/mol	molar proportion

25.1 Introduction

The working group "Technik biologischer Prozesse" (Technology of Biological Processes) (chairmen: Prof. Dr. PAUL PRÄVE, Hoechst AG; Prof. Dr. HEINZ BRAUER, Berlin) of the DECHEMA technical committee "Biotechnology" has extended recommendations on the design of a reference reactor to include specifications of comparative tests for fermentation processes. Following proposals from members of the working group, process guidelines were developed and experimentally tested at GBF – Gesellschaft für Biotechnologische Forschung mbH, Braunschweig.

It will not be attempted to give a detailed discussion of the biological, physiological, and biochemical aspects of the processes used in the tests, since the processes are applied industrially, and extensive literature is already available.

Table 25.1 lists the process types, products, and strains of microorganism used in the processes considered.

Biological, chemical, and physical parameters exert a complex influence on the progress of biological processes. The reproducibility of results is particularly difficult when the processes are carried out in different test equipment and at different locations. Under such conditions, identical test conditions are practically impossible to guarantee.

The recommendations on the technical design of a reference reactor which have been laid down by DECHEMA (Deutsche Gesellschaft für chemisches Apparatewesen, Frankfurt am Main) represent the first step towards solving this problem. The comparative tests for fermentation processes described in the following represent a second step towards the ability to reproduce the results of biological processes. The guidelines not only assist the comparison of tests carried out at different locations, but also the comparison of results achieved with changed test conditions.

The processes chosen for the purposes of comparison are relevant to particular fields of microbial production, and consider biological, chemical, physical, and physiological aspects. All results were obtained with a fermentation reactor of 30 L working volume, constructed in accordance with the DECHEMA recommendations.

Both reference reactor design and design of the test series should also take account of the requirements of producers and operators of bioreactors when transferring and comparing the results of biological processes. In order to ensure reproducible initial conditions, only those organisms were used which are available from readily accessible strain collections.

Table 25.1. Types of Process, Product, and Strain of Microorganism Used in the Comparative Tests (References see under Sect. 25.10)

Type	Product	Strain
Biomass production	Baker's yeast	*Saccharomyces cerevisiae* DSM 2155
Biosynthesis of antibiotics	Pentalenolactone	*Streptomyces arenae* TÜ 469 DSM 40734
Polysaccharide production	Xanthan	*Xanthomonas campestris* DSM 1526
Biotransformation	Gluconic acid	*Aspergillus niger* DSM 2466

The results described were obtained under reference conditions. The processes are not intended to be optimized. However, they are suitable for the purpose of comparison.

25.2 Reference Reactor and Test Conditions

The DECHEMA standards concerning the construction of bioreactors include recommendations on reactor geometry, type and arrangement of stirring devices and supports, power requirements for drives and aeration, materials used, gaskets, and limiting values for temperature and pressure.

No recommendations are made for the parameters to be measured or types of measuring equipment; this should be a matter of individual decision. A series of data obtained from measurements both within and outside the reactor is needed in order to compare the results. The following describes the required measuring devices.

Process monitoring, control, and analysis require a reference reactor, which must be equipped with the following devices:

1. A scale for liquid level;
2. calibrated containers for acids, alkalis, and antifoaming agents;
3. a device for temperature measurement and control;
4. a device for pH-measurements and control;
5. a device for measurement and control of agitator speed;
6. a device for measuring gas flow rates;
7. a device for measuring p_{O_2};
8. a device for measuring O_2 and CO_2 in the exhaust gas;
9. arrangements for measuring and controlling pressure.

25.3 Reference Conditions

25.3.1 Process Conditions

Within the framework of these investigations, both batch and fed-batch processes have been carried out.

25.3.1.1 Batch Process

For this method, all necessary nutrients are initially provided. The rotary speed, aeration rate, temperature, and fermenter pressure are not changed during the test. All concentrations are allowed to change freely, and are not controlled. The process is terminated after a certain period of time, or when a predetermined concentration is reached.

25.3.1.2 Fed-Batch Process

This method also starts by providing all necessary nutrients. However, additional quantities of nutrients, etc. can be added during the process according to a time schedule or in order to adjust concentrations to predetermined levels. The process is terminated after a certain period of time or at a predetermined concentration.

25.3.2 Documentation

The listed test specifications cover the following aspects:

1. Initial cultivation of vegetative cells from the permanent forms;
2. preparation of the medium, sequence of the processes, sterilization requirements;
3. initial conditions for the seed culture(s) and main culture;
4. process parameters of the main culture;
5. time schedule for addition of nutrients, etc., or control parameters on which additions are to be based;
6. analysis and calculating methods.

25.3.3 Control Parameters

In the test series, the following parameters were recorded:

1. Fermenter total volume;
2. fermenter working volume;
3. volume of inoculum;
4. agitator speed;
5. temperature;
6. pH value, controlled or not;
7. gas flow rates;
8. p_{O_2} value;
9. fermenter pressure;
10. time required for completion of cultivation, or for a limiting concentration to be achieved.

25.4 Process Monitoring

For all the tests described in the following, process monitoring was based on the volume-related oxygen uptake rate Q_{O_2}, the volume-related CO_2 evolution rate Q_{CO_2}, and the respiration quotient RQ. Q_{O_2} and Q_{CO_2} were calculated from material balance equations from the inflowing and outflowing gas quantities. The parameters for the material balance equations are schematically illustrated in Fig. 25.1.

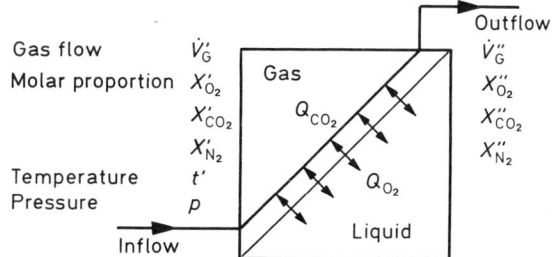

Figure 25.1. Parameters required for calculating the volume-related oxygen uptake rate and the CO_2 formation rate via a material balance.

25.4.1 Volume-related Oxygen Uptake Rate

The following assumptions were made when calculating the oxygen uptake rate Q_{O_2}:

1. Inflowing and outflowing gas mixtures are ideal gases.
2. Inflowing and outflowing gases are dry. In accordance with this requirement, inflowing gases must be measured before reaching the gas humidifier (if used), and the partial stream used for exhaust gas determination must be dried.
3. Inflowing gases consist of the components O_2, CO_2, and N_2; where the term N_2 encompasses all inert components.
4. Relative concentrations (X) of the components of the inflowing gases correspond to that of dry air, i.e.:

$$X'_{O_2} = 0.2095 \; \frac{\text{mol } O_2}{\text{mol gas}}$$

$$X'_{CO_2} = 0.0003 \; \frac{\text{mol } CO_2}{\text{mol gas}}$$

$$X'_{N_2} = 0.7902 \; \frac{\text{mol } N_2}{\text{mol gas}}$$

The volume-related oxygen uptake rate can then be determined from the following relationship:

$$Q_{O_2} = \frac{32}{22.4} \cdot 10^6 \frac{\dot{V}'_G}{V_R} \cdot \left[X'_{O_2} - X''_{O_2} \cdot \frac{X'_{N_2}}{1 - X''_{O_2} - X''_{CO_2}} \right]$$

where:

Q_{O_2}	$\dfrac{\text{mg } O_2}{\text{L} \cdot \text{h}}$	volume-related oxygen uptake rate
$\dot{V}'_G$	m^3/h	flow rate of the inflowing air at NTP
X'_{O_2}, X''_{O_2}	$\dfrac{\text{mol } O_2}{\text{mol gas}}$	molar proportion of oxygen in the inflowing and outflowing gases
X'_{CO_2}, X''_{CO_2}	$\dfrac{\text{mol } CO_2}{\text{mol gas}}$	molar proportion of CO_2 in the inflowing and outflowing gases
X'_{N_2}	$\dfrac{\text{mol } N_2}{\text{mol gas}}$	molar proportion of all inert components in the inflowing gases
V_R	L	momentaneous liquid volume in the fermenter.

25.4.2 Volume-related CO_2 Evolution Rate

Q_{CO_2} can be calculated in an analogous way according to the relationship:

$$Q_{CO_2} = \frac{44.01}{22.4} \cdot 10^6 \frac{\dot{V}'_G}{V_R} \cdot \left[X''_{CO_2} \cdot \frac{X'_{N_2}}{1 - X''_{O_2} - X''_{CO_2}} - X'_{CO_2} \right]$$

with the units:

Q_{CO_2}	$\dfrac{\text{mg } O_2}{\text{L} \cdot \text{h}}$	volume-related CO_2 evolution rate

25.4.3 Respiratory Quotient

The quotient of the CO_2 evolution rate and the oxygen uptake rate is the respiratory quotient RQ:

$$RQ = \frac{Q_{CO_2}}{Q_{O_2}} \cdot \frac{32.00}{44.01} \; \frac{\text{mol } CO_2}{\text{mol } O_2}.$$

Changes in the metabolic processes are usually indicated very clearly by changes in RQ.

25.5 Comparative Tests

25.5.1 Baker's Yeast

25.5.1.1 Cultivation

Strain: Cultivation is carried out with commercial baker's yeast, which in most German baker's yeast production plants is cultivated from the strain *Saccharomyces cerevisiae* VdH2. This strain is identical with *Saccharomyces cerevisiae* strains DSM 2155, ATCC 7754, and CBS 1368.

Seed culture propagation: Baker's yeast is generally supplied with the moisture content H_{27}, i.e., the yeast dry matter has 27% of the wet substance. The *Saccharomyces cerevisiae* strain is first mixed into a sterile medium, then 5 mg of tetracycline is added to each liter of working volume (see below).

Medium: The medium is based on a yeast dry mass (YDM) concentration of 60 g YDM/L.

Salts:
- 16 g/L $(NH_4)_2SO_4$, Merck 1216
- 1.75 g/L $(NH_4)H_2PO_4$, Merck 1126
- 1.16 g/L $MgSO_4 \cdot 7 H_2O$, Merck 5886
- 5 g/L KCl, Merck 4936
- 0.1 mL/L antifoaming agent, UCON-N 38

The salts are dissolved in tap water and adjusted to pH 5.2 with 2 M $Na_2CO_3 \cdot H_2O$ solution, and sterilized in the fermenter for 20 min at 121 °C.

Vitamins:
- 0.0833 mg/L D-biotin, Merck 24514
- 8.33 mg/L D-pantothenic acid, SIGMA No. P 2250
- 0.1 g/L meso-inosit, Merck 4731

The above are dissolved in 50 mL tap water and sterilized for 20 min at 121 °C.

Process: Cultivation is carried out as a fed-batch process, the substrate feed rate being exponential-linear as depicted in Table 25.2.

Table 25.2. Logarithmic-linear Substrate Feed Schedule for Baker's Yeast Production

Culture Time (in h)	Substrate Feed Rate (in mL/min)	Cell Density (in g YDM/L[a])	Total Yeast Quantity (in g YDM[a])	Hourly Increase (in h^{-1})
0	6.25	18.0	337.5	1.15
1	6.25	20.26	387.5	1.15
2	7.75	22.97	450.0	1.62
3	9.41	26.05	525.0	1.69
4	11.83	29.83	622.4	1.182
5	14.41	33.95	737.7	1.187
6	17.0	42.64	1012.7	1.154
7	16.66	42.64	1012.7	1.154
8	16.66	46.37	1147.7	1.137
9	16.66	49.82	1282.9	1.137
10	16.66	53.00	1417.8	1.105
11	16.66	55.96	1552.9	1.095
12	16.66	58.72	1687.9	1.087

[a] YDM = yeast dry mass

Culture conditions

Reactors: Reference fermenter "Biostat" U 30 D (Braun, Melsungen). 20 L medium are sterilized at 121 °C for 20 min in the fermenter. After sterilizing, excess liquid is drained to 15 L working volume.

Inoculum: 1250 g of baker's yeast H_{27} are dissolved in 2 L of sterile medium, and filled up to 3.7 L. 50 mL vitamin solution is added together with the antibiotic. The inoculum is transferred to the reactor under sterile conditions.

Nutrient: 10 L 25% saccharose solution is added in accordance with the dosing schedule described in Table 25.2. The required flow rate can be provided with, e.g., a metering pump from B. Braun, type FE 211.

Temperature:	32 °C
pH:	5.2, adjusted with 2 M $Na_2CO_3 \cdot H_2O$, Merck 6386
Aeration:	starting at 1 vvm = 1.8 N m³/h and adjusted to maintain p_{O_2} > 10% air saturation
Agitator speed:	600 rpm
Fermenter pressure:	not controlled (gas sterile filter)
Fermentation time:	12 h

25.5.1.2 Analysis

Determination of dry mass

A 20 mL duplicate sample is removed and centrifuged for 10 min at 12 000 g. The supernatant is carefully removed, and the cell pellet dried at 80 °C for 24 h under vacuum.

25.5.1.3 Results

It is recommended that the substrate feed rate is not solely based on the logarithmic-linear schedule described in Table 25.2. As often observed with commercial baker's yeast there is a risk of ethanol formation at the start of cultivation as a result of a saccharose overfeeding. It is thus recommended that the feed rate is also adjusted in accordance with the RQ, which should be around 1.2. If necessary, the substrate feed can be temporarily interrupted. Fig. 25.2

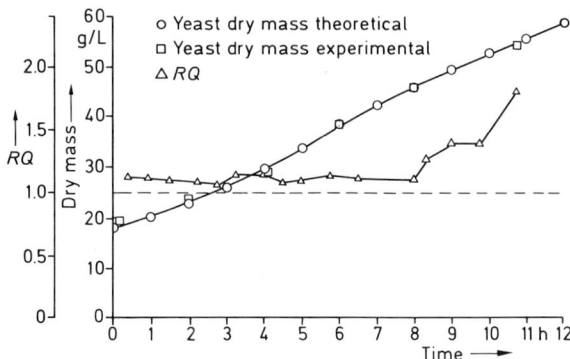

Figure 25.2. The cultivation of baker's yeast from *Saccharomyces cerevisiae* according to the substrate feed schedule in Table 25.2.

shows the results obtained. The yield corresponded to the theoretical value, i.e., no loss of yield occurred.

25.5.2 Biosynthesis of Antibiotics

25.5.2.1 Pentalenolactone Cultivation

Strain: Streptomyces arenae TÜ 469 DSM 40734

Seed culture propagation: In accordance with the guidelines in the DSM catalog for slant agar

Medium: Same medium for all cultures (reagent numbers as used by Merck)

40 g/L	D-Mannite, 5980
1 g/L	NaCl, 6404 E
0.4 g/L	$CaCl_2 \cdot 2 H_2O$, 2582

3 g/L K$_2$HPO$_4$·3 H$_2$O, 5099
2 g/L L-Asparagine monohydrate, 1565
1 g/L MgSO$_4$·7 H$_2$O, 5886
20 mg/L FeSO$_4$·7 H$_2$O, 3965
10 mg/L ZnSO$_4$·7 H$_2$O, 8883

The above are dissolved in deionized water, adjusted to pH 6.2 with H$_2$SO$_4$, and sterilized for 20 min at 121 °C.

Process: Seed culture
Stage 1: 250 mL medium are inoculated into a 1 L shake flask with 1 loop of cell material (slant agar). Incubation occurs for 100 h at 30 °C at 80 rpm on a shaker.

Stage 2: 250 mL medium are inoculated into a 1 L shake flask with 5 mL of the first seed culture. Incubation occurs for 50 h at 30 °C at 80 rpm on a shaker.

Process: Main culture
The main culture is inoculated with 10% of its volume of the liquid from stage 2. Cultivation of the required quantity of inoculant takes place in twelve separate flasks.

Culture conditions: Reference fermenter "Biostat" U 30 D (Braun, Melsungen)
Volume of medium before
    sterilizing: 27.5 L
    Volume of inoculant: 3 L
    Temperature: 30 °C
    pH: adjusted to 6.2 with 12.5% ammonia solution
    Aeration: 0.1 vvm
    Agitator speed: 850 rpm
    Fermenter pressure: not regulated
    Fermentation time: 50 h
    Final concentration:
        approx. 3 mg/L pentalenolactone.

25.5.2.2 Analysis

Determination of dry mass
Centrifugation of 2 × 20 mL samples at 12 000 *g* for 10 min. The supernatant liquid is used for product determination. The cell pellet is washed once with 2 N H$_3$PO$_4$ and twice with distilled water, and dried for 24 h at 80 °C.

Pentalenolactone determination
Pentalenolactone inhibits the activity of glyceraldehyde-3-phosphate dehydrogenase. The inhibition is used for the enzyme-based determination of pentalenolactone.
The following equipment is required for the determination:
    Enzyme pipettes (20 µL, 50 µL, 100 µL, 200 µL)
    1 photometer, adjusted to 334 nm
    1 cm cuvette
    1 pen-recorder, adjusted to 5 cm/min.

The following chemicals are required:
1. NAD: β-Nicotinamide adenine dinucleotide free acid, grade II 98%, Boehringer 127981
2. GAPDH: Glyceraldehyde-3-phosphate dihydrogenase, from rabbit muscle, Boehringer 105694
3. GAP: D-Glyceraldehyde-3-phosphate diethylacetate, as the crystallized dicyclohexylammonium salt, Boehringer 105759
4. Disodium hydrogen arsenate-7-hydrate, Merck 6284
5. Tetrasodium diphosphate-10-hydrate, Merck 6591
6. Titriplex III, Merck 8418
7. 2 M Sulfuric acid
8. 2-Mercaptoethanol, Merck-Schuchardt 805740
9. Sodium hydrogen carbonate
10. 2 N Phosphoric acid.

Solutions
All solutions are made with twice distilled water.
1. Buffer solution: 22.3 g tetrasodium diphosphate-10-hydrate dissolved in 950 mL H$_2$O and adjusted to pH 8.5 with 2 N H$_3$PO$_4$. Then, 0.3 mL mercaptoethanol and 1.86 g titriplex III are added and filled up to 1 L.
2. NAD solution: 332 mg NAD dissolved in water and filled up to 10 mL. To be stored at −18 °C.
3. Arsenate solution: 0.2 M (=624 mg) arsenate dissolved in water and filled up to 10 mL.

4. GAPDH solution: The commercial solution contains 10 mg/mL. A 1:400 dilution is made with twice distilled water. Attention! Only keeps for 6 hours.

5. GAP solution: 50 mg D-GAP-diethylacetate dissolved in water, filled up to 18 mL, and heated to 80°C; 0.5 mL 2 M H_2SO_4 is then added and maintained at 80°C for 4 min. The solution is then cooled in an ice bath, and adjusted to pH 5 with an approx. 2 M $NaHCO_3$ solution, and subsequently frozen in portions. The solution is stable for ten days at $-18°C$, or about 6 hours at room temperature.

Monitoring

The enzyme activity can be monitored with a photometer at 334 nm. The change of extinction of a reacting sample should be measured with a pen-recorder, at a sensitivity where an extinction of 0.25 corresponds to approx. 20 cm deflection.

Recording of the zero curve

1630 µL buffer, 100 µL NAD, 50 µL arsenate, and 20 µL GAPDH solution are mixed in a cuvette. In the photometer the temperature of the sample is allowed to stabilize for 5 min at room temperature. The extinction is adjusted to 0.02; 200 µL of GAP solution is added, and the paper feed of the pen-recorder activated. The extinction curve is recorded for approx. 1 min.

Recording of the sample curve

20 µL of the supernatant liquid are mixed with 1610 µL buffer solution in a new cuvette, and the remaining reagents added as for zero curve registration. The chart paper is reset back to the origin in order to record the extinction curve of the sample parallel to the zero curve.

Fig. 25.3 shows typical results obtained. The residual activity can be calculated from the quotient of curve sections a and b as shown. The pentalenolactone concentration in the sample can then be determined from the measured residual activity with the help of the calibration chart in Fig. 25.4.

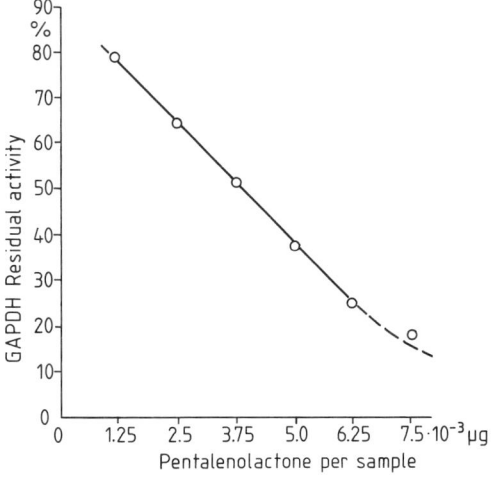

Figure 25.4. Calibration curve: GAPDH residual activity vs. pentalenolactone in the sample.

25.5.2.3 Results

Figs. 25.5 and 25.6 show the results obtained with the reference conditions described above.

The maximum pentalenolactone concentration of 3.2 mg/L is reached after 58 hours, as seen from Fig. 25.5. Under the reference conditions, the growth of the culture is approx. exponential. The increases in cell density, oxygen uptake rate, and CO_2 evolution rate are exponential. The morphology of the mycelia is flocculent.

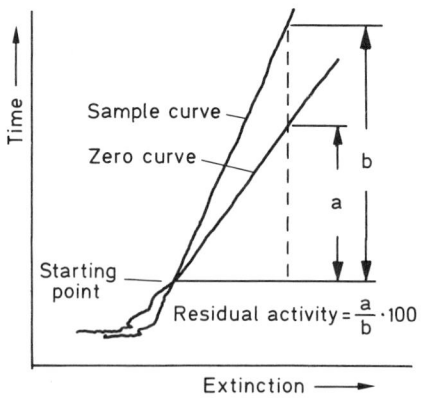

Figure 25.3. Typical extinction curve obtained in the enzymatic determination of pentalenolactone.

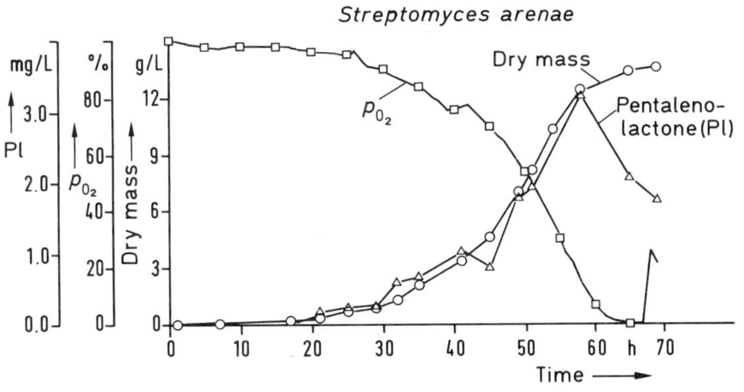

25.5 Comparative Tests

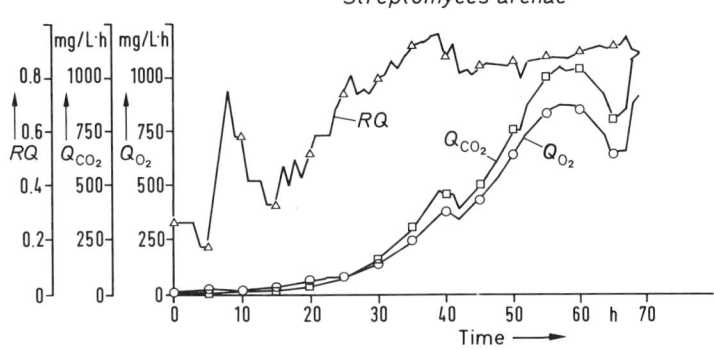

Figure 25.5. Dry mycelium mass, pentalenolactone (Pl), and oxygen partial pressure in % air saturation when cultivating *Streptomyces arenae* under reference conditions.

Figure 25.6. Volume-related oxygen uptake rate Q_{O_2} and CO_2 formation rate Q_{CO_2} when cultivating *Streptomyces arenae* under reference conditions.

Pentalenolactone standard

The standard substance can be obtained from Prof. Dr. MECKE (University of Tübingen, Federal Republic of Germany).

25.5.3 Production of Xanthan Polysaccharide

25.5.3.1 Cultivation

Strain: Xanthomonas campestris DSM 1526.

Seed culture propagation: Procedure on slant agar according to the regulations in the DSM-catalog.

Medium: Slant agar

20 g/L glucose, Merck 8342
10 g/L yeast extract, DIFCO 0127-01
20 g/L $CaCO_3$, Merck 2066
20 g/L agar, Merck 1614

Medium: Seed culture

10 g/L glucose (added under sterile conditions as a 50% solution after autoclaving)
3 g/L yeast extract
5 g/L bacto-peptone, DIFCO 0118-01
3 g/L malt extract, DIFCO 0186-1

The above are dissolved in tap water, adjusted to pH 7 with KOH, Merck 5033, and sterilized for 20 min at 121°C.

Medium: Main culture

50 g/L glucose and 0.4 mL antifoaming agent UCON-N 38 are dissolved in tap water, adjusted to pH 4-4.5 with H_2SO_4, Merck 731, and sterilized for 20 min at 121 °C in the fermenter before adding:

5 g/L K_2HPO_4, Merck 5101
8 g/L   yeast extract
0.2 g/L $MgSO_4 \cdot 7 H_2O$, Merck 5886
0.4 g/L   urea, Merck 8487

as a mixture dissolved in distilled water, adjusted to pH 7.0 ± 0.1 with KOH, and sterilized for 20 min at 121 °C.

Process: Seed culture

Stage 1: 14 mL of seed culture medium inoculated into a 100 mL shake flask with 2 loops of cell material from a slant agar culture, which is not older than one week, incubated for 24 h at 25 °C and 100 rpm on a shaker.

Stage 2: 86 mL of seed culture medium inoculated into a 500 mL shake flask with 14 mL of liquid from stage 1; conditions as in stage 1.

Stage 3: 200 mL of seed culture medium inoculated into a 1 L shake flask with 100 mL of liquid from stage 2. Incubation is performed as in stage 1.

Process: Main culture

The main culture is inoculated with 10% of its volume of the liquid from stage 3. The necessary quantity of inoculant is cultivated in 10 separate flasks.

Culture conditions: Reference fermenter "Biostat" U 30 (Braun, Melsungen)

Glucose is dissolved in 24 L tap water, adjusted to pH 4.0–4.5, and sterilized in the fermenter. Yeast extract, K_2HPO_4, $MgSO_4 \cdot 7 H_2O$, and urea are dissolved in distilled water, filled up to 3 L, adjusted to pH 7.0 with KOH, sterilized separately, and transferred to the fermenter.

Volume of inoculant: 3.0 L
Temperature: 28 °C
pH: 7.0 ± 0.1
Adjusted with: 20 weight % KOH
Aeration: 1 vvm
Agitator speed: 600 rpm
Fermenter pressure: not controlled, gas exhaust through sterile filter
Fermentation time: 100 h.

25.5.3.2 Analysis

Determination of dry mass

A 10 mL sample is centrifuged at 12 000 g for 20 min. If viscosity is too high, 10 g of the sample should be diluted to 50 g with distilled water. The supernatant liquid is used to determine xanthan and glucose. The cell pellet is washed with distilled water, 2 N HCl, and distilled water, respectively, centrifuging each time at 12 000 g for 10 min. The cell mass is dried for 24 h at 80 °C under vacuum.

Xanthan determination

15 mL of pure ethanol is added to 5 mL cell-free sample while stirring. Precipitated xanthan is separated by centrifuging and dried for 1–2 h at 100 °C, followed by 12 h at room temperature in a desiccator.

Glucose determination

Measurement to be carried out with a Beckman Glucose Analyzer.

25.5.3.3 Results

Due to the rapidly increasing viscosity, oxygen transport limitation occurs after 20 h as indicated by a reduction of p_{O_2} to zero. The pH value drops simultaneously. Control starts at pH 7. Due to the nitrogen limitation in the medium the cell mass no longer increases after 20 h (see Fig. 25.7).

The xanthan concentration continues to increase although the O_2 transport limitation increases due to the increase in the viscosity, and thus the transferred oxygen quantity Q_{O_2} drops steadily. Unstirred zones appear in the reactor, reducing the rate of xanthan accumulation. The total glucose quantity is consumed after approxi-

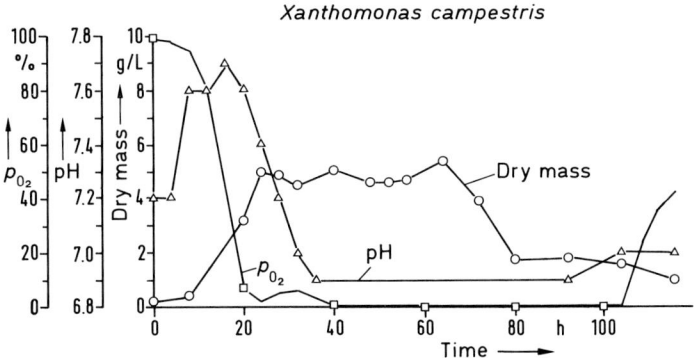

Figure 25.7. Variation of dry cell mass, pH, and oxygen partial pressure p_{O_2} when cultivating *Xanthomonas campestris* for xanthan production.

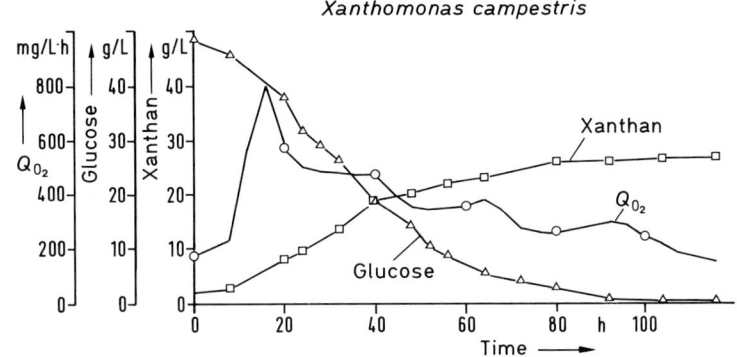

Figure 25.8. Glucose, xanthan, and Q_{O_2} when cultivating *Xanthomonas campestris* for xanthan production.

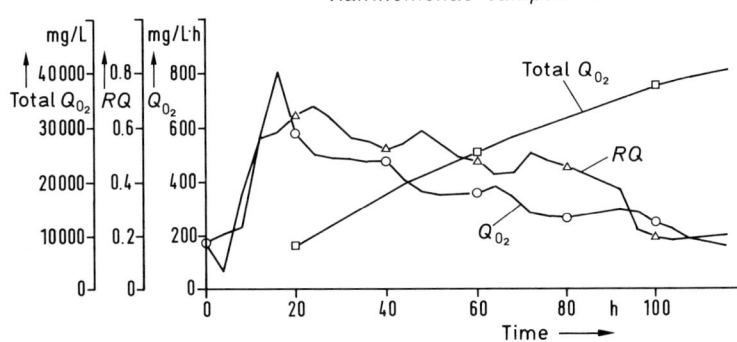

Figure 25.9. Variation of Q_{O_2}, RQ, and the integral oxygen uptake rate per liter (total O_2) when cultivating *Xanthomonas campestris* for xanthan production.

mately 100 h as shown by a sudden increase in p_{O_2} and a reduction in Q_{O_2} (see Fig. 25.8).

Fig. 25.9 shows the change in Q_{O_2}, the total oxygen uptake per liter at any given time, and the RQ value. RQ drops with increasing oxygen limitation. The glucose exhaustion can be particularly clearly observed after the 92nd hour. The total yield was 0.53 g xanthan/g glucose. 1.2 kg O_2 were required to convert 1.5 kg glucose. The apparent viscosity of the xanthan solution was measured with a cylindrical rotary viscosimeter at a shear velocity of $D = 28.8 \text{ s}^{-1}$.

Fig. 25.10 shows the apparent viscosity of the xanthan solutions produced in tests with different reactors. No reactor-specific differences were observed.

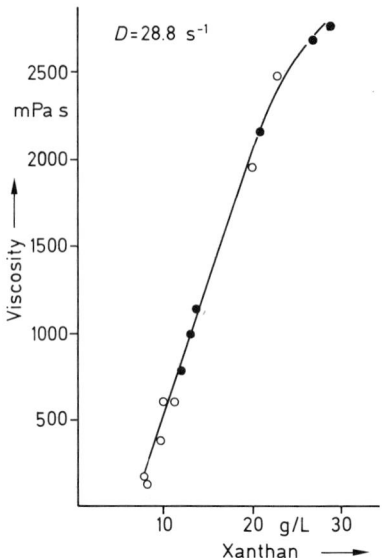

Figure 25.10. Viscosity of aqueous xanthan medium. Measured with a cylindrical rotary viscosimeter, $D = 28.8 \text{ s}^{-1}$. Reactor U 30 D (Braun) ○, reactor b 50 (Giovanola) ●.

25.5.4 Biotransformation of Gluconic Acid

25.5.4.1 Cultivation

Strain: Aspergillus niger, DSM 2466

Seed culture propagation: Procedure carried out on agar following the guidelines in the DSM-catalog.

Medium: Media for seed culture and main culture are identical.

Substrate: 170 g/L glucose
2 g/L corn steep liquor (45% dry matter)
Salts: 0.2 g/L MgSO$_4 \cdot$ 7 H$_2$O
0.5 mL/L H$_3$PO$_4$ 85%
0.5 mL/L conc. ammonia solution
1 g/L sodium borate
50 g/L CaCO$_3$ (sterilized for 2 h at 160°C in drying cabinet)
0.4 mL/L antifoaming agent UCON-N 38

The above are dissolved in deionized water, adjusted to pH 6.7 ± 0.1, and sterilized for 1 h at 105°C in the fermenter while stirring at 200 rpm.

Process: Seed culture: 100 mL medium inoculated into 300 mL shake flasks with one loop of cell material (agar) for 24 h at 30°C and 100 rpm.

Process: Main culture: The main culture is inoculated with 5% of its volume. The required quantity of inoculant is cultivated in 5 flasks.

Culture conditions: Reference fermenter "Biostat" U 30 D (Braun, Melsungen)

Medium volume before
sterilizing: 27.5 L
Volume of inoculant: 3 L
Temperature: 30°C
pH: uncontrolled
Aeration: 0.25 vvm
Agitator speed: 600 rpm
Fermenter pressure: uncontrolled
Fermentation time: the end of fermentation is indicated by a 100% increase of the oxygen saturation as indicated by the oxygen electrode.

25.5.4.2 Analysis

Determination of dry mass
A 20 mL sample is centrifuged at 12 000 g for 20 min. The supernatant liquid is used to determine glucose and calcium gluconate. The pellet is washed twice with 2 N HCl and once with distilled water, each time centrifuging for 20 min. The cell mass is dried at 80°C for 24 h under vacuum.

Glucose determination
The glucose content is determined with an autoanalyzer using dinitrosalicylic acid

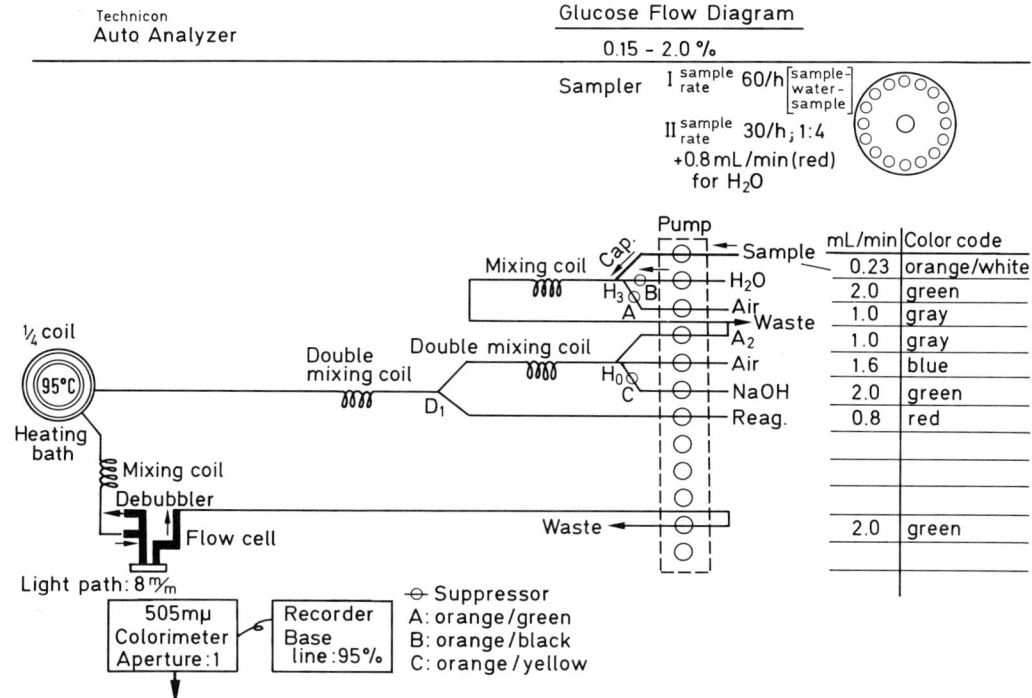

Figure 25.11. Autoanalyzer for glucose determination (according to K. BUCHTA, Boehringer Ingelheim).

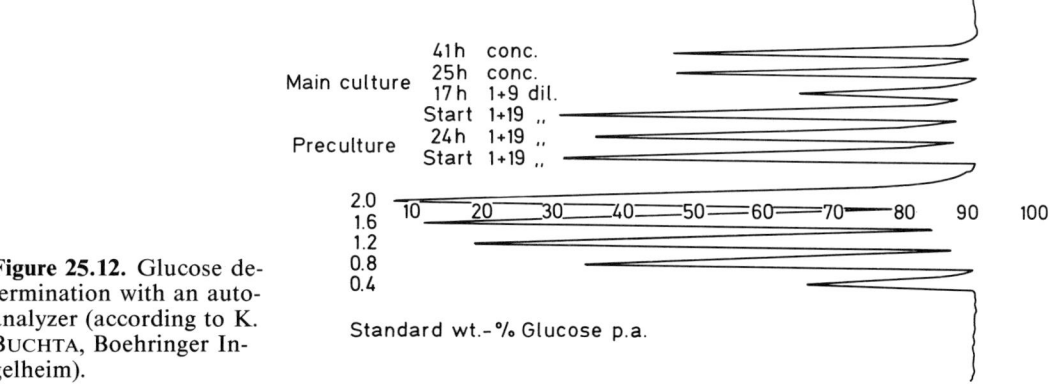

Figure 25.12. Glucose determination with an autoanalyzer (according to K. BUCHTA, Boehringer Ingelheim).

as a reagent. The analyzer is schematically shown in Fig. 25.11.

Reagents: 1.0 g 3.5-dinitrosalicylic acid + 300 mL water + 7.0 mL 1 N NaOH, filled up with water to 500 mL and filtered.

NaOH: Prepare 2 N solution.

Ca-gluconate determination

The formed gluconic acid can be determined via the dissolved calcium, using titriplex and eriochrome-red as indicators in an ammoniacal solution with an indicator buffer tablet: Merck 8418. 5 mL conc. ammonia solution, one buffer tablet and 2 mL

Table 25.3. Results – Gluconate Production (according to K. Buchta, Boehringer Ingelheim)

Exp. No.	Initial glucose (in g/L)	Time (in h)	Residual glucose (in g/L)	Calcium (in g/L)	Calcium-gluconate (in g/L)	Gluconic acid yield (in %)	pH	Aeration (in vvm)	Agitation (in rpm)	Reactor
1	154	0	—	0.4	—	—	6.55	0.16	550	"Biostat" U 30 D
		17	76.0	11.4	118.1	70.4	5.00			
		25	10.4	16.4	171.8	102.4	5.60			
2	174	0	—	0.4	—	—	6.60	0.10	550	"Biostat" U 30 D
		17	86.0	10.0	103.0	54.4	5.30			
		25	13.4	17.2	180.0	95.2	5.30			
		41	7.1	18.6	195.0	103.1	5.55			
3	174	0	—	0.4	—	—	6.50	0.25	500	Self Construction
		17	120.0	9.2	94.0	49.8	5.50			
		25	20.0	15.6	163.0	86.1	5.30			
		41	9.0	18.2	191.0	100.8	5.50			
4	168	0	—	0.4	—	—	6.50	0.1	900	"Biostat" U 30 D
		17	41.0	14.4	150.0	82.1	5.30			
		26	9.6	17.6	184.0	100.9	5.60			
5	170	0	—	0.4	—	—	6.60	0.25	550	"Biostat" U 30 D
		17	40.0	14.0	146.0	78.8	5.40			
		25	5.8	17.2	180.0	97.4	5.60			
		41	5.9	18.8	197.0	106.7	5.60			

of the supernatant liquid is added to approximately 300 mL distilled water.

10 mL broth is centrifuged and heated if required; 2 mL of the supernatant liquid are added to 600 mL distilled water. After adding 5 mL conc. ammonia solution, a pinch of eriochrome-red B-powder [1 g dye + 99 g NaCl (p.a.) ground in a mortar], and 0.1 M titriplex solution, the sample is titrated until the color changes to yellow-green.

1 mL titriplex III = 4.008 mg Ca
or   = 2.004 g Ca/L broth
(with respect to the sample).

Calculations
a) Gluconate formation:

$$g \frac{\text{Ca-gluconate formed}}{\text{L broth}} =$$

$$= 10.74 \left[\frac{g\ Ca}{L\ sample} - \frac{g\ Ca}{L\ initial\ broth} \right]$$

b) Gluconic acid yield:

$$\%\ yield = 986.21 \left[\frac{\dfrac{g\ Ca}{L\ sample} - \dfrac{g\ Ca}{L\ initial\ broth}}{\dfrac{g\ glucose}{L\ initial\ broth}} \right]$$

25.5.4.3 Results

The formed gluconic acid is determined by the concentration of Ca-gluconate in the medium. Table 25.3 shows values obtained by K. BUCHTA, Boehringer Ingelheim (compare with Fig. 25.12). Results obtained at GBF are shown in Figs. 25.13 and 25.14. After an initial lag of approximately 5 h, culture growth and Ca-gluconate production started simultaneously. Antifoaming agent must be added as required. Cell growth deteriorates after about 20 h. Gluconate formation continues at a constant

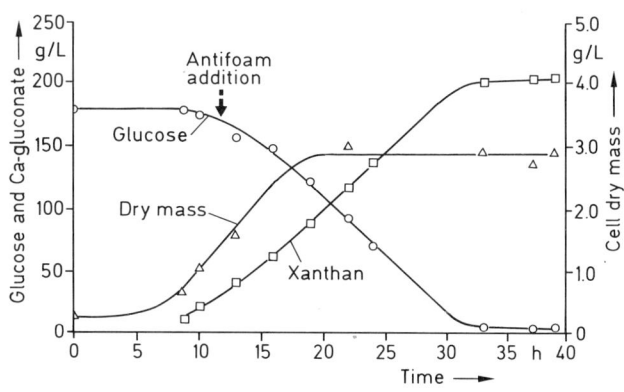

Figure 25.13. Cultivation of *Aspergillus niger* for the production of gluconic acid. Variation of Ca-gluconate concentration, cell dry mass, and glucose in the medium.

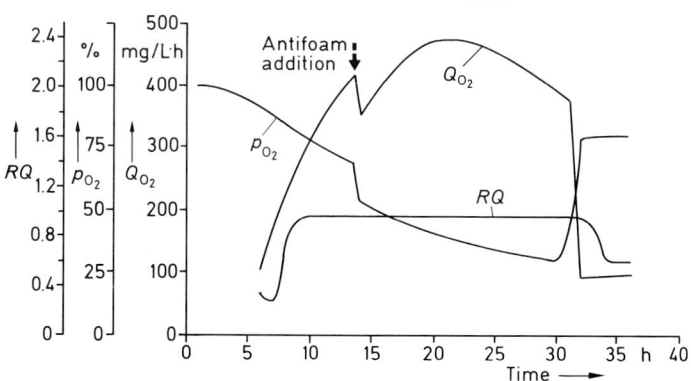

Figure 25.14. Change of Q_{O_2}, oxygen partial pressure p_{O_2}, and respiration quotient RQ during the cultivation of *Aspergillus niger* for gluconic acid production.

speed until the glucose is exhausted after 32 h.

As shown in Fig. 25.13, oxygen transport limitation does not take place. The more or less constant oxygen transfer rate with linear increase in the gluconate and 30% oxygen saturation shows that the O_2 transport capacity of the reactor was not exhausted.

During the formation of Ca-gluconate, the CO_2, which is released by the reaction of gluconic acid with $CaCO_3$, is removed by the aeration, resulting in a high, apparent respiration quotient.

At a constant cell density, the conversion rate of the glucose – and thus gluconate – production is obviously constant.

The gluconate yield was identical with the results obtained by K. BUCHTA.

25.6 Addresses of Strain Collections

1. DSM – Deutsche Sammlung von Mikroorganismen
   Grisebachstr. 8
   D-3400 Göttingen

2. VdH – Versuchsanstalt der Hefeindustrie
   Institut für Gärungsgewerbe und Biotechnologie
   Seestr. 13
   D-1000 Berlin 65

3. ATCC – American Type Culture Collection
   12301 Parklawn Drive
   Rockville, Maryland 20852
   U.S.A.

4. CBS – Centraalbureau voor Schimmelcultures
   P.O. Box 273
   NL-3740 Baarn (Netherlands)

25.7 Equipment List

Type	Model	Manufacturer (all Germany)
Fermenter	BIOSTAT U 30 D	B. Braun, Melsungen
O_2-analyzer	OXYGOR 3	MAIHAK AG, Hamburg
CO_2-analyzer	UNOR N 4	MAIHAK AG, Hamburg
Oxygen electrode	(sterilizable)	INGOLD KG, Frankfurt
Substrate dosing	Precision metering pump FE 211	B. Braun, Melsungen
	Microprocessor controlled programming device D-95-600	M. K. Juchheim GmbH, Fulda
Centrifuge	JUNIOR	Heraeus/Christ, Osterode
Spectrophotometer	PM 2 D	ZEISS
Rotary viscosimeter	Rotovisko RV 3, Rotor MV 2, Stator MV ST	HAAKE, Karlsruhe

25.8 Reagents Required

If not stated otherwise in the text, the tests were carried out with the reagents listed here.

Type	Manufacturer
Agar Bacto	DIFCO 0140-01
Ammonia solution, concentrated	MERCK 5432
Antifoaming agent	BRENNTAG UCON N 38
Borax (disodium tetraborate-10-hydrate)	MERCK 6308
Calcium carbonate	MERCK 2066
Corn steep liquor (with approx. 50% dry substrate)	HENKEL
Dodecyl hydrogen sulfate (sodium salt)	MERCK 2969
Glucose	MERCK 8342
Hydrochloric acid, min. 37%	MERCK 317
2-Hydroxy-3,5-dinitrobenzoic acid	MERCK-SCHUCHARDT 800 141
Indicator buffer tablets	MERCK 8418
Magnesium sulfate $\cdot 7\,H_2O$	MERCK 5586
Phosphoric acid, 85%	MERCK 563
Sodium hydroxide	MERCK 6498
Sulfuric acid, 95–97%	MERCK 731
Titriplex III (ethylenedinitrilotetraacetic acid)	MERCK 8418

25.9 References

Baker's yeast:
W. K. BRONN: "Standardmethoden der Versuchsanstalt der Hefeindustrie 1974" (Standard Methods of the Experimental Institute of the Yeast Industry). Institut für Gärungsgewerbe und Biotechnologie, Berlin.

Pentalenolactone:
J. LEMKE; Doctoral Thesis, Universität Tübingen 1971.
S. HARTMANN, J. NAEFF, U. HEER, and D. MECKE: "Arenaemycine (pentalenolactone): a specific inhibitor of glycolysis"; FEBS Lett. *93*, 2 (1978), 9.

Xanthan:
R. A. MORAINE and P. ROGOVIN; Biotechnol. Bioeng. *12* (1971), 281–391.
G. W. PACE and R. C. RIGHELATO: "Production of extracellular microbial polysaccharides" in "Advances in Biochemical Engineering", Vol. 15. Springer Verlag, Heidelberg–New York–Berlin 1980.

Gluconic acid:
H.-J. REHM: "Industrielle Mikrobiologie". Springer Verlag, Heidelberg–New York–Berlin 1980.
K. Z. ZETELAKI: The role of aeration and agitation in the production of glucose oxidase in submerged culture, II"; Biotechnol. Bioeng. *12* (1971), 379–397.

4 Selected Unit Operations

Chapter 26
Preparation of Media

Alena Cejka

Schering AG
Bergkamen, Federal Republic of Germany

26.1 Introduction
26.2 Components of Industrial Fermentation Media
26.2.1 Carbon Sources
26.2.1.1 Carbohydrates
  Monosaccharides
    Xylose
    Glucose
  Disaccharides
    Sucrose
    Molasses
    Lactose
    Maltose
  Polysaccharides
    Starch
    Dextrin
    Inulin
    Cellulose
26.2.1.2 Alcohols
  Methanol
  Ethanol
  Polyalcohols
26.2.1.3 Carboxylic Acids
26.2.1.4 Fats
26.2.1.5 Hydrocarbons
  Methane
  n-Butane
  n-Pentane
  n-Paraffins
26.2.1.6 Gaseous Substrates
26.2.1.7 Coal

26.2.2 Technical Substrates Mainly Serving as Nitrogen Sources
26.2.2.1 Mineral Salts
26.2.2.2 Organic Nitrogen Sources
    Urea
    Meals
        Soybean Meal
        Cottonseed Meal
        Rapeseed Meal
        Corn Meal
        Corn Gluten Meal
        Fish Meal
    Potato Proteins
    Cornsteep Liquor
    Yeast Autolysates and Dried Yeast
    Protein Hydrolysates
    Distillation Residues
    Gelatin
26.2.3 Inorganic Components
26.2.3.1 Phosphorus
26.2.3.2 Sulfur
26.2.3.3 Other Elements
    Macroelements
    Microelements
26.2.4 Vitamins
26.2.5 Other Medium Additives
26.2.5.1 Amino Acids
26.2.5.2 Precursors and Inducers
26.2.5.3 Detergents
26.2.5.4 Antifoaming Agents
26.2.5.5 Solids
26.2.5.6 Antiseptics
26.2.5.7 Toxins
26.2.5.8 Water
26.2.5.9 Enzyme Preparations
26.3 Storage and Handling of Raw Materials
26.4 Design of Industrial Nutrient Media
26.5 Processing of Wastes
26.6 Commercial References
26.7 References

26.1 Introduction

A great variety of raw materials is currently available as nutrients for industrial fermentations. The spectrum includes, apart from traditional carbon and nitrogenous substances, also various agricultural and industrial by-products and waste materials, which in most countries previously had not been considered as substrates.

In 1969, SOLOMONS [26.1] compiled a detailed survey for England on industrially employed nutrient substances including a list of suppliers. ZABRISKIE et al. [26.2] published a corresponding survey for the USA.

In the following an attempt will be made to describe presently used raw materials for industrial fermentations in Europe, while devoting special attention to publications which have not been treated adequately in review articles of the past.

26.2 Components of Industrial Fermentation Media

Fermentation substrates are classified according to their predominant function within a nutrient medium as carbon-containing and nitrogenous substances, minerals, and vitamins. Carbohydrates, for instance, serve as carbon sources, but also supply oxygen and hydrogen. Complex raw materials, such as molasses or whey, additionally contain nitrogen, minerals, and vitamins. Proteins are an important nitrogen source and are employed, e.g., in the form of meals, which simultaneously supply carbon, oxygen, hydrogen, and minerals.

In order to develop a complex medium for industrial fermentations it is necessary to determine the following criteria:

1. Specific nutrient requirement of selected microorganism and procedure.
2. Exact composition of industrial nutrients, possible modifications during storage, and influence of technological procedure in their fabrication.
3. Properties of nutrients in terms of storage and handling, as well as their behavior as components of complex media during media preparation (sterilization), fermentation, and product isolation.
4. Cost of nutrients.

26.2.1 Carbon Sources

26.2.1.1 Carbohydrates

Monosaccharides

Xylose

On a technical scale, xylose nutrient media have only been employed for the production of glucose isomerase; it may either serve as sole carbon source (LEE et al. [26.3]; POPOV et al. [26.4]) or as an inductor in complex nutrient media containing potato starch (IIZUKA et al. [26.5]), corn starch (MILES LABORATORIES [26.6]; OUTTRUP [26.7]), glucose (WEBER [26.8]; SHIEH et al. [26.9]), or glycerol (VAHERI and KAUPPINEN [26.10]).

Fermentations of D-xylose to ethanol have been carried out on a laboratory scale with *Kluyveromyces marxianus* (MARGARITIS and BAJPAI [26.11]) as well as with mutants of *Candida* and *Saccharomyces* (GONG [26.12]).

A further procedure involving pure, crystalline xylose is the isomerization of D-xylose to D-xylulose by means of glucose isomerase (Sweetzyme® Type Q, Novo Industri, Inc.) and subsequent fermentative degradation of the D-xylulose to ethanol

Glucose

Glucose is frequently used in fermentations intended to yield highly purified products, especially where colored carbohydrate-containing substrate mixtures would require economically inefficient further processing.

Glucose is formed by acid or enzymatic hydrolysis of corn or potato starch. Various companies offer glucose either as a powder, a paste, or a syrup (Maizena GmbH; Roquette Frères, SA). Characteristics of the different brands are summarized in Tables 26.1 and 26.2.

Dextrose monohydrate is the most frequently employed form of glucose ("Cerelose" by Maizena GmbH, Roferose® by Roquette Frères, SA). It is used in the production of antibiotics, steroids, amino acids, butyric, itaconic and tartaric acid, yeastlytic enzyme, xanthan gum, and heteropolysaccharides (Table 26.3).

Starch dextrose is utilized, for instance, in steroid conversions (SCHERING AG [26.33]).

with *Saccharomyces cerevisiae* or *Candida tropicalis* (CHIANG et al. [26.13]).

Table 26.1. Composition of Commercial Glucose Substrates (Courtesy of Maizena GmbH) (in %)

Constituents	Type of Product		
	Dextrose Hydrate	Starch Dextrose	Liquid Dextrose
	Cerelose 02001	Glucodex® 15050	Ceredex® 02761
Dry matter	91	61	71
Saccharides (d.m.)	100	99	99.9
Spectrum of sugars (% of saccharides):			
D-Glucose	100	65	96
D-Fructose	0	3	0
Maltose	0	10	2
Higher dextrins	0	22	2
Ash	0.02	1.0	0.1

Table 26.2. Composition of Commercial Glucose Syrups (Courtesy of Roquette Frères, SA)

Constituents	Type of Product		
	L 99	74/968	74/904
Dry matter (%)	70 ± 0.5	74 ± 0.5	74 ± 0.5
DE[a] Value	98 – 99	96 – 98	90 – 94
Composition of dry matter (%)			
Monosaccharides	min. 96	92 – 96	80 – 88
Disaccharides	max. 4	4 – 8	12 – 20
pH in solution	—	3.5–5.0	3.5–5.0
Ash (%)	max. 0.1	max. 0.2	max. 0.2

[a] DE Dextrose Equivalent

Table 26.3. Dextrose as Carbon-Source in Processes of Commercial Importance

Product	Microorganisms	Reference
Antibiotic U 60.394	*Streptomyces wolensis*	DOLAK and JOHNSON [26.14]
Antibiotic	*Streptomyces cattleya*	KEMPF and WILSON [26.18]
Lysolipin	*Streptomyces violaceoniger*	ZAEHNER et al. [26.15]
Cephalosporin	*Streptomyces rochii*	BROWN et al. [26.19]
Daunorubicin	*Streptomyces* sp.	McGUIRE et al. [26.16]
Gentamicin derivatives	*Micromonospora purpurea*	YAMAMOTO and DAUM [26.17]
Steroids	*Mycobacterium fortuitum*	WOVCHA and BROOKS [26.20]
	Mycobacterium sp.	MITSUBISHI Chem. Ind. Ltd. [26.21a]
Butyric acid	*Clostridium butyricum*	SHARPEL and STEGMAN [26.25]
Itaconic acid	*Aspergillus terreus*	COURTAULDS, Ltd. [26.26]
Tartaric acid	*Gluconobacter suboxydans*	KODAMA et al. [26.27]
Xanthan gum	*Xanthomonas* sp.	PACE and COOTE [26.29]
Heteropolysaccharides	*Alcaligenes* sp.	KANG and VEEDER [26.31]
Polysaccharides	Review	LAWSON and SUTHERLAND [26.30]
D-Biotin	*Sporobolomyces carniocolor*	SHIBATA et al. [26.23]
Yeast-lytic enzyme	*Cytophaga* sp.	ASENJO et al. [26.28]
Amino acids	Review	KINOSHITA and NAKAYAMA [26.24]
Alcohols	Review	MICHAELS and BLANCH [26.32]

Table 26.4. Composition of Hydrol (KLAUSHOFER [26.34])

Constituents	
Dry matter (refr.) (%)	50.80–55.70
DE[a] (d. m.)	78.8–81.6
Glucose (d. m.) (%)	68.3–78.0
Maltose (d. m.) (%)	2.4–5.2
Non-fermentable oligosaccharides and dextrins (d. m.) (%)	5.2–8.0
pH	3.32–3.60

[a] DE Dextrose Equivalent

Hydrol is a syrupy residue of dextrose production from corn starch containing 68–78% dextrose (dry weight), as well as considerable amounts of non-fermentable substances, which mainly are dextrins, reversion products, and gentobiose (Table 26.4; KLAUSHOFER [26.34]).

Hydrol has been used in the production of proteases with *Aspergillus awamori* (POPOVA [26.35]) and was found to be a potential carbon source for the synthesis of valine (POPOVA and MURGOV [26.36]). It has also been used in penicillin production (MATELOVA [26.37]), but its use for this purpose is limited because of low yields (NEWMAN et al. [26.38]). Baker's yeast has been made from a mixture of hydrol with yeast extract or molasses (KLAUSHOFER [26.34]).

Disaccharides

Sucrose and lactose are used in industrial fermentations as pure substances and in the form of raw materials containing these sugars, i.e., molasses and whey, respectively. Maltose is used in the form of malt, malt extract, and wort.

Sucrose

Pure, commercial-grade sucrose has been the main substrate for industrial-scale, sub-

Table 26.5. Composition of European Beet Molasses

Type of Molasses	Raw Sugar and Refinery Sugar			"Lime Separation" (Steffen)	Quentin
	OLBRICH [26.47]	MALANOWSKA and LABENDZINSKI [26.48]	LANGPAULOVA and JANDA [26.49]	TERTYSHNYI et al. [26.50]	WIELAND and KOVACS [26.51a]
	1952–1954[a]	1966–1967[b]	1975–1976[c]	1979[d]	1978/79[e]
Dry solids (%)	72.4–86.0	78.9–86.2	79.5–82.2	71.0–80.6	82.0–84.0
Sucrose (%)	55.7–69.0	43.8–54.4	50.0–54.6	43.0–56.2	45.0–47.6
Invert sugar (%)	0–3.41	1.19–4.5	0.12–1.56	4.0–5.4	—
Raffinose (%)	0–4.20	0–2.75	1.24–1.54	—	—
Ash (%)	9.83–19.46	5.3–9.6	11.14–14.33	4.59–8.32	11.43–11.78
Total nitrogen (%)	1.170–2.946	0.82–1.90	1.45–1.65	0.38–2.21	1.78–2.11
Purity quotient	55.2–67.1	51.8–65.1	62.3–66.5	58.9–76.0	53.5–58.0
pH	5.51–10.12	6.4–8.1	7.5–8.5	5.54–6.57	7.2–7.9

Analysis of: [a] 11 European countries   [b] 5 factories   [c] 8 factories   [d] 9 factories   [e] 3 factories

merged citric acid production in England (J. E. Sturge) and Austria (Jungbunzlauer Spiritus, Chem. Fab.) (MIALL [26.39]). Lactic acid has also been produced from pure sucrose; this minimizes lactic acid purification (KRUMPHANZL and DYR [26.40]). Furthermore, sucrose serves as a carbon source in complex nutrient media for the production of, e.g., L-tryptophan (MAKSIMOVA et al. [26.41]), ergot alkaloids (WACK et al. [26.42]), glucose-6-phosphate dehydrogenase (NAKAJIMA et al. [26.43]), L-lysine (WALCZAK and OBERMAN [26.44]), chlortetracycline (WELWARD and HALAMA [26.45]), and xanthan gum (PACE and COOTE [26.29]).

The most diverse use of sucrose is in the form of molasses.

Molasses

Molasses are among the most important raw materials of the fermentation industry. They are by-products of sugar-beet and sugarcane extraction procedures and are correspondingly called beet or cane molasses.

Molasses are the major raw material for the production of baker's yeast, citric acid, feed yeasts, acetone/butanol, organic acids, and amino acids. They are often added as a supplementary carbohydrate source to nutrient media for the production of antibiotics and enzymes. Citric acid is mainly produced with beet molasses.

Beet molasses are syrupy, brownish liquids with a content of approximately 50% sucrose. The texture and composition of beet molasses is determined by the technology of the sugar extraction process and by the quality of the crop (agricultural technology, climate, storage of beets, etc.). The fermentation industry predominantly uses raw sugar or refiner's molasses; Steffen and

Table 26.6. Composition of Carbonate-free Ash of Beet Molasses (DYR et al. [26.62]; OLBRICH [26.63], [26.64])

Constituent	% of Ash
K_2O	66.15–72.74
Na_2O	9.42–15.86
CaO	4.37–7.09
MgO	0–0.78
Fe_2O_3	0.01–0.45
P_2O_5	0.23–0.80
SiO_2	0–1.45
Cl	8.51–11.32
SO_3	0.59–2.56

Table 26.7. Organic Acids in Beet Molasses

Acid	SCHIWECK and HABERL [26.46] g/100 g Non-Sugar	MALANOWSKA and LABENDZINSKI [26.48] g/100 g Molasses
Formic acid	0.681–1.557	—
Acetic acid	0.942–1.843	0.38–0.99
Propionic acid	—	0.05–0.33
Butyric acid	—	0.10–0.85
Isobutyric acid	—	0.04–0.16
Valeric acid	—	0.01–0.11
Capronic acid	—	0–0.10
Lactic acid	3.930–12.900	—
Citric acid	0.472–1.742	—
Malic acid	0.610–1.837	—

Table 26.8. Content of Total, α-Amino, and Betaine Nitrogen in Beet Molasses

Constituent	OLBRICH [26.64]	CEJKOVA[a] [26.56]	GINTEROVA[b] [26.65]
Total nitrogen (%)	1.80	1.23–1.74	0.99–1.48
α-Amino nitrogen (%)	0.58	0.12–0.28	0.11–0.19
Betaine nitrogen (%)	1.12	0.37–0.80	0.44–0.69

[a] 13 factories were examined during the sugar campaigns 1962/63 and 1963/64
[b] 16 factories were examined during five sugar campaigns 1962–1967.

Quentin molasses are rarely employed. While sugar raw and refiner's molasses hardly show any difference in their chemical composition, Steffen molasses have a high content of raffinose and Quentin molasses a high concentration of magnesium. Examples of the remarkably different compositions of European molasses are given in Table 26.5.

In addition to sucrose, molasses contain the following mono- and oligosaccharides: glucose, fructose, raffinose, fructosyl saccharoses, and galactosyl m-inosite (SCHIWECK and HABERL [26.46]; SCHIWECK [26.53]). The non-sugar components, making up about 30% of the dry matter, are composed of inorganic (Table 26.6) and organic, nitrogen-free (Table 26.7) and nitrogenous compounds (Table 26.8). Glutamic acid accounts for 40–55% of the assimilable nitrogen (OLBRICH [26.47]).

The following concentrations of growth factors have been determined: biotin 17.7–21.8 µg/kg, Ca-pantothenate 23–70 mg/kg, m-inositol 0.69–1.83 g/kg (SCHIWECK and HABERL [26.46]), and thiamine 26–75 mg/kg (TERTYSHNYI et al. [26.50]). Quentin molasses have only 2/3 the abovementioned concentration of biotin (KOVACS and WIELAND [26.51]). Furthermore, volatile nitrogenous pyrazines and pyrroles, as well as furans and phenols have been detected, all in concentrations in the ppm range (ZAUNER et al. [26.52]; SCHIWECK [26.53]).

The suitability of molasses for industrial fermentations cannot be evaluated according to their origin and chemical composition. The only effective means for such an evaluation is the biological test.

Over the years repeated systematic analyses of various types and charges of beet molasses have been carried out in order to find the most important criteria determining productivity and quality. The critical parameters in the production of baker's yeast (STROS and SYHOROVA [26.54]; GINTEROVA [26.55]), citric acid (CEJKOVA [26.56]), and ethanol (GINTEROVA [26.57]) were determined to be: α-amino and total nitrogen content, betaine nitrogen, the degree of purity, volatile organic acids, lime, and the buffering capacity. Undesirable compounds are: betaine, volatile acids, and lime, as well as colloids formed from pectin and caramelized pigments at concentrations reaching 1.4–12.9% (MALANOWSKA and LABENDZINSKI [26.48]), or caramelan at concentrations of 0.4–1.3% (SHVETS et al. [26.58]). Other unwelcome contaminants are heavy metals (Pb 1.0–68.0 ppm; Cu 1.6–20.0 ppm; Fe 56.0–290.0 ppm), nitrates and nitrites at, respectively, 233.7–1208.2 and 7.3–17.0 mg/100 g non-carbohydrate (SCHIWECK and HABERL [26.46]), SO_2 (OLBRICH [26.47]), and pesticide residues at concentrations in the ppb and ppm range (KUBACKI and KASPROWICZ [26.59]; HUNCIKOVA [26.60]; SCHIWECK [26.53]).

The quality of the molasses varies during the sugar campaign depending on the time of the onset of beet extraction. For the fermentation industry the most suitable molasses are those usually obtained during the months of October and November from freshly harvested beets with a high content of amino acids (MALANOWSKA and LABENDZINSKI [26.48]; GINTEROVA [26.55], [26.57]; HUNCIKOVA [26.60]; LABENDZINSKI [26.61]).

Storage before processing causes changes in the beets, which lead to a decrease of the quality of the molasses; the contents of amino compounds decreases in December and January, with an increase of betaine, nitrates, nitrites, and volatile acids (Table 26.9).

During the last years the quality of molasses has been changed due to optimization of the technology of beet sugar extraction and more intense and effective fertilization of crops. With higher sugar extraction, the sucrose content of the molasses decreased between 1954 and 1979 from 63.3 to 59.9% (EMMERICH [26.66]) and between 1971 and 1977 from 64.1 to 60.9% (BESPALIJ and SAVSHUK [26.67]). Changes of agricultural technology led to a higher content of total nitrogenous compounds, ash, calcium, and electrolytes, thus reducing the usefulness of the molasses for yeast and ethanol production (EMMERICH [26.66]; BESPALIJ and SAVSHUK [26.67]; STUDNICKY [26.68]).

In processes in which they are the sole carbon source, the molasses are pretreated and the inhibitors are partly removed. For the production of yeast and ethanol the molasses are often merely neutralized with $CaCO_3$ (REHM [26.69]). For many of these processes the molasses are only boiled in an

Table 26.9. Changes in the Composition of Beet Molasses During the Sugar Campaign

Constituent	Campaign		Reference
	Begin	End	
Amino acids (g/100 g non-sugar)	4.37–4.91	2.53	BELOVA [26.70]
α-Amino nitrogen (%)	1.148–1.170	0.455	GINTEROVA [26.57]
Total nitrogen (%)	1.74	1.41	FARAGO et al. [26.72]
Betaine nitrogen (% total N)	36.37	69.34	GINTEROVA [26.57]
Invert sugar (%)	0.313	2.043	GINTEROVA [26.71]
Raffinose (%)	0.9	2.0	EMMERICH [26.66]
Volatile acid (%)	0.91	1.36	GINTEROVA [26.57]
P_2O_5 (%)	0.036	0.008	GINTEROVA [26.55]
Buffering capacity mL 1 N H_2SO_4 pH 6.5→5.5	5.12	0.99	GINTEROVA [26.55]
pH	7.6	5.8	FARAGO et al. [26.72]

acidic or alkaline medium and, after settling out, separated from the precipitate (DYR et al. [26.62]). For citric acid production the molasses are boiled with potassium ferrocyanide and fermented together with the precipitate (MEZZADROLI [26.73]; LEOPOLD and VALTR [26.74]; LEOPOLD et al. [26.75]; review: MIALL [26.39]). A possible inhibitory effect of free potassium ferrocyanide can be eliminated by adding sodium dithionite ($Na_2S_2O_3$), potassium pyrosulfite ($K_2S_2O_5$), or sodium bisulfite ($NaHSO_3$) (ILCZUK [26.76]).

Phosphorus must be added to the medium in the form of superphosphate or phosphoric acid, and nitrogen in the form of ammonium sulfate. In baker's yeast production, Quentin and Steffen molasses are enriched with cornsteep liquor (KOVACS and WIELAND [26.51]; BELOVA and ERMAKOVA [26.385]).

For the preparation of nutrient media containing other carbohydrate nutrients in addition to molasses, a pretreatment of the molasses is usually not necessary.

Sugarcane molasses are employed in industrial fermentations as refiner's and as raw cane sugar molasses, the latter also called blackstrap molasses. Another kind of cane molasses are the high-test molasses, a concentrate of purified, partly inverted sugarcane juice. They are not the by-product of the sugar crystallization process.

High-test molasses are produced mainly in the Republic of South Africa, Cuba, and Australia, but there only if specially ordered. The composition of cane molasses is shown in Table 26.10.

Cane molasses are characterized by a high content of invert sugar and biotin: in blackstrap molasses the biotin content is 800–1000 µg/kg (WATANABE and SATO [26.57a]; YANCHEVSKII et al. [26.77]). They contain SO_2, volatile acids, nitrates, and colloids which are undesirable in microbial fermentations. Microbial conversion of easily assimilable carbohydrates and the low lime content are responsible for changes in the composition of the molasses during storage, which has a negative effect on its usefulness for fermentations.

Table 26.10. Composition of Cane Molasses

Constituent (%)	Blackstrap Molasses		Refinery Molasses		High-Test Molasses	
	OLBRICH [26.64]	YANCHEVSKII et al. [26.77]	WHITE [26.78]	TERTYSHNYI et al. [26.50]	OLBRICH [26.63]	PATURAU [26.79]
Dry substance	77–84	80.8	78–85	72.9	81–86	86
Polarization	—	32.60	—	—	—	—
Invert sugar	52–65	19.07	50–58	4.6	72–79	77
Total sugar	—	55.08	—	43.0	—	—
Ash	7–11	11.15	3.5–7.5	8.32	2–3	2.5
Total nitrogen	0.4–1.5	0.5	0.08–0.5	1.52	0.07–0.20	0.15
pH	4.5–6.0	6.0	—	5.98	6.0	—
P_2O_5	0.6–2.0	0.17	0.009–0.07	—	0.2–0.6	—
CaO	0.1–1.1	1.38	0.15–0.8	—	0.08–0.30	—
MgO	0.03–0.10	—	0.25–0.8	—	0.01–0.08	—
K_2O	2.6–5.0	—	0.8–2.2	—	0.7–1.4	—

Blackstrap molasses serve as the main raw material for the production of ethanol, baker's and feed yeast, acetone, and butanol (LYNESS and DOELLE [26.80]; YANCHEVSKII et al. [26.77]; SAITO [26.81]; QADEER et al. [26.82]; FOUAD et al. [26.83]; SPIVEY [26.84]; VOLESKY et al. [26.85]); in Europe, however, it is not used for production of baker's yeast (OLBRICH [26.47]). In Brazil cane molasses and sugarcane juice are employed in ethanol production (AMORIM and CAMPOS [26.86]). The molasses are reported to be utilized in combination with other nutrients to produce antibiotics (BAGHLAFF et al. [26.87]; SHEHATA et al. [26.88]), amino acids (MISRA et al. [26.89]; TOSAKA et al. [26.90]; SAITO [26.81]), and organic acids (BASILICO et al. [26.91]; YAMADA and HIDAKA [26.92]; HAMISSA et al. [26.93]; TAURO et al. [26.94]).

High-test molasses are used in ethanol production (PATURAU [26.79]) and in the fermentation of citric acid (MILES LABORATORIES [26.95]).

Prior to the preparation of a medium from blackstrap molasses, colloids and heavy metals are removed by clarification with superphosphate, boiling, and centrifuging (MISRA et al. [26.89]), or by treating the diluted molasses with bentonites (NASIM et al. [26.96]), potassium ferrocyanide, EDTA, methylene blue, and subsequent centrifuging (ABOU-ZEID et al. [26.97]). Refinery molasses are only centrifuged as a diluted solution (HAMISSA et al. [26.93]). High-test molasses are treated by ion exchange (MILES LABORATORIES [26.95]).

High-ash sucrose-syrup, resulting from crystalline sucrose refining and containing 3–25% ash and 40–83% sucrose in dry matter, is used for the production of penicillin and cephalosporins (NEWMAN et al. [26.38]; RIZZUTO et al. [26.98]).

In recent years, sugar-beet pulp has been introduced for the industrial production of ethanol (MICHAELS and BLANCH [26.32]; KIRBY and MARDON [26.99]; LARSEN et al. [26.100]); pectinases (VORONOVA et al. [26.101]; KALUNYANTS et al. [26.102]; SHINKARENKO et al. [26.103]), fungal cellulase (FENIKSOVA et al. [26.104]), and protein feed (JANICKI et al. [26.105]; ATAMANYUK et al. [26.106]). The pulp is either extracted and the sugarbeet juice used as a nutrient medium or a kind of slurry is produced and fermented. Sugarcane juice is also converted to ethanol in a similar manner (LYNESS and DOELLE [26.80]). A new technology with the above procedure is called the Ex-Ferm process (ROLZ [26.107]; ER-EL et al. [26.108]).

Lactose

As a carbon source in nutrient media, lactose exerts only a minimal repressive effect and thus can be used in high concentration (DEMAIN et al. [26.109]). Despite its slow assimilation, leading to extended fermentation periods (SEHGAL et al. [26.110]), lactose is often employed in antibiotics and especially in penicillin production (EL-SAIED et al. [26.111]; EL-MARSAFY et al. [26.112]; RYU and HOSPODKA [26.113]; NEWMAN et al. [26.38]; MATELOVA [26.37]). Lactose is also used to produce alkaline proteases from *Bacillus subtilis* (KASZAB and KOVACS [26.114]).

Thermo-sterilization partly converts the lactose contained in more complex media into non-assimilable lactulose. In the bacterial production of α-amylase 24 and 36% of the lactose contained in media with originally 8 and 10% lactose, respectively, were 'lost' by sterilization (THAYANITHY et al. [26.115]).

Whey is the 'serum' obtained from milk, during the production of cheese, after removing fats and casein. We generally differentiate between sweet and sour whey. Sweet whey is formed in the preparation of Cheddar, Swiss, and Italian cheese; acid whey from Cottage and Cream cheese (BERNSTEIN and TZENG [26.116]). The pH values are in the range of pH 6.3–6.5 for sweet and pH 4.1 for acid whey (REHM [26.69]).

On an industrial scale, whey is mainly processed into protein feed and, in part, employed in the preparation of ethanol. Whey is used in the following forms:

1. liquid or dried, sweet or acid whey,
2. deproteinized whey, via thermal or chemical treatment (DE SANCHEZ

Table 26.11. Composition of Whey and Permeate

Constituent %	Whey		Permeate Dry
	Raw BERNSTEIN and TZENG [26.116]	Dry BEAUSEJOUR et al. [26.123]	VANANUVA and KINSELLA [26.119]
Dry matter	5.6–6.3	—	96
Protein	11.9–14.4	11.4	3
Lactose	69.0–78.0	78.0	75
Fat	—	1.4	—
Ash	5.0–8.2	9.2	—
Minerals	8.0–9.0	—	12

and CASTILLO [26.117]; CHEN and ZALL [26.118]),
3. permeate obtained by ultrafiltration with or without subsequent reverse-osmosis (VANANUVA and KINSELLA [26.119]; FRIEND et al. [26.131]; MOON et al. [26.129]; REESEN and STRUBE [26.120]),
4. demineralized permeate obtained via ion exchanger (TENNEY [26.121]).

The composition of whey and permeate is shown in Table 26.11.

The type of cheese as well as production procedures determine the composition of the whey. The content of essential amino acids varies largely in the different types of whey. The principal amino acids found were histidine, aspartic acid, alanine, and proline (SKVORTSOVA [26.122]) and leucine, threonine, isoleucine, and valine (ZABRISKIE et al. [26.2]). Concentrations of vitamin A and choline are high (ZABRISKIE et al. [26.2]). The processing of whey and permeate is summarized in Table 26.12.

Maltose

Maltose is present in malt, malt extract, and brewer's wort. Malt is obtained from germinated barley after drying and milling, and is characterized by its high amylase activity. In beer brewing the malt is mashed in water in order to obtain a malt extract. Hops extract is then added, thus yielding the wort.

Malt extract is seldom used as nutrient for industrial production other than in brewing. However, it is an important adjunct to nutrient media designed for the storage of strains of yeasts and fungi (HAYNES et al. [26.136]; BEECH and DAVENPORT [26.137]; ANTHEUNISSE [26.138]; KUDRYAVTSEV et al. [26.139]; REHBERG [26.140]; HALL [26.141]). Malt extract is commercially offered as a syrup (HALL [26.141]) or in dry form. The composition of wort and malt extract is given in Table 26.13.

Polysaccharides

Starch, inulin, and cellulose are frequently employed carbon sources for industrial fermentations.

Starch

Starch is either used in pure form as an amorphous powder or as a component of potatoes, wheat, oats, barley, sorghum, corn, tapioca, and sago (Table 26.14). These are the main raw materials for the preparation of ethanol, acetone, and butanol.

Starch is the most significant reserve carbohydrate of plants and is synthesized in the form of small granules with a typical structure. These starch granules are insoluble in water and very slowly decomposed by microorganisms. In heated water the starch granules swell, become gelatinous, and lose their firmness. The different types

Table 26.12. Whey and Permeate Used as Fermentation Substrates

Product	Substrate	Microorganism	Reference
Single Cell Protein[a] (SCP)	Whey	*Saccharomyces fragilis*	BERNSTEIN and TZENG [26.116]
	Whey	*Kluyveromyces fragilis*	BEAUSEJOUR et al. [26.123]
	Whey	*Chlorella* sp.	AL'BITSKAYA et al. [26.125]
	Deproteinized whey	*Kluyveromyces fragilis*	DE SANCHEZ and CASTILLO [26.117]
	Permeate	*Saccharomyces cerevisiae*	VANANUVA and KINSELLA [26.119]
SCP + carotene	Whey + starch + malt	*Saccharomyces cerevisiae*	MOEBUS and TEUBER [26.126]
SCP + vitamin B 12	Whey + beet molasses	*Rhodotorula gracilis* K1	ATAMANYUK and VAKAR [26.127]
SCP + fatty acids	Whey	Propionic acid bacteria	LUTSKOVA [26.128]
	Whey + permeate	*Candida curvata*	
		Trichosporon cutaneum	MOON et al. [26.129]
Ethanol	Whey	*Kluyveromyces fragilis*	MOULIN and GALZY [26.130]
	Deproteinized whey	*Saccharomyces fragilis*	CHEN and ZALL [26.118]
	Permeate	*Kluyveromyces fragilis*	REESEN and STRUBE [26.120]
	Permeate + grain	*Kluyveromyces fragilis* + *Saccharomyces cerevisiae*	FRIEND et al. [26.131]
Porphyrins + corrinoids	Whey + Co^{++} + Fe^{++}	*Propionibacterium shermanii*	GIEC et al. [26.132]
Nisin	Whey + potato syrup	*Streptococcus* sp.	MELLER et al. [26.133]
Citric acid	Permeate	*Aspergillus niger*	SOMKUTI and BENCIVENGO [26.134]
Brewer's yeast	Demineralized, hydrolyzed permeate + wort	Brewer's yeast	TENNEY [26.121]
Baker's yeast	Whey + beet molasses	Baker's yeast	HROBONI et al. [26.135]

[a] Review: MORESI [26.124]

Table 26.13. Composition of Wort and Malt Extract

Wort (MOSTEK [26.142])
Reducing sugar as maltose (%)	6.07
Dextrins (%)	1.27
Total nitrogen (mg/100 mL)	82.9
α-Amino nitrogen (mg/100 mL)	12.89

Amino acids (mg/100 mL)
Lysine	10.68	Alanine	8.00
Histidine	Trace	Valine	12.93
Arginine	7.03	Methionine	0.56
Aspartic acid	6.29	Isoleucine	7.98
Threonine	6.58	Leucine	16.58
Glutamic acid	8.35	Tyrosine	7.12
Proline	32.32	Phenylalanine	12.35
Glycine	11.71	γ-Amino-butyric acid	9.68

Malt extract (Maltzin Dry PT) (Courtesy of Diamalt AG)
Moisture (%)	2.4
Reducing sugar as maltose (%)	70.1
Ash (%)	1.4
Acidity as lactic acid (%)	1.44
pH, 20% sol.	1.7

pletely. Gelatinized starch may serve many microorganisms as a carbon source.

For ethanol production gelatinized starch is saccharified by acid or enzymatic hydrolysis into the assimilable maltose or dextrose. The classical saccharification with malt has largely been replaced by microbial enzyme processes. According to MAISCH et al. [26.143] the only process in which malt saccharification has been maintained without an addition of enzymes is the manufacturing of Bourbon whiskey.

A direct saccharification of sweet potato starch for ethanol production without prior gelatinization has been reported by a treatment with glucoamylase obtained from a *Rhizopus* species (MATSUOKA et al. [26.144]).

The saccharification of potato mash by means of the enzymatic preparations Thermamyl® 60 and Novo L 120 (Novo Industri A/S) has been described by ROSEN [26.145]. Cassava and corn are saccharified by Thermamyl and Fungamyl (NIELSEN

Table 26.14. Composition of Starch Crops Employed in Nutrient Media (DYR et al. [26.62])

Constituent (%)	Raw Materials						
	Potato	Jerusalem Artichoke	Barley	Wheat	Oat	Corn	Rye
Water	76.64	79.12	13.4	13.0	13.0	12–14	13.0
Ash	1.08	1.16	3.32	1.6	3.4	1.4	1.9
Protein (N × 6.25)	1.97	1.80	10.06	14.0	10.9	11.0	12.5
Nitrogen-free substances	19.17	16.40	65.21	67.6	58.5	69.0	68.3
Starch	17.02	—	53.04	—	—	60–63	50–60
Inulin	—	16.0	—	—	—	—	—
Pentosans	—	—	—	—	—	—	10.0
Fat	0.15	0.18	1.94	1.8	4.7	5.0	2.1
Cellulose	0.98	1.26	6.07	2.2	9.5	2.0	2–3

of starch gelatinize at different temperatures: potato starch 65 °C, corn starch 75 °C, oat starch 85 °C, and rye, wheat, barley, and rice starch 80 °C (DYR et al. [26.62]). Complete gelatinization is achieved at a minimal ratio of starch to water of 1:4 and a temperature of 110 °C. At temperatures of 120–130 °C starch liquifies com-

and ROSENDAL [26.146]); saccharification of cassava tubers and rye have been described by KOSARIC et al. [26.147] and OFFER et al. [26.148], respectively. Next to the industrial products from isolated enzymes, crude microbial enzyme preparations have been reported, as e.g., malt in conjunction with a crude xylanase preparation from *As*-

pergillus awamori (RODZEVITSCH et al. [26.149]); wheat bran koji (ZIFFER and JOSIF [26.150]); a mixture of amylases extracted from *Aspergillus oryzae* with glucoamylase from *Aspergillus awamori*, cellulase from *Trichoderma konigii* (LOSYAKOVA et al. [26.151a]); saccharifying enzymes from *Endomycopsis bispora* with glucoamylases from *Aspergillus batatae* (TÄUFEL et al. [26.152]). Enzymatic preparations used to replace malt have led in all cases to a complete saccharification of the raw materials and to higher ethanol yields. Such preparations rapidly hydrolyze dextrins to fermentable sugars, thus preventing limitation of the rate of yeast fermentation. The quantity of dextrins within the mash saccharified with malt varies greatly depending on the quality of the malt.

In acetone/butanol production the gelatinized starch is often added directly to cultures of *Clostridium acetobutylicum* (FOUAD et al. [26.83]). Raw potato starch can be added to certain *Streptomyces* cultures to yield an alkaline proteinase (YAMAMOTO et al. [26.153]). Investigations have been reported on a direct ethanol fermentation employing raw starch by means of a selected amylolytic, homofermentative culture of *Schwanniomyces alluvius* (CALLEJA et al. [26.154]).

Saccharified starch is industrially used in the preparation of 1. β-mannanase (BASKIS and FIRANTAS [26.155]) and α-amylase (WINDISH and MHATRE [26.156]) from *Bacillus subtilis*, 2. pullulan polymers from *Aureobasidium pullulans* (SHIPMAN and FAN [26.157]), 3. proteins from *Candida utilis* (DESCHAMPS [26.158]), 4. ribonuclease from *Bacillus subtilis* (BASKIS and RAGAVICIUS [26.159]), 5. baker's yeast (EDELMAN and PENTTILA [26.160]), 6. inositol and guanosine from *Bacillus subtilis* and *Bacillus megaterium* (SHIBAI et al. [26.161]), and 7. antibiotics (BHADRA et al. [26.162]).

Amylaceous raw materials, such as bran, are employed as solids, e.g., for many fermented foods in the orient (ROSE [26.163]), in the synthesis of ochratoxin (LINDENFELSER and CIEGLER [26.164]), the production of α-amylases from *Bacillus subtilis* (QADEER et al. [26.164a]), of fungal amylases (WINDISH and MHATRE [26.156]), and pectinases from *Aspergillus foetidus* (LOSYAKOVA et al. [26.151]).

Dextrin

Dextrin is the spray-dried product of α-amylase degradation of starch. The components of a commercial dextrin are listed in Table 26.15.

Dextrin has been used either as sole carbon source or in combination with dextrose in nutrient media to produce some antibiotics (ASHY et al. [26.165]; FLICKINGER and PERLMAN [26.166]; REUSSER and SLECHTA [26.167]; TOYO JOZO [26.168]).

Inulin

Inulin is the reserve substance of Jerusalem artichokes, chicory, and other com-

Table 26.15. Composition of Maltodextrins (Courtesy of Roquette Frères, SA)

Constituent	Type of Product				
	MD 05	MD 25	MD 40	MD 50	MD 63
Moisture (% max.)	5	5	5	5	6
Ash (% max.)	0.3	0.3	0.3	0.3	0.3
Protein (% max.)	0.15	0.15	0.15	0.15	0.15
pH in sol.	4.8–5.2	4.8–5.2	4.8–5.2	4.8–5.2	4.8–5.2
DE[a] Value	18–20	28–31	38–41	45–48	60–64
Spectrum of sugars (in %):					
Glucose	2–3	8–11	3	5	30–32
Maltose	6–8	8–10	37	50	34–36
Polysaccharides	92–89	84–79	60	45	36–32

[a] DE – Dextrose Equivalent

posites (Asteraceae). It is a linear polymer of $\beta(2\rightarrow 1)$D-fructose and $\alpha(1\rightarrow 2)$D-glucose. The ratio of fructose to glucose is seldom lower than 80:20 (KIERSTAN [26.169]).

Jerusalem artichokes contain about 25% inulin and have been used in France in ethanol production. Inulin was first hydrolyzed to levulin and fructose by inulinase from *Aspergillus niger* and subsequently fermented by *Saccharomyces cerevisiae* (DYR et al. [26.62]). A direct fermentation of a tuber extract without prior hydrolysis has been carried out by employing inulinase-active cultures of *Kluyveromyces* (GUIRAUD et al. [26.170]) and *Kluyveromyces marxianus* ATCC 12708 (DUVNIAK et al. [26.171]).

In the ethanol fermentation of chicory by means of *Kluyveromyces* cultures, the hydrolysis of inulin has been supported by adding the cellulose-degrading enzyme preparations Celluclast® or Pectinex® (Novo Industri A/S) thereby increasing the yields of ethanol (ANONYMOUS [26.172]).

Cellulose

Cellulose exists as a crystalline substance in plant cell walls together with hemicellulose and lignin at a ratio of approximately 4:3:3. It is a homogeneous polymer of glucose, whereas hemicellulose is a mixed polymer of pentoses (xylose and arabinose), hexoses (mannose), and various glyco acids. Lignin is a macromolecular polyphenol. The crystalline cellulose in cell walls is resistant to hydrolysis, also by the 'shielding' effect of the surrounding lignin. Hemicellulose is easily hydrolyzed into mono- and oligosaccharides by acids, bases, and enzymes (TSAO et al. [26.173]).

Cellulose is contained in agricultural and industrial wastes of plant origin. Such agricultural wastes are straw of all kinds, corn cobs and stalks, bagasse, and other plant residues; industrial wastes of this kind result from paper and wood processing. Cellulose may also be obtained from peat.

Microbial conversions of cellulose into protein feed, ethanol, and other products have been considered by many authors (reviews: HAN [26.174]; GHOSE [26.175]). Industrial plant waste materials are now processed on a commercial scale, while for agricultural plant residues this only occurs sporadically.

One of the crucial steps in cellulose processing is the decomposition of the association between lignin and cellulose. Different pretreatment techniques of cellulose-containing raw materials permit removal of the lignin 'barrier' around the cellulose, e.g.: intense mechanical degradation followed by steam treatment (DATTA [26.176]), γ-irradiation of high intensity (KAMAKURA and KAETSU [26.177]; HAN and CIEGLER [26.178]; HAN et al. [26.179]), contacting with ozone, SO_2 (MIRON and BEN-GHEDALIA [26.180]), or NH_4OH (HAN [26.181]), steeping in alkaline solutions (DATTA [26.176]), or treatment with Cadoxen (anaqueous, alkaline solution of ethylenediamine) (TSAO [26.182]; GHOSE and GHOSCH [26.183]; review: FAN et al. [26.184]). Bagasse is often delignified by mere alkaline treatment.

The pretreated substrate is either saccharified by a mild acid hydrolysis or by enzymes. Plant material degraded by acid hydrolysis contains mainly pentose monosaccharides from hemicellulose, e.g., 8.3% xylose, 1.9% glucose, and 1.7% arabinose (dry weight) (ROSENBERG et al. [26.185]). These hydrolysates may either be directly fermented into ethanol by *Bacillus macerans* (ROSENBERG et al. [26.185]) or used as a substrate for the preparation of protein feed by means of certain strains of *Candida* (GONZALES-VALDES and MOO-YOUNG [26.186]; VOLFOVA et al. [26.187]). Furfural is generally present in such hydrolysates; thus, for yeast production it is necessary to decrease the content of furfural, if in excess of 0.01%, by aeration (BELENKIJ and KOMAROVA [26.188]) or by UV irradiation (BELENKIJ et al. [26.189]). Most frequently a direct processing of the hydrolysate into protein feed is accomplished by employing such fungi as, e.g., *Chaetomium cellulolyticum* (CHAHAL et al. [26.190]; MOO-YOUNG et al. [26.191]). Delignified bagasse has mainly been processed into protein feed with *Cellulomonas* bacteria (PEROTTI DE GALVEZ and MOLINA [26.192]; PLANES et al. [26.193]; ENRIQUEZ and RODRIGUEZ

[26.194]) or with *Aspergillus terreus* (GARG and NEELAKANTAN [26.195]).

Plant residues enzymatically saccharified by cultures of *Cellulomonas, Trichoderma reesei, Aspergillus wentii*, or isolated enzyme preparations are treated with *Saccharomyces cerevisiae* for ethanol production (GHOSE [26.175]; SAVARESE and YOUNG [26.196]), with *Clostridium saccharoperbutylacetonicum* for butanol production (SONI et al. [26.197]), and with *Candida* (ARAUJO and D'SOUZA [26.198]) or *Torula* species (HAN [26.181]) for yeast production for feed purposes.

Waste paper, such as newsprint, is first grated and then hydrolyzed with dilute acids under mild conditions. Cellulase is industrially obtained from *Trichoderma reesei* with paper hydrolysates (NYSTROM and ALLEN [26.199]).

Peat, the most extensive reserves of which are located in the USSR, is used in that country on an industrial scale to produce protein feed and ethanol. The hemicellulose of peat contains mainly hexoses. High-moor peat consists of 63–78% hydrolyzable components (per organic mass), 45–59% of which are reducing compounds (SHIMANSKIJ [26.200]). Peat is processed by grating, hydrolyzing with dilute sulfuric acid, and subsequently treating with water glass, lime, and charcoal (LEDUY [26.201]). In order to remove furfural the hydrolyzed peat is intensely aerated at 60–80 °C for 2 hours and neutralized thereafter with aqueous ammonia (MUCENIECE and ABELE [26.202]). Adding aqueous ammonia to the hydrolyzed peat pulp and aerating at 80–85 °C during neutralization with lime enhances the filtration properties, the solubility of humic acids, and reduces the loss of reducing substances (SHISHKOVA et al. [26.203]). The neutralized and filtered extract then contains, e.g., 3.7% reducing compounds, 1.6% humic acids, 0.4% uronic acid, 0.06% volatile and 0.2% non-volatile organic acids, 0.048% P_2O_5, and 0.152% iron (SHISHKOVA et al. [26.204]).

Peat hydrolysates are processed into feed by adding mineral salts and inoculating with osmophilic strains of *Candida tropicalis* (SHISHKOVA et al. [26.203]) or other *Candida* species (BOGDANOVSKAYA and ZALASHKO [26.205]). Mixtures of peat hydrolysates with molasses have been used for L-lysine production (BEKER [26.206]), with molasses residues for methane fermentation (ZEILE and KALEJA [26.207]), and with plant residue hydrolysates (SHISHKOVA et al. [26.208], [26.204]) or wood hydrolysates in the production of protein feed (ERMACHENKO et al. [26.209]; LOJKO et al. [26.210]). Amylases and proteases have been produced from hydrolyzed peat by means of *Bacillus subtilis* (EKA and FOGARTY [26.211]), pullulan by *Aureobasidium pullulans* (LEDUY and BOA [26.212]). Peat oxydants, by-products of peat hydrolysis, contain mono- and dicarboxylic acids and can be employed for feed production (BEL'KEVICH and MINKEVICH [26.213]; review: LEDUY [26.214]) or to yield fats with *Candida tropicalis* and *Lipomyces lipoferus* (ANDREEVSKAYA and ZALASHKO [26.215]).

Wood hydrolysates. Waste materials from wood processing are industrially hydrolyzed and used in the production of feed (KUZNETSOVA et al. [26.216]; KADLEC et al. [26.217]; GLUSHCHENKO et al. [26.218]) and ethanol (GAUSS et al. [26.219]; HAJNY [26.220]).

Wood hydrolysates, however, have a high content of furfural, which inhibits yeast growth. The ratio of reducing compounds to furfural can be taken as a measure of the utility of the hydrolysate for yeast conversion procedures (SEMUSHINA et al. [26.221]). Furfural concentrations can be lowered by different types of treatment: γ-irradiation in the presence of KCl, which specifically catalyzes furfural radiolysis (KLIMENTOV et al. [26.222]), vacuum freezing of the hydrolysate (KRASAVINA et al. [26.223]; POKINTSOKHA [26.224]), distillation with lignosulfonic acid (KIN and WITKOWSKI [26.225]), removal of furfural by steam distillation (SEMUSHINA et al. [26.221]), oxidation at 45 °C with activated carbon exposed to air oxygen (GLUSHCHENKO et al. [26.218]).

Hydrolysates of the wood of coniferous trees contain 80.3% hexoses and 19.7% pentoses per total content of monosaccharides; beechwood hydrolysates yield 71.5% hexoses and 28.5% pentoses (SEMUSHINA et al. [26.221]). A typical composition of a liquid wood hydrolysate is given by the following figures (in %) (PELECHOVA et al. [26.226]): reducing compounds 2.1; dextrin 0.2; furfural 0.04; acidity 0.94; mineral acids 0.63; organic acids 0.31; colloids 0.44; sediment 3.24.

In order to reduce the concentration of furfural, wood hydrolysates are often diluted or only small quantities added to the nutrient medium, then the low carbon content is augmented by adding other carbon sources, e.g., acetic acid for L-lysine (SMEKAL et al. [26.227]), ethanol for yeast (PELECHOVA et al. [26.228]), and molasses for ethanol and butanol production (MADDOX [26.229]). Wood hydrolysates have been successfully used as raw materials for producing itaconic acid (KOBAYASHI [26.230]), L-lysine, and L-glutamic acid (GLUSHCHENKO et al. [26.218]).

Sulfite waste liquors from wood pulping are significant carbohydrate-containing by-products of the paper manufacturing industries. During cellulose extraction the sulfite waste liquors remove about half of the pulped wood in dissolved form. During the 1940s these were industrially employed in many countries to produce ethanol and, some years later, protein feed with, e.g., *Paecilomyces varioti* (Pekilo process, ROMANTSCHUK and LEHTOMÄKI [26.231]), *Cryptococcus diffluens* (BARTA [26.232]), *Candida utilis,* and *Candida tropicalis* (BOLOTIN et al. [26.233]).

Wood pulping by means of the sulfite process involves the boiling of wood chips with bisulfite, hydrogen, and Ca-ions in aqueous solution. The composition of sulfite waste liquors is determined by the type of wood and technology employed and may vary drastically: reducing compounds 19.8–39.0 g/L, free SO_2 0.096–0.912 g/L, total SO_2 0.416–1.888 g/L, free furfural 0.144–0.768 g/L, volatile acids (as acetic acid) 0.537–6.15 g/L, pentoses 5.5–20.0 g/L (KARCZEWSKA [26.234]).

More advanced technologies substitute Mg- and Na- for Ca-ions; simultaneously the cellulose yield has been increased and the fraction of assimilable reducing compounds for microorganisms reduced to a maximum of 40 g/L (RYCHTERA et al. [26.235]).

The quality of sulfite waste liquors for microbial procedures is controlled by the amount of volatile compounds, i.e., volatile acids, SO_2 and furfural, because of their growth-inhibiting effects (SESTAKOVA [26.236]). The concentrations of volatile acids and SO_2 for different types of sulfite waste liquors are given in Table 26.16 (SESTAKOVA and STROS [26.237]).

Table 26.16. Content of Volatile Acids and SO_2 in Sulfite Waste Liquors (SESTAKOVA and STROS [26.237])

Sulfite Liquor	Volatile Acids		SO_2	
	Formic	Acetic	Free	Bound
	(mg/100 mL)			
Na	6.3	302.0	339.4	987.6
Ca	0.95	507.5	121.4	718.8
Mg	0.62	485.5	6.4	67.2

Before sulfite waste liquors are used for media preparation they are, e.g., treated with hot air and water vapor and neutralized with lime and NH_4OH. This decreases the content of SO_2, furfural, and formic acid, and reduces the inhibitory effect (GRUZINOVA and KOZLOVA [26.238]). In sodium sulfite liquors, however, the high SO_2 content persists (SESTAKOVA and STROS [26.237]). Its conversion into highly soluble sodium bisulfite or monosulfite may enhance the formation of undesirable glycerol during alcoholic fermentation (KRIZANOVA [26.239]).

There is only a low content of assimilable substances in sulfite waste liquors, but it may be increased for microbial conversion by the following methods:

a) The main component of sulfite waste liquors is lignosulfonate at concentrations of 50–65% (based on solids); when treated with ozone it is cleaved into compounds which are assimilable by *Torula* strains (PATTON et al. [26.240]).

b) The content of assimilable carbohydrates could be raised by 4–22% through additional acid hydrolysis (7.6% H_2SO_4 for 2 h at 80–85 °C) (IVANOVA [26.241]).

c) Sulfite waste liquors are combined with other nutrients such as molasses and plant residue hydrolysates in the Pekilo process (ROMANTSCHUK and LEHTOMÄKI [26.231]), with synthetic ethanol in the cultivation of mixed cultures of *Candida tropicalis* and *Candida utilis* (RYCHTERA et al. [26.242]; VERNEROVA et al. [26.243]), or with sulfite liquor residue and ethanol for protein feed production (MOSTECKY et al. [26.244]).

d) For yeast cultivation, sulfite waste liquors are enriched with yeast autolysates or cornsteep liquor (KVASNIKOV et al. [26.245]).

Sulfite waste liquors have also been employed as a xylose source in the production of glucose isomerase with *Aerobacter levanicus* (SHIEH et al. [26.246]).

26.2.1.2 Alcohols

Methanol

Methanol is produced from natural gas and naphtha at a purity exceeding 99.8%. It is employed as a nutrient for the biosynthesis of protein feed (COONEY and LEVINE [26.274]; ANONYMOUS [26.275], [26.276]). For this purpose *Methylophilus methylotrophus* is continuously grown in aerated media containing a mixture of ammonia and methanol. The proteinaceous feed product is commercially sold as Pruteen (Imperial Chem. Ind., Billigham, U. K.; ANONYMOUS [26.277]). The Provesteen process produces yeasts (Phillips Petroleum Co.; KRIEGER [26.278]), the protein products of which are also supposed to be used in human nutrition. Other microorganisms employed for SCP production are: *Candida boidinii* (CARDINI et al. [26.279]; ZOTTI and CARDINI [26.280]; TANI et al. [26.281]) and *Hansenula polymorpha* (BABEL et al. [26.282]).

Methanol also serves in the biosynthesis of vitamin B_{12} by *Pseudomonas* (KAMIKUKO et al. [26.283]), L-serine by *Arthrobacter globiformis* (TANI et al. [26.284]), L-leucine and valine (OGATA et al. [26.285]), and polyhydroxy butyrates by *Alcaligenes eutrophus* (HOWELLS [26.286]). Methanol is further used as a carbon source in combination with xylose for growth of *Hansenula polymorpha* (BABEL et al. [26.282]).

Ethanol

Ethanol traditionally has been employed for vinegar production, i.e., acetic acid fermentation. The type of ethanol varies locally and may either be wine, cider, a distillate, or ethanol from fermented malt saccharified by diastase (with or without additives of grain) (GREENSHIELDS [26.287]). Increasing quantities of ethanol were used for fermentation when it became synthetically available from ethylene. Several commercial procedures were reported on ethanol fermentations to produce livestock feed by *Candida* sp. (ANONYMOUS [26.275], [26.276]; MADRON et al. [26.288]) and to produce protein additives for human nutrition (ANONYMOUS [26.276]; KRIEGER [26.278]).

Synthetic, technical-grade ethanol employed for microbial conversions contains various by-products, of which the following may be toxic for microorganisms: crotonaldehyde, allyl alcohol, acroleine, isobutyl alcohol, 1-butanol, isoamyl alcohol, 1-pentanol (*n*-amyl alcohol), and 1-hexanol (SESTAKOVA et al. [26.289]; PELECHOVA et al. [26.290]). Concentrations of crotonaldehyde have been determined as 905.0 µg/kg ethanol, of isobutyl alcohol 237.5 µg/kg, and of 1-butanol 35.0 µg/kg ethanol (ROSAK et al. [26.291]). In batch cultures toxic concentrations are not reached; however, in processes using recycled, centrifuged nutrient media the growth of yeast cultures is inhibited.

Prior to the preparation of a medium the synthetic, technical-grade ethanol can be purified by extractive distillation (ROSAK et al. [26.292]), or the crotonaldehyde, which is most toxic for *Candida utilis,* can be converted with ammonia into the non-toxic trimeric crotonaldehyde-ammonia (SESTAKOVA and STROS [26.293]).

Table 26.17. Glycerol as Carbon Source in Commercially Important Nutrient Media

Product	Microorganism	Reference
Glycerol dehydrogenase	*Aerobacter aerogenes*	STAHL et al. [26.307]
Glycerol kinase	*Streptomyces canus*	TOYO JOZO [26.168]
Nuclease	*Serratia marcescens*	BELYAEVA et al. [26.308]
Erythromycin	*Arthrobacter* sp.	FRENCH et al. [26.309]
Tobramycin	*Streptomyces cremeus*	MOTKOVA et al. [26.310]
Antibiotics	*Streptomyces wadayamensis*	IMANAKA et al. [26.311]
Steroids	*Mycobacterium* sp.	WIX et al. [26.312]
Steroids	*Mycobacterium fortuitum*	WOVCHA et al. [26.313]
Steroids	*Mycobacterium fortuitum*	WOVCHA et al. [26.314]
L-Glutamic acid	*Brevibacterium*	LINDE et al. [26.315]
L-Threonine	*Escherichia* mutant	AJINOMOTO CO [26.316]
Dihydroxyacetone	*Acetobacter suboxydans*	OVERCHENKO et al. [26.317]

In the Republic of South Africa a SASOL-alcohol mixture is produced as by-product in the synthetic fuel procedure and contains 90.7% ethanol, 8.3% isopropyl alcohol, 0.5% cyclohexane, 0.13% benzene, 0.014% diethyl ketone, and 0.1% non-identified compounds. Attempts have been made to convert this product with *Candida krusei* (KILIAN et al. [26.294]) or a mixture of ethanol, isopropyl alcohol, and water (22:2:1) with *Candida utilis* to biomass (PRIOR et al. [26.295]).

Furthermore, ethanol is employed in the production of L-lysine (NAKAYAMA and HAGINO [26.296]; KYOWA HAKKO KOGYO CO [26.297]) and penicillin V and G (MATELOVA and CULIK [26.298]).

Polyalcohols

Glycerol is often used for antibiotics and steroid synthesis; sorbitol and mannitol only rarely serve as carbon sources. However, mannitol is suitable for the synthesis of the antibiotic papulacandin (TRAXLER et al. [26.299]), the insecticide nikkomycin (HAGEMAIER et al. [26.300]), and of ergot alkaloids from *Aspergillus fumigatus* and *Claviceps fusiformis*, respectively (NARAYAN and RAO [26.301]; SOCIC et al. [26.302]). D-Sorbitol is used with *Acetobacter suboxydans* to obtain L-sorbose for vitamin C synthesis (WELLS et al. [26.303]; PRESCOTT and DUNN [26.304]); furthermore, it serves as an ideal carbon source for the synthesis of orotic acid by a mutant of *Arthrobacter paraffineus* (KAWAMOTO et al. [26.305]). It is also suitable for the production of proteases from *Bacillus thuringiensis* (EGOROV et al. [26.306]).

The different applications of glycerol as a carbon source in nutrient media are listed in Table 26.17.

26.2.1.3 Carboxylic Acids

Acetic acid has recently been used as an inexpensive carbon source in microbial processes. Microbial amino acid syntheses with acetic acid have been extensively reported. Mixtures of acetic acid with each of the following compounds have been employed to produce L-lysine: NH_4-acetate (KABUSHIKI KAISHI TEKKOSHA CO [26.256]) at an ideal ratio of 1:0.25 (OKUMURA et al. [26.257]), glucose (KUBOTA et al. [26.258]), sucrose (RUKLISA et al. [26.259]), wood hydrolysates (SMEKAL et al. [26.227], [26.260]; PELECHOVA et al. [26.261]), yeast extract, and cornsteep liquors (BACHINA et al. [26.262]). L-Glutamic acid is synthesized with media containing only acetic acid (HARADA et al. [26.263]; KITANO et al. [26.264]; KANZAKI et al. [26.265]) or mixtures of acetic acid with glucose, sugarcane molasses, or NH_4-acetate (TOSAKA et al. [26.90]). L-Isoleucine (IKEDA et al. [26.266]) and threonine (AKASHI et al. [26.267]) also have been produced on acetic acid nutrient media. The microorganisms for all above cases are strains of *Brevibacterium*, *Corynebacterium*, and *Micrococcus*. Protein feed has been obtained with *Candida utilis* on the basis of acetic acid/synthetic ethanol mixtures at a ratio of acetic acid carbon to ethanol carbon of 22:1 (VERNEROVA et al. [26.243]).

In the Republic of South Africa SASOL oil is formed as a by-product in the production of synthetic fuel from coal. This oil contains approximately 1% fatty acids (0.7% acetic acid, 0.15% propionic acid, 0.1% butyric acid, and 0.5% valerianic acid). SASOL oil has been employed to produce protein feed with *Candida* and *Acinetobacter* (LATEGAN and DU FREEZ [26.268]).

For L-glutamic acid synthesis unsaturated fatty acids such as oleic, vaccenic, and linoleic acid have been added in concentrations of 50–100 mg/L to glucose nutrient media (OKUMURA et al. [26.269]; TAKEDA CHEM. IND. [26.270]). Oleic acid is employed as a carbon source for prodigiosine production with *Serratia marcescens* (NAKAMURA and KITAMURA [26.271]).

Of the hydroxy carboxylic acids succinic acid is used with *Penicillium digitatum* to synthesize pectases (LOBANOK et al. [26.272]), and fumaric acid or α-keto-glutaric acid are employed on an industrial scale to yield creatinine iminohydroxylase (MASUREKAR [26.273]).

26.2.1.4 Fats

Animal and vegetable oils and fats are often employed as additional carbon sources in conjunction with carbohydrates. They either are added directly to the nutrient media in form of pure glycerides (NISHIKAWA et al. [26.247]) or as oils (LURIE et al. [26.248]); and they may serve as metabolizable antifoaming agents (WELWARD et al. [26.249]; CHERKASOVA and MAKAREVICH [26.250]).

The utility of an oil for microbial procedures is determined by its acid value and its content of peroxides and glycerides. During storage the acid value of the oil increases. An acid value higher than 10 inhibits tetracycline synthesis (CHERKASOVA and MAKAREVICH [26.250]). The content of peroxides also rises during storage and may adversely affect the synthesis of penicillin, chlorotetracycline, and L-lysine (WELWARD et al. [26.249]). Glycerides are used in steroid production in the form of oils or non-defatted meal (MITSUBISHI [26.21a]; NISHIKAWA et al. [26.247]). Non-defatted soybean meal, e.g., Soyapan or Nurupan, contains 20–22% oil, of which 2.2–2.5% is lecithin (courtesy of Edelsoja GmbH, Hamburg). Standardized, refined soybean lecithins are also employed, such as M-C-Thin AF-1 and P-1, Emulbesto 100-A with 60–65% phospholipids, or Emulpur N with at least 95% phospholipids (Lucas Meyer GmbH, Hamburg). The content of glycerides of several commercially available oils is shown in Ta-

Table 26.18. Glyceride Content and Iodine Values of Some Commercially Available Oils

Oil	Glyceride (%) (MITSUBISHI [26.21])	Iodine Value (SPECTOR [26.251])
Coconut	65–75	10.4
Palm	51–67	54.2
Olive	40–70	81.1
Peanut	29–39	93.4
Rapeseed	22–50	98.6
Cottonseed	16–35	105.7
Soybean	15–23	130.0
Corn	4–6	122.6

ble 26.18. Their iodine value specifically relates to the content of unsaturated fatty acids.

Besides their application in the synthesis of antibiotics and steroids, oils are also used in the production of lipase (PAL et al. [26.252]), amylase (YAMAMOTO et al. [26.253]), and proteases (LEUCHTENBERGER and RUTTLOFF [26.254]). Yields of alcoholic fermentations of sugarcane molasses at 40 °C can be increased by adding 0.5% linseed oil (SAIGAL and VISWANATHAN [26.255]).

26.2.1.5 Hydrocarbons

Methane
Natural gas contains 90–92% methane as well as 1.5% ethane, 1–2% butane, and traces of CO_2, argon, and N_2. Often it is contaminated with sulfur compounds, which must be removed prior to its use for microbiological processes (REHM [26.69]). *Methylomonas* has been grown industrially on methane for the production of feed protein (ANONYMOUS [26.275]; WENDLAND et al. [26.318]; HILGER et al. [26.319]).

n-Butane
n-Butane can be utilized as a carbon source by various microorganisms (MCLEE et al. [26.320]). *Pseudomonas butanovora*, e.g., converts *n*-butane into single cell protein (ICHIKAWA et al. [26.321]). The oxidation products of *n*-butane have gained economic importance; they are formed, e.g., in the production of acetic acid at Gelanese Canada, Ltd. and consist of a mixture of ketones, aldehydes, acids, and esters. *Candida* and *Saccharomyces* yeasts are used to convert these substrate mixtures to SCP (WAYMAN [26.322]).

n-Pentane
n-Pentane has been used as a carbon source for biomass synthesis with *Corynebacterium hydrocarboclastus* (UEDA et al. [26.323]).

n-Paraffins
These hydrocarbons have been extensively employed as carbon sources for the synthesis of microbial biomass, organic acids, and amino acids. Today, commercial media mainly contain C_{10}–C_{20} paraffins, while crude oil is only very rarely used for fermentative purposes. Microbial biomass was industrially produced in various countries during the 1970s (ANONYMOUS [26.275], [26.277], [26.276]; KNECHT et al. [26.323a]; BIRKENSTEDT et al. [26.324]; BENNET et al. [26.325]). Yeast production has been achieved with *n*-alkanes with chain lengths around C_{10}–C_{14} and C_{15}–C_{18} (KNECHT et al. [26.323a]) or C_{13}–C_{21} (CEJKOVA et al. [26.326]); highest yields were reached with the C_{15}–C_{17} fraction (FEDOROV et al. [26.327]). In order to avoid carbon depletion in fermentations with hydrocarbons several carbon-source additives have been used: CO_2 (gas) via aeration (VESELOV et al. [26.328]; ERICKSON [26.329]), or an extract of rice bran (IIZUKA [26.330]), still residues (IRISKHANOV [26.331]), or autolyzed brewer's yeast (OLTARZHEVSKAYA et al. [26.332]) directly added to the medium.

Other than for protein feed *n*-paraffins are employed for the production of citric acid (HUSTEDE and SIEBERT [26.333]; FURUKAWA et al. [26.334a]), citric acid and isocitric acid (AKIYAMA et al. [26.335]; BEHRENS et al. [26.336]; NARA et al. [26.337]; PUKLOWSKI et al. [26.338]; KIMURA and NAKANISHI [26.339]), 2-ketoglutaric acid (TSUGAWA and OKUMURA [26.340]), succinic acid (SATO et al. [26.341]), fumaric acid (FURUKAWA et al. [26.334]), L-glutamic acid (NAKAO et al. [26.342]), L-lysine (KYOWA HAKKO KOGYO CO [26.343]), L-threonine and L-valine (KYOWA HAKKO KOGYO CO [26.344]), orotic acid and orotidine (KAWAMOTO et al. [26.305]), erythritol (HATTORI and SUZUKI [26.345]), and wax esters (DEWITT et al. [26.346]). For L-lysine production a mixture of *n*-paraffins with ethanol (5:1) has been reported (KYOWA HAKKO KOGYO CO [26.343]).

26.2.1.6 Gaseous Substrates

Gaseous substrates such as CO, CO_2, and H_2 hold a special position in microbial procedures. These are converted into protein feed, for instance, by *Hydrogenomonas* bacteria (LAFFERTY et al. [26.347]) or the not further characterized microorganism *Zimmet* B 612 (KARBAUM et al. [26.348]). Several anaerobic cultures of *Clostridium* utilize CO, CO_2, and H_2 to form different organic acids (LEVY et al. [26.349]; KERBY and ZEIKUS [26.350]).

26.2.1.7 Coal

Aqueous extracts of coal and its oxidation products, such as fatty acids, alcohols, ketones, hydrocarbons, and humic acids, obtained by alkaline extraction, can be used for biomass production with *Actinomyces, Penicillium, Pseudomonas* (FIEDLER et al. [26.350a]), *Candida tropicalis* (ROSE et al. [26.351]; KUCHER et al. [26.352]), *Bacillus paracoccus,* and others (PACA and GREGR [26.353]). The composition of the extracts depends on the quality of the coal and the type of oxidation procedure applied. A typical composition is, e.g.: water-soluble organic acids 5–26 g/L (KUCHER et al. [26.352]), volatile acids 5.8–9.2 g/L (PACA and GREGR [26.353]), and a mixture of organic acids composed of benzoic acids (0.7–14.2%), naphthoic acids (2.88–16.67%), and aliphatic dicarboxylic acid (15.3–19.8%) (BAZAROVA et al. [26.354]).

26.2.2 Technical Substrates Mainly Serving as Nitrogen Sources

26.2.2.1 Mineral Salts

Ammonium salts and nitrates serve as nitrogen sources for nutrient media in industrial fermentations. Aqueous ammonia and ammonium sulfate are almost exclusively used as nitrogen sources for the production of yeasts on molasses, potatoes, whey, bagasse, and other natural carbon sources, and for ethanol and acetone/butanol fermentation. Ammonium ions are necessary in penicillin synthesis (EL-MARSAFY et al. [26.112]) and are used in the form of ammonium sulfate (Table 26.41).

The pH values of media with high cellulose concentrations can be effectively controlled with ammonia which may simultaneously serve as a nitrogen source for *Trichoderma reesei* (STERNBERG and DORWAL [26.355]). Nitrogenous mineral salts are the sole nitrogen source used for the conversion of n-alkanes to erythritol by *Candida lipolytica* and *Candida zeylaroides* (HATTORI and SUZUKI [26.345]).

Other substances employed as nitrogen sources are ammonium chloride, ammonium nitrate, ammonium phosphate, sodium nitrate, and potassium nitrate (commercial grade). The suitability of mineralic nitrogen sources cannot be defined for individual procedures.

It generally applies that mineralic nitrogen sources are present in complex media either as a part of a carbon-containing substrate, such as molasses, whey, etc. or they are added separately.

26.2.2.2 Organic Nitrogen Sources

Urea

Urea is a highly suitable nitrogen source for nutrient media, specifically because of its buffering capacity. It is employed together with synthetic ethanol for the growth of *Candida utilis* (SESTAKOVA and STROS [26.356]) and is added in various mixtures with other nitrogen-containing raw materials (Table 26.41). However, urea is not stable during thermal sterilization, thus limiting its use on a technical scale.

Meals

Soybean meal, cottonseed meal, rapeseed meal, corn meal, and fish meal are frequently used as nitrogen source in nutrient media for industrial purposes. Fish meal is dried, comminuted fish or fish waste. The other meals are residues of oil extraction

Table 26.19. Composition of Soybean Meal for Fermentation Purposes[a]

Constituent (%)	Soybean Meal					
	Defatted		Partly defatted	Full fat		
	Edel-Soja GmbH[b]	Akzo AG[c]	Hafen-Mühlen-Werke GmbH		Edel-Soja GmbH	
					Soyapan[d]	Nurupan[e]
Protein	48.56–52.80	46.53–51.80	46.31	40.63	33.75–38.00	40.00–42.44
Fat	0.43–1.59	1.61–2.79	3.99	14.40	19.50–20.46	18.20–26.00
Ash	5.88–7.20	4.50–7.33	7.34	5.1	4.52–4.81	4.40–4.21
Humidity	6.17–9.48	4.13–7.32	11.00	4.7	8.84–9.29	4.55–6.82

[a] Analyses by Schering AG
[b] 25 charges analyzed in 1981
[c] 24 charges analyzed in 1977
[d] 2 charges analyzed in 1982–1983
[e] 11 charges analyzed in 1982–1984

processes and contain proteins, carbohydrates, and small quantities of residual fat.

Meals are good and inexpensive sources of amino acids, mineral salts, and vitamins. The application of meal on a technical scale is often complicated because of difficulties in preparing the media and in isolating the products from the medium. When suspended in water soy meal and corn meal form conglomerates of different form and size (from 400–600 µm) (EREMINA and MOTINA [26.357]). These conglomerates must be sterilized at a temperature of 122–127 °C for a minimum of 16–21 minutes. During product extraction with solvents the proteins contained within the meals often form emulsions and the remaining fats are extracted as waxy substances along with the product.

Soybean Meal

Soybean meal is most frequently used. It is commercially available in three different grades according to the fat content: full-fat meal with a minimum of 18% fat, low-fat meal with 4.5–9% fat, and defatted meal with a maximum of 2% fat (CIRCLE and SMITH [26.358]). An analysis of industrially used soybean meals is given in Table 26.19.

Defatted soybean meal on the average contains 0.26% calcium, 0.44% magnesium, 2.18% potassium, 0.007% zinc, 0.73% phosphorus, 0.02% iron, and 0.07% silicon (own analysis, Schering AG).

It contains all essential amino acids, glutamic acid being the main component with little methionine and cystine. 7–10 Proteinase inhibitors are present in soymeal; these, however, can be inactivated by thermal treatment at 100 °C (RACKIS [26.259]). They are undesirable only in enzymatically pretreated nutrient media. Soybean meal also contains thermo-resistant proteins, which inhibit the growth of *Clostridium perfringens* (BUSTA and SCHRODER [26.360]). Of nine proteinaceous fractions extracted with water at different pH values (between 1 and 12) only one – extracted at pH 11.5–12 – stimulated the formation of gibberellic acid by *Gibberella fujikuroi*. The other fractions led to the formation of gibberellins (FUSKA et al. [26.361]).

Our (unpublished) investigations with defatted soybean meal, its aqueous extract (pH 6.6), and percolated soybean meal residues showed that only the extract gave higher conversions of 6α-fluor-16α-methyl-11β-hydrodesoxycorticosterone by Δ^1-dehydration with *Bacillus lentus*.

Soybean meal appears to be the best nitrogen source for the synthesis of streptomycin by *Streptomyces griseus* (AHARONOWITZ [26.362]) and it is also used for various other antibiotics, as, e.g.: neomycin (BHADRA et al. [26.162]), capreomycin (BRISTOL MYERS CO [26.363]), and antimycin (VEZINA et al. [26.364]). Furthermore, soybean meal is employed for steroid conversions, as in the 11α-hydroxylation of 3-oxo-$\Delta^{1,4}$-androstane (PETZOLD et al. [26.365]) or in the side-chain degradation with *Mycobacterium* (WOVCHA et al. [26.313], [26.314]; WOVCHA and BROOKS [26.20]), for which it may also serve as a glyceride donor (IMADA and TAKAHASHI [26.366]). A further application of soybean meal is found in the production of microbial rennet (AUNSTRUP [26.367]).

Cottonseed Meal

Cottonseed meal is commercially offered as Pharmamedia and Proflo (Traders Oil Mills, Ltd.). Both are made from the embryo of cottonseed after the linters, hull and oil have been removed at a lower (Pharmamedia) or at a higher temperature (Proflo). An analysis of these products is shown in Tables 26.20 and 26.21.

Gossypol is an important constituent of cottonseed flour. It stimulates the synthesis of amphotericin B (LINKE et al. [26.368]) and acts as an antioxidant, eliminating the inhibitory effect of oxidized fats which are employed as antifoaming agents in the production of antibiotics and amino acids (WELWARD et al. [26.249]). It also increases overall yields. Both mentioned products are frequently employed nutrients for industrial antibiotics and vitamin production (Table 26.41).

Table 26.20. Typical Analyses of Proflo and Pharmamedia (Courtesy of Traders Oil Mills, Ltd.)

Constituent (%)	Proflo	Pharmamedia
Total solids	98.75	99.0
Protein (N × 6.25)	61.06	59.20
Amino nitrogen	4.19	4.67
Ammonium nitrogen	1.16	1.27
Carbohydrates	23.18	24.13
Reducing sugars	1.17	1.18
Non-reducing sugars	1.30	1.16
Fat	4.10	4.02
Ash	6.73	6.71
Gossypol	0.031	0.03

Table 26.21. Amino Acids Composition of Proflo and Pharmamedia[a] (Courtesy of Traders Oil Mills, Ltd.)

Amino Acid (%)	Proflo	Pharmamedia
Glutamic acid	15.99	21.77
Aspartic acid	8.01	9.86
Arginine	7.15	12.28
Leucine	6.46	6.11
Phenylalanine	4.17	5.82
Alanine	4.13	3.88
Serine	3.77	4.58
Lysine	3.30	4.49
Valine	3.25	4.57
Proline	2.68	3.94
Glycine	2.75	3.78
Tyrosine	2.53	3.42
Threonine	2.46	3.31
Isoleucine	2.42	3.29
Histidine	1.98	2.96
Cystine	1.45	1.52[b]
Methionine	1.40	1.52
Tryptophan	0.86	0.95

[a] Moore-Stein Technique – calculated on 16% nitrogen basis
[b] ½ Cystine

Rapeseed Meal

Because of its toxicity rapeseed meal, after oil extraction, cannot be used as a livestock feed. It is either disposed of as a waste or employed as a fertilizer. Rapeseed meal, however, may also serve – with or without pretreatment – as a nitrogen source for fermentations (PHILIPCHUK and JACKSON [26.369]). It is employed in nutrient media for steroid conversion (MITSUBISHI CHEM. IND. [26.21a]; NISHIKAWA et al. [26.247]) and its hydrolysates are used to grow *Candida utilis* (PHILIPCHUK and JACKSON [26.369]).

Corn Meal

Corn meal serves as a nitrogen source in the biosynthesis of chlorotetracycline (MAL'TSEV and CHALENKO [26.370]) and biomycin (FEDOROVA et al. [26.371]). It may substitute for cornsteep liquor in the formation of amylase by *Bacillus subtilis* (DZAVAKHIYA et al. [26.372]). Various processes employ corn meal as a carbon source in complex nutrient media, e.g., in the production of glucoamylase by *Endomycopsis* (MOSKVICHEVA et al. [26.373]; BEKERS et al. [26.374]; DREIMANE [26.375]), proteases by *Aspergillus* (POPOVA [26.35]), and tetracycline (VECHER et al. [26.376a]).

Corn Gluten Meal

This is a by-product of the corn wet milling process. It is used in conjunction with oat meal in antibiotics synthesis (REUSSER and SLECHTA [26.167]). It serves as a substitute for cornsteep liquor in griseofulvin production (LEWIS and MITCHELL [26.377]; KUZNETSOVA et al. [26.378]), and in steroid transformation (BAHN et al. [26.379]). The composition of a corn gluten meal is shown in Table 26.22.

lus terricola which is used in the fish industry to yield fish hydrolysates (POPOVA [26.35]). Commercial fish meal (Lohmann Tierernährung GmbH) contains 55–64% raw protein, 12% fat, and 4–5% NaCl.

Potato Proteins

Potato proteins are obtained during the process of starch extraction and offered under the trade name Alburex® (Roquette Frères, SA). Industrially potato proteins are used as an alternate to soybean meal in nutrient media for the production of enzymes and antibiotics (HUCHETTE and DEVOS [26.380]). Alburex® contains 75–85% protein (dry weight). It is completely water soluble, and highly emulsive.

Cornsteep Liquor

Cornsteep liquor, a by-product of cornstarch extraction, frequently serves as a nitrogen source in some industrial culture media. It contains essential amino acids, vitamins, and mineral salts (Tables 26.23 to 26.25).

A disadvantage of cornsteep liquor is the great fluctuation in composition (CEJKA

Table 26.22. Composition of Rapeseed and Corn Meal

Constituent (%)	Rapeseed Meal	Corn Meal		Corn Gluten Meal
	Hafen-Mühlen-Werke GmbH[a]	Traders Oil Mills, Ltd.[b]		KUZNETSOVA et al. [26.378]
Moisture	8.20	7.0	8	5.55–8.60
Total nitrogen	5.42	—	—	9.20–10.90
Protein (N × 6.25)	33.88	22.6	42.0	—
Fat	4.41	1.6	2.5	—
Carbohydrate	16.97	53.2	40.2	11.56–21.84
Ash	6.52	3.3	2.0	—
Phosphorus	1.05	0.5	0.46	0.11–0.15

[a] Analysis by Schering AG
[b] Courtesy of Traders Oil Mills, Ltd.

Fish Meal

Fish meal is only rarely used in nutrient media for fermentation procedures. It has been successfully used instead of cornsteep liquor in protease production with *Aspergil-* [26.381]; SKVORTSOVA et al. [26.122]). This often necessitates the choice of other nitrogenous raw materials (DZAVAKHIYA et al. [26.372]; KUZNETSOVA et al. [26.382]). The quality of the cornsteep liquor depends on

Table 26.23. Composition of Commercial Cornsteep Liquor (CSL) Solulys L® (Roquette Frères, SA)

Constituent (g/100 g original CSL)	Range of Value	Mean Value	S. D. of Sample[a]
Dry matter	46.80–49.60	48.47	1.71
Ash	8.04–10.43	8.81	0.61
Total nitrogen	3.33–3.67	3.64	0.11
pH	4.00–4.70	4.26	0.15
Total sugar (as glucose)	0.74–4.39	2.21	0.94
Lactic acid	11.60–19.30	13.21	1.71
Acidity (mL 0.1 N NaOH/10 g orig. CSL)	108–144	125.62	8.92
Volatile acids (mL 0.1 N NaOH/10 g orig. CSL)	0.1–1.1	0.56	0.31
Iron	0.009–0.027	0.016	0.006
Phosphorus	1.5–1.9	1.72	0.098
Calcium	0.02–0.07	0.036	0.015
Zinc	0.005–0.012	0.008	0.002
Potassium	2.0–2.5	2.25	0.143
SO_2	Trace–0.02	0.07	0.003
Sedimentary solids	38.4–52.9	45.1	3.501

50 charges of CSL were analyzed at Schering AG
[a] S. D. standard deviation

the grade and germinating power of corn and on the technology of cornsteep liquor production (KUZNETSOVA et al. [26.382]). For steroid transformation the best suitable cornsteep liquors are from corn having a germinating power of 70%. These contain ash, phosphorus, calcium, zinc, SO_2, and proline in sufficiently high or non-inhibitory concentrations for steroid conversions (CEJKA [26.381]). The undesirable reduction of the 20β-group accompanying the Δ^1-dehydrogenation of prednisolone is enhanced by cornsteep liquor, probably due to the presence of lactic acid (GOTOVTSEVA et al. [26.383]). For the production of the insecticide entobacterine by *Bacillus thuringiensis* (SKVORTSOVA et al. [26.122]) cornsteep liquors are needed which contain the following minimum concentrations of amino acids: alanine 40 µg/mL, aspartic acid 10 µg/mL, and glutamic acid 3 µg/mL. In glucose isomerase synthesis by *Actinoplanes misguricusis,* cornsteep liquor with 3–7% solids gives best results (SHIEH et al. [26.246]).

Table 26.24. Composition of Cornsteep Powder (CSP) (Roquette Frères, SA)

Constituent (g/100 g CSP)	Range of Value
Dry matter	90.37–91.74
Ash	12.89–13.74
Total nitrogen	7.36–8.33
Total sugar as glucose	5.71–5.88
Acidity (mL 0.1 N NaOH/10 g CSP)	169.0–198.0
Volatile acids (mL 0.1 N NaOH/10 g CSP)	0.5–1.35
Lactic acid	15.06–17.72
pH	3.80–4.40
Phosphorus	2.60–2.80
Nitrogen	3.0–3.60

5 charges of CSP were analyzed at Schering AG

A constant quality of cornsteep liquor is available in the product Solulys L® (Roquette Frères, SA). It is produced from

Table 26.25. Amino Acid Content of Cornsteep Liquor and Powder (Courtesy of Roquette Frères, SA)

Amino Acid	% Total Nitrogen	
	Liquor	Powder
Aspartic acid	5.50	5.65
Threonine	4.0	4.5
Serine	4.6	3.6
Glutamic acid	17.5	14.0
Proline	8.0	10.0
Glycine	5.1	6.1
Alanine	11.2	11.7
Valine	5.8	6.2
Cystine	1.9	2.2
Methionine	1.9	2.1
Isoleucine	3.6	3.6
Leucine	11.3	9.5
Tyrosine	3.4	2.8
Phenylalanine	4.4	3.75
Lysine	2.5	3.0
Histidine	2.8	3.7
Arginine	3.3	4.0

high-quality corn by a controlled lactic acid fermentation (DEVOS et al. [26.384]). Here, fermentation conditions are maintained by temperature regulation. During corn steeping the temperature is decreased from 55 °C to 35 °C, which permits complete fermentation of free sugars and degradation of free amino acids, which are undesirable, e.g., in penicillin production. The composition of Solulys L® is given in Tables 26.23 and 26.25.

A dried product is used in order to avoid great fluctuations in the quality of cornsteep liquors, e.g., when used together with beet molasses in baker's yeast production (BELOVA and ERMAKOVA [26.385]). The dried cornsteep liquor is offered as a powder (Roquette Frères, SA). Its composition is given in Tables 26.24 and 26.25. Other applications of cornsteep liquor are listed in Table 26.41.

Yeast Autolysates and Dried Yeast

These products frequently serve as a nitrogen source for nutrient media. Yeast autolysates are rich in protein and amino nitrogen as well as important vitamins. They are obtained from different types of brewer's and baker's yeast *Saccharomyces cerevisiae*, and their composition depends on the type of yeast used (BRIDSON and BRECKER [26.386]), as shown in Table 26.26. Yeast autolysates are offered either as liquids, pastes, or dried products with different solubilities. A survey of commercial yeast autolysates is shown in Tables 26.27 and 26.28. Analytical data of five industrially employed yeast autolysates (Ohly GmbH) are summarized in Table 26.29 (review: PEPPLER [26.387]).

Table 26.26. Composition of Yeast Autolysates (BRIDSON and BRECKER [26.386])

Constituent[a] (%)	Baker's Yeast	Brewer's Yeast
	Yeast Extract	
Moisture	30	30
Ash	8.35	10.0
Acid soluble ash	0.26	0.13
P as P_2O_5	1.98	4.0
Total nitrogen	9.97	7.6
Formol nitrogen	2.91	4.5

[a] % in original product

Next to commercial yeast autolysate products, several other types of autolysates are being employed, e.g., derived from brewer's yeast wastes (OLTARZHEVSKAYA et al. [26.332]) and feed yeast cultivated with wood hydrolysates or *n*-paraffins (EGOROV et al. [26.388]; MESHINYA et al. [26.389]). Feed yeast has been hydrolyzed with acids (VAGAVICIUS et al. [26.390]), with a lytic enzyme complex from *Actinomyces cinerosus* (SHKLYAR et al. [26.391]), or with protosubtilin and lysosubtilin (KUPLETSKAYA et al. [26.392]). These hydrolysates are used as nitrogen or vitamin sources in various industrial nutrient media, often in place of cornsteep liquor.

Applications of yeast autolysates are listed in Table 26.41.

Table 26.27. Composition of Some Commercial Yeast Autolysates (Courtesy of the Producers)

Product	Characteristics	Total N (%)	Amino N (% of total N)	Dry Matter (%)	NaCl (%)
Producer: Amber Laboratories					
BYF Series 100	Spray-dried, autolyzed	8.3	25	96	—
BYF Series 300	Brewer's yeast, contained some insoluble yeast	6.8	23	96	—
Amberex 510	Spray-dried, autolyzed yeast	10.4	65	95	—
Amberex 1003	water soluble	8.0	32.4	97	2.5
Producer: Bovril Food Ingredients, Ltd.					
Yeatex C Paste	Brewer's yeast autolysate	6.8	3.8	75	13
Producer: Ohly GmbH					
KAV	*Saccharomyces cerevisiae*	8.8	—	70	1.0
NAV	autolysate paste	7.2	—	80	15.0
M1V		6.4	—	70	20.0
KAT	Spray-dried, autolysates	11.7	—	97	<1
NAT		7.2	—	97	18.0
M1T		6.1	—	97	38.0

Table 26.28. Amino Acid Composition of Some Yeast Autolysates

Amino Acid (% protein basis)	Product		
	Amber BYF Series 100	Amberex 510	Yeatex C Paste
Arginine	3.7	3.9	1.5
Lysine	6.0	6.3	6.1
Tryptophan	1.0	1.0	1.5
Methionine	2.0	2.1	1.3
Cystine	0.8	0.8	—
Histidine	2.3	2.4	2.0
Tyrosine	3.8	4.0	2.6
Phenylalanine	4.0	4.2	3.3
Threonine	3.2	3.4	0.4
Leucine	6.0	6.3	6.1
Isoleucine	4.2	4.4	4.1
Valine	4.0	4.2	5.2
Glutamic acid	8.5	8.8	10.0
Glycine	5.0	5.3	4.8

Dried yeast for industrial nutrient media is commercially available under the trade names Amber DBY Series 1 (brewer's yeast), Nutrex 540 *(Kluyveromyces fragilis)* – both by Universal Foods Corp. –, Yesta A (wine yeast), Yesta B (brewer's yeast) – both by Bovril Food Ingredients –, and yeast powder (Hefe-Pulver) by Auxoferm GmbH. These contain approximately 6–7% total nitrogen and are rich in amino acids and vita-

Table 26.29. Composition of Commercial Yeast Autolysates (Ohly GmbH)

Constituent (%)	Range of Value[a]
Moisture	32.6–40.84
Ash	10.19–17.70
Total nitrogen (d. m.)	8.54–9.19
Amino nitrogen (d. m.)	2.84–4.57
Glucose	1.7–2.7
Iron	0.002–0.014
Calcium	0.06–0.09
Potassium	1.32–1.40
Zinc	0.001–0.011
Phosphorus	0.82–1.05
Natrium chloride	14.54–15.30
pH	5.0–5.2
Acidity (mL 0.1 N NaOH/ 10 g yeast autolysate)	49.0–66.0

5 charges were analyzed at Schering AG
[a] % in original product

Table 26.30. Composition of Dried *Torula* Yeast (Courtesy of Cellulose Attisholz AG)

Typical Analysis (%)			
Moisture	7.0	Ash	7
Protein	54–57	Fat	6–8
Carbohydrates	28–35	Fiber	3–7
Minerals			
P_2O_5 (%)	3.9	MgO (%)	0.3
K_2O (%)	1.8	Zn (ppm)	100
CaO (%)	0.7	Cu (ppm)	8
Amino acids (%)			
Lysine	3.94	Cystine	0.58
Methionine	0.63	Alanine	3.72
Leucine	3.56	Arginine	2.74
Isoleucine	2.26	Aspartic acid	5.05
Phenylalanine	2.18	Glutamic acid	9.71
Threonine	2.40	Glycine	2.27
Tryptophan	0.72	Proline	2.02
Valine	2.61	Serine	2.27
Histidine	1.02	Tyrosine	1.73
Vitamins (ppm)			
Thiamine	35	Pyridoxine	9
Riboflavin	53	Nicotinic acid	347

mins. A Swiss company (Cellulose Attisholz AG) offers a *Torula* feed yeast product; its analysis is given in Table 26.30. Dried yeast applications for nutrient media are exemplified in Table 26.41.

Protein Hydrolysates

Plant and animal protein is hydrolyzed by acid or enzymes to yield commercial protein hydrolysates, which are offered either as spray-dried or liquid products. Casein and soybean protein are most frequently employed for this purpose. Enzymatically hydrolyzed and acid-hydrolyzed products differ in their composition (BRIDSON and BRECKER [26.386]). Trypsin-hydrolyzed casein contains 12.9% total nitrogen and 6.6% amino nitrogen, whereas acid-hydrolyzed casein has 8.3% total nitrogen and 6.4% amino nitrogen. The ratio of amino nitrogen to total nitrogen is 51.2 and 74.0, respectively. Analytical data of some technically employed hydrolysates are presented in Table 26.31.

A West German company (Diamalt AG) also offers an acid-hydrolyzed product (Table 26.32).

Commercially available protein hydrolysates mainly serve in the production of pharmaceuticals, e.g., in steroid transformation (SCHERING AG [26.393]; MARSHECK et al. [26.394]), for L-asparaginase (HO and BOECK [26.395]), diphtheria toxin (STAINER and SCHOLTE [26.396]), antibiotics (VEZINA et al. [26.364]; KUPLETSKAYA et al. [26.397]), and amino acids (OKUMURA et al. [26.369]; CHATTOPADHYAY and BANERJEE [26.398]; AKASHI et al. [26.399]). Soybean protein hydrolysates are mainly applied in these processes.

Other, non-commercial protein hydrolysates are made from different types of meal. These may be used as nutrients in conjunction with waste mycelia for erythromycin production (having served in primary erythromycin synthesis) (RUCZAJ and PASS [26.400]). Furthermore, hydrolysates formed via *Rhizopus delemar* from corn meal are employed in synthesis of the insecticide dendrobacillin (KOKIN [26.401]), acid-hydrolyzed peanut meal serves in L-lysine production (WELWARD et al. [26.402]).

Inexpensive protein hydrolysates are obtained from waste mycelia of *Penicillium chrysogenum* following penicillin extraction (KIRSANOV [26.403]). The mycelia of *Penicillium chrysogenum, Streptomyces aureofaciens,* and brewer's yeast are dried, pulver-

Table 26.31. Composition of Protein Hydrolysates (Courtesy of the Producers)

Product	Primary Protein Source	Digestion	Total Nitrogen (%)	$\frac{\text{Amino N}}{\text{Total N}} \times 100$	Solubility (g/L)	Moisture (%)
Producer: Sheffield Products						
N-Z-Amine A	Casein	Pancreatic	13.1	49.6	25	3.4
N-Z-Amine AT	Casein	Pancreatic	13.0	49.2	a)	3.5
N-Z-Amine AS	Casein	Pancreatic	13.0	50.0	210	3.5
N-Z-Amine B	Casein	Pancreatic	13.2	42.4	35	3.4
N-Z-Amine BT	Casein	Pancreatic	13.0	43.1	a)	3.5
N-Z-Amine E	Casein	Pancreatic	13.2	34.1	35	3.7
N-Z-Amine ET	Casein	Pancreatic	13.1	34.4	a)	3.5
N-Z-Amine YT	Casein	Pancreatic	13.2	47.7	50	3.7
N-Z-Amine YTT	Casein	Pancreatic	13.0	47.0	a)	3.8
Edamin	Lactalbumin	Pancreatic	12.3	56.1	20	3.2
Edamin T	Lactalbumin	Pancreatic	12.3	55.3	a)	3.4
PMN[b]	Non-fat dry milk	Pancreatic	5.4	31.5	425	3.0
PMN II	Non-fat dry milk	Pancreatic	5.5	29.1	a)	2.0
Hy – Soy	Soybean meal	Papaic	9.5	20.0	120	2.8
Hy – Soy T	Soybean meal	Papaic	7.7	20.8	a)	3.6
Primaton RL	Meat	Peptic	11.9	47.9	120	1.7
Amicase	Casein	Mineral acid	13.4	71.0	100	1.8
Amisoy	Soy protein	Mineral acid	12.6	69.0	100	3.0
Amiton	Animal protein	Mineral acid	12.7	77.1	100	1.8
Cottonseed peptone	Cottonseed flour	Enzymatic	8.7	33.3	a)	3.4
Hydrolyzed corn gluten	Corn gluten	Enzymatic	10.1	25.7	a)	—
Producer: Amber Laboratories						
Amber EHC	Casein	Enzymatic	13.5	30.0	—	4.0

[a] Technical grade products. They are intended for large-scale, low-cost usage where the presence of insoluble material is unimportant. They are unfiltered mixtures of hydrolyzed protein and deactivated enzyme.
[b] Peptonized milk nutrient

Table 26.32. Composition of Commercial Acid Protein Hydrolysates Technical Grade (Courtesy of Diamalt AG)

Constituent[a] (%)	Diamin N 9	Diamin N 9 Paste	Product Diamin ML II	Diamin HE	Diamin SFD (without salt)
Dry matter	50	81	28	29	50
Total nitrogen	9	15	6	4.1	8
Amino acids	30	53	10	21	45
NH_4Cl	18	21	18	8	<2
Alanine	1.8	2.8	0.4	1.1	2.9
Arginine	3.5	5.9	0.8	1.4	4.0
Aspartic acid	2.7	4.7	0.6	1.5	4.0
Cystine	0.4	–	0.3	2.2	0.3
Glutamic acid	5.6	10.6	1.4	4.2	11.0
Glycine	1.5	3.4	0.4	1.0	3.6
Histidine	0.4	0.9	0.1	0.2	0.8
Isoleucine	0.5	0.8	0.6	0.5	0.4
Leucine	1.0	1.7	1.4	1.2	0.4
Lysine	1.3	4.3	0.3	0.5	1.9
Methionine	0.1	0.2	0.1	0.2	0.1
Phenylalanine	0.6	1.3	0.4	0.2	0.9
Proline	3.0	3.8	0.5	2.8	4.1
Serine	3.5	6.1	0.8	2.3	5.3
Threonine	2.7	3.8	0.6	1.3	3.3
Tyrosine	0.4	0.7	0.3	0.4	0.2
Valine	1.7	2.4	0.7	0.8	1.7

[a] % in original product

ized, and subsequently submitted to an alkaline hydrolysis; the product is then used in nutrient media (KIRSANOV [26.404]). For L-lysine production with *Brevibacterium lactofermentum,* a hydrolysate from *Pseudomonas* is incorporated into the nutrient medium (NITELEA et al. [26.406]).

Distillation Residues

These are the still residues of the fermented starch-containing mash or molasses from ethanol and acetone/butanol distillation. In many countries they are used as livestock feed or fertilizer; however, they often also serve as a nitrogen source and are used as a supplement to molasses in the production of ethanol and feed yeast.

Depending on the type of raw material used we distinguish between molasses, grain, potato distillation residues, and mixtures thereof. All contain non-assimilable rests of nutrients and metabolites of yeasts or *Clostridium* as well as their autolyzed cell fragments. The composition of these distillation residues is summarized in Table 26.33. The betaine, glutamic acid, and aspartic acid concentration ranges between 6-8% and that of leucine and isoleucine between 1-2% of the total nitrogen (MACKO [26.407]). Other constituents may be 11% carbohydrates, 9% protein, 1.5% volatile acids, 21% gelatinous substances, 4.5% lactic acid, and 5.5% glycerol (KUJALA et al. [26.408]).

The components of the ash of a molasses distillation residue are listed in Table 26.34.

Torula yeast is produced either from a mixture of 12–14% molasses distillation residues (syrup) and molasses (MACKO [26.407]) or the molasses distillation residue itself may be used as the main substrate (KUJALA et al. [26.408]).

Table 26.33. Composition of Corn, Grain, and Molasses Distillation Residues

Constituent (% d.m.)	Corn (OFFER and LAUE [26.405])	Distillation Residue Grain[a] (Strothmann Brennereien)	Molasses (MACKO [26.407])
Dry matter	5.9	85.99	75.32–77.46
Ash	5.1	6.87	24.11–24.80
Organic matter	94.1	79.10	50.52–52.93
Total nitrogen	4.41	6.40	3.27– 3.74
Fat	10.00	3.86	—
Total potassium	—	—	14.41–14.46

[a] analyzed at Institut für Gärungsgewerbe und Biotechnologie, West-Berlin. Courtesy of Strothmann Brennereien.

Table 26.34. Composition of Non-soluble Ash of Molasses Distillation Residues (RYBAROVA [26.409])

Component	% of Distillation Residue
Potassium	0.204–0.322
Natrium	0.198–0.406
Calcium	51.6–57.8
Magnesium	1.11–6.98
Iron	0.822–2.11
Zinc	0.108–0.283
Manganese	0.021–0.096
Aluminum	0.132–0.147

data from three different factories

Table 26.35. Amino Acid Composition of GEWO® Gelatin (Courtesy of Geistlich Wolhusen)

Amino Acid	(%)	Amino Acid	(%)
Glycine	26.4	Serine	4.1
Alanine	10.7	Methionine	0.9
Valine	2.8	Cysteine	—
Leucine	3.3	Cystine	—
Isoleucine	1.4	Phenylalanine	2.6
Threonine	2.2	Tyrosine	0.6

Furthermore, molasses distillation residues can replace cornsteep liquor and sugar in catalase production by *Penicillium vitale* and serve as a nitrogen source in the processing of plant hydrolysates by *Candida scottii* (ZABRODSKIJ et al. [26.410]; review: BRAUN and MEYRATH [26.411]; RIBEIRO and BRANCO [26.412]).

Gelatin

Gelatin is extracted from bones and rind after fats and minerals, chiefly phosphates, have been removed. It serves as a nitrogen source in nutrient media for collagenase synthesis by *Flavobacterium* species (LABADIE and GOUET [26.413]) and for gelatinase formation by *Bacillus mesentericus* (VASKIVNYUK et al. [26.376]). The composition of amino acids in GEWO® gelatin is given in Table 26.35.

26.2.3 Inorganic Components

26.2.3.1 Phosphorus

This element, in the form of $H_2PO_4^-$, is essential for cell growth and the regulation of several metabolic processes. Phosphates are the most frequently employed substrates in nutrient media to meet this requirement. Cornsteep liquor is a suitable phosphorus source, the phosphorus content being proportional to the ash content (CEJKA [26.381]). Molasses nutrient media employed in baker's yeast, ethanol, and acetone/butanol production require phosphorus in the form of phosphates or superphosphates. This also applies to media from sugar-beet-pulp or sugarcane (JONES et al. [26.414]; FENIKSOVA et al. [26.104]; ER-EL et al. [26.108]) and media containing cellulose, n-paraffins, and ethanol (see Table 26.41).

26.2.3.2 Sulfur

Equally important as phosphorus, sulfur usually is employed in the form of sulfates. In cephalosporin synthesis, calcium sulfate may be used in place of methionine (NISS [26.415]).

26.2.3.3 Other Elements

Macroelements

K, Mg, Ca, Zn, Fe, and Cl are required in nutrient media in concentrations of about 0.1–1 mM. They are either added in the form of pure phosphates or sulfates or as constituents of raw materials.

Microelements

Co, B, Cd, Cr, Cu, Mo, and Ni requirements range between 0.1–100 µM (JONES et al. [26.414]).

For many complex nutrient media the supply of these elements is covered by the water and raw materials. In certain cases, depending on the procedure and types of raw materials used, it may be necessary to add the above elements directly. This may be achieved, as for instance in SCP production by *Candida utilis,* by adding the ash of molasses distillation residues or sludge from potash plants to the ethanol substrate (RYBAROVA and PECKA [26.416]; RYBAROVA [26.417]). In n-paraffin media these elements must also be supplemented (BENNET et al. [26.325]). The synthesis of chlorotetracycline by *Streptomyces aureofaciens* is enhanced by Cu^{2+} (AMERICAN CYANAMID CO [26.418]), pectinase formation in sugar-beet-pulp nutrient media is increased by Ti^{4+} (SHINKARENKO et al. [26.103]) and the synthesis of glucose isomerase by *Streptomyces flavoriceus* is improved by adding Mg- and Co-ions (VAHERI and KAUPPINEN [26.419]). The yield of baker's yeast can be raised in molasses nutrient media by adding carboxylin (a mixture of $NaHCO_3$, $MgSO_4$, $MnSO_4$, and $ZnSO_4$) (BELOVA et al. [26.420]). In citric acid fermentation of beet molasses by *Aspergillus niger,* the acid yield can be optimized by controlling and regulating the concentration of ferrocyanid ions (CEJKOVA et al. [26.421]).

26.2.4 Vitamins

Many of the raw materials used in fermentations contain sufficient concentrations of vitamins for microbial growth. The vitamin content of several products incorporated in nutrient media for industrial fermentations is shown in Table 26.36.

Biotin and thiamine, the most common essential vitamins, frequently do not occur in sufficient concentrations and thus must be separately added. These two vitamins as well as inositol and pantothenic acid are required in cultivating *Saccharomyces cerevisiae*. Beet molasses lack sufficient amounts of biotin and, when used in baker's yeast production, must be supplemented with biotin in the form of desthiobiotin, corn-steep liquor (ZAPARA and BALYBERDINA [26.422]), whey (HROBONI et al. [26.135]), a product of lactose production containing 80% of the milk's biotin (DERKANOSOV and NAZINTSEVA [26.423]), or a mixture of beet and cane molasses (OLBRICH [26.47]). Several strains of *Candida,* when grown on methanol, ethanol, and n-paraffins, require additional biotin (VIKHANSKY et al. [26.424]) or thiamine (KNECHT et al. [26.323a]; BIRKENSTEDT et al. [26.324]; ERMAKOVA et al. [26.425]). These are supplemented either as pure vitamins (thiamine as its hydrochloride) or as a yeast extract (CEJKOVA et al. [26.326]; TSUGAWA and OKUMURA [26.340]; SUZUKI et al. [26.426]; a.o.). Both

Table 26.36. Vitamin Content of Medium Ingredients for Industrial Fermentations

Medium Ingredient	Vitamin (μg/g)							
	Thiamine	Riboflavin	Pyridoxin	Biotin	Pantothenic acid	Niacin	Inositol	Choline
Beet molasses[a]	—	—	—	0.0204	36	—	1280	—
Cane molasses[b]	8.3	2.5	6.5	1.2	214	210	—	—
Soybean meal expeller[c]	—	3.08	—	—	14.08	30.36	—	2420
Cottonseed meal								
Proflo[c]	4.36	5.11	0.885	0.794	12.51	83.90	3640	33.48
Pharmamedia[c]	3.99	4.82	16.40	1.52	12.40	83.30	10800	3270
Cornsteep liquor[d]	5.4	11.00	15	—	28	170	4950	5400
Yeast autolysates								
BYF Series 100[e]	50	35	25	2	100	550	3000	2000
Yeatex C Paste[f]	10	20	25	1	50	400	1500	1500
KAT[g]	30–40	60–80	34–40	—	300	800	—	—
NAT 38[g]	10–15	40–60	14–20	—	300	100	—	—
Dried distiller's solubles[c]	5.5	15.4	—	2.86	19.8	110	—	4400

[a] SCHIWECK and HABERL [26.46], [b] OLBRICH [26.64], [c] courtesy of Traders Oil Mills, Ltd., [d] courtesy of Maizena GmbH, [e] courtesy of Amber Laboratories, [f] courtesy of Bovril Ingredients, [g] courtesy of Ohly GmbH

vitamins are also supplied in cellulose conversions by *Thermomonospora* sp. (MOREIRA et al. [26.427]).

Biotin influences the bacterial synthesis of amino acids (review: KINOSHITA and NAKAYAMA [26.24]) and is added in L-lysine production in concentrations of 15–50 µg/L to nutrient media containing acetate or ethanol as carbon source (KABUSHIKI KAISHA TEKKOSHA CO [26.256]; KYOWA HAKKO KOGYO CO [26.297]; NAKAYAMA and HAGINO [26.296]). Biotin is also an essential growth factor for bacteria synthesizing L-glutamic acid; here, the biotin concentration may not exceed 5 µg/L, as higher quantities inhibit L-glutamic acid formation (KINOSHITA and NAKAYAMA [26.24]). The biotin content of nutrient media of sugarcane molasses containing up to 800 µg/L of biotin can be reduced by treatment with activated carbon, ion exchange (BALITSKAYA et al. [26.428]), or γ-irradiation (WATANABE and SATO [26.57a]).

26.2.5 Other Medium Additives

26.2.5.1 Amino Acids

Free amino acids are added to nutrient media for the production of certain rare amino acids by deficient strains, or they may serve as precursors (examples given in Table 26.41). Technical grade amino acids are commercially available in L- and D,L-form (Diamalt AG).

26.2.5.2 Precursors and Inducers

Media employed to synthesize certain extracellular enzymes and antibiotics must contain precursors or inducers (Table 26.37). To reduce the high costs of pure precursor/inducer some procedures have been adapted to operate with raw materials containing the required substrate, e.g., corn cobs instead of pure xylose or industrial waste meat instead of collagen.

26.2.5.3 Detergents

Detergents serve as emulsifiers in preparing water-insoluble nutrients and substrates such as *n*-alkanes, oils, and steroid-containing raw materials. Their use leads to increased yields of biomass from *n*-alkanes (TANAKA and FUKUJI [26.439]; NAKAHARA et al. [26.440]; KUCHER et al. [26.441]), amino acids (WATANABE et al. [26.442]), and organic acids (FURUKAWA et al. [26.334]). The conversion of β-sitosterol by *Nocardia* is enhanced by Tween 40 and Tween 60, though it also serves as a source of fatty acids for the synthesis of surface-active lipids in cell walls (SCHOEMER and WAGNER [26.443]). Detergents in nutrient media alter the permeability of cell membranes and activate various cellular enzymes. Therefore, their stimulating effect has also been observed in complex nutrient media with carbohydrate carbon sources, e.g., for obtaining protease from *Aspergillus* (BHUMIBHAMON and EKLUND [26.444]), lipase from *Alcaligenes* (KOKUSHO et al. [26.445]), antibiotics from *Melanconis flavovirens* (SASEK and GUPTA [26.446]), and vaccines from *Leptospira* (JOHNSON and BEY [26.447]).

The most frequently employed commercially available detergents for nutrient media are listed in Table 26.38.

26.2.5.4 Antifoaming Agents

Antifoams are mostly vegetable or animal oils and fats and are often used as carbon sources (CHERKASOVA and MAKAREVICH [26.250]; FRANZKE et al. [26.448]). The type of industrially used antifoams depends greatly on local conditions. The USSR and Eastern Block states utilize animal oils, especially whale oil, and vegetable oils, while western fermentation industries mainly employ synthetic products (Table 26.39).

26.2.5.5 Solids

Solids and substances increasing the viscosity of nutrient media may intensify certain fermentation processes. The yield of

Table 26.37. Inductors and Precursors Employed in Nutrient Media

Product	Inductor or Precursor	Microorganism	Reference
Glucosoisomerase	Xylose (corn cobs)	*Streptomyces*	SANCHEZ and QUINTO [26.429]
β-Glucosidase	Amygdalin, salicyn	*Aspergillus niger*	WOODWARD and WISEMAN [26.430]
Cholinesterase	Lecithin (soyalecithin)	*Pseudomonas*	BEAUCAMP et al. [26.431]
Collagenase	Collagen (waste from meat industry)	*Flavobacterium* sp.	LABADIE and GOUET [26.413]
Erythromycin	n-Propanol	*Streptomyces erythreus*	GRUZINA et al. [26.432]
Penicillin V	Phenoxy acetic acid	*Penicillium chrysogenum*	MATELOVA and CULIK [26.298]
Penicillin G	Phenylacetate	*Penicillium chrysogenum*	MATELOVA and CULIK [26.298]
D-Biotin	Azelaic acid	*Sporobolomyces carnicolor*	SHIBATA et al. [26.23]
Ergot alkaloids	α-Ketobutyric acid	*Claviceps purpurea*	UDVARY-NAGY et al. [26.433]
Xanthan	Na-desoxycholate	*Xanthomonas campestris*	WEISROCK [26.434]
Neoviridogriseins	L-Proline	*Streptomyces griseoviridus*	OKUMURA et al. [26.435]
L-Serine	Glycine	*Sarcina albida*	EMA et al. [26.436]
Formycin	Lysine, glutamate	*Streptomyces* sp.	OCHI et al. [26.437]
Desacetoxycephalosporin C	L-Cysteine	*Paecilomyces carneus*	NAKAO et al. [26.342a]
Vitamin B_{12}	$CoCl_2$ + 2,6-dimethylbenzimidazol	*Pseudomonas thermophila* K-2	KATRUK [26.438]

Table 26.38. Some Industrial Emulsifiers Applied in Fermentations

Producer	Trade Name	Chemical Basis	Appearance[a]
Akzo Chemie GmbH	Armotan® PM 0 20	Polyoxy-ethylene-sorbate-mono-oleate	L
	Armotan® MS	Sorbate-mono-stearate	S
Atlas Chemie	Arlacel® 186	Glycerol-mono-di-oleate	L
	Atmos® 150	Glycerol-mono-di-stearate	S
	Tween® 20	Polyoxy-ethylene-(20)-sorbate-mono-laurate	L
	Tween® 40	Polyoxy-ethylene-(20)-sorbate-mono-palmitate	L
	Tween® 61	Polyoxy-ethylene-(4)-sorbate-mono-stearate	S
	Tween® 65	Polyoxy-ethylene-(20)-sorbate-stearate	S
	Tween® 80	Polyoxy-ethylene-(20)-sorbate-mono-oleate	L
Chemische Werke Grünau GmbH	Lamecreme® V 35600 P	Glycerol-mono-di-stearate	S
Goldschmidt AG	Tagat® O	Polyoxy-ethylene-glycerol-mono-oleate	L
	Tagat® S	Polyoxy-ethylene-glycerol-mono-stearate	S part. L
	Tegin®	Palmitate-stearate-mono-glyceride	S
	Tegin® G	Ethylene-glycol-mono-di-stearate	S
	Tegin® M	Palmitate-stearate-mono-di-glyceride	S
	Tegin® P	1,2-propylene-glycol-mono-di-glyceride	W
	Tegin® 90	Palmitate-stearate-mono-di-tri-glyceride	S
	Emulgator® BTO 2	Blends of non-ionics	L
	Emulgator® E 2155	Blends of non-ionics	W
	Axol® C 62	Citric acid fatty acid glyceride	S
	Lisat® C	Calcium stearoyl-2-lactylate	S
Lucas Meyer GmbH	Emulpur N	Defatted soybean lecithin	S
	Epikuron 100 G	Blends of isolated soybean phospholipids	S
	Emulthin M-35	Spray-dried soybean lecithin concentrate	S
	Emulfluid A	Water-dispersed soybean lecithins	V
Ugine Kuhlmann GmbH	Emkapol® 200, 300, 400	Polyethylene-glycols	V
	Ukamil® 190	Polyoxy-ethylene with linear synthetic alcohol	L

[a] Appearance: L Liquid, S Solid, W Waxy, V Viscous

Table 26.39. Some Industrial Antifoams Employed in Fermentation Procedures

Producer	Trade Name	Chemical Basis	Appearance[a]
Bayer AG	Entschäumer 7800 neu	Blends of higher hydrocarbons and sulfonic acid derivatives	L
	Baysilon Entschäumer E	Blends of dimethylpolysiloxan and non-ionics; 40% (weight) emulsion	V
Dow Corning Int. Ltd.	Entschäumer E 100 konz.	Blends of glycerides and higher fatty acids	V
	Dow Corning® 1520	Silicon antifoams, emulsion	L
Goldschmidt AG	Entschäumer EH 7365	Concentrate of organic antifoams	L
Hoechst AG	Entschäumer FN	Blends of non-silicon and non-mineral surfactants	V
Imperial Chemical Industries Ltd.	Silcolapse® 5000	Silicon antifoam emulsion; 30% (weight)	L
Schill und Seilacher GmbH & Co	Struktol® J 647	Blends of non-ionics	L
	Struktol® SB 2020	Blends of higher vegetable esters and alcohols	L
Ugine Kuhlmann GmbH	Pluronic® F 68	Polyethylene-oxide-propylene-glycol	S
	Pluronic® L 81		L
	Tetronic	Polyethylene-propylene-oxide-diamine	L
Union Carbide Co.	SAG® 100	Silicon antifoams	V
	SAG® 471	Silicon antifoams	V
	SAG® 5693	Silicon antifoams	V
	Sentry® Simethicone NF	100% silicon antifoams	L
	SAG® 10	10% (weight) emulsion	L
	SAG® 30	30% (weight) emulsion	V
	SAG® 4130	Silicon emulsion	V
	Sentry® Simethicone Emulsion	Silicon emulsion	L
Wacker-Chemie GmbH	Silicon antifoam SH	Silicon antifoams	V

[a] Appearance: L Liquid, V Viscous, S Solid

neomycin could be significantly increased by adding either agar, paper pulp, or asbestos to the nutrient medium containing glucose as carbon source (LEVITOV and MESHKOV [26.449]). Equally, submerged citric acid fermentation with *Aspergillus niger* could be improved by pellet formation, which increased by adding starch, agar, algenic acid, and carrageenin in concentrations of maximally 0.5% (CIMERMAN et al. [26.450]). Sulfite waste liquors have been enriched with cellulose fibers adsorbing certain yeasts capable of adsorption, thus raising the ethanol yield (KALIUZHNYJ [26.451], [26.452]). Presently, particles of, e. g., stainless steel are being tested for their ability to serve as mechanical supports of growth in media with *Streptomyces*, yeasts, and fungi. Such products already have been successfully employed in tower fermentations in so-called fluidized-bed procedures of different kinds (ATKINSON et al. [26.453]).

NH_4-ions contained in complex nutrient media inhibit leucomycin synthesis by *Streptomyces kitasatoensis* as well as the formation of cerulenin by *Cephalosporium coerulens*. By adding insoluble magnesium phosphate (TANAKA et al. [26.454]) or natural zeolites (MASUMA et al. [26.455]) the NH_4-ions are trapped by the latter, thus increasing the antibiotics synthesis.

26.2.5.6 Antiseptics

Addition of antiseptics to fermentation media has rarely been reported. Surface citric acid fermentations with beet molasses have been carried out with formalin (LEOPOLD et al. [26.75]), 5-nitro-2-furyl-acrylate sodium and potassium salts (LEOPOLD et al. [26.456]), or furacycline (KARKLIN [26.457]). Furazolidin has been used in the production of an insecticide with *Bacillus thuringiensis*. This compound is not toxic to the microorganism and inhibits the growth of phages (FAIBICH et al. [26.458]).

26.2.5.7 Toxins

Aflatoxin, patulin, zearalenone, and other toxins are formed by such fungi as *Aspergillus, Penicillium, Fusarium,* and others. They may become accidentally incorporated into nutrient media when fungus-infected substrates/raw materials are employed. Corn is the best substrate for aflatoxin production by *Aspergillus* (LILLEHOJ et al. [26.459]; SANCHIS et al. [26.460]). Aflatoxins may also pass into cornsteep liquors during corn steeping in cornstarch production. We have found that in all cornsteep liquors, obtained from corn with a 20% germinating power, aflatoxin B_1, G_1, and M occurred in concentrations up to 200 µg/kg cornsteep liquor (unpublished results). These cornsteep liquors all inhibited steroid conversions (CEJKA [26.381]). Solulys L (Roquette Frères SA), produced from standardized, commercial corn, was free of aflatoxins (DEVOS and HUCHETTE, personal communication). WOOD [26.461] found that the presence of 0.15 and 0.3% SO_2 during corn steeping decreases aflatoxin formation by 17 and 56%, respectively. Corn may also contain deoxynivalenol and zearalenone (MIROCHA et al. [26.462]). Concentrations of the latter have been found to reach 120-6400 µg/kg corn.

Aflatoxins inhibit the growth of *Bacillus, Clostridium,* and *Streptomyces* (BURMEISTER and HESSELTINE [26.463]), and also alcoholic fermentations with *Saccharomyces cerevisiae*. The latter process is also inhibited by patulin, diacetoxyscipenol, pennicollic acid, and Toxin T 2 (LAFONT et al. [26.464]).

26.2.5.8 Water

Water serves to dissolve or suspend nutrient substrates and forms the basis of all liquid media. Thus, water quality is of crucial importance in fermentation procedures. Different technologies may require different water qualities. The food industry generally employs drinking water for production of human nutrients. The same applies to pharmaceutical products, though demineralized water occasionally may be required.

Drinking water contains metals and trace elements, increasing the yield of product in organic acid syntheses, in contrast to the use of distilled water (FURUKAWA et al. [26.334]).

The production of feed yeast from synthetic ethanol does not impose special requirements on water quality. Thus, surface water, treated with aluminum salts and chlorinated, has been employed for this purpose. However, the concentrations of aluminum and active chlorine may not exceed 1 mg/L and 0.3 mg/L, respectively (RYBAROVA and ADAMEK [26.465]). River water has been used in alcoholic fermentations (TOMCZYNSKA [26.466]): Organic contaminants were removed in chloroform and ethanol; limited amounts of this extract are occasionally added back to the water. Yeast fermentation was thereby unaltered, though yeast growth was inhibited.

The use of highly salty seawater is especially interesting for producing glycerol by means of the alga *Dunaliella* sp. (AVRON and BEN-AMOTZ [26.467]; CHEN and CHI [26.468]): The high osmotic pressure of the seawater is counteracted by the formation of intercellular glycerol.

Several modern technologies employ the spent, fermented, and centrifuged media liquids as a substitute for water. This recycling reduces the amount of waste liquids and effluents and is increasingly used in feed yeast production and alcoholic fermentation. The industrial recycling of media liquids was carried out in the CSSR in 1959 in *Torula* yeast production. The media were reemployed ten times (BARTA and GREGR [26.469]; GREGR et al. [26.470]). The composition of the media changed during recycling: the content of non-assimilable compounds, such as betaine, increased from 0.312 to 2.82%, ash from 0.494 to 4.75%, potassium from 24.3 to 230 mmol/L, accompanied by a change of density from 1.0138 to 1.0765 kg/m^2 (4.2 to 19 °Bg) and an increase of the osmotic pressure from 5.8×10^5 to 53.95×10^5 Pa (CEJKOVA [26.471]). In commercial yeast production by the continuous procedure the media could be recycled up to an osmotic pressure of $93.2 \cdot 10^5$ Pa (FENCL et al. [26.472]).

The applicability of a recycling procedure, however, depends on the adaptability of the microorganism strain to the conditions of the recycled, fermented media (CEJKOVA [26.473]), especially to the high salt concentrations (MURPHY et al. [26.474]). Certain yeast strains have converted beet molasses, diluted with distillation residues instead of water, into ethanol (DAHIYA et al. [26.475]).

The recycling of nutrient media was then rediscovered in the 1970s and especially tested for biomass production (review: HAMER [26.476]). Media with the following substrates were investigated: synthetic ethanol (SIMEK and STROS [26.477], [26.478]), hexadecane (DAINIPPIN INK CHEM. [26.479]), methanol (TOPIWALA and KHOSROVI [26.480]), wood hydrolysates (KUNDEV [26.481]), and synthetic ethanol supplemented with molasses distillation residues (STROS et al. [26.482]). Recycling has also been applied in commercial alcoholic fermentations (ROSEN [26.483]) and has been proposed for ethanol, acetone, and butanol production (VOLESKY et al. [26.85]).

26.2.5.9 Enzyme Preparations

Commercial enzyme preparations have been increasingly applied in the fermentation industries in recent years. They are mainly used for the degradation of polysaccharides, such as starch, inulin, cellulose, and pectin, but also for breaking down foam-producing proteins during medium preparation. Furthermore, enzyme preparations serve in isolating products from complex media containing microbial cells and their metabolites.

Various enzyme preparations and mixtures are commercially available (Kali Chemie AG; Novo Industri A/S). Several enzyme preparations are characterized by the data in Table 26.40.

Table 26.40. Enzyme Preparations Used for Pretreatment of Industrial Nutrient Fermentation Media (Courtesy of Novo Industri A/S)

Enzyme Common Name Systematic Name EC Number	Trade Name Origin	Reaction which is Catalyzed Typical Application	Operating Conditions	
			pH	Temperature (°C)
α-Amylase 1,4-α-D-Glucan glucanohydrolase 3.2.1.1	BAN *Bacillus subtilis* Termamyl® *Bacillus licheniformis* Fungamyl® *Aspergillus oryzae*	Endohydrolytic cleavage of α-1,4-glycosidic linkages in saccharides with three or more glucose units Decomposition of starch to dextrins in processing of starch-containing products	6–6.5 6–8 5–7	70–90 105–110 50
Amyloglucosidase 1,4-α-D-Glucanohydrolase 3.2.1.3	AMG *Aspergillus niger* SAN *Aspergillus niger*	Hydrolytic cleavage of 1,4-glycosidic linkage in polysaccharides by splitting off glucose units from the non-reducing end Saccharification of polysaccharides in brewing and spirit industry	4–5	60 50–60
β-Glucanase 1,3-(1,3; 1,4)-β-D-Glucan 3(4)-glucanohydrolase 3.2.1.39 3.2.1.6	Cereflo® *Bacillus subtilis*	Endohydrolytic cleavage of 1,3- or 1,4-linkages in β-D-glucans to 3–5 glucose units Prevention of filtration problems in barley processing	5	50–70
Glucose isomerase D-Xylose ketol isomerase 5.3.1.5	Sweetzyme® *Bacillus coagulans* Immobilized	Transformation of D-xylose to D-xylulose; D-glucose to D-fructose	7.0–8.0	55–58
Cellulase 1,4-(1,3; 1,4)-β-D-Glucan glucanohydrolase 3.2.1.4	Celluclast™ *Trichoderma reesei*	Hydrolytic breakdown of β-1,4-glucosidic linkages in cellulose to glucose, cellobiose and higher glucose polymers	4.5–6.0	50–60
Cellobiase β-D-Glucoside glucohydrolase 3.2.1.21	Novozym 188 *Aspergillus niger*	Hydrolysis of terminal non-reducing β-D-glucoside compounds in cellobiose by liberating β-glucose	4.8	50–60

Table 26.40. (continued)

Enzyme Common Name Systematic Name EC Number	Trade Name Origin	Reaction which is Catalyzed Typical Application	Operating Conditions	
			pH	Temperature (°C)
Hemicellulase 1,4-β-D-Mannan mannanohydrolase 3.2.1.78	Gamanase® Aspergillus niger	Hydrolytic cleavage of β-1,4-D-mannosidic bonds in mannans, galactomannan and glucomannans for reduction of viscosity	3–6	70–80
Inulinase β-2,1-D-Fructosidase 3.2.1.7	Novozym 230	Endo- and exohydrolytic breakdown of β-2,1-linkages in inulin to fructose up to a degree of hydrolysis of 98%	4–5	60–65
Pullulanase Pullulan 6-glucanohydrolase 3.2.1.41	Promozyme™ Bacillus sp.	Endohydrolysis of 1,6-α-D-glucosidic bonds in pullulan, amylopectin and glycogen. Degradation of starch in combined application with amyloglucosidase after pretreatment with α-amylase	4.5–5.5	60
Pectinase Poly-(1,4-α-D-galacturonide) glycanohydrolase 3.2.1.15	Pectinex® Aspergillus niger	Hydrolysis of 1,4-α-D-galacturonic linkages in pectate and other galacturonans. Depolymerization of pectin for the macerating of plant tissues in vegetable processing	4.5	15–50
Proteases Peptide hydrolases 3.4	Alcalase® 0.6 L Bacillus licheniformis Neutrase® Bacillus subtilis	Hydrolysis of proteins and peptides of vegetable and animal origin	6.5–9.5 5.5–7.5	60–65 45–55
Mixture of enzymes	Ceremix™ Bacillus subtilis	Amylo-, glucano- and proteolytic activity in processing of barley	5.7	70–90

26.3 Storage and Handling of Raw Materials

Industrial fermentations often require tonnage quantities of raw materials per individual batch: e.g., a 100 m^3 production fermenter is supplied with a medium consisting of 2% soybean meal and 2% cornsteep liquor; 2 tons of each nutrient are filled into the fermenter – under stirring to avoid clumping – and sterilized. Storage and handling of such large quantities of raw material is of crucial importance.

Powdered materials must be stored under dry conditions in order to avoid caking as, e.g., in phosphates and lime. Modern drying techniques allow uncomplicated storage of highly hygroscopic, spray-dried proteinaceous yeast autolysates and cornsteep liquors. Enzymes required for pretreatment, e.g., saccharification of amylaceous raw materials, are stored at 4–10 °C and are handled (i.e., weighed and portioned) in ventilated environments. Skin contact and inhalation especially of powdered products must be avoided. Paste-type raw materials such as glucose paste and glucose syrup must be immediately processed, ideally at a close fermentation plant.

Mold and fungus infections are especially frequent during summer. Many liquids such as molasses and cornsteep liquors with a high content of solids must be tempered and stirred during storage. The type and conditions of storage tanks is also highly important. The interior walls of plastic-lined tanks must be undamaged and free of microbes. Most suitable are storage tanks without lining which tolerate steam treatment. Regular cleaning and disinfecting with volatile disinfectants, such as perchloracetic acid and steam, are simple methods to guarantee high quality of stored raw materials. Oily raw materials must also be stored under temperate conditions to simplify pumping and bottling. Industrial steroid conversion equipment includes huge solvent storage tanks, which must be installed and surveyed under strict precautions, especially in terms of possible charge buildup.

26.4 Design of Industrial Nutrient Media

The general discrepancy between design of industrial nutrient media and knowledge of the primary and secondary metabolism of organisms has been discussed extensively by CORBETT [26.484].

The composition of industrial nutrient media is initially the result of empirical trials aimed at guaranteeing growth and metabolic activity of a microorganism. Optimization for commercial production is usually kept secret.

Most nutrient media are complex formulas of several nutrient substrates. Synthetic media have been employed in antibiotics production (PERLMAN [26.485]) and applications of synthetic ethanol for producing biomass have been patented (ADAMEK et al. [26.487]).

Important examples of industrially utilized nutrient media are listed in Table 26.41.

Optimization methods greatly depend on the type of product: for biomass a continuous cultivation scheme is suggested (MATELES and BATTAT [26.530]; SUMMERS et al. [26.531]), while for secondary metabolites certain countries have mostly applied statistical planning methods, such as those by ROSENBROCK, by BOX-WILSON, and others.

Statistical evaluations rely mainly on the most important parameter, which is microbial activity, i.e., product yield. However, other criteria such as foaming and filtration properties are also taken into account.

Nutrient media occasionally contain unidentifiable raw materials which may cause

Table 26.41. Production Media from Natural Ingredients
(Product – Microorganism – Reference)

Antibiotics. *Dactylosporangium thailandense*
G 367. Toyo Jozo Kabushiki Kaisha [26.484a].
5.0% Dextrins   0.7% CaCO$_3$
0.5% Glucose   1.3 ppm CoCl$_2$
3.0% Soybean meal, defatted pH 7.0

Antibiotics WS 3442. *Streptomyces wadayamensis*
ATCC 21948. Imanaka et al. [26.311].
3.0% Glycerol   1.0% Cottonseed meal
2.0% Soybean meal   0.2% D,L-Methionine
1.0% Gluten meal

Anthracycline antibiotics. *Streptomyces galilaeus*
KE 303. Oki et al. [26.485a].
1.5% Soluble starch   0.1% MgSO$_4$·7H$_2$O
1.0% Glucose   0.0007% CuSO$_4$·5H$_2$O
3.0% Soybean meal   0.0001% FeSO$_4$·7H$_2$O
0.2% Yeast extract   0.0008% MnCl·4H$_2$O
0.3% NaCl   0.0002% ZnSO$_4$·7H$_2$O
0.1% K$_2$HPO$_4$ pH 7.4

Bactobolin. *Pseudomonas yoshitomiensis* Y 12278.
Munakata et al. [26.486].
2.0% Mannitol   0.02% K$_2$HPO$_4$
0.25% Dried yeast   0.8% CaCO$_3$
0.5% (NH$_4$)$_2$SO$_4$ 0.1% Antifoam agent
0.4% KCl   pH 7.6

Cephalosporins. *Streptomyces rochei* S 3907 C.
Brown et al. [26.19].
25 g/L Cerelose   5 g/L Pharmamedia
15 g/L Cornsteep   0.01 g/L CoCl$_2$·6H$_2$O
  liquor   pH 7.3
10 g/L Distiller's
  solubles

Cephalosporin C. *Cephalosporium acremonium*
mutant BC 2116 derived from *C. acremonium*
M 8650. Pan et al. [26.487a].
3.0% Methyl oleate   0.6% D,L-Methionine
3.0% Fish meal   0.5% CaCO$_3$
3.0% Peanut meal   0.1% Silicone SAG 471
0.5% (NH$_4$)$_2$SO$_4$

Capreomycin. *Dactylosporangium variesporum*
D 409 (ATCC 31203). Bristol-Myers Co
[26.363].
2.0% Corn starch   0.33% MgSO$_4$·7H$_2$O
3.0% Soybean meal   1.0% CaCO$_3$

Chlortetracycline. *Streptomyces aureofaciens*.
Welward and Halama [26.45].
4.0% Sucrose   0.4% (NH$_4$)$_2$SO$_4$
2.0% Soybean meal   0.2% Beet molasses
0.7% Peanut flour   0.2% Cornsteep liquor
0.25% NaCl   1 drop/100 mL Soybean oil
0.4% CaCO$_3$ pH 6.3–6.4

Gentamicin derivatives. *Micromonospora purpurea*
mutants. Yamamoto and Daun [26.17].
2.0% Glucose   0.0001% CoCl$_2$·6H$_2$O
0.75% Soybean meal   0.01% Antifoam
0.75% Yeast extract   0.02% Streptamine
0.4% CaCO$_3$

Macrolidic antibiotics. *Streptomyces antibioticus*
ATC 31771. Cappelletti et al. [26.488].
5.0% Dextrose   3.0% CaCO$_3$
2.0% Soybean meal   0.3% NaCl
0.4% Dried baker's yeast 1.0% Fat oil
1.6% Corn meal   Tap water

Narasin. *Streptomyces granuloruber* NRRL 12389.
Kastner and Hamill [26.489].
1.0% Glucose   0.2% CaCO$_3$
2.0% Molasses   0.02% Silicon anti-
0.5% Peptone (Difco)   foam agent
                       Deionized water

Neomycin. *Streptomyces fradiae*. Penick & Co
[26.490].
2% Soy flour   1% Cottonseed protein
1% (NH$_4$)$_2$SO$_4$ 1% Corn gluten
7–13% Amylopectin   1% CaCO$_3$

Neomycin. *Streptomyces fradiae* 3535. Bhadra et al. [26.162].
5% Tapioca starch   1% (NH$_4$)$_2$SO$_4$
1% Soy flour   1% CaCO$_3$
1% Yeast powder   pH 7.5

Neplanocin A. *Ampullariella regularis* A 11079.
Yaginuma et al. [26.491].
4.0% Glucose   0.1% Yeast extract
1.0% Soybean flour   0.25% NaCl
0.4% Meat extract   0.25% CaCO$_3$
0.4% Peptone   pH 6.5

Oxytetracycline as an animal feed supplement.
Streptomyces rimosus 12907. Baghlaff et al.
[26.87].
30 g/L Blackstrap molasses   10 g/L Rice bran
20 g/L Fodder yeast   0.2 g/L KH$_2$PO$_4$

Penicillin. *Penicillium chrysogenum* KAIS 12690.
Ryu and Hospodka [26.113].
2.0% Lactose   0.5% (NH$_4$)$_2$SO$_4$
1.0% Pharmamedia   0.02% MgSO$_4$·7H$_2$O
0.5% CaCO$_3$ 0.025% Polypropylene
0.3% KH$_2$PO$_4$ glycol
                       0.2% Soybean oil

Prodigiosin. *Serratia marcescens* ATCC 31453.
Nakamura and Kitamura [26.271].
2.0% Na-oleate   0.05% MgSO$_4$
0.4% (NH$_4$)$_2$SO$_4$ pH 8.0
0.3% KH$_2$PO$_4$

Table 26.41. (continued)

Streptomycin. *Streptomyces griseus albus* mutant. MERCK & Co [26.492].
Seed medium:
1.0% Dextrose
0.6% Meat extract
1.0% Enzymatic digest of casein
Distilled water

Production medium:
3.5% Soybean meal
2.75% Dextrose
0.5% Distiller's dried solubles
0.25% NaCl
0.4 mL/100 mL Soybean oil
Distilled water

Tetracycline antibiotics. *Streptomyces galilaeus* OBB 111610. FUJIWARA et al. [26.493].
2.0% D-Glucose   0.1% $MgSO_4 \cdot 7 H_2O$
2.0% Starch   0.3% NaCl
1.0% Pharmamedia   0.3% $CaCO_3$
0.1% K_2SO_4 Tap water

Tunicamycin. *Streptomyces chartreusis* NRRL 12338. HAMILL et al. [26.494].
2.5% Glucose   0.05% $MgSO_4 \cdot 7 H_2O$
1.0% Corn starch   0.2% $CaCO_3$
1.0% Meat peptone   Tap water
0.4% Enzyme-hydro-   pH 6.6
    lyzed casein
0.5% Blackstrap molasses

Insecticides. *Streptomyces avermitilis* MA 4848. ARISON et al. [26.495].
Seed medium:
2.0% Lactose   0.5% Autolyzed yeast
1.5% Distiller's   0.32 mL/L Polygly-
    solubles   col 2000
pH 7.0

Production medium:
4.5% Dextrose   0.25% Autolyzed yeast
2.4% Peptonized milk   2.5 mL/L Polygly-
pH 7.0   col 2000

Insecticides B 41 D. *Streptomyces* B 41-146. SANKYO Co [26.496].
8.0% Glucose   1.0% Skim milk
1.0% Soybean flour   0.2% Cornsteep liquor
0.5% Corn starch   0.3% NaCl

Pesticides (Nikkomycins). *Streptomyces tendae* ATCC 31160. HAGEMAIER et al. [26.300].
2.0% Mannitol
2.0% Soybean meal
pH 7.5

Pesticides (Botrycidin A). *Bacillus subtilis* NRRL B 12231. MISATO et al. [26.497].
2.0% Starch
2.0% Glucose
2.0% Soybean meal
0.05% Antifoam   pH 6.2

Steroid transformation. *Curvularia lunata* NRRL 2380. SCHERING AG [26.498].
0.7% Glucose
3.0% Cornsteep liquor   pH 5.5

Steroid transformation. *Mycobacterium fortuitum* NRRL B-11359. WOVCHA and BROOKS [26.20].
5.0 g/L Cerelose   2.0 g/L $MgSO_4 \cdot 7 H_2O$
3.0 g/L NH_4Cl 0.5 g/L Urea
0.5 g/L KH_2PO_4 2.0 g/L Tween 80
3.0 g/L $CaCO_3$ 1.0 g/L Soy flour
3.0 g/L Na-citrate $\cdot H_2O$ 8.0 g/L Ucon
pH 7.0

Steroid transformation. *Arthrobacter, Nocardia, Fusarium, Microbacterium, Mycobacterium* a.o. MITSUBISHI CHEM. IND. [26.21a].
Seed medium:
1.0% Glucose   1.0% Peptone
0.3% Meat extract   pH 7   0.5% NaCl

Fermentation medium:
2.0% Glucose   0.2% Asparagine
2.0% Cornsteep liquor   0.2% KH_2PO_4
0.4% Na-glutamate   pH 7   1.0% Rapeseed oil

Steroid transformation. *Botryoplodia malorum* CBS 13450. SCHERING AG [26.33]; PETZOLD et al. [26.499].
1% Starch dextrose
1% Soybean meal   pH 6.2

Steroid transformation. Mutant *Mycobacterium fortuitum* NRRL B-8119. WOVCHA et al. [26.314].
1.0% Glycerol   0.15% NH_4Cl
0.1% Soy flour   0.005% $FeCl_3 \cdot 6 H_2O$
0.05% K_2HPO_4 Distilled water
0.05% $MgSO_4 \cdot 7 H_2O$ pH 7.0

Steroid transformation. *Arthrobacter simplex* ATCC 6946. WEBER et al. [26.500].
Seed medium:
0.5% Cornsteep liquor   0.1 mL/L Silicon SH
0.05% Glucose   0.1 mL/L Pluronic
0.1% Yeast extract   pH 7.0

Production medium:
0.5% Cornsteep liquor
0.05% Glucose
0.3% Protein hydrolysate
0.1 mL/L Silicon SH
0.1 mL/L Pluronic   pH 7.0

Table 26.41. (continued)

Steroid transformation. *Mycobacterium fortuitum* NRRL B-8119. SALMOND et al. [26.501].
0.5% Cerelose	0.1% Soybean meal
0.3% NH$_4$Cl	0.05% KH$_2$PO$_4$
0.3% CaCO$_3$	0.05% Urea
0.3% Na-citrate	0.8% Ucon
0.2% Tween 80	pH 7.0

Steroid transformation. *Curvularia lunata* NRRL 2380. ALIG et al. [26.502].
1.0% Cornsteep liquor	0.005% Soybean oil
1.25% Soybean meal	pH 6.2

Steroid transformation. Bacterial mutant DSM 1444. BAHN et al. [26.503].
0.5% Glucose	1.8% Corn gluten
0.3% Yeast extract	0.15% K$_2$HPO$_4$
0.8% Cornsteep liquor	0.09% (NH$_4$)$_2$HPO$_4$
	pH 7.0

Steroid transformation. *Mycobacterium fortuitum* NRRL B-12433. WOVCHA et al. [26.504].
0.5% Cerelose	0.1% Soybean meal
0.3% NH$_4$Cl	0.05% KH$_2$PO$_4$
0.3% CaCO$_3$	0.05% Urea
0.3% Na-citrate · 2 H$_2$O	0.8% Ucon
0.2% Tween 80	pH 7.0 Tap water

Steroid transformation. *Rhodotorula equi* MCI-1416. MITSUBISHI CHEM. IND. [26.21].
4.0% Soybean meal	0.2% NaNO$_3$
2.0% Yeast	0.2% K$_2$HPO$_4$
1.0% Glycerol	0.1% MgSO$_4$ · 7 H$_2$O
	pH 7.0

Clavine ergot alkaloids. *Claviceps fusiformis*. SOČIČ et al. [26.302].
5.0% Peptone
10.0% Mannitol

Ergot alkaloids. *Claviceps purpurea* MNG 00186. WACK et al. [26.42].
100 g/L Sucrose	1.0 g/L NH$_4$NO$_3$
10 g/L Succinic acid	1.0 g/L CaCl$_2$
0.25 g/L KH$_2$PO$_4$	pH 5.2–5.3
0.25 g/L MgSO$_4$	

Feeding with valine and isoleucine total 100 and 200 g/L.

Ergot alkaloids. *Aspergillus fumigatus* NCIM 902. NARAYAN and RAO [26.301].
25.0 g/L Glucose	20 mg/L ZnSO$_4$ · 7 H$_2$O
25.0 g/L Mannitol	2.5 mg/L CuSO$_4$ · 5 H$_2$O
12.0 g/L NH$_4$-citrate	15 mg/L MnSO$_4$ · H$_2$O
	1 mg/100 mL Sodium lauryl sulfate
2.0 g/L MgSO$_4$ · 7 H$_2$O	
0.25 g/L KH$_2$PO$_4$	
pH 5.5	

Guanosine. *Bacillus pumilus* No. 148-S-16 (FERM BP-6). SUMINO et al. [26.505].
Seed medium:
4.0% Glucose	0.2% CaCl$_2$ · 2 H$_2$O
0.2% Urea	2.0% Cornsteep liquor
0.5% Na-L-glutamate	0.25% Ribonucleic acid
0.2% MgSO$_4$ · 7 H$_2$O	0.03% Histidine
0.1% KCl	200 µg/L Biotin
0.003% MnSO$_4$ · 4 H$_2$O	

Production medium:
22.0% Glucose	0.4% CaCl$_2$ · 2 H$_2$O
0.8% (NH$_4$)$_2$SO$_4$	0.006% MnSO$_4$ · 4 H$_2$O
1.0% Na-L-glutamate	0.03% Histidine
0.4% MgSO$_4$ · 7 H$_2$O	400 µg/L Biotin
0.2% KCl	0.1% Inosine
0.03% Ribonucleic acid	5.0% Cornsteep liquor

(about 10% of the required amount of the adenine-containing material)

Inosine and/or guanosine. *Bacillus pumilus* No. 158-A-17 (FERM BP-7). DOI et al. [26.506].
Seed medium:
2.0% Sorbitol
0.1% KH$_2$PO$_4$
0.3% K$_2$HPO$_4$
2.0% Dried yeast

Production medium:
4.0% Corn starch saccharification liquor (CSSL)
0.4% (NH$_4$)$_2$SO$_4$
0.5% Na-glutamate	0.03% Histidine
0.05% KH$_2$PO$_4$	200 µg/L Biotin
0.1% KCl	0.2% MgSO$_4$ · 7 H$_2$O
0.2% CaCl$_2$ · 2 H$_2$O	0.25% Ribonucleic acid
0.003% MnSO$_4$ · 4 H$_2$O	(purity 70.3%)
pH 6.4	1.0% Cornsteep liquor

Starting from 20 h CSSL was intermittently added. 20% based on the starting amount of the medium was consumed.

Polysaccharide S-53. *Klebsiella pneumoniae* ATCC 31488. KANG and VEEDER [26.31].
3.0% Hydrolyzed starch	0.01% MgSO$_4$ · 7 H$_2$O
0.19% NaNO$_3$	0.05% K$_2$HPO$_4$
pH 6.0–7.0	

Biopolymers. *Xanthomonas campestris* NRRL-B-1459 A. BAHN et al. [26.507].
2.8% Glucose	0.02% MgSO$_4$ · 7 H$_2$O
0.6% Soybean meal	0.08% (NH$_4$)$_2$HPO$_4$
1.0% Cornsteep liquor	0.09% Na$_2$HPO$_4$
pH 6.8	

2.8% Glucose was added at 24 h; 30% oil phase (sorbitan monooleate and isoparaffin mixture) was added at 40 h.

Table 26.41. (continued)

Biosurfactans. *Corynebacterium salvinicum* SFC. ZAJIC et al. [26.508].
3.0% Hexadecane 0.02% $MgSO_4 \cdot 7 H_2O$
0.2% $(NH_4)_2SO_4$ 0.0001% $CaCl_2 \cdot 2 H_2O$
0.4% KH_2PO_4 0.0001% $FeSO_4 \cdot 7 H_2O$
0.6% Na_2HPO_4 pH 7.0

Biotin. *Sporobolomyces carnicolor* ATCC 16407. SHIBATA et al. [26.23].
5.0% Glucose 0.3% K_2HPO_4
5.0% Sucrose 0.01% $MgSO_4$
1.0% Soybean flour 1.0% $CaCO_3$
0.1% Na-aspartate 0.005–0.02 mg/L Azelaic acid

Asparaginase. *Escherichia coli.* HO and BOECK [26.395].
0.75% Glucose 0.3% Soy peptone
1.7% Casamino acid 0.1% L-Asparagine
pH 7.3 0.25% KH_2PO_4

Cellulase. *Trichoderma reesei* QM 9414. ALLEN et al. [26.509].
50.4 g/L Cellulose 0.6 g/L $CaCl_2$
11.7 g/L $(NH_4)_2SO_4$ 0.6 g/L Urea
3.7 g/L KH_2PO_4 2.8 g/L Proteose
0.6 g/L $MgSO_4 \cdot 7 H_2O$ peptone
1.4 mg/L $ZnSO_4 \cdot 7 H_2O$ 5.0 mg/L $FeSO_4 \cdot 7 H_2O$
2.0 mg/L $CoCl_2$ 1.56 mg/L $MnSO_4$

Cellulase. *Trichoderma reesei.* FENIKSOVA et al. [26.104].
5.0% Beet sugar pulp 0.2% KH_2PO_4
1.0% $(NH_4)_2SO_4$

Cholesterol oxidase. *Oudemansiella mucida* FERM-P 5778. MATSUI et al. [26.510].
2.0% Soluble starch 0.1% $MgSO_4 \cdot 7 H_2O$
0.3% Yeast extract 0.5% Soybean oil
1.0% Polypeptone

β-Glucanase. *Bacillus subtilis* Cohn 1872. HUBER et al. [26.511].
1.5% Wheat grits 0.05% K_2HPO_4
0.9% Soybean grits 0.3% Cornsteep liquor
1.5% Starch 0.2% Wheat wax
0.5% $(NH_4)_2HPO_4$ 0.05% $MgSO_4 \cdot 7 H_2O$
0.2% Na_2HPO_4 pH 7.0

Glucoamylase. *Endomycopsis fibuliger* R-313. BEKERS et al. [26.374].
2.0% Corn meal 0.15% KH_2PO_4
2.0% Cornsteep liquor 0.033% $CaCl_2$
0.33% $(NH_4)_2SO_4$ pH 7.5–8.0

Glucose-isomerase. *Arthrobacter* nov. sp. NRRL B-3724. LEE et al. [26.3].
20 g/L Xylose 6 g/L $(NH_4)_2HPO_4$
5 g/L Bacto-tryptone 0.2 g/L KH_2PO_4
1 g/L Bacto-yeast 0.25 g/L $MgSO_4 \cdot 7 H_2O$
extract
pH 6.9

Glucose-isomerase. *Arthrobacter levanicum* NRRL B-1678. SHIEH et al. [26.9].
0.8% Xylose 0.18% KCl
0.2% Glucose 0.009% $CoSO_4 \cdot 7 H_2O$
2.0% Yeast extract pH 7.5
or
4.0% Birchwood sulfite liquor (as xylose)
0.0036% $CoSO_4 \cdot 7 H_2O$
0.024% KCl pH 7.5

Glucose-isomerase. *Streptomyces flavoricens.* VAHERI and KAUPPINEN [26.10].
0.6% Xylose 0.1% $MgSO_4 \cdot 7 H_2O$
0.5% Glycerol 0.02% $CoCl_2 \cdot 6 H_2O$
1.0% Cornsteep liquor 0.1% K_2HPO_4
1.0% Peptone

Glucose-isomerase. *Bacillus coagulans.* OUTTRUP [26.7].
4.0 g/L Xylose 5.0 g/L $(NH_4)_2SO_4$
80.0 g/L Corn mash 0.2 g/L $MgSO_4 \cdot 7 H_2O$
5.0 g/L Yeast extract 0.02 g/L $MnSO_4 \cdot H_2O$
1.0 g/L K_2HPO_4

Glycerinkinase. *Streptomyces canus* A 2408 FERM-P Nr. 4977. TOYO JOZO KABUSHI KAISHA [26.512].
1.0% Peptone 0.20% KCl
0.1% K_2HPO_4 1.5% Glycerol
0.05% $MgSO_4$ pH 7.0

Inulinase. *Aspergillus ficuum* A 524. ZITTAN et al. [26.513].
2.0% Cornsteep liquor 0.01% $FeSO_4 \cdot 7 H_2O$
1.2% $(NH_4)H_2PO_4$ 2.5% Sucrose
0.07% KCl 0.005% $CaCl_2$
0.05% $MgSO_4 \cdot 7 H_2O$ 0.01%/Antifoamer
1.0% K_2HPO_4 pH 4.5

β-Mannase. *Bacillus subtilis* G 78. BASKIS and FIRANTAS [26.155].
63.0 g/L Hydrolyzed 3.0 g/L K_2SO_4
starch 0.2 g/L $MgSO_4$
17.5 g/L Cornsteep 1.0 g/L $MnSO_4 \cdot 5 H_2O$
liquor 0.2 g/L $CaCl_2$
2.6 g/L $(NH_4)_2SO_4$
12.0 g/L $Na_2HPO_4 \cdot$
$\cdot 12 H_2O$

Table 26.41. (continued)

Pectinase. *Aspergillus foetidus.* LOSYAKOVA et al. [26.151].
30 g Beet sugar pulp 1.0 g $(NH_4)_2SO_4$
60 g Wheat bran 60–62% Humidity

Protease. Mixture of *Actinomyces rimosus* and *Actinomyces violaceus* 5 G; 5 and 1.25% in inoculum, resp. EGOROV et al. [26.514].
1.0% Glycerol 0.011% K_2HPO_4
0.002% NH_4Cl 0.4% $CaCO_3$
0.002% $NaNO_3$ Tap water

Rennet. *Mucor miehei.* AUNSTRUP [26.367].
4% Potato starch 10% Ground barley
3% Soybean meal 0.5% $CaCO_3$

Rennet. *Phellinus chrysoloma* (Fr.) Donk. HYLMAR et al. [26.515].
1.5% Glucose
1.5% Cornsteep liquor
0.1% $MgSO_4$
Distilled water pH 5.5

Yeast lytic enzyme. *Cytophaga* sp. ASENJO et al. [26.28].
1% Yeast extract (Yatex Bovril)
1% Glucose
pH 7.2

L-Glutamic acid. *Micrococcus glutamicus* ATCC 13032. BALITSKAYA et al. [26.428].
20–25% Treated beet molasses
0.8% Urea 0.1–0.2% CSL
0.05% KH_2PO_4 strictly limited biotin
0.025% $MgSO_4 \cdot 7H_2O$ content 2.5 µg/L
pH 7.0–7.2

L-Glutamic acid. Mutant *Brevibacterium lactofermentum* AJ 11637 FERM-P 3811. HIRAGA et al. [26.516].
10.0% Cane molasses (as sugar)
0.1% KH_2PO_4 1 µg/ml Thiamine·HCl
0.1% $MgSO_4 \cdot 7H_2O$ pH 7.0

L-Isoleucine. Mutant *Brevibacterium thiogenitalis* ATCC 19240. UPDIKE and CALTON [26.517].
10% Glucose 5.0% $CaCO_3$
5% $(NH_4)_2SO_4$ 0.3% Soyton
0.3% KH_2PO_4 100 µg/L Biotin
0.04% $MgSO_4 \cdot 7H_2O$ 1 µg/L Thiamine
Trace element mixture: 30 g/L Na-α-Hydroxybutyrate
8.8 g/L $ZnSO_4 \cdot 7H_2O$
10.0 g/L $FeSO_4 \cdot 7H_2O$ 1 mL/L Trace element mixture
0.06 g/L $CaSO_4 \cdot 5H_2O$
0.088 g/L $Na_2B_4O_7 \cdot 10H_2O$ pH 7.8
0.053 g/L $Na_2MoO_4 \cdot 2H_2O$
7.5 g/L $MnSO_4 \cdot H_2O$
0.12 g/L $CoCl_2 \cdot 6H_2O$
0.055 g/L $CaCl_2$
pH 2.0

L-Lysine. *Nocardia* sp. 258-NP-4 (ATCC 21336). KYOWA HAKKO KOGYO CO [26.343].
5–10% *n*-Paraffin mixture
1% Ethanol 0.001% $MnSO_4 \cdot 3H_2O$
0.2% KH_2PO_4 1.0% NH_4NO_3
0.2% Na_2HPO_4 0.5% Casamino acid
0.1% $MgSO_4 \cdot 7H_2O$ 10 mg/L Thiamine
0.001% $FeSO_4 \cdot 7H_2O$ 250 µg/mL L-Homoserine
0.001% $ZnSO_4 \cdot 7H_2O$ 2% $CaCO_3$
pH 7.0

L-Lysine. *Corynebacterium acetophilum* hom-73 (NRRL B-3671). KABUSHIKI KAISHA TEKKOSHA CO [26.256].
5.0% Na-acetate 0.001% $MnCl_2 \cdot 4H_2O$
1.9% NH_4-acetate 0.05% Homoserine
0.1% K_2PO_4 2 γ/100 mL Biotin
0.05% $MgSO_4 \cdot 7H_2O$
0.0001% $FeCl_3 \cdot 6H_2O$
The pH of the medium was adjusted to 7.4 by feeding an aqueous solution containing 40% acetic acid and 10% ammonium acetate.

L-Lysine. *Corynebacterium glutamicum* ATCC 13287 mutants. PELECHOVA et al. [26.226].
85 mL Wood hydrolysate 0.1 g KH_2PO_4
15 ml Peanut meal 3.0 g $CaCO_3$
  hydrolysate 0.01 g $MgSO_4 \cdot 7H_2O$
1 g CSL 2.0 g Na-acetate
1 g $(NH_4)_2SO_4$
pH 7.0–7.2

L-Lysine. *Corynebacterium glutamicum.* MISRA et al. [26.89].
150 g/L Pretreated blackstrap molasses as sugar
20 g/L CSL
20 g/L NH_4Cl
1 g/L K_2HPO_4
0.4 g/L $MgSO_4 \cdot 7H_2O$
Arsenic free superphosphate (25% P_2O_5) was added to the diluted 1:1 molasses in the quantity of 1% of the sugar content, the mixture was boiled, neutralized and kept at 10 °C. The heavy precipitate was removed by centrifugation.

Table 26.41. (continued)

L-Lysine. *Brevibacterium lactofermentum* ICCF 3. NITELEA et al. [26.406].
10.0% Beet molasses
0.27% Ca-superphosphate
3.0% NH$_4$Cl
1.0% CaCO$_3$
1.2–2.4 g dry matter/100 mL hydrolysate of *Pseudomonas* sp.

L-Lysine. *Corynebacterium glutamicum* ATCC 13287. WALCZAK and OBERMAN [26.44].
15.0% Sucrose   0.001% FeSO$_4 \cdot$ 7 H$_2$O
0.8% Na-acetate   0.001% CaCl$_2$
4.6% (NH$_4$)$_2$SO$_4$ 50 µg/L D(+)-Biotin
3.5% CaCO$_3$ 500 µg/L Thiamine·HCl
0.1% KH$_2$PO$_4$ 700 µg/L D,L-Homoserine
0.02% MgSO$_4 \cdot$ 7 H$_2$O

L-Lysine. *Brevibacterium flavum* ATCC 13326. HIRAKAWA et al. [26.518].
8.0% Whey (as lactose)   0.05% K$_2$HPO$_4$
2.0% Glucose   0.05% KH$_2$PO$_4$
3.5% (NH$_4$)$_2$SO$_4$ 0.002% MgSO$_4 \cdot$ 7 H$_2$O
0.75% Cornsteep liquor   0.002% FeSO$_4 \cdot$ 7 H$_2$O
5.0% CaCO$_3$ 0.001% ZnSO$_4 \cdot$ 7 H$_2$O
0.02% Homoserine   0.001% MnSO$_4 \cdot$ 7 H$_2$O
pH 6.7

L-Threonine. *Brevibacterium, Corynebacterium, Arthrobacter, Microbacterium.* AJINOMOTO CO [26.519].
0.3 g/L Ethanol (by feeding total 5.0 g/L)
5.0 g/L (NH$_4$)$_2$SO$_4$ 2 mg Fe^{2+}
2.5 g/L Urea   2 mg Mn^{2+}
3.0 g/L KH$_2$PO$_4$ 20 mL/L Soybean
0.4 g/L MgSO$_4 \cdot$ 7 H$_2$O   hydrolysate
                          50 γ/L Biotin
                          1000 γ/L Thiamine

L-Tryptophan. *Bacillus subtilis* strain 3557. MAKSIMOVA et al. [26.41].
10.0% Sucrose   0.1% MgSO$_4$
2.0% Cornsteep liquor   0.05% NaCl
0.06% KH$_2$PO$_4$ 0.5% Urea
0.14% K$_2$HPO$_4$

L-Tryptophan. *Bacillus subtilis.* SEMENOVA et al. [26.520].
10.0% Glucose   0.14% K$_2$HPO$_4$
0.5% Urea   0.1% MgSO$_4$
2.0% CSL   0.05% NaCl
0.06% KH$_2$PO$_4$ pH 5

L-Valine. *Brevibacterium flavum.* POPOVA and MURGOV [26.36].
10–25% Hydrol   1.5% Cornsteep liquor
4% (NH$_4$)$_2$SO$_4$ 200.0 µg/L Thiamine
3% CaCO$_3$ 30.0 µg/L Biotin

Butyric acid. *Clostridium butyricum.* SHARPELL and STEGMANN [26.25].
3% Cerelose
1% Yeast autolysate Amber BYF 300
1% CaCO$_3$

Citric acid. *Aspergillus niger.* LEOPOLD and VALTR [26.74].
15.0% Beet molasses
0.045 g/L H$_3$PO$_4$
0.75–2.1 g/L K$_4$Fe(CN)$_4$ pH 6.6
The mixture was boiled, sterilized, and used for fermentation with precipitate.

Citric acid. *Candida oleophila* ATCC 20122 mutant strain AC 7. HUSTEDE and SIEBERT [26.333].
10.0% *n*-Dodecane (wgt/vol)
0.05% KH$_2$PO$_4$ 0.05% Cornsteep
0.02% MgSO$_4 \cdot$ H$_2$O   powder
0.025% MnSO$_4 \cdot$ 4 H$_2$O   8.0% CaCO$_3$
                         1 × 10^{-4} mol 2,4-dinitrophenol was added after 24 h of cultivation

Lactic acid. *Lactobacillus delbruckii* B-70. KRUMPHANZL and DYR [26.40].
100.0 g/L Sucrose
20.0 g/L Yeast extract
2.5 g/L KH$_2$PO$_4$

Acetone-Butanol. *Clostridium acetobutylicum.* BARBER et al. [26.521].
134 g/L Cane molasses   1 g/L Starch
2 g/L (NH$_4$)$_2$SO$_4$ 1 g/L CaCO$_3$
pH 7.0–7.3

Ethanol. Flocculant strain *Zymomonas mobilis* ZM 401 (ATCC 31822). ROGERS and TRIBE [26.522].
15.0% Glucose   0.1% KH$_2$PO$_4$
1.0% Yeast extract   0.05% MgSO$_4 \cdot$ 7 H$_2$O
0.1% (NH$_4$)$_2$SO$_4$ pH 5.0

Table 26.41. (continued)

Ethanol. *Saccharomyces cerevisiae.* GONG et al. [26.523].
8.8% Xylulose, chemically isomerized in mixture with
2.2% Xylose
0.85% Yeast extract
0.13% NH_4Cl
0.011% $MgSO_4 \cdot 7 H_2O$
0.006% $CaCl_2$
pH 5.5

Fats and oils (as cocoa butter substitutes). *Rhodosporidium toruloides.* GIERHART [26.524].
0.5% Peptone   10 mg/ml Antibiotic
0.1% Yeast extract   0.01% Tween 20
2.0% Glucose   1.0% Stearic acid
0.1% K_2HPO_4 pH 5.5–6.0

Baker's yeast. *Saccharomyces cerevisiae.* POPOVA and KALYUZHNYI [26.525].
25–60 g/L Hydrolysate of agricultural waste as reducing substances
5–15 g/L $(NH_4)_2SO_4$
2.6–7.8 g/L Superphosphate
0.25–0.75 g/L KCl
0.15–0.60 g/L $MgSO_4$
0.3–0.5 g/L Cornsteep liquor

Fodder yeast. *Torulopsis utilis.* GREGR et al. [26.470].
1750 kg Beet molasses 41° Bg.
1520 kg Water
  10 kg H_2SO_4
   4 kg Formaldehyde   pH 4.2–4.8
The mixture was boiled 1 h, cooled 1 h and used without filtration.

SCP. *Candida lipolytica.* CEJKOVA et al. [26.326].
10 g/L n-Paraffin (C_{13}–C_{21})
7.0 g/L KH_2PO_4 0.1 g/L NaCl
0.4 g/L NH_4-nitrogen   0.1 g/L Yeast extract
0.4 g/L Urea-nitrogen
0.2 g/L $MgSO_4$

SCP. *Candida curvata.* MOON et al. [26.129].
1 L Whey permeate   2.5 mg/L $MnSO_4$
1.0 mL NH_4OH 5.3 mg/L K_2HPO_4

SCP. *Saccharomyces fragilis.* BERNSTEIN and TZENG [26.116].
12.0% Whey solids
 0.1% Phophoric acid
 0.13% Yeast extract (Amber BYF Series 100)
 0.3–0.5% Ammonia
 0.5–0.2% Hydrochloric acid (pH 4.5)

SCP. *Candida tropicalis.* SHISHKOVA et al. [26.208].
1 L Wood hydrolysate
5.0 g peat hydrolysate with 40.8% dry matter
1.2 g $(NH_4)H_2PO_4$
0.8 g KCl   pH 4.2–4.4

High protein product. *Kluyveromyces fragilis.* BEAUSEJOUR et al. [26.123].
7.0% Cheese whey powder   0.25% KH_2PO_4
0.5% $(NH_4)_2SO_4$ pH 4.0 with H_3PO_4

SCP. An unidentified newly isolated acidophilic fungus. BOA and LEDUY [26.526].
5.3% Hydrolysate of Fafard peat (as total carbohydrate)
0.5% $(NH_4)_2SO_4$
0.5% K_2HPO_4
0.04% $MgSO_4$ pH 2.5

SCP. *Candida intermedia.* BAYER [26.527].
5.7% Sweet whey (as dry matter) deproteinized
0.75% $(NH_4)_2SO_4$
0.1% Yeast extract   pH 4.0

Culture for flax retting. *Clostridium felsineum* 13. VOZNYAKOVSKAYA and KULIKOVSKAYA [26.528].
1.35% Potato pulp (as dry biomass weight)
0.4% Cornsteep liquor
0.2% $(NH_4)_2SO_4$
0.02% $FeCl_3$
0.4% Chalk

Plant pathogen culture to control hemp. *Fusarium oxysporum* f. sp. *cannabis.* HILDEBRAND and MCCAIN [26.529].
800 g Barley straw
  2 L Distilled water
160 g Cottonseed meal

Table 26.42. Optimization of Nutrient Medium Using Mathematical Experiment Design

Optimization Criterion	Microorganism	Parameter	Reference
Glucoamylase activity	*Endomycopsis bispora*	Cornmeal, soybean meal, CSL, KH$_2$PO$_4$	MOSKVITCHEVA et al. [26.373]
	Endomycopsis fibuliger R-313	Cornmeal, CSL, whale oil, (NH$_4$)$_2$SO$_4$, KH$_2$PO$_4$, CaCl$_2$	DREIMANE [26.375]
Amylase activity	*Bacillus subtilis* strain 103	Starch, CSL, cornmeal, vitamin extract, NH$_4$HPO$_4$, Na$_2$SO$_4$	DZHAVAKHIYA et al. [26.372]
β-Fructofuranosidase activity	*Aspergillus awamori*	Molasses, soybean meal, NaNO$_3$, MgSO$_4$, KH$_2$PO$_4$	SEROVA et al. [26.536]
β-Mannanase activity	*Bacillus subtilis* G-78	Hydrolyzed starch, CSL, (NH$_4$)$_2$SO$_4$, Na$_2$HPO$_4$, MgSO$_4$, MnSO$_4$·5H$_2$O, CaCl$_2$	BASKIS and FIRANTAS [26.155]
Cellulase activity	*Trichoderma reesei* QM 9414	CaCl$_2$, MgSO$_4$·7H$_2$O, KH$_2$PO$_4$	HAULY et al. [26.537]
Esterase activity	*Mycobacterium album* Sohngen 726	Carbon sources: citrate, sucrose, sorbit, glycerol	MATVEEVA and LESTROVAYA [26.538]
Pectinase activity	*Aspergillus awamori* 44-2 b	Sugar-beet pulp, KH$_2$PO$_4$, (NH$_4$)$_2$SO$_4$, MgSO$_4$	VORONOVA et al. [26.101]
Carboxymethylcellulase activity	*Schizophyllum commune* Nr. 13 Delmer ATCC 38548	Cellulose, peptone, glycerol	DESROCHERS et al. [26.539]
Protease activity	*Bacillus subtilis* 2 M	Mineral components	VASKIVNYUK and SEMENOVA [26.540]
Exoprotease activity	*Bacillus thuringiensis* var. *finitimus*	Glucose, glycerol, sorbit, CoCl$_2$, K$_2$HPO$_4$, ZnSO$_4$	EGOROV et al. [26.306]
Gelatinase acitivity	*Bacillus mesentericus* 316 m	KH$_2$PO$_4$, (NH$_4$)$_2$SO$_4$, ZnSO$_4$	VASKIVNYUK et al. [26.540a]
Rennin activity	*Mucor pusillus* 917	Wheat bran, dry defatted milk, peptone, KH$_2$PO$_4$, MgSO$_4$·7H$_2$O	KRAYUSHKINA et al. [26.541]
Yeast-lytic enzyme activity	*Streptomyces* sp.	Baker's yeast, lactose, cornmeal	GALAS et al. [26.542]

Table 26.42. (continued)

Optimization Criterion	Microorganism	Parameter	Reference
Rennin activity	*Endothia parasitica* CBS 25051	Extracted soybean meal, glucose, Ca-salts	ERDELYI and KISS [26.543]
Ribonuclease activity	*Bacillus subtilis* KR 349	Carbon- and nitrogen-sources	BASKIS and RAGAVICIUS [26.159]
Tobramycin production	*Streptomyces cremeus* var. *tobramycini* var. nov. 2242	Maltose, NH_4Cl, K_2HPO_4, $CaCO_3$; glucose, whale oil, soybean meal	MOTKOVA [26.544]; MOTKOVA et al. [26.545]
Bleomycin production	*Streptoverticillium griseocarneum* var. *bleomycini* 1129/2646	Soybean meal, glucose, lactose, K_2HPO_4, $ZnSO_4$	KOROBKOVA et al. [26.546]
Albofungin productivity	*Streptomyces tumemacerans*	CSL, $(NH_4)_2SO_4$, NaCl, $CaCO_3$, starch, glucose	SABIROV et al. [26.547]
Gramicidin C productivity	*Bacillus brevis* var. G-B	Aeration, glycerol, lactic acid, yeast autolysate, casein, minerals	KUPLETSKAYA et al. [26.397]
Streptomycin productivity, foam and filtration ability of medium	—	CSL, soybean meal, hydrol, chalk, minerals	IVANKOVA et al. [26.535]
Nisin productivity	*Streptococcus lactis* strain MGU	Molasses, fodder yeast autolysate, $NH_4H_2PO_4$, K_2SO_4	EGOROV et al. [26.388]
Nisin productivity	*Streptococcus lactis* strain MGU	Na-acetate, Na-oxalate, Na-citrate, NaH_2PO_4, $NH_4H_2PO_4$, glucose	EGOROV et al. [26.548]
L-Lysine productivity	*Brevibacterium flavum* 178	Sucrose, L-threonine, $(NH_4)_2SO_4$, KH_2PO_4	MURGOV and ZAITSEVA [26.549]
L-Tryptophan productivity	*Torulopsis utilis*	Sucrose, anthranilic acid, yeast inoculum	SOBCZAK and MAJCHRZAK [26.550]
Ethanol productivity	Yeast	Glucose syrup concentration, yeast concentration, $NH_4H_2PO_4$, $ZnSO_4$	CHEN [26.551]

Table 26.42. (continued)

Optimization Criterion	Microorganism	Parameter	Reference
Ethanol productivity	Compressed distiller's yeast	Corn mash concentration; yeast concentration; saccharification time	CHEN and GUTMANIS [26.552]
Ethanol productivity	Distiller's yeast	Temperature and duration of starch hydrolysis; diameter of starch raw particles	USTINNIKOV et al. [26.553]
SCP on ethanol	Candida utilis	O_2-content in air (21–99.5%)	EDERER et al. [26.554]
SCP on permeate of milk and whey	Trichosporon cutaneum	Urea, permeate	KÄPPELI et al. [26.555]
SCP on pretreated bagasse	Cellulomonas sp., II bc and GiIII	$Na_2HPO_4 \cdot 12 H_2O/KH_2PO_4$; $NaCl$, NH_4Cl, $MgSO_4 \cdot 7 H_2O$; Trace element solution; $CaCl_2$, thiamine	RODRIGUEZ et al. [26.556]
SCP productivity on whey	Kluyveromyces fragilis IMAT 1872	Lactose concentration	MORESI et al. [26.557]
SCP productivity on methanol	Candida boidinii 11 bh	Yeast extract, nitrogen concentration, KH_2PO_4, $MgSO_4 \cdot 7 H_2O$, $ZnSO_4 \cdot 7 H_2O$ biotin, thiamine	VOTRUBA et al. [26.558] PILAT et al. [26.559], [26.560]
SCP productivity on ethanol	Candida utilis A 49	15 Elements	PROKOP et al. [26.561]
Growth intensity for flax retting	Clostridium felsineum strain 13	Potato pulp, CSL, $FeCl_3$, $(NH_4)_2SO_4$	VOZNYAKOVSKAYA and KULIKOVSKAYA [26.528]

foaming during fermentation. Strongly foaming substrates are molasses, soybean meal, cornsteep liquor, hydrol, and sulfite waste liquors. Adding chalk and ammonium sulfate to media containing soybean meal and cornsteep liquor reduces foaming (SOIFER et al. [26.532], [26.533]; KRISTAPSON [26.534]). Statistical optimization for streptomycin production has shown, however, that both mentioned compounds are unimportant in terms of foaming, whereas the ratio of soybean meal to cornsteep liquor and hydrol is a major factor determining foaming and antibiotic activity (IVANKOVA et al. [26.535]).

Many combinations of media components have been reported as well as various parameters in statistical design. Evaluation of the obtained data should lead to the design of a nearly minimal medium permitting the highest possible microbial activity with the lowest nutrient concentration. Examples of media optimization are shown in Table 26.42.

26.5 Processing of Wastes

Constant efforts are being made to employ industrial by-products and wastes as raw materials for fermentation processes. Some aspects of such procedures are compiled in Table 26.43.

Table 26.43. Applications of Industrial Wastes in the Fermentation Industry

Waste Product	Fermentation	Reference
Potato processing waste	SCP	FORNEY and REDDY [26.562]
		LEMMEL et al. [26.563]
		SENEZ et al. [26.564]
		BLOCH et al. [26.565]
Sauerkraut waste	SCP	HANG [26.566]
Canning industry waste	SCP	GHAI et al. [26.567]
Waste of dehydrated-onion industry	SCP	GHONAIM et al. [26.568]
Coffee waste	Ethanol	BHAT and SINGH [26.569]
Effluent of starch plant	SCP	NOJIRI et al. [26.570]
Waste from manufacture of soups	L-Lysine	BRECKA et al. [26.571]
Keratin waste of the meat industry	SCP	BALABUSHEVICH and KRASNOBRIZHII [26.572]
		SCHERBAKOV et al. [26.573]
Ammonia waste from pressure gas works	SCP	BARTA et al. [26.574]
Ammoniacal extract produced during reduction of nucleic acid content of yeast	SCP	RUT and STROS [26.575]
Waste from production of polyvinylacetate	SCP	ERZINKIAN et al. [26.576]
Saponified solid residues from production of palm oil	SCP	MARTINET et al. [26.577]

Agricultural waste and urban sewage products and their utilization are the issue of Volume 8 of "Biotechnology".

Acknowledgement

I would like to express my gratitude to Schering AG for permitting the compilation of this chapter.

26.6 Commercial References

1. Akzo Chemie GmbH, Postf. 641, D-5160 Düren.
2. Amber Laboratories, 6101 N. Teutonia Ave., Milwaukee, Wisconsin 53209, USA.
3. Atlas Chemie, Niederlassung der Deutschen ICI GmbH, Goldschmidtstr. 100, D-4300 Essen 1.
4. Auxoferm GmbH, Wandsbeker Zollstr. 59, D-2000 Hamburg 70.
5. Bayer AG, D-5090 Leverkusen.
6. Bovril Ingredients, Ltd., Burton upon Trent, Staffordshire DE14 2AB, England. Distributor: Georg Fles GmbH, Postf. 520128, D-2000 Hamburg 52.
7. Cellulose Attisholz AG, CH-4708 Luterbach. Distributor: Fritz Köster Handelsges. mbH Co, Rothenbaumchaussee 59, D-2000 Hamburg 13.
8. Chemische Fabrik Grünau GmbH, Postf. 120, D-7918 Illertissen.
9. Diamalt AG, Postf. 400469, D-8000 München 40.
10. Dow Corning Int. Ltd., Chaussée de la Hulpe 154, B-1170 Bruxelles, Belgium. West Germany: Dow Corning GmbH, Pelkovenstr. 152, D-8000 München 50.
11. Edelsoja GmbH, Postf. 280246, D-2000 Hamburg 28.
12. Ed. Geistlich Söhne AG, CH-6110 Wolhusen.
13. Th. Goldschmidt AG, Goldschmidtstr. 100, D-4300 Essen 1.
14. Hafen-Mühlen-Werke GmbH, Postf. 105246, D-2800 Bremen 1.
15. Hoechst AG, D-6230 Frankfurt am Main 80.
16. Imperial Chemical Industries PLC, Organic Division, P.O. Box 42, Blacklay, Manchester M9 3DA, England. West Germany: Deutsche ICI GmbH, Lyoner Str. 36, D-6000 Frankfurt am Main.
17. Kali Chemie AG, Postf. 220, D-3000 Hannover 1.
18. Lohmann Tierernährung GmbH, Postf. 446, D-2190 Cuxhaven.
19. Lucas Meyer GmbH & Co, Ausschläger Elbteich 62-72, D-2000 Hamburg 28.
20. Maizena Industrieprodukte GmbH, Postf. 104320, D-2000 Hamburg 1.
21. Novo Industri A/S, Enzyme Division, DK-2880 Bagsvaerd. West Germany: Kantstr. 2, D-6500 Mainz 1.
22. Ohly GmbH, Wandsbeker Zollstr. 59, D-2000 Hamburg 70. Since 1984: Chemische Werke Hüls AG, D-4690 Herne.
23. Roquette Frères SA, 4, rue Patou, F-59022 Lille. West Germany: Schaumainkai 45, D-6000 Frankfurt 70.
24. Sheffield Products Kraft, Inc., P.O. Box 398 Memphis, Tennessee 38101, USA. Distributor: Otto Aldac, Curslacker Neuerdeich 66, D-2000 Hamburg 80.
25. Schill und Seilacher GmbH & Co, Moorfleeter Str. 28, D-2000 Hamburg 74.
26. Strothmann-Brennereien, Postf. 1140, D-4950 Minden.
27. Traders Oil Mills Co, P.O. Box 1837, Forth Worth, Texas 76101, USA.
28. Ugine Kuhlmann, Deutschland GmbH, Postf. 8208, D-4030 Ratingen 8.
29. Union Carbide Corp., 270 Park Ave., New York, N.Y. 10017, USA. West Germany: Postf. 300945, D-4000 Düsseldorf 30.
30. Universal Foods Corp., 433 E Michigan St., Milwaukee, Wisconsin 53201, USA.
31. Wacker-Chemie GmbH, Prinz-Regenten-Str. 22, D-8000 München 22.

26.7 References

[26.1] G. L. SOLOMONS: "Material and Methods in Fermentation". Academic Press, London-New York, 1969.

[26.2] D. W. ZABRISKIE, W. B. ARMIGER, D. H. PHILLIPS, and P. A. ALBANO: "Traders Protein". 2nd Ed. Memphis, Tennessee (USA), 1982.

[26.3] C. K. LEE, L. E. HAYES, and M. E. LONG: US Pat. 3 645 848, 1972.

[26.4] M. POPOV, G. DZHEDIEVA, I. TODOROV, and N. STOEVA: Brit. Pat. 2 063 885, 1981.

[26.5] H. IIZUKA, M. SUEKANE, and M. KANNO: West Ger. Pat. 1 934 461, 1971.

[26.6] MILES LABORATORIES Inc.: Brit. Pat. 1 376 983, 1971.

[26.7] H. OUTTRUP: West Ger. Pat. 2 400 323, 1974.

[26.8] P. WEBER: Brit. Pat. 1 410 579, 1973.

[26.9] K. K. SHIEH, H. A. LEE, and B. J. DONNELLY: West Ger. Pat. 2 351 443, 1974.

[26.10] M. VAHERI and V. KAUPPINEN: Process Biochem. *12* (1977) July/August, 5.

[26.11] A. MARGARITIS and P. BAJPAI: Appl. Environ. Microbiol. *44* (1982), 1039.

[26.12] C. S. GONG: Eur. Pat. 66 396, 1982.

[26.13] L. C. CHIANG, H. Y. HSIAO, P. P. UENG, L. F. CHEN, and G. T. TSAO: Biotechnol. Bioeng. Symp. *11* (1981), 263.

[26.14] L. A. DOLAK and L. E. JOHNSON: US Pat. 4 306 021, 1981.

[26.15] H. ZAEHNER, H. DRAUTZ, and W. KELLER: US Pat. 4 277 478, 1981.

[26.16] J. C. MCGUIRE, B. K. HAMILTON, and R. J. WHITE: Process Biochem. *14* (1979) December, 2.

[26.17] H. YAMAMOTO and S. J. DAUM: West Ger. Pat. 3 042 075, 1981.

[26.18] A. J. KEMPF and K. E. WILSON: US Pat. 4 247 640, 1981.

[26.19] D. BROWN, A. F. GILES, H. W. CRAMER, H. M. NOBLE, L. J. NISBET, M. E. BUSHELL, G. WEARE, and I. Y. CALDWELL: Eur. Pat. 28 511, 1981.

[26.20] M. G. WOVCHA and K. E. BROOKS: US Pat. 4 293 646, 1981.

[26.21] MITSUBISHI CHEM. IND., Ltd.: Jap. Pat. 82 125 696, 1982.

[26.21a] MITSUBISHI CHEM. IND., Ltd.: Brit. Pat. 1 539 233, 1979.

[26.22] J. P. ROLLS: US Pat. 4 088 537, 1978.

[26.23] M. SHIBATA, T. HASEGAVA, E. HIGASHIDA, K. MIZUNO, and S. KAMEDA: Jap. Pat. 67 03 074, 1967.

[26.24] S. KINOSHITA and K. NAKAYAMA: In "Economic Microbiology", Vol. 2, p. 209 (A. H. ROSE, ed.). Academic Press, London-New York, 1978.

[26.25] F. SHARPELL and C. STEGMANN: In "Adv. Biotechnol." (Proc. 6th Int. Ferment. Symp. 1980), Vol. 2, p. 71 (M. MOO-YOUNG and C. W. ROBINSON, eds.). Pergamon Press, Toronto, Ontario, 1980.

[26.26] COURTAULDS, Ltd.: Brit. Pat. 1 085 901, 1967.

[26.27] T. KODAMA, U. KOTERA, and K. YAMADA: Agric. Biol. Chem. *36* (1972), 1299.

[26.28] J. A. ASENJO, P. DUNNIL, and M. D. LILLY: Biotechnol. Bioeng. *23* (1981), 97.

[26.29] G. W. PACE and S. D. J. COOTE: Eur. Pat. 32 293, 1981.

[26.30] C. J. LAWSON and I. W. SUTHERLAND: In "Economic Microbiology", Vol. 2 (A. H. ROSE, ed.). Academic Press, London-New York, 1978.

[26.31] K. S. KANG and G. T. VEEDER: Fr. Pat. 2 465 000, 1981.

[26.32] S. L. MICHAELS and H. W. BLANCH: Acta Biotechnol. *1* (1981), 351.

[26.33] SCHERING AG: Eur. Pat. 51 143, 1980.

[26.34] H. KLAUSHOFER: In "Proceeding Problem with Molasses in the Yeast Industry", p. 67 (E. SINDA and E. PARKKINEN, eds.). Kauppakirjapamo, Helsinki, 1980.

[26.35] N. V. POPOVA: Fermentn. Spirt. Promst. *31* (1965) 5, 4.

[26.36] ZH. POPOVA and I. MURGOV: Nauchni Tr. Vissh. Inst. Khranit. Vkusova Promst. *23* (1976), 269.

[26.37] V. MATELOVA: Kvasny Prum. *27* (1981), 21.

[26.38] H. NEWMAN, R. D. SKOLE, J. HOGU, and A. B. RIZZUTO: Dev. Ind. Microbiol. *21* (1980), 375.

[26.39] L. M. MIALL: In "Economic Microbiology", Vol. 2, p. 48 (A. H. ROSE, ed.). Academic Press, London-New York, 1978.

[26.40] V. KRUMPHANZL and J. DYR: Kvasny Prum. *7* (1961), 81.

[26.41] E. A. MAKSIMOVA, L. E. SEMENOVA, L. A. MUZYCHENKO, V. V. SHCHERBACHEV, and V. I. VALUEV: Khim. Farm. Zh. *11* (1977), 75.

[26.42] G. WACK, J. KISS, L. NAGY, E. UDVARDY-NAGY, K. ZALAI, and S. E. ZSOKA: Fr. Pat. 2 475 573, 1981.

[26.43] H. NAKAJIMA, K. NAGATA, M. KAGEYAMA, T. SUGA, and K. MOTOSUGI: West Ger. Pat. 3 041 744, 1981.

[26.44] P. WALCZAK and H. OBERMAN: Zesz. Nauk Politech. Lodz. Ser. Technol. Chem. Spozyw. *361* (1980), 115.

[26.45] L. WELWARD and D. HALAMA: Folia Microbiol. *23* (1978), 12.

26.7 References

[26.46] H. SCHIWECK and L. HABERL: Branntweinwirtschaft *113* (1973), 76.
[26.47] H. OLBRICH: Branntweinwirtschaft *113* (1973), 53.
[26.48] J. MALANOWSKA and S. LABENDZINSKI: Pr. Inst. Lab. Badaw. Przem. Spozyw. *19* (1969), 27.
[26.49] J. LANGPAULOVA and A. JANDA: Kvasny Prum. *24* (1978), 9.
[26.50] V. N. TERTYSHNYI, N. Y. KRASNOBRIZHII, A. D. POLOTNYAK, and M. I. BALABUSHEVICH: Mikrobiol. Zh. Kiev *43* (1981), 316.
[26.51] B. KOVACS and A. WIELAND: Szeszipar *45* (1979), 92.
[26.51a] A. WIELAND and B. KOVACS: Szeszipar, January/March (1979), 27.
[26.52] E. ZAUNER, W. K. BRONN, H. DELLWEG, and R. TRESSL: Branntweinwirtschaft *119* (1979), 154.
[26.53] H. SCHIWECK: In "Proceeding Problem with Molasses in the Yeast Industry", p. 21 (E. SINDA and E. PARKKINEN, eds.). Kauppakirjapamo, Helsinki, 1980.
[26.54] F. STROS and V. SYHOROVA: Listy Cukrov. *81* (1965), 285.
[26.55] A. GINTEROVA: Kvasny Prum. *19* (1973), 6.
[26.56] A. CEJKOVA: Kvasny Prum. *11* (1965), 250.
[26.57] A. GINTEROVA: Kvasny Prum. *19* (1973), 37.
[26.57a] H. WATANABE and T. SATO: J. Ferment. Technol. *59* (1981), 169.
[26.58] V. N. SHVETS, P. A. KULISH, and E. I. KNOGOTKOVA: Izv. Vyssh. Uchebn. Zaved. Pishch. Tekhnol. *1* (1982), 72.
[26.59] S. KUBACKI and W. KASPROWICZ: Prace Inst. Lab. Bad. Przemysl. Spoz. *2* (1972), 295.
[26.60] S. HUNCIKOVA: Kvasny Prum. *22* (1976), 58.
[26.61] S. LABENDZINSKI: In "Proceeding Problem with Molasses in the Yeast Industry", p. 39 (E. SINDA and E. PARKKINEN, eds.). Kauppakirjapamo, Helsinki, 1980.
[26.62] J. DYR, V. GREGR, Z. KUTTELVASER, A. SEILER, J. TOMASEK, and S. ZELENKA. "Lihovarstvi", Vol. 1. Statni Nakladatelstvi Techn. Lit. Prague (Czech.), 1955.
[26.63] H. OLBRICH: "Die Melasse". Verlag Institut für Gärungsgewerbe, Berlin, 1956.
[26.64] H. OLBRICH: In "Principles of Sugar Technology", Vol. 3, p. 511 (P. HONIG, ed.). Elsevier, Amsterdam, 1963.
[26.65] A. GINTEROVA: Kvasny Prum. *14* (1968), 18.
[26.66] A. EMMERICH: Branntweinwirtschaft *120* (1980), 234.
[26.67] M. N. BESPALIJ and M. I. SAVSHUK: Ferment. Spirt. Promst. *5* (1978), 28.
[26.68] J. STUDNICKY: Kvasny Prum. *21* (1975), 9.
[26.69] H.-J. REHM: "Industrielle Mikrobiologie". Springer Verlag, Berlin-Heidelberg-New York, 1980.
[26.70] O. I. BELOVA: Sakh. Promst. *42* (1968), 37.
[26.71] A. GINTEROVA: Kvasny Prum. *18* (1972), 224.
[26.72] A. FARAGO, O. GOTTLASZ, F. PANDI, and I. TOTH: Szeszipar *21* (1973), 20.
[26.73] G. MEZZADROLI: Fr. Pat. 833631, 1938.
[26.74] J. LEOPOLD and Z. VALTR: Czech. Pat. 100941, 1961.
[26.75] H. LEOPOLD, M. BURIAM, V. CERNY, and J. PASEK: Nahrung Chem. Biochem. Mikrobiol. Technol. *21* (1977), 655.
[26.76] Z. ILCZUK: Eur. J. Appl. Microbiol. Biotechnol. *17* (1983), 69.
[26.77] V. K. YANCHEVSKII, A. D. KOVALENKO, and L. V. LEVANDOVSKII: Fermentn. Spirt. Promst. *1* (1982), 9.
[26.78] J. WHITE: "Yeast Technology." Chapman & Hall, London, 1954.
[26.79] J. M. PATURAU: "By-products of the Cane Sugar Industry. An Introduction to their Industrial Utilization." Elsevier, Amsterdam, 1969.
[26.80] E. LYNESS and H. W. DOELLE: Biotechnol. Lett. *3* (1981), 257.
[26.81] T. SAITO: Process Biochem. *12* (1977) March, 17.
[26.82] M. A. QADEER, F. M. CHOUDRY, S. AHMAD, S. RASHID, and M. A. AKHTAR: Pak. J. Sci. Res. *32* (1980), 157.
[26.83] M. FOUAD, A. A. ABOU-ZEID, and M. YASSEIN: Acta Biol. Acad. Sci. Hung. *27* (1976), 107.
[26.84] M. J. SPIVEY: Process Biochem. *13* (1978) November, 2.
[26.85] B. VOLESKY, A. MULCHANDANI, and J. WILLIAMS: Ann. N.Y. Acad. Sci. *369* (1981), 205.
[26.86] H. V. AMORIM and H. CAMPOS: In "Adv. Biotechnol." (Proc. 6th Int. Ferment. Symp. 1980), p. 201 (M. MOO-YOUNG and C. W. ROBINSON, eds.).

Pergamon Press, Toronto (Canada), 1981.

[26.87] A. D. BAGHLAF, A.-Z. A. ABOU-ZEID, A. I. EL-DEWANY, A. E.-W. I. EISSA, M. FOUAD, and M. YASSEIN: Zentralbl. Bakteriol. Parasitenkd. Infektionskr. Hyg. Abt. 2, *135* (1980), 427.

[26.88] Y. M. SHEHATA, A. A. ABOU-ZEID, and M. M. ABDEL-HAMID: Microbiol. Esp. *30* (1978), 29.

[26.89] A. K. MISRA, J. DASGUPTA, and V. C. VORA: J. Chem. Technol. Biotechnol. *30* (1980), 453.

[26.90] O. TOSAKA, Y. MURAKAMI, S. IKEDA, and H. YOSHII: Fr. Pat. 2 472 610, 1981.

[26.91] J. C. BASILICO, F. T. POMAR, and C. A. MEINARDI: Rev. Fac. Ing. Quim. Univ. Nac. Litoral *43* (1979), 25.

[26.92] K. YAMADA and H. HIDAKA: Agric. Biol. Chem. *28* (1964), 876.

[26.93] F. A. HAMISSA, S. S. MABROUK, and A. F. ABDEL-FATTAH: J. Gen. Appl. Microbiol. *23* (1977), 23.

[26.94] P. TAURO, K. CHAUDHURY, and S. ETHIRAJ: In "Abstr. 5th Int. Ferment. Symp.", p. 407 (H. DELLWEG, ed.). Verlag Versuchs- und Lehranstalt für Spiritusfabrikation und Fermentationstechnologie, Berlin, 1976.

[26.95] MILES LABORATORIES Inc.: Brit. Pat. 669 733, 1952.

[26.96] S. NASIM, J. DASGUPTA, A. W. KHAN, and V. C. VORA: Indian J. Microbiol. *21* (1981), 343.

[26.97] A. A. ABOU-ZEID, A. I. EISSA, A. I. EL-DEWANY, F. FOUAD, M. FAHMI, and M. YASSEIN: Indian J. Technol. *16* (1978), 161.

[26.98] A. B. RIZZUTO, R. D. SKOLE, H. H. NEWMAN, J. N. B. HOGU, and V. A. TOSCANO: US Pat. 4 003 791, 1977.

[26.99] K. D. KIRBY and C. J. MARDON: Biotechnol. Bioeng. *22* (1980), 2425.

[26.100] D. H. LARSEN, D. L. DONEY, and H. A. ORIEN: Dev. Ind. Microbiol. *22* (1981), 719.

[26.101] L. I. VORONOVA, V. L. YAROVENKO, Y. P. GRACHEV, and L. A. KHERSONOVA: Fermentn. Spirt. Promst. *4* (1974), 16.

[26.102] K. A. KALUNYANTS, B. A. VELICHKO, G. A. CHERNOVA, E. D. FALALEEVA, P. A. STANKEVICH, Y. T. STEPANOV, M. A. DOROFEEVA, B. VALTERE, I. A. BRUNS, and S. M. MALEI: USSR Pat. 345 198, 1972.

[26.103] N. T. SHINKARENKO, K. A. KALUNYANTS, L. I. GOLGER, B. A. VELICHKO, and R. P. KAVTORINA: USSR Pat. 558 041, 1977.

[26.104] R. V. FENIKSOVA, K. A. KALUNYANTS, M. S. VAGANOVA, G. V. SHCHUS, and V. G. RYZHAKOVA: USSR Pat. 590 957, 1978.

[26.105] J. JANICKI, K. SZEBIOTKO, R. WOJTAL, and W. GRAJEK: Przem. Ferment. Rolny. *15* (1971), 7.

[26.106] D. I. ATAMANYUK, T. A. BORISOVA, and T. E. TSYGULYA: Izv. Akad. Nauk Mold. SSR, Ser. Biol. Khim. Nauk *5* (1978), 87.

[26.107] C. ROLZ: Process Biochem. *15* (1980) August/September, 2.

[26.108] Z. ER-EL, E. BATTAT, U. SHECHTER, and I. GOLDBERG: Biotechnol. Lett. *3* (1981), 385.

[26.109] A. L. DEMAIN, Y. M. KENNEL, and Y. AHARONOWITZ: Symp. Soc. Gen. Microbiol. *29* (1979), 163.

[26.110] S. N. SEHGAL, R. SAUCIER, and C. VEZINA: J. Antibiot. *29* (1976), 265.

[26.111] H. M. EL-SAIED, S. B. EL-DIN, and M. A. AKHER: Process Biochem. *12* (1977) October, 31.

[26.112] M. EL-MARSAFY, M. ABDEL-AKHER, and H. EL-SAIED: Zentralbl. Bakteriol. Parasitenkd., Infektionskr. Hyg. Abt. 2 *132* (1977), 117.

[26.113] D. D. Y. RYU and J. HOSPODKA: Biotechnol. Bioeng. *22* (1980), 289.

[26.114] I. KASZAB and A. KOVACS: In "Abstr. 5th Int. Ferment. Symp." (H. DELLweg, ed.). Verlag Versuchs- und Lehranstalt für Spiritusfabrikation und Fermentationstechnologie, Berlin, 1976.

[26.115] K. THAYANITHY, G. HARDING, and D. A. J. WASE: Biotechnol. Lett. *4* (1982), 423.

[26.116] S. BERNSTEIN and C. H. TZENG: Environ. Prot. Technol. Ser. 600/Z-77-1 (1977), 33, 42.

[26.117] S. B. DE SANCHEZ and F. J. CASTILLO: Acta Cient. Venez. *31* (1980), 24.

[26.118] H. C. CHEN and R. R. ZALL: Process Biochem. *17* (1982) January/February, 20.

[26.119] P. VANANUVA and J. E. KINSELLA: J. Food Sci. *40* (1975), 336.

[26.120] L. REESEN and R. STRUBE: Process Biochem. *13* (1978) November, 21.

[26.121] R. I. TENNEY: J. Am. Soc. Brew. Chem. *38* (1980), 74.

[26.122] M. M. SKVORTSOVA, Y. G. TOLCHINA, and N. L. GURGUTSA: Tr. Biol. Inst.

Akad. Nauk SSSR Sib. Otd. *27* (1974), 100.
[26.123] D. BEAUSEJOUR, A. LEDUY, and R. S. RAMALHO: Can. J. Chem. Eng. *59* (1981), 522.
[26.124] M. MORESI: Chim. Ind. Milan *63* (1981), 593.
[26.125] O. N. AL'BITSKAYA, O. I. GORONKOVA, L. P. NOSOVA, A. P. KIRICHKOV, and L. N. LUNIN: USSR Pat. 678065, 1979.
[26.126] O. MOEBUS and M. TEUBER: Kiel. Milchwirtsch. Forschungsber. *31* (1979), 297.
[26.127] D. I. ATAMANYUK and L. I. VAKAR: Izv. Akad. Nauk Mold. SSR, Ser. Biol. Khim. Nauk *4* (1976), 48.
[26.128] M. LUTSKOVA: Proc. Int. Dairy Congr. *5* (1966), 75.
[26.129] N. J. MOON, E. G. HAMMOND, and B. A. GLATZ: J. Dairy Sci. *61* (1978), 1537.
[26.130] G. MOULIN and P. GALZY: In "Adv. Biotechnol." (Proc. 6th Int. Ferment. Symp. 1980), Vol. 2, p. 181 (M. R. MOO-YOUNG and W. CAMPBELL, eds.). Pergamon Press, Toronto (Canada), 1981.
[26.131] B. A. FRIEND, M. L. CUNNINGHAM, and K. M. SHAHANI: Agric. Wastes *4* (1982), 55.
[26.132] A. GIEC, H. WEKER, and J. SKUPIN: Bull. Acad. Pol. Sci. Ser. Sci. Biol. *24* (1976), 497.
[26.133] J. MELLER, H. WOSKO, B. RAKOWSKI, K. BRONIKOWSKI, E. LIPINSKA, E. JAKUBCZYK, and M. SZADKOWSKA: West Ger. Pat. 2000818, 1970.
[26.134] G. A. SOMKUTI and M. M. BENCIVENGO: Dev. Ind. Microbiol. *22* (1981), 557.
[26.135] L. HROBONI, B. KAROLUS, M. PAWLOWSKI, J. MALANOWSKA, B. JANUSZKIEWICZ, H. ZIAJA, and S. TOMCZYK: Pol. Pat. 75050, 1975.
[26.136] W. C. HAYNES, L. J. WICKERHAM, and C. W. HESSELTINE: Appl. Microbiol. *3* (1955), 361.
[26.137] F. W. BEECH and R. P. DAVENPORT: In "Methods in Microbiology", Vol. 4, p. 153 (C. BOOTH, ed.). Academic Press, London-New York, 1971.
[26.138] J. ANTHEUNISSE: Antonie van Leeuwenhoek; J. Microbiol. Serol. *38* (1972), 617.
[26.139] V. I. KUDRYAVTSEV, M. V. FATEEVA, and T. N. NIKITINA: Mikrobiologiya *41* (1972), 903.

[26.140] R. REHBERG: Monatsschr. Brau. *30* (1977), 222.
[26.141] J. A. D. HALL: Process Biochem. *13* (1978) July, 20.
[26.142] J. MOSTEK: Kvasny Prum. *20* (1974), 193.
[26.143] W. F. MAISCH, M. SOBOLOV, and A. J. PETRICOLA: In "Microbiology Technology", Vol. 2, 2nd Ed., p. 79 (H. J. PEPPLER and D. PERLMAN, eds.). Academic Press, New York-San Francisco, 1979.
[26.144] H. MATSUOKA, Y. KOBA, and S. UEDA: J. Ferment. Technol. *60* (1982), 599.
[26.145] K. ROSEN: In "Abstr. 5th Int. Ferment. Symp.", p. 375 (H. DELLWEG, ed.). Verlag Versuchs- und Lehranstalt für Spiritusfabrikation und Fermentationstechnologie, Berlin, 1976.
[26.146] B. H. NIELSEN and P. ROSENDAL: Proc. 6th Int. Symp. Alcohol Fuels Technol. Guaruja *1* (1980), 51.
[26.147] N. KOSARIC, D. C. M. NG, I. RUSSELL, and G. S. STEWART: Adv. Appl. Microbiol. *26* (1980), 147.
[26.148] G. OFFER, V. GOSLICH, and M. HALDENWANGER: Branntweinwirtschaft *110* (1970), 151.
[26.149] V. I. RODZEVITSCH, V. L. YAROVENKO, B. A. USTINNIKOV, N. S. MAZUR, G. M. DOBROLINSKAYA, and T. A. TSCHEREMNOVA: USSR Pat. 467929, 1975.
[26.150] J. ZIFFER and M. C. JOSIF: Biotechnol. Lett. *4* (1982), 573.
[26.151] L. S. LOSYAKOVA, K. A. KALUNYANTS, L. N. MUSHNIKOVA, T. R. STRUNNIKOVA, and T. P. RUDENKO: USSR Pat. 310929, 1971.
[26.151a] L. S. LOSYAKOVA, S. N. ROMANOVA, A. A. SHILOVA, and G. I. FERTMAN: USSR Pat. PCT, WO 81/00, 857, 1981.
[26.152] A. TÄUFEL, H. RUTTLOFF, V. JAROVENKO, P. LIETZ, B. USTINNIKOV, and P. STEFFEN: Lebensmittelind. *26* (1979), 167.
[26.153] T. YAMAMOTO, T. NAKANISHI, and N. HASEGAWA: In "Abstracts 5th Intern. Ferment. Symp.", p. 264 (H. DELLWEG, ed.). Verlag Versuchs- und Lehranstalt für Spiritusfabrikation und Fermentationstechnologie, Berlin, 1976.
[26.154] G. B. CALLEJA, S. LEVY-RICK, C. V. LUSENA, A. NASIM, and F. MORANELLI: Biotechnol. Lett. *4* (1982), 543.
[26.155] E. BASKIS and S. FIRANTAS: Fermentn. Spirt. Promst. *4* (1981), 35.

[26.156] W. WINDISCH and N. S. MHATRE: Adv. Appl. Microbiol. *7* (1965), 273.
[26.157] R. H. SHIPMAN and L. T. FAN: Process Biochem. *13* (1978) March, 19.
[26.158] F. DESCHAMPS: Tec. Molitoria *32* (1981), 685.
[26.159] E. BASKIS and A. RAGAVICIUS: Proizvod. Primen. Mikrobn. Ferment. Prep. *3* (1976), 24.
[26.160] K. EDELMANN and L. PENTTILA: In "Util. Enzymes Technol. Aliment., Int. Symp." (P. DUPUY, ed.). Tech. Doc. Lavoisier, Paris, 1982.
[26.161] H. SHIBAI, H. ENEI, and Y. HIROSE: Process Biochem. *13* (1978) November, 6.
[26.162] R. BHADRA, S. K. GOSWAMI, and S. K. MAJUMDAR: Folia Microbiol. Prague *18* (1973), 300.
[26.163] A. H. ROSE: In "Fermented Foods. Economic Microbiology", Vol. 7, p. 1 (A. H. ROSE, ed.). Academic Press, London-New York, 1982.
[26.164] L. A. LINDENFELSER and A. CIEGLER: Appl. Microbiol. *29* (1975), 323.
[26.164a] M. A. QADEER, J. I. ANJUM, and R. AKHTAR: Pak. J. Sci. Ind. Res. *23* (1980), 25.
[26.165] M. A. ASHY, A. A. ABOU-ZEID, A. I. EL-DIWANY, and M. R. GAD: Enzyme Microbiol. Technol. *4* (1982), 20.
[26.166] M. C. FLICKINGER and D. PERLMAN: J. Appl. Biochem. *2* (1980), 280.
[26.167] F. REUSSER and L. SLECHTA: US Pat. 4267112, 1981.
[26.168] TOYO JOZO KABUSHIKI KAISHA: Belg. Pat. 884925, 1980.
[26.169] M. P. J. KIERSTAN: Process Biochem. *15* (1980), No. 4, 2.
[26.170] J. P. GUIRAUD, J. DAURELLES, and P. GALZY: Biotechnol. Bioeng. *23* (1981), 1461.
[26.171] Z. DUVNIAK, N. KOSARIC, and S. KLIZA: Biotechnol. Bioeng. *24* (1982), 2297.
[26.172] ANONYMOUS: Res. Discl. *212* (1981), 456.
[26.173] G. T. TSAO, M. LADISCH, C. LADISCH, T. A. HSU, B. DALE, and T. CHOU: Annu. Rep. Ferment. Proc. *2* (1978), 1.
[26.174] Y. W. HAN: Adv. Appl. Microbiol. *23* (1978), 119.
[26.175] T. K. GHOSE: In "Adv. Food-Prod. Syst. Arid Semiarid Lands" (Proc. Symp. 1980), pp. 225–266, 331–332 (J. T. MANASSAH and E. J. BRISKEY, eds.). Academic Press, New York, 1981.
[26.176] R. DATTA: Process Biochem. *16* (1981), 16, 42.
[26.177] M. KAMAKURA and I. KAETSU: Biotechnol. Bioeng. *24* (1982), 991.
[26.178] Y. W. HAN and A. CIEGLER: Process Biochem. *17* (1982) January/February, 32.
[26.179] Y. W. HAN, J. TIMPA, and A. CIEGLER: Biotechnol. Bioeng. *23* (1981), 2525.
[26.180] J. MIRON and D. BEN-GHEDALIA: Biotechnol. Bioeng. *23* (1981), 2863.
[26.181] Y. W. HAN: Appl. Microbiol. *29* (1975), 510.
[26.182] G. T. TSAO: Process Biochem. *13* (1978) October, 12.
[26.183] T. K. GHOSE and P. GHOSH: Process Biochem. *14* (1979) November, 20.
[26.184] L. T. FAN, Y. H. LEE, and M. M. CHARPURAY: Adv. Biochem. Eng. *23* (1982), 157.
[26.185] S. L. ROSENBERG, T. R. BATTER, H. W. BLANCH, and C. R. WILKE: AIChE Symp. Ser. *77* (1981), 107.
[26.186] A. GONZALES-VALDES and M. MOO-YOUNG: Biotechnol. Lett. *3* (1981), 143.
[26.187] O. VOLFOVA, E. KYSLIKOVA, B. SIKYTA, and J. PANOS: Folia Microbiol. Prague *24* (1979), 163.
[26.188] S. I. BELENKIJ and L. I. KOMAROVA: Gidroliz. Lesokhim. Promst. *18* (1965), 20.
[26.189] S. I. BELENKIJ, K. P. ZAITSEV, N. F. BAKHIREV, V. S. JELISEEV, V. M. KOZLOVA, M. I. KURZIM, V. V. TSCHUDAEV, V. V. RASSOCHIN, A. A. GORELOV, N. D. EREMINA, and L. K. ALFEROVA: USSR Pat. 261341, 1970.
[26.190] D. S. CHAHAL, M. MOO-YOUNG, and D. VLACH: Biotechnol. Bioeng. *23* (1981), 2417.
[26.191] M. MOO-YOUNG, A. J. DAUGULIS, D. S. CHAHAL, and D. G. MACDONALD: Process Biochem. *14* (1979) October, 38.
[26.192] N. I. PEROTTI DE GALVEZ and O. E. MOLINA: Biotechnol. Lett. *3* (1981), 717.
[26.193] R. L. PLANES, G. IGLESIAS, and L. M. HERNANDEZ: Cellulose Chem. Technol. *15* (1981), 17.
[26.194] A. ENRIQUEZ and H. RODRIGUEZ: Biotechnol. Bioeng. *25* (1983), 877.
[26.195] S. K. GARG and S. NEELAKANTAN: Biotechnol. Bioeng. *24* (1982), 2407.
[26.196] J. J. SAVARESE and S. D. YOUNG: Biotechnol. Bioeng. *20* (1978), 1291.

[26.197] B. K. SONI, K. DAS, and T. K. GHOSE: Biotechnol. Lett. *4* (1982), 19.
[26.198] A. ARAUJO and J. D'SOUZA: J. Ferment. Technol. *58* (1980), 399.
[26.199] J. M. NYSTROM and A. L. ALLEN: Biotechnol. Bioeng. Symp. *6* (1976), 55.
[26.200] V. S. SHIMANSKIJ: "Zapasy Torfa." Nauka i Tekhnika, Minsk, USSR, 1977.
[26.201] A. LEDUY: Process Biochem. *14* (1979), 3, 5.
[26.202] R. MUCENIECE and K. ABELE: USSR Pat. 422 765, 1974.
[26.203] Z. P. SHISHKOVA, J. GAILITIS, U. Y. SHMIT, N. A. VEDERNIKOV, and V. KRASTINS: In "Nov. Protsessy Prod. Pererab. Torfa", p. 152 (I. I. LISHTVAN, ed.). Nauka i Tekhnika, Minsk, USSR, 1982.
[26.204] Z. P. SHISHKOVA, A. KAININS, J. GAILITIS, U. SMIT, N. VEDERNIKOV, V. KRASTINS, D. A. KALINKIN, V. R. VAAKH, M. A. ZININA, V. D. BELYAEV, N. S. MAXIMENKO, and Y. V. EPSHTEIN: US Pat. 4 178 214, 1979.
[26.205] ZH. N. BOGDANOVSKAYA and M. V. ZALASHKO: In "Nov. Protsessy Prod. Pererab. Torfa", p. 174 (I. I. LISHTVAN, ed.). Nauka i Tekhnika, Minsk, USSR, 1982.
[26.206] M. BEKER: Adv. Microbiol. Eng. Part 1, *4* (1971), 233.
[26.207] M. ZEILE and E. KALEJA: Latv. PSR Zinat. Akad. Vestis. *2* (1969), 79.
[26.208] Z. P. SHISHKOVA, A. KAININS, J. GAILITIS, U. SMIT, N. A. VEDERNIKOV, V. KRASTINS, D. A. KALINKIN, V. VAAKS, M. A. ZININA, V. D. BELYAEV, N. S. MAXIMENKO, and Y. V. EPSHTEIN: West Ger. Pat. 2 740 785, 1979.
[26.209] V. A. ERMACHENKO, D. A. KALINKIN, and M. V. ZALASHKA: Vestsi Akad. Navuk B. SSR Ser. Biyal. Navuk *5* (1976), 58.
[26.210] M. N. LOJKO, V. S. SHIMANSKII, and R. T. BRATISHKO: Ref. Gidroliz. Proizv. *1* (1978), 9.
[26.211] O. U. EKA and W. M. FOGARTY: West Afr. J. Biol. Appl. Chem. *20* (1977), 8.
[26.212] A. LEDUY and J. M. BOA: Can. J. Microbiol. *29* (1983), 143.
[26.213] P. I. BEL'KEVICH and M. I. MINKEVICH: Vestsi Akad. Nauk Bel. SSR Ser. Khim. Nauk *4* (1979), 118.
[26.214] A. LEDUY: In "Proc. Int. Peat Symp.", p. 89 (C. H. FUCHSMAN, ed.). Bemidji State University, Bemidji, Minnesota, USA, 1981.
[26.215] V. D. ANDREEVSKAYA and M. V. ZALASHKO: Prikl. Biokhim. Mikrobiol. *15* (1979), 522.
[26.216] V. I. KUZNETSOVA, L. S. MAKSIMOVA, T. N. MANKOVA, L. K. STAKHORSKAYA, V. F. GITERMAN, and A. A. NEOKESARIISKII: USSR Pat. 311 956, 1971.
[26.217] K. KADLEC, J. PELECHOVA, V. KRUMPHANZL, and T. SOKOLOV: Kvasny Prum. *25* (1979), 82.
[26.218] N. V. GLUSHCHENKO, V. N. BUKIN, M. E. BEKER, L. V. DMITRENKO, V. A. UTENKOVA, M. A. KUZMINA, L. S. KUTSEVA, N. M. BAZDYREVA, G. K. LIEPINSH, E. B. TRUSLE, and T. A. PAVLOVA: US Pat. 4 286 060, 1981.
[26.219] W. F. GAUSS, S. SUZUKI, and M. TAKAGI: US Pat. 3 990 944, 1976.
[26.220] G. F. HAJNY: Report, FSRP-FPL-385, Order AD-AO98945 Avail. NTIS from Gov. Rep. Announce, Index (U.S.) *81* (1981), 4197.
[26.221] T. N. SEMUSHINA, N. L. MONAKHOVA, N. V. GLUSHCHENKO, and B. I. TOKAREV: Gidroliz. Lesokhim. Promst. *4* (1981), 26.
[26.222] A. S. KLIMENTOV, B. G. ERSHOV, I. F. VYSOTSKAYA, and L. N. KRAEV: Gidroliz. Proizvod. *1* (1978), 10.
[26.223] E. P. KRASAVINA, L. I. SOKOLOVA, and R. P. MARTYNOVA: Gidroliz. Lesokhim. Promst. *19* (1966), 24.
[26.224] G. S. POKINTSOKHA: Gidroliz. Lesokhim. Prom. *4* (1973), 21.
[26.225] Z. KIN and C. WITKOWSKI: Pol. Pat. 104 986, 1979.
[26.226] J. PELECHOVA, F. SMEKAL, V. BULANT, and V. KRUMPHANZL: Kvasny Prum. *25* (1979), 174.
[26.227] F. SMEKAL, J. PELECHOVA, E. KINDLOVA, and V. KRUMPHANZL: Kvasny Prum. *26* (1980), 200.
[26.228] J. PELECHOVA, K. KADLEC, J. STANEK, V. KRUMPHANZL, and T. SOKOLOV: Kvasny Prum. *25* (1979), 56.
[26.229] I. S. MADDOX: Biotechnol. Lett. *4* (1982), 23.
[26.230] T. KOBAYASHI: Process Biochem. *13* (1978) May, 15.
[26.231] H. ROMANTSCHUK and M. LEHTOMÄKI: Process Biochem. *13* (1978) March, 16.
[26.232] J. BARTA: Kvasny Prum. *17* (1971), 145.
[26.233] D. B. BOLOTIN, L. D. ZONOVA, V. V. OVCHINNIKOVA, and G. L. AKIM: USSR Pat. 722 238, 1981.

[26.234] H. KARCZEWSKA: Przem. Ferment. Rolny (1967) May, 180.

[26.235] M. RYCHTERA, G. PANUSCHKA, J. VERNEROVA, and H. FISCHER: Kvasny Prum. 25 (1979), 270.

[26.236] M. SESTAKOVA: Kvasny Prum. 24 (1978), 149.

[26.237] M. SESTAKOVA and F. STROS: Kvasny Prum. 24 (1978), 103.

[26.238] V. V. GRUZINOVA and V. C. KOZLOVA: Gidroliz. Lesokh. Promst. 28 (1976), 20.

[26.239] M. KRIZANOVA: Kvasny Prum. 12 (1966), 183.

[26.240] J. T. PATTON, M. F. JURGENSEN, and B. J. DELANEY: Process Biochem. 14 (1979) June, 16.

[26.241] V. M. IVANOVA: Gidroliz. Lesokhim. Promst. 28 (1976), 21.

[26.242] M. RYCHTERA, J. BARTA, A. FIECHTER, and A. A. EINSELE: Process Biochem. 12 (1977) March, 26.

[26.243] J. VERNEROVA, J. BARTA, M. RYCHTERA, and J. MOSTECKY: Czech. Pat. 199950, 1979.

[26.244] J. MOSTECKY, J. BARTA, M. NEMEC, R. SARA, O. SLIVA, F. STROS, and J. VERNEROVA: Czech. Pat. 174431, 1976.

[26.245] E. I. KVASNIKOV, M. B. TEVELEVICH, L. P. PANTJUSHKINA, V. SADIAKHMATOV, O. A. BURAKOVA, M. I. ZJUZINA, R. N. ZUDOVA, and E. V. KOZUBENKO: Mikrobiol. Zh. 38 (1976), 160.

[26.246] K. K. SHIEH, B. J. DONNELLY, and H. A. LEE: US Pat. 3813320, 1974.

[26.247] D. NISHIKAWA, Y. IMADA, M. KINOSHITA, K. TAKAHASHI, H. MACHIDA, and M. NAGASAWA: US Pat. 4101378, 1978.

[26.248] L. M. LURIE, N. E. STEPANOVA, Y. E. BARTOSHEVICH, and M. M. LEVITOV: Antibiotiki 23 (1978), 86.

[26.249] L. WELWARD, J. RAKYTA, R. FRIMM, R. KOSALKO, and L. LACKO: Czech. Pat. 185858, 1980.

[26.250] G. N. CHERKASOVA and V. G. MAKAREVICH: Khim. Farm. Zh. 11 (1977), 87.

[26.251] W. S. SPECTOR (ed.): "Handbook of Biological Data." W. B. Sanders Co., Philadelphia-London, 1961.

[26.252] N. PAL, S. DAS, and A. K. KUNDU: J. Ferment. Technol. 56 (1978), 593.

[26.253] T. YAMAMOTO, F. HATTORI, and H. TAKATSU: Bull. Inst. Chem. Res. Kyoto 42 (1964), 252.

[26.254] A. LEUCHTENBERGER and H. RUTTLOFF: Abh. Akad. Wiss. DDR, Abt. Math. Naturwiss. Tech. (3N, Mikrob. Enzymprod.) (1981), p. 347.

[26.255] D. SAIGAL and L. VISWANATHAN: Enzyme Microbiol. Technol. 6 (1984), 78.

[26.256] KABUSHIKI KAISHA TEKKOSHA CO: Brit. Pat. 1300654, 1972.

[26.257] S. OKUMURA, F. YOSHINAGA, Y. YOSHINARA, A. KAMIMURA, and T. KAJIWARA: West Ger. Pat. 2100159, 1972.

[26.258] K. KUBOTA, Y. YOSHIHARA, H. HIRAKAWA, H. KAMIJO, S. NOSAKI, F. YOSHINAGA, S. OKUMURA, and H. OKADA: West Ger. Pat. 2321461, 1973.

[26.259] M. RUKLISA, A. SAKSE, U. VIESTURE, J. SVINKA, and D. MARAUSKA: In "Vlijanije Uslovij Kultivirovanija", p. 23 (Y. O. YAKOBSON, ed.). Zinatne, Riga (USSR), 1980.

[26.260] F. SMEKAL, J. PELECHOVA, and V. KRUMPHANZL: Kvasny Prum. 27 (1981), 276.

[26.261] J. PELECHOVA, F. SMEKAL, V. KOURA, J. PLACHY, and V. KRUMPHANZL: Folia Microbiol. Prague 25 (1980), 341.

[26.262] T. A. BACHINA, V. P. ANTIPOV, L. N. MARKIZOVA, A. K. SOKOLOV, and M. D. FRANK-KAMENETSKAYA: Khim. Farm. Zh. 11 (1977), 69.

[26.263] T. HARADA, K. SETO, and Y. MUROOKA: J. Ferment. Technol. 46 (1968), 169.

[26.264] K. KITANO, Y. SUGIYAMA, and T. KANZAKI: J. Ferment. Technol. 50 (1972), 182.

[26.265] T. KANZAKI, K. KITANO, Y. SUMINO, and H. OKAZAKI: J. Agric. Chem. Soc. 46 (1972), 95.

[26.266] S. IKEDA, I. FUJITA, and Y. HIROSE: Agric. Biol. Chem. 40 (1976), 517.

[26.267] K. AKASHI, H. SHIBAI, and Y. HIROSE: Agric. Biol. Chem. 43 (1979), 1563.

[26.268] P. M. LATEGAN and J. C. DU FREEZ: In "Abstr. 5th Int. Ferment. Symp.", p. 138 (H. DELLWEG, ed.). Verlag Versuchs- und Lehranstalt für Spiritusfabrikation und Fermentationstechnologie, Berlin, 1976.

[26.269] S. OKUMURA, R. TSUGAWA, T. TSUNODA, N. MIYACHI, A. KITAI, and A. OZAKI: West Ger. Pat. 1279585, 1968.

[26.270] TAKEDA CHEM. IND. Ltd: Fr. Pat. 1580214, 1969.

[26.271] K. NAKAMURA and K. KITAMURA: US Pat. 4266028, 1981.

[26.272] A. G. LOBANOK, R. V. MIKHAILOVA, I. V. STAKHEEV, and S. G. LATYSHEVA: USSR Pat. 591498, 1978.

[26.273] P. S. MASUREKAR: Eur. Pat. 29382, 1981.
[26.274] C. L. COONEY and D. W. LEVINE: Adv. Appl. Microbiol. *15* (1972), 337.
[26.275] ANONYMOUS: Chem. Eng. *82* (1975) December, 87.
[26.276] ANONYMOUS: Eur. Chem. News, December 17/24 (1976), 8.
[26.277] ANONYMOUS: Chem. Eng. News *54* (1976), 42, 25.
[26.278] J. KRIEGER: Chem. Ind. London, August 1 (1983), 21.
[26.279] G. CARDINI, L. DI FIORE, and A. ZOTTI: In "Abstr. 5th Int. Ferment. Symp.", p. 202 (H. DELLWEG, ed.). Verlag Versuchs- und Lehranstalt für Spiritusfabrikation und Fermentationstechnologie, Berlin, 1976.
[26.280] A. ZOTTI and G. CARDINI: US Pat. 4288554, 1981.
[26.281] Y. TANI, N. KATO, and H. YAMADA: Adv. Appl. Microbiol. *24* (1978), 165.
[26.282] W. BABEL, R. SCHOLZ, F. GLOMBITZA, U. ISKE, E. FEILER, U. HILGER, J. SCHNEIDER, Ch. GWENNER, and K. RICHTER: East Ger. Pat. 139870, 1980.
[26.283] T. KAMIKUKO, M. HAYASHI, and N. NISHIO: In "Abstr. 5th Int. Ferment. Symp.", p. 401 (H. DELLWEG, ed.). Verlag Versuchs- und Lehranstalt für Spiritusfabrikation und Fermentationstechnologie, Berlin, 1976.
[26.284] Y. TANI, T. KANAGAWA, A. HANPONGKITTIKUN, K. OGATA, and H. YAMADA: Agric. Biol. Chem. *42* (1978), 2275.
[26.285] K. OGATA, Y. IZUMI, M. KAWAMORI, Y. ASANO, and Y. TANI: J. Ferment. Technol. *55* (1977), 444.
[26.286] E. R. HOWELLS: Chem. Ind. London *15* (1982), 508.
[26.287] R. N. GREENSHIELDS: In "Economic Microbiology", Vol. 2, p. 121 (A. H. ROSE, ed.). Academic Press, London-New York, 1978.
[26.288] F. MADRON, K. EDERER, and F. STROS: Kvasny Prum. *25* (1979), 249.
[26.289] M. SESTAKOVA, L. ADAMEK, and F. STROS: Folia Microbiol. Prague *21* (1976), 444.
[26.290] J. PELECHOVA, J. UHER, J. ROSAK, and V. KRUMPHANZL: Kvasny Prum. *26* (1980), 10.
[26.291] J. ROSAK, J. PELECHOVA, and V. KRUMPHANZL: Kvasny Prum. *26* (1980), 219.
[26.292] J. ROSAK, V. KRUMPHANZL, and J. PELECHOVA: Kvasny Prum. *26* (1980), 274.
[26.293] M. SESTAKOVA and F. STROS: Kvasny Prum. *26* (1980), 126.
[26.294] S. G. KILIAN, B. A. PRIOR, P. M. LATEGAN, and M. C. J. KRUGER: Biotechnol. Bioeng. *23* (1981), 267.
[26.295] B. PRIOR, S. KILIAN, and P. LATEGAN: Arch. Microbiol. *125* (1980), 133.
[26.296] K. NAKAYAMA and H. HAGINO: Sw. Pat. 492783, 1970.
[26.297] KYOWA HAKKO KOGYO Co Ltd.: Brit. Pat. 1200587, 1970.
[26.298] V. MATELOVA and K. CULIK: Czech. Pat. 181397, 1980.
[26.299] P. TRAXLER, J. GRUNER, and J. A. L. AUDEN: J. Antibiotik. *30* (1977), 289.
[26.300] H. P. HAGEMAIER, W. KOENIG, H. ZAEHNER, H. P. FIEDLER, W. DEHLER, A. KECKEISEN, H. HOLST, and G. ZOEBELEIN: West Ger. Pat. 2928137, 1981.
[26.301] V. NARAYAN and K. K. RAO: Indian J. Microbiol. *21* (1981), 104.
[26.302] H. SOČIČ, E. PERTOT, and M. DIDEK-BRUMEC: Vest. Slov. Kem. Drus. *20* (1982), 323.
[26.303] P. A. WELLS, L. B. LOCKWOOD, J. J. STUBBS, E. T. ROE, N. PORGES, and E. A. GASTROK: Ind. Eng. Chem. *31* (1939), 1518.
[26.304] S. C. PRESCOTT and C. G. DUNN: "Industrial Microbiology." McGraw-Hill, New York, 1959.
[26.305] I. KAWAMOTO, T. NARA, M. MISAWA, and S. KINOSHITA: Agric. Biol. Chem. *34* (1970), 1142.
[26.306] N. S. EGOROV, T. G. YUDINA, Zh. L. LORIYA, and R. N. ZELENEVA: Prikl. Biokhim. Mikrobiol. *17* (1981), 676.
[26.307] P. STAHL, H. SEIDEL, and H. BRUNNER: West Ger. Pat. 2952410, 1981.
[26.308] M. I. BELAYEVA, D. V. YUSUPOVA, and A. Z. GAREISHINA: USSR Pat. 340691, 1972.
[26.309] J. C. FRENCH, J. D. HOWELLS, and L. E. ANDERSON: US Pat. 3551294, 1970.
[26.310] M. O. MOTKOVA, E. G. GLADKIKH, and T. P. KOROBKOVA: Antibiotiki Moscow *26* (1981), 492.
[26.311] H. IMANAKA, J. HOSODA, K. JOMON, H. SAKAI, I. UEDA, and D. MORINO: US Pat. 4283492, 1981.
[26.312] G. WIX, K. G. BÜKI, E. TÖMÖRKENY, and G. AMBRUS: Steroids *11* (1968), 401.

[26.313] M. G. Wovcha, F. J. Antosz, J. C. Knight, L. A. Kominek, and T. R. Pyke: Biochim. Biophys. Acta 53 (1978), 308.

[26.314] M. G. Wovcha, F. J. Antosz, J. M. Beaton, A. B. Garcia, and L. A. Kominek: US Pat. 4 221 868, 1980.

[26.315] E. Linde, T. Sturis, and J. Jakobson: In "Aminokysloty Mikrobnogo Sinteza", p. 119 (M. Bekers, ed.). Izd. Zinatne, Riga (USSR), 1968.

[26.316] Ajinomoto Co. Inc.: Fr. Pat. 1 580 549, 1969.

[26.317] M. B. Overchenko, V. A. Yanson, and G. M. Dobrolinskaya: Prikl. Biokhim. Mikrobiol. 17 (1981), 81.

[26.318] D. Wendland, E. Bruehl, J. Ambrosius, E. Kranz, K. Karbaum, and G. Lingner: East Ger. Pat. 148 465, 1981.

[26.319] U. Hilger, H. Quarder, R. Scholz, E. Hagen, K. D. Wendlandt, A. N. Grigoryan, I. L. Silant'ev, and N. V. Osokina: East Ger. Pat. 149 873, 1981.

[26.320] A. G. McLee, A. C. Kormendy, and M. Wayman: Can. J. Microbiol. 18 (1972), 1191.

[26.321] Y. Ichikawa, S. Sato, and J. Takahashi: J. Ferment. Technol. 59 (1981), 269.

[26.322] M. Wayman: Can. J. Microbiol. 20 (1974), 1675.

[26.323] K. Ueda, J. Takahashi, and N. Uemura: West Ger. Pat. 1 921 886, 1972.

[26.323a] R. Knecht, P. Präve, R. Seipenbusch, and D. A. Sukatsch: Process Biochem. 12 (1977) May, 11.

[26.324] J. W. Birkenstedt, U. Faust, and W. Sambeth: Process Biochem. 12 (1977) November, 7.

[26.325] I. C. Bennett, J. C. Hondermarck, and J. R. Todd: Hydrocarbon Process. 3 (1969), 104.

[26.326] A. Cejkova, F. Stros, J. Rybarova, and K. Hauser: Czech. Pat. 135 820, 1970.

[26.327] V. V. Fedorov, R. V. Katrush, G. I. Vorob'eva, V. A. Garbalinskii, and Y. A. Belogortsev: Khim. Tekhnol. Topl. Masel 5 (1981), 32.

[26.328] I. Y. Veselov, A. D. Gologobov, E. R. Davidov, L. G. Eliseeva, N. N. Latysheva, V. V. Rachinskii, and I. P. Uvarov: Prikl. Biokh. Mikrobiol. 10 (1974), 697.

[26.329] L. E. Erickson: Biotechnol. Bioeng. 23 (1981), 793.

[26.330] H. Iizuka: West Ger. Pat. 1 517 785, 1971.

[26.331] K. A. Iriskhanov: USSR Pat. 241 372, 1970.

[26.332] T. N. Oltarzhevskaya, V. F. Semenov, and V. T. Vaskivnyuk: Mikrobiol. Zh. 42 (1980), 247.

[26.333] H. Hustede and D. Siebert: US Pat. 3 843 465, 1974.

[26.334] T. Furukawa, T. Nakahara, and K. Yamada: Agric. Biol. Chem. 34 (1970), 1402.

[26.334a] T. Furukawa, T. Ogino, and T. Matsuyoshi: J. Ferment. Technol. 60 (1982), 281.

[26.335] S. Akiyama, T. Suzuki, Y. Sumino, Y. Nakao, and H. Fukuda: Agric. Biol. Chem. 37 (1973), 879.

[26.336] U. Behrens, R. Hellmig, U. Stottmeister, and E. Weissbrodt: East Ger. Pat. 148 347, 1981.

[26.337] K. Nara, K. Ohta, H. Fukuda, H. Katamoto, and O. Yamazaki: West Ger. Pat. 2 046 576, 1971.

[26.338] W. D. Puklowski, I. Reiff, and H. J. Rehm: In "Abstr. 5th Int. Ferment. Symp.", p. 408 (H. Dellweg, ed.). Verlag Versuchs- und Lehranstalt für Spiritusfabrikation und Fermentationstechnologie, Berlin, 1976.

[26.339] K. Kimura and T. Nakanishi: Brit. Pat. 1 332 180, 1973.

[26.340] R. Tsugawa and S. Okumura: Agric. Biol. Chem. 33 (1969), 676.

[26.341] M. Sato, T. Nakahara, and K. Yamada: Agric. Biol. Chem. 36 (1972), 1969.

[26.342] Y. Nakao, M. Kikuchi, M. Suzuku, and M. Doi: Agric. Biol. Chem. 34 (1970), 1875.

[26.342a] Y. Nakao, K. Kitano, K. Kintaka, S. Suzuki, K. Katamoto, and K. Nara: West Ger. Pat. 2 428 957, 1975.

[26.343] Kyowa Hakko Kogyo Co Ltd.: Brit. Pat. 1 241 901, 1971.

[26.344] Kyowa Hakko Kogyo Co Ltd.: Brit. Pat. 1 190 546, 1970.

[26.345] K. Hattori and T. Suzuki: Agric. Biol. Chem. 38 (1974), 581.

[26.346] S. Dewitt, J. L. Ervin, D. Howes-Orchison, D. Dalietos, and S. L. Neidleman: J. Am. Oil Chem. Soc. 59 (1982), 69.

[26.347] R. M. Lafferty, A. Moser, and W. Steiner: In "1st Symp. Mikrobielle Proteingewinnung", p. 87 (F. Wagner, ed.). Verlag Chemie, Weinheim, 1975.

[26.348] K. Karbaum, K. D. Wendlandt, J.

AMROSIUS, and Ch. GWENNER: East Ger. Pat. 149810, 1981.
[26.349] P. F. LEVY, G. W. BARNARD, D. V. GARCIA-MARTINEZ, J. E. SANDERSON, and D. L. WISE: Biotechnol. Bioeng. 23 (1981), 2293.
[26.350] R. KERBY and J. G. ZEIKUS: Curr. Microbiol. 8 (1983), 27.
[26.350a] B. FIEDLER, J. KORANDA, V. FOUSEK, V. KUDELA, and J. KOLAR: Czech. Pat. 139031, 1970.
[26.351] M. J. ROSE, J. M. GARROSELLA, J. D. CORRICK, and J. A. SUTTON: US Pat. 3540983, 1970.
[26.352] R. V. KUCHER, A. A. TUROVSKII, N. V. DZUMEDZEI, O. V. BAZAROVA, M. I. PAVLJUK, and D. I. KHMELNITSKAYA: Mikrobiologiya 46 (1977), 583.
[26.353] J. PACA and V. GREGR: Process Biochem. 12 (1977) April, 14.
[26.354] O. V. BAZAROVA, N. V. DZUMEDZEI, L. V. MOTOVILOVA, L. N. EKATERINA, and R. V. KUCHER: Mikrobiologiya 49 (1980), 331.
[26.355] D. STERNBERG and S. DORVAL: Biotechnol. Bioeng. 21 (1979), 181.
[26.356] M. SESTAKOVA and F. STROS: Kvasny Prum. 23 (1977), 8.
[26.357] V. A. EREMINA and G. L. MOTINA: Khim. Farm. Zh. 15 (1981), 91.
[26.358] S. J. CIRCLE and A. K. SMITH: In "Soybeans: Chemistry and Technology", Vol. 1, p. 294. AVI Publ. Co. Inc., Westport, Connecticut (USA), 1972.
[26.359] J. J. RACKIS: In "Soybeans: Chemistry and Technology", Vol. 1, p. 158 (A. K. SMITH and S. J. CIRCLE, eds.). AVI Publ. Co. Inc., Westport, Connecticut (USA), 1972.
[26.360] F. F. BUSTA and D. J. SCHRODER: Appl. Microbiol. 22 (1971), 177.
[26.361] I. FUSKA, N. KUGR, and I. ZAICHEK: Mikrobiologiya 33 (1964), 783.
[26.362] Y. AHARONOWITZ: Annu. Rev. Microbiol. 34 (1980), 209.
[26.363] BRISTOL-MYERS CO: US Pat. 4026766, 1977.
[26.364] C. VEZINA, C. BOLDUC, A. KUDELSKI, and S. N. SEHGAL: J. Antibiot. 29 (1976), 248.
[26.365] K. PETZOLD, K. KIESLICH, and A. WEBER: West Ger. Pat. 2940285, 1981.
[26.366] Y. IMADA and K. TAKAHASHI: US Pat. 4223091, 1980.
[26.367] K. AUNSTRUP: In "Biotechnology and Fungal Differentiation", p. 157 (J. MEYRATH and J. D. BU'LOCK, eds.).

Academic Press, London-New York, 1977.
[26.368] H. A. B. LINKE, W. MECHLINSKI, and C. P. SCHAFFNER: J. Antibiot. 27 (1974), 155.
[26.369] G. E. PHILIPCHUK and H. JACKSON: J. Gen. Appl. Microbiol. 25 (1979), 117.
[26.370] P. M. MAL'TSEV and N. V. CHALENKO: Ferment. Spirt. Promst. 36 (1970) 6, 18.
[26.371] N. Y. FEDOROVA, E. N. PISARCHUK, and I. N. FEDORENKO: Tr. Ukr. Naucho Issled. Inst. Spirt. Likero Vodoch. Promsti. 12 (1969), 260.
[26.372] G. Y. DZAVAKHIYA, K. A. KALUNYANTS, and M. S. VAGANOVA: Ferment. Spirt. Promst. 3 (1978), 16.
[26.373] E. P. MOSKVITCHEVA, B. A. USTINNIKOV, V. L. YAROVENKO, P. A. BELOZEROV, E. A. MAKSIMOVA, and V. N. MAKSIMOV: Prikl. Biochim. Mikrobiol. 10 (1974), 797.
[26.374] M. BEKERS, M. DREIMANE, V. VIESTURE, A. OZOLS, I. KRAUZE, A. APSITE, A. VECOZOLA, K. A. KALUNYANTS, and L. I. ORESHCHENKO: USSR Pat. 361199, 1972.
[26.375] M. DREIMANE: In "Mikrobiologicheskie Preparaty", pp. 56–64 (Y. O. YAKOBSON, ed.). Izd. Zinatne, Riga (USSR), 1976 (Russian).
[26.376] V. T. VASKIVNYUK, T. M. LUGININA, and A. M. PASICHNIK: Mikrobiol. Zh. 39 (1977), 144.
[26.376a] A. S. VECHER, I. I. PAROMCHIK, E. N. SKACHKOV, V. N. RESHETNIKOV, Y. U. ZABORONOK, L. N. AKIMOVA, and I. S. TSARENKOVA: Antibiotiki Moscow 23 (1978), 963.
[26.377] L. LEWIS and S. MITCHELL: Brit. Pat. 934528, 1963.
[26.378] N. A. KUZNETSOVA, E. N. BOLSHAKOVA, and E. B. PETROVA: Antibiotiki 13 (1968), 1063.
[26.379] M. BAHN, R. SCHMID, and R. WAGNER: Eur. Pat. 33439, 1981.
[26.380] M. HUCHETTE and F. DEVOS: Fr. Pat. 2496689, 1982.
[26.381] A. CEJKA: Eur. J. Appl. Microbiol. 3 (1976), 145.
[26.382] N. A. KUZNETSOVA, E. B. PETROVA, and E. N. BOLSHAKOVA: Antibiotiki 13 (1968), 820.
[26.383] V. A. GOTOVTSEVA, L. F. SKVOTSOVA, and A. S. KOROVKINA: Mikrobiologiya 48 (1979), 833.
[26.384] F. DEVOS, P. BEUQUE, and M. HUCHETTE: Eur. Pat. 26125, 1981.

[26.385] L. D. BELOVA and T. V. ERMAKOVA: Khlebopek. Konditer. Promst. *11* (1981), 43.

[26.386] E. Y. BRIDSON and A. BRECKER: In "Methods in Microbiology", Vol. 3 A, p. 229 (J. R. NORRIS and D. V. RIBBONS, eds.). Academic Press, London-New York, 1970.

[26.387] H. J. PEPPLER: In "Fermented Foods. Economic Microbiology", Vol. 7, p. 293 (A. H. ROSE, ed.). Academic Press, London-New York, 1982.

[26.388] N. S. EGOROV, I. P. BARANOVA, Y. I. KOZLOVA, A. G. VOLKOV, V. A. GRUSHINA, E. I. ISAI, P. P. ISAI, and A. T. SIDORENKO: Antibiotiki Moscow *25* (1980), 260.

[26.389] G. R. MESHINYA, E. V. KAMINSKA, V. V. SAVIENKO, M. F. KALNINYA, and E. A. DOMBROVSKA: In "Fermentatsya", p. 85 (D. Y. KRESLINYA, ed.). Zinatne, Riga (USSR), 1974.

[26.390] A. VAGAVICIUS, A. UZKURENAS, M. JAMONTIENE, O. BAGDONATTE, R. MUNIC, and R. TRINKA: In "Dostizheniya i Zadachi v Oblasti Mikrobiologii v Sovetskoj Litve", pp. 64–66, 1977. Ref. Zh. Biol. Khim. *1978,* Abstr. Nr. 13 Kh 150.

[26.391] B. C. SHKLYAR, A. P. SHPOKENE, A. B. RAGAVICIUS, A. G. LOBANOK, A. P. UZKURENAS, and V. V. GREBENKO: USSR Pat. 649 745, 1979.

[26.392] M. B. KUPLETSKAYA, I. V. AVSYUK, G. P. GOLOVKINA, A. N. GRIGORYAN, and N. S. EGOROV: Prikl. Biokhim. Mikrobiol. *17* (1981), 389.

[26.393] SCHERING AG: Eur. Pat. 54 810, 1981.

[26.394] W. J. MARSHEK, S. KRAYCHY, and R. D. MUIR: Appl. Microbiol. *23* (1972), 72.

[26.395] P. P. K. HO and L. D. BOECK: South Afr. Pat. 69 05 427, 1970.

[26.396] D. W. STAINER and M. J. SCHOLTE: Biotechnol. Bioeng. Symp. *4,* Part 1 (1974), 283.

[26.397] M. B. KUPLETSKAYA, V. N. MAKSIMOV, and T. B. KASATKINA: Prikl. Biokhim. Mikrobiol. *5* (1969), 541.

[26.398] S. P. CHATTOPADHYAY and A. K. BANERJEE: Folia Microbiol. *23* (1978), 469.

[26.399] K. AKASHI, H. SHIBAI, and Y. HIROSE: J. Ferment. Technol. *57* (1979), 321.

[26.400] Z. RUCZAJ and L. PASS: Przem. Chem. *56* (1977), 360.

[26.401] V. K. KOKIN: In "Fiziko-Chimicheskie Osnovy Pishchevoj Technologii.", p. 61 (A. V. ZUBCHENKO, ed.). Voronezh. Tekhnol. Institut, Voronezh (USSR), 1973.

[26.402] L. WELWARD, R. FRIMM, and R. KOSALKO: Kvasny Prum. *21* (1975), 63.

[26.403] G. P. KIRSANOV: Lab. Delo *7* (1967), 431.

[26.404] G. P. KIRSANOV: Prikl. Biokhim. Mikrobiol. *5* (1969), 211.

[26.405] G. OFFER and F. LAUE: Branntweinwirtschaft *122* (1982), 18.

[26.406] I. NITELEA, C. IONESCU, C. VLADEANU, and N. MUSAT: Roman. Pat. 72 350, 1980.

[26.407] J. MACKO: Kvasny Prum. *16* (1970), 197.

[26.408] P. KUJALA, R. HULL, F. ENGSTRÖM, and E. JACKMAN: Sugar Azucar, March (1976), 28.

[26.409] J. RYBAROVA: Kvasny Prum. *26* (1980), 180.

[26.410] A. G. ZABRODSKII, A. N. OSOVIK, V. J. PSCHEVORSKAYA, M. Y. KALYUZHNYI, and D. D. TARASYUK: Fermentn. Spirt. Promst. *3* (1975), 26.

[26.411] R. BRAUN and J. MEYRATH: Branntweinwirtschaft *121* (1981), 102.

[26.412] C. C. RIBERIO and J. R. C. BRANCO: Process Biochem. *17* (1981) April/May, 8.

[26.413] L. LABADIE and P. GOUET: Fr. Pat. 2 474 052, 1981.

[26.414] R. P. JONES, N. PAMMENT, and P. F. GREENFIELD: Process Biochem. *16* (1981), 42.

[26.415] H. F. NISS: US Pat. 3 539 694, 1970.

[26.416] J. RYBAROVA and K. PECKA: Kvasny Prum. *24* (1978), 224.

[26.417] J. RYBAROVA: Kvasny Prum. *26* (1980), 156.

[26.418] AMERICAN CYANAMID CO: US Pat. 3 050 446, 1960.

[26.419] M. E. O. VAHERI and V. KAUPPINEN: In "Abstr. 5th Int. Ferment. Symp.", p. 267 (H. DELLWEG, ed.). Verlag Versuchs- und Lehranstalt für Spiritusfabrikation und Fermentationstechnologie, Berlin, 1976.

[26.420] L. D. BELOVA, N. M. SEMIKHATOVA, T. V. ERMAKOVA, and L. N. VOROB'EVA: Khlebopek. Konditer. Promst. *7* (1981), 41.

[26.421] A. CEJKOVA, J. RYBAROVA, and M. SESTAKOVA: Czech. Pat. 121 742, 1967.

[26.422] E. M. ZAPARA and L. M. BALYBERDINA: Khlebopekarn. Konditor. Promst. *2* (1980), 39.

[26.423] N. I. DERKANOSOV and T. G. NAZINTSEVA: USSR Pat. 547469, 1977.
[26.424] Y. D. VIKHANSKY, V. A. MIRONOV, and G. KAPULTSEVICH: Mikrobiologiya 49 (1979), 628.
[26.425] I. T. ERMAKOVA, P. MÜLLER, T. V. FINOGENOVA, and A. B. LOZINOV: Mikrobiologiya 48 (1979), 349.
[26.426] M. SUZUKI, A. BERGLUND, A. UNDEN, and C. G. HEDEN: J. Ferment. Technol. 55 (1977), 466.
[26.427] A. R. MOREIRA, J. A. PHILLIPS, and A. E. HUMPHREY: Biotechnol. Bioeng. 23 (1981), 1325.
[26.428] R. M. BALITSKAYA, L. A. MINEEVA, Z. M. ZAITSEVA, N. I. ZHDANOVA, E. S. MOROZOVA, and S. I. ALIKHANYAN: Prikl. Biokhim. Mikrobiol. 5 (1969), 12.
[26.429] S. SANCHEZ and C. M. QUINTO: Appl. Microbiol. 30 (1975), 750.
[26.430] J. WOODWARD and A. WISEMAN: Enzyme Microbiol. Technol. 4 (1982), 73.
[26.431] K. BEAUCAMP, M. NELBOECK, H. GAUHL, H. SEIDEL, W. GRUBER, and H. BRUNNER: West Ger. Pat. 2933648, 1981.
[26.432] V. D. GRUZINA, L. D. SHISHCHENKO, and N. T. TROFIMOVA: Antibiotiki Moscow 26 (1981), 407.
[26.433] E. UDVARY-NAGY, M. BUDAI, G. FEKETE, S. GOROG, B. HERENYI, G. WACK, and K. ZALAI: Belg. Pat. 887392, 1981.
[26.434] W. P. WEISROCK: US Pat. 4301247, 1981.
[26.435] Y. OKUMURA, K. OKAMURA, T. TAKEI, K. KOUNO, J. LEIN, T. ISHIKURA, and Y. FUKAGAWA: J. Antibiot. 32 (1979), 584.
[26.436] M. EMA, T. KAKIMOTO, and I. CHIBATA: Appl. Environ. Microbiol. 37 (1979), 1053.
[26.437] K. OCHI, S. IWAMOTO, E. HAYASE, S. YASHIDA, and Y. OKAMI: J. Antibiot. 27 (1974), 909.
[26.438] E. A. KATRUK: Izv. Akad. Nauk Mold. SSR, Ser. Biol. Khim. Nauk 4 (1982), 59.
[26.439] A. TANAKA and S. FUKUI: J. Ferment. Technol. 49 (1971), 809.
[26.440] T. NAKAHARA, K. HISATSUKA, and Y. MINODA: J. Ferment. Technol. 59 (1981), 415.
[26.441] R. V. KUCHER, N. V. DZUMEDZEI, and D. L. KHMELNITSKAYA: Mikrobiologiya 50 (1981), 1105.
[26.442] K. WATANABE, T. TANAKA, T. KIRAKAWA, and M. SEZAK: Ger. Pat. 2136317, 1972.
[26.443] U. SCHOEMER and F. WAGNER: Eur. J. Appl. Microbiol. Biotechnol. 10 (1980), 99.
[26.444] O. BHUMIBHAMON and E. EKLUND: In "Abstr. 5th Int. Ferment. Symp.", p. 154 (H. DELLWEG, ed.). Verlag Versuchs- und Lehranstalt für Spiritusfabrikation und Fermentationstechnologie, Berlin, 1976.
[26.445] Y. KOHUSHO, H. MACHIDA, and S. IWASAKI: US Pat. 4283494, 1981.
[26.446] V. SASEK and A. R. GUPTA: Folia Microbiol. Prague 26 (1981), 124.
[26.447] R. C. JOHNSON and R. F. BEY: US Pat. 4133717, 1979.
[26.448] C. FRANZKE, R. GÖBEL, J. KROLL, P. LIEBS, G. RITTER, and M. SCHULTZE: Nahrung 23 (1979), 283.
[26.449] M. M. LEVITOV and A. N. MESHKOV: Mikrobiologiya 22 (1963), 717.
[26.450] A. CIMERMAN, V. JOHANIDES, and S. SKAFAR: In "Abstr. 5th Int. Ferment. Symp.", p. 405 (H. DELLWEG, ed.). Verlag Versuchs- und Lehranstalt für Spiritusfabrikation und Fermentationstechnologie, Berlin, 1976.
[26.451] M. Y. KALIUZHNYI: Tr. Inst. Mikrobiol. Akad. Nauk SSR 6 (1959), 172.
[26.452] M. Y. KALIUZHNYI: Mikrobiologiya 31 (1962), 720.
[26.453] B. ATKINSON, G. M. BLACK, and A. PINCHES: Process Biochem. 15 (1980), 24.
[26.454] Y. TANAKA, Y. TAKAHASHI, R. MASUMA, Y. IWAI, H. TANAKA, and S. OMURA: Agric. Biol. Chem. 45 (1981), 2475.
[26.455] R. MASUMA, Y. TANAKA, and S. OMURA: J. Antibiot. 35 (1982), 1184.
[26.456] J. LEOPOLD, M. LINHART, and L. TREFIL: Czech. Pat. 191011, 1981.
[26.457] R. Y. KARKLIN: In "Mikrobnyj Sintez Biologicheki Vazhnych Vechchestv", pp. 63-77. Zinatne, Riga (USSR), 1968.
[26.458] M. M. FAIBICH, V. A. SHIRSHOV, V. P. MOTORNAYA, and L. L. SINITSYNA: USSR Pat. 509642, 1976.
[26.459] E. B. LILLEHOJ, W. J. GARCIA, and M. LAMBROW: Appl. Microbiol. 28 (1974), 763.
[26.460] V. SANCHIS, I. VINAS, M. JIMENEZ, M. A. CALVO, and E. HERNANDEZ: Mycopathologia 80 (1982), 89.
[26.461] G. M. WOOD: Chem. Ind. London, December (1982), 972.

[26.462] C. J. Mirocha, S. V. Pathre, B. Schauerhamer, and C. M. Christensen: Appl. Environ. Microbiol. *32* (1976), 553.

[26.463] H. R. Burmeister and C. W. Hesseltine: Appl. Microbiol. *14* (1966), 403.

[26.464] J. Lafont, A. Romand, and P. Lafont: Mycopathologia *74* (1981), 119.

[26.465] J. Rybarova and L. Adamek: Kvasny Prum. *21* (1975), 126.

[26.466] J. Tomczynska: Pr. Inst. Lab. Badaw. Przem. Spozyw. *19* (1969), 707.

[26.467] M. Avron and A. Ben-Amotz: US Pat. 4 115 949, 1978.

[26.468] B. J. Chen and C. H. Chi: Biotechnol. Bioeng. *23* (1981), 1267.

[26.469] J. Barta and V. Gregr: Kvasny Prum. *6* (1960), 35.

[26.470] V. Gregr, J. Dyr, and J. Barta: Czech. Pat. 96 374, 1960.

[26.471] A. Cejkova: Folia Microbiol. *31* (1966), 439.

[26.472] Z. Fencl, J. Barta, K. Beran, J. Hospodka, and A. Cejkova: Czech. Pat. 121 277, 1966.

[26.473] A. Cejkova: Kvasny Prum. *11* (1965), 155.

[26.474] T. K. Murphy, H. W. Blanch, and C. R. Wilke: Process Biochem. November/December (1982), 6.

[26.475] D. S. Dahiya, M. Koshy, S. S. Dhamija, B. S. Yadav, and P. Tauro: Int. Sugar J. *84* (1982), 232.

[26.476] G. Hamer: Biotechnol. Bioeng. *24* (1982), 511.

[26.477] V. Simek and F. Stros: Kvasny Prum. *20* (1974), 217.

[26.478] V. Simek and F. Stros: Kvasny Prum. *21* (1975), 226.

[26.479] Dainippon Ink and Chemicals, Inc.: Fr. Pat. 2 219 224, 1974.

[26.480] H. H. Topiwala and B. Khosrovi: Biotechnol. Bioeng. *20* (1978), 73.

[26.481] K. Kundev: Nauchni Tr. Vissh Inst. Khranit. Vkusova Promst. Plovdiv *26* (1979), 267.

[26.482] F. Stros, L. Adamek, M. Rut, J. Rybarova, and Z. Aunicky: Czech. Pat. 187 982, 1978.

[26.483] K. Rosen: Process Biochem. *13* (1978) May, 26.

[26.484] K. Corbett: Br. Mycol. Soc. Symp. Ser. *3,* Fungal Biotechnol. (1980), p. 25.

[26.484a] Toyo Jozo Kabushiki Kaisha: Belg. Pat. 884 926, 1980.

[26.485] D. Perlman: Ann. N. Y. Acad. Sci. *139* (1966), 258.

[26.485a] T. Oki, A. Yoshimoto, V. Matsuzawa, T. Ishikura, T. Takeuchi, and H. Umezawa: Eur. Pat. 62 327, 1982.

[26.486] T. Munakata, Y. Ikeda, H. Matsuki, and K. Isagai: Agric. Biol. Chem. *47* (1983), 929.

[26.487] L. Adamek, F. Stros, M. Svojgr, K. Hauser, and A. Prokop: Czech. Pat. 158 991 (1975).

[26.487a] C. H. Pan, S. V. Speth, E. McKillip, and C. H. III Nash: Dev. Ind. Microbiol. *23* (1982), 313.

[26.488] L. M. Cappelletti, R. Spagnoli, and L. Toscano: Eur. Pat. 56 290, 1982.

[26.489] R. E. Kastner and R. L. Hamill: US Pat. 4 342 829, 1982.

[26.490] S. B. Penick & Co: US Pat. 3 352 761, 1967.

[26.491] S. Yaginuma, N. Muto, M. Tsujino, Y. Sudate, M. Hayashi, and M. Otan: J. Antibiot. *34* (1981), 359.

[26.492] Merck & Co, Inc.: US Pat. 2 571 693, 1951.

[26.493] A. Fujiwara, T. Hoshino, and M. Tazoe: Fr. Pat. 2 492 386, 1982.

[26.494] R. L. Hamill, M. M. Hoehn, and L. V. D. Boeck: US Pat. 4 336 333, 1982.

[26.495] B. H. Arison, R. T. Goegelman, and V. P. Gullo: US Pat. 4 285 963, 1981.

[26.496] Sankyo Co, Ltd.: Jap. Pat. 82 120 589, 1982.

[26.497] T. Misato, K. Ko, Y. Kobayashi, Y. Matsuzawa, T. Watanabe, and H. Mizuno: Fr. Pat. 2 467 215, 1981.

[26.498] Schering AG: Eur. Pat. 54 768, 1980.

[26.499] K. Petzold, R. Wiechert, H. Laurent, K. Nickisch, and D. Bittler: Eur. Pat. 51 143, 1982.

[26.500] A. Weber, M. Kennecke, and R. Müller: Eur. Pat. 55 832, 1982.

[26.501] W. G. Salmond, C. E. Sacks, and M. G. Wovcha: West Ger. Pat. 3 225 747, 1983.

[26.502] L. Alig, A. Fürst, M. Müller, U. Kerb, K. Kieslich, and R. Wiechert: Swiss Pat. 629 828, 1982.

[26.503] M. Bahn, J. Schindler, and R. Schmidt: Eur. Pat. 61 688, 1982.

[26.504] M. G. Wovcha, J. C. Knight, and A. B. Garcia: Fr. Pat. 2 505 360, 1982.

[26.505] Y. Sumino, K. Sonoi, and M. Doi: Brit. Pat. 2 101 131, 1983.

[26.506] M. Doi, K. Sonoi, and Y. Sumino: Brit. Pat. 2 100 731, 1983.

[26.507] M. Bahn, K. Engelskirchen, L. Schieferstein, J. Schindler, R.

[26.507 cont.] SCHMIDT, and W. STEIN: West Ger. Pat. 3 105 556, 1982.
[26.508] J. E. ZAJIC, D. F. GERSON, R. K. GERSON, and C. PANCHAL: Can. Pat. 1 125 683, 1982.
[26.509] A. L. ALLEN, C. R. BLODGETT, and J. M. NYSTROM: AIChE Symp. Ser. 75 (1979), 20.
[26.510] S. MATSUI, K. NAKAJIMA, and T. TANIGUCHI: Fr. Pat. 2 497 228, 1982.
[26.511] J. HUBER, R. PIPPIG, M. WOELBING, S. RICHTER, and K. H. MANGOLD: East Ger. Pat. 154 546, 1982.
[26.512] TOYO JOZO KABUSHIKI KAISHA: Austrian Pat. 368 187, 1982.
[26.513] L. E. ZITTAN, I. V. DIERS, K. M. OXENBOELL, and B. HOEJER-PEDERSEN: Fr. Pat. 2 504 939, 1982.
[26.514] N. S. EGOROV, N. S. LANDAU, V. I. GESHEVA, and V. N. MAKSIMOV: USSR Pat. 971 877, 1982.
[26.515] B. HYLMAR, L. POKORNA, L. PETERKOVA, V. HUSEK, M. DEDEK, V. MUSILEK, and V. ZALABAK: Czech. Pat. 188 789, 1981.
[26.516] H. HIRAGA, M. YOSHIMURA, S. IKEDA, and H. YOSHII: Fr. Pat. 2 497 232, 1982.
[26.517] M. H. UPDIKE and G. J. CALTON: Fr. Pat. 2 491 493, 1982.
[26.518] K. HIRAKAWA, R. TAKAKUMA, K. NOMURA, M. KATOH, and K. WATANABE: Fr. Pat. 2 491 495, 1982.
[26.519] AJINOMOTO CO. INC.: Brit. Pat. 1 286 208, 1972.
[26.520] L. E. SEMENOVA, L. A. MINEEVA, N. I. ZHDANOVA, and G. A. VELIKZHANINA: Khim. Farm. Zh. 12 (1978), 94.
[26.521] J. M. BARBER, F. T. ROBB, J. R. WEBSTER, and D. R. WOODS: Appl. Environ. Microbiol. 37 (1979), 433.
[26.522] P. L. ROGERS and D. E. TRIBE: Fr. Pat. 2 495 637, 1982.
[26.523] C. S. GONG, L. F. CHEN, M. C. FLICKINGER, and G. T. TSAO: Eur. Pat. 38 723, 1981.
[26.524] D. L. GIERHART: Brit. Pat. 2 091 286, 1982.
[26.525] V. A. POPOVA and M. Y. KALYUZHNYJ: USSR Pat. 521 308, 1976.
[26.526] J. BOA and A. LEDUY: Can. J. Chem. Eng. 60 (1982), 532.
[26.527] K. BAYER: J. Dairy Sci. 66 (1983), 214.
[26.528] Y. M. VOZNYAKOVSKAYA and O. K. KULIKOVSKAYA: Mikrobiologiya 41 (1972), 739.
[26.529] D. C. HILDEBRAND and A. H. McCAIN: Phytopathology 68 (1978), 1099.
[26.530] R. I. MATELES and E. BATTAT: Appl. Microbiol. 28 (1974), 901.
[26.531] R. J. SUMMERS, D. P. BOUDREAUX, and V. R. SRINIVASAN: Appl. Environ. Microbiol. 38 (1979), 66.
[26.532] R. D. SOIFER, T. A. GORSKAYA, T. A. IVANKOVA, and N. F. BELIKOVA: Antibiotiki Moscow 13 (1968), 120.
[26.533] R. D. SOIFER, S. V. GORSKAYA, and T. A. IVANKOVA: Biotechnol. Bioeng. Symp. 4, Part 2 (1974), 755.
[26.534] M. Zh. KRISTAPSON: In "Poluchenie i Primenenie Aminokislot." (R. A. KUKAIN, ed.). Zinatne, Riga (USSR), 1970.
[26.535] T. A. IVANKOVA, R. D. SOIFER, N. F. BELIKOVA, S. V. GORSKAYA, S. A. ZHUKOVSKAYA, and V. M. FISHMAN: Khim. Farm. Zh. 4 (1970), 32.
[26.536] Y. Z. SEROVA, G. M. DOBROLINSKAYA, and Y. P. GRACHEV: Izv. Vyssh. Uchebn. Zaved. Pishch. Tekhnol. 4 (1978), 52.
[26.537] M. C. O. HAULY, R. S. F. DA SILVA, and C. S. RAO: Rev. Bras. Tecnol. 11 (1980), 1.
[26.538] N. I. MATVEEVA and N. N. LESTROVAYA: Mikrobiologiya 44 (1975), 42.
[26.539] M. DESROCHERS, L. JURASEK, and M. G. PAICE: Dev. Ind. Microbiol. 22 (1981), 675.
[26.540] V. T. VASKIVNYUK and T. V. SEMENOVA: Mikrobiol. Zh. Kiev 43 (1981), 380.
[26.540a] V. T. VASKIVNYUK, T. M. LUGININA, and A. M. PASICHNIK: Mikrobiol. Zh. 39 (1977), 144.
[26.541] E. A. KRAYUSHKINA, N. A. ZHEREBTSOV, and V. I. ZVYAGINTSEV: Prikl. Biokhim. Mikrobiol. 9 (1973), 883.
[26.542] E. GALAS, S. BIELECKI, T. ANTCZAK, A. WIECZOREK, and R. BLASZCZYK: In "Adv. Biotechnol." (Proc. 6th Int. Ferment. Symp. 1980), Vol. 3, p. 301 (C. VEZINA and K. SINGH, eds.). Pergamon Press, Toronto (Canada), 1981.
[26.543] A. ERDELYI and E. KISS: Acta Aliment. Acad. Sci. Hung. 7 (1978), 155.
[26.544] M. O. MOTKOVA: Antibiotiki Moscow 23 (1978), 1068.
[26.545] M. O. MOTKOVA, T. P. KOROBKOVA, and E. G. GLADKIKH: Antibiotiki Moscow 26 (1981), 651.
[26.546] T. P. KOROBKOVA, T. S. MAKSIMOVA, O. L. OL'KHOVATOVA, M. S. YURINA, and V. A. ZENKOVA: Antibiotiki Moscow 23 (1978), 1065.

[26.547] S. SABIROV, V. D. KUZNETSOV, S. N. FILIPPOVA, and V. M. FISHMAN: Antibiotiki Moscow 23 (1978), 590.

[26.548] N. S. EGOROV, O. A. ALESHINA, I. P. BARANOVA, T. I. GOLIKOVA, and Y. I. KOZLOVA: Deposited Doc. 1982, VINITI 1238-82.

[26.549] I. D. MURGOV and Z. M. ZAITSEVA: Prikl. Biokhim. Mikrobiol. 9 (1973), 845.

[26.550] E. SOBCZAK and R. MAJCHRZAK: Przem. Ferment. Rolny. 14 (1970), 10, 11.

[26.551] S. L. CHEN: Biotechnol. Bioeng. 23 (1981), 1927.

[26.552] S. L. CHEN and F. GUTMANIS: Process Biochem. 17, November/December (1982), 2.

[26.553] B. A. USTINNIKOV, S. I. GROMOV, L. A. ROVINSKII, N. B. PCHELINA, and A. Z. BASYROV: Fermentn. Spirt. Promst. 1 (1983), 29.

[26.554] K. EDERER, F. MADRON, and F. STROS: Kvasny Prum. 28 (1982), 248.

[26.555] O. KÄPPELI, N. HALTER, and Z. PUHAN: GBF (Gesellschaft für Biotechnologische Forschung). Monograph. Ser. 6 (1982), 75.

[26.556] H. RODRIGUEZ, A. ENRIQUEZ, and O. VOLFOVA: Folia Microbiol. 28 (1983), 163.

[26.557] M. MORESI, C. NACCA, R. NARDI, and C. PALLESCHI: Appl. Microbiol. Biotechnol. 8 (1979), 49.

[26.558] J. VOTRUBA, P. PILAT, and A. PROKOP: Biotechnol. Bioeng. 17 (1975), 1833.

[26.559] P. PILAT, J. VOTRUBA, A. PROKOP, O. KSANDROVA, M. RYCHTERA, and V. GREGR: Sb. Vys. Sk. Chem. Technol. Praze Potraviny E46 (1976), 117.

[26.560] P. PILAT, A. PROKOP, J. VOTRUBA, O. KSANDROVA, M. RYCHTERA, and V. GREGR: Sb. Vys. Sk. Chem. Technol. Praze Potraviny E46 (1976), 133.

[26.561] A. PROKOP, J. VOTRUBA, M. RYCHTERA, and H. MOJOVA-MÜLLEROVA: Sb. Vys. Sk. Chem. Technol. Praze Potraviny E49 (1977), 187.

[26.562] L. J. FORNEY and C. A. REDDY: Dev. Ind. Microbiol. 119 (1977), 135.

[26.563] S. A. LEMMEL, R. C. HEIMSCH, and L. L. EDWARDS: Environ. Microbiol. 37 (1979), 227.

[26.564] J. C. SENEZ, M. RAIMBAULT, and F. DESCHAMPS: World Anim. Rev. 35 (1980), 36.

[26.565] F. BLOCH, G. E. BROWN, and D. F. FARKAS: Am. Potato J. 50 (1973), 357.

[26.566] Y. D. HANG: In "Abstr. 5th Int. Ferment. Symp.", p. 350 (H. DELLWEG, ed.). Verlag Versuchs- und Lehranstalt für Spiritusfabrikation und Fermentationstechnologie, Berlin, 1976.

[26.567] S. K. GHAI, S. S. KAHLON, and D. S. CHAHAL: Indian J. Exp. Biol. 17 (1979), 789.

[26.568] S. A. GHONAIM, A. A. ABOU-ZEID, A. F. A. EL-FETTAH, and M. A. FARID: Zentralbl. Bakteriol. Parasitenkd. Infektionskr. Hyg. Abt. 2 135 (1980), 82.

[26.569] P. K. BHAT and M. B. D. SINGH: J. Coffee Res. 5 (1975), 71.

[26.570] M. NOJIRI, K. KAKUTANI, S. UEDONO, K. UENAKAI, and M. MATSUMOTO: Brit. Pat. 1 579 632, 1980.

[26.571] A. BRECKA, A. BENDA, M. BUCKO, J. DASEK, and J. ZAJICEK: Czech. Pat. 131 199, 1969.

[26.572] M. I. BALABUSHEVICH and N. Y. KRASNOBRIZHII: Mikrobiol. Zh. Kiev 43 (1981), 312.

[26.573] A. A. SCHERBAKOV, L. N. KRAEV, M. Y. KALYUZHNYI, and D. D. TARASJUK: Gidroliz. Lesokhim. Prom. 5 (1977), 10.

[26.574] J. BARTA, F. STROS, and R. ZABOJNIK: Kvasny Prum. 10 (1964), 256.

[26.575] M. RUT and F. STROS: Kvasny Prum. 26 (1980), 57.

[26.576] L. A. ERZINKIAN, R. M. AKHIAN, F. G. SARUKHANIAN, R. S. KARIMIAN, L. G. PETROSIAN, G. P. MOVSESIAN, and R. A. ARAKELIAN: USSR Pat. 438 680, 1974.

[26.577] F. MARTINET, A. BA, R. RATOMAHENINA, J. GRAILLE, and P. GALZY: Oleagineux 37 (1982), 193.

Chapter 27

Sterilization

Karl Heinz Wallhäusser

Hoechst Aktiengesellschaft
Frankfurt am Main 80
Federal Republic of Germany

27.1 Introduction
27.1.1 Sterility Requirements for Fermentation Processes
27.1.2 Definition of Sterility
27.2 Suitable Sterilization Processes
27.2.1 Classification of Sterilization Processes
27.2.2 Sterilization with Moist Heat
27.2.2.1 Influence of Type of Microorganism and Functional State
27.2.2.2 Influence of Initial Germ Count and Acceptable Final Concentration
27.2.2.3 Influence of Treatment Temperature and Contact Time
27.2.2.4 Influence of Other Environmental Conditions
27.2.2.5 Sterilization of Lab Fermenters and Air Filters in the Autoclave
27.2.2.6 Sterilization of Stationary Large-scale Fermenters
27.2.2.7 Sterilization in the Instantaneous Heater
27.2.3 Sterilization with Dry Heat
27.2.4 Sterilization by Germ Removal Filtration
27.2.4.1 Definition of Germ Removal Filtration
27.2.4.2 Germ Removal Filtration for Air
27.2.4.3 Germ Removal Filtration for Additions Forming Clear Solutions
27.2.4.4 Germ Removal by Ultrafilters
27.2.4.5 Aseptic Work under Laminar Flow
27.2.5 Plant Sterilization with Microbicidal Gases
27.2.5.1 Formaldehyde/Water Vapor Mixtures
27.3 Disinfection of Equipment
27.3.1 Definition and Areas of Application
27.3.2 Available Active Substances
27.3.3 Suitable Disinfectants
27.4 References

27.1 Introduction

27.1.1 Sterility Requirements for Fermentation Processes

In contrast with natural fermentation processes, i.e. spontaneous alcoholic fermentation and lactic fermentation, industrial fermentation is as a rule a controlled process using pure cultures of highly cultivated production strains in a carefully balanced nutrient medium. Here, it is rarely possible to use selective media to prevent the growth of contaminating microorganisms so that the production strain can grow freely. For fermentation, in the original meaning of the term, it is a fact that the reactions take place under anaerobic conditions, which exercise a selective influence and prevent the growth of undesirable microorganisms. A further restrictive effect is the low pH value, which is optimal for the development of yeast but in addition suppresses the growth of anaerobic bacteria. In a few industrial fermentation processes, as in the production of organic acids (acetic acid, citric acid), the low pH can be used as a protective function. However, in a large majority of industrial fermentation processes employing complex natural nutrients (see Sect. 27.2.2.3), usually with an initially high level of microbial contamination, it is necessary to use sterilization in an effort to eliminate all microorganisms present and thus enable the production strain to develop freely. This requires not only initial sterile conditions; during the entire period of fermentation the whole process must be protected by a variety of measures from contaminating microorganisms.

27.1.2 Definition of Sterility

Sterilization is understood as the elimination (by removal or killing) of all microorganisms and the inactivation of viruses present in or on a product.

According to this definition sterility is an absolute concept. Day-to-day practice has shown, however, that this theoretical goal cannot always be achieved, depending on

Table 27.1. Requirements for the Reduction of the Germ Count in Various Processes (WALLHÄUSSER [27.3])

Process	Conditions	Exposure Time Contact Time	Reduction in Germ Count	
			by powers of ten	by percent
Autoclave	121 °C 2 bar	15 min	12	99.9999999999
Hand disinfection	21 °C	0.5 min	5	99.999
Surface disinfection	21 °C	5 min	5	99.999
Preservation: Ophthalmic solution				
B.P. 1980	20–25 °C	1 day	6	99.9999
FIP	20–25 °C	1 day	2	99
		7 days	3	99.9
USP XX	20–25 °C	14 days	3	99.9
Chemotherapy: for antibiotics	37 °C	1–2 hours	3	99.9

the respective conditions. Consequently, one accepts today, above all for the calculation of the required lethality of a sterilization process (see Sect. 27.2.2.3), a contamination probability of 10^{-6} [27.1], [27.2].

In other processes, such as disinfection or preservation to prevent microbial spoilage, one is prepared to accept lower killing and inactivation rates (Table 27.1).

27.2 Suitable Sterilization Processes

27.2.1 Classification of Sterilization Processes

For the sterilization processes used in the pharmaceutical and food industries we distinguish between processes employing

- moist heat (steam)
- dry heat
- microbicidal gases
  - ethylene oxide
  - formaldehyde
- ionizing radiation
- germ removal by filtration (WALL-HÄUSSER [27.4], [27.5]).

A suitable process is expected to achieve the desired degree of sterility without impairing the treated product. Of the above-mentioned processes, however, only filtration can fulfill this objective, and its application is restricted to the treatment of air and other gases and clear liquids.

Depending on the process temperature and contact time, the application of heat frequently leads to changes in certain starting materials, which may later adversely affect the course of fermentation. The most important effects of this kind are:

- the caramelization of sugar solutions;
- the denaturing of proteins, normally used as sources of nitrogen;
- the inactivation of numerous vitamins and other substances essential to growth;
- reaction of aldo sugars with amino acids and other materials containing amino groups;
- polymerization processes in unsaturated aldehydes;
- hydrolytic cleavages.

Some of these reactions can be avoided by sterilizing the individual components of a nutrient medium separately or occasionally by regulating the pH value.

The use of dry heat for sterilization purposes, which requires considerably higher temperatures than for moist heat treatment, is restricted in industrial fermentation processes to the sterilization of defoamers based on oils (sperm oil, lard oil, etc.).

Microbicidal gases have never played a dominant role in fermentation technology. They can be employed only for the sterilization, or more properly, disinfection of plant installations (fermentation tanks, pipe systems, filtration plant, etc.), but not for treating the entire fermentation batch, i.e., including the culture medium, because in aqueous media they immediately react with other substances and can thus lead to toxic reaction products such as ethylene chlorohydrin (from ethylene oxide and chlorine containing compounds). With ethylene oxide and its mixtures with inert gases such as CO_2 and methyl formate there is additionally the risk of explosion and inhalation toxicity, which by themselves prohibit the use of this gas. The microbicidal action of these gases depends to a large degree on the concentration used, the relative humidity, temperature, and contact time.

The above factors also seriously limit the use of formaldehyde/water vapor mixtures, particularly the long contact time (6–16 hours) required for low concentrations. Furthermore, the polymerization products formed (paraformaldehyde) can interfere with the subsequent fermentation process (antimicrobial action) or contaminate the isolated fermentation product during the processing stage (see Sect. 27.2.5.1).

The microbicidal action of ionizing radiation cannot be used in the actual fermentation process, but is limited to the treatment of a few fermentation end products, such as crude enzymes (enzyme powder for the washing detergents industry), to reduce the germ count. No effort is made here to achieve true sterilization, since in the event of a high initial germ count this would bring about serious impairment of enzyme activity (WALLHÄUSSER [27.4]). For this reason only a mild treatment with a radiation dose of less than 1 Mrad ($<1 \cdot 10^4$ Gy) is possible and justifiable.

Germ removal filtration is used primarily for the sterilization of the considerable amounts of air required for fermentation, for the treatment of clear precursor solutions, for all important steps in the aseptic processing of the fermentation products, and above all for the antibiotics that are to be administered parenterally such as numerous penicillin and cephalosporin derivatives.

27.2.2 Sterilization with Moist Heat

Processes employing superheated, saturated steam are the most reliable sterilization processes, but their results are decisively affected by the following factors:

1. the type of microorganisms present and their functional state (vegetative forms or spores);
2. the initial germ count and the acceptable final concentration (probability of contamination of 10^{-6}, i.e., sterility by definition, or a reduction in the germ count with a few isolated germs being tolerated in the end product),
3. the temperature at the coldest point in the product and the treatment time at the lethal temperature, and finally,
4. the environmental conditions (e.g., microbes protectively encased in dirt or enclosed in crystals or granules, the pH of the solution, the synergistic or antagonistic influences of other substances present, etc.).

An optimal, material-preserving sterilization process using saturated live steam cannot be planned without knowledge of the above points. Consequently, for each fermenter and each culture medium these questions must be clarified already during the development phase.

27.2.2.1 Influence of Type of Microorganism and Functional State

The choice of the treatment temperature, which can, indeed, considerably influence the product to be sterilized, is governed primarily by the types of microorganisms found in the raw materials and water and by their functional state. With respect to the functional state we distinguish the vegetative forms, which in most cases are killed or inactivated in a few minutes at temperatures as low as 65–100 °C, just as are bacteriophages and streptophages. Difficulties are always encountered with survival forms, such as spores. Among these, the fungal spores (conidiospores) are relatively easy to kill (100–105 °C). It is primarily the *Clostridium* and *Bacillus* spores that cause more severe difficulties and make it necessary to use higher treatment temperatures. Table 27.2 summarizes the heat sensitivity of different types of microorganisms, their functional state being taken into account.

27.2.2.2 Influence of Initial Germ Count and Acceptable Final Concentration

Evaluation of this factor is made with the most resistant microorganism in the fermenter culture medium, preferably with the most heat-resistant form, the spore. Upon

Table 27.2. Heat Resistance Levels for Various Microorganisms (WALLHÄUSSER [27.5]; KONRICH and STUTZ [27.6])

Resistance Level	Effective Process	Temperature Range and Contact Time		Microorganisms in each Category
I	pasteurization	61.5°C	30 min	*Mycobacterium tuberculosis,* Abortus Bang, path. streptococci *Listeria,* polio viruses
		72°C	15 s	*Mycobacterium tuberculosis* Rickettsia
				Coxiella (Q fever) polio viruses
II	gentle heating to 80°C	80°C	30 min	most vegetative bacteria, yeasts and molds, all viruses except hepatitis viruses
III	boiling	100°C	5 min	hepatitis viruses
			15–30 min	*Bacillus anthracis* spores, most fungal spores
	boiling with 0.5% soda		15 min	*Bacillus anthracis* spores
IV	superheated steam 105°C	105°C	5 min	*Bacillus anthracis* spores
V	superheated steam 121°C	121°C	8–12 min	*Bacillus stearothermophilus* spores and other bacilli and *Clostridium* spores
VI	superheated steam 134°C	134°C	up to 6 h	highly thermo-resistant spores from agar agar (occasionally present)

Table 27.3. D, z, and k Values for Some Spore-forming Bacteria (drawn from various sources)

Microorganism (spores in aqueous suspension)	D Value (in min) $D_{115°}$	$D_{121°}$	z Value (in °C)	k Value (for 121°C)
Bacillus stearothermophilus	10–24	1.5–4.0	6–7	1.5–0.56
Bacillus subtilis	2.2	0.4–0.7	8–13	5.75–3.29
Bacillus megaterium	0.025	0.04	7	57.57
Clostridium sporogenes	2.8–3.6	0.8–1.4	13	2.87–1.6

heating to an appropriate lethal temperature individual spores begin to die. In analogy to a monomolecular reaction, with temperature and other environmental conditions held constant, the number of microorganisms killed per time unit depends on the number of viable microorganisms originally present, N_0 (number of colony-forming units, CFU). The number of surviving microorganisms (N) after time unit (t) can be determined by the plate count method. If the log of the number of surviving microorganisms is plotted against time, a survival curve is obtained (Fig. 27.1).

The D value, the decimal reduction time or destruction value, is the time in minutes required to reduce the initial germ count of a particular variety of microorganism in a particular functional state (e.g., spores) under the selected process conditions by one power of ten, which corresponds to a killing rate of 90%.

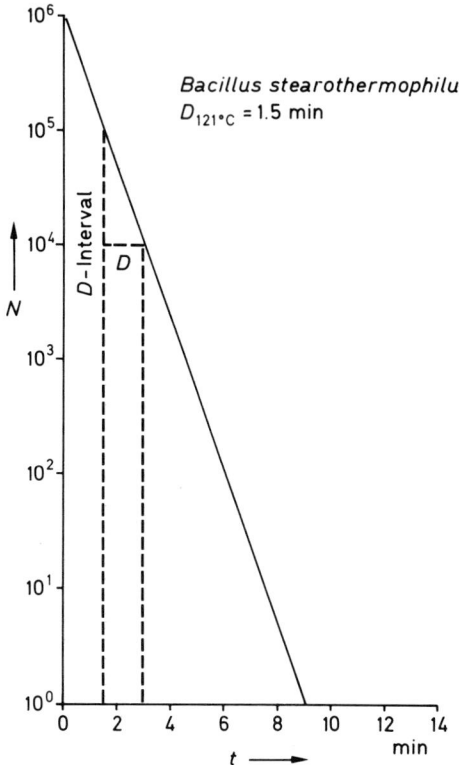

Figure 27.1. Relationship between the number N of surviving spores of *Bacillus stearothermophilus* per trial vessel and the treatment time t in saturated, superheated steam at 121 °C (WALLHÄUSSER [27.5]). – The D value can be read from the survival curve.

The relationship between the survival rate N/N_0 and the heat treatment time t is expressed by:

$$\ln\left(\frac{N}{N_0}\right) = -kt. \quad (27.1)$$

The constant k (see Table 27.3) depends on the type of microorganism and the other conditions of the environment.

Eq. (27.1) can be used to calculate the D value from any pair of N_0 and N values. For this purpose 1/10 is substituted for the survival rate N/N_0 and D for the time t, which yields:

$$\ln \frac{1}{10} = -kD \quad \text{or}$$
$$2.303 \log(0.1) = -kD$$
$$-2.303 = -kD$$
$$D = \frac{2.303}{k}. \quad (27.2)$$

27.2.2.3 Influence of Treatment Temperature and Contact Time

The temperature dependence of the D value can be well represented as a straight line by plotting the log of D versus the temperature T (°C or F; see Fig. 27.2).

From the microorganism-specific thermal killing time curve one can obtain two further important reference variables for the sterilization process in question, namely the z value and the F value.

z value: The microorganism-specific z value indicates the temperature increase or decrease (in °C or F) required to change the D value by a factor of 10. From Fig. 27.2 the z value found for *Clostridium botulinum* is 9.8 °C or 17.6 F. Further z values are shown in Table 27.3.

F value: Also from Fig. 27.2 one can determine for any lethal temperature from 100 °C to 130 °C the F value required to achieve a sterilization effect.

The F value is the time in minutes required to kill all spores in a suspension at a specific temperature (e.g., 121 °C). For the F value the following relation applies:

$$F_T^z = n \cdot D. \quad (27.3)$$

That is, the sterilization time F depends on the microorganism-specific z and D values, the chosen lethal temperature of the process, the initial germ count and the accepta-

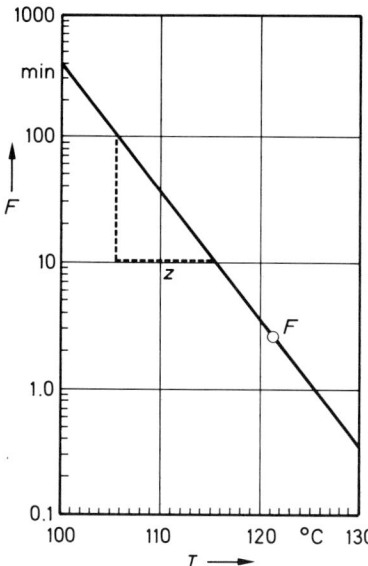

Figure 27.2. The thermal killing time curve for spores of *Clostridium botulinum* with F and z values indicated (PFLUG [27.7]).

ble final concentration in the nutrient medium to be sterilized. n is equal to the number of log-phases of the total germ count.

In respect of the initial germ count a distinction must be made, however, between spores and vegetative forms. Since the vegetative forms are killed at temperatures below 100°C, they need not be considered in calculating for the actual sterilization process.

Determination of the initial germ count in the nutrient medium to be sterilized is made immediately after the medium is prepared by the Koch plate count method.

Example
Initial ($CFU_{20°C}$) $5 \cdot 10^5$/mL
  in total batch $4 \cdot 10^{13}/80\ m^3$

After heating
30 min at 80°C ($CFU_{80°C}$) $1.25 \cdot 10^3$/mL
                               $10^{11}/80\ m^3$

Determination of the $D_{121°C}$ value
In the overkill process, which is recommended by most pharmacopoeias, one starts with $D_{121°C} = 1.5$ min, pertaining to the spores of *Bacillus stearothermophilus*, which are frequently used as bioindicators. These spores, which are classified as highly thermo-resistant, occur in the raw materials used in fermentation processes, particularly when these raw materials derive from waste products of the sugar industry. In all other raw materials (soybean meal, etc.) the microorganisms usually found are mesophilic spore-forming types with $D_{121°C}$ values between 0.5 and 1.0 min. Several batches of the raw materials used should always be examined for their germ profiles before the required D value is determined.

Calculation of the required F value
This calculation requires that, in addition to the initial germ count, the acceptable final concentration be specified. In the pharmaceutical field, as previously mentioned (Sect. 27.1.2), a probability of contamination of 10^{-6} is taken as the base, that is, of 1 million batches only one should be contaminated and all the others sterile. Frequently in the fermentation industry, as in aseptic manufacturing processes, a higher probability of contamination of 10^{-3} is accepted. On the basis of the initial, experimentally determined germ count (example: 10^{11}/fermenter) and a desired probability of final contamination equal to 10^{-6} resp. 10^{-3}/fermenter, n, the number of log-phases of total germ count, yields $n = 11 + 6$ resp. $n = 11 + 3$. The F values are calculated then as follows:

$F_T^z = n \cdot D_{121°C}^z$,
here taking $D_{121°C}^6 = 1.5$ min and
$n = 17$ $n = 14$
$F_{121°C}^6 = 17 \cdot 1.5$ $F_{121°C}^6 = 14 \cdot 1.5$
$F\ \ \ \ = 25.5$ min $F\ \ \ \ = 21$ min

If the calculation is, however, based on a D value of 1.0 min, the sterilization times turn out to be 17 and 14 minutes.

The discrepancy between theory and practice
The calculated sterilization time leads to the desired effect, however, only if the lethal temperature (here 121°C) can be

achieved at every point in the material being treated. Difficulties arise when the raw materials for fermentation include crumbly granular materials such as soybean meal, in which spores can become enclosed and insulated from the sterilization process and thus survive. Such products should be soaked for an extended period and vigorously stirred, if necessary, before the other nutrients are added in order to avoid the above-mentioned difficulties.

Equivalent processes

If the material to be sterilized cannot tolerate higher temperatures without suffering thermal damage, then, as shown in Fig. 27.2, equivalent processes can be used.

On the basis of an existing process (yielding the reference temperature, for example, 120 °C) the required equivalent treatment time can be calculated for a temperature (T) that is tolerable for the material by use of the following equation (27.4):

$$F^z_{121°C}/F^z_T = 10^{\frac{T-121}{z}}, \tag{27.4}$$

$$F^z_T = \frac{F^z_{121°C}}{10^{\frac{T-121}{z}}}. \tag{27.5}$$

Inclusion of the heating and cooling times in the sterilization process: In the course of the sterilization process every lethal temperature in a given time unit exerts a lethal effect:

$$F^z_T = \int L \cdot dt. \tag{27.6}$$

The lethality rate (L) is defined as the reciprocal of the sterilization time F required at temperature T:

$$L = \frac{1}{F}, \tag{27.7}$$

$$L = 10^{\frac{T-121.1}{z}}.$$

Numerical procedures

To work with $F^z_T = \int L \cdot dt$, the range covered by the lethality curve, whose graph

Table 27.4. Lethal Rates for Different Temperatures T per Time Unit (min) Calculated for a Reference Temperature of 120 °C/min (PFLUG 27.7]).

$$F^z_{120}/F^z_T = 10^{\frac{T-120}{z}} \quad \text{for } z = 6°C \text{ and } 10°C$$

T (in °C)	Bacillus stearothermophilus ($z = 6°C$)	Clostridium botulinum ($z = 10°C$)
95	0.000	0.003
96	0.000	0.004
97	0.000	0.005
98	0.000	0.006
99	0.000	0.008
100	0.000	0.010
101	0.000	0.012
102	0.001	0.016
103	0.001	0.020
104	0.002	0.025
105	0.002	0.032
106	0.005	0.040
107	0.007	0.051
108	0.010	0.063
109	0.015	0.079
110	0.022	0.100
111	0.032	0.126
112	0.046	0.159
113	0.068	0.200
114	0.100	0.251
115	0.147	0.316
116	0.215	0.398
117	0.316	0.501
118	0.464	0.631
119	0.681	0.794
120	1.000	1.000
121	1.468	1.259
122	2.154	1.585
123	3.162	1.995
124	4.642	2.512
125	6.813	3.162
126	10.000	3.981
127	14.680	5.012
128	21.540	6.310
129	31.620	7.943
130	46.420	10.000

yields an irregular geometrical figure, is divided into narrow parallel strips ($y_0 \to y_n$) with a width of Δt (the time interval of the "unit process", i.e., 1 min) for successive temperature measurements. According to PATASHNIK [27.8] the area under the curve

($=A$) can be represented in a simplified form as:

$$A = \Delta t(y_1 + y_2 + \cdots + y_{n-1}), \quad (27.8)$$

where the y values stand for the lethal rates

$$y = 10^{\frac{T-121.1}{z}}.$$

as required by most pharmacopoeias for sterilization in the autoclave, the equivalent value $F_{121\,°C}^{z=10} = 22.885$ min when the heating time, coming-up time (to allow for heat penetration to reach the equilibrium temperature), and cooling time are taken into account. That means, the treatment lasts longer than necessary. Thus, when sensitive

Table 27.5. Calculation of the Lethality of Autoclave Sterilization after the Method of PATASHNIK [27.8]

Time (in min)	Temperature (in °C)	Lethal Rate $\left(\frac{\text{min at } 120°C}{\text{min at } T}\right)$	
		for $z = 6°C$	for $z = 10°C$
5	95	0.000	0.003
6	101	0.000	0.012
7	107	0.006	0.051
8	110	0.022	0.100
9	112	0.046	0.158
10	114	0.100	0.251
11	116	0.215	0.398
12	117	0.316	0.501
13	118	0.464	0.631
14	119	0.681	0.794
15	120	1.000	1.000
16–30	121	15·1.468	15·1.259
31	110	0.022	0.100
	$F = \Delta t (\sum L) =$	1·24.892 min	1·22.885 min

To integrate the area (A) under the lethality curve, one first should compile in tabular form the temperatures measured in the sterile material at 1 min intervals, enter the corresponding lethal rate to be read from Table 27.4 for a temperature of 120°C under the required z value (z value for *Clostridium botulinum* is 10°C, for *Bacillus stearothermophilus* 6°C), and add them.

The sum of the lethal rates ($\sum L$) in the example shown in Table 27.5 is 22.885 min for $z = 10°C$.

$$F = \Delta t (\sum L) = 1 \cdot 22.885 \text{ min}$$
corresponding to 121°C.

In the example chosen here with a sterilization time (holding time) of 15 min at 121°C, products are being sterilized, it is possible to shorten the total process time by including the equivalent effect of the coming-up time and cooling time.

Digital-computer method

Steam sterilization can be optimally regulated with a computer program. The sterilization temperature can be monitored at several critical points and a running record of the calculated F values is printed out. By taking into consideration the heating time, coming-up time, and cooling time with their contribution to overall lethality, overheating of the material is avoided. This applies especially to the sterilization of laboratory fermenters in autoclaves (see Sect. 27.2.2.5).

27.2.2.4 Influence of Other Environmental Conditions

Influence of Air

The D and F values cited above are valid for saturated, superheated steam. Although a steam-air mixture has the same pressure as pure steam, the temperature of the gas mixture is lower than that of pure steam. Consequently, it is essential to monitor both pressure and temperature simultaneously. In fermenter sterilization, in which as a rule the steam enters through air filters and the water inlet line, a pressure of up to 2.2 bar is required to maintain a sterilization temperature of 121°C, depending on the design, the mixer, and the associated air displacement.

Influence of pH

The F value of a sterilization process is very strongly influenced by the pH of the nutrient medium being treated. If, for example, the pH is reduced from 6.5 to 4.5, the treatment time is shortened to 1/5 of its original duration.

27.2.2.5 Sterilization of Lab Fermenters and Air Filters in the Autoclave

Small laboratory fermenters with a capacity of 5, 10, or 30 liters, usually equipped with glass connection pipes, can only be sterilized in the autoclave. After cooling to about 40°C they are installed in the fermentation plant. Disadvantages of this process are that the air cannot be completely removed from these fermenters during sterilization (see Sect. 27.2.2.4) and that, due to a lack of mixing in the sediment of culture-medium constituents consisting of large, insoluble particles (e.g., soybean meal), spore-forming microorganisms can escape destruction.

To specify the sterilization (as a rule about 30 min at 121°C), the critical points, the "coldest point" in the fermenter and the maximum temperature in the sediment, must be determined with the aid of thermocouples.

Small computers such as the MIC-F_o computer from MMM can be used for precise control of the F value. The process can also be validated with any test ampoules such as Sterikon test ampoules [27.9], which can be attached to a wire, inserted through the inoculum pipe, and positioned at the coldest point in the fermenter. The indicator spores *(Bacillus stearothermophilus)* are destroyed with an $F_{121°C}$ value of 15–17 minutes.

Air filters should also be sterilized in the autoclave along with the fermenter. At the end of the cooling phase it is advisable to install the fermenter immediately and to dry the air filter with forced air. If left to stand for a longer period, microorganisms can grow through the filter medium and contaminate the fermenter contents (WALLHÄUSSER [27.10]).

27.2.2.6 Sterilization of Stationary Large-scale Fermenters

The stationary fermenters of a large-scale plant, frequently having a capacity of 0.4, 4.0, 40, and 120 m^3, are designed as pressure vessels with a stirring apparatus, an aeration system using air from special air filters, and other auxiliary equipment (see Fig. 27.3).

One aim of the stepwise transfer of the inoculum from the prestages to the final production stage is to reduce the considerable financial risks of microbial contamination. In the manufacture of β-lactam antibiotics, contamination with β-lactamase-producing organisms can destroy the entire amount of the active principle in a few hours.

The sterilization of stationary fermenters along with their nutrient media is as a rule carried out by the introduction of saturated steam through air filters and steam jacket in the smaller stages and for the large-scale fermenters also through air filters and water

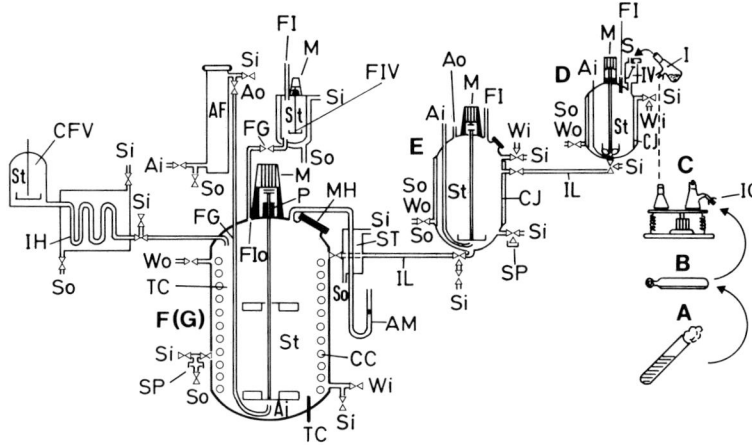

Figure 27.3. Schematic diagram of an antibiotic fermentation plant (WALLHÄUSSER and SCHMIDT [27.11]). – **A** Test tube inoculated with a production strain in sterile, dry sand-loam mixture for inoculating the Roux bottle **B**. **B** Spore suspension for inoculating **C**. **C** Shake flask with IC inoculum connection. **D** Inoculation stage, 400 L fermenter, IL inoculum line. **E** 4000 L fermenter. **F** 40 000 L fermenter. **G** 120 000 L fermenter. FIV Foam inhibitor vessel. IH Instantaneous heater. CFV Continuous feed vessel.
Further abbreviations: A air, AF air filter, AM air meter (rotameter), CC cooling coil, CJ cooling jacket, FG foam gauge to regulate foam inhibitor feed, FI foam inhibitor, HJ heating jacket, IV inoculation vessel, M motor, MH manhole (sealed with steam), P packing (seal), S steam, SP sampling point, St stirrer, ST steam trap, TC thermocouple, i inlet, o outlet.

inlet lines, and always with a running agitator. Circulation of the nutrient medium with the stirrer primarily prevents the settling of the coarse constituents (e.g., soybean meal) and simultaneously effects uniform heat distribution.

Coarse constituents of the nutrient medium, particularly when they tend to clump, frequently contain encased microorganisms, which under such a protective mantle escape destruction and can later interfere with the process of fermentation.

In most fermentation processes the carbon source, often a sugar (unrefined sugar, starch, glucose, etc.), is not added all at once, but is fed in continuously at a precisely metered rate after being sterilized in an instantaneous heater (see Sect. 27.2.2.7). This also applies to most of the other additions, especially the precursor substances, which are built into the molecule of the target product by the production strain and thus contribute to increasing the yield. Although the foam inhibitor (see Sect. 27.2.3) is heated in its storage vessel by means of a steam jacket (see Fig. 27.3), the steam in this case serves only as a heat transfer medium; the sterilization process itself is subject to the laws of hot-air sterilization (see Sect. 27.2.3).

The entire process is monitored by built-in thermocouples and pressure transducers, and at the same time an F-value computer can be hooked up to ensure optimal performance. As soon as the F value attains the precalculated level (the effect of the subsequent cooling phase is included), the cooling phase begins under automatic control. This is carried out in smaller fermenters by means of a cooling jacket and in larger vessels by cooling coils, which are frequently mounted in the fermenter, but also occasionally around its outer wall. The cooling equipment also regulates the temperature during the fermentation process. It is important that condensation formed during the sterilization process is taken into account.

27.2.2.7 Sterilization in the Instantaneous Heater

Certain nutrients such as the carbon source (sugar) and also the precursor solutions are fed in continuously during the fermentation process and must be sterilized before they reach the fermenter. This is accomplished with the help of instantaneous heaters, which are designed with such dimensions that the desired lethal effect is achieved at a tolerable temperature in an acceptable time. Since much value is placed on avoiding as much time loss as possible, usually HTSH (high temperature short time heating) processes are used, which work with temperatures in the range from 130 to 140 °C and use saturated steam (steam injection).

If the calculation of the F value is based on the thermo-resistance of *Bacillus stearothermophilus*, then for an initial germ count of $10^9/m^3$ and a desired probability of final contamination equal to $10^{-6}/m^3$ ($n = 9 + 6$) a treatment temperature of 127 °C, and a D value of 9 s (Table 27.6), the formula:

$F = n \cdot D$

$F_{127°C} = 15 \cdot 9 \text{ s}$

yields an F value of 135 s.

If for an instantaneous heater with steam injection, heating, and cooling times of 20 s each are added in the calculation, the throughput or treatment time is 175 s or roughly 3 min.

Table 27.6. D and k Values for Spores of *Bacillus stearothermophilus*

Temperature (in °C)	D Value		k Value
115	15.0 min		0.1535
121	1.5 min	90 s	1.535
127		9 s	15.35
133		0.9 s	153.5

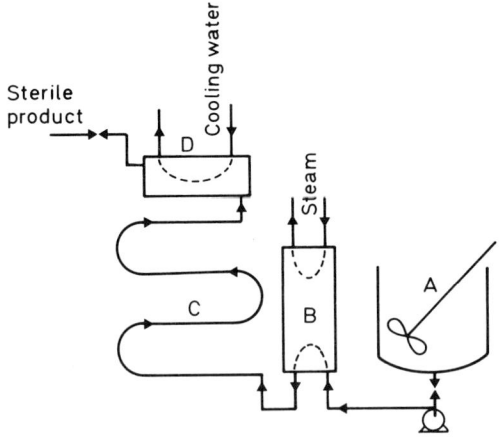

Figure 27.4. Schematic diagram of an instantaneous heater based on a heat exchanger (SIMON and MEUNIER [27.12]). – A Mixing vessel for the product to be sterilized, B steam treatment vessel, C pipe section at sterilizing temperature, D cooling section.

27.2.3 Sterilization with Dry Heat

This method of sterilization is only of secondary importance in industrial fermentations, but it is used for the sterilization of non-aqueous media, such as foam inhibitors based on animal and vegetable oils, fats, and silicone oils. In comparison with the processes employing moist heat (superheated steam), considerably higher temperatures are required to achieve sterilization. Foam inhibitor vessels are double-walled, steam-heated through a jacket, have built-in stirrers, and a capacity of 100–200 L.

Sterilization with dry heat presents difficulties if heat is transferred by hot air; cold air pockets interfere with heat transfer.

Table 27.7. D and z Values for Spores of *Bacillus* and *Clostridium* sp. under the Influence of Dry Heat (from various authors)

Organism (spores of)	D Value (in min or s) at					z Value (in °C)	
	120°C	140°C	150°C	160°C	180°C		
Bacillus subtilis	30′	2.7′	2′	80″	30″	23.3	18.3
B. stearothermophilus				20″	10″	14.4	24.4
Clostridium sporogenes			6′	2′	15″	33.0	18–22

These difficulties do not exist where heat is transferred directly from the heating jacket to the material to be sterilized. This material (oils) is, furthermore, normally low in germ content, and microorganisms cannot multiply in this water-free environment. The initial germ count of natural oils lies in the range of about 200/mL, so that a 500 L-vessel of foam inhibitor would contain 10^8 microorganisms. If the desired state is a probability of contamination equal to 10^{-6}, this yields $n=14$. $D_{150°C} \approx 1$ min (cf. Table 27.7).

Using $F = n \cdot D$
$F = 14 \cdot 1$ min,

we obtain a sterilization time of 14 min at a temperature of 150°C. At 120°C, i.e., the process temperature for steam sterilization, dry heat sterilization, with a D value of 30 min, requires 420 min or 7 h.

27.2.4 Sterilization by Germ Removal Filtration

Germ removal filtration plays a decisive role in fermentation processes. It is the method of choice for the "sterilization" of the air required for the process and is also frequently used for the germ removal treatment of water and other constituents of the nutrient medium and additions that yield a clear solution.

27.2.4.1 Definition of Germ Removal Filtration

Germ removal filtration is the separation of microorganisms present in liquids (aqueous and oil solutions: melted fats, paraffins, and solvents) and gases by means of suitable filter media. In certain cases special adsorption filters and ultrafilters are suitable for the removal of viruses. The aim of germ removal filtration is a sterile filtrate. Here too, as with other sterilization processes, there is no absolutely efficient germ removal and consequently, as with autoclave sterilization (residual germ count less than 10^{-6} where initial germ count is 10^6), a certain low removal rate may be acceptable. Thus, for example, HIMAC [27.14] expects no detectable germ passage with a contamination level of $10^7/cm^2$ filter area.

Filter Types
Filters are classified either as:
- surface filters (e.g., membrane filters) or
- deep-bed filters,

depending on whether germ removal occurs preferentially at the surface or within the inner structure of the filter material. Both types of filter function mainly as screens. Whereas in the first case the germs are retained at the surface of the filter medium and consequently block it after a certain time in operation, in the case of the deep-bed filters (in the area of fermentation usually glass-wool filters for air filtration) the microorganisms penetrate into the twisted paths of the labyrinthine region, where they are retained. In addition to the pure screening effect, occasionally adsorption effects also come into play, and in some filters are enhanced by charging the fibers with a zeta-plus potential. Because of the possible reversal of charge of bacteria due to a coat of soil, however, the process must rely primarily on the screening effect.

contamination, such as 10^{-6}). The higher the initial germ count, the greater is the probability, and hence the risk, of germ passage through the filter medium.

3. Filter specifications (rating) for membrane filters (WALLHÄUSSER [27.15])

Today standard germ removal filters are validated by the manufacturer before they

Table 27.8. Destruction of Microorganisms Through Drying on Smooth Surfaces (WALLHÄUSSER [27.4])

Microorganism	Initial Germ Count per 25 cm^2	Surviving Microbes/25 cm^2 after			
		1	2	4	6 hours
Staphylococcus aureus	1 000	445	396	161	145
Pseudomonas aeruginosa	10 000	325	150	120	29

In the filtration of air it is advantageous that vegetative bacteria very quickly dry out in the steady flow of air and die, whereas the long-lived forms (spores of bacilli) survive for a long period (Table 27.8).

Factors Influencing Germ Filtration
1. Particle size – removal rate

Particularly in the filtration of liquids, but also with gases, the size of the organism to be removed plays an important role. Coccoid bacteria have as a rule a cell diameter of 0.8–1.2 µm. For rod-shaped bacteria with a length of 1–3 µm it is the frequently very small cell diameter of only 0.3 µm (in various *Pseudomonas* species) which enables their passage through membrane filter layers and even through deep-bed filters. The removal rate should amount to at least 10^7 test bacteria per cm^2 of filter area.

2. Initial germ count

As in other methods of sterilization the F value ($F = n \cdot D$; see Sect. 27.2.2.3) is decisively affected by n, the number of log-phases of the total count, which is a function of the initial germ count and the acceptable final concentration (probability of

are approved for sale. They are characterized by the following parameters:

1. Flow-rate (throughput of air or water under a certain pressure)
2. Maximum "pore diameter" (calculated by the bubble point method) (WALLHÄUSSER [27.16])
3. Mean pore diameter (determined by the mercury intrusion method)
4. Germ removal test with specified test bacteria, such as *Serratia marcescens* for 0.45 µm filters and *Pseudomonas diminuta* for 0.2 µm filters.

On the basis of these data the filter material is assigned a nominal pore size.

A filter suitable to remove bacteria from liquids has a nominal pore size of 0.2 µm and smaller, but for air, in addition to the hydrophobic membrane filters with 0.2 µm, suitable filters are also those with a pore size of 0.3 and 0.45 µm. They are currently in use for the "sterilization" of the air required for fermentation.

Filters designed specifically for air filtration are best tested with a suitable germ aerosol (for example, *Bacillus* spores in absolute alcohol).

4. Ratings of glass-wool filters

In large-scale industrial fermentation, glass wool filters were used from the start for providing sterile air. Depending on the required throughput rate certain container dimensions have become established (see Table 27.9) and proved successful in more than 40 years of practical experience. The germ removal performance of such a filter depends on certain variables, such as the diameter and length of the glass fibers, the filter dimension (length and breadth), and the weight and active surface area of the fibers. A factor that must not be ignored is the method of packing, which must be such as to avoid larger voids that raise the likelihood of germ passage. Theoretical calculations must be omitted here, but can be found elsewhere (SIMON and MEUNIER [27.12]). The best practical test is always the aeration of a dilute nutrient broth through the freshly packed and sterilized air filter. These large-dimension filters are far more reliable than the modern filter cartridges, which are now replacing them because of their ease of handling and lower space requirements.

5. In-process controls of the leak-tightness of filtration systems

Particularly in the filtration of liquids but also of gases the results of germ removal filtration depend on the integrity of the filter material and the container or support on which it is mounted. Repeated sterilization with steam must be expected to damage the membrane or cause a leak around the O-ring used to seal off the sterile from the non-sterile side.

The integrity of a filtration system can be tested with the aid of a pressure hold test. Various companies supply suitable equipment for testing the integrity of membrane filter cartridges. Frequently the testing equipment also permits a check of the filter material with the bubble-point test in addition to the pressure-hold test.

Different instruments are available as a filtration system control, such as the Sarto-

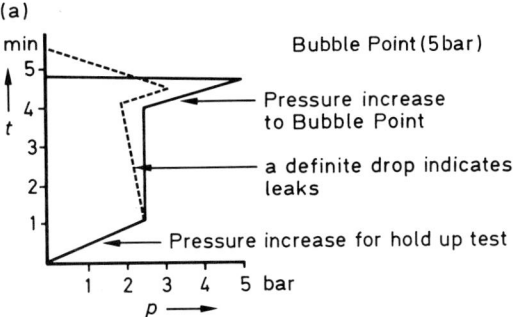

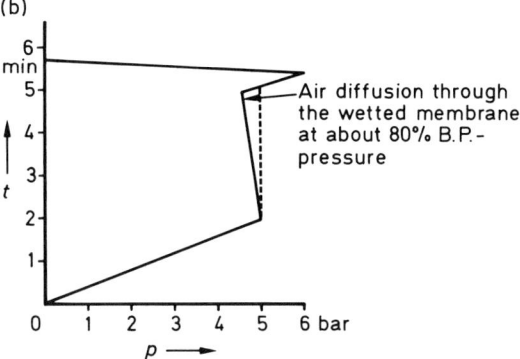

Figure 27.5. Filtration system control. The pressure hold test for integrity testing, the 3-phase test procedure with Sartocheck (WALLHÄUSSER [27.13]). – (a) Test under normal working conditions at 60% of Bubble Point pressure. A straight line at 2.5 bar for 3 min indicates that the filter is intact. (b) The pressure hold test near Bubble Point (Forward Flow Test at about 80% of Bubble Point pressure).

check (Fig. 27.5), which permit the integrity of the system to be tested without affecting the sterile side. The filter must, however, be wetted beforehand, which can be done with water (in the filtration of aqueous solutions) or with a special wetting agent (for hydrophobic filters). The functional control is performed by the pressure hold test, in which an intact filter will maintain the pressure specified by the manufacturer for a specified time (about 3–5 min). In smaller systems this test is not difficult to perform, but it is problematical in larger systems (WALLHÄUSSER [27.13]).

27.2.4.2 Germ Removal Filtration for Air

Germ Content of the Air

The germ content of the air is subject to seasonal fluctuations and generally lies between 100 and 500 CFU/m^3. The air required for the aeration of fermenters is usually cleaned in prefilters (e.g., carbon filters) before it is fed through the actual germ removal filters to the process. The germ content of the pre-cleaned air is on the average below 100 CFU/m^3.

Glass-wool Filters for Germ Removal
a) Use in laboratory fermenters

In the beginnings of fermentation technology, glass-wool filtering tubes (made of glass with a rubber stopper and glass tubing) (Fig. 27.6a) were always used for the sterile filtration of air (4 cm in diameter, length of part packed with glass wool about 15 cm, sufficient for a 20 L-fermenter). These are sterilized in an autoclave along with the fermenter. Today instead of filtering tubes, frequently membrane filter units are used for incorporation into pipe systems (see Table 27.11).

b) Use for large-scale fermenters (Fig. 27.6b)

Stationary, steam-sterilizable glass-wool filters are cylindrical containers of stainless steel with a cover flange and connection pipes for steam and for inlet and outlet lines for the gas (air) to be filtered. Inside the container is a tray with a perforated plate and handle to hold the glass wool.

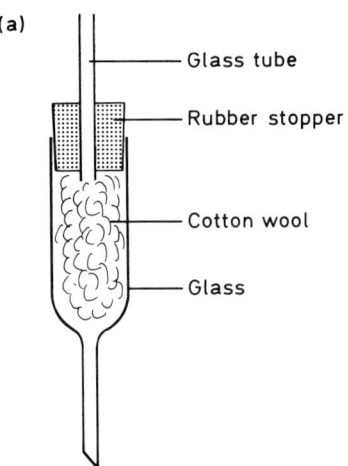

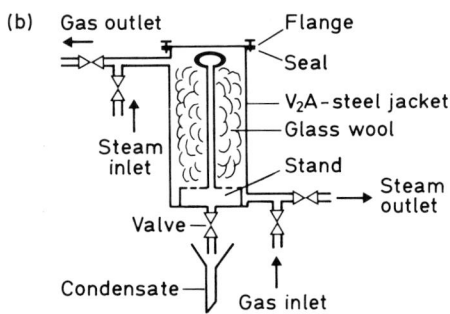

Figure 27.6. Two types of filters. (a) Filtering tube with glass-wool packing; (b) glass-wool filter for large-scale fermenters (WALLHÄUSSER and SCHMIDT [27.11]).

The latter is packed in such a way that the amount by weight indicated in Table 27.9 is uniformly distributed and lies everywhere tight against the container wall. Protective

Table 27.9. Dimensions of Glass-wool Filters for the Sterile Air Supply of Fermenters

Fermenter Volume			Air Filter			
(in L)	Filling (in m^3)	Flow Rate (in m^3/h)	Diameter (in cm)	Height (in cm)	Glass-wool Filling (in kg)	Volume (in L)
400	0.3		20	150	1.5	47.1
4 000	2.4		30	300	4.5	212.0
40 000	32	600–1800	120	150	50.0	1695.6
120 000	90		150	250	150.0	4415.6

Air throughput/filter: 0.3–1.0 parts by vol. air/parts by vol. culture medium/min

gloves and a face mask should be worn during packing.

The handle facilitates the emptying of the container when the glass wool is renewed, which in larger filters can only be managed with block and tackle. The valve on the outlet for condensate is first opened somewhat wider after the conclusion of sterilization of the filter, which is steam-treated together with the fermenter in order to blow out the condensed water and air, and then closed again.

Membrane Cartridge Filter for Air Sterilization
a) Use in laboratory fermenters

For supplying laboratory fermenters with sterile air we have available today a whole range of disposable and reusable membrane filter devices, which are attached with flexible hose to the air inlet connection pipes of the fermenter and are sterilized together with the latter in the autoclave. For small fermenters up to 10 liters capacity, disk filters with a diameter of 47–50 mm are adequate. The filter material used for this purpose consists of a hydrophobic membrane with a nominal pore size of 0.2–0.45 µm, which is mounted in a suitable V2A steel housing. For larger laboratory filters (capacity about 20 liters) disposable cartridge filters with 0.025–0.05 m² filter area (for example, Millipak, DFA, or FLF Pall) or reusable types are recommended (see Fig. 27.7a).

b) Use for large-scale fermenters
(see Fig. 27.7b)

The "fiber problem" associated with glass fibers, just as with asbestos fibers, has created a situation which suggests the use of alternate materials. For the HEPA filters (see Sect. 27.2.4.5), however, no replacement for glass fibers has yet been found. In the fermentation field, on the other hand, the traditional glass-wool filter is being replaced more and more by hydrophobic membrane cartridge filters, which can be combined on the modular principle, depending on the air volume flow rate required, in a horizontal or vertical arrangement. Although difficulties appeared at first

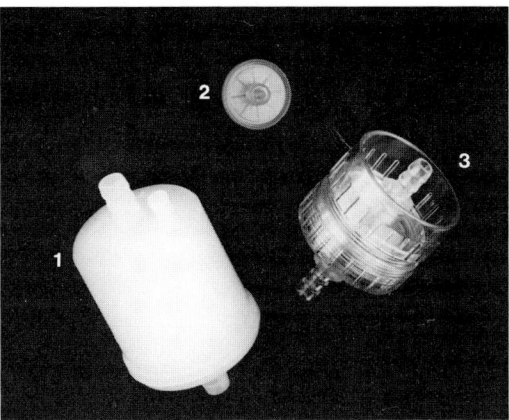

Figure 27.7a. Membrane filter attachment for aeration of lab fermenters with sterile air. (1) Sartofluor capsule (filter area 0.1 m²), (2) Minisart (filter area 5.3 cm²), (3) polycarbonate filter (filter area 12.5 cm²; ⌀ 50 mm).

Figure 27.7b. Membrane cartridge housing for large-volume air filtration.

Table 27.10. Cartridge Housings for Fermenter Aeration

Type	Manufacturer	Cartridge Length (in cm)	Removal Rate with PTFE (in µm)	Number	Arrangement Vertical	Arrangement Horizontal	Filter Area (in m²)	Max. Operat. Pressure (in bar)	Max. Throughput (in N m³/h)	Pressure (in bar)	Sterilizable at 121 °C (+)
DFA 3001 FRP	Pall	6.5	0.2	1	1		0.14	5 for liquids 30 °C 3.5 for gases	25	0.3	+
MSS 4463 G 08/G 09	Pall	13.3	0.2	1	1		0.29	10 at 95 °C	80	0.3	+
MDS 4463 G 08/G 09	Pall	13.3	0.2	1	1		0.29	16 at 120 °C	80	0.3	+
MSS 5001 G 716/GNW 25	Pall	25.4	0.2	1	1		0.72	10 for liquids 6 for gases	250	0.3	+
MSS 5002 G 716/GNW 25	Pall	50.8	0.2	1	2		1.44	10 for liquids 6 for gases	300	0.3	+
ECS 3000 GNW 25	Pall	13.5	0.2	1	1		0.36	10	600	0.3	+
ECS 6001 GNW 50	Pall	25.4	0.2	1	1		0.72	10	1000	0.3	+
ECS 6002 GNW 50	Pall	50.8	0.2	1	2	1	1.44	10	1400	0.3	+
ECS 7003 G 1 F 080 H4	Pall	25.4	0.2	3	1	3	2.16	6 [a]	1800	0.3	+
ECS 7006 G 1 F 080 H4	Pall	50.8	0.2	3	2	3	4.32	6 [a]	2100	0.3	+
ECS 7005 G 1 F 100 H4	Pall	25.4	0.2	5	1	5	3.6	6 [a]	2600	0.3	+
ECS 7010 G 1 F 100 H4	Pall	50.8	0.2	5	2	5	7.2	6 [a]	4000	0.3	+
ECS 7018 G 1 F 150 B4	Pall	50.8	0.2	9	2	9	12.96	6 [a]	7500	0.3	+
ECS 7030 G 1 F 200 H4	Pall	50.8	0.2	15	2	15	21.6	10 [a]	12000	0.3	+
ECS 7032 G 1 F 300 H4	Pall	25.4	0.2	32	1	32	23.04	10 [a]	17000	0.3	+
ECS 7064 G 1 F 300 H4	Pall	50.8	0.2	32	2	32	46.08	10 [a]	28000	0.3	+
YY 14131-00ST1	Millipore	80.0	0.22	3	3		2.1	7	150	0.2	+
YY 14331-20ST3	Millipore	80.0	0.22	9	3	3	6.3	7	450	0.2	+
YY 1403101ST12	Millipore	80.0	0.22	36	12	3	25.2	7	1800	0.2	+
5181206T8	Sartorius	8.2	0.45	1			0.1	5	4	0.5	+
5181306T9	Sartorius	13.5	0.45	1			0.2	5	4	0.5	+
5180406T1	Sartorius	25.0	0.45	1			0.4	5	20	0.5	+
5180407T1	Sartorius	25.0	0.2	1			0.4	5	13	0.5	+

[a] Executions of the ECS-housing in 2.5, 6, 10, and 16 bar   PTFE: polytetrafluoroethylene

Table 27.11. Hydrophobic Membrane Cartridge Filters for the Filtration of Air

Designation (Type)	Manufacturer	Material	Removal Rate (in μm)	Cartridge Length (in cm)	Cartridge Area (in m²)	Flow Rate Air (in L/min) at Δp (in mbar)						
						7	21	35	50	70	100
SLK 7001 FRP	PALL	PTFE	0.2	6.6	0.14	22.05	66.15	110.25		220.5	
SLK 7002 FRP	PALL	PTFE	0.2	13.3	0.29	44.1	132.3	220.5		441.0	
AB 1 FR 7 PV	PALL	PTFE	0.2	25.4	0.72	116.67	350.0	583.3		1166.7	
AB 2 FR 7 PV	PALL	PTFE	0.2	50.8	1.44	233.3	700.0	1166.7		2333.3	
AB 3 FR 7 PV	PALL	PTFE	0.2	76.2	2.16	350.0	1050.0	1750.0		3500.1	
CVGB 01TP1	MILLIPORE	PVDF	0.22	25	0.67	25.0	75.0	125.0		250.0	
CVGB 03TP1	MILLIPORE	PVDF	0.22	75	2.1	75.0	225.0	375.0		750.0	
518 0407 T1	SARTORIUS	PTFE	0.2	25	0.4				500.0		900.0
T2	SARTORIUS	PTFE	0.2	50	0.8				1000.0		1800.0
T3	SARTORIUS	PTFE	0.2	75	1.2				1500.0		2700.0
518 0406 T1	SARTORIUS	PTFE	0.45	25	0.4				700.0		1200.0
T2	SARTORIUS	PTFE	0.45	50	0.8				1400.0		2400.0
T3	SARTORIUS	PTFE	0.45	75	1.2				2100.0		3600.0
518 1207 T8	SARTORIUS	PTFE	0.2	8.2	0.1				83.3		166.6
T9	SARTORIUS	PTFE	0.2	13.5	0.2				125.0		250.0
518 9206 T8	SARTORIUS	PTFE	0.45	8.2	0.1				108.3		216.6
T9	SARTORIUS	PTFE	0.45	13.5	0.2				166.6		333.3
518 1307 T8	SARTORIUS	PTFE	0.2	15	0.1				96.0		175.0
T9	SARTORIUS	PTFE	0.2	18.5	0.2				150.0		250.0
518 1306 T8	SARTORIUS	PTFE	0.45	15	0.1				125.0		225.0
T9	SARTORIUS	PTFE	0.45	18.5	0.2				191.6		308.3

PTFE: polytetrafluoroethylene; PVDF: polyvinylidenedifluoride

after repeated sterilization (leaks at the caps of the cartridge filters), they have now been largely eliminated by the use of suitable membrane materials (based on PTFE = polytetrafluoroethylene). Suitable cartridge housings for one or several filter cartridges are listed in Table 27.10. According to PALL [27.17] these filters will withstand steam treatment at 125°C for 100 h and at 138°C for 50 h, which corresponds to about 100 sterilization cycles (cf. Tables 27.10 and 27.11).

27.2.4.3 Germ Removal Filtration for Additions Forming Clear Solutions

Numerous filterable additions that form clear solutions and are fed in continuously during the course of the fermentation process can be sterilized by germ removal filtration. As a rule one uses suitable membrane filters (disk filters or cartridge filters, depending on the flow rate), which are stable to the material being filtered. Suitable nominal pore sizes for this purpose are not larger than 0.2 µm. In filtration processes that operate rapidly and continuously there is little risk of accidental germ passage. This risk increases, however, with increasing contact time and particularly in intermittent processes. During the stationary phase when the filtration process is interrupted, single germ cells can grow into the filter labyrinth and then, when operation is resumed, be flung out by the pressure buildup (blowpipe effect, WALLHÄUSSER [27.9]). Double-layer cartridges afford greater safety; they can delay the blowpipe effect, but not permanently prevent it. In any case, the complete filtration system should be checked by means of suitable integrity tests before the start and the end of filtration. Suitable instruments with automatic recording, such as the Sartocheck, can control performance by the pressure hold test and also the bubble-point test.

27.2.4.4 Germ Removal by Ultrafilters (see Table 27.12)

Only ultrafilters with an asymmetrical pore structure and a nominal separation capacity for nominal molecular weight limits from 10 000 to 160 000 are suitable for production purposes. Through a special flow technique, usually with turbulent tangential flow across the membrane, premature clogging of these very fine-pore membranes is prevented. Compared with the true membrane filters, ultrafilters permit very much higher germ passage rates. Whereas the removal rate in the use of membrane filters is more than seven powers of ten per cm^2, systems using ultrafilters achieve a rate of only 2 to 5 powers of ten per cm^2 filter area. Germ removal, according to our own studies, is most efficient with spiral and cassette modules and falls off markedly with hollow-fiber (tube) modules (WALLHÄUSSER [27.18]).

The use of ultrafilters in the fermentation field is found mainly in the processing and concentration of certain metabolic products such as enzymes, "pyrogens", and other high molecular weight compounds. The fact that a sterile permeate may not be formed in the process must always be borne in mind.

27.2.4.5 Aseptic Work under Laminar Flow

Transportable laminar flow (LF) boxes are today indispensable aids in the performance of aseptic operations, particularly in the fermentation field, and especially when changes or repairs have to be carried out on the running laboratory fermenter. When HEPA filters of glass fiber wadding (which remove more than 99.97% of all particles >0.5 µm) are used as flow alignment devices, it is possible to achieve low-turbulence distribution of the purified air. With small LF units practically ideal conditions are obtained in the region near the flow source, where microbial contamination is quickly eliminated by the displacement

Table 27.12. Germ Removal with Ultrafilters (WALLHÄUSSER [27.18])

Ultrafilter Type	Manufacturer	Filter Area (in m^2)	Through-put (in L)	Starting Material CFU/mL	Total Count (CFU) Total	Germ-passage CFU/cm^2	CFU-Reduction Index per cm^2 Filter Area
Sartocon®	SARTORIUS	1.5	40	$1.0 \cdot 10^6$	>10000	~0.67	$2.7 \cdot 10^6$
		0.5	35	$3.0 \cdot 10^3$	720	0.14	$1.5 \cdot 10^5$
		0.5	35	$3.0 \cdot 10^3$	1376	0.27	$7.8 \cdot 10^5$
		0.5	35	$3.6 \cdot 10^3$	840	0.17	$1.5 \cdot 10^5$
		0.5	35	$3.5 \cdot 10^3$	1760	0.35	$7.0 \cdot 10^4$
Spiral-Modul	MILLIPORE	1.2	35	$5.5 \cdot 10^5$	>5000	~0.42	$5.9 \cdot 10^6$
HTSS		1.2	40	$5.5 \cdot 10^5$	>10000	~0.83	$2.2 \cdot 10^6$

flow. An effort should always be made to avoid disturbance zones in the form of deflection surfaces, since otherwise eddies will form and the objective will not be achieved.

27.2.5 Plant Sterilization with Microbicidal Gases

Only ethylene oxide mixed with inert gases and formaldehyde/water vapor mixtures can be considered for this application. Ethylene oxide, however, cannot be used for safety reasons, since it forms explosive mixtures with air and, as a strong protoplasm poison, it irritates the skin and respiratory organs. The threshold limit value (= TLV; MAK = maximale Arbeitsplatzkonzentration in Germany) laid down in the Federal Republic of Germany is 10 ppm (= 0.018 mg/L air).

Nor can formaldehyde be used without drawbacks, but where steam sterilization is not possible it is the only microbicidal gas that can be used to a limited extent for the "sterilization" of surfaces of piping systems and equipment (centrifuges), storage vessels, etc. and also for room disinfection.

27.2.5.1 Formaldehyde/Water Vapor Mixtures

As early as the turn of the century, dilute formaldehyde solutions have been used for room disinfection by evaporation or atomizer application.

In the last few years this technique has come under heavy attack, because formaldehyde in a 15 ppm concentration through daily contact for 6 hours on 5 days a week over a period of 2 years causes tumor formation in 43.2% of rats, but in only 2.4% of mice. Years ago the TLV (see Sect. 27.2.5) of formaldehyde was established at 1 ppm, so it is apparent that such an experimental design as that used to obtain the above figures ignores the actual conditions. For humans the threshold above which eye irritation occurs in 0.05-2.0 ppm and for nasal irritation 0.05 to 1.0 ppm. In a room having a volume of 30 m^3 smoke from about 22 cigarettes already exceeds the TLV of 1 ppm (CIIT Conference [27.19]). In view of the fact that formaldehyde is the only substance that, with sufficiently long contact times, is capable of killing *Bacillus* spores, the German Federal Office of Public Health in the preliminary recommendations on laboratory safety [27.20] stated the following: "As the substance of choice only formaldehyde can be recommended: evaporated or atomized dilute formaldehyde solutions with a suitable apparatus. Dosage rate: 5 g formaldehyde per m^3 room volume with at least 70% relative humidity. Other substances are either highly toxic, explosive, carcinogenic, or require conditions of application that can only rarely be met".

Since at present no other and better substance is available, formaldehyde vapors are used in fermentation technology and also to ensure aseptic conditions in the pharmaceutical industry in a concentration of a few hundred ppm for the sterilization of surfaces in equipment, where the problem is also to eliminate any spore-forming organisms that are not normally destroyed under the conditions of desinfection (peracetic acid, see Sects. 27.3.2 and 27.3.3, is too corrosive).

Factors Influencing the Sterilization Process

The sterilization process is influenced by the following factors:

1. The formaldehyde concentration. It must be borne in mind that formaldehyde is available on the market in the form of a 35-40% stabilized solution under the name formalin or formol. Often one starts with these commercial solutions and forgets to calculate on the basis of the actual formaldehyde content, so that the concentration then used is only 1/3 the prescribed concentration.

2. The existing humidity. It is optimum at 80-90% relative humidity. If the air is too dry, the formaldehyde polymerizes to paraformaldehyde and settles on the surfaces being treated.

3. Temperature. It must not lie below 18°C. As with numerous other antimicrobial processes, the effect increases with rising temperature.

4. The contact time. It depends largely on the concentration present and the other processing conditions mentioned above. At lower concentrations (about 100-140 ppm) it exceeds 6 hours. It is advantageous to allow the process to run overnight.

The Killing of Microorganisms

The D value (see Sect. 27.2.2.2) is affected by the above-mentioned conditions, the types of organisms present, their functional state, the initial germ count and acceptable final concentration. For a formaldehyde concentration of 100-140 ppm at a relative humidity of 80-90% and a temperature of 20°C it is about 30 minutes.

If it is admissible to base the calculation of plant decontamination on the very resistant spores of *Bacillus subtilis* var. *niger* with a total of 10^6 spores and the target level is a 10^{-6} probability of contamination, then

$F = n \cdot D$; $n = 12$, $D = 30$ min
$F = 12 \cdot 30$ min
$F = 6$ h.

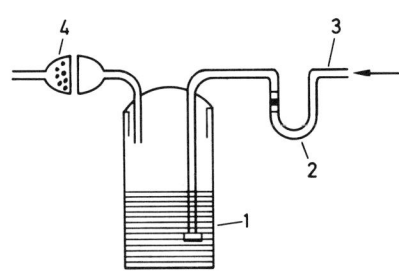

Figure 27.8. Apparatus for sterilization with formaldehyde (WALLHÄUSSER [27.21]). – Survival count on CSA-medium with 0.5% histidine after a contact time of 1, 2, 3, 4, 5, and 6 hours. – 1 Glass bottle (500 mL) with 300 mL 8.0% formaldehyde solution, 2 rotameter, 3 air stream (100 L/h), 4 sintered glass filter type G1 (∅ 60 mm) with bioindicator-carrier (glass beads). CSA: caseinpeptone-soypeptone-agar.

Method

In a suitable V2A steel container with a built-in atomizer (Fig. 27.8 shows an experimental unit), and with air flowing through a 7.5% formaldehyde solution, an amount of substance is entrained such that the formaldehyde content of the outlet air (for treating the plant) is about 100–140 ppm.

Residual Formaldehyde

It is always to be expected that certain residual amounts of formaldehyde will persist in the system, where especially in fermentation processes they can affect the growth of the culture organism. The minimum inhibitory concentration (MIC) for formaldehyde for bacteria lies in the range of 50–100 µg/mL and for molds in the range of 500–1000 µg/mL. Consequently, care must be taken that the formaldehyde concentration in a culture medium lies below these values.

In the processing of fermentation products such as antibiotics, in aseptic plants that can be sterilized only with formaldehyde, the FDA (USA) accepts a residual content of 10 ppm in the end product.

27.3 Disinfection of Equipment

27.3.1 Definition and Areas of Application

Disinfection, which derives from medical practice, is restricted to a selective reduction of the initial germ count on the surface of objects and on the skin to prevent the transmission of certain pathogens.

Here one knowingly accepts the fact that under the circumstances no "more" can be achieved, and numerous saprophytes, particularly the sporiferous forms, survive the disinfection treatment. In the treatment of installations in the fermentation field it is frequently of great importance, however, that all microorganisms present be eliminated. Consequently, the aim is to achieve "cold sterilization", one form of which is gasing with formaldehyde (see Sect. 27.2.5.1).

27.3.2 Available Active Substances

As Fig. 27.9 shows, the number of active substances from which a choice can be made is very small. Frequently commercial preparations consist of combinations of these disinfectants.

27.3.3 Suitable Disinfectants

In the disinfection of medical instruments, where one assumes a maximum contact time (frequently immersion time in the disinfectant bath) of one hour and an applied concentration of 2–4%, the predominant disinfectants are based on aldehydes, preferably formaldehyde and/or glutaraldehyde, frequently in combination with quaternary compounds or phenol derivatives. Also available on the market are preparations containing peracetic acid in stabilized form. These preparations exhibit broad-spectrum activity (applied concentration 1%), particularly against spores (see Fig. 27.9), but they are extremely corrosive, and, unfortunately, in the final applied form often unstable. Where these compounds are handled for use on larger equipment, their effect on the skin and their inhalation toxicity must be pointed out for the safety of personnel.

In any case, as has been pointed out in connection with formaldehyde (see Sect. 27.2.5.1), in fermentation units the final concentration of the active substance or substances used must remain below the

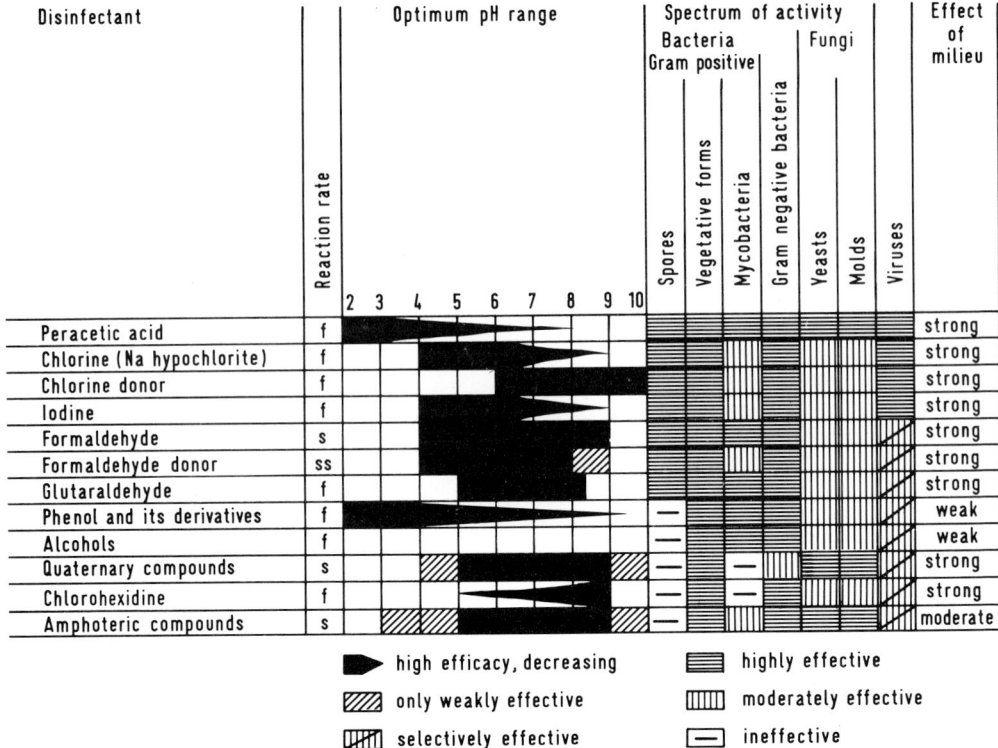

Figure 27.9. Spectrum of activity and pH dependence of the most important disinfectants (WALLHÄUSSER [27.5]). – f fast acting, s slowly acting, ss very slowly acting.

MIC (in µg/mL) for the microorganism to be cultured. Where necessary this condition can be approached by subsequent rinsing with sterile water.

27.4 References

[27.1] USP XX. United States Pharmacopeia 1980.
[27.2] Pharmacopeia Nordica ADD. (1974).
[27.3] K. H. WALLHÄUSSER: "Moderne Aspekte zur Produktionshygiene". Parfuem. Kosmet. *64* (1983), 677.
[27.4] K. H. WALLHÄUSSER: "Sterilisation-Desinfektion-Konservierung", 2nd Ed., p. 258. Thieme Verlag, Stuttgart, 1978.
[27.5] K. H. WALLHÄUSSER: "Praxis der Sterilisation-Desinfektion-Konservierung", 3rd Ed., p. 178. Thieme Verlag, Stuttgart, 1984.
[27.6] K. F. KONRICH and L. STUTZ: "Die bakterielle Keimtötung durch Wärme", 2nd Ed., Enke Verlag, Stuttgart, 1963.
[27.7] I. J. PFLUG: In "Industrial Sterilization" (G. B. PHILLIPS, ed.). Duke University Press, Durham, N.C., 1973.
[27.8] M. V. PATASHNIK; Food Technol. *7* (1953), 1.
[27.9] E. MERCK, Darmstadt: "Sterikon-Bioindikatoren".
[27.10] K. H. WALLHÄUSSER: "Durchwachs- und Durchblaseffekte bei Langzeit-Sterilfiltrationsprozessen". Pharm. Ind. *45* (1983), 527.
[27.11] K. H. WALLHÄUSSER and H. SCHMIDT: "Sterilisation-Desinfektion-Konservie-

rung-Chemotherapie". Thieme Verlag, Stuttgart, 1967.

[27.12] P. SIMON and R. MEUNIER: "Microbiologie industrielle et génie biochemique". Masson, Paris, 1970.

[27.13] K. H. WALLHÄUSSER: "Validierungsverfahren zur Überwachung der Entkeimungsfiltration". Pharm. Ind. *44* (1982), 401.

[27.14] HIMAC Health Industry Manufacturers Association: Document No. 3, Vol. 4 (1982) April, p. 1.

[27.15] K. H. WALLHÄUSSER: "Sécurité programmée lors de la filtration sterilisante". Inf. Chim. *230* (1982) October, 201.

[27.16] K. H. WALLHÄUSSER: "Germ removal filtration". Adv. Pharm. Sci. *5* (1982), 1–116.

[27.17] PALL GmbH Dreieich (Germany): "Emflon-Membranfilter", Druckschrift SD 900 G, 1982.

[27.18] K. H. WALLHÄUSSER: "Wasseraufbereitung mit Ultrafiltern". Pharm. Ind. *45* (1983), 1157.

[27.19] CIIT Conference on Formaldehyde Toxicity. Raleigh, N.C., November 20–21, 1980.

[27.20] BGA (Bundesgesundheitsamt, Federal Republic of Germany), Bundesgesundheitsblatt *24* (1981), 357.

[27.21] K. H. WALLHÄUSSER: "Disinfectants as an aid for good manufacturing practice in the pharmaceutical industry". J. Pharm. Belg. *36* (1981), 283.

Chapter 28
Recovery Operations

Maria-Regina Kula

Gesellschaft für Biotechnologische Forschung mbH
Braunschweig-Stöckheim
Federal Republic of Germany

28.1 Introduction
28.2 Harvesting
28.2.1 General Aspects
28.2.2 Centrifugation
28.2.2.1 Tubular and Chamber Centrifuges
28.2.2.2 Disc-stack Centrifuges
28.2.2.3 Solid Discharge Systems
28.2.3 Filtration
28.2.3.1 Filtration Equipment
28.3 Cell Disintegration
28.3.1 General Aspects
28.3.2 High-pressure Homogenization
28.3.3 Industrial Bead Mills
28.4 Clarification of Crude Extracts
28.4.1 Mechanical Separation
28.4.2 Extraction of Proteins from Cell Homogenates
28.5 Product Enrichment
28.5.1 General Aspects
28.5.2 Precipitation
28.5.2.1 Salting out
28.5.2.2 Isoelectric Precipitation
28.5.2.3 Precipitation by Organic Solvents
28.5.2.4 Precipitation by Non-ionic Polymers
28.5.2.5 Precipitation by Polyelectrolytes
28.5.3 Ultrafiltration
28.5.4 Extraction
28.5.5 Batch Adsorption
28.6 Chromatography
28.7 Concentration and Final Processing
28.8 Conclusion
28.9 References

List of Symbols

A	m²	area equivalent of centrifuges
a	—	exponent of pressure in Eq. (28.12)
C	kg/m³, kmol/m³	concentration
D_i	m²/s	diffusion coefficient
D	m	diameter of centrifuge bowl
d	m	diameter of particle or droplet
F	m²	filter area
f	—	protein-protein interaction coefficient, Eq. (28.18)
g	m/s²	earth acceleration
I	kmol/m³	ionic strength
J	m/s	flux (area-specific volumetric flow rate)
k	—	proportionality factor, Eqs. (28.8), (28.16), (28.17)
k_1	mol/kg·s	first order rate constant
K_d	m/s	mass transfer coefficient
K	—	partition coefficient
l	m	length
L	m	height of filter cake
M	kg/kmol	molar mass
N	—	number of passages
p	Pa = N/m²	pressure
Q	m³/s	volumetric feed flow
q	—	protein-polymer interaction coefficient, Eq. (28.18)
R	kg/kg cell mass	protein released
R_{max}	—	maximal amount of protein available for release
R_c	Pa·s/m	hydraulic resistance of filter cake
R_M	Pa·s/m	hydraulic resistance of filter medium
Re	—	Reynolds number
r	m	radius
r_1, r_2	m	outer and inner radius of disc stack
r_u, r_l	m	radius position of discharge of overflow, underflow
r_s	m	radius position of interface
S	kg/m³	solubility
Sc	—	Schmidt number
Sh	—	Sherwood number, function of Re, Sc, d_t/l_t
t	s	time
T	K, °C	temperature
$\bar{u}$	m/s	average linear velocity
V	m³	volume
V_a	m³	accumulated volume
v	m/s	speed of centrifugal sedimentation
v_s	m/s	speed of sedimentation due to gravity
W_m	Pa·s/m	hydraulic resistance of membrane
W_g	Pa·s/m	hydraulic resistance of gel layer
X	—	thermodynamic value $=(\mu_i-\mu_i^0/RT)$ in Eq. (28.18)
$x = \dfrac{\eta \cdot R_c \cdot w}{F^2 p}$		[in Eq. (28.11)]
$y = \dfrac{\eta \cdot R_M}{F p}$		[in Eq. (28.11)]
z	—	number of discs in stack

Greek Letters

α	m²	surface area of particles (or molecules)
$\alpha 1, \alpha 2, \alpha-\beta$	—	internal angle of centrifuge bowls, see Fig. 28.4
β'	—	constant in solubility equations
β	—	factor in Eq. (28.15)
Δ	—	difference
δ	C	total surface charge
γ	J/m²	surface free energy
ε	—	porosity
φ	—	angle between centrifuge axis and disc stack
Φ	V	electrostatic potential
η	Pa·s	dynamic viscosity
ν	m²/s	kinematic viscosity
μ^0	J/mol	standard chemical potential
σ_p^2	m²	variance of pore diameter
σ_i^2	m²	variance of particle diameter
ω	rad/s	angular velocity
τ	s	mean residence time
ϱ_s	kg/m³	density of solids
ϱ_l	kg/m³	liquid density
Σ	m²	factor for comparison of centrifuges
ζ	—	acceleration number = $\dfrac{r \cdot \omega^2}{g}$

Indices

b	bulk
c	convective

f	filter
g	gel
i	referred to component i (except D_i)
l	lower phase (except ϱ_l)
m	membrane
p	pore
t	tube
u	upper phase

28.1 Introduction

The products of biotechnological processes are usually found in complex mixtures of rather dilute solutions and must be concentrated and purified in order to be used. A general scheme of downstream processing is presented in Fig. 28.1. For products with limited stability recovery operations often encounter constraints with regard to temperature, pH, salt or solvent composition. This is especially true for proteins with catalytic or biological activities which depend on the structural integrity of the macromolecule as well as its secondary, tertiary, and quaternary structure. Unit operations in downstream processing, e.g., solid-liquid separation or chromatography use similar technical devices for low-molecular-weight products such as antibiotics or proteins; the actual performance of the procedures is quite different, however, because of different properties of the two product classes. This chapter concentrates on the recovery of enzymes or biologically active proteins from biological sources only. Efficient production of proteins requires the use of the metabolic machinery of living cells. Microorganisms are a major source of enzymes, and only few enzymes and protein hormones are still extracted from plant and animal tissue. The rapid development of genetic engineering has led to the production of mammalian proteins, especially human proteohormones and interferons in microorganisms, predominantly by *Escherichia coli*. This development is likely to continue.

Thus, in the future microorganisms will become the most important producers of enzymes and biologically active proteins by means of recombinant DNA technology. In this chapter we will therefore concentrate on the recovery of proteins from microbial sources.

There are numerous interrelationships between production and recovery of proteins which are poorly understood. For instance, the influence of a particular strain of microorganism as well as the effects of varying growth conditions on the spectrum of proteins and undesirable by-products are rarely investigated. These aspects, however, will not be covered here. In the future integrated process development is needed. The optimization of single steps alone is not sufficient to achieve optimal results in a complex biotechnological process. The complexity of the subject and the limited space make it necessary to give only a general overview of the topic. For details the reader is referred to current monographs and reviews.

28.2 Harvesting

28.2.1 General Aspects

Depending on the scale of operation, the location, and the sensitivity of the product one of the operations in Table 28.1 will provide a most convenient and/or economic solution for product separation.

Flocculation and flotation are currently employed on a large scale in waste water treatment and in single-cell protein production. For the recovery of biologically active proteins other methods are preferred. In practice centrifugation is often employed to recover the product within cells, while filtration is used when extracellular, excreted

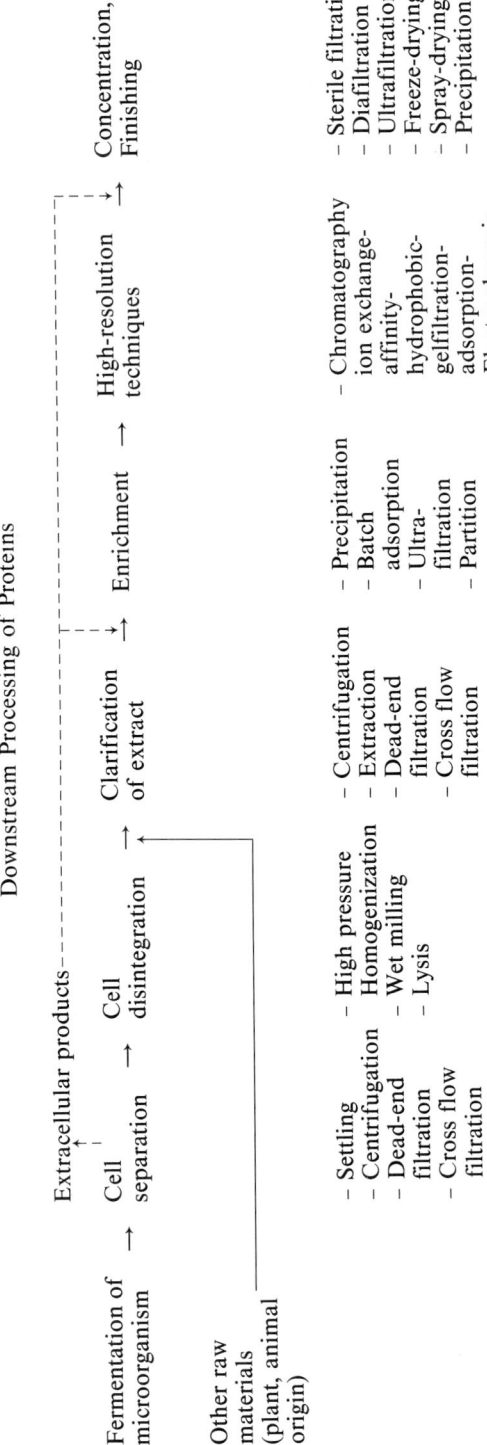

Figure 28.1. Flow diagram and common operations in downstream processing of proteins.

Table 28.1. Unit Operations in Solid-liquid Separation

Operation	Separation Criteria	Additional Limiting Parameters
Centrifugation	Density difference	Size of particle, viscosity of broth
Filtration	Size	Shape and compressability of particle, viscosity of broth
Flocculation	Surface charge	Shear forces
Flotation	Surface potential	Influence of other components in the broth

proteins are collected from the broth. To improve the separation step, flocculation is often carried out first to increase particle size, a dominating parameter in centrifugation and filtration.

28.2.2 Centrifugation

Different types of centrifuges are employed in biotechnology, mainly for solid/liquid separation; liquid/liquid/solid and liquid/liquid separations are less common (HEMFORT [28.1] and SOKOLOV [28.2]). There are four different designs (see Fig. 28.2). The tubular, chamber, and disc centrifuge, rotated in vertical position for continuous operation with regard to the liquid phases. The solids may be discharged discontinuously or semicontinuously, but continuous discharge is also possible. The horizontally rotating decanter or scroll centrifuge discharges liquids and solids continuously. Table 28.2 summarizes general criteria for the choice of a separator with regard to particle size and concentration of solids in the feed. The following parameters are similarly important for separation: shape and density of particles and the viscosity of the liquid medium. A solid particle suspended in a liquid will experience differential acceleration and therefore differential movement when centrifuged as a result of the density difference between solid and liquid. The acceleration determines the sedimentation period for density differences, as is usually found in biotechnological processes. The acceleration factor is given by Eq. (28.1) and summarized in Table 28.3 for the most commonly employed centrifuges. Ultracentrifuges have by far the highest accelerations leading to sequential sedimentation according to molecular mass, even for soluble components like proteins. Ultracentrifuges are important tools for the analysis of proteins and nucleic acids and in genetic engineering for the purification of plasmids. However, a typical ultracentrifuge operates discontinuously and has a restricted processing capacity. A continuously operating ultracentrifuge is commercially available with a tubular rotor driven by an air turbine at 35 000 rpm or $90 000 \cdot g$, (ROUND et al. [28.3]). During high speed centrifugation large forces act on the rotor demanding high mechanical safety. The rotor experiences forces due to its own rotating mass as well as to the pressure of the liquid column, which depends in turn on the speed of rotation, the density of the suspension, and the height of the liquid column. In a first approximation it can be assumed that the tension in the rotor wall is proportional to $D \cdot \zeta$, where D denotes the diameter

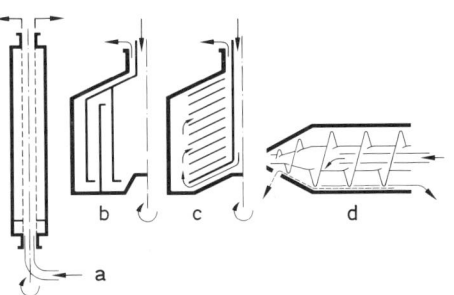

Figure 28.2. Basic design of separator bowls. (a) Tubular, (b) chamber, (c) disc stack, (d) scroll. – The arrows indicate the flow pattern within the bowl.

Table 28.2. Selection Criteria of Continuous Separators with Regard to Particle Size and Solid Content of Feed

Type of Separator	Range of Particle Size (in μm)	Range of Solid Concentration in Feed (in %)
Purifier bowl	0.5–500	<3
Intermittant discharge	0.5–500	1–10
Nozzle	0.5–500	5–25
Scroll	3–30 000	3–40

Table 28.3. Acceleration Factors for Common Centrifuges

Ultracentrifuges	10^5–$10^6\,g$
Tubular centrifuges	13 000–17 000 g
Chamber centrifuges	6 000–11 000 g
Disc-stack centrifuges	5 000–15 000 g
Scroll centrifuges	1 500–4 500 g

of the rotor and ζ the acceleration factor, which is equal to $r \cdot \omega^2/g$, [see Eqs. (28.1) and (28.2)]. This means that high acceleration and large volumes cannot be realized simultaneously beyond certain limits. Very high accelerations are only reached in rotors with a small diameter. This led to the design of a tubular centrifuge with a rather high ratio of height to diameter of approximately 5–7.

To describe sedimentation in a centrifugal field the following equation is generally used:

$$v = \frac{d^2 \cdot (\varrho_s - \varrho_1)}{18\eta} \cdot r \cdot \omega^2, \quad (28.1)$$

where v is speed of sedimentation, d particle diameter, ϱ_s density of the solids, ϱ_1 density of the liquid, r radius of rotation, ω angular velocity, and η dynamic viscosity.

A volumetric flow is superimposed on the sedimentation in the gravitational field. Only particles transported out of the flow-path by sedimentation or acceleration will be removed from the feed. The ratio of volumetric flow to effective clarifying surface is important for the performance of a centrifuge or sedimentation tank, respectively. From the relations between the load per unit clarifying surface and the radial sedimentation velocity in a centrifugal field, the limit particle size to be separated can be calculated according to the following equation:

$$Q = \frac{d^2 \cdot \Delta\varrho \cdot g}{18\eta} \cdot \zeta \cdot A, \quad (28.2)$$

with Q being the volumetric feed flow, g gravitational constant, ζ acceleration factor, and A the area equivalent of the rotor.

The same equation can be used to approximate maximal volumetric flow through a rotor from independently determined properties of the suspension to be separated. The maximal volumetric flow then is:

$$Q = v_s \cdot \zeta \cdot A, \quad (28.3)$$

with v_s as the sedimentation velocity due to gravity only.

The product of the acceleration number ζ and the effective clarifying surface A is called the sigma factor Σ, which is used for comparison of different centrifuges and for scale-up. The clarifying surface has been improved in the chamber centrifuge and disc-stack centrifuge (Fig. 28.2). Several options in the basic design of certain centrifuges are offered on the market which may also meet unusual demands. Certain equipment for biotechnological procedures can be operated under sterile conditions, with or without containment, under cooling and temperature control of the bowl within a narrow range (BRUNNER [28.4]).

28.2.2.1 Tubular and Chamber Centrifuges

Due to the limited capacity of the bowl, tubular centrifuges are used only for difficult separations for which high acceleration numbers have certain advantages. Since the mass of a tubular rotor filled with a suspension is comparatively low, this type of centrifuge can be more rapidly accelerated and decelerated. This allows to reduce the down time operating with a spare bowl. The rotor is driven by a belt from the top via a flexible spindle. The product enters the bowl from below and, in general, is discharged under gravity into a recipient at the top. The radius position of the overflow is fixed. Once the bowl is filled, the main movement of the feed is in a thin cylindrical layer determined by the radius position of the overflow. The maximal clarifying surface is defined by this cylindrical layer and not by the rotor wall. When the feed rate is increased, the thickness of the separating layer is enlarged, thereby limiting the performance by increasing the distance a particle must travel towards the wall. Separation efficiency is fairly constant until the sediment has filled the outer part of the bowl completely, up to the radius position of the overflow. The rather low sediment capacity limits economic operation to suspensions with a low solid concentration. In a chamber centrifuge the equivalent clarifying surface is enlarged by installing several concentric cylinders within a bowl. Each chamber thus created operates in principle like a tubular centrifuge as discussed above. For a given feed rate, the thickness of the axial cylindrical layer decreases with increasing distance from the axis of rotation. This minimizes the path a particle must travel from the inner to the outermost chamber. Since the centrifugal forces are increasing in the same direction, a certain classification of larger and smaller particles within the chambers is possible. The enforced flow through the chambers also increases the residence time within the centrifuge. Volumetric flows up to 15000 L/h are possible. Bowls with 2, 4, or 6 chambers are available with a solid holdup volume up to 65 L. Chamber centrifuges show a relatively constant and effective performance until the chambers are filled with solids. However, they can only be used in batch operations for the clarification of liquids. Once the chambers are filled, the separator has to be stopped. The bowl has to be opened manually and the chambers emptied one after the other. For this reason downtime is considerable and much time and effort is needed to recover the solids and clean the bowl. For the harvest of microorganisms, continuously operating disc-stack centrifuges are therefore successfully competing with chamber centrifuges. The overflow is, in general, discharged from the bowl of the chamber centrifuge under pressure by inserting a centripedal pump into the bowl top. The pump is stationary and immersed in the spinning liquid. This arrangement converts some rotational energy into pressure, which can be utilized for the transport of the liquid phase. At the same time a liquid seal is formed, which eliminates formation of foam and aerosols. Since the sediment may collect for a long time in the rotor of a chamber centrifuge, direct bowl cooling systems as well as cooling of the centripedal pump are necessary for the recovery of labile products.

28.2.2.2 Disc-stack Centrifuges

For continuous operations disc-stack bowls have been developed as a special design. In this bowl a continuous liquid flow is possible as well as periodical or continuous removal of the sludge. The effective clarifying surface is increased by installing a large number of conical, superimposed discs within the bowl. The interspaces are as small as 0.3 mm. In the disc stack, the liquid stream is divided into many thin layers and the settling path is considerably shortened. Fig. 28.3 illustrates the flow pattern in

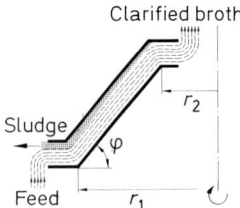

Figure 28.3. Solid-liquid separation in a single segment of a disc-stack centrifuge.

a separation space within a conical annular compartment. Once a solid particle has reached the surface of the next disc by travelling under the influence of the centrifugal force, it can slide down continuously in a cohesive layer into the sediment holding space towards the rotor wall. The clarified liquid flows towards the center of the bowl and leaves the disc stack near the axis of rotation. Discharge from the bowl can be accomplished under gravity or under pressure, as discussed above. The equivalent clarifying surface of a disc-stack separator can be roughly calculated from the following equation:

$$\Sigma = \frac{2\pi}{3g} \cdot \omega^2 \cdot \text{tg}\varphi \cdot z \cdot (r_1^3 - r_2^3), \quad (28.4)$$

with $\text{tg}\varphi$ the tangens of the angle φ (see Fig. 28.3), z the number of discs, and r_1 and r_2 the outer and inner radii of the discs in the stack, respectively.

The minimal density difference for successful separation in a disc-stack separator is approximately 0.01–0.03 kg/m³. For appropriate flow rates the limiting particle diameter in a disc-stack separator is approximately 0.5 m^{-6}. For a given bowl diameter the radius of the disc stack is determined by the necessary solid collecting space. The angle φ should be such that the sediments are gliding smoothly. For solid discharging separators the upper and lower parts of the bowl are conical (Fig. 28.4). The resulting angles $\alpha 1$ and $\alpha 2$, should also allow gliding of the sediments to the point of discharge. In general, continuous discharge of solids requires a certain plasticity of the solids for effective operation, which, however, is usually the case for microorganisms. Residual solid components from the nutrient broth may give rise to some problems if compacted in the solid sedimenting zone of a continuously operating separator.

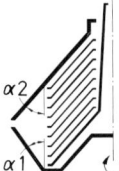

Figure 28.4. Design of a disc-stack centrifuge with intermittent discharge of sludge.

The distance between individual discs depends on the solid concentration in the feed, on the particle size, but especially on the properties of the sludge, which must glide along the disc in order to avoid clogging of the stack. In continuously operated separators the disc stack may contain along the same radius a number of channels, which serve to distribute the suspension within the stack. In order to achieve maximal separation for two immiscible liquids, the interfacial line should be positioned within the rising channels. This can be accomplished by variation of the radius position of the discharge point of the underflow, keeping the radius position of the overflow constant. From consideration of the hydrostatic equilibrium:

$$\varrho_u(r_s^2 - r_u^2) = \varrho_l(r_s^2 - r_l^2), \quad (28.5)$$

it follows:

$$r_l = \sqrt{\frac{r_s^2(\varrho_l - \varrho_u) + \varrho_u r_u^2}{\varrho_l}}, \quad (28.6)$$

with ϱ_u and ϱ_l being the density of the upper and lower phase, and r_s, and r_u, r_l the radii of the interface, the overflow, and underflow discharge point, respectively.

28.2.2.3 Solid Discharge Systems

Two different constructions are used for discharge of sludge from a rotating centrifuge bowl. In intermittent discharge centrifuges a hydraulic system opens a slit around the edge of the bowl (see Fig. 28.4). This is either controlled manually or by electronic timers, setting the interval as well as the duration of the bowl opening. If a complete discharge of the bowl content is intended, the feed pump is stopped during the opening. In order to avoid excessive dilution and loss of product in the sediment, fully automatic operations can be installed. The optical properties of the clarified overflow are analyzed and the discharge activated when a preset value is reached. In this manner, fluctuations of solid concentration and feed rate can be compensated for and a more constant quality of the discharged sediment obtained. When high solid concentrations in the feed require very short intervals between discharges, a nozzle separator may be used to accomplish the desired separation.

Nozzle separators are disc-stack centrifuges operating with a continuous discharge of solids. By their concentrating effect they can handle solid concentrations up to 30% in the feed; the concentration of solids achieved by a nozzle separator may be as high as 20fold. The solids sedimented and concentrated along the periphery of the bowl are continuously discharged through the nozzle openings. For suspensions of microorganism producing sludges of suitable plastic behavior, nozzles can be placed at a smaller radius than the periphery of the bowl (see Fig. 28.4). This reduces the pressure at the nozzle and allows larger diameters of the nozzle orifice. Such bowls require less electric energy for operation and have been successfully employed, e.g., in the production of baker's yeast. The amount of sludge delivered by a nozzle separator during continuous operation depends on the speed of revolution, the number of nozzles, and on the square of the diameter of the orifice. In addition, the dimension of the mean radius between the position of the nozzle and the overflow has to be considered. The maximal concentration of solids in the discharged sludge depends to some extent on the construction of the bowl, especially on the angles $\alpha 1$ and $\alpha 2$ and the sediment angle $\alpha\beta$ in the solids recipient (Fig. 28.5). High solid concentrations can be achieved in rotors with large diameters and a large number of nozzles. The final concentration is also influenced by the feed rate and the total solid concentration in the feed. For this reason, partial recycling of the sludge improves the performance of a nozzle separator when a high final solid concentration in the sludge is required (Fig. 28.6). For a given rotor, the operator can change the number of nozzles and the diameter of the orifice to accommodate various kinds of feedstock. A minimal diameter of 0.4 mm is recommended to avoid clogging. Lowering the number of nozzles and changing the diameter has to be

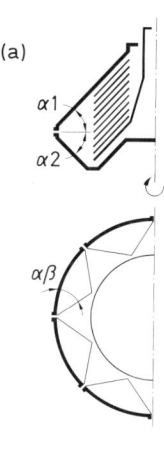
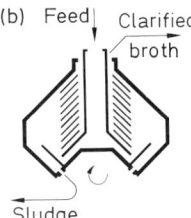

Figure 28.5. Design of nozzle separators for continuous discharge of sludge.

carried out maintaining a symmetrical flow pattern in the bowl and ensuring a balanced operation.

Intermittent discharge centrifuges and special designs of nozzle bowls allow complete emptying of the bowl and can therefore be incorporated in product lines requiring cleaning in place (CIP).

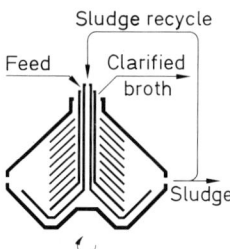

Figure 28.6. Nozzle separator with central recirculation of sludge.

Decanter or scroll centrifuges can operate with feed suspensions containing up to 60% solids. In biotechnology this type of centrifuge is used in waste water treatment to reduce the water content of the sludge or in SCP production for the final concentration of microorganisms. The special feature of the centrifuge is the scroll, which rotates with a different speed than the rotor; it transports the collected solids out of the centrifuge and assists in dewatering the sludge (ALT and GÖSELE [28.5]). The sigma factor of scroll centrifuges is generally not as high as for disc-stack centrifuges. Therefore, the particle size of the sedimented microorganisms is of primary concern and in general has to be improved by flocculation or coagulation. In recent years special attention has been drawn to the forces in the centrifuge acting on microbial flocs. Attempts have been made to improve the hydrodynamic flow and to avoid excessive disintegration of preformed flocs in the machine (BELL and BRUNNER [28.6]).

28.2.3 Filtration

In contrast to centrifuging, filtration as a unit operation for solid/liquid separation does not depend on a density difference. Since the density difference between microorganism and broth is fairly small, filtration is a valuable alternative and should always be considered (SVAROVSKY [28.8]). In biotechnological processes we mainly encounter cake filtration for the clarification of microbial cultures. Deep-bed filtration is a specialty for sterile filtration of labile substances in the cold or is used at the end of a biotechnological process to reduce or eliminate microbial contamination of pure products in order to improve stability and shelf life. In such cases, there will be only low concentrations of microorganisms. Ultrafiltration techniques are applied to solutions in order to concentrate and separate biopolymers from small molecules. This process will be considered in Sect. 28.5.3. A recent development is the application of crossflow filtration in biotechnology (KRONER et al. [28.9]). This technique prevents the formation of a cake on top of a filter medium by suitable hydrodynamic measures. The importance of this approach is evident, considering the following equations describing the filtration.

$$Q = \frac{k \cdot p}{L}, \quad (28.7)$$

with Q as the flux, k the proportionality factor, p pressure, and L height of the filter bed.

As early as 1856 DARCY recognized that the flux is proportional to the pressure applied and inversely proportional to the height of the filter bed. A closer inspection of the process leads to the equation:

$$\frac{dV}{dt} \cdot \frac{1}{F} = \frac{k \cdot P}{\eta \cdot L}, \quad (28.8)$$

where V is the volume, t the time, dV/dt the flux rate, and F the filter area.

The filter cake thickness, however, is not constant during a dead-end filtration. Therefore, the ratio of k/L, describing the resistance of the filter cake, has to be divided into two terms, one relating to the filter medium and the other to the filter cake. As a first approximation, it can be assumed that the resistance of the filter medium will be constant during operation, while the resistance of the filter cake depends on the nature of the particles to be filtered. The solid content of the feed and the accumulated volumes during operation decrease with increasing filter area. The flux can therefore be described by:

$$\frac{dV}{dt} \cdot \frac{1}{F} = \frac{P}{\eta \left(\dfrac{R_c \cdot w \cdot V_a}{F} + R_M \right)}, \quad (28.9)$$

where R_M is the resistance of the filter medium, R_c the specific resistance of the cake, V_a the accumulated volume filtered, and w the solid content per unit volume of suspension.

For filtrations at constant pressure, as realized in the vacuum drum filter, Eq. (28.9) can be transformed into:

$$\frac{dt}{dV} = \frac{\eta \cdot R_c \cdot w}{F^2 \cdot p} \cdot V_a + \frac{\eta \cdot R_M}{F \cdot p}, \quad (28.10)$$

or

$$\frac{dt}{dV} = x V_a + y. \quad (28.11)$$

Eq. (28.10) can be used in simple experiments with bench filters to determine the values of R_c and R_M by plotting the reciprocal of the flux at time t versus V_a and calculating the unknown from the slope and intercept, respectively, according to Eqs. (28.11) and (28.10).

There is a relation between the resistance of the filter cake and the size and morphology of the microorganisms. Mycelia, in general, can be filtered quite easily, the resulting filter deposit has a sufficiently large porosity to pose little resistance to solvent flux. Yeast and bacteria are much more difficult to filter effectively because of their small size. According to HAGEN-POISEUILLE, the flow rate is proportional to the fourth power of the diameter of a capillary. The smaller the particles to be separated, the smaller are the interstitial spaces, which act similarly to a capillary. In addition, the resulting cake shows poor mechanical stability and is easily compressed by moderate pressures, which leads to a further reduction of the porosity of the cake and eventually blocks the filter completely. A conventional strategy to avoid or circumvent the problem of high resistance of microbial filter cakes is to add filter aids such as diatomeous earth, e.g., celite. A filter aid mixed with a microbial suspension improves the mechanical stability of the developing cake and allows larger accumulative volumes to be filtered. As an alternative, vacuum drum filters can be precoated with similar materials and the microbial deposit on the surface can be kept very small by removing the outermost layer during each revolution by a blade or string. Addition of filter aids can only be recommended for extracellular products; in the case of intracellular compounds, the microorganism would become contaminated by a comparatively high concentration of filter aid, which would lead to severe problems in later processing.

In crossflow filtration the buildup of a filter cake is prevented either by high tangential flow along the membrane or by mechanical means in so-called dynamic filters. If the cake buildup on the filter can be prevented, all problems related to the high resistance of filter cakes formed from bacteria and yeasts can be avoided over a considerably large concentration range. The success of such operations then depends more on the rheological properties of the microbial suspension with increasing concentration. Rather low flux rates show that interactions between the microporous membranes used in crossflow filtration and components of the suspension lead to a very pronounced decrease in flux rate, even in the absence of a visible filter cake (KRONER et al. [28.10]). Improvements of the membranes employed for microfiltration may eventually overcome this problem.

Also, in the early days of ultrafiltration fouling of the membrane was a major problem; this has been largely solved by recent developments. For example, the use of crossflow filtration in an industrial process to separate an antibiotic from microbial cells by using an ultrafiltration membrane has been reported (GRAVATT and MOLNAR [28.11]). The cells are concentrated from the broth only threefold, and the residual product is then washed out through the membrane. Surprisingly high retentions of proteins are encountered even on microporous membranes when filtering microbial cultures and cell homogenates, thus presently limiting the application of crossflow techniques in the recovery of enzymes.

28.2.3.1 Filtration Equipment

The three most commonly used types of filters are shown in Fig. 28.7: the vacuum drum filter, the filter press, and the horizontal leaf filter. The vacuum filter is often operated by first depositing a filter aid as a precoat on the filter medium covering the drum. The suspension is fed into the trough and may or may not be mixed with additional filter aid. The thickness of the developing filter cake can be influenced by changing the liquid level in the trough or the rotational speed of the drum. As soon as the cake emerges from the suspension dewatering begins. The filtrate is collected in channels inside the drum and discharged. If necessary, a washing step can be added by spraying water or a suitable solution on the cake near the top. The filter cake is finally removed by a blade, which exposes a new surface of the precoat for the next revolution. Other designs for removing the filter cake are available if the vacuum drum filter is operated without precoat. The maximal pressure differential in the vacuum drum filter is one bar. Vacuum drum filters are commercially available up to 100 m² filter area. The end volume of the suspension can be kept very small, which is important when the culture broth contains the compound of interest. Depending on the diameter of the drum and the desired

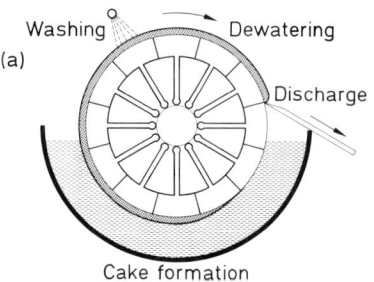

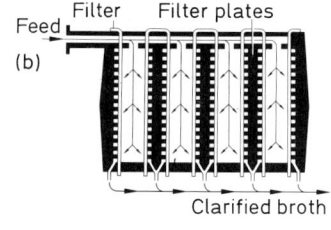

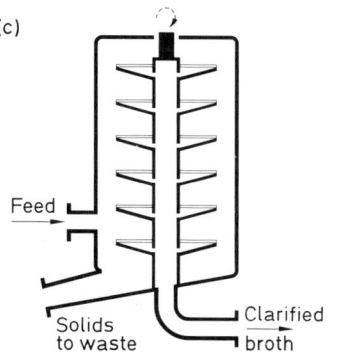

Figure 28.7. Schematic diagram of different filters. (a) Rotory drum filter, (b) filter press, (c) leaf filter.

residual water content of the cake, the drum is operated between 0.1 and 3 rpm. Vacuum drum filters are extensively used in the production of extracellular enzymes.

Leaf filters may be operated as vacuum or pressure filters. The general construction allows one or both sides of the leaves to be covered by filter medium. The suspension is delivered from the outside and the filtrate recovered through the internal pipe system of the leaves. Single leaves are combined into larger units. Fig. 28.7 shows a vertical arrangement of filter discs, exposing the filter media only on the upper side. The hollow shaft collects the filtrate from the leaves into a central line. During filtration the leaf arrangement is static, but it may be rotated at the end of filtration for the removal of solids. In this manner, the operation of such filters can be automated. Closed operation is also possible. Addition of filter aid to the suspension is generally required to insure sufficiently high flow rates. Precoating of the filter leaves with filter aid is also possible, but is only recommended for polishing of solutions, since the surface cannot be renewed as with a vacuum drum filter.

Plate and frame filters are usually operated as pressure filters. During start-up of the operation, the resistance of the filter medium is relatively small so that the pressure differential is not large. During this initial phase, a constant volume of filtrate is obtained. However, the pressure usually develops rapidly with increasing height of the cake until the operating pressure of the pump is reached. Then, filtration will be conducted with constant pressure resulting in a decreasing filtration rate. Filter presses are operated with pressures up to 20 bars and may therefore have relatively higher flow rates. These are used for suspensions which are difficult to filter and require less filter aid. The main disadvantage is a discontinuous operation and relatively long off-times for removal of the filter cake, cleaning, and reassembly. For all filtration processes, extensive trials are required during process development and scale-up (BENDER and REDEKER [28.12]).

28.3 Cell Disintegration

28.3.1 General Aspects

The interior of a cell is separated from the environment by a complex wall and membrane structure. Fig. 28.8 gives a schematic view of the most commonly encountered types of Gram positive and Gram negative bacteria, yeasts, and fungi. The polysaccharide and peptidoglycan structures are responsible for the mechanical strength of the envelope, while membranes composed of lipids and proteins act as the permeation barrier. In order to isolate intracellular proteins, the cell wall has to be disintegrated. Biological, chemical, or physical procedures are summarized in Table 28.4. Mechanical methods are mainly used for large-scale applications. Two methods dominate: high pressure homogenization and wet milling.

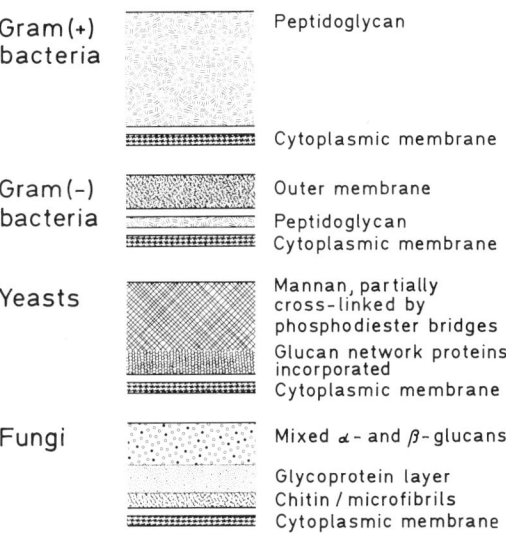

Figure 28.8. Schematic and simplified illustration of cell envelopes of different classes of microorganisms used in biotechnological processes.

Table 28.4. Methods of Cell Disintegration

1. Non-mechanical Methods	a) Chemical treatment (acids, bases, solvents, or detergents)
	b) Physical treatment (freeze-thawing, osmotic shock)
	c) Enzymatic digestion (lytic enzymes, phages)
2. Mechanical Methods	a) Wet milling
	b) High-pressure homogenization
	c) Pressure extrusion
	d) Sonification

28.3.2 High-pressure Homogenization

The flow path of a microbial suspension through a homogenizer is shown in Fig. 28.9. The suspension is transported by a positive displacement pump and discharged through an adjustable, spring-loaded, restricted orifice valve. The operating pressure is controlled by altering the pressure of the spring on the valve piston by means of a large handwheel. Differently designed valve units are available. The disruption of yeast in a high pressure homogenizer may be described as a first order rate process according to HETHERINGTON et al. [28.13]:

$$\log \frac{R_{max}}{R_{max} - R} = k \, N p^a , \qquad (28.12)$$

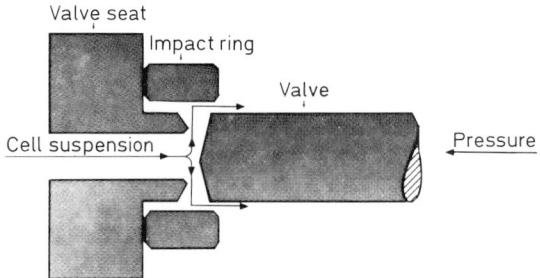

Figure 28.9. Cell disintegration in a high-pressure homogenizer. Detail of the functional parts in a Manton-Gaulin high-pressure homogenizer (Gaulin Corporation, Everett, Mass.).

where R_{max} is the maximally released amount of protein per unit weight and R the amount of protein released at time t, k is a first order rate constant, N the number of passages, p the pressure applied, and a an exponent.

Protein release and enzyme solubilization are influenced mainly by the operating pressure and the design of the valve unit and are independent of yeast concentration up to 600 g of compressed yeast per liter of suspension. The efficiency of disintegration depends on the number of passages and on the pressure. For commercial baker's yeast, the exponent a in Eq. (28.12) has been determined as 2.9; however, the exponent depends on the organism itself, as well as on medium composition, and on growth conditions (ENGLER and ROBINSON [28.14]). The first order rate constant is also a function of the temperature. Due to the extremely short residence time, temperature control within the homogenizer is impossible. The temperature is linearly correlated to the pressure and increases by 2.2–2.4 °C/100 bar, irrespective of the valve design (SCHÜTTE et al. [28.15]). Therefore, a temperature rise of 14–15 °C occurs after a single passage through the homogenizer at an operating pressure of 550 bars. If more than one passage is necessary for complete cell disintegration, the product must be cooled in order to keep the accumulated temperature low enough to ensure high yield. The power consumption of the machine is a linear function of the operating pressure and corresponds to 0.35 kW per 100 bar applied.

The homogenizer is usually operated at a constant through-put. The suspension has to be well strained to avoid malfunctioning of the valve. The valve unit is also sensitive to stalling by air or gas cushions; therefore, suitable precautions have to be taken. The valve unit is the focal point of erosion and has to be well maintained and precision-manufactured to insure a constant high efficiency.

Cell disintegration has been studied in detail only with *Saccharomyces cerevisiae*. An increasing number of microorganisms has been successfully disintegrated in a high pressure homogenizer, including *Escherichia coli*, *Aspergillus* and *Bacillus* species. However, there are some reports that the release of enzymes from certain microorganisms by high pressure homogenization is very small, as e.g., for the release of fumarase from *Brevibacterium* (SCHÜTTE et al. [28.16], [28.17]). Also other small bacteria like *Nocardia* and *Micrococcus* seem to be very resistant to high pressure homogenization.

preferred. Since the residence time in the mill is longer than in the homogenizer and an excess of heat is produced by wet milling, an efficient cooling system through a jacket and at the seal is mandatory.

The agitator normally consists of a central shaft onto which discs or rings of different geometry or pins are mounted, as shown in Fig. 28.10. The agitator accelerates the microbeads and transfers the energy necessary for cell breakage. The separation of the glass beads from the suspension is usually performed by a self-cleaning slit mechanism. The slit has to be sufficiently small to retain the beads completely within the reactor. For the release of cytoplasmic enzymes, beads of <1 mm diameter are required for optimal operation.

Cell disintegration in a high-speed mill is a complex process with a large number of contributing parameters pertaining to the biological system and its history, the opera-

28.3.3 Industrial Bead Mills

Cell disintegration has been studied in different types of bead mills. In recent years, mills with horizontal grinding chambers and variable speed agitators have been

Table 28.5. Process Variables of a Bead Mill

- Size of beads
- Weight of beads
- Packing density of beads
- Stirrer speed
- Design of stirrer blades
- Feed rate of suspension
- Cell concentration in suspension
- Temperature

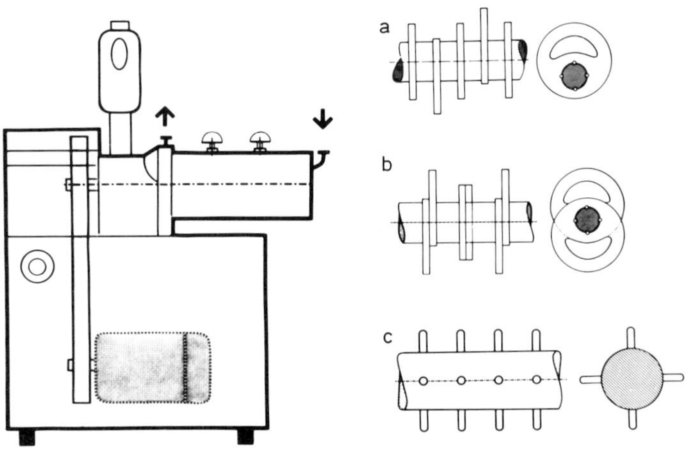

Figure 28.10. High-speed glass bead mill for cell disintegration. Different designs of stirrer blades or pins, respectively, used on the horizontally positioned agitator shaft (Netzsch Mill, Netzsch-Feinmahltechnik, Selb).

tion, and the machine. The most important ones are summarized in Table 28.5. Cell disruption in high-speed mills was found to be a first-order reaction and can be described by:

$$\log \frac{R_{max}}{R_{max} - R} = k \cdot t, \qquad (28.13)$$

according to CURRIE et al. [28.18] and MARFFY and KULA [28.19]. The first-order rate constant k was shown to depend on the agitator speed, on the flow rate of the suspension, the concentration of microorganism, the size of the beads, the load volume, the density of the beads, and on the temperature during disruption. However, the geometry of the grinding chamber and the design of the agitator are important for the flow path of the suspension in the mill and the degree of dispersion and mixing and influence the performance of the mill (SCHÜTTE et al. [28.20], LIMON-LASON et al. [28.21]).

Insufficiently understood parameters are the energy dissipation per unit volume, the influence of the material and surface properties of the stirrer blades, and the effect of the beads participating in the energy transfer on the efficiency of cell disruption. A better knowledge of these factors appears necessary in order to optimize high-speed bead mills for the disruption of microorganisms. Such mills have been originally designed for the disintegration of pigments, and are commercially available with capacities between 0.6 and 200 liters. For the larger models, heat removal and continuous separation of the glass beads can be difficult. Detailed studies are reported for a 22 L mill, indicating that efficient disruption of yeast as well as bacteria is possible with through-puts of 40–70 kg per h, for several strains of yeast with enzyme solubilization >90% and up to 200 kg/h with yields of around 80% (SCHÜTTE et al. [28.16]). High activity yields of bacterial enzymes require repeated passages through the mill and reduce the through-put to 10–20 kg/h (SCHÜTTE et al. [28.17]). The higher price of bacteria requires higher yields (>90%) of disintegration.

28.4 Clarification of Crude Extracts

28.4.1 Mechanical Separation

The separation of insoluble fragments from solubilized proteins after cell disintegration is a very difficult task. For the utilization of certain intracellular enzymes, a removal of cell debris can be avoided by immobilizing the cell homogenate. This has been described for glucose isomerase (HEMMINGSEN [28.22]). However, in most cases further purification and concentration of the relevant product will require the removal of cell debris. In principle, solid-liquid separation techniques such as centrifugation (MOSQUEIRA et al. [28.23]) or filtration (GRAY et al. [28.24]) can be employed. These operations are discussed in Sect. 28.2 together with aspects of the harvest of microorganisms. Mechanical clarification of cell homogenates is much more difficult to perform. This is due to the fact that cell fragments are smaller than intact microorganisms, the range of particle sizes is much wider – extending into the submicron range –, the density difference is small and the viscosity increases during homogenization. Therefore, a compromise often must be sought in terms of operating times, yield, and purity in performing mechanical separations (NAEHER and THUM [28.25]).

During centrifugation the through-put has to be decreased considerably. A longer residence time of the product in the rotor may lead to a pronounced temperature rise, which may adversely affect the quality of biological materials. Therefore, separators with direct bowl cooling are required. Centrifugation is the method of choice for the processing of inclusion bodies formed during expression of mammalian proteins in recombinant *Escherichia coli*.

Removing cell debris by filtration requires the use of filter aids. This is permissible for dissolved product; however, adsorption of proteins may occur on porous materials and has to be investigated during process development. The temperature rise is not significant during filtration, while the longer processing times may be critical, e.g., with regard to proteolytic product degradation. First experiments with crossflow filtration techniques to remove cell debris show that retention becomes a limiting factor. Retentions as high as 90% on microporous membranes have been reported for total protein present in a cell homogenate (KRONER et al. [28.9]). The secondary membranes formed are similar to ultrafiltration membranes having a cut-off around 50000 dalton. Depending on the molecular weight of the desired protein, prolonged diafiltration is required to obtain reasonable yields by this method. An undesired side-effect is an increase of the process volume and simultaneously the protein concentration is lowered which also may have a negative effect on protein stability.

In summary, it can be said that mechanical separation steps for removing cell debris are possible but performance is severely limited by inherent suspension and separation characteristics. One of the new developments of recent years has been the introduction of a thermal separation technique for removing cell debris, based on the partition of proteins, cells, and cell-wall fragments in aqueous phase systems. This will be discussed in detail in the following section.

28.4.2 Extraction of Proteins from Cell Homogenates

Extraction methods are highly advanced in the pharmaceutical and chemical industry. They are used in the downstream processing of antibiotics but have only recently been introduced for recovering biologically active proteins (KULA et al. [28.26]). Commonly employed organic solvents are unsuitable for protein recovery. However, aqueous two-phase systems formed by adding hydrophilic polymer(s) to aqueous salt solutions can be used for this purpose (KULA et al. [28.27], HUSTEDT et al. [28.28]). In such systems a high water content is maintained in both phases despite the fact that the liquids are no longer miscible. Such aqueous two-phase systems provide favorable conditions for protecting the biological activities of cells, organelles, and proteins (ALBERTSON [28.29]). By selection of proper conditions, cell fragments can be confined to one phase and the protein of interest partitioned preferentially in the other phase.

The partition coefficient K can be defined as the ratio of concentrations of the compound in the top and bottom phase:

$$K = \frac{C_{iu}}{C_{il}}. \qquad (28.14)$$

The partition coefficient is a constant over a wide range of concentrations – also for proteins – as long as the molecular properties in the two phases are not changed. Surface properties are the guiding forces for partition as indicated by Eq. (28.15) (GERSON [28.30]):

$$-\log K = \alpha \Delta \gamma + \delta \Delta \Phi + \beta, \qquad (28.15)$$

where K is the partition coefficient, α the surface area, $\Delta \gamma$ the difference of surface free energy in the top and bottom phase, δ the surface charge, $\Delta \Phi$ the electrostatic potential difference between the phases, and β a value comprising the standard chemical potential and activity coefficients.

The partition coefficient is exponentially related to the surface free energy γ of a particle or macromolecule and the electrostatic potential Φ between the phases, or a linear combination of both. The surface area increases with increasing molecular weight. Therefore, it is expected that large "molecules" such as cells and cell fragments show one-sided partition; small differences of the surface free energy in the two phases should then lead to large changes of the partition coefficient.

For technical reasons, cell debris and residual cells are most favorably collected in the lower phase in order to facilitate later

Table 28.6. Parameters Affecting Partition in Aqueous Two-phase Systems

- Choice of the hydrophilic polymer(s)
- Average molecular weight of the polymer(s) selected and the molecular-weight distribution
- pH in the system selected
- Kind of salt selected to induce phase formation in one-polymer systems
- Concentration of salting-in ions in relation to the phase forming salt in one-polymer systems
- Phase potential in two-polymer systems
- Length of the tie-line (a complex function of the concentration of phase forming components)
- Concentration of disintegrated cells

separation. This can be achieved as discussed below. The separation efficiency is also influenced by the volume ratio V_u/V_l. Successful extractions of enzymes require partition coefficients of ≥ 3 in order to achieve high yields in one step. Table 28.6 summarizes important parameters which influence partition coefficients in aqueous two-phase systems. The interdependence of many of these parameters, however, makes it presently impossible to predict a partition coefficient from molecular properties of a given protein. Suitable systems must be experimentally designed.

Obviously, the choice of the polymer and its average molecular weight mainly influences the surface free energy. This enables subtle changes of the hydrophobic character of a solution. The choice of ions in two polymer systems mainly affects the potential, while the pH affects the potential and the surface charge. Alterations of the surface charge, in turn, affect the surface free energy. In systems with a higher salt content, especially in polyethylene glycol-salt systems, protein solubilization seems important, as shown by the unique properties of systems containing a mixture of chloride and phosphate. For economic reasons, high concentrations of cell homogenates are desirable for initial extractions; this lowers the processing volume and the expenses for polymers. However, for large concentrations of cell homogenate in the extraction systems, the concentration of biopolymers will reach levels at which they are no longer negligible with regard to the hydrophilic polymers employed to induce phase formation. Perturbations of the carrier system are first noticeable by a changing volume ratio and subsequently by a changing partition coefficient. At still higher concentrations, multiple liquid and solid phases may be formed. The useful working range is normally limited by the yield of a single-step extraction. Nevertheless, 20-40% of a cell homogenate (based on the packed volume of intact cells) can be extracted. Partition coefficients in such complex systems are highly reproducible. The mode of cell disintegration does not seem to influence the results. One can even take advantage of the contribution of cell constituents on phase formation. Several extraction procedures have been reported to operate below the binodal line of the carrier system. Large-scale extractions are usually performed at room temperature without temperature control in the separating equipment. The partition coefficients are not very sensitive to temperature provided the systems are sufficiently remote from the binodal to ensure phase formation. A temperature increase of 1-2°C usually does not affect the yield and purity of the desired product. Therefore, temperature control in the separators is unnecessary. The polymers support the stability of the proteins and enable operation at ambient temperatures. This also improves the separation of the phase systems by lowering the viscosity. Operation at room temperature and without cooling contributes significantly to the energy balance of the process and the ease of operation.

For the industrial application of the proposed method, it is important to ensure a suitable partition coefficient, but equally significant to consider the time to reach equilibrium in phase formation and partition, and the mechanical separation of phases. Partition is diffusion-dependent and, therefore, one initially would expect rather long periods until equilibrium is

Table 28.7. Technical Data and Performance of Open and Closed Separators for Liquid-liquid Separation of Different Phase Systems in the Absence of Solids (data from HUSTEDT et al. [28.28])

Separator	Σ-Value (m^2)	Speed (rpm)	Phase System	Q_{max} (L/h)	τ (s)	Se[c]
Gyrotester B[a]	700	12 000	10% PEG 4000 2% Dextran T500	12	123	>0.99
LAPX-202[a]	970	9 300	13% PEG 4000 11% KPi	90	13	>0.99
SAOH-205[b]	1 460	9 700	10% PEG 4000 12.5% KPi	420	2	~0.99
SA-7[b]	7 000	8 400	10% PEG 4000 12.5% KPi	2 200	5	~0.99

[a] open separators, α-Laval-Industrietechnik GmbH, D-2000 Hamburg 80
[b] closed separators, Westfalia Separator AG, D-4740 Oelde
[c] Se separation efficiency [28.32]

reached. However, the low interfacial tension characteristic of aqueous two-phase systems leads to the formation of very small droplets in the dispersion, even with a low energy input during mixing.

The approach to equilibrium in partition seems to be governed by the short distance a single particle or molecule has to travel. Combined with a large surface area for exchange and fast coalescence and redispersion, equilibrium is achieved rapidly under optimal mixing conditions, even if the bottom-phase in the initial extraction step is very viscous (up to 3000 cp). Equilibrium is reached in less than one minute, which makes it very difficult to resolve the time course and establish mass transfer rates (FAUQUEX et al. [28.31]).

The low interfacial tension also minimizes denaturation of unstable proteins at the interface and contributes to the observed high activity yields. The low interfacial tension also has to be considered with regard to mechanical phase separation. There, redispersion may occur in flow systems, especially during acceleration of the feed in a centrifugal separator and its flow through narrow channels in a disc stack. Such problems, however, do not represent major difficulties and can be controlled by a suitable feed rate and careful adjustment in the radial position of the underflow.

In general, commercially available liquid/liquid separators have been successfully applied in the separation of aqueous phase systems and the extraction of enzymes from cell homogenates. Table 28.7 summarizes performance data for differently designed equipment. It should be noted that nozzle separators are employed for this purpose when the viscosity of the bottom phase leads to a higher flow resistance in the rotor (KRONER et al. [28.32], [28.33]). The high feed rates and the corresponding small residence time limit the temperature rise of the product-containing phase and permit the use of small equipment for industrial purposes.

Appropriate operating conditions can ensure a high phase purity of over- and underflow. Mechanical losses of product can thus be kept generally below 2%. This is much lower than the losses in the interstitial spaces of a soft pellet in case of a solid-liquid separation of a comparable sample. The importance of high yields in the initial steps of downstream processing increases with the price of the feedstock.

Table 28.8 presents data for the extraction of different enzymes from a variety of microorganisms; on the average there is >90% yield for a single extraction step. A recent economic analysis (KRONER et al. [28.34]) compared extraction to other methods of separation for clarifying crude extracts, demonstrating that extraction shows a high spacetime yield at a comparatively low cost. The price for chemicals needed to

Table 28.8. Extraction of Enzymes from Microbial Cells[a] (data from HUSTEDT et al. [28,28])

Enzyme	Organism	Kind of Phase System	Biomass Concentration (%)	Yield (%)	Purification Factor
Isoleucyl-tRNA synthetase	Escherichia coli	PEG/salt	20	93	2.3
Fumarase		PEG/salt	25	93	3.4
Aspartase		PEG/salt	25	96	6.6
Penicillin acylase		PEG/salt	20	90	8.2
β-Galactosidase		PEG/salt	12	87	9.3
α-Glucosidase	Saccharomyces cerevisiae	PEG/salt	30	95	3.2
Glucose-6-phosphate dehydrogenase		PEG/salt	30	91	1.8
Alcohol dehydrogenase		PEG/salt	30	96	2.5
Hexokinase		PEG/salt	30	92	1.6
Glucose isomerase	Streptomyces sp.	PEG/salt	20	86	2.5
Pullulanase	Klebsiella pneumoniae	PEG/dextran	25	91	2
Phosphorylase		PEG/dextran	16	85	1
Leucine dehydrogenase	Bacillus sphaericus	PEG/crude dextran	20	98	2.4
D-Lactate dehydrogenase	Lactobacillus confusus	PEG/salt	20	95	1.5
L-2-Hydroxyisocaproate dehydrogenase	Lactobacillus confusus	PEG/salt	20	94	16
D-2-Hydroxyisocaproate dehydrogenase	Lactobacillus casei	PEG/salt	20	95	4.9
Leucine dehydrogenase	Bacillus cereus	PEG/salt	20	98	1.3
Fumarase	Brevibacterium ammoniagenes	PEG/salt	20	83	7.5
Glucose-6-phosphate dehydrogenase	Leuconostoc sp.	PEG/salt	35	94	1.3
Formate dehydrogenase	Candida boidinii	PEG/salt	33	90	2.0
Formate dehydrogenase		PEG/crude dextran	20	91	n.d.
Formaldehyde dehydrogenase		PEG/dextran	20	94	n.d.
Isopropanol dehydrogenase		PEG/salt	20	98	2.6

[a] Before liquid extraction the cells were disrupted by high-pressure homogenization or wet milling – with the exception of the extraction of pullulanase

n.d. not determined

establish a phase system is more than compensated for by savings in labor, energy, and equipment. In addition, the quality of the product extracted from a cell homogenate is much higher than a clarified extract obtained by mechanical separation. The major difference is the better purification by extracting preferentially the desired protein, reflected by the purification factors in Table 28.8. Even a moderate purification factor of 2 or 3 indicates that respectively 50 or 75% of the contaminating proteins have been removed from the desired product together with the cell debris. Since surface free energy, charge, and molecular weight are individual properties of proteins, selective removal of interfering activities is possible by designing the process accordingly. Scale-up of an extraction process can be predicted from laboratory data with high accuracy. Detailed investigations at the 50 kg level have been reported (KRONER et al. [28.33]) and the scale-up appears to be straightforward and simple.

28.5 Product Enrichment

28.5.1 General Aspects

A volume reduction and concentration of the product becomes necessary, e.g., for extracellular enzymes obtained in the culture broth, but also for enzymes in crude microbial extracts before final high-resolution methods are applied. At the same time undesirable by-products like nucleic acids, pigments, and residual components from the broth should be eliminated. For this purpose four different methods are applicable: precipitation (BELL et al. [28.7]), ultrafiltration (FLASCHEL et al. [28.35]), batch adsorption, and partition (KULA et al. [28.27], HUSTEDT et al. [28.36]).

28.5.2 Precipitation

For soluble proteins the majority of polar, hydrophilic groups in the side-chains of amino acids are found on the surface and exposed to the solvent, while the hydrophobic residues are buried inside the macromolecule. Charged and dipolar groups form a layer of counter-ions closely associated with the surface of the protein. In this boundary there are changing levels of tightly and loosely bound water forming a hydration shell, which acts as a stabilizing barrier and prevents aggregation under normal conditions. Protein precipitation will result by altering the hydration shell in adding salt in high concentrations, or hydrophilic polymers, by changing the surface charge via pH, by increasing the electrostatic interaction between proteins through lowering of the dielectric constant, or by bridging protein molecules through flocculating agents or multivalent ions. The resulting precipitate is not a crystal but a protein aggregate of mostly undetermined structure.

Precipitation depends on the concentration of the desired protein but also on the total concentration of proteins. Especially, fractionating by precipitation requires that environmental conditions, e.g., pH, temperature, ionic strength, and composition of the feed be maintained as closely as possible in order to obtain reproducible results. Since solid/liquid separation must follow precipitation, the average size of the formed particle and the size distribution is important. The yield in separation is also influenced by the density of the resulting aggregate and its resistance to disintegration due to shear forces in agitated systems or centrifuges. These factors are affected by the type of precipitating agent and by the contacting conditions and mixing behavior of the reactor. Suitable aging conditions positively influence the result. The complex interaction between these factors has recently been studied in detail, mainly with soybean proteins. These investigations have been reviewed by BELL et al. [28.7]. The albumin precipitation process developed originally by COHN et al. [28.37] about 40 years ago demonstrates that it is possible to effectively fractionate protein mixtures by precipitation. However, purification factors obtained by differential precipitation are only moderate and normally do not exceed 2 or 3, since they rely on general physicochemical properties of proteins.

Recent developments aim at a more specific precipitation of desired products utilizing affinity interactions. Two concepts have been evaluated in the laboratory: the use of bifunctional reagents carrying two head groups providing biospecific interaction. A three-dimensional network may result under proper conditions leading to precipitation of the desired product (LARSSON and MOSBACH [28.38]). In addition to the effort to synthesize the reagent, care has to be taken during operation, which requires stoichiometric reactions for the formation of the network. Another approach utilizes the properties of a special polymeric carrier covalently attached to an affinity ligand, which may or may not be soluble in aqueous solutions, depending on the pH and the charge of carboxyl groups in the polymer (SCHNEIDER et al. [28.39]).

Generally, precipitation reactions carried out in the past only to reduce volume have been replaced by ultrafiltration processes. However, precipitation processes are still important in cases where fractional separation can be achieved or where the high viscosity of the solution makes ultrafiltration troublesome. The different precipitating agents are discussed below.

28.5.2.1 Salting out

Proteins can be precipitated by adding high concentrations of inorganic salts, e.g., ammonium sulfate, sodium sulfate, or magnesium sulfate. The relative effectiveness of neutral salts is described in the so-called Hofmeister's series of lyotropic ions. Sulfates are preferred to phosphates because of their better solubility. Especially ammonium sulfate is highly soluble at low temperatures and is widely used in the biochemical laboratory. Its large-scale application, however, is problematic since it is corrosive and releases ammonia at higher pH values. COHN developed Eq. (28.16) to describe the precipitation of proteins in quantitative terms:

$$\log S = \beta' - kI, \qquad (28.16)$$

with S as the solubility, I the ionic strength, and β' and k as characteristic constants for each protein. β' is strongly dependent on the pH and temperature, while k depends on the salt employed. A precipitate is formed in a protein solution if the ionic strength exceeds a limiting value:

$$I = \frac{\beta' - \log S}{k}. \qquad (28.17)$$

The onset of precipitation of a given protein is particularly important for fractional precipitations. The equilibrium between solid and liquid phases is only slowly established and results depend critically on the mode of operation, e.g., whether a crystalline salt or a concentrated solution is used for precipitation (FOSTER et al. [28.40]).

28.5.2.2 Isoelectric Precipitation

For the recovery of biologically active proteins, pH-dependent precipitations at low, constant ionic strength are rarely attempted since the biological activity is often lost at low pH values. In principle, Eq. (28.16) can be used to describe the process, since β is a function of pH and normally reaches a minimum value at the isoelectric point. pH treatment may be used to remove contaminating proteins while keeping the desired product in solution. For this purpose, mineral acids with anions positioned high in Hofmeister's series should be used, as these tend to stabilize solubilized proteins.

28.5.2.3 Precipitation by Organic Solvents

Water-miscible organic solvents such as ethanol, methanol, isopropylalcohol, or acetone can be used for the precipitation of proteins. Such solvents lower the dielectric constant of the medium and therefore increase intermolecular electrostatic interactions. These are also modified by the pH, by temperature, ionic strength, and protein concentration, which have to be controlled in order to achieve reproducible results. Since organic solvents tend to inactivate biologically active proteins, operations generally have to be performed at a low temperature, e.g., fractionating of human plasma is carried out at $-10\,°C$; in this case the temperature has to be maintained within $\pm 0.5\,°C$.

Precautions against fire hazard and expenses for explosion-proof equipment have to be considered for large-scale operation. The advantage of using organic solvents as precipitating agents is their availability in high quality and the ease of recovery and recycling.

28.5.2.4 Precipitation by Non-ionic Polymers

To overcome some of the disadvantages of organic solvents, such as the danger of protein denaturation and the fire hazard, hydrophilic polymers have been studied as protein precipitating agents, in particular polyethylene glycol (FOSTER et al. [28.41], HÖNIG and KULA [28.42]). It is assumed that the polymers exclude proteins from part of the solution thereby increasing local concentrations and promoting protein-protein interactions. In addition, the hydrophilic polymers and the proteins compete for the water available for solvation. Precipitation by polyethylene glycol can be described by the equation:

$$\ln S + fS = X - qC, \qquad (28.18)$$

with S as the protein solubility, C the polyethylene glycol concentration, and f the protein-protein interaction coefficient; X is related to the standard chemical potential μ^0 by $X = (\mu_i - \mu_i^0)/RT$, q is the protein-polymer interaction coefficient.

The solubility of proteins in the presence of hydrophilic polymers is influenced by temperature, ionic strength, pH value, and protein concentration. Other parameters are the molecular weight of the precipitating agents as well as the molecular weight of the desired protein (HÖNIG and KULA [28.42]). The protein-protein interacting term already becomes significant at protein concentrations of about 10 mg/mL and limits the efficiency of any protein fractionation (FOSTER et al. [28.41]). Non-ionic polymers tend to stabilize proteins. Precipitation and separation, therefore, can be performed at room temperature, which is also desirable in view of the rising viscosity of the solutions. It should be noted that at higher ionic strengths two liquid phases may result when polyethylene glycol is added. Partition and precipitation can occur simultaneously in such a complex mixture. The operating mechanism for the desired compound can be analyzed by following the concentration dependence, which is distinctly different for the two possible reactions.

28.5.2.5 Precipitation by Polyelectrolytes

Certain polyelectrolytes are used as flocculating agents. These also act on soluble proteins by forming networks of polyelectrolytes and proteins. The degree to which this occurs depends on the number of counter-ions available on the protein surface and, therefore, shows a minimum near the isoelectric point. In addition, polyelectrolytes compete for water in the solution and may exhibit molecular exclusion as discussed above. Natural acidic polysaccharides, as e.g., alginate, pectate, and carrageenan, but also carboxymethyl cellulose, and polyacrylic and polymetacrylic acids have been used to precipitate proteins. Anionic polyelectrolytes must be used at pH values below the isoelectric point of proteins. Thus, they are mainly applied under acidic conditions, which limits their application for the purpose of purifying enzymes. In contrast, cationic polyelectrolytes such as polyethylene imine are used above the isoelectric point, i.e., in a range which in general is more compatible with enzyme stability. Polyethylene imine with a molecular weight of 40000–60000 is widely used as a protein precipitant (BERGMEYER et al. [28.43]).

The precipitation reaction by polyelectrolytes is also strongly influenced by the type and concentration of counter-ions. The complexes formed between protein and polyelectrolytes can be dissociated at high salt concentrations, which permits recovery of the product.

28.5.3 Ultrafiltration

Ultrafiltration had been introduced into the laboratory by 1965 and has ever since expanded to large-scale applications in biotechnology (FLASCHEL et al. [28.35]). This is especially the case in the production of extracellular enzymes from dilute culture broth and in numerous processes where reducing the volume of a protein solution is necessary for operation, transport, or storage. Similarly important is the removal of low-molecular-weight components from a protein solution by a process called diafiltration. Here, water or a buffer of defined composition is fed through an ultrafiltration device for solvent exchange; the protein remains in solution. The development of this procedure became possible by the introduction of hydrophilic, anisotropic membranes, which are composed of an ultra-thin skin mounted on a fibrous support. Such membranes show sufficient hydraulic permeability for pressure-driven operations. Solvent permeability through a pore or capillary can be described, according to HAGEN-POISEUILLE, by the equation:

$$J = \varepsilon_m \frac{d_p^2}{32 \eta l_p} \cdot \Delta P, \qquad (28.19)$$

where J denotes the flux, ε_m the membrane porosity, d_p the pore diameter, l_p the pore length, η the dynamic viscosity, and ΔP the trans-membrane pressure difference. For a given membrane, the hydraulic resistance W_m can be defined:

$$J = \frac{1}{W_m} \Delta P. \qquad (28.20)$$

The hydraulic resistance is a characteristic value for a given membrane and is defined for pure solvent flux only.

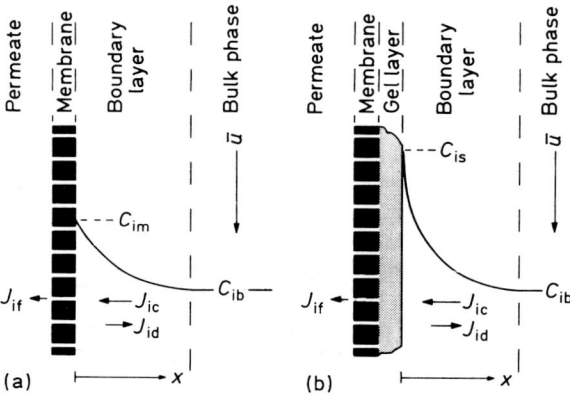

Figure 28.11. Schematic presentation of concentration polarization without gel formation (a) and with gel formation (b) (reprinted from FLASCHEL et al. [28.35]).

For real solutions, when solutes are retained by the membrane, the hydraulic resistance will rise due to adsorption. At higher solute concentrations the hydraulic resistance rises further by the so-called concentration polarization. This phenomenon is illustrated in Fig. 28.11. Any retained solute accumulates on the surface of the membrane, while the solvent passes through. Therefore, a higher solute concentration results in front of the membrane from where it diffuses back into the bulk solution. If the concentration and/or the flux is high enough, saturating concentrations of the solute will be reached, giving rise to a secondary membrane as a gel layer of solutes, which exhibits a higher hydraulic resistance described by the equation:

$$J = \frac{l}{W_m + W_g} \Delta P, \qquad (28.21)$$

where W_g is the hydraulic resistance of the gel layer.

In contrast to the hydraulic resistance of the membrane itself, the resistance of the secondary membrane increases with flux. At steady state the following equation holds:

$$J_{ic} - J_{if} - J_{id} = 0, \qquad (28.22)$$

where J_{ic} is the molar flux of solute i by convection, J_{if} the molar flux of solute in the filtrate, and J_{id} the molar flux of solute by back-diffusion.

The concentration profile of solute in the boundary layer can be integrated and solved for the flux:

$$J = k_d \ln \frac{C_{im} - C_{if}}{C_{ib} - C_{if}}. \qquad (28.23)$$

where k_d is a mass transfer coefficient, C_{im} the solute concentration at the membrane, C_{ib} the solute concentration in the bulk phase, and C_{if} the solute concentration in the filtrate. In case of quantitative retention, the concentration in the filtrate approaches zero and Eq. (28.23) can be simplified to:

$$J = k_d \ln \frac{C_{im}}{C_{ib}}. \qquad (28.24)$$

Plotting the flux versus the natural logarithm of the solute concentration in the bulk phase yields the mass transfer coefficient. It is essential to measure it at constant trans-membrane pressure and under identical hydrodynamic conditions.

As a consequence of gel polarization, the flux is determined by the hydraulic resistance of the gel layer, which in turn depends on the nature of the proteins or other compounds forming the gel layer and the flow rate along the membrane. The model predicts that flux becomes independent of

Table 28.9. Main Parameters Which Influence the Performance of Ultrafiltration (FLASCHEL et al. [28.35])

	Membrane	Module	System	Operation
Basic parameters	d_p σ_p^2	d_t l_t	M_i σ_i^2	Δp J
	cut-off	F_m	D_i v	$\bar{u}$
	l_p ε_m		C_i (saturation) C_{ib}	T
Derived quantities	W_m apparent, intrinsic retention	F_m/V	$Sc = v/D_i$	$Re = \bar{u} d_t / v$ Sh $k_d = Sh\, D_i / d_t$

pressure since any increase in the transmembrane pressure will immediately increase the gel layer thickness, which re-establishes the former flux by increasing the hydraulic resistance. All measures which enhance the back-transport of the solute into the bulk solution will influence the mass-transfer coefficient and improve the flux.

Ultrafiltration, therefore, presents itself as a complex process which depends on properties of the membrane, the geometry of the module, molecular properties of the solutes involved, and operational conditions. The main parameters are summarized in Table 28.9. The geometry of the module and operating variables are the main means for improving reverse transport of retained species. Since the diffusion coefficient of the solute as well as the viscosity of the medium depend on temperature, the flux will increase at higher temperatures. Operational temperature, however, is limited by the thermal stability of the product, but may also be governed by requirements for control of microbial contamination, which is reduced at lower temperatures. On an industrial scale, the concentration of extracellular enzymes is reportedly carried out at 10°C. An increase in the capacity of 1–3% can be expected for 1°C temperature increase.

The mathematical description and engineering aspects of ultrafiltration processes are well established and have led to improved equipment and process design for the concentration and diafiltration of proteins. The final concentration depends on the hydrodynamics of the system and the maximal applicable pressure. In the case of human albumin, final concentrations of 40% have been achieved in thin channel and cassette systems, and 25% in hollow fiber cartridges. If ultrafiltration is carried out prior to a drying step, it may be uneconomic to attempt to reach such high protein concentrations in solution, even when the solubility of the product would allow such treatment. The simultaneous removal of low-molecular-weight constituents of the culture broth, e.g., sugars, amino acids, and peptides from extracellular enzymes during ultrafiltration improves the product considerably, since these components cause discoloration and odors during drying and lead to hygroscopic, gumming, and caking enzyme powders. Table 28.10 illustrates the scale of industrial operation and the results achieved.

While low-molecular-weight components have been successfully separated by ultrafiltration, it has not been possible to fractionate proteins according to molecular weight or by their Stoke's-Einstein radius, except at extremely low protein concentrations which are unrealistic for practical purposes. Fractionation is impaired by the polarization layer which cannot be completely avoided. Minor fractionations can be achieved by selecting suitable membranes, but only when the molecular weights are sufficiently different.

During the last decade, ultrafiltration has become established in biotechnology, since reliable, non-fouling, high-flux membranes have become commercially available. The concentrating process is generally a batch operation, which recycles the concentrate at

Table 28.10. Concentration of Enzymes from Culture Broths (data from NEUBECK [28.44])

Enzyme	Source	Initial Mass (tons)	Final Mass (tons)	Final Solids (%)	Final Volume (%)	Yield (%)
α-Amylase	Bacillus subtilis	14	3.2	33	22.9	82
α-Amylase	Bacillus subtilis	18.6	3.9	25	21.0	88
Neutral protease	Aspergillus flavus	1.6	0.2	15	12.5	90
Acid protease	Aspergillus oryzae	59	4.4	13	7.5	90
Neutral protease	Bacillus subtilis	~2.6	0.4	16	15.4	95

a constant membrane pressure, thereby increasing the solute concentration as slowly as possible and maintaining high flux. In very large-scale procedures, ultrafiltration is carried out operating continuously with a cascade design. The design is based on layout data using experimental results combined with theoretical models to account for the polarization of the membranes. The improved, chemically resistant membranes can be periodically cleaned and disinfected. Under normal conditions these membranes may be used for one year or longer.

28.5.4 Extraction

As discussed in Sect. 28.4.2, proteins can be extracted from cell homogenates in aqueous two-phase systems. A substantial purification can already be achieved by the initial extraction (HUSTEDT et al. [28.28]). By further exploiting the same principle of partition in aqueous phase systems, it is possible to purify the desired product. By these means, polyethylene glycol as well as salts employed during the first extraction can be reused.

In general, the first extraction step is designed to remove the desired protein in the polyethylene glycol-rich top phase. Addition of salt to the top phase will generate a secondary phase system. If nucleic acids and polysaccharides must be removed in order to improve subsequent chromatographic separations, appropriate conditions are chosen to retain the desired protein in the polyethylene glycol-rich top phase. Nucleic acids and polysaccharides are more hydrophilic and can be preferentially extracted into the salt-rich phase. The second and/or third extraction step can be designed to efficiently separate interfering activities. This has been demonstrated for the separation of fumarase and aspartase from *Escherichia coli* (KULA et al. [28.27]) and D-lactate dehydrogenase and L-2-hydroxyisocaproate dehydrogenase from *Lactobacillus confusus* (SCHÜTTE et al. [28.20]). More hydrophobic proteins and other contaminants, such as pigments, stay in the polyethylene glycol-rich phase, while the protein of interest is extracted into the salt-rich bottom phase. Phase separation is easily performed under gravity; separating times of 60–90 min are common. Otherwise liquid/liquid separators can be used achieving high separation efficiency with residence times of less than 10 s (Table 28.7). In the final extraction step, the protein should be shifted to the salt-rich phase, thereby recovering a larger proportion of the initial supply of polyethylene glycol. Depending on circumstances this may either be during the second or third extraction step. A general process scheme is presented in Fig. 28.12. Overall purification factors up to 30 have been reported for two to three consecutive single-step extractions with overall recoveries of ~70%. The majority of these enzymes, summarized in Table 28.11, has been successfully tested for application as industrial catalysts.

A detailed economic analysis for the fumarase process has been published (KRONER et al. [28.34]) indicating a relatively low cost of the secondary extraction step, combined with an exceptionally high space-time yield. The relative cost for the chemicals used to build up the aqueous two-phase system amounts to 13%, including expenses for waste water treatment. Especially for high-priced raw materials, the comparatively high yields combined with the ease of large-scale operation make extraction systems very attractive. First attempts to recycle polyethylene glycol have been successful. If generally applicable, this would reduce the relative cost of the chemicals even further.

Furthermore, there are two aspects to be considered for further development of protein partitioning in aqueous two-phase systems and their technical application. First, the specificity of extraction in aqueous phase systems may be improved by affinity partitioning. If, for example, an affinity ligand is covalently bound to polyethylene glycol, it will be largely confined to the polyethylene glycol-rich phase. Biospecific complexing can be used to carry the desired

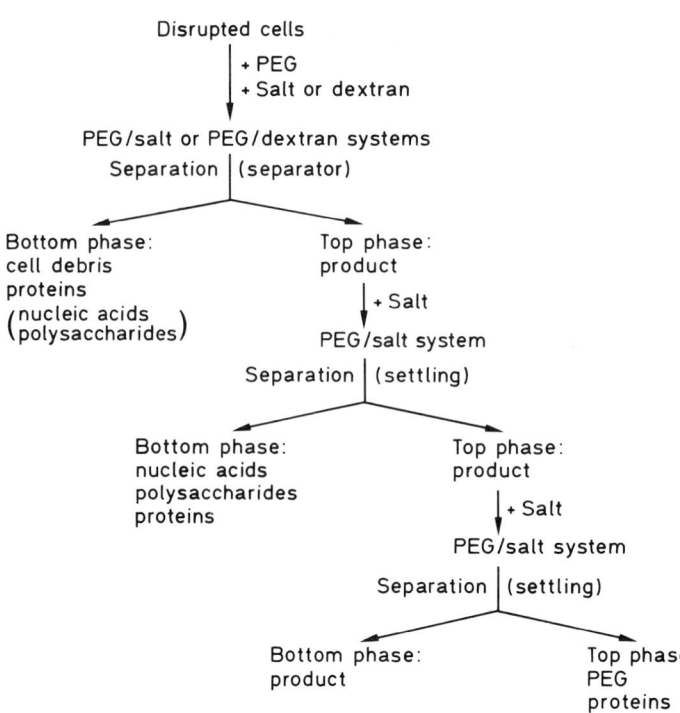

Figure 28.12. General flow diagram of an extractive enzyme purification (three-step process). PEG = Polyethylene glycol.

Table 28.11. Proteins Isolated by Several Subsequent Single-stage Extraction Steps (data from HUSTEDT et al. [28.28])

Protein	Organism	No. of Extraction Steps	Overall Purification Factor	Overall Yield
L-2-Hydroxyisocaproate dehydrogenase	Lactobacillus confusus	2	24	80
D-2-Hydroxyisocaproate dehydrogenase	Lactobacillus casei	2	7	85
D-Lactate dehydrogenase	Lactobacillus confusus	2	1.9	91
Glucose-6-phosphate dehydrogenase	Leuconostoc sp.	2	5	80
Fumarase	Brevibacterium ammoniagenes	2	22	75
Aspartase	Escherichia coli	3	18	82
Penicillin acylase	Escherichia coli	2	10	78
α-1,4-Glucan phosphorylase	Klebsiella pneumoniae	2	2.5	81
Pullulanase	Klebsiella pneumoniae	4	6.3	70
Glucose dehydrogenase	Bacillus sp.	3	33	83
Leucine dehydrogenase	Bacillus cereus	2	2.4	89
Leucine dehydrogenase	Bacillus sphaericus	2	3.1	87
Aspartate β-decarboxylase	Pseudomonas dacunhae	3	6	78
Formate dehydrogenase	Candida boidinii	3	4.2	78

The first extraction step always aimed at removal of cell debris

protein into the polyethylene glycol-rich phase under selective conditions. This has been studied especially by using triazine dyes as general ligands (KRONER et al. [28.45], KOPPERSCHLÄGER and JOHANSSON [28.46]). Affinity partitioning is sensitive to the phase potential and requires two-polymer systems and low ionic strength unless zero potential systems are used. Affinity complexes dissociate by adding salt to a separated top phase, which extracts the protein into the salt phase, generally in high yield. The ligand density per unit volume in a phase system is expected to be higher than on solid support. The complexes are formed in the homogeneous phase. Therefore, mass transfer resistance should be negligible and equilibrium conditions should be obtained rapidly. Quantitative recovery of the soluble modified ligand, however, is more difficult than for a solid affinity resin.

Other efforts are directed to continuous processes to extract and purify intracellular enzymes. Since final equilibrium conditions are obtained surprisingly fast, even for cell debris and macromolecules like proteins, in-line mixing has been tested. This can be successfully accomplished with static mixers under appropriate conditions. In-line mixing has only to be combined with a continuous phase separation in a separator to accomplish continuous processing (HUSTEDT et al. [28.36], [28.47]). Further developments are somewhat hampered by the high capacity of the method, which cannot be scaled down sufficiently. Extraction of 10 kg of disintegrated cells per hour has been reported with a mean residence time in the system for a three-step-extraction process of less than 7 min (KULA [28.48]). Continuous operation for 5 h has been accomplished and had to be terminated not for technical reasons but for lack of starting material. With improved in-line controls, continuous processes appear feasible and advantageous for the industrial production of intracellular enzymes or other proteins.

28.5.5 Batch Adsorption

Batch adsorption is carried out if the properties of the adsorbent, e.g., particle size or swelling behavior or if the properties of the suspension, e.g., viscosity or presence of particulate matter, make operation of a chromatographic column not advisable. In batch adsorption, usually coarse-grade materials of chromatographic resins are used, as described in Sect. 28.6. In addition, some inorganic materials like bentonite are used only in batch procedures for the adsorption of proteases or nucleases. Contact usually occurs in a stirred tank, and the adsorbent is separated by filtration in a basket centrifuge or similar device, allowing repeated washing and complete recovery of the solution containing the product. Resolution is not as high as by chromatography, but may be sufficient for certain applications, especially if contaminants can thus be removed quickly.

28.6 Chromatography

Chromatography is a high resolution technique and therefore preferred if proteins of high purity are required. The technique has been extensively used, improved, and diversified ever since PETERSON and SOBER [28.49] introduced cellulosic ion exchangers for protein purification. However, it has only fairly recently been established as an industrial process in the pharmaceutical industry, e.g., for the production of highly purified insulin (COONEY [28.50]) and in plasma fractionation (CURLING [28.51]). Presently, a large number of new products are being developed by using recombinant DNA technology to produce rare proteins. Most of these proteins have to be purified to homogeneity. Therefore, it can be expected that chromatography will

Table 28.12. Methods for Large-scale Protein Chromatography

Protein Property Used for Separation	Method	Resolution	Capacity
Size and shape	Gelfiltration	moderate	low, volume limited
Charge	Ion-exchange	high	very high
Isoelectric point	Chromatofocusing	very high	high, but volume limited
Surface free energy	Hydrophobic chromatography	high	high
Bio-specific interaction	Affinity chromatography	excellent	very high
	Immuno adsorption	excellent	high

be increasingly applied in the new biotechnological industry (COONEY [28.50]).

Several separation criteria can be employed as summarized in Table 28.12. Ideally, chromatographic media should be insoluble and chemically stable during operation and cleaning procedures. For a sufficient capacity, they should be permeable and have a large internal surface area, while at the same time they should exhibit high mechanical strength and should show no or minimal non-specific adsorption. The last requirement excludes the majority of resins developed for fractionating inorganic ions and other low-molecular-weight products. High recovery of biologically active proteins can only be obtained with sufficiently hydrophilic resins. The basis for these materials are natural, semisynthetic, or synthetic polymers such as cellulose, dextran, agarose, polyacrylamide, or polyvinyl derivatives. These polymers can be derivatized to introduce special groups available for differential interaction with a mixture of proteins, e.g., ion exchange groups, hydrophobic chains, biospecific ligands, or antibodies. An inherent drawback of hydrophilic gels is their low rigidity, at least with porosities suitable for protein chromatography. This fact has hampered large-scale application of chromatography to protein separation. Recently, macroporous gels of higher rigidity (CURLING [28.52]) and composite materials (TAYOT et al. [28.53], [28.54]) have become available. This indicates the increased effort to improve mechanical properties of chromatographic resins. In addition, column design has been improved (JANSON and HEDMAN [28.55]) to cope with the limiting mechanical properties of the media. Performance of chromatographic columns can also be improved by producing resins of uniform sizes (COONEY [28.50]). This could dramatically increase the resolution and speed of operation of chromatographic separations, since particles with a much smaller diameter can be employed.

Chromatography is a batch operation. However, chromatographic systems can be well monitored and controlled and therefore easily automated and operated in repetitive cycles. Automation has to include the necessary washing steps to remove all remaining material from the column; re-equilibration must then be performed. Cleaning in place is very difficult. Depending on the nature and concentration of impurities present in the sample, rather harsh conditions such as high salt and/or dilute sodium hydroxide solutions are required, which may affect the swelling of gel beads, and therefore influence the packing of a column. Regeneration in place is necessary, since the packing of large-size columns is a laborious and time-consuming task. Repeated use of chromatographic media is mandatory in view of the high cost. Affinity chromatography and immune adsorbance techniques lead to highest purification, but such resins are very expensive, considering the fact that specific ligands have to be provided. Costs are especially high for monoclonal antibodies. In addition, considering the very restricted application, the sales volume is low. Affinity resins with group-specific ligands, e.g., coenzymes, triazine dyes, lectins, heparin, etc. have a wider range of application and, therefore, are less expensive. Elution from affinity columns may also be difficult. If harsh conditions are required, the product and resins may become damaged. The service time of columns has to be carefully considered and columns protected to insure long-lasting performance. Feed solutions applied to such expensive columns must be pretreated to remove the bulk of contaminants which may become potentially dangerous to the packing material. In addition, precautions should be taken to avoid microbial contamination. Clogging can be prevented by use of an in-line filter with a pore size of 0.22 μm and 0.9 μm to free all feed solutions from insoluble matter. Properly maintained, chromatographic columns can be repeatedly used, e.g., DEAE-Sepharose has been used more than 100 times to separate albumin from human serum (CURLING [28.52]) and Heparin-Sepharose more than 50 times in 1000 L-scale preparations of antithrombin III (EKETORP [28.56]). Microbial extracts are more difficult to process than plasma or serum due to the presence of nucleic acids and colloidal materials. The former especially affects the performance of anion exchangers. Partition, precipitation, or batch adsorption can be employed as a pretreatment before chromatography.

In principle, chromatography has provided the means of separating single proteins from very complex mixtures. The large variety of physical properties and functions of proteins makes it impossible to predict the most favorable sequence of operations to handle an essentially unknown mixture. Generally, the attempt is made to arrive at the lowest number of steps necessary to achieve the desired product quality and yield. An integration of different chromatographic steps can be achieved by applying the eluate of column A to column B without change in ionic composition, thus saving time and material. This consideration would place, e.g., a hydrophobic chromatography step after ion exchange or affinity chromatography, since hydrophobic interactions increase in high salt concentrations, often encountered in the eluate of such columns. Development of a chromatographic separation process has to be carried out in the laboratory, keeping the special requirements of large-scale operation in mind. This applies especially to the choice of suitable media, when mechanical properties, size, and size distribution, swelling behavior, and stationary regeneration have to be considered. The potential height of the gel bed depends on the physical properties. There is a general tendency to lower the height of the gel bed in order to achieve fast flow rates. Therefore, the design of end plates becomes very important, which are crucial for a uniform distribution of the inlet stream over the entire cross-sectional area and for collecting the zones without distortion at the other end of the column.

The capacity of chromatographic media varies and depends, among others, on the starting conditions, the pH, and the ionic strength of the feed solution. The capacity should be optimized with regard to load by operating columns slightly below the maximum load compatible with separation. Adsorption and ionic exchange chromatography are not limited by the volume of the applied sample and therefore have higher capacities than gel filtration, where the sample volume should not exceed a fraction of the column volume. For separations by gel filtration, the sample volume should be below 5% of the column volume, while up to 25% may be used for desalting. The capacity of gel filtration is also limited by the viscosity and the specific density of the solutions. These may cause channelling in large columns when the difference in viscosity between sample and elution buffer becomes too large. Flow rates are presently more limited by physical properties of the media, though eventually the kinetics of the adsorption and desorption steps and partition between gel and mobile phase determines the ultimate values. Separation seems to depend on the flow rate only in case of gel filtration (see also Fig. 28.13).

In large-scale operation, adsorption and ion-exchange chromatography are con-

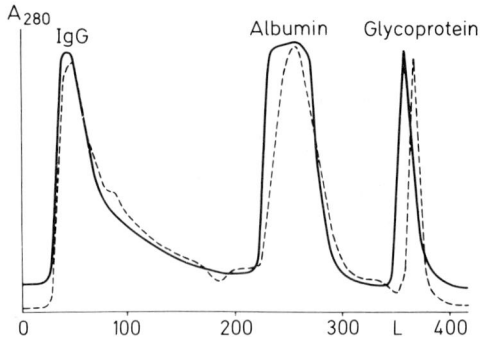

Figure 28.13. Separation of albumin from bovine plasma by high performance liquid chromatography on DEAE-Sepharose; fast flow. Column: stainless steel, 75 L total volume (80 cm Ø × 15 cm height), flow rate of elution buffer: 10 L/min, (2 mL/cm²·min at 75 kPa, solid line) and 5 L/min, (1 mL/cm²·min at 35 kPa, broken line); sample: desalted bovine plasma; capacity: 4.5 kg/h protein processed, 2.3 kg/h albumin separated. IgG = Immunoglobulin.

ducted at the highest possible flow rate with regard to the pressure differential of the resin and column. Elution can be performed by isocratic operation, by continuous or stepwise change in the composition of the eluting buffer. In contrast to laboratory practice where gradient elution for ion-exchange and adsorption chromatography is preferred, process chromatography is conducted preferentially by stepwise elution. This minimizes time and cost of operation. For the selection of a suitable elution buffer it should also be considered – apart from financial aspects – whether buffer components can be tolerated in small amounts in the end product. Otherwise, additional purification steps may become necessary. Alternate strategies have to be evaluated in the laboratory before scale-up is attempted. Since the gel material for large-scale columns represents a considerable investment, later changes in a process will be very costly.

Once the basic data have been established in the laboratory, scale-up of a single or a sequence of chromatographic steps to production scale is comparatively easy. One must increase the diameter of the column and keep most other parameters constant, such as chromatographic media, height of column, linear flow-rate (the ratio of volumetric flux to cross-sectional area), composition of feed, equilibration conditions, ionic strength, and pH of elution buffers. The sample volume and the volume of the elution buffers are increased in linear proportions with the column volume. Thus, the load per unit column volume is kept constant during scale-up. Performance and capacity of large-scale chromatography is demonstrated in Fig. 28.13 illustrating the purification of albumin from plasma. For successful operation it is necessary to have auxiliary equipment, e.g., pumps, valves, and monitoring systems – besides resins and columns. In addition to physical parameters of the eluate, such as pH, conductivity, and UV-adsorption it is also valuable to know the flow rate, operating time, and accumulated volume. These are especially needed to identify the desired product and control the collection of samples (JEFFERIS and KULA [28.57]). The available information can serve to automate the process by also considering a number of control and safety functions. One should not neglect that process-scale protein chromatography requires large quantities of high-quality water (CURLING and COONEY [28.58]), which must be continuously available and supplied at the chromatography plant.

28.7 Concentration and Final Processing

Depending on the intended application of the product, the final formulation may vary considerably. Usually, after purification, a concentration and a sterile-filtration step are performed to reduce the volume and the microbial contamination of the sample. Products intended for intravenous injection may require additional operations – not considered here – in order to obtain safe, sterile, and pyrogen-free preparations. Ultrafiltration is the method of choice if a liquid concentrate is required. Some products are stored as precipitates in ammonium sulfate- or polyethylene glycol-containing mother liquors, which are obtained by a final precipitation step.

Often a dry product is requested for storage and transport. For biologically active proteins, freeze-drying is the preferred method, where special ingredients, like sugar, substrates, or dextrans are added to protect the labile protein during processing. Freeze-drying is a rather energy-consuming and therefore expensive operation, which restricts its application to high-value products (GOLDBLITH et al. [28.59]). For larger amounts of product, spray-drying under optimized conditions may also be successful (LUYBEN et al. [28.60]). Care has to be taken to prevent uncontrolled release of biologically active and potent proteins into the atmosphere. Containment may be necessary for one or several of the above mentioned unit operations and production steps. To avoid possible allergic reactions, special formulations have been developed for proteases in dry form which prevent dust formation (AUNSTRUP et al. [28.61]). Similar techniques may be adapted to handle other dry proteins in highly concentrated form.

28.8 Conclusion

Considering the fact that the first enzyme was crystallized in 1926 by SUMNER and the chemical nature of proteins was established unequivocally not even 40 years ago by F. SANGER, it is astonishing to observe the pace of scientific and technological developments with regard to proteins. In addition to the numerous new enzymes discovered in nature, new methods to increase production of proteins by recombinant DNA technology and the potential to improve characteristic properties by protein engineering has opened new horizons. However, in order to be applied as a catalyst, drug, or analytical tool, all these proteins have to be purified to a greater or lesser extent, a challenge for the downstream processing of this class of substances in the near and distant future.

28.9 References

[28.1] H. HEMFORT: "Separatoren: Zentrifugen für Klärung, Trennung, Extraktion". Westfalia Separator AG, Oelde, 1979.

[28.2] V. I. SOKOLOV: "Moderne Industriezentrifugen". Verlag Technik, Berlin, 1967.

[28.3] J. J. ROUND, R. A. LIPTAK, and W. C. MC GREGOR: "Continuous-flow ultracentrifugation in preparative biochemistry"; in "Biochemical engineering II"; Ann. N.Y. Acad. Sci. *369* (1981), 265–274.

[28.4] K. H. BRUNNER: "Separatoreneinsatz in der Biotechnologie"; Chemie-Technik *1983*, Nr. 4.

[28.5] C. ALT and W. GÖSELE: "Einsatzkriterien für Dekanter"; Chem.-Ing.-Tech. *54* (1982), 425–430.

[28.6] D. J. BELL and K. H. BRUNNER: "A method for the evaluation of floc break up in centrifuges"; Filtr. Sep. July/Aug. (1983), 274–278.

[28.7] D. J. BELL, M. HOARE, and P. DUNNILL: "The formation of protein precipitates and their centrifugal recovery"; Adv. Biochem. Eng. Biotechnol. *26* (1983), 1–7.

[28.8] L. SVAROVSKY: "Solid-liquid Separation"; Chem. Eng. Series. Butterworth, London, 1977.

[28.9] K. H. KRONER, H. SCHÜTTE, H. HUSTEDT, and M. R. KULA: "Crossflow filtration in the downstream processing of enzymes"; Process Biochem. *19* (1984), 67–74.

[28.10] K. H. Kroner, W. Hummel, J. Völkel, and M. R. Kula: "Effect of antifoams on crossflow filtration of microbial suspensions"; Proceedings of the Europe-Japan Congress on Membrane Science and Technology, Stresa, Italy, 1984. Plenum Press, New York, in press.

[28.11] D. P. Gravatt and E. T. Molnar: "Recovery of an extracellular antibiotic by ultrafiltration"; Lecture presented at Engineering Foundation Conference "Recovery of Fermentation Products". Sea Island, Georgia, 1984.

[28.12] W. Bender and D. Redeker: "Fortschritte bei der mechanischen Flüssigkeitsabtrennung durch Filtration"; Chem.-Ing.-Tech. 53 (1981), 227–236.

[28.13] P. J. Hetherington, M. Follows, P. Dunnill, and M. D. Lilly: "Release of protein from baker's yeast in an industrial homogenizer"; Trans. Inst. Chem. Eng. 49 (1971), 142–148.

[28.14] C. R. Engler and C. W. Robinson: "Effects of organism type and growth conditions on cell disruption by impingement"; Biotechnol. Lett. 3 (1981), 83–88.

[28.15] H. Schütte, K. H. Kroner, and M. R. Kula: "Recent studies of mechanical disintegrators for the large scale disruption of microorganisms"; Proc. 4th Eur. Congr. Biotechnology, Vol. I, pp. 621–628. Verlag Chemie, Weinheim – Deerfield Beach/Florida – Basel, 1984.

[28.16] H. Schütte, K. H. Kroner, H. Hustedt, and M. R. Kula: "Experiences with a 20 l industrial bead mill for the disruption of microorganism"; Enzyme Microbiol. Technol. 5 (1983), 143–148.

[28.17] H. Schütte, K. H. Kroner, H. Hustedt, and M. R. Kula: "Disintegration of microorganisms in a 20 l industrial bead mill"; in "3. Rothenburger Fermentationssymposium, Enzyme Technology", pp. 115–124. (R. M. Lafferty, ed.). Springer Verlag, Berlin–Heidelberg–New York, 1983.

[28.18] A. J. Currie, P. Dunnill, and M. D. Lilly: "Release of proteins from baker's yeast by disruption in an industrial agitator mill"; Biotechnol. Bioeng. 14 (1972), 725–736.

[28.19] F. Marffy and M. R. Kula: "Enzyme yields from cells of brewer's yeast disrupted by treatment in a horizontal disintegrator"; Biotechnol. Bioeng. 16 (1974), 623–634.

[28.20] H. Schütte, W. Hummel, and M. R. Kula: "L-2-Hydroxyisocaproate dehydrogenase – a new enzyme from Lactobacillus confusus for the stereospecific reduction of 2-ketocarboxylic acids"; Appl. Microbiol. Biotechnol. 19 (1984), 167–176.

[28.21] J. Limon-Lason, M. Hoare, C. B. Orsborn, D. J. Doyle, and P. Dunnill: "Reactor properties of a high speed bead mill for microbial cell rupture"; Biotechnol. Bioeng. 21 (1979), 745–774.

[28.22] S. H. Hemmingsen: "Development of an immobilized glucose isomerase for industrial application"; in "Applied Biochemistry and Bioengineering" (L. B. Wingard, Jr., E. Katchalski-Katzir, and L. Goldstein, eds.), Vol. 2, pp. 157–183. Academic Press, New York, 1979.

[28.23] F. G. Mosqueira, J. J. Higgins, P. Dunnill, and M. D. Lilly: "Characteristics of mechanically disrupted baker's yeast in relation to its separation in industrial centrifuges"; Biotechnol. Bioeng. 23 (1981), 335–343.

[28.24] P. P. Gray, P. Dunnill, and M. D. Lilly: "The clarification of mechanically disrupted yeast suspensions by rotory vacuum precoat filtration"; Biotechnol. Bioeng. 15 (1973), 309–320.

[28.25] G. Naeher and W. Thum: "Production of enzymes for research and clinical use"; in "Industrial Aspects of Biochemistry" (B. Spencer, ed.), Part I, pp. 47–64. North Holland Publishing Co, Amsterdam, 1974.

[28.26] M. R. Kula, K. H. Kroner, H. Hustedt, S. Grandja, and W. Stach: "Process for the Separation of Enzymes"; Ger. Patent 26 39 129, US Patent 4 144 130, 1976.

[28.27] M. R. Kula, K. H. Kroner, and H. Hustedt: "Purification of enzymes by liquid-liquid extraction"; Adv. Biochem. Eng. Biotechnol. 24 (1982), 73–118.

[28.28] H. Hustedt, K. H. Kroner, and M. R. Kula: "Applications of phase partitioning in biotechnology"; in "Partitioning in Aqueous Two-Phase Systems" (H. Walter, D. E. Brooks, and D. Fisher, eds.). Academic Press, New York, in press.

[28.29] P. Å. Albertsson: "Partition of Cell Particles and Macromolecules"; 2nd Ed. John Wiley & Sons, New York, 1971.

[28.30] D. F. Gerson: "Cell surface energy,

contact angles and phase partition I, lymphocytic cell lines in biphasic aqueous mixtures"; Biochim. Biophys. Acta *602* (1980), 269–280.

[28.31] P. F. FAUQUEX, H. HUSTEDT, and M. R. KULA: "Phase equilibration in agitated vessels during extractive enzyme recovery"; J. Chem. Tech. Biotechnol., in press, 1985.

[28.32] K. H. KRONER, H. HUSTEDT, and M. R. KULA: "Evaluation of crude dextran as phase forming polymer for the extraction of enzymes in aqueous two-phase systems in large scale"; Biotechnol. Bioeng. *24* (1982), 1015–1045.

[28.33] K. H. KRONER, H. SCHÜTTE, W. STACH, and M. R. KULA: "Scale-up of formate dehydrogenase isolation by partition"; J. Chem. Tech. Biotechnol. *32* (1982), 130–137.

[28.34] K. H. KRONER, H. HUSTEDT, and M. R. KULA: "Extractive enzyme recovery: Economic considerations"; Process Biochem. *19* (1984), 170–179.

[28.35] E. FLASCHEL, Ch. WANDREY, and M. R. KULA: "Ultrafiltration for the separation of biocatalysts"; Adv. Biochem. Eng. Biotechnol. *26* (1983), 73–142.

[28.36] H. HUSTEDT, K. H. KRONER, and M. R. KULA: "Continuous enzyme purification by crosscurrent extraction"; Proceedings 3rd Eur. Congr. Biotechnology, Vol. I, pp. 597–605. Verlag Chemie, Weinheim – Deerfield Beach/Florida – Basel, 1984.

[28.37] E. J. COHN, L. E. STRONG, W. L. HUGHES, D. J. MULFORD, J. N. ASHWORTH, M. MELIN, and H. L. TAYLOR: "Preparation and properties of serum and proteins. A system for separation into fractions of the protein and lipoprotein components of biological tissues and fluids". J. Am. Chem. Soc. *68* (1946), 459–475.

[28.38] P. O. LARSSON and K. MOSBACH: "Affinity precipitation of enzymes"; FEBS-Lett. *98* (1979), 333–338.

[28.39] M. SCHNEIDER, C. GUILLOT, and B. LAMY: "The affinity precipitation technique"; in "Biochemical engineering II"; Ann. N.Y. Acad. Sci. *369* (1981), 257–263.

[28.40] P. R. FOSTER, P. DUNNILL, and M. D. LILLY: "Salting out of enzymes with ammonium sulphate"; Biotechnol. Bioeng. *13* (1971), 713–718.

[28.41] P. R. FOSTER, P. DUNNILL, and M. D. LILLY: "The precipitation of enzymes from cell extracts of *Saccharomyces cerevisiae* by polyethylene glycol"; Biochim. Biophys. Acta *317* (1973), 505–516.

[28.42] W. HÖNIG and M. R. KULA: "Selectivity of protein precipitation with polyethylene glycol fractions of various molecular weights"; Anal. Biochem. *72* (1976), 502–512.

[28.43] H. U. BERGMEYER, G. NAEHER, W. THUM, and G. WEIMANN: "Verfahren zur Anreicherung von Proteinen"; DP 2 001 902 7, 1970.

[28.44] C. E. NEUBECK: US Patent 4 233 405, Nov. 11, 1980.

[28.45] K. H. KRONER, A. CORDES, A. SCHELPER, M. MORR, A. F. BÜCKMANN, and M. R. KULA: "Affinity partition studied with glucose-6-phosphate dehydrogenase in aqueous two-phase systems in response to triazine dyes"; in "Affinity Chromatography and Related Techniques", pp. 491–501 (T. C. GRIBNAU, J. VISSER, and R. J. NIVARD, eds.). Elsevier, Amsterdam, 1982.

[28.46] G. KOPPERSCHLÄGER and G. JOHANSSON: "Affinity partitioning with polymer-bound cibacron blue F3G-A for rapid, large-scale purification of phosphofructokinase from baker's yeast"; Anal. Biochem. *124* (1982), 117–124.

[28.47] H. HUSTEDT, K. H. KRONER, H. SCHÜTTE, and M. R. KULA: "Extractive purification of intracellular enzymes"; in "3rd Rothenburger Fermentationssymposium, Enzyme Technology", pp. 135–145. (R. M. LAFFERTY, ed.) Springer Verlag, Berlin, 1983.

[28.48] M. R. KULA: "Continuous extraction of enzymes in aqueous phase systems"; Lecture presented at Engineering Foundation Conference "Recovery of Fermentation Products"; Sea Island, Georgia, 1984.

[28.49] E. A. PETERSON and H. A. SOBER: "Chromatography of proteins I. Cellulose ion-exchange adsorbents"; J. Am. Chem. Soc. *78* (1956), 751–755.

[28.50] J. M. COONEY: "Chromatographic gel media for large-scale protein purification"; Biotechnology *2* (1984), 41–43, 46–51, 54, 55.

[28.51] J. M. CURLING (ed.): "Methods in Plasma Fractionation". Academic Press, New York, 1980.

[28.52] J. M. CURLING: "Use of preparative chromatography in the purification and isolation of proteins"; Lecture presented at Novo Biotechnology Symposium, Copenhagen, April 1984.

[28.53] J. L. TAYOT, M. TARDY, P. GATTEL, R. PLAN, and M. ROUMIANTZEFF: "Industrial ion-exchange chromatography of proteins on DEAE dextran derivatives of porous silica beads"; in "Chromatography of Synthetic and Biological Polymers" (R. EPTON, ed.). Vol. 2, pp. 95–110. Ellis Horwood, Chichester, 1978.

[28.54] J. L. TAYOT, M. TARDY, and P. GATTEL: "Ion-exchange and affinity chromatography on silica derivatives"; in "Methods of Plasma Fractionation" (J. CURLING, ed.). Academic Press, New York, 1980.

[28.55] J. C. JANSON and P. HEDMAN: "Large-scale chromatography of proteins"; Adv. Biochem. Eng. 25 (1982), 43–99.

[28.56] R. EKETORP: "Affinity chromatography in industrial blood plasma fractionation"; in "Affinity Chromatography and Related Techniques" (T. C. J. GRIBNAU, J. VISSER, and R. J. F. NIVARD, eds.), pp. 263–273. Elsevier, Amsterdam, 1982.

[28.57] R. P. JEFFERIS, III, and M. R. KULA: "Application of computers to enzyme recovery"; in "Enzyme Engineering" (E. K. PYE and H. H. WEETALL, eds.), Vol. 3, pp. 241–248. Plenum Press, New York, 1978.

[28.58] J. M. CURLING and J. M. COONEY: "Operation of large-scale gel filtration and ion-exchange systems"; J. Parent. Sci. Technol. March/April (1982), 59–63.

[28.59] S. A. GOLDBLITH, R. REY, and W. W. ROTHMAYR (eds.): "Freeze-drying and Advanced Food Technology". Academic Press, London, 1975.

[28.60] K. C. LUYBEN, J. K. LIOU, and S. BRUIN: "Enzyme degradation during drying"; Biotechnol. Bioeng. 24 (1982), 533–552.

[28.61] K. AUNSTRUP, O. ANDRESEN, E. A. FALCH, and T. K. NIELSEN: "Production of microbial enzymes"; in "Microbial Technology" (H. J. PEPPLER and D. PERLMAN, eds.), 2nd Ed., Vol. 1, pp. 281–309. Academic Press, New York, 1979.

5 Measurement and Control

Chapter 29

Measurement and Instrumentation

Ulfert Onken
Rainer Buchholz

Abteilung Chemietechnik
Universität Dortmund
Dortmund, Federal Republic of Germany

Wolfgang Sittig

Abteilung Biotechnik
Hoechst AG
Frankfurt am Main, Federal Republic of Germany

29.1   Introduction
29.2   Physical Process Variables
29.2.1   Temperature
29.2.2   Pressure
29.2.3   Reactor Hold-up
29.2.4   Flow Rates (Gas and Liquid)
29.2.4.1 Floating Body Flowmeters
29.2.4.2 Differential Pressure Flowmeters
29.2.4.3 Rotating Flowmeters
29.2.4.4 Electromagnetic Flowmeters
29.2.5   Impeller Speed and Power Input
29.2.6   Liquid Viscosity
29.2.7   Foam
29.2.8   Gas Hold-up
29.2.9   Bubble Size and Interfacial Area
29.2.10 Liquid Turbulence
29.3   Chemical Process Variables
29.3.1   Exhaust Gas Analysis
29.3.2   pH
29.3.3   Dissolved Gases and Volatiles

29.3.3.1 Electrochemical Methods
29.3.3.2 Mass Spectrometry
29.3.3.3 Fluorescence Quenching
29.3.4 Analysis of Fermenter Broth by Sampling
29.3.5 Physical Methods for Biomass Determination
29.3.5.1 Turbidity Measurement
29.3.5.2 Light Scattering
29.3.5.3 Other Methods
29.3.6 Redox Potential
29.3.7 Enzymatic Analysis of Substrates and Metabolites
29.3.8 Ion-specific Electrodes
29.4 References

29.1 Introduction

The carrying out of a process requires knowledge of its state; this applies to chemical processes in general as well as to biochemical processes in particular. To obtain the necessary information, measurements can be carried out by using a wide range of methods employing specific sensors. The gained information can then be used for either manual or automatic process control and can serve as the basis for obtaining better knowledge of the process for its development and optimization.

For the majority of variables, measuring methods used in biochemical processes are the same as used for other chemical processes. However, in biochemical processes a number of requirements have to be fulfilled which necessitate the use of special methods of specific precautions, as e.g., with respect to sterility. In highly complex biochemical reaction systems, containing microorganisms or isolated enzymes, it is difficult to define the state of the system in terms of a few variables. Therefore, on average, the amount of instrumentation for biochemical processes is generally more extensive than for other chemical processes.

In biochemical processes one can differentiate between methods measuring physical and chemical variables. The physical variables comprise general process engineering variables, such as temperature, pressure, mass, mass-flow rates, liquid level, and impeller speed. The chemical variables are mainly the concentrations of substrates and metabolites (organic compounds, O_2 and CO_2 in the gas and liquid phase), also including biomass and pH.

As an example for the instrumentation of biochemical processes, Fig. 29.1 shows the basic measuring equipment of a standard fermenter. The sensors for this outfit do not affect sterility nor disturb sterilization, with the exception of probes for pH and dissolved oxygen. For these latter two variables, probes must be specially designed to allow sterilization by steam at 121°C. This also applies to the measurement of most other chemical variables, whether performed by probes or by sampling.

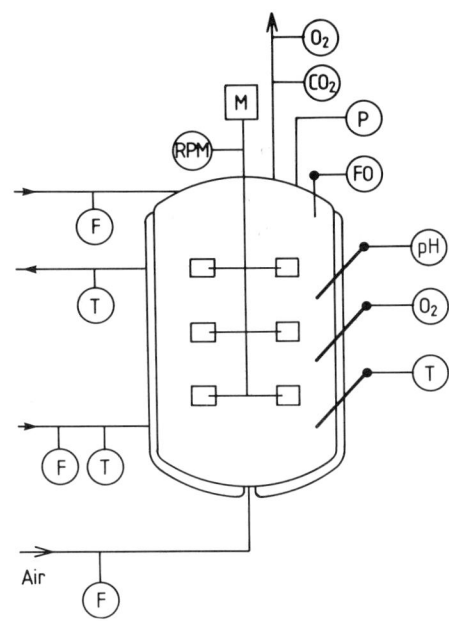

Figure 29.1. Basic measuring equipment of a standard fermenter. – F flow, FO foam, P pressure, RPM revolutions per minute, T temperature, M motor.

The measuring methods used in chemical processes have been, in general, thoroughly treated in monographs and textbooks (e.g., HENGSTENBERG et al. [29.1]; JOHNSON [29.2]) as well as in encyclopedias and handbooks (HASSELBACH [29.3]; MELZER and JAENICKE [29.4]; CONSIDINE [29.5]; PEINKE [29.6]). The reader should refer to these sources for detailed information on standard measuring techniques. In this chapter emphasis will be given to measuring methods specific for biochemical processes and to special procedures and precautions.

29.2 Physical Process Variables

29.2.1 Temperature

Temperature is an important parameter in biochemical processes; this is true not only for the reaction itself, but also for auxiliary operations, such as sterilization and downstream processing. Temperatures are usually monitored by one of the following standard devices (HASSELBACH [29.3]; JOHNSON [29.2]; WILLIAMS and ANDREW [29.7]):

- resistance thermometers
- thermocouples
- liquid expansion thermometers.

The risk of infection is minimal in all these methods. Resistance thermometry prevails because of its accuracy and reliability; sensors usually are encased platinium wires (100 ohms electrical resistance at 0 °C). The use of thermocouples (iron/constantan) is less frequent. Electric measuring signals from both resistance thermometers and thermocouples can be transferred to control boards. This is not possible with liquid expansion thermometers (e.g., mercury or ethanol in glass), which occasionally are employed for direct on-the-spot measurement.

Another method of on-the-spot temperature indication is by means of thermocolors, which change their color at a certain temperature. They can be applied in the form of thermofoils attached at critical spots, e.g., fermenters during sterilization.

29.2.2 Pressure

Pressure is measured, e.g., inside a fermenter vessel, by means of conventional pressure gauges (ANDREW and MILLER [29.8]; MORRISON [29.9]; HASSELBACH [29.3]). Since the manometer is not in direct contact with the fermenter contents (pressure transmitted via membrane), no sterility problems arise. Often, measurement of pressure is not included in the standard equipment, though it may yield valuable information, especially with laboratory glass vessels. Here, any clogging of the exhaust pipe may cause a build-up of a pressure head, and thereby, apart from the danger of cracking the glass, other parameters, such as solubility of gases, will be affected. In fermenters containing cultures which tend to form wall growth, deposits of microbial mass on the membrane may lead to errors in the monitored pressure.

29.2.3 Reactor Hold-up

The reactor liquid hold-up can be measured by weight or hydrostatic pressure (ELFERS [29.10]; ANDREW and RHEA [29.11]). Knowing the liquid hold-up of vessels is necessary to avoid overfilling and to calculate mass balances. Weight measurements can be carried out in two ways:

1. by measuring the hydrostatic pressure difference between the bottom of the tank and the headspace using conventional pressure gauges. The pressure difference is proportional to the weight of the contents of the vessel;
2. by measuring the total weight. Here, the vessel is suspended on mechanical scales or on pressure cells. The empty vessel is first weighed and the obtained value subtracted from the weight of the filled vessel. With large, stirred tank fermenters of high power input, oscillation resonance may cause problems in weight indication.

In the case of glass vessels or steel tanks with sight glasses, the liquid volume can be measured by direct observation or by means of a light barrier. Another possibility is to install a number of probes which make use of one of the following measuring effects:

- heat conductance
- electric capacitance
- electric conductivity.

In mass measurements, gas hold-up must be accounted for.

29.2.4 Flow Rates (Gas and Liquid)

Gas flow rate is one of the most important parameters in aerobic fermentations; likewise, the rate of gas production is of interest for cultures producing biogas.

Liquid flow rates must be known for continuous and fed-batch processes, where the rate of nutrient feed is an essential variable for efficient operation of the process, by means of mass balancing and control. Furthermore, knowing the liquid flow rate is necessary to control the addition of corrective liquid feed streams, such as the amount of base or acid consumed for pH control or the amount of antifoam input.

Flow rates are mainly measured by the following devices (ANDREW et al. [29.12]; ERICSON [29.13]; JOHNSON [29.2]):

- floating body flowmeter
- differential pressure flowmeter
- rotating flowmeter
- electromagnetic flowmeter.

29.2.4.1 Floating Body Flowmeters

The most commonly used type of flowmeter is the floating body type, also known as rotameter (Fig. 29.2). In a conical tube (a), the upstreaming fluid (gas or liquid) exerts a lifting force onto the floating body (b, cone or ball). The lifting force is balanced by the weight of the floating body minus its buoyancy. Since the lifting force is produced by the differential pressure across the slot between the tube wall and floating body, the position to which the float rises in the tube is a function of the flow rate, also depending on the properties of the fluid. Therefore, the equipment must be calibrated under defined conditions.

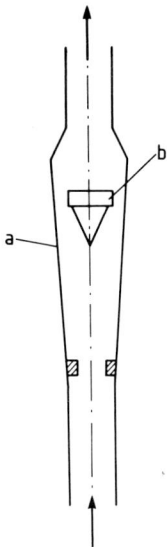

Figure 29.2. Floating body flowmeter (rotameter). – a tapered column, b floating body.

The accuracy of rotameters ($\pm 3\%$ of full scale) is not exceedingly high. Another disadvantage is their nonlinear response, which has to be accounted for by calibration. They are mainly used for measuring small and medium flow rates. When the conical measuring tube is made of glass, as is usually the case in laboratory and pilot plant installations, flow rates can be read by direct observation. Such glass rotameters have the advantages of low cost and low maintenance requirement. However, when the measuring value is to be transferred to a remote control board, it has to be converted by a transducer. For rotameters with measuring pipes made of metal, a transducer is necessary in any case.

29.2.4.2 Differential Pressure Flowmeters

Differential pressure flowmeters are based on the determination of the pressure drop caused by an artificial restriction in the pipe. The three most common types of this kind of flowmeters are shown in Fig. 29.3. As a rule, standard designs of these devices are used, which do not need to be calibrated if the properties of the fluid are known; temperature and pressure corrections have to be applied.

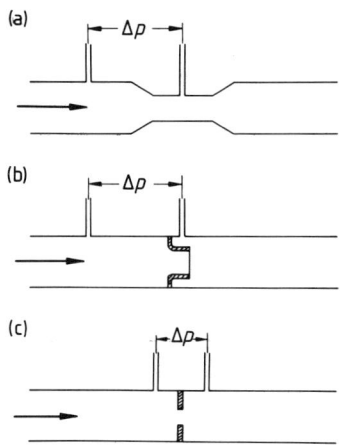

Figure 29.3. Differential pressure flowmeters. – (a) Venturi restriction, (b) nozzle restriction, (c) orifice plate restriction, Δp differential pressure.

The differential pressure from these restriction flowmeters is proportional to the square root of the flow rate if the flow is turbulent. Therefore, accuracy at low flow rates is rather poor. Accuracy is also reduced by uneven distribution of fluid velocity, which may be caused by flow obstacles, such as valves, bends, or probes, located shortly before the restriction. Faulty readings may also result from deposition of solids on the measuring restriction. On an average, accuracy of restriction flowmeters will be ± 1 to 2%.

29.2.4.3 Rotating Flowmeters

The measuring principle of rotating flowmeters (turbine flowmeters) is based on the transfer of flow energy onto rotating bodies. The flow impact can be directed either tangentially (turbine wheel) or radially (impeller wheel). For liquids, both types are used, whereas for gases, only turbine wheel flowmeters are employed, mainly for rather high flow rates. The number of revolutions of the rotating body is proportional to the rate of flow. The accuracy of the method is about $\pm 1\%$.

Higher accuracies up to $\pm 0.5\%$ can be achieved by special types of rotating flowmeters, which function by periodically filling and discharging defined volumes. Devices of this kind are the oval gear meter with two counter-rotating wheels and the rotary piston meter. They are used especially for accurate volumetric measurement of liquid flow (MCMAHON [29.14]).

29.2.4.4 Electromagnetic Flowmeters

Electrically conductive liquids passing a magnetic field will induce a voltage between two electrodes positioned rectangular to the direction of flow. The voltage is proportional to the flow velocity. The method does not require the installation of any parts which could disturb the measured flow. It can be applied to fermenting liquids without difficulty, since these contain electrolytes, and used on corrosive fluids. Results are not affected by changes in fluid density or viscosity, and the measuring device does not cause any pressure loss, in contrast to other methods. The procedure is, however, rather costly. With its accuracy of $\pm 1\%$, this method is preferably used to reliably measure liquid flow rates.

29.2.5 Impeller Speed and Power Input

For stirred tank fermenters, impeller speed is an important operating variable, which is very often kept constant. It is usually measured by monitoring the number of revolutions per time unit outside the aseptic area of the impeller shaft. This is accomplished by one of the following conventional devices: generator-type tachometers, in which a direct electric current proportional to the number of revolutions is generated, and electronic tachometers, producing an alternating current, the frequency of which is determined by a digital counter. Still another device makes use of light-chopping disks which are installed on the agitator shaft, where the intermittent light signals are recorded via a photocell and counted digitally.

The power consumption of agitators depends on stirrer speed (number of revolutions per second) and physical properties of the stirred fluid, especially on its viscosity, which may change drastically during batch fermentations, e.g., in some processes for the production of antibiotics, such as penicillin.

In large-scale fermenters, the consumption of electric energy, as determined by a wattmeter, yields useful information on the input of agitating power when friction losses in stuffing box, seals, and motor are accounted for. Since in laboratory fermenters this fraction of energy consumption is rather high and also not reproducible, no meaningful results on power input can be obtained by this method. More direct measurements of agitator power are possible by using torsion dynamometers or strain gauges. The latter method is more accurate, because, in the case of measuring elements mounted inside the fermenter, measuring signals are not affected by frictional losses in seals and bearings. Sterility is not restricted, since the measuring in strain gauges consists in changes of electric resistance monitored in a bridge circuit.

29.2.6 Liquid Viscosity

Viscosity is a physical property which is defined by Newton's law:

$$\tau = -\eta \frac{dw}{dy}, \qquad (29.1)$$

with τ shear stress, η dynamic viscosity, and dw/dy shear rate.

The viscosity η is independent of the shear rate. Fluids obeying Eq. (29.1) are called Newtonian. Most liquids of higher molecular weight and many suspensions show more complex flow behavior. For these non-Newtonian liquids, an apparent viscosity can be defined as a function of shear rate (METZNER and OTTO [29.15]). Depending on the type of flow characteristics, various classes of non-Newtonian liquids can be distinguished, such as pseudoplastic or dilatant liquids. Many fermentation broths, e.g., mycelial fermentation

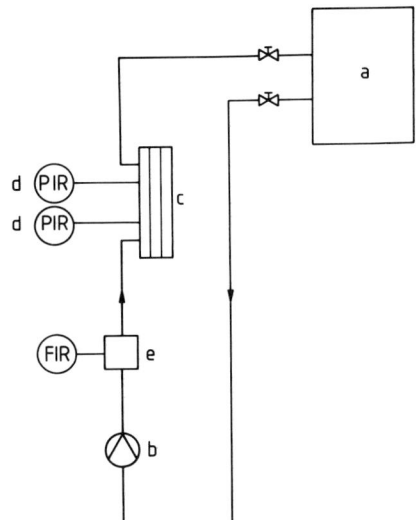

Figure 29.4. On-line measurement of viscosity. - a fermenter, b pump, c slot viscosimeter, d pressure pick-ups, e flowmeter. - PIR pressure indication and registration, FIR flow indication and registration.

broths in antibiotics production or polysaccharide solutions, show non-Newtonian behavior. Since viscosity is dependent on broth composition and cell morphology, it can be used as an indicator of the state of a culture, especially in the case of non-Newtonian flow behavior. Unfortunately, to date, not much attention has been paid to the development of devices for the on-line determination of viscosity in fermenters. One installation, described by LANGER and WERNER [29.16] should be suitable for this purpose. It is essentially a slot-type viscosimeter (Fig. 29.4) through which a sampling stream is pumped. Pressure differences are measured by pressure pick-ups mounted in the wall of the slot. This installation can be used in two ways:

a) At constant rate of flow through the slot, viscosity is determined from the pressure difference of a constant shear gradient as a function of time.
b) The flow rate is changed in steps, yielding the shear diagram; this is repeated at intervals during the process.

For both these procedures, an on-line computer is needed. All parts of the liquid circuit have to be sterilized together with the fermenter before inoculation. The system has been successfully tested in an antibiotics plant (NEUHAUS et al. [29.17]).

Viscosity can, of course, be determined by taking samples to be measured off-line (for measurement methods see: GERTH [29.18]; MUMMA [29.19]; DINSDALE and MOORE [29.20]). For highly viscous broths from aerobic cultures, one problem remains to be mentioned which exists for both on-line and off-line measurement; that is, proper degassing. Because of the low rising velocity of small gas bubbles in viscous liquids, complete separation of gas may be practically impossible, thus leading to faulty readings.

29.2.7 Foam

Foaming is a nuisance occurring in most fermentation broths. It may be caused by surface-active metabolites (e.g., proteins, polysaccharides), by components of the medium (e.g., starch, molasses), or by the cells. Two types of foams are to be distinguished: soft foams, which are rather unstable, and hard ones, which are stable. Foaming must be suppressed in order to prevent contamination of the culture from wetted exit filters in the first place, but also to avoid clogging of the exhaust system including its measuring devices and loss of culture broth.

Foam destruction can be achieved either mechanically (mostly by centrifugal systems) or chemically, i.e., by antifoaming agents. Often both methods are combined. In this case, the mechanical system is operated permanently, and the chemical antifoam is added, when mechanical defoaming does not suffice. In any case, foam control necessitates foam detection. This can be performed by one of the following types of sensors mounted inside the fermenter above the liquid level: electric conductivity probe, capacitance probe, heat conductivity probe. For further information see HALL et al. [29.21].

29.2.8 Gas Hold-up

For aerobic submerged fermentations, gas hold-up is an important parameter, because the interfacial area for gas-liquid mass transfer in the aerated broth depends largely on gas content. However, fermenters are usually not equipped with devices for measuring gas hold-up. It is determined mainly when investigating mass transfer and process modelling, though knowledge of gas content might well be used to define the state of aerobic fermentation processes, and this so much more, since the measurement can be performed without affecting sterility. From the various methods for the determination of gas hold-up, only those will be treated here which are frequently used. For other methods as well as for more detailed information see HEWITT [29.22, 29.23].

The *mean gas hold-up* can easily be measured by simply installing a stand pipe in which the liquid level is viewed through a sight glass. In addition, the liquid level inside the aerated fermenter must also be known. Because levels will oscillate, great accuracy can not be achieved by this method, especially with foaming media.

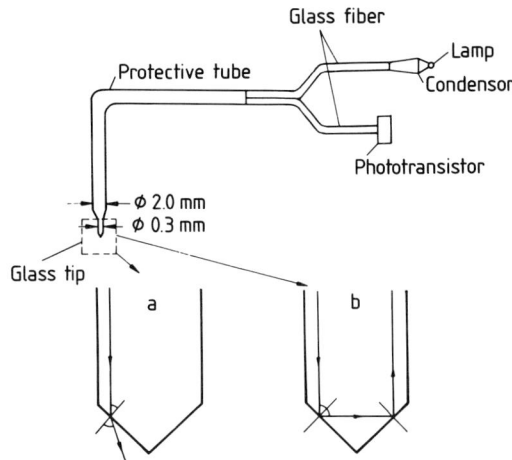

Figure 29.5. Optical probe for measuring the gas hold-up. − a liquid: $n=n_L>1.15$, b gas: $n=n_G<1.15$.

More information can be obtained by measuring *local gas hold-up*. Here, two methods shall be mentioned. The first makes use of the electric conductivity of fermentation media (SERIZAWA et al. [29.24]). A probe for this method essentially consists of the blank tip of an insulated needle. The electric resistance between this needle and its metal jacket is recorded; it will increase by orders of magnitude, when the tip is passed by a gas bubble. An important measure in the application of the method is hydrophobizing the probe to exclude erroneous signals caused by an adhering waterfilm (cf. bubble size measurement in the next section).

The second method for measuring gas content employs an optical probe, which uses the difference in the optical density of liquid and gas to determine gas content (MILLER and MITCHIE [29.25]; GALAUP and DELHAYE [29.26]). In this case, the intrinsic part of the probe is a light-conducting glass fiber. As the sensitive part of the probe is very small, it shows great sensitivity to mechanical stress (Fig. 29.5).

29.2.9 Bubble Size and Interfacial Area

As already mentioned, the interfacial area between gas bubbles and broth is decisive for oxygen transfer in aerobic fermentations. Its overall description is given by the following equation:

$$\dot{n}_{O_2} = \beta_L a (c_L^* - c_L) V_R \qquad (29.2)$$

with $\dot{n}_{O_2}$ molar oxygen transfer rate, β_L mass transfer coefficient, a specific interfacial area, c_L^*, c_L dissolved oxygen concentration in phase boundary and in the bulk of the liquid, respectively, and V_R reactor volume.

In the term for mass transfer resistance $1/\beta_L a$, the interfacial area a is much more dependent on operating parameters than β_L. While β_L varies usually by a factor of two at the most, the interfacial area may change by more than a factor of ten when aeration rate or stirring speed are altered. This is due to changes in gas hold-up and mean bubble size.

For determining the interfacial area in culture media, chemical methods, such as sulfite oxidation or CO_2 absorption, cannot be used for a number of reasons (e.g., lysis of microbes under the conditions of the chemical reactions; O_2 and CO_2, respectively, are substrate and metabolite of microbial metabolism). Therefore, physical methods have to be applied. These are based on the following equation:

$$a = \frac{6\varepsilon_G}{d_S}, \qquad (29.3)$$

with ε_G relative gas hold-up and d_S Sauter diameter.

The latter quantity is the mean bubble diameter which can be evaluated from experimental data of bubble size distribution. Hence, physical methods for interfacial area measurement are essentially methods for determining bubble-size distributions (HEWITT [29.22]). Table 29.1 gives a survey of these methods.

tically fails with turbid media, which are frequent in fermentations.

The optical probe, or more precisely *light reflecting probe*, utilizes the difference in optical refraction of gas and liquid in the same way as the optical probe for measuring gas hold-up. Its disadvantage is its great sensitivity to mechanical stress (cf. Sect. 29.2.8).

Electric conductivity probes for measuring bubble sizes also should be very small (SERIZAWA et al. [29.24]; BUCHHOLZ et al. [29.34]). The design of one type of such a microprobe is given in Fig. 29.6. It consists

Table 29.1. Physical Methods of Determining Bubble Size Distribution

Method	Bubble Size Range (in mm)	Limitations and Problems	Reference
Photographic	0.1–20.0	only in vicinity of wall, glass windows required	SIEMES and BORCHERS [29.27]
Electric conductivity probe	0.6–20.0	only electrically conducting liquids	SERIZAWA et al. [29.24], BURGESS and CALDERBANK [29.28], BUCHHOLZ et al. [29.29]
Optical probe	1.0–20.0	sensitive to mechanical stress	MILLER and MITCHIE [29.25], GALAUP and DELHAYE [29.26], CALDERBANK and PEREIRA [29.30]
Photoelectric bypass method	0.5–6.0[a]	limited to low viscosities, not at high turbulence	TODTENHAUPT [29.31], PILHOFER et al. [29.32], WEILAND et al. [29.33]

[a] range depending on capillary diameter of probe

In order to assess the applicability of the different methods, it has to be remembered that in fermenters bubble size spectra are ranging from 0.1 to 10 mm diameter; sometimes even larger bubbles (up to 20 mm diameter) may be found. Microbubbles with diameters of less than 0.5 mm contribute only negligibly to mass transfer.

The *photographic method* suffers from the fact that a plane sight glass is required and that only bubbles in the vicinity of the wall can be registered; besides, the method prac-

of four 20 µm diameter platinum wires fused in glass. The total diameter of the probe amounts to 300 to 500 µm. The signals from this four-point microprobe yield information not only on size and rising velocity of the bubbles, but also on their shape and direction of ascent (STEINEMANN and BUCHHOLZ [29.35]). To prevent disturbance of the signals by adhering waterfilms, the probe must be hydrophobized. In contrast to the other physical methods for bubble size determination, both optical and electri-

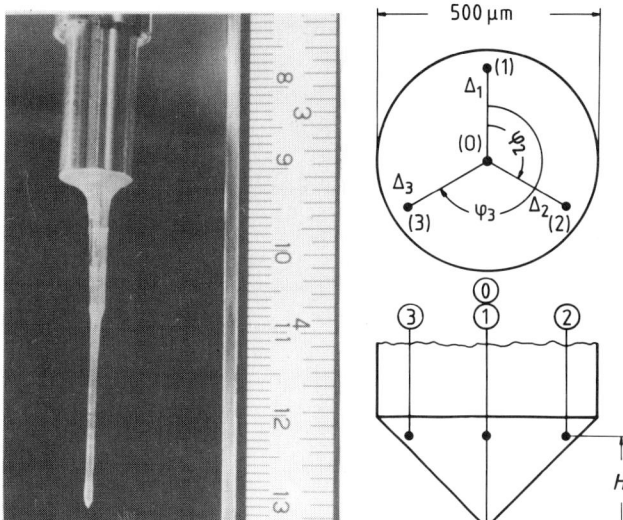

Figure 29.6. Electric conductivity microprobe for measuring bubble sizes. – Spatial arrangement of the probe points (0)-(3).

cal conductivity probes provide reliable information on relative gas hold-up, which together with mean bubble size is necessary for the evaluation of the interfacial area, according to Eq. (29.3).

The *photoelectric bypass method* (PILHOFER et al. [29.32]; WEILAND et al. [29.33]) has the disadvantage that external circulation of a sample stream from the fermenter is required with additional precautions for sterility. Besides, one must ensure that flow conditions at the location of withdrawal are not disturbed. This is practically impossible at the high turbulence which prevails in stirred tanks. The essential feature of the method consists in conducting the sample stream through a glass capillary, where the gas bubbles are stretched to form plugs to be monitored by two light beams. Apart from the disadvantages already mentioned, the method is not suited for highly viscous broths (BUCHHOLZ et al. [29.36]).

29.2.10 Liquid Turbulence

Turbulence and flow behavior of the medium are essential for mass transfer and mixing in fermenters. Relevant information on these processes can be gained by measuring the quasi-turbulence structure of the liquid phase. Moreover, turbulence measurement supplies information on shear effects in fermentations, which may cause severe reduction of the performance of shear-sensitive microorganisms (e.g., KIM et al. [29.37]). For measuring of turbulence properties, Laser-Doppler or hot-film anemometry can be employed.

Laser-Doppler anemometry (LDA) does not affect flow conditions; but because gas bubbles and turbidity prevent penetrating

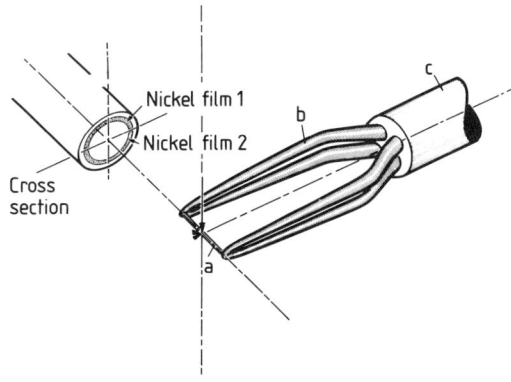

Figure 29.7. Split-film probe. – a quartz fiber, b prongs, c ceramic tube.

of the laser beam to a greater depth of the gas-liquid dispersion, the use of LDA in fermentation media is limited. Hot-film anemometry (HFA) does not show such problems (FRANZ et al. [29.38]). Using simple probes for this method with one electrically heated metal film, yields absolute values of instantaneous local liquid velocities only. With so-called split-film probes (Fig. 29.7), both magnitude and direction of these velocities are obtained (BUCHHOLZ et al. [29.34]). Such a split-film probe consists of a glass fiber coated by two half-cylindrical nickel films. Probes of this type have successfully been used for the determination of flow structure in tower fermenters for biological effluent treatment.

29.3 Chemical Process Variables

29.3.1 Exhaust Gas Analysis

The determination of gaseous compounds in the off-gas leaving the fermenter can be performed by conventional gas analysis (MELZER and JAENICKE [29.4], MCCULLOUGH and ANDREW [29.39]), e.g., by using paramagnetism (O_2), infrared absorption (CO_2), or flame ionization (hydrocarbons). Of these, the devices for measuring O_2 and CO_2 are part of the standard instrumentation of fermenters. Mass spectrometry can be used to analyze all gaseous components of the exhaust gas stream together with organic volatiles, such as lower alcohols (REUSS et al. [29.40]; cf. Sect. 29.3.3.2).

Another possibility for analyzing organic volatiles in the off-gas is enzymatic analysis. Thus, VAN STEENBERGEN and WILLIAMS [29.41] used alcohol dehydrogenase together with nicotinamid dinucleotide (NAD^+) as oxidant for the determination of ethanol concentrations in a sampling stream withdrawn from the off-gas line. The increase of NADH due to oxidized ethanol was photometrically monitored at 340 nm.

The sampling stream for exhaust gas analysis should be purified carefully from drops and dust particles, especially when the results are needed for mass-balance computations and process modelling. Separation of aerosols is particularly important when the sampling gas is removed from the fermenter head space. In industrial installations with a set of several fermenters, exhaust gas streams and fresh air are often measured in a single set of analyzers via a multiplexer.

29.3.2 pH

The pH value is an important indicator of the state of a biochemical process. Measurement is performed electrochemically as is the rule in industrial chemical processes (MELZER and JAENICKE [29.4], MCCUL-

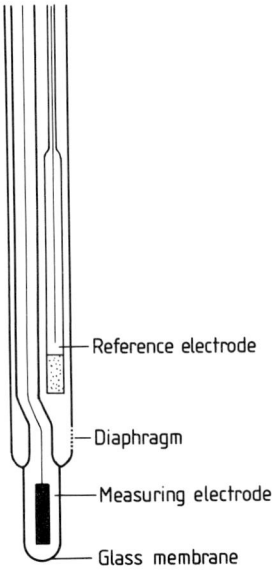

Figure 29.8. Schematic set-up of a sterilizable pH electrode.

LOUGH and ANDREW [29.39]). Sterilizable glass electrodes are offered as compact units by several producers; they belong to the standard equipment of fermenters. Fig. 29.8 shows a schematic set-up of such a sterilizable pH electrode.

For employment under aseptic conditions, the electrodes are usually steam-sterilized together with the fermenter vessel (20–60 min at 121 °C). During sterilization, a pressure difference builds up on both sides of the glass membrane. In order to avoid its destruction, a counter-pressure is superimposed. For long-time measurement with sterilizable electrodes, repeated calibration is necessary, because the glass membrane might be blocked by deposition of solids (e.g., proteins), and, hence, its resistance to H^+-ion diffusion will increase. Reliable calibration can only be guaranteed when parallel samples are analyzed from time to time.

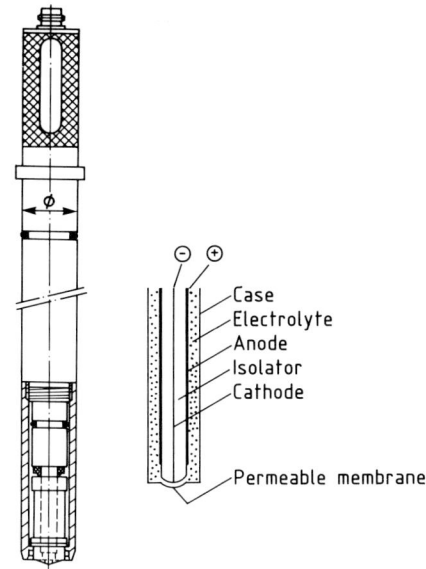

Figure 29.9. Oxygen electrode.

29.3.3 Dissolved Gases and Volatiles

Dissolved oxygen and carbon dioxide, both of which are important variables in fermentations, are normally determined by electrochemical methods. Other methods for measuring these two variables are mass spectrometry and fluorescence quenching. Furthermore, mass spectrometry can be used for analyses of other dissolved gases and volatile organic compounds.

29.3.3.1 Electrochemical Methods

The electrochemical determination of oxygen and carbon dioxide in fermentation media is performed by means of special sterilizable electrodes. As a rule, oxygen electrodes are part of the standard outfit for laboratory fermenters, but not as much for industrial fermenters, because of rather high maintenance and calibration requirements.

Analysis of oxygen by electrochemical probes is based on detecting the amount of oxygen diffusing from the liquid to be analyzed through a membrane into a amperometric or polarographic measuring cell. More frequently employed are electrodes using the amperometric principle (JOHNSON et al. [29.42], BUEHLER and INGOLD [29.43], SCHINDLER and SCHINDLER [29.43a], MELZER and JAENICKE [29.4]). The construction of this type of probe is shown in Fig. 29.9. Between cathode and anode a constant voltage (ca. 650 mV) is applied. At the cathode, the oxygen that has diffused into the cell is reduced to hydroxide ions according to the following reaction scheme:

Cathode: $O_2 + 2H_2O + 4e^- \rightarrow 4OH^-$

Anode: $4Ag + 4Cl^- \rightarrow 4AgCl + 4e^-$

Response of the probe is proportional to oxygen activity in the liquid.

Since at equilibrium, that is, for a saturated liquid, the activity of a solute is directly related to its partial pressure (more precisely to its fugacity), the readings of electrochemical probes are commonly given as % saturation or "partial pressure". In order to obtain mass concentrations from data for degree of saturation, as is required for mass-balance calculations, solubility data given as mass concentration must be available for the actual medium at the prevailing temperature. For electrochemical oxygen probes, a temperature coefficient of up to 3%/K is to be expected. This is caused by changes in diffusivity and solubility in the membrane (LEE and TSAO [29.44]). Furthermore, oxygen solubility in water changes about 2%/K at 298 K. In order to achieve a good temperature control of the probe, the output of commercial devices is often compensated by use of a thermistor circuit.

Accuracy of oxygen electrodes is about ±5% of the total measuring range. In large-scale reactors, local dissolved oxygen concentrations depend on measuring location and on flow and mixing conditions. Therefore, the position of the probe must be carefully chosen in order to obtain representative data. For dynamic response of oxygen electrodes, see LEE and TSAO [29.44].

Sterilization of oxygen electrodes is critical because of the extremely thin membranes. An electrode can be sterilized a few times only, sometimes only once or twice; that means that its application is accompanied by considerable uncertainties. Often, the response of the probe exhibits a drastical change after sterilization. Therefore, recalibration after sterilizing is necessary. Finally, it should be mentioned that for determining oxygen profiles in microscopic ranges, e.g., in cell agglomerates, microprobes are available (TSAO and LEE [29.45]; MAASS et al. [29.46]).

Electrochemical analysis of dissolved carbon dioxide in sterile environments has been extremely difficult for a long time, since the electrodes exhibited unsatisfactory stability after sterilization. Improvement of membranes and the use of temperature-proof pH electrodes now allow standard application of CO_2 electrodes in steam-sterilized bioreactors (PUHAR et al. [29.47];

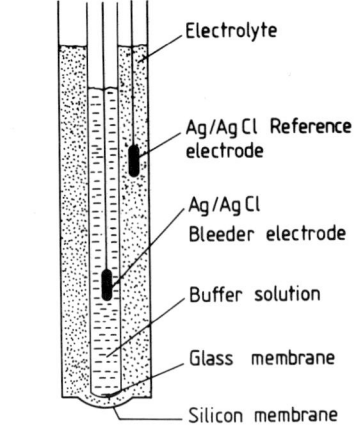

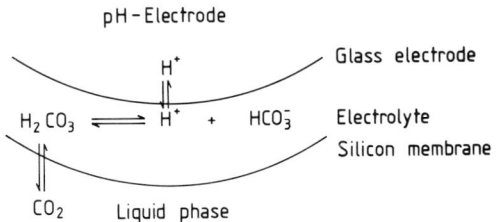

Figure 29.10. Principal design of a CO_2-probe.

SHODA and ISHIKAWA [29.48]). Fig. 29.10 shows the principal design of a CO_2 probe (type Ingold). Measurement is based on the diffusion of CO_2 into aqueous bicarbonate solution. The shift in pH is recorded by means of a glass electrode. The problems of reliable measurement are the same as for the oxygen electrode. The overall dimensions of this electrode coincide with the oxygen electrode, so this probe can be mounted in bioreactors into the same type of plug.

29.3.3.2 Mass Spectrometry

An alternative to electrochemical determination of dissolved gases is mass spectrometry which can be applied to exhaust gas analysis as well (REUSS et al. [29.49]; HEINZLE and LAFFERTY [29.50]; cf. Sect. 29.3.1). Simple, cheap mass spectrometers, e.g., quadrupole mass spectrometers, can be used. The evaluation of concentrations of dissolved gases, such as oxygen or carbon dioxide, and of volatile substrates, such as methanol or ethanol, differs from the measurement of the components of gaseous mixtures only by a modified entrance system. Accordingly, the principles of the application of mass spectrometry in fermentations shall be treated here for both cases. The use of the method for on-line analysis is particularly economic, because with one apparatus alone exhaust gas measurement as well as the determination of dissolved compounds can be performed; moreover, several fermenters can be connected to one apparatus via a multiplexer.

Fig. 29.11 gives the set-up of a process mass spectrometer including two types of sampling systems for the determination of dissolved compounds. One type of probe (Fig. 29.11b) is essentially a steel capillary of which one end, which is sealed by a membrane, is immersed into the medium to be analyzed. The membrane made from silicon or PTFE (polytetrafluoro ethylene) should show low water permeability in order to enrich the gaseous and volatile compounds in the sample stream. Because pressure in the mass spectrometer must be extremely low ($<10^{-6}$ mbar), the sample stream is regulated by a valve. To reduce time delay in the sampling system, the flow through the capillary can be accelerated by introducing a stream of inert gas, e.g., helium, into the sampling system (see Fig. 29.11c) (HEINZLE et al. [29.51]). Depending on their affinity to the membrane material, the different components will exhibit different migration velocities, leading to different enrichment in the ultrahigh vacuum of the mass spectrometer. This effect must be compensated by calibration. Errors may be induced by attachment of bubbles on the membrane and by varying flow velocities of the broth. Agitation may significantly influence the response by changing the thickness of the liquid film on the membrane, particularly at low stirrer speed and with highly viscous fluids. Other system-dependent errors have been discussed by MILLARD [29.52].

Mass spectrometric analysis of fermentation media by means of sampling systems allows determination not only of dissolved gases, such as O_2, CO_2, and CH_4, but also of volatile compounds, e.g., methanol, ethanol, butanol, acetone. When a mass spectrometer system has been installed, it can, of course, be employed also for analyzing waste-gas streams. In this application, accuracy is much higher than for analysis

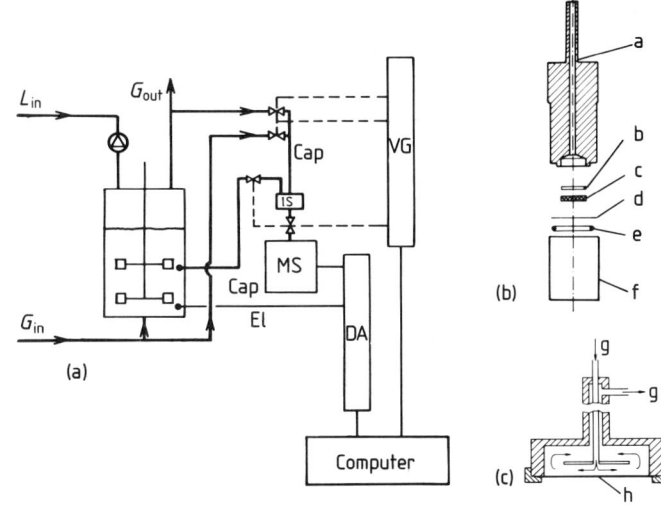

Figure 29.11. (a) Set-up of a process mass spectrometer. – MS mass spectrometer, L_{in} nutrient feed rate, $G_{in,\,out}$ gas flow rate, Cap capillary inlet, Mem membrane inlet, IS inlet system, El oxygen electrode, IEEE-bus standard bus of computer, VG valve gear, DA data acquisition. (b) Sampling system. – a steel capillary, b distance holder (ring), c glass frit, d membrane, e O-ring, f jacket tube. (c) Sampling system (with inert gas stream). – g helium stream, h membrane.

from liquids. Thus, oxygen may be determined at an accuracy of better than 0.1%, which is of interest for oxygen balances in fermentations with low oxygen consumption and high aeration rates.

29.3.3.3 Fluorescence Quenching

For medical investigations, so-called optodes for O_2 and CO_2 have been developed (LUEBBERS and OPITZ [29.53]). They make use of the quenching of fluorescence. Recently, the application of these optodes to fermentation media has been proposed (OPITZ and LUEBBERS [29.53a]). The sensitive element in the measuring probe (Fig. 29.12) is a membrane into which a fluorescence indicator has been incorporated. This membrane is brought into contact with the culture broth. As fluorescence indicators, pyrene-butyric acid (for O_2, fluorescence wavelength $\lambda = 395$ nm) or β-methylbelliferon purin (for CO_2, $\lambda = 445$ nm) may be used. After calibration, the method yields mass concentrations. In contrast to electrochemical probes for oxygen measurement, fluorescence quenching does not consume oxygen.

29.3.4 Analysis of Fermenter Broth by Sampling

Although on-line (or *in vitro*) measuring methods have been greatly improved in the last decade, many parameters can only be determined by sampling and subsequent off-line analysis. This procedure for obtaining process information has lately been more frequently used for a number of reasons. On the one hand, microbiological and biochemical research has brought about numerous improved methods of enzyme and protein identification requiring laboratory

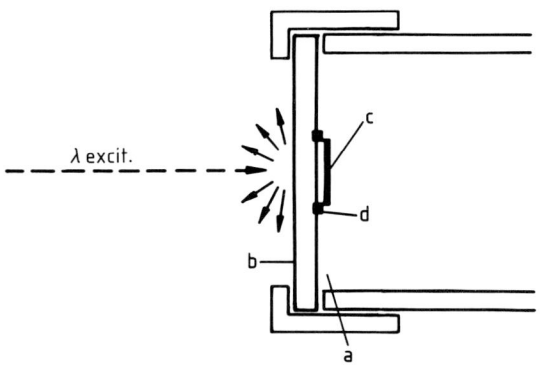

Figure 29.12. Fluorescence – optical sensor (optode) for oxygen. – a sensor, b plexiglass, c membrane, d O-ring.

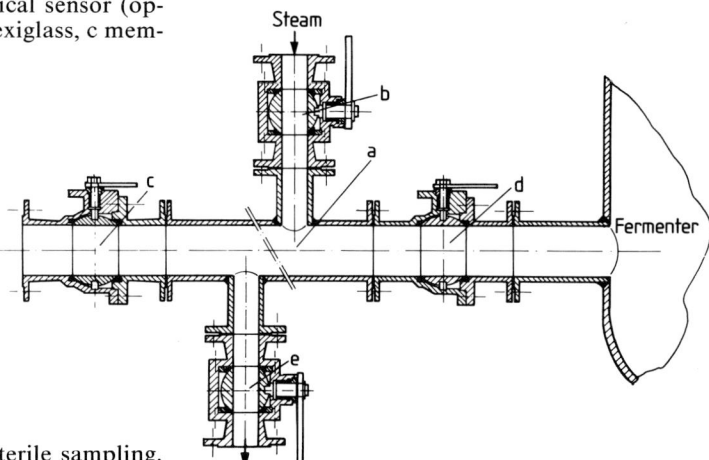

Figure 29.13. Steam sealing for sterile sampling. – a steam-sealed chamber, b–e valves.

preparation of the sample. On the other hand, for the purpose of computer application and mathematical modelling, more ample and more accurate process information is needed. As a rule, samples have to be taken whenever an on-line measurement is either impossible, too complicated or expensive, or when sterility is critical. Sampling requires sterile equipment and skill. To keep the sampling ports sterile, they are usually protected by a steam seal (Fig. 29.13), between the open (and hence polluted) atmosphere (c) and the valve or stopcock to the fermenter proper (d) a chamber (a) is formed. This is rinsed by steam which is collected by a condensing trap as long as the sampling device is not in action. Often, computerized automatic sampling devices are used. In fully automated systems, the samples are automatically deep-frozen to be processed later.

Samples are analytically evaluated by various different procedures (see Table 29.2).

Table 29.2. Measurement Procedures for Samples from Fermenters

Type of Investigation	Method
Sterility test	Agar plate
Morphology of culture	Agar plate
Biological and enzymatic activity	Enzymatic methods
Analysis of substrates, products, and metabolites	Various methods
Recalibration of probes	Various methods
Physical properties of culture medium	Various methods

29.3.5 Physical Methods for Biomass Determination

Biomass concentration is a crucial point for the state of any fermentation process. The most frequently used methods for determining biomass concentration are the microscopic measurement of the cell number, the number of cells in a defined volume, and the measurement of dry cell weight. Both methods require an aseptic sample from the fermenter. A grave disadvantage of both methods is the time requirement: the values are available only a considerable time (up to several hours) after sampling.

29.3.5.1 Turbidity Measurement

With increasing biomass concentration, the optical transparency of the culture broth decreases, and hence its turbidity increases. Since some substrates, such as starch and whey, as well as gas bubbles affect optical transparency in a similar way, the interpretation of turbidity measurements cannot be straightforward. Disturbances due to gas bubbles may be eliminated by inserting the sensor into a gas-free zone of the fermenter or better into a recycled sample stream with a degassing chamber. The light absorption by quasi-homogeneous fermenter media is measured by using either mono- or dual-beam photometers, of which the latter type is less susceptible to fluctuations in the intensity of the light source (McCullough and Andrew [29.39], Melzer and Jaenicke [29.4]).

The different systems for measuring turbidity in fermenter media have their inherent advantages and disadvantages:

- Measurement in a bypass after degassing
  The fermentation broth is pumped continuously or discontinuously through a bypass. Gas bubbles are allowed to disengage in a degassing chamber, from which the liquid is conducted through a conventional flow cuvette, where transparency is measured.
  Advantage: continuous real-time signal.
  Disadvantage: possible cell damage, danger of cell sedimentation or wall growth in the bypass, low reproducibility.

- Turbidity probes
  Built-in probes for use in fermenters have been developed applying both mono-beam (OHASHI et al. [29.54]) and dual-beam systems (METZ [29.55]) with light source and photocells incorporated in a stainless cage.

  Advantage: continuous real-time signal, little danger of contamination since neither sampling nor bypassing are required.

  Disadvantage: danger of microbial wall growth, possible error by gas bubbles.

29.3.5.2 Light Scattering

The measurement of the light scattered by the cells can be correlated with biomass contents (HANCHER et al. [29.56]). In contrast to the measurement of optical transparency, the determination of the intensity of scattered light is proportional to cell density.

into the photocell (e). As in the measurement system for turbidity, the culture broth is pumped through a degassing chamber (i) and measuring cell (c), and then recirculated. An error of $\pm 10\%$ has to be expected; but the simplicity of the method and the prompt response have rendered it very useful.

29.3.5.3 Other Methods

Particle counters can determine cell densities of 10^3 to 10^6 cells/mL (KUBITSCHEK [29.57]). Here, a laminar sampling stream passes a measuring chamber to which an electric field is applied. The electrical resistance is recorded; it correlates with the number of cells.

Another method makes use of the fluorescence of active biomass, which is caused by inherent components. Thus, NADH, upon UV radiation at 366 nm, emits fluorescence at about 460 nm. The concentration of NADH is directly related to biomass density; however, this also applies to other

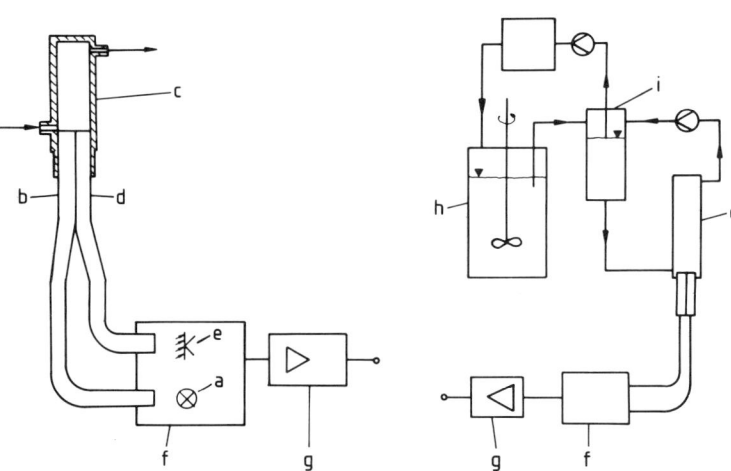

Figure 29.14. Biomass concentration by light scattering. – a light source, b, d light conductors, c measuring cell, e photo resistor, f transmitter and receiver, g amplifier, h fermenter, i degassing chamber.

Fig. 29.14 shows a device for determining biomass from light scattering. Light from a low-voltage lamp (a) is conducted through a light conductor (b) into the measuring cell (c), from where the scattered light passes through a second light conductor (d) back

components, such as oxygen and organic substrates. When these latter quantities are known or can be kept constant, significant information on biomass density can be obtained (ZABRISKIE and HUMPHREY [29.58], ZABRISKIE [29.59]).

The schematic arrangement of the fluorescence probe developed by BEYELER et al. [29.60] is shown in Fig. 29.15. In contrast to the fluorimeters formerly used, this probe does not require the installation of a sight glass but can be inserted into a 25 mm standard flange of the fermenter. With this set-up, no special problems of sterility arise. The probe operates as follows:

A UV light source at 366 nm induces fluorescence at about 460 nm. This emitted light is registered via a filter system (M, F_2 in Fig. 29.15) and amplified by a photomultiplier (D, A). In general, fluorescence measurements can directly monitor biochemical reaction steps taking place in the fermenter.

Accordingly, an exact interpretation of signals from redox potential measurements cannot be given. It is for that reason that JACOB [29.61] suggested to rather name this potential 'platinum-electrode potential' instead of 'redox'. The competing donors oxygen and glucose, which are simultaneously present in a fermentation, may serve as a typical example. In spite of the difficulty of interpreting the results, measurements of redox potentials permit an important insight into the course of fermentations (JACOB [29.62], KJAERGAARD [29.63], KJAERGAARD and JOERGENSEN [29.64]).

Sterilizable platinum electrodes are commercially available; they either contain built-in reference electrodes (Ag/AgCl)

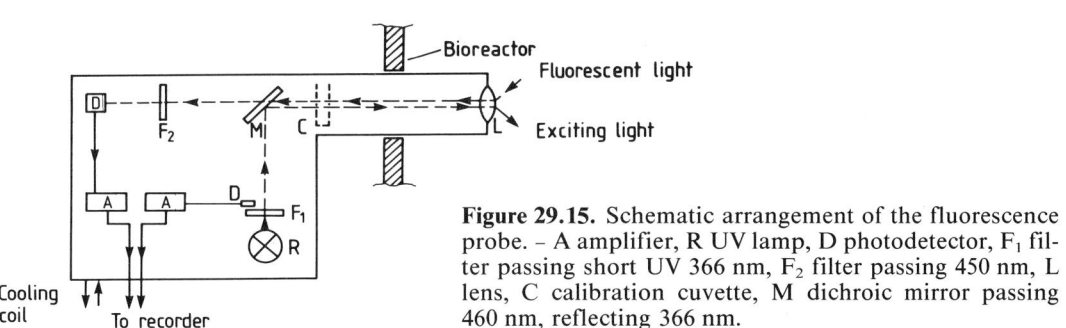

Figure 29.15. Schematic arrangement of the fluorescence probe. – A amplifier, R UV lamp, D photodetector, F_1 filter passing short UV 366 nm, F_2 filter passing 450 nm, L lens, C calibration cuvette, M dichroic mirror passing 460 nm, reflecting 366 nm.

29.3.6 Redox Potential

Another parameter obtained with electrochemical methods is the redox potential. Every redox system consists of two components, the one is oxidized by electron donation, the other is reduced by electron acceptance. In such a system, an electrochemical potential can be measured by means of an unprotected electrode consisting of a noble metal (Au/Pt), the composition of which is chosen on the basis of the relation of donor to acceptor (MELZER and JAENICKE [29.4]).

In a fermenter culture, a great number of redox systems are present simultaneously.

similiar to pH electrodes, or they are used in combination with pH measurement making use of the same reference electrode. The amplifier for redox measurements is of the same type as in conventional pH meters. As in the latter case, the built-in electrode is sterilized together with the fermenter. Since no membrane is required, no special sterilization problems exist. The signal of the redoxmeter is influenced by pH. This can be tolerated because fermentations are usually run at a constant pH. As mentioned above, the results from redox measurements cannot be assigned to a single parameter; they are used rather to obtain direct information on trends of a running process.

29.3.7 Enzymatic Analysis of Substrates and Metabolites

Enzymatic analysis allows very specific determination of many organic compounds. This group of methods takes advantage of the ability of enzymes to react selectively with well-defined compounds, or rather chemical structures: Organic compounds present in fermentation media (substrates and metabolites) can be analyzed (GUIL-BAULT [29.65], [29.66]; BOWERS and CARR [29.67]). Usually, this is performed off-line with samples taken from the fermenter; but methods for on-line enzymatic analysis have also been developed, which shall be described here.

Table 29.3 gives a number of examples of enzymatic analyses. The actual procedure consists of determining the conversion of the enzyme-catalyzed reaction with the respective substrate by analyzing one of the products either calorimetrically or electrochemically. For instance, the amount of oxygen produced in an enzymatic reaction may be monitored by an oxygen probe, as in the first example of Table 29.3, where β-D-glucose is converted into gluconic acid and hydrogen peroxide by means of glucose oxidase. The product hydrogen peroxide can be measured in various ways. e.g., with the aid of another enzyme (catalase) which generates oxygen.

Mostly immobilized enzymes on a carrier (adsorbent, ion-exchange resin, membrane) are used. Immobilization of enzymes is performed by physical or chemical methods (BOWERS and CARR [29.67]). Using immobilized enzymes offers two advantages: the valuable enzymes can be reused, and immobilized enzymes are less susceptible to inhibition by other components.

In applying on-line enzymatic analysis to sterile fermentations, special difficulties

Table 29.3. Systems for Enzymatic Analysis

Substrate	Enzyme	Product
β-D-Glucose	Glucose oxidase (+catalase)	H_2O_2 (O_2)
Saccharose	Invertase (+ mutarotase + glucose oxidase)	(H_2O_2)
Lactose	β-Galactosidase (+ glucose oxidase)	(H_2O_2)
Alcohols (primary, short-chain)	Alcohol oxidase	H_2O_2
L-Amino acid	L-Amino acid oxidase	NH_4^+, H_2O_2
Urea	Urease	NH_4^+, CO_2
Penicillin	Penicillinase	H^+

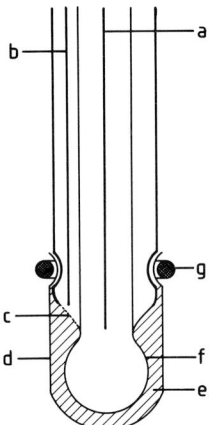

Figure 29.16. Principle of glass electrode with enzyme system. – a sensor electrode, b reference electrode, c diaphragm, d polymer membrane, e enzyme layer, f glass membrane, g O-ring.

arise because conventional sterilization techniques apply steam at 120°C which will destroy the enzymes. One solution is to employ a dialysis system through which a recycled sample stream of broth is conducted; components diffusing through the dialysis membrane can then be monitored continuously by means of enzymatic analysis. Systems of this type have been developed, for instance, for measuring glucose, saccharose, and lactose (LEEDS & NORTHRUP [29.68]), and glucose and glycerol.

Another possibility has been described by HEWETSON et al. [29.69] who assembled their enzyme electrode for penicillin (see Fig. 29.16) under sterile conditions and installed it after the fermenter had been steam-sterilized. The enzymatic reaction in this electrode is the hydrolytic cleavage of the β-lactam ring, by which H-ions are produced; they are monitored via a glass electrode.

Still another way of applying enzymatic analysis to sterile fermentations has been shown by MATTIASSON et al. [29.70]. Their procedure is based on the heat effect of enzymatic reactions. In order to monitor this effect, a continuous sample stream from the

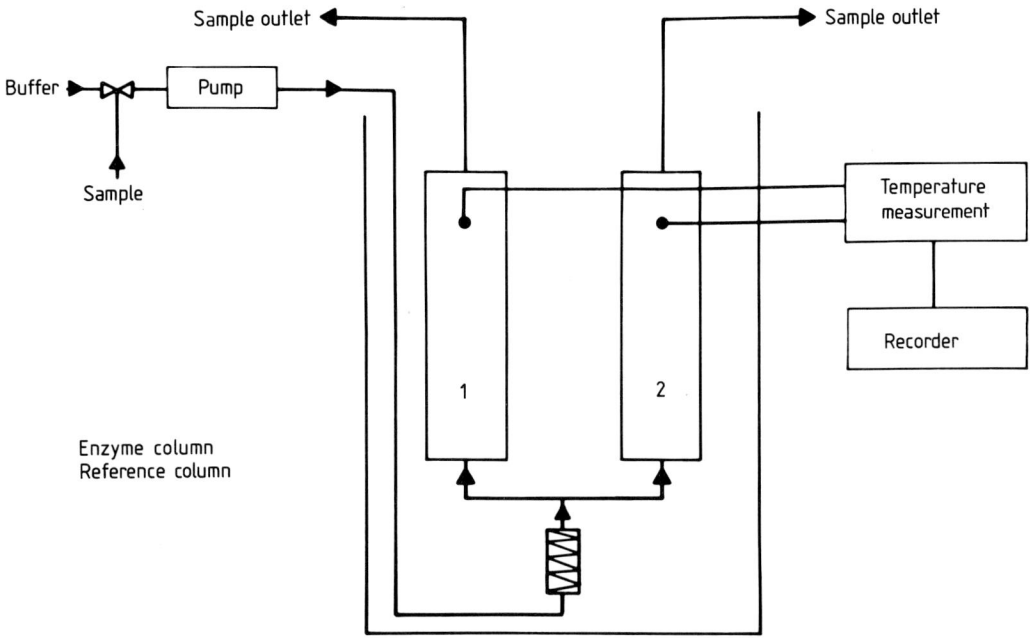

Figure 29.17. Enzyme thermistor unit. – 1 enzyme column, 2 reference column.

fermenter is fed through an enzyme-thermistor unit (see Fig. 29.17). This consists of two small parallel columns, one of which is packed with immobilized enzyme; the other column contains the same packing as the former except the enzyme. Thermistors in each of the columns indicate the temperature difference produced by the enzyme reaction, which is proportional to the concentration of the substrate, i.e., the component to be determined in the sample stream. MANDENIUS et al. [29.71] have used such an enzyme-thermistor unit with invertase to determine and control saccharose concentrations in a continuous culture producing ethanol from saccharose.

29.3.8 Ion-specific Electrodes

Not only carbon sources and other organic compounds, but also inorganic salts containing, e.g., N, P, S, K, Mg, Ca, Na, Fe are necessary constituents of fermentation broths. Ion-specific electrodes have been proposed for a number of these and for other ions (EBEL [29.72]). Some electrodes of this type are commercially available for off-line measurement; examples have been compiled in Table 29.4. There is little known about on-line measurement using ion-sensitive electrodes; but these probes will find wider use in fermentation processes in the future when more reliable, sterilizable electrode systems will have been developed.

Ion-sensitive electrodes are, in fact, potentiometric electrodes applying different principles. For the measurement of Na^+, glass electrodes with glass membranes, especially sensitive to Na^+, are employed; other modifications of glass electrodes are sensitive to NH_4^+ and K^+. Another type of ion-sensitive electrodes makes use of organic membranes, which, in some cases, may contain enzymes; here, sterilization is a special problem (see Sect. 29.3.7). Other types of ion-sensitive electrodes make use of the selectivity of membranes from certain crystals or are based on simple metal electrodes. The latter type has the disadvantage of simultaneously reacting to the redox potential and the specific metal. A detailed description of special measuring principles has been compiled by SCHINDLER and SCHINDLER [29.43a].

Table 29.4. Ion-sensitive Electrodes

Ion	Lowest Concentration Range (mol/L) of Application	Interference by Other Ions
NO_3^-	10^{-5}	Cl^-, Br^-, F^-, NO_2^-, SO_4^{2-}, CH_3COO^-
S^{2-}	10^{-15}	Hg^{2+}
Cl^-	10^{-5}	Br^-, J^-, CN^-, S^{2-}
J^-	10^{-7}	S^{2-}
K^+	10^{-5}	Na^+, NH_4^+, Cs^+, H^+, Li^+
Ca^{2+}	10^{-5}	Na^+, Mg^{2+}, Pb^{2+}, Fe^{3+}, Zn^{2+}, Cu^{2+}
Cu^{2+}	10^{-10}	S^{2-}, Ag^+, Hg^{2+}, Cl^-, Br^-, J^-, Fe^{3+}, Cd^{2+}

29.4 References

[29.1] J. HENGSTENBERG, B. STURM, and O. WINKLER; "Messen und Regeln in der chemischen Technik". 3rd Ed. Springer Verlag, Berlin 1980/81.

[29.2] C. D. JOHNSON; "Process Control Instrumentation Technology", 2nd Ed. Wiley, New York 1982.

[29.3] J. HASSELBACH; "Betriebsmesstechnik" in "Ullmanns Encyklopaedie der technischen Chemie", 4th Ed., Vol. 5, pp. 797–890. Verlag Chemie, Weinheim–Deerfield Beach, Florida–Basel, 1980.

[29.4] W. MELZER and D. JAENICKE; "Prozessanalytik" in "Ullmanns Encyklopaedie der technischen Chemie". 4th Ed., Vol. 5, pp. 891-944. Verlag Chemie, Weinheim–Deerfield Beach, Florida–Basel, 1980.

[29.5] D. M. CONSIDINE; in "Chemical Engineer's Handbook", 5th Ed., Sect. 22. McGraw-Hill, New York, 1973.

[29.6] W. PEINKE; In "Chemische Technologie" (K. WINNACKER, H. HARNISCH, and R. STEINER, eds.) 4th Ed., Vol. 1, pp. 504-598. Hanser, München, 1984.

[29.7] B. WILLIAMS and W. G. ANDREW; "Temperature measurement" in "Applied Instrumentation in Process Industries" (W. G. ANDREW and B. WILLIAMS, eds.), 2nd Ed., Vol. 1, pp. 168-195. Gulf Publishing Comp., Houston, 1979.

[29.8] W. G. ANDREW and W. F. MILLER; "Pressure measurement" in "Applied Instrumentation in the Process Industries" (W. G. ANDREW and O. WILLIAMS, eds.), 2nd Ed., Vol. 1, pp. 142-167. Gulf Publishing Company, Houston, 1979.

[29.9] J. H. MORRISON; "Pressure and vacuum measurement" in "Handbook of Applied Instrumentation" (D. M. CONSIDINE and S. D. ROSS, eds.), Sect. 4, pp. 10-23. McGraw-Hill, New York, 1964.

[29.10] P. A. ELFERS; "Liquid level" in "Handbook of Applied Instrumentation" (D. M. CONSIDINE and S. D. ROSS, eds.), Sect. 5, pp. 55-67. McGraw-Hill, New York, 1964.

[29.11] W. G. ANDREW and K. G. RHEA; "Level measurement" in "Applied Instrumentation in the Process Industries" (W. G. ANDREW and B. WILLIAMS, eds.), 2nd Ed., Vol. 1, pp. 99-141. Gulf Publishing Company, Houston, 1979.

[29.12] W. G. ANDREW, B. J. NORMAND, and F. E. EDMONDSON; "Flow measurement" in "Applied Instrumentation in the Process Industries" (W. G. ANDREW and B. WILLIAMS, eds.), 2nd Ed., Vol. 1, pp. 46-98. Gulf Publishing Company, Houston, 1979.

[29.13] F. A. ERICSON; "Pumps and flow measurement" in "Laboratory Engineering and Manipulations" (E. S. PERRY and A. WEISSBERGER ed.), 3rd Ed., pp. 182-220. John Wiley & Sons, New York, 1979.

[29.14] J. B. MCMAHON; "Flow of fluids" in "Handbook of Applied Instrumentation" (D. M. CONSIDINE and S. D. ROSS eds.), Vol. 5, pp. 5-21. McGraw-Hill, New York, 1964.

[29.15] A. B. METZNER and R. E. OTTO; AIChE J. *3*(1957), 3.

[29.16] G. LANGER and U. WERNER; Rheolog. Acta *14* (1975), 237.

[29.17] O. NEUHAUS, G. LANGER, and U. WERNER; Chem. Ing. Tech. *54* (1982), 1188.

[29.18] C. GERTH; "Rheometrie" in "Ullmanns Encyklopaedie der technischen Chemie", 4th Ed., Vol. 5, pp. 755-777. Verlag Chemie, Weinheim-Deerfield Beach, Florida-Basel, 1980.

[29.19] R. MUMMA; "Viscosity and consistency" in "Handbook of Applied Instrumentation" (D. M. CONSIDINE and S. D. ROSS, eds.), Vol. 7, pp. 39-50. McGraw-Hill, New York, 1964.

[29.20] A. DINSDALE and F. MOORE; in "Viscosity and its Measurement". Chapman and Hall, London, and Reinhold, New York, 1962.

[29.21] M. J. HALL, S. D. DICKINSON, R. PRITCHARD, and J. I. EVENS; "Foams and foam control in fermentation processes". Prog. Ind. Microbiol. *12* (1973), 170-234.

[29.22] G. F. HEWITT; in "Measurement of Two Phase Flow Parameters". Academic Press, London, 1978.

[29.23] G. F. HEWITT; "Measurement techniques" in "Handbook of Multiphase Systems" (G. HETSRONI, ed.), Vol. 10. Hemisphere Publishing Corp., Washington, 1982.

[29.24] A. SERIZAWA, I. KATAOKA, and I. MICHIYOSHI; Int. J. Multiphase Flow *2* (1975), 221.

[29.25] N. MILLER and R. E. MITCHIE; J. Br. Nucl. Energy Soc. *2* (1970), 94.

[29.26] J.-P. GALAUP and J.-M. DELHAYE; Houille Blanche *1* (1976), 17.

[29.27] W. SIEMES and E. BORCHERS; Chem. Eng. Sci. *12* (1960), 77.

[29.28] J. M. BURGESS and P. H. CALDERBANK; Chem. Ing. Sci. *30* (1975), 743, 1107, and 1511.

[29.29] R. BUCHHOLZ, K. FRANZ, and U. ONKEN; Chem. Ing. Tech. *54* (1982), 608.

[29.30] P. H. CALDERBANK and J. PEREIRA; Chem. Ing. Sci. *32* (1977), 1427.

[29.31] E. K. TODTENHAUPT; Chem. Ing. Tech. *43* (1971), 337.

[29.32] Th. PILHOFER, H. JEKAT, H. D. MILLER, and H. J. MÜLLER; Chem. Ing. Tech. *46* (1974), 913.

[29.33] P. WEILAND, L. BRENTRUP, and U. ONKEN; Ger. Chem. Eng. *3* (1980), 296.

[29.34] R. BUCHHOLZ, J. TSEPETONIDES, J. STEINEMANN, and U. ONKEN; Ger. Chem. Eng. *6* (1982), 105.

[29.35] J. STEINEMANN and R. BUCHHOLZ; Part. Charact. *3* (1984), 102.

[29.36] R. BUCHHOLZ, W. ZAKRZEWSKI, and K. SCHÜGERL; Chem. Ing. Tech. *51* (1979), 568.

[29.37] J. H. KIM, S. M. LEBEAULT, and M. REUSS; Eur. J. Appl. Microb. Biotechnol. *18* (1983), 11.

[29.38] K. FRANZ, TH. BOERNER, H. J. KANTOREK, and R. BUCHHOLZ; Chem. Ing. Tech. *56* (1984), 154 and Ger. Chem. Eng. *7* (1984), 365.

[29.39] R. L. MCCULLOUGH and W. G. ANDREW; "Analytical instruments" in "Applied Instrumentation in the Process Industries" (W. G. ANDREW and B. WILLIAMS, eds.), 2nd Ed., Vol. 1, p. 209. Gulf Publishing Comp., Houston, 1979.

[29.40] M. REUSS, J. GNIESER, H. G. RENGG, and F. WAGNER; Eur. J. Appl. Microbiol. *1* (1975), 295.

[29.41] W. VAN STEENBERGEN-HORROCKS and J. H. S. WILLIAMS; Z. Lebensm. Unters. Forsch. *168* (1979), 112.

[29.42] M. D. JOHNSON, J. BORKOWSKI, and C. ENGBLOM; Biotechnol. Bioeng. *6* (1964), 457.

[29.43] H. BUEHLER and W. INGOLD; Process Biochem. *11* (1976), 19.

[29.43 a] J. G. SCHINDLER and M. M. SCHINDLER; In "Bioelektrochemische Membranelektroden". Walter de Gruyter, Berlin–New York, 1983.

[29.44] Y. H. LEE and G. T. TSAO; Adv. Biochem. Eng. *13* (1979), 35.

[29.45] G. T. TSAO and D. D. LEE; AIChE J. *21* (1975), 979.

[29.46] B. MAASS, H. BAUMGAERTL, and D. W. LUEBBERS; Arch. Oto. *214* (1976), 109.

[29.47] E. PUHAR, A. EINSELE, H. BUEHLER, and W. INGOLD; Biotechnol. Bioeng. *22* (1980), 2411.

[29.48] M. SHODA and Y. ISHIKAWA; Biotechnol. Bioeng. *23* (1981), 461.

[29.49] M. REUSS, H. PIEHL, and F. WAGNER; Eur. J. Appl. Microbiol. *1* (1975), 323.

[29.50] E. HEINZLE and R. M. LAFFERTY; 1st Eur. Congr. Biotechnol. 1978, Interlaken, Switzerland.

[29.51] E. HEINZLE, O. BOLZERN, I. J. DUNN, and J. R. BOURNE; Adv. Biotechnol. *1* (1980), 439.

[29.52] B. J. MILLARD; In "Quantitative Mass Spectrometry". Heyden, London, 1978.

[29.53] D. W. LUEBBERS and N. OPITZ; Sensors and Actuators *4* (1983), 641.

[29.53 a] N. OPITZ and D. W. LUEBBERS; Chem. Ing. Tech. *56* (1984), 248.

[29.54] M. OHASHI, T. WATABE, T. ISHIKAWA, Y. WATANABA, K. MIWA, M. SHODA, Y. ISHIKAWA, T. ANDO, T. SHIBATA, T. KITSUNAI, N. KAMIYAMA, and Y. OIKAWA; Biotechnol. Bioeng. Symp. *9* (1979), 103.

[29.55] H. METZ; Chem. Tech. *10* (1981), 691.

[29.56] C. W. HANCHER, C. H. THACKER, and E. F. PHARES; Biotechnol. Bioeng. *16* (1974), 475.

[29.57] H. E. KUBITSCHEK; "Counting and sizing microorganismns with the coulter counter" in "Methods in Microbiology" (NORRIS and RIBBONS, eds.), p. 593. Academic Press, London, 1969.

[29.58] D. W. ZABRISKIE and A. E. HUMPHREY; Appl. Environ. Microbiol. *35* (1978), 337.

[29.59] D. W. ZABRISKIE; Biotechnol. Bioeng. Symp. *9* (1979), 117.

[29.60] W. BEYELER, A. EINSELE, and A. FIECHTER; Eur. J. Appl. Microbiol. Biotechnol. *13* (1981), 10.

[29.61] H. E. JACOB; Z. Allg. Mikrobiol. *11* (1971), 691.

[29.62] H. E. JACOB; in "Methods in Microbiology" (NORRIS and RIBBONS, eds.), Vol. 2, p. 91. Academic Press, London, 1970.

[29.63] L. KJAERGAARD; Adv. Biochem. Eng. *7* (1977), 131.

[29.64] L. KJAERGAARD and B. B. JOERGENSEN; Biotechnol. Bioeng. Symp. *9* (1979), 85.

[29.65] G. G. GUILBAULT; Methods Enzymol. *44* (1976), 579.

[29.66] G. G. GUILBAULT; Enzyme Microbiol. Technol. *2* (1980), 258.

[29.67] L. D. BOWERS and P. W. CARR; Adv. Biochem. Eng. *15* (1980), 90.

[29.68] LEEDS & NORTHRUP; Ger. Pat. (DOS) 2520417.5.

[29.69] J. W. HEWETSON, T. H. JONG, and P. R. GRAY; Biotechnol. Bioeng. Symp. *9* (1979), 125.

[29.70] B. MATTIASSON, B. DANIELSSON, and K. MOSBACH; Anal. Lett. *9* (1976), 867.

[29.71] C. F. MANDENIUS, B. DANIELSSON, and B. MATTIASSON; Acta Chem. Scand. Ser. B *34* (1980), 463.

[29.72] S. EBEL; "Elektrochemische Analysenverfahren" in "Ullmanns Encyklopaedie der technischen Chemie", 4th Ed., Vol. 5, pp. 651–683. Verlag Chemie, Weinheim–Deerfield Beach, Florida–Basel, 1980.

Chapter 30
Control and Optimization

Ulfert Onken

Abteilung Chemietechnik
Universität Dortmund
Dortmund, Federal Republic of Germany

Peter Weiland

Institut für Technologie
Bundesforschungsanstalt für Landwirtschaft
Braunschweig, Federal Republic of Germany

30.1 Introduction
30.2 Conventional Control
30.2.1 General Considerations
30.2.2 Physical Parameters
30.2.2.1 Temperature
30.2.2.2 Pressure
30.2.2.3 Gas Flow Rate
30.2.2.4 Liquid Feed Rates
30.2.2.5 Foam
30.2.3 Chemical Parameters
30.2.3.1 pH
30.2.3.2 Dissolved Oxygen
30.3 Computers in Biochemical Processes
30.3.1 General Remarks
30.3.2 Data Acquisition
30.3.3 Parameter Evaluation
30.3.4 Process Control
30.3.5 Process Modelling and Optimization
30.3.5.1 Process Modelling
30.3.5.2 Optimization Procedures
30.3.6 Computer Configurations
30.4 References

30.1 Introduction

Microbial processes are regulated by the biochemical activities of the microbes and by the conditions of their environment. These two types of factors do interact; in fermentation technology, however, only environmental conditions can be controlled directly. The individual environmental variables, e.g., temperature, pH, partial pressure of oxygen, and concentrations of nutrients (carbon and nitrogen sources, salts, vitamins), create an environment which influences the metabolic activity of the cells. The objective of process control, therefore, is to create the most favorable environment for the culture in order to produce an optimum of the desired metabolic activity.

Table 30.1. Main Controllable Process Parameters

Physical Parameters	Chemical Parameters
Temperature	pH
Pressure	Dissolved oxygen
Gas flow rate	
Substrate feed rate	
Foam	

Control parameters can be grouped into two major categories: physical and chemical parameters. Table 30.1 lists the variables which can be controlled in industrial fermentation. The physical parameters are largely independent of the cellular activity, whereas the chemical parameters reflect the enzymatic activity inducing the appearance or disappearance of compounds (cell mass, substrates, metabolites such as acids, alcohols, surface-active compounds).

Just as in chemical processes, control of fermentation processes can be performed in two entirely different ways: conventionally and computer-aided. With conventional methods, only those variables are controlled which can be monitored directly. Usually the set-point values of the variables are held constant during the process, independent of the physiological state of the culture, which is changing during the fermentation. Control devices are mostly of the feedback type, well known from process control in chemical plants (BECKER and MUCKLI [30.1], CONSIDINE [30.2], HENGSTENBERG et al. [30.3], JOHNSON [30.4], SCHÖNE [30.5], SHINSKEY [30.6]).

In contrast to these classical techniques, computer-aided control offers additional and more advanced possibilities besides maintaining individual variables at constant preset values. For this latter purpose, a computer has the advantage that it can employ highly sophisticated control algorithms adapted specifically to the response behavior of the system. This type of so-called Direct Digital Control (DDC) will, however, not pay for the cost of a computer. Only the use of additional capabilities is profitable. One of these is data logging, that is employing the computer for registration of data from the process. Here, another capability of the computer comes in: it can be utilized for the original purpose of computers, i.e., for performing calculations. Thus, parameters which are characteristic for the state of process can be evaluated, e.g., oxygen uptake rate. The next step is employing the relevant data for process control, which means altering the set-points of control parameters in order to improve the conditions for the culture. This is the first stage of computer-aided process optimization. This type of control will be more efficient when a model for the process is available. The development of such models is not easy; for this purpose the process computer is a necessary tool.

30.2 Conventional Control

30.2.1 General Considerations

Classical control techniques primarily utilize automatic feedback control with a closed loop configuration. The control variable is measured and compared with a reference set-point. If there is a difference between actual and desired level of the controlled variable, a control action will be started. In its simplest form the control loop consists of a measurement sensor, a measurement transmitter, the controller, and the control element (Fig. 30.1). The transmitter transforms the analog output of the sensor into a standard output signal which is compared by the controller with the desired set-point. Depending on the difference between the signal from the set-point and from the sensor, the controller produces an output signal to the control element.

In the simplest case, the controller acts only for a predetermined time interval if the measured parameter is beyond a fixed lower or upper limit. These on-off or two-point controllers do not take into account the degree of the deviation, and therefore this often leads to an unsteady state of the controlled parameter. These simple but inexpensive controllers are suitable only for those applications where close control is not essential, e.g., for foam control. Better control can be achieved when the controller takes into account the degree of deviation and the rate of its change. There are three principal types of control actions which are generally employed, viz., proportional (P), integral (I) and differential (D). Depending on the time behavior of the controlled variable P, PI, PD, or PID controllers are used.

In the first case (P control), the controller produces an output signal which is proportional to the difference between the set-point and the sensor signal actually trans-

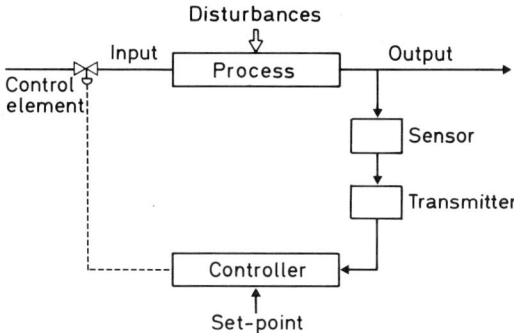

Figure 30.1. Conventional feedback control of a process variable.

mitted. Hence, with proportional control, the greater the deviation of the sensor signal from its set-point, the stronger is the corrective action. *Integral action* (frequently called "automatic reset") gives an output signal of the control element which is proportional to the time integral of the error, whereas *derivative action* (often termed "rate control") gives an output which is proportional to the derivative of the error. The integral and differential modes are normally used in combination with the proportional mode. Characteristic for proportional control alone, is a permanent difference between the desired process parameter value and its set-point, which is called offset. With PI control the controller output increases as long as a difference between desired value and control point exists. Therefore, the PI controller does not provide a permanent deviation (no offset). The disadvantages of PI control are higher maximum deviations, longer response times, and longer periods of oscillations compared to mere proportional action. Therefore, PI control action is used, when its shortcomings can be tolerated, and when an exact value of the controlled parameter is essential for the microorganisms. For these reasons, PI control is the method most frequently used in fermentations. In the case of differential mode (D control), control action will take place only when the deviation is changing. With D control action the controlled variable shows smallest oscillations and lowest maximum deviations. The offset is the same as with proportional control alone. PID control is a compromise between the advantages and disadvantages of PI and PD. Offset and the derivative mode reduce maximum deviations and oscillation times, although the latter are still larger than with PD control.

The controllers used in fermentation technology are of the same types as in chemical processes. In most cases, conventional control of fermentation processes is performed by single-loop control systems. They are sometimes, however, not able to maintain constant favorable conditions for the culture. This may happen, e.g., in large industrial fermenters, where some variables may show distinctly differing values at different locations of the reactor. Examples of this behavior with so-called distributed parameters are fed-batch or semicontinuous processes with intermittant substrate feeding. Here, different levels of substrate concentration are observed not only varying with time but also with distance from feed location. This poses a problem especially when the culture is sensitive to substrate poisoning. In cases of this kind, multiple loop control will be necessary, where several sensors and control actions are combined. An example of multiloop control is given in the scheme of Fig. 30.2, where an application of the so-called cascade control is shown. The process is a fed-batch fermentation in which substrate concentration shall be kept at an optimum. Because mixing of fresh substrate feed in the fermenter will take some time, the concentrations of sensors A and B will differ, especially just after an amount of substrate has been fed into the fermenter, with B showing a lower substrate concentration than A. Whereas controller A is acting directly on the control element, controller B ("master controller") alters the set-point of controller A ("slave controller") according to the information coming from sensor B.

There are various types of multi-loop control techniques which are commonly used in chemical processes; for more detailed information see BECKER and MUCKLI [30.1], CONSIDINE [30.2], HENGSTENBERG et al. [30.3], JOHNSON [30.4], SCHÖNE [30.5], SHINSKEY [30.6].

30.2.2 Physical Parameters

30.2.2.1 Temperature

Biochemical reactions are as dependent on temperature as any other type of chemical reactions, especially with regard to reaction rate. Besides, the activity of biochemical catalysts, the so-called enzymes, is most sensitive to temperature within a rather narrow range for optimum activity. In particular, temperatures 10–20 °C above the optimum may cause irreversible damage to the enzymes as well as to proteins in general and to microorganisms. Therefore, an efficient temperature control is required for the majority of biochemical processes. As temperature sensors, platinum resistance thermometers are used in most cases (cf. Sect. 29.2.1).

Temperature control is usually effected in single-loop mode as shown in principle in Fig. 30.3. For the steam sterilization cycle the set-point of the controller will be changed automatically or manually to the sterilization temperature. In case of instrument power failure, the cooling valve will open and the heating valve will close for safe operation. In small fermenters, control is frequently accomplished with two electri-

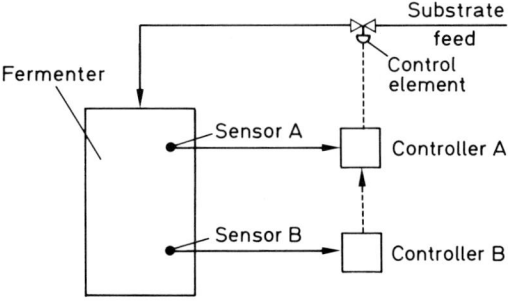

Figure 30.2. Control of substrate feed rate as an example of cascade control.

cal on-off valves. Cold water is added to lower the temperature, whereas heating of the thermostat medium is performed via a steam-water mixer, a steam heat exchanger, or an electric heater.

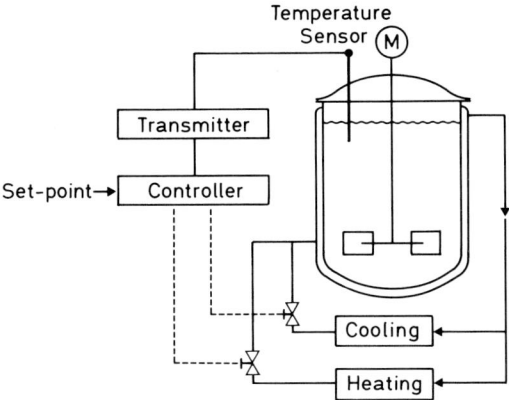

Figure 30.3. Principle of a temperature control system.

Sometimes, a cascade type of control may be preferable, e.g., when the cooling medium shows greater changes in temperature than the temperature difference between bioreactor and cooling circuit. This is the case, when river water is used for cooling a fermenter operating at 35 °C; river water may easily have a maximum temperature of up to 25 °C in summer and a minimum of 3–4 °C in winter.

30.2.2.2 Pressure

Pressure should be kept constant throughout the whole process especially in aerobic fermentations for various reasons, e.g., because of the pressure dependence of gas solubilities and of the effect of pressure on mass flow rates. In order to ensure sterility, operating pressure should be about 0.2 bars higher than atmospheric pressure. Pressure control is usually achieved by regulating the exit gas stream via a control valve.

30.2.2.3 Gas Flow Rate

In aerobic fermentations the flow of fresh air should be controlled. This can be done by monitoring the flow rate with a floating body flowmeter, e.g., a rotameter (cf. Sect. 29.2.4), and converting the position of the float in the meter into an electric signal which is used to regulate a valve in the gas-feeding pipe. In a similar way, the signals from other flow metering devices can be employed.

30.2.2.4 Liquid Feed Rates

Control of liquid feed rates is important for continuous processes as well as for fed-batch cultures. In continuous processes, liquid feed rates may be either kept constant or adapted to the state of the process.

In the first case, standard flow-control devices can be used with flowmeters for monitoring, such as rotameters or electromagnetic flowmeters (cf. Sect. 29.2.4). Another possibility is keeping the reactor hold-up constant by monitoring either weight or liquid level of the reactor (cf. Sect. 29.2.3).

If it is necessary, however, to adapt liquid feed rates to the state in the reactor, other control techniques must be used. One such case is substrate feeding into cultures which are sensitive both to higher and lower concentrations of substrate. An example is single cell protein (SCP) production from methanol, where the microorganisms cannot tolerate a high level of methanol in the medium on the one side, and where, on the other side, decreasing methanol concentrations will soon lead to a lowering of growth rates. Here, the logical procedure is to control substrate feed rates via methanol concentration in the fermenter broth using a rather fast sensor for methanol. In other cases, when cultures are not as sensitive to substrate poisoning, the more time-consuming procedure of taking probes at definite intervals and analyzing them off-line will be sufficient and the only possible way, because suitable and reliable sensors are not available for most substrates. In order to obtain the analytical data without loss of time, some type of autoanalyzer should be used. This procedure is commonly applied in fed-batch cultures, e.g., in several processes for secondary metabolites such as antibiotics.

30.2.2.5 Foam

Foaming is an annoying phenomenon accompanying most aerobic and many anaerobic fermentations. With foam entering the waste-gas line, contamination of the culture will be the consequence. Foam control can be achieved by mechanical devices or chemically by adding antifoam agents (e.g., polypropylene glycol). Installations for mechanical foam destruction will be in operation throughout the fermentation process. They may, however, prove partially or totally inefficient during a certain stage of the fermentation. Therefore, chemical defoaming must always be kept available.

For this purpose a foam detector at a suitable position in the fermenter will monitor whether the mechanical defoamer is not able to destroy the foam sufficiently (cf. Sect. 29.2.7). When this happens, the signal from the foam sensor will open the control valve in the antifoamer feeding line for a limited period, thus adding a predetermined amount of antifoaming agent. Because this substance may be consumed by the microbes, which after some time will lead to increased foaming again, the control will have to work once more. Sporadic splashing of droplets carried in the waste gas onto the foam sensor may also produce signals. A time delay in the control circuit will prevent these false signals from causing a control action.

30.2.3 Chemical Parameters

30.2.3.1 pH

In order to achieve high growth rates and optimum product formation in fermentation processes, the pH must be kept constant within a narrow range depending on the individual type of organism and on the actual phase of the culture. For many organisms the optimum pH is close to 7. However, there are a number of cultures where optimum pH is quite different. For instance, yeasts and lactobacilli prefer a lower pH of about 4 to 5, whereas for thiobacilli and some fungi a pH as low as 2 is possible. Without pH-control, microbial metabolism causes a shift in pH due to the formation of acid intermediates and products and to the consumption of certain substrates. In order to maintain a constant pH, base or acid is fed into the fermenter according to the deviation between the pH signal from the sensor (glass-electrode, cf. Sect. 29.3.2) and the set-point. This difference acts upon one of the two control valves for base or acid feeding.

30.2.3.2 Dissolved Oxygen

In aerobic fermentations, oxygen usually is a limiting substrate because of its low solubility in the fermentation broth. In order to secure a sufficient oxygen supply, a surplus of air is often fed into the fermenter with the consequence of a low oxygen yield. However, even then a high aeration rate will not always lead to sufficiently high concentrations of dissolved oxygen and to a satisfying supply of the microbes with oxygen, because resistance to oxygen transfer into the medium is limiting. For a stirred fermenter, a possible mode of acting in such a situation would be an increase in impeller speed, which enhances mass transfer and consequently oxygen concentration.

This shows that in a stirred fermenter the concentration of dissolved oxygen can be manipulated by at least two variables, i.e., aeration rate and impeller speed. The dissolved oxygen concentration chosen for a specific fermentation can be attained by individually, sequentially, or simultaneously manipulating inlet air flow and impeller speed. Because of the large variations in oxygen demand throughout a batch-fermentation and the considerable differences in oxygen demand for different cultures, the configuration of such a control system should allow free selection of one of these three control strategies. In most cases, the controller first increases impeller speed, and only when the desired dissolved oxygen content cannot be reached at the upper limit of impeller speed, the controller increases gas throughput. Such an interactive control system is shown schematically in Fig. 30.4. The p_{O_2} master (PID type) controller compares the deviation of the signal from the polarographic oxygen sensor from set-point; the result is used to adjust the set-point of the impeller speed controller and to manipulate the valve for air supply. Another possibility of dissolved oxygen control, which is sometimes used, consists in varying the oxygen concentration of the feed gas. In a typical procedure for this method, a group of three proportional valves controlling the flows of air, oxygen, and nitrogen is used. The oxygen and nitrogen valves allow automatic enrichment or depletion of the oxygen concentration of the feed gas, depending on the actual requirements of the culture. As this method permits higher oxygen partial pressures than when only air is used, greater oxygen transfer rates can be achieved. This is important for cultures with high oxygen consumption.

Because of the problems connected with the presently available methods for measuring dissolved oxygen (durability and reproducibility after sterilization, see Sect. 29.3.3.1), dissolved oxygen control is primarily employed industrially in aerobic cultures with high oxygen demand, when productivity depends greatly on oxygen supply, e.g., in some of the processes for antibiotics. On the other hand, laboratory and pilot plant fermenters are often equipped with an optional dissolved oxygen control.

30.3 Computers in Biochemical Processes

30.3.1 General Remarks

Biochemical processes, especially fermentations, are generally much more complex than chemical processes. First, biochemical processes are highly sensitive even to small changes in operating conditions, such as temperature, pH, or concentration of substrate. Biochemical catalysts, the so-called enzymes, usually have a very narrow range for their optimum activity; besides, at environmental conditions not very far from optimum, enzymes may be deactivated irreversibly. Deviations from optimum may have detrimental consequences for microbial cultures in particular, as they constitute systems with many different enzymes. These systems may adapt to environmental changes, but this will not be equivalent to optimum process conditions. In any case,

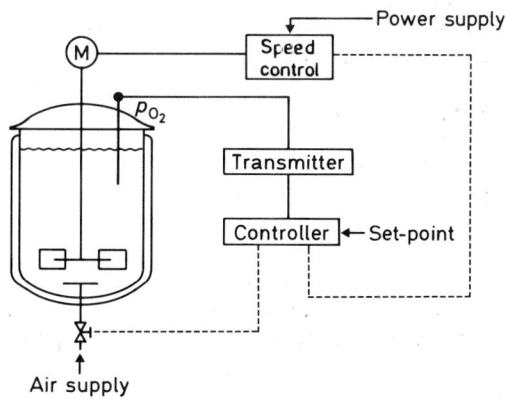

Figure 30.4. Principle of a dissolved oxygen control system.

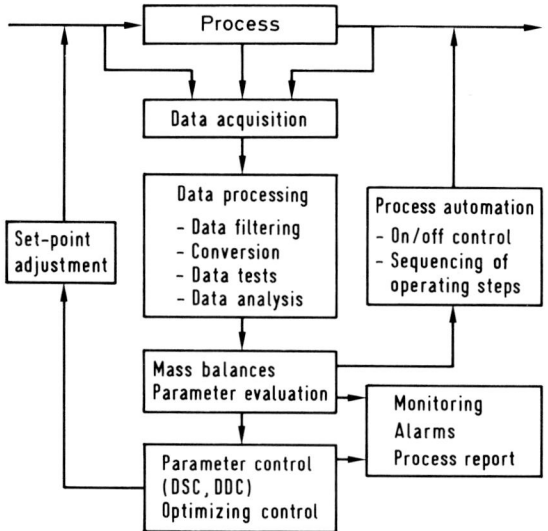

Figure 30.5. Functions of the computer in fermenter control.

whether it is a process with microbes or with isolated enzymes, the reaction system is characterized by a large number of parameters, most of them interacting in various ways.

Computers are very useful in controlling and optimizing biochemical processes. Principally, a computer can fulfill several tasks in process development in industrial production in general, and in biochemical processes in particular. Fig. 30.5 gives a schematical survey of computer functions for fermenter control.

The first and most simple computer task is mere data logging, which may be done for the purpose of data collection and retrieval alone. At the same time, data acquisition is a prerequisite for the more advanced uses of a computer, namely data analysis, process control, and process optimization. All of these tasks require suitable processing of the information from the process, such as data reduction.

Before use of the data for process control and optimization, it is necessary to check them for possible errors. Besides, process parameters not directly accessible by measurement, such as substrate consumption or oxygen uptake rate, can be easily calculated.

With all the data from the different sensors available, the computer can be used as a controller system. In this step of application it replaces the various classical control devices which are used in conventional fermenter control. Here, another advantage of computer utilization is at hand: the computer can employ more advanced and specific control algorithms than are used in conventional controllers (P, PI, PD, PID control; see Sect. 30.2.1). This special type of computer control is known as Direct Digital Control (DDC).

Finally, computers are suitable and necessary tools for process optimization. Any type of optimization is based on reliable process data, as they are available on a computer memory from measurements. With these data, the actual optimization is performed either off-line by developing a model for process control or on-line using search techniques together with informations on the actual state of the process. Whichever procedure is chosen, it has to be performed on a computer.

30.3.2 Data Acquisition

For the purpose of data acquisition the computer has to be connected with the various sensors of the biochemical reactor. At this point it should be mentioned that only part of the process parameters can be measured directly without problems, which means reliably in real-time mode. In fermentations, these are temperature, pressure, gas and liquid flow rates, exhaust gas composition, stirrer speed and power consumption, and, with precaution, pH, and dissolved oxygen.

The sensors for these variables yield primary signals of different kind, e.g., pneumatic, voltage, current, or digital signals. Before these signals can be fed into the computer, they must be converted into a form which is accepted by the data-processing unit. This conversion is performed by the so-called interface. Since computers can process only digital information, the analog signals are transformed by converting them into standardized voltage (mV) or current signals in a first step and into digital information thereafter. The transducing steps may include amplication and noise filtering of the signals, if necessary. Because analog-to-digital (A/D) converters are rather costly, the electric signals from the various sensors are scanned sequentially by a computer-controlled multiplexer, which is also part of the interface (for more detailed information see HAMPEL [30.7]; RAY [30.8]; JOHNSON [30.4]).

The digital signals from the interface are fed into the computer where they are processed in various ways depending on the nature of the data and their further utilization. In general, the data will be first transformed into true parameter values in standard engineering units (e.g., °C for the temperature) via calibration functions and correction factors. Together with the time of measurement taken from the digital clock of the computer, the parameter value is now written on a mass storage device, where the recorded data are available for subsequent calculations, e.g., computations of mean values and deviations or of characteristic process parameters not accessible by direct measurement (cf. Sect. 30.3.3). Before storing, the data may be checked for reliability by comparing them with former data or tolerance limits. Deviations can produce an error message or an alarm, if necessary.

For biochemical processes it is as a rule not necessary to work with high scanning rates; intervals in the order of one minute will be sufficient for data acquisition in most cases. At regular intervals, e.g., every 30 or 60 minutes, the computer prints out mean values of the recorded parameters either as charts or graphs. Besides, nowadays most computers are provided with a video screen, on which actual and former process data may be shown upon request. In this way, process reports are obtained to be stored for future reference (BOWSKI et al. [30.9]; GREINER [30.10]).

Besides process variables from on-line measurements, data obtained off-line may also be fed into the computer system and used as additional information for the actual process and for historical data retrieval (LEE and BEERENBACH [30.11]).

30.3.3 Parameter Evaluation

One of the advantages of computer utilization, which is of special interest for fermentation processes, is the evaluation of process parameters not directly accessible by measuring sensors. Several important parameters of this kind are given in Table 30.2. Some of these are essential for characterizing the state of the process, such as cell density or substrate utilization rate. Without an on-line computer these data would have to be determined off-line, which means that they would be available only some time. This on-line determination of derived process parameters has become common practice by now (e.g., COONEY et

Table 30.2. Process Parameters to be Determined with On-line Computers

Chemical	Physical
Cell density	Volumetric mass-transfer coefficient
Oxygen uptake rate	
Carbon dioxide evolution rate	Apparent viscosity (in non-Newtonian media)
Respiratory quotient	
Substrate utilization rate	
Heat evolution rate	

al. [30.12]; HUMPHREY [30.13]; BRAVARD et al. [30.14]; BOWSKI et al. [30.9]; BULL [30.15]).

An example of the evaluation of derived process parameters is the respiratory quotient (RQ = moles CO_2 produced/moles O_2 consumed). For its determination, data are needed for the following quantities: concentrations of oxygen and carbon dioxide in the exhaust gas, temperature, and pressure. With this information the computer first determines the moles of produced CO_2 and of consumed O_2 per unit of time, and subsequently the respiratory coefficient. RQ values obtained in this way may be used for process analysis; e.g., in yeast fermentation, the stages of ethanol formation (RQ > 1.0), oxidative growth (0.9 < RQ < 1.0), and endogenous metabolism (0.7 < RQ < 0.8) can be distinguished (WANG et al. [30.16]). Besides, calculated parameters, such as RQ or the volumetric mass-transfer coefficient, may serve as input information for process control. In this way, RQ data have been used by several investigators (WHAITE et al. [30.17]; SPRUYTENBERG et al. [30.18]; cf. Sect. 30.3.4).

Other parameters for which on-line determination is difficult or impossible, due to the lack of suitable sensors, are cell density, cell growth rate, and substrate consumption rate. Without knowledge of these parameters more sophisticated control strategies and on-line optimization are not possible. Their determination is often an essential step in the use of on-line computers in fermentation processes. In contrast to parameters calculated from data directly available from sensors, the evaluation of these other process variables requires additional information on cell growth and metabolism, stoichiometry of substrate conversion, composition of biomass, etc.

One example of this type of parameter determination is the estimation of cell mass concentration which is of central importance in fermentation processes. Here, several procedures have been proposed. One of these starts with the basic assumption that the different nutrients are converted into biomass and other metabolic products in definite stoichiometric ratios, i.e.:

$$\text{a substrate} + \text{b } O_2 + \text{c } NH_3 \rightarrow \text{cell mass} + \text{d by-product} + \text{e } CO_2 + \text{f } H_2O \quad (30.1)$$

with a, b, c, d, e, f being the number of moles per 1 kg of cell mass.

When the stoichiometric relations between nutrient consumption, cell mass and by-product formation are known, cell mass concentration and cell growth rate can be calculated from oxygen uptake, carbon dioxide production, or ammonia consumption rates (COONEY et al. [30.12]). If stoichiometric coefficients do not change during the fermentation, computer results for cell mass will be in good agreement with actual values, as was shown by WANG et al. [30.16] for baker's yeast, by PARK et al. [30.19] for a glutamic acid fermentation system, and by SWARTZ and COONEY [30.20] for continuous cultures of *Hansenula polymorpha* on methanol.

Another procedure of determining cell density uses the material balance for oxygen as a basis. Its application is limited to aerobic processes; but it has the advantage that the key component, oxygen, is measured directly and continuously in both the gas and the liquid phase. Another basis of the method is the close relation between oxidative phosphorylation and growth energetics. With the assumption that oxygen is utilized for cell growth and cell maintenance only, the oxygen uptake rate (OUR) can be determined by a two-parameter model (PIRT [30.21]):

$$\text{OUR} = \frac{1}{Y_{X/O}} \cdot \frac{dX}{dt} + m \cdot X \quad (30.2)$$

with X = cell density,
$Y_{X/O}$ = biomass yield coefficient based on oxygen,
and m = maintenance coefficient.

Rearrangement of Eq. (30.2) leads to the relation for calculating the cell growth rate:

$$\frac{dX}{dt} = Y_{X/O} \cdot (\text{OUR}) - m \cdot Y_{X/O} \cdot X. \quad (30.3)$$

Integration of Eq. (30.3) yields the equation for the cell mass concentration $X(t)$ at the time t:

$$X(t) = \exp(-m \cdot Y_{X/O} \cdot t) \cdot \left\{ \int_0^t Y_{X/O} \cdot \exp(m \cdot Y_{X/O} \cdot t) \cdot (\text{OUR}) \, dt + X_0 \right\} \quad (30.4)$$

with X_0 as cell mass concentration at $t=0$.

For the calculation of both cell density $X(t)$ and growth rate dX/dt, only the yield coefficient $Y_{X/O}$, the maintenance coefficient m, and the cell density X_0 at zero time are required. ZABRISKIE et al. [30.22] tested this method with different microorganisms. They extended the model to cultures with production of secondary metabolites and with diauxic growth. In the first case, a product yield coefficient has to be included in Eqs. (30.2), (30.3), and (30.4), whereas in the second case a metabolic correction factor is introduced.

Another way of determining cell density and growth makes use of the fact that all microbial processes evolve heat. The heat generated by the microbial process can be evaluated from a heat balance for the fermenter, using data for the heat streams into and out of the fermenter and information on the heat production of the stirrer. The result correlates with microbial activity. Together with data on the enthalpies of biochemical reactions, derived from heats of combustion, and with oxygen uptake, the evolution of heat by the microbial process can be fed into correlations yielding growth rate and cell density (WANG et al. [30.23]; BAYER and FUEHRER [30.24]; LUONG and VOLESKY [30.25]).

With procedures similar to those described for cell density, other process parameters, such as substrate concentration or product formation, can also be evaluated. Such data are valuable tools for characterizing process behavior and for recognizing deviations from regular operation. This data analysis is not so much a goal in itself but aids in process development and operation.

30.3.4 Process Control

With the full amount of information from all sensors of the process and with additional input and calculated data stored, a computer can fulfill the following tasks in process control, given in the order of increasing level of quality and complexity (ARMIGER and MORAN [30.26]; HAMPEL [30.7]):

1. simple monitoring, on/off-control of operating steps (e.g., filling, discharging, sterilizing), alarms;
2. sequencing of operating steps in batch and fed-batch processes (process automation);
3. control of individual process parameters;
4. control of the entire process (process optimization).

The last function requires special control strategies and detailed knowledge of process behavior, usually in the form of a process model. This area will be treated in a chapter of its own.

The first two items in the list of computer control tasks include alarm signalling, instrument recalibration, fail-safe shutdown procedures, sequencing of the various process steps from startup to shutdown, such as feeding the medium, heating for sterilization, cooling, taking samples, discharging the vessel, isolating biomass and products via manipulating valves or starting and stopping pumps. This level of computer control poses no special problems; it is important mainly on an industrial scale, where process automation has the advantage of safer and more reliable operation (METZ and WENZEL [30.27]). In combination with computerized data management, even this rather limited application of computers may be profitable in biochemical production plants (GREINER [30.10]).

The next more advanced level of computer use in process control involves manipulation of process parameters in individual loops. Whereas in conventional control the controlled parameter is maintained at a

Table 30.3. Examples of Computer-controlled Fermentations

Culture	Main Control Task	Process[a]	Reference
Yeast, aerobic	Constant dissolved O_2 concentration via aeration rate and agitation	B, Lab.	Nyiri et al. [30.29]
		B, Lab.	Kobayashi et al. [30.30]
		F-B, Lab.	Pons et al. [30.31]
	Constant RQ via substrate feed rate	F-B, Lab.	Aiba et al. [30.32]
			Wang et al. [30.16]
			Whaite et al. [30.17]
			Pons et al. [30.31]
	Constant RQ via dilution rate	C, Lab.	Spruytenberg et al. [30.18]
	Optimum of cell yield and growth rate	F-B, Lab.	Wang et al. [30.33]
	Maximum product formation via substrate feed rate	F-B, Lab.	Nelligan and Calam [30.34]
Yeast on sulfite liquor, aerobic	Optimum growth rate	C, Lab.	Cooney and Swartz [30.35]
		C, Prod.	Halme and Tiussa [30.36]
SCP on methanol	Prevention of methanol accumulation	C, Lab.	Swartz and Cooney [30.20]
			Cooney and Swartz [30.35]
Penicillium chrysogenum	Cell growth rate (for maximum penicillin production)	F-B, Lab.	Mou and Cooney [30.37]
Aspergillus niger for production of β-galactosidase	Maximum production of β-galactosidase	B, Prod.	Lundell [30.38]

[a] Abbreviations: B batch process, Lab. laboratory, F-B fed-batch process, C continuous process, Prod. production

previously defined value (set-point), in computerized control the value for this parameter can be adapted to the state of the process. For this control function two basic concepts are available: "Direct Set-point Control" (DSC) and "Direct Digital Control" (DDC).

In the first case (DSC), the control loop is equipped with a conventional analog controller; the set-point of this controller is adjusted according to the information supplied by the computer at definite time intervals. In the case of DDC, the conventional electric or pneumatic controller is replaced by the computer which now acts as a controller, receiving information from sensors for the controlled variable, comparing the data with the desired values, and initiating corrective action according to the deviation between actual and desired value. In contrast to conventional controllers, with DDC the computer is not limited to classical control algorithms (P, PI, PD, PID; cf. Sect. 30.2.1), but can choose some other more suitable response. Besides higher quality of control, DDC has, compared to DSC, the advantage of lower investment cost, when at least about ten control loops need to be equipped with conventional controllers. The higher requirement of computer time for DDC does not have much effect on cost at today's low and still decreasing computer prices. It is, however, not advisable, to run a process installation completely with DDC with only one computer, because a computer failure would result in total loss of control. Therefore, either a second computer must be installed for backup, which may be suitable when the computer is controlling several fermenters, as is the case in production plants, or a few important control loops must be equipped with analog controllers to which control can be returned in case of computer breakdown (DOBRY and JOST [30.28]; HAMPEL [30.7]).

Table 30.3 gives examples of computer-controlled fermentations reported in the literature. With two exceptions all of these are coming from university laboratories and research institutes. The major part of industrial applications, however, will certainly not be published. Thus, for SCP production from methanol, accumulation of methanol with a maximum tolerable concentration of about 0.8% must be prevented at any location of the fermenter, possibly by a strategy suggested by SWARTZ and COONEY [30.20]; but nothing is known about how this problem has been solved in industrial practice.

In some of the examples given in Table 30.3, the main control task consists in obtaining the optimum of a controlled variable. Such an optimizing control requires quantitative knowledge of process behavior in the form of a mathematical model (cf. Sect. 30.3.5).

30.3.5 Process Modelling and Optimization

30.3.5.1 Process Modelling

Optimization is a much-used word with different meanings. In chemical engineering and bioengineering, when used with respect to processes, it means operating a process at optimum conditions including the search for this optimum. With industrial processes, this optimum will be an economic optimum in the end, though very often an objective more directly connected with the process is used instead, such as maximum yield or maximum productivity. The selection of such an objective function implies that it is closely related to the profitability of the process. Thus, operating a process at the optimum means operation at an extreme of the objective function. For this purpose, the dependence of this objective function on process variables must be known. This knowledge is what we call process model. When it is represented in the form of mathematical equations – the usual and most convenient way of doing it – a mathematical model results.

Since real systems, and in particular biochemical systems, are extremely complex, our knowledge of them and the models for their description will always be incomplete. Besides, in many cases it is neither neces-

Table 30.4. Types of Process Models

Model	Describes
Black-box	Over-all behavior of process (e.g., by regression functions)
Mechanistic	Mechanism of process
Physical	Physical processes (e.g., heat and mass transfer, mixing)
Biochemical	Kinetics of biochemical reactions (e.g., Michaelis-Menten equation)
System	Physical processes and biochemical kinetics

sary nor suitable to use a complicated model, because a more simple one will be sufficient. Often the process model is composed of only a few material balances (e.g., WANG et al. [30.16]; BRAVARD et al. [30.14]; SWARTZ and COONEY [30.20]; PARK et al. [30.19]). Models of biochemical processes show a wide range of complexity, depending on available process information and on the purpose of their application.

Table 30.4 gives a survey of the different types of models in the order of increasing complexity. The two main groups – *black-box models* and *mechanistic models* – can be characterized shortly by stating that black-box models say only how a process behaves, whereas mechanistic models also say why. The explanations in Table 30.4 apply only to ideal types of models; most practical models will be somewhere between these ideal types.

Up to this point, there is no essential difference between models for chemical and biochemical processes. Thus, multiphase system models can be used to simulate processes with living cells by considering the cells as a separate phase, which is sufficiently specified by the number of cells or by the dry weight of biomass. This means that the properties of the biomass (e.g., protein content, metabolic activity) are assumed to be constant throughout the process. This approach does not take into account that the cells in a culture will generally be different, e.g., in age, size, content of protein, RNA and DNA, and correspondingly, in metabolic activity. When these properties cannot be neglected, so-called *structured models* must be used (HARDER and ROELS [30.51]; ROELS [30.52]). This will be necessary for quick

Table 30.5. Examples of Models for Computer Control of Fermentation Processes

Culture	Model	Reference
Yeast, aerobic	B [a] Material balances (C, O, N) S [b] O_2 transfer and material balances (C, O, N)	BRAVARD et al. [30.14] NYIRI et al. [30.43]
Citric acid	B Material balances based on metabolic pathway B Material balance (O) and heat balance	AIBA and MATSUOKA [30.47] HO [30.48]
Whey fermentation	S O_2 transfer and lactose inhibition	MORESI et al. [30.49]
Pseudomonas on glucose and 2,4-dichlorophenoxy-acetic acid	B (structured) Biomass and concurrent substrate utilization	PAPAGEORGAKOPOULOU and MAIER [30.50]

[a] B Biochemical   [b] S System

changes of biomass composition (e.g., lag phase of growth) or processes for the production of extracellular compounds. For structural models, as for any other model, complexity must be limited according to the general guideline, that a model shall contain only significant parameters. A rather simple version of structural models is based on the distinction of two or three sections of biomass of different properties (two- and three-compartment models). In setting up mechanistic models it is necessary to find out which steps of a proposed mechanism are relevant to the process on the basis of accuracy and reliability of process data. This model identification is performed by simulating the possible models on a computer (AIBA [30.53]), using suitable numerical techniques, such as linearization (REUSS et al. [30.54]) or statistical methods (MORESI et al. [30.49]).

In Table 30.5 some examples of models for optimizing control are given. As can be seen from the second column, most of the models are rather simple, though all of them are mechanistic (biochemical and system models). Apparently black-box models are not used; this is also true for physical models. The reason is clear: without minimum information on the chemistry of the process, models will be too unrealistic. For applications outside of process control, the situation may be different. Thus, in fermenter design meaningful physical models are required to describe flow and mixing behavior and its interaction with mass transfer between gas and liquid phases and cells.

30.3.5.2 Optimization Procedures

In on-line optimization the computer acts as a master-controller which calculates optimum process conditions with a model and compares them with actual data from the process. The resulting deviations are used to effect corrective actions (controlling variables, set-points) in order to force the system to the optimum. Several relatively simple cases of this optimizing control have already been given in Table 30.3. Examples for more sophisticated control strategies are shown in Table 30.6 for fed-batch cultures of yeast. In two cases time-optimal control policies have been applied: KALOGERAKIS and BOYLE [30.45] used time varying constraints, whereas DAIRAKU et al. [30.46] employed a model describing dynamic process behavior. An example of optimizing control in continuous culture is reported by VERES et al. [30.39] for gluconic acid fermentation;

Table 30.6. Optimizing Control of Fed-batch Cultures of Yeast (Aerobic)

Control Tasks	Control Strategy (simplified)	Reference
Optimal cell growth, suppression of ethanol formation	Control of nutrient feed (molasse, NH_3) via RQ based on constant C/N-ratio in biomass	NYIRI et al. [30.43]
Maximum productivity, suppression of ethanol formation	Control of substrate feed via RQ; constraints for RQ and oxygen uptake rate	RAMIREZ et al. [30.44]
Maximum yield of biomass	Constant cell and substrate concentrations via time-optimal substrate feed	KALOGERAKIS and BOYLE [30.45]
Time-optimal substrate feed during start-up and exponential growth	Model for dynamic behavior	DAIRAKU et al. [30.46]

they used the statistical *Box-Wilson* technique for the determination of optimum model parameters.

To this point we have not considered an important feature of process behavior, in particular of fermentation processes, i.e., the uncertainty of process information and consequently of process parameters. In more advanced optimization strategies, the model parameters can be updated on the basis of actual process information. An example of such an adaptive control is described by YOUSEFPOUR and WILLIAMS [30.40], who applied a control system of this type to yeast fed-batch fermentation. An interesting and promising possibility to cope with the problem of uncertainty of process data is the application of statistical methods. For this stochastic control, on-line state estimation techniques have been developed in control theory (cf. RAY [30.8]). They consist of filtering of input data and subsequent prediction. The expression "filtering" stems from the idea that actual process data are disturbed by a noise from which signals have to be extracted for prediction. One such filtering technique is the so-called *Kalman* filter (KALMAN [30.41]; KALMAN and BUCY [30.42]). It has been used in simulation studies of continuous fermentations (FAWZY and HINTON [30.55]; STEPHANOPOULOS and SAN [30.56]). The results show that the technique is both practicable and efficient. Another technique of adaptive control of continuous fermentations has been simulated by PRAUSE et al. [30.57].

Further examples of on-line optimization of fermentation processes are cited in several review articles (DOBRY and JOST [30.28]; WEIGAND [30.58]; BULL [30.15]). An important point must not be forgotten: for complete utilization of the capacities of modern computers for optimizing control, actual process data are required, especially for the biochemical parameters. The fact that for many of them reliable sensors are not yet available, limits computer applications.

30.3.6 Computer Configurations

Figure 30.6 shows an example of a computer-controlled fermenter, where pressure, temperature, pH, dissolved oxygen, air flow rate, exhaust air composition, and foam formation are measured with continuous control, using fresh and exhaust air flow rates, cooling water and steam, acid and base dosage, nutrient feed, and antifoam dosage as controlling variables. Filling and sterilizing the fermenter is effected by sequential control. The diagram is not complete, e.g., it does not contain discharging facilities and alarm monitoring; but it shows the essential features of a computer-controlled fermenter in a production plant.

Besides sensors, amplifiers, and control elements, a computer configuration for process control comprises the following elements:

- the computer interface with signal conditioning, analog-to-digital and digital-to-analog conversion, and a real time multiplexer or scanner;
- a processing unit with memory and mass storage devices;
- peripheral devices for data storage (disk or tapes), data input (keyboard), data output (printer, plotter, display), alarm monitoring;
- a real time clock.

Whereas for computer control of laboratory fermenters usually one computer is dedicated to a single fermenter, in pilot and production plants with a larger number of fermenters systems composed of several computers are preferred. This type of installation has become of special interest with the development of efficient and low-cost microcomputers. These can perform data processing and lower-level control tasks, thus relieving the process computer, usually some type of minicomputer, of most of its real-time burden.

An example of such an hierarchic micro-minicomputer system is shown in Fig. 30.7. The microprocessors which are operating at the lower level of the control systems are

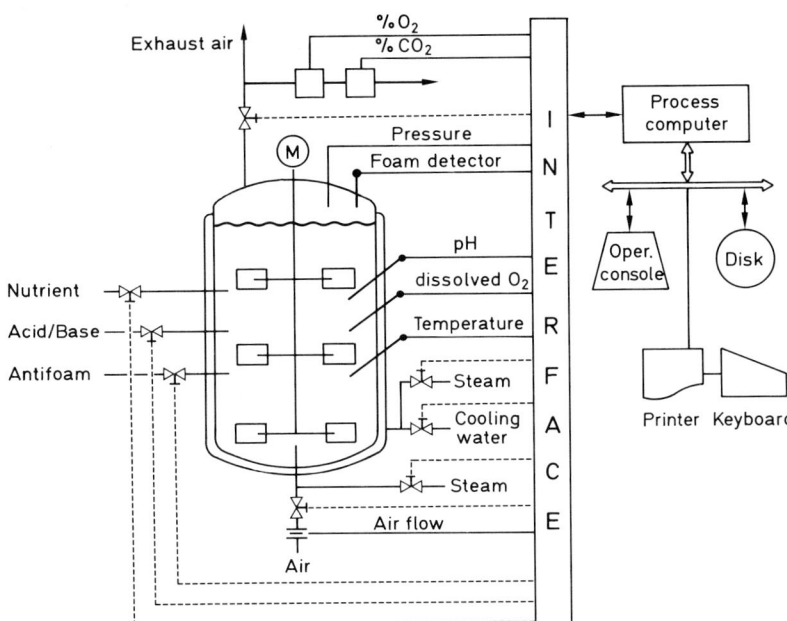

Figure 30.6. Example of a computer controlled fermenter.

connected directly with the process interface. One of them is dedicated to instrumental analyzers, such as process gas chromatograph and mass spectrometer. The others are servicing the fermenter units, whereby each of these microprocessors may operate and control one to about ten fermenters or other process units at the same time, depending on demand of computer capacity. In case of failure of one of the microprocessors, its function can be taken over by one of the other microprocessors or by the minicomputer operating at the higher level of the system.

Typical functions of microprocessors within an hierarchical control system are:

- input and output of analog and digital signals including encoding and decoding;
- intermediate storage of data;
- data processing;
- routine calculations, such as determination of process parameters;
- process monitoring, sequencing, and alarms;
- operating DSC and DDC control loops.

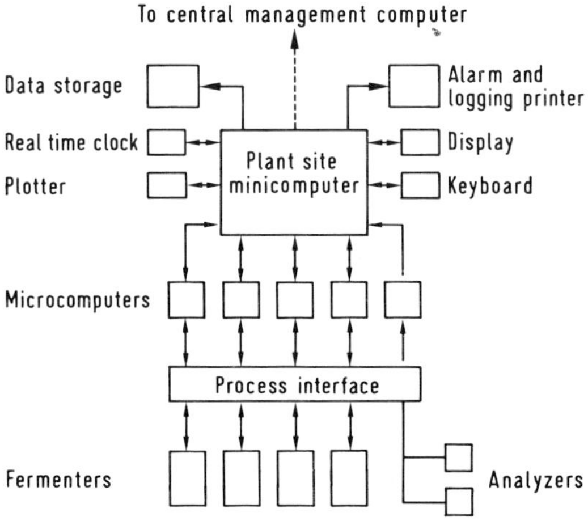

Figure 30.7. Hierarchic computer system for fermenter control.

With much of the routine work being done by the microcomputers, the plant minicomputer is free to perform control tasks and computations which require longer CPU times, such as:

- supervisory control (e.g., trend graphs of current fermentations, comparison of current and historical process data);
- reports of plant operation;
- data analysis (e.g., error calculations, statistical tests, recalculation of parameters);
- advanced control and optimizing strategies, e.g., adaptive control;
- model and program development.

In cases where a large central computer is available in the plant, it may be advantageous to connect the control system with it via the plant minicomputer. Thus, highly complex problems of process analysis and modelling can be investigated with the aid of larger program packages; besides, production data can be transferred directly to central management.

Examples of micro-minicomputer systems for fermenter control in pilot plants have been described for both university (ROLF et al. [30.59]; HENNIGAN et al. [30.60]) and industrial research (ALFORD [30.61]). Coupled computer systems have also been used in production plants, though publications are rather scarce. One such case has been reported, where a plant with 63 fermenters has been equipped with a micro-minicomputer system (OINAS [30.62]).

Besides coupled micro-minicomputer systems, minicomputer systems are also in use for process control in fermentation pilot plants (LEE and BEERENBACH [30.11]). BOWSKI et al. [30.9] describe such an installation which is connected with a remote computer-controlled production plant by means of a telephone for bidirectional information transfer which allows direct comparison of process data from production and pilot plant.

General aspects of computer systems for fermenter control have been discussed by several authors (JEFFERIS [30.63]; REUSS et al. [30.54]; DOBRY and JOST [30.28]; BLACHÈRE et al. [30.64]; HAMPEL [30.7]; JEFFERIS et al. [30.65]; ROLF et al. [30.59]). They point out that systems built up of more than one computer show better flexibility and greater safety of operation. An important requirement for successful implementation of computer control systems is user-orientated software, including a high-level easy-to-learn and easy-to-use programming language such as FORTRAN IV or BASIC with extensions for process control. Since these possibilities of easy programming have become available for desk calculators and personal computers, control of single laboratory fermenters using such a small computer can be realized at low cost (NELLIGAN and CALAM [30.34]; NYESTE et al. [30.66]).

In the future, the application of computers in fermentation control will definitely become more and more wide-spread. By way of example, at present all manufacturers of laboratory fermenters are offering versions with computer control. However, full use of the possibilities of computerized control and optimization is still hampered by the lack of adequate sensors. Even for the measurement of standard variables, for which accurate methods are available, inadequate sensors are widely used, such as rotameters for the measurement of gas and liquid flow rates. It is hoped that the development and application of better sensors will attract the same interest in the future, as has been given to the use of computers.

30.4 References

[30.1] W. BECKER and W. MUCKLI: In "Ullmanns Encyklopädie der technischen Chemie", Vol. 4, p. 173–268, 4th Ed. Verlag Chemie, Weinheim-Deerfield Beach/Florida-Basel, 1974.

[30.2] D. M. CONSIDINE: In "Chemical Engineer's Handbook", Sect. 22, 5th Ed. McGraw-Hill, New York, 1973.

[30.3] J. HENGSTENBERG, B. STURM, and O. WINKLER: "Messen und Regeln in der chemischen Technik"; 3rd Ed. Springer Verlag, Berlin-Heidelberg-New York, 1980/81.

[30.4] C. D. JOHNSON: "Process Control Instrumentation Technology", 2nd Ed. Wiley, New York, 1982.
[30.5] A. SCHÖNE: In "Ullmanns Encyklopädie der technischen Chemie", Vol. 4, p. 269–312, 4th Ed. Verlag Chemie, Weinheim–Deerfield Beach/Florida–Basel, 1974.
[30.6] F. G. SHINSKEY: "Process-Control Systems". McGraw-Hill, New York, 1979.
[30.7] W. HAMPEL: Adv. Biochem. Eng. *13* (1979), 1.
[30.8] W. H. RAY: "Advanced Process Controll". McGraw-Hill, New York, 1981.
[30.9] L. BOWSKI, C. R. PERLEY, and J. M. WEST: Biotechnol. Bioeng. *25* (1983), 1237.
[30.10] B. GREINER: Chem. Ing. Tech. *46* (1974), 680.
[30.11] S. E. LEE and B. A. BEERENBACH: In "Computer Applications in Fermentation Technology", p. 57. Internat. Conference London, 1981. Society of Chemical Industry, Manchester, 1982.
[30.12] C. L. COONEY, H. Y. WANG, and D. I. C. WANG: Biotechnol. Bioeng. *19* (1977), 55.
[30.13] A. E. HUMPHREY: Process Biochem. March (1977), 19.
[30.14] J. P. BRAVARD, M. CORDONNIER, J. P. KERVENEZ, and M. LEBEAULT: Biotechnol. Bioeng. *21* (1979), 1239.
[30.15] D. N. BULL: Annu. Rep. Ferment. Proc. *6* (1983), 359.
[30.16] H. Y. WANG, C. L. COONEY, and D. I. C. WANG: Biotechnol. Bioeng. *19* (1977), 69.
[30.17] P. WHAITE, S. ABORHEY, E. HONG, and P. L. ROGERS: Biotechnol. Bioeng. *20* (1978), 1459.
[30.18] R. SPRUYTENBERG, I. J. DUNN, and J. R. BOURNE: Biotechnol. Bioeng. Symp. *9* (1979), 359.
[30.19] S. H. PARK, K. T. HONG, J. H. LEE, and J. C. BAE: Eur. J. Appl. Microbiol. Biotechnol. *17* (1983), 168.
[30.20] J. R. SWARTZ and C. L. COONEY: Biotechnol. Bioeng. Symp. *9* (1979), 95.
[30.21] S. J. PIRT: "Principles of Microbe and Cell Cultivation". Blackwell Scientific Publications, Oxford, 1975.
[30.22] W. D. ZABRISKIE, W. B. ARMIGER, and A. E. HUMPHREY: "Workshop on Computer Applications in Fermentation Technology, Braunschweig-Stöckheim 1976" (R. P. JEFFERIS, ed.), p. 59. Verlag Chemie, Weinheim–Deerfield Beach/Florida–Basel, 1977.
[30.23] H. Y. WANG, D. I. C. WANG, and C. L. COONEY: Eur. J. Appl. Microbiol. *5* (1978), 207.
[30.24] K. BAYER and F. FUEHRER: Process Biochem. July/Aug. (1982), 42.
[30.25] J. H. LUONG and B. VOLESKY: Adv. Biochem. Eng. Biotechnol. *28* (1983), 1.
[30.26] W. B. ARMIGER and D. M. MORAN: Biotechnol. Bioeng. Symp. *9* (1979), 215.
[30.27] H. METZ and F. WENZEL: 5th Internat. Ferment. Symp. Berlin, 1976 (H. DELLWEG, ed.), p. 34. Institut f. Gärungsgewerbe u. Biotechnologie, Berlin, 1976.
[30.28] D. D. DOBRY and J. L. JOST: Annu. Rep. Ferment. Proc. *1* (1977), Chap. 5.
[30.29] L. K. NYIRI, R. P. JEFFERIS, and A. E. HUMPHREY: Biotechnol. Bioeng. Symp. *4* (1974), 613.
[30.30] T. KOBAYASHI, T. YANO, and S. SHIMUZU: Adv. Biotechnol. *1* (1981), 413.
[30.31] M. N. PONS, J. BORDET, and J. M. ENGASSER: In "Modelling and Control of Biotechnical Processes", p. 259. IFAC, Helsinki, 1982.
[30.32] S. AIBA, S. NAGAI, and Y. NISHIZAWA: Biotechnol. Bioeng. *18* (1976), 1001.
[30.33] H. Y. WANG, C. L. COONEY, and D. I. C. WANG: Biotechnol. Bioeng. *21* (1979), 975.
[30.34] J. NELLIGAN and C. T. CALAM: Biotechnol. Lett. *2* (1980), 531.
[30.35] C. L. COONEY and J. R. SWARTZ: In "Modelling and Control of Biotechnical Processes", p. 243. IFAC, Helsinki, 1982.
[30.36] A. HALME and E. TIUSSA: In "Modelling and Control of Biotechnical Processes", p. 267. IFAC, Helsinki, 1982.
[30.37] D.-G. MOU and C. L. COONEY: Biotechnol. Bioeng. *25* (1983), 225.
[30.38] R. LUNDELL: In "Computer Applications in Fermentation Technology", p. 181. Internat. Conference London, 1981. Society of Chemical Industry, Manchester, 1982.
[30.39] A. VERES, L. NYESTE, I. KURUCZ, L. KIRCHKNOPF, L. SZIGETI, and J. HOLLÓ: Biotechnol. Bioeng. *23* (1981), 391.
[30.40] P. YOUSEFPOUR and D. WILLIAMS: Biotechnol. Lett. *3* (1981), 249.
[30.41] R. E. KALMAN: J. Basic Eng. (1960), 35.
[30.42] R. E. KALMAN and R. S. BUCY: J. Basic Eng. (1961), 95.
[30.43] L. K. NYIRI, G. M. TOTH, C. S. KRISHNASWAMI, and D. V. PARMENTER: In

[30.43] "Workshop on Computer Applications in Fermentation Technology, Braunschweig-Stöckheim 1976" (R. P. JEFFERIS, ed.), p. 37. Verlag Chemie, Weinheim–Deerfield Beach/Florida–Basel, 1977.
[30.44] A. RAMIREZ, A. DURAND, and H. T. BLACHÈRE: Biotechnol. Lett. *3* (1981), 555.
[30.45] N. KALOGERAKIS and T. J. BOYLE: Biotechnol. Bioeng. *23* (1981), 921.
[30.46] K. DAIRAKU, Y. YAMASAKI, H. MORIKAWA, S. SHIOYA, and T. TAKAMATSU: J. Ferment. Technol. *60* (1982), 67.
[30.47] S. AIBA and M. MATSUOKA: Eur. J. Appl. Microbiol. *5* (1978), 247.
[30.48] L. HO: Biotechnol. Bioeng. *21* (1979), 1289.
[30.49] M. MORESI, A. COLLICCHIO, and F. SANSOVINI: Eur. J. Appl. Microbiol. Biotechnol. *9* (1980), 173.
[30.50] H. PAPAGEORGAKOPOULOU and W. J. MAIER: Biotechnol. Bioeng. *26* (1984), 275.
[30.51] A. HARDER and J. A. ROELS: Adv. Biochem. Eng. *21* (1982), 55.
[30.52] J. A. ROELS: "Energetics and Kinetics in Biotechnology". Elsevier, Amsterdam, 1983.
[30.53] S. AIBA: Biotechnol. Bioeng. Symp. *9* (1979), 269.
[30.54] M. REUSS, R. P. JEFFERIS, and J. LEHMANN: In "Workshop on Computer Applications in Fermentation Technology, Braunschweig-Stöckheim 1976" (R. P. JEFFERIS, ed.), p. 107. Verlag Chemie, Weinheim–Deerfield Beach/Florida–Basel, 1977.
[30.55] A. S. FAWZY and O. R. HINTON: J. Ferment. Technol. *58* (1980), 61.
[30.56] G. STEPHANOPOULOS and K. Y. SAN: Adv. Biotechnol. *1* (1981), 399.
[30.57] M. PRAUSE, H.-J. SCHULZ, and D. WAGLER: Acta Biotechnol. *4* (1984), 143.
[30.58] W. A. WEIGAND: Annu. Rep. Ferment. Proc. *2* (1978), Chap. 3.
[30.59] M. J. ROLF, P. J. HENNIGAN, and R. D. MOHLER: Biotechnol. Bioeng. *24* (1982), 1191.
[30.60] P. J. HENNIGAN, M. J. ROLF, W. A. WEIGAND, and C. H. LIM: In "Computer Applications in Fermentation Technology", p. 99. Internat. Conference London, 1981. Society of Chemical Industry, Manchester, 1982.
[30.61] J. S. ALFORD: In "Computer Applications in Fermentation Technology". Internat. Conference London, 1981. Society of Chemical Industry, Manchester, 1982.
[30.62] R. OINAS: Process Biochem. *6,* Nov./Dec. (1982), 10.
[30.63] R. P. JEFFERIS: In "Workshop on Computer Applications in Fermentation Technology, Braunschweig-Stöckheim 1976" (R. P. JEFFERIS, ed.), p. 21. Verlag Chemie, Weinheim–Deerfield Beach/Florida–Basel, 1977.
[30.64] H. T. BLACHÈRE, P. PERINGER, and A. CHERUY: Proc. 1st Eur. Congr. Biotechnol. DECHEMA-Monogr. *85.* Verlag Chemie, Weinheim–Deerfield Beach/Florida–Basel, 1978.
[30.65] R. J. JEFFERIS, S. S. KLEIN, and J. DRAKEFORD: Biotechnol. Bioeng. Symp. *9* (1979), 231.
[30.66] L. NYESTE, L. SZIGETI, A. VERES, E. PUNGOR, JR., I. KURUCZ, and J. HOLLÓ: Biotechnol. Bioeng. *23* (1981), 405.

Index

A

Acetic acid in media 647
Actinomyces rimosus growth media for 676
Actinomyces violaceus growth media for 676
Aeration devices sorption characteristics determination of 555
Aflatoxin in media 667
Agitator speed effect on release of intracellular material 371
— influence on mold growth 371
Air bubbles *see* Bubbles
Air sparger *see* Gas sparger
Alcohols in media 646
Allosteric control of enzyme activity 222
Amensalism of microbial populations 337
Amino acids in media 663
Ammonia in media 650
— solubility in water 162
Ammonium salts in media 650
Ampullariella regularis growth media for 672
Anacystis nidulans free jet experiments 381
Antifoam agents components of 666
— effect on bubble coalescence 554
Antifoaming agents in media 663
Antiseptics in media 667
Arrhenius equation temperature dependence of enzyme kinetics 219
Arthrobacter sp. growth media for 673, 677
Arthrobacter levanicum growth media for 675
Arthrobacter simplex growth media for 673
Aspergillus awamori media optimization 679
Aspergillus ficuum growth media for 675
Aspergillus foetidus growth media for 676
Aspergillus fumigatus growth media for 674
Aspergillus niger formation of gluconate 620
— growth media for 677
— influence of medium viscosity 389
— oxygen solubility in media for 168
— release of protoplasts caused by mechanical forces 372
Aspergillus oryzae influence of medium viscosity 389

B

Bacillus brevis media optimization 680
Bacillus cereus enzyme extraction 744
Bacillus coagulans growth media for 675
Bacillus mesentericus media optimization 679
Bacillus pumilus growth media for 674
Bacillus sphaericus enzyme extraction 744
Bacillus subtilis growth media for 673, 675, 677
— media optimization 679f
Bacillus thuringiensis media optimization 679
Bacteria reaction to mechanical stress 524
Bagasse in media 643
Baker's yeast growth under standard conditions 613
— substrate feed schedule 613
Balancing methods in bioprocessing 240
Batch adsorption of enzymes 753
Batch fermentation diauxic growth 265
— effect of salinity 264
— effect of temperature 264
— endogenous metabolism 262
— heat production 275
— kinetics 243 ff
— multi-substrate kinetics 264, 266
— multi-substrate uptake interactive model 267
— — non-interactive model 268
— rate equations differential evaluation 250
— — integral evaluation 250
— specific growth rate 247
— transient growth kinetic models for 269
Batch growth kinetics death phase 261
— inhibition-free substrate-limiting model 253
— inhibition kinetics 256
— lag time 260
— product inhibition 258
— stationary phase 261
— substrate inhibition 257
Batch process standardization 611
Bayer injector 544
Bead mill for cell disintegration 739
Beer wort sugar uptake sequential 266

Beet juice in media 638
Beet molasses composition 634
— in media 634
Bingham fluids 7
Biochemical reactions at the surface of solid spheres effect on mass transfer 93
— effect on mass transfer 73
Biocoenosis 340
Biofilm growth transport limitation 272
Biofilm reactors 341
Biofilms biological rate equation 360
Biokinetics transport phenomena interaction 177
Biological oxygen demand see BOD
Biological rate equation biofilm processing 361
Biomass measurement by fluorescence 780
— — by particle counters 780
— molecular weight 233
Biomass determination by light scattering 780
— by turbidity measurement 779
Biomass production in air lift fermenter 510
Biomass yield effect of specific growth rate 237
Bioprocesses 227 ff
— sterility 700
— unstructured modelling 239
Bioprocessing modelling 248
Bioprocess kinetic analysis strategy for different processes 180
Bioprocess kinetics mathematical modelling 178
Bioreactor for membrane processes hollow fiber fermenter 326
— — plug fermenter 326
— — rotor fermenter 327
— single-stage continuous stirred tank 287
Bioreactor operation balanced growth 189
— homogeneity 185
— pseudohomogeneity 185
— unbalanced growth 189
Bioreactors see also Loop reactors
— bench scale standardization 610
— estimation of process kinetic parameters 181
— heat transfer 431
— loop reactors 465 ff
— modelling 571 ff
— models for biokinetic analysis 187
— operation modes 183
— optimization 190
— reactor balances 183
— recycle operation 192
— scale-up 571 ff
— stirred flooding in 436
— — mass transfer in 440
— stirred tank versus plug flow reactor 191

— stirred vessels 395 ff
— types 184
Biotin in media 661
Bisubstrate kinetics of enzymatic reactions 211
Bodenstein number 300
BOD in waste water treatment 528
Bond number 460
Botryoplodia malorum growth media for 673
Boundary layer hypothesis in fluid flow 63, 146
Boundary layer thickness in fluid flow 66
Bran in media 642
Brevibacterium ammoniagenes enzyme extraction 744
Brevibacterium flavum growth media for 677
— media optimization 680
Brevibacterium lactofermentum growth media for 676 f
Brevibacterium thiogenitalis growth media for 676
Bubble break-up in bioreactors 452
Bubble coalescence effect of alcohols 553
— effect of antifoam agents 554
— effect of geometric parameters 554
— effect of inorganic salts 552
— in bioreactors 452
Bubble column geometry effect on sorption number 459
Bubble column reactor 445 ff
Bubble columns mathematical models 461
— modified 448
Bubble formation energy transfer 438
Bubble generation propeller stirrer 434
— turbine stirrer 433
Bubble movement in stirred bioreactors 435
Bubble properties in bubble columns 452
Bubble size distribution measurement of 771 f
Bubble swarm dynamics in bubble columns 452
Bubbles coalescence effect on sorption characteristics 547
— deformation turbulence 87, 89
— fluid flow within 85
— non-spherical mass transfer through the interface 97
— spherical mass transfer through the interface 96
Bubbles surrounding fluid flow 85
Bunsen coefficient gas solubility 161

C

Candida boidinii enzyme extraction 744
— media optimization 681
— oxygen solubility in media for 168
Candida curvata growth media for 678

Candida intermedia growth media for 678
Candida lipolytica biomass production 510
— growth media for 678
Candida oleophila growth media for 677
Candida tropicalis growth media for 678
Candida utilis media optimization 681
— oxygen solubility in media for 168
Cane juice in media 638
Cane molasses composition 637
— in media 637
Carbohydrates in media 631
Carbon dioxide evolution rate determination 612
— in media 649
— solubility in water 162
Carbon mole biomass 232
Carbon monoxide in media 649
— solubility in water 162
Carbon sources in media 631
Cascade control 790
Casein hydrolysates in media 657
Cell cycle in synchronous culture 333
Cell debris removal 740
Cell disintegration 737 ff
— by bead mills 739
— high pressure homogenization 737
— methods 738
— wet milling 737
Cell synchrony 331
Cellulomonas sp. media optimization 681
Cellulose hydrolysis 643
— in media 643
Centrifugation as a means of separation 729 f
Centrifuge chamber 731
— disc-stack 731 f
— tubular 731
Cephalosporium acremonium growth media for 672
Chaetomium cellulolyticum oxygen solubility in media for 168
Chemical control parameters 792
Chemical oxygen demand *see* COD
Chemical process variables measurement of 774 ff
Chemostat *see* Continuous stirred tank
Chicory in media 643
Chlamydomonas reinhardii critical stress values in free jet experiments 391
— — in stirring experiments 391
— free jet experiments 381
— stirring experiments 384
Chlorella vulgaris free jet experiments 381
— photosynthetic culture influence of mechanical forces 377
Chromatography in recovery of proteins 753 ff
Citric acid production effect of mechanical forces 373

Clarification of crude extracts 740 ff
Claviceps fusiformis growth media for 674
Claviceps purpurea growth media for 674
Clostridium acetobutylicum growth media for 677
Clostridium butyricum growth media for 677
Clostridium felsineum growth media for 678
— media optimization 681
Coal extracts in media 650
COD in waste water treatment 528
Commensalism of microbial populations 337
Competition of microbial populations 335
Computer data acquisition 794
— for process control configuration 802
— in biochemical processes 793
— parameter evaluation 795
— process control 797
Computer-controlled fermentations 798
Continuous cultivation 285 ff
— multi-stage operation 294 ff
— — applications 297
— multi-stream multi-stage operation 296
— plug flow reactor 298
— — applications 301
— recycling operation 302
— residence time distribution 300
— single-stream multi-stage operation 295
Continuous plug flow reactor with recycling 303
Continuous stirred tank cascade with recycling 303
— comparison with plug flow reactor 304
— effect of incomplete mixing 289
— effect of lysis 289
— effect of maintenance 289
— effect of mixing-process dynamics 292
— product formation kinetics without product inhibition 292
— — with product inhibition 293
— product inhibition kinetics 288
— simple kinetics 287
— with recycling 302
Control 761 ff, 787 ff
— cascade- 790
— conventional 789 ff
— feedback- 789
Control parameters chemical 788
— physical 788
Convective transport in fluid flow 17
Conventional control 789
Corn gluten meal in media 653
Corn meal composition 653
— in media 653
Cornsteep liquor composition 654
— in media 653
Corynebacterium acetophilum growth media for 676

Corynebacterium glutamicum growth media for 676f
Corynebacterium salvinicum growth media for 675
Cottonseed meal composition 652
— in media 652
Crossflow filtration 734
Culture collections 624
Culture media oxygen solubility in 169
Curvularia lunata growth media for 673f
Cyanobacteria hydromechanical stress capacity 380
Cytophaga sp. growth media for 676

D

Dactylosporangium thailandense growth media for 672
Dactylosporangium variesporum growth media for 672
Damage mechanisms 376
Damköhler number 22
Deborah number 460
Deep-bed filtration 734
Deformation turbulence hypothesis in fluid flow 151
Deformation turbulence 117
— in liquid films 125
Detergents in media 663
Dextrin in media 642
Dextrose in media 632
Dialysis culture comparison with non-dialysis culture 328
— performance equations 327
Dialysis cultures 325ff
Diauxie in batch fermentation 265
Differential pressure flowmeter 768
Dilatant fluids 7
Dilution rate critical 287
— — effect of recycling ratio 304
— effect on steady state biomass concentration 291, 299
Dimensional analysis in scale-up 590
Dimensionless numbers in differential equations 28
Direct Digital Control (DDC) 788, 794, 799
Direct Set-point Control (DSC) 799
Disc-stack centrifuge 732
Disk stirrer 404
Dissolved gases determination 775ff
Dissolved oxygen control of 792
Distillation residues composition 660
— in media 659

E

Eadie-Hofstee plot effect of internal mass transfer resistance 362
Electrochemical determination of carbon dioxide 775
— of oxygen 775
Electrodes ion-specific 784
Electromagnetic flowmeter 768
Emulsifiers components of 665
— in media 663
Endogenous metabolism kinetic modelling in batch fermentation 262
Endomycopsis bispora media optimization 679
Endomycopsis fibuliger growth media for 675
— media optimization 679
Endothia parasitica media optimization 680
Energetic yield energy efficiency coefficient 233
— of growth 233
Energy dissipation rate in bubble columns 451
Energy efficiency factor for oxygen 233
Energy input in bubble columns 451
Enzymatic analysis of metabolites 782
— of substrates 782
Enzyme activity kinetic models of control 223
— regulation by allosteric control 222
— regulation by enzyme synthesis 222
— regulation by mass action law 222
Enzyme inhibition kinetics competitive 214
— Dixon plots 216
— non-competitive 214
— substrate inhibition 216
— uncompetitive 214
Enzyme kinetics complex reactions 209
— control and regulation 222
— differential method of analysis 204
— enthalpy/entropy-compensation 221
— Hill 203
— integral method of analysis 204
— Langmuir 204
— Langmuir-Hinshelwood 210
— linearization plot 205
— Lineweaver-Burk plot 206
— Michaelis-Menten 204, 210
— model equations 210
— Monod 204
— pH effect 217
— rate equations 199ff
— reaction mechanisms 208
— saturation type kinetics 204
— simple 204
— stationary state 202
— steady state 202
— temperature dependence 219
— transient state 202
— Walker diagram 206
Enzyme preparations commercial 669

Enzyme reactions inhibition kinetics 213
— rate equations for complex kinetics 213
Enzyme synthesis kinetic models 224
Enzymes in media preparation 668
Escherichia coli enzyme extraction 744
— growth media for 675
— oxygen solubility in media for 168
Ethane solubility in water 162
Ethanol effect on oxygen solubility 167
— in media 646
Ethanol production in air lift reactor 509
— *Zymomonas* sp. membrane process 327
Exhaust gas analysis 774
Extraction for product enrichment 751
— of enzymes from microbial cells 744
— of proteins from cell homogenates 741 ff

F
Fats in media 648
Fatty acids in media 648
Fed-batch cultivation classification 316
Fed-batch process 318
— applications 324
— computer simulation 318
— kinetics structured models 321
— — unstructured models 319
— standardization 611
— variable-volume continuous operation 317
Feedback control 789
Fermentation fluids fluid-mechanic properties 400
Fermentation processes yield coefficents 236
Fermentations comparative tests 607 ff
Fermenter broth analysis 778
Fermenters *see* Bioreactors
Fick's law 14
Filamentous microorganisms free jet experiments 385
— influence of mechanical forces 370
Film hypothesis concentration profiles 148
— in fluid flow 148
Films *see* Liquid films
Filter design 736
Filtration as a means of separation 734 f
— equipment 736
Fish meal in media 653
Fixed-bed reactors 342
Floating body flowmeter 767
Flocs microbial 342
Flowmeters 767 ff
Flow rate measurement of 767
Flow regimes in bubble columns 449
Fluid flow *see also* Turbulent flow
— *see also* Laminar flow

— analogy of momentum, heat and mass transfer 153 ff
— around solid spheres velocity field 81
— boundary layer thickness 67
— energy content 408
— in loop reactors 482
— laminar friction factor 68
— — velocity field 65
— Newtonian fluids in jet loop reactors 488
— non-Newtonian in jet loop reactors 490
— of Newtonian fluids 33 ff
— of non-Newtonian fluids 49 ff
— parallel to flat plates 147
— parallel to plates 63
— primary in stirred vessels 406
— secondary in stirred vessels 406
— transport phenomena 1 ff
— transport processes in 61 ff
— turbulent friction factor 68
— volumetric flow rates 409
Fluidized bed bioreactors 342
Fluid mechanics concentration fields differential equations 23 ff
— temperature fields differential equations 23 ff
— velocity differential equations 23 ff
Fluid particles transport processes 115
Fluids *see also* Non-Newtonian fluids
— *see also* Newtonian fluids
— in motion temperature field 427
— quiescent temperature field 426
Fluorescence measurement for biomass determination 781
Fluorescence quenching for determination of dissolved gases 778
Foam control of 792
— destruction chemical 770
— — mechanical 770
Formaldehyde for sterilization 719
Free jet experiments 378
Friction factor laws for plates 68
Froude number 460
Fusarium oxysporum growth media for 678
Fusarium sp. growth media for 673

G
Galilei number 460
Gas bubbles *see* Bubbles
Gas dispersion energy transfer 438
— in stirred vessels 432
Gas flow rate control of 791
— measurement of 767
Gas hold-up in bubble columns 451
— measurement of 770
— volumetric in jet loop reactors 493

Gas/liquid interface transport limitation oxygen transfer 358
Gas-liquid mass transfer in bioreactors 456
Gas phase dispersion in bubble columns 455
Gas solubility determination of 161
— effect of anions 165
— effect of cations 164
— effect of electrolytes 163
— effect of organic solutes 165
— effect of pressure 162
— effect of temperature 162
— in biomedia 159 ff
Gas sparger 448
— effect of hole diameter 458
Gelatin in media 660
Germ removal by ultrafilters 718
Germ removal filtration definition 711
— for air 714 ff
Gluconic acid formation by *Aspergillus niger* 620
Glucose effect on oxygen solubility 165
— in media 632
Glycerol in media 647
Green algae hydromechanical stress capacity 380
Growth balanced steady-state 315
— synchrony 330
Growth media *see* Media
Growth rate oxygen transfer rate limitation 272
— specific critical O_2-concentration 271
— — diffusion limitation 270
— — in batch fermentation 247

H

Haldane relationship enzyme kinetics 209
Hansenula polymorpha oxygen solubility in media for 168
Harvest of microorganisms 727 ff
Heat convective transport 17
Heat and mass transfer analogy of 153 ff
Heat conductivity molecular 12
Heat diffusivity 12
Heat exchange in bubble columns 448
Heat production in batch fermentation 275
Heat resistance of microorganisms 703
Heat transfer at flat plates 72
— deformation turbulence hypothesis 151
— in bubble columns 460
— in laminar tube flow 38
— in power law fluids 57 ff
— — effect of internal frictional heat 57
— in stirred vessels 424
— in turbulent tube flow 42
— steady-state at solid spheres 92

Heat transfer coefficient 17
Heat transport in fluid flow 3 ff
— molecular 11
— turbulent 12
High-pressure homogenization 738
Hill enzyme kinetics 203, 223
Hoechst waste water treatment aerobic 561
Hydrazine method absorbed oxygen 546
Hydrocarbons in media 649
Hydrodynamic forces effect on biomass production 373
Hydrodynamic stress in a free jet 378
Hydrodynamics in bubble columns 448
Hydrogen in media 649
— solubility in water 162
Hydrogen sulfide solubility in water 162
Hydrol in media 633

I

Immobilized enzymes as reactants in mass transport 44
Impeller speed measurement of 769
Industrial effluent *see* Waste water
Industrial fermentation media *see* Media
Inhibition kinetics in batch fermentation 256
— in bioreactor operation 193
Injector design data 540
— pressure drop characteristics 540
— slot 545
— sorption characteristics 540, 546
Inorganic components in media 660
Instrumentation 763 ff
Inulin in media 642
Isoelectric precipitation of proteins 746
Isokinetic model enthalpy/entropy-compensation 221

J

Jerusalem artichokes in media 643
Jet bioreactor reciprocating 519 ff, 522
— — application 534
— volumetric power input 533
— waste water specific conversion rate 533
Jet loop reactors 473
— liquid circulation velocity 496
— power input 486

K

Kinetic equations for unstructured modelling 239
Kinetics of batch fermentation 243 ff

King-Altman model of enzyme bisubstrate reactions 211
Klebsiella pneumoniae enzyme extraction 744
— growth media for 674
Kluyveromyces fragilis growth media for 678
— media optimization 681
Kuenen coefficient gas solubility 161

L

Lactobacillus casei enzyme extraction 744
Lactobacillus confusus enzyme extraction 744
Lactobacillus delbruckii growth media for 677
Lag time batch growth kinetics 260
Laminar flow friction factor 36
— velocity profiles 35
Langmuir enzyme kinetics 204
Langmuir plot effect of internal transport limitation 362
Leuconostoc sp. enzyme extraction 744
Light scattering measurement 780
Linear growth phenomena macrokinetics 271
Lineweaver-Burk double reciprocal plot 206
Lineweaver-Burk plot effect of internal mass transfer resistance 363
Liquid feed rate control of 791
Liquid films fluid mechanic properties 116
— mass transfer 120
— — with biochemical reactions 127
— — with first order reactions 138
— — with second order reactions 130
— transport processes 113
— wavy film surface 116
— with smooth surfaces mass transfer 118
— with wavy surfaces mass transfer 120
Liquid flow rate measurement of 767
Liquid/gas interface transport limitation 358
Liquid hold-up measurement of 766
Liquid jets transport processes 115
Liquid phase micromixing 358
Liquid phase dispersion in bubble columns 453
Liquid-solid bioprocess heterogeneous concentration/space profile 359
Liquid/solid interface transport limitation 357
Liquid-solid mass transfer in bioreactors 456
Liquid turbulence measurement of 773
Liquid viscosity measurement of 769
Loop reactors 465 ff
— air-lift 472
— fluid dynamics 481
— improved productivity 508
— industrial applications 509
— jet 473
— mass transfer 502
— mixing behavior 499

— parts 478
— plunging liquid jet 477
— pressure cycle fermenter 471
— propeller 471
— residence time behavior 502
— types 470

M

Macrobalances stoichiometric 229
Macroelements in media 661
Maintenance coefficient 237
— in batch fermentation 262
Malt extract composition 641
Maltose in media 639
Mass convective transport 17
Mass and heat transfer analogy of 153 ff
Mass conversion biochemical 20
Mass spetrometry for determination of dissolved gases 776 f
Mass transfer at flat plates 69
— combined external and internal 363
— deformation turbulence hypothesis 151
— hypotheses 143 ff
— in bubble columns 455
— in laminar tube flow 38
— in liquid films 113
— in power law fluids 55
— in turbulent tube flow 42
— steady-state at bubbles 96
— — at solid spheres 90
— unsteady-state through the interface of spherical particles 99, 107
— with biochemical reactions 43, 72
— — local concentration 45
— — mean concentration 45
Mass tranfer and reaction interaction 355
— interactions in microbial systems 349 ff
Mass transfer coefficient 17
— volumetric 457
Mass transport in fluid flow 3 ff
— in laminar flow fields 16
— in turbulent flow fields 16
Mathematical models batch growth kinetics 253
Measurement 761 ff
— of physical process variables 766
Mechanical forces change in metabolism 373
— effect on microbial cultures 370
Mechanical stress damage effects 375
— effect on bacteria 524
— in microbial production 369 ff
Mechanistic models of microkinetics in batch fermentation 249
Media carbon sources 631
— design of industrial 671

— inorganic components 660
— nitrogen sources 650
— preparation of 629 ff
— raw material storage 671
— recycling of 668
— vitamin sources 661
Membrane processing in bioreactors 325
Methane in media 649
— solubility in water 162
Methanol effect on oxygen solubility 167
— in media 646
Methylomonas clara biomass production 510
— change of metabolism by mechanical forces 372
Methylotrophs yield stoichiometric coefficient 238
Michaelis-Menten enzyme kinetics 204
Microbacterium sp. growth media for 673, 677
Microbalances stoichiometric 229
Microbial biofilms 341
Microbial flocs biological rate equation 360
Microbial interactions kinetic analysis 335
Microbial populations interactions 334
Micrococcus glutamicus growth media for 676
Microelements in media 661
Micromonospora purpurea growth media for 672
Microorganisms regulatory mechanism 269
Minerals in media 661
Mixed microbial fermentations 334
Mixed microbial populations waste water treatment applications 340
Mixing in bubble columns 453
— macro in bioreactor operation 191
— micro in bioreactor operation 194
Mixing time in bubble columns 454
Modelling batch fermentation 248
Modelling of biokinetics mathematical 178
— unstructured elemental balances 239
— — kinetic equation 239
— — of bioprocesses 239
Modelling of bioprocesses computer use 578
— mathematical complexity 577
— parameter optimization 580
— parameter sensitivity 579
Models *see also* Mathematical models
— for computer control of fermentation processes 800
— for process control 800
Models of bioreactors black box models 575
— classification 575
— continuous time 576
— discrete time 576
— gray box models 575
— types 574
Molasses in media 634
Molds influence of mechanical stress 387

— morphology influence of mechanical forces 370
Molecular mass transport in fluid flow 13
"Mole" microorganism 232
Momentum analogy with heat and mass transfer 153 ff
— convective transport of 19
Momentum transfer deformation turbulence hypothesis 151
— local wall shear stress 67
— tube flow 35
Momentum transport in fluid flow 3 ff
— turbulent 9
Monod enzyme kinetics 204
Monod kinetics in batch fermentation 247
Monod volumentric reaction rate equation 20
Mucor javanicus influence of impeller speed 388
Mucor miehei growth media for 676
Mucor pusillus media optimization 679
Multi-phase system transport limitation 357
Multi-substrate kinetics in batch fermentation 266
Mutualism of microbial populations 338
Mycelia hydrolysates in media 659
Mycelial growth kinetic modelling 263
Mycobacterium sp. growth media for 673
Mycobacterium album media optimization 679
Mycobacterium fortuitum growth media for 673 f

N

Nabla operator 27
Navier-Stokes equations 24
n-Butane in media 649
Newtonian fluids molecular momentum transport in 6
— transport processes through tubes 33 ff
Nitrogen solubility in water 162
Nocardia sp. growth media for 673, 676
Non-Newtonian fluids *see also* Power law fluids
— classification 7
— in stirred vessels 401
— molecular momentum transport in 7
— power law 51
— transport processes through tubes 49 ff
Non-Newtonian power law fluids velocity profiles 52
Nozzle propulsion 541
— radial flow type 563
— two-phase *see* Injector
Nozzle separator 733
n-Paraffins in media 649
n-Pentane in media 649

Nusselt number 18, 57
Nutrient media see Media

O

Optimization 787ff, 799ff
Optimizing control of fed-batch cultures of yeast 801
Ostwald coefficient gas solubility 160
Ostwald-de Waele power law 460
Ostwald fluids 8
Oudemansiella mucida growth media for 675
Oxygen dissolved control of 792
— solubility in water 162
Oxygen consumption rate effect of transport phenomena 364
Oxygen demand proportionality to biomass yield 239
Oxygen requirement prediction from mass balance equation 240
Oxygen solubility in culture media 169
Oxygen transfer enhancement of gas/liquid 365
— measuring electrodes 354
— transport enhancement 354
Oxygen transfer efficiency effect of temperature 551
— of injectors 549
Oxygen transfer rate of various bioreactors 478
Oxygen transport in waste water from air to microbe 525
Oxygen uptake rate determination 612
Ozone solubility in water 162

P

Paddle stirrer 404
Particle interface transport processes 77
Particles spherical mass transfer 99
Peat hydrolysis 644
— in media 644
Pellet growth mass transfer limitation 272
Pellets microbial 342
Penetration hypothesis concentration profile 150
— in fluid flow 149
Penicillin production effect of mechanical forces 374
Penicillin synthesis enthalpy balances 240
— reaction stoichiometry 240
Penicillium chrysogenum effect of mechanical forces 374
— growth media for 672

— growth rate of dependence on stirrer speed 371
— impeller speed influence 389
— oxygen solubility in media for 167
Pentalenolactone formation from *Streptomyces arenae* 614
Periodic feed bioprocessing 317
pH control of 792
— effect on growth in batch fermentation 264
— enzyme kinetics effect on 217
— measurement of 774
pH-optimum of enzyme reactions 218
Phellinus chrysoloma growth media for 676
Phosphorus in media 660
Physical control parameters 790
Physical process variables measurement of 766ff
Plug flow reactor comparison with continuous stirred tank 304
— practical properties 298
Potato proteins in media 653
Power input measurement of 769
Power law 7
Power law fluids 401
— dilatant 52
— friction factors 53
— heat transfer in 57
— mass transfer in 55
— pseudoplastic fluids 52
— velocity profiles 51
Power number 458
Prandtl boundary layer hypothesis 63
Prandtl number 13
Precipitation by non-ionic polymers 747
— by organic solvents 747
— by polyelectrolytes 747
— for product enrichment 745
— isoelectric of proteins 746
Predation of microbial populations 338
Predator-prey interactions 338
Pressure control of 791
— measurement of 766
Pressure drop in bubble columns 450
Process computer 788
Process control 797
Process development pragmatic empiric 181
— systematic empiric 182
Process mass spectrometer 777
Process modelling 799
Process optimization 799
Process parameters evaluation 795
Product enrichment 745ff
— ultrafiltration 748
Product formation growth-associated 275
— kinetic models 273
— non-growth-associated 275
— pseudohomogeneous rates of 274

— rate equations for 277
Product inhibition kinetics of batch fermentation 259
Propeller stirrer 404
— bubble generation 434
Propulsion jet nozzle *see* Injector
Protein hydrolysates composition 658
— in media 657
Proteins chromatography 754 ff
— downstream processing 728
— isoelectric precipitation 746
— isolation by cell disintegration 737
— — by extraction 741, 752
— precipitation by non-ionic polymers 747
— — by organic solvents 747
— — by polyelectrolytes 747
— recovery 727 ff
Pseudomonas methylotrophus biomass production 510
Pseudomonas yoshitomiensis growth media for 672
Pseudoplastic fluids 7

Q

Quasi-steady state periodic operation 315

R

Rapeseed meal composition 653
— in media 652
Raw materials for media storage 671
Reaction and mass transfer interaction modelling 355
— interactions in microbial systems 349 ff
Reaction engineering methodology 175
— microbial fundamentals 171 ff
Reaction flux density 20
Reaction rate integration with transport process 21
Reaction rate equation 20
Reactor hold-up measurement of 766
Recovery of proteins from microbial sources 727 ff
Recovery operations 725 ff
— product enrichment 745
— removal of by-products 745
Recycle operation in bioreactors 192
Recycling in continuous cultivation 302
— — applications 305
Recycling ratio optimization 304
Redox potential in fermentations 781
Regime analysis experimental methods 600
— theoretical methods 600

Removal of cell debris by centrifugation 740
— by filtration 740
Residence time in bioreactors 192
Residence time distribution in continuous cultivation 300
Resistance factor for spherical bubbles 86
— in fluid flow around solid spheres 83
Resistance thermometer 766
Respiratory quotient determination 612
Reynolds number 29
Rhizopus javanicus influence of impeller speed 388
Rhodosporidium toruloides growth media for 678
Rhodotorula equi growth media for 674
Rotating disk fermenters 342
Rotating flowmeter 768

S

Saccharomyces cerevisiae enzyme extraction 744
— growth media for 678
— growth under standard conditions 613
— oxygen solubility in media for 168
— rate of heat production 277
Saccharomyces fragilis growth media for 678
Salinity effect on growth in batch fermentation 264
Salting out for product enrichment 746
Sampling for process information 778
Scale-up aeration rate 589
— dimensional analysis 586
— fundamental method 585
— heterogeneous systems 594
— laboratory culture to production plant 181
— methodology 585
— of bioreactors 581 ff
— pragmatic-empiric 182
— regime analysis 599
— rules of thumb 587
— sludge digester 596
— stirred vessel 587
— systematic-empiric 183
Schizophyllum commune media optimization 679
Schmidt number 16
Seawater in media 668
Sechenov equation gas solubility 163
Separation solid-liquid 729
Separator for liquid-liquid separation 743
Serratia marcescens growth media for 672
Sherwood number 18
— in transport processes Newtonian fluids 46

Slug flow in bubble columns 450
Sodium chloride effect on bubble coalescence 548
Solid/liquid interface transport limitation 357
Solids in media 663
Sorption number 458
Soybean meal composition 651
— in media 651
Spheres solid surrounding fluid flow 81
Spirulina platensis critical stress values in free jet experiments 391
— — in stirring experiments 391
— free jet experiments 385
— stirring experiments 386
Sporobolomyces carnicolor growth media for 675
Stanton number 364
Starch in media 639
— saccharification 641
Starch crops composition 641
Steady state biomass concentration effect of dilution rate 291
Steady-state operation relaxed 315
Stefan diffusion 15
Sterile sampling 778
Sterility definition 700
— requirements for fermentation processes 700
Sterilization 699 ff
— by germ removal filtration 711
— disinfection of equipment 721
— influence of air 708
— influence of contact time 704 ff
— influence of pH 708
— influence of temperature 704 ff
— in the instantaneous heater 710
— laminar flow boxes 718
— of air filters in the autoclave 708
— of air for fermenters 714
— of lab fermenters 708
— of plants with microbicidal gases 719
— of stationary large-scale fermenters 708
— with dry heat 710
— with formaldehyde/water vapor mixtures 719
— with moist heat 702
Sterilization processes 701 ff
Still residues in media 659
Stirred vessels *see also* Bioreactors
— fluid flow 405
— gas content 437
— gas dispersion 432
— heat transfer 424, 431
Stirrer speed effect on release of intracellular material 371
— influence on mold growth 371
Stirrers arrangement in bioreactors 403
— energy transfer to fluids 410

— energy transfer to Newtonian fluids 414
— energy transfer to non-Newtonian fluids 422
— functions of 402
— mechanical energy transfer 413
— shapes 403
Stoichiometric balance of semi-discontinuous reactor operation 236
Stoichiometry 227 ff
— application to bioprocessing 236
— definition 229
— macrobalances 229
— microbalances 229
— of complex reaction systems 235
— of complex reactor operation 236
— of non-stationary reactor operation 236
— of technical processes 230
Stream lines in fluid flow 64
Streptococcus lactis media optimization 680
Streptomyces sp. enzyme extraction 744
Streptomyces antibioticus growth media for 672
Streptomyces arenae growth under standard conditions 614
Streptomyces aureofaciens growth media for 672
Streptomyces avermitilis growth media for 673
Streptomyces canus growth media for 675
Streptomyces chartreusis growth media for 673
Streptomyces cremeus media optimization 680
Streptomyces flavoricens growth media for 675
Streptomyces fradiae growth media for 672
Streptomyces galilaeus growth media for 672 f
Streptomyces granuloruber growth media for 672
Streptomyces griseus albus growth media for 673
Streptomyces hygroscopicus growth kinetics 264
Streptomyces rimosus growth media for 672
Streptomyces rochei growth media for 672
Streptomyces tendae growth media for 673
Streptomyces tumemacerans media optimization 680
Streptomyces wadayamensis growth media for 672
Streptoverticillium griseocarneum media optimization 680
Substrate degree of reduction 233, 235
Substrate inhibition Dixon plots 216
Substrate inhibition kinetics batch fermentation 257
Sucrose in media 633
Sulfite waste liquor in media 645
Sulfur in media 661
Surface active agents in media 663
Synchronous culture 330 ff
— phased-continuous 333

T

Temperature control of 790
— dependence of enzyme kinetics 219
— effect on growth in batch fermentation 264
— measurement of 766
Tetrahymena pyriformis hydromechanical stress capacity 379
Thermal death rate of microbial cells 221
Thermal inactivation rate of enzymes 221
Thermodynamic efficiency of aerobic growth 234
— of growth 233
Thiamine in media 661
Thiobacillus ferrooxidans oxygen solubility in media for 162
TOC in waste water treatment 528
Torulopsis utilis growth media for 678
— media optimization 680
Total organic carbon *see* TOC
Tower fermenter *see* Bubble column reactor
Tower-shaped reactors aerobic waste water treatment 537 ff
— — plant description 557
Trace elements in media 661
Transient operation techniques 314
Transport enhancement in bioprocessing 353 f, 365
Transport limitation external 353
— internal 354, 359
— — theory 356
Transport phenomena biokinetics interaction 177
Transport processes in fluid flow parallel to plates 61 ff
— in liquid films 113 ff
— Newtonian fluids flowing through tubes 33 ff
— non-Newtonian fluids flowing through tubes 49 ff
— scale dependence 582
— through the interface of particles 77 ff
— through the interface of spherical particles 104
— integration with reaction rate 21
Trichoderma reesei growth media for 675
— media optimization 679
— oxygen solubility in media for 168
Trichosporon cutaneum media optimization 681
Trickling filters 342
Tube flow momentum transfer 35
— rough tubes 42
Turbidity measurement of 779
Turbine stirrer 404
— bubble generation 433
Turbulence of liquid measuring 773
Turbulent flow friction factor 36
— velocity profiles 35

U

Ultrafiltration 734
— germ removal 718
— in protein recovery 748 ff
Urea in media 650

V

Velocity field around non-spherical bubbles 88
— around spherical bubbles 85
— in fluid flow around solid spheres 81
Viscosity measurement of 769
Vitamin content of medium ingredients 662
Vitamins in media 661
Vortex field in fluid flow around solid spheres 82

W

Walker linearization 206
Wall shear stress momentum transfer 67
Wash-out in continuous fermentation 287
Waste treatment biological biokinetic relationships 240
— — stoichiometric relationships 240
Waste water aerobic treatment plant description 557
Waste water treatment aerobic tower-shaped fermenter 537 ff
— biological jet bioreactor 519
— bioreactor development 521
— conversion efficiency 530
— in bubble columns 458
— in loop bioreactors 511
— pilot plant 526
— sedimentation system 530
— substrate uptake simultaneous 266
Water activity effect on growth in batch fermentation 264
Water quality in media 667
Wavy liquid films mass transfer in 125
Whey composition 639
— in media 638
Wood hydrolysate in media 644

X

Xanthan formation from *Xanthomonas campestris* 617
Xanthomonas campestris growth media for 674
— growth under standard conditions 617
Xylose in media 631

Y

Yeast *Candida utilis* composition 657
— dried in media 655
— fed-batch cultures optimizing control 801
— molecular formula 232
Yeast autolysates composition 656
— in media 655
Yield expression in molar units 232
— in g/mol ATP 232
Yield coefficient effect of maintenance coefficient 237
— effect of specific growth rate 237
— estimation graphic method 231
— in bioprocessing 231
— in fermentation processes 236
Yield factor effect of temperature 238
— in absence of product formation 238
— relation to stoichiometric coefficients 238
— Y_{ATP} 231
Yield factors concept of 230